Mirror Symmetry V

AMS/IP
Studies in
Advanced
Mathematics

Volume 38

Mirror Symmetry V

Proceedings of the BIRS Workshop on
Calabi-Yau Varieties and Mirror Symmetry
December 6–11, 2003
Banff International Research Station
for Mathematics Innovation & Discovery

Noriko Yui,
Shing-Tung Yau,
James D. Lewis, Editors

American Mathematical Society • International Press

Shing-Tung Yau, General Editor

2000 *Mathematics Subject Classification.* Primary 14-XX, 11-XX, 32-XX, 81-XX.

Library of Congress Cataloging-in-Publication Data

Mirror symmetry V : proceedings of the BIRS workshop, Calabi-Yau varieties and mirror symmetry, December 6–11, 2003 / Noriko Yui, Shing-Tung Yau, and James D. Lewis, editors.
p. cm. — (AMS/IP studies in advanced mathematics, ISSN 1089-3288 ; v. 38)
Includes bibliographical references.
ISBN-13: 978-0-8218-4251-5 (alk. paper)
ISBN-10: 0-8218-4251-X (alk. paper)
1. Mirror symmetry—Congresses. 2. Geometry, Differential—Congresses. 3. Geometry, Analytic—Congresses. I. Yui, Noriko. II. Yau, Shing-Tung, 1949– III. Lewis, James Dominic, 1953– IV. Title: Mirror symmetry 5.

QC174.17.S9M5634 2006
516.3′62—dc22 2006047914

∞ The paper used in this book is acid-free and falls within the guidelines established to ensure permanence and durability.
Visit the AMS home page at http://www.ams.org/
Visit the International Press home page at URL: http://www.intlpress.com/
Visit the Centre de Recherches Mathématiques home page at URL: http://www.crm.umontreal.ca

10 9 8 7 6 5 4 3 2 1 11 10 09 08 07 06

CONTENTS

Part IV: Geometric Analytic Aspects: Picard–Fuchs Equations

PREFACE

In 2003, the Pacific Institute for Mathematical Sciences (PIMS), in cooperation with the Mathematical Sciences Research Institute (MSRI) Berkeley, inaugurated the Banff International Research Station for Mathematical Innovation and Discovery (BIRS) as one of the North American counterparts for the research institutes in Europe, e.g., Oberwolfach and Luminy. Our proposed workshop "Calabi–Yau Varieties and Mirror Symmetry" was selected to run in the period of December 6 to 11, 2003.

The organizers of the workshop were Victor Batyrev (Tübingen), Shinobu Hosono (Tokyo), James D. Lewis (Alberta), Bong H. Lian (Brandeis), Shing-Tung Yau (Harvard), Noriko Yui (Queen's) and Don Zagier (Max-Planck-Institüt, Bonn/College de France).

This workshop was the second in the series, following the first one held at the Fields Institute in the summer of 2001. The Proceedings of the first workshop has been published in the Fields Communication Series Volume **38** under the title "Calabi–Yau Varieties and Mirror Symmetry", American Mathematical Society/Fields Institute, in the fall of 2003.

We have witnessed tremendous progress since our first workshop, and the present volume is the testimony to some of the particularly noteworthy recent developments on Calabi–Yau varieties and mirror symmetry.

Since its discovery in the early 1990s, mirror symmetry, or more generally, string theory, has literally exploded onto the mathematical scene, encompassing many branches of mathematics and mathematical physics. It has pointed to deep connections between many subjects previously thought unrelated, for instance, provocatively put, number theory might be considered one aspect of mirror symmetry!

The participants were a mixture of mathematicians and mathematical physicists actively working on mirror symmetry and related fields.

This volume is a record of our five-day workshop at BIRS, and it contains twenty-four articles about and around Mirror Symmetry from mathematical and physics perspectives. Some articles in this volume are based on the talks presented at the workshop, while others report on results obtained subsequently. Also, the editors have solicited articles from a number of people who could not, for one or another reason, physically attend the workshop, but the subject matters are in perfect accordance with the theme of the workshop.

The papers in this volume are roughly classified into four sub-categories:

- Part I : Arithmetic Aspects;
- Part II : Geometric Aspects;
- Part III : Differential Geometric and Mathematical Physics Aspects, and
- Part IV : Geometric Analytic Aspects (Picard–Fuchs Differential Equations).

In the sub-category of Arithmetic Aspects, a large number of articles are devoted to discussions on mirror symmetry from arithmetic points of view. The topics include calculation of zeta-functions and L-series of Calabi–Yau varieties, modularity of Calabi–Yau varieties, especially modularity of non-rigid Calabi–Yau threefolds, arithmetic mirror symmetry (conjectures), as well as Mahler measures of Calabi–Yau varieties.

Under Geometric Aspects, a number of articles address transcendental methods in the study of algebraic cycles and Chow groups of zero-cycles, while others are concerned with variations of Hodge structures, structure of families, and Calabi–Yau varieties with complex multiplication. Also, the homological mirror symmetry (conjecture) and derived categories are explored in some of the articles. Mirror symmetry for toric varieties is another topic discussed here, for instance, computations of the integral cohomology groups of Calabi–Yau threefolds from hypersurfaces in 4-dimensional Gorenstein toric Fano varieties. Also the aspects of the even/odd cohomology exchange under mirror symmetry for Calabi–Yau threefolds are addressed.

The sub-category of Differential Geometric and Mathematical Physics Aspects contains articles which address boundary renormalization group flows of Landau-Ginzburg models with an unbroken $N = 2$ supersymmetry, local mirror symmetry, special Lagrangian torus fibrations, Gromov–Witten invariants and relative Gromov–Witten invariants, Gromov–Witten theory for conifolds and orientifolds, among others.

Under Geometric Analytic Aspects, the majority of the papers are concerned with Picard–Fuchs differential equations from various points of view. The topics include computation of Picard–Fuchs differential equations, mirror maps, monodromy, structure of families, open strings, and geometric applications of Picard–Fuchs equations. Also, mirror symmetry and its relation to integral variations of Hodge structure, as well as the geometric realization problem are discussed here.

On behalf of the organizing committee, we thank the BIRS director at the time of our workshop, Robert V. Moody, and the BIRS staff members for their warm welcome and superlative support.

The editors wish to express their gratitude to all the contributors for their effort in preparing their manuscripts for this volume.

All papers in this volume have gone through a rigorous refereeing process. We are deeply grateful to all our referees for their immense efforts and discipline in evaluating the papers.

The editors are especially grateful to Arthur Greenspoon of Mathematical Reviews for his expert editing of the manuscript. He read through all the papers of this volume and made numerous comments and suggestions for their improvements.

Finally, the editors would like to thank Debbie Iscoe and Carl Riehm of the Fields Institute for their technical help in producing the final version of the manuscript.

J. D. Lewis, S.-T. Yau and Noriko Yui, the Editors

PART I

ARITHMETIC ASPECTS

AMS/IP Studies in Advanced Mathematics
Volume 38, 2006

Mahler's Measure and L-Series of $K3$ Hypersurfaces

Marie José Bertin

ABSTRACT. We express the logarithmic Mahler measure of three polynomials defining $K3$ hypersurfaces in $\mathbb{P}^3$ in terms of the L-series of the corresponding hypersurface.

1. Introduction

Let $\mathbb{T}^2 = \{(x, y) \in \mathbb{C}^2 : |x| = |y| = 1\}$ be the 2-torus. Let $P(x, y)$ be a polynomial in two variables with integer coefficients. Suppose that P does not vanish on $\mathbb{T}^2$. If P is associated to an elliptic curve E, such that the faces P_F of P (defined in terms of the Newton polygon of P) are cyclotomic, the following relation between the logarithmic Mahler measure of P, $m(P)$, and the L-series of the elliptic curve E,

$$m(P) \doteq \frac{N}{4\pi^2} L(E, 2) = L'(E, 0)$$

is conjectured to hold. ($\doteq$ means equality up to a rational coefficient and N is the conductor of E [**5**] [**11**] [**12**]).

Some of these formulae have been proved by Rodriguez-Villegas [**11**], for some curves E with complex multiplication. His proof uses the expression of the logarithmic Mahler measure of a polynomial in terms of Eisenstein-Kronecker series. For more examples see [**3**].

We seek the analog of these formulae for P defining a $K3$ hypersurface. For this purpose, we give some examples of formulae relating the logarithmic Mahler measure of P and the L-series of the $K3$ hypersurface. The polynomials P in question belong to two families of $K3$ hypersurfaces whose logarithmic Mahler measures have been studied in [**4**]. As explained in [**4**], the derivative of the logarithmic Mahler measure with respect to the parameter is a period of the corresponding $K3$ hypersurface; thus, only the transcendental lattice is relevant for the Mahler measure. The generic member of both families has Picard number 19. Special members of the families have Picard number 20, and thus a 2-dimensional transcendental lattice. The three polynomials below are among these special members. The L-functions

---

2000 *Mathematics Subject Classification.* Primary 11G40; Secondary 14J28, 14G10, 11R09, 11F23.

Key words and phrases. Modular Mahler's Measure, Eisenstein-Kronecker Series, K3 hypersurface, L-series, modular form.

correspond to just a 2-dimensional piece of the 22-dimensional H^2 of the $K3$ hypersurfaces. Since these $K3$ hypersurfaces have Picard number 20, this 2-dimensional piece is their transcendental lattice.

We first recall the expression of the Mahler measure of two families of polynomials defining $K3$ hypersurfaces, in terms of Eisenstein–Kronecker series, together with the results of Shioda–Inose giving the computation of the determinant of the transcendental lattice. Then, we give the L-series of some $K3$ hypersurfaces and prove the following theorem.

THEOREM 1. *Let Y_2, S_0 and S_{12} be the $K3$ hypersurfaces associated to the corresponding polynomials P_2, Q_0 and Q_{12},*

$$P_2 = x^2yz + xy^2z + xyz^2 + t^2(xy + xz + yz) - 2xyzt$$

$$Q_0 = (x + y + z + t)[t(xy + xz + yz) + xyz] - 4xyzt$$

$$Q_{12} = (x + y + z + t)[t(xy + xz + yz) + xyz] - 16xyzt.$$

If $L(Y, s)$ denotes the L-series of the hypersurface Y, one gets the following relations.

(1)

$$L(Y_2, s) = L(f, s)$$

where $f(z) = \eta(z)^2\eta(2z)\eta(4z)\eta(8z)^2 \in S_3(\Gamma_0(8), \epsilon_8)$, ϵ_8 being some Dirichlet character modulo 8,

$$L(S_0, s) = L(S_{12}, s) = L(g, s)$$

where $g(z) = \eta(2z)^3\eta(6z)^3 \in S_3(\Gamma_0(12), \epsilon_{12})$, ϵ_{12} being some Dirichlet character modulo 12, $L(f, s)$ and $L(g, s)$ denote the Mellin transform of the modular forms f and g.

(2)

$$m(P_2) = 4\frac{|\det T_{Y_2}|^{3/2}}{4\pi^3}L(Y_2, 3)$$

$$m(Q_0) = 2\frac{|\det T_{S_0}|^{3/2}}{4\pi^3}L(S_0, 3)$$

$$m(Q_{12}) = \frac{1}{2}\frac{|\det T_{S_{12}}|^{3/2}}{4\pi^3}L(S_{12}, 3)$$

where T_Y denotes the transcendental lattice of the surface Y.

2. Definitions and previous results

2.1. Logarithmic Mahler measure.

DEFINITION 1. *The logarithmic Mahler measure $m(P)$ of a Laurent polynomial $P \in \mathbb{C}[X_1^{\pm}, ..., X_n^{\pm}]$ is defined by*

$$m(P) := \frac{1}{(2\pi i)^n}\int_{\mathbb{T}^n} \log|P(x_1, ..., x_n)|\frac{dx_1}{x_1}\cdots\frac{dx_n}{x_n}$$

where $\mathbb{T}^n$ is the n-torus $\{(x_1, ..., x_n) \in \mathbb{C}^n : |x_1| = ... = |x_n| = 1\}$.

From the definition, one can derive the following. If P is a homogeneous polynomial of degree d in the variables X_0, X_1, $\cdots$, X_n, defining an hypersurface in $\mathbb{P}^n$, one can associate to P the Laurent polynomial $\tilde{P}$ in the variables $x_1 = \frac{X_1}{X_0}$, $\cdots$,

$x_n = \frac{X_n}{X_0}$. Note that here we are dealing with a variety together with an embedding in $\mathbb{P}^n$. Isomorphic but differently embedded varieties are likely to get different Mahler measures.

Let P_k and Q'_k denote the following Laurent polynomials:

$$P_k = X + \frac{1}{X} + Y + \frac{1}{Y} + Z + \frac{1}{Z} - k$$

and

$$Q'_k = X + \tfrac{1}{X} + Y + \tfrac{1}{Y} + Z + \tfrac{1}{Z} + XY + \frac{1}{XY} + ZY + \frac{1}{ZY} + XYZ + \frac{1}{XYZ} - k.$$

We recall the following theorem [**4**].

THEOREM 2. *(1) Let us write* $k = t + \frac{1}{t}$ *for*

$$t = (\frac{\eta(\tau)\eta(6\tau)}{\eta(2\tau)\eta(3\tau)})^6,$$

where η *is the Dedekind eta function*

$$\eta(\tau) = e^{\frac{\pi i \tau}{12}} \prod_{n\geq 1}(1 - e^{2\pi i n\tau}).$$

Then

$$\begin{aligned} m(P_k) = &\frac{\Im\tau}{8\pi^3}\{\Sigma'_{m,n}(-4(2\Re\frac{1}{(m\tau+n)^3(m\bar{\tau}+n)} + \frac{1}{(m\tau+n)^2(m\bar{\tau}+n)^2}) \\ &+ 16(2\Re\frac{1}{(2m\tau+n)^3(2m\bar{\tau}+n)} + \frac{1}{(2m\tau+n)^2(2m\bar{\tau}+n)^2}) \\ &- 36(2\Re\frac{1}{(3m\tau+n)^3(3m\bar{\tau}+n)} + \frac{1}{(3m\tau+n)^2(3m\bar{\tau}+n)^2}) \\ &+ 144(2\Re\frac{1}{(6m\tau+n)^3(6m\bar{\tau}+n)} + \frac{1}{(6m\tau+n)^2(6m\bar{\tau}+n)^2}))\}. \end{aligned}$$

(2) Similarly, if $k = -(t + \frac{1}{t}) - 2$ *and*

$$t = \frac{\eta(3\tau)^4\eta(12\tau)^8\eta(2\tau)^{12}}{\eta(\tau)^4\eta(4\tau)^8\eta(6\tau)^{12}}$$

then

$$\begin{aligned} m(Q'_k) = &\frac{\Im\tau}{8\pi^3}\{\Sigma'_{m,n}(2(2\Re\frac{1}{(m\tau+n)^3(m\bar{\tau}+n)} + \frac{1}{(m\tau+n)^2(m\bar{\tau}+n)^2}) \\ &- 32(2\Re\frac{1}{(2m\tau+n)^3(2m\bar{\tau}+n)} + \frac{1}{(2m\tau+n)^2(2m\bar{\tau}+n)^2}) \\ &- 18(2\Re\frac{1}{(3m\tau+n)^3(3m\bar{\tau}+n)} + \frac{1}{(3m\tau+n)^2(3m\bar{\tau}+n)^2}) \\ &+ 288(2\Re\frac{1}{(6m\tau+n)^3(6m\bar{\tau}+n)} + \frac{1}{(6m\tau+n)^2(6m\bar{\tau}+n)^2}))\}. \end{aligned}$$

The summation Σ' *means that the sum is for rational integers* $(m, n) \neq (0, 0)$.

2.2. Some known facts concerning the L-series of a $K3$-surface. For definitions concerning $K3$-surfaces, see for example [**4**].

Let V be a projective variety of dimension d over a finite field k with q elements. Suppose moreover that V is geometrically irreducible (i.e. it is irreducible over some algebraic closure of k). Let N_n denote the number of points of V in a k-extension of degree n. Then the zeta function attached to V is defined by

$$Z_V(T) := exp(\sum_{n\geq 1} N_n \frac{T^n}{n}).$$

By Weil's conjectures (1949) proved by Dwork (1960) we know that $Z_V(T)$ is a rational function with coefficients in $\mathbb{Q}$ satisfying the functional equation

$$Z_V(\frac{1}{q^d T}) = \pm (q^{d/2}T)^c Z_V(T)$$

for some $c \in \mathbb{N}$.

If V is an algebraic $K3$-surface defined over $\mathbb{Q}$, then, for almost all primes p, its reduction modulo p is, after desingularization, again a $K3$-surface, which we denote by V_p. Moreover, one can show that $Z_{V_p}(T)$ is of the form

$$Z_{V_p}(T) = \frac{1}{(1-T)(1-p^2T)P_2(T)}$$

where $P_2(T)$ is a polynomial of degree 22 satisfying $P_2(0) = 1$ and such that

$$P_2(T) = Q_p(T)R_p(T)$$

where $R_p(T)$ comes from the algebraic cycles with $R_p(\frac{T}{p}) \in \mathbb{Z}[T]$ and $Q_p(T)$ comes from the transcendental cycles. Under some conditions, the $H^{1,1}$ part of the Hodge decomposition of H^2 has dimension 20. In that case, the $K3$-surface is singular, that is to say of Picard number $\rho = 20$, and its transcendental lattice is of dimension 2. Hence $Q_p(T)$ is of degree 2 for all primes p not dividing an integer N. The Langlands philosophy says that one can define the $Q_p(T)$ for p dividing N in such a manner that

$$Z(V,s) := \prod_p \frac{1}{Q_p(p^{-s})} = \sum_{n\geq 1} \frac{a_n}{n^s}$$

is the Dirichlet series of a cusp form.

The number N is related to the absolute value of the determinant of the transcendental lattice of the $K3$-surface.

For example [**14**], if the affine equation of the surface is given by

$$t^2(x+y)(x+z)(y+z) + xyz = 0,$$

the weight of the cusp form is 3 and the level is determined by the determinant of the transcendental lattice. In that case, the cusp form is

$$f(z) = \eta(z)^2\eta(2z)\eta(4z)\eta(8z)^2 \in S_3(\Gamma_0(8), \epsilon_8)$$

where ϵ_8 is some character modulo 8.

2.3. Two results of Shioda. (I) For the first result used here, we refer the reader to Shioda–Inose's paper [**15**].

Let $\Phi : X \to \mathbb{P}^1$ be an elliptic surface with a section and consider the sections of this elliptic fibration, i.e. determined by the rational points of the corresponding elliptic curve defined over the field of rational functions in s.

Denote by $r(\Phi)$ the rank of the group of sections. Then the Picard number $\rho(X)$ satisfies the equation

$$\rho(X) = r(\Phi) + 2 + \sum_{\nu=1}^{h}(m_\nu - 1) \tag{1}$$

where h is the number of singular fibers and m_ν the number of irreducible components of the corresponding singular fiber.

In addition, if $r(\Phi) = 0$ and if $n(\Phi)$ is the order of the group of sections of Φ, one gets

$$\mid \det T_X \mid = \mid \det S_X \mid = (\prod_{\nu=1}^{h} m_\nu^{(1)})/n(\Phi)^2 \tag{2}$$

where $m_\nu^{(1)}$ denotes the number of simple components of the divisor $D_\nu = \Phi^{-1}(t_\nu)$ of a singular fiber, S_X the sublattice of algebraic cycles in $H_2(X, \mathbb{Z})$ and $T_X = S_X^\perp$ the orthogonal complement of S_X in $H_2(X, \mathbb{Z})$, called the transcendental lattice.

(II) For the second result we need, we refer the reader to Shioda's article [**13**].

Let $(S, \Phi, \mathbb{P}^1)$ be an elliptic surface with a section Φ, without exceptional curves of the first kind. Denote by $NS(S)$ the group of algebraic equivalence classes of divisors of S. Let u be the generic point of $\mathbb{P}^1$ and $\Phi^{-1}(u) = E$ the elliptic curve defined over $K = \mathbb{C}(u)$ with a K-rational point $o = o(u)$. Then $E(K)$ is an abelian group of finite type provided that $j(E)$ is transcendental over $\mathbb{C}$. Let r be the rank of $E(K)$ and $s_1, ..., s_r$ be generators of $E(K)$ modulo torsion. Also, the torsion group $E(K)_{tors}$ is generated by at most two elements t_1 of order e_1 and t_2 of order e_2 such that $1 \leq e_2$, $e_2|e_1$ and $\mid E(K)_{tors} \mid = e_1 e_2$. The group $E(K)$ of K-rational points of E is canonically identified with the group of sections of S over $\mathbb{P}^1(\mathbb{C})$. For $s \in E(K)$, we denote by (s) the image curve in S of the section corresponding to s. Let us define

$$D_\alpha := (s_\alpha) - (o) \quad 1 \leq \alpha \leq r$$

$$D'_\beta := (t_\beta) - (o) \quad \beta = 1, 2.$$

Consider now the singular fibers of S over $\mathbb{P}^1$. We set

$$\Sigma := \{v \in \mathbb{P}^1 \ : \ C_v = \Phi^{-1}(v) \ \text{ is a singular fiber}\}$$

and for each $v \in \Sigma$, we denote by $\Theta_{v,i}$ $(0 \leq i \leq m_v - 1)$ the irreducible components of C_v. Let $\Theta_{v,0}$ be the unique component of C_v passing through $o(v)$. One gets

$$C_v = \Theta_{v,0} + \sum_{i \geq 1} \mu_{v,i}\Theta_{v,i}, \qquad \mu_{v,i} \geq 1.$$

Let A_v be the matrix of order $m_v - 1$ whose entry of index (i, j) is $(\Theta_{v,i}\Theta_{v,j})$, $i, j \geq 1$, where (DD') is the intersection number of the divisors D and D' along S. Finally f will denote a non-singular fiber, i.e. $f = C_{u_0}$ for $u_0 \notin \Sigma$.

THEOREM 3. *The Néron–Severi group $NS(S)$ of the elliptic surface S is generated by the following divisors:*

$$f, \Theta_{v,i} \quad (1 \leq i \leq m_v - 1, \quad v \in \Sigma)$$
$$(o), D_\alpha \quad (1 \leq \alpha \leq r), \quad D'_\beta \quad (\beta = 1, 2).$$

The only relations between these divisors are at most the two relations

$$e_\beta D'_\beta \approx e_\beta(D'_\beta(o))f + \sum_{v \in \Sigma} (\Theta_{v,1}, ..., \Theta_{v,m_v-1}) e_\beta A_v^{-1} \begin{pmatrix} (D'_\beta \Theta_{v,1}) \\ . \\ . \\ . \\ (D'_\beta \Theta_{v,m_v-1}) \end{pmatrix}$$

where $\approx$ stands for algebraic equivalence.

3. Proof of Theorem 1

3.1. Determinant of the transcendental lattice.

THEOREM 4.

(1) *The determinant of the transcendental lattice of Y_2 is equal to -8.*
(2) *For S_0 and S_{12} the determinant is equal to -12.*

PROOF. To compute the determinant of the transcendental lattice of Y_2 and S_0 we can find a good elliptic fibration so that 2.3 (I) applies. For S_{12}, we need 2.3 (II). The proof, however involves, several lemmas which we will prove later on.

The polynomials defining respectively Y_2, S_0 and S_{12} are related to Beauville's famous list [**1**] , more precisely to the elliptic pencil

$$(x+y)(x+z)(y+z) + uxyz = 0.$$

Let us recall the singular fibers of this pencil

at	$u = \infty$	$xyz = 0$	of type	I_6
at	$u = 0$	$(x+y)(x+z)(y+z) = 0$	of type	I_3
at	$u = 1$	$(x+y+z)(xy+xz+yz) = 0$	of type	I_2
at	$u = -8$	$(x+y)(y+z)(z+x) - 8xyz = 0$	of type	I_1.

Now cutting Y_2, (resp. S_0 and S_{12}) by the hyperplane $x + y + z = st$ (resp. $t = s(x+y+z)$), one gets, after simplification, the corresponding elliptic fibration

$$\begin{array}{ll} (A) & (x+y)(x+z)(y+z) + (s-1)^2 xyz = 0 \\ (B) & s(s+1)(x+y)(x+z)(y+z) + (s-1)^2 xyz = 0 \\ (C) & s(s+1)(x+y)(x+z)(y+z) + (s^2 - 14s + 1)xyz = 0. \end{array}$$

(i) The surface Y_2

From above, taking $u = (s-1)^2$, we deduce from (A) the singular fiber structure, that is

at	$s = \infty$	$xyz = 0$	$t = 0$	of type	I_{12}
at	$s = 1$	$(x+y)(x+z)(y+z) = 0$	$x+y+z = t$	of type	I_6
at	$s = 0$	$(x+y+z)(xy+xz+yz) = 0$	$x+y+z = 0$	of type	I_2
at	$s = 2$	$(x+y+z)(xy+xz+yz) = 0$	$x+y+z = 2t$	of type	I_2
at	$s = \alpha$	$(x+y)(x+z)(y+z) - 8xyz = 0$	$x+y+z = \alpha t$	of type	I_1
at	$s = \beta$	$(x+y)(x+z)(y+z) - 8xyz = 0$	$x+y+z = \beta t$	of type	I_1,

where α and β are the roots of the polynomial $x^2 - 2x + 9$. From 2.3 (I), there follows $r = 0$ and with a Weierstrass model of Beauville's pencil

$$Y^2 + uXY - u(u-1)Y = X^3 - (2u-1)X^2 + u(u-1)X,$$

we find that the origin is a point of 6-torsion. According to Miranda [**8**], $n(\Phi) = 6$ and using 2.3 (I)

$$|\det T_X| = \frac{12 \times 6 \times 2 \times 2}{6^2} = 8.$$

Since $\det T_X$ is negative, $\det T_X = -8$. Taking, as in Shioda–Inose [**15**], a suitable basis of the transcendental lattice, one deduces that the surface Y_2 corresponds to the unique class modulo $GL_2(\mathbb{Z})$ of the matrix

$$\begin{pmatrix} 2 & 0 \\ 0 & 4 \end{pmatrix}$$

belonging to the set of matrices with integer coefficients

$$\begin{pmatrix} 2a & b \\ b & 2c \end{pmatrix}$$

such that $b^2 - 4ac = -8$.

We have also proved the following lemma.

LEMMA 1. *In the reduction modulo p, $p \neq 2$, $p \neq 3$, the singular fibers are of the same type as in the complex case. Their irreducible components not passing through the zero section are defined over the base field.*

(ii) The surface S_0

From (B), with $u = \frac{(s-1)^2}{s(s+1)}$, we deduce the singular fiber structure

at	$s = 0$	$xyz = 0$	$t = 0$	of type	I_6
at	$s = -1$	$xyz = 0$	$x + y + z + t = 0$	of type	I_6
at	$s = 1$	$(x+y)(x+z)(y+z) = 0$	$x + y + z - t = 0$	of type	I_6
at	$s = \infty$	$xy + xz + yz = 0$	$x + y + z = 0$	of type	I_2
at	$s = -\frac{1}{3}$	$(x+y)(x+z)(y+z) - 8xyz = 0$	$x + y + z + 3t = 0$	of type	I_2
at	$s = \frac{1}{3}$	$(x+y+z)(xy+xz+yz) = 0$	$x + y + z - 3t = 0$	of type	I_2.

From 2.3 (I), there follows $r = 0$. Now using the previous Weierstrass model for $u = \frac{(s-1)^2}{s(s+1)}$ and MAPLE, we find for the elliptic curve over $\mathbb{Q}(s)$ a torsion subgroup of order 12 and with the help of Miranda [**8**], $n(\Phi) = 12$; so as above

$$|\det S_0| = \frac{6^3 \times 2^3}{12^2} = 12$$

and $\det S_0 = -12$.

Moreover, we can state the following lemma.

LEMMA 2. *Modulo p and for $p \neq 2$, $p \neq 3$, the singular fibers and their components are of the same type as in the complex case. They are always defined over the base field.*

(iii) The surface S_{12}

We apply the result 2.3 (II). The surface S_{12} contains eleven double points:

$$\begin{array}{llll} P_{01}=(1,0,0,0), & P_{02}=(0,1,0,0), & P_{03}=(0,0,1,0), & P_{04}=(0,0,0,1), \\ P_{12}=(1,-1,0,0), & P_{13}=(1,0,-1,0), & P_{14}=(1,0,0,-1), & P_{23}=(0,1,-1,0), \\ P_{24}=(0,1,0,-1), & P_{34}=(0,0,1,-1), & E=(1,1,1,1). & \end{array}$$

As above, with $u=\frac{s^2-14s+1}{s(s+1)}$, we get from (C) the singular fiber structure of S_{12}.

$$\begin{array}{llll} \text{at } s=0 & xyz=0 & t=0 & \text{of type } I_6 \\ \text{at } s=-1 & xyz=0 & x+y+z+t=0 & \text{of type } I_6 \\ \text{at } s=\alpha & (x+y)(x+z)(y+z)=0 & \alpha(x+y+z)-t=0 & \text{of type } I_3 \\ \text{at } s=\beta & (x+y)(x+z)(y+z)=0 & \beta(x+y+z)-t=0 & \text{of type } I_3 \\ \text{at } s=\infty & (x+y+z)(xy+xz+yz)=0 & x+y+z=0 & \text{of type } I_2 \\ \text{at } s=\frac{1}{15} & (x+y+z)(xy+xz+yz)=0 & x+y+z-15t=0 & \text{of type } I_2 \\ \text{at } s=\frac{1}{3} & (x+y)(x+z)(y+z)-8xyz=0 & x+y+z-3t=0 & \text{of type } I_2. \end{array}$$

Here, α and β are the roots of $x^2-14x+1=0$.

The elliptic fibration (C) has a Weierstrass model obtained from a projective transform

$$Y^2+dXY=X^3-s(s+1)(s^2-16s-1)X^2+s^3(s+1)^3(15s-1)X$$

with

$$\begin{cases} x=(Xd+Y)(s^2-14s+1) \\ y=-Y(s^2-14s+1) \\ z=s(s+1)d[s(s+1)(15s-1)Z-X] \end{cases}$$

Thanks to MAPLE-APECS, we find a torsion group isomorphic to $\mathbb{Z}/6\mathbb{Z}$ and a 6-torsion point $(s(s+1)(15s-1),-s(15s-1)(s+1)d,1)$. Using Miranda [**8**], we deduce $n(\Phi)=6$ and from formula 2.3 (I), $r=1$. So, to apply the Theorem 3, we need a divisor D_1. Instead of finding a point of infinite order on the elliptic curve, we look for a line of infinite order on the surface. We denote by S the following line on S_{12}:

$$S=(-\lambda,\alpha\mu,\mu,\lambda).$$

Thus, $D_1=(S)-(0)$. It can be shown as in [**10**] that it is a section of infinite order. Let us denote by (s_6) the section defined by the 6-torsion point

$$s_6=(0,1,0,s).$$

Thus, $D_1'=(s_6)-(0)$.

Let us now describe each singular fiber and its intersection with s_6 and S.

For $s=0$, the irreducible components of the special fiber I_6 are $\Theta_{0,0}$ defined by $z=0$ and $t=0$, $\Theta_{0,2}$ defined by $x=0$ and $t=0$, $\Theta_{0,4}$ defined by $y=0$ and $t=0$, $\Theta_{0,1}$ the blow-up of the point $P_{0,2}$, $\Theta_{0,3}$ the blow-up of the point $P_{0,3}$, $\Theta_{0,5}$ the blow-up of the point $P_{0,1}$. One can easily see that S cuts $\Theta_{0,2}$ and s_6 cuts $\Theta_{0,1}$.

For $s=-1$, the irreducible components of the special fiber I_6 are $\Theta_{-1,0}$ defined by $z=0$ and $t=-(x+y)$, $\Theta_{-1,2}$ defined by $x=0$ and $t=-(x+y)$, $\Theta_{-1,4}$ defined by $y=0$ and $t=-(x+y)$, $\Theta_{-1,1}$ the blow-up of the point $P_{2,4}$, $\Theta_{-1,3}$ the blow-up of the point $P_{3,4}$, $\Theta_{-1,5}$ the blow-up of the point $P_{1,4}$. Now, S cuts $\Theta_{-1,5}$ and s_6 cuts $\Theta_{-1,1}$.

For $s = \alpha$, one gets a fiber of type I_3 with $\Theta_{\alpha,0}$ defined by $x + y = 0$ and $t = \alpha(x + y + z)$, $\Theta_{\alpha,1}$ defined by $y + z = 0$ and $t = \alpha(x + y + z)$ and $\Theta_{\alpha,2}$ defined by $x + z = 0$ and $t = \alpha(x + y + z)$. Thus, S cuts $\Theta_{\alpha,0}$ at $(-\alpha, \alpha, 1, \alpha)$ and s_6 cuts $\Theta_{\alpha,2}$.

For $s = \beta$, there is a fiber of type I_3 with $\Theta_{\beta,0}$ defined by $x + y = 0$ and $t = \beta(x + y + z)$, $\Theta_{\beta,1}$ defined by $y + z = 0$ and $t = \beta(x + y + z)$ and $\Theta_{\beta,2}$ defined by $x + z = 0$ and $t = \beta(x + y + z)$. Thus, S cuts $\Theta_{\beta,2}$ at $(-\beta, 1, \beta, \beta)$ and s_6 cuts $\Theta_{\alpha,2}$.

For $s = \infty$, there is a fiber of type I_2 with $\Theta_{\infty,1}$ the blow-up of $P_{0,4}$ and $\Theta_{\infty,0}$ the rational curve $(1+\tau+\tau^2, \tau(1+\tau+\tau^2), -(1+\tau)(1+\tau+\tau^2), 15\tau(1+\tau))$. Now, S cuts $\Theta_{\infty,0}$ at $(-(\alpha+1), \alpha, 1, \frac{15\alpha(\alpha+1)}{\alpha^2+\alpha+1})$, a point corresponding to $\tau = \frac{-\alpha}{\alpha+1}$. But, s_6 cuts $\Theta_{\infty,1}$.

For $s = 1/15$, the irreducible components of the special fiber I_2 are $\Theta_{1/15,0}$ defined by $x+y+z = 0$ and $t = 0$, $\Theta_{1/15,1}$ the rational curve defined by $15t = x+y+z$ and $xy + xz + yz = 0$. Thus, S cuts $\Theta_{1/15,1}$ along $(-(\alpha+1)/16, \alpha, 1, (\alpha+1)/16)$ and s_6 cuts also $\Theta_{1/15,1}$.

For $s = 1/3$, the irreducible components of the special fiber I_2 are $\Theta_{1/3,0}$ the blow-up of E, $\Theta_{1/3,1}$ the rational curve defined by $3t = x + y + z$ and $t(xy + xz + yz) - 3xyz = 0$. It follows that S cuts $\Theta_{1/3,1}$ at $(-(\alpha+1)/4, \alpha, 1, (\alpha+1)/4)$ and s_6 cuts also $\Theta_{1/3,1}$.

Now let us consider the 20 divisors in the following order

$$(o),\ D_1,\ f,\ \Theta_{0,i}\,(1 \le i \le 5),\ \Theta_{-1,i}\,(1 \le i \le 5),\ \Theta_{\alpha,1},\ \Theta_{\alpha,2},$$
$$\Theta_{\beta,1},\ \Theta_{\beta,2},\ \Theta_{\infty,1},\ \Theta_{1/15,1},\ \Theta_{1/3,1}.$$

From the above facts, the intersection matrix is the following:

$$\begin{pmatrix}
-2 & 0 & 1 & 0 & 0 & 0 & 0 & 0 & 0 & 0 & 0 & 0 & 0 & 0 & 0 & 0 & 0 & 0 & 0 & 0\\
0 & -2 & 1 & 0 & 1 & 0 & 0 & 0 & 0 & 0 & 0 & 0 & 1 & 0 & 0 & 0 & 1 & 0 & 1 & 1\\
1 & 1 & 0 & 0 & 0 & 0 & 0 & 0 & 0 & 0 & 0 & 0 & 0 & 0 & 0 & 0 & 0 & 0 & 0 & 0\\
0 & 0 & 0 & -2 & 1 & 0 & 0 & 0 & 0 & 0 & 0 & 0 & 0 & 0 & 0 & 0 & 0 & 0 & 0 & 0\\
0 & 1 & 0 & 1 & -2 & 1 & 0 & 0 & 0 & 0 & 0 & 0 & 0 & 0 & 0 & 0 & 0 & 0 & 0 & 0\\
0 & 0 & 0 & 0 & 1 & -2 & 1 & 0 & 0 & 0 & 0 & 0 & 0 & 0 & 0 & 0 & 0 & 0 & 0 & 0\\
0 & 0 & 0 & 0 & 0 & 1 & -2 & 1 & 0 & 0 & 0 & 0 & 0 & 0 & 0 & 0 & 0 & 0 & 0 & 0\\
0 & 0 & 0 & 0 & 0 & 0 & 1 & -2 & 0 & 0 & 0 & 0 & 0 & 0 & 0 & 0 & 0 & 0 & 0 & 0\\
0 & 0 & 0 & 0 & 0 & 0 & 0 & 0 & -2 & 1 & 0 & 0 & 0 & 0 & 0 & 0 & 0 & 0 & 0 & 0\\
0 & 0 & 0 & 0 & 0 & 0 & 0 & 0 & 1 & -2 & 1 & 0 & 0 & 0 & 0 & 0 & 0 & 0 & 0 & 0\\
0 & 0 & 0 & 0 & 0 & 0 & 0 & 0 & 0 & 1 & -2 & 1 & 0 & 0 & 0 & 0 & 0 & 0 & 0 & 0\\
0 & 0 & 0 & 0 & 0 & 0 & 0 & 0 & 0 & 0 & 1 & -2 & 1 & 0 & 0 & 0 & 0 & 0 & 0 & 0\\
0 & 1 & 0 & 0 & 0 & 0 & 0 & 0 & 0 & 0 & 0 & 1 & -2 & 0 & 0 & 0 & 0 & 0 & 0 & 0\\
0 & 0 & 0 & 0 & 0 & 0 & 0 & 0 & 0 & 0 & 0 & 0 & 0 & -2 & 1 & 0 & 0 & 0 & 0 & 0\\
0 & 0 & 0 & 0 & 0 & 0 & 0 & 0 & 0 & 0 & 0 & 0 & 0 & 1 & -2 & 0 & 0 & 0 & 0 & 0\\
0 & 0 & 0 & 0 & 0 & 0 & 0 & 0 & 0 & 0 & 0 & 0 & 0 & 0 & 0 & -2 & 1 & 0 & 0 & 0\\
0 & 1 & 0 & 0 & 0 & 0 & 0 & 0 & 0 & 0 & 0 & 0 & 0 & 0 & 0 & 1 & -2 & 0 & 0 & 0\\
0 & 0 & 0 & 0 & 0 & 0 & 0 & 0 & 0 & 0 & 0 & 0 & 0 & 0 & 0 & 0 & 0 & -2 & 0 & 0\\
0 & 1 & 0 & 0 & 0 & 0 & 0 & 0 & 0 & 0 & 0 & 0 & 0 & 0 & 0 & 0 & 0 & 0 & -2 & 0\\
0 & 1 & 0 & 0 & 0 & 0 & 0 & 0 & 0 & 0 & 0 & 0 & 0 & 0 & 0 & 0 & 0 & 0 & 0 & -2
\end{pmatrix}$$

This matrix has determinant $-6^2 \times 12$.

We shall prove the following lemma.

LEMMA 3. *The Néron–Severi group $NS(S_{12})$ is generated by the classes of the following divisors:*

$$(0), D_1, D_1', f, \Theta_{0,i}, \Theta_{-1,i}\ (1 \le i \le 5), \Theta_{\alpha,1}, \Theta_{\alpha,2}, \Theta_{\beta,1}, \Theta_{\beta,2}, \Theta_{\infty,1}, \Theta_{1/15,1}, \Theta_{1/3,1}.$$

The only relation between these classes is:

$$-6D_1' = 12f + 5(\Theta_{0,1} + \Theta_{-1,1}) + 4(\Theta_{0,2} + \Theta_{-1,2}) + 3(\Theta_{0,3} + \Theta_{-1,3}) + 2(\Theta_{0,4} + \Theta_{-1,4})$$
$$+ \Theta_{0,5} + \Theta_{-1,5} + 3\Theta_{1/15,1} + 2(\Theta_{\alpha,1} + \Theta_{\beta,1}) + 4(\Theta_{\alpha,2} + \Theta_{\beta,2}).$$

The determinant of the Picard lattice is equal to -12.

PROOF. Let Λ be the lattice generated by the classes of the divisors given in the theorem and Λ' the sublattice generated by $(0), D_1, f$ and the components of the singular fibers. From the computation above

$$\det \Lambda' = -6^2 \times 12.$$

By Theorem 1, the only relation between these classes is the one given in the theorem. Therefore, the determinant of Λ is -12. Thus, we only need to show that Λ is not of index 2 in the Néron–Severi group. Otherwise, there would be an element of $S+ks_6,\ \ k=0,...,5$ belonging to $2\,MW(S_{12})$. So, taking the restrictions to the fibers over 0 and -1, one should get homomorphisms from $MW(S_{12})$ into $\mathbb{Z}/6\mathbb{Z} \times \mathbb{C}^*$. But, S (resp. s_6) cuts $\Theta_{0,2}$ (resp. $\Theta_{0,1}$). Similarly, S (resp. s_6) cuts $\Theta_{-1,5}$ (resp. $\Theta_{-1,1}$). And it is easy to see that the sections $S+ks_6$ do not restrict to elements which are simultaneously 2-divisible. Hence $\det NS(S_{12}) = -12$. □

From the above, we can also deduce the following lemma.

LEMMA 4.

1. *Over the complex field, all the singular fibers are of type* I_n.
2. *Under the reduction modulo p,* $p \neq 3$, $p \neq 5$, *the singular fibers are of the same type. Their components are defined over* $\mathbb{F}_p$ *if* $(\frac{3}{p}) = 1$. *If* $(\frac{3}{p}) = -1$, *the Frobenius exchanges the two fibers of type* I_3 *component by component. The components of the other singular fibers are defined over the base field.*
3. *Under the reduction modulo p, the cycle S and its conjugate* S' *are defined over the base field if* $(\frac{3}{p}) = 1$ *and over* $\mathbb{F}_{p^2}$ *otherwise.*
4. *If* $p = 5$, *we get the same conclusions except for* $s = \infty$ *and* $s = 1/15$ *replaced by a singular fiber of type* I_4 *at* $s = \infty$.

3.2. The L**-series.** If X denotes a $K3$ hypersurface, the L^*-series of X is defined by

$$L^*(X, s) := \prod {}^*Z(X/\mathbb{F}_p, p^{-s}) = \sum_{n \geq 1} \frac{a(n)}{n^s},$$

where the product is taken over the primes p of good reduction. Giving a suitable value to the local factors corresponding to the primes p dividing N, N being the absolute value of the determinant of the transcendental lattice, the L-series of the surface X can be expressed in terms of the Mellin transform of a modular form. Let us recall the formula

$$Z(X/\mathbb{F}_p, T) = \frac{1}{(1-T)(1-p^2T)P_2(T)}$$

with $P_2(T) \in \mathbb{Z}[T]$ of degree 22 and $P_2(0) = 1$.

LEMMA 5. *(i) For* Y_2 *and* $p \neq 2$,

$$P_2(T) = (1-pT)^{20} Q_2(T).$$

(ii) For S_0 *and* $p \neq 2$, $p \neq 3$,

$$P_2(T) = (1-pT)^{20} Q_2(T).$$

(iii) For S_{12} and $p \neq 2$, $p \neq 3$,

$$P_2(T) = (1 - pT)^{17}(1 - (\frac{3}{p})pT)^3 Q_2(T),$$

where $(\frac{3}{p})$ is the Legendre symbol.

PROOF. We deduce (i) and (ii) from Lemmas 1, 2.

For (iii), it follows from Lemma 4, if $(\frac{3}{p}) = -1$, that the components

$$\Theta_{\alpha,1},\ \Theta_{\alpha,2},\ \Theta_{\beta,1},\ \Theta_{\beta,2}$$

are not defined over the base field and that the Frobenius exchanges them. Similarly, S is exchanged with its conjugate S'. Therefore, $\{\Theta_{\alpha,1}, \Theta_{\alpha,2}, \Theta_{\beta,1}, \Theta_{\beta,2}\}$ is a space of dimension 6 invariant under the Frobenius F_p. In this subspace, we have $\det(1 - tF_p) = (1 - p^2t^2)^3$. In its orthogonal space of dimension 14 (for the intersection form) in the Néron–Severi lattice, the action of the Frobenius is the multiplication by p, since there is a basis of cycles defined over the base field. □

If

$$P_2(T) = (1 - pT)^{k_1}(1 + pT)^{k_2} Q_2(T)$$

with

$$Q_2(T) = (1 - \lambda_1 T)(1 - \lambda_2 T) \in \mathbb{Z}[T],$$

then the number N_q of $\mathbb{F}_q$-rational points of X for $q = p^r$ is given by the formula

$$N_q = 1 + q^2 + q(k_1 + (-1)^r k_2) + A_q$$

with $A_q = \lambda_1^r + \lambda_2^r$.

Now we deduce, from Lemma 5, the following corollary.

COROLLARY 1. *Let $p \neq 2,\ 3$.*

$$\begin{aligned} N_q(Y_2) &= 1 + q^2 + 20q + A_q(Y_2) \\ N_q(S_0) &= 1 + q^2 + 20q + A_q(S_0) \\ N_q(S_{12}) &= 1 + q^2 + 17q + 3q((\tfrac{3}{p}))^r + A_q(S_{12}). \end{aligned}$$

The computation of N_q will finally gives A_q and the L-series.

Let us recall the following lemma.

LEMMA 6. [**7**] *The $K3$ hypersurface Y_2 is a double covering of the elliptic pencil*

$$(x + y)(xy + z^2) + vxyz = 0,$$

isomorphic over $\mathbb{C}$ to another Beauville pencil [**1**]. *It corresponds to the modular elliptic surface for the modular group $\Gamma_1(8)$.*

There exists on some open set of Y_2 an automorphism σ of order 8 given by

$$\sigma : (x, y, z, t) \mapsto (-t\frac{yt + xz}{xt + zy}, \frac{(yt + xz)(t^2 + zy)}{(y + z)(xt + zy)}, z, t).$$

PROOF. For the proof we refer to [**7**]. Just notice that cutting by the hyperplane $z = st$, we get after simplification

$$s(x + y)(xy + t^2) + (s - 1)^2 xyt = 0.$$

□

LEMMA 7. *Let $q = p^r$ and $p \neq 2, 3$.*

$$\begin{array}{llcll}
(1) & N_q(Y_2) & \equiv & 5q + 3((\frac{2}{p})^r + 1) + 5 & \mod 8 \\
 & A_q(Y_2) & \equiv & q - q^2 + 3((\frac{2}{p})^r + 1) + 4 & \mod 8 \\
(2) & N_q(S_0) & \equiv & 0 & \mod 6 \\
 & A_q(S_0) & \equiv & -1 - q^2 + 4q & \mod 6 \\
(3) & N_q(S_{12}) & \equiv & 3q - 3 & \mod 6 \\
 & A_q(S_{12}) & \equiv & 2 + 4q - 3q(\frac{3}{p})^r - q^2 & \mod 6
\end{array}$$

PROOF. (1) Let us define

$$Y := \{(x, y, z, t) \in \mathbb{P}^3 \,:\, xyz(x + y + z) + t^2(xy + xz + yz) - 2xyzt = 0\}.$$

The $K3$ hypersurface Y_2 is obtained from Y by blowing up the 10 double points $P_{01}, P_{02}, P_{03}, P_{04}, P_{12}, P_{13}, P_{23}, A = (1, -1, 1, 1), B = (1, 1, -1, 1), C = (-1, 1, 1, 1)$. Let $\mathbb{P}(Y')$ be the projective variety corresponding to

$$Y' := \{(x, y, z, 1) \in \mathbb{P}^3 \,:\, (x, y, z, 1) \in Y, xyz \neq 0\}.$$

First we count the points in Y_2 with at least one zero coordinate. We get:

- $q - 2$ points with only one zero coordinate of the form $(x, y, z, 0)$ with $x + y + z = 0$,
- $6(q-1)$ points with only two zero coordinates (without intersection points) on the double lines $t = 0 \quad x = 0$, $t = 0 \quad y = 0$, $t = 0 \quad z = 0$,
- $3(q-1)$ points with only two zero coordinates (without intersection points) on the lines $x = 0 \quad y = 0$, $x = 0 \quad z = 0$, $y = 0 \quad z = 0$,
- 4 points with three zero coordinates P_{01}, P_{02}, P_{03}, P_{04},
- 3 points P_{01}, P_{02}, P_{03}, the intersection points of the double lines,
- $13q$ points coming from the blowing up of the 10 double points and P_{01}, P_{02} and P_{03}.

So we find $23q - 4$ points with at least one zero coordinate and

$$N_q(Y_2) = N_q(Y') + 23q - 4.$$

Now fix z, $z \neq 0, \pm 1$ and consider for x and $y \neq 0, \pm 1$

$$\sigma' : (x, y, z) \mapsto (-\frac{y + xz}{x + zy}, \frac{(y + xz)(1 + zy)}{(y + z)(x + zy)}, z).$$

Then we see that σ' is an automorphism of Y' such that

$$\sigma'^2 : (x, y, z) \mapsto (\frac{1}{y}, x, z).$$

And we can show that, if $y + xz \neq 0$, $x + yz \neq 0$, $z + y \neq 0$ and $1 + zy \neq 0$, then σ' is an automorphism of order 8 on Y'. Otherwise, the corresponding points, for x and $y \neq 0, \pm 1$, are of the form $(1, y, -y, 1)$, $(x, 1, -\frac{1}{x}, 1)$, $(x, -x, 1, 1)$, $(1, y, -\frac{1}{y}, 1)$, $(x, 1, -x, 1)$.

Now we consider the cases $z = \pm 1$, still with x and $y \neq \pm 1$. If $z = 1$, we get $(x + y)(xy + 1) = 0$; so, furthermore, points of the form $(x, -\frac{1}{x}, 1, 1)$.

If $z = -1$, one finds $x^2 y + x(y^2 - 4y + 1) + y = 0$ with discriminant $(y^2 - 6y + 1)(y - 1)^2$. The case $(\frac{2}{p}) = 1$ corresponds to 2 solutions in $\mathbb{F}_p$ of the form $(-1, y_1, -1, 1)$. But that case is discarded.

If $y^2 - 6y + 1$ is a square modulo p, combining σ'^2 and permutation of the two first variables, there are 8 solutions of the type $(x, y, -1, 1)$, $(\frac{1}{x}, y, -1, 1)$, $(y, x, -1, 1)$,

$(\frac{1}{y}, x, -1, 1)$, $(\frac{1}{y}, \frac{1}{x}, -1, 1)$, $(\frac{1}{x}, \frac{1}{y}, -1, 1)$, $(y, \frac{1}{x}, -1, 1)$ and $(x, \frac{1}{y}, -1, 1)$, since for $x = y$, $x = 1$.

Finally there remain the points with coordinates $x = \pm 1$, $y = \pm 1$. Among them we find the three double points $(1, 1, -1)$, $(1, -1, 1)$ and $(-1, 1, 1)$. There remain thus the points of the type $(-1, -1, z_1)$ with z_1 root of $z^2 - 6z + 1$ and their permuted. That give $3((\frac{2}{p})^r + 1)$ solutions. So,

$$N_q(Y') = 8t + 6(q-3) + 3((\frac{2}{p})^r + 1) + 3$$

and

$$N_q(Y_2) = 23q - 4 + 8t + 6(q-3) + 3((\frac{2}{p})^r + 1) + 3.$$

Thus

$$N_q(Y_2) \equiv 5q + 3((\frac{2}{p})^r + 1) + 5 \mod 8$$

and by corollary 1

$$A_q(Y_2) \equiv q - q^2 + 3((\frac{2}{p})^r + 1) + 4 \mod 8.$$

(2) We embed S_0 in $\mathbb{P}^4$ and set X_0 where X_0 is defined by

$$\{(x, y, z, t, u) \in \mathbb{P}^4 \ : \ x + y + z + t + u = 0, \ txyu + txzu + tyzu + xyzu + 4xyzt = 0\}.$$

We denote by $\mathbb{P}(X_0')$ the projective variety corresponding to

$$X_0' ; = \{(x, y, z, t) \in X_0 \ : \ xyztu \neq 0\}.$$

Notice that X_0 inherits 14 double points from the double points of S_0:

$$P_{01}, P_{02}, P_{03}, P_{04}, P_{12}, P_{13}, P_{14}, P_{23}, P_{24}, P_{34}, A, B, C, D = (-1, -1, -1, 1).$$

By blowing up these 14 double points one gets S_0.

First we count the points in X_0 with at least one zero coordinate.

If a point in X_0 has one zero coordinate, automatically it has a second zero coordinate but no more than three zero coordinates. Thus

$$N_q(X_0) = N_q(\mathbb{P}(X_0')) + 10q - 10$$

and

$$N_q(S_0) = N_q(\mathbb{P}(X_0')) + 24q - 10.$$

Since X_0' is fixed by any permutation of x, y, z, t, the number of points of X_0' for which x, y, z, t are all different is congruent to 0 modulo 6.

If only 2 coordinates are equal we must count the points of the form $(x, x, z, 1, u)$, $xz \neq 0$, $x \neq z$ and the points derived from them by permuting the first 3 coordinates. Now they satisfy

$$(z+1)(2x^2 + x(z+1) + 2z) = 0.$$

So one checks easily that the number of such points is

$$3(2q - 5) \equiv 3 \mod 6.$$

If the first two coordinates are equal, one gets the point D. So,

$$N_q(\mathbb{P}(X_0')) \equiv 4 \mod 6$$

By corollary 1 we deduce

$$A_q(S_0) \equiv -1 - q^2 + 4q \mod 6.$$

(3) Similarly, we define X_{12} as the set

$\{(x,y,z,t,u) \in \mathbb{P}^4 \,:\, x+y+z+t+u=0;\ txyu+txzu+tyzu+xyzu+16xyzt=0\}$.

Since X_{12} has 11 double points, one gets

$$N_q(S_{12}) = N_q(\mathbb{P}(X'_{12}) + 21q - 10,$$

if

$$X'_{12} := \{(x,y,z,t,u) \in X_{12} \,:\, xyztu \neq 0\}.$$

Since X'_{12} is fixed by any permutation of x, y, z, t, the number of points of X'_{12} for which x, y, z, t are all different is congruent to 0 modulo 6. If 2 coordinates are equal for points of the form $(x, x, z, 1, u)$ and points deduced by permutation of the three first coordinates, then

$$2(z+1)x^2 + x(z^2+14z+1) + 2z(z+1) = 0.$$

For $z \in \mathbb{F}_q$, in general, it gives two different x, but only one if

$$(z^2+14z+1)^2 - 16z(z+1)^2 = 0,$$

that is, for an even number of z, if $p \neq 2$, $p \neq 3$.

Thus the number of such solutions is congruent to 0 modulo 6.

If the three coordinates are equal, we get the point D. So

$$N_q(X'_{12}) \equiv 1 \mod 6.$$

Hence

$$N_q(S_{12}) \equiv 3q - 3 \mod 6$$

and by corollary 1

$$A_q(S_{12}) \equiv -1 - q^2 + q - 3q((\frac{3}{p}))^r \mod 6.$$

□

From Shioda–Inose [**15**], since we know the lattice of transcendental cycles, the zeta function of Y_2 (resp. S_0, S_{12}) on $\mathbb{F}_q$ can be expressed in terms of the zeta function of a model of the elliptic curve $\mathbb{C}/\mathbb{Z}+\mathbb{Z}\sqrt{-2}$ (resp. $\mathbb{C}/\mathbb{Z}+\mathbb{Z}\omega$ with ω a cube root of unity). Following now Stienstra and Beukers [**14**], denoting $A_p := \lambda_1 + \lambda_2$, one finds the following possibilities.

(1) For Y_2,

$$\begin{array}{llll} \text{if} & p = a^2 + 2b^2 (p \equiv 1,\ 3 \bmod 8), & A_p = \pm 2(a^2 - 2b^2) & \lambda_1\lambda_2 = \pm p^2 \\ \text{if} & p \equiv 5,\ 7 \bmod 8, & \text{either } A_p = 0 & \lambda_1\lambda_2 = \pm p^2 \\ & & \text{or } A_p = \pm 2p, \pm p & \lambda_1\lambda_2 = p^2. \end{array}$$

But $(\frac{2}{p}) = (-1)^{\frac{p^2-1}{8}}$; so by Lemma 7, if $p \equiv 1$ or $3 \mod 8$, then $A_p(Y_2) \equiv 2p \mod 8$, $A_{p^2} \equiv 2 \mod 8$ hence $A_p(Y_2) = 2(a^2 - 2b^2)$ and $\lambda_1\lambda_2 = p^2$.

Furthermore, if $p \equiv 5$ or $7 \mod 8$, then $A_p(Y_2) \equiv 0 \mod 8$, $A_{p^2} \equiv 2 \mod 8$ hence $A_p(Y_2) = 0$ and $\lambda_1\lambda_2 = -p^2$.

Thus we obtain,

$$\begin{array}{llll} \text{if} & p = a^2 + 2b^2 (p \equiv 1,\ 3 \bmod 8), & A_p(Y_2) = 2(a^2 - 2b^2) & \lambda_1\lambda_2 = p^2 \\ \text{if} & p \equiv 5,\ 7 \bmod 8, & A_p(Y_2) = 0 & \lambda_1\lambda_2 = -p^2. \end{array}$$

(2) For S_0 and S_{12}, we have the following possibilities

$$\begin{array}{llll} \text{if} & p = a^2 + 3b^2 (p \equiv 1 \bmod 6), & A_p = \pm 2(a^2 - 3b^2), & \lambda_1\lambda_2 = \pm p^2 \\ & & \pm(a^2 \pm 6ab - 3b^2) & \\ \text{if} & p \equiv 5 \bmod 6, & \text{either } A_p = 0 & \lambda_1\lambda_2 = \pm p^2 \\ & & \text{or } A_p = \pm 2p, \pm p & \lambda_1\lambda_2 = p^2. \end{array}$$

Thus using Lemma 7 , if $p \equiv 5 \mod 6$, then $A_p = 0$ and $A_{p^2} \equiv 2 \mod 6$, hence $\lambda_1\lambda_2 = -p^2$.

Furthermore, if $p \equiv 1 \mod 6$, then $A_p = 2(a^2 - 3b^2)$ and $A_{p^2} \equiv 2 \mod 6$, hence $\lambda_1\lambda_2 = p^2$. Thus we obtain, for $S = S_0$ or $S = S_{12}$,

$$\begin{array}{llll} \text{if} & p = a^2 + 3b^2 (p \equiv 1 \bmod 6), & A_p(S) = 2(a^2 - 3b^2) & \lambda_1\lambda_2 = p^2 \\ \text{if} & p \equiv 5 \bmod 6, & A_p(S) = 0 & \lambda_1\lambda_2 = -p^2. \end{array}$$

Now, with a good choice of local factors for $p = 2$, $p = 3$, we deduce the result as in Stienstra-Beukers [**14**]. This achieves the proof of Theorem 1.

The proof of Theorem 2 comes from the following relations given in [**4**]:

$$m(P_2) = \frac{16\sqrt{2}}{\pi^3}\frac{1}{2}\sum_{l,m}{}' \frac{l^2 - 2m^2}{(l^2 + 2m^2)^3}$$

$$m(Q'_{12}) = 4m(Q'_0)$$

$$m(Q'_0) = \frac{12\sqrt{3}}{\pi^3}\frac{1}{2}\sum_{l,m}{}' \frac{l^2 - 3m^2}{(l^2 + 3m^2)^3}.$$

These relations can be deduced from Theorem 2. The τ corresponding to polynomials P_2, Q'_0 and Q'_{12} are respectively $\frac{-2+\sqrt{-2}}{6}$, $\frac{3+\sqrt{-3}}{12}$ and $\frac{3+\sqrt{-3}}{6}$. Moreover, using the transformation $x = tX$, $y = tXY$, $z = tXYZ$, we observe that $m(Q'_0) = m(Q_0)$ and $m(Q'_{12}) = m(Q_{12})$. $\square$

Acknowledgments

I would like to thank particularly J. D. Lewis for his invitation to the Workshop Calabi-Yau Varieties and Mirror Symmetry held in Banff International Research Station (Canada), from 6 to 12 December 2003, and Noriko Yui for asking me to write this paper.

I am also grateful to the referee for his stimulating and relevant suggestions and remarks.

References

[1] A. Beauville, *Les familles stables de courbes elliptiques sur* $\mathbb{P}^1$ *admettant quatre fibres singulières* C.R.Acad.Sci. Paris Sér. I Math. t.294 (1982), 657-660.

[2] A. Beilinson, *Higher regulators of modular curves*, Application of algebraic K-theory to algebraic geometry and number theory, Part I,II (Boulder, Colo., 1983), Contemp. Math., **55**, Amer. Math. Soc., Providence, R. I., (1986), 1-34.

[3] M.J. Bertin, *Mesure de Mahler d'une famille de polynômes*, J. reine angew. Math. 569 (2004), 175-188.

[4] M.J. Bertin, *Mesure de Mahler d'hypersurfaces* $K3$, Preprint (2004), http://arxiv.org/pdf/math.NT/0501153.

[5] D.W. Boyd, *Mahler's measure and special values of L-functions*, Experiment. Math. **7** (1998), 37-82.

[6] J.W.S. Cassels, Lectures on Elliptic Curves, Cambridge University Press (1991)

[7] O. Lecacheux, *Points rationnels sur la surface elliptique modulaire associée au groupe modulaire* $\Gamma_1(8)$, Preprint (2004).
[8] R. Miranda & U. Persson, *Torsion groups of elliptic surfaces*, Comp. Math. 72 (1989), 249-267.
[9] C. Peters & J. Stienstra, *A pencil of $K3$ surfaces related to Apéry's recurrence for $\zeta(3)$ and Fermi surfaces for potential zero*, Arithmetic of Complex Manifolds (Erlangen, 1988) (W.-P. Barth and H. Lange, eds.), Lecture Notes in Math., vol. 1399, Springer, Berlin 1989, 110-127.
[10] C. Peters, J. Top & M. van der Vlugt, *The Hasse zeta function of a $K3$ surface related to the number of words of weight 5 in the Melas codes*, J. reine angew. Math. 432 (1992), 151-176.
[11] F. Rodriguez-Villegas, *Modular Mahler Measures*, http://www.ma.utexas.edu/users/villegas/research.html, Preprint (1996).
[12] F. Rodriguez-Villegas, *Modular Mahler measures I*, Topics in Number Theory (S.D. Ahlgren, G.E. Andrews & K. Ono, ed.), Kluwer, Dordrecht (1999), 17-48.
[13] T. Shioda, *On elliptic modular surfaces*, J. Math. Soc. Japan, vol. 24, n1, (1972), 20-59.
[14] J. Stienstra & F. Beukers, *On the Picard-Fuchs Equation and the Formal Brauer Group of Certain Elliptic $K3$-Surfaces*, Math. Ann. 271, (1985), 269-304.
[15] T. Shioda & H. Inose, *On singular $K3$ surfaces* in Complex analysis and algebraic geometry (Baily, Shioda T. ed.) , Cambridge (1977), 119-135.
[16] H. Verrill & N. Yui, *Thompson Series, and the Mirror Maps of Pencils of $K3$ Surfaces*, CRM Proceedings and Lecture Notes, **24** (2000), 399-432.
[17] N. Yui, *Arithmetic of Certain Calabi-Yau Varieties and Mirror Symmetry*, IAS/Park City Mathematics Series, Vol. 9, (2001), 509-569.

Université Pierre et Marie Curie (Paris 6), Institut de Mathématiques, 175 rue du Chevaleret, 75013 PARIS, France

E-mail address: bertin@math.jussieu.fr

AMS/IP Studies in Advanced Mathematics
Volume 38, 2006

On the Modularity of Calabi-Yau Threefolds Containing Elliptic Ruled Surfaces

K. Hulek and H. Verrill

ABSTRACT. In this paper we prove that (not necessarily rigid) Calabi-Yau threefolds defined over $\mathbb{Q}$ which contain sufficiently many elliptic ruled surfaces are modular (under certain mild restrictions on the primes of bad reduction). This is an application of the results of Dieulefait and Manoharmayum [**DM03**], [**Die05**] who showed modularity for rigid Calabi-Yau threefolds.

0. Introduction

In the last two years different authors have found several examples of non-rigid Calabi-Yau varieties defined over $\mathbb{Q}$ for which modularity could be proved (see [**LY03**], [**HV05**] and [**Sch05**]). The examples of Livné and Yui arise by taking a quotient of a product of a $K3$ surface with an elliptic curve, whereas the examples constructed by Schütt and the authors have the structure of a fibre product of elliptic surfaces. In each case the motive splits into low-dimensional pieces which can be explained geometrically.

In this brief note we want to give a criterion which implies modularity if the Calabi-Yau variety contains sufficiently many elliptic ruled surfaces. This applies to the examples given in [**HV05**] and [**Sch05**] and can also be used to explain the part of the motive in the examples in [**LY03**] which does not come from the transcendental lattice of the $K3$ surface. The main ingredient of this paper is an application of a modularity result proved by Dieulefait and Manoharmayum in their paper [**DM03**] (and its extension by Dieulefait [**Die05**]) where they establish modularity criteria for rigid Calabi-Yau threefolds.

Finally we shall show how to use our criterion to construct numereous examples of modular (birational) Calabi-Yau varieties. In some of these examples we have strong numerical evidence as to the modular forms which appear. The levels include 10,12,16,17,21,27,28,32,35,48,55,60,64,68,90,480 and the corresponding varieties are listed in Tables 1,2, 3 and 4.

2000 *Mathematics Subject Classification.* 11G40, 11G35, 14G10, 14J32.

Key words and phrases. Calabi-Yau manifolds, modularity.

Both authors are grateful to the DFG for financial support through the Schwerpunktprogramm *Globale Methoden in der komplexen Geometrie.* The second author was partly supported by Louisiana Board of Regents Grant LEQSF (2004-7) RD-A-16.

Acknowledgements. We are very grateful to Professors Dieulefait and Manoharmayum for the exchange of some e-mails in which they explained some of their results and details about l-adic Galois representations to us.

We would like to thank the referee for computing the levels of the weight 4 modular forms appearing in Tables 2, and 3, and the level 480 example in Table 1. We also thank Matthias Schütt for reading through our manuscript and making several suggestions for improvements as well as for computational help with one example.

1. A modularity criterion

Let X be a smooth projective variety with $h^{30}(X) = h^{03}(X) = 1$. (In our applications this will be a Calabi-Yau variety or a variety birational to a nodal Calabi-Yau variety.) A (birational) elliptic ruled surface Y is a smooth projective surface which is birational to a $\mathbb{P}^1$-bundle over an elliptic curve. In this case $h^{21}(Y) = h^{12}(Y) = 1$ and $h^{30}(Y) = h^{03}(Y) = 0$.

DEFINITION 1.1. Let X be smooth projective and define $b := h^{12}(X)$. Let $Y_j, j = 1, \ldots, b$, be birational elliptic ruled surfaces contained in X. We say that these surfaces *span* $H^{21}(X) \oplus H^{12}(X)$ if the natural map

$$H^3(X, \mathbb{C}) \to \oplus_{j=1}^{b} H^3(Y_j, \mathbb{C})$$

is surjective.

REMARK 1.2. If the surfaces Y_j are as in the previous definition, then we have an exact sequence

$$0 \to H^{30}(X) \oplus H^{03}(X) \to H^3(X, \mathbb{C}) \to \oplus_{j=1}^{b} H^3(Y_j, \mathbb{C}) \to 0.$$

This follows since the restriction of a $(3,0)$ or a $(0,3)$ form to an elliptic ruled surface must vanish, and together with the assumption that the surfaces Y_j span $H^{21}(X) \oplus H^{12}(X)$ this gives the claim. In particular we obtain an isomorphism $H^{21}(X) \oplus H^{12}(X) \cong \oplus_{j=1}^{b} H^3(Y_j, \mathbb{C}) = \oplus_{j=1}^{b}(H^{21}(Y_j) \oplus H^{12}(Y_j))$.

REMARK 1.3. If X is a smooth Calabi-Yau threefold and Y is a smooth elliptic ruled surface contained in X, then a straightforward argument from intersection theory (cf. the proof of Proposition 2.1) using the adjunction formula shows that the map $H^3(X, \mathbb{C}) \to H^3(Y, \mathbb{C})$ is surjective. If $h^{12}(X) = 1$ then this shows that we are automatically in the situation described above. In fact this observation is often useful when one wants to prove modularity of Calabi-Yau threefolds with $h^{12}(X) = 1$.

We now assume that we are in an arithmetic situation, i.e. that the variety X as well as the subvarieties Y_j are defined over $\mathbb{Q}$. When we speak about a birational elliptic surface Y defined over $\mathbb{Q}$ we mean that there is a birational map defined over $\mathbb{Q}$ from Y to $E \times \mathbb{P}^1$. In what follows the assumption that all Y_j are defined over $\mathbb{Q}$ can often be weakened to the assumption that the union of the surfaces Y_j is defined over $\mathbb{Q}$, but for the sake of simplicity we shall assume that all the Y_j themselves are defined over $\mathbb{Q}$. Then the map $H^3(X) \to \oplus_{j=1}^{b} H^3(Y_j)$ can be interpreted in the l-adic cohomology and gives rise to a morphism of $\mathrm{Gal}(\overline{\mathbb{Q}}/\mathbb{Q})$-modules $H^3_{\text{ét}}(X, \mathbb{Q}_l) \to \oplus H^3_{\text{ét}}(Y_j, \mathbb{Q}_l)$. (Strictly speaking we have a representation of the Galois group $\mathrm{Gal}(\overline{\mathbb{Q}}/\mathbb{Q})$ on the dual vector spaces, but we shall, by misuse of language, refer to the cohomology groups as Galois representations.) If the surfaces Y_j span $H^{21}(X) \oplus H^{12}(X)$ this gives rise to an exact sequence of $\mathrm{Gal}(\overline{\mathbb{Q}}/\mathbb{Q})$-modules

$$0 \to U \to H^3_{\text{ét}}(X, \mathbb{Q}_l) \to \oplus H^3_{\text{ét}}(Y_j, \mathbb{Q}_l) \to 0 \tag{1}$$

where U has dimension 2.

We shall want to make use of a criterion proved by Dieulefait and Manoharmayum [**DM03**] and further extended by Dieulefait [**Die05**] which was developed in order to prove the modularity of rigid Calabi-Yau varieties.

PROPOSITION 1.4. Let l be a prime and let

$$\rho : \mathrm{Gal}(\overline{\mathbb{Q}}/\mathbb{Q}) \to \mathrm{GL}_2(\overline{\mathbb{Q}}_l)$$

be an irreducible continuous Galois representation, unramified outside finitely many primes, and having determinant ε_l^3 where ε_l is the cyclotomic character. Consider the following conditions:

(1) ρ is crystalline at l with Hodge-Tate weights $\{0, 3\}$.
(2) The residual representation

$$\overline{\rho} : \mathrm{Gal}(\overline{\mathbb{Q}}/\mathbb{Q}) \to \mathrm{GL}_2(\overline{\mathbb{F}}_l)$$

takes values in $\mathrm{GL}_2(\mathbb{F}_l)$.
(3) The l-adic representation ρ has the field of coefficients $\mathbb{Q}$.

Moreover, assume that one of the following assumptions holds:

(DM1) $l = 7$, ρ is unramified at 3 and conditions (1) and (2) hold.
(DM2) $l = 5$ and there is a prime $p \equiv \pm 2 (\mathrm{mod}\, 5)$ where ρ is unramified and the trace of Frobenius $t_3(p)$ at p is not divisible by 5, where we define $t_3(p) := \mathrm{trace}(\rho(\mathrm{Frob}_p))$. Moreover, conditions (1) and (2) hold.
(D3) $l > 5$, ρ is unramified at 3, $t_3(3)$ is not divisible by 3 and conditions (1) and (3) hold.
(D4) $l > 5$, ρ is unramified at 5, $t_5(5)$ is not divisible by 5 and conditions (1) and (3) hold.

Then ρ is modular.

PROOF. Dieulefait and Manoharmayum proved that either of the conditions (DM1) or (DM2) are sufficient for modularity (see [**DM03**, Remark 3.2]). The fact that condition (D3) or (D4) are sufficient is shown by Dieulefait in the appendix to this paper ([**Die05**, Theorem A.1], resp. [**Die05**, Section A.2]). □

We now want to apply this to our situation.

PROPOSITION 1.5. Let X be a smooth projective threefold defined over $\mathbb{Q}$ with $h^{30}(X) = h^{03}(X) = 1$ and assume that X contains birational ruled surfaces $Y_j, j = 1, \ldots, b$ which are defined over $\mathbb{Q}$ and which span $H^{21}(X) \oplus H^{12}(X)$. Let ρ be the 2-dimensional Galois representation given by the kernel U from the exact sequence

$$0 \to U \to H^3_{\text{ét}}(X, \mathbb{Q}_l) \to \oplus H^3_{\text{ét}}(Y_j, \mathbb{Q}_l) \to 0.$$

Assume that one of the following conditions holds:

(i) X has good reduction at 3 and 7 or
(ii) X has good reduction at 5 and at some prime $p \equiv \pm 2 (\mathrm{mod}\, 5)$ and the trace of Frob_p on U is not divisible by 5 or

(iii) X has good reduction at 3 and the trace of Frob_3 on U is not divisible by the prime 3.
(iv) X has good reduction at 5 and the trace of Frob_5 on U is not divisible by the prime 5.

Then X is modular. More precisely

$$L(X,s) \doteq L(f_4,s)\prod_{j=1}^{b} L(g_2^j, s-1) \tag{2}$$

where f_4 is a weight 4 form and the g_2^j are the weight 2 forms associated to the base curves E_j of the birational ruled surfaces Y_j. (As usual $\doteq$ denotes equality of the L-series up to possibly finitely many factors.)

PROOF. Since the surfaces Y_j are birational over $\mathbb{Q}$ to a product $E_j \times \mathbb{P}^1$ we have

$$L(Y_j,s) \doteq L(g_2^j, s-1)$$

where g_2^j is the weight 2 form associated to the elliptic curve E_j. After passing to the semi-simplification we have an isomorphism

$$H^3_{\text{ét}}(X,\mathbb{Q}_l) \cong U \oplus \bigoplus_{j=1}^{b} H^3_{\text{ét}}(Y_j,\mathbb{Q}_l).$$

Our claim follows if we can show that we can apply the results of Dieulefait and Manoharmayum, resp. Dieulefait to deduce the modularity of the 2-dimensional $\mathrm{Gal}(\overline{\mathbb{Q}}/\mathbb{Q})$ representation U. Clearly U is a continuous representation which is unramified outside finitely many primes, namely the primes of bad reduction. It follows from the Weil conjectures as proved by Deligne that U is irreducible [**DM03**, pp. 1,2]. Moreover, if U is unramified at l then its determinant is ε_l^3. Condition (2), resp. (3) of 1.4 follows since U is a subspace of $H^3_{\text{ét}}(X,\mathbb{Q}_l)$ for which (2) and (3) hold. Similarly we can deduce that U is crystalline and has Hodge-Tate weights $\{0,3\}$. The first follows e.g. from the fact that $H^3_{\text{ét}}(X,\mathbb{Q}_l)$ is crystalline for $l>3$ (see [**FM87**, Theorem 2.2 A (ii)]) and the fact that this property is preserved for submodules and quotients (see [**FM87**, Proposition 1.7]). The statement about the Hodge-Tate weight follows from standard comparison theorems (see [**FM87**, Section 3.1]). The conditions (DM1), (DM2) or (D) follow immediately from the hypothesis of the theorem. □

REMARK 1.6. The weight 4 form which is associated to the 2-dimensional representation U can in many cases be determined explicitly by the method of Faltings, Livné and Serre.

2. Fibre products

We shall now investigate a certain class of Calabi-Yau threefolds which often (but not always) contain many elliptic ruled surfaces in the sense of our discussion in the previous section.

Let $r : Y \to \mathbb{P}^1$ and $r' : Y' \to \mathbb{P}^1$ be semi-stable rational elliptic fibrations with a section. Then the fibre product

$$W = Y \times_{\mathbb{P}^1} Y'$$

is a nodal Calabi-Yau variety. We denote by $\hat{W}$ a small and by $\tilde{W}$ a big resolution of W. Note that small resolutions need not be projective, in fact in many cases there exists no projective small resolution. However, $\hat{W}$ is a Moishezon manifold and hence we still have a Hodge decomposition and thus can speak about the Hodge numbers $h^{pq}(\hat{W})$, cf. [**Uen75**]. Let S and S' be the sets of critical values of r and r'. We define $S'' = S \cap S'$ and $\tilde{S} = S \cup S' \setminus S''$. Then all fibres over points in $\tilde{S}$ are of type $I_n \times I_0$ or $I_0 \times I_n$. For any $t \in \mathbb{P}^1$ we denote by $b(t)$, resp. $b'(t)$ the number of components of $r^{-1}(t)$, resp. $r^{-1}(t')$. Then it is well known [**Sch88**, p. 190] that

$$h^{12}(\tilde{W}) = h^{12}(\hat{W}) = 1 + \operatorname{rank}\operatorname{Pic}(W_\eta) - \#S'' + \sum_{t\in\tilde{S}}(b(t)b'(t) - 1).$$

Here $\operatorname{Pic}(W_\eta)$ is the Picard group of the general fibre. Its rank may be computed using the formula

$$\operatorname{rank}\operatorname{Pic}(W_\eta) = \operatorname{rank}\operatorname{Pic}(Y_\eta) + \operatorname{rank}\operatorname{Pic}(Y'_\eta) + d,$$

where $d = 1$ if Y_η and Y'_η are isogenous and 0 otherwise. By [**Sch88**, p. 190] this can be rewritten as

$$\operatorname{rank}\operatorname{Pic}(W_\eta) = d + 18 - \sum_{t\in S} b(t) - \sum_{t'\in S'} b'(t') + \#S + \#S'.$$

We define the *fibre defect*

$$\delta = \delta(r, r') := 1 + \operatorname{rank}\operatorname{Pic}(W_\eta) - \#S''$$

which can be rewritten as

$$\delta = d + 19 - \sum_{t\in S} b(t) - \sum_{t'\in S'} b'(t') + \#S + \#S' - \#S''. \tag{3}$$

When we discuss examples in § 3.2, we will restrict consideration to the case where S and S' are rational elliptic fibrations. Since by Noether's theorem the Euler number of these surfaces is 12 it follows that $\sum_{t\in S} b(t) = \sum_{t'\in S'} b'(t') = 12$, and so in this situation (3) becomes

$$\delta = d - 5 + \#S + \#S' - \#S''. \tag{4}$$

Using its definition, the fibre defect can be computed explicitly for all pairs (r, r') of elliptic fibrations. Note that this number is not an invariant of the variety W but depends on its representation as a fibre product (as we shall see in examples). We call δ the *fibre defect*, as this measures the size of $H^{12}(\hat{W}) \oplus H^{21}(\hat{W})$ which does not come from fibres of type $I_n \times I_0$.

We shall now consider the fibres over the points in $\tilde{S}$. As we have already remarked these are of type $I_n \times I_0$ or $I_0 \times I_n$, i.e. a cycle of elliptic ruled surfaces whose components are isomorphic to $\mathbb{P}^1_i \times E_s$ where $i = 1, \ldots, b(t)$, resp. $i = 1, \ldots, b'(t)$ and $t \in \tilde{S}$. We shall write $\gamma(t)$ for $b(t)$ or $b'(t)$ in order to simplify the notation. Let $\alpha_t, \beta_t \in H_1(E_t, \mathbb{Z})$ be generators with $\alpha_t^2 = \beta_t^2 = 0$ and $\alpha_t \cdot \beta_t = 1$. Then $A_t^i = \mathbb{P}^1_i \times \alpha_t$ and $B_t^i = \mathbb{P}^1_i \times \beta_t$ are generators of $H_3(\mathbb{P}^1_i \times E_t)$. Finally let $\gamma_{i,t} : \mathbb{P}^1_i \times E_t \to \hat{W}$, resp. $\gamma_{i,t} : \mathbb{P}^1_i \times E_t \to \tilde{W}$ be the inclusion.

PROPOSITION 2.1. There is an exact sequence

$$0 \to U \to H^3(\hat{W}, \mathbb{C}) \xrightarrow{\oplus\gamma^*_{i,t}} \bigoplus_{t\in\tilde{S}} \bigoplus_{i=1}^{\gamma(t)-1} H^3(\mathbb{P}^1_i \times E_t, \mathbb{C}) \to 0$$

where $\dim_{\mathbb{C}} U = 2\delta + 2$. A similar exact sequence exists for $\tilde{W}$ instead of $\hat{W}$.

PROOF. We have to compute the rank of the intersection matrix of the cycles A_t^i and B_t^i. Clearly cycles in different fibres are independent, hence it suffices to look at one fibre of type $I_n \times I_0$. We assume that the enumeration of the components is such that $\mathbb{P}_i^1 \times E_t$ intersects $\mathbb{P}_j^1 \times E_t$ if and only if $|i - j| \leq 1$ where the index i has to be read cyclically, i.e. we can assume $i \in \mathbb{Z}/\gamma(t)\mathbb{Z}$. Then it follows immediately from the geometry that

$$A_t^i \cdot A_t^j = B_t^i \cdot B_t^j = 0 \quad \text{for all } i, j$$

and

$$A_t^i \cdot B_t^j = -B_t^j \cdot A_t^i = \begin{cases} 0 & \text{if} \quad |i - j| \geq 2 \\ 1 & \text{if} \quad |i - j| = 1. \end{cases}$$

It remains to determine $A_t^i \cdot B_t^i$. Recall that the normal bundle of a component $\mathbb{P}_i^1 \times E_t$ in $\hat{W}$, resp. $\tilde{W}$ has degree -2 on the fibres $\mathbb{P}_i^1 \times \{pt\}$. It then follows from [**Ful98**, 19.2.2] that

$$A_t^i \cdot B_t^i = -B_t^i \cdot A_t^i = -2.$$

Altogether the intersection matrix of the cycles $A_t^i, B_t^i, i = 1, \ldots, \gamma(t) - 1$, is given by

$$\begin{pmatrix} \begin{pmatrix} 0 & -2 \\ 2 & 0 \end{pmatrix} & \begin{pmatrix} 0 & 1 \\ -1 & 0 \end{pmatrix} & & & & \\ \begin{pmatrix} 0 & 1 \\ -1 & 0 \end{pmatrix} & \begin{pmatrix} 0 & -2 \\ 2 & 0 \end{pmatrix} & \begin{pmatrix} 0 & 1 \\ -1 & 0 \end{pmatrix} & & & \\ & \begin{pmatrix} 0 & 1 \\ -1 & 0 \end{pmatrix} & \begin{pmatrix} 0 & -2 \\ 2 & 0 \end{pmatrix} & & & \\ & & & \ddots & & \\ & & & & \begin{pmatrix} 0 & -2 \\ 2 & 0 \end{pmatrix} & \begin{pmatrix} 0 & 1 \\ -1 & 0 \end{pmatrix} \\ & & & & \begin{pmatrix} 0 & 1 \\ -1 & 0 \end{pmatrix} & \begin{pmatrix} 0 & -2 \\ 2 & 0 \end{pmatrix} \end{pmatrix}.$$

It is obvious that this matrix has rank $2\gamma(t) - 2$ and this proves the proposition. $\square$

REMARK 2.2. It is actually not necessary that we start with semi-stable fibrations provided $\hat{W}$ is a Calabi-Yau variety and we obtain sufficiently many contributions from the fibres of type $I_n \times I_0$.

We now want to consider the modularity of such fibre products.

THEOREM 2.3. *Assume that* $r : Y \to \mathbb{P}^1$ *and* $r' : Y' \to \mathbb{P}^1$ *are semi-stable, rational elliptic fibrations with a section, which are defined over* $\mathbb{Q}$. *Let* $\tilde{W}$ *be a (big) resolution of* $W = Y \times_{\mathbb{P}^1} Y'$. *Moreover assume that*

(1) *the fibre defect* $\delta = \delta(r, r') = 0$,
(2) *if* $t \in \tilde{S}$ *and the fibre* W_t *is isomorphic to* $E_t \times I_n$ *for some elliptic curve* E_t *and an* n*-cycle* $I_n, n \geq 2$, *then all components of this fibre are defined over* $\mathbb{Q}$.

Then there is an exact sequence of $\mathrm{Gal}(\overline{\mathbb{Q}}/\mathbb{Q})$*-modules*

$$0 \to U \to H^3_{\acute{e}t}(\tilde{W}, \mathbb{Q}_l) \to \bigoplus_{t \in \tilde{S}} \bigoplus_{i=1}^{\gamma(t)-1} H^3_{\acute{e}t}(\mathbb{P}^1_i \times E_t) \to 0$$

with 2*-dimensional kernel* U*.*

Moreover assume that one of the following assumptions holds:

(3) $\tilde{W}$ *has good reduction at* 3 *and* 7 *or*

(4) $\tilde{W}$ *has good reduction at* 5 *and at some prime* $p \equiv \pm 2 (\mathrm{mod}\, 5)$ *with the trace of* Frob_p *on the kernel* U *not divisible by* 5 *or*

(5) $\tilde{W}$ *has good reduction at* 3 *and the trace of* Frob_3 *on* U *not divisible by* 3*.*

Then $\tilde{W}$ *is modular and the same is true for a small projective resolution defined over* $\mathbb{Q}$ *if it exists. Moreover*

$$L(H^3_{\acute{e}t}(\tilde{W}), s) \doteq L(f, s) \prod_{t \in \tilde{S}} (g_{E_t}, s-1)^{\gamma(t)-1}$$

where f *is a modular form of weight* 4 *and* g_{E_t} *is the weight* 2 *form associated to* E_t*.*

PROOF. By Proposition 2.1 the condition $\delta = 0$ assures that the components of the fibres $E_t \times \mathbb{P}^1$ over the points in $\tilde{S}$ span $H^{21}(\tilde{W}) \oplus H^{12}(\tilde{W})$ and that the kernel U has, therefore, dimension 2. The result is now an immediate application of Proposition 1.5. □

3. Examples

In this section we want to discuss various examples where the method described above can be applied. We use (4) to find the fibre defect δ in these examples.

3.1. Relation with previous examples. In [**HV05**] we considered the following examples. Let $\mathcal{E}_{(\alpha,\beta,\gamma)}$ be the elliptic fibration which is the resolution of the surface $\mathcal{E}'_{(\alpha,\beta,\gamma)} \subset \mathbb{P}^2 \times \mathbb{P}^1$ given by

$$\mathcal{E}'_{(\alpha,\beta,\gamma)}: \qquad (x+y+z)(\alpha xy + \beta yz + \gamma zx)t_0 = t_1 xyz, \tag{5}$$

where $(x:y:z) \in \mathbb{P}^2, (t_0:t_1) \in \mathbb{P}^1$. There is a projection $p: \mathcal{E}_{(\alpha,\beta,\gamma)} \to \mathbb{P}^1$, with fibre $\mathcal{E}_{(\alpha,\beta,\gamma),t} := p^{-1}(1:t)$ for generic t. We write $\mathcal{E}_\alpha := \mathcal{E}_{(1,1,\alpha)}$ and $\mathcal{E}_{\alpha,t} := \mathcal{E}_{(1,1,\alpha),t}$. The singular fibres of these families can be read off from [**HV05**, Section 4, Table 2].

Taking fibre products and resolving singularities one obtains smooth varieties

$$X_{\mathbf{a}} = X_{(\alpha_1:\alpha_2:\alpha_3:\alpha_4:\alpha_5:\alpha_6)} \sim_{bir} \mathcal{E}_{(\alpha_1:\alpha_2:\alpha_3)} \times_{\mathbb{P}^1} \mathcal{E}_{(\alpha_4:\alpha_5:\alpha_6)}$$

which are resolutions of nodal Calabi-Yau threefolds. The examples discussed in [**HV05**] were $X_{25} = X_{(1:1:1:1:1:25)}$, $X_{(1:1:1:9:9:9)}$ and $X_{(1:1:1:4:4:9)}$ where $h^{12} = 4, 2$ and 1 respectively. In the last case

$$X_{(1:1:4:4:4:16)} \sim_{bir} \mathcal{E}_{(1:1:16)} \times_{\mathbb{P}^1} \mathcal{E}_{(4:4:4)}$$

and we have $\#S = \#S' = 4$ and $\#S'' = 3$ and hence, since these families are not isogenous, we have $\delta = 0$. In this case $h^{12} = 1$ and it is sufficient to consider the $I_2 \times I_0$-fibre over the point 16. We can then derive modularity from Theorem 2.3.

In the second example we can either write

$$X_{(1:1:1:9:9:9)} \sim_{bir} \mathcal{E}_{(1:1:1)} \times_{\mathbb{P}^1} \mathcal{E}_{(9:9:9)}$$

or

$$X_{(1:1:1:9:9:9)} \sim_{bir} \mathcal{E}_{(1:1:9)} \times_{\mathbb{P}^1} \mathcal{E}_{(1:9:9)}.$$

In the first case we have again $\#S = \#S' = 4$ and $\#S'' = 3$ and hence $\delta = 0$. We have a fibre of type $I_3 \times I_0$ over the point 1 and since $h^{12} = 2$ this spans all of $H^{21} \oplus H^{12}$ and our method applies. In the second description we have $\#S = \#S' = 5$ and $\#S'' = 4$ and this gives $\delta = 1$. This is also an example which shows that δ is not a birational invariant but depends on the chosen fibration.

Finally, in the first example

$$X_{25} \sim_{bir} \mathcal{E}_{(1:1:1)} \times_{\mathbb{P}^1} \mathcal{E}_{(1:1:25)}$$

and in this case $\#S = 4$, $\#S' = 5$, $\#S'' = 3$ and thus $\delta = 1$. Hence we cannot invoke Theorem 2.3. However, we showed in [**HV05**] that one can still find sufficiently many elliptic ruled surfaces in X_{25} in order to be able to apply Proposition 1.5.

The weight 4 forms corresponding to $H^{30} \oplus H^{03}$ were computed in [**HV05**].

A further class of examples where this method works was discussed by Schütt [**Sch05**]. In his examples $\#S = 5$ or 4 and $\#S' = 4$ with $\#S'' = 4$ and 3 respectively. In all cases this gives $\delta = 0$ and hence one is in a situation where Theorem 2.3 applies.

3.2. New examples. In this section we give a large number of examples of (birational) Calabi-Yau threefolds for which Theorem 2.3 implies modularity. In many cases we shall also conjecture the modular forms which appear. We have strong numerical evidence for these conjectures, i.e. we computed some of the Fourier coefficients modulo p for all primes of good reduction less than 100. It seems very likely that, using the technique of Faltings, Livné and Serre, these conjectures can be verified.

We use the following notation. If $\mathcal{F}^t$ is an elliptic fibration with parameter t, and if M is a fractional linear transformation, then $\mathcal{F}^{Mt}$ denotes the family obtained by replacing t with Mt.

We denote by $\mathcal{E}(\Gamma)$ Beauville's modular elliptic surfaces [**Bea82**], where Γ is one of $\Gamma(3)$ $\Gamma_1(4) \cap \Gamma(2)$, $\Gamma_1(6)$, $\Gamma_1(5)$, $\Gamma_0(8) \cap \Gamma_1(4)$ or $\Gamma_0(9) \cap \Gamma_1(3)$. Note that $\mathcal{E}(\Gamma_1(6)) = \mathcal{E}_{(1:1:1)}$. The others may be given by equations as in the table below. The values of t for which the fibres are singular are tabulated, and the last column shows the number of components of the singular fibres, all of which are of I_n type.

Γ	equation for $\mathcal{E}(\Gamma)$	singular t	fibres
$\Gamma(3)$	$x^3 + y^3 + z^3 = 3txyz$	$\infty, 1, \frac{-1\pm\sqrt{-3}}{2}$	$3, 3, 3, 3$
$\Gamma_1(4) \cap \Gamma(2)$	$x(x^2 + z^2 + 2zy) = tz(x^2 - y^2)$	$\infty, 0, 1, -1$	$4, 4, 2, 2$
$\Gamma_1(5)$	$x(x - z)(y - z) = tzy(x - y)$	$\infty, 0, \frac{-11\pm 5\sqrt{5}}{2}$	$5, 5, 1, 1$
$\Gamma_1(6)$	$(x + y + z)(xy + yz + zx) = txyz$	$\infty, 0, 1, 9$	$6, 2, 3, 1$
$\Gamma_0(8) \cap \Gamma_1(4)$	$(x + y)(xy + z^2) = 4txyz$	$\infty, 0, 1, -1$	$8, 2, 1, 1$
$\Gamma_0(9) \cap \Gamma_1(3)$	$x^2y + y^2z + z^2x = 3txyz$	$\infty, 1, \frac{-1\pm\sqrt{-3}}{2}$	$9, 1, 1, 1$

If Y and Y' are **not** isogenous then the condition $\delta = 0$ in Theorem 2.3 translates into

$$\#S + \#S' = \#S'' + 5.$$

This shows that we cannot have examples where at least 1 elliptic fibration has 6 singular fibres. In order to produce examples with $\delta = 0$ we have three possibilities:

	$\#S$	$\#S'$	$\#S''$	possible examples
Case A	5	5	5	$\mathcal{E}_{(1:\alpha^2:\alpha^2)} \times_{\mathbb{P}^1} \mathcal{E}^{Mt}_{(1:\beta^2:\beta^2)}$
Case B	5	4	4	$\mathcal{E}_{(1:\alpha^2:\alpha^2)} \times_{\mathbb{P}^1} \mathcal{E}^{Mt}_{(1:1:1)}$
				$\mathcal{E}_{(1:\alpha^2:\alpha^2)} \times_{\mathbb{P}^1} \mathcal{E}^{Mt}(\Gamma)$
Case C	4	4	3	$\mathcal{E}_{(1:1:1)} \times_{\mathbb{P}^1} \mathcal{E}^{Mt}_{(1:1:1)}$
				$\mathcal{E}(\Gamma_1) \times_{\mathbb{P}^1} \mathcal{E}^{Mt}(\Gamma_2)$

We now construct explicit examples for each case.

Case A: $\#S = \#S' = \#S'' = 5$ (and the two surfaces are not isogenous). Notice that this can never give non-rigid examples since we have no fibres of type $I_n \times I_0$.

Given a fibration $\mathcal{E}_{(\alpha,\beta,\gamma)}$ and a fractional linear transformation M, we consider

$$\mathcal{E}^{Mt}_{(\alpha,\beta,\gamma)} : (x+y+z)(axy+byz+czx) = (Mt)xyz.$$

We shall take two families given by $\mathcal{E}_{(1,\alpha^2,\alpha^2)}$ and $\mathcal{E}_{(1,\gamma^2,\gamma^2)}$, with $1, \alpha, 2\alpha, 0$ distinct, and $1, \gamma, 2\gamma, 0$ distinct, and analyse possible fibre products. Recall from [**HV05**, Section 4, Table 2] that the singular fibres of $\mathcal{E}_{(1,\alpha^2,\alpha^2)}$ lie over the points $\{\infty, 0, 1, (1 \pm 2\alpha)^2\}$, with fibre types I_6, I_2, I_2, I_1, I_1 respectively, provided $2\alpha \neq 0, 1, 2$. If there is a fractional linear transformation M with $M\{\infty, 0, 1, (1 \pm 2\alpha)^2\} = \{\infty, 0, 1, (1 \pm 2\gamma)^2\}$, then the resolution of

$$\mathcal{E}_{(1,\alpha^2,\alpha^2)} \times \mathcal{E}^{Mt}_{(1,\gamma^2,\gamma^2)}$$

is a rigid (birational) Calabi-Yau threefold, provided that $\#\{\infty, 0, 1, (1 \pm 2\alpha)^2\} = 5$ and that this is not an isogenous pairing. There are 60 such choices (this can be reduced by symmetry considerations) for which three of the points of $\infty, 0, 1, (1 \pm 2\gamma)^2$ map to 0, 1 and ∞. This determines M, and the remaining two points β, δ then map to $M(\beta)$, $M(\delta)$, which are given by rational functions in α and γ. The requirement that

$$M(\beta) = (1+2\alpha)^2 \text{ and } M(\delta) = (1-2\alpha)^2, \tag{6}$$

implies $\alpha = (M(\beta) - M(\delta))/8$. This expression for α in terms of γ can be substituted into (6) to give polynomials satisfied by γ, and from this we can determine whether α^2 and $\gamma^2 \in \mathbb{Q} \setminus \{0\}$. One must also check that $(1 \pm 2\gamma)^2 \neq 0, 1, \infty$. In this case, we get a new example of a rigid Calabi-Yau threefold.

We list all possible cases in Table 1. The singular fibres of the first family have types I_6, I_2, I_2, I_1, I_1, in this order, so we only list the fibre types of the second family. Note that $\mathcal{E}^t_{(1,1/9,1/9)} \times \mathcal{E}^{9t}_{(1,4,4)}$ is the same as $\mathcal{E}_{(9,1,1)} \times \mathcal{E}_{(1,4,4)} \sim_{bir} X_{(1:1:1:4:4:9)}$, which we considered in [**HV05**], so this case gives only four new cases.

Computational evidence indicates that the L-series of the middle cohomology of $\mathcal{E}^t_{(1,-1/8,-1/8)} \times \mathcal{E}^{1-t}_{(1,-1/8,-1/8)}$ is a weight 4 level 32 modular form, with q-expansion

$$q - 8q^3 - 10q^5 - 16q^7 + 37q^9 + 40q^{11} + \cdots.$$

			fibre types of $\mathcal{E}^{Mt}_{(1,\gamma^2,\gamma^2)}$									level
α^2	γ^2	Mt	∞	1	0	$(1\pm 2\alpha)^2$		$(1\pm 2\alpha)^2$		$(1\pm 2\gamma)^2$		
$\frac{-1}{8}$	$\frac{-1}{8}$	$1-t$	I_6	I_2	I_2	I_1	I_1	$\frac{1\pm\sqrt{-8}}{2}$		$\frac{1\pm\sqrt{-8}}{2}$		32
$\frac{1}{2}$	$\frac{1}{2}$	$1/t$	I_2	I_6	I_2	I_1	I_1	$3\pm 2\sqrt{2}$		$3\pm 2\sqrt{2}$		64
$\frac{1}{9}$	4	$9t$	I_6	I_2	I_1	I_2	I_1	$1/9$	$25/9$	9	25	60
$\frac{9}{4}$	$\frac{9}{4}$	$16/t$	I_2	I_6	I_1	I_2	I_1	16	4	16	4	480
$\frac{1}{64}$	$\frac{1}{64}$	$\frac{25}{16}-t$	I_6	I_1	I_1	I_2	I_2	$\frac{9}{16}$	$\frac{25}{16}$	$\frac{9}{16}$	$\frac{25}{16}$	35

TABLE 1. Rigid Calabi-Yau threefolds of the form $\mathcal{E}_{(1,\alpha^2,\alpha^2)} \times \mathcal{E}^{Mt}_{(1,\gamma^2,\gamma^2)}$

Similarly, computations indicate that $\mathcal{E}^t_{(1,\frac{1}{2},\frac{1}{2})} \times \mathcal{E}^{1/t}_{(1,\frac{1}{2},\frac{1}{2})}$ is associated with a weight 4 level 64 modular form, with q-expansion

$$q - 22q^5 - 27q^9 + 18q^{13} + \cdots,$$

$\mathcal{E}^t_{(1,9/4,9/4)} \times \mathcal{E}^{16/t}_{(1,9/4,9/4)}$ is associated with a weight 4 level 480 modular form,

$$q + 3q^3 + 5q^5 + 4q^7 + 9q^9 - 40q^{11} - 90q^{13} + \cdots,$$

and $\mathcal{E}^t_{(1,\frac{1}{64},\frac{1}{64})} \times \mathcal{E}^{25/16-t}_{(1,\frac{1}{64},\frac{1}{64})}$ is associated with the weight 4 level 35 form

$$q + q^2 - 8q^3 - 7q^4 - 5q^5 - 8q^6 + \cdots.$$

Case B: $\#S = 5$, $\#S' = \#S'' = 4$.

i. Fibre products of $\mathcal{E}_{(1:a:a)}$ fibrations. We first consider products of the form

$$\mathcal{E}_{(1:\alpha^2:\alpha^2)} \times_{\mathbb{P}^1} \mathcal{E}^{Mt}_{(1:1:1)}$$

for some fractional linear transformation M, and $\alpha \neq 0, 1, 1/2$. Since we are more interested in non-rigid cases, we will not investigate the rigid possibilities. Recall that the singular fibres of $\mathcal{E}_{(1:1:1)}$ lie over the points $\{\infty, 0, 1, 9\}$ and that the singular fibres are of type I_6, I_2, I_3, I_1 (in this order). For a non-rigid example, we are therefore looking for matrices M with

$$\{\infty, 0, 1\} \not\subset M\{\infty, 0, 1, 9\} \subset \{\infty, 0, 1, (1\pm 2\alpha)^2\}.$$

Denote the cross ratio by $C(a,b,c,d) = (a-c)(b-d)/(a-d)(b-c)$, and set $j(a,b,c,d) = j(C(a,b,c,d))$, where

$$j(\lambda) = \frac{(\lambda^2 - \lambda + 1)^3}{(\lambda-1)^2\lambda^2}.$$

Possible values of α^2 are given by solutions to $73^3 2^{-6} 3^{-4} = j(\infty, 0, 1, 9) = J$, for $J = j(0, 1, (1\pm 2\alpha)^2)$, $j(\infty, 1, (1\pm 2\alpha)^2)$ or $j(0, \infty, (1\pm 2\alpha)^2)$. The first possibility,

$$j\left(\left(\frac{2\alpha+1}{2\alpha-1}\right)^2 \left(\frac{\alpha+1}{\alpha-1}\right)\right) = 73^3 2^{-6} 3^{-4}$$

has no rational solutions. Up to sign, the second,

$$j\left(\frac{\alpha+1}{\alpha-1}\right) = \frac{(\alpha^2+3)^3}{4(\alpha^2-1)^2} = 73^3 2^{-6} 3^{-4}$$

α	Mt	fibre types							level	modular form
		∞	1	0	$(1 \pm 2\alpha)^2$		$(1 \pm 2\alpha)^2$			
17	$\frac{(t-1)}{(t-33^2)}$	I_3	I_2	–	I_6	I_1	33^2	35^2	17	$q - 3q^2 - 8q^3 \cdots$
$\frac{5}{4}$	$\frac{45}{4(t-1)}$	I_2	I_6	–	I_1	I_3	$\frac{9}{4}$	$\frac{49}{4}$	90	$q - 2q^2 + 4q^4$ $-5q^5 - 4q^7 - 8q^8$ $+10q^{10} - 12q^{11} \cdots$
$\frac{7}{9}$	$\frac{(81t-25)}{56}$	I_6	I_3	–	I_2	I_1	$\frac{25}{81}$	$\frac{23^2}{81}$	21	$q - 3q^2 - 3q^3 \cdots$
$\frac{1}{4}$	$\frac{9}{4t}$	I_2	–	I_6	I_1	I_3	$\frac{1}{4}$	$\frac{9}{4}$	90	$q - 2q^2 + 4q^4$ $+5q^5 - 4q^7 - 8q^8$ $-10q^{10} + 48q^{11} \cdots$

TABLE 2. Modular Calabi-Yau varieties of the form $\mathcal{E}_{(1:\alpha^2:\alpha^2)} \times_{\mathbb{P}^1} \mathcal{E}^{Mt}_{(1:1:1)}$

α^2	Γ	M	level	modular form
-3	$\Gamma(3)$	$(t+7)/8$	27	$q - 3q^2 + q^4 - 15q^5 \cdots$
9	$\Gamma_1(4) \cap \Gamma(2)$	$(25-t)/24$	12	$q + 3q^3 - 18q^5 + 8q^7 \cdots$
121/125	$\Gamma_1(5)$	$\frac{125}{88}(1-t)$	55	$q + q^2 - 3q^3 - 7q^4 + \cdots$
9	$\Gamma_0(8) \cap \Gamma_1(4)$	$24/(25-t)$	12	$q + 3q^3 - 18q^5 + 8q^7 \cdots$
-3	$\Gamma_0(9) \cap \Gamma_1(3)$	$(t+7)/8$	27	$q - 3q^2 + q^4 - 15q^5 \cdots$

TABLE 3. Modular Calabi-Yau threefolds of the form $\mathcal{E}_{(1:\alpha^2:\alpha^2)} \times_{\mathbb{P}^1} \mathcal{E}(\Gamma)^{Mt}$

gives $\alpha = 17, 5/4, 7/9$, and the third,

$$j\left(\left(\frac{2\alpha+1}{2\alpha-1}\right)^2\right) = 73^3 2^{-6} 3^{-4}$$

gives $\alpha = 1, 1/4$ (plus irrational solutions), but $\alpha = 1$ is a degenerate case. We obtain the examples given in Table 2. As before, the first family, $\mathcal{E}_{(1:\alpha^2:\alpha^2)}$, has fibre types I_6, I_2, I_2, I_1, I_1 in this order, so we only list the fibre types for the second family, $\mathcal{E}^{Mt}_{(1:1:1)}$. In all cases $h^{12} = 1$. Note that the two level 90 modular forms in this table are not twists of each other.

ii. $\mathcal{E}_{(1:a:a)}$ with Beauville modular fibrations. The same procedure may be applied to fibrations of the form

$$\mathcal{E}_{(1:\alpha^2:\alpha^2)} \times_{\mathbb{P}^1} \mathcal{E}(\Gamma)^{Mt}.$$

Table 3, which is not exhaustive, gives a few examples, all of which have $h^{12} = 1$. Note that the level 55 example is the same as Schütt's W_2 in [**Sch05**]. There is a 3 : 1 isogeny between the two cases in Table 3 where level 27 occurs. This isogeny is described by Schoen [**Sch95**].

Case C: $\#S = \#S' = 4$, $\#S'' = 3$.

i. Fibre products of $\mathcal{E}_{(1:1:1)}$ fibrations. We consider non-rigid possibilities of the form

$$\mathcal{E}_{(1:1:1)} \times_{\mathbb{P}^1} \mathcal{E}^{Mt}_{(1:1:1)}.$$

We must take M such that

$$\#(\{0,1,\infty,9\} \cup M\{0,1,\infty,9\}) = 3.$$

For example, there are 6 possible ways to map $0,1,9$ to $0,1,9$, only one of which (the identity) maps ∞ to ∞. The other 5 cases are given by transformations given by matrices

$$M := \begin{pmatrix} 81 & -81 \\ 17 & -81 \end{pmatrix}, \begin{pmatrix} 81 & -81 \\ 73 & -9 \end{pmatrix}, \begin{pmatrix} 9 & -81 \\ 73 & -81 \end{pmatrix}, \begin{pmatrix} 9 & 0 \\ 10 & -9 \end{pmatrix}, \text{ or } \begin{pmatrix} -1 & 9 \\ 7 & 1 \end{pmatrix}.$$

The resulting fibrations all have $h^{12} = 10$.

For each of the 10 choices of pairings, we have 5 maps for which $S \neq S'$ (there are 6 maps (the number of permutations of three objects), but this includes the identity, or a map such as $t \mapsto 9/t$ which leaves the cross ratio of $0,1,9,\infty$ invariant (the cross-ratio is invariant under $a \leftrightarrow b, c \leftrightarrow d$)). Explicitly, the maps are given by fractional linear transformations defined by matrices $m_i T^j (RT)^k m_l^{-1}$ where $1 \leq i \leq l \leq 4$, $j = 0,1$, $k = 0,1,2$, and $T = \begin{pmatrix} -1 & 1 \\ 0 & 1 \end{pmatrix}$, $R = \begin{pmatrix} 0 & 1 \\ 1 & 0 \end{pmatrix}$, $m_1 = \begin{pmatrix} 9 & 0 \\ 1 & 8 \end{pmatrix}$, $m_2 = \begin{pmatrix} 9 & 0 \\ 0 & 1 \end{pmatrix}$, $m_3 = \begin{pmatrix} 8 & 1 \\ 0 & 1 \end{pmatrix}$, $m_4 = I$. This gives 50 fibrations, (six of which are given in Table 4), which may not all be new, for example,

$$\mathcal{E}_{(1:1:1)} \times \mathcal{E}^{9t}_{(1:1:1)} \sim_{bir} X_{(1:1:1:9:9:9)},$$

and may not all be distinct, for example, $\mathcal{E}_{(1:1:1)} \times \mathcal{E}^{Mt}_{(1:1:1)}$ is the same variety as $\mathcal{E}_{(1:1:1)} \times \mathcal{E}^{M^{-1}t}_{(1:1:1)}$. The only rigid examples are those with $M\{0,1,\infty\} = \{0,1,\infty\}$, (two of which are the same) so we have 4 rigid cases, and at most 45 non-rigid cases. All of the rigid cases (given as the first 4 examples in Table 4) have already been studied in detail by Schütt [**Sch04**]. Schütt also pointed out to us that the level 90 example in Table 4 is a twist of the level 10 example in this table. Note that this level 90 form is the same as the first level 90 example in Table 2, but for that case we had $h^{12} = 1$. We believe that those examples in Table 4 which have the same modular form are isogenous.

ii. Beauville case.

We can obtain examples from the $\mathcal{E}(\Gamma)$ in exactly the same way as for $\mathcal{E}_{(1:1:1)} = \mathcal{E}(\Gamma_1(6))$ above. However, in general, unless all the singular t are rational, we have fewer cases. Also in some examples, many of the possible M give rise to isomorphic cases. Table 4 gives a number of examples of modular fibrations of the form

$$\mathcal{E}(\Gamma_1) \times_{\mathbb{P}^1} \mathcal{E}^{Mt}(\Gamma_2),$$

but is far from exhaustive. The level means the conjectural level of the corresponding weight 4 modular form, indicated by computations. Note that there are some isogenous pairs in this table, for which we expect that the weight 4 modular forms are the same. However, h^{12} is not preserved by the map corresponding to the isogeny. Several cases involving $\Gamma_1(4) \cap \Gamma(2)$, with $h^{12} = 1$, could be replaced by rigid cases with $\Gamma_0(8) \cap \Gamma_1(4)$ replacing $\Gamma_1(4) \cap \Gamma(2)$.

The isogenous case: Finally, If Y and Y' **are** isogenous then $\#S = \#S' = \#S''$, and the condition $\delta = 0$ in Theorem 2.3 translates into $\#S = 4$.

Γ_1	Γ_2	M	h^{12}	level	modular form
$\Gamma_1(6)$	$\Gamma_1(6)$	$t/(t-1)$	0	21	$q-3q^2-3q^3\cdots$
$\Gamma_1(6)$	$\Gamma_1(6)$	$1/t$	0	10	$q+2q^2-8q^3\cdots$
$\Gamma_1(6)$	$\Gamma_1(6)$	$1-t$	0	17	$q-3q^2-8q^3\cdots$
$\Gamma_1(6)$	$\Gamma_1(6)$	$(t-1)/t$	0	73	$q+3q^2-8q^3\cdots$
$\Gamma_1(6)$	$\Gamma_1(6)$	$9t$	2	90	$q-2q^2+4q^4$ $-5q^5-4q^7-8q^8\cdots$
$\Gamma_1(6)$	$\Gamma_1(6)$	$9-t$	4	21	$q-3q^2-3q^3\cdots$
$\Gamma_0(8)\cap\Gamma_1(4)$	$\Gamma_0(8)\cap\Gamma_1(4)$	$1-t$	0	12	$q+3q^3-18q^5\cdots$
$\Gamma_1(4)\cap\Gamma(2)$	$\Gamma_1(4)\cap\Gamma(2)$	$1-t$	2	12	$q+3q^3-18q^5\cdots$
$\Gamma_0(8)\cap\Gamma_1(4)$	$\Gamma_0(8)\cap\Gamma_1(4)$	$\frac{-2}{t+1}$	1	48	$q-3q^3-18q^5\cdots$
$\Gamma_0(8)\cap\Gamma_1(4)$	$\Gamma_1(6)$	t	0	80	$q-2q^3-5q^5\cdots$
$\Gamma_1(4)\cap\Gamma(2)$	$\Gamma_1(6)$	t	1	80	$q-2q^3-5q^5\cdots$
$\Gamma_1(4)\cap\Gamma(2)$	$\Gamma_1(6)$	$1/t$	1	80	$q-2q^3-5q^5\cdots$
$\Gamma_1(4)\cap\Gamma(2)$	$\Gamma_1(6)$	$1-t$	1	28	$q-10q^3-8q^5\cdots$
$\Gamma_1(4)\cap\Gamma(2)$	$\Gamma_1(6)$	$1/(1-t)$	1	28	$q-10q^3-8q^5\cdots$
$\Gamma_1(4)\cap\Gamma(2)$	$\Gamma_1(6)$	$1-1/t$	1	68	$q-2q^3-8q^5\cdots$
$\Gamma_0(9)\cap\Gamma_1(3)$	$\Gamma_0(9)\cap\Gamma_1(3)$	$1/t$	16	27	$q-3q^2+q^4\cdots$

TABLE 4. Modular Calabi-Yau threefolds of the form $\mathcal{E}(\Gamma_1)\times\mathcal{E}(\Gamma_2)^{Mt}$

All examples of this type are rigid. This case includes all self fibre products of Beauville examples, given in the appendix to [**Yui03**]. Another example is

$$\mathcal{E}(\Gamma_0(8)\cap\Gamma_1(4))\times_{\mathbb{P}^1}\mathcal{E}^{(t-1)/(t+1)}(\Gamma_0(8)\cap\Gamma_1(4))$$

The fact that $t\mapsto(t-1)/(t+1)$ is an isogeny is explained in more detail in Example 3.5 in [**Ver01**], which deals with the same elliptic pencil, but with parameter t replaced by $\frac{1}{4t}$.

We have computationally verified that the Mellin transform of the L-series of the middle cohomology of this variety is the weight 4 level 16 modular form,

$$q+4q^3-2q^5-24q^7-11q^9+44q^{11}\cdots.$$

Appendix A. A Modularity Criterion for Integral Galois Representations and Calabi-Yau Threefolds[1]

Luis V. Dieulefait
Department of Mathematics
Universitat de Barcelona
E-mail address: ldieulefait@mat.ub.es

ABSTRACT. In a previous article, we have proved a result asserting the existence of a compatible family of Galois representations containing a given crystalline irreducible odd two-dimensional representation. We apply this result, combined with the potential modularity results of Taylor, to prove modularity for any irreducible crystalline ℓ-adic odd 2-dimensional Galois representation (with finite ramification set) unramified at 3 verifying an "ordinarity at 3" easy to check condition, with Hodge-Tate weights $\{0, w\}$ such that $2w < \ell$ (and $\ell > 3$) and such that the traces a_p of the images of Frobenius verify $\mathbb{Q}(\{a_p\}) = \mathbb{Q}$. This result applies in particular to any motivic compatible family of odd two-dimensional Galois representations of $\mathrm{Gal}(\bar{\mathbb{Q}}/\mathbb{Q})$ if the motive has rational coefficients, good reduction at 3, and the "ordinarity at 3" condition is satisfied. As a corollary, this proves that all rigid Calabi-Yau threefolds defined over $\mathbb{Q}$ having good reduction at 3 and such that a_3 is not divisible by 3 are modular, and it can also be used to obtain a modularity criterion for non-rigid Calabi-Yau threefolds containing elliptic ruled surfaces. The results are also valid if the local condition is imposed at 5 instead of at 3.

A.1. The result and its proof. The main tools in our proof are the existence of families proved in [**Die04**], together with the potential modularity results proved in [**Tay02, Tay01**]. We will also need a result from [**Wil88**] which controls ordinarity for Hilbert modular forms.

We will call the field of coefficients of a Galois representation the field generated by the traces of the images of Frobenius elements.

THEOREM A.1. Let $\ell > 3$ be a prime. Let σ_ℓ be a two-dimensional odd irreducible ℓ-adic Galois representation (of the absolute Galois group of $\mathbb{Q}$, continuous) ramified only at ℓ and at a finite set of primes S not containing 3, with field of coefficients $\mathbb{Q}$. Assume that σ_ℓ is crystalline at ℓ, with Hodge-Tate weights $\{0, w\}$ (w odd). Assume also that $\ell > 2w$. Then, if $a_3 := \mathrm{trace}(\sigma_\ell(\mathrm{Frob}_3))$ is not divisible by 3, the representation σ_ℓ is modular.

COROLLARY A.2. If an odd, two-dimensional, compatible family of Galois representations is attached to a motive defined over $\mathbb{Q}$, having rational coefficients, good reduction at 3 and such that 3 does not divide $a_3 := \mathrm{trace}(\sigma_\ell(\mathrm{Frob}_3))$, $\ell \neq 3$, then the family (thus, the motive) is modular. In particular, any rigid Calabi-Yau threefold defined over $\mathbb{Q}$ with good reduction at 3 and 3 not dividing a_3 is modular.

REMARK A.1 (Remark). From the theorem there can also be deduced a similar modularity criterion for non-rigid Calabi-Yau threefolds containing elliptic ruled surfaces, as shown in the accompanying article of Hulek and Verrill.

REMARK A.2 (Remark). This modularity criterion for rigid Calabi-Yau threefolds is different from those obtained in [**DM03**]. In particular, the criteria in

---

[1]2000 Mathematics Subject Classification: Primary 11F11, 11F80, 14J32; Secondary 11G40, 14F32

[**DM03**] required good reduction at 3 and 7 or at 5 and some other prime.

PROOF OF COROLLARY. Just apply the theorem to the ℓ-adic representation in the family for a sufficiently large prime ℓ where the motive has good reduction. □

PROOF OF THEOREM. First, recall that from the existence of a family result in [**Die04**] we can insert σ_ℓ in a (strongly) compatible family $\{\sigma_q\}$ which has rational coefficients and is unramified at 3 for any $q \neq 3$. Also, from the results in [**Tay01**], we know that this family is potentially modular, i.e., that when restricted to some totally real number field F all representations in the family agree with those attached to some Hilbert modular form h over F.

Now, we will translate the "easy to check condition" on the trace of σ_ℓ at Frob_3 into ordinarity of the 3-adic representation in the compatible family. We apply a result of Wiles [**Wil88**], which tells us that we can read off from the corresponding eigenvalue of a Hilbert modular form that the attached Galois representation is ordinary. Our family of representations becomes modular when restricted to the Galois group of a totally real number field F, which can be assumed (using solvable base change) to be such that 3 is totally split in $F/\mathbb{Q}$ (cf. [**Die04**]), so that ordinarity at 3 of the restriction to the Galois group of F is equivalent to ordinarity of the full 3-adic representation. The condition 3 does not divide a_3 implies (cf. [**Wil88**]) that when we restrict σ_3 to the Galois group of F we get a modular Galois representation which is ordinary at (every prime of F dividing) 3, therefore we conclude that σ_3 is ordinary.

The proof finishes using Skinner-Wiles results: since the residual mod 3 representation has coefficients in $\mathbb{F}_3$, it is known that it is either modular or reducible (by results of Langlands and Tunnell). Knowing that σ_3 is ordinary, an application of [**SW99**] and [**SW01**] gives the modularity of σ_3, thus of σ_ℓ (because they are compatible). □

A.2. An extension of the result. The theorem and the corollary remain true if we replace the prime 3 by the prime 5 everywhere in their statements, and in particular this implies that any rigid Calabi-Yau threefold defined over $\mathbb{Q}$ with good reduction at 5 and 5 not dividing a_5 is modular. The proof given above extends automatically to this case; just recall that (and this was also recalled and used in [**DM03**]) since the mod 5 residual representation has values in $\mathbb{F}_5$ it is known that it is either modular or reducible.

References

[Bea82] Arnaud Beauville. Les familles stables de courbes elliptiques sur $\mathbf{P}^1$ admettant quatre fibres singulières. *C. R. Acad. Sci. Paris Sér. I Math.*, 294(19):657–660, 1982.

[Die04] Luis V. Dieulefait. Existence of families of Galois representations and new cases of the Fontaine-Mazur conjecture. *J. Reine Angew. Math.*, 577:147–151, 2004.

[Die05] Luis V. Dieulefait. A modularity criterion for integral Galois representations and Calabi–Yau threefolds, Appendix to the article of Hulek and Verrill, in this volume.

[DM03] Luis Dieulefait and Jayanta Manoharmayum. Modularity of rigid Calabi-Yau threefolds over **Q**. In *Calabi-Yau varieties and mirror symmetry (Toronto, ON, 2001)*, volume 38 of *Fields Inst. Commun.*, pages 159–166. Amer. Math. Soc., Providence, RI, 2003.

[FM87] Jean-Marc Fontaine and William Messing. p-adic periods and p-adic étale cohomology. In *Current trends in arithmetical algebraic geometry (Arcata, Calif., 1985)*, volume 67 of *Contemp. Math.*, pages 179–207. Amer. Math. Soc., Providence, RI, 1987.

[Ful98] William Fulton. *Intersection theory*, volume 2 of *Ergebnisse der Mathematik und ihrer Grenzgebiete. 3. Folge. A Series of Modern Surveys in Mathematics [Results in Mathematics and Related Areas. 3rd Series. A Series of Modern Surveys in Mathematics]*. Springer-Verlag, Berlin, second edition, 1998.

[HV05] Klaus Hulek and Helena Verrill. On modularity of rigid and non-rigid Calabi-Yau varieties associated to the root lattice A_4. *Nagoya Math. J.*, 2005. math.AG/0304169. To appear.

[LY03] R. Livné and Noriko Yui. The modularity of certain non-rigid Calabi-Yau threefolds. math.AG/0304497, 2003. J. Math. Kyoto Univ., to appear

[Sch88] Chad Schoen. On fiber products of rational elliptic surfaces with section. *Math. Z.*, 197(2):177–199, 1988.

[Sch95] Chad Schoen. On the computation of the cycle class map for nullhomologous cycles over the algebraic closure of a finite field. *Ann. Sci. École Norm. Sup. (4)*, 28(1):1–50, 1995.

[Sch04] Matthias Schütt. New examples of modular rigid Calabi-Yau threefolds. *Collect. Math.*, 55(2):219–228, 2004.

[Sch05] Matthias Schütt. On the modularity of three Calabi-Yau threefolds with bad reduction at 11. *Canad. Math. Bull.*, 2005. math.AG/0405450. To appear.

[SW99] C. M. Skinner and A. J. Wiles. Residually reducible representations and modular forms. *Inst. Hautes Études Sci. Publ. Math.*, (89):5–126 (2000), 1999.

[SW01] C. M. Skinner and A. J. Wiles. Nearly ordinary deformations of irreducible residual representations. *Ann. Fac. Sci. Toulouse Math. (6)*, 10(1):185–215, 2001.

[Tay01] Richard Taylor. On the meromorphic continuation of degree two L-functions. available at http://abel.math.harvard.edu/~rtaylor/, 2001.

[Tay02] Richard Taylor. Remarks on a conjecture of Fontaine and Mazur. *J. Inst. Math. Jussieu*, 1(1):125–143, 2002.

[Uen75] Kenji Ueno. *Classification theory of algebraic varieties and compact complex spaces.* Springer-Verlag, Berlin, 1975. Notes written in collaboration with P. Cherenack, Lecture Notes in Mathematics, Vol. 439.

[Ver01] H. A. Verrill. Picard-Fuchs equations of some families of elliptic curves. In *Proceedings on Moonshine and related topics (Montréal, QC, 1999)*, volume 30 of *CRM Proc. Lecture Notes*, pages 253–268, Providence, RI, 2001. Amer. Math. Soc.

[Wil88] A. Wiles. On ordinary λ-adic representations associated to modular forms. *Invent. Math.*, 94(3):529–573, 1988.

[Yui03] Noriko Yui. Update on the modularity of Calabi-Yau varieties. In *Calabi-Yau varieties and mirror symmetry (Toronto, ON, 2001)*, volume 38 of *Fields Inst. Commun.*, pages 307–362. Amer. Math. Soc., Providence, RI, 2003. With an appendix by Helena Verrill.

Institut für Mathematik (C), Universität Hannover, Welfengarten 1, 30060 Hannover, Germany

E-mail address: hulek@math.uni-hannover.de

Department of Mathematics, Louisiana State University, Baton Rouge, LA 70803-4918, USA

E-mail address: verrill@math.lsu.edu

AMS/IP Studies in Advanced Mathematics
Volume 38, 2006

Arithmetic Mirror Symmetry for a Two-parameter Family of Calabi–Yau Manifolds

Shabnam Kadir

ABSTRACT. We study mirror symmetric pairs of Calabi–Yau manifolds over finite fields. In particular we compute the number of rational points of the manifolds as a function of the complex structure parameters. The data of the number of rational points of a Calabi–Yau $X/\mathbb{F}_q$ can be encoded in a generating function known as the congruent zeta function. The Weil Conjectures (proved in the 1970s) show that, for smooth varieties, these functions take a very interesting form in terms of the Betti numbers of the variety. This has interesting implications for mirror symmetry, as mirror symmetry exchanges the odd and even Betti numbers. Here the zeta function for a two-parameter family of octics in weighted projective space, $\mathbb{P}_4{}^{(1,1,2,2,2)}$ [8] is computed. The form of the zeta function at points in the moduli space of complex structures where the manifold is singular (where the Weil conjectures, apart from rationality, are not applicable), is investigated. The zeta function appears to be sensitive to monomial and non-monomial deformations of complex structure (or, equivalently on the mirror side, toric and non-toric divisors). Various conjectures about the form of the zeta function for mirror symmetric pairs are made in light of the results of this calculation.

1. Introduction

In this paper we are interested in studying arithmetic properties of Mirror symmetry and present some of the results obtained in the author's thesis [**K**]. Mirror symmetry is a conjecture in string theory according to which certain 'mirror pairs' of Calabi–Yau manifolds give rise to isomorphic physical theories. A Calabi–Yau manifold is a complex variety of dimension d with trivial canonical bundle and vanishing Hodge numbers $h^{i,0}$ for $0 < i < d$.

Physicists concerned with mirror symmetry usually deal with Calabi–Yau manifolds defined over $\mathbb{C}$; here however, in order to study the arithmetic, we shall reduce these algebraic Calabi–Yau varieties over discrete finite fields, $\mathbb{F}_q$, $q = p^r$, where p is prime and $r \in \mathbb{N}$; these are the extensions of degree r of the finite field $\mathbb{F}_p$. The data of the number of rational points of the reduced variety, $N_{r,p}(X) = \#(X/\mathbb{F}_{p^r})$,

---

2000 *Mathematics Subject Classification.* Primary 14J32, 14G10; Secondary 14G15, 14D05, 11F23, 11G25.

Key words and phrases. Calabi–Yau manifolds, mirror symmetry, zeta-functions, finite ground fields.

This research was conducted when the author was supported by an E.P.S.R.C. studentship at the Mathematical Institute, University of Oxford.

can be encoded in a generating function known as the congruent zeta function, which takes the form

$$Z(X/\mathbb{F}_{p^r}, t) \equiv \exp\left(\sum_{r\in\mathbb{N}} N_{r,p}(X)\frac{t^r}{r}\right). \tag{1.1}$$

The motivation for choosing the above type of generating function is that this expression leads to rational functions in the formal variable t. This result is part of the famous Weil conjectures (no longer conjectures for at least the last 30 years [**Dw, Del**]). The Weil conjectures show that the zeta function for a smooth variety is a rational function determined by the cohomology of the variety; in particular, that the degree of the numerator is the sum of the odd Betti numbers of the variety and the denominator, the sum of the even Betti numbers. As mirror symmetry interchanges the odd and even Betti numbers of the Calabi–Yau variety, it is a natural question to investigate the zeta functions of pairs of mirror symmetric families. Hence there is speculation as to whether a 'quantum modification' to the congruent zeta function can be defined such that the zeta functions of mirror pairs of Calabi–Yau varieties are inverses of each other. Notice that the above conjecture cannot hold using the 'classical definition' (1.1) above because this would mean that for a pair of manifolds, (X, Y), we would have to have $N_{r,p}(X) = -N_{r,p}(Y)$, which is not possible.

In order to study these questions, we study the two-parameter family of Calabi–Yau threefolds, octic hypersurfaces in weighted projective space $\mathbb{P}_4{}^{(1,1,2,2,2)}$ [8] and use methods similar to [**CdOV1, CdOV2**]. The (non-arithmetic) mirror symmetry for this model was studied in detail in [**CdOFKM**]. We shall be concerned with Calabi–Yau manifolds which are hypersurfaces in toric varieties, as this provides a powerful calculational tool, and it enables one to use the Batyrev formulation of mirror symmetry [**Bat**].

We shall be particularly interested in the behaviour of zeta functions of singular members of such families, as for singular varieties there is no guiding principle similar to the Weil conjectures for the smooth case. In the computations involving the two-parameter family of octic Calabi–Yau manifolds, it is observed that the zeta function degenerates in a consistent way at singularities. The nature of the degeneration depends on the type of singularity. In this case there are two types of singularity: a one-dimensional locus (in the base space) of conifold points and another one-dimensional locus of points where the Calabi–Yau manifolds are birational to a one-parameter family, whose conifold locus has previously been studied by Rodriguez-Villegas [**R-V**].

The method of computation involves use of Gauss sums, that is the sum over an additive character times a multiplicative character. For the purpose of counting rational points the Dwork character is a very suitable choice for the additive character along with the Teichmüller character as multiplicative character. The Dwork character was first used by Dwork [**Dw**] to prove the rationality of the congruent zeta function for varieties (this is one of the Weil conjectures). The number of rational points can be written in terms of these Gauss sums, thus enabling computer-aided computation of the zeta function.

The mirror of the Calabi–Yau manifolds can be found using the Batyrev mirror construction [**Bat**]. This is a construction in toric geometry in which a dual pair of reflexive polytopes can be related via toric geometry to mirror symmetric pairs of Calabi–Yau manifolds. Using this construction and using Gauss sums it is possible to also find the zeta function of the mirror Calabi–Yau manifold. In **[CdOV2]**, the mirror zeta function was found to have some factors in common with the original zeta function, namely a contribution to the number of points associated to the unique interior lattice point of the polyhedron. We shall find the same phenomenon for the octic where there is a sextic, $R_{(0,0,0,0,0)}$, that appears in both the original family and the mirror. This phenomenon will be explained in Section 7.7 for the special case of Fermat hypersurfaces in weighted projective space (these admit a Greene–Plesser description). For the octic threefolds, on the mirror side, there was also a contribution which may be thought of as related to a 'zero-dimensional Calabi–Yau manifold' (studied in **[CdOV1]**) which was sensitive to a particular type of singularity where upon resolution there was birationality with a one-parameter model. This contrasts with the the fact that the zeta function of the original family of octics has a contribution related to a particular monomial $(4,0,3,2,1)$ (not in the Newton polyhedron because it has degree 16 as opposed to 8) which is sensitive to the presence of conifold points. This phenomenon of a special contribution at conifold points was also found for the quintic threefolds in **[CdOV2]**.

We verify that the number of rational points of the octic Calabi–Yau manifolds over $\mathbb{F}_p$ can be given in terms of its periods. The periods satisfy a system of differential equations known as the Picard–Fuchs equations, with respect to the parameters. The Picard–Fuchs equations for the quintics (and also the octics) simplify considerably because of the group of automorphisms, $\mathcal{A}$, of the manifold. The elements of the polynomial ring can be classified according to their transformation properties under the automorphisms, that is, into representations of $\mathcal{A}$, and owing to the correspondence with the periods, the periods can be classified accordingly. Analogous results were found for the quintic threefold in **[CdOV1]**.

The toric data for a Calabi–Yau manifold can be used to calculate many important invariants, e.g. Hodge numbers. It can also be used to calculate the number of monomial and non-monomial deformations of complex structure **[Bat]**, or equivalently, on the mirror side, toric and non-toric divisors (by the monomial-divisor map **[AGM]**). The zeta functions appear to be sensitive to the type of deformation/divisor.

Arithmetic geometers have been interested in the cohomological L-series of Calabi–Yau varieties over $\mathbb{Q}$ or number fields **[Yui]**. One important question is the modularity of Calabi–Yau varieties, i.e. are their cohomological L-series completely determined by certain modular cusp forms? A brief review of these ideas is presented in Section 2.3, see **[Yui]** for a thorough overview . In particular, rigid Calabi–Yau threefolds defined over the field of rational numbers are equipped with 2-dimensional Galois representations, which are conjecturally equivalent to modular forms of one variable of weight 4 on some congruence subgroup of $PSL(2,\mathbb{Z})$ **[Yui]**. For not necessarily rigid Calabi–Yau threefolds over the rationals, the Langlands Program predicts that there should be some automorphic forms attached to them. Modularity was observed for the octic family of Calabi–Yau threefolds at the special values in the moduli space where there was birationality with the one-parameter family and, simultaneously, a conifold singularity, which reproduced the results in **[R-V]**. However, other indications of potential modularity were observed for which related cusp forms were found using the tables of **[St]**.

It is thought that higher dimensional Galois representations will correspond to L-functions associated with Siegel modular forms. In **[K]** we find evidence of connections to the spinor L-function of genus 2 Siegel modular forms, and possibly to the standard L-function of some genus 3 modular forms.

The structure of this paper is as follows: Section 2 gives a brief introduction to L-functions, outlining what is expected for the zeta functions of mirror symmetric manifolds by only considering the Weil Conjectures. Section 3 provides an introduction to the Batyrev procedure and introduces the family of Calabi–Yau manifolds we shall be considering. Section 4 outlines the Dwork–Katz–Griffiths method for finding the Picard–Fuchs equations. The relation between periods and rational points can be explained by comparing the classes of monomials obtained using the Dwork–Katz–Griffiths method with the classes of monomials obtained by calculating the number of rational points in terms of Gauss sums in Section 5. This section includes an introduction to the Gauss sums employed in the computation, and the derivation of formulae for the number of rational points for both the original family of manifolds and its mirror. In Section 7 we analyse the zeta functions obtained from these derived formula, observing many instances of probable modularity, and make conjectures concerning the structure of the zeta functions corresponding to Calabi–Yau manifolds. Particular interest attaches to the form of the zeta function for values of the parameters for which the manifold is singular. We conclude with a mirror theorem; other examples of arithmetic mirror theorems appear in the recent work of Daqing Wan [**Wn**], also in this volume. The explicit zeta functions for $p = 3, 5, 7, 11, 13, 17$ for the octic Calabi–Yau threefolds as a function of the parameters can be found in the author's thesis [**K**]. The results for $p = 19, 23$ are also available and support the conjectures that we make regarding the general form of the zeta function, but were too large to display in print. In this paper only the results for the primes 3 and 5 are given.

2. L-functions

In this section we review some general properties of zeta functions and existing conjectures about modularity; throughout this paper we shall often refer to some of the known results presented here.

2.1. The congruent zeta function. The arithmetic structure of Calabi–Yau varieties can be encoded in the congruent zeta function. The Weil Conjectures (finally proven by Deligne) show that the zeta function is a rational function determined by the cohomology of the variety.

DEFINITION 2.1 (Congruent Zeta Function). The congruent zeta function for a smooth projective variety X defined over $\mathbb{F}_p$ is defined as follows:

$$Z(X/\mathbb{F}_p, t) := \exp\left(\sum_{r\in\mathbb{N}} \#(X/\mathbb{F}_{p^r})\frac{t^r}{r}\right). \tag{2.1}$$

2.2. Weil conjectures and mirror symmetry.

THEOREM 2.1 (Weil Conjectures).
The three Weil conjectures are:

(1) *Functional Equation*
The zeta function of an algebraic variety of dimension d satisfies a functional equation

$$Z\left(X/\mathbb{F}_p, \frac{1}{p^d t}\right) = (-1)^{\chi+\mu} p^{\frac{d\chi}{2}} t^{\chi} Z(X/\mathbb{F}_p, t), \tag{2.2}$$

where χ is the Euler characteristic of the variety X over $\mathbb{C}$. Furthermore μ is zero when the dimension d of the variety is odd, and when d is even, μ is the multiplicity of $-p^{d/2}$ as an eigenvalue of the action induced on the cohomology by the Frobenius automorphism:

$$F : X \to X, \quad x \mapsto x^p \ . \tag{2.3}$$

(2) *Rationality*
The zeta function is rational $Z(X/\mathbb{F}_p, t) \in \mathbb{Q}(t)$. There is a factorization:

$$Z(X/\mathbb{F}_p, t) = \frac{\prod_{j=1}^{d} P_{2j-1}^{(p)}(t)}{\prod_{j=0}^{d} P_{2j}^{(p)}(t)} \ , \tag{2.4}$$

with each $P_j^{(p)}(t) \in \mathbb{Z}[t]$. Further, $P_0^{(p)}(t) = 1 - t$, $\quad P_{2d}^{(p)}(t) = 1 - p^d t$, and for each $1 \leq j \leq (2d-1)$, $P_j^{(p)}(t)$ factors (over $\mathbb{C}$) as

$$P_j^{(p)}(t) = \prod_{l=1}^{b_j} (1 - \alpha_l^{(j)}(p) t) \ , \tag{2.5}$$

where the b_j are the Betti numbers of the variety, $b_j = \dim H^j_{\mathrm{DeRham}}(X)$. This was first proven by Dwork [**Dw**].

(3) *Riemann Hypothesis*
The coefficients $\alpha_l^{(j)}(p)$ are algebraic integers which satisfy the Riemann hypothesis:

$$\|\alpha_l^{(j)}(p)\| = p^{j/2} \ . \tag{2.6}$$

This was proved by Deligne.

For Calabi–Yau threefolds with finite fundamental group, i.e. $h^{1,0} = 0 = h^{2,0}$, the expressions above simplify in the following way:

$$Z(X/\mathbb{F}_p, t) = \frac{P_3^{(p)}(t)}{(1-t) P_2^{(p)}(t) P_4^{(p)}(t) (1 - p^3 t)} \ , \tag{2.7}$$

with

$$\deg(P_3^{(p)}(t)) = 2 + 2h^{2,1}, \quad \deg(P_2^{(p)}(t)) = \deg(P_4^{(p)}(t)) = h^{1,1}. \tag{2.8}$$

as the Hodge diamond in this case is of the form

$$\begin{array}{ccccccc} & & & 1 & & & \\ & & 0 & & 0 & & \\ & 0 & & h^{1,1} & & 0 & \\ 1 & & h^{1,2} & & h^{2,1} & & 1 \\ & 0 & & h^{2,2} & & 0 & \\ & & 0 & & 0 & & \\ & & & 1 & & & \end{array}$$

where $h^{1,1} = h^{2,2}$ and $h^{2,1} = h^{1,2}$.

The Weil conjectures suggest that some residue of Mirror Symmetry survives in the form of the zeta function for Calabi–Yau threefolds. This is because mirror symmetry interchanges the Hodge numbers, i.e. for a mirror symmetric pair, $\mathcal{M}$, $\mathcal{W}$:

$$h^{1,1}_{\mathcal{M}} = h^{2,1}_{\mathcal{W}}, \quad h^{2,1}_{\mathcal{M}} = h^{1,1}_{\mathcal{W}} . \tag{2.9}$$

Section 3 provides more details about (2.9).

2.3. Hasse–Weil L-functions. The concept of the Hasse–Weil L-Function of an elliptic curve can be generalized for a Calabi–Yau d-fold. It can be shown that the *ith* polynomial $P_i(t)$ is associated to the action induced by the Frobenius morphism on the ith cohomology group $H^i(X)$. It is useful to decompose the zeta function in order to find out the arithmetic information encoded in the Frobenius action.

DEFINITION 2.2 (Local L-Function). Let $P_i(t)$ be the polynomials in the zeta function over the field $\mathbb{F}_p$. The *ith* L-function of the variety X over $\mathbb{F}_p$ is defined as:

$$L^{(i)}(X/\mathbb{F}_p, s) = \frac{1}{P_i^{(p)}(p^{-s})} . \tag{2.10}$$

The Hasse–Weil L-Function of a Calabi–Yau d-fold X over $\mathbb{Q}$ (assume $h^{i,0} = 0$ for $1 < i < (d-1)$) is defined as follows:

$$L_{HW}(X, s) = \prod_{p\, good} L^{(d)}(X/\mathbb{F}_p, s) = \prod_{p\, good} \frac{1}{\prod_{j=1}^{b_d}(1 - \alpha_j^{(d)}(p)p^{-s})} , \tag{2.11}$$

where the product is taken over the primes of good reduction for the variety.

2.4. Modularity conjectures for Calabi–Yau threefolds. Rigid Calabi–Yau threefolds have the property that $h^{2,1} = 0$. In a certain sense, rigid Calabi–Yau varieties are the natural generalization of elliptic curves to higher dimensions.

CONJECTURE 2.1 (Yui). *Every rigid Calabi–Yau threefold over $\mathbb{Q}$ is modular, i.e. its L-series coincides with the Mellin transform of a modular cusp form, f, of weight 4 on $\Gamma_0(N)$. Here N is a positive integer divisible by primes of bad reduction. More precisely:*

$$L(X, s) = L(f, s) \quad \text{for } f \in S_4(\Gamma_0(N)). \tag{2.12}$$

At certain points in the complex structure moduli space for the mirror family of octic Calabi–Yau threefolds that we shall consider, the manifold is rigid. Here we observe indications of elliptic modularity as expected by the above conjecture. Our calculations not only verify the results of [**R-V**], but also point to connections with another modular cusp form. This is described in detail in Section 7.5.

For not necessarily rigid Calabi–Yau threefolds over the rationals, the Langlands Program predicts that there should be some automorphic forms attached to them. In this paper, we shall be considering a family of threefolds which is non-rigid for all but a few sub-cases. In [**K**] possible connections with Siegel modular forms were observed.

3. Toric geometry, mirror symmetry and Calabi–Yau manifolds

In this section we shall first provide a summary of Batyrev's construction for mirror pairs of toric Calabi–Yau hypersurfaces given in [**Bat, CK, Voi**] as well as numerous other sources. We shall make use of this construction in our calculations for the zeta function for the two-parameter mirror symmetric pair of octic threefolds. A precise geometric description of the two-parameter family that we study is given in Section 3.3.

3.1. Batyrev's construction. To describe the toric variety $\mathbb{P}_\Delta$, let us consider an n-dimensional convex integral polyhedron $\Delta \in \mathbb{R}^n$ containing the origin $\mu_0 = (0, \ldots, 0)$ as an interior point. An integral polyhedron is a polyhedron with integral vertices, and is called reflexive if its dual, defined by

$$\Delta^* = \{(x_1, \ldots, x_n) | \sum_{i=1}^{n} x_i y_i \geq -1 \quad \forall (y_1, \ldots, y_n) \in \Delta\}, \tag{3.1}$$

is again an integral polyhedron. It is clear that if Δ is reflexive, then Δ^* is also reflexive, since $(\Delta^*)^* = \Delta$.

We associate to Δ a complete rational fan, $\Sigma(\Delta)$, in the following way: For every l-dimensional face, $\Theta_l \subset \Delta$, we define an n-dimensional cone $\sigma(\Theta_l)$ by

$$\sigma(\Theta_l) := \{\lambda(p' - p) | \lambda \in \mathbb{R}_+, \ p \in \Delta, \ p' \in \Theta_l\}. \tag{3.2}$$

The fan, $\Sigma(\Delta)$, is then given as the collection of $(n-l)$-dimensional dual cones $\sigma^*(\Theta)$ for $l = 0, 1, \ldots, n$ for all faces of Δ. The toric variety, $\mathbb{P}_\Delta$, is the toric variety associated to the fan $\Sigma(\Delta)$, i.e. $\mathbb{P}_\Delta := \mathbb{P}_{\Sigma(\Delta)}$ (See [**F**] for a nice description of how this is done).

Denote by μ_i $(i = 0, \ldots, s)$ the integral points of Δ and consider an affine space $\mathbb{C}^{s+1}$ with coordinates $(a_0, \ldots, a_s)$. Consider the zero locus Z_f of the Laurent polynomial:

$$f(a, X) = \sum_{i=0}^{s} a_i X^{\mu_i}, \quad f(a, X) \in \mathbb{C}[X_1^{\pm 1}, \ldots, X_n^{\pm 1}], \tag{3.3}$$

in the algebraic torus $(\mathbb{C}^*)^n \in \mathbb{P}_\Delta$, and its closure $\bar{Z}_f$ in $\mathbb{P}_\Delta$. For conciseness, we have used the notation $X^\mu := X_1^{\mu_1} \ldots X_n^{\mu_n}$.

Let f_{Θ_l} denote the polynomial consisting of all the monomials in the face Θ_l, and let $f \equiv f_\Delta$. f and Z_f are called Δ-regular if for all $l = 1, \ldots, n$, the f_{Θ_l} and $X_i \frac{\delta}{\delta X_i} f_{\Theta_l}, \forall i = 1, \ldots, n$ are not zero simultaneously in $(\mathbb{C}^*)^n$. This is equivalent to the transversality condition for the quasi-homogeneous polynomials W_i.

We shall now assume Δ to be reflexive. When we vary the parameters a_i under the condition of Δ-regularity, we will have a family of Calabi–Yau varieties. In general the ambient space $\mathbb{P}_\Delta$ is singular, and so in general $\bar{Z}_f$ inherits some of the singularities of the ambient space. The Δ regularity condition ensures that the only singularities of $\bar{Z}_f$ are the inherited ones. $\bar{Z}_f$ can be resolved to a Calabi–Yau manifold $\hat{Z}_f$ if and only if $\mathbb{P}_\Delta$ only has Gorenstein singularities, which is the case if Δ is reflexive [**Bat**].

The families of Calabi–Yau manifolds $\hat{Z}_f$ will be denoted $\mathcal{F}(\Delta)$. Of course, the above definitions also hold for the dual polyhedron Δ^* with its integral points $\mu_i^* (i = 0, 1, \ldots, s^*)$.

Batyrev observed that a pair of reflexive polyhedra (Δ, Δ^*) naturally gives rise to a pair of mirror Calabi–Yau families $(\mathcal{F}(\Delta), \mathcal{F}(\Delta^*))$ as the following identities on the Hodge number hold when $n \geq 4$ ($(n-1)$ is the dimension of the Calabi–Yau varieties):

$$\begin{aligned} h^{1,1}(\hat{Z}_{f,\Delta}) &= h^{n-2,1}(\hat{Z}_{f,\Delta^*}) \\ &= l(\Delta^*) - (n+1) - \sum_{\operatorname{codim} \Theta^*=1} \acute{l}(\Theta^*) + \sum_{\operatorname{codim} \Theta^*=2} \acute{l}(\Theta^*)\acute{l}(\Theta) , \end{aligned}$$

(3.4)

$$\begin{aligned} h^{1,1}(\hat{Z}_{f,\Delta^*}) &= h^{n-2,1}(\hat{Z}_{f,\Delta}) \\ &= l(\Delta) - (n+1) - \sum_{\operatorname{codim} \Theta=1} \acute{l}(\Theta) + \sum_{\operatorname{codim} \Theta=2} \acute{l}(\Theta)\acute{l}(\Theta^*) . \end{aligned}$$

Here $l(\Theta)$ and $\acute{l}(\Theta)$ are the number of integral points on a face Θ of Δ and in its interior, respectively. An l-dimensional face can be represented by specifying its vertices $v_{i_1}, \ldots, v_{i_k}$. Then the dual face defined by $\Theta^* = \{x \in \Delta^* | (x, v_{i_1}) = \ldots = (x, v_{i_k}) = -1\}$ is an $(n-l-1)$-dimensional face of Δ^*. By construction $(\Theta^*)^* = \Theta$, and we thus have a natural pairing between l-dimensional faces of Δ and $(n-l-1)$-dimensional faces of Δ^*.The last sum in (3.5) is over pairs of dual faces. Their contribution cannot be associated with a monomial in the Laurent polynomial.

We shall denote by

$$h^{1,1}_{\text{toric}}(\hat{Z}_{f,\Delta}) = h^{n-1,1}_{\text{poly}}(\hat{Z}_{f,\Delta^*}) \text{ and } h^{1,1}_{\text{poly}}(\hat{Z}_{f,\Delta^*}) = h^{n-2,1}_{\text{toric}}(\hat{Z}_{f,\Delta})$$

the expressions (3.5) without the last terms which sum over codimension 2 faces. These are the dimensions of the spaces

$$H^{n-1}_{\text{poly}}(\hat{Z}_{f,\Delta^*}) \text{ and } H^{1,1}_{\text{poly}}(\hat{Z}_{f,\Delta^*}).$$

$H^{n-1}_{\text{poly}}(\hat{Z}_{f,\Delta^*})$ is isomorphic to the space of first-order polynomial deformations of $\hat{Z}_{f,\Delta^*}$ and can be generated by monomials. The space $H^{1,1}_{\text{poly}}(\hat{Z}_{f,\Delta^*})$ is the part of the second cohomology of $\hat{Z}_{f,\Delta^*}$ which comes from the ambient toric variety, and can be generated by toric divisors. There is a one-to-one correspondence between monomials and toric divisors called the *monomial-divisor map* explained in [**AGM**].

The formulae (3.5) are invariant under interchange of Δ with Δ^* and $h^{1,1}$ with $h^{n-2,1}$, and are thus manifestly mirror symmetric. For Calabi–Yau threefolds, $n = 4$ and we get the interchange of $h^{1,1}$ with $h^{2,1}$.

3.2. Cox variables. Global coordinates, akin to the homogeneous coordinates on projective space, can be defined for a toric variety. Hence we shall define the homogeneous coordinate ring of a toric variety. It turns out, as we shall see later, that using these coordinates is extremely convenient.

DEFINITION 3.1. If $X = X_\Sigma$ is given by a fan Σ in $N_{\mathbb{R}}$, then we can introduce a variable x_ρ for each $\rho \in \Sigma(1)$ (where $\Sigma(1)$ are the one-dimensional cones), and consider the polynomial ring

$$S = \mathbb{C}[x_\rho : \rho \in \Sigma(1)]. \tag{3.5}$$

A monomial in S is written $x^D = \prod_\rho x_\rho^{a_p}$, where $D = \sum_\rho a_\rho D_\rho$ is an effective torus-invariant divisor on X (this uses the monomial divisor correspondence $(\rho \leftrightarrow D_\rho)$), and we say that x^D has degree

$$\deg(x^D) = [D] \in A_{n-1}(X). \tag{3.6}$$

Here A_{n-1} is the Chow group. The ring S is graded by the Chow group and together they give us the homogeneous coordinate ring. Given a divisor class $\alpha \in A_{n-1}(X)$, let S_α denote the graded piece of S is degree α. For $\mathbb{P}^n$ it can be shown that the homogeneous coordinate ring defined above is the usual one. Hence, for weighted projective space we can associate a coordinate to each point in the Newton polyhedron Δ.

3.3. Calabi–Yau manifolds with two parameters. We shall consider Calabi–Yau threefolds which are obtained by resolving singularities of degree eight (octic) hypersurfaces in the weighted projective space $\mathbb{P}_4^{(1,1,2,2,2)}$. A typical defining polynomial for such a hypersurface is

$$P = c_1 x_1^8 + c_2 x_2^8 + c_3 x_3^4 + c_4 x_4^4 + c_5 x_5^4 + \alpha x_1^6 x_3 + \beta x_2^4 x_3^4 + \gamma x_1^7 x_2 \, , \tag{3.7}$$

though in general an octic may contain many more terms. There are 105 degree eight monomials, but using the freedom of homogeneous coordinate redefinitions, which provide 22 parameters, we are left with 83 possible monomial deformations. The Newton polyhedron for this family has 7 points, one important two-face of which is illustrated in Figure 1. There are two other types of two-face: a triangle with vertices $(0,0,4,0,0)$, $(0,0,0,4,0)$ and $(0,0,0,0,4)$, and six triangles with a vertex of the form $(8,0,0,0,0)$ and two vertices of the form $(0,0,4,0,0)$. The mirror family has a Newton polyhedron (the dual) containing 105 points, two-faces of which are illustrated in Section 6 in Figures 5, 6 and 7.

We can define a 2-dimensional sub-family of the above 83-dimensional family, namely degree eight hypersurfaces with two parameters of deformations defined by the equation

$$P(x) = x_1^8 + x_2^8 + x_3^4 + x_4^4 + x_5^4 - 2\phi x_1^4 x_2^4 - 8\psi x_1 x_2 x_3 x_4 x_5 \, . \tag{3.8}$$

We shall be comparing the zeta functions of this 2-dimensional sub-family of octics with that of the mirror family of the full 83-dimensional family. This is a special family because it only contains monomials which are invariant under the group G of automorphisms by which one quotients when using the Greene–Plesser construction.

The singularities occur along $x_1 = x_2 = 0$, where there is a curve C of singularities of type A_1. In our particular example, the curve C is described by

$$x_1 = x_2 = 0, \quad x_3^4 + x_4^4 + x_5^4 = 0 \, . \tag{3.9}$$

It only containing monomials which are invariant under the automorphism group. In general it will just be a smooth quartic plane curve, which always has genus 3. To resolve these singularities we blow up the locus $x_1 = x_2 = 0$; replacing the curve of singularities by a (exceptional divisor) E, which is a ruled surface over the curve C. That is, each point of C is blown up to a $\mathbb{P}^1$. We point out the existence of this exceptional divisor because later it contributes the the number of rational points in a way not dependent on the parameters (ϕ, ψ).

However, as we wish to exploit the Batyrev mathod for finding the mirror, it is best to represent the manifolds torically:

DEFINITION 3.2. The family of octic Calabi–Yau threefolds can be defined over the toric variety $\frac{\mathbb{C}^6 - F}{(\mathbb{C}^*)^2}$ (where the rôle of F is to restrict the coordinates that are allowed by toric geometry to vanish simultaneously), as a hypersurface defined by the following polynomial:

$$P(\mathbf{x}, x_6) = x_1^8 x_6^4 + x_2^8 x_6^4 + x_3^4 + x_4^4 + x_5^4 - 8\psi x_1 x_2 x_3 x_4 x_5 x_6 - 2\phi x_1^4 x_2^4 x_6^4 \, , \tag{3.10}$$

where the x_i are Cox variables associated to the monomials in the Newton polyhedron as follows (we have shifted back the monomials by $(1,1,1,1,1)$ for ease of manipulation):

$$x_1 = x_{(8,0,0,0,0,1,4)}, \quad x_2 = x_{(0,8,0,0,0,1,4)}, \quad x_3 = x_{(0,0,4,0,0,1,0)},$$
$$x_4 = x_{(0,0,0,4,0,1,0)}, \quad x_5 = x_{(0,0,0,0,4,1,0)}, \quad x_6 = x_{(4,4,0,0,0,1,4)}. \tag{3.11}$$

The two-faces of the triangulated Newton polyhedron which consists of 7 points (the dual polyhedron corresponding to the mirror contains 105 points) are shown below. The triangulation imposes restrictions on the allowable simultaneous vanishing of the coordinates. It should be noted that for this model the Hodge numbers

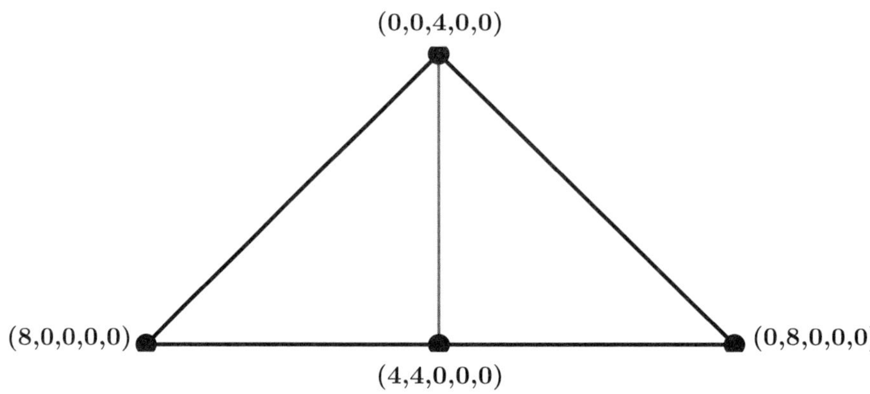

Figure 1: A two-face triangulated to resolve singularities

for values of (ϕ, ψ) for which the variety is smooth (which can be calculated from (3.5)) are $h^{1,1} = 2$, and $h^{2,1} = 86$. Hence we expect a zeta function of the form:

$$Z(X/\mathbb{F}_p, t) = \frac{R_{174}(t)}{(1-t)R_2(t)R_2'(t)(1-p^3t)}, \tag{3.12}$$

where $R_{174}(t)$ is of degree 174; $R_2(t)$ and $R_2'(t)$ are both of degree 2. We shall see in Section 5 that $R_2(t) = (1-pt)^2$ and $R_2'(t) = (1-p^2t)^2$.

It should be noted that using Batyrev's formula for the Hodge numbers, (3.5), for $n = 4$ and leaving out the last terms, we obtain $h^{2,1}_{\text{poly}} = 83$ (for the original family). (Using all of (3.5) we obtain $h^{2,1} = 86$). Thinking about the octics as hypersurfaces in weighted projective space, this corresponds to the fact that we can only incorporate 83 complex structure deformations as monomial perturbations of the homogeneous polynomial

$$P_0(x) = x_1^8 + x_2^8 + x_3^4 + x_4^4 + x_5^4\,. \tag{3.13}$$

For the mirror family, $H^{1,1}_{\text{toric}}(Y) = 83$, whereas $H^{1,1}(Y) = 86$, indeed we shall see in Section 5 that in the calculation of the number of rational points for the mirror there is a splitting in the denominator of the zeta function.

3.4. K3 fibration. The octic family of Calabi–Yau threefolds is K3-fibred. A K3 can be given by a hypersurface in $\mathbb{P}_3^{\mathbf{k}}$, if $\mathbf{k} = (1, k_2, k_3, k_4)$ and $1 + k_2 + k_3 + k_4 = 4$. A K3 fibration can thus be formed in $\mathbb{P}_4^{(1,1,2k_2,2k_3,2k_4)}$: the $\mathbb{P}_1$ base can be given by the ratio of the two coordinates with weight one, and the K3 are quartic hypersurfaces in $\mathbb{P}_3^{(1,k_2,k_3,k_4)}$, i.e. letting $x_2 = \lambda x_1$, $y_1 = x_1^2$ and $y_i = x_{i+1}$ for $i = 2, \dots, 4$, we get the pencil of K3 surfaces (for fixed ψ and ϕ):

$$(3.14) \qquad (1 + \lambda^8)y_1^4 + y_2^4 + y_3^4 + y_4^4 - 2\phi\lambda^4 y_1^4 - 8\psi\lambda y_1 y_2 y_3 y_4 \,,$$

hence, in our case $\mathbf{k} = (1, 1, 1, 1)$.

3.5. Singularities. This two-parameter family of varieties was extensively studied in [**CdOFKM**]; some essential features are summarized here.

For the purposes of examining the singularities it is easier to think of the family once again as hypersurfaces in weighted projective space. Using the Greene–Plesser construction [**GP**] (the very first construction for finding mirror manifolds), the mirror of $\mathbb{P}_4^{(1,1,2,2,2)}[8]$ can be found by the process of 'orbifolding', i.e. it may be identified with the family of Calabi–Yau threefolds of the form $\{p = 0\}/G$ where

$$(3.15) \qquad p(x) = x_1^8 + x_2^8 + x_3^4 + x_4^4 + x_5^4 - 2\phi x_1^4 x_2^4 - 8\psi x_1 x_2 x_3 x_4 x_5 \,,$$

and where $G \cong \mathbb{Z}_4^3$ is the group with generators

$$(3.16) \qquad (\mathbb{Z}_4; 0, 3, 1, 0, 0), \quad (\mathbb{Z}_4; 0, 3, 0, 1, 0), \quad (\mathbb{Z}_4; 0, 3, 0, 0, 1).$$

We have used the notation $(\mathbb{Z}_k; r_1, r_2, r_3, r_4, r_5)$ for a $\mathbb{Z}_k$ symmetry with the action:

$$(3.17) \quad (x_1, x_2, x_3, x_4, x_5) \to (\omega^{r_1}x_1, \omega^{r_2}x_2, \omega^{r_3}x_3, \omega^{r_4}x_4, \omega^{r_5}x_5), \quad \text{where } \omega^k = 1.$$

For a good description of the moduli space, it is useful to enlarge G to $\hat{G}$ consisting of elements $g = (\alpha^{a_1}, \alpha^{a_2}, \alpha^{2a_3}, \alpha^{2a_4}, \alpha^{2a_5})$ acting as

$$(x_1, x_2, x_3, x_4, x_5) \mapsto (\alpha^{a_1}x_1, \alpha^{a_2}x_2, \alpha^{2a_3}x_3, \alpha^{2a_4}x_4, \alpha^{2a_5}x_5; \alpha^{-a}\psi, \alpha^{-4a}\phi) \,,$$

where $a = a_1 + a_2 + 2a_3 + 2a_4 + 2a_5$, and where α^{a_1} and α^{a_2} are 8th roots of unity and $\alpha^{2a_3}, \alpha^{2a_4}, \alpha^{2a_5}$ are 4th roots of unity.

If we quotient the family of weighted projective hypersurfaces $\{p = 0\}$ by the full group $\hat{G}$ we must quotient the parameter space (ϕ, ψ) by a $\mathbb{Z}_8$ with a generator g_0 acting as follows:

$$(3.18) \qquad g_0 : (\psi, \phi) \to (\alpha\psi, -\phi).$$

The quotiented parameter space has a singularity at the origin and can be described by three functions:

$$(3.19) \qquad \tilde{\xi} := \psi^8, \quad \tilde{\eta} := \psi^4\phi, \quad \tilde{\zeta} := \phi^2,$$

subject to the relation $\tilde{\xi}\tilde{\zeta} = \tilde{\eta}^2$. The above equations describe an affine quadric cone in $\mathbb{C}^3$; this can be compactified to the projective quadric cone in $\mathbb{P}_3$. This is isomorphic to the weighted projective space $\mathbb{P}^{(1,1,2)}$. We embed $\mathbb{C}^3$ in $\mathbb{P}_3$ by sending the point of $\mathbb{C}^3$ with the coordinates $(\tilde{\xi}, \tilde{\eta}, \tilde{\zeta})$ to the point in $\mathbb{P}_3$ with homogeneous coordinates $[\xi, \eta, \zeta, 1]$.

Note that the square of the generator g_0^2 acts trivially on ϕ, and so fixes the entire line $\psi = 0$. This means that the quotiented family $\{p = 0\}/\hat{G}$ will have new singularities along the $\psi = 0$ locus (which becomes the $\xi = \eta = 0$ locus in the quotient).

We now locate the parameters for which the original family of hypersurfaces is singular, and study the behaviour of these singularities under quotienting. We obtain the following:

(1) C_{con}: Along the locus $(8\psi^4 + \phi)^2 = \pm 1$, the original family of threefold acquires a collection of conifold points. For the mirror, these are identified under the G-action, giving only one conifold point per threefold on the quotient.
(2) C_1: Along the locus $\phi^2 = 1$, the threefold acquires 4 isolated singularities, leading to only one singular point on the quotient.
(3) C_∞ If we let (ϕ, ψ) approach infinity, we obtain a singular threefold

$$2\phi x_1^4 x_2^4 + 8\psi x_1 x_2 x_3 x_4 x_5 \ . \tag{3.20}$$

(4) C_0: $\psi = 0$ - this locus requires special treatment , as it leads to additional singularities on the quotient by $\hat{G}$.

The points of intersection of the above singular loci are described in Figure 2.

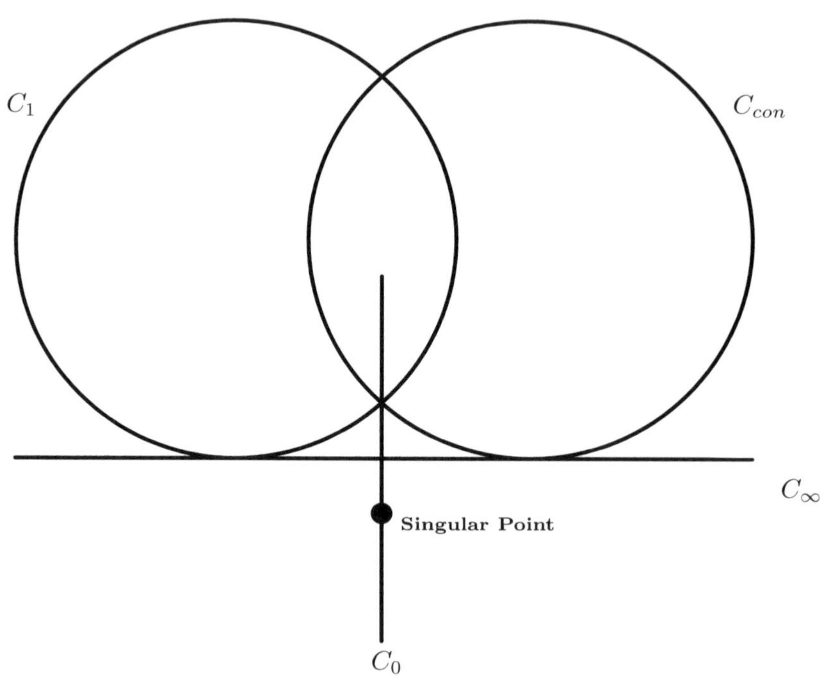

Figure 2: The Compactified Moduli Space

3.5.1. *The locus* $\phi^2 = 1$. When we restrict $\mathbb{P}_4{}^{(1,1,2,2,2)}[8]$ to the locus $\phi^2 = 1$, the resulting family of singular manifolds is birationally equivalent to the mirror family of $\mathbb{P}_5[2,4]$, see [**CdOFKM**] for an explicit map. The equations defining the complete intersection can have the form

$$y_0^2 + y_1^2 + y_2^2 + y_3^2 \quad = \quad \eta y_4 y_5 \ ,$$

$$y_4^4 + y_5^4 = y_0 y_1 y_2 y_3 ,$$

where η is the single parameter. This complete intersection needs to be quotiented by the group H of coordinate rescalings which preserve both hypersurfaces as well as the holomorphic three-form. H can be described as follows:

(3.21)

$$\{(\alpha^{4a}, \alpha^{4b}, \alpha^{4c}, \alpha^{4a}, \alpha^e, \alpha^f) | e + f \equiv 0 \mod 8,\ 4a + 4b + 4c + 4d \equiv 4e \mod 8\} .$$

The Hodge numbers of the mirror of this complete intersection are as follows: [**CK**] $h^{1,1} = 1$ and $h^{2,1} = 89$. When there is both a $\phi^2 = 1$ singularity and a conifold point the (mirror) threefold upon resolution becomes rigid, i.e. $h^{2,1} = 0$. The modularity of the zeta functions at these points was established by Rodriguez-Villegas in [**R-V**].

The Hodge numbers above can be understood by means of an extremal transition. This class of extremal transitions generalises the notion of a conifold transition; for details see [**KMP**], where a genus g curve of A_{N-1} singularities was considered, where $N \geq 2$. A Calabi–Yau hypersurface will have this type of singularity precisely when Δ^* has an edge joining 2 vertices v_0, v_N of Δ^* with $N - 1$ equally spaced lattice points $v_1, \ldots, v_{N-1}$ in the interior of the edge. As explained previously, the Calabi–Yau threefold, M, is a hypersurface in the toric variety whose fan is a suitable refinement of the fan consisting of the fan over the faces of Δ^*. In particular, there are edges corresponding to the lattice points $v_1, \ldots, v_{N-1}$. These correspond to toric divisors, which resolves a surface S of A_{N-1} singularities in V. Restricting to the hypersurface M we see that there are $N - 1$ divisors in M which resolve a curve C of A_{N-1} singularities. The genus g of the curve C can be found by finding the dual to the edge $D_1^* =< v_0, v_N >$ which determines a 2-dimensional face, Δ_2, of Δ. The number of integral interior points of the triangle Δ_2 is equal to the genus g.

In our case $g = 3$, $N = 2$ and the relevant edge is that shown in Figure 1.

Overall, in smoothing the singularities, the change in Hodge numbers is given by the following formula [**KMP**]:

$$(3.22)\quad h^{1,1} \to h^{1,1} - (N-1) , \quad h^{2,1} \to h^{2,1} + (2g-2)\binom{N}{2}(N-1) .$$

The change above is due to the fact that the transition 'kills' $(N - 1)$ independent homology cycles. Homotopically, the transition is obtained by replacing $(2g-2)\binom{N}{2}$ two-spheres by $(2g - 2)\binom{N}{2}$ three-spheres. This results in a change in the Euler characteristic of $-2(2g - 2)\binom{N}{2}$, hence the change in the 3D-homology has to be $2(2g-2)\binom{N}{2} - 2(N-1)$, which is equally shared between $H^{2,1}$ and $H^{1,2}$, giving us (3.22).

4. Periods and Picard–Fuchs equations

In our calculation for the octic threefolds we show that the zeta function factorizes into pieces that can be associated with certain set of periods. In this section we give a review of a method of computing the Picard–Fuchs equations, the Dwork–Katz–Griffiths method, and then exhibit some of the the Picard–Fuchs equations that we computed for the octic threefolds in Section 4.3; the complete set of diagrams can be found in [**K**].

4.1. Periods and Picard–Fuchs equations. The dimension of the third cohomology group for a Calabi–Yau is $\dim H^3 = b_3 = 2(h^{2,1}+1)$. Furthermore, the unique holomorphic three-form Ω depends only on the complex structure. If we take derivatives with respect to the complex structure moduli, we get elements in $H^{3,0} \bigoplus H^{2,1} \bigoplus H^{1,2} \bigoplus H^{0,3}$. Since b_3 is finite, there must be linear relations between derivatives of Ω of the form $\mathcal{L}\Omega = d\eta$ where $\mathcal{L}$ is a differential operator with moduli-dependent coefficients. If we integrate this equation over an element of the third homology group H_3, we will get a differential equation $\mathcal{L}\Pi_i = 0$ satisfied by the periods of Ω. They are defined as

$$\Pi_i(a) = \int_{\Gamma_i} \Omega(a), \quad \Gamma_i \in H_3(X, \mathbb{Z}). \tag{4.1}$$

where a represents the complex structure moduli. These are the Picard–Fuchs equations, and in general we obtain a set of coupled linear partial differential equations (in the case of one-parameter models (i.e. $b_3 = 4$) we just get a single ordinary differential equation of order 4).

4.2. Dwork, Katz and Griffiths' method. One procedure for determining the Picard–Fuchs equations for hypersurfaces in $\mathbb{P}[\vec{w}]$ is the Dwork, Katz and Griffiths reduction method [**CK**]. For hypersurfaces in weighted projective space $\mathbb{P}_4^{(w_1,w_2,w_3,w_4,w_5)}$ the periods $\Pi_i(a)$ of the holomorphic three-form $\Omega(a)$ can be written as

$$\Pi_i(a) = \int_{\Gamma_i} \Omega(a) = \int_\gamma \int_{\Gamma_i} \frac{\omega}{P(a)}, \quad \Gamma_i \in H_3(X, \mathbb{Z}), \quad \text{i.e. } i = 1, \dots, 2(h^{2,1}+1)\,. \tag{4.2}$$

Here $\omega = \sum_{i=1}^5 (-1)^i w_i z_i dz_1 \wedge \dots \wedge \hat{dz_i} \wedge \dots \wedge dz_5$ and γ is a small curve around the hypersurface $P = 0$ in the 4-dimensional embedding space. The numbers a_i are the coefficients of the perturbations of the quasi-homogeneous polynomial P.

Observe that $\frac{\partial}{\partial z_i}\left(\frac{f(z)}{P^r}\right)\omega$ is exact if $f(z)$ is homogeneous with degree such that the whole expression has degree zero. This leads to the partial integration rule $\left(\partial_i = \frac{\partial}{\partial z_i}\right)$:

$$\frac{f\partial_i P}{P^r} = \frac{1}{r-1}\frac{\partial_i f}{P^{r-1}}. \tag{4.3}$$

In practice one chooses a basis $\{\varphi_k(z)\}$ for the elements of the local ring $\mathcal{R}$. For hypersurfaces $\mathcal{R} = \mathbb{C}[z_1, \dots, z_{n+1}]/(\partial_i P)$. From the Poincaré polynomial associated to P one sees that there are $(1, \tilde{h}^{2,1}, \tilde{h}^{2,1}, 1)$ basis elements with degrees $(0, d, 2d, 3d)$ respectively. The elements of degree d are the perturbing monomials.

One then takes derivatives of the expressions

$$\Pi_j^k = \int_{\Gamma_j} \frac{\varphi_k(z)}{P^{n+1}} \quad (n = \deg(\varphi_k)/d) \tag{4.4}$$

with respect to the moduli. If one produces an expression such that the numerator in the integrand is not one of the basis elements, one relates it, using the equations $\partial_i P = \dots$, to the basis and uses (4.3). This leads to a system of first order differential equations (known as 'Gauss–Manin equations') for the Π_j^k which can be rewritten as a system of partial differential equations for the period (the Picard–Fuchs equations): $\partial_{a_k}\Pi = M^{(k)}(a)\Pi$, $k = 1, \dots, \tilde{h^{2,1}}$. The Picard–Fuchs equations reflect the structure of the local ring and express relations between its elements (modulo the ideal).

Note that the method above only gives us the Picard–Fuchs equations for which there exists a monomial perturbation. Also, the above method only works for manifolds embedded in projective spaces. In [**CdOV1**] it was shown that the number of rational points of the quintic Calabi–Yau manifolds over $\mathbb{F}_p$ can be given in terms of the periods; our calculations verify this relation for the octic threefolds, as we have 15 classes of monomials each providing its own Picard–Fuchs equations. The following table lists the monomials (the meaning of the $\lambda_{\mathbf{v}}$ will be explained in Section 5):

Monomial $\mathbf{v}$	degree	Permutations $\lambda_{\mathbf{v}}$
(0,0,0,0,0)	0	1
(0,2,1,1,1)	8	2
(6,2,0,0,0)	8	1
(0,0,0,2,2)	8	3
(3,1,2,0,0)	8	6
(4,0,2,0,0)	8	3
(0,0,2,1,1)	8	3
(6,0,1,0,0)	8	6
(5,1,1,0,0)	8	3
(4,0,1,1,0)	8	3
(2,0,3,0,0)	8	6
(1,1,0,0,3)	8	3
(0,0,3,1,0)	8	6
(2,0,2,1,0)	8	12
(7,3,2,1,0)	16	6

Later on we shall see that the zeta function for the octic threefolds can be decomposed into pieces corresponding to certain monomial classes which turn out to be identical to the above monomial classes. This establishes the correspondence between periods and the number of rational points.

Mirror symmetric pairs of threefolds (X, Y) often have a certain number of periods in common; these periods are associated to the monomials that the two reflexive polyhedra have in common. For our two-parameter model, 6 periods are 'shared'. The shared periods are those associated to the monomial $(0,0,0,0,0)$ (the G- invariant ones).

We are now able to list the 15 different Picard–Fuchs Equations obtained. Using the notation of [**CdOV1**], we associate a period to each monomial:

$$\varpi_{\mathbf{v}}(\phi,\psi) = \int_{\gamma_{\mathbf{v}}} \Omega(\phi,\psi) = C \int_{\Gamma} \frac{x^{\mathbf{v}}}{P^{w(\mathbf{v})+1}}\,, \tag{4.5}$$

where C is a convenient normalization constant. In going from the first equality to the second, we have used the residue formula for the holomorphic 3-form [**BCdOFHJG**]:

$$\Omega := \operatorname{Res}\left[\frac{\omega}{P}\right]. \tag{4.6}$$

In affine space $\mathbb{C}_5^{(w_1,w_2,w_3,w_4,w_5)}$, we may take ω to be given by $\omega = \prod_{i=0}^{5} dx_i$. The contour Γ is chosen so as to reproduce the integral of the residue over the cycle $\gamma_{\mathbf{v}}$.

Differentiating with respect to each parameter we obtain (suppressing constants):

$$\frac{d}{d\psi}\varpi_{\mathbf{v}} = \int_\Gamma \mathrm{d}^5x \, \frac{x^{\mathbf{v}+(1,1,1,1,1)}}{P^{w(\mathbf{v})+2}} 8\psi(w(\mathbf{v})+1) \; , \tag{4.7}$$

$$\frac{d}{d\phi}\varpi_{\mathbf{v}} = \int_\Gamma \mathrm{d}^5x \, \frac{x^{\mathbf{v}+(4,4,0,0,0)}}{P^{w(\mathbf{v})+2}} 2\phi(w(\mathbf{v})+1) \; . \tag{4.8}$$

The Picard–Fuchs equation can be found by repeated use of the following operations:
For $i = 1, 2$:

$$\begin{aligned} D_i\left(\frac{x^{\mathbf{v}}}{P^{w(\mathbf{v})+1}}\right) \\ &:= \frac{\partial}{\partial x_i}\left(x_i \frac{x^{\mathbf{v}}}{P^{w(\mathbf{v})+1}}\right) \\ &= (1+v_i)\frac{x^{\mathbf{v}}}{P^{w(\mathbf{v})+1}} + \frac{\partial P}{\partial x_i}(w(\mathbf{v})+1)\frac{x^{\mathbf{v}}}{P^{w(\mathbf{v})+2}} \\ &= (1+v_i)\frac{x^{\mathbf{v}}}{P^{w(\mathbf{v})+1}} + (w(\mathbf{v})+1)\frac{x^{\mathbf{v}}}{P^{w(\mathbf{v})+2}} 8(x_i^8 - \psi x_1x_2x_3x_4x_5 - \phi x_1^4x_2^4) \; ; \end{aligned} \tag{4.9}$$

for $i = 3, 4, 5$:

$$\begin{aligned} D_i\left(\frac{x^{\mathbf{v}}}{P^{w(\mathbf{v})+1}}\right) \\ &:= \frac{\partial}{\partial x_i}\left(x_i \frac{x^{\mathbf{v}}}{P^{w(\mathbf{v})+1}}\right) \\ &= (1+v_i)\frac{x^{\mathbf{v}}}{P^{w(\mathbf{v})+1}} + \frac{\partial P}{\partial x_i}(w(\mathbf{v})+1)\frac{x^{\mathbf{v}}}{P^{w(\mathbf{v})+2}} \\ &= (1+v_i)\frac{x^{\mathbf{v}}}{P^{w(\mathbf{v})+1}} + (w(\mathbf{v})+1)\frac{x^{\mathbf{v}}}{P^{w(\mathbf{v})+2}} 4(x_i^4 - 2\psi x_1x_2x_3x_4x_5) \; . \end{aligned} \tag{4.10}$$

Now

$$\int_\Gamma \mathrm{d}^5x \; D_i\left(\frac{x^{\mathbf{v}}}{P^{w(\mathbf{v})+1}}\right) = 0 \; , \tag{4.11}$$

because $D_i\left(\frac{x^{\mathbf{v}}}{P^{w(\mathbf{v})+1}}\right)$ corresponds to an exact differential form and hence (4.9) and (4.10) establish identities between the differential form associated to $x^{\mathbf{v}}$ and the differential forms associated to $x^{\mathbf{v}+(1,1,1,1,1)}$, $x^{\mathbf{v}+(4,4,0,0,0)}$, and either $x^{\mathbf{v}+(8,0,0,0,0)}$ or $x^{\mathbf{v}+(0,0,4,0,0)}$ (the last two up to permutation). Using 4.7 and 4.8 the Picard–Fuchs equation can be derived. These identities are summarized in a straightforward way using arrows in a diagramatic representation of each Picard–Fuchs equation.

4.3. Picard–Fuchs diagrams. In this section we present diagramatic representations of the Picard–Fuchs equations. The notation used is as follows: the monomials are written in the form:

$$\left(v_1, v_2 | v_3, v_4, v_5 | \frac{v_1+v_2+2(v_3+v_4+v_5)}{8}, \frac{v_1+v_2}{2}\right), \tag{4.12}$$

the last two digits denoting the degree and the values of the Cox variable $x_{(\mathbf{4,4,0,0,0})}$, respectively. Unbroken arrows labelled by D_i relate periods associated to $x^{\mathbf{v}}$ to those associated to either (up to permutation of the added monomial) $x^{\mathbf{v}+(8,0,0,0,0)}$ or $x^{\mathbf{v}+(0,0,4,0,0)}$. Horizontal unbroken arrows denote the correspondence between periods associated to $x^{\mathbf{v}}$ and $x^{\mathbf{v}+(1,1,1,1,1)}$. Curvy broken lines relate, in the same way, $x^{\mathbf{v}}$ with $x^{\mathbf{v}+(4,4,0,0,0)}$. Hence it is possible, although rather involved, to derive the Picard–Fuchs equations from these diagrams by using (4.9), and (4.10). The pairs of identical monomials encased in boxes are to be identified, thus making the diagrams 'closed'. Hence these diagrams are pictorial representations of the Picard–Fuchs equations. In this paper we only list two, in Figures 3 and 4; for the full set of diagrams, consult the author's thesis [**K**].

The monomial $(7,3|0,2,1|2,5)$ is related to itself via the Jacobian ideal; it is not in the ideal on the conifold locus, $(8\psi^4+\phi)^2=1$. This is illustrated by its corresponding equation, derived from the diagram in Figure 3:

$$\begin{aligned}16[(8\psi^4+\phi)^2-1]\frac{x_1^7x_2^3x_4^2x_5}{P^3} &= 8\psi^3[\partial_3(\frac{x_1^2x_2^6x_4}{P^2})+(8\psi^4+\phi)\partial_3(\frac{x_1^6x_2^2x_4}{P^2})]\\ &+4\psi^2[\partial_5(\frac{x_1^5x_2x_3^2}{P^2})+(8\psi^4+\phi)\partial_5(\frac{x_1x_2^5x_3^2}{P^2})]\\ &+2\psi[\partial_4(\frac{x_2^4x_3x_5^2}{P^2})+(8\psi^4+\phi)\partial_4(\frac{x_1^4x_3x_5^2}{P^2})]\\ &+[\partial_1(\frac{x_2^3x_4^2x_5}{P^2})+(8\psi^4+\phi)\partial_2(\frac{x_1^3x_4^2x_5}{P^2})]\,.\end{aligned} \tag{4.13}$$

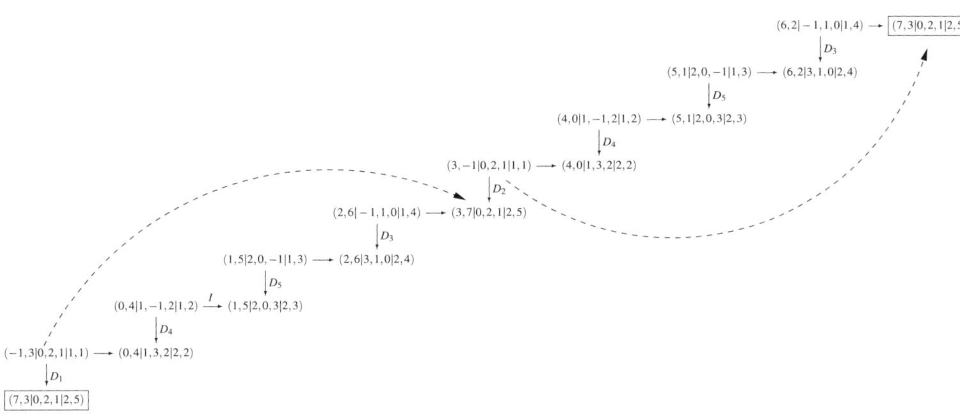

Figure 3: Picard–Fuchs Equation for $(4,0,2,3,1)$

Equivalently, the monomial $(7,3,0,2,1)$ corresponds to a differential form which is exact except at the conifold locus, when $(8\psi^4+\phi)^2=1$. Hence, as long as $(8\psi^4+\phi)^2\neq 1$, the period associated to $(7,3,0,2,1)$ is zero. It will later be seen that, generically, each class of monomials contributes to the zeta function except $(7,3,0,2,1)$. The monomial $(7,3,0,2,1)$ only contributes when there is a conifold singularity. This is similar to the case for the quintic threefolds [**CdOV1**], where there was a 'period' associated to the monomial $(4,3,2,1,0)$ which was only non-zero on the conifold locus.

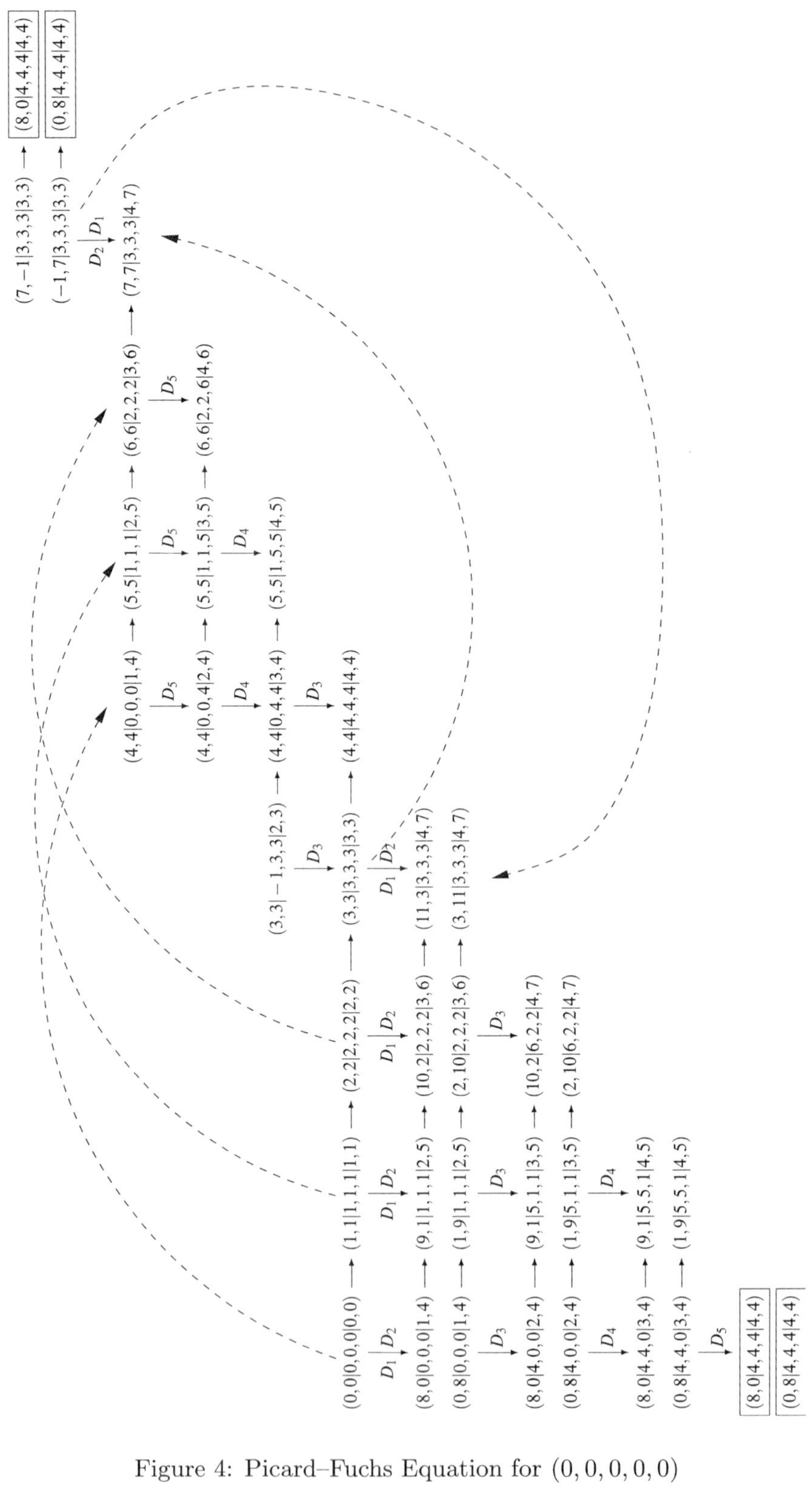

Figure 4: Picard–Fuchs Equation for $(0, 0, 0, 0, 0)$

5. Exact calculation of the number of rational points

In this section we describe in detail the derivation of our results for the octic threefolds. In the first few sections, we review some of the machinery (much of which was used by Dwork to prove the rationality of the zeta function [**Dw**]) that we use in the main computation, which begins in Section 5.2. The formulae we obtained for the number of rational points are summarized in Section 5.4; Mathematica code was written to compute the number of rational points using these formulae. Section 6 describes the computation of the number of points for the mirror octic threefolds.

5.1. Dwork's character and Gauss sums. We record the expressions for the number of points, which we shall denote $\nu(\phi, \psi)$, in terms of Gauss sums. We use machinery originally invented by Dwork to prove the rationality of the zeta function - part of the Weil Conjectures. We begin with a quick review of the properties of Dwork's character [**CdOV1**]. The character is defined by

$$\Theta(x) = F(\pi \mathrm{Teich}(x)) \,, \tag{5.1}$$

where π is a number in $\mathbb{C}_p$ such that $\pi^{p-1} = -p$, Teich is the Teichmüller character for the multiplicative group $\mathbb{F}_p^*$. The Teichmüller character is a homomorphism of the multiplicative groups Teich : $\mathbb{F}_p^* \to \mathbb{Z}_p^*$, such that Teich$(a)$ is the unique $(p-1)$-th root of unity in $\mathbb{Z}_p^*$, which is congruent to a modulo p. F is the function defined by

$$F(X) = \exp\left(X + \frac{X^p}{p}\right) = \sum_{n=0}^{\infty} c_n X^n \,. \tag{5.2}$$

Θ is a nontrivial additive character, i.e. $\Theta(x+y) = \Theta(x)\Theta(y)$. This construction was adapted by Dwork to give an additive character $\Theta_s : \mathbb{F}_q \to \mathbb{C}_p$ where $q = p^s$:

$$\Theta_s(x) = \Theta(\mathrm{tr}(x)) \,, \tag{5.3}$$

where tr : $\mathbb{F}_q \to \mathbb{F}_p$ is the trace map $\mathrm{tr}(x) = x + x^p + x^{p^2} + \cdots + x^{p^{s-1}}$. The character Θ_s has a remarkable splitting property:

$$\Theta_s(x) = \prod_{l=0}^{s-1} \Theta_{p^l}(x) \,. \tag{5.4}$$

A Gauss sum is the sum of an additive character, Θ, times a multiplicative character, $\mathrm{Teich}^n(x)$. Define:

$$G_n = \sum_{x \in \mathbb{F}_p^*} \Theta(x) \mathrm{Teich}^n(x); \tag{5.5}$$

thus, in particular, we have $G_0 = \sum_{x \in \mathbb{F}_p^*} \Theta(x) = -1$.

The following is a standard properties of Gauss sums:

$$G_n G_{-n} = (-1)^n p \quad \text{for} \quad p-1 \nmid n \,. \tag{5.6}$$

The Gross–Koblitz formula gives G_n for the case $p-1 \nmid n$:

$$G_n = p(-p)^{-\langle \frac{n}{p-1} \rangle} \Gamma_p\left(1 - \left\langle \frac{n}{p-1} \right\rangle\right); \quad p-1 \nmid n \,, \tag{5.7}$$

where $\langle x \rangle$ denotes the fractional part of x.The following is a very useful identity, which can be derived from (5.5):

$$\Theta(x) = \frac{1}{p-1} \sum_{m=0}^{p-2} G_{-m} \mathrm{Teich}^m(x) \ . \tag{5.8}$$

Similarly, for finite extensions $\mathbb{F}_{p^s}$ with $q = p^s$ points, we define:

$$G_{s,n} = \sum_{x \in \mathbb{F}_q^*} \Theta_s(x) \mathrm{Teich}^n(x), \tag{5.9}$$

where now Teich is the Teichmüller character for $\mathbb{F}_q^*$. The Gross–Koblitz formula is now

$$G_{s,n} = (-1)^{s+1} q \pi^{-S(n)} \prod_{l=0}^{s-1} \Gamma_p \left(1 - \left\langle \frac{p^l n}{q-1} \right\rangle \right); \quad q - 1 \nmid n \ , \tag{5.10}$$

where $S(n)$ is the is the sum of the p-adic digits of n.

5.2. Number of points in terms of Gauss sums. When a variety is defined as the vanishing set of a polynomial $P \in k[X_1, \ldots, X_n]$, where k is a field, a non-trivial additive character like Dwork's character can be exploited to count points over k. Since $\Theta(x)$ is a character:

$$\sum_{y \in k} \Theta(yP(x)) = \begin{cases} 0 & \text{if } P(x) \neq 0, \\ q := \mathrm{Card}(k) & \text{if } P(x) = 0 \ ; \end{cases} \tag{5.11}$$

hence

$$\sum_{x_i \in k} \sum_{y \in k} \Theta(yP(x)) = q\nu \ , \tag{5.12}$$

where ν is the number of rational points of the variety.

5.3. The octic family of threefolds. We shall now apply the identity of the previous section to the octic threefolds as described in terms of Cox variables in Section 3.3, i.e. where,

$$P(\mathbf{x}, x_6) = x_1^8 x_6^4 + x_2^8 x_6^4 + x_3^4 + x_4^4 + x_5^4 - 8\psi x_1 x_2 x_3 x_4 x_5 x_6 - 2\phi x_1^4 x_2^4 x_6^4. \tag{5.13}$$

We obtain:

$$\begin{aligned}
q\nu(\psi, \phi) &= \sum_{y, x_6 \in \mathbb{F}_q} \sum_{\mathbf{x} \in (\mathbb{F}_q)^5} \Theta(yP(\mathbf{x}, x_6)) - q\nu_F \\
&= \sum_{\substack{y=0, \\ x_6 \neq 0}} \sum_{\mathbf{x} \in (\mathbb{F}_q)^5} \Theta(yP(\mathbf{x}, x_6)) + \sum_{\substack{y=0, \\ x_6 = 0}} \sum_{\mathbf{x} \in (\mathbb{F}_q)^5} \Theta(yP(\mathbf{x}, 0)) \\
&\quad + \sum_{\substack{y\neq 0, \\ x_6 \neq 0}} \sum_{\mathbf{x} \in (\mathbb{F}_q)^5} \Theta(yP(\mathbf{x}, x_6)) + \sum_{\substack{y\neq 0, \\ x_6 = 0}} \sum_{\mathbf{x} \in (\mathbb{F}_q)^5} \Theta(yP(\mathbf{x}, 0)) - q\nu_F \\
&= (q-1)(q^2-1)(q^3-1) + (q^2-1)(q^3-1) + \sum_{\substack{y\neq 0, \\ x_6 \neq 0}} \sum_{\mathbf{x} \in (\mathbb{F}_q)^5} \Theta(yP(\mathbf{x}, x_6)) \\
&\quad - \sum_{\substack{y\neq 0, \\ x_6 \neq 0}} \sum_{\substack{x_1 = x_2 = 0, \\ x_3, x_4, x_5 \in \mathbb{F}_q}} \Theta(y(x_3^4 + x_4^4 + x_5^4)) + \sum_{\substack{y\neq 0, \\ x_6 = 0}} \sum_{\mathbf{x} \in (\mathbb{F}_q)^5} \Theta(y(x_3^4 + x_4^4 + x_5^4)) \\
&= (q-1)^2(q+1)q(1+q+q^2) + \sum_{\substack{y\neq 0, \\ x_6 \neq 0}} \sum_{\mathbf{x} \in (\mathbb{F}_q)^5} \Theta(yP(\mathbf{x}, x_6))
\end{aligned}$$

$$- \sum_{\substack{y\neq 0,\\ x_6\neq 0}} \sum_{\substack{x_1=x_2=0,\\ x_3,x_4,x_5\in\mathbb{F}_q}} \Theta(y(x_3^4+x_4^4+x_5^4)) + \sum_{\substack{y\neq 0,\\ x_6=0}} \sum_{\mathbf{x}\in(\mathbb{F}_q)^5} \Theta(y(x_3^4+x_4^4+x_5^4)) \,, \tag{5.14}$$

where ν_F is the number of points that needed to be subtracted from the total sum due to the combinations of coordinates which are not allowed to vanish simultaneously. The rule given by toric geometry is that a subset of the Cox variables can vanish simultaneously if and only if the corresponding monomials lie in the same cone.

In this case the following vanishings are forbidden by the triangulation (see Figure 1):

$$\begin{aligned} y=0, \quad & x_6\neq 0, \quad & x_1=x_2=0,\\ & & x_3=x_4=x_5=0\ ;\\ y=0, \quad & x_6=0, \quad & x_1=x_2=0,\\ & & x_3=x_4=x_5=0\ . \end{aligned} \tag{5.15}$$

Thus the first two terms (minus the above forbidden vanishings) are given by

$$\begin{aligned} \sum_{\substack{y=0,\\ x_6\neq 0}} \sum_{\mathbf{x}\in(\mathbb{F}_q)^5\setminus F} \Theta(yP(\mathbf{x},x_6)) &= (q-1)(q^2-1)(q^3-1)\,,\\ \sum_{\substack{y=0,\\ x_6=0}} \sum_{\mathbf{x}\in(\mathbb{F}_q)^5\setminus F} \Theta(yP(\mathbf{x},0)) &= (q^2-1)(q^3-1)\,. \end{aligned} \tag{5.16}$$

We shall use the relation $N=\nu(q-1)^2$, where $N=N_{q=p^s}=\#(X/\mathbb{F}_{p^s})$ i.e. the number of points on the Calabi–Yau threefold reduced over the finite field $\mathbb{F}_{p^s}$. This relation comes from the fact that we are looking at hypersurfaces in $\frac{\mathbb{C}^6-F}{(\mathbb{C}^*)^2}$, the extra factor of $(q-1)^2$ comes from taking the quotient by $(\mathbb{C}^*)^2$. Now:

$$\begin{aligned} &\left(\sum_{\substack{y\neq 0,\\ x_6=0}} \sum_{\substack{(x_1,x_2)\neq(0,0)\\ x_3,x_4,x_5\in(\mathbb{F}_q)^3}} - \sum_{\substack{y\neq 0,\\ x_6\neq 0}} \sum_{\substack{x_1=x_2=0,\\ x_3,x_4,x_5\in\mathbb{F}_q}} \right) \Theta(y(x_3^4+x_4^4+x_5^4))\\ &\quad = ((q^2-1)(q-1)-(q-1)^2)\sum_{\mathbf{v}} \prod_{i=3,4,5} G_{-\frac{v_i(q-1)}{4}}\\ &\quad = q(q-1)^2 \sum_{\mathbf{v}} \prod_{i=3,4,5} G_{-\frac{v_i(q-1)}{4}}\ . \end{aligned} \tag{5.17}$$

Where the sum $\sum_{\mathbf{v}}$ is over monomials (and their permutations) of the form $\mathbf{v}:=(v_3,v_4,v_5)$ of degree $d=0 \mod 4$ with $0\leq v_i\leq 3$, namely $(2,1,1)$ and $(3,3,2)$. This is a very simple case of the use of Lemma (5.1), which we shall come to soon. The procedure shall be described in detail for the more complicated term. However, note that the contribution above can be thought of as the contribution from the ruled surface of $x_3^4+x_4^4+x_5^4$ over a $\mathbb{P}^1$ described in Section 3.3 and from now on it shall be denoted $N_{\text{exceptional}}$. Note that $N_{\text{exceptional}}$ is only non-zero when $4|q-1$.

$$N(\psi,\phi) = 1+2q+2q^2+q^3+N_{\text{exceptional}}+\frac{1}{q(q-1)^2}\sum_{\substack{y\neq 0,\\ x_6\neq 0}}\sum_{\mathbf{x}\in(\mathbb{F}_q)^5}\Theta(yP(\mathbf{x},x_6)) \tag{5.18}$$

It appears that (5.18) will give us the terms $(1-t)(1-pt)(1-p^2t)(1-p^3t)$ in the denominator of the zeta function. We shall see at the end of the calculation that this is indeed the case.

The term in (5.18): $\frac{1}{q(q-1)^2}\sum_{\substack{y\neq 0,\\ x_6\neq 0}}\sum_{\mathbf{x}\in(\mathbb{F}_q)^5}\Theta(yP(\mathbf{x},x_6))$, shall from now on be denoted N_{mon}.

$$
\begin{aligned}
&q(q-1)^2N_{\text{mon}}\\
&\quad= q(q-1)^2(N_{\text{mon}}(x_i\neq 0\,\forall i)+N_{\text{mon}}(x_1x_2\neq 0, x_3x_4x_5=0)+N_{\text{mon}}(x_1x_2=0, x_3x_4x_5=0))\\
&\quad\equiv \sum_{y\in\mathbb{F}_p^*} g_\phi(y)g_\psi(y)g_8(y)^2g_4(y)^3+\sum_{y\in\mathbb{F}_p^*} g_\phi(y)g_8(y)^2[(g_4(y)+1)^3-g_4(y)^3]\\
&\qquad+\sum_{y\in\mathbb{F}_p^*}[(g_8(y)+1)^2-g_8(y)^2](g_4(y)+1)^3\,,
\end{aligned}
\tag{5.19}
$$

where $g_8(y)=\Theta(yx_i^8x_6^4)$, $g_4(y)=\Theta(yx_i^4)$, $g_\phi(y)=\Theta(-y2\phi x_1^4x_2^4x_6^4)$, and $g_\psi(y)=\Theta(-y8\psi x_1x_2x_3x_4x_5x_6)$. The above expression is obtained by simply expanding out in Θ over all the combinations of vanishing x_i (as we have already subtracted out the number of points due to the 'forbidden vanishings', ν_F).

An important observation is that $g_8(y)+1=\sum_{s=1}^{p-2}G_{-s}\text{Teich}^s(yx^8x_6^4)$. Hence whenever there are expressions involving $(g_8(y)+1)$ or $(g_4(y)+1)$, we can restrict the argument of the Gauss sum, e.g. to be non-zero. As we shall see, this shall be immensely useful in dealing with, for instance, the second and third terms in (5.19).

The expression, (5.19), is for the case $\phi,\psi\neq 0$, the general case. When either ϕ or ψ are zero, the expressions for N_{mon} are as follows:

(1) For $\phi=0$

$$
\begin{aligned}
q(q-1)^2N_{\text{mon}} &= q(q-1)^2(N_{\text{mon}}(\prod_{i=1}^5 x_i\neq 0)+N_{\text{mon}}(\prod_{i=1}^5 x_i=0))\\
&\equiv \sum_{y\in\mathbb{F}_p^*} g_\psi(y)g_8(y)^2g_4(y)^3\\
&\quad+\sum_{y\in\mathbb{F}_p^*}[(g_8(y)+1)^2(g_4(y)+1)^3-g_8(y)^2g_4(y)^3]\,.
\end{aligned}
$$

(2) For $\psi=0$

$$
\begin{aligned}
q(q-1)^2N_{\text{mon}} &= q(q-1)^2(N_{\text{mon}}(x_1x_2\neq 0)+N_{\text{mon}}(x_1x_2=0))\\
&\equiv \sum_{y\in\mathbb{F}_p^*} g_\phi(y)g_8(y)^2(g_4(y)+1)^3\\
&\quad+\sum_{y\in\mathbb{F}_p^*}((g_8(y)+1)^2-g_8(y)^2)(g_4(y)+1)^3\,.
\end{aligned}
$$

(3) For $\psi=\phi=0$

$$
q(q-1)^2N_{\text{mon}} = \sum_{y\in\mathbb{F}_p^*}(g_8(y)+1)^2(g_4(y)+1)^3\,.
$$

First we shall calculate $N_{\rm mon}$ for the case when $x_i \neq 0\,\forall i$ and $\phi\psi \neq 0$:

$$\begin{aligned} N_{\rm mon}(x_i \neq 0) &= \frac{1}{q(q-1)^2} \sum_{y,x_6\in\mathbb{F}_p^*} \sum_{\mathbf{x}\in(\mathbb{F}_p^*)^5} \Theta(yx_1^8x_6^4)\Theta(yx_2^8x_6^4)\Theta(yx_3^4)\Theta(yx_4^4) \\ &\quad \times\Theta(yx_5^4)\Theta(-2y\phi x_1^4x_2^4x_6^4)\Theta(-8y\psi x_1x_2x_3x_4x_5x_6) \\ &= \frac{1}{q(q-1)^2} \sum_{s_\phi,s_\psi,s_i=0}^{q-2} \prod_{i=1}^{5} G_{-s_i}G_{-s_\psi}G_{-s_\phi}\,\mathrm{Teich}^{s_\psi}(-8\psi)\mathrm{Teich}^{s_\phi}(-2\phi) \\ &= \frac{1}{q(q-1)^2} \sum_{s_\phi,s_\psi,s_i=0}^{q-2} \prod_{i=1}^{5} G_{-s_i}G_{s_\psi}G_{s_\phi}\,\mathrm{Teich}^{-s_\psi}(-8\psi)\mathrm{Teich}^{-s_\phi}(-2\phi) \\ &= \frac{1}{q(q-1)^2} \sum_{s_\phi,s_\psi}^{q-2} \sum_{\mathbf{v}} \prod G_{-\left(\frac{v_i(q-1)}{8}+\frac{s_\phi}{2}+\frac{s_\psi}{8}\right)} \prod G_{-\left(\frac{v_i(q-1)}{4}+\frac{s_\psi}{4}\right)} \\ &\quad \times G_{s_\psi}G_{s_\phi}\,\mathrm{Teich}^{-s_\psi}(-8\psi)\mathrm{Teich}^{-s_\phi}(-2\phi)\,. \end{aligned} \tag{5.20}$$

In our calculations repeated use shall be made of the following lemma:

LEMMA 5.1. *Let $i_1, \ldots, i_n$ be non-negative integers. Then*

$$\sum_{(a_1,\ldots,a_n)\in(\mathbb{F}_q)^n} \mathrm{Teich}^{i_1}(a_1)\ldots\mathrm{Teich}^{i_n}(a_n) = 0\,, \tag{5.21}$$

unless each i_j is non-zero and divisible by $q-1$.

PROOF. Suppose first that $n = 1$. If $i = 0$, then $\sum_{a\in\mathbb{F}_q} \mathrm{Teich}^0(a) = q = 0$ in $\mathbb{F}_q$. Assume that $i > 0$, and let b denote a generator of the finite cyclic group $\mathbb{F}_q^*$. If $(q-1) \nmid i$, then

$$\sum_{a\in\mathbb{F}_q} \mathrm{Teich}^i(a) = \sum_{j=0}^{q-2} (\mathrm{Teich}^j(b))^i = \frac{(\mathrm{Teich}^i(b))^{q-1}-1}{\mathrm{Teich}^1(b)-1} = 0. \tag{5.22}$$

If $i = (q-1)j$, then

$$\sum_{a\in\mathbb{F}_q} \mathrm{Teich}^{(q-1)j}(a) = \sum_{a\in\mathbb{F}_q^*} 1^j = q-1. \tag{5.23}$$

In general,

$$\begin{aligned} &\sum_{(a_1,\ldots,a_n)\in(\mathbb{F}_q)^n} \mathrm{Teich}^{i_1}(a_1)\ldots\mathrm{Teich}^{i_n}(a_n) \\ &= \left(\sum_{a_1\in\mathbb{F}_q} \mathrm{Teich}^{i_1}(a_1)\right) \cdots \left(\sum_{a_n\in\mathbb{F}_q} \mathrm{Teich}^{i_n}(a_n)\right) \\ &= 0, \end{aligned} \tag{5.24}$$

unless all the exponents i_j are non-zero and divisible by $q-1$. □

Using Lemma 5.1 on expressions of the form $\sum_{x_i\in\mathbb{F}_{p^s}} \mathrm{Teich}^n(x_i)$, we obtain the following simultaneous set of constraints on the values of the s_i:

$$\begin{array}{lll} q-1| & 8s_i - 4s_\phi - s_\psi & i = 1,2 \\ q-1| & 4s_i - s_\psi & i = 3,4,5 \\ q-1| & 4s_1 + 4s_2 - 4s_\phi - s_\psi & \\ q-1| & \sum_{i=1}^5 s_i - s_\psi - s_\phi & \\ 0 \leq & s_i, s_\phi, s_\psi \leq (q-2) & i = 1,2,3,4,5. \end{array}$$

Hence define a vector of integers, $\mathbf{v^t} = (v_1, v_2, v_3, v_4, v_5, v_\phi, v_\psi)$, such that

$$\begin{array}{rcll} v_i(q-1) & = & 8s_i - 4s_\phi - s_\psi & i = 1,2 \\ v_i(q-1) & = & 4s_i - s_\psi & i = 3,4,5 \\ v_\phi(q-1) & = & 4s_1 + 4s_2 - 4s_\phi - s_\psi & \\ v_\psi(q-1) & = & \sum_{i=1}^5 s_i - s_\psi - s_\phi \,. & \end{array}$$

It is easy to see that we need to consider vectors $\mathbf{v}$ such that

$$\begin{aligned} v_1 + v_2 + 2(v_3 + v_4 + v_5) = 8v_\psi \\ \frac{v_1+v_2}{2} = v_\phi \\ 0 \le v_i \le 7,\ i = 1,2 \\ 0 \le v_i \le 3,\ i = 3,4,5. \end{aligned} \tag{5.25}$$

Every vector of the form (5.25) comes from a set of integers (s_i, s_ϕ, s_ψ) which satisfies the set of constraints. From now on we shall only write down the first five independent entries of $\mathbf{v}$. Notice that given one solution of (5.25), $(v_1, v_2, v_3, v_4, v_5)$, another solution can be obtained by adding $(1,1,1,1,1)$ or $(4,4,0,0,0)$ and reducing $v_1, v_2 \bmod 8$, $v_3, v_4, v_5 \bmod 4$. In (5.20), $\sum_{\mathbf{v}}$, denotes summing over all valid monomials, $\mathbf{v}$. We can define an equivalence, $\sim$, on the set of monomials: two monomials are equivalent if they differ by a factor of $(1,1,1,1,1) \mod 8$ in the first two coordinates and $\mod 4$ in the last three coordinates. We can thus form equivalence classes of $\sim$. Notice that permutations of the form $P_2 \times P_3$, where P_2, and P_3 are the permutations of the first two and the last three coordinates, respectively, leave products of Gauss sums unchanged, e.g. $\prod_{i=1,2} G_{-\left(\frac{v_i(q-1)}{8}+\frac{s_\phi}{2}+\frac{s_\psi}{8}\right)} \prod_{i=3,4,5} G_{-\left(\frac{v_i(q-1)}{4}+\frac{s_\psi}{4}\right)}$.
We define a concept of a 'length of orbit' of classes, which is the number of times the vector $(1,1,1,1,1)$ needs to be added to a monomial until you return to a permutation of the monomial. For the class of monomials labelled by $(4,0,1,1,0)$, for example, we see that:

$$(4,0,1,1,0) \to (5,1,2,2,1) \to (6,2,3,3,2) \to (7,3,0,0,3) \to (0,4,1,1,0)\,. \tag{5.26}$$

Since we return to (a permutation of) the same monomial after 4 additions of $(1,1,1,1,1)$, we can regard this class as having an orbit of length 4.

Because permutations of monomials leave products of Gauss sums unchanged, we can simply do the sum using one representative of each class, and place a 'permutation' factor $\lambda_{\mathbf{v}}$ in front, where it is clear that we need

$$\lambda_{\mathbf{v}} = \#\textit{Permutations of } \mathbf{v} \times \textit{'Length of orbit'}/8. \tag{5.27}$$

Overall, there are 15 classes of monomials:

Monomial $\mathbf{v}$	degree	Permutations $\lambda_{\mathbf{v}}$	Length of orbit
(0,0,0,0,0)	0	1	8
(0,2,1,1,1)	8	2	8
(0,4,2,2,2)	16	1	4
(0,0,0,2,2)	8	3	8
(2,0,1,3,3)	16	6	8
(4,0,2,0,0)	8	3	4
(0,0,2,1,1)	8	3	8
(6,0,1,0,0)	8	6	8
(0,4,0,3,3)	16	3	4
(4,0,1,1,0)	8	3	4
(2,0,3,0,0)	8	6	8
(2,2,1,1,0)	8	3	8
(0,0,3,1,0)	8	6	8
(2,0,2,1,0)	8	12	8
(4,0,2,3,1)	16	6	4

Recall that we remarked that $N_{\text{exceptional}}$ is only non-zero when $4|q-1$ indeed $N_{\text{exceptional}}$ could have been combined with N_{mon} as part of the contributions from the monomials $(0,0,2,1,1)$ and $(1,1,0,0,3)$.

We shall continue the calculation of N_{mon}, but for the case $q \equiv 3 \mod 8$, as this is a typical non-trivial case. The essential consideration at each step is to ensure that the arguments n of the Gauss sums G_n are integers. Thus the first step would be to define $m = \frac{q-1}{2} \in \mathbb{Z}$, and $t = s_\phi/4$. We need to ensure $\frac{v_i}{4}m + \frac{s_\phi+t}{2} \in \mathbb{Z}$ and $\frac{v_i}{2}m + t \in \mathbb{Z}$. Notice that for the case we are considering, $\frac{3m+1}{4} = \frac{3q-1}{8} \in \mathbb{Z}$. First we split the sums over $\mathbf{v}$ and t into two parts: $(\mathbf{v}, t)$ and $\left(\mathbf{v} + (3,3,3,3,3), t + \frac{1}{2}\right)$. For the second step, notice that the transformation on the integer variable t of

$$t \to t + \frac{3m+1}{2} \iff \begin{cases} \mathbf{v} \to \mathbf{v} + (3,3,3,3,3)\ , \\ t \to t + \frac{1}{2}\ . \end{cases} \tag{5.28}$$

So we replace the second sum over $\mathbf{v} + (3,3,3,3,3)$, by just extending the range of t in the first sum by $\frac{3m+1}{2}$:

$$\begin{aligned} N_{\text{mon}}(x_i \neq 0) &= \frac{1}{q(q-1)^2} \sum_{\mathbf{v}} \sum_{s_\phi=0}^{q-2} \sum_{0 \le t \le \lfloor \frac{q-2}{4} \rfloor = \frac{q-3}{4}} \prod G_{-\left(\frac{v_i}{4}m + \frac{s_\phi+t}{2}\right)} \\ &\quad \times \prod G_{-\left(\frac{v_i}{2}m+t\right)} G_{4t} G_{s_\phi} \text{Teich}^{-4t}(-8\psi) \text{Teich}^{-s_\phi}(-2\phi) \\ &\quad + \frac{1}{q(q-1)^2} \sum_{\mathbf{v}} \sum_{s_\phi=0}^{q-2} \sum_{0 \le t \le \frac{q-7}{4}} \prod G_{-\left(\frac{(v_i+3)}{4}m + \frac{(s_\phi+t)}{2} + \frac{1}{4}\right)} \\ &\quad \times \prod G_{-\left(\frac{(v_i+3)}{2}m + t + \frac{1}{2}\right)} G_{4t+2} G_{s_\phi} \text{Teich}^{-4t-2}(-8\psi) \text{Teich}^{-s_\phi}(-2\phi) \\ &= \frac{1}{q(q-1)^2} \sum_{s_\phi=0}^{q-2} \sum_{\substack{0 \le t \le \lfloor \frac{q-2}{4} \rfloor = \frac{q-3}{4} \\ \frac{3q-1}{4} \le t \le (q-2)}} \prod G_{-\left(\frac{v_i}{4}m + \frac{s_\phi+t}{2}\right)} \\ &\quad \times \prod G_{-\left(\frac{v_i}{2}m+t\right)} G_{4t} G_{s_\phi} \text{Teich}^{-4t}(-8\psi) \text{Teich}^{-s_\phi}(-2\phi)\ . \end{aligned} \tag{5.29}$$

Now we undertake an analogous step, this time introducing a new variable, $u = \frac{s_\phi+t}{2}$, and allowing u to be either integer or half-integer. We again split the sum over the monomials into two equal sized sets: $\mathbf{v}$ and $\mathbf{v}+(2,2,2,2,2)$. For the second step, notice that $\frac{m+1}{2} = \frac{q+1}{4} \in \mathbb{Z}$ and that the transformation this time on both t and u:

$$\begin{cases} t \to t + m \\ u \to u + \frac{m+1}{2} \end{cases} \iff \begin{cases} \mathbf{v} \to \mathbf{v} + (2,2,2,2,2) \\ u \to u + \frac{1}{2} \ . \end{cases} \tag{5.30}$$

However the transformation in u follows from its dependence on t. Hence again we just need to extend the sum over t:

$$\begin{aligned} 0 \le t \le \frac{q-3}{4} &\to \begin{cases} 0 \le t \le \frac{q-3}{4}, \\ \frac{q-1}{2} \le t \le \frac{3q-5}{4} \ . \end{cases} \\ \frac{3q-1}{4} \le t \le q-2 &\to \begin{cases} \frac{q+1}{4} \le t \le \frac{q-3}{2}, \\ \frac{3q-1}{4} \le t \le q-2 \ . \end{cases} \end{aligned} \tag{5.31}$$

which means having $0 \le t \le q-2$, which is rather satisfying:

$$\begin{aligned} N_{\mathrm{mon}}(x_i \neq 0) &= \frac{1}{q(q-1)^2} \sum_{\substack{0 \le t \le \frac{q-3}{4} \\ \frac{3q-1}{4} \le t \le (q-2)}} \sum_{\lceil \frac{t}{2} \rceil \le u \le \lfloor \frac{q-2+t}{2} \rfloor} \prod G_{-(\frac{v_i}{4}m+u)} \\ &\quad \times \prod G_{-(\frac{v_i}{2}m+t)} G_{4t} G_{2u-t} \mathrm{Teich}^{-4t}(-8\psi) \mathrm{Teich}^{-s_\phi}(-2\phi) \\ &\quad + \frac{1}{q(q-1)^2} \sum_{\substack{0 \le t \le \frac{q-3}{4} \\ \frac{3q-1}{4} \le t \le (q-2)}} \sum_{\lceil \frac{t}{2} - 1 \rceil \le u \le \lfloor \frac{q-3+t}{2} \rfloor} \prod G_{-(\frac{(v_i+2)}{4}m+u+\frac{1}{2})} \\ &\quad \times \prod G_{-(\frac{(v_i+2)}{2}m+t)} G_{4t} G_{2u+1-t} \mathrm{Teich}^{-4t-2}(-8\psi) \mathrm{Teich}^{-s_\phi}(-2\phi) \\ &= \frac{1}{q(q-1)^2} \sum_{t=0}^{q-2} \sum_{\lceil \frac{t}{2} \rceil \le u \le \lfloor \frac{q-2+t}{2} \rfloor} \prod G_{-(\frac{v_i}{4}m+u)} \\ &\quad \times \prod G_{-(\frac{v_i}{2}m+t)} G_{4t} G_{2u-t} \mathrm{Teich}^t \left(\frac{-2\phi}{(8\psi)^4} \right) \mathrm{Teich}^u \left(\frac{1}{(2\phi)^2} \right) . \end{aligned} \tag{5.32}$$

Note that we have also carried out the following simple steps on the Teichmüller characters:

$$\begin{aligned} \mathrm{Teich}^{-s_\psi}(-8\psi)\mathrm{Teich}^{-s_\phi}(-2\phi) &= \mathrm{Teich}^{-4t}(-8\psi)\mathrm{Teich}^{-2u+t}(-2\phi) \\ &= \mathrm{Teich}^t \left(\frac{-2\phi}{(8\psi)^4} \right) \mathrm{Teich}^{-2u}(2\phi) \\ &= \mathrm{Teich}^u \left(\frac{1}{(2\phi)^2} \right) \mathrm{Teich}^t \left(\frac{-2\phi}{(8\psi)^4} \right) \\ &= \mathrm{Teich}^u(\mu)\mathrm{Teich}^t(\Psi) , \end{aligned} \tag{5.33}$$

where Ψ and μ happen to be the large complex structure coordinates:

$$\mu = \frac{1}{(2\phi)^2}, \quad \Psi = \frac{-2\phi}{(8\psi)^4} \ . \tag{5.34}$$

The final step involves extending the range of u. We can split the sum over the $\mathbf{v}$s into two pieces $\mathbf{v}$ and $\mathbf{v} + (4,4,0,0,0)$, but

$$\mathbf{v} \to \mathbf{v} + (2,2,2,2,2) \iff u \to u + m \ ; \tag{5.35}$$

hence we extend the range of u from $\lceil \frac{t}{2} \rceil \le u \le \lfloor \frac{q-2+t}{2} \rfloor$ to $\lceil \frac{t}{2} \rceil \le u \le \lfloor \frac{2q-3+t}{2} \rfloor$

giving

$$N_{\rm mon}(x_i \neq 0) = \sum_{\mathbf{v}} \lambda_{\mathbf{v}} \frac{1}{q(q-1)^2} \sum_{t=0}^{q-2} \sum_{\lceil \frac{t}{2} \rceil \leq u \leq \lfloor \frac{2q-3+t}{2} \rfloor} \prod G_{-(\frac{v_i}{4}m+u)}$$

$$\times \prod G_{-(\frac{v_i}{2}m+t)} G_{4t} G_{2u-t} {\rm Teich}^t \left(\frac{-2\phi}{(8\psi)^4}\right) {\rm Teich}^u \left(\frac{1}{(2\phi)^2}\right) . \tag{5.36}$$

A similar analysis for the case $q \equiv 7 \mod 8$ gives an identical expression. Note that for these cases the allowed monomials have to have the following properties: $4|v_i$, $i = 1, 2$, and $2|v_i$, $i = 3, 4, 5$. Hence, only four out of the fifteen classes of monomial are allowed to contribute; these are:

Monomial $\mathbf{v}$	degree	Permutations $\lambda_{\mathbf{v}}$	Length of orbit
(0,0,0,0,0)	0	1	8
(0,4,2,2,2)	8	1	4
(0,0,0,2,2)	8	3	8
(4,0,2,0,0)	8	3	4

Now consider the case when only $x_1x_2 \neq 0$, $x_3x_4x_5 = 0$:

$$\begin{aligned} N_{\rm mon}(x_1x_2 \neq 0,\ x_3x_4x_5 = 0) &= \frac{1}{q(q-1)} \sum_{y \in \mathbb{F}_p^*} g_\phi(y) g_8(y)^2 [(g_4(y)+1)^3 - g_4(y)^3] \\ &= \frac{1}{q(q-1)} \sum_{y, x_6 \in \mathbb{F}_p^*} \sum_{\mathbf{x} \in (\mathbb{F}_p^*)^5} \Theta(y x_1^8 x_6^4) \Theta(y x_2^8 x_6^4) \Theta(y x_3^4) \\ &\quad \times \Theta(y x_4^4) \Theta(y x_5^4) \Theta(-2y\phi x_1^4 x_2^4 x_6^4) \\ &= \frac{1}{q(q-1)} \sum_{s_\phi, s_i = 0}^{q-2} \prod_{i=1}^{5} G_{-s_i} G_{-s_\phi} {\rm Teich}^{s_\phi}(-2\phi) \\ &= \frac{1}{q(q-1)} \sum_{s_\phi, s_i = 0}^{q-2} \prod_{i=1}^{5} G_{-s_i} G_{s_\phi} {\rm Teich}^{-s_\phi}(-2\phi) \\ &= \frac{-1}{q(q-1)} \sum_{s_\phi = 0}^{q-2} \sum_{\substack{\mathbf{v}\, s.t. \\ v_3v_4v_5 = 0}} \prod G_{-(\frac{v_i(q-1)}{8} + \frac{s_\phi}{2})} \\ &\quad \times \prod G_{-(\frac{v_i(q-1)}{4})} G_{s_\phi} {\rm Teich}^{-s_\phi}(-2\phi) . \end{aligned} \tag{5.37}$$

Notice that we have again introduced monomials $\mathbf{v}$. These also have to satisfy (5.25); however, in addition, the constraint, $v_3v_4v_5 = 0$ is required, as the expression $\sum_{y \in \mathbb{F}_p^*} [(g_4(y)+1)^3 - g_4(y)^3]$ can be considered as:

$$\begin{aligned} &(\textit{Non-zero arguments of Gauss sums}) - (\textit{All Gauss sums}) \\ &= -(\textit{Gauss sums with at least one zero argument}). \end{aligned}$$

Hence we need to restrict the allowable monomials with the constraint $\prod v_3v_4v_5 = 0$ and there is an overall minus sign.
Now for $q \equiv 3 \mod 4$ we can write

$$N_{\rm mon}(x_1x_2 \neq 0,\ x_3x_4x_5 = 0)$$

$$= \frac{-1}{q(q-1)} \sum_{s_\phi = 0}^{q-2} \sum_{\substack{\mathbf{v}\, s.t. \\ \prod v_3v_4v_5 = 0}} \prod G_{-(\frac{v_i}{4}m + \frac{s_\phi}{2})} \prod G_{-(\frac{v_i}{2}m)} G_{s_\phi} {\rm Teich}^{-s_\phi}(-2\phi)$$

$$= \frac{-1}{q(q-1)} \sum_{u=0}^{m-1} \sum_{\substack{\mathbf{v} s.t. \\ v_3v_4v_5=0}} \prod G_{-(\frac{v_i}{4}m+u)} \prod G_{-(\frac{v_i}{2}m)} G_{2u} \mathrm{Teich}^{-u}(-2\phi)$$

$$+ \frac{-1}{q(q-1)} \sum_{u=0}^{m-1} \sum_{\substack{\mathbf{v} s.t. \\ v_3v_4v_5=0}} \prod G_{-\left(\frac{(v_i+2)}{4}m+u+\frac{1}{2}\right)} \prod G_{-\left(\frac{v_i+2}{2}m\right)} G_{2u+1} \mathrm{Teich}^{-(2u+1)}(-2\phi).$$

Here we have defined $u = \frac{s_\phi}{2}$ and have split the sum over the **v**s.(Notice that we could not have done this had $8|q-1$ and in that case we only get the first term). Now we can extend the sum over u by noticing that

$$\mathbf{v} \to \mathbf{v} + (4,4,0,0,0) \Longleftrightarrow u \to u+m; \tag{5.38}$$

hence we can extend the range of u from $0 \le u \le m-1$ to $0 \le u \le q-2$:

$$\begin{aligned} & N_{\mathrm{mon}}(x_1x_2 \neq 0,\, x_3x_4x_5 = 0) \\ &= -\frac{1}{q(q-1)} \sum_{u=0}^{q-2} \sum_{\substack{\mathbf{v} s.t. \\ v_3v_4v_5=0}} \prod G_{-(\frac{v_i}{4}m+u)} \prod G_{-(\frac{v_i}{2}m)} \\ &\quad \times G_{2u} \mathrm{Teich}^u\left(\frac{1}{(2\phi)^2}\right) \\ &\quad -\frac{1}{q(q-1)} \sum_{u=0}^{q-2} \sum_{\substack{\mathbf{v} s.t. \\ v_3v_4v_5=0}} \prod G_{-\left(\frac{(v_i+2)}{4}m+u+\frac{1}{2}\right)} \\ &\quad \times \prod G_{-(\frac{v_i+2}{2}m)} G_{2u+1} \mathrm{Teich}^u\left(\frac{1}{(2\phi)^2}\right) \mathrm{Teich}\left(-\frac{1}{2\phi}\right). \end{aligned} \tag{5.39}$$

Finally the contribution from $N_{\mathrm{mon}}(x_1x_2 = 0, x_3x_4x_5 = 0)$ needs to be found and added:

$$\begin{aligned} & N_{\mathrm{mon}}(x_1x_2 = 0, x_3x_4x_5 = 0) \\ &= \sum_{y \in \mathbb{F}_p^*} [(g_8(y)+1)^2 - g_8(y)^2](g_4(y)+1)^3 \\ &= -\frac{1}{q} \sum_{y \in \mathbb{F}_q^*} \sum_{b=0}^{1} \sum_{\substack{\mathbf{v} s.t. v_3v_4v_5 \neq 0 \\ (v_i+4b) \neq 0}} \lambda_{\mathbf{v}} \prod_{i=1,2} G_{-\frac{(v_i+4b)}{4}m} \prod_{i=3,4,5} G_{-\frac{v_i}{2}m}. \end{aligned} \tag{5.40}$$

In total we obtain for $q \equiv 3 \mod 8$:

$$\begin{aligned} N_{\mathrm{mon}} &= \frac{1}{q(q-1)^2} \sum_{t=0}^{\frac{q-3}{4}} \sum_{u=\lceil \frac{t}{2} \rceil}^{\lfloor \frac{2q-3+t}{2} \rfloor} \sum_{\mathbf{v}} \lambda_{\mathbf{v}} \prod_{i=1,2} G_{-(\frac{v_i}{4}m+u)} \prod_{i=3,4,5} G_{-(\frac{v_i}{2}m+t)} \\ &\quad \times G_{2u-t} G_{4t} \mathrm{Teich}^u\left(\frac{1}{(2\phi)^2}\right) \mathrm{Teich}^t\left(\frac{-2\phi}{(8\psi)^4}\right) \\ &\quad -\frac{1}{q(q-1)} \sum_{u=0}^{q-2} \sum_{\substack{\mathbf{v} s.t. \\ v_3v_4v_5=0}} \lambda_{\mathbf{v}} \prod_{i=1,2} G_{-(\frac{v_i}{4}m+u)} \prod_{i=3,4,5} G_{-\frac{v_i}{2}m} G_{2u} \mathrm{Teich}^u\left(\frac{1}{(2\phi)^2}\right) \\ &\quad -\frac{1}{q(q-1)} \sum_{u=0}^{q-2} \sum_{\substack{\mathbf{v} s.t. \\ v_3v_4v_5=0}} \lambda_{\mathbf{v}} \prod_{i=1,2} G_{-(\frac{v_i+2}{4}m+u+\frac{1}{2})} \prod_{i=3,4,5} G_{-\frac{v_i+2}{2}m} G_{2u+1} \\ &\quad \times \mathrm{Teich}^u \frac{1}{(2\phi)^2} \mathrm{Teich}^{-1}(-2\phi) \end{aligned}$$

$$-\frac{1}{q}\sum_{y\in\mathbb{F}_q^*}\sum_{b=0}^{1}\sum_{\substack{\mathbf{v}\\ v_3v_4v_5\neq 0\ \&\ (v_i+4b)\neq 0}}\lambda_{\mathbf{v}}\prod_{i=1,2}G_{-\frac{(v_i+4b)}{4}m}\prod_{i=3,4,5}G_{-\frac{v_i}{2}m}\,. \tag{5.41}$$

The computation for the other cases is similar and the results are summarized in the next four sections.

5.4. Summary of results.

5.4.1. *Summary for $\phi=\psi=0$.*

(1) When $8|q-1$:

$$N_{\mathrm{mon}} = \frac{1}{q}\sum_{b=0}^{1}\sum_{a=0}^{3}\sum_{\mathbf{v}}\lambda_{\mathbf{v}}\prod_{i=1,2}G_{-(v_i+a+4b)k}\prod_{i=3,4,5}G_{-2(v_i+a)k}\,. \tag{5.42}$$

(2) When $4|q-1$, $8\nmid q-1$:

$$N_{\mathrm{mon}} = \frac{1}{q}\sum_{a=0}^{3}\sum_{\mathbf{v}}\lambda_{\mathbf{v}}\prod_{i=1,2}G_{-\frac{(v_i+2a)}{2}l}\prod_{i=3,4,5}G_{-(v_i+2a)l}\,. \tag{5.43}$$

(3) When $4\nmid q-1$:

$$N_{\mathrm{mon}} = \frac{1}{q}\sum_{b=0}^{1}\sum_{\mathbf{v}}\lambda_{\mathbf{v}}\prod_{i=1,2}G_{-\frac{(v_i+4b)}{4}m}\prod_{i=3,4,5}G_{-\frac{v_i}{2}m}\,. \tag{5.44}$$

5.4.2. *Summary for $\phi=0$, $\psi\neq 0$.*

(1) When $8|q-1$:

$$\begin{aligned} N_{\mathrm{mon}} &= \frac{1}{q(q-1)}\sum_{w=0}^{q-2}\sum_{\mathbf{v}}\lambda_{\mathbf{v}}\prod_{i=1,2}G_{-(v_ik+w)}\prod_{i=3,4,5}G_{-(2v_ik+2w)}G_{8w}\mathrm{Teich}^{w}\left(\frac{1}{(8\psi)^8}\right)\\ &\quad -\frac{1}{q}\sum_{b=0}^{1}\sum_{a=0}^{3}\sum_{\mathbf{v}}\lambda_{\mathbf{v}}\prod_{i=1,2}G_{-(v_i+a+4b)k}\prod_{i=3,4,5}G_{-2(v_i+a)k}\,. \end{aligned} \tag{5.45}$$

(2) When $4|q-1$, $8\nmid q-1$:

$$\begin{aligned} N_{\mathrm{mon}} &= \frac{1}{q(q-1)}\sum_{w=0}^{q-2}\sum_{\mathbf{v}}\lambda_{\mathbf{v}}\prod_{i=1,2}G_{-(\frac{v_i}{2}l+w)}\prod_{i=3,4,5}G_{-(v_il+2w)}G_{8w}\mathrm{Teich}^{w}\left(\frac{1}{(8\phi)^8}\right)\\ &\quad -\frac{1}{q}\sum_{a=0}^{3}\sum_{\mathbf{v}}\lambda_{\mathbf{v}}\prod_{i=1,2}G_{-\frac{(v_i+2a)}{2}l}\prod_{i=3,4,5}G_{-(v_i+2a)l}\,. \end{aligned} \tag{5.46}$$

(3) When $4\nmid q-1$:

$$\begin{aligned} N_{\mathrm{mon}} &= \frac{1}{q(q-1)}\sum_{w=0}^{q-2}\sum_{\mathbf{v}}\lambda_{\mathbf{v}}\prod_{i=1,2}G_{-(\frac{v_i}{4}m+w)}\prod_{i=3,4,5}G_{-(\frac{v_i}{2}m+2w)}G_{8w}\mathrm{Teich}^{w}\left(\frac{1}{(8\phi)^8}\right)\\ &\quad -\frac{1}{q}\sum_{b=0}^{1}\sum_{\mathbf{v}}\lambda_{\mathbf{v}}\prod_{i=1,2}G_{-\frac{(v_i+4b)}{4}m}\prod_{i=3,4,5}G_{-\frac{v_i}{2}m}\,. \end{aligned} \tag{5.47}$$

5.4.3. *Summary for $\psi = 0$, $\phi \neq 0$.*

(1) When $8|q-1$:

$$N_{\text{mon}} = \frac{1}{q(q-1)}\sum_{u=0}^{q-2}\sum_{\mathbf{v}}\lambda_{\mathbf{v}}\sum_{a=0}^{3}\prod_{i=1,2}G_{-((v_i+a)k+u)}\prod_{i=3,4,5}G_{-2(v_i+a)k}G_{2u}\text{Teich}^{u}\left(\frac{1}{(2\phi)^2}\right)$$
$$-\frac{1}{q}\sum_{b=0}^{1}\sum_{a=0}^{3}\sum_{\mathbf{v}}\lambda_{\mathbf{v}}\prod_{i=1,2}G_{-(v_i+a+4b)k}\prod_{i=3,4,5}G_{-2(v_i+a)k}\,. \tag{5.48}$$

(2) When $4|q-1$, $8\nmid q-1$:

$$N_{\text{mon}} = \frac{1}{q(q-1)}\sum_{u=0}^{q-2}\sum_{\mathbf{v}}\lambda_{\mathbf{v}}\sum_{a=0}^{1}\prod_{i=1,2}G_{-(\frac{v_i+2a}{2}l+u)}\prod_{i=3,4,5}G_{-(v_i+2a)l}G_{2u}\text{Teich}^{u}\left(\frac{1}{(2\phi)^2}\right)$$
$$+\frac{1}{q(q-1)}\sum_{u=0}^{q-2}\sum_{\mathbf{v}}\lambda_{\mathbf{v}}\sum_{a=0}^{1}\prod_{i=1,2}G_{-(\frac{v_i+2a+1}{2}l+u+\frac{1}{2})}\prod_{i=3,4,5}G_{-(v_i+2a+1)l}G_{2u+1}$$
$$\times\text{Teich}^{u}\frac{1}{(2\phi)^2}\text{Teich}^{-1}(-2\phi)$$
$$-\frac{1}{q}\sum_{b=0}^{1}\sum_{a=0}^{1}\sum_{\mathbf{v}}\lambda_{\mathbf{v}}\prod_{i=1,2}G_{-\frac{(v_i+2a+4b)}{2}l}\prod_{i=3,4,5}G_{-(v_i+2a)l}\,. \tag{5.49}$$

(3) When $4\nmid q-1$:

$$N_{\text{mon}} = \frac{1}{q(q-1)}\sum_{u=0}^{q-2}\sum_{\mathbf{v}}\lambda_{\mathbf{v}}\prod_{i=1,2}G_{-(\frac{v_i}{4}m+u)}\prod_{i=3,4,5}G_{-\frac{v_i}{2}m}G_{2u}\text{Teich}^{u}\left(\frac{1}{(2\phi)^2}\right)$$
$$+\frac{1}{q(q-1)}\sum_{u=0}^{q-2}\sum_{\mathbf{v}}\lambda_{\mathbf{v}}\prod_{i=1,2}G_{-(\frac{v_i+2}{4}m+u+\frac{1}{2})}\prod_{i=3,4,5}G_{-\frac{v_i+2}{2}m}G_{2u+1}$$
$$\times\text{Teich}^{u}\frac{1}{(2\phi)^2}\text{Teich}^{-1}(-2\phi)$$
$$-\frac{1}{q}\sum_{b=0}^{1}\sum_{\mathbf{v}}\lambda_{\mathbf{v}}\prod_{i=1,2}G_{-\frac{(v_i+4b)}{4}m}\prod_{i=3,4,5}G_{-\frac{v_i}{2}m}\,. \tag{5.50}$$

5.4.4. *Summary for $\phi\psi \neq 0$.*

(1) When $8|q-1$:

$$N_{\text{mon}} = \frac{1}{q(q-1)^2}\sum_{t=0}^{q-2}\sum_{u=\lceil\frac{t}{2}\rceil}^{\lfloor\frac{2q-3+t}{2}\rfloor}\sum_{\mathbf{v}}\lambda_{\mathbf{v}}\prod_{i=1,2}G_{-(v_ik+u)}\prod_{i=3,4,5}G_{-(2v_ik+t)}$$
$$\times G_{2u-t}G_{4t}\text{Teich}^{u}\left(\frac{1}{(2\phi)^2}\right)\text{Teich}^{t}\left(\frac{-2\phi}{(8\psi)^4}\right)$$
$$-\frac{1}{q(q-1)}\sum_{u=0}^{q-2}\sum_{a=0}^{3}\sum_{\substack{\mathbf{v}\,s.t.\\ v_3v_4v_5=0}}\lambda_{\mathbf{v}}\prod_{i=1,2}G_{-((v_i+a)k+u)}\prod_{i=3,4,5}G_{-2(v_i+a)k}G_{2u}$$

$$\times \mathrm{Teich}^u\left(\frac{1}{(2\phi)^2}\right)$$
$$-\frac{1}{q}\sum_{b=0}^{1}\sum_{a=0}^{3}\sum_{\substack{\mathbf{v}\\ v_3v_4v_5\neq 0\ \&\ (v_i+a+4b)\neq 0}}\lambda_{\mathbf{v}}\prod_{i=1,2}G_{-(v_i+a+4b)k}\prod_{i=3,4,5}G_{-2(v_i+a)k}\cdot \tag{5.51}$$

(2) When $4|q-1$, $8\nmid q-1$:

$$N_{\mathrm{mon}} = \frac{1}{q(q-1)^2}\sum_{t=0}^{q-2}\sum_{u=\lceil\frac{t}{2}\rceil}^{\lfloor\frac{2q-3+t}{2}\rfloor}\sum_{\mathbf{v}}\lambda_{\mathbf{v}}\prod_{i=1,2}G_{-(\frac{v_i}{2}l+u)}\prod_{i=3,4,5}G_{-(v_il+t)}$$
$$\times G_{2u-t}G_{4t}\mathrm{Teich}^u\left(\frac{1}{(2\phi)^2}\right)\mathrm{Teich}^t\left(\frac{-2\phi}{(8\psi)^4}\right)$$
$$-\frac{1}{q(q-1)}\sum_{u=0}^{q-2}\sum_{a=0}^{1}\sum_{\substack{\mathbf{v} s.t.\\ v_3v_4v_5=0}}\lambda_{\mathbf{v}}\prod_{i=1,2}G_{-(\frac{(v_i+2a)}{2}l+u)}$$
$$\times\prod_{i=3,4,5}G_{-(v_i+2a)l}G_{2u}\mathrm{Teich}^u\left(\frac{1}{(2\phi)^2}\right)$$
$$-\frac{1}{q(q-1)}\sum_{u=0}^{q-2}\sum_{a=0}^{1}\sum_{\substack{\mathbf{v} s.t.\\ v_3v_4v_5=0}}\lambda_{\mathbf{v}}\prod_{i=1,2}G_{-(\frac{(v_i+2a+1)}{2}l+u+\frac{1}{2})}$$
$$\times\prod_{i=3,4,5}G_{-(v_i+2a+1)l}G_{2u+1}\mathrm{Teich}^u\frac{1}{(2\phi)^2}\mathrm{Teich}^{-1}(-2\phi)$$
$$-\frac{1}{q}\sum_{b=0}^{1}\sum_{a=0}^{1}\sum_{\substack{\mathbf{v}\\ v_3v_4v_5\neq 0\ \&\ (v_i+2a+4b)\neq 0}}\lambda_{\mathbf{v}}\prod_{i=1,2}G_{-\frac{(v_i+2a+4b)}{2}l}\prod_{i=3,4,5}G_{-(v_i+2a)l}\cdot \tag{5.52}$$

(3) When $4\nmid q-1$:

$$N_{\mathrm{mon}} = \frac{1}{q(q-1)^2}\sum_{t=0}^{\frac{q-3}{4}}\sum_{u=\lceil\frac{t}{2}\rceil}^{\lfloor\frac{2q-3+t}{2}\rfloor}\sum_{\mathbf{v}}\lambda_{\mathbf{v}}\prod_{i=1,2}G_{-(\frac{v_i}{4}m+u)}\prod_{i=3,4,5}G_{-(\frac{v_i}{2}m+t)}$$
$$\times G_{2u-t}G_{4t}\mathrm{Teich}^u\left(\frac{1}{(2\phi)^2}\right)\mathrm{Teich}^t\left(\frac{-2\phi}{(8\psi)^4}\right)$$
$$-\frac{1}{q(q-1)}\sum_{u=0}^{q-2}\sum_{\substack{\mathbf{v} s.t.\\ v_3v_4v_5=0}}\lambda_{\mathbf{v}}\prod_{i=1,2}G_{-(\frac{v_i}{4}m+u)}\prod_{i=3,4,5}G_{-\frac{v_i}{2}m}G_{2u}\mathrm{Teich}^u\left(\frac{1}{(2\phi)^2}\right)$$
$$-\frac{1}{q(q-1)}\sum_{u=0}^{q-2}\sum_{\substack{\mathbf{v} s.t.\\ v_3v_4v_5=0}}\lambda_{\mathbf{v}}\prod_{i=1,2}G_{-(\frac{v_i+2}{4}m+u+\frac{1}{2})}\prod_{i=3,4,5}G_{-\frac{v_i+2}{2}m}G_{2u+1}$$
$$\times\mathrm{Teich}^u\frac{1}{(2\phi)^2}\mathrm{Teich}^{-1}(-2\phi)$$
$$-\frac{1}{q}\sum_{y\in\mathbb{F}_q^*}\sum_{b=0}^{1}\sum_{\substack{\mathbf{v}\\ v_3v_4v_5\neq 0\ \&\ (v_i+4b)\neq 0}}\lambda_{\mathbf{v}}\prod_{i=1,2}G_{-\frac{(v_i+4b)}{4}m}\prod_{i=3,4,5}G_{-\frac{v_i}{2}m}\cdot \tag{5.53}$$

The number of points over $\mathbb{F}_q$ depends on the value of $q \mod 8$. Note that $8|p^2-1$ for all primes p. Hence it can be concluded that:

(1) If $p-1 \equiv 2 \mod 4$, then $p^{2r}-1 \equiv 0 \mod 8$ and $p^{2r-1}-1 \equiv 2 \mod 4$ $\forall r \in \mathbb{N}$,
(2) If $p-1 \equiv 0 \mod 4$, then $p^{2r}-1 \equiv 0 \mod 8$ and $p^{2r-1}-1 \equiv 0 \mod 4$ $\forall r \in \mathbb{N}$,
(3) If $p-1 \equiv 0 \mod 8$, then $p^{r}-1 \equiv 0 \mod 8$ $\forall r \in \mathbb{N}$.

Hence the zeta function can be split into odd and even powers:

$$\begin{aligned}\zeta(\phi,\psi,p,t) = &\equiv \exp\left(\sum_{r\in\mathbb{N}} N_{r,p}(X)\frac{t^r}{r}\right)\\ &= \exp\left(\sum_{r\in\mathbb{N}} N_{2r+1,p}(X)\frac{t^{2r+1}}{2r+1}\right)\exp\left(\sum_{r\in\mathbb{N}} N_{2r,p}(X)\frac{t^{2r}}{2r}\right). \qquad (5.54)\end{aligned}$$

We conclude that for the three cases above the zeta function takes the form:
(1) $\zeta(\phi,\psi,p,t) = R_m(p,t)R_{k'}(p,t)$,
(2) $\zeta(\phi,\psi,p,t) = R_l(p,t)R_{k'}(p,t)$,
(3) $\zeta(\phi,\psi,p,t) = R_k(p,t)R_{k'}(p,t)$,

where $R_{k'}(p,t)$ is the contribution from the odd powers and $R_m(p,t)$, $R_l(p,t)$, $R_k(p,t)$ are the contributions from the even powers.

6. The mirror octics

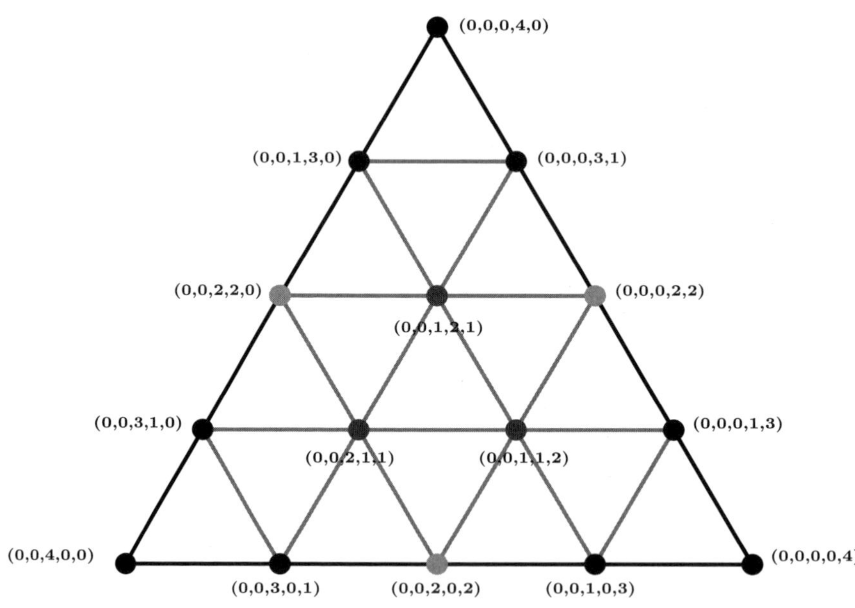

Figure 5: The two-face with vertices $(0,0,4,0,0)$, $(0,0,0,4,0)$ and $(0,0,0,0,4)$.

The Batyrev mirror of the octic is given by a hypersurface in the toric variety $\frac{\mathbb{C}^{104}-F}{(\mathbb{C}^*)^{100}}$. We now introduce 104 Cox variables $\tilde{x}_{\mathbf{m}}$ for the mirror, one for each monomial $\mathbf{m}$ in the Newton polyhedron (i.e. all the degree 8 monomials

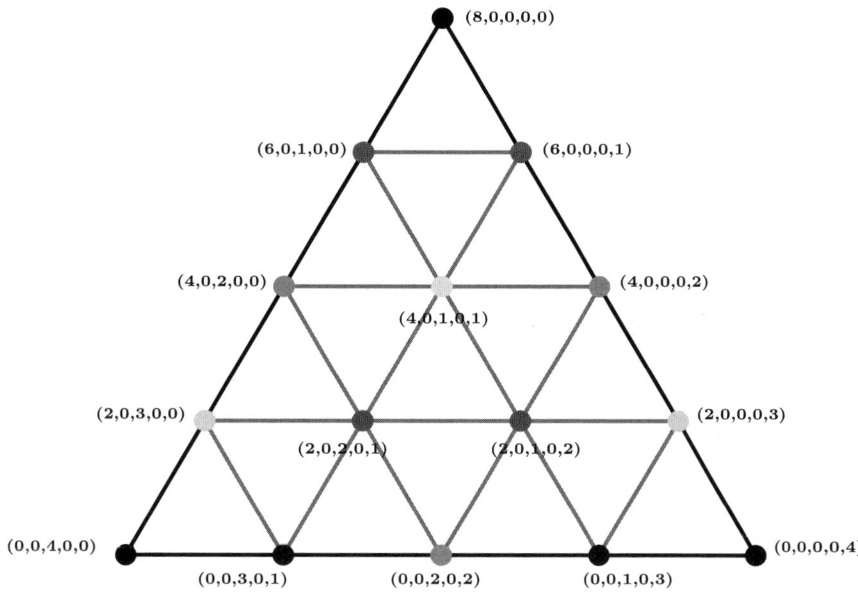

Figure 6: A two-face with vertices $(8,0,0,0,0)$, $(0,0,4,0,0)$ and $(0,0,0,0,4)$.

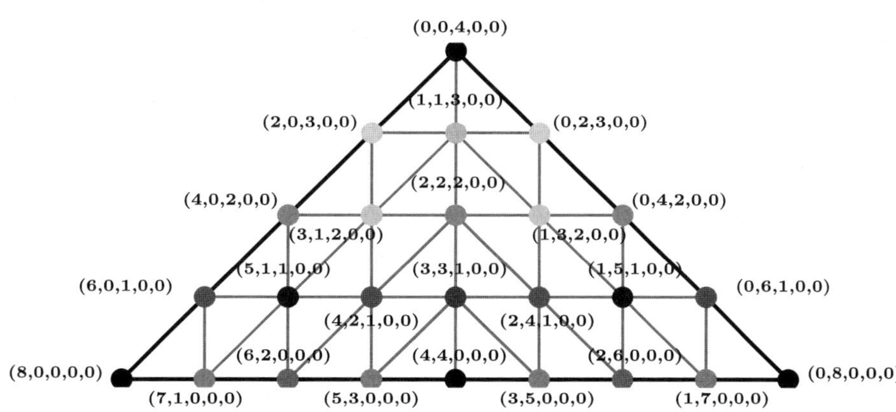

Figure 7: A two-face with vertices $(8,0,0,0,0)$, $(0,8,0,0,0)$ and $(0,0,4,0,0)$.

apart from the interior point $(1,1,1,1,1)$). Let us write each monomial in the following way: $(m_1, m_2, m_3, m_4, m_5, m_\psi, m_\phi)$, where $m_\phi = \frac{m_1+m_2}{2}$ and $m_\psi = \frac{m_1+m_2+2(m_3+m_4+m_5)}{8}$. Let us stack all these monomials (with $(1,1,1,1,1,1,1)$ on the top) to form a 105×7 matrix:

$$\begin{pmatrix} 1 & 1 & 1 & 1 & 1 & 1 & 1 \\ 8 & 0 & 0 & 0 & 0 & 1 & 4 \\ 0 & 8 & 0 & 0 & 0 & 1 & 4 \\ 0 & 0 & 4 & 0 & 0 & 1 & 0 \\ 0 & 0 & 0 & 4 & 0 & 1 & 0 \\ 0 & 0 & 0 & 0 & 4 & 1 & 0 \\ 4 & 4 & 0 & 0 & 0 & 1 & 4 \\ 7 & 1 & 0 & 0 & 0 & 1 & 4 \\ \vdots & \vdots & \vdots & \vdots & \vdots & \vdots & \vdots \end{pmatrix}$$

Let the columns of this matrix be denoted $(\mathbf{w}_j, w_\psi, w_\phi)$, for $j = 1, \ldots, 5$; then the mirror polynomial takes the form

$$\tilde{P}(\tilde{x}) = \tilde{x}^{w_1} + \tilde{x}^{w_2} + \tilde{x}^{w_3} + \tilde{x}^{w_4} + \tilde{x}^{w_5} - 2\phi\tilde{x}^{w_\phi} - 8\psi\tilde{x}^{w_\psi} \ . \tag{6.1}$$

where $\tilde{x}^{w_j} := \prod_{\mathbf{m}} \tilde{x}_{\mathbf{m}}^{m_j}$. We can also define an extra coordinate $\tilde{y}$ to be associated to the interior point.
We need to evaluate:

$$\begin{aligned} q(q-1)^{100}\tilde{N} &= \sum_{(\tilde{y},\tilde{\mathbf{x}})\in(\mathbb{F}_q^{105}\setminus F)} \Theta\left(\tilde{y}(\tilde{P}(\tilde{x})\right) \\ &= q(q-1)^{100}(\tilde{N}(\tilde{y}=0) + \tilde{N}(\tilde{y}\neq 0)) \ . \end{aligned} \tag{6.2}$$

A symmetric triangulation of the Newton polyhedron has been chosen, as illustrated in Figures 5, 6, and 7 (each shade represents one of the 14 classes to which a monomial in the Newton polyhedron can belong). The rule provided by toric geometry dictates that a subset of the coordinates $\tilde{x}_{\mathbf{m}}$ can be zero simultaneously if and only if the corresponding $\mathbf{m}$'s lie in the same cone. Hence, a maximum of 3 Cox variables are allowed to vanish at the same time, and they must lie within the same 'small triangle'.

First consider the case when none of the 105 Cox variables vanish:

$$\begin{aligned} &\sum_{\substack{(\tilde{y},\tilde{\mathbf{x}})\\ \in(\mathbb{F}_q^*)^{105}}} \Theta\left(\tilde{y}(\tilde{P}(\tilde{x})\right) \\ &= \sum_{(\tilde{y},\tilde{\mathbf{x}})\in(\mathbb{F}_q^*)^{105}} \Theta\left(\tilde{y}\left(\tilde{x}^{w_1} + \tilde{x}^{w_2} + \tilde{x}^{w_3} + \tilde{x}^{w_4} + \tilde{x}^{w_5} - 2\tilde{\phi}\tilde{x}^{w_\phi} - 8\tilde{\psi}\tilde{x}^{w_\psi}\right)\right) \\ &= \sum_{(\tilde{y},\tilde{\mathbf{x}})\in(\mathbb{F}_q^*)^{105}} \Theta\left(\tilde{y}\tilde{x}^{w_1}\right)\Theta\left(\tilde{y}\tilde{x}^{w_2}\right)\Theta\left(\tilde{y}\tilde{x}^{w_3}\right)\Theta\left(\tilde{y}\tilde{x}^{w_4}\right)\Theta\left(\tilde{y}\tilde{x}^{w_5}\right) \\ &\quad \times\Theta\left(-2\phi\tilde{y}\tilde{x}^{w_\phi}\right)\Theta\left(-8\psi\tilde{y}\tilde{x}^{w_\psi}\right) \\ &= \frac{(q-1)^{105}}{(q-1)^7} \sum_{\substack{0\leq s_i\leq q-2,\\ q-1\mid\sum_i s_i m_i}} \prod_{i=1}^{5} G_{-s_i} G_{s_\phi} G_{s_\psi} \mathrm{Teich}^{-s_\psi}(-8\psi)\mathrm{Teich}^{-s_\phi}(-2\phi) \ . \end{aligned} \tag{6.3}$$

Notice that here we are required to solve the constraints

$$q-1 | (s_1 m_1 + s_2 m_2 + s_3 m_3 + s_4 m_4 + s_5 m_5 - s_\phi m_\phi - s_\psi m_\psi) \quad \forall\mathbf{m} \ , \tag{6.4}$$

where $\mathbf{m} \equiv (m_1, m_2, m_3, m_4, m_5, m_\psi, m_\phi)$, and $m_\psi = 1$ and $m_\phi = \frac{m_1+m_2}{2}$. Hence define a vector $(v)_{\mathbf{m}}$ of length 105 indexed by $\mathbf{m}$, such that:

$$v_{\mathbf{m}}(q-1) = \mathbf{m}.\mathbf{s}\ , \tag{6.5}$$

where $s \equiv (s_1, s_2, s_3, s_4, s_5, -s_\psi, -s_\phi)$. We shall rename the v's associated to the five vertex monomials as follows:

$$\begin{aligned} &v_{(8,0,0,0,0,1,4)} \equiv v_1\ , \quad v_{(0,8,0,0,0,1,4)} \equiv v_2\ , \quad v_{(0,0,4,0,0,1,0)} \equiv v_3\ , \\ &v_{(0,0,0,4,0,1,0)} \equiv v_4\ , \quad v_{(0,0,0,0,4,1,0)} \equiv v_5\ . \end{aligned} \tag{6.6}$$

It is clear that from applying (6.5) to the monomials $(8,0,0,0,0,1,4)$, $(0,8,0,0,0,1,4)$, $(0,0,4,0,0,1,0)$, $(0,0,0,4,0,1,0)$, and $(0,0,0,4,0,1,0)$ we obtain:

$$\begin{aligned} s_1 &= \frac{v_1(q-1)}{8} + \frac{s_\phi}{2} + \frac{s_\psi}{8}\ , \quad s_2 = \tfrac{v_2(q-1)}{8} + \tfrac{s_\phi}{2} + \tfrac{s_\psi}{8}\ , \\ s_3 &= \frac{v_3(q-1)}{4} + \frac{s_\psi}{4}\ , \qquad s_4 = \tfrac{v_4(q-1)}{4} + \tfrac{s_\psi}{4}\ , \qquad s_5 = \frac{v_5(q-1)}{4} + \frac{s_\psi}{4}\ . \end{aligned} \tag{6.7}$$

By using that fact that, by the definition of the Cox variables for this family of Calabi–Yau manifolds, $m_1 + m_2 + 2(m_3 + m_4 + m_5) = 8$, and (6.5), it is easy to show that

$$v_{\mathbf{m}} = \frac{(m_1 v_1 + m_2 v_2 + 2(m_3 v_3 + m_4 v_4 + m_5 v_5))}{8}, \quad \forall \mathbf{m}\ . \tag{6.8}$$

It is easy to check that the only solutions to these numerous constraints is to have

$$(v_1, v_2, v_3, v_4, v_5) = (1,1,1,1,1)a, \quad a = 0, 1, \ldots, 7; \tag{6.9}$$

where the last three digits are to be considered mod 4; hence,

$$\begin{aligned} \sum_{(\tilde{y},\tilde{\mathbf{x}})\in(\mathbb{F}_q^*)^{105}} \Theta\left(\tilde{y}(\tilde{P}(\tilde{x})\right) &= (q-1)^{98} \sum_{a=0}^{7} G^2_{-\left(\frac{a(q-1)}{8}+\frac{s_\phi}{2}+\frac{s_\psi}{8}\right)} G^3_{-\left(\frac{a(q-1)}{4}+\frac{s_\psi}{4}\right)} G_{s_\phi} G_{s_\psi} \\ &\quad \times \mathrm{Teich}^{-s_\psi}(-8\psi)\mathrm{Teich}^{-s_\phi}(-2\phi)\ . \end{aligned} \tag{6.10}$$

Notice that the above expression is almost equal to the contribution to the number of points from the class of monomials $(0,0,0,0,0)$ in the original family of octic threefolds.

Now we need to consider the case when only monomials on the two-face in Figure 5 with vertices $(0,0,4,0,0)$, $(0,0,0,4,0)$, and $(0,0,0,0,4)$ are allowed to vanish.

To begin with, consider the case $\tilde{x}_{(0,0,4,0,0)} = 0$. Here the defining equation reduces to

$$\tilde{P} = \tilde{x}^{w_1} + \tilde{x}^{w_2} + \tilde{x}^{w_4} + \tilde{x}^{w_5} - 2\phi\tilde{x}^{w_\phi}\ , \tag{6.11}$$

and so

$$\begin{aligned} &\sum_{\substack{(\tilde{y},\tilde{\mathbf{x}})\in(\mathbb{F}_q^{104}\setminus F),\\ \tilde{x}_{(0,0,4,0,0)}=0}} \Theta\left(\tilde{y}(\tilde{P}(\tilde{x})\right) \\ &= (q-1)^{103} + (q-1)^{99} \sum_{\substack{0\le s_i\le q-2,\\ q-1|\sum_i s_i m_i}} \prod_{i=1}^{5} G_{-s_i} G_{s_\phi} \mathrm{Teich}^{-s_\phi}(-2\phi)\ . \end{aligned} \tag{6.12}$$

The first term on the right hand side is the contribution to the sum from $\tilde{y} = 0$. The second term was obtained from the straightforward expansion of

$$(6.13) \qquad \sum_{(\tilde{y},\tilde{\mathbf{x}})\in(\mathbb{F}_q^*)^{104}} \Theta\left(\tilde{y}\tilde{x}^{w_1}\right)\Theta\left(\tilde{y}\tilde{x}^{w_2}\right)\Theta\left(\tilde{y}\tilde{x}^{w_4}\right)\Theta\left(\tilde{y}\tilde{x}^{w_5}\right)\Theta\left(-2\phi\tilde{y}\tilde{x}^{w_\phi}\right)$$

in terms of Gauss sums.

We have constraints identical to (6.5), except that now there is no s_ϕ, s_3 or m_ϕ, m_3:

$$(6.14) \qquad \begin{gathered} v_{\mathbf{m}}(q-1) = \mathbf{m}.\mathbf{s} \quad \forall \mathbf{m} \\ q-1|\left(m_1 s_1 + m_2 s_2 + m_4 s_4 + m_5 s_5 - m_\phi s_\phi\right). \end{gathered}$$

This time the only solutions to the above constraints are

$$(6.15) \qquad (v_1, v_2, v_3, v_4, v_5) = (0,0,0,0,0) \& (4,4,0,0,0).$$

Hence we can rewrite the second term:

$$(6.16) \qquad (q-1)^{99} \sum_{a=0,4} G^2_{-\left(\frac{(q-1)}{8}a+\frac{s_\phi}{2}\right)} G_0^2 G_{s_\phi} \mathrm{Teich}^{-s_\phi}(-2\phi);$$

using the fact that $G_0 = -1$, we obtain

$$(6.17) \qquad \sum_{\substack{(\tilde{y},\tilde{\mathbf{x}})\in(\mathbb{F}_q^{104}\setminus F),\\ \tilde{x}_{(0,0,4,0,0)}=0}} \Theta\left(\tilde{y}(\tilde{P}(\tilde{x})\right) = (q-1)^{103} + (q-1)^{99} \sum_{a=0,1} G^2_{-\left(\frac{(q-1)}{2}a+\frac{s_\phi}{2}\right)} G_{s_\phi} \mathrm{Teich}^{-s_\phi}(-2\phi).$$

Now we can add up all the contributions from the various different combinations of allowable vanishings on this two-face. In each case, the contribution will have the form

$$(6.18) \qquad C(q-1)^A \pm C(q-1)^B \sum_{a=0,1} G^2_{-\left(\frac{(q-1)}{2}a+\frac{s_\phi}{2}\right)} G_{s_\phi} \mathrm{Teich}^{-s_\phi}(-2\phi),$$

where C can be thought of as the multiplicity of the type of points under consideration, e.g. for the case when there is only one variable vanishing and these are vertices, $C = 3$, as there are 3 vertices.

The values of A, B, C are tabulated below:

Number of Points	Subsidiary condition	A	B	Sign	Multiplicity C
Point	Vertex	103	99	+	3
	Interior to Edge	103	100	−	9
	Interior to 2-Face	103	101	+	3
Link	Edge Link	102	99	−	12
	Interior Links	102	100	+	18
Triangle		101	99	+	16

We now have to consider cases where vanishing can take place on the other two-faces.

# of Points	Subsidiary condition	Contribution
Point	Vertex, e.g. $(8,0,0,0,0)$	$2(q-1)^{103}+2(q-1)^{100}\sum G_{-s_1}G_{-s_3}G_{-s_4}G_{-s_5}$
	Interior to Edge	$18(q-1)^{103}+2(q-1)^{101}\sum G_{-s_1}G_{-s_3}G_{-s_4}$
	e.g. $(7,1,0,0,0)$	$7(q-1)^{103}+7(q-1)^{101}\sum G_{-s_3}G_{-s_4}G_{-s_5}$
	Interior to 2-Face (Type 2)	$18(q-1)^{103}+2(q-1)^{102}\sum G_{-s_1}G_{-s_3}$
	e.g. $(1,1,3,0,0)$	$27(q-1)^{103}+27(q-1)^{102}\sum G_{-s_3}G_{-s_4}$
Link	Edge Link	$24(q-1)^{102}+24(q-1)^{100}\sum G_{-s_1}G_{-s_3}G_{-s_4}$
	Interior Link	$108(q-1)^{102}+108(q-1)^{101}\sum G_{-s_1}G_{-s_3}$
		$8(q-1)^{102}+108(q-1)^{100}\sum G_{-s_3}G_{-s_4}G_{-s_5}$
		$120(q-1)^{102}+120(q-1)^{101}\sum G_{-s_3}G_{-s_4}$
Triangle		$96(q-1)^{101}+96(q-1)^{100}\sum G_{-s_1}G_{-s_3}$
		$96(q-1)^{101}+96(q-1)^{100}\sum G_{-s_3}G_{-s_4}$

It turns out that all the s_i's in the above expressions have to vanish (again due to a large number of constraints similar to (6.5)). Altogether we get the following contribution:

$$\begin{aligned}&72(q-1)^{103}+26(q-1)^{102}+192(q-1)^{101}+2(q-1)^{100}\sum G_{-s_1}G_{-s_3}G_{-s_4}G_{-s_5}\\&+6(3q+1)(q-1)^{100}\sum G_{-s_1}G_{-s_3}G_{-s_4}+(7q+1)(q-1)^{100}\sum G_{-s_3}G_{-s_4}G_{-s_5}\\&+3(9q^2+22q+1)(q-1)^{100}\sum G_{-s_3}G_{-s_4}\\&+6(1+12q+3q^2)(q-1)^{100}\sum G_{-s_1}G_{-s_3}\ .\end{aligned} \tag{6.19}$$

Hence, using the fact that $G_0=-1$ we obtain

$$\begin{aligned}\tilde{N} &= \frac{1}{(q-1)^2}\sum_{a=0}^{7}G^2_{-\left(\frac{a(q-1)}{8}+\frac{s_\phi}{2}+\frac{s_\psi}{8}\right)}G^3_{-\left(\frac{a(q-1)}{4}+\frac{s_\psi}{4}\right)}G_{s_\phi}G_{s_\psi}\\&\quad\times\mathrm{Teich}^{-s_\psi}(-8\psi)\mathrm{Teich}^{-s_\phi}(-2\phi)\\&\quad+\frac{1}{(q-1)}(1+3q+3q^2)\sum_{a=0,1}G^2_{-\left(\frac{(q-1)}{2}a+\frac{s_\phi}{2}\right)}G_{s_\phi}\mathrm{Teich}^{-s_\phi}(-2\phi)\\&\quad-2+80q+83q^2+q^3\ .\end{aligned} \tag{6.20}$$

Perusal of Section (5.4.4) shows that the contribution to the number of points from $(0,0,0,0,0)$ to the expression for the original family of octics $N_{(0,0,0,0,0)}$ is given by:

$$\begin{aligned}N_{(0,0,0,0,0)} &= \frac{1}{(q-1)^2}\sum_{a=0}^{7}G^2_{-\left(\frac{a(q-1)}{8}+\frac{s_\phi}{2}+\frac{s_\psi}{8}\right)}G^3_{-\left(\frac{a(q-1)}{4}+\frac{s_\psi}{4}\right)}G_{s_\phi}G_{s_\psi}\\&\quad\times\mathrm{Teich}^{-s_\psi}(-8\psi)\mathrm{Teich}^{-s_\phi}(-2\phi)\\&\quad+\frac{1}{(q-1)}\sum_{a=0,1}G^2_{-\left(\frac{(q-1)}{2}a+\frac{s_\phi}{2}\right)}G_{s_\phi}\mathrm{Teich}^{-s_\phi}(-2\phi)\ ;\end{aligned} \tag{6.21}$$

hence we can rewrite the total in the form

$$\begin{aligned}\tilde{N} &= -2+80q+83q^2+q^3+N_{(0,0,0,0,0)}\\&\quad+\frac{3(q+1)}{(q-1)}\sum_{a=0,1}\sum_{s_\phi=0}^{q-2}G^2_{-\left(\frac{(q-1)}{2}a+\frac{s_\phi}{2}\right)}G_{s_\phi}\mathrm{Teich}^{-s_\phi}(-2\phi)\\&= -2+80q+83q^2+q^3+N_{(0,0,0,0,0)}\end{aligned}$$

$$+\frac{3(q+1)}{(q-1)}\sum_{u=0}^{q-2} G_{-u}^2 G_{2u}\mathrm{Teich}^u\left(\frac{1}{(2\phi)^2}\right). \tag{6.22}$$

It turns out that the final term in (6.22) is related to the number of points on a Calabi–Yau manifold of dimension zero, which is described in [**CdOV1**]. This Calabi–Yau is defined as the points which satisfy the following equation in $\mathbb{F}_p\mathbb{P}^1$ (projective space with coordinates in $\mathbb{F}_p$):

$$\begin{aligned} P(x) &= x_1^2 + x_2^2 - 2\phi x_1 x_2 \\ &= (x_1 - \phi x_2)^2 - (\phi^2 - 1)x_2^2 . \end{aligned} \tag{6.23}$$

The rewriting of the equation in the second line makes it clear that the number of points over $\mathbb{F}_p\mathbb{P}^1$ is given by

(6.24)

$$N(\phi) = 1 + \left(\frac{\phi^2-1}{p}\right) = \begin{cases} 1, & \text{if } \phi^2 - 1 \equiv 0 \mod p\,, \\ 2, & \text{if } \phi^2 - 1 \text{ is a non-zero square residue of p}\,, \\ 0, & \text{if } \phi^2 - 1 \text{ is a non-square residue of p}\,, \end{cases}$$

where $\left(\frac{a}{p}\right)$ is the Legendre symbol.

Now computing N in the usual way gives

$$N = 1 - \frac{1}{q} + \frac{1}{q(q-1)}\sum_{u=0}^{q-2} G_u^2 G_{2u}\mathrm{Teich}^u\left(\frac{1}{(2\phi)^2}\right). \tag{6.25}$$

Equating (6.24) and (6.25) gives us

$$\sum_{u=0}^{q-2} G_u^2 G_{2u}\mathrm{Teich}^u\left(\frac{1}{(2\phi)^2}\right) = (q-1)\left(1 + q\left(\frac{\phi^2-1}{p}\right)\right). \tag{6.26}$$

Generalizing to $\mathbb{F}_{p^s}$ we obtain

$$\tilde{N} = 1 + 83q + 83q^2 + q^3 + N_{(0,0,0,0,0)} + 3q(q+1)\left(\frac{\phi^2-1}{p}\right)^s. \tag{6.27}$$

This leads to a zeta function of the form

(6.28)

$$\tilde{\zeta}(\phi,\psi) = \frac{R_{(\mathbf{0},\mathbf{0},\mathbf{0},\mathbf{0},\mathbf{0})}}{(1-t)(1-pt)^{83}(1-p^2t)^{83}\left(1-\left(\frac{\phi^2-1}{p}\right)pt\right)^3\left(1-\left(\frac{\phi^2-1}{p}\right)p^2t\right)^3(1-p^3t)}.$$

In summary, the form of the zeta function for the mirror is one of the following:

(1) When $\phi^2 - 1 \neq 0 \mod p$, we get either:

$$\tilde{\zeta}(\phi,\psi) = \begin{cases} \frac{R_{(\mathbf{0},\mathbf{0},\mathbf{0},\mathbf{0},\mathbf{0})}}{(1-t)(1-pt)^{86}(1-p^2t)^{86}(1-p^3t)} & \text{if } \left(\frac{\phi^2-1}{p}\right) = 1, \\ \\ \frac{R_{(\mathbf{0},\mathbf{0},\mathbf{0},\mathbf{0},\mathbf{0})}}{(1-t)(1-pt)^{83}(1-p^2t)^{83}(1+pt)^3(1+p^2t)^3(1-p^3t)} & \text{if } \left(\frac{\phi^2-1}{p}\right) = -1. \end{cases}$$

In this case $R_{(\mathbf{0},\mathbf{0},\mathbf{0},\mathbf{0},\mathbf{0})}$ has degree 6 or (if in addition $(8\psi^4+\phi)^2-1 \equiv 0 \mod p$) degree 5.

(2) When $\phi^2 - 1 \equiv 0 \mod p$ we get

$$\tilde{\zeta}(\phi,\psi) = \frac{R_{(0,0,0,0,0)}}{(1-t)(1-pt)^{83}(1-p^2t)^{83}(1-p^3t)}\,; \tag{6.29}$$

in this case $R_{(0,0,0,0,0)}$ either has degree 4 or (if in addition $(8\psi^4+\phi)^2-1 \equiv 0 \mod p$) degree 3.

The splitting of the denominator according to the value of ϕ^2 may have a deeper significance. It was noted in Section 3 that $h^{1,1}_{\text{toric}} = 83$, whereas $h^{1,1} = 86$ for the mirror octics. For the original family, $h^{1,1} = h^{1,1}_{\text{toric}} = 2$, and correspondingly, no splitting was observed in the denominator. This prompts Conjecture 7.2 described in the next section.

7. Zeta functions, results, conjectures and open questions

In this section we shall present and discuss the zeta functions obtained from the formulae derived in the previous section. The Gross–Koblitz formula (5.10) was used repeatedly to expand the Gauss sums in terms of p-adic Gamma functions. The computation being extremely heavy, this task was undertaken by writing extensive Mathematica code.

For the octic threefolds results for the number of points over $\mathbb{F}_{p^r}$ are known for up to $r = 2$ for the primes up to 73. For the primes up to 23 the number of points could be computed up to $r = 3$. For smaller primes higher values of r were computable, e.g. for $p = 3$, we computed up to $r = 8$. To compute the zeta function fully for larger primes would require considerably more computing power and memory than is currently typically available in top of the range desktop machines. It turns out that calculation up to $r = 3$ is usually sufficient for determining the full zeta function. In [**K**] the author has provided full tables of the zeta functions found for primes from 3 to 17.The prime 2 is the only prime of bad reduction.

7.1. Zeta function of the octic Calabi–Yau threefolds. It was found that the number of points can be written in the following way:

$$N_r(\psi) = N_{(0,0,0,0,0)} + \sum_{\mathbf{v}} \lambda_{\mathbf{v}} N_{\mathbf{v},r}(\psi)\,, \tag{7.1}$$

where the $N_{\mathbf{v},r}$ are not in general either positive or integral. Hence, for each (ϕ,ψ) the zeta function is given by

$$\zeta(\phi,\psi,p,t) = \frac{R_{\text{excep}}(t)\prod_{\mathbf{v}} R_{\mathbf{v}}(t)^{\lambda_{\mathbf{v}}}}{(1-t)(1-pt)^2(1-p^2t)^2(1-p^3t)}\,, \tag{7.2}$$

where $R_{\mathbf{v}}(t)$ is the contribution to the zeta function associated to the monomial $\mathbf{v}$.

$R_{\text{excep}}(t)$, a sextic, which always factorizes into the cube of a quadratic, can be thought of as a contribution from the exceptional divisor, which as explained in (3.5) was a ruled surface over a curve of genus 3. A curve of genus 3 has a Hodge diamond of the form

$$\begin{matrix} & 1 & \\ 3 & & 3 \\ & 1 & \end{matrix}$$

and thus, its zeta function would also have a sextic in the numerator. The factor $R_{\text{excep}}(t)$ is independent of the parameters (ϕ,ψ). It could, however, be included as part of the contribution from $(0,0,2,1,1)$ and $(2,2,1,1,0)$ combined. The presence of $R_{\text{excep}}(t)$ arises from having singularities in the ambient space, in this case

$\mathbb{P}^4_{(1,1,2,2,2)}$, which need to be resolved. It also reflects the fact that not all complex structure deformations can be described as monomial deformations, since for this case $h^{2,1}_{\text{poly}} = 83$, $h^{2,1} = 86$. This prompts the following conjecture:

CONJECTURE 7.1 (Non-monomial splitting of the Numerator). *For a family of three-dimensional Calabi–Yau hypersurfaces, if $h^{2,1} > h^{2,1}_{\text{poly}}$, the numerator of the zeta function has a factor that is independent of the parameters labelling each member of the family. This factor, denoted $R_{excep}(t)$, will have degree $2h'$, where $h' = h^{2,1} - h^{2,1}_{\text{poly}}$. and may exhibit modularity connected with the modularity properties of certain curves which are singular sets of the Calabi-Yau varieties.*

For the family of quintic threefolds[**CdOV2**] no such parameter independent factor was found in the zeta function, in keeping with the above conjecture, as in that case, $h^{2,1}_{\text{poly}} = h^{2,1} = 101$.

For various values of p, $R_{\text{excep}}(t)$ takes the form:

p	$R_{\text{excep}}(t)$				
3	$(1+3^3t^2)^3$	19	$(1+19^3t^2)^3$	47	$(1+47^3t^2)^3$
5	$(1-2.5t+5^3t^2)^3$	23	$(1+23^3t^2)^3$	53	$(1+14.53t+53^3t^2)^3$
7	$(1+7^3t^2)^3$	29	$(1-10.29t+29^3t^2)^3$	59	$(1+59^3t^2)^3$
11	$(1+11^3t^2)^3$	31	$(1+31^3t^2)^3$	61	$(1-10.61t+61^3t^2)^3$
13	$(1+6.13t+13^3t^2)^3$	37	$(1-2.37t+37^3t^2)^3$	67	$(1+67^3t^2)^3$
17	$(1-2.17t+17^3t^2)^3$	41	$(1-10.41t+41^3t^2)^3$	71	$(1+71^3t^2)^3$
		43	$(1+43^3t^2)^3$	73	$(1+6.73t+73^3t^2)^3$

When $p \not\equiv 1 \mod 4$ it seems that $R_{\text{excep}}(t) = (1+p^3t^2)^3$.
When $p \equiv 1 \mod 4$, $R_{\text{excep}}(t) = (1+b(p)pt+p^3t^2)^3$.
The coefficients $b(p)$ appear to be related to a normalized Hecke cusp eigenform of weight 2 and level 64. The appearance of this factor was shown for the case $\phi = \psi = 0$ in [**LS**]; our analysis shows that this factor of the zeta function is present at all points in the moduli space.

The calculation of the mirror octic in Section 6 hinted that the arithmetic may 'know' about the monomial-divisor map, i.e. the monomial and non-monomial deformations of the defining polynomial for the original family and equivalently the toric and non-toric divisors of the mirror. This leads to the following conjecture:

CONJECTURE 7.2 (Toric Splitting of Denominator). *For Calabi–Yau threefolds for which $h^{1,1}(X) \neq h^{1,1}_{\text{toric}}(X)$. Let $\tilde{h} = h^{1,1}(X) - h^{1,1}_{\text{toric}}(X)$ the denominator of the zeta function takes the form*

$$(1-t)(1-pt)^{h^{1,1}_{\text{toric}}}(1-p^2t)^{h^{1,1}_{\text{toric}}}(1-\chi pt)^{\tilde{h}}(1-\chi p^2 t)^{\tilde{h}}(1-p^3t)\ , \tag{7.3}$$

where χ is a character that depends on the defining equation of a singular locus. At this singular locus, the denominator will take the form

$$(1-t)(1-pt)^{h^{1,1}_{\text{toric}}}(1-p^2t)^{h^{1,1}_{\text{toric}}}(1-p^3t)\ . \tag{7.4}$$

REMARK 7.1. Indeed, the above two conjectures may really be viewed as a single mirror symmetry conjecture viewed from different sides, because from Batyrev's formulae, $h' = \tilde{h}$, i.e. $h^{2,1}(X) - h^{2,1}_{\text{poly}}(X) = h^{1,1}(Y) - h^{1,1}_{\text{toric}}(Y)$, for a mirror symmetric pair of manifolds (X, Y).

7.2. Behaviour at singularities. The degree of the rational function $R_{\mathbf{v}}(t)$ obtained from the contribution to the rational points associated to each monomial class is dependent on the point in complex structure moduli space, which is specified by the values of ϕ and ψ. By 'degree' of a rational function (f/g) we mean $\deg(f/g) = \deg(f) - \deg(g)$.

It is apparent that not only does the zeta function change dramatically at the discriminant locus, but the change appears to be characterized by the type of singularity. Hence we propose the following conjecture:

CONJECTURE 7.3 (Degeneration due to Singularities). *The degeneration of the degree of each piece of the zeta function* $\deg(R_v)$ *of a Calabi–Yau manifold at a singular point in the moduli space is determined by the nature of the singularity.*

The following table summarizes this degeneration at various types of singularity:

Degree of Contribution $R_{\mathbf{v}}(t)$ According to Singularity				
Monomial $\mathbf{v}$	Smooth	Conifold	$\phi^2 = 1$	Conifold and $\phi^2 = 1$
(0,0,0,0,0)	6	5	4	3
(0,2,1,1,1)	4	3	2	1
(6,2,0,0,0)	4	3	2	1
(0,0,0,2,2)	4	3	3	2
(2,0,1,3,3)	2	1	1	0
(4,0,2,0,0)	4	3	3	2
(0,0,2,1,1)	3	2	2	1
(6,0,1,0,0)	3	2	2	1
(0,4,0,3,3)	4	3	3	2
(4,0,1,1,0)	4	3	3	2
(2,0,3,0,0)	3	2	2	1
(2,2,1,1,0)	3	2	2	1
(0,0,3,1,0)	2	1	2	1
(2,0,2,1,0)	2	1	2	1
(4,0,2,3,1)	0	−1	0	−1

It can be shown that the contribution from $(4,0,3,2,1)$ is only non-trivial (i.e. not 1) at conifold points:

$$R_{(4,0,3,2,1)}(t) = \delta((8\psi^4+\phi)^2-1)\begin{cases} \frac{1}{[(1-p^2t)(1+p^2t)]^{\frac{1}{2}}}; & 4 \nmid (p-1) \\ \frac{1}{1+p^2t}; & 4|(p-1),\ 8\nmid(p-1) \\ \frac{1}{1-p^2t}; & 8|(p-1) \end{cases} \tag{7.5}$$

CONJECTURE 7.4 (Degenerations due to Isolated A_1 Singularities). *For isolated A_1 singularities (ordinary double points or conifold points), the degree of each contribution* $\deg(R_v)$ *goes down by exactly* 1.

CONJECTURE 7.5 (Degeneration Conservation). *Where two types of singularity coincide, the total degeneration in degree of each contribution* $\deg(R_v)$ *is the sum of the degenerations due to each type of singularity.*

For instance, Table (7.2) shows that when there is both a conifold singularity and a $\phi^2 = 1$ singularity, the total degeneration is indeed as specified by Conjecture (7.5).

It would be interesting to classify the types of degeneration that occur at singular points in the moduli space. This 'arithmetic classification' of singularities may differ from geometric classifications; comparisons would be illuminating. The establishment of a procedure for relating the zeta function of a singular variety to the zeta functions of its resolutions would also be desirable.

One way to proceed which is likely to yield quick results would be to study K3 surfaces (in [**K**] the zeta functions for a family of quartics in $\mathbb{P}_3$ was found). Mirror symmetry of K3 surfaces has a slightly different flavour: the Hodge diamond remains unchanged. Here the significant objects are the Picard lattices. The Picard number of a K3 and its mirror add up to 20, by Dolgachev's construction of mirror symmetry [**Dol**]. K3 manifolds often admit an elliptic Weierstrass fibrations. Given such a fibration, the Picard number can be calculated by using the Shioda–Tate formula. For K3 manifolds with both a toric description and a Weierstrass description, the behaviour of the zeta function at singularities can be explored in a more systematic manner due to the Kodaira classification of singular fibres.

7.3. Functional equation. It shall be seen that the rational functions obtained for the various pieces of the zeta function all satisfy a functional equation. In fact it would be the functional equation desired if it were possible to 'break up' χ in (2.2) into its 'components', based on the formula for the Euler characteristic in terms of the Betti numbers, i.e. up to sign, the degree of the polynomial can replace χ in (2.2):

$$\chi = \sum_{i=0}^{2d} (-1)^i b_i \ . \tag{7.6}$$

So for sextics, it would be desirable for the following functional equation to hold:

$$R_{\mathbf{v}}\left(\frac{1}{p^3 t}\right) = \frac{1}{p^9 t^6} R_{\mathbf{v}}(t) \ , \tag{7.7}$$

for all values of (ϕ, ψ). Hence, if $R_{\mathbf{v}}$ is a sextic one would expect it to have the following form:

$$1 + a_1 t + a_2 t^2 + a_3 t^3 + a_2 p^3 t^4 + a_1 p^6 t^5 + p^9 t^6 \ . \tag{7.8}$$

This is indeed the form of the sextics observed. It can be observed that the $R_{\mathbf{v}}$ always decompose into (fractional powers of) even degree polynomials satisfying an analogous functional equation, and some linear terms which are $(1 \pm p^u t)$, $u = 1, 2$.

Similarly, all quartics encountered are of the form

$$1 + a_1 t + a_2 t^2 + a_1 p^3 t^3 + p^6 t^4 \ . \tag{7.9}$$

7.4. Interesting pairings. So far our calculations indicate that for $4|p-1$, (e.g. for the primes for which we have calculated this is $p = 5, 13, 17$) the contributions from the pairs of monomials $(0,0,1,1,2)$ & $(2,2,1,1,0)$, $(6,0,1,0,0)$ & $(2,0,3,0,0)$ and $(0,4,3,3,0)$ & $(4,0,1,1,0)$ multiply together to give polynomials or fractional powers of polynomials (individually they gives infinite series). Notice that the value of $\lambda_{\mathbf{v}}$ of the monomials in each pair are the same. In addition it can be observed that, if you add up the various pairs (mod 8(4) for the first two (last three) digits) you always obtain monomials that are equivalent to $(0,0,0,0,0)$. It has been 'experimentally' observed that for the cases where $4 \nmid p-1$, these monomials individually contribute polynomials. This hints at analogues of the $\mathcal{A}$ and $\mathcal{B}$ Euler curves found in [**CdOV2**]; these curves could be associated to certain pairs of monomials (namely, $(4,1,0,0,0)$ & $(2,3,0,0,0)$ and $(3,1,1,0,0)$ & $(2,2,1,0,0)$)

whose contributions multiplied together to give fractional powers of polynomials. In that case, the relatively simple form of the solutions to the Picard–Fuchs equations, namely hypergeometric functions, enabled them to be associated to Euler's integral. This led directly to the presence of genus 4 curves. In our case, there appears to be no such fortuitous correspondence, as the associated Picard–Fuchs equations are complicated partial differential equations, as can be seen from the diagrams in Section 4.

Hence the largest polynomials that need to be found in order to fully determine the zeta function are octics, namely those corresponding to $R_{(0,4,0,3,3)} \times R_{(4,0,0,1,1)}$. However, it was found that in all cases this octic factorises. An octic with the appropriate reflection formula is fully defined by only 4 parameters, (a_1, a_2, a_3, a_4):

$$1 + a_1 t + a_2 t^2 + a_3 t^3 + a_4 t^4 + a_3 p^3 t^5 + a_2 p^6 t^6 + a_1 p^9 t^7 + p^{12} t^8 \ . \tag{7.10}$$

The Weil Conjectures alone only indicate a zeta function of the form (3.3), with a polynomial of degree 174 in the numerator of the form $P_j^{(p)}(t) = \prod_{l=1}^{b_j}(1 - \alpha_l^{(j)}(p)t)$, with $\|\alpha_j^{(i)}(p)\| = p^{i/2}$. This appears to indicate that we would need to compute $\mathcal{N}_{p,r}$ up to $r = 174$, however, as a consequence of the decomposition, the zeta function can be fully determined by computing only up to $r = 4$, and in practise $r = 3$.

The sextics and quartics which constitute parts of the zeta function take the same form as the denominator of either the standard or the spinor L-function for a Siegel modular forms of the symplectic groups $Sp(4, \mathbb{Z})$ and $Sp(6, \mathbb{Z})$. As remarked previously in 2.4, a connection to Siegel modular forms is expected according to the Langlands program. At the moment this identification is preliminary, owing to the fact that there are currently no readily available published tables akin to those for elliptic modular forms [**St**], consisting of eigenvalues of Siegel modular forms (although there is an online modular forms database under construction by Nils-Peter Skoruppa [**Sk**], which may be of use in the future). This is because, unlike for the elliptic modular case, there is no obvious connection between the known invariants of the Hecke algebra (the Satake p-parameters) and the Fourier coefficients of a Hecke eigenform; see [**A, V**] for a good introduction to these concepts. The 'problem of multiplicity one'(the eigenvalues determining a modular form) is still a hard problem. The author is grateful to Valery Gritsenko for clarifying this point.

A natural question to ask would be how would Siegel modular forms be related to the variety in question. In fact, they arise very naturally when studying the modular group which arises from considering changes of basis in the Special Geometry description of the space of complex structures [**CdO**], [**Str**]. The arithmetic properties of this modular group warrant future study. Unfortunately, as stressed above, identification of any particular Siegel modular form with the arithmetic has not yet been possible. Preliminary observations are discussed in detail in [**K**].

7.5. The locus $\phi^2 = 1$ and modularity. As expected from the fact that on the singular locus $\phi^2 = 1$ the family is birationally equivalent to a one-parameter family of Calabi–Yau threefolds after resolution of singularities. Indeed when $\phi^2 = 1$ and there is no conifold singularity, we always obtain a zeta function of the form:

$$\frac{R_{128}}{(1-t)(1-p^2t)^2(1-p^3t)} \ , \tag{7.11}$$

where R_{128} is a polynomial of degree 128. The quartic denominator is thus in keeping with the fact that we are now dealing with a family birationally equivalent to a one-parameter family, i.e. $h^{1,1} = 1$.

Recall that $h^{2,1} = 89$ for this one-parameter birational model (after smoothing). Hence, after smoothing, we would expect the numerator to be a polynomial of degree 180, and the denominator to be of the form $(1-t)(1-pt)(1-p^2t)(1-p^3t)$.

When there is also a conifold singularity, the form of the zeta function agrees with the results of [**R-V**]. In that paper it is found that at the conifold points, the factor $R_0(t)$ in the numerator of the zeta function (which corresponds to our $R_{(0,0,0,0,0)}$) takes the following form:

$$R_0(t) = \left(1 - \left(\frac{D}{p}\right) pt\right)\left(1 - a_p t + p^3 t^2\right). \tag{7.12}$$

Generically for $\phi^2 = 1$, $R_0(t)$ is of degree four, but at conifold points, it degenerates to the above cubic. One can verify that a_p is the p-th coefficient of a modular newform of weight 4 for a congruence subgroup for the form $\Gamma_0(N)$of $SL_2(\mathbb{Z})$. The modular forms were identified by computing a_p using p the p-adic formulae employed here and in [**CdOV1, CdOV2**], and by comparison with the tables of W.Stein [**St**]. For this case, $D = 2^3, N = 2^4$; the value of a_p is listed below for the first few primes:

p	2	3	5	7	11	13	17	19	23	29	31	37
a_p	0	4	−2	−24	44	22	50	−44	56	198	160	−162

p	41	43	47	53	59	61	67	71	73
a_p	−198	−52	−528	−242	668	550	−188	−728	154

The values so far have been confirmed up to $p = 73$. It should be noted that the coefficients of these forms are always $0 \mod 2$.

The cusp form above, denoted $V_{2,4}(q)$, admits an η-product expression:

$$V_{2,4}(q) = \eta(q^2)^{-4}\eta(q^4)^{16}\eta(q^8)^{-4}. \tag{7.13}$$

Our calculations verify the results of [**R-V**]. The cases of ours which correspond to those in [**R-V**] are when both singularities coincide and $\psi \neq 0$. When $\psi = 0$, we also observed some factorization reminiscent of modularity, a possible cusp form is of level $N = 32 = 2^5$ (a higher power of the bad prime, 2):

p	2	3	5	7	11	13	17	19	23	29	31
a_p	0	0	22	0	0	−18	−94	0	0	−130	0

p	37	41	43	47	53	59	61	67	71	73
a_p	214	−230	0	0	518	0	830	0	0	1098

Notice that our results predict that this modular form will only have coefficients a_p for p such that $4|p-1$.

There was found to be agreement so far, for all the primes computed, that is, up to $p = 73$. We are using the notation of Section A.1, later employed extensively in the tables. Here, $(a)_1 \equiv (1 + at)$ and $(a)_2 \equiv (1 + at + p^3t^2)$.

Prime	3	5	7	11	13
$\psi \neq 0$	$(3)_1(-4)_2$	$(5)_1(2)_2$	$(-7)_1(24)_2$	$(11)_1(-44)_2$	$(13)_1(-22)_2$
$\psi = 0$	$(-3)_1(0)_2$	$(-5)_1(-22)_2$	$(-7)_1(0)_2$	$(-11)_1(0)_2$	$(-13)_1(18)_2$

Prime	17	19	23	29	31
$\psi \neq 0$	$(-17)_1(-50)_2$	$(19)_1(44)_2$	$(-23)_1(-56)_2$	$(29)_1(-198)_2$	$(-31)_1(-160)_2$
$\psi = 0$	$(-17)_1(94)_2$	$(-19)_1(0)_2$	$(-23)_1(0)_2$	$(-29)_1(130)_2$	$(-31)_1(0)_2$

Prime	37	41	43	47	53
$\psi \neq 0$	$(37)_1(162)_2$	$(-41)_1(198)_2$	$(43)_1(52)_2$	$(-47)_1(528)_2$	$(53)_1(242)_2$
$\psi = 0$	$(-37)_1(-214)_2$	$(-41)_1(230)_2$	$(-43)_1(0)_2$	$(-47)_1(0)_2$	$(-53)_1(-518)_2$

Prime	59	61	67	71	73
$\psi \neq 0$	$(59)_1(-668)_2$	$(61)_1(-550)_2$	$(67)_1(188)_2$	$(-71)_1(728)_2$	$(-73)_1(-154)_2$
$\psi = 0$	$(-59)_1(0)_2$	$(-61)_1(-830)_2$	$(-67)_1(0)_2$	$(-71)_1(0)_2$	$(-73)_1(-1098)_2$

The level N of the modular form is always a product of powers of the bad primes; this satisfies a well known 'folk' conjecture:

CONJECTURE 7.6 (Modularity Folklore). *For a d-dimensional Calabi–Yau variety, the powers α_p that occur in the expansion in the bad primes, $N = \prod_{p\ bad} p^{\alpha_p}$, of the level N of the weight $d+1$ modular form for some $\Gamma_0(N)$ increase with the 'badness' of the singularity.*

Hence in the case above, when the manifolds had a singularity of the type $C_{\rm con} \cap C_1$ (see Section 3.5), $N = 2^4$, however when the singularity was at the intersection of the loci $C_{\rm con} \cap C_1 \cap C_0$, $N = 2^5$.

This is a generalization of the case for modular forms related to elliptic curves ($d = 1$): An elliptic curve E has bad reduction when it has a singularity modulo p. The type of singularity determines the power of p that occurs in the conductor. If the singularity is a 'node' corresponding to a double root of the polynomial, the curve is said to have 'multiplicative reduction' and p occurs to the first power in the conductor. If the singularity is a 'cusp' corresponding to a triple root, E is said to have 'additive reduction', and p occurs in the conductor with a power of 2 or more. The conductor is equal to the level of the associated elliptic modular form.

7.6. $\psi = 0$. Generally when $\psi = 0$ (a special locus in the moduli space), there seemed to be more factorization of the $R_{(0,0,0,0,0)}$'s; in fact it can be factorized into a product of quadratic terms with coefficients in $\mathbb{Q}(\sqrt{2})$ if $4 \nmid (p-1)$ and $\mathbb{Q}(\sqrt{p})$, when $4|p-1$ but $8 \nmid p-1$:

Prime p	$\phi = 0$
3	$(0)_2(-3.2\sqrt{2})_2(3.2\sqrt{2})_2$
5	$(-10)_2(-2^3\sqrt{5})_2(2^3\sqrt{5})_2$
7	$(0)_2^3$
11	$(0)_2(-11.2\sqrt{2})_2(11.2\sqrt{2})_2$
13	$(6.13)_2(-2^3.3\sqrt{13}t)_2(2^3.3\sqrt{13})_2$
17	$(2(45-16\sqrt{2}))_2(2(45+16\sqrt{2}))_2$

In general, factorization may indicate decomposition of Galois representations into those of lower dimension.

7.7. Mirror theorem. To begin, we recall the result of our main calculation. A priori, for arbitrary octic hypersurfaces in $\mathbb{P}_4{}^{(1,1,2,2,2)}$, one would expect a zeta function of the form

$$Z(X/\mathbb{F}_p, t) = \frac{R_{174}(t)}{(1-t)R_2(t)\acute{R}_2(t)(1-p^3t)}, \tag{7.14}$$

however, restricting to a two-parameter family containing only monomials which are invariant under the group of automorphisms, G, we find a factorization of the zeta function of the form

$$\zeta(\phi,\psi,p,t)=\frac{R_{\rm excep}(t)\prod_{\mathbf{v}}R_{\mathbf{v}}(t)^{\lambda_{\mathbf{v}}}}{(1-t)(1-pt)^2(1-p^2t)^2(1-p^3t)}\,, \tag{7.15}$$

where each **v** labels a classes of representations of G. We would not expect such a beautiful decomposition in general. The mirror is obtained by quotienting out this special sub-family by the group of automorphisms, G [**GP**]. The zeta function of the mirror shares a factor $R_{(0,0,0,0,0)}$ which has degree equal to the number of shared periods. Hence, in the case of the quintic threefolds, $R_{(0,0,0,0,0)}$ was a quartic, and for the octic threefolds $R_{(0,0,0,0,0)}$ was a sextic. This gives rise to the following theorem:

THEOREM 7.1 (Mirror Theorem for Hypersurfaces of Fermat Type). *For mirror pairs of Calabi–Yau threefolds which can be described using the Greene–Plesser construction (i.e. hypersurfaces of Fermat type in weighted projective space), the G-invariant part, i.e. $R_{(0,0,0,0,0)}$ of the zeta function of the Calabi–Yau threefold is preserved. The degree of this factor is equal to the number of shared periods.*

This was observed for the quintic in [**CdOV2**], and for the octic here. It would be interesting to compute the zeta functions for mirror symmetry pairs of Calabi–Yau manifolds that are not of Fermat type. This should be possible using existing methods, i.e. using Cox variables and the Batyrev procedure.

PROOF. For Fermat hypersurfaces the dual of (the larger) Newton polyhedron consists of a subset of the points of the original, this is because the mirror is formed by taking the quotient by the Greene–Plesser group of automorphisms. All monomials in this 'smaller' polyhedron are invariant under the Greene–Plesser group of automorphisms. Each point corresponds to a period, as was shown in Section 4.

The number of shared periods for Fermat hypersurfaces is equal to the number of monomials in the smaller of the Newton polyhedra excluding the unique interior point. For the family of manifolds that have defining equations given by the monomials in the smaller polyhedron (in our example, for instance, this is the mirror octic threefolds defined by (6.1)), the shared periods are all the periods, and the number of periods is equal to the dimension of the third cohomology group. Let us denote this family the 'mirror family', for convenience.

The method of calculation of the number of rational points by using Gauss sums consisting of Dwork's character gives the number of points in terms of the monomials or, equivalently, the periods. In both families there is a contribution to the number of rational points coming from the G-invariant monomials. However, applying the Weil conjectures for the 'mirror family' gives us that the degree of the numerator of the zeta function has to be equal to the third Betti number, b_3. The contribution to the number of rational points that gives the denominator is not dependent on the monomials/periods, and hence the numerator of the zeta function comes from the contribution to the number of points from the G-invariant monomials. Let this numerator be denoted $R_{(0,0,0,0,0)}$.

The special G-invariant sub-family of the original family that we choose also has a contribution to the zeta function in terms of the G-invariant monomials/periods. Hence, the numerator of the zeta function of this family contains the factor $R_{(0,0,0,0,0)}$. □

Acknowledgements

The author would like to thank Philip Candelas, her DPhil supervisor, for introducing her to this interesting new area. She would also like to thank Xenia de la Ossa for many fruitful conversations.

Appendix A. Results for the octic threefolds

A.1. New notation. Occasionally for clarity and conciseness, the following notation shall be used in the tables:

Notation	Polynomial, for prime p
$(a)_1$	$(1+at)$
$(a)_2$	$(1+at+p^3t^2)$
$(a,b)_4$	$(1+at+bt^2+ap^3t^3+p^6t^4)$
$(a,b,c)_6$	$(1+at+bt^2+ct^3+bp^3t^4+ap^6t^5+p^9t^6)$
$(a,b,c,d)_8$	$(1+at+bt^2+ct^3+dt^4+cp^3t^5+bp^6t^6+ap^9t^7+p^{12}t^8)$

A.2. Numerical data obtained. In the table below we give the maximum r for which the number of rational points over $\mathbb{F}_{p^r}$ was computed for each prime p.

Prime p	Maximum r
3	8
5	5
7	4
11	3
13	3
17	3

The zeta functions for $p = 19, 23$ were also computed (using data up to $r = 3$), but the results were too large to display in this paper. For primes up to 73, the numerical data up to $r = 2$ has been obtained and this data was used to check modularity, but again there is insufficient space to display the corresponding zeta functions.

A.3. p=3 complete tables. For $p=3$, $R_{\text{excep}}(t) = (1+27t^2)^3$.
For $\phi = 0, \psi = 0$:

Monomial $\mathbf{v}$	Contribution	Power $\lambda_{\mathbf{v}}$
(0,0,0,0,0)	$(1+27t^2)(1-18t^2+729t^4)$	1
(0,2,1,1,1)	$(1+27t^2)[(1-6t+27t^2)(1+6t+27t^2)]^{\frac{1}{2}}$	2
(6,2,0,0,0)	$(1+27t^2)^2$	1
(0,0,0,2,2)	$(1-18t^2+729t^4)$	3
(2,0,1,3,3)	$(1+27t^2)$	6
(4,0,2,0,0)	$(1+27t^2)^2$	3
(0,0,2,1,1)	$(1+27t^2)^{\frac{1}{2}}(1-18t^2+729t^4)^{\frac{1}{2}}$	3
(6,0,1,0,0)	$(1+27t^2)^{\frac{1}{2}}(1-27t^2)$	6
(0,4,0,3,3)	$(1+27t^2)^2$	3
(4,0,1,1,0)	$(1+27t^2)^2$	3
(2,0,3,0,0)	$(1+27t^2)^{\frac{1}{2}}(1-27t^2)$	6
(2,2,1,1,0)	$(1+27t^2)^{\frac{1}{2}}(1-18t^2+729t^4)^{\frac{1}{2}}$	3
(0,0,3,1,0)	$(1+27t^2)$	6
(2,0,2,1,0)	$[(1-6t+27t^2)(1+6t+27t^2)]^{\frac{1}{2}}$	12
(4,0,2,3,1)	1	6

For $\phi = 0, \psi = 1, 2$ which is a conifold point:

Monomial $\mathbf{v}$	Contribution	Power $\lambda_{\mathbf{v}}$
(0,0,0,0,0)	$(1-3t)(1-6t+27t^2)(1+8t+27t^2)$	1
(0,2,1,1,1)	$[(1-3t)(1+3t)(1-6t+27t^2)(1+6t+27t^2)]^{\frac{1}{2}}$	2
(6,2,0,0,0)	$(1+3t)(1+6t+27t^2)$	1
(0,0,0,2,2)	$(1-3t)(1+6t+27t^2)$	3
(2,0,1,3,3)	$[(1-3t)(1+3t)]^{\frac{1}{2}}$	6
(4,0,2,0,0)	$(1+3t)(1-6t+27t^2)$	3
(0,0,2,1,1)	$[(1-3t)(1+3t)(1+27t^2)]^{\frac{1}{2}}$	3
(6,0,1,0,0)	$[(1-3t)(1+3t)(1+27t^2)]^{\frac{1}{2}}$	6
(0,4,0,3,3)	$[(1-3t)(1+3t)(1-18t^2+729t^4)]^{\frac{1}{2}}$	3
(4,0,1,1,0)	$[(1-3t)(1+3t)(1-18t^2+729t^4)]^{\frac{1}{2}}$	3
(2,0,3,0,0)	$[(1-3t)(1+3t)(1+27t^2)]^{\frac{1}{2}}$	6
(2,2,1,1,0)	$[(1-3t)(1+3t)(1+27t^2)]^{\frac{1}{2}}$	3
(0,0,3,1,0)	$[(1-3t)(1+3t)]^{\frac{1}{2}}$	6
(2,0,2,1,0)	$[(1-3t)(1+3t)]^{\frac{1}{2}}$	12
(4,0,2,3,1)	$\frac{1}{((1-9t)(1+9t))^{\frac{1}{2}}}$	6

For $\phi = 1, \psi = 1, 2$, here $\phi^2 = 1$:

Monomial $\mathbf{v}$	Contribution	Power $\lambda_{\mathbf{v}}$
(0,0,0,0,0)	$(1 + 2t - 18t^2 + 54t^3 + 729t^4)$	1
(0,2,1,1,1)	$(1 + 27t^2)$	2
(6,2,0,0,0)	$(1 + 6t + 27t^2)$	1
(0,0,0,2,2)	$(1 + 3t)(1 - 6t + 27t^2)$	3
(2,0,1,3,3)	$[(1 - 3t)(1 + 3t)]^{\frac{1}{2}}$	6
(4,0,2,0,0)	$(1 + 3t)(1 - 6t + 27t^2)$	3
(0,0,2,1,1)	$(1 - 18t^2 + 729t^4)^{\frac{1}{2}}$	3
(6,0,1,0,0)	$[(1 - 6t + 27t^2)(1 + 6t + 27t^2)]^{\frac{1}{2}}$	6
(0,4,0,3,3)	$(1 + 27t^2)^{\frac{1}{2}}(1 - 18t^2 + 729t^4)^{\frac{1}{2}}$	3
(4,0,1,1,0)	$(1 + 27t^2)^{\frac{1}{2}}(1 - 18t^2 + 729t^4)^{\frac{1}{2}}$	3
(2,0,3,0,0)	$[(1 - 6t + 27t^2)(1 + 6t + 27t^2)]^{\frac{1}{2}}$	6
(2,2,1,1,0)	$(1 - 18t^2 + 729t^4)^{\frac{1}{2}}$	3
(0,0,3,1,0)	$(1 + 27t^2)$	6
(2,0,2,1,0)	$[(1 - 6t + 27t^2)(1 + 6t + 27t^2)]^{\frac{1}{2}}$	12
(4,0,2,3,1)	1	6

Intersection of both a conifold point and $\phi^2 = 1$:

Monomial $\mathbf{v}$	$\phi = 1, \psi = 0$	$\phi = 2, \psi = 0$	$\phi = 2, \psi = 1, 2$
(0,0,0,0,0)	$(1 - 3t)(1 + 27t^2)$	$(1 + 3t)(1 + 27t^2)$	$(1 + 3t)(1 - 4t + 27t^2)$
(0,2,1,1,1)	$[(1 - 3t)(1 + 3t)]^{\frac{1}{2}}$	$[(1 - 3t)(1 + 3t)]^{\frac{1}{2}}$	$[(1 - 3t)(1 + 3t)]^{\frac{1}{2}}$
(6,2,0,0,0)	$\frac{(1+27t^2)}{(1-9t)}$	$\frac{(1+27t^2)}{(1-9t)}$	$(1 + 3t)$
(0,0,0,2,2)	$(1 + 27t^2)$	$(1 + 27t^2)$	$(1 - 3t)(1 + 3t)$
(2,0,1,3,3)	1	1	$\left(\frac{(1-3t)(1+3t)}{(1-9t)(1+9t)}\right)^{\frac{1}{2}}$
(4,0,2,0,0)	$(1 + 27t^2)$	$(1 + 27t^2)$	$(1 + 3t)^2$
(0,0,2,1,1)	$(1 + 27t^2)^{\frac{1}{2}}$	$(1 + 27t^2)^{\frac{1}{2}}$	$[(1 - 3t)(1 + 3t)]^{\frac{1}{2}}$
(6,0,1,0,0)	$(1 + 27t^2)^{\frac{1}{2}}$	$(1 + 27t^2)^{\frac{1}{2}}$	$[(1 - 3t)(1 + 3t)]^{\frac{1}{2}}$
(0,4,0,3,3)	$(1 + 27t^2)$	$(1 + 27t^2)$	$[(1 + 3t)(1 - 3t)(1 + 27t^2)]^{\frac{1}{2}}$
(4,0,1,1,0)	$(1 + 27t^2)$	$(1 + 27t^2)$	$[(1 + 3t)(1 - 3t)(1 + 27t^2)]^{\frac{1}{2}}$
(2,0,3,0,0)	$(1 + 27t^2)^{\frac{1}{2}}$	$(1 + 27t^2)^{\frac{1}{2}}$	$[(1 - 3t)(1 + 3t)]^{\frac{1}{2}}$
(2,2,1,1,0)	$(1 + 27t^2)^{\frac{1}{2}}$	$(1 + 27t^2)^{\frac{1}{2}}$	$[(1 - 3t)(1 + 3t)]^{\frac{1}{2}}$
(0,0,3,1,0)	$[(1 - 3t)(1 + 3t)]^{\frac{1}{2}}$	$[(1 - 3t)(1 + 3t)]^{\frac{1}{2}}$	$[(1 - 3t)(1 + 3t)]^{\frac{1}{2}}$
(2,0,2,1,0)	$[(1 - 3t)(1 + 3t)]^{\frac{1}{2}}$	$[(1 - 3t)(1 + 3t)]^{\frac{1}{2}}$	$[(1 - 3t)(1 + 3t)]^{\frac{1}{2}}$
(4,0,2,3,1)	$\frac{1}{((1-9t)(1+9t))^{\frac{1}{2}}}$	$\frac{1}{((1-9t)(1+9t))^{\frac{1}{2}}}$	$\frac{1}{((1-9t)(1+9t))^{\frac{1}{2}}}$

A.4. p=5 complete tables. For $p = 5$, $R_{\text{excep}}(t) = (1 - 2.5t + 125t^2)^3$.

Monomial $\mathbf{v}$	$\phi = 0, \psi = 0$	$\phi = 0, \psi \neq 0$
(0,0,0,0,0)	$(-10)_2(0,-70)_4$	$(0)_2(-4,-90)_4$
(0,2,1,1,1)	$(0)_2(10)_2$	$(-20)_2(10)_2$
(4,0,2,2,2)	$(-20)_2(20)_2$	$(0)_2(10)_2$
(0,0,0,2,2)	$(0)_2$	$(20)_2(-10)_2$
(2,0,1,3,3)	$(10)_2$	$(10)_2$
(4,0,2,0,0)	$(-20)_2(20)_2$	$(0)_2(-10)_2$
$(0,0,2,1,1) \times (2,2,1,1,0)$	$(0)_2(-10)_2$	$(-20)_2(10)_2(20)_2$
$(6,0,1,0,0) \times (2,0,3,0,0)$	$(-20)_2(-10)_2(20)_2$	$(-10,75,-1500)_6$
$(0,4,0,3,3) \times (4,0,1,1,0)$	$(-20)_2(10)_2(20)_2$	$(-20)_2(-10)_2(10)_2(20)_2$
(0,0,3,1,0)	$(10)_2$	$(-10)_2$
(2,0,2,1,0)	$(0)_2$	$(-10)_2$
(4,0,2,3,1)	1	1

Monomial $\mathbf{v}$	$\phi = 1, \psi = 0$	$\phi = 1, \psi \neq 0$
(0,0,0,0,0)	$(-10)_2(0,-70)_4$	$(0)_2(-4,-90)_4$
(0,2,1,1,1)	$(0)_2(10)_2$	$(-20)_2(10)_2$
(4,0,2,2,2)	$(-20)_2(20)_2$	$(0)_2(10)_2$
(0,0,0,2,2)	$(0)_2$	$(20)_2(-10)_2$
(2,0,1,3,3)	$(10)_2$	$(10)_2$
(4,0,2,0,0)	$(-20)_2(20)_2$	$(0)_2(-10)_2$
$(0,0,2,1,1) \times (2,2,1,1,0)$	$(0)_2(-10)_2$	$(-20)_2(10)_2(20)_2$
$(6,0,1,0,0) \times (2,0,3,0,0)$	$(-20)_2(-10)_2(20)_2$	$(-10,75,-1500)_6$
$(0,4,0,3,3) \times (4,0,1,1,0)$	$(-20)_2(10)_2(20)_2$	$(-20)_2(-10)_2(10)_2(20)_2$
(0,0,3,1,0)	$(10)_2$	$(-10)_2$
(2,0,2,1,0)	$(0)_2$	$(-10)_2$
(4,0,2,3,1)	1	1

Monomial $\mathbf{v}$	$\phi = 1, \psi = 0$	$\phi = 1, \psi \neq 0$
(0,0,0,0,0)	$(1-5t)(1-22t+125t^2)$	$(1+5t)(1+2t+125t^2)$
(0,2,1,1,1)	$(1-5t)$	$(1-5t)$
(6,2,0,0,0)	$\frac{(1+10t+125t^2)}{(1-25t)}$	$1\ (1-5t)$
(0,0,0,2,2)	$(1+10t+125t^2)$	$3\ (1-5t)(1+5t)$
(2,0,1,3,3)	1	6 1
(4,0,2,0,0)	$(1+10t+125t^2)$	$3\ (1-5t)(1+5t)$
$(0,0,2,1,1) \times (2,2,1,1,0)$	$(1-10t+125t^2)$	$3(1+5t) \times (1+5t)$
$(6,0,1,0,0) \times (2,0,3,0,0)$	$(1-10t+125t^2)$	$(1-5t) \times (1-5t)\ 6$
$(0,4,0,3,3) \times (4,0,1,1,0)$	$(1-10t+125t^2) \times (1-10t+125t^2)$	$(1+5t)^2(1-10t+125t^2)$
(0,0,3,1,0)	$(1+5t)$	$(1+5t)$
(2,0,2,1,0)	$(1+5t)$	$(1-5t)$
(4,0,2,3,1)	$\frac{1}{(1+25t)}$	$\frac{1}{(1+25t)}$

Monomial $\mathbf{v}$	$\phi = 2, \psi = 0$	$\phi = 2, \psi \neq 0$
(0,0,0,0,0)	$(-10)_2(10)_2(-4)_2$	$(0,-5,1200)_6$
(0,2,1,1,1)	$(-10)_2(-20)_2$	$(0,-50)_4$
(4,0,2,2,2)	$(-10)_2(20)_2$	$(0)_2(10)_2$
(0,0,0,2,2)	$(-10)_2(20)_2$	$(10)_2(-20)_2$
(2,0,1,3,3)	$(20)_2$	$(10)_2$
(4,0,2,0,0)	$(-10)_2(-20)_2$	$(0)_2(10)_2$
$(0,0,2,1,1) \times (2,2,1,1,0)$	$(10)_2^2(-20)_2$	$(10)_2^2(-20)_2$
$(6,0,1,0,0) \times (2,0,3,0,0)$	$(-10)_2(10)_2(-20)_2$	$(0,-25,-2000)_6$
$(0,4,0,3,3) \times (4,0,1,1,0)$	$(-10)_2(10)_2^2(20)_2$	$(0,-2.125)_4(10)_2(-20)_2$
(0,0,3,1,0)	$(-10)_2$	$(10)_2$
(2,0,2,1,0)	$(10)_2$	$(0)_2$
(4,0,2,3,1)	1	1

$\phi = 3, \psi = 0$	$\phi = 3, \psi \neq 0$	$\phi = 4, \psi = 0$	$\phi = 4, \psi \neq 0$
$(4)_2(10)_2^2$ $(-20)_2(10)_2$ $(-20)_2(10)_2$	$(5)_1(-12, 190)_4$ $(5)_1(-10)_2$ $(5)_1(10)_2$	$(-5)_1(22)_2$ $(5)_1$ $\frac{(-10)_2}{(-25)_1}$	$(-14)_2(10)_2$ $(-10)_2$ $(10)_2$
$(-20)_2(10)_2$ $(20)_2$ $(10)_2(20)_2$	$(5)_1(-10)_2$ $(5)_1$ $(5)_1(-10)_2$	$(-10)_2$ 1 $(-10)_2$	$(-5)_1(10)_2$ $(5)_1$ $(-5)_1(10)_2$
$(-10)_2(10)_2(-20)_2$	$(5)_1^2(-20)_2$	$(10)_2$	$(1 - 125t^2) \times (1 - 125t^2)$
$(-10)_2(10)_2(-20)_2$	$(5)_1^2(-20)_2$	$(10)_2$	$(-10)_2 \times (-10)_2$
$(-20)_2(-10)_2^2(10)_2$	$(5)_1^2(20)_2(-10)_2$	$(-10)_2(10)_2$	$(0, -2.125)_4(-10)_2$
$(-10)_2$ $(-10)_2$ 1	$(5)_1$ $(5)_1$ $1/(25)_1$	$(5)_1$ $(-5)_1$ $\frac{1}{(25)_1}$	$(-10)_2$ $(10)_2$ 1

References

[A] A. N. Andrianov, *Euler Products Coresponding to Siegel Modular Forms of Genus 2*, Russian Math. Surveys **29**:3(1974), 45-116.

[AGM] D. Aspinwall, B. Greene and D. Morrison, *The Monomial-Divisor Mirror Map*, arXiv:alg-geom/9309007, Internat. Math. Res. Notices 1993, no. 12, 319-337.

[Bat] V. Batyrev, *Dual Polyhedra and Mirror Symmetry for Calabi–Yau Hypersurfaces in Toric Varieties* J. Algebraic Geom. **3** (1994), 493-535..

[BCdOFHJG] P. Berglund, P. Candelas, X. de la Ossa, A. Font, T. Hübsch, D. Jančić, F. Quevedo, *Periods for Calabi–Yau and Landau–Ginzburg Vacua*, arXiv:hep-th/9308005, Nuclear Phys. B. **419** (1994), no.2, 352-403.

[BCDT] C. Breuil, B. Conrad, F. Diamond, and R. Taylor, *On the Modularity of Elliptic Curves over* $\mathbb{Q}$, J.A.M.S. **14** (2001), 843-939.

[CdO] P.Candelas and X. de la Ossa, *Moduli Space of Calabi–Yau Manifolds*, Nuclear Physics **B355** (1991), 455-481.

[CdOFKM] P. Candelas, X. de la Ossa, A. Font, S. Katz and D. Morrison, *Mirror Symmetry for two-parameter Models I* arXiv:hep-th/9308083, Nuclear Phys. B. **416** (1994), no. 2, 481-538.

[CdOK] P. Candelas, X. de la Ossa, S. Katz, *Mirror Symmetry for Calabi–Yau Hypersurfaces in Weighted* $\mathbb{P}^4$ *and an Extension of Landau-Ginzburg Theory*, hep-th/9412117, Nuclear Phys. B. **450** (1995), no. 1-2, 267-290.

[CdOV1] P. Candelas, X. de la Ossa, and F.Rodriguez Villegas, *Calabi–Yau manifolds over finite fields I*, arXiv:hep-th/0012233.

[CdOV2] P. Candelas, X. de la Ossa, and F.Rodriguez Villegas, *Calabi–Yau manifolds over finite fields II*, Fields Institute Communications Volume **38**, (2003), 121-159.

[C] H. Cohen, *Elliptic Curves*, From Number Theory to Physics, Springer-Verlag.

[CK] David A. Cox and Sheldon Katz, *Mirror Symmetry and Algebraic Geometry* Mathematical Surveys and Monographs, Volume **68**, American Mathematical Society.

[Del] P. Deligne, *La Conjecture de Weil I*, Publ.Math.IHES **43** (1974) 273-307.

[Dol] I. V. Dolgachev, *Mirror Symmetry for Lattice Polarized K3 Surfaces*, Algebraic Geometry 4, J. Math. Sci. 1996, **81**(3), 2599-2630.

[Dw] B. M.Dwork, *On the rationality of the Zeta Function of an Algebraic Variety*, Amer.J.Math. **82** (1960) 631-648.

[F] W. Fulton, *Introduction to Toric Varieties*, Princeton University Press, 1993.

[GP] B. Greene, M. Plesser, *Duality in Calabi–Yau moduli space*, Nuclear Physics **B338** (1990), 15-37.

[HKT] S. Hosono, A. Klemm and S. Theisen, *Lectures on Mirror Symmetry* arXiv:hep-th/9403096, Lectures Notes in Phys., 436, Springer, Berlin, 1994.

[K] S. N. Kadir, *The Arithmetic of Calabi–Yau Manifolds and Mirror Symmetry*, University of Oxford, 2004, arXiv: hep-th/0409202.

[KMP] S. Katz, D.R. Morrison and M. R. Plesser *Enhanced Gauge Symmetry in Type II String Theory* , Nuclear Phys. B477 (1996), no.1, 105-140.

[LS] M. Lynker and R. Schimmrigk, *Geometric Kac-Moody Modularity*, arXiv: hep-th/0410189.

[M] J. S. Milne, *Lectures on Étale Cohomology* **www.math.lsa.umich.edu/ jmilne/**.
[R-V] Fernando Rodriguez-Villegas, *Hypergeometric Familites of Calabi–Yau Manifolds* Fields Institute Communications Volume **38**,(2003), 223-231.
[Sk] N.-P. Skoruppa, *MODI* **http://wotan.algebra.math.uni-siegen.de/ modi/**.
[St] W. Stein, **www.math.harvard.edu/~was**.
[Str] A. Strominger, *Special Geometry*, Commun. Math. Phys. **133** (1990), 163-180.
[TW] A. Wiles and R. Taylor, *Ring Theoretic Properties of Certain Hecke Algebras*, Ann. of Math. (2) 141 (1995), no.3, 553-572.
[V] T. B. Veenstra ,*Siegel Modular Forms, L-functions, and Satake Parameters*, Journal of Number Theory **87**, 15-30 (2001).
[Voi] C. Voisin, *Mirror Symmetry* Société Mathemématique de France/ American Mathematical Society Texts and Monographs, Volume **1**.
[Wn] D. Wan *Mirror Symmetry for Zeta Function* in this volume and arXiv;math.AG/0411464.
[W] A. Wiles, *Ring Theoretic Properties of Certain Hecke Algebras* Ann. Math. **141**, 553-572, (1995).
[Yui] N. Yui, *Update on the Modularity of Calabi–Yau Varieties*, Fields Institute Communications Volume **38**,(2003), 309-362.

Fields Institute for Research in Mathematical Sciences, 222 College Street, Toronto M5T 3J1, Canada
E-mail address: skadir@fields.utoronto.ca

AMS/IP Studies in Advanced Mathematics
Volume 38, 2006

A Rational Map Between Two Threefolds

Kenichiro Kimura

Let $V_{33} \subset \mathbb{P}^5$ be the threefold defined by the following 2 cubic equations:

$$X_0^3 + X_1^3 + X_2^3 + X_3^3 = 0$$
$$X_2^3 + X_3^3 + X_4^3 + X_5^3 = 0.$$

Let $\widetilde{V}_{33}$ be the blow-up of V_{33} along its singular locus. In [**13**] it is shown that the third Betti number of $\widetilde{V}_{33}$ is 2 and the L-series $L(H^3(\widetilde{V}_{33}), s) = L(f, s)$ with f the newform of weight 4 on $\Gamma_0(9)$. This is also the L-series of a Hecke character of the field $\mathbb{Q}(\sqrt{-3})$.

Let $E \subset \mathbb{P}^2$ be the curve defined by

$$X_0^3 + X_1^3 + X_2^3 = 0.$$

In Remark 4.5 of [**4**] it is mentioned that there is a piece of the cohomology $H^3(E^3)$ which has the same L-series as that of $H^3(\widetilde{V}_{33})$. According to the Tate conjecture there should be a correspondence between V_{33} and E^3. We give such a correspondence explicitly. Other relatives of these varieties are the fibered squares of the universal curves for $\Gamma(3)$ and $\Gamma_0(9)$. In [**9**] and in section 13 of [**10**] Schoen constructs correspondences between these varieties and E^3.

We refer the reader to Remark 4.5 of [**4**] for a list of threefolds which have a two-dimensional Galois representation in H^3 whose L-series is associated to an elliptic modular form of weight 4, and known correspondences among those varieties. We include the relevant works in the reference.

THEOREM 1. *There is a dominant rational map from E^3 to V_{33} of degree 3.*

PROOF. Consider the affine piece $\{X_3 \neq 0\}$ of V_{33} which is equal to

$$\operatorname{Spec} \mathbb{Q}[X_0, X_1, X_2, X_4, X_5]/(X_0^3 + X_1^3 + X_2^3 + 1,\ X_2^3 + 1 + X_4^3 + X_5^3).$$

On the other hand, the affine open $(E - \{X_2 = 0\})^3$ of E^3 is given by

$$\operatorname{Spec} \mathbb{Q}[x_1, y_1, x_2, y_2, x_3, y_3]/(x_1^3 + y_1^3 + 1, x_2^3 + y_2^3 + 1, x_3^3 + y_3^3 + 1).$$

2000 *Mathematics Subject Classification.* Primary 14F20; Secondary 14G10.

Key words and phrases. the Tate conjecture, L-series.

[1]After this paper was written, the author was informed that it is mentioned in [**6**] that one can construct a Calabi-Yau threefold from a quotient of E^3.

Then there is a map from $(E - \{X_2 = 0\})^3$ to V_{33} induced by the ring homomorphism

$$X_0 \mapsto -x_1y_3,\ X_1 \mapsto -y_1y_3,\ X_2 \mapsto x_3,\ X_4 \mapsto -x_2y_3,\ X_5 \mapsto -y_2y_3.$$

We denote by $\mathbb{Q}(E^3)$ and $\mathbb{Q}(V_{33})$ the function fields of E^3 and of V_{33} respectively. Then one can see that the field $\mathbb{Q}(E^3)$ is generated over $\mathbb{Q}(V_{33})$ by y_3 with the equation $y_3^3 + X_2^3 + 1 = 0$. So this map is dominant of degree 3. □

Acknowledgment. The author is grateful to the referee for comments on an earlier version of this paper which helped improve the exposition.

References

[1] Ekedahl,T.; van Geemen, B., *An exceptional isomorphism between modular varieties*, Progr. Math. **89** (1991), 51–74.

[2] van Geemen, B.; Nygaard, N., *On the Geometry and Arithmetic of Some Siegel Modular Threefolds*, J. of Number Theory **53** (1995), 45–87.

[3] van Geemen, B.; Top, J., *An Isogeny of K3 surfaces*, mathAG/0309272 (2003).

[4] Hulek, K.; Spandaw, J.; van Geemen, B.; van Straten, D., *The modularity of the Barth-Nieto quintic and its relatives*, Adv. Geom. **1** (2001), 263-289 also available as AG/0010049.

[5] Livne, R. , *Cubic exponential sums and Galois representations*, Contemp. Math. **67** (1987), 247–261.

[6] Paranjape, K.; Ramakrishnan, D., *Quotients of E^n by $\mathfrak{a}_{n+1}$ and Calabi-Yau manifolds*, preprint (2004).

[7] Saito, M-H.; Yui, N., *The modularity conjecture for rigid Calabi-Yau threefolds over $\mathbb{Q}$*, J. Math. Kyoto Univ. **41** (2001), 403–419.

[8] Schoen, C., *On the geometry of a special determinantial hypersurface associated to the Mumford-Horrocks vector bundle*, J. Reine Angew. Math. **364** (1986), 85–111.

[9] Schoen, C., *Zero cycles modulo rational equivalence for some varieties over fields of transcendence degree one*, Proc. Sympos. Pure Math. **46**, Part 2 (1987), 463–473.

[10] Schoen, C., *On the computation of the cycle class map for nullhomologous cycles over the algebraic closure of a finite field*, Ann. Sci. Ecole Norm. Sup. (4) **28**, (1995), 1–50.

[11] van Straten, D., *A quintic hypersurface in $\mathbb{P}^4$ with 130 nodes*, Topology **32**, (1993), 857–864.

[12] Verrill, H., *The L-series of certain rigid Calabi-Yau threefolds*, J. Number Theory **81**, (2000), 509–542.

[13] Werner, J.; van Geemen, B., *New examples of threefolds with $c_1 = 0$*, Math. Z. **203**, (1990), 211-225.

[14] Yui, N., *The arithmetic of certain Calabi-Yau varieties over number fields*, NATO Sci. Ser. C Math. Phys. Sci. **548**, (2000), 515–560.

Institute of Mathematics, University of Tsukuba, Tsukuba, Ibaraki, 305-8571, Japan
E-mail address: kimurak@math.tsukuba.ac.jp

AMS/IP Studies in Advanced Mathematics
Volume 38, 2006

A Modular Non-Rigid Calabi-Yau Threefold

Edward Lee

ABSTRACT. We construct an algebraic variety by resolving singularities of a quintic Calabi-Yau threefold. The middle cohomology of the threefold is shown to contain a piece coming from a pair of elliptic surfaces. The resulting quotient is a two-dimensional Galois representation. By using the Lefschetz fixed-point theorem in étale cohomology and counting points on the variety over finite fields, this Galois representation is shown to be modular.

1. Introduction

In this paper we investigate the geometry and arithmetic of a Calabi-Yau threefold $X \subset \mathbb{P}^4 \times \mathbb{P}^4$ given as a complete intersection of five hypersurfaces of bidegree $(1,1)$. X is a partial desingularization common to a pair of quintic threefolds F and G in $\mathbb{P}^4$; we will in fact be interested in a big resolution $\tilde{X}$ of X. So $\tilde{X}$ is not actually a Calabi-Yau threefold, but it is birational to one.

A conjecture of Fontaine and Mazur [**8**] predicts that two-dimensional l-adic Galois representations coming from geometry should be modular. More precisely, the statement is that a continuous irreducible two-dimensional l-adic representation of the absolute Galois group $G_{\mathbb{Q}}$ that is isomorphic to a Tate twist of a subquotient of an étale cohomology group of a variety $X/\mathbb{Q}$ should be modular. This is a higher-dimensional generalization of the Taniyama-Shimura conjecture on the modularity of elliptic curves over $\mathbb{Q}$, as proved by Wiles, Taylor et al. [**29**], [**2**].

A rigid Calabi-Yau threefold X defined over $\mathbb{Q}$ has two-dimensional middle cohomology, and is thus expected to be modular. We expect the L-series of the Galois action on the l-adic cohomology to be, up to factors associated to the primes of bad reduction of X, the Mellin transform of a weight 4 modular form. Recently Dieulefait and Manoharmayum [4] proved that rigid Calabi-Yau threefolds that have good reduction at 3 and 7, or at 5 and another suitable prime, are modular. A handful of explicit examples of modular Calabi-Yau threefolds are known. Some examples are given in [**24**], [**30**], [**16**], [**31**], [**32**].

Some nonrigid Calabi-Yau threfolds have been shown to be modular by Hulek-Verrill [**16**] and Schütt [**25**]. Here the middle cohomology group of X has dimension greater than 2, so one must figure out how to extract a 2-dimensional piece on which $G_{\mathbb{Q}}$ acts.

2000 *Mathematics Subject Classification.* Primary 14J15; Secondary 11F23, 14J32, 11G40.

Key words and phrases. Calabi–Yau threefold, non-rigid Calabi–Yau threefold, a two-dimensional Galois representation, modular variety, Horrocks–Mumford vector bundle.

Hulek-Verrill found threefolds in a toric variety with $h^3 = 4, 6$ and 10; in each case they showed that the semisimplification of H^3 was a direct sum $\oplus_i W_i \oplus V$, where the W_i came from elliptic surfaces defined over $\mathbb{Q}$ and V was the remaining quotient. The W_i were thus isomorphic to cohomology groups of elliptic curves twisted by (-1), and V was shown to correspond to a modular form of weight 4.

Schütt constructed some fiber products of rational elliptic surfaces; he showed that each of their middle cohomology groups also broke up into a sum of two-dimensional pieces coming from elliptic curves and a leftover two-dimensional piece corresponding to a modular form of weight 4.

Our threefold $\tilde{X}$ is constructed by resolving singularities of a pair of Horrocks-Mumford quintic threefolds F and G. The Horrocks-Mumford bundle HM is a stable indecomposable rank 2 vector bundle over $\mathbb{P}^4$, and zero sets of its sections are abelian surfaces. We take zero sets of a pair of sections of the line bundle $\wedge^2(HM) \cong \mathcal{O}(5)$ as our quintic threefolds F and G; they are thus pencils of abelian surfaces.

First we take a common partial resolution X of F and G as mentioned at the outset; it is a Calabi-Yau threefold given as a complete intersection in $\mathbb{P}^4 \times \mathbb{P}^4$. We then blow up the singularities of X to obtain $\tilde{X}$. Unfortunately, there exists no model for a small resolution of X over $\mathbb{Q}$.

By studying the geometry of $\tilde{X}$ and by exploiting the Weil conjectures, we are able to show that $h^3(\tilde{X}) = 6$. We show that the semisimplification of the Galois representation $H^3(\tilde{X})$ is a direct sum of a two-dimensional piece V and a four-dimensional piece W. The four-dimensional piece W arises from the cohomology of a pair of elliptic surfaces E_1 and E_2 that are complex conjugates of each other. Thus their union is defined over $\mathbb{Q}$, and as a Galois representation W is induced from a representation of the subgroup $G_{\mathbb{Q}(i)}$. We then show that the two-dimensional piece V is modular; by using a theorem of Faltings-Serre-Livné [**19**], we are able to prove this by studying the reduction of $\tilde{X}$ modulo a finite set of primes. In practice this amounts to counting the points on $\tilde{X}$ over $\mathbb{F}_p$, a task which can easily be done by computer.

This paper is organized as follows: In section 2 we review the construction and key properties of the Horrocks-Mumford vector bundle. In section 3 we construct the Horrocks-Mumford quintic threefolds F and G as pencils of abelian surfaces. In section 4 we construct the common resolution $\tilde{X}$ of the threefolds F and G and study its geometry. In section 5 we find the elliptic surfaces E_1 and E_2 in $\tilde{X}$, count points and apply Livné's method to show that $\tilde{X}$ is modular.

This paper is based upon doctoral research conducted at Harvard University. The author would like to thank Shing-Tung Yau for his guidance and encouragement, Richard Taylor and Klaus Hulek for valuable discussions, and the referee for some helpful comments.

2. The Horrocks-Mumford vector bundle

2.1. Construction of the bundle. The Horrocks-Mumford vector bundle HM is a stable, indecomposable rank 2 bundle over the complex projective space $\mathbb{P}^4$. It is essentially the only known bundle satisfying these properties; all other such bundles that are currently known are derived from HM by twisting by powers of the sheaf $\mathcal{O}(1)$ or by taking pullbacks to branched covers of $\mathbb{P}^4$. It was first discovered by Horrocks and Mumford in [**14**], and has been further studied by many other authors (see for example [**5**], [**6**], [**15**], [**24**]). In this section we will describe the construction of HM and explain some of its properties that we will use later.

The following exposition of the Horrocks-Mumford bundle is taken from [**13**]. A *monad* is a three-term complex

$$A \xrightarrow{\ p\ } B \xrightarrow{\ q\ } C$$

of vector bundles where p is injective and q is surjective. The cohomology of the monad

$$E = \ker q/\operatorname{im} p$$

is also a vector bundle. To construct the Horrocks-Mumford bundle using a monad we fix a vector space

$$V \cong \mathbb{C}^5.$$

Denote its standard basis by $e_i, i \in \mathbb{Z}/5$. On the projective space $\mathbb{P}^4 = \mathbb{P}(V)$, we have the Koszul complex

$$\begin{aligned} 0 \longrightarrow \mathcal{O} \xrightarrow{\wedge s} V \otimes \mathcal{O}(1) \xrightarrow{\wedge s} \wedge^2 V \otimes \mathcal{O}(2) \\ \xrightarrow{\wedge s} \wedge^3 V \otimes \mathcal{O}(3) \xrightarrow{\wedge s} \wedge^4 V \otimes \mathcal{O}(4) \longrightarrow \mathcal{O}(5) \longrightarrow 0. \end{aligned}$$

Now the quotient $\mathcal{O}(1) \otimes V/\mathcal{O}$ is isomorphic to the tangent sheaf T, and in the Koszul complex the sheaf of cycles $\operatorname{im}(\mathcal{O}(i) \otimes \wedge^i V) \subset \mathcal{O}(i+1) \otimes \wedge^{i+1} V$ is isomorphic to $\wedge^i T$. Thus from the map

$$\mathcal{O}(2) \otimes \wedge^2 V \longrightarrow \mathcal{O}(3) \otimes \wedge^3 V$$

we obtain the sequence of maps

$$\mathcal{O}(2) \otimes \wedge^2 V \xrightarrow{p_0} \wedge^2 T \xrightarrow{q_0} \mathcal{O}(3) \otimes \wedge^3 V$$

where the first map is surjective and the second is injective.

Horrocks and Mumford defined the following maps

$$\begin{aligned} f^+ : V \longrightarrow \wedge^2 V, \quad f^+(\sum v_i e_i) = \sum v_i e_{i+2} \wedge e_{i+3}, \\ f^- : V \longrightarrow \wedge^2 V, \quad f^-(\sum v_i e_i) = \sum v_i e_{i+1} \wedge e_{i+4}. \end{aligned}$$

Using these maps one can define

$$\begin{aligned} p : V \otimes \mathcal{O}(2) \xrightarrow{(f^+, f^-)(2)} 2 \wedge^2 V \otimes \mathcal{O}(2) \xrightarrow{2p_0} 2 \wedge^2 T \\ q : 2 \wedge^2 T \xrightarrow{2q_0} 2 \wedge^3 V \otimes \mathcal{O}(3) \xrightarrow{(-f^{-*}, f^{+*})(3)} V^* \otimes \mathcal{O}(3). \end{aligned}$$

One easily checks that $q \circ p = 0$. Hence we obtain a monad

$$V \otimes \mathcal{O}(2) \xrightarrow{p} 2 \wedge^2 T \xrightarrow{q} V^* \otimes \mathcal{O}(3).$$

Its cohomology

$$HM = \ker q/\operatorname{im} p$$

is the Horrocks-Mumford bundle. It is a rank 2 bundle, and its total Chern class $c(HM)$ equals $c(\wedge^2 T)^2 c(V^* \otimes \mathcal{O}(3))^{-1} c(V \otimes \mathcal{O}(2))^{-1}$. Using the splitting principle, one computes this class to be $1 + 5H + 10H^2$. Therefore, zero sets of sections of HM are surfaces of degree 10; Horrocks and Mumford showed that the generic zero set is a smooth abelian surface.

2.2. Symmetries of HM and invariant quintics. The study of HM has been greatly expedited by the fact that it admits a large group of discrete symmetries. Consider the Heisenberg group of rank 5, which we denote by H_5. We present it as a subgroup of $GL_5(\mathbb{C})$ generated by the matrices

$$\sigma = \begin{pmatrix} & 1 & & & \\ & & 1 & & \\ & & & 1 & \\ & & & & 1 \\ 1 & & & & \end{pmatrix}, \tau = \begin{pmatrix} 1 & & & & \\ & \epsilon & & & \\ & & \epsilon^2 & & \\ & & & \epsilon^3 & \\ & & & & \epsilon^4 \end{pmatrix},$$

where $\epsilon = e^{\frac{2\pi i}{5}}$ is a primitive fifth root of unity. H_5 is a central extension

$$1 \to \mu_5 \to H_5 \to \mathbb{Z}/5 \times \mathbb{Z}/5 \to 1$$

where σ is sent to $(1,0)$ and τ to $(0,1)$. Here μ_5 is the multiplicative group of fifth roots of unity.

In fact, the normalizer N_5 of H_5 in $SL_5(\mathbb{C})$ preserves HM. N_5 is a semidirect product of H_5 with the binary icosahedral group $SL(2,\mathbb{Z}_5)$. We will need the following elements of N_5:

$$\iota = \begin{pmatrix} 1 & & & & \\ & & & & 1 \\ & & & 1 & \\ & & 1 & & \\ & 1 & & & \end{pmatrix}, \mu = \begin{pmatrix} 1 & & & & \\ & & 1 & & \\ & & & & 1 \\ & 1 & & & \\ & & & 1 & \end{pmatrix}, \nu = \begin{pmatrix} 1 & & & & \\ & \epsilon & & & \\ & & \epsilon^4 & & \\ & & & \epsilon^4 & \\ & & & & \epsilon \end{pmatrix}.$$

These matrices act on $N_5/\mu_5 \simeq \mathbb{Z}/5\times\mathbb{Z}/5$ by conjugation; Horrocks and Mumford showed that the action is unimodular. Their images in $SL_2(\mathbb{Z}_5)$ are

$$\overline{\iota} = \begin{pmatrix} -1 & \\ & -1 \end{pmatrix}, \overline{\mu} = \begin{pmatrix} 2 & \\ & 3 \end{pmatrix}, \overline{\nu} = \begin{pmatrix} 1 & 2 \\ & 1 \end{pmatrix}.$$

En route to determining the sections of HM, Horrocks and Mumford determined the N_5/H_5-module $\Gamma_{H_5}(\mathcal{O}(5))$ of H-invariants of $\Gamma(\mathcal{O}(5))$, i.e. Heisenberg-invariant quintics in $\mathbb{P}^4$. It is six-dimensional, spanned by the polynomials

$$\sum x_i^5, \sum x_i^3 x_{i+1} x_{i+4}, \sum x_i x_{i+1}^2 x_{i+4}^2,$$
$$\sum x_i^3 x_{i+2} x_{i+3}, \sum x_i x_{i+2}^2 x_{i+3}^2, x_0 x_1 x_2 x_3 x_4$$

where the sums are taken over powers of σ. The base locus of this space of quintics is the set of 25 lines L_{ij}, where

$$L_{00} = \{x \in \mathbb{P}^4 : x_0 = x_1 + x_4 = x_2 + x_3 = 0\},$$

$$L_{ij} = \sigma^i \tau^j L_{00}.$$

Since $c(HM) = 1+5H+10H^2$, $c_1(\wedge^2(HM)) = 5H$ and thus $\wedge^2(HM) \cong \mathcal{O}(5)$. Hence if s_1 and s_2 are sections of HM, the zero set of the section $s_1 \wedge s_2$ of $\wedge^2 HM$ is a (singular) quintic Calabi-Yau threefold that has the structure of a pencil of abelian surfaces.

PROPOSITION 2.1. *For generic sections s_1 and s_2 of HM, the singularities of the resulting threefold are the 100 nodes coming from the intersection of $Z(s_1)$ and $Z(s_2)$.*

PROOF. For a generic choice of s_1 and s_2, $Z(s_1)$ and $Z(s_2)$ intersect transversely in 100 points; these points form the base locus of the pencil.

Let p be a point at which $Z(s_1)$ and $Z(s_2)$ intersect transversely. Choose a trivialization of HM near p, and put $s_1 = (s_{11}, s_{12})$ and $s_2 = (s_{21}, s_{22})$ relative to this trivialization. Since $Z(s_1)$ and $Z(s_2)$ intersect transversely, we may then use s_{11}, s_{12}, s_{21} and s_{22} as local coordinates on $\mathbb{P}^4$ centered at p. The local equation for the threefold is then

$$s_{11}s_{22} - s_{12}s_{21} = 0.$$

Hence p is a node. □

Nodes on threefolds result from the vanishing of an S^3 cycle on a a smooth family of threefolds. One expects that degenerating the S^3 cycles and then resolving the singularities will cause the Betti number h^3 to drop. We will be interested in birationally equivalent Calabi-Yau threefolds with low Betti number. Taking one-parameter families of abelian surfaces in $\mathbb{P}^4$ gives us a quick way of manufacturing nodal Calabi-Yau threefolds, whose singularities can then be resolved. In [**24**], Schoen studied the Fermat quintic Q defined by the equation

$$x_0^5 + x_1^5 + x_2^5 + x_3^5 + x_4^5 - 5x_0x_1x_2x_3x_4 = 0.$$

Schoen showed that it was a Horrocks-Mumford quintic with 125 nodes instead of the usual 100, and he proved that the blowup $\tilde{Q}$ of Q was rigid and modular. Other nodal Calabi-Yau threefolds whose resolutions are modular were studied in [**30**].

REMARK. Instead of manufacturing nodal Calabi-Yaus in $\mathbb{P}^4$, one can also use as the ambient space other Fano fourfolds such as $\mathbb{P}^3 \times \mathbb{P}^1$; we then want to consider a rank 2 bundle whose determinant bundle is anticanonical. As above, we can then consider surfaces cut out by sections of the bundle and take pencils of these surfaces to obtain other nodal Calabi-Yau threefolds. In [**17**] and [**18**], Lange has proven the existence of abelian surfaces in $\mathbb{P}^1 \times \mathbb{P}^3$ and by the Serre construction found the rank 2 bundle V whose zero sections yield these surfaces.

3. Abelian surfaces in $\mathbb{P}^4$

3.1. Sections of HM. In the previous section we mentioned that an abelian surface in $\mathbb{P}^4$ is projectively equivalent to $Z(s)$ for some section s of HM. Since $h^0(HM)$ is 4, $\mathbb{P}^3$ is a parameter space of (possibly degenerate) abelian surfaces in $\mathbb{P}^4$.

Any vector bundle over $\mathbb{P}^1$ splits into a direct sum of line bundles; for most lines in $\mathbb{P}^4$, the restriction of HM is isomorphic to $\mathcal{O}(2) \oplus \mathcal{O}(3)$. Lines L such that $HM|_L$ is isomorphic to $\mathcal{O}(2-a) \oplus \mathcal{O}(3+a)$ are called *jumping lines* of order a. It is well known that the 25 lines L_{ij} are jumping lines of order 3; the restriction of HM to these lines is isomorphic to $\mathcal{O}(-1) \oplus \mathcal{O}(6)$. Barth, Hulek and Moore ([**5**]) proved that the restriction map $\Gamma_{HM} \longrightarrow \Gamma_{HM|_{L_{00}}}$ is injective, and they were able to determine the sections of $\Gamma_{HM|_{L_{00}}}$:

PROPOSITION 3.1. *Let λ and μ be the restrictions of the coordinates x_1 and x_2 to L_{00}. Then the image of Γ_{HM} in $\Gamma_{HM|_{L_{00}}}$ is spanned by the sections $t_0 = \lambda^6 + 2\mu^5\lambda, t_1 = \mu^6 - 2\mu\lambda^5, t_2 = 5\lambda^4\mu^2, t_3 = 5\lambda^2\mu^4$ of $\mathcal{O}(-1) \oplus \mathcal{O}(6)$.* □

Given a section s of HM, we can associate to it the vector (c_0, c_1, c_2, c_3) representing the coordinates of $s_{L_{00}}$ with respect to the basis t_0, t_1, t_2, t_3. The coordinates

c_i are then homogeneous coordinates on the moduli space $\mathbb{P}^3$ of abelian surfaces in $\mathbb{P}^4$. We will call these coordinates *BHM coordinates.*

The polynomial $c_0t_0 + c_1t_1 + c_2t_2 + c_3t_3$ determines the singularities of $Z(s)$:

THEOREM 3.2. *Let s be a section of HM, and let $f = c_0t_0+c_1t_1+c_2t_2+c_3t_3$ be its restriction to L_{00}. The degeneracies of $Z(s)$ are determined by the multiplicities of the roots $(\lambda : \mu)$ of f:*

Multiplicities of roots	Degeneracy of $X(s)$
$(1,1,1,1,1,1)$	smooth
$(2,1,1,1,1,1)$	translation scroll of elliptic normal curve
$(3,1,1,1)$	tangent scroll of elliptic normal curve
$(2,2,1,1)$	union of five quadrics
$(2,2,2)$	doubled elliptic quintic scroll
$(4,2)$	union of five double planes

The automorphisms μ, ν and δ of $\mathbb{P}^4$ induce automorphisms, also denoted μ, ν and δ, of the moduli space $\mathbb{P}^3$. In BHM coordinates, they have the following form:

$$(1)\quad \mu = \begin{pmatrix} & 1 & & \\ 1 & & & \\ & & & 1 \\ & & 1 & \end{pmatrix}, \nu = \begin{pmatrix} \epsilon^4 & & & \\ & \epsilon & & \\ & & \epsilon^3 & \\ & & & \epsilon^2 \end{pmatrix}, \delta = \frac{1}{\sqrt{5}} \begin{pmatrix} -1 & 1 & \eta' & -\eta \\ 1 & -1 & -\eta & \eta' \\ \eta' & -\eta & 1 & -1 \\ -\eta & \eta' & -1 & 1 \end{pmatrix}$$

where $\eta = \epsilon + \epsilon^4$ and $\eta' = \epsilon^2 + \epsilon^3$.

3.2. A pencil of abelian surfaces. One can ask if the abelian surfaces corresponding to fixed points of these automorphisms have any interesting properties; this is how we found the threefold X. Let us find the fixed points of the automorphism μ; these correspond to the eigenspaces of the matrix

$$\begin{pmatrix} & 1 & & \\ 1 & & & \\ & & & 1 \\ & & 1 & \end{pmatrix}.$$

We find the eigenspace V_{-1} spanned by the vectors $(1,-1,0,0)$ and $(0,0,1,-1)$ and the eigenspace V_1 spanned by $(1,1,0,0)$ and $(0,0,1,1)$ These correspond to lines in the moduli space $\mathbb{P}^3$, also denoted V_{-1} and V_1. Our threefold X will be derived from the threefolds swept out by the abelian surfaces in V_{-1} and V_1.

As in [**6**], we will use the Clebsch diagonal cubic

$$(2)\qquad X_3 = \{c \in \mathbb{P}^3 : c_0^2c_3 + c_1^2c_2 - c_0c_2^2 - c_1c_3^2 = 0\}.$$

X_3 is the image of the $SL(2,\mathbb{F}_5)$-equivariant rational map $p : \mathbb{P}^2 \to \mathbb{P}^3$ sending $(y_1 : y_2 : y_3)$ to

$$(3)\qquad x = (y_1y_3^2 - y_2^3 : y_3^3 - y_1y_2^2 : y_2^2y_3 - y_2y_1^2 : y_3y_1^2 - y_2y_3^2).$$

The rational map p is undefined at the points $(1 : 0 : 0)$ and $(1 : \epsilon^k : \epsilon^{-k})$; these six points correspond to the six exceptional divisors when we exhibit X_3 as $\mathbb{P}^2$ blown up in six points.

Recall the configuration of 27 lines on the cubic surface: the cubic surface is isomorphic to $\mathbb{P}^2$ blown up in six points $p_1, p_2, \ldots, p_6$. The lines E_m, $m = 1, 2, \ldots, 6$

are the exceptional divisors. The lines F_{mn}, $1 \leq m < n \leq 6$ are the proper transforms of the lines through p_m and p_n. The lines G_n, $n = 1, 2, \ldots, 6$ are the proper transforms of the conics through the five points other than p_n.

To locate the six exceptional divisors in X_3, temporarily dehomogenize y by setting $y_1 = 1$. Also set $y_2 = 0$ and consider what happens when we let y_3 approach 0; we see that $p((1:0:0))$ approaches the point $(0:0:1:0)$. Now set $y_3 = 0$ and consider what happens when y_2 approaches 0; $p((1:0:0))$ approaches the point $(0:0:0:1)$. Hence the line E_0 in X_3 is spanned by the points $(0:0:0:1)$ and $(0:0:1:0)$. Repeating the same local analysis at the other five points, we find that the lines E_k, $k = 1, 2, \ldots, 5$ are spanned by the points $(-3\epsilon^{2k} : -2\epsilon^{k} : 1 : -\epsilon^{-2k})$ and $(2\epsilon^{-k} : 3\epsilon^{-2k} : \epsilon^{2k} : -1)$.

LEMMA 3.3. *The line V_{-1} is exactly the line F_{56}.*

PROOF. One checks that the line V_{-1} intersects the E-lines E_5, E_6 and no others. □

Let F be the threefold swept out by the abelian surfaces parametrized by $F_{56} = V_{-1}$. Let G be the threefold swept out by V_1. We can determine what all the fibres of F are:

PROPOSITION 3.4. *The singular fibres of F are as follows: there are two fibres of type (2,2,1,1) corresponding to unions of five quadrics and two fibres of type (2,2,2) corresponding to doubled elliptic quintic scrolls.*

The singular fibres of G are as follows: there are two fibres of type (2,2,1,1) corresponding to unions of five quadrics and two fibres of type (3,1,1,1) corresponding to tangent scrolls of elliptic normal curves.

PROOF. For $(a : b) \in \mathbb{P}^1$ and $(\lambda : \mu) \in \mathbb{P}^1$, consider the correspondence in $\mathbb{P}^1 \times \mathbb{P}^1$ defined by the condition that $(\lambda : \mu)$ be a root of the polynomial $a(t_0 - t_1) + b(t_2 - t_3)$. This is a $(1, 6)$ correspondence and thus has genus zero. Projection to the first copy of $\mathbb{P}^1$ is a map between rational curves of degree 6. By the Riemann-Hurwitz theorem, there are 10 ramification points (counted with multiplicity).

The line F_{56} intersects E_5 and E_6 at $(2 : -2 : 4 : 4)$ and $(0 : 0 : 1 : -1)$ respectively. The corresponding sextic polynomials in λ, μ have multiplicities $(2, 2, 1, 1)$, so the corresponding fibres are unions of five quadrics each.

Four branch points have already been accounted for. We found that the line F_{56} intersects the curve C_6 in the points $(5 : -5 : 2i+1 : -2i-1)$ and $(5 : -5 : -2i+1 : 2i-1)$. These points correspond to polynomials of type $(2, 2, 2)$. They account for the other six ramification points. We have therefore found all the singular fibres of F.

For G, we look at the correspondence on $\mathbb{P}^1 \times \mathbb{P}^1$ where $(\lambda : \mu)$ is a root of $a(t_0+t_1)+b(t_2+t_3)$. This time, the correspondence has two horizontal components because $(1 : i)$ and $(1 : -i)$ are common roots of $t_0 + t_1$ and $t_2 + t_3$. The remaining part of the correspondence is a $(1, 4)$ curve that maps surjectively onto the first copy of $\mathbb{P}^1$, so there are 6 ramification points.

By inspection, the polynomial $a(t_0 + t_1) + b(t_2 + t_3)$ has repeated roots when $(a : b) = (1 : 0)$, $(0 : 1)$, $(1 : \frac{3+4i}{5})$ and $(1 : \frac{3-4i}{5})$ of type $(2, 2, 1, 1)$, $(2, 2, 1, 1)$, $(3, 1, 1, 1)$ and $(3, 1, 1, 1)$ respectively. Eliminating the common roots $(1 : \pm i)$, the remaining roots have multiplicity $(2, 2)$, $(2, 2)$, $(2, 1, 1)$ and $(2, 1, 1)$ respectively. This accounts for the 6 ramification points. □

Once we have the defining equation for F, simple calculations will show that the quadric surface $T_0 = \{x : x_0 = x_1x_4 - x_2x_3 = 0\}$ and its translates $T_i = \sigma^i T_0$

are contained in F; the T_i are τ-invariant. Put $U_0 = \delta T_0$; we then have $U_0 = \{x : \Sigma_i x_i = \Sigma_{i \neq j} x_i x_j = 0\}$. The quadric surface U_0 and its translates $U_i = \tau^i W_0$ are contained in F as well; the U_i are σ-invariant. Since the surfaces $\cup_i T_i$ and $\cup_i U_i$ are of degree 10 and invariant under H_5, they must be two of the singular fibres in the abelian surface fibration of F.

Similarly, one checks that the quadric surface $Q_0 = \{z : z_0 = z_1 z_4 + z_2 z_3 = 0\}$ and its translates $Q_i = \sigma^i Q_0$ are in G, and that the quadric surface $R_0 = \delta Q_0 = \{z : \Sigma_i z_i = \Sigma_{i \neq j} z_i z_j = 0\}$ and its translates $Q_i = \tau^i Q_0$ are in G. These two unions of five quadrics are two of the singular fibres in the fibration of G.

We can identify the other fibres of F; first we need to identify certain elliptic curves in $\mathbb{P}^4$. Aure, Decker, Hulek, Popescu and Ranestad [**1**] have shown that the set of H_5-equivariant elliptic normal curves in $\mathbb{P}^4$ is parametrized by $\mathbb{P}^1$: to the point $(\lambda : \mu)$ corresponds the elliptic normal curve $E_{(\lambda:\mu)}$ defined by the set of equations

$$q_i^{(\lambda:\mu)}(x) = -\lambda\mu x_i^2 - \mu^2 x_{i+1} x_{i+4} + \lambda^2 x_{i+2} x_{i+3}. \tag{4}$$

By inspection, we see that the curves corresponding to $(1 : i)$ and $(1 : -i)$ are in F and G. Denote these curves by E_1 and E_2 respectively. Proposition 4.3 in [**1**] shows that the elliptic quintic scrolls Q_1 and Q_2 contain E_1 and E_2 respectively, where Q_1 is defined by the equations

$$x_i^3 + x_i x_{i+1} x_{i+4} + x_i x_{i+2} x_{i+3} - i(x_{i+1}^2 x_{i+3} + x_{i+2} x_{i+4} + x_{i+1} x_{i+2}^2 + x_{i+3}^2 x_{i+4}) = 0 \tag{5}$$

and Q_2 is defined by replacing i above with its complex conjugate $-i$. Simple calculations show that Q_1 and Q_2 are contained in F. Being H_5-invariant, they must be the elliptic quintic scroll fibers of F.

4. The threefold $\tilde{X}$

4.1. Definition of $\tilde{X}$. Although we have defined the Horrocks-Mumford bundle only over $\mathbb{C}$, the pencils of abelian surfaces it defines are quintics in $\mathbb{P}^4$, and the quintics we are interested in have integer coefficients and can thus be studied over arbitrary fields k.

In particular, we have the Horrocks-Mumford quintics F and G defined by the lines V_{-1} and V_1 in $\mathbb{P}^3$. In [**20**], Manolache was able to determine the equation of the Horrocks-Mumford quintic in terms of the BHM-coordinates of the parametrizing line:

THEOREM 4.1. *Suppose the Horrocks-Mumford quintic X is determined by the line passing through the points $(a_0 : a_1 : a_2 : a_3)$ and $(b_0 : b_1 : b_2 : b_3)$. Then the defining equation for X is*

$$\begin{aligned} &25(a_3 b_4 - a_4 b_3)(\Sigma x_0 x_1 x_2 x_3 x_4) \\ &+ 5(a_2 b_3 - a_3 b_2)(\Sigma x_0 x_2^2 x_3^2) \\ &- 5(a_1 b_3 - a_3 b_1)(\Sigma x_0^3 x_2 x_3) \\ &+ 5(a_2 b_4 - a_4 b_2)(\Sigma x_0^3 x_1 x_4) \\ &- 5(a_1 b_4 - a_4 b_1)(\Sigma x_0 x_1^2 x_4^2) \\ &+ (a_1 b_2 - a_2 b_1)(\Sigma(x_0^5 - x_0 x_1 x_2 x_3 x_4) \end{aligned} \tag{6}$$

where the sums are taken over cyclic permutations of the indices.

An easy calculation shows that F is defined by the equation

$$\Sigma_i(x_i^3x_{i+1}x_{i+4} + x_i^3x_{i+2}x_{i+3} - x_ix_{i+1}^2x_{i+4}^2 - x_ix_{i+2}^2x_{i+3}^2) = 0 \tag{7}$$

and that G is defined by the equation

$$\Sigma_i(z_i^3z_{i+1}z_{i+4} - z_i^3z_{i+2}z_{i+3} - z_iz_{i+1}^2z_{i+4}^2 + z_iz_{i+2}^2z_{i+3}^2) = 0. \tag{8}$$

where the summations are taken over cyclic permutations of the indices. As before, F and G are both invariant under the action of the matrices σ and τ.

Consider now the complete intersection threefold X in $\mathbb{P}^4(x) \times \mathbb{P}^4(z)$ given by the matrix equation

$$M(x)z = 0, \tag{9}$$

where

$$M(x) = \begin{pmatrix} & -x_3 & x_1 & x_4 & -x_2 \\ -x_3 & & -x_4 & x_2 & x_0 \\ x_1 & -x_4 & & -x_0 & x_3 \\ x_4 & x_2 & -x_0 & & -x_1 \\ -x_2 & x_0 & x_3 & -x_1 & \end{pmatrix}. \tag{10}$$

Note that this is equivalent to the matrix equation

$$L(z)x = 0, \tag{11}$$

where

$$L(z) = \begin{pmatrix} & z_2 & -z_4 & -z_1 & z_3 \\ z_4 & & z_3 & -z_0 & -z_2 \\ -z_3 & z_0 & & z_4 & -z_1 \\ -z_2 & -z_4 & z_1 & & z_0 \\ z_1 & -z_3 & -z_0 & z_2 & \end{pmatrix}. \tag{12}$$

Note also that $\det M(x)$ and $\det L(z)$ give us the equations for F and G respectively (up to a factor of 2). Hence the projections π_1 and π_2 of X onto each factor give us F and G. In [**22**], Moore first considered the matrices

$$M(x,y) = \begin{pmatrix} x_0y_0 & x_3y_2 & x_1y_4 & x_4y_1 & x_2y_3 \\ x_3y_3 & x_1y_0 & x_4y_2 & x_2y_4 & x_0y_1 \\ x_1y_1 & x_4y_3 & x_2y_0 & x_0y_2 & x_3y_4 \\ x_4y_4 & x_2y_1 & x_0y_3 & x_3y_0 & x_1y_2 \\ x_2y_2 & x_0y_4 & x_3y_1 & x_1y_3 & x_4y_0 \end{pmatrix}, \tag{13}$$

$$L(z,y) = \begin{pmatrix} z_0y_0 & z_2y_4 & z_4y_3 & z_1y_2 & z_3y_1 \\ z_4y_1 & z_1y_0 & z_3y_4 & z_0y_3 & z_2y_2 \\ z_3y_2 & z_0y_1 & z_2y_0 & z_4y_4 & z_1y_3 \\ z_2y_3 & z_4y_2 & z_1y_1 & z_3y_0 & z_0y_4 \\ z_1y_4 & z_3y_3 & z_0y_2 & z_2y_1 & z_4y_0 \end{pmatrix}. \tag{14}$$

For a generic choice of $y \in \mathbb{P}^4$, the threefold determined by $\det M(x,y) = 0$ and the threefold determined by $\det L(z,y) = 0$ are both Horrocks-Mumford quintics with the expected 100 nodes. For our threefolds F and G, we have taken $y = (0 : 1 : -1 : -1 : 1)$.

4.2. Singularities of X. We need to know the singularities of X:

PROPOSITION 4.2. *Over a field of characteristic not equal to 2 or 5, X has 60 singular points. The 60 singular points of X are all ordinary double points (nodes).*

REMARK. In characteristic 0, Gross and Popescu [**11**] have studied the two-parameter family of threefolds X_y given by perfoming the construction above for y a generic point of the plane

$$\mathbb{P}^2_+ = \{y : y_1 - y_4 = y_2 - y_3 = 0\}.$$

Thus X_y is a common partial resolution of the quintic threefolds F_y and G_y. Our threefold X is thus a special member of this family. Over $\mathbb{C}$, Gross and Popescu proved the following statements for a generic choice of y:

(1) F_y is singular along the union of two elliptic curves $D_{1,y}$ and $D_{2,y}$. These curves are the base curves of the elliptic quintic scrolls $Q_{1,y}$ and $Q_{2,y}$ appearing in the abelian surface fibration of F_y, and they intersect in the 25 points comprising the H_5 orbit of y. These 25 points comprise the base locus of the pencil of abelian surfaces.
(2) G_y is singular along a disjoint union of two elliptic curves $E_{1,y}$ and $E_{2,y}$.
(3) X_y is singular along 50 nodes, 2 of which lie over each point of the Heisenberg orbit of y when projecting via the map π_1. We call these 50 nodes *regular nodes*.

Essentially our threefold X is a special member of the family X_y where $y = (0 : 1 : -1 : -1 : 1)$. Over $\mathbb{C}$, it is easy to check that X has the 50 expected nodes and 10 others. Over fields of arbitrary characteristic, we resort to brute-force computation. A basic trick we use is to compute Gröbner bases of ideals over $\mathbb{Z}$ in order to obtain results valid in fields of unknown characteristic.

We will begin the proof of the proposition after we dispense with some preliminary results. If x is a point in $\mathbb{P}^4(x)$, we say that x is a rank n point if rank $M(x) = n$. Similarly, if z is a point in $\mathbb{P}^4(z)$, we say that z is a rank n point if rank $L(z) = n$.

LEMMA 4.3. *If x is a point of F, then x is a rank 4 point or a rank 3 point.*

PROOF. If x is a point of F, then $\det M(x) = 0$. Hence the rank of $\det M(x)$ is at most 4.

The 3 by 3 matrices

$$\begin{pmatrix} & -x_3 & x_1 \\ -x_3 & & -x_4 \\ x_1 & -x_4 & \end{pmatrix} \begin{pmatrix} & -x_4 & x_2 \\ -x_4 & & -x_0 \\ x_2 & -x_0 & \end{pmatrix} \begin{pmatrix} & -x_0 & x_3 \\ -x_0 & & -x_1 \\ x_3 & -x_1 & \end{pmatrix}$$

$$\begin{pmatrix} & x_1 & -x_2 \\ x_1 & & x_3 \\ -x_2 & x_3 & \end{pmatrix} \begin{pmatrix} & -x_3 & -x_2 \\ -x_3 & & x_0 \\ -x_2 & x_0 & \end{pmatrix} \begin{pmatrix} & x_1 & x_4 \\ x_1 & & -x_0 \\ x_4 & -x_0 & \end{pmatrix}$$

$$\begin{pmatrix} & x_2 & x_0 \\ x_2 & & -x_1 \\ x_0 & -x_1 & \end{pmatrix} \begin{pmatrix} & x_1 & -x_2 \\ x_1 & & x_3 \\ -x_2 & x_3 & \end{pmatrix} \begin{pmatrix} & -x_3 & x_4 \\ -x_3 & & x_2 \\ x_4 & x_2 & \end{pmatrix}$$

$$\begin{pmatrix} & -x_4 & x_0 \\ -x_4 & & x_3 \\ x_0 & x_3 & \end{pmatrix}$$

are all 3 by 3 minors of $M(x)$. Their determinants give us $\pm 2x_ix_jx_k$ for all subsets $\{i,j,k\}$ of $\{0,1,2,3,4\}$. Suppose these determinants are all zero. Then three of the x_i must be zero.

Without loss of generality, we can assume either that x_0, x_1, x_2 are 0 or that x_0, x_2, x_4 are zero. Suppose the former holds. One sees then that the matrix

$$\begin{pmatrix} & -x_3 & & x_4 & \\ -x_3 & & -x_4 & & \\ & -x_4 & & & x_3 \\ x_4 & & & & \\ & & x_3 & & \end{pmatrix}$$

has at least three linearly independent columns. A similar result holds of x_0, x_2, x_4 are zero. Hence the rank of $M(x)$ is at least 3 for all x. □

LEMMA 4.4. *If z is a point of G, then z is a rank 4 point or a rank 3 point.*

PROOF. Again, if z is a point of G, then $\det L(z) = 0$. Hence the rank of $\det L(z)$ is at most 4.

The polynomials $\pm(z_iz_{i+1}^2 + z_{i+2}^2z_{i+3})$ and $\pm(z_i^2z_{i+3} + z_{i+2}^2z_{i+4})$ are all determinants of 3 by 3 minors of $L(z)$. Suppose that these minors are all zero. Without loss of generality, assume $z_0 = 1$.

Since $z_0z_1^2 + z_2^2z_3 = 0$, we have $z_1^2 = -z_2^2z_3$. From $z_4z_0^2 + z_1^2z_2 = 0$, we get $z_4 = -z_1^2z_2 = z_2^3z_3$. From $z_2z_3^2 + z_4^2z_0 = 0$, we get $z_2z_3^2(1 + z_2^5) = 0$.

On the other hand, we also have $z_4^2z_2 + z_1^2z_3 = 0$. From this we see that $z_2^7z_3^2 - z_2^2z_3^2 = 0$, or $(z_2^5 - 1)z_2^2z_3^2 = 0$. Thus we must have either $z_2 = 0$ or $z_3 = 0$. In either case, we have $z = (1:0:0:0:0)$, which is a rank 4 point. Hence there are no points of rank lower than 3. □

LEMMA 4.5. *Points of F or G of rank less than 4 are singular points.*

PROOF. More generally, let A be an n by n matrix whose ij entry is the indeterminate x_{ij}. Using the Laplace expansion formula, we see that $\frac{\partial \det(A)}{\partial x_{ij}}$ is equal to $(-1)^{i+j}\det A_{ij}$, where A_{ij} is the ij minor of A.

Now suppose that the x_{ij} are functions of some other indeterminates y_k. By the chain rule, $\frac{\partial \det(A)}{\partial y_k} = \sum_{i=1}^n \sum_{j=1}^n (-1)^{i+j} \det A_{ij} \frac{\partial x_{ij}}{\partial y_k}$. If y is a point of rank less than n, then the determinants of all the $(n-1)$ by $(n-1)$ minors are zero, and hence $\frac{\partial \det(A)}{\partial y_k}$ is zero for all k. □

LEMMA 4.6. *The projections $\pi_1 : X \to F$ and $\pi_2 : X \to G$ are isomorphisms outside the rank 3 loci of F and G respectively.*

PROOF. If x is a rank 4 point of $\mathbb{P}^4(x)$, this means that the kernel of $M(x)$ is 1-dimensional. Hence the kernel of $M(x)$ defines a unique point z in $\mathbb{P}^4(z)$. We thus have a regular map $f_1 : F - F_4 \to \pi^{-1}(F - F_4)$.

Now given a point (x, z) in X such that x is a rank 4 point, the coordinates of z are given by the determinants of 4 by 4 minors of $M(x)$. Hence the map $\pi_1 : \pi^{-1}(F - F_4) \to (F - F_4)$ is also regular, proving the result for F. A similar result holds for G. □

LEMMA 4.7. *Suppose the characteristic of the base field is not 2. Then the rank 3 locus of G is the union E of E_1 and E_2, where E_1 is cut out by the polynomials $iz_i^2 + z_{i+1}z_{i+4} + z_{i+2}z_{i+3}$ and E_2 is cut out by the polynomials $-iz_i^2 + z_{i+1}z_{i+4} + z_{i+2}z_{i+3}$.*

PROOF. This is proven by using Macaulay2. One shows that the radicals of the ideals defining the two varieties have identical Gröbner bases, up to constant factors of 2 and 5. □

QUESTION. Is there a geometric explanation for this result?

LEMMA 4.8. *If the characteristic of the base field is not 2 or 5, E_1 and E_2 are elliptic normal curves.*

PROOF. This was proven by Fisher in Chapter 4 of [7]. Assuming the characteristic of the base field is not 5, Fisher constructs the universal curve $\mathcal{X}(5) \subset X(n) \times \mathbb{P}^4$ as the closure of the scheme defined by the 4 by 4 Pfaffians of the 5 by 5 matrix

$$\begin{pmatrix} & -a_1x_1 & -a_2x_2 & a_2x_3 & a_1x_4 \\ a_1x_1 & & -a_1x_3 & -a_2x_4 & a_2x_0 \\ a_2x_2 & a_1x_3 & & -a_1x_0 & -a_2x_1 \\ -a_2x_3 & a_2x_4 & a_1x_0 & & -a_1x_2 \\ -a_1x_4 & -a_2x_0 & a_2x_1 & a_1x_2 & \end{pmatrix}$$

where $a = (0 : a_1 : a_2 : -a_2 : -a_1)$. One checks that E_1 and E_2 are the curves obtained when $a = (0 : 1 : -i : i : -1)$ and $(0 : 1 : i : -i : 1)$ respectively. By considering the $SL_2(\mathbb{Z}/5\mathbb{Z})$ action on $\mathcal{X}(5)$, Fisher shows that the fibers are smooth elliptic normal curves when $a_1a_2(a_1^{10} - 11a_1^5a_2^5 - a_2^{10}) \neq 0$, in analogy with the case where the base field is $\mathbb{C}$. □

Note that $S_1 = \pi_2^{-1}(E_1)$ and $S_2 = \pi_2^{-1}(E_2)$ are elliptic ruled surfaces in X. We can now classify the singularities of X.

PROOF OF PROPOSITION 4.2. A point (x, z) of X is a singular point if and only if the kernel of the matrix $\begin{pmatrix} L(z) & M(x) \end{pmatrix}$ has dimension at least 6, i.e. if the rank is at most 4. This is equivalent to saying that the rank of the transposed matrix $\begin{pmatrix} L^T(z) \\ M(x) \end{pmatrix}$ is at most 4, which is equivalent to saying that the kernel of $\begin{pmatrix} L^T(z) \\ M(x) \end{pmatrix}$ has dimension at least 1, i.e. the kernel of $L^T(z)$ and the kernel of $M(x)$ have nontrivial intersection.

CASE 1. x is a rank 4 point.

Since (x, z) is a point of X, z is in the kernel of $M(x)$. Therefore z must span the space $\ker L^T(z) \cap M(x)$. Hence $L^T(z)z = 0$. Simple algebra shows that the only singular points (x, z) in this case are $((1 : 0 : 0 : 0 : 0), (1 : 0 : 0 : 0 : 0))$ and $((1 : 1 : 1 : 1 : 1), (1 : 1 : 1 : 1 : 1))$ and their orbits under (σ, σ) and (τ^3, τ). We will call these nodes in X $\sigma - nodes$ and $\tau - nodes$ respectively. We will also use the same nomenclature for the images of these nodes in F and G.

CASE 2. x is a rank 3 point.

We will suppose that z must then be a rank 3 point. Using Macaulay2, we show that if (x, z) is a singular point of X and x is a rank 3 point of F, then some x_i or some z_i is zero. (The code can be found in the Appendix.)

Suppose that some x_i is zero. Without loss of generality, suppose x_0 is zero. The polynomials $(x_1x_2+x_4^2)^2$, $(x_2x_4+x_3^2)^2$, $(x_1x_3+x_2^2)^2$, $(x_1x_4+x_3^2)^2$ and $(x_2x_3-x_1x_4)^2$ are all 4×4 minors of $M(x)$ when x_0 is zero.

If x_1 is also zero, we quickly see that all x_i are zero, a contradiction. So suppose $x_1 = 1$. We quickly see that $x_4 = \epsilon^k$, $x_2 = -\epsilon^{2k}$ and $x_3 = -\epsilon^{4k}$. Solving for z gives us the singular points

$$(x, z) = ((0 : 1 : -\epsilon^k : -\epsilon^{2k} : \epsilon^{3k}), (0 : 1 : \pm i\epsilon^{2k} : \mp i\epsilon^{4k} : -\epsilon^k)),$$

whose orbits under (σ, σ) will be the 50 regular nodes of X.

Now suppose instead that some z_i is zero. Without loss of generality, we may assume $z_0 = 0$.. If we assume that z is a rank 3 point of G, then z is a point of $E = E_1 \cup E_2$ with $z_0 = 0$. We have equations defining E, and using these equations we recover the 50 points above.

If z is a rank 4 point, then z must itself be a rank 4 node of G, since the map $\pi_2 : X \to G$ is an isomorphism on the rank 4 locus. Again using Macaulay2, we can find generators of an ideal that defines the set of singular points on G with $z_0 = 0$. We eventually find one of two things: either z is a σ-node, all of whose corresponding x are not rank 3 points, or z is of the form $(0 : 1 : \pm i\epsilon^k : \mp i\epsilon^{2k} : -\epsilon^{3k})$, which are not rank 4 points of G. So we have found no new nodes.

We have assumed throughout that our nodes have $x_0 = 0$ or $z_0 = 0$; taking the orbits of (x, z) under (σ, σ), we get all possible values of z for which there is a node (x, z) with x and z rank 3 points.

There are 60 singular points on X: the σ-nodes; the τ-nodes; and the 50 regular nodes given by $((0 : 1 : -1 : -1 : 1), (0 : 1 : \pm i : -\pm i : -1))$ and their orbits under (σ, σ) and (τ^3, τ).

To prove that the 60 singular points are indeed nodes, we need to construct local coordinates around each singular point.

$((1 : 0 : 0 : 0 : 0), (1 : 0 : 0 : 0 : 0))$: Dehomogenize by fixing $x_0 = 1$ and $z_0 = 1$. X is then defined by the five equations

$$\begin{aligned} -x_3z_1 + x_1z_2 + x_4z_3 - x_2z_4 &= 0, \\ -x_3 - x_4z_2 + x_2z_3 + z_4 &= 0, \\ x_1 - x_4z_1 - z_3 + x_3z_4 &= 0, \\ x_4 + x_2z_1 - z_2 - x_1z_4 &= 0, \\ -x_2 + z_1 + x_3z_2 - x_1z_3 &= 0. \end{aligned} \tag{15}$$

Let I be the ideal generated by these five polynomials.

Notice that $z_4 = x_3$ plus terms of higher order after we complete the coordinate ring at the ideal $(x_1, \ldots, x_4, z_1, \ldots, z_4)$. Similarly $z_3 = x_1 + h.o.t.$, $z_2 = x_4 + h.o.t.$ and $z_1 = x_2 + h.o.t.$.

Replacing the z's in this manner, we see that X is defined by the single equation

$$-x_3x_2 + x_1x_4 + x_4x_1 - x_2x_3 + h.o.t. = 0 \tag{16}$$

in the variables x_1, x_2, x_3, x_4. Let J be the ideal generated by this formal power series. We see that the completed local ring $\overline{k}[[x_1, \ldots, x_4, z_1, \ldots, z_4]]/I$ is isomorphic to the completed local ring $\overline{k}[[x_1, \ldots, x_4]]/J$.

The second-order piece of equation (16) reads

$x^T Bx = 0$ where $B = \begin{pmatrix} & & & 1 \\ & & -1 & \\ & -1 & & \\ 1 & & & \end{pmatrix}$. The determinant of B is 1. Therefore, we can find a matrix C such that $B = C^T C$. Put $y = Cx$; we now have an isomorphism

$$\overline{k}[[x_1, \ldots, x_4]]/J \longrightarrow \overline{k}[[y_1, \ldots, y_4]]/L$$

where L is generated by some formal power series of the form $y_1^2+y_2^2+y_3^2+y_4^2+h.o.t.$ and $Cx \to y$.

Finally, we show that there is an isomorphism

$$\overline{k}[[y_1 \ldots y_4]]/L \to \overline{k}[[t_1, \ldots, t_4]]/M$$

where M is generated by $t_1^2+t_2^2+t_3^2+t_4^2$. Put $t_i = y_i+P_{i2}(y)+P_{i3}(y)+\ldots$, where P_{ik} is a monomial of order k in the y's. L is generated by $y_1^2+y_2^2+y_3^2+y_4^2+L_3(y)+\ldots$. We can find P_{i2} such that $\sum_i 2y_i P_{i2} = L_3(y)$, and we can solve for higher P_{ik} recursively. The resulting expressions for the t_i define the isomorphism. At last we see that our singular point is a node.

$((1:1:1:1:1),(1:1:1:1:1))$: Dehomogenize by setting $x_0 = 1$ and $z_0 = 1$. Now put $x_i = u_i + 1$ and $y_i = w_i + 1$ for $i = 1,2,3,4$. The defining equations for X are now

(17)
$$\begin{aligned} u_1 - u_2 - u_3 + u_4 - w_1 + w_2 + w_3 - w_4 + (-u_3w_1 + u_1w_2 + u_4w_3 - u_2w_4) &= 0, \\ u_2 - u_3 - u_4 - w_2 + w_3 + w_4 + (-u_4w_2 + u_2w_3) &= 0, \\ u_1 + u_3 - u_4 - w_1 - w_3 + w_4 + (-u_4w_1 + u_3w_4) &= 0, \\ -u_1 + u_2 + u_4 + w_1 - w_2 - w_4 + (u_2w_1 - u_1w_4) &= 0, \\ -u_1 - u_2 + u_3 + w_1 + w_2 - w_3 + (u_3w_2 - u_1w_3) &= 0. \end{aligned}$$

Adding them up, we get

$$\begin{aligned} &u_1w_2 - u_1w_3 - u_1w_4 + u_2w_1 + u_2w_3 - u_2w_4 \\ &- u_3w_1 + u_3w_2 + u_3w_4 - u_4w_1 - u_4w_2 + u_4w_3 = 0. \end{aligned} \tag{18}$$

Again we can eliminate the w_i; we then have

$$\begin{aligned} &u_1(u_2 - u_3 - u_4) + u_2(u_1 + u_3 - u_4) \\ &+ u_3(-u_1 + u_2 + u_4) + u_4(-u_1 - u_2 - u_3) + h.o.t. = 0. \end{aligned} \tag{19}$$

after passing to the completion of the coordinate ring.

The second-order piece of (19) can be written as

$$2u^T Bu = 0$$

with $B = \begin{pmatrix} & 1 & -1 & -1 \\ 1 & & 1 & -1 \\ -1 & 1 & & 1 \\ -1 & -1 & 1 & \end{pmatrix}$. The determinant of B is 5. Therefore, $((1:1:1:1:1),(1:1:1:1:1))$ is a node.

$((1 : -1 : -1 : 1 : 0), (1 : i : -i : -1 : 0))$: Again dehomogenize by setting $x_0 = z_0 = 1$. Put $x = u + (-1, -1, 1, 0), z = w + (i, -i, 1, 0)$. In the (u, w) coordinates, the defining equations for X are

$$
\begin{aligned}
-iu_1 - iu_3 - u_4 - w_1 - w_2 + w_4 + (u_1w_2 - u_2w_4 - u_3w_1 + u_4w_3) &= 0,\\
-u_2 - u_3 + iu_4 - w_3 + w_4 + (u_2w_3 - u_4w_2) &= 0,\\
u_1 - iu_4 - w_3 + w_4 + (u_3w_4 - u_4w_1) &= 0,\\
iu_2 + u_4 - w_1 - w_2 + w_4 + (-u_1w_4 + u_2w_1) &= 0,\\
u_1 - u_2 - iu_3 + w_1 + w_2 + w_3 + (-u_1w_3 + u_3w_2) &= 0.
\end{aligned} \tag{20}
$$

From these equations, we can obtain the equation

$$
\begin{aligned}
&- u_1w_2 + u_2w_4 + u_3w_1 - u_4w_3 + iu_2w_3\\
&- iu_4w_2 - iu_3w_4 + iu_4w_1 - u_1w_4 + u_2w_1 = 0.
\end{aligned} \tag{21}
$$

This time we want to eliminate u_1, u_2, u_3 and w_1. We have

$$
\begin{aligned}
u_1 &= iu_4 + w_3 - w_4 + h.o.t.,\\
u_2 &= iu_4 + iw_3 + \frac{2-i}{5}w_4 + h.o.t.,\\
u_3 &= (-1-i)w_3 + \frac{3+i}{5}w_4 + h.o.t.,\\
w_1 &= -w_2 - w_3 + \frac{6+2i}{5}w_4 + h.o.t.
\end{aligned} \tag{22}
$$

We then get the equation

$$
\begin{aligned}
&- 4iu_4w_2 - (2+2i)u_4w_3\\
&- \frac{4-12i}{5}u_4w_4 - (4-2i)w_3w_4 + \frac{14-2i}{5}w_4^2 + h.o.t. = 0.
\end{aligned} \tag{23}
$$

The second-order part of this equation can be written in matrix form:

$$
v^TBv = 0 \tag{24}
$$

with $v = (u_4, w_2, w_3, w_4)$ and

$$
B = \begin{pmatrix} & -2i & -1-i & \frac{-2+6i}{5}\\ -2i & & & \\ -1-i & & & -2+i\\ \frac{-2+6i}{5} & & -2+i & \frac{14-2i}{5} \end{pmatrix}.
$$

The determinant of this matrix is $-12+16i$. So this point is a node. By symmetry, all of our 60 singular points are nodes. □

Note that the surface S_1 contains the 25 nodes

$$
(\sigma^i\tau^3j, \sigma^i\tau^j)((0 : 1 : -1 : -1 : 1), (0 : 1 : -i : i : -1))
$$

and that S_2 contains the 25 nodes

$$
(\sigma^i\tau^3j, \sigma^i\tau^j)((0 : 1 : -1 : -1 : 1), (0 : 1 : i : -i : -1).
$$

Furthermore, the surfaces $\pi_1^{-1}T_i$ contain the 5 σ-nodes and the surfaces $\pi_1^{-1}U_i$ contain the 5 τ-nodes. Hence we can blow up these surfaces in some order to obtain a projective small resolution $\hat{X}$ of X over $\mathbb{Q}(\epsilon)$. However, we cannot obtain a model for any small resolution over $\mathbb{Q}$ because the determinant of the local quadratic equation at the node $((1:1:1:1:1),(1:1:1:1:1))$, which is not a square in $\mathbb{Q}$.

Assume from now on that the characteristic of our base field is different from 2 and 5. Since the singularities of X are 60 nodes, we can blow up the nodes to get a smooth threefold $\tilde{X}$. (Note that the equations defining the union of the 60 points are defined over $\mathbb{Z}$, so the blowup is defined over $\mathbb{Z}$.) Each exceptional divisor is isomorphic to $\mathbb{P}^1 \times \mathbb{P}^1$, and we will need to know when the rulings on each $\mathbb{P}^1 \times \mathbb{P}^1$ are defined over $\mathbb{F}_p$:

PROPOSITION 4.9. *If $p \equiv 1 \pmod{20}$, then all 60 nodes are defined over $\mathbb{F}_p$. If $p \equiv 11 \pmod{20}$, then only the 10 σ-nodes and τ-nodes are defined over $\mathbb{F}_p$. If $p \equiv 9, 13, 17 \pmod{20}$, then the σ-nodes $((1:1:1:1:1),(1:1:1:1:1))$ and 10 of the regular nodes are defined over $\mathbb{F}_p$. Otherwise only the σ-nodes and $((1:1:1:1:1),(1:1:1:1:1))$ are defined over $\mathbb{F}_p$.*

PROOF. Whether or not our nodes are defined over $\mathbb{F}_p$ depends on the values of p modulo 4 and 5. If $p \equiv 1 \pmod 4$ and $p \equiv 1 \pmod 5$, then both i and ϵ are in $\mathbb{F}_p$, and hence all 60 nodes are defined over $\mathbb{F}_p$. If $p \equiv 3 \pmod 4$ and $p \equiv 1 \pmod 5$, then the σ-nodes and τ-nodes are all defined over $\mathbb{F}_p$. If $p \equiv 1 \pmod 4$ and $p \equiv 2, 3, 4 \pmod 5$, then the σ-nodes, $((1:1:1:1:1),(1:1:1:1:1))$ and the nodes $(0:1:-1:-1:1),(0:1:\pm i:\mp i:-1))$ and their orbits under (σ,σ) are defined over $\mathbb{F}_p$. Finally, if $p \equiv 3 \pmod 4$ and $p \equiv 2,3,4 \pmod 5$ then only the σ-nodes and $((1:1:1:1:1),(1:1:1:1:1))$ are defined over $\mathbb{F}_p$. Using the Chinese Remainder Theorem gives us the result. □

PROPOSITION 4.10. *If $p \equiv 1 \pmod{20}$, all 60 nodes have blowups whose rulings are defined over $\mathbb{F}_p$. If $p \equiv 11 \pmod{20}$, then the σ-nodes and τ-nodes have rulings defined over $\mathbb{F}_p$. If $p \equiv 9 \pmod{20}$, then the σ-nodes, $((1:1:1:1:1),(1:1:1:1:1))$ and the 10 regular nodes have rulings defined over $\mathbb{F}_p$. If $p \equiv 13, 17 \pmod{20}$, then the σ-nodes and the 10 regular nodes have rulings defined over $\mathbb{F}_p$. If $p \equiv 19 \pmod{20}$, then the σ-nodes and $((1:1:1:1:1),(1:1:1:1:1))$ have rulings defined over $\mathbb{F}_p$. Otherwise only the σ-nodes have rulings defined over $\mathbb{F}_p$.*

PROOF. Given a node defined over $\mathbb{F}_p$, the rulings of the exceptional divisor are defined over $\mathbb{F}_p$ if and only if the determinant of the symmetric matrix defining the node is a square in $\mathbb{F}_p$. We calculated these determinants to be 1, 5 and $-12+16i$ (up to square factors). Now 1 is always a square, and $-12+16i$ is a square in $\mathbb{F}_p$ if i is defined over $\mathbb{F}_p$. So the σ-nodes and 50 regular nodes have rulings defined over $\mathbb{F}_p$ as long as the nodes themselves are defined over $\mathbb{F}_p$. Now 5 is a square in $\mathbb{F}_p$ if and only if $p \equiv \pm 1 \pmod 5$. Using the Chinese Remainder Theorem again gives us our result. □

4.3. Topology of $\tilde{X}$. We need to compute the topological invariants of $\tilde{X}$. First let X' be a smooth deformation of X; that is, let X' be a smooth threefold obtained by intersecting $\mathbb{P}^4 \times \mathbb{P}^4$ with five divisors of type $(1,1)$. Topologically, $\tilde{X}$ is obtained from X' by contracting 60 copies of S^3 and replacing them with 60 copies of $\mathbb{P}^1 \times \mathbb{P}^1$. Therefore, we have

$$\chi((\tilde{X})) = \chi(X') + 60\chi(\mathbb{P}^1 \times \mathbb{P}^1) = \chi(X') + 240.$$

For $0 \leq i \leq 5$, let X^i denote the intersection of $\mathbb{P}^4 \times \mathbb{P}^4$ with i generic divisors of type $(1,1)$. So we have $\mathbb{P}^4 \times \mathbb{P}^4 = X^0$ and $X' = X^5$. We have the exact sequences

$$
\begin{array}{lllll}
0 \longrightarrow & T(X^1) \longrightarrow & T(\mathbb{P}^4 \times \mathbb{P}^4)|_{X^1} \longrightarrow & N|_{X^1/\mathbb{P}^4\times\mathbb{P}^4} \longrightarrow 0 \\
0 \longrightarrow & T(X^i) \longrightarrow & T(X^{i-1})|_{X^i} \longrightarrow & N|_{X^i/X^{i-1}} \longrightarrow 0 \\
0 \longrightarrow & TX' \longrightarrow & T(X^4)|_{X'} \longrightarrow & N|_{X'/X^4} \longrightarrow 0
\end{array} \tag{25}
$$

Let X and Y denote the hyperplane classes on the two copies of $\mathbb{P}^4$. Then the Chern class of the third term in each sequence is equal to $(1 - X - Y)$. We thus have

$$c(X') = c(\mathbb{P}^4 \times \mathbb{P}^4) \cdot (1 - X - Y)^5. \tag{26}$$

Taking the terms of order three, we have

$$
\begin{aligned}
c_3(X') &= (10X^3 + 50X^2Y + 50XY^2 + 10Y^3) \\
&\quad - 5(X+Y)(10X^2 + 25XY + 10Y^2) \\
&\quad + 10(X+Y)^2(5X + 5Y) \\
&\quad - 10(X+Y)^3.
\end{aligned} \tag{27}
$$

The Euler characteristic of X' is

$$\chi(X') = \int_{\mathbb{P}^4\times\mathbb{P}^4} c_5(N|_{X'/\mathbb{P}^4\times\mathbb{P}^4}) \cdot c_3(X') \tag{28}$$

Now only the terms of degree 8 in the characteristic class contribute to the integral, and the orientation class of $\mathbb{P}^4 \times \mathbb{P}^4$ is X^4Y^4. The integral is thus equal to the coefficient of the X^4Y^4 term in $c_3(X')(1 + X + Y)^5$, which we compute to be -100. We thus have $\chi(X') = -100$ and $\chi(\tilde{X}) = 140$.

We can also compute most of the Hodge numbers of $\tilde{X}$. Since $\tilde{X}$ is obtained from X' by the surgery procedure explained above, we have $h^{0,0} = h^{3,3} = 1, h^{1,0} = h^{0,1} = h^{2,0} = h^{0,2} = 0$, and $h^{3,0} = h^{0,3} = 1$. The only unknown Hodge numbers are $h^{1,1} = h^{2,2}$ and $h^{1,2} = h^{2,1}$, and we know that $2h^{1,1} - h^{2,1} = 140$.

REMARK. In section 1, we remarked that Lange [**18**] had found a rank 2 vector bundle over $\mathbb{P}^1 \times \mathbb{P}^3$ whose zero sections yielded abelian surfaces. The Chern class of this bundle is $1 + (2H_1 + 4H_3) + (8H_1H_3 + 6H_3^2)$, where H_1 and H_3 denote the pullbacks to $\mathbb{P}^1 \times \mathbb{P}^3$ of the hyperplane class on $\mathbb{P}^1$ and $\mathbb{P}^3$ respectively. Since $c_1(\mathbb{P}^1 \times \mathbb{P}^3) = (2H_1 + 4H_3)$, a pencil of such abelian surfaces would yield a Calabi-Yau threefold. A generic pencil of such abelian surfaces would have a base locus consisting of $2 \cdot 8 \cdot 6 = 96$ nodes. One computes the Euler characteristic of the resulting threefold to be 176. Since $176 < 4 \cdot 96$, it is within the realm of possibility that resolving the singularities would yield a threefold with $h^{1,2}$ small or zero.

5. Proof that $\tilde{X}$ is modular

5.1. Modular forms. This brief review of modular forms is taken from [**3**]. Recall that the special linear group $SL_2(\mathbb{Z})$ acts on points z in the upper half plane $\mathcal{H}$: if $\gamma = \begin{pmatrix} a & b \\ c & d \end{pmatrix}$, $\gamma z = \frac{a\tau+b}{c\tau+d}$.

Let $\Gamma(N)$ denote the subgroup of matrices γ in $SL_2(\mathbb{Z})$ such that

$$\gamma \equiv \begin{pmatrix} 1 & 0 \\ 0 & 1 \end{pmatrix} (\text{mod } N).$$

Call a subgroup Γ of $SL_2(\mathbb{Z})$ a *congruence subgroup* if it contains $\Gamma(N)$ for some N. The *level* of N is the smallest N such that Γ contains $\Gamma(N)$. The most important congruence subgroups are

$$\Gamma_1(N) = \left\{ \begin{pmatrix} a & b \\ c & d \end{pmatrix} : \begin{pmatrix} a & b \\ c & d \end{pmatrix} \cong \begin{pmatrix} 1 & * \\ 0 & 1 \end{pmatrix} (\text{mod } N) \right\},$$

$$\Gamma_0(N) = \left\{ \begin{pmatrix} a & b \\ c & d \end{pmatrix} : \begin{pmatrix} a & b \\ c & d \end{pmatrix} \cong \begin{pmatrix} * & * \\ 0 & * \end{pmatrix} (\text{mod } N) \right\}.$$

DEFINITION 5.1. A *modular function* f of weight $2k$ and level N is a holomorphic function on the upper half plane $\mathcal{H}$ such that $f(\gamma(z)) = (cz + d)^{-2k} f(z)$ for all γ in some congruence subgroup Γ of level N.

DEFINITION 5.2. A *modular form* f is a modular function satisfying the following property: for all γ in $SL_2(\mathbb{Z})$, the function $(c\tau + d)^{-2k} f(\gamma\tau)$ has a Puiseux series expansion $\sum_{n \geq 0} a_n q^{\frac{n}{h}}$ in fractional powers of $q = e^{2\pi i\tau}$. We call this series the Fourier expansion of f at the cusp $\gamma^{-1}(i\infty)$, since the limit $q \to 0$ corresponds to the limit $z \to i\infty$. A *cusp form* is a modular form such that the Fourier expansion at each cusp has vanishing constant term.

We will focus on congruence subgroups Γ lying between $\Gamma_1(N)$ and $\Gamma_0(N)$. In this case, the matrix $\begin{pmatrix} 1 & 1 \\ 0 & 1 \end{pmatrix}$ is in Γ. If f is a modular form with respect to Γ, we must then have $f(\gamma z) = f(z)$ for all z in $\mathcal{H}$. Since f is periodic with period 1, its expansion at the cusp at infinity can be written as a power series in $q = e^{2\pi i\tau}$.

5.2. Galois theory. This quick review of Galois theory is taken from [**28**]. Recall the p-adic absolute value on $\mathbb{Q}$ given by $|x|_p = p^{-k}$ if x can be expressed as $p^k \frac{a}{b}$ with a and b both coprime to p. The completion of the field $\mathbb{Q}$ under this absolute value gives us the field of p-adic numbers $\mathbb{Q}_p$, and $| \quad |_p$ extends uniquely to $\mathbb{Q}_p$. Under this metric, $\mathbb{Q}_p$ is a locally compact, totally disconnected field. Let $G_{\mathbb{Q}_p}$ denote the absolute Galois group $\mathrm{Gal}(\overline{\mathbb{Q}}_p/\mathbb{Q}_p)$; it is identified with the group of continuous automorphisms of $\mathbb{Q}_p$. $G_{\mathbb{Q}_p}$ is a topological group; a basis at the identity is given by the collection of subgroups of finite index.

Given an embedding of $\overline{\mathbb{Q}}$ into $\overline{\mathbb{Q}}_p$, we obtain a closed embedding of $G_{\mathbb{Q}_p}$ into $G_{\mathbb{Q}}$; this embedding varies by conjugation as the embedding varies.

Let $\mathcal{O}_{\overline{\mathbb{Q}}_p}$ denote the ring of integers of $\overline{\mathbb{Q}}_p$; it is the ring of elements with absolute value less than or equal to 1. It is a local ring with maximal ideal $\mathfrak{m}_{\overline{\mathbb{Q}}_p}$, which is the ideal of elements with absolute value strictly less than 1. The residue field $\mathcal{O}_{\overline{\mathbb{Q}}_p}/\mathfrak{m}_{\overline{\mathbb{Q}}_p}$ is an algebraic closure of $\mathbb{F}_p$, which we denote by $\overline{\mathbb{F}}_p$. We thus obtain a continuous map $G_{\mathbb{Q}_p} \to G_{\mathbb{F}_p}$ which is surjective. Its kernel $I_{\mathbb{Q}_p}$ is called the inertia subgroup of $G_{\mathbb{Q}_p}$. The group $G_{\mathbb{F}_p}$ is procyclic, being the inverse limit $\lim_{\leftarrow} G(\mathbb{F}_{p^k}/\mathbb{F}_p)$. The Frobenius element $Frob_p$ defined by

$$\mathrm{Frob}_p(x) = x^p$$

generates a dense subgroup of $G_{\mathbb{F}_p}$.

We will be looking at representations $\rho : G_{\mathbb{Q}} \to GL_d(K)$ coming from algebraic geometry in the next section. Given an embedding $\overline{\mathbb{Q}} \hookrightarrow \overline{\mathbb{Q}}_p$, we have $I_{\mathbb{Q}_p} \subset G_{\mathbb{Q}_p} \subset G_{\mathbb{Q}}$. We say that ρ is unramified at a prime p if it is trivial on the inertia group I_p. In this case, $\rho(\mathrm{Frob}_p)$ is well-defined given the choice of embedding $\overline{\mathbb{Q}} \hookrightarrow \overline{\mathbb{Q}}_p$, and $\mathrm{tr}\,\rho(\mathrm{Frob}_p)$ is well-defined independent of the embedding.

5.3. Modularity of an algebraic variety. Let X be an projective algebraic variety defined over $\mathbb{Z}$. We can then consider the reduction of X modulo a prime p, i.e. the scheme $X \times_{\mathbb{Z}} \overline{\mathbb{F}}_p$.

Fixing once and for all an embedding $\overline{\mathbb{Q}} \subset \mathbb{C}$, the Galois group $G_{\mathbb{Q}}$ acts on X and thus induces an action on the cohomology groups

$$H^i(X(\mathbb{C}), \mathbb{Q}_l) \cong H^i_{et}(X \times_{\mathbb{Z}} \overline{\mathbb{Q}}, \mathbb{Q}_l).$$

where the left-hand side is the cohomology of X in the analytic topology.

Given a prime $p \neq l$ and an embedding $\overline{\mathbb{Q}} \subset \overline{\mathbb{Q}}_p$, we have the decomposition subgroup $G_{\mathbb{Q}_p} \subset G_{\mathbb{Q}}$. In addition, we have the surjective map $G_{\mathbb{Q}_p} \to G_{\mathbb{F}_p}$. It is well-known that if p is a prime of good reduction for X, then the representation $\rho : G_{\mathbb{Q}} \to GL(H^i(X \times_{\mathbb{Z}} \overline{\mathbb{Q}}, \mathbb{Q}_l))$ is unramified at p.

If p is a prime of good reduction for X, then we can assign a well-defined value to $\rho(\mathrm{Frob}_p)$ by choosing a lifting of Frob_p from $G_{\mathbb{F}_p}$ to $G_{\mathbb{Q}_p}$. By abuse of notation, such a lifting will also be denoted by Frob_p. Furthermore, by passing from $X \times_{\mathbb{Z}} \overline{\mathbb{Q}}_p$ to the reduction $X \times_{\mathbb{Z}} \overline{\mathbb{F}}_p$, we can consider the action of Frob_p on the cohomology of $X \times_{\mathbb{Z}} \overline{\mathbb{F}}_p$.

We say that an n-dimensional algebraic variety X is *modular* if for some subquotient V of $H^i(X, \mathbb{Q}_l)$, the numbers $\mathrm{tr}\,\mathrm{Frob}_p(V)$ are equal to the coefficients of a cusp form f for all but finitely many primes p.

En route to proving Fermat's Last Theorem, Wiles and Taylor [**29**] proved that for any elliptic curve E with semistable reduction at 3 and 5, the numbers $\mathrm{tr}\,\mathrm{Frob}_p(H^1(X))$ are the coefficients of a modular form; Breuil, Conrad, Diamond and Taylor [**2**] later proved the statement for all elliptic curves E.

5.4. Lefschetz, Weil and counting points. In étale cohomology, we have the Lefschetz theorem [**9**]:

THEOREM 5.1. *If f is an automorphism of the variety X, then the number of fixed points of f is given by the following formula:*

$$Fix(f, X) = \sum_{i=0}^{2n} (-1)^i \mathrm{tr}\, f^*(H^i(X)).$$

In the case X is defined over $\mathbb{F}_p$ and $f = \mathrm{Frob}_p$, $Fix(f, X)$ is simply the number of points of X over $\mathbb{F}_p$. We see that the number of points of X over $\mathbb{F}_p$ is related to the action of Frob_p on the étale cohomology of X.

PROPOSITION 5.2. *For p congruent to 1 modulo 20, the semisimplification of the action of the Frobenius map* Frob_p *on $H^2(\tilde{X} \times_{\mathbb{Z}} \overline{\mathbb{F}}_p, \mathbb{Q}_l)$ is multiplication by p.*

PROOF. We prove this statement in two steps.

STEP 1. The Frobenius map Frob_p acts on $H^2(X \times_{\mathbb{Z}} \overline{\mathbb{F}}_p, \mathbb{Q}_l)$ by multiplication by p.

Consider the embedding $i : \mathbb{P}^4 \times \mathbb{P}^4 \to \mathbb{P}^{24}$ that sends (x, z) to y where $y_{5i+j} = x_i z_j$. Recall that X is a section of $\mathbb{P}^4 \times \mathbb{P}^4$ by five divisors of type $(1, 1)$; correspondingly $i(X)$ is a section of $i(\mathbb{P}^4 \times \mathbb{P}^4)$ by five hyperplanes; as in section 3 let X^i denote successive sections of $\mathbb{P}^4 \times \mathbb{P}^4$ by these hyperplanes, with the exception that here X^i is X, instead of a deformation of X. Moreover, by Bertini's theorem these sections can be chosen such that $X^{i-1} - X^i$ is smooth for all i; for $p = 101$ a specific choice of hyperplane sections is given in a Macaulay2 program in the Appendix.

By the Lefschetz theorem in étale cohomology,

$$i^* : H^2(\mathbb{P}^4 \times \mathbb{P}^4, \mathbb{Q}_l) \to H^2(X \times_{\mathbb{Z}} \overline{\mathbb{F}}_p, \mathbb{Q}_l)$$

is an isomorphism that preserves the Frobenius action. (The proof of the Lefschetz theorem depends only on the fact that the $i(X^{i-1} - X^i)$ are smooth and affine; see [**9**], p. 106.)

Since all of $H^2(\mathbb{P}^4 \times \mathbb{P}^4, \mathbb{Q}_l)$ can be represented by divisors defined over $\mathbb{F}_{101}$, the Frobenius map acts by multiplication by p on $H^2(\mathbb{P}^4 \times \mathbb{P}^4, \mathbb{Q}_l)$. Thus the Frobenius map acts likewise on $H^2(X \times_{\mathbb{Z}} \overline{\mathbb{F}}_p, \mathbb{Q}_l)$. Note that Step 1 is valid for any prime p, not just those congruent to 1 modulo 20.

STEP 2. The semisimplification of the action of the Frobenius map Frob_p on $H^2(\tilde{X} \times_{\mathbb{Z}} \overline{\mathbb{F}}_p, \mathbb{Q}_l)$ is multiplication by p.

Recall that $\pi : \tilde{X} \to X$ is a blowup of 60 ordinary double points. From the Leray spectral sequence for π, we obtain an exact sequence

$$0 \longrightarrow H^2(X \times_{\mathbb{Z}} \overline{\mathbb{F}}_p, \mathbb{Q}_l) \longrightarrow H^2(\tilde{X} \times_{\mathbb{Z}} \overline{\mathbb{F}}_p, \mathbb{Q}_l) \longrightarrow \oplus_{i=1}^{60} H^2(Q_i, \mathbb{Q}_l), \tag{29}$$

where the Q_i are the exceptional divisors. For $p \equiv 1 \pmod{20}$, the rulings on all the Q_i are defined over $\mathbb{F}_p$. Hence the Frobenius map acts by multiplication by p on the Q_i. From the exact sequence, we see that the semisimplification of the Frobenius map acts by multiplication by p on $H^2(\tilde{X} \times_{\mathbb{Z}} \overline{\mathbb{F}}_p, \mathbb{Q}_l)$. $\square$

We will now concentrate our attention on the prime $p = 101$. Let us collect the information we have so far about the cohomology of $\tilde{X} \times_{\mathbb{Z}} \overline{\mathbb{F}}_{101}$:

(1) $h^0 = h^6 = 1$.
(2) $h^1 = h^5 = 0$.
(3) The semisimplification of the Frob_{101} action on H^2 is multiplication by 101. By Poincaré duality, the semisimplification of the Frob_{101} action on H^4 is multiplication by 101^2.
(4) $2h^2 - h^3 = 138$.
(5) $\#X(\mathbb{F}_{101}) = 1 + 101h^2 + 101^2h^2 + 101^3 - \mathrm{tr}\,\mathrm{Frob}_{101}(H^3(\tilde{X} \times_{\mathbb{Z}} \overline{\mathbb{F}}_{101}, \mathbb{Q}_l))$.
(6) For primes p of good reduction, $|\mathrm{tr}\,\mathrm{Frob}_p(H^3(\tilde{X} \times_{\mathbb{Z}} \overline{\mathbb{F}}_p, \mathbb{Q}_l))| \le h^3 p^{\frac{3}{2}} = (138 - 2h^2)p^{\frac{3}{2}}$ by the Weil conjectures.

The number of points in $\tilde{X}$ is easily computed by the following procedure:

(1) For a given prime p, count the number of points in G using a computer. (The code is in the Appendix.) We start by counting points in G instead of X because it is faster, the running time of the program being an exponential function of the number of variables.
(2) Add p times the number of points in E, since each point in E is replaced by a copy of $\mathbb{P}^1$ upon passage to X. Note that E has points defined over $\mathbb{F}_p$ only when $p \equiv 1 \pmod 4$.

(3) Add the number of points arising from the blowup of the nodes.

The σ-nodes are defined for all $\mathbb{F}_p$, and the rulings over the exceptional divisors exist over all $\mathbb{F}_p$. Hence each of these five nodes adds $p^2 + p$ points to the total.

The node $((1:1:1:1:1),(1:1:1:1:1))$ is defined over all $\mathbb{F}_p$, but the other τ-nodes in its orbit are defined only if $p \equiv 1 \pmod 5$. The rulings over the exceptional divisors exist over $\mathbb{F}_p$ only if $\sqrt{5}$ is defined in $\mathbb{F}_p$, i.e. if $p \equiv \pm 1 \pmod 5$. If the rulings are defined over $\mathbb{F}_p$, we add $p^2 + p$ points. Otherwise we add only p^2 points.

The node $((0:1:-1:-1:1),(0:1:i:-i:-1))$ is defined over $\mathbb{F}_p$ if $p \equiv 1 \pmod 4$, as are the other nodes in its σ-orbit. However, the other nodes in its H_5-orbit are defined over $\mathbb{F}_p$ only if i and ϵ are both defined. If the nodes are defined, then the rulings over the exceptional divisors are as well. Hence we add $p^2 + p$ times the number of these nodes.

For $p = 101$, we obtain $\#X(\mathbb{F}_{101}) = 1770940$. Thus

$$\operatorname{tr} \operatorname{Frob}_{101}(H^3(\tilde{X})) = 1 + 101h^2 + 101^2h^2 + 101^3 - 1770940, \tag{30}$$

so we must have

$$|1 + 101h^2 + 101^2h^2 + 101^3 - 1770940| \leq (138 - 2h^2)101^{\frac{3}{2}}. \tag{31}$$

Separating the absolute value inequality into two inequalities, we have

$$\begin{aligned} h^2(101 + 101^2 + 2 \cdot 101^{\frac{3}{2}}) &\leq 138 \cdot 101^{\frac{3}{2}} - 1 - 101^3 + 1770940, \\ h^2(101 + 101^2 - 2 \cdot 101^{\frac{3}{2}}) &\geq -138 \cdot 101^{\frac{3}{2}} - 1 - 101^3 + 1770940. \end{aligned} \tag{32}$$

Since h^2 must be an integer, the two inequalities force h^2 to be equal to 72. We then get $h^3 = 6$. (This trick for computing h^2 is due to Werner and van Geemen in [**30**].)

For $p \not\equiv 1 \pmod{20}$, we no longer know that the semisimplification of Frob_p acts by multiplication by p on H^2. However, we know that Frob_p acts on $H^2(X)$ by multiplication by p. We also know that $\oplus_{i=1}^{60} H^2(Q_i, \mathbb{Q}_l)$ is spanned by algebraic cycles, so the eigenvalues of Frob_p acting on this space are all p times roots of unity. Using the exact sequence (29) again, the eigenvalues of the semisimplification of Frob_p acting on $H^2(\tilde{X})$ are all p times roots of unity. By the Weil conjectures, the trace of Frob_p is a rational integer. Hence the trace of Frob_p must be p times an integer h.

Suppose the eigenvalues of Frob_p acting on $H^2(\tilde{X})$ are $p\zeta_i$, with the ζ_i being roots of unity. Choosing a basis of $H^2(\tilde{X})$ and a Poincaré dual basis of $H^4(\tilde{X})$, the action of Frob_p on $H^2(\tilde{X})$ can be represented as a matrix. This matrix is similar to a matrix of upper Jordan blocks having diagonal entries $p\zeta_i$. By Poincaré duality, the action of Frob_p acting on $H^4(\tilde{X})$ will be p^2 times the contragredient of the action on $H^2(\tilde{X})$. Thus the matrix of Frob_p acting on $H^4(\tilde{X})$ will be similar to a matrix of lower triangular blocks having diagonal entries $p^2\overline{\zeta_i}$ with the same multiplicities as in $H^2(\tilde{X})$. Hence the trace of Frob_p acting on H^4 is hp^2.

Furthermore, we have the additional piece of information that $h^2 = 72$. So we can use the Weil conjectures again. We now have

$$|1 + ph + p^2h + p^3 - \#(\tilde{X})| \leq 6p^{\frac{3}{2}}. \tag{33}$$

This is equivalent to the two inequalities

$$h(p+p^2) \leq \#(\tilde{X}) - 1 - p^3 - 6p^{\frac{3}{2}} \tag{34}$$

$$h(p+p^2) \geq 1 + p^3 + 6p^{\frac{3}{2}} - \#(\tilde{X}). \tag{35}$$

It turns out that for $p = 59, 67, 71$ and $p \geq 79$, the inequalities determine h exactly. Our calculations are listed in Appendix B.

CONJECTURE 5.1. For primes p of good reduction, the trace of Frob_p acting on $H^2(\tilde{X})$ is $12p$, $20p$, $24p$ or $72p$ depending on whether i and ϵ are defined over $\mathbb{F}_p$.

There is a unique cusp form f of level 5 and weight 4, whose Fourier coefficients a_p can be found in William Stein's Modular Forms Database [**26**]. The first few coefficients of the q-expansion of f are as follows:

$$f = q - 4q^2 + 2q^3 + 16q^4 - 5q^5 - 8q^6 + 6q^7 - 64q^8 + \ldots \tag{36}$$

Comparing the values of $\mathrm{tr}\,\mathrm{Frob}_p(H^3)$, we notice that for primes $p \equiv 3 \pmod 4$, $a_p = \mathrm{tr}\,\mathrm{Frob}_p(H^3)$, whereas for primes $p \equiv 1 \pmod 4$, $\mathrm{tr}\,\mathrm{Frob}_p(H^3) - a_p = p(2p + 2 - \#(E))$.

5.5. Proof of the main theorem. We can finally prove the main theorem of this paper:

THEOREM 5.3. *The third cohomology group H^3 of $\tilde{X}$ is modular in the following sense: as a Galois representation, its semisimplification is a direct sum of a modular rank 2 motive V and a 4-dimensional piece $W \cong H^1(\tilde{S}, \mathbb{Q}_l)(-1)$.*

PROOF. First we note that $\tilde{S}$ and $\tilde{X}$ are defined over $\mathbb{Z}$. Since we will only consider the coefficient group $\mathbb{Q}_l$, we can use the proper-smooth base change theorem (see for example [**21**]) to pass from $\tilde{X}$ to $\tilde{X} \times_{\mathbb{Z}} \overline{\mathbb{F}}_p$ (for $p \neq 2, 5, l$) and to $\tilde{X} \times_{\mathbb{Z}} \mathbb{C}$. In addition, we can pass from étale cohomology on $\tilde{X} \times_{\mathbb{Z}} \mathbb{C}$ to analytic cohomology by the comparison theorem.

We present the proof in several steps.

STEP 1. $H^1(\tilde{S} \times_{\mathbb{Z}} \overline{\mathbb{F}}_p, \mathbb{Q}_l)(-1) \cong \mathrm{Ind}_{G_{\mathbb{Q}(i)}}^{G_{\mathbb{Q}}} H^1((E_1 \times_{\mathbb{Z}} \overline{\mathbb{Q}}_p) \times_{\overline{\mathbb{Q}}_p} \overline{\mathbb{F}}_p, \mathbb{Q}_l)(-1)$.

For now, we use the analytic topology. Recall that the map $\pi : \tilde{S} \longrightarrow S$ is just the blowup at 50 points. A standard Mayer-Vietoris argument (see for example [**10**], p. 473) shows that $H^1(\tilde{S} \times_{\mathbb{Z}} \overline{\mathbb{F}}_p, \mathbb{Q}_l) \cong H^1(S \times_{\mathbb{Z}} \overline{\mathbb{F}}_p, \mathbb{Q}_l)$.

The map $\pi : S \longrightarrow E$ is the projectivization of the rank 2 bundle $\ker L \longrightarrow E$. Using the analytic topology for now, the Leray-Hirsch theorem tells us that $H^*(S, \mathbb{Q}_l)$ is a truncated polynomial ring over $H^*(E, \mathbb{Q}_l)$ generated by the single element $c_1(\ker L)$, which has dimension 2. Hence $\pi^* : H^1(E, \mathbb{Q}_l) \longrightarrow H^1(S, \mathbb{Q}_l)$ is an isomorphism. This statement also holds in étale cohomology.

Finally, recall that $E = E_1 \cup E_2$, where E_1 and E_2 are complex conjugates of each other. Thus $H^1(E \times_{\mathbb{Z}} \overline{\mathbb{F}}_p, \mathbb{Q}_l) \cong \mathrm{Ind}_{G_{\mathbb{Q}(i)}}^{G_{\mathbb{Q}}} H^1((E_1 \times_{\mathbb{Z}} \overline{\mathbb{Q}}_p) \times_{\overline{\mathbb{Q}}_p} \overline{\mathbb{F}}_p, \mathbb{Q}_l)$. Putting everything together completes Step 1.

STEP 2. $H^1(\tilde{S} \times_{\mathbb{Z}} \overline{\mathbb{F}}_p, \mathbb{Q}_l)(-1)$ is a subrepresentation of $H^3(\tilde{X} \times_{\mathbb{Z}} \overline{\mathbb{F}}_p, \mathbb{Q}_l)$.

Note that $H^1(\tilde{S} \times_{\mathbb{Z}} \overline{\mathbb{F}}_p, \mathbb{Q}_l)(-1)$ is isomorphic to $H^3_{\tilde{S}}(\tilde{X} \times_{\mathbb{Z}} \overline{\mathbb{F}}_p, \mathbb{Q}_l)$ (see [**21**], p. 98). So it is sufficient to show that inclusion induces an injection

$$j^* : H^3_{\tilde{S}}(\tilde{X} \times_{\mathbb{Z}} \overline{\mathbb{F}}_p, \mathbb{Q}_l) \longrightarrow H^3(\tilde{X} \times_{\mathbb{Z}} \overline{\mathbb{F}}_p, \mathbb{Q}_l).$$

By base change and the comparison theorem, it suffices to prove the same statement in the complex analytic topology.

In the analytic topology, we have the Lefschetz duality diagram (see [**23**], p. 429)

$$\begin{array}{ccc} H^3_{\tilde{S}}(\tilde{X} \times_{\mathbb{Z}} \mathbb{C}, \mathbb{Q}_l) & \xrightarrow{j^*} & H^3(\tilde{X} \times_{\mathbb{Z}} \mathbb{C}, \mathbb{Q}_l) \\ \downarrow \cong & & \downarrow \cong \\ H_3(\tilde{S} \times_{\mathbb{Z}} \mathbb{C}, \mathbb{Q}_l) & \xrightarrow{j_*} & H_3(\tilde{X} \times_{\mathbb{Z}} \mathbb{C}, \mathbb{Q}_l) \end{array}$$

where the vertical arrows are isomorphisms. Hence it is sufficient to show that inclusion induces an injective map $j_* : H_3(\tilde{S} \times_{\mathbb{Z}} \mathbb{C}, \mathbb{Q}_l) \longrightarrow H_3(\tilde{X} \times_{\mathbb{Z}} \mathbb{C}, \mathbb{Q}_l)$. We do this by computing intersection classes of cycles in $H_3(\tilde{S} \times_{\mathbb{Z}} \mathbb{C}, \mathbb{Q}_l)$.

The 3-cycles in $\tilde{S}$ are of the form $\alpha_i \times \mathbb{P}^1$ and $\beta_j \times \mathbb{P}^1$, where α_i and β_j generate $H_1(E_i)$. Note that the α_i and β_j can be chosen to miss the exceptional divisors.

We have the fundamental result

$$j_*(\alpha) \cap j_*(\beta) = j_*(PD[\tilde{S}]|_{\tilde{S}} \cap \alpha \cap \beta)$$

for any homology cycles α and β.

Note that

$$[\tilde{S}]|_{\tilde{S}} = (K_{\tilde{X}} - K_{\tilde{X}} + \tilde{S})|\tilde{S} = K_{\tilde{S}} - K_{\tilde{X}}|\tilde{S}.$$

But X has trivial canonical bundle. Hence $K_{\tilde{X}}$ is supported on its exceptional fibers. Therefore the restriction of $K_{\tilde{X}}$ to $\tilde{S}$ is supported on the exceptional fibers of $\tilde{S}$. Let γ_i and δ_i be 1-cycles generating the first homology of E_i, and put $\alpha_i = \gamma_i \times C_i$ and $\beta_i = \delta_i \times C_i$, where C_i is a ruling of S_i.

Since the cycles α_i and β_j can be chosen to miss the exceptional fibers, we have

$$PD[\tilde{S}]|_{\tilde{S}} \cap \alpha_i \cap \beta_j = PD[K_{\tilde{S}}] \cap \alpha_i \cap \beta_j.$$

The canonical bundle of $\tilde{S}$ is well-known; it is simply $-2(D) + \Sigma E_i$, where D is a horizontal section and the E_i are the exceptional divisors. We also have

$$\gamma_i \cap \gamma_j = \delta_i \cap \delta j = 0,$$

$$\gamma_1 \cap \delta_2 = \gamma_2 \cap \delta_1 = 0,$$

$$\gamma_i \cap \delta_i = C_i,$$

where C_i is a line belonging to the ruling of S_i.

Therefore we have

$$j_*(\alpha_i) \cap j_*(\alpha_j) = j_*(\beta_i) \cap j_*(\beta_j) = 0$$

$$j_*(\alpha_i) \cap j_*(\beta_j) = -2\delta_{ij}.$$

Putting these results together, we see that the intersection matrix of the 3-cycles is

$$\begin{pmatrix} & -2 & & \\ 2 & & & \\ & & & -2 \\ & & 2 & \end{pmatrix},$$

which is nonsingular. Hence $H_3(\tilde{S})$ injects into $H_3(\tilde{X})$.

Upon passing to the semisimplification, we now see that as Galois representations,

$$H^3(\tilde{X} \times_{\mathbb{Z}} \overline{\mathbb{F}}_p, \mathbb{Q}_l) = V \oplus \operatorname{Ind}_{G_{\mathbb{Q}(i)}}^{G_{\mathbb{Q}}} H^1(E \times_{\mathbb{Z}} \overline{\mathbb{F}}_p, \mathbb{Q}_l)(-1),$$

with V some undetermined 2-dimensional piece. This observation was first made by Taylor [**27**].

STEP 3. Away from the primes of bad reduction, the traces of V coincide with the coefficients of the unique modular form f of level 5 and weight 4.

For Step 3, we invoke a result of Faltings-Serre-Livné [**19**] which says essentially that it is enough to check the equality at a suitably chosen finite set of primes:

THEOREM 5.4. *(Faltings-Serre-Livné) Suppose that ρ_1 and ρ_2 are two 2-adic 2-dimensional Galois representations unramified outside a set of primes S. Let K_S be the smallest field containing all quadratic extensions of $\mathbb{Q}$ ramified at primes in S, and let T be a set of primes disjoint from S. Then if*

$\operatorname{tr}\rho_1 \equiv \operatorname{tr}\rho_2 \equiv 0$ *and* $\det\rho_1 \equiv \det\rho_2 \pmod 2$*;*

$\{\mathrm{Frob}_p|_{K_S} : p \in T\}$ *is equal to the set* $\mathrm{Gal}(K_S/K) - Id$*;*

for all $p \in T$, $\operatorname{tr}\rho_1\mathrm{Frob}_p = \operatorname{tr}\rho_2\mathrm{Frob}_p$ *and* $\det\rho_1\mathrm{Frob}_p = \det\rho_2\mathrm{Frob}_p$*;*

then ρ_1 and ρ_2 have isomorphic semisimplifications.

We will apply this result in the case $\rho_1 = V$ and ρ_2 is the modular Galois representation coming from the cusp form f of level 5 and weight 4. Thus we must do the following:

Check that $\operatorname{tr}\mathrm{Frob}_p(H^3(\tilde{X} \times_{\mathbb{Z}} \overline{\mathbb{F}}_p, \mathbb{Q}_l))$ is even for primes p of good reduction.

Check that the coefficients a_p of the modular form f are even for $p \neq 5$.

Observe that the determinants of both representations are given by χ^3, where χ is the cyclotomic character.

Find a suitable set of primes T_s.

Compute $\operatorname{tr}\mathrm{Frob}_p H^3(\tilde{X} \times_{\mathbb{Z}} \overline{\mathbb{F}}_p, \mathbb{Q}_l)$ for all p in T_s and check that these are equal to the corresponding coefficients in the modular form f.

LEMMA 5.5. $\operatorname{tr}\mathrm{Frob}_p H^3(\tilde{X} \times_{\mathbb{Z}} \overline{\mathbb{F}}_p, \mathbb{Q}_l)$ *is even for $p \neq 2, 5$.*

PROOF. Recall that

$$\operatorname{tr}\mathrm{Frob}_p(H^3(\tilde{X})) = p^3 + p(p+1)h + 1 - \#\tilde{X}(\mathbb{F}_p). \tag{37}$$

Thus we need only check that $\#\tilde{X}(\mathbb{F}_p)$ is even.

Recall the automorphism of order 4

$$\mu = \begin{pmatrix} 1 & & & & \\ & & 1 & & \\ & & & & 1 \\ & 1 & & & \\ & & & 1 & \end{pmatrix}$$

acting on F, and notice that it lifts to an automorphism (μ, μ) of X. Each point (x, z) of X lies in an orbit of size 1, 2 or 4. The points in orbits of size 2 or 4 obviously contribute an even number of points to the total. If any of these points are nodes, then so are the other points in their orbits. Blowing up each node adds either p^2 or $p^2 + 2p$ points to the total. Since there must be an even number of such nodes, blowing up adds another even number of points to the total. So we are left with counting the number of points in $\tilde{X}$ coming from fixed points of X.

The fixed points of μ in F are of the form $(1 : x : x : x : x)$, $(0 : 1 : 1 : 1 : 1)$, $(0 : 1 : -1 : -1 : 1)$, and $(0 : 1 : \pm i : \mp i : -1)$; this is a total of $p + 3$ points, which is even. Of these points, the only ones that are singular points in F are $(1 : 1 : 1 : 1 : 1)$, $(1 : 0 : 0 : 0 : 0)$ and $(1 : \pm\frac{1}{2i} : \pm\frac{1}{2i} : \pm\frac{1}{2i} : \pm\frac{1}{2i})$. The other $p - 1$ points are smooth points in F and thus lift uniquely to points in X fixed by (μ, μ).

The points $(1 : \pm\frac{1}{2i} : \pm\frac{1}{2i} : \frac{1}{2i} : \frac{1}{2i})$ (if they exist over $\mathbb{F}_p$) are singular; their inverse images under π_1 are lines. In these lines, the only fixed points of (μ, μ) are

$((1 : \frac{1}{2i} : \frac{1}{2i} : \frac{1}{2i} : \frac{1}{2i}), (1 : \frac{1}{2i} : \frac{1}{2i} : \frac{1}{2i} : \frac{1}{2i}))$,
$((1 : \frac{1}{2i} : \frac{1}{2i} : \frac{1}{2i} : \frac{1}{2i}), (0 : 1 : -i : i : -1))$,
$((1 : -\frac{1}{2i} : -\frac{1}{2i} : -\frac{1}{2i} : -\frac{1}{2i}), (1 : -\frac{1}{2i} : -\frac{1}{2i} : -\frac{1}{2i} : -\frac{1}{2i}))$,
$((1 : -\frac{1}{2i} : -\frac{1}{2i} : -\frac{1}{2i} : -\frac{1}{2i}), (0 : 1 : i : -i : -1))$,

and none of these points are nodes. So we have added another even number of points to the total.

Finally, we look at the contribution coming from the points $(1 : 1 : 1 : 1 : 1)$ and $(1 : 0 : 0 : 0 : 0)$ in F. The fixed points of (μ, μ) in X lying over these points are $((1 : 1 : 1 : 1 : 1), (1 : 1 : 1 : 1 : 1))$ and $((1 : 0 : 0 : 0 : 0), (1 : 0 : 0 : 0 : 0))$. These points are nodes; blowing them up adds either p^2 or $p^2 + 2p$ points for each node. So again we end up adding an even number of points to the total. Hence the number of points in $\tilde{X}$ is even. □

LEMMA 5.6. *The coefficients of f are all even.*

PROOF. f is a modular form of level 5. The corresponding Galois representation $\rho : \mathrm{Gal}(\overline{\mathbb{Q}}/\mathbb{Q}) \to GL_2(\mathbb{Q}_2)$ is unramified outside 5. If the trace of ρ were not congruent to $0 \pmod 2$, then the image of the reduction $\overline{\rho} : \mathrm{Gal}(\overline{\mathbb{Q}}/\mathbb{Q}) \to GL_2(\mathbb{Z}/2)$ would contain one of the elements $\begin{pmatrix} 1 & 1 \\ 1 & \end{pmatrix}$ or $\begin{pmatrix} & 1 \\ 1 & 1 \end{pmatrix}$. These matrices are of order 3. Therefore, if L is the extension of $\mathbb{Q}$ cut out by $\ker \overline{\rho}$, L is a C_3 or S_3 extension of $\mathbb{Q}$ unramified outside 5. No such field exists. □

LEMMA 5.7. *Let T be the set $\{67, 71, 101, 103, 113, 131, 157\}$. Then as p runs over this set,* Frob_p *runs over all the non-identity members of* $\mathrm{Gal}(\mathbb{Q}_S/\mathbb{Q})$.

PROOF. The field $\mathbb{Q}_S$ is the compositum of all quadratic extensions of $\mathbb{Q}$ unramified outside $\{2, 5\}$, which is $\mathbb{Q}(i, \sqrt{2}, \sqrt{5})$. Now $\mathrm{Gal}(\mathbb{Q}_S/\mathbb{Q})$ is isomorphic to $(\mathbb{Z}/2)^3$, and the coordinates of Frob_p are just the values of $(\frac{-1}{p})$, $(\frac{2}{p})$ and $(\frac{5}{p})$. These are controlled by the behavior of p modulo 40, as illustrated in Table 5.1. The set of primes T gives us representatives of every element of $\mathrm{Gal}(\mathbb{Q}_S/\mathbb{Q})$ other than the identity. □

Finally, we show that for primes p in the set T, $\operatorname{tr} V = \operatorname{tr} \rho$.

The trace of V is equal to the trace of $\mathrm{Frob}_p(H^3(\tilde{X} \times_{\mathbb{Z}} \overline{\mathbb{F}}_p, \mathbb{Q}_l))$ minus the trace of $\mathrm{Frob}_p(H^1(\tilde{S} \times_{\mathbb{Z}} \overline{\mathbb{F}}_p, \mathbb{Q}_l)(-1))$.

We obtain $\operatorname{tr} \mathrm{Frob}_p(H^3(\tilde{X} \times_{\mathbb{Z}} \overline{\mathbb{F}}_p, \mathbb{Q}_l))$ by counting points on $\tilde{X}$ over $\mathbb{F}_p$ and by knowing what h is by using the Weil conjectures, as discussed above. For p

$p \pmod{40}$	$\left(\frac{-1}{p}\right)$	$\left(\frac{2}{p}\right)$	$\left(\frac{5}{p}\right)$
3, 27	-1	-1	-1
7, 23	-1	1	-1
11, 19	-1	-1	1
13, 37	1	-1	-1
17, 33	1	1	-1
17, 33	1	1	-1
21, 29	1	-1	1
31, 39	-1	1	1
1, 9	1	1	1

Table 5.1: Images of Frobenius elements in $\mathrm{Gal}(\mathbb{Q}_s/\mathbb{Q})$.

in the set S, the bounds provided by the Weil conjectures determine h and thus $\operatorname{tr}\mathrm{Frob}_p(H^3(\tilde{X} \times_{\mathbb{Z}} \overline{\mathbb{F}}_p, \mathbb{Q}_l))$. Calculations are in the Appendix.

Recall that $H^1(\tilde{S} \times_{\mathbb{Z}} \overline{\mathbb{F}}_p, \mathbb{Q}_l)(-1)$ is isomorphic to $\mathrm{Ind}_{G_{\mathbb{Q}}}^{G_{\mathbb{Q}(i)}} H^1(E_1 \times_{\mathbb{Z}} \overline{\mathbb{F}}_p, \mathbb{Q}_l)(-1)$. For p congruent to $1 \pmod 4$, i is in $\mathbb{F}_p$. Hence Frob_p acts on $H^1(E_1 \times_{\mathbb{Z}} \overline{\mathbb{F}}_p, \mathbb{Q}_l)$ and on $H^1(E_2 \times_{\mathbb{Z}} \overline{\mathbb{F}}_p, \mathbb{Q}_l)$, so the the trace of Frob_p acting on $\mathrm{Ind}_{G_{\mathbb{Q}(i)}}^{G_{\mathbb{Q}}} H^1(\tilde{S} \times_{\mathbb{Z}} \overline{\mathbb{F}}_p, \mathbb{Q}_l)(-1)$ is equal to p times the sum of the traces on $H^1(E_1 \times_{\mathbb{Z}} \overline{\mathbb{F}}_p, \mathbb{Q}_l)$ and $H^1(E_2 \times_{\mathbb{Z}} \overline{\mathbb{F}}_p, \mathbb{Q}_l)$. In each case the trace of Frob_p acting on $H^1(E_i \times_{\mathbb{Z}} \overline{\mathbb{F}}_p, \mathbb{Q}_l)$ is just $p + 1 - \#E_i(\mathbb{F}_p)$, so the trace of Frob_p acting on $H^1(\tilde{S} \times_{\mathbb{Z}} \overline{\mathbb{F}}_p, \mathbb{Q}_l)(-1)$ is $p(2p + 2 - \#E(\mathbb{F}_p))$.

For p equal to $3 \pmod 4$, i is not in $\mathbb{F}_p$. Hence Frob_p interchanges the two representations $H^1(E_1 \times_{\mathbb{Z}} \overline{\mathbb{F}}_p, \mathbb{Q}_l)$ and $H^1(E_2 \times_{\mathbb{Z}} \overline{\mathbb{F}}_p, \mathbb{Q}_l)$, so the trace of

$$\mathrm{Ind}_{G_{\mathbb{Q}(i)}}^{G_{\mathbb{Q}}} H^1(\tilde{S} \times_{\mathbb{Z}} \overline{\mathbb{F}}_p, \mathbb{Q}_l)(-1)$$

is equal to zero.

Taking the differences of these two traces gives us the traces of V, which are seen to be equal to the coefficients of the modular form f. □

Appendix A. C++ and Macaulay2 code

In this appendix we put the C++ and Macaulay2 programs we used for our computations.

```
#include<iostream.h> // Count points on G over Fp
int psols;

long long G(long long z0, long long z1, long long z2, long long z3,
long long z4)
// the equation for the determinantal quintic
{
  return z0 * z0 * z0 * z1 * z4 + z1 * z1 * z1 * z2 * z0
  + z2 * z2 * z2 * z3 * z1 + z3 * z3 * z3 * z4 * z2 + z4 * z4 * z4 * z0 * z3
  + z0 * z2 * z2 * z3 * z3 + z1 * z3 * z3 * z4 * z4 + z2 * z4 * z4 * z0 * z0
  + z3 * z0 * z0 * z1 * z1 + z4 * z1 * z1 * z2 * z2 - (z0 * z1 * z1 * z4 * z4
  + z1 * z2 * z2 * z0 * z0 + z2 * z3 * z3 * z1 * z1 + z3 * z4 * z4 * z2 * z2
  + z4 * z0 * z0 * z3 * z3 + z0 * z0 * z0 * z2 * z3 + z1 * z1 * z1 * z3 * z4
  + z2 * z2 * z2 * z4 * z0 + z3 * z3 * z3 * z0 * z1 + z4 * z4 * z4 * z1 * z2);
}

void counter(long long z0, long long z1, long long z2, long long z3, long long z4, int p) // count points on G
{
  if (G(z0, z1, z2, z3, z4) % p == 0)
    ++psols;
}

int main()
{
  long long j0, j1, j2, j3, j4;
  int p;
  cout << ``Enter a prime p: ``;
  cin >> p;
  j0 = 1; // Affine open set with j0 != 0
  for (j1 = 0; j1 < p; ++j1) {
    for (j2 = 0; j2 < p; ++j2) {
      for (j3 = 0; j3 < p; ++j3) {
        for (j4 = 0; j4 < p; ++j4) {
          counter(j0, j1, j2, j3, j4, p);
        }
      }
    }
  }
  j0 = 0;
  j1 = 1; // Affine A^3 with j0 = 0, j1 != 0
  for (j2 = 0; j2 < p; ++j2) {
    for (j3 = 0; j3 < p; ++j3) {
      for (j4 = 0; j4 < p; ++j4) {
        counter(j0, j1, j2, j3, j4, p);
      }
    }
  }
  j1 = 0;
  j2 = 1; // Affine A^2 with j0 = j1 = 0, j2 != 0
  for (j3 = 0; j3 < p; ++j3) {
    for (j4 = 0; j4 < p; ++j4) {
     counter(j0, j1, j2, j3, j4, p);
    }
  }
  j2 = 0;
  j3 = 1; // Affine A^1 with j0 = j1 = j2 = 0, j3 != 0
  for (j4 = 0; j4 < p; ++j4) {
    counter(j0, j1, j2, j3, j4, p);
  }
  counter(j0, j1, j2, 0, 1, p);
  cout << ``\n''
    ``The number of solutions mod p is `` << psols << ``\n'';
}
```

Program A.1: Counting points on G.

```
#include <iostream.h> // Count points on E modulo a prime p

int psols;

long long E0(long long z0, long long z1, long long z2, long long z3, long long z4, long long y)
{
  return (-y * z0 * z0 - z1 * z4 + y * y * z2 * z3);
}

long long E1(long long z0, long long z1, long long z2, long long z3, long long z4, long long y)
{
  return (-y * z1 * z1 - z2 * z0 + y * y * z3 * z4);
}

long long E2(long long z0, long long z1, long long z2, long long z3, long long z4, long long y)
{
  return (-y * z2 * z2 - z3 * z1 + y * y * z4 * z0);
}

long long E3(long long z0, long long z1, long long z2, long long z3, long long z4, long long y)
{
  return (-y * z3 * z3 - z4 * z2 + y * y * z0 * z1);

long long E4(long long z0, long long z1, long long z2, long long z3, long long z4, long long y)
{
  return (-y * z4 * z4 - z0 * z3 + y * y * z1 * z2);
}

void counter(long long z0, long long z1, long long z2, long long z3,
long long z4, long long y, int p)
// count points
{
  if (E0(z0, z1, z2, z3, z4, y) % p == 0) {
    if (E1(z0, z1, z2, z3, z4, y) % p == 0) {
      if (E2(z0, z1, z2, z3, z4, y) % p == 0) {
        if (E3(z0, z1, z2, z3, z4, y) % p == 0) {
          if (E4(z0, z1, z2, z3, z4, y) % p == 0) {
            ++psols;
          }
        }
      }
    }
  }
}

int main()
{
  long long j0, j1, j2, j3, j4, y;
  int p;
  cout << ``Enter a parameter y: ``; // usually y is a square root of -1
  cin >> y;
  cout << ``Enter a prime p: ``;
  cin >> p;
  j0 = 1; // Affine open set with j0 != 0
  for (j1 = 0; j1 < p; ++j1) {
    for (j2 = 0; j2 < p; ++j2) {
      for (j3 = 0; j3 < p; ++j3) {
        for (j4 = 0; j4 < p; ++j4) {
          counter(j0, j1, j2, j3, j4, y, p);
        }
      }
    }
  }
  j0 = 0;
  j1 = 1; // Affine A^3 with j0 = 0, j1 != 0
  for (j2 = 0; j2 < p; ++j2) {
    for (j3 = 0; j3 < p; ++j3) {
      for (j4 = 0; j4 < p; ++j4) {
        counter(j0, j1, j2, j3, j4, y, p);
      }
    }
  }
  j1 = 0;
  j2 = 1; // Affine A^2 with j0 = j1 = 0, j2 != 0
  for (j3 = 0; j3 < p; ++j3) {
    for (j4 = 0; j4 < p; ++j4) {
      counter(j0, j1, j2, j3, j4, y, p);
    }
  }
  j2 = 0;
  j3 = 1; // Affine A^1 with j0 = j1 = j2 = 0, j3 != 0
  for (j4 = 0; j4 < p; ++j4) {
    counter(j0, j1, j2, j3, j4, y, p);
  }
  counter(j0, j1, j2, 0, 1, y, p);
  cout << ``\n''
    ``The number of solutions to Ey mod p is `` << psols << ``\n'';
```

Program A.2: Counting points on E_1 and E_2.

```
F = ZZ;

R = F[x0,x1,x2,x3,x4,z0,z1,z2,z3,z4];

f0 = 0 * 0  -x3*z1 + x1*z2 + x4*z3 - x2*z4;
f1 = -x3*z0 +0 * 0 - x4*z2 + x2*z3 + x0*z4;
f2 =  x1*z0 -x4*z1 + 0 * 0 - x0*z3 + x3*z4;
f3 =  x4*z0 +x2*z1 - x0*z2 + 0 * 0 - x1*z4;
f4 = -x2*z0 +x0*z1 + x3*z2 - x1*z3 + 0 * 0;

L = matrix{{0,z2,-z4,-z1,z3},{z4,0,z3,-z0,-z2},{-z3,z0,0,z4,-z1},{-z2,-z4,z1,0,z0},{z1,-z3,-z0,z2,0}};
G = minors(4,L);

M = matrix{{0,-x3,x1,x4,-x2},{-x3,0,-x4,x2,x0},{x1,-x4,0,-x0,x3},{x4,x2,-x0,0,-x1},{-x2,x0,x3,-x1,0}};
F = minors(4,M);

S0 = diff f0;
S1 = S0 || diff f1;
S2 = S1 || diff f2;
S3 = S2 || diff f3;
S = S3 || diff f4;

X = ideal(f0,f1,f2,f3,f4);
sing = minors(5,S);
slocus = X + sing;
slocus2 = slocus + G + F;
k = 100*x0*x1*x2*x3*x4*z0*z1*z2*z3*z4;

--now check that k is contained in slocus2
```

Program A.3: Proving that nodes on X have $x_0 = 0$ or $z_0 = 0$.

```
-- Let G be the quintic threefold as in our thesis.  E1 U E2 is a union of two elliptic curves.
-- We show that over any field of characteristic not equal to 2,
-- the set of points in E1 U E2 is precisely the set of rank 3 points of G.

-- I is the ideal of the singular locus of G.

F = ZZ;
R = F[z0,z1,z2,z3,z4];
L = matrix{{0,-z4,z3,z2,-z1},{-z2,0,-z0,z4,z3},{z4,-z3,0,-z1,z0},{z1,z0,-z4,0,-z2},{-z3,z2,z1,-z0,0}};
I = minors(4,L);

-- the ideal of E1 is generated by the elements I ri + si, i = 0,1,2,3,4.
-- the ideal of E2 is generated by the elements -I ri + si, i = 0,1,2,3,4.

r0 = z0^2;
r1 = z1^2;
r2 = z2^2;
r3 = z3^2;
r4 = z4^2;
s0 = (z1*z4 + z2*z3);
s1 = (z2*z0 + z3*z4);
s2 = (z3*z1 + z4*z0);
s3 = (z4*z2 + z0*z1);
s4 = (z0*z3 + z1*z2);

-- if the characteristic of our base field is not 2, then the ideal of E1 * E2 is generated by the elements
-- pi = (I ri + si)*(-I ri + si) = (ri^2 + si^2)
-- and the elements (I ri + si)*(-I rj + sj) = ri*rj + si*sj + I(ri*sj - si*rj).
-- But since the characteristic is not 2, the generators ri*rj + si*sj + I(ri*sj - si*rj)
-- and ri*rj + si*sj - I(ri*sj - si*rj)
-- can be replaced by ri*rj + si*sj and ri*sj - si*rj.

p0 = r0^2 + s0^2;
p1 = r1^2 + s1^2;
p2 = r2^2 + s2^2;
p3 = r3^2 + s3^2;
p4 = r4^2 + s4^2;

q01 = r0 * r1 + s0 * s1;
q10 = -r0 * s1 + r1 * s0;
q02 = r0 * r2 + s0 * s2;
q20 = -r0 * s2 + r2 * s0;
q03 = r0 * r3 + s0 * s3;
q30 = -r0 * s3 + r3 * s0;
q04 = r0 * r4 + s0 * s4;
q40 = -r0 * s4 + r4 * s0;
q12 = r1 * r2 + s1 * s2;
q21 = -r1 * s2 + r2 * s1;
q13 = r1 * r3 + s1 * s3;
q31 = -r1 * s3 + r3 * s1;
q14 = r1 * r4 + s1 * s4;
q41 = -r1 * s4 + r4 * s1;
q23 = r2 * r3 + s2 * s3;
q32 = -r2 * s3 + r3 * s2;
q24 = r2 * r4 + s2 * s4;
q42 = -r2 * s4 + r4 * s2;
q34 = r3 * r4 + s3 * s4;
q43 = -r3 * s4 + r4 * s3;

E = ideal(p0,p1,p2,p3,p4,q01,q10,q02,q20,q03,q30,q04,q40,q12,q21,q13,q31,q14,q41,q23,q32,q24,q42,q34,q43);

-- I is the ideal of the rank 3 locus of G.
-- E is the ideal of E1 U E2.
-- we show that the radicals of these ideals are the same.

Erad = radical E;
Irad = radical I;
Egens = transpose gens gb Erad;
Igens = transpose gens gb Irad;

-- one checks that the elements of Egens and the elements of Igens are identical, up to factors of 2.
-- Hence over a base field of characteristic not equal to 2, Egens and Igens cut out the same points.
-- The singular locus of G thus consists of E and some extra rank 4 points.
```

Program A.4: Proof that $E_1 \cup E_2$ is the rank 3 locus of G.

```
-- We want to find the singular points of G.
-- with z0 = 0

F = ZZ;
R = F[z0,z4,z2,z3,z1];

L = matrix{{0,z2,-z4,-z1,z3},{z4,0,z3,-z0,-z2},{-z3,z0,0,z4,-z1},{-z2,-z4,z1,0,z0},{z1,-z3,-z0,z2,0}};

-- here G is actually 2 times what it should be but that's ok
-- if we are not in characteristic 2

G = det L;

H = diff(G);

-- the ideal slocus defines the singular locus of G

slocus = ideal H + ideal (G);
slocusgens = transpose gens gb slocus;

-- the ideal slocusz0 defines the set of singular points of G
-- with z0 = 0

slocusz0 = slocus + ideal(z0);
slocusz0gens = transpose gens gb slocusz0;

-- checking the Grobner basis slocusz0gens, we find that either some
-- other zi is 0, in which case z is a sigma-node, or
-- z1 z4 + z2 z3 = 0.  In this case,

slocus2z0 = slocusz0 + ideal(z1 * z4 + z2 * z3);
slocus2z0gens = transpose gens gb slocus2z0;

-- checking the Grobner basis slocus2z0gens,
-- and setting z1 = 1, one concludes that
-- z0 = 0, z1 = 1, z2 = z3^3, z4 = -z3^4,
-- and z3^10 = -1.
```

Program A.5: Finding nodes of G that satisfy $z_0 = 0$.

```
-- consider the segre variety P4 x P4 in P24.  our threefold X is a section of P4 by five
-- hyperplanes.  we show that one can choose five hyperplanes such that no singularities appear until
-- the final slice.  we can then apply the Lefschetz theorem to compute Hodge numbers of X.

F = ZZ/101;
R = F[x0, x1, x2, x3, z0, z1, z2, z3];
x4 = 1;
z4 = 1;
f0 = -x3 * z1 + x1 * z2 + x4 * z3 - x2 * z4;
f1 = -x3 * z0 - x4 * z2 + x2 * z3 + x0 * z4;
f2 = x1 * z0 - x4 * z1 - x0 * z3 + x3 * z4;
f3 = x4 * z0 + x2 * z1 - x0 * z2 - x1 * z4;
f4 = -x2 * z0 + x0 * z1 + x3 * z2 - x1 * z3;
X0 = ideal(f0+3*f1+7*f2+2*f3+10*f4);
X1 = X0 + ideal(f0+5*f1+27*f2+53*f3+18*f4);
X2 = X1 + ideal(f0+54*f1+33*f2+42*f3+20*f4);
X3 = X2 + ideal(f0+9*f2+38*f2+19*f3+64*f4);
X4 = X3 + ideal(f0);
J0 = diff (f0+3*f1+7*f2+2*f3+10*f4);
J1 = J0 || diff (f0+5*f1+27*f2+53*f3+18*f4);
J2 = J1 || diff (f0+54*f1+33*f2+42*f3+20*f4);
J3 = J2 || diff (f0+9*f2+38*f2+19*f3+64*f4);
J4 = J3 || diff f0;
sing0 = minors(1,J0);
sing1 = minors(2,J1);
sing2 = minors(3,J2);
sing3 = minors(4,J3);
sing4 = minors(5,J4);
slocus0 = X0 + sing0;
slocus1 = X1 + sing1;
slocus2 = X2 + sing2;
slocus3 = X3 + sing3;
slocus4 = X4 + sing4;

-- check that dim slocus0 = dim slocus1 = dim slocus2 = dim slocus 3 = -1
-- and dim slocus4 = 0
```

Program A.6: Proof that $\mathbb{P}^4 \times \mathbb{P}^4$ can be successively sliced to obtain X in such a manner that singularities only appear at the end.

Appendix B. Calculating traces of Frob_p

p	59	67	71	79	83	89
$\#G(\mathbb{F}_p)$	225766	327706	407910	529886	613006	751756
σ-nodes defined over $\mathbb{F}_p$	5	5	5	5	5	5
τ-nodes defined over $\mathbb{F}_p$	1	1	5	1	1	1
Other nodes defined over $\mathbb{F}_p$	0	0	0	0	0	10
Points on $E_1 \cup E_2$	0	0	0	0	0	180
i in $\mathbb{F}_p$?	0	0	0	0	0	1
$\sqrt{5}$ in $\mathbb{F}_p$?	1	0	1	1	0	1
ϵ in $\mathbb{F}_p$?	0	0	1	0	0	0
$\#X(\mathbb{F}_p)$	247360	355310	459740	568280	655170	897360
$p^3 + 1 - \#X(\mathbb{F}_p)$	-41980	-54546	-101828	-75240	-83382	-192390
$p^2 + p$	3540	4556	5112	6320	6972	8010
h	12	12	20	12	12	24
$\mathrm{tr}\,\mathrm{Frob}_p$ on H^3	500	126	412	600	282	-150
$6p^{\frac{3}{2}}$	2719.2	3290.6	3589.6	4213.1	4537.0	5037.8
a_p	500	126	412	600	282	-150
$\mathrm{tr}\,\mathrm{Frob}_p - a_p$	0	0	0	0	0	0
$(\mathrm{tr}\,\mathrm{Frob}_p - a_p)/p$	0	0	0	0	0	0
$2p + 2 - \#(E_1 \cup E_2)(\mathbb{F}_p)$						0

Table B.1: Counting points on X and calculating traces of Frob_p.

p	97	101	103	107	109	113
$\#G(\mathbb{F}_p)$	967966	1126560	1157186	1295146	1365776	1517046
σ-nodes defined over $\mathbb{F}_p$	5	5	5	5	5	5
τ-nodes defined over $\mathbb{F}_p$	1	5	1	1	1	1
Other nodes defined over $\mathbb{F}_p$	10	50	0	0	10	10
Points on $E_1 \cup E_2$	170	200	0	0	220	230
i in $\mathbb{F}_p$?	1	1	0	0	1	1
$\sqrt{5}$ in $\mathbb{F}_p$?	0	1	0	0	1	0
ϵ in $\mathbb{F}_p$?	0	1	0	0	0	0
$\#X(\mathbb{F}_p)$	1137910	1770940	1221870	1364910	1583340	1750730
$p^3 + 1 - \#X(\mathbb{F}_p)$	-225236	-740638	-129142	-139866	-288310	-307832
$p^2 + p$	9506	10302	10712	11556	11990	12882
h	24	24	72	12	12	24
$\mathrm{tr}\,\mathrm{Frob}_p$ on H^3	2908	1106	-598	-1194	-550	1336
$6p^{\frac{3}{2}}$	5732.1	6090.3	3589.6	6272.1	6640.9	6828.0
a_p	386	702	-598	-1194	-550	1562
$\mathrm{tr}\,\mathrm{Frob}_p - a_p$	2522	404	0	0	0	-226
$(\mathrm{tr}\,\mathrm{Frob}_p - a_p)/p$	26	4	0	0	0	-2
$2p + 2 - \#(E_1 \cup E_2)(\mathbb{F}_p)$	26	4			0	-2

Table B.2: Counting points on X and calculating traces of Frob_p, cont.

p	127	131	137	139	149	151
$\#G(\mathbb{F}_p)$	2143566	24219190	2685206	2802246	3437616	3669110
σ-nodes defined over $\mathbb{F}_p$	5	5	5	5	5	5
τ-nodes defined over $\mathbb{F}_p$	1	5	1	1	1	4
Other nodes defined over $\mathbb{F}_p$	0	0	10	0	10	0
Points on $E_1 \cup E_2$	0	0	290	0	300	0
i in $\mathbb{F}_p$?	0	0	1	0	1	0
$\sqrt{5}$ in $\mathbb{F}_p$?	0	1	0	1	1	1
ϵ in $\mathbb{F}_p$?	0	1	0	0	0	1
$\#X(\mathbb{F}_p)$	2241610	2596140	3029350	2919840	3842300	3900140
$p^3 + 1 - \#X(\mathbb{F}_p)$	-193226	-348048	-457966	-234220	-534350	-457188
$p^2 + p$	16256	17292	18906	19460	22350	22952
h	12	20	24	12	24	20
$\operatorname{tr}\operatorname{Frob}_p$ on H^3	1846	-2208	-4252	-700	2050	1852
$6p^{\frac{3}{2}}$	8587.4	8996.2	9621.3	9832.8	10912.7	11133.2
a_p	1846	-2208	-2334	-700	2050	1852
$\operatorname{tr}\operatorname{Frob}_p - a_p$	0	0	-1918	0	0	0
$(\operatorname{tr}\operatorname{Frob}_p - a_p)/p$	0	0	-14	0	0	0
$2p + 2 - \#(E_1 \cup E_2)(\mathbb{F}_p)$			-14		0	

Table B.3: Counting points on X and calculating traces of Frob_p, cont.

p	157
$\#G(\mathbb{F}_p)$	4019026
σ-nodes defined over $\mathbb{F}_p$	5
τ-nodes defined over $\mathbb{F}_p$	1
Other nodes defined over $\mathbb{F}_p$	10
Points on $E_1 \cup E_2$	350
i in $\mathbb{F}_p$?	1
$\sqrt{5}$ in $\mathbb{F}_p$?	0
ϵ in $\mathbb{F}_p$?	0
$\#X(\mathbb{F}_p)$	4473070
$6p^{\frac{3}{2}}$	11803.3
a_p	-2494
$\operatorname{tr}\operatorname{Frob}_p - a_p$	-5338
$(\operatorname{tr}\operatorname{Frob}_p - a_p)/p$	-34
$2p + 2 - \#(E_1 \cup E_2)(\mathbb{F}_p)$	-34

Table B.4: Counting points on X and calculating traces of Frob_p, cont.

References

[1] A. Aure, W. Decker, K. Hulek, S. Popescu and K. Ranestad, Syzygies of Abelian and Bielliptic Surfaces in $\mathbb{P}^4$, *Internat. J. Math.* **8**, 1997, 849-919.
[2] C. Breuil, B. Conrad, F. Diamond and R. Taylor, On the Modularity of Elliptic Curves over $\mathbb{Q}$: Wild 3-adic Exercises, *J. Amer. Math. Soc.* **14**, 2001, 843-939.
[3] H. Darmon, F. Diamond and R. Taylor, Fermat's Last Theorem, *Elliptic Curves, Modular Forms and Fermat's Last Theorem*, International Press (Boston), 1999, 2-140.
[4] L. Dieulefait and J. Manoharmayum, Modularity of Rigid Calabi-Yau Threefolds over $\mathbb{Q}$, *Calabi-Yau Varieties and Mirror Symmetry*, American Mathematical Society (Providence, RI), 2003, 159-166.
[5] W. Barth, K. Hulek and R. Moore, Degenerations of Horrocks-Mumford Surfaces, *Math. Ann.* **277**, 1987, 735-755.
[6] W. Barth and R. Moore, Geometry in the Space of Horrocks-Mumford Surfaces, *Topology* **28**, 1989, 231-245.
[7] T. Fisher, *On 5 and 7 Descents for Elliptic Curves*, Thesis, University of Cambridge (2000). `http://www.dpmms.cam.ac.uk/~taf1000/thesis.html`
[8] J.-M. Fontaine and B. Mazur, Geometric Galois Representations, *Elliptic Curves, Modular Forms and Fermat's Last Theorem*, International Press (Boston), 1999, 190-227.
[9] E. Freitag and R. Kiehl, *Étale Cohomology and the Weil Conjecture*, Springer-Verlag (New York), 1987.
[10] P. Griffiths and J. Harris, *Principles of Algebraic Geometry*, Wiley-Interscience (New York), 1978.
[11] M. Gross and S. Popescu, Calabi-Yau Threefolds and Moduli of Abelian Surfaces I, *Comp. Math.* **127**, 2001, 169-228.
[12] R. Hartshorne, Algebraic Geometry, Springer-Verlag (New York), 1977.
[13] K. Hulek, The Horrocks-Mumford Bundle, *Vector Bundles and Algebraic Geometry: Durham 1993*, Cambridge University Press (New York), 1993, 139–178.
[14] G. Horrocks and D. Mumford, A Rank 2 Vector Bundle on $\mathbb{P}^4$ with $15,000$ Symmetries, *Topology* **12**, 1973, 63-81.
[15] K. Hulek and H. Lange, The Hilbert Modular Surface for the Ideal $(\sqrt{5})$ and the Horrocks-Mumford Bundle, *Math. Z.* **198**, 1988, 95-116.
[16] K. Hulek and H. Verrill, *On modularity of rigid and nonrigid Calabi-Yau varieties associated to the root lattice A_4*, `math.AG/0304169`.
[17] H. Lange, Abelian Surfaces in $\mathbb{P}^1 \times \mathbb{P}^3$, *Arch. Math.* **63**, 1994, 80-84.
[18] H. Lange, A Vector Bundle of Rank 2 on $\mathbb{P}^1 \times \mathbb{P}^3$, *J. London Math. Soc.* (2) **57**, 1998, 583-598.
[19] R. Livné, Cubic Exponential Sums and Galois Representations, *Contemporary Mathematics* **67**, 1987, 247–261.
[20] N. Manolache, The Equations of the Abelian Surfaces Embedded in $\mathbb{P}^4(\mathbb{C})$, *J. Reine. Angew. Math.* **394**, 1989, 196-202.
[21] J. Milne, *Lectures on Étale Cohomology.* `http://www.jmilne.org/math/CourseNotes/`, 1998.
[22] R. Moore, Heisenberg-invariant quintic threefolds and sections of the Horrocks-Mumford bundle, preprint, Canberra 1985.
[23] J. Munkres, *Elements of Algebraic Topology*, Addison-Wesley (New York), 1984.
[24] C. Schoen, On the Geometry of a Special Determinantal Hypersurface Associated to the Mumford-Horrocks Vector Bundle, *J. Reine. Angew. Math.* **364**, 1986, 85-111.
[25] M. Schütt, On the Modularity of Three Calabi-Yau Threefolds with Bad Reduction at 11. To appear in *Bull. Canadian Math. Soc.*
[26] W. Stein, The Modular Forms Database, `http://modular.fas.harvard.edu/Tables/`
[27] R. Taylor, private communication.
[28] R. Taylor, Galois Representations, *Ann. Fac. Sci. Toulouse Math. (6)* **13**, 2004, 73-119.
[29] R. Taylor and A. Wiles, Ring-theoretic Properties of Certain Hecke Algebras, *Ann. Math. (2)* **141**, 1995, 553-572.
[30] J. Werner and B. van Geemen, New Examples of Threefolds with $c_1 = 0$, *Math. Z.* **203**, 1990, 211-225.
[31] N. Yui, The Arithmetic of Certain Calabi-Yau Varieties over Number Fields, *The Arithmetic and Geometry of Algebraic Cycles* (Banff, AB, 1998), NATO Sci. Ser. C Math. Phys. Sci. 548, Kluwer Acad. Publ., Dordrecht, 200, pp. 515-560.
[32] N. Yui, Update on the Modularity of Calabi-Yau Varieties, *Proceedings of the Workshop on Calabi-Yau Varieties and Mirror Symmetry at the Fields Institute*, Toronto, July 2001.

Department of Mathematics, Harvard University, Cambridge, MA 02138
E-mail address: `edward.lee@post.harvard.edu`

AMS/IP Studies in Advanced Mathematics
Volume 38, 2006

Arithmetic of Algebraic Curves and the Affine Algebra $A_1^{(1)}$

Monika Lynker and Rolf Schimmrigk

ABSTRACT. Arithmetic methods provide tools for the construction of a relation between algebraic varieties and the affine Lie algebra $A_1^{(1)}$. Such a relation is useful in string theory as a link between the geometry of spacetime and the physics on the world sheet. We describe these methods in the context of algebraic curves.

1. Introduction

1.1 The problem of string compactification can mathematically be viewed, roughly, as the goal to construct a map which relates Kac-Moody vertex algebras to the geometry of Calabi-Yau varieties. At present the complete structure of this map remains conjectural, but certain ingredients have been identified. Part of the data of the conformal field theory on the string world sheet are the conformal weights of the fields of the theory. These scaling dimensions are rational numbers that appear in the correlation functions of the quantum field theory. It turns out that it is more useful to think about these values in terms of the number field generated by their associated quantum dimensions, defined via the characters of the Kac-Moody algebra. A relation between these quantum dimensions and the arithmetic of Calabi-Yau varieties of Brieskorn-Pham type has been established in [**21**] by considering the Hecke theoretic nature of the Hasse-Weil L-functions of these varieties. A more conceptual framework of this relation has been formulated in [**18**].

A second aspect of conformal field theory is the modular invariance of the string partition function. It has been an open question for a long time how the string theoretic building blocks on the world sheet are reflected in the geometry of spacetime, if at all. One way to make this question more precise is by asking how the characters that appear in the partition function of the string emerge from spacetime, and

2000 *Mathematics Subject Classification.* Primary 11F23, 11F22, 11G30, 11G40, 17B67; Secondary 14J32, 81T30, 81R10 .

Key words and phrases. algebraic curves, affine Kac–Moody algebra, Hasse–Weil L-function, quantum dimension.

M.L. and R.S. were supported in part by the NSF under Grant No. PHY99-07949. They are grateful to the Kavli Institute for Theoretical Physics, Santa Barbara, for hospitality and support through KITP Scholarships during the course of part of this work.

how in turn the spacetime geometry can be constructed from string theoretic quantities. Questions of this type have a long history in arithmetic algebraic geometry, in particular in the context of the Shimura-Taniyama conjecture [**32, 1**], and the Langlands program, but string theory provides a somewhat different point of view than the one encountered in the broader framework of arithmetic geometry. Modularity of geometric L-functions associated to algebraic varieties along the lines of Langlands' program is not sufficient. In order to be of string theoretic significance, the geometric modular forms should allow an interpretation in terms of modular forms that arise in the description of the physics on the Riemann surfaces defined by the propagating string.

It was shown in [**23**] in the context of string compactifications on elliptic curves that the Mellin transform of the Hasse-Weil L-function of the elliptic curve

$$C_3 = \{(x : y : z) \in \mathbb{P}_2 \mid x^3 + y^3 + z^3 = 0\} \tag{1.1}$$

factors into a product of modular forms that arise from the characters of the underlying two-dimensional theory. More precisely, the following result was obtained. Define $q = e^{2\pi i \tau}$ and let $\eta(\tau) = q^{1/24}\prod_{n=1}^{\infty}(1-q^n)$ denote the Dedekind η-function, $c^k_{\ell,m}(\tau)$ be the affine $A_1^{(1)}$ string functions at affine level $k \in \mathbb{N}$, and $\Theta^k_{\ell,m}(\tau) = \eta^3(\tau)c^k_{\ell,m}(\tau)$ be the Hecke indefinite modular forms associated to the string functions.

Theorem 1.1. *The Mellin transform of the Hasse-Weil L-function $L_{\rm HW}(C_3, s)$ of the cubic elliptic curve $C_3 \subset \mathbb{P}_2$ is a modular form $f_{\rm HW}(C_3, q) \in S_2(\Gamma_0(27))$ which factors into the product*

$$f_{\rm HW}(C_3, q) = \Theta^1_{1,1}(q^3)\Theta^1_{1,1}(q^9). \tag{1.2}$$

Here $\Theta^1_{1,1}(\tau)$ is the unique Hecke modular form at affine level $k = 1$, associated to the quadratic extension $\mathbb{Q}(\sqrt{3})$.

1.2 Elliptic curves are somewhat degenerate examples of Calabi-Yau manifolds, and the question arises whether the results above can be extended to more general algebraic curves, e.g.

$$C_n = \{(x : y : z) \in \mathbb{P}_2 \mid x^n + y^n + z^n = 0\}. \tag{1.3}$$

The genus of these Riemann surfaces is given by $g(C_n) = (n-1)(n-2)/2$, hence except for the cubic curve C_3 just discussed these curves are not elliptic. It is therefore not obvious whether one should be able to find a string theoretic interpretation, generalizing the result described above for C_3. For such general Riemann surfaces there exists in particular no a priori statement, like the proof of the Shimura-Taniyama conjecture, which would ensure the modularity of the $q-$expansion derived from the Hasse-Weil L-function.

An argument in favor of a conformal field theoretic interpretation of the curves C_n is provided by the fact that they appear as singular loci of higher dimensional Calabi-Yau varieties that are expected to be exactly solvable. For the cubic curve C_3 such higher dimensional manifolds are abundant, examples being provided by the elliptic K3 surface defined by the degree 6 polynomial in weighted projective space $\mathbb{P}_{(1,1,2,2)}$

$$S_6 = \{(z_1 : \cdots : z_4) \in \mathbb{P}_{(1,1,2,2)} \mid z_1^6 + z_2^6 + z_3^3 + z_4^3 = 0\}, \tag{1.4}$$

or the degree twelve Calabi-Yau threefold

$$X_{12} = \{(z_0 : \cdots : z_4) \in \mathbb{P}_{(1,1,2,4,4)} \mid z_0^{12} + z_1^{12} + z_2^6 + z_3^3 + z_4^3 = 0\}. \tag{1.5}$$

The variety X_{12} is an elliptic fibration and the curve C_3 is the generic elliptic fiber which can be identified by an iterative application of the construction described in [**8**]. First we can construct the elliptic K3 surface S_6 via the twist map as

$$\mathbb{P}_{(2,1,1)}[6] \times \mathbb{P}_2[3] \longrightarrow \mathbb{P}_{(1,1,2,2)}[6], \tag{1.6}$$

and then the hypersurface X_{12} can be constructed by applying the twist map again in the form

$$\mathbb{P}_{(2,1,1)}[12] \times \mathbb{P}_{(1,1,2,2)}[6] \longrightarrow \mathbb{P}_{(1,1,2,4,4)}[12]. \tag{1.7}$$

We therefore see that the twist map allows us to construct the threefold X_{12} as an iterated orbifold of products of curves and makes explicit the origin of the two minimal factors at $k = 1$. There is a large number of varieties of this type among the Calabi-Yau hypersurface threefolds constructed in the mirror paper [**2**], and in ref. [**14**], which describe the complete construction of this class of varieties.

For higher genus curves of type C_n similar embeddings can be obtained. Examples are provided by the quartic plane curve C_4, which is embedded in the octic hypersurface in weighted projective space $\mathbb{P}_{(1,1,2,2,2)}$, the quintic curve C_5, embedded in $\mathbb{P}_{(1,3,2,2,2)}$, or the septic curve C_7 in $\mathbb{P}_{(1,7,2,2,2)}$. The class of higher dimensional varieties of this type is large. Because they form part of an exactly solvable variety we might expect these curves to inherit the information of the underlying string theory. In this way we are led to ask whether more general, non-Calabi-Yau varieties are related to affine Kac-Moody algebras similar to the cubic.

Theorem 1.2 *The Hasse-Weil L-function $L_{\mathrm{HW}}(C_4, s)$ of the quartic plane curve C_4 factors into a triple product $L_{\mathrm{HW}}(C_4, s) = L(E_4, s)^3$, where $L(E, s)$ is the Hasse-Weil L-functions of the weighted elliptic curve*

$$E_4 = \{(x_0 : x_1 : x_2) \in \mathbb{P}_{(1,1,2)} \mid x_0^4 + x_1^4 + x_2^2 = 0\}. \tag{1.8}$$

The Hasse-Weil modular form $f(E_4, q) \in S_2(\Gamma_0(64))$ associated to $L(E_4, s)$ factors into a twisted product

$$f(E_4, q) = \Theta_{1,1}^2(4\tau)^2 \otimes \chi_2. \tag{1.9}$$

Here the twist character is the Legendre symbol $\chi_2(\cdot) = \left(\frac{2}{\cdot}\right)$ and $\Theta_{1,1}^2(\tau)$ is the affine $A_1^{(1)}$ theta function of the string function $c_{1,1}^2(\tau)$ at affine level $k = 2$.

This result shows that the modular forms derived from the quartic curve admits a Kac-Moody theoretic interpretation via an $A_1^{(1)}$ theta function.

1.3 We can turn this result around and read it as an arithmetic interpretation of a conformal field theoretic object, the string theoretic parafermionic string functions.

Corollary 1.3 *The $A_1^{(1)}$ string function $c_{1,1}^2(\tau)$ at level $k = 2$ is determined by the twisted Hasse-Weil modular form*

$$c_{1,1}^2(\tau) = \frac{1}{\eta^3(\tau)}\sqrt{f(E_4) \otimes \chi_2(q^{1/4})}. \tag{1.10}$$

1.4 The weighted elliptic curve E_4 which emerges as the arithmetic building block of the quartic curve also appears as the singular locus of many higher dimensional varieties and it is possible to repeat the conformal field theoretic analysis indicated above for Fermat curves. This can be illustrated by considering the

Calabi-Yau threefold X_{16} of degree 16 embedded in the weighted projective space $\mathbb{P}_{(1,1,2,4,8)}$. This variety is a K3 fibration with a generic fiber described by a degree 8 hypersurface S_8 embedded in $\mathbb{P}_{(1,1,2,4)}$, which is elliptic. The generic fiber of this elliptic fibration is precisely the curve E_4 and we can iteratively construct the degree sixteen threefold in $\mathbb{P}_{(1,1,2,4,8)}$ with the methods of [**8**] by first considering the map

$$\mathbb{P}_{(2,1,1)}[8] \times \mathbb{P}_{(1,1,2)}[4] \longrightarrow \mathbb{P}_{(1,1,2,4)}[8], \tag{1.11}$$

and then applying the map

$$\mathbb{P}_{(2,1,1)}[16] \times \mathbb{P}_{(1,1,2,4)}[8] \longrightarrow \mathbb{P}_{(1,1,2,4,8)}[16]. \tag{1.12}$$

Alternatively, E_4 appears as the $\mathbb{Z}_2$−singular curve on this threefold. As in the case of Theorem 1.1 for the cubic curve, Theorem 1.2 provides a geometric interpretation of the string theoretic modular form of the minimal factor at affine level $k = 2$ which is part of the Gepner model corresponding to this threefold.

It becomes clear by an iterative application of the twist map construction [**8**] that elliptic threefolds which are K3 fibrations can be built from the generic elliptic fibre and other curves. This reduces the question of modularity for these higher dimensional varieties to the problem of modularity for curves.

1.5 This paper is organized as follows. In Section 2 we compute the Hasse-Weil L-function of the quartic. We will see that arithmetically the basic building block of this curve is given by the elliptic Brieskorn-Pham curve of degree four. In Section 2 we briefly review the relevant aspects of non-twisted affine Kac-Moody algebras. In Section 4 the modular form defined by the twisted Mellin transform of the Hasse-Weil L-function is related to a modular form derived from the character of an affine Kac-Moody algebra. A relation between the character defining this twist and the quantum dimensions of the affine Kac-Moody algebra is described in Section 5. In the final Section 6 we show that the factorization behavior of the quartic is typical for Fermat curves, indicating that other curves can be treated in a similar way. As an example we briefly indicate results for the Fermat curve of degree six.

2. Hasse-Weil L-function

2.1 For algebraic varieties X the congruent zeta function at a prime number p can be defined as the generating function

$$Z(X/\mathbb{F}_p, t) \equiv \exp\left(\sum_{r\in\mathbb{N}} \#\left(X/\mathbb{F}_{p^r}\right)\frac{t^r}{r}\right). \tag{2.1}$$

It was first shown by F.K. Schmidt [**24, 7**] that for algebraic curves X the zeta function $Z(X/\mathbb{F}_p, t)$ is a rational function which takes the form

$$Z(X/\mathbb{F}_p, t) = \frac{P^{(p)}(t)}{(1-t)(1-pt)}, \tag{2.2}$$

where

$$P^{(p)}(t) = \sum_{i=0}^{2g} \beta_i(p)t^i, \tag{2.3}$$

is a polynomial whose degree is given by the genus $g(X) = (2 - \chi(X))/2$ of the curve, where $\chi(X)$ is the Euler characteristic of the curve. More important is the global zeta function obtained by setting $t = p^{-s}$ and taking the product over all rational primes at which the variety has good reduction. Let S denote the set of rational primes at which X becomes singular and denote by P_S the set of primes that are not in S. The global zeta function then becomes

$$Z(X,s) = \prod_{p \in P_S} \frac{P^{(p)}(p^{-s})}{(1-p^{-s})(1-p^{1-s})} = \frac{\zeta(s)\zeta(s-1)}{L_{\rm HW}(X,s)}, \tag{2.4}$$

with the Hasse-Weil L-function

$$L_{\rm HW}(X,s) = \prod_{p \in P_S} \frac{1}{P^{(p)}(p^{-s})}, \tag{2.5}$$

and the Riemann zeta function $\zeta(s) = \prod_p (1-p^{-s})^{-1}$.

The most direct way to compute the expansion for the lower primes involves the comparison of the definition of the congruent zeta function with the rational expression found by Schmidt. This results in the expansion

$$L_{\rm HW}(C_4,s) = 1 + \frac{6}{5^s} - \frac{9}{9^s} - \frac{18}{13^s} + \frac{6}{17^s} + \frac{9}{25^s} + \frac{30}{29^s} + \frac{6}{37^s} + \frac{30}{41^s} + \cdots \tag{2.6}$$

with the associated $q-$series given by

$$f_{\rm HW}(C_4,q) = q + 6q^5 - 9q^9 - 18q^{13} + 6q^{17} + 9q^{25} + 30q^{29} + 6q^{37} + 30q^{41} + \cdots \tag{2.7}$$

Finite expansions like this often turn out to be useful because of theorems by Faltings and Serre, which show that such functions are determined uniquely by a finite number of terms.

2.2 The behavior of the coefficients a_n of the Hasse-Weil $q-$expansion under Hecke operators indicates that this series does not describe a Hecke eigenform. Hence the above result for the Hasse-Weil L-function is not particularly illuminating except for the fact that all the coefficients are divisible by 3. This suggests that the Hasse-Weil L-function of the quartic plane curve can be viewed as the cubic power of a more basic L-series. We can write $L_{\rm HW}(C_4,s) = L^3(s)$ with

$$L(s) = 1 + \frac{2}{5^s} - \frac{3}{9^s} - \frac{6}{13^s} + \frac{2}{17^s} - \frac{1}{25^s} + \frac{10}{29^s} + \frac{2}{37^s} + \frac{10}{41^s} + \cdots \tag{2.8}$$

with the corresponding $q-$expansion

$$f(q) = q + 2q^5 - 3q^9 - 6q^{13} + 2q^{17} - q^{25} + 10q^{29} + 2q^{37} + 10q^{41} + \cdots \tag{2.9}$$

This result indicates that $f(q)$ is the basic building block of the Hasse-Weil L-form $f_{\rm HW}(C_4,q)$, providing a sort of cube root of it via its L-series.

2.3 The question therefore becomes whether it is possible to give the expansion $f(q)$ a geometric origin as the Hasse-Weil modular form of some other geometric object, and whether this geometric object admits a string theoretic interpretation. This turns out to be the case. The first step is to factorize the polynomials $P^{(p)}(t)$ in a more systematic way in terms of Jacobi sums.

Theorem 2.3 [29] *For the plane curve*

$$C_n = \{z_0^n + z_1^n + z_2^n = 0\} \subset \mathbb{P}_2 \tag{2.10}$$

defined over the finite field $\mathbb{F}_q$ *set* $d = (n, q-1)$ *and define the set* $A_2^{q,n}$ *of triplets* $\alpha = (\alpha_0, \alpha_1, \alpha_2)$ *of rational numbers as*

$$A_2^{q,n} = \left\{ \alpha \in \mathbb{Q}^3 \mid 0 < \alpha_i < 1,\ d\alpha_i \equiv 0 \bmod 1,\ \sum_{i=0}^{2} \alpha_i \equiv 0 \bmod 1 \right\}. \tag{2.11}$$

For such triplets define the Jacobi sums

$$j_q(\alpha_0,\alpha_1,\alpha_2) = \frac{1}{q-1} \sum_{\substack{u_0,u_1,u_2\in\mathbb{F}_q \\ u_0+u_1+u_2=0}} \chi_{\alpha_0}(u_0)\chi_{\alpha_1}(u_1)\chi_{\alpha_2}(u_2) \tag{2.12}$$

where $\chi_{\alpha_i}(u_i) = e^{2\pi i\alpha_i m_i}$ and the integers m_i are determined via $u_i = g^{m_i}$, where $g \in \mathbb{F}_q$ is a generator. Then the cardinality of $C_n/\mathbb{F}_q$ is given by

$$\#(C_n/\mathbb{F}_q) = 1 + q + \sum_{\alpha\in A_2^p} j_q(\alpha). \tag{2.13}$$

With these Jacobi sums one can express the F.K. Schmidt polynomial as

$$P^{(p)}(t) = \prod_{\alpha\in A_2^n} \left(1 + j_{p^f}(\alpha)t^f\right)^{1/f}, \tag{2.14}$$

where for any α in the set A_2^n defined by

$$A_2^n = \left\{\alpha \in \mathbb{Q}^3 \mid \alpha_i \in (0,1),\ n\alpha_i \in \mathbb{N},\ \sum_{i=0}^{2} \alpha_i \in \mathbb{N}\right\}, \tag{2.15}$$

f is the smallest positive integer such that $(p^f - 1)\alpha_i \in \mathbb{N}$ for all i. The sums $j_q(\alpha)$ are constant over the two permutation orbits of A_2^4, and with $j_p\left(\frac{1}{2},\frac{3}{4},\frac{3}{4}\right) = \overline{j_p\left(\frac{1}{2},\frac{1}{4},\frac{1}{4}\right)}$ one finds

$$L_{\mathrm{HW}}(C_4,s) = \prod_p \frac{1}{(1+2a_p\cdot p^{-s} + p\cdot p^{-2s})^3}, \tag{2.16}$$

where $a_p = \mathrm{Re}\ j_p\left(\frac{1}{2},\frac{1}{4},\frac{1}{4}\right)$.

Hence $L_{\mathrm{HW}}(C_4,s) = L^3(s)$ in terms of an L-function

$$L(s) = \prod_p \frac{1}{1+2a_p\cdot p^{-s} + p\cdot p^{-2s}} \tag{2.17}$$

which has the form of the Hasse-Weil series of an elliptic curve. The quartic curve C_4 has genus $g = 3$, hence this factorization indicates that the L-function of this curve splits into a product of g elliptic contributions. Such a factorization into elliptic factors will not happen for general Fermat curves.

2.4 The above decomposition allows us to return to the question asked above of whether one can determine an elliptic curve whose Hasse-Weil series has been determined above. The quartic curve C_4 becomes singular when reduced at $p = 2$. Hence we would expect the purported elliptic curve to have a conductor that is divisible by two, suggesting an elliptic curve of degree four. Such a curve is the well known weighted torus, described by

$$E_4 = \{(x_0 : x_1 : x_2) \in \mathbb{P}_{(1,1,2)} \mid x_0^4 + x_1^4 + x_2^2 = 0\}. \tag{2.18}$$

The Hasse-Weil modular form resulting from the solutions $E_4/\mathbb{F}_p$ starts out as

$$f_{\mathrm{HW}}(E_4,q) = q + 2q^5 - 3q^9 - 6q^{13} + 2q^{17} - q^{25} + 10q^{29} + 2q^{37} + 10q^{41} + \cdots \tag{2.19}$$

in agreement with the expansion $f(q)$ associated to the 'cube root' $L(s)$ discussed above.

We are therefore in the same situation as in [**23**] and we can ask whether the modular form $f(q) = f_{\mathrm{HW}}(E_4,q)$ can be related to modular forms that are induced by conformal field theoretic characters.

3. Affine Kac-Moody algebras

3.1 A construction of non-twisted affine Kac-Moody algebras, also called affine Lie algebras, is provided by an extension of the loop algebra [**9**]. In the present context it is sufficient to consider the algebra $A_1^{(1)}$. For fixed level k of the theory the characters χ_ℓ, $0 \leq \ell \leq k$ can be expressed in terms of the string functions $c_{\ell,m}^k$ introduced by Kac-Peterson [**10**].

3.2 Supersymmetric string models can be constructed in terms of conformal field theories with $N = 2$ supersymmetry. It turns out that for the present discussion the important structure is determined by the Hecke indefinite modular forms, which can be defined as

$$\Theta_{\ell,m}^k(\tau) = \sum_{\substack{-|x|<y\leq|x| \\ (x,y) \text{ or } (\frac{1}{2}-x,\frac{1}{2}+y) \\ \in\mathbb{Z}^2+\left(\frac{\ell+1}{2(k+2)},\frac{m}{2k}\right)}} \operatorname{sign}(x) e^{2\pi i\tau((k+2)x^2-ky^2)} \tag{3.1}$$

and which are related to the string functions via the Dirichlet eta-function $\eta(\tau)$ as $\Theta_{\ell,m}^k(\tau) = \eta^3(\tau)c_{\ell,m}^k(\tau)$.

3.3 It was suggested by Gepner some time ago that exactly solvable string compactifications obtained by tensoring several copies of $N = 2$ supersymmetric conformal field theories derived from affine Kac-Moody algebras should yield, after performing appropriate projections, theories that correspond in some limit to geometric compactifications described by Brieskorn-Pham type Calabi-Yau varieties [**5**]. The evidence for this conjecture was initially based mostly on spectral information for all models in the Gepner class of solvable string compactifications and the agreement of certain types of intersection numbers which allow an interpretation as Yukawa couplings, as well as Landau-Ginzburg type arguments [**15**]. In the case of the Fermat cubic curve these results suggest that there is an underlying conformal field theory of this elliptic curve that is described by the GSO projection of a tensor product of three models at affine level $k = 1$.

In the case of higher genus curves no such direct relation is expected from a string theory perspective. There is, however, a weaker embedding argument that suggests a possible modular interpretation. Higher genus curves can be embedded in higher dimensional Calabi-Yau varieties which in turn are conjectured to be exactly solvable. Examples of such embeddings for the quartic curve are provided by threefolds such as Brieskorn-Pham hypersurfaces $\mathbb{P}_{(1,1,2,2,2)}$[8], $\mathbb{P}_{(1,2,3,3,3)}$[12], $\mathbb{P}_{(1,4,5,5,5)}$[20], and others. Hence we might expect that conformal field theoretic aspects are encoded in the quartic curve as well.

We can ask whether there are relations between the modular objects determined by the affine algebra and the modular objects determined by the variety. The results of [**23**] suggest that interesting quantities to consider are the theta functions associated to string functions of the affine Lie algebra $A_1^{(1)}$ at affine level k. The expansion of these functions at affine level $k = 2$ follows from those of the string function expansions. It turns out that relevant for the present discussion is the theta function

$$\Theta_{1,1}^2(q) = q^{1/8}(1 - q - 2q^2 + q^3 + 2q^5 + O(q^6)). \tag{3.2}$$

It will become clear below that $\Theta_{1,1}^2(q)$ emerges as the building block of the Hasse-Weil modular form $f_{\rm HW}(C_4, q)$ of the quartic plane Fermat curve.

4. Geometric modularity

4.1 From a physical perspective it is not clear a priori which conformal field theoretic quantities should be the correct building blocks of the Hasse-Weil function, if any. Possibilities include the twists of the affine or parafermionic characters, or some elements of the $N = 2$ superconformal model. The coset construction shows that the basic part of the $N = 2$ theory is given by the affine $A_1^{(1)}$ Kac-Moody algebra. The string functions $c_{\ell,m}^k(\tau)$ of the $N = 2$ characters would appear to be natural candidates because they capture the essential interacting nature of the field theory. Furthermore, associated to string functions are natural number theoretic theta functions.

The main result of [**23**] shows that the Hasse-Weil L-function of the cubic plane curve C_3 is determined by the Kac-Moody string function of the affine $A_1^{(1)}$ algebra at affine level $k = 1$. At $k = 1$ there is only one string function, denoted by $c_{1,1}^1(\tau)$, which can be computed to lead to the expansion

$$c_{1,1}^1(\tau) = q^{-1/24}(1 + q + 2q^2 + 3q^3 + 5q^4 + 7q^5 + O(q^6)). \tag{4.1}$$

It turns out that more important than the string functions are the associated $A_1^{(1)}$ theta functions $\Theta_{\ell,m}^k(\tau)$. These objects are associated to quadratic number fields determined by the level of the affine theory [**10**]. At level $k = 1$ the unique theta function $\Theta_{1,1}^1(\tau)$ is associated to the real quadratic extension $\mathbb{Q}(\sqrt{3})$ of the rational field $\mathbb{Q}$. Its expansion follows from the string function expansion, resulting in

$$\Theta_{1,1}^1(q) = q^{1/12}(1 - 2q - q^2 + 2q^3 + q^4 + 2q^5 + O(q^6)). \tag{4.2}$$

It is this modular form of weight one which emerged in [**23**] as the building block of the Hasse-Weil modular form $f_{\mathrm{HW}}(C_3, q)$ of the cubic elliptic curve, as established in Theorem 1.1.

4.2 The case of the cubic elliptic curve is special because it is a Calabi-Yau curve and therefore defines a consistent string theory background. For higher genus curves this is not the case, and it is therefore not a priori clear whether one should expect a relation to the string theoretic Kac-Moody algebra. As mentioned above, an argument that is encouraging is that higher genus algebraic curves such as the quartic, and others, appear in higher dimensional Calabi-Yau varieties as singular curves which have to be resolved.

4.3 The quartic curve is singular at the prime $p = 2$, therefore we expect the conductor of the elliptic curves involved to be divisible by 2, and perhaps by powers of two. Using Weil's conductor observation [**30**] for elliptic curves, we further expect the modular conductor of the corresponding modular form to be divisible by some power of 2. The arithmetic conductor of the elliptic curve E can be determined via Tate's algorithm [**28**] and turns out to be $N = 64$.

4.5 Combining the conductor argument with the string theoretic embedding of the quartic into higher dimensional varieties we are led to consider the form

$$\Theta_{1,1}^2(q^4)^2 = q - 2q^5 - 3q^9 + 6q^{13} + 2q^{17} - q^{25} - 10q^{29} + \cdots \tag{4.3}$$

Comparing this with the Mellin transform of the cube root of the Hasse-Weil L-form of the quartic C_4 shows agreement except for the signs in the terms q^n with $n = 1 \bmod 8$. The necessary sign flip can be produced by the Legendre symbol

$$\chi_2(r) = \left(\frac{2}{r}\right) = \left\{ \begin{array}{rl} 1 & r \equiv 1 \bmod 8 \\ -1 & r \equiv 5 \bmod 8 \end{array} \right\} \tag{4.4}$$

We therefore see that if we denote by $\Theta^2_{1,1}(4\tau)^2 \otimes \chi_2$ the modular form obtained by twisting the form $\Theta^2_{1,1}(4\tau)^2$ by the character $\chi_2(\cdot)$, we can write the Hasse-Weil L-function as

$$L_{\mathrm{HW}}(C_4, s) = (L(\Theta^2_{1,1}(4\tau)^2 \otimes \chi_2, s)^3. \tag{4.5}$$

This concludes the proof of the Theorem stated in the Introduction.

4.6 The remaining question is whether the theta function input is unique. This follows from the Eichler-Shimura theory [**3, 25, 26**]. $\Theta^2_{1,1}(4\tau)^2$ is an element in the space of cusp forms $S_2(\Gamma_0(32))$. For arbitrary modular level N, the dimension of the space $S_2(\Gamma_0(N))$ can be computed as the genus of the modular curve $X_0(N)$ via

$$g(X_0(N)) = 1 + \frac{\mu(N)}{12} - \frac{\nu_2(N)}{4} - \frac{\nu_3(N)}{3} - \frac{\nu_\infty}{2}, \tag{4.6}$$

where $\mu(N)$ is the index of $\Gamma_0(N)$ in $\Gamma(1) = \mathrm{SL}(2,\mathbb{Z})$, $\nu_2(N)$ and $\nu_3(N)$ are the number of elliptic points of order 2 and 3, and $\nu_\infty(N)$ is the number of $\Gamma_0(N)$ inequivalent cusps. For $N = 32$ this implies that the space of cusp forms is one-dimensional, hence $\Theta^2_{1,1}(4\tau)^2$ is its unique generator, up to multiplication by a constant.

5. Quantum dimensions

5.1 The character $\chi_2(\cdot)$ which appears above in the context of providing a string theoretic interpretation of the Hasse-Weil modular form furthermore points toward the field of quantum dimensions, thereby providing a link between the problem of a geometric explanation of the string theoretic modularity and the problem of providing a geometric explanation of the string theoretic spectrum.

5.2 For general square-free n the Legendre characters $\chi_n(\cdot)$ afford a characterization of the factorization behavior of rational primes p when viewed as elements in a quadratic extension $\mathbb{Q}(\sqrt{n})$. For $n = 2$ the character χ_2 is associated to the quadratic extension $\mathbb{Q}(\sqrt{2})$, which is precisely the field determined by the conformal weights of the affine theory when mapped into the quantum dimensions. This can be seen as follows.

5.3 It was noted in [**21**] that a link between the geometry and the conformal field theory can be obtained by using a translation of the conformal weights into algebraic integers via the Rogers dilogarithm. In the simple case of the affine Lie algebra $A_1^{(1)}$ at affine level k the modular transformations of the characters χ associated to the primary fields Φ_i with conformal weights

$$\Delta_\ell^{(k)} = \frac{\ell(\ell+2)}{4(k+2} \tag{5.1}$$

are given by

$$\chi_\ell\left(-\frac{1}{\tau}, \frac{u}{\tau}\right) = e^{\pi i k u^2/2} \sum_m S_{\ell m} \chi_m(\tau, u) \tag{5.2}$$

with modular S-matrix

$$S_{\ell m} = \sqrt{\frac{2}{k+2}} \sin\left(\frac{(\ell+1)(m+1)\pi}{k+2}\right), \quad 0 \le \ell, m \le k. \tag{5.3}$$

With these matrices one can define the generalized quantum dimensions as $Q_{\ell m} = \frac{S_{\ell m}}{S_{0m}}$ for the affine $A_1^{(1)}$ algebra at level k. The importance of these numbers derives from the fact that even though they do not directly provide the scaling behavior of the correlation functions, they do contain the complete information about the conformal weights as well as the central charge. The first step in this direction was the realization by Kirillov and Reshetikhin that the central charge can be expressed in terms of the quantum dimensions. Earlier mathematical results had been obtained by Lewin. Denote by L Rogers' dilogarithm

$$L(z) = Li_2(z) + \frac{1}{2}\log(z)\ \log(1-z) \tag{5.4}$$

and Li_2 is Euler's classical dilogarithm

$$Li_2(z) = \sum_{n\in\mathbb{N}} \frac{z^n}{n^2}. \tag{5.5}$$

Then one has the following result.

Theorem 5.3[**19, 11, 12**]. *For the generalized quantum dimensions $Q_{\ell m}$ one finds the following relations*

$$\frac{1}{L(1)}\sum_{\ell=1}^{k} L\left(\frac{1}{Q_{\ell m}^2}\right) = \frac{3k}{k+2} - 24\Delta_m^{(k)} + 6m. \tag{5.6}$$

For $m=0$ this theorem reduces to the central charge result in terms of the quantum dimensions $Q_\ell = S_{\ell 0}/S_{00}$

$$\frac{1}{L(1)}\sum_{\ell=1}^{k} L\left(\frac{1}{Q_{\ell}^2}\right) = \frac{3k}{k+2}, \tag{5.7}$$

obtained earlier in [**13, 16**].

5.4 It follows that the quantum dimensions contain the essential information about the spectrum of the conformal field theory and Rogers' dilogarithm provides, via Euler's dilogarithm, the map from the quantum dimensions to the central charge and the conformal weights. A review of these results and references to the original literature can be found in [**12**].

5.5 In the case of the elliptic curve E_4 the exactly solvable model is a tensor product of two $A_1^{(1)}$ theories at affine level $k=2$, equipped with the diagonal affine invariant, indicated as usual by the index A,

$$\mathbb{P}_{(1,1,2)} \supset E_4 \ \sim\ \left((A_1^{(1)})^{\otimes 2}_{k=2,A}\right)_{\rm GSO}. \tag{5.8}$$

The quantum dimensions of the $A_1^{(1)}$ theory at level $k=2$ take values in the quadratic extension $\mathbb{Q}(\sqrt{2})$ and therefore we find that the field of quantum dimensions provides a physical explanation of the emergence of the Legendre character in the modularity relation above. Hence Theorems 1.1 and 1.2 can be interpreted as a geometric derivation of a conformal field theoretic objects – the conformal field theoretic modular forms $\Theta^1_{1,1}(3\tau)\Theta^1_{1,1}(9\tau)$ and $\Theta^2_{1,1}(4\tau)^2$ are determined by the geometric modular forms defined by the Hasse-Weil L-functions. The interpretation of the second form involves the twist by the quadratic character associated to the number field of the quantum dimensions.

This physical explanation of the origin of the twist makes it clear why no character was necessary in the discussion in [**23**] of the plane cubic elliptic curve C_3 – the quantum dimensions of the affine theory $A_1^{(1)}$ at conformal level $k=1$ take values in the field of rational numbers $\mathbb{Q}$.

6. Generalizations

6.1 For more general curves of higher genus it is possible to decompose the Jacobians in a way similar to the analysis in previous sections. In general the factorization behavior can be quite involved and in the following we briefly describe what is known about the splitting behavior. For more details we refer to the original literature [**4, 20**] (see also [**17, 18**]). The reason why this factorization behavior is relevant in the present context is that the splitting occurs on the level of isogenies, and isogenous varieties have the same L-functions. Also important is that the varieties that emerge in this factorization admit complex multiplication and therefore their L-functions are determined by algebraic Hecke characters (see e.g. [**17, 18**]). Hence the factorization of Jacobian varieties allows one to determine modular forms associated to these higher genus curves.

6.2 It was shown by Faddeev [**4**] that the Jacobian variety $J(C_d)$ of Fermat curves $C_d \subset \mathbb{P}_2$ of prime degree splits into a product of abelian factors A_{O_i}

$$J(C_d) \cong \prod_{O_i \in I/(\mathbb{Z}/d\mathbb{Z})^\times} A_{O_i}, \tag{6.1}$$

where the set I provides a parametrization of the cohomology of C_d, and the sets O_i are orbits in I of the group $(\mathbb{Z}/d\mathbb{Z})^\times$. More precisely, it was shown that there is an isogeny

$$i : J(C_d) \longrightarrow \prod_{O_i \in I/(\mathbb{Z}/d\mathbb{Z})^\times} A_{O_i}. \tag{6.2}$$

Explicitly, I is the set of triplets (r,s,t) parametrizing a basis of the cohomology

$$\mathrm{H}^1(C_d) = \left\{\omega_{r,s,t} = x^{r-1}y^{s-d}dx \mid r,s,t \in \mathbb{N}, 0 < r,s,t < d,\ r+s+t = 0 \bmod d\right\}, \tag{6.3}$$

and the abelian varieties $A_d^{[(r,s,t)]}$ are associated to orbits $[(r,s,t)]$ of the triplets (r,s,t) with respect to the group $(\mathbb{Z}/d\mathbb{Z})^\times$.

The periods of the Fermat curve have been computed by Rohrlich [**20**] to be

$$\int_{A^jB^k\kappa} \omega_{r,s,t} = \frac{1}{d}B\left(\frac{s}{d},\frac{t}{d}\right)(1-\xi^s)(1-\xi^t)\xi^{js+kt}, \tag{6.4}$$

where ξ is a primitive $d-$th root of unity, and

$$B(u,v) = \int_0^1 t^{u-1}(1-v)^{v-1}dt \tag{6.5}$$

is the classical beta function. A, B are the two automorphism generators

$$\begin{aligned} A(1,y,z) &= (1,\xi y, z)\\ B(1,y,z) &= (1,y,\xi z) \end{aligned} \tag{6.6}$$

and κ is the generator of $\mathrm{H}_1(C_d)$ as a cyclic module over $\mathbb{Z}[A,B]$. The period lattice of the Fermat curve is therefore the span of

$$\left(\dots,\xi^{jr+ks}(1-\xi^r)(1-\xi^s)\frac{1}{d}B\left(\frac{r}{d},\frac{s}{d}\right),\dots\right)_{\substack{1\le r,s,t\le d-1\\ r+s+t=d}},\quad \forall 0 \le j,k \le d-1. \tag{6.7}$$

The abelian factor $A_{[(r,s,t)]}$ associated to the orbit $O_{r,s,t} = [(r,s,t)]$ can be obtained as the quotient

$$A_{[(r,s,t)]} = \mathbb{C}^{\varphi(d_0)/2}/\Lambda_{r,s,t}, \tag{6.8}$$

where $d_0 = d/\gcd(r,s,t)$ and the lattice $\Lambda_{r,s,t}$ is generated by elements of the form

$$\sigma_a(z)(1-\xi^{as})(1-\xi^{at})\frac{1}{d}B\left(\frac{\langle as\rangle}{d},\frac{\langle at\rangle}{d}\right), \tag{6.9}$$

where $z \in \mathbb{Z}[\mu_{d_0}]$, $\sigma_a \in \mathrm{Gal}(\mathbb{Q}(\mu_{d_0})/\mathbb{Q})$ runs through subgroups of the Galois group of the cyclotomic field $\mathbb{Q}(\mu_{d_0})$ and $\langle x\rangle$ is the smallest integer $0 \leq x < d$ congruent to x mod d.

6.3 Alternatively, the abelian variety $A_d^{r,s,t}$ can be constructed in a more geometric way as follows. Consider the orbifold of the Fermat curve C_d with respect to the group defined as

$$G_d^{r,s,t} = \left\{(\xi_1,\xi_2,\xi_3) \in \mu_d^3 \mid \xi_1^r\xi_2^s\xi_3^t = 1\right\}. \tag{6.10}$$

The quotient $C_d/G_d^{r,s,t}$ can be described algebraically via projections

$$\begin{aligned} T_d^{r,s,t}: C_d &\longrightarrow C_d^{r,s,t} \\ (x,y) &\mapsto (x^d, x^r y^s) =: (u,v), \end{aligned} \tag{6.11}$$

which map C_d into the curves

$$C_d^{r,s,t} = \left\{v^d = u^r(1-u)^s\right\}. \tag{6.12}$$

For prime degrees the abelian varieties $A_d^{r,s,t}$ can be defined simply as the Jacobians $J(C_d^{r,s,t})$ of the projections $C_d^{r,s,t}$. When d has nontrivial divisors $m|d$, this definition must be modified as follows. Consider the projected Fermat curves

$$\begin{aligned} C_d &\longrightarrow C_m \\ (x,y) &\mapsto (\bar{x},\bar{y}) := \left(x^{\frac{d}{m}}, y^{\frac{d}{m}}\right), \end{aligned} \tag{6.13}$$

whose Jacobians can be embedded as $e: J(C_m) \longrightarrow J(C_d)$. Composing the projection $T_d^{r,s,t}$ as

$$J(C_m) \xrightarrow{e} J(C_d) \xrightarrow{T_d^{r,s,t}} J(C_d^{r,s,t}) \tag{6.14}$$

for all proper divisors $m|d$ leads to a collection of subvarieties $\cup_{m|d} T_d^{r,s,t}(e(J(C_m)))$. The abelian variety of interest is then defined as

$$A_d^{r,s,t} = J(C_d^{r,s,t})/\cup_{m|d} T_d^{r,s,t}(e(J(C_m))). \tag{6.15}$$

The abelian varieties $A_d^{r,s,t}$ are not necessarily simple but it can happen that they in turn can be factored. This question can be analyzed via a criterion of Shimura-Taniyama, described in [**27**].

6.4 We briefly summarize in a further example how the factorization of Fermat curve Jacobians guides the modular analysis described in the case of the quartic Fermat curve. The curve C_4 is particularly simple because its Jacobian decomposes into a triple product of the Jacobian of a single curve. In general the factorization will involve different abelian factors, which is illustrated by the curve C_6 of degree 6. The L-function of C_6 is given by (here $\doteq$ indicates that we consider only the terms involving the good primes)

$$L_{\mathrm{HW}}(C_6,s) \doteq 1 + \frac{8}{7^s} - \frac{4}{13^s} - \frac{52}{19^s} + \frac{32}{31^s} + \cdots \tag{6.16}$$

and it is not as obvious as in the case of C_4 how one might try to decompose this series.

Analyzing the structure of the Jacobian shows that it factors into three different products associated to the orbits of $(\mathbb{Z}/6\mathbb{Z})^\times$ and their permutations, leading to the orbits

$$O_{1,1,4},\ O_{1,2,3},\ O_{2,2,2}. \tag{6.17}$$

All three orbits define elliptic curves, and one finds that the Jacobian of the genus ten curve C_6 decomposes into a product of Jacobians of elliptic curves of conductor 432, 144, and 27 respectively. The elliptic factor of conductor 27 is precisely the curve C_3, analyzed in Theorem 1.1, providing one of the three known solvable Brieskorn-Pham curves. The elliptic curve E_4 discussed in the context of the Fermat curve C_4 provides the second exactly solvable curve, and the elliptic factor of conductor 144 appearing here is particularly interesting because its L-function

$$L_{\mathrm{HW}}(E_6, s) = 1 + \frac{4}{7^s} + \frac{2}{13^s} - \frac{8}{19^s} - \frac{5}{25^s} + \frac{4}{31^s} + \cdots \tag{6.18}$$

is that of the degree six Brieskorn-Pham curve

$$E_6 = \left\{(x_0, x_1, x_2) \in \mathbb{P}_{(1,2,3)} \mid x_0^6 + x_1^3 + x_2^3 = 0\right\}, \tag{6.19}$$

which is the third and final elliptic Brieskorn-Pham curve conjectured to be exactly solvable. This class of exactly solvable elliptic curves is described in detail in [**22**]. It is shown there that the Mellin transform $f_{\mathrm{HW}}(E_6, q) \in S_2(\Gamma_0(144))$ of the L-function of E_6 can be factored in terms of the theta function $\Theta_{1,1}^1$ as

$$f_{\mathrm{HW}}(E_6, q) = \Theta_{1,1}^1(q^6)^2 \otimes \chi_3, \tag{6.20}$$

where χ_3 is the usual Legendre character.

These results for E_3, E_4, and E_6 show that for all elliptic curves of Brieskorn-Pham type the geometric modular form derived from the Hasse-Weil L-function factors into a twisted product of Hecke indefinite modular forms derived from the string functions of the affine algebra $A_1^{(1)}$.

Acknowledgements. Part of this work was completed while RS was supported as a Scholar at the Kavli Institute for Theoretical Physics in Santa Barbara. It is a pleasure to thank the KITP for hospitality. This work was supported in part by the National Science Foundation under Grant No. PHY99-07949, an IUSB Faculty Research Grant (ML), and a KSU Incentive Grant for Scholarship (RS).

References

[1] C. Breuil, B. Conrad, F. Diamond, R. Taylor, *On the modularity of elliptic curves over* $\mathbb{Q}$*, or wild 3-adic exercises*, J. Am. Math. Soc. **14** (2001) 843 – 939

[2] P. Candelas, M. Lynker and R. Schimmrigk, *Calabi-Yau Manifolds in Weighted* $\mathbb{P}_4$, Nucl. Phys. **B341** (1990) 383 – 402

[3] M. Eichler, *Quaternäre quadratische Formen und die Riemannsche Vermutung für die Kongruenzzetafunktion*, Arch. Math. **5** (1954) 355 – 366

[4] D. K. Faddeev, *On the Divisor Groups of some Algebraic Curves*, Dokl. Tom **136** (1961) 296-298 [Sov. Math **2**(1961) 67-69]
Invariants of Divisor Classes for the Curves $y^\ell = x^k(1-x)$ *in an* ℓ*-adic Cyclotomic Field*, Trudy Mat. Inst. Steklova **64** (1961) 284 – 293 (In Russian)

[5] D. Gepner, *Spacetime Supersymmetry in Compactified String Theory and Superconformal Models*, Nucl.Phys. **B296**(1988)757

[6] B. H. Gross, *On the Periods of Abelian Integrals and a Formula of Chowla and Selberg*, Inv. math. **45** (1978) 193

[7] H. Hasse, *Über die Kongruenzzetafunktionen. Unter Benutzung von Mitteilungen von Prof. Dr. F.K. Schmidt und Prof. Dr. E. Artin*, S. Ber. Preuß. Akad. Wiss. H. **17** (1934) 250 – 263

[8] B. Hunt and R. Schimmrigk, *K3-fibered Calabi-Yau Threefolds I, The Twist Map*, Int. J. Math. **10** (1999) 833 – 869, [arXiv: math.AG/9904059]

[9] V.G. Kac, *Infinite Lie Algebras*, Cambridge University Press, 1990s

[10] V.G. Kac and D.H. Peterson, *Infinite-Dimensional Lie Algebras, Theta Functions and Modular Forms*, Adv. Math. **53** (1984) 125 – 264

[11] A.N. Kirillov, *Dilogarithm Identities, Partitions, and Spectra in Conformal Field Theory*, Algebra i Analiz **6** (1994), no.2, 152–175; translation in St. Petersburg Math. J. **6** (1995), no.2, 327–348, [arXiv: hep-th/9212150]

[12] A.N. Kirillov, *Dilogarithm Identities*, Progr. Theor. Phys. Suppl **118** (1995) 61, [arXiv: hep-th/9408113]

[13] A.N. Kirillov and N.Yu. Reshetikhin, *Exact solution of the integrable XXZ Heisenberg model with arbitrary spin, I, II*, J. Phys. **A20** (1987) 1565 – 1586, 1587 – 1597

[14] A. Klemm and R. Schimmrigk, *Landau-Ginzburg String Vacua*, Nucl. Phys. **B411** (1994) 559 – 583, [arXiv: hep-th/9204060];
M. Kreuzer and H. Skarke, *No Mirror Symmetry in Landau-Ginzburg Spectra!*, Nucl. Phys. **B388** (1992) 113 – 130, [arXiv: hep-th/9205004]

[15] E.Martinec, *Algebraic Geometry and Effective Lagrangians*, Phys.Lett. **B217**(1989)431;
C.Vafa and N.Warner, *Catastrophes and the Classification of Conformal Field Theories*, Phys.Lett. **B218**(1989)51;
W.Lerche, C.Vafa and N.Warner, *Chiral Rings in N=2 Superconformal Theories*, Nucl. Phys. **B324** (1989) 427

[16] L. Lewin, *Polylogarithms and Associated Functions*, North-Holland Publishing Co. 1981.

[17] M. Lynker, V. Periwal and R. Schimmrigk, *Complex Multiplication and Black Hole Attractor Varieties*, Nucl. Phys. **B667** (2003) 484 – 504, [hep-th/0303111]

[18] M. Lynker, R. Schimmrigk and S. Stewart, *Complex Multiplication of Exactly Solvable Calabi-Yau Varieties*, Nucl. Phys. **B700** (2004) 463 – 489, [hep-th/0312319]

[19] W. Nahm, A. Recknagel and M. Terhoeven, *Dilogarithm Identities in Conformal Field Theory*, Mod. Phys. Letts. **A8** (1993) 1835 – 1848, [arXiv: hep-th/9211034]

[20] D. E. Rohrlich, *The Periods of the Fermat Curve*, Appendix to Gross 1978, Inv. math. **45** (1978) 208

[21] R. Schimmrigk, *Arithmetic of Calabi-Yau Varieties and Rational Conformal Field Theory*, J. Geom. Phys. **44** (2003) 555 – 565, [arXiv: hep-th/0111226]

[22] R. Schimmrigk, *Arithmetic Spacetime Geometry from String Theory*, to appear

[23] R. Schimmrigk and S. Underwood, *The Shimura-Taniyama Conjecture and Conformal Field Theory*, J. Geom. Phys. **48** (2003) 169 – 189, [arXiv: hep-th/0211284]

[24] F. K. Schmidt, *Analytische Zahlentheorie in Körpern der Charakteristik p*, Math. Z. **33** (1931) 1 – 32

[25] G. Shimura, *Correspondences modulaires et les fonction ζ de courbes algebrique*. J. Math. Soc. Japan **10** (1958) 1 – 28

[26] G. Shimura, *Sur les intégrales attachés aux formes automorphes*, J. Math. Soc. Japan **11** (1959) 291 – 311

[27] G. Shimura and Y. Taniyama, *Complex Multiplication of Abelian Varieties and its Application to Number Theory*, 1961

[28] J. Tate, *Algorithm for Determining the Type of a Singular Fiber in on Elliptic Pencil*, Letter to J. Cassels, in MODULAR FUNCTIONS OF ONE VARIABLE IV, eds. B.J. Birch and W. Kuyk,, Springer Verlag, LNM 476, 1975

[29] A. Weil, *Jacobi sums as "Grössencharaktere"*, Trans. Amer. Math. Soc. **73** (1952) 487

[30] A. Weil, *Über die Bestimmung Dirichletscher Reihen durch Funktionalgleichungen*, Math. Ann. **168** (1967) 149–156.

[31] A. Weil, *Sur les périodes des intégrales abeliennes*, Comm. Pure Appl. Math **29** (1976) 813 – 819.

[32] A. Wiles, *Modular elliptic curves and Fermat's Last Theorem*, Ann. Math. **141** (1995) 443 – 551;
R. Taylor and A. Wiles, *Ring-theoretic properties of certain Hecke algebras*, Ann. of Math. **141** (1995) 553 – 572.

INDIANA UNIVERSITY SOUTH BEND, SOUTH BEND, IN 46634
E-mail address: mlynker@iusb.edu

KENNESAW STATE UNIVERSITY, KENNESAW, GA 31044
E-mail address: netahu@yahoo.com

AMS/IP Studies in Advanced Mathematics
Volume 38, 2006

Mahler Measure Variations, Eisenstein Series and Instanton Expansions

Jan Stienstra

Abstract. This paper points at an intriguing inverse function relation with on the one hand the coefficients of the Eisenstein series in Rodriguez Villegas' paper on "Modular Mahler Measures" and on the other hand the instanton numbers in papers on "Non-Critical Strings" by Klemm-Mayr-Vafa and Lerche-Mayr-Warner. In a companion paper Mahler measures will be related to dimer models. Thus, if it is not just a lucky coincidence in a few examples, our inverse function relation may be an incarnation of the duality between string models and dimer models proposed by Okounkov, Reshetikhin and Vafa in their paper "Quantum Calabi-Yau and Classical Crystals".

Introduction

The *logarithmic Mahler measure* $\mathsf{m}(F)$ and the *Mahler measure* $\mathsf{M}(F)$ of a Laurent polynomial $F(x,y)$ with complex coefficients are:

$$\mathsf{m}(F) := \frac{1}{(2\pi i)^2}\oint\oint_{|x|=|y|=1} \log|F(x,y)|\,\frac{dx}{x}\frac{dy}{y}\,, \tag{1}$$

$$\mathsf{M}(F) := \exp(\mathsf{m}(F))\,. \tag{2}$$

Wondering how the Mahler measure depends on the coefficients of the Laurent polynomial, and more in particular on the constant term, we will look at the Mahler measure $\mathsf{M}(F_t)$ of Laurent polynomials

$$F_t(x,y) := t - (xy)^{-1}\widetilde{P}(x,y,1)$$

where $\widetilde{P}(X,Y,Z)$ is a fixed weighted homogeneous polynomial and t is a complex variable. Geometrically speaking, we look at the Mahler measure of the fibres of the map $\mathbb{P}^2 \longrightarrow \mathbb{P}^1$, $(X,Y,Z) \mapsto [\widetilde{P}(X,Y,Z) : (XYZ)]$ on the (weighted) projective plane $\mathbb{P}^2$. In the examples these fibres are elliptic curves. The variation in the Mahler measure is then related to the variation in the periods, or rather the period ratio, of these elliptic curves (see §1). This is what Rodriguez Villegas was investigating when he noticed that some of his formulas looked similar to formulas in the literature on Mirror Symmetry; see the final remarks in [**11**]. The modest goal of the present paper is to point out that actual instanton expansions matching his

---

2000 *Mathematics Subject Classification.* Primary 11R99, 11F11.

Key words and phrases. Mahler measure, Eisenstein series, Instanton expansion, dimers.

examples do occur in the physics literature [**8, 6, 7, 4**]. The relation is, however, more intriguing and less straightforward than one might expect from the remarks in [**11**]:

Observation:[1] *The expansions* $1-\sum b_n \frac{n^2\mathsf{q}^n}{1-\mathsf{q}^n}$ *of the weight* 3 *modular forms in Examples* 1, 2, 3 *of* [**11**] *are related to the instanton expansions* $\kappa(1-\sum a_n \frac{n^3\mathsf{Q}^n}{1-\mathsf{Q}^n})$ *in Examples* E_6, E_5, E_8, *respectively, in* [**6, 8**], *via the inverse function relations*

$$-\mathsf{Q} = \mathsf{q}\prod_{n\geq 1}(1-\mathsf{q}^n)^{nb_n}\,, \qquad -\mathsf{q} = \mathsf{Q}\prod_{n\geq 1}(1-\mathsf{Q}^n)^{n^2a_n}\,. \tag{3}$$

The constant κ *is* $-3, -4, -1$ *for Examples* E_6, E_5, E_8, *respectively.*

The Mahler measure of the elliptic curve with modular parameter q *is* $|\mathsf{Q}|^{-1/\nu}$ *with* $\nu = 3,\ 2,\ 6$ *for Examples* E_6, E_5, E_8, *respectively.* □

That Examples 1, 2, 3 of [**11**] line up with Examples E_6, E_5, E_8 in [**6, 8**] is immediately obvious from the formulas for the elliptic pencils and their Picard-Fuchs equations which are explicitly given in [**6, 8, 11**]. Looking at how the various authors get from the Picard-Fuchs equations to their expansions directly leads to Relation (3); for more details see Section 1.

In Section 2 we present six such examples including the above ones and the E_7 case of [**6, 8**]. In each case the computer output suggests that the numbers a_n and b_n are integers. Except for the b_n in four examples we have no proof for this suggested integrality.

Discussion and new directions. While the second product in (3) summarizes some of the results (instanton expansions) in the String Theory literature of the 1990's, Mahler measure and an analogue of the first product in (3) appear in the String related literature of the 2000's. Theorem 3.5 of [**5**], for instance, gives a formula for the partition function of a dimer model which is exactly the same as the defining formula (2) for the Mahler measure of the characteristic polynomial of that model. In [**14**] we show that the polynomials of Examples ♯ 3, ♯ 4 and ♯ 6 of the present paper are characteristic polynomials of dimer models. The partition functions of these dimer models can therefore be given as the multiplicative inverses of the first products in (3). For Example ♯ 3, for instance, this product is

$$\mathsf{q}\prod_{n\geq 1}(1-\mathsf{q}^n)^{9n\chi_{-3}(n)}\,, \qquad \chi_{-3}(n) = 0, 1, -1 \quad \text{for} \quad n \equiv 0, 1, 2 \bmod 3.$$

There is here some similarity with MacMahon's function

$$M(q) := \prod_{n\geq 1}(1-\mathsf{q}^n)^{-n}$$

which appears in partition functions in [**9, 10**]. The fact that MacMahon's function is the generating function for plane partitions (i.e. with 3-dimensional Young diagrams) led to unexpected new directions in String Theory research: *the melting crystal and random surface pictures* [**5, 9, 10**].

[1] I call this an observation and not a theorem, since it can be 'proved' by just pointing out how pieces of existing literature fit together (which, however, as far as I know, were never presented together in this form).

Acknowledgements. I want to thank Don Zagier for bringing [**11**] to my attention (somewhen around 1998) and for keeping me well informed about his work in progress [**15**]. I thank Bogdan Florea for discussing with me in November 2001 some parts of his work, thereby also providing references to [**6, 8**]. I want to thank Noriko Yui for the invitation to the Banff Workshop on Mirror Symmetry with its stimulating atmosphere to report on this work.

1. Details of the observation

We investigate Laurent polynomials of the form

$$F_t(x,y) := t - (xy)^{-1}\widetilde{P}(x,y,1)$$

with t a complex parameter and $\widetilde{P}(X,Y,Z)$ as in Table 1. Writing $\xi = t^{-1}$ and assuming $\max_{|x|=|y|=1} |\widetilde{P}(x,y,1)||\xi| < 1$ one then finds

$$\mathsf{M}(F_t) = |Q|^{-1} \tag{4}$$

with

$$\begin{aligned} Q \;&:=\; \beta \exp\left(-\frac{1}{(2\pi i)^2}\oint\!\!\oint_{|x|=|y|=1} \log(t - \frac{\widetilde{P}(x,y,1)}{xy})\,\frac{dx}{x}\frac{dy}{y}\right) \\ &=\; \beta\,\xi \exp\left(\sum_{n=1}^{\infty}\frac{\xi^n}{n}\left(\frac{1}{(2\pi i)^2}\oint\!\!\oint_{|x|=|y|=1}\frac{\widetilde{P}(x,y,1)^n}{(xy)^n}\frac{dx}{x}\frac{dy}{y}\right)\right) \end{aligned} \tag{5}$$

and β an as yet undetermined complex number of absolute value 1. The number β is, of course, irrelevant for (4), but for the correct match with [**8, 6**] it will be necessary to take here an appropriate ξ-independent constant.

Thus

$$Q = \beta\,\xi \exp\left(\sum_{n=1}^{\infty} c_n \frac{\xi^n}{n}\right) \tag{6}$$

with

$$c_n = \text{coefficient of } X^nY^nZ^n \text{ in } \widetilde{P}(X,Y,Z)^n. \tag{7}$$

We refer to Table 2 for concrete examples. Now let $\overline{\mathcal{E}}_t$ denote the closure of

$$\mathcal{E}_t = \{(x,y) \in \mathbb{C}^* \times \mathbb{C}^* \mid \widetilde{P}(x,y,1) = txy\}$$

in the (appropriately weighted) projective plane. Restricting to values of $t = \xi^{-1}$ for which the curve $\overline{\mathcal{E}}_t$ is non-singular, differentiating (5) and using the residue theorem one finds

$$\xi\frac{d}{d\xi}\log Q \;=\; \frac{1}{(2\pi i)^2}\oint\!\!\oint_{|x|=|y|=1}\frac{t\,dx\,dy}{t\,xy - \widetilde{P}(x,y,1)} \;=\; \frac{1}{2\pi i}\oint_{\gamma}\omega_t \tag{8}$$

where γ is a closed loop and ω_t is a holomorphic 1-form on the curve $\overline{\mathcal{E}}_t$,

$$\omega_t := \frac{t dx}{tx - \frac{\partial}{\partial y}\widetilde{P}(x,y,1)}$$

on the coordinate patch where the denominator does not vanish.

Thus $\xi\frac{d}{d\xi}\log Q$ is a period of ω_t. The periods of ω_t are functions of $\xi = t^{-1}$ which satisfy a second order linear differential equation: the *Picard-Fuchs equation* of the elliptic pencil. In our examples there is only one solution $g_1(\xi)$ of the Picard-Fuchs equation which is holomorphic in a neighborhood of $\xi = 0$ and has $g_1(0) = 1$. Differentiating (6) shows that this solution is

$$g_1(\xi) \;=\; \xi \frac{d}{d\xi} \log Q \;=\; \sum_{n=0}^{\infty} c_n \xi^n . \tag{9}$$

A second solution has the form

$$g_2(\xi) \;=\; g_1(\xi) \log \xi \;+\; h(\xi),$$

where $h(\xi)$ is a holomorphic function of ξ which vanishes at $\xi = 0$. The classical procedure in analyzing elliptic pencils is to consider the period ratio and to set

$$q := \exp\left(\frac{g_2(\xi)}{g_1(\xi)}\right) \;=\; \xi \exp\left(\frac{h(\xi)}{g_1(\xi)}\right). \tag{10}$$

Formulas (6) and (10) give Q and q as functions of ξ, while

$$q = \xi + O(\xi^2)\,, \qquad Q = \beta\xi + O(\xi^2).$$

One can eliminate ξ and directly express Q as a function of q and vice versa. Although one could now express this inverse function relationship through product expansions as in (3), it is better to wait: in the examples there is a positive integer ν (see Table 2) such that Q^ν is a function of q^ν and vice versa. It is this latter inverse function relationship which is used in (3) with

$$\mathsf{Q} := Q^\nu\,, \qquad \mathsf{q} := q^\nu\,. \tag{11}$$

Here are more details about the role of ν. One solution of the Picard-Fuchs differential equation is:

$$g_1 = \sum_{m=0}^{\infty} u_m \psi^m = \sum_{n=0}^{\infty} c_n \xi^n . \tag{12}$$

with $\psi = \xi^\nu$, u_m and c_n as in Table 2. The numbers u_m and (A, B, λ) in Table 2 satisfy the recurrence relation

$$(m+1)^2 u_{m+1} - (Am^2 + Am + \lambda) u_m + Bm^2 u_{m-1} = 0 \qquad (m \geq 0)\,. \tag{13}$$

This can easily be checked by hand for the hypergeometric cases ♯ 1–♯ 4. The recurrence in case ♯ 5 is due to Apéry; the numbers u_m appear in his famous irrationality proof for $\zeta(2)$. The recurrence in case ♯ 6 is due to Cusick. Proofs for the recurrences in cases ♯ 5 and ♯ 6 (which in fact first derive the Picard-Fuchs equation) can be found in, for instance, [**13**] §11. The (A, B, λ)-notation was introduced by Zagier [**15**], who made an extensive search for all such triples for which all u_m in the recurrence (13) are integers.

The recurrence relation (13) is equivalent to the differential equation

$$\psi \frac{d}{d\psi}\left((1 - A\psi + B\psi^2)\psi \frac{d}{d\psi} g\right) + \psi(-\lambda + B\psi) g = 0 \tag{14}$$

for the generating function g_1 in (12). A second solution of the differential equation (14) can be obtained by the method of Frobenius. For this we make the Ansatz

$$g(\psi) = \sum_{m=0}^{\infty} u_m(\epsilon) \psi^{m+\epsilon}\,, \qquad u_0(\epsilon) = 1.$$

This is a solution of (14) if and only if $\epsilon^2 = 0$ and for all $m \geq 0$

(15) $(m+\epsilon)^2 u_m(\epsilon) = (A(m+\epsilon)^2 - A(m+\epsilon)+\lambda)u_{m-1}(\epsilon) - B(m-1+\epsilon)^2 u_{m-2}(\epsilon).$

We set $u_0(\epsilon) = 1$, solve the recurrrence and then expand

(16) $$g(\psi) = g_1(\psi) + g_2(\psi)\epsilon.$$

Then $g_1(\psi)$ and $g_2(\psi)$ are solutions of (14), $g_1(\psi) = \sum_{m=0}^{\infty} u_m \psi^m$ as before and $g_2(\psi) = g_1(\psi)\log\psi + h(\psi)$ where $h(\psi)$ is a power series in ψ with $h(0) = 0$. Now set, with ν as in Table 2,

(17) $$q := \exp\left(\frac{g_2(\psi)}{\nu\, g_1(\psi)}\right) = \xi \exp\left(\frac{h(\xi^\nu)}{\nu\, g_1(\xi^\nu)}\right).$$

This definition of q agrees with (10), but note that $\mathsf{q} := q^\nu$ is actually a function of $\psi = \xi^\nu$ and $\mathsf{q} = \psi + O(\psi^2)$.

Now recall (9) and (12):

(18) $$\psi\frac{d}{d\psi}\log Q^\nu \;=\; \xi\frac{d}{d\xi}\log Q \;=\; \sum_{n=0}^{\infty} c_n \xi^n \;=\; \sum_{m=0}^{\infty} u_m \psi^m .$$

This shows that $\mathsf{Q} := Q^\nu$ is actually a function of ψ and that $\mathsf{Q} = \alpha\psi + O(\psi^2)$ for some constant α. One can eliminate ψ and directly express Q as a function of q and vice versa. This is the inverse function relationship used in (3), with $\alpha = -1$.

By comparing the above formulas for Q and for q with the formulas in n°13 of [**11**] one sees that our $\mathsf{q}\frac{d}{d\mathsf{q}}\log\mathsf{Q} = \left(\psi\frac{d}{d\psi}\log\mathsf{Q}\right)\left(\mathsf{q}\frac{d}{d\mathsf{q}}\log\psi\right)$ equals Rodriguez Villegas' $e(\tau)$, which in the examples has an expansion like an Eisenstein series of weight 3:

(19) $$\mathsf{q}\frac{d}{d\mathsf{q}}\log\mathsf{Q} = 1 - \sum_{n\geq 1} b_n \frac{n^2\mathsf{q}^n}{1-\mathsf{q}^n}.$$

Note that (9) determines Q only up to a multiplicative constant. This constant is irrelevant for the expansion (19). On the other hand, for [**8**] fixing this constant is very relevant. It is called fixing the B-field. Equations (4.10) and (4.13) of [**8**] essentially boil down to

(20) $$\mathsf{Q}\frac{d}{d\mathsf{Q}}\log\mathsf{q} \;=\; 1 - \sum_{n\geq 1} a_n \frac{n^3\mathsf{Q}^n}{1-\mathsf{Q}^n};$$

to really get the expansions of [**8**] Equation (4.10) the right hand side of (20) has to be multiplied by an appropriate constant (see the numerical results in the next section).

With (19) and (20) one can express Q as a function of q and vice versa, both up to a multiplicative constant. These constants are fixed by

$$\mathsf{Q} = \alpha\psi + O(\psi^2), \qquad \mathsf{q} = \psi + O(\psi^2)$$

Thus one arrives at the following functional relation between Q and q:

(21) $$\alpha^{-1}\mathsf{Q} = \mathsf{q}\prod_{n\geq 1}(1-\mathsf{q}^n)^{nb_n}, \qquad \alpha\mathsf{q} = \mathsf{Q}\prod_{n\geq 1}(1-\mathsf{Q}^n)^{n^2 a_n},$$

Incidentally, the relation between the α here and the β in (5) is $\alpha = \beta^\nu$. In the examples $\alpha = \pm 1$ and all a_n and b_n turn out to be integers. Except for some of the b_n we have no proof for this integrality.

	$\widetilde{P}(X,Y,Z)$
$\sharp$ 1	$X^2+Y^3+Z^6$
$\sharp$ 2	$X^2+Y^4+Z^4$
$\sharp$ 3	$X^2Y+Y^2Z+Z^2X$
$\sharp$ 4	$(X+Y)(XY+Z^2)$
$\sharp$ 5	$(X+Y+Z)(X+Z)(Y+Z)$
$\sharp$ 6	$(X+Y)(Y+Z)(Z+X)$

TABLE 1

2. Numerical results for six examples

Our examples deal with pencils of elliptic curves in a (weighted) projective plane $\mathbb{P}^2$ described by the polynomials in Table 1:

$$\overline{\mathcal{E}}_t := \{(X,Y,Z) \in \mathbb{P} \mid \widetilde{P}(X,Y,Z) = tXYZ\}. \tag{22}$$

Beauville [**2**] showed that there are exactly six semi-stable families of elliptic curves over $\mathbb{P}^1$ with four singular fibres. His list consists of cases $\sharp$ 3 - $\sharp$ 6 in our Table 1 plus the polynomials $(X+Y)((X+Y+Z)^2-4YZ)$ and $X^3+Y^3+Z^3$. We do not explicitly discuss the latter two, since the results coincide with those of $\sharp$ 4 resp. $\sharp$ 3.

Table 2 shows that cases $\sharp$ 1 - $\sharp$ 4 are of hypergeometric origin; more precisely they result by a substitution $t \mapsto t^6, t^4, t^3, t^2$, respectively, from elliptic pencils with exactly three singular fibres. Families of elliptic curves with exactly three singular fibres have been classified in [**12**]. From the tables in [**12**] one can see that our examples come from the singular fibre configurations $\sharp 1 : (II^*\ I_1\ I_1)$, $\sharp 2 : (III^*\ I_2\ I_1)$, $\sharp 3 : (IV^*\ I_1\ I_3)$, $\sharp 4 : (I_1^*\ I_1\ I_4)$. Kodaira's notation for the singular fibres corresponds as follows with Dynkin diagrams: $II^* = E_8$, $III^* = E_7$, $IV^* = E_6$, $I_1^* = E_5 = D_5$. This agrees with the computations hereafter according to which examples $\sharp$ 1 –$\sharp$ 4 give cases E_8, E_7, E_6, E_5 in [**6, 8**].

We now come to the actual calculations (using PARI). From the input in the form of the triple (A, B, λ) we compute g_1 and $\mathsf{q} = q^\nu$ as functions of ψ using Formulas (15)-(17). Inverting the latter functional relation we get ψ and g_1 as functions of q. The numbers b_n are then computed from the expansion (cf. (19), (18))

$$g_1(\mathsf{q})\,\frac{\mathsf{q}}{\psi(\mathsf{q})}\,\frac{d\psi(\mathsf{q})}{d\mathsf{q}} = 1 - \sum_{n\geq 1} b_n \frac{n^2\mathsf{q}^n}{1-\mathsf{q}^n} = 1 - \sum_{n\geq 1}\left(\sum_{k\mid n} k^2 b_k\right)\mathsf{q}^n\,. \tag{23}$$

Having the numbers b_n we compute Q and the numbers a_n from (21):

$$\alpha^{-1}\mathsf{Q} = \mathsf{q}\prod_{n\geq 1}(1-\mathsf{q}^n)^{nb_n}\,, \qquad \alpha\mathsf{q} = \mathsf{Q}\prod_{n\geq 1}(1-\mathsf{Q}^n)^{n^2 a_n}\,. \tag{24}$$

Typically $\alpha = -1$ leads to the numbers which are reported in [**8**] Equations (4.10); only for the conifold case loc. cit. uses $\alpha = 1$.

We report the results of the calculations in tables for the numbers a_n and b_n and as η-products or Eisenstein series for $\psi(\mathsf{q})$, $g_1(\psi(\mathsf{q}))$ and $\mathsf{q}\frac{d}{d\mathsf{q}}\log\mathsf{Q}$. The elliptic pencils and Picard-Fuchs equations of Examples $\sharp$ 3-$\sharp$ 6 are well-known and well documented in the literature [**15, 11**]; also $\sharp$ 1 appears in [**11**] as example 3. So

	u_m	$c_n = u_m$	(A,B,λ)	ν
$\sharp$ 1	$\dfrac{(6m)!}{m!(2m)!(3m)!}$	if $6\|n$, $m=\frac{n}{6}$	$(432,0,60)$	6
$\sharp$ 2	$\dfrac{(4m)!}{m!^2(2m)!}$	if $4\|n$, $m=\frac{n}{4}$	$(64,0,12)$	4
$\sharp$ 3	$\dfrac{(3m)!}{m!^3}$	if $3\|n$, $m=\frac{n}{3}$	$(27,0,6)$	3
$\sharp$ 4	$\dfrac{(2m)!^2}{m!^4}$	if $2\|n$, $m=\frac{n}{2}$	$(16,0,4)$	2
$\sharp$ 5	$\sum_{k=0}^{m}\binom{m}{k}^2\binom{m+k}{k}$	$m=n$	$(11,\text{-}1,3)$	1
$\sharp$ 6	$\sum_{k=0}^{m}\binom{m}{k}^3$	$m=n$	$(7,\text{-}8,2)$	1

TABLE 2. $c_n = 0$ if n does not satisfy the condition in the third column

we have taken the expressions for $\psi(\mathsf{q})$ and $g_1(\psi(\mathsf{q}))$ directly from [**15, 11**]. Zagier [**15**] gives an argument for "recognizing modularity" which actually proves that the expressions are correct. However, as an extra security check we have checked that the q-expansions of these expressions coincide with our PARI calculations to order at least 100.

$$\boxed{\sharp\ \mathbf{1}\text{:}\ (A,B,\lambda) = (432,0,60)\,,\ \alpha = -1}$$

This corresponds to the E_8 example in [**8, 6**] and Example 3 in [**11**].

n	a_n	b_n
1	252	252
2	-9252	-13374
3	848628	1253124
4	-114265008	-151978752
5	18958064400	21255487740
6	-3589587111852	-3255937602498
7	744530011302420	531216722607876
8	-165076694998001856	-90773367805541376
9	38512679141944848024	16069733941012586748

The numbers of a_n are the instanton numbers of [**8, 6**] Example E_8. They also appear in [**4**] Table 7 column $X_6(1123)$. More precisely the expansion for the E_8 case in [**8**] Eq. (4.10) is:

$$-1\cdot\left(1-\sum_{n\geq 1} a_n \frac{n^3\mathsf{Q}^n}{1-\mathsf{Q}^n}\right)$$

Rodriguez Villegas [**11**] Example 3 gives the following expressions for, in our notation, $\psi(\mathsf{q})$, $g_1(\psi(\mathsf{q}))$ and $\mathsf{q}\frac{d}{d\mathsf{q}}\log \mathsf{Q}$:

$$\begin{aligned}\psi(\mathsf{q}) &= \frac{1}{864}\left(1 - E_6(\mathsf{q})E_4(\mathsf{q})^{-3/2}\right)\\ g_1(\psi(\mathsf{q})) &= E_4(\mathsf{q})^{1/4}\\ \mathsf{q}\frac{d}{d\mathsf{q}}\log \mathsf{Q} &= 1 - \sum_{n\geq 1} b_n \frac{n^2\mathsf{q}^n}{1-\mathsf{q}^n} = \frac{1}{2}E_4(\mathsf{q})^{3/4}\left(1 + E_6(\mathsf{q})E_4(\mathsf{q})^{-3/2}\right)\end{aligned}$$

where $E_4(\mathsf{q})$ and $E_6(\mathsf{q})$ are the Eisenstein series

$$E_4(\mathsf{q}) := 1 + 240\sum_{n=1}^{\infty}\sum_{d|n} d^3\mathsf{q}^n\,, \qquad E_6(\mathsf{q}) := 1 - 504\sum_{n=1}^{\infty}\sum_{d|n} d^5\mathsf{q}^n\,.$$

$\sharp$ **2:** $(A, B, \lambda) = (64, 0, 12)\,,\ \alpha = -1$

This corresponds to the E_7 example in [**8, 6**], but has no explicitly worked out counterpart in [**11**].

n	a_n	b_n
1	28	28
2	−136	−134
3	1620	996
4	−29216	−10720
5	651920	139292
6	−16627608	−2019450
7	465215604	31545316
8	−13927814272	−520076672
9	439084931544	8930941980

The numbers of $2a_n$ are the instanton numbers of [**8, 6**] Example E_7. They also appear in [**4**] Table 7 column $X_4(1112)$. More precisely the expansion for the E_7 case in [**8**] Eq. (4.10) is:

$$-2\cdot\left(1 - \sum_{n\geq 1} a_n \frac{n^3\mathsf{Q}^n}{1-\mathsf{Q}^n}\right)$$

Here we can only offer (without reference to proofs in the literature) expressions for $\psi(\mathsf{q})$, $g_1(\psi(\mathsf{q}))$ and $\mathsf{q}\frac{d}{d\mathsf{q}}\log \mathsf{Q}$, which are correct to at least $O(\mathsf{q}^{100})$:

$$\begin{aligned}\psi(\mathsf{q})^{-1} &= 64 + \frac{\eta(\mathsf{q})^{24}}{\eta(\mathsf{q}^2)^{24}}\\ g_1(\psi(\mathsf{q}) &= (2E_2(\mathsf{q}^2) - E_2(\mathsf{q}))^{\frac{1}{2}}\\ \mathsf{q}\frac{d}{d\mathsf{q}}\log \mathsf{Q} &= 1 - \sum_{n\geq 1} b_n \frac{n^2\mathsf{q}^n}{1-\mathsf{q}^n} = (2E_2(\mathsf{q}^2) - E_2(\mathsf{q}))^{\frac{3}{2}}\left(1 + 64\frac{\eta(\mathsf{q}^2)^{24}}{\eta(\mathsf{q})^{24}}\right)^{-1}\end{aligned}$$

where

$$E_2(\mathsf{q}) := 1 - 24\sum_{n=1}^{\infty}\sum_{d|n} d\mathsf{q}^n\,, \qquad \eta(\mathsf{q}) := \mathsf{q}^{\frac{1}{24}}\prod_{n\geq 1}(1-\mathsf{q}^n)\,.$$

♯ 3: $(A, B, \lambda) = (27, 0, 6)\,,\ \alpha = -1$

This corresponds to the E_6 example in [**8, 6**] and Example 1 in [**11**].

n	a_n	b_n
1	9	9
2	−18	−9
3	81	0
4	−576	9
5	5085	−9
6	−51192	0
7	565362	9
8	−6684480	−9
9	83246697	0

The numbers of $3a_n$ are the instanton numbers of [**8, 6**] Example E_6. The numbers $\frac{1}{3}a_n$ appear in [**4**] Table 1 and in [**7**] Fig. 2. More precisely the expansion for the E_6 case in [**8**] Eq. (4.10) is:

$$-3 \cdot \left(1 - \sum_{n \geq 1} a_n \frac{n^3 \mathsf{Q}^n}{1 - \mathsf{Q}^n}\right)$$

Zagier [**15**] and Rodriguez Villegas [**11**] give the following expressions for, in our notation, $\psi(\mathsf{q})$, $g_1(\psi(\mathsf{q}))$ and $\mathsf{q}\frac{d}{d\mathsf{q}} \log \mathsf{Q}$:

$$\begin{aligned}
\psi(\mathsf{q})^{-1} &= 27 + \frac{\eta(\mathsf{q})^{12}}{\eta(\mathsf{q}^3)^{12}} \\
g_1(\psi(\mathsf{q})) &= E_{1,\chi_{-3}}(\mathsf{q}) := 1 + 6 \sum_{n \geq 1} \chi_{-3}(n) \frac{\mathsf{q}^n}{1 - \mathsf{q}^n} \\
\mathsf{q}\frac{d}{d\mathsf{q}} \log \mathsf{Q} &= 1 - \sum_{n \geq 1} b_n \frac{n^2 \mathsf{q}^n}{1 - \mathsf{q}^n} = E_{3,\chi_{-3}}(\mathsf{q}) := 1 - 9 \sum_{n \geq 1} \chi_{-3}(n) \frac{n^2 \mathsf{q}^n}{1 - \mathsf{q}^n}
\end{aligned}$$

where χ_{-3} denotes the quadratic Dirichlet character $\left(\frac{-3}{\cdot}\right)$, i.e.

$$\chi_{-3}(n) = \left(\tfrac{-3}{n}\right) = 0,\ 1,\ -1 \qquad \text{if} \quad n \equiv 0,\ 1,\ 2 \bmod 3.$$

According to [**1**] one also has $E_{3,\chi_{-3}}(\mathsf{q}) \;=\; \eta(\mathsf{q})^9 \eta(\mathsf{q}^3)^{-3}$.

$\sharp$ **4:** $(A,B,\lambda)=(16,0,4)\,,\ \alpha=-1$

This corresponds to the E_5 example in [**8, 6**] and Example 2 in [**11**].

n	a_n	b_n
1	4	4
2	−5	0
3	12	−4
4	−48	0
5	240	4
6	−1359	0
7	8428	−4
8	−56000	0
9	392040	4

The numbers of $4a_n$ are the instanton numbers of [**8, 6**] Example E_5. More precisely the expansion for the E_5 case in [**8**] Eq. (4.10) is:

$$-4\cdot\left(1-\sum_{n\geq 1}a_n\frac{n^3\mathsf{Q}^n}{1-\mathsf{Q}^n}\right)$$

Zagier [**15**] and Rodriguez Villegas [**11**] give the following expressions for, in our notation, $\psi(\mathsf{q})$, $g_1(\psi(\mathsf{q}))$ and $\mathsf{q}\frac{d}{d\mathsf{q}}\log\mathsf{Q}$:

$$\begin{aligned}
\psi(\mathsf{q}) &= \frac{\eta(\mathsf{q})^8\eta(\mathsf{q}^4)^{16}}{\eta(\mathsf{q}^2)^{24}}\\
g_1(\psi(\mathsf{q})) &= E_{1,\chi_{-4}}(\mathsf{q}) \;:=\; 1+4\sum_{n\geq 1}\chi_{-4}(n)\frac{\mathsf{q}^n}{1-\mathsf{q}^n}\\
\mathsf{q}\frac{d}{d\mathsf{q}}\log\mathsf{Q} &= 1-\sum_{n\geq 1}b_n\frac{n^2\mathsf{q}^n}{1-\mathsf{q}^n}=E_{3,\chi_{-4}}(\mathsf{q})\;:=\;1-4\sum_{n\geq 1}\chi_{-4}(n)\frac{n^2\mathsf{q}^n}{1-\mathsf{q}^n}
\end{aligned}$$

where χ_{-4} denotes the quadratic Dirichlet character $\left(\frac{-4}{\cdot}\right)$, i.e.

$$\chi_{-4}(n)=\left(\tfrac{-4}{n}\right)=0,\,1,\,0,\,-1\qquad\text{if}\quad n\equiv 0,\,1,\,2,\,3\bmod 4.$$

According to [**1**] one also has $E_{3,\chi_{-4}}(\mathsf{q})\;=\;\eta(\mathsf{q})^4\eta(\mathsf{q}^2)^6\eta(\mathsf{q}^4)^{-4}$.

Remark: Taking $(A,B,\lambda)=(16,0,4)\,,\ \alpha=1$ leads to the instanton numbers of [**8**] Example Conifold.

$\sharp$ **5:** $(A,B,\lambda)=(11,-1,3)\,,\ \alpha=-1$

This example does not appear in [**8, 6**] or [**11**].

n	a_n	b_n
1	2	2
2	−2	1
3	3	−1
4	−8	−2
5	27	0
6	−102	2
7	420	1
8	−1856	−1
9	8649	−2

Zagier [**15**] gives for $\psi(\mathsf{q})$ and $g_1(\psi(\mathsf{q}))$ the expressions

$$\begin{aligned}\psi(\mathsf{q}) &= \mathsf{q}\prod_{n=1}^{\infty}(1-\mathsf{q}^n)^{5(\frac{n}{5})}\\ g_1(\psi(\mathsf{q})) &= 1+\frac{1}{2}\sum_{n\geq 1}((3-i)\chi(n)+(3+i)\overline{\chi(n)}\,)\frac{\mathsf{q}^n}{1-\mathsf{q}^n}\end{aligned}$$

where $(\frac{\cdot}{5})$ is the quadratic character and χ is the *quartic character* $\left(\frac{\cdot}{2-i}\right)_4$, i.e.

$$(\tfrac{n}{5})=(-1)^j\,,\qquad \chi(n)=i^j\qquad \text{if}\quad n\equiv 2^j \bmod 5;$$

the latter congruence can also be written as $n\equiv i^j \bmod (2-i)$ in $\mathbb{Z}[i]$.

From this we infer

$$\mathsf{q}\frac{d}{d\mathsf{q}}\log \mathsf{Q}=1-\sum_{n\geq 1}b_n\frac{n^2\mathsf{q}^n}{1-\mathsf{q}^n}=1+\frac{1}{2}\sum_{n\geq 1}((2-i)\chi(n)+(2+i)\overline{\chi(n)})\frac{n^2\mathsf{q}^n}{1-\mathsf{q}^n}\,.$$

$\sharp$ **6:** $(A,B,\lambda)=(7,-8,2)\,,\ \alpha=-1$

This example does not appear in [**8, 6**] or [**11**].

n	a_n	b_n
1	1	1
2	−1	1
3	1	0
4	−2	−1
5	5	−1
6	−14	0
7	42	1
8	−136	1
9	465	0

Zagier [**15**] gives for $\psi(\mathsf{q})$ and $g_1(\psi(\mathsf{q}))$ the expressions

$$\begin{aligned}\psi(\mathsf{q}) &= \frac{\eta(\mathsf{q})^3\eta(\mathsf{q}^6)^9}{\eta(\mathsf{q}^2)^3\eta(\mathsf{q}^3)^9}\\ g_1(\psi(\mathsf{q})) &= \frac{\eta(\mathsf{q}^2)\eta(\mathsf{q}^3)^6}{\eta(\mathsf{q})^2\eta(\mathsf{q}^6)^3} = \frac{1}{3}(E_{1,\chi_{-3}}(\mathsf{q}) + 2E_{1,\chi_{-3}}(\mathsf{q}^2))\,.\end{aligned}$$

From this one then obtains

$$\mathsf{q}\frac{d}{d\mathsf{q}}\log\mathsf{Q} = 1 - \sum_{n\geq 1} b_n \frac{n^2\mathsf{q}^n}{1-\mathsf{q}^n} = \frac{1}{9}(E_{3,\chi_{-3}}(\mathsf{q}) + 8E_{3,\chi_{-3}}(\mathsf{q}^2))\,.$$

Here $E_{1,\chi_{-3}}(\mathsf{q})$ and $E_{3,\chi_{-3}}(\mathsf{q})$ are the same as in Example ♯ 3.
Note that $b_n = (-1)^{n-1}\chi_{-3}(n)$.

References

[1] Ahlgren, S., B. Berndt, A. Yee, A. Zaharescu, *Integrals of Eisenstein series and derivatives of L-functions*, International Math. Research Notices 32 (2002) 1723–1738.

[2] Beauville, A., *Les familles stables de courbes elliptiques sur* $\mathbf{P}^1$ *admettant quatre fibres singulières*, C. R. Acad. Sc. Paris, t. 294 (1982) 657-660

[3] Boyd, D., *Mahler's measure and special values of L-functions*, Experimental Math. vol. 7 (1998) 37–82

[4] Chiang, T-M., A. Klemm, S-T. Yau, E. Zaslow, *Local mirror symmetry: Calculations and interpretations*, hep-th/9903053

[5] Kenyon, R., A. Okounkov, S. Sheffield, *Dimers and Amoebae*, arXiv:math-ph/0311005

[6] Klemm, A., P. Mayr, C. Vafa, *BPS states of exceptional non-critical strings*, arXiv:hep-th/9607139

[7] Klemm, A., E. Zaslow, *Local Mirror Symmetry at Higher Genus*, arXiv:hep-th/9906046

[8] Lerche, W., P. Mayr, N.P. Warner, *Non-critical strings, Del Pezzo singularities and Seiberg-Witten curves*, arXiv:hep-th/9612085

[9] Okounkov, A., *Random surfaces enumerating algebraic curves*, arXiv:math-ph/0412008

[10] Okounkov, A., N. Reshetikin, C. Vafa, *Quantum Calabi-Yau and Classical Crystals*, arXiv:hep-th/0309208

[11] Rodriguez Villegas, F., *Modular Mahler measures I*, Topics in number theory (University Park, PA, 1997), Ahlgren, S., G. Andrews, K. Ono (eds) 17–48, Math. Appl., 467, Kluwer Acad. Publ., Dordrecht, 1999. See also: http://www.ma.utexas.edu/users/villegas/research.html

[12] Schmickler-Hirzebruch, U., *Elliptische Flächen über* $\mathbb{P}_1\mathbb{C}$ *mit drei Ausnahmefasern und die hypergeometrische Differentialgleichung*, Schriftenreihe des Mathematischen Instituts der Universität Münster, 2. Serie, Heft 33 (1985)

[13] Stienstra, J., F. Beukers, *On the Picard-Fuchs equation and the formal Brauer group of certain elliptic K3-surfaces*, Math. Ann. 271 (1985) 269-304

[14] Stienstra, J., *Mahler Measure, Eisenstein Series and Dimers*, these proceedings

[15] Zagier, D., *Integral solutions of Apéry-like recurrence equations*, preprint

Mathematisch Instituut, Universiteit Utrecht, Postbus 80.010, 3508 TA Utrecht, the Netherlands

E-mail address: stien@math.uu.nl

AMS/IP Studies in Advanced Mathematics
Volume 38, 2006

Mahler Measure, Eisenstein Series and Dimers

Jan Stienstra

ABSTRACT. This note reveals a mysterious link between the partition function of certain dimer models on 2-dimensional tori and the L-function of their spectral curves. It also relates the partition function in certain families of dimer models to Eisenstein series.

Introduction

The *logarithmic Mahler measure* $\mathsf{m}(F)$ and the *Mahler measure* $\mathsf{M}(F)$ of a Laurent polynomial $F(x,y)$ with complex coefficients are:

$$\text{(1)} \quad \mathsf{m}(F) := \frac{1}{(2\pi i)^2}\oint\oint_{|x|=|y|=1} \log|F(x,y)|\,\frac{dx}{x}\frac{dy}{y}\,, \qquad \mathsf{M}(F) := \exp(\mathsf{m}(F))\,.$$

Boyd [**2**] gives a survey of many Laurent polynomials for which $\mathsf{m}(F)$ equals (numerically to many decimal places) a 'simple' non-zero rational number times the derivative at 0 of the L-function of the projective plane curve Z_F defined by the vanishing of F:

$$\text{(2)} \qquad \mathsf{m}(F)\cdot\mathbb{Q}^* \;=\; \mathrm{L}'(Z_F,0)\cdot\mathbb{Q}^*\,.$$

Tables 2, 5, 1 in [**2**] give the following families of cubic polynomials $\tilde{F}(X,Y,Z)$ for which (2) was checked numerically with $F(x,y)=(xy)^{-1}\tilde{F}(x,y,1)$ and with integer values of the parameter s,

$$\text{(3)} \qquad X^2Y+XY^2+X^2Z+XZ^2+Y^2Z+YZ^2-sXYZ$$

$$\text{(4)} \qquad X^2Y+Y^2Z+Z^2X-sXYZ$$

$$\text{(5)} \qquad X^2Y+XY^2+XZ^2+YZ^2-sXYZ$$

Deninger [**3**] and Rodriguez Villegas [**8**] showed that the experimentally observed relations (2) agree with predictions from the Bloch-Beilinson conjectures. Rodriguez Villegas [**8**] gave a proof of (2) for some values of s in Examples (3), (4), (5).

On the other hand, Kenyon, Okounkov and Sheffield gave a formula for the *partition function per fundamental domain* of a dimer model which is exactly the same as Formula (1) for the Mahler measure of the characteristic polynomial of

2000 *Mathematics Subject Classification.* Primary 11R99, 11F11.
Key words and phrases. Dimers, Mahler measure, Eisenstein series.

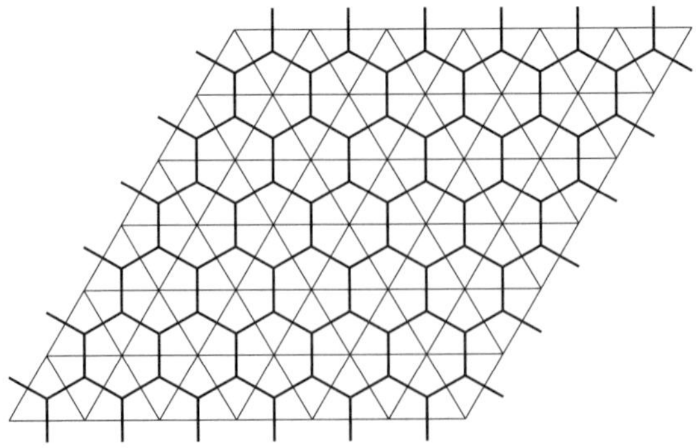

FIGURE 1

that dimer model (see [**5**] Theorem 3.5). In this paper we show examples of dimer models with characteristic polynomials (3), (4), (5).

This suggests that there may be some mysterious link between the partition function of a dimer model and the L-function of its spectral curve.

Our dimer models come in families with the parameter s explicitly related to the weights in the dimer model. The results for the Mahler measures of the polynomials (3), (4), (5) presented in [**9**] now imply that, for $|s|$ sufficiently large, the partition functions in these families of dimer models are $\mathsf{Q}_6^{-1}, \mathsf{Q}_3^{-\frac{1}{3}}, \mathsf{Q}_4^{-\frac{1}{2}}$ respectively, with

$$\begin{aligned} \mathsf{Q}_6 &= \mathsf{q}\prod_{n\geq 1}(1-\mathsf{q}^n)^{(-1)^{n-1}n\chi_{-3}(n)}, \quad \mathsf{Q}_3 = \mathsf{q}\prod_{n\geq 1}(1-\mathsf{q}^n)^{9n\chi_{-3}(n)}, \\ \mathsf{Q}_4 &= \mathsf{q}\prod_{n\geq 1}(1-\mathsf{q}^n)^{4n\chi_{-4}(n)}, \end{aligned} \tag{6}$$

where $\chi_{-3}(n) = 0, 1, -1$ if $n \equiv 0, 1, 2 \bmod 3$ and $\chi_{-4}(n) = 0, 1, 0, -1$ if $n \equiv 0, 1, 2, 3 \bmod 4$ and where q is explicitly related to s.

The Eisenstein series in the title are the logarithmic derivatives of the products in (6):

$$1+\sum_{n\geq 1}\chi_{-3}(n)\frac{n^2(-\mathsf{q})^n}{1-\mathsf{q}^n}, \quad 1-9\sum_{n\geq 1}\chi_{-3}(n)\frac{n^2\mathsf{q}^n}{1-\mathsf{q}^n}, \quad 1-4\sum_{n\geq 1}\chi_{-4}(n)\frac{n^2\mathsf{q}^n}{1-\mathsf{q}^n}.$$

The products and Eisenstein series for Q_3 and Q_4 were also studied by Ramanujan (see [**1**]). We find the similarity with MacMahon's function

$$M(\mathsf{q}) := \prod_{n\geq 1}(1-\mathsf{q}^n)^{-n}$$

very intriguing, in particular because MacMahon's function appears in the partition functions in [**6, 7**].

1. Dimers

For general theory and background information on dimer models we refer to [**4, 5, 6, 7**]. We restrict ourselves to dimer models on the standard hexagonal graph Γ, which is dual to the tessellation of the plane by regular triangles; see Figure 1.

A *dimer configuration*, or perfect matching, on Γ is a collection $\mathcal{M}$ of its edges such that each vertex of Γ is incident to exactly one edge from $\mathcal{M}$. Each edge

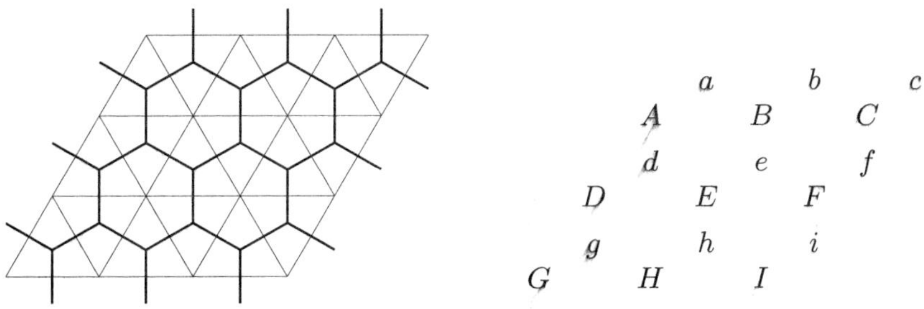

$$\mathsf{K} := \begin{bmatrix} Aa & Ba & 0 & 0 & 0 & 0 & Ga*x & 0 & 0 \\ 0 & Bb & Cb & 0 & 0 & 0 & 0 & Hb*x & 0 \\ Ac*y & 0 & Cc & 0 & 0 & 0 & 0 & 0 & Ic*x \\ Ad & 0 & 0 & Dd & Ed & 0 & 0 & 0 & 0 \\ 0 & Be & 0 & 0 & Ee & Fe & 0 & 0 & 0 \\ 0 & 0 & Cf & Df*y & 0 & Ff & 0 & 0 & 0 \\ 0 & 0 & 0 & Dg & 0 & 0 & Gg & Hg & 0 \\ 0 & 0 & 0 & 0 & Eh & 0 & 0 & Hh & Ih \\ 0 & 0 & 0 & 0 & 0 & Fi & Gi*y & 0 & Ii \end{bmatrix}.$$

FIGURE 2
The matrix entries give an abstract, tautological weighting of the graph edges. The $*x$ (resp. $*y$) marks the graph edges for which the top-bottom (resp. left-right) sides of the fundamental parallellogram must be identified.

of Γ intersects exactly one edge in the triangle tessellation and vice versa. Given a dimer configuration $\mathcal{M}$ one can remove from the triangle tessellation all edges which intersect a Γ-edge belonging to $\mathcal{M}$. The two triangles adjacent to such an edge—one like ∇, the other one like Δ—glue together to form a rhombus and as a result one gets a tessellation of the plane by rhombi (see Figure 5). Conversely every tessellation of the plane by rhombi, each of which is the union of two adjacent triangles in the triangle tessellation, comes from a dimer configuration on Γ.

View the plane as the complex plane $\mathbb{C}$ and let $\omega := e^{2\pi i/3}$. Then the vertices of the triangulation are precisely the elements of the lattice

$$\Lambda := \{a + b\omega \mid a, b \in \mathbb{Z}\}.$$

We first look at dimer configurations which are invariant under translations by vectors from the lattice 3Λ. One may also say that these are dimer configurations on the graph $\Gamma/3\Lambda$, embedded in the torus $\mathbb{C}/3\Lambda$. Figure 2 shows the fundamental domain and a labeling scheme for the vertices of the graph and the triangles of the tessellation. The matrix K in Figure 2 gives the same adjacency and labeling structure. K is the matrix of the *Kasteleijn operator* of the dimer model on $\Gamma/3\Lambda$. The determinant

$$P(x,y) := \det \mathsf{K} \tag{7}$$

is called the *characteristic polynomial* and the zero locus of $P(x,y)$ in $\mathbb{C}^{*2}$ is called the *spectral curve* of the dimer model (see [**4**] §2.1). A straightforward computation gives the characteristic polynomial shown in Figure 3.

$$\widetilde{P}(X,Y,Z) := Z^3 P(\tfrac{X}{Z},\tfrac{Y}{Z}) =$$

$$X^3\begin{bmatrix} i&h&g&f&e&d&c&b&a\\ F&E&D&C&B&A&I&H&G\end{bmatrix}+Y^3\begin{bmatrix} i&h&g&f&e&d&c&b&a\\ G&I&H&D&F&E&A&C&B\end{bmatrix}+Z^3\begin{bmatrix} i&h&g&f&e&d&c&b&a\\ I&H&G&F&E&D&C&B&A\end{bmatrix}$$

$$+XZ^2(\begin{bmatrix} i&h&g&f&e&d&c&b&a\\ F&H&G&C&E&D&I&B&A\end{bmatrix}+\begin{bmatrix} i&h&g&f&e&d&c&b&a\\ I&E&G&F&B&D&C&H&A\end{bmatrix}+\begin{bmatrix} i&h&g&f&e&d&c&b&a\\ I&H&D&F&E&A&C&B&G\end{bmatrix})$$

$$+X^2Z(\begin{bmatrix} i&h&g&f&e&d&c&b&a\\ F&E&G&C&B&D&I&H&A\end{bmatrix}+\begin{bmatrix} i&h&g&f&e&d&c&b&a\\ F&H&D&C&E&A&I&B&G\end{bmatrix}+\begin{bmatrix} i&h&g&f&e&d&c&b&a\\ I&E&D&F&B&A&C&H&G\end{bmatrix})$$

$$+YZ^2(\begin{bmatrix} i&h&g&f&e&d&c&b&a\\ G&I&H&F&E&D&C&B&A\end{bmatrix}+\begin{bmatrix} i&h&g&f&e&d&c&b&a\\ I&H&G&D&F&E&C&B&A\end{bmatrix}+\begin{bmatrix} i&h&g&f&e&d&c&b&a\\ I&H&G&F&E&D&A&C&B\end{bmatrix})$$

$$+Y^2Z(\begin{bmatrix} i&h&g&f&e&d&c&b&a\\ G&I&H&D&F&E&C&B&A\end{bmatrix}+\begin{bmatrix} i&h&g&f&e&d&c&b&a\\ G&I&H&F&E&D&A&C&B\end{bmatrix}+\begin{bmatrix} i&h&g&f&e&d&c&b&a\\ I&H&G&D&F&E&A&C&B\end{bmatrix})$$

$$+X^2Y(\begin{bmatrix} i&h&g&f&e&d&c&b&a\\ G&E&D&C&F&A&I&H&B\end{bmatrix}+\begin{bmatrix} i&h&g&f&e&d&c&b&a\\ F&E&H&D&B&A&I&C&G\end{bmatrix}+\begin{bmatrix} i&h&g&f&e&d&c&b&a\\ F&I&D&C&B&E&A&H&G\end{bmatrix})$$

$$+XY^2(\begin{bmatrix} i&h&g&f&e&d&c&b&a\\ G&E&H&D&F&A&I&C&B\end{bmatrix}+\begin{bmatrix} i&h&g&f&e&d&c&b&a\\ G&I&D&C&F&E&A&H&B\end{bmatrix}+\begin{bmatrix} i&h&g&f&e&d&c&b&a\\ F&I&H&D&B&E&A&C&G\end{bmatrix})$$

$$-XYZ(\begin{bmatrix} i&h&g&f&e&d&c&b&a\\ G&E&H&C&F&D&I&B&A\end{bmatrix}+\begin{bmatrix} i&h&g&f&e&d&c&b&a\\ G&H&D&C&F&E&I&B&A\end{bmatrix}+\begin{bmatrix} i&h&g&f&e&d&c&b&a\\ G&E&H&F&B&D&I&C&A\end{bmatrix}$$

$$+\begin{bmatrix} i&h&g&f&e&d&c&b&a\\ F&H&G&D&B&E&I&C&A\end{bmatrix}+\begin{bmatrix} i&h&g&f&e&d&c&b&a\\ G&H&D&F&B&E&I&C&A\end{bmatrix}+\begin{bmatrix} i&h&g&f&e&d&c&b&a\\ F&I&G&D&B&E&C&H&A\end{bmatrix}$$

$$+\begin{bmatrix} i&h&g&f&e&d&c&b&a\\ G&I&D&F&B&E&C&H&A\end{bmatrix}+\begin{bmatrix} i&h&g&f&e&d&c&b&a\\ F&H&G&D&E&A&I&C&B\end{bmatrix}+\begin{bmatrix} i&h&g&f&e&d&c&b&a\\ G&H&D&F&E&A&I&C&B\end{bmatrix}$$

$$+\begin{bmatrix} i&h&g&f&e&d&c&b&a\\ F&I&G&C&E&D&A&H&B\end{bmatrix}+\begin{bmatrix} i&h&g&f&e&d&c&b&a\\ I&E&G&C&F&D&A&H&B\end{bmatrix}+\begin{bmatrix} i&h&g&f&e&d&c&b&a\\ F&I&G&D&E&A&C&H&B\end{bmatrix}$$

$$+\begin{bmatrix} i&h&g&f&e&d&c&b&a\\ G&I&D&F&E&A&C&H&B\end{bmatrix}+\begin{bmatrix} i&h&g&f&e&d&c&b&a\\ I&E&G&D&F&A&C&H&B\end{bmatrix}+\begin{bmatrix} i&h&g&f&e&d&c&b&a\\ F&I&H&C&E&D&A&B&G\end{bmatrix}$$

$$+\begin{bmatrix} i&h&g&f&e&d&c&b&a\\ I&E&H&C&F&D&A&B&G\end{bmatrix}+\begin{bmatrix} i&h&g&f&e&d&c&b&a\\ I&H&D&C&F&E&A&B&G\end{bmatrix}+\begin{bmatrix} i&h&g&f&e&d&c&b&a\\ F&I&H&D&E&A&C&B&G\end{bmatrix}$$

$$+\begin{bmatrix} i&h&g&f&e&d&c&b&a\\ I&E&H&D&F&A&C&B&G\end{bmatrix}+\begin{bmatrix} i&h&g&f&e&d&c&b&a\\ I&E&H&F&B&D&A&C&G\end{bmatrix}+\begin{bmatrix} i&h&g&f&e&d&c&b&a\\ I&H&D&F&B&E&A&C&G\end{bmatrix})$$

FIGURE 3

Here $\begin{bmatrix} i&h&g&f&e&d&c&b&a\\ F&E&D&C&B&A&I&H&G\end{bmatrix} = Fi * Eh * Dg * Cf * Be * Ad * Ic * Hb * Ga$, the product of the weights. It also shows the dimer configuration as a map from ∇- to Δ-triangles.

Next we want to specify a 3Λ-invariant weight function W on the edges of Γ so that the characteristic polynomial of the dimer model with these weights is a constant times one of the polynomials (3), (4), (5).

Thus, the coefficients of certain monomials must be 0 and the coefficients of all other monomials different from XYZ must be equal and non-zero. Moreover, the ratio between the coefficient of XYZ and the other non-zero coefficient should be $-s$. Altogether these conditions constitute 9 equations on the weights. Exactly which 9 equations depends on the polynomial one wants to reconstruct.

The graph $\Gamma/3\Lambda$ has 27 edges and 18 vertices. If one multiplies the weights of the three edges incident to a vertex by the same non-zero number k, the whole characteristic polynomial gets multiplied by k. So there is a torus $\mathbb{C}^{*18}$ acting on the solutions of the 9 equations, but the 1-dimensional subtorus given by multiplication factors k for the nine Δ-vertices and k^{-1} for the nine ∇-vertices ($k \in \mathbb{C}^*$) acts trivially. So, for fixed s, the solution space of the 9 equations modulo the action of $\mathbb{C}^{*18}$ has dimension at least $1 = 27 - 9 - 17$.

Without further delving into general solutions we now exhibit just three examples from a larger collection we found by trial and error. Correctness of these examples can easily be checked by direct inspection of the general characteristic polynomial $\widetilde{P}(X,Y,Z)$ in Figure 3.

As a common structure in these examples there are only four edge weights (0, 1, w, m) and s is a function of m. The parameter w confirms that for fixed s the solution space modulo $\mathbb{C}^{*18}$ has dimension ≥ 1. Edges with weight 0 cannot occur in a dimer configuration. Thus if the dimer configuration is presented as a tessellation by rhombi the 0-weight edges of Γ turn into edges of the tessellation which *must be present*; it is like a boundary condition. This is shown for Example 2 in Figure 4. Any tessellation by rhombi which respects the boundary condition is allowed. Figure 5 shows the 3Λ-periodic dimer configurations for Example 2.

EXAMPLE 1. *Setting* $W(Hg) = W(Dd) = W(Eh) = 0$, $W(Hb) = W(Ee) = W(Df) = m$, $W(Ii) = W(Ic) = W(Gi) = W(Aa) = W(Ac) = W(Ga) = 1$ *and all remaining weights* $= w$ *yields a characteristic polynomial of type (3):*

$$mw^6(X^2Y + XY^2 + X^2Z + XZ^2 + Y^2Z + YZ^2) - (4 + 3m + 3m^2 + m^3)w^6XYZ$$

with $s = (4 + 3m + 3m^2 + m^3)m^{-1}$.

EXAMPLE 2. *Setting* $W(Hg) = W(Dd) = W(Eh) = W(Be) = W(Cb) = W(Ff) = 0$, $W(Hb) = W(Ee) = W(Df) = m$, $W(Ii) = W(Ic) = W(Gi) = W(Aa) = W(Ac) = W(Ga) = 1$ *and all remaining weights* $= w$ *yields a characteristic polynomial of type (4):*

$$mw^6(X^2Z + XY^2 + YZ^2) - (2 + m^3)w^6XYZ$$

with $s = (2 + m^3)m^{-1}$.

EXAMPLE 3. *Setting* $W(Aa) = W(Bb) = W(Cf) = W(Gi) = W(Fe) = 0$, $W(Cb) = W(Ff) = W(Be) = 1$, $W(Cc) = W(Fi) = W(Ba) = m$ *and all remaining weights* $= w$ *yields a characteristic polynomial of type (5):*

$$mw^6(X^2Y + XY^2 + XZ^2 + YZ^2) - (2 + m^2 + m^3)w^6XYZ$$

with $s = (2 + m^2 + m^3)m^{-1}$.

Remark: Among the examples in [**5**] there is a dimer model on the square-octagon graph with characteristic polynomial of type (5): $x + x^{-1} + y + y^{-1} - s$.

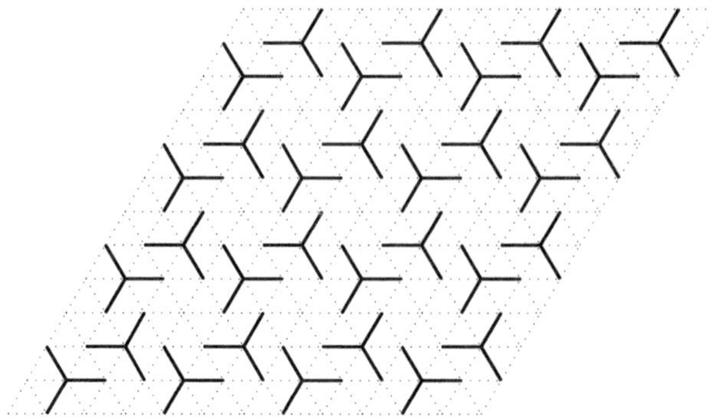

FIGURE 4

As in [**5**] §3.2, we now fix a 3Λ-invariant $\mathbb{R}$-valued weight function W on the edges of Γ and consider for a positive integer n the dimer configurations on (Γ, W) which are invariant under translations by vectors in the lattice $3n\Lambda$. Thus, the preceding discussion and examples concern the case $n = 1$. Theorem 3.3 of [**5**] expresses the characteristic polynomial $P_n(x, y)$ of the dimer model on $(\Gamma/3n\Lambda, W)$ in terms of the characteristic polynomial $P(x, y)$ of the dimer model on $(\Gamma/3\Lambda, W)$:

$$P_n(x,y) = \prod_{u^n=x} \prod_{v^n=y} P(u,v)\,.$$

Corollary 3.4 of [**5**] uses this to compute the *partition function* $Z(\Gamma/3n\Lambda, W)$ for the dimer model on $(\Gamma/3n\Lambda, W)$:

$$Z(\Gamma/3n\Lambda, W) = \frac{1}{2}(-Z_n^{(0,0)} + Z_n^{(0,1)} + Z_n^{(1,0)} + Z_n^{(1,1)}) \qquad \text{where}$$

$$Z_n^{(a,b)} = \prod_{u^n=(-1)^a} \prod_{v^n=(-1)^b} P(u,v)\,. \tag{8}$$

Finally, [**5**] Thm. 3.5 computes the *partition function per fundamental domain* Z for periodic dimer models on (Γ, W) with period lattice $3n\Lambda$, $n \in \mathbb{N}$:

$$\begin{aligned} \log Z &:= \lim_{n\to\infty} \frac{1}{n^2} \log Z(\Gamma/3n\Lambda, W) \\ &= \frac{1}{(2\pi i)^2} \oint\!\!\oint_{|x|=|y|=1} \log|P(x,y)|\, \frac{dx}{x}\frac{dy}{y} \quad =: \quad \mathsf{m}(P)\,. \end{aligned} \tag{9}$$

Note: $|P(x,y)| = P(x,y)$ in (9), because the polynomial $P(x,y)$ has real coefficients and (8) contains with every factor also its complex conjugate.

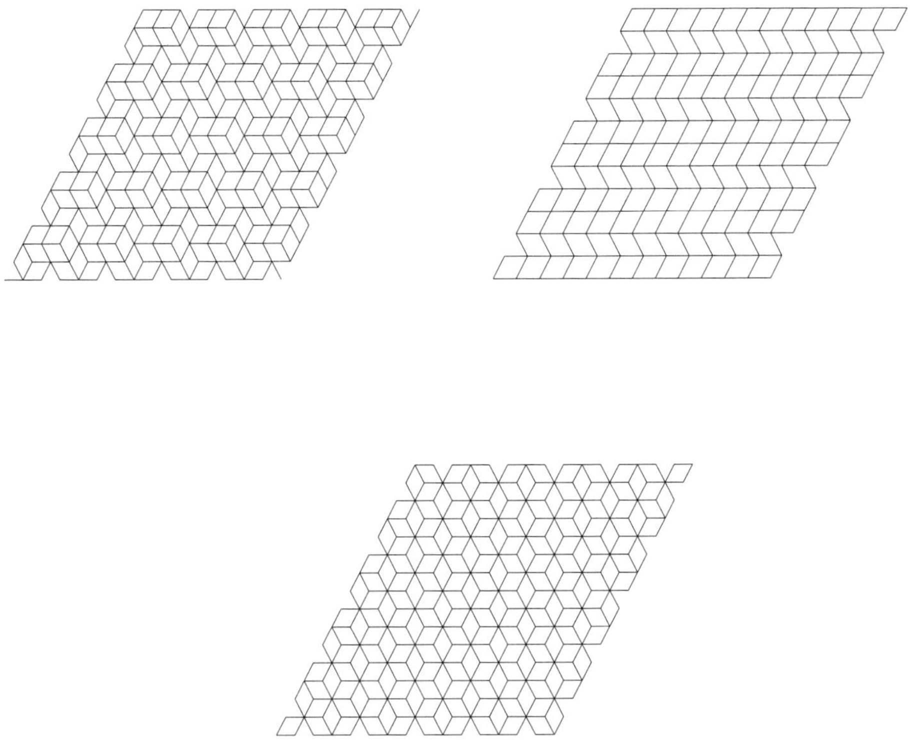

FIGURE 5
180° degree rotation of the top left picture and $\pm 60°$ degree rotations of the top right picture give also legitimate tessellations.

COROLLARY 1. *The partition function per fundamental domain is equal to the Mahler measure of the characteristic polynomial:*

$$Z = \mathsf{M}(P)\,.$$

□

We now combine the above examples with Examples ♯ 6, ♯ 3, ♯ 4 of [**9**].

EXAMPLE 1. **(continued)** *According to* [**9**] *Example ♯ 6 the Mahler measure of the polynomial*

$$(X+Y)(Y+Z)(Z+X) - tXYZ$$

with $|t|$ *sufficiently large, is* Q_6^{-1} *where*

$$\mathsf{Q}_6 = \mathsf{q}\prod_{n\geq 1}(1-\mathsf{q}^n)^{(-1)^{n-1}n\chi_{-3}(n)} \quad \textit{and} \quad t = \frac{\eta(\mathsf{q}^2)^3\eta(\mathsf{q}^3)^9}{\eta(\mathsf{q})^3\eta(\mathsf{q}^6)^9}$$

and $\chi_{-3}(n) = 0,\,1,\,-1$ *if* $n \equiv 0,\,1,\,2 \bmod 3$. *Thus, if we rescale the dimer model in Example 1 by multiplying all edge-weights by* $(mw^6)^{-\frac{1}{9}}$*, the partition function per fundamental domain for the rescaled dimer model is* Q_6^{-1} *with*

$$t - 2 = s = (4 + 3m + 3m^2 + m^3)m^{-1}\,.$$

EXAMPLE 2. **(continued)** *According to* [**9**] *Example ♯ 3 the Mahler measure of the polynomial*

$$X^2Y + Y^2Z + Z^2X - tXYZ$$

with $|t|$ *sufficiently large, is* $\mathsf{Q}_3^{-\frac{1}{3}}$ *where*

$$\mathsf{Q}_3 = \mathsf{q}\prod_{n\geq 1}(1-\mathsf{q}^n)^{9n\chi_{-3}(n)} \quad \textit{and} \quad t^3 = 27 + \frac{\eta(\mathsf{q})^{12}}{\eta(\mathsf{q}^3)^{12}}\,.$$

Thus, if we rescale the dimer model in Example 2 by multiplying all edge-weights by $(mw^6)^{-\frac{1}{9}}$*, the partition function per fundamental domain for the rescaled dimer model is* $\mathsf{Q}_3^{-\frac{1}{3}}$ *with*

$$t = s = (2 + m^3)m^{-1}\,.$$

EXAMPLE 3. **(continued)** *According to* [**9**] *Example ♯ 4 the Mahler measure of the polynomial*

$$(X+Y)(XY+Z^2) - tXYZ$$

with $|t|$ *sufficiently large, is* $\mathsf{Q}_4^{-\frac{1}{2}}$ *where*

$$\mathsf{Q}_4 = \mathsf{q}\prod_{n\geq 1}(1-\mathsf{q}^n)^{4n\chi_{-4}(n)} \quad \textit{and} \quad t = \frac{\eta(\mathsf{q}^2)^{12}}{\eta(\mathsf{q})^4\eta(\mathsf{q}^4)^8}$$

and $\chi_{-4}(n) = 0,\,1,\,0,\,-1$ *if* $n \equiv 0,\,1,\,2,\,3 \bmod 4$. *Thus, if we rescale the dimer model in Example 3 by multiplying all edge-weights by* $(mw^6)^{-\frac{1}{9}}$*, the partition function per fundamental domain for the rescaled dimer model is* $\mathsf{Q}_4^{-\frac{1}{2}}$ *with*

$$t = s = (2 + m^2 + m^3)m^{-1}\,.$$

References

[1] Ahlgren, S., B. Berndt, A. Yee, A. Zaharescu, *Integrals of Eisenstein series and derivatives of L-functions*, International Math. Research Notices 32 (2002) 1723–1738.

[2] Boyd, D., *Mahler's measure and special values of L-functions*, Experimental Math. vol. 7 (1998) 37–82

[3] Deninger, D. *Deligne periods of mixed motives, K-theory and the entropy of certain* $\mathbb{Z}^n$*-actions*, J. Amer. Math. Soc. 10 (1997) 259–281

[4] Kenyon, R., A. Okounkov, *Planar dimers and Harnack curves*, arXiv:math.AG/0311062

[5] Kenyon, R., A. Okounkov, S. Sheffield, *Dimers and Amoebae*, arXiv:math-ph/0311005

[6] Okounkov, A., *Random surfaces enumerating algebraic curves*, arXiv:math-ph/0412008

[7] Okounkov, A., N. Reshetikin, C. Vafa, *Quantum Calabi-Yau and Classical Crystals*, arXiv:hep-th/0309208

[8] Rodriguez Villegas, F., *Modular Mahler measures I*, Topics in number theory (University Park, PA, 1997), Ahlgren, S., G. Andrews, K. Ono (eds) 17–48, Math. Appl., 467, Kluwer Acad. Publ., Dordrecht, 1999. See also: http://www.ma.utexas.edu/users/villegas/research.html

[9] Stienstra, J., *Mahler Measure Variations, Eisenstein Series and Instanton Expansions*, these proceedings

MATHEMATISCH INSTITUUT, UNIVERSITEIT UTRECHT, POSTBUS 80.010, 3508 TA UTRECHT, THE NETHERLANDS

E-mail address: `stien@math.uu.nl`

AMS/IP Studies in Advanced Mathematics
Volume 38, 2006

Mirror Symmetry for Zeta Functions

Daqing Wan

ABSTRACT. In this paper, we study the relation between the zeta function of a Calabi-Yau hypersurface and the zeta function of its mirror. Two types of arithmetic relations are discovered. This motivates us to formulate two general arithmetic mirror conjectures for the zeta functions of a mirror pair of Calabi-Yau manifolds.

CONTENTS

1. Introduction

In this section, we describe two mirror relations between the zeta function of a Calabi-Yau hypersurface in a projective space and the zeta function of its mirror manifold. Along the way, we make comments and conjectures about what to expect in the general case.

Let d be a positive integer. Let X and Y be two d-dimensional smooth projective Calabi-Yau varieties over $\mathbb{C}$. A necessary condition (the topological mirror

2000 *Mathematics Subject Classification.* Primary 11G25; Secondary 14J32, 14G05, 14G15, 11G25.

Key words and phrases. strong mirror pairs, mirror congruences, Calabi-Yau hypersurfaces, generic mirror pairs, rational points, finite fields, zeta functions, slope zeta functions, arithmetic mirror symmetry.

Partially supported by NSF. The auhor thanks V. Batyrev, P. Candelas, H. Esnault, L. Fu, K.F. Liu, Y. Ruan, S.T. Yau for helpful discussions. The paper was motivated by some questions in my lectures at the 2004 Arizona Winter School.

test) for X and Y to be a mirror pair is that their Hodge numbers satisfy the Hodge symmetry:

$$h^{i,j}(X) = h^{d-i,j}(Y), \ 0 \le i,j \le d. \tag{1}$$

In particular, their Euler characteristics are related by

$$e(X) = (-1)^d e(Y). \tag{2}$$

In general, there is no known rigorous algebraic geometric definition for a mirror pair, although many examples of mirror pairs are known at least conjecturally. Furthermore, it does not make sense to speak of "the mirror" of X as the mirror variety usually comes in a family. In some cases, the mirror does not exist. This is the case for rigid Calabi-Yau 3-folds X, since the rigidity condition $h^{2,1}(X) = 0$ would imply that $h^{1,1}(Y) = 0$, which is impossible.

We shall assume that X and Y are a given mirror pair in some sense and are defined over a number field or a finite field. We are interested in how the zeta function of X is related to the zeta function of Y. Since there is no algebraic geometric definition for X and Y to be a mirror pair, it is difficult to study the possible symmetry between their zeta functions in full generality. On the other hand, there are many explicit examples and constructions which at least conjecturally give a mirror pair, most notably in the toric hypersurface setting as constructed by Batyrev [**1**]. Thus, we shall first examine an explicit example and see what kind of relations can be proved for their zeta functions in this case. This would then suggest what to expect in general.

Let $n \ge 2$ be a positive integer. We consider the universal family of Calabi-Yau complex hypersurfaces of degree $n+1$ in the projective space $\mathbb{P}^n$. Its mirror family is a one-parameter family of toric hypersurfaces. To construct the mirror family, we consider the one parameter subfamily X_λ of complex projective hypersurfaces of degree $n+1$ in $\mathbb{P}^n$ defined by

$$f(x_1, \cdots, x_{n+1}) = x_1^{n+1} + \cdots + x_{n+1}^{n+1} + \lambda x_1 \cdots x_{n+1} = 0,$$

where $\lambda \in \mathbb{C}$ is the parameter. The variety X_λ is a Calabi-Yau manifold when X_λ is smooth. Let μ_{n+1} denote the group of $(n+1)$-st roots of unity. Let

$$G = \{(\zeta_1, \cdots, \zeta_{n+1}) \mid \zeta_i^{n+1} = 1, \zeta_1 \cdots \zeta_{n+1} = 1\}/\mu_{n+1} \cong (\mathbb{Z}/(n+1)\mathbb{Z})^{n-1},$$

where μ_{n+1} is embedded in G via the diagonal embedding. The finite group G acts on X_λ by

$$(\zeta_1, \cdots, \zeta_{n+1})(x_1, \cdots, x_{n+1}) = (\zeta_1 x_1, \cdots, \zeta_{n+1} x_{n+1}).$$

The quotient X_λ/G is a projective toric hypersurface Y_λ in the toric variety $\mathbb{P}_\Delta$, where $\mathbb{P}_\Delta$ is the simplex in $\mathbb{R}^n$ with vertices $\{e_1, \cdots, e_n, -(e_1 + \cdots + e_n)\}$ and the e_i's are the standard coordinate vectors in $\mathbb{R}^n$. Explicitly, the variety Y_λ is the projective closure in $\mathbb{P}_\Delta$ of the affine toric hypersurface in $\mathbb{G}_m^n$ defined by

$$g(x_1, \cdots, x_n) = x_1 + \cdots + x_n + \frac{1}{x_1 \cdots x_n} + \lambda = 0.$$

Assume that X_λ is smooth. Then Y_λ is a (singular) mirror of X_λ. It is an orbifold. If W_λ is a smooth crepant resolution of Y_λ, then the pair (X_λ, W_λ) is called a mirror pair of Calabi-Yau manifolds. Such a resolution exists for this example but is not unique if $n \ge 3$. The number of rational points and the zeta function are independent of the choice of the crepant resolution. We are interested in understanding how the arithmetic of X_λ is related to the arithmetic of W_λ, in particular how the zeta function of X_λ is related to the zeta function of W_λ. Our main concern in this paper is to consider Calabi-Yau manifolds over finite fields, although we shall mention some conjectural implications for Calabi-Yau manifolds defined over number fields.

In this example, we see two types of mirror pairs. The first one is the generic mirror pair $\{X_\Lambda, W_\lambda\}$, where X_Λ is the generic member in the moduli space of smooth projective Calabi-Yau hypersurfaces of degree $(n+1)$ in $\mathbb{P}^n$ and W_λ is the generic member in the above one-parameter family of Calabi-Yau manifolds. Note that X_Λ and Y_λ are parameterized by different parameter spaces (of different dimensions). The possible zeta symmetry in this case would then have to be a relation between certain generic property of their zeta functions.

The second type of mirror pairs is the one-parameter family of mirror pairs $\{X_\lambda, W_\lambda\}$ parameterized by the same parameter λ. This is a stronger type of mirror pair than the first type. For $\lambda \in \mathbb{C}$, we say that W_λ is a **strong mirror** of X_λ. For such a strong mirror pair $\{X_\lambda, W_\lambda\}$, we can really ask for the relation between the zeta function of X_λ and the zeta function of W_λ. If $\lambda_1 \neq \lambda_2$, W_{λ_1} would not be called a strong mirror for X_{λ_2}, although they would be a usual **weak mirror** pair. In fact, we do not have a definition for a strong mirror pair in general, as there is not even a definition for a generic or weak mirror pair in general.

Let $\mathbb{F}_q$ be a finite field of q elements, where $q = p^r$ and p is a prime. For a scheme X of finite type of dimension d over $\mathbb{F}_q$, let $\#X(\mathbb{F}_q)$ denote the number of $\mathbb{F}_q$-rational points on X. Let

$$Z(X,T) = \exp(\sum_{k=1}^{\infty} \frac{T^k}{k} \#X(\mathbb{F}_{q^k})) \in 1 + T\mathbb{Z}[[T]]$$

be the zeta function of X. It is well known that $Z(X,T)$ is a rational function in T whose reciprocal zeros and reciprocal poles are Weil q-integers. Factor $Z(X,T)$ over the p-adic numbers $\mathbb{C}_p$ and write

$$Z(X,T) = \prod_i (1 - \alpha_i T)^{\pm 1}$$

in reduced form, where the algebraic integers $\alpha_i \in \mathbb{C}_p$. One knows that the slope $\mathrm{ord}_q(\alpha_i)$ is a rational number in the interval $[0, d]$. For two real numbers $s_1 \leq s_2$, we define the slope $[s_1, s_2]$ part of $Z(X,T)$ to be the partial product

$$Z_{[s_1,s_2]}(X,T) = \prod_{s_1 \leq \mathrm{ord}_q(\alpha_i) \leq s_2} (1 - \alpha_i T)^{\pm 1}. \tag{3}$$

For a half-open and half-closed interval $[s_1, s_2)$, the slope $[s_1, s_2)$ part $Z_{[s_1,s_2)}(X,T)$ of $Z(X,T)$ is defined in a similar way. These are rational functions with coefficients in $\mathbb{Z}_p$ by the p-adic Weierstrass factorization. It is clear that we have the decomposition

$$Z(X,T) = \prod_{i=0}^{d} Z_{[i,i+1)}(X,T).$$

Our main result of this paper is the following arithmetic mirror theorem.

THEOREM 1.1. *Assume that $\lambda \in \mathbb{F}_q$ is such that (X_λ, W_λ) is a strong mirror pair of Calabi-Yau manifolds over $\mathbb{F}_q$. For every positive integer k, we have the congruence formula*

$$\#X_\lambda(\mathbb{F}_{q^k}) \equiv \#Y_\lambda(\mathbb{F}_{q^k}) \equiv \#W_\lambda(\mathbb{F}_{q^k}) \pmod{q^k}.$$

Equivalently, the slope $[0,1)$ *part of the zeta function is the same for the mirror varieties* $\{X_\lambda, Y_\lambda, W_\lambda\}$*:*

$$Z_{[0,1)}(X_\lambda, T) = Z_{[0,1)}(Y_\lambda, T) = Z_{[0,1)}(W_\lambda, T).$$

We now discuss a few applications of this theorem. In terms of cohomology theory, this suggests that the semi-simplifications of the De Rham-Witt cohomology (in particular, the p-adic etàle cohomology) for $\{X_\lambda, Y_\lambda, W_\lambda\}$ are all the same. A corollary of the above theorem is that the unit root parts (slope zero parts) of their zeta functions are the same:

$$Z_{[0,0]}(X_\lambda, T) = Z_{[0,0]}(Y_\lambda, T) = Z_{[0,0]}(W_\lambda, T).$$

The p-adic variation of the rational function $Z_{[0,0]}(X_\lambda, T)$ as λ varies is closely related to the mirror map which we do not discuss here, but see [**4**] for the case $n \leq 3$. From the arithmetic point of view, the p-adic variation of the rational function $Z_{[0,0]}(X_\lambda, T)$ as λ varies is explained by Dwork's unit root zeta function [**5**]. We briefly explain the connection here.

Let B be the parameter variety of λ such that (X_λ, W_λ) form a strong mirror pair. Let $\Phi : X_\lambda \to B$ (resp. $\Psi : W_\lambda \to B$) be the projection to the base by sending X_λ (resp. W_λ) to λ. The pair (Φ, Ψ) of morphisms to B is called a **strong mirror pair of morphisms to** B. Each of its fibres gives a strong mirror pair of Calabi-Yau manifolds. Recall that Dwork's unit root zeta function attached to the morphism Φ is defined to be the formal infinite product

$$Z_{\text{unit}}(\Phi, T) = \prod_{\lambda \in |B|} Z_{[0,0]}(X_\lambda, T^{\deg(\lambda)}) \in 1 + T\mathbb{Z}_p[[T]],$$

where $|B|$ denotes the set of closed points of B over $\mathbb{F}_q$. This unit root zeta function is no longer a rational function, but was conjectured by Dwork in [**5**] and proved by the author in [**11**][**12**][**13**] to be a p-adic meromorphic function in T. The above theorem immediately implies

COROLLARY 1.2. *Let* (Φ, Ψ) *be the above strong mirror pair of morphisms to the base* B*. Then their unit root zeta functions are the same:*

$$Z_{\text{unit}}(\Phi, T) = Z_{\text{unit}}(\Psi, T).$$

If λ is in a number field K, then Theorem 1.1 implies that the Hasse-Weil zeta functions of X_λ and Y_λ differ essentially by the L-function of a pure motive $M_n(\lambda)$ of weight $n-3$. That is,

$$\zeta(X_\lambda, s) = \zeta(Y_\lambda, s) L(M_n(\lambda), s-1).$$

In the quintic case $n = 4$, the pure weight 1 motive $M_4(\lambda)$ would come from a curve. This curve has been constructed explicitly by Candelas, de la Ossa and F. Rodriguez-Villegas [**3**]. The relation between the Hasse-Weil zeta functions of X_λ and W_λ are similar, differing by a few more factors consisting of Tate twists of the Dedekind zeta function of K.

Theorem 1.1 motivates the following more general conjecture.

CONJECTURE 1.3 (Congruence mirror conjecture). *Suppose that we are given a strong mirror pair* $\{X, Y\}$ *of Calabi-Yau manifolds defined over* $\mathbb{F}_q$*. Then, for every positive integer* k*, we have*

$$\#X(\mathbb{F}_{q^k}) \equiv \#Y(\mathbb{F}_{q^k}) \pmod{q^k}.$$

Equivalently,

$$Z_{[0,1)}(X, T) = Z_{[0,1)}(Y, T).$$

Equivalently (by the functional equation),

$$Z_{(d-1,d]}(X,T) = Z_{(d-1,d]}(Y,T).$$

The condition in the congruence mirror conjecture is vague since one does not know at present an algebraic geometric definition of a strong mirror pair of Calabi-Yau manifolds, although one does know many examples such as the one given above. Thus, a major part of the problem is to make the definition of a strong mirror pair mathematically precise. For an additional piece of evidence for the congruence mirror conjecture, see Theorem 6.2, which can be viewed as a generalization of Theorem 1.1. As indicated before, this conjecture implies that Dwork's unit root zeta functions for the two families forming a strong mirror pair are the same p-adic meromorphic function. This means that under the strong mirror family involution, Dwork's unit root zeta function stays the same.

Just like the zeta function itself, its slope $[0,1)$ part $Z_{[0,1)}(X_\lambda, T)$ depends heavily on the algebraic parameter λ, not just on the topological properties of X_λ. This means that the congruence mirror conjecture is really a continuous type of arithmetic mirror symmetry. This continuous nature requires the use of a strong mirror pair, not just a generic mirror pair.

Assume that $\{X, Y\}$ forms a mirror pair, not necessarily a strong mirror pair. A different type of arithmetic mirror symmetry reflecting the Hodge symmetry, which is discrete and hence generic in nature, is to look for a suitable quantum version $Z_Q(X,T)$ of the zeta function such that

$$Z_Q(X,T) = Z_Q(Y,T)^{(-1)^d},$$

where $\{X, Y\}$ is a mirror pair of Calabi-Yau manifolds over $\mathbb{F}_q$ of dimension d. This relation cannot hold for the usual zeta function $Z(X,T)$ for obvious reasons, even for a strong mirror pair, as it contradicts the congruence mirror conjecture for odd d. No non-trivial candidate for $Z_Q(X,T)$ has been found. Here we propose a p-adic quantum version which would have the conjectural properties for most (and hence generic) mirror pairs. We will call our new zeta function the slope zeta function, as it is based on the slopes of the zeros and poles.

DEFINITION 1.4. For a scheme X of finite type over $\mathbb{F}_q$, write as before

$$Z(X,T) = \prod_i (1 - \alpha_i T)^{\pm 1}$$

in reduced form, where $\alpha_i \in \mathbb{C}_p$. Define the slope zeta function of X to be the two-variable function

$$S_p(X,u,T) = \prod_i (1 - u^{\mathrm{ord}_q(\alpha_i)} T)^{\pm 1}. \tag{4}$$

Note that

$$\alpha_i = q^{\mathrm{ord}_q(\alpha_i)} \beta_i,$$

where β_i is a p-adic unit. Thus, the slope zeta function $S_p(X,u,T)$ is obtained from the p-adic factorization of $Z(X,T)$ by dropping the p-adic unit parts of the roots and replacing q by the variable u. This is not always a rational function in u and T. It is rational if all slopes are integers. Note that the definition of the slope zeta function is independent of the choice of the ground field $\mathbb{F}_q$ where X is defined. It depends only on $X \otimes \bar{\mathbb{F}}_q$ and thus is also a geometric invariant. It would be interesting to see if there is a Diophantine interpretation of the slope zeta function. If we have a smooth proper family of varieties, the Grothendieck specialization theorem implies that the generic Newton polygon on each cohomology exists and hence the generic slope zeta function exists as well.

If X is a scheme of finite type over $\mathbb{Z}$, then for each prime number p the reduction $X \otimes \mathbb{F}_p$ has the p-adic slope zeta function $S_p(X \otimes \mathbb{F}_p, u, T)$. At first glance, one might think that this gives infinitely many discrete invariants for X, as the set of prime numbers is infinite. However, it can be shown that the set $\{S_p(X \otimes \mathbb{F}_p, u, T) | p \text{ prime}\}$ contains only finitely many distinct elements. In general, it is a very interesting but difficult problem to determine this set

$$\{S_p(X \otimes \mathbb{F}_p, u, T) | p \text{ prime}\}.$$

Suppose that X and Y form a mirror pair of d-dimensional Calabi-Yau manifolds over $\mathbb{F}_q$. For simplicity and for comparison with the Hodge theory, we always assume in this paper that X and Y can be lifted to characteristic zero (to the Witt ring of $\mathbb{F}_q$). In this good reduction case, the mod p Hodge numbers equal the characteristic zero Hodge numbers. Taking $u = 1$ in the definition of the slope zeta function, we see that the specialization $S_p(X, 1, T)$ already satisfies the desired relation

$$S_p(X, 1, T) = (1 - T)^{-e(X)} = (1 - T)^{-(-1)^d e(Y)} = S_p(Y, 1, T)^{(-1)^d}.$$

This suggests that there is a chance that the slope zeta function might satisfy the desired slope mirror symmetry

$$S_p(X, u, T) = S_p(Y, u, T)^{(-1)^d}. \tag{5}$$

In section 7, we shall show that the slope zeta function satisfies a functional equation. Furthermore, the expected slope mirror symmetry does hold if both X and Y are ordinary. If either X or Y is not ordinary, the expected slope mirror symmetry is unlikely to hold in general.

If $d \leq 2$, the congruence mirror conjecture implies that the slope zeta function does satisfy the expected slope mirror symmetry for a strong mirror pair $\{X, Y\}$, whether X and Y are ordinary or not. For $d \geq 3$, we believe that the slope zeta function is still a little bit too strong for the expected symmetry to hold in general, even if $\{X, Y\}$ forms a strong mirror pair. And it should not be too hard to find a counter example, although we have not done so. However, we believe that the expected slope mirror symmetry holds for a generic mirror pair of 3-dimensional Calabi-Yau manifolds.

CONJECTURE 1.5 (Slope mirror conjecture). *Suppose that we are given a generic mirror pair $\{X, Y\}$ of 3-dimensional Calabi-Yau manifolds defined over $\mathbb{F}_q$. Then we have the slope mirror symmetry for their generic slope zeta functions:*

$$S_p(X, u, T) = \frac{1}{S_p(Y, u, T)}. \tag{6}$$

A main point of this conjecture is that it holds for all prime numbers p. For arbitrary $d \geq 4$, the corresponding slope mirror conjecture might be false for some prime numbers p, but it should be true for all primes $p \equiv 1 \pmod{D}$ for some positive integer D depending on the mirror family, if the family comes from the reduction modulo p of a family defined over a number field. In the case $d \leq 3$, one could take $D = 1$ and hence get the above conjecture.

Again the condition in the slope mirror conjecture is vague as an algebraic geometric definition of a mirror family is not presently known, although many examples are known in the toric setting. In a future paper, using the results in **[10][14]**, we shall prove that the slope mirror conjecture holds in the toric hypersurface case if

$d \leq 3$. For example, if X is a generic quintic hypersurface, then X is ordinary by the results in [**8**][**10**] for every p and thus one finds

$$S_p(X \otimes \mathbb{F}_p, u, T) = \frac{(1-T)(1-uT)^{101}(1-u^2T)^{101}(1-u^3T)}{(1-T)(1-uT)(1-u^2T)(1-u^3T)}.$$

This is independent of p. Note that we do not know if the one-parameter subfamily X_λ is generically ordinary for every p. The ordinarity property for every p was established only for the universal family of hypersurfaces, not for a one-parameter subfamily of hypersurfaces such as X_λ. If Y denotes the generic mirror of X, then by the results in [**10**] [**14**], Y is ordinary for every p and thus we obtain

$$S_p(Y \otimes \mathbb{F}_p, u, T) = \frac{(1-T)(1-uT)(1-u^2T)(1-u^3T)}{(1-T)(1-uT)^{101}(1-u^2T)^{101}(1-u^3T)}.$$

Again, it is independent of p. The slope mirror conjecture holds in this example.

Remark: The slope zeta function is completely determined by the Newton polygon of the Frobenius acting on cohomologies of the variety in question. The converse is not true, as there may be cancellations coming from different cohomology dimensions in the slope zeta function.

For a mirror pair over a number field, we have the following harder conjecture.

CONJECTURE 1.6 (Slope mirror conjecture over $\mathbb{Z}$). *Let $\{X, Y\}$ be two schemes of finite type over $\mathbb{Z}$ such that their generic fibres $\{X \otimes \mathbb{Q}, Y \otimes \mathbb{Q}\}$ form a usual (weak) mirror pair of d-dimensional Calabi-Yau manifolds defined over $\mathbb{Q}$. Then there are infinitely many prime numbers p (with positive density) such that*

$$S_p(X \otimes \mathbb{F}_p, u, T) = S_p(Y \otimes \mathbb{F}_p, u, T)^{(-1)^d}.$$

Remarks. If one uses the weight $2\log_q|\alpha_i|$ instead of the slope $\mathrm{ord}_q\alpha_i$, where $|\cdot|$ denotes the complex absolute value, one can define a two-variable weight zeta function in a similar way. It is easy to see that the resulting weight zeta function does not satisfy the desired symmetry, as the weight has nothing to do with the Hodge symmetry, while the slopes are related to the Hodge numbers, as the Newton polygon (slope polygon) lies above the Hodge polygon.

In practice, one is often given a mirror pair of singular Calabi-Yau orbifolds, where there may not exist a smooth crepant resolution. In such a case, one could define an orbifold zeta function, which would be equal to the zeta function of the smooth crepant resolution whenever such a resolution exists. Similar results and conjectures should carry over to such orbifold zeta functions.

In the appendix, D. Haessig (my student at UC Irvine) proves some additional congruence results for the strong mirror pair (X_λ, Y_λ), some of which are used in Section 7.

2. A counting formula via Gauss sums

Let $V_1, \cdots, V_m$ be m distinct lattice points in $\mathbb{Z}^n$. For $V_j = (V_{1j}, \cdots, V_{nj})$, write

$$x^{V_j} = x_1^{V_{1j}} \cdots x_n^{V_{nj}}.$$

Let f be the Laurent polynomial in n variables written in the form:

$$f(x_1, \cdots, x_n) = \sum_{j=1}^{m} a_j x^{V_j}, a_j \in \mathbb{F}_q,$$

where not all a_j are zero. Let M be the $n \times m$ matrix

$$M = (V_1, \cdots, V_m),$$

where each V_j is written as a column vector. Let N_f^* denote the number of $\mathbb{F}_q$-rational points on the affine toric hypersurface $f = 0$ in $\mathbb{G}_m^n$. If each $V_j \in \mathbb{Z}_{\geq 0}^n$, we let N_f denote the number of $\mathbb{F}_q$-rational points on the affine hypersurface $f = 0$ in $\mathbb{A}^n$. We first derive a well known formula for both N_f^* and N_f in terms of Gauss sums.

For this purpose, we now recall the definition of Gauss sums. Let $\mathbb{F}_q$ be the finite field of q elements, where $q = p^r$ and p is the characteristic of $\mathbb{F}_q$. Let χ be the Teichmüller character of the multiplicative group $\mathbb{F}_q^*$. For $a \in \mathbb{F}_q^*$, the value $\chi(a)$ is just the $(q-1)$-st root of unity in the p-adic field $\mathbb{C}_p$ such that $\chi(a)$ modulo p reduces to a. Define the $(q-2)$ Gauss sums over $\mathbb{F}_q$ by

$$G(k) = \sum_{a \in \mathbb{F}_q^*} \chi(a)^{-k} \zeta_p^{\mathrm{Tr}(a)} \quad (1 \leq k \leq q-2),$$

where ζ_p is a primitive p-th root of unity in $\mathbb{C}_p$ and Tr denotes the trace map from $\mathbb{F}_q$ to the prime field $\mathbb{F}_p$.

LEMMA 2.1. *For all $a \in \mathbb{F}_q$, the Gauss sums satisfy the following interpolation relation:*

$$\zeta_p^{\mathrm{Tr}(a)} = \sum_{k=0}^{q-1} \frac{G(k)}{q-1} \chi(a)^k,$$

where

$$G(0) = q-1, \ G(q-1) = -q.$$

Proof. By the Vandermonde determinant, there are numbers $C(k)$ $(0 \leq k \leq q-1)$ such that for all $a \in \mathbb{F}_q$, one has

$$\zeta_p^{\mathrm{Tr}(a)} = \sum_{k=0}^{q-1} \frac{C(k)}{q-1} \chi(a)^k.$$

It suffices to prove that $C(k) = G(k)$ for all k. Take $a = 0$; then one finds that $C(0)/(q-1) = 1$. This proves that $C(0) = q-1 = G(0)$. For $1 \leq k \leq q-2$, one computes that

$$G(k) = \sum_{a \in \mathbb{F}_q^*} \chi(a)^{-k} \zeta_p^{\mathrm{Tr}(a)} = \frac{C(k)}{q-1}(q-1) = C(k).$$

Finally,

$$0 = \sum_{a \in \mathbb{F}_q} \zeta_p^{\mathrm{Tr}(a)} = \frac{C(0)}{q-1} q + \frac{C(q-1)}{q-1}(q-1).$$

This gives $C(q-1) = -q = G(q-1)$. The lemma is proved.

We also need to use the following classical theorem of Stickelberger.

LEMMA 2.2. *Let $0 \leq k \leq q-1$. Write*

$$k = k_0 + k_1 p + \cdots + k_{r-1} p^{r-1}$$

in p-adic expansion, where $0 \leq k_i \leq p-1$. Let $\sigma(k) = k_0 + \cdots + k_{r-1}$ be the sum of the p-digits of k. Then,

$$\mathrm{ord}_p G(k) = \frac{\sigma(k)}{p-1}.$$

Now we turn to deriving a counting formula for N_f in terms of Gauss sums. Write $W_j = (1, V_j) \in \mathbb{Z}^{n+1}$. Then

$$x_0 f = \sum_{j=1}^{m} a_j x^{W_j} = \sum_{j=1}^{m} a_j x_0 x_1^{V_{1j}} \cdots x_n^{V_{nj}},$$

where x now has $n+1$ variables $\{x_0, \cdots, x_n\}$. Using the formula

$$\sum_{t \in \mathbb{F}_q} t^k = \begin{cases} 0, & \text{if } (q-1) \not| k, \\ q-1, & \text{if } (q-1)|k \text{ and } k > 0, \\ q, & \text{if } k = 0, \end{cases}$$

one then calculates that

$$\begin{aligned}
qN_f &= \sum_{x_0,\cdots,x_n \in \mathbb{F}_q} \zeta_p^{\mathrm{Tr}(x_0 f(x))} \\
&= \sum_{x_0,\cdots,x_n \in \mathbb{F}_q} \prod_{j=1}^{m} \zeta_p^{\mathrm{Tr}(a_j x^{W_j})} \\
&= \sum_{x_0,\cdots,x_n \in \mathbb{F}_q} \prod_{j=1}^{m} \sum_{k_j=0}^{q-1} \frac{G(k_j)}{q-1} \chi(a_j)^{k_j} \chi(x^{W_j})^{k_j} \\
&= \sum_{k_1=0}^{q-1} \cdots \sum_{k_m=0}^{q-1} (\prod_{j=1}^{m} \frac{G(k_j)}{q-1} \chi(a_j)^{k_j}) \sum_{x_0,\cdots,x_n \in \mathbb{F}_q} \chi(x^{k_1 W_1 + \cdots + k_m W_m}) \\
&= \sum_{\sum_{j=1}^m k_j W_j \equiv 0 (\mathrm{mod}\ q-1)} \frac{(q-1)^{s(k)} q^{n+1-s(k)}}{(q-1)^m} \prod_{j=1}^{m} \chi(a_j)^{k_j} G(k_j), \qquad (7)
\end{aligned}$$

where $s(k)$ denotes the number of non-zero entries in $k_1 W_1 + \cdots + k_m W_m$.

Similarly, one calculates that

$$\begin{aligned}
qN_f^* &= \sum_{x_0 \in \mathbb{F}_q, x_1,\cdots,x_n \in \mathbb{F}_q^*} \zeta_p^{\mathrm{Tr}(x_0 f(x))} \\
&= (q-1)^n + \sum_{x_0,\cdots,x_n \in \mathbb{F}_q^*} \prod_{j=1}^{m} \zeta_p^{\mathrm{Tr}(a_j x^{W_j})} \\
&= (q-1)^n + \sum_{\sum_{j=1}^m k_j W_j \equiv 0 (\mathrm{mod}\ q-1)} \frac{(q-1)^{n+1}}{(q-1)^m} \prod_{j=1}^{m} \chi(a_j)^{k_j} G(k_j). \qquad (8)
\end{aligned}$$

We shall use these two formulas to study the number of $\mathbb{F}_q$-rational points on certain hypersurfaces in the next two sections.

3. Rational points on Calabi-Yau hypersurfaces

In this section, we apply formula (7) to compute the number of $\mathbb{F}_q$-rational points on the projective hypersurface X_λ in $\mathbb{P}^n$ defined by

$$f(x_1, \cdots, x_{n+1}) = x_1^{n+1} + \cdots + x_{n+1}^{n+1} + \lambda x_1 \cdots x_{n+1} = 0,$$

where λ is an element of $\mathbb{F}_q^*$. We shall handle the easier case $\lambda = 0$ separately. Let M be the $(n+2) \times (n+2)$ matrix

$$(9) \qquad M = \begin{pmatrix} 1 & 1 & 1 & \cdots & 1 & 1 \\ n+1 & 0 & 0 & \cdots & 0 & 1 \\ 0 & n+1 & 0 & \cdots & 0 & 1 \\ \vdots & \vdots & \vdots & \cdots & \vdots & \vdots \\ 0 & 0 & 0 & \cdots & n+1 & 1 \end{pmatrix}$$

Let $k = (k_1, \cdots, k_{n+2})$ written as a column vector. Let N_f denote the number of $\mathbb{F}_q$-rational points on the affine hypersurface $f = 0$ in $\mathbb{A}^{n+1}$. By formula (7), we deduce that

$$qN_f = \sum_{Mk \equiv 0 (\text{mod } q-1)} \frac{(q-1)^{s(k)} q^{n+2-s(k)}}{(q-1)^{n+2}} (\prod_{j=1}^{n+2} G(k_j)) \chi(\lambda)^{k_{n+2}},$$

where $s(k)$ denotes the number of non-zero entries in $Mk \in \mathbb{Z}^{n+2}$. The number of $\mathbb{F}_q$-rational points on the projective hypersurface X_λ is then given by the formula

$$\frac{N_f - 1}{q-1} = \frac{-1}{q-1} + \sum_{Mk \equiv 0 (\text{mod } q-1)} \frac{q^{n+1-s(k)}}{(q-1)^{n+3-s(k)}} (\prod_{j=1}^{n+2} G(k_j)) \chi(\lambda)^{k_{n+2}}.$$

If $k = (0, \cdots, 0, q-1)$, then $Mk = (q-1, \cdots, q-1)$ and $s(k) = n+2$. In this case, the corresponding term in the above expression is $-(q-1)^n$, which is $(-1)^{n-1}$ modulo q. If $k = (0, ..., 0)$, then $s(k) = 0$ and the corresponding term is $q^{n+1}/(q-1)$, which is zero modulo q.

Thus, we obtain the congruence formula modulo q:

$$\frac{N_f - 1}{q-1} \equiv 1 + (-1)^{n-1} + \sum^*_{Mk \equiv 0 (\text{mod } q-1)} \frac{q^{n+1-s(k)}}{(q-1)^{n+3-s(k)}} (\prod_{j=1}^{n+2} G(k_j)) \chi(\lambda)^{k_{n+2}},$$

where $\sum^*$ means summing over all the solutions $k = (k_1, \cdots, k_{n+2})$ with $0 \leq k_i \leq q-1$, $k \neq (0, \cdots, 0)$, and $k \neq (0, \cdots, 0, q-1)$.

LEMMA 3.1. *If $k \neq (0, \cdots, 0)$, then $\prod_{j=1}^{n+2} G(k_j)$ is divisible by q.*

Proof. Let k be a solution of $Mk \equiv 0 (\text{mod } q-1)$ such that $k \neq (0, \cdots, 0)$. Then there are positive integers $\ell_0, \cdots, \ell_{r-1}$ such that

$$k_1 + \cdots + k_{n+2} = (q-1)\ell_0,$$
$$< pk_1 > + \cdots + < pk_{n+2} > = (q-1)\ell_1,$$
$$\cdots$$
$$< p^{r-1}k_1 > + \cdots + < p^{r-1}k_{n+2} > = (q-1)\ell_{r-1},$$

where $< pk_1 >$ denotes the unique integer in $[0, q-1]$ congruent to pk_1 modulo $(q-1)$ and which is 0 (resp. $q-1$) if $pk_1 = 0$ (resp., if pk_1 is a positive multiple of $q-1$). By the Stickelberger theorem, we deduce that

$$\text{ord}_p \prod_{j=1}^{n+2} G(k_j) = \frac{\sum_j \sigma(k_j)}{p-1} = \frac{1}{q-1} \sum_{i=0}^{r-1} (q-1)\ell_i = \sum_{i=0}^{r-1} \ell_i.$$

Since $\ell_i \geq 1$, it follows that

$$\text{ord}_q \prod_{j=1}^{n+2} G(k_j) = \frac{1}{r} \sum_{i=0}^{r-1} \ell_i \geq 1$$

with equality holding if and only if all $\ell_i = 1$. The lemma is proved.

Using this lemma and the previous congruence formula, we deduce

LEMMA 3.2. *Let $\lambda \in \mathbb{F}_q^*$. We have the congruence formula modulo q:*

$$\#X_\lambda(\mathbb{F}_q) \equiv 1 + (-1)^{n-1} + \sum_{\substack{Mk\equiv 0(\text{mod } q-1)\\ s(k)=n+2}}^{*} \frac{1}{q(q-1)}(\prod_{j=1}^{n+2} G(k_j))\chi(\lambda)^{k_{n+2}}.$$

4. Rational points on the mirror hypersurfaces

In this section, we apply formula (8) to compute the number of $\mathbb{F}_q$-rational points on the affine toric hypersurface in $\mathbb{G}_m^n$ defined by the Laurent polynomial equation

$$g(x_1, \cdots, x_n) = x_1 + \cdots + x_n + \frac{1}{x_1 \cdots x_n} + \lambda = 0,$$

where λ is an element of $\mathbb{F}_q^*$. Let N be the $(n+1) \times (n+2)$ matrix

$$N = \begin{pmatrix} 1 & 1 & \cdots & 1 & 1 & 1 \\ 1 & 0 & \cdots & 0 & -1 & 0 \\ 0 & 1 & \cdots & 0 & -1 & 0 \\ \vdots & \vdots & \cdots & \vdots & \vdots & \vdots \\ 0 & 0 & \cdots & 1 & -1 & 0 \end{pmatrix} \tag{10}$$

Let $k = (k_1, \cdots, k_{n+2})$ written as a column vector. By formula (8), we deduce that

$$qN_g^* = (q-1)^n + \sum_{Nk\equiv 0(\text{mod } q-1)} \frac{1}{(q-1)}(\prod_{j=1}^{n+2} G(k_j))\chi(\lambda)^{k_{n+2}},$$

where $k = (k_1, \cdots, k_{n+2})$ with $0 \le k_i \le q-1$.

The contribution of the trivial terms k (where each k_i is either 0 or $q-1$) is given by

$$\frac{1}{q-1}\sum_{s=0}^{n+2}(-q)^s(q-1)^{n+2-s}\binom{n+2}{s} = \frac{(-1)^n}{q-1}.$$

Since

$$(q-1)^n + \frac{(-1)^n}{q-1} = \frac{(q-1)^{n+1} + (-1)^n}{q-1} \equiv q(n+1)(-1)^{n-1}(\text{mod } q^2),$$

we deduce

LEMMA 4.1. *For $\lambda \in \mathbb{F}_q^*$, we have the following congruence formula modulo q:*

$$N_g^* \equiv (n+1)(-1)^{n-1} + \sum_{Nk\equiv 0(\text{mod } q-1)}' \frac{1}{q(q-1)}(\prod_{j=1}^{n+2} G(k_j))\chi(\lambda)^{k_{n+2}},$$

where $\sum'$ means summing over all the non-trivial solutions k.

5. The mirror congruence formula

THEOREM 5.1. *For $\lambda \in \mathbb{F}_q^*$, we have the congruence formula*

$$\#X_\lambda(\mathbb{F}_q) \equiv N_g^* + 1 - n(-1)^{n-1} \ (\text{mod } q).$$

Proof. If k is a non-trivial solution of $Nk \equiv 0 \,(\mathrm{mod}\ q-1)$, then we have

$$k_1 \equiv k_2 \equiv \cdots \equiv k_n \equiv k_{n+1} \,(\mathrm{mod}\ q-1)$$

and

$$k_1 + \cdots + k_{n+1} + k_{n+2} \equiv 0 \,(\mathrm{mod}\ q-1).$$

Since k is non-trivial, we must have

$$0 < k_1 = k_2 = \cdots = k_{n+1} < q-1,$$

$$k_1 + \cdots + k_{n+2} = (n+1)k_1 + k_{n+2} = (n+1)k_2 + k_{n+2} = \cdots \equiv 0 \ (\mathrm{mod}\ q-1).$$

This gives all solutions of the equation $Mk \equiv 0 \,(\mathrm{mod}\ q-1)$ with $k_1 = \cdots = k_{n+1}$, $0 < k_1 < q-1$ and $s(k) = n+2$. The corresponding terms for these k's in $(N_f - 1)/(q-1)$ and N_g^* are exactly the same.

A solution of $Mk \equiv 0 \,(\mathrm{mod}\ q-1)$ is called **admissible** if $s(k) = n+2$ and its first $k+1$ coordinates $\{k_1, \cdots, k_{n+1}\}$ contain at least two distinct elements. The above results show that we have

$$\frac{N_f - 1}{q-1} - 1 - (-1)^{n-1} - (N_g^* - (n+1)(-1)^{n-1})$$

$$\equiv \sum_{\text{admissible } k} \frac{1}{q(q-1)} (\prod_{j=1}^{n+2} G(k_j)) \chi(\lambda)^{k_{n+2}} \ (\mathrm{mod}\ q).$$

This congruence together with the following lemma completes the proof of the theorem.

LEMMA 5.2. *If k is an admissible solution of $Mk \equiv 0 \,(\mathrm{mod}\ q-1)$, then*

$$\mathrm{ord}_q(\prod_{j=1}^{n+2} G(k_j)) \geq 2.$$

Proof. If k is an admissible solution, then $< pk >, \cdots, < p^{r-1}k >$ are also admissible solutions. For each $1 \leq i \leq n+1$, write

$$(n+1)k_i + k_{n+2} = (q-1)\ell_i,$$

where ℓ_i is a positive integer. Adding these equations together, we get

$$(n+1)(k_1 + \cdots + k_{n+1}) + (n+1)k_{n+2} = (q-1)(\ell_1 + \cdots + \ell_{n+1}).$$

Thus, the integer

$$\frac{k_1 + \cdots + k_{n+2}}{q-1} = \frac{\ell_1 + \cdots + \ell_{n+1}}{n+1} = \ell \in \mathbb{Z}_{>0}.$$

It is clear that $\ell = 1$ if and only if each $\ell_i = 1$ which would imply that $k_1 = \cdots = k_{n+1}$, contradicting the admissibility of k. Thus, we must have that $\ell \geq 2$.

Similarly, for each $0 \leq i \leq r-1$, we have

$$< p^i k_1 > + \cdots + < p^i k_{n+2} > = (q-1)j_i,$$

where $j_i \geq 2$ is a positive integer. We conclude that

$$\mathrm{ord}_q(\prod_{j=1}^{n+2} G(k_j)) = \frac{j_0 + \cdots + j_{r-1}}{r} \geq 2.$$

The lemma is proved.

6. Rational points on the projective mirror

Let Δ be the convex integral polytope associated with the Laurent polynomial g. It is the n-dimensional simplex in $\mathbb{R}^n$ with the following vertices:

$$\{e_1, \cdots, e_n, -(e_1 + \cdots + e_n)\},$$

where the e_i's are the standard unit vectors in $\mathbb{R}^n$.

Let $\mathbb{P}_\Delta$ be the projective toric variety associated with the polytope Δ, which contains $\mathbb{G}_m^n$ as an open dense subset. Let Y_λ be the projective closure in $\mathbb{P}_\Delta$ of the affine toric hypersurface $g = 0$ in $\mathbb{G}_m^n$. The variety Y_λ is then a projective toric hypersurface in $\mathbb{P}_\Delta$. We are interested in the number of $\mathbb{F}_q$-rational points on Y_λ.

The toric variety $\mathbb{P}_\Delta$ has the following disjoint decomposition:

$$\mathbb{P}_\Delta = \bigcup_{\tau \in \Delta} \mathbb{P}_{\Delta,\tau},$$

where τ runs over all non-empty faces of Δ and each $\mathbb{P}_{\Delta,\tau}$ is isomorphic to the torus $\mathbb{G}_m^{\dim \tau}$. Accordingly, the projective toric hypersurface Y_λ has the corresponding disjoint decomposition

$$Y_\lambda = \bigcup_{\tau \in \Delta} Y_{\lambda,\tau}, \; Y_{\lambda,\tau} = Y_\lambda \cap \mathbb{P}_{\Delta,\tau}.$$

For $\tau = \Delta$, the subvariety $Y_{\lambda,\Delta}$ is simply the affine toric hypersurface defined by $g = 0$ in $\mathbb{G}_m^n$. For zero-dimensional τ, $Y_{\lambda,\tau}$ is empty. For a face τ with $1 \le \dim \tau \le n-1$, one checks that $Y_{\lambda,\tau}$ is isomorphic to the affine toric hypersurface in $\mathbb{G}_m^{\dim \tau}$ defined by

$$1 + x_1 + \cdots + x_{\dim \tau} = 0.$$

For such a τ, the inclusion-exclusion principle shows that

$$\#Y_{\lambda,\tau}(\mathbb{F}_q) = q^{\dim \tau - 1} - \binom{\dim \tau}{1} q^{\dim \tau - 2} + \cdots + (-1)^{\dim \tau - 1} \binom{\dim \tau}{\dim \tau - 1}.$$

Thus,

$$\#Y_{\lambda,\tau}(\mathbb{F}_q) = \frac{1}{q}((q-1)^{\dim \tau} + (-1)^{\dim \tau + 1}).$$

This formula holds even for zero-dimensional τ as both sides would then be zero.

Putting these calculations together, we deduce that

$$\#Y_\lambda(\mathbb{F}_q) = N_g^* - \frac{(q-1)^n + (-1)^{n+1}}{q} + \sum_{\tau \in \Delta} \frac{1}{q}((q-1)^{\dim \tau} + (-1)^{\dim \tau + 1}),$$

where τ runs over all non-empty faces of Δ including Δ itself. Since Δ is a simplex, one computes that

$$\sum_{\tau \in \Delta} ((q-1)^{\dim \tau} + (-1)^{\dim \tau + 1}) = \frac{q^{n+1} - 1}{q - 1} + (-1) = \frac{q(q^n - 1)}{q - 1}.$$

This implies that

$$\#Y_\lambda(\mathbb{F}_q) = N_g^* - \frac{(q-1)^n + (-1)^{n+1}}{q} + \frac{q^n - 1}{q - 1}. \tag{11}$$

This equality holds for all $\lambda \in \mathbb{F}_q$, including the case $\lambda = 0$. Reducing modulo q, we get

$$\#Y_\lambda(\mathbb{F}_q) \equiv N_g^* + 1 - n(-1)^{n-1} \pmod{q}. \tag{12}$$

This and Theorem 5.1 prove the case $\lambda \neq 0$ of the following theorem.

THEOREM 6.1. *For every finite field* $\mathbb{F}_q$ *with* $\lambda \in \mathbb{F}_q$, *we have the congruence formula*

$$\#X_\lambda(\mathbb{F}_q) \equiv \#Y_\lambda(\mathbb{F}_q) \pmod q.$$

If furthermore, $\lambda \in \mathbb{F}_q$ *such that* g *is* Δ*-regular and* W_λ *is a mirror manifold of* X_λ, *then*

$$\#Y_\lambda(\mathbb{F}_q) \equiv \#W_\lambda(\mathbb{F}_q) \pmod q.$$

Proof. For the first part, it remains to check the case $\lambda = 0$. The proof is similar and in fact somewhat simpler than the case $\lambda \neq 0$. We give an outline. Since $\lambda = 0$, we can take $k_{n+2} = 0$ in the calculations of N_f and N_g^*. One finds then

$$\#X_0(\mathbb{F}_q) \equiv 1 + \sum_{\substack{Mk\equiv 0 (\mathrm{mod}\ q-1) \\ s(k)=n+2}}^{*} \frac{1}{q(q-1)}\left(\prod_{j=1}^{n+2} G(k_j)\right),$$

where $\sum^*$ means summing over all those solutions $k = (k_1, \cdots, k_{n+1}, 0)$ with $0 \leq k_i \leq q-1$ and $k \neq (0, \cdots, 0)$.

Similarly, one computes that

$$N_g^* \equiv n(-1)^{n-1} + \sum_{Nk\equiv 0 (\mathrm{mod}\ q-1)}' \frac{1}{q(q-1)}\left(\prod_{j=1}^{n+2} G(k_j)\right),$$

where $\sum'$ means summing over all those non-trivial solutions k with $k_{n+2} = 0$. By (12), we deduce

$$\#Y_0(\mathbb{F}_q) \equiv 1 + \sum_{Nk\equiv 0 (\mathrm{mod}\ q-1)}' \frac{1}{q(q-1)}\left(\prod_{j=1}^{n+2} G(k_j)\right).$$

As before, one checks that

$$\sum_{\substack{Mk\equiv 0 (\mathrm{mod}\ q-1) \\ s(k)=n+2}}^{*} \left(\prod_{j=1}^{n+2} G(k_j)\right) \equiv \sum_{Nk\equiv 0 (\mathrm{mod}\ q-1)}' \left(\prod_{j=1}^{n+2} G(k_j)\right) (\mathrm{mod}\ q^2).$$

The first part of the theorem follows.

To prove the second part of the theorem, let Δ^* be the dual polytope of Δ. One checks that Δ^* is the simplex in $\mathbb{R}^n$ with the vertices

$$(n+1)e_i - \sum_{j=1}^{n} e_j \ (i = 1, ..., n), \ -\sum_{j=1}^{n} e_j.$$

This is the $(n+1)$-multiple of a basic (regular) simplex in $\mathbb{R}^n$. In particular, the codimension 1 faces of Δ^* are $(n+1)$-multiples of a basic simplex in $\mathbb{R}^{n-1}$. By the parallel hyperplane decomposition in [**7**], one deduces that the codimension 1 faces of Δ^* have a triangulation into basic simplices. Fix such a triangulation which produces a smooth crepant resolution $\phi : W_\lambda \to Y_\lambda$. One checks [**2**] that for each point $y \in Y_\lambda(\mathbb{F}_q)$, the fibre $\phi^{-1}(\lambda)$ is stratified by affine spaces over $\mathbb{F}_q$. Since the fibres are connected, it follows that the number of $\mathbb{F}_q$-rational points on $\phi^{-1}(\lambda)$ is congruent to 1 modulo q. Thus, modulo q, we have the congruence

$$\#W_\lambda(\mathbb{F}_q) \equiv \sum_{y\in Y_\lambda(\mathbb{F}_q)} \phi^{-1}(\lambda)(\mathbb{F}_q) \equiv \sum_{y\in Y_\lambda(\mathbb{F}_q)} 1 = \#Y_\lambda(\mathbb{F}_q).$$

The proof is complete.

In terms of zeta functions, the above theorem says that the slope $[0,1)$ part of the zeta function for X_λ equals the slope $[0,1)$ part of the zeta function for Y_λ.

The above elementary calculations can be used to treat some other examples of toric hypersurfaces and complete intersections. In a joint work [**6**] with Lei Fu, we have proved the following partial generalization.

THEOREM 6.2. *Let X be a smooth connected Calabi-Yau scheme defined over the ring $W(\mathbb{F}_q)$ of Witt vectors of $\mathbb{F}_q$. Let G be a finite group of W-morphisms acting on X. Assume that G fixes the non-zero global section of the canonical bundle of X. Then, for each positive integer k, we have the congruence formula*

$$\#(X \otimes \mathbb{F}_q)(\mathbb{F}_{q^k}) \equiv \#(X/G \otimes \mathbb{F}_q)(\mathbb{F}_{q^k})(\mathrm{mod}\ q^k).$$

7. Applications to zeta functions

In this section, we compare the two zeta functions $Z(X_\lambda, T)$ and $Z(Y_\lambda, T)$, where $\{X_\lambda, Y_\lambda\}$ is our strong mirror pair.

First, we recall what is known about $Z(X_\lambda, T)$. Let $\lambda \in \mathbb{F}_q$ such that X_λ is smooth projective. By the Weil conjectures, the zeta function of X_λ over $\mathbb{F}_q$ has the following form:

$$Z(X_\lambda, T) = \frac{P(\lambda, T)^{(-1)^n}}{(1-T)(1-qT)\cdots(1-q^{n-1}T)}, \tag{13}$$

where $P(\lambda, T) \in 1 + T\mathbb{Z}[T]$ is a polynomial of degree $n(n^n - (-1)^n)/(n+1)$, pure of weight $n-1$. By the results in [**8**][**10**], the universal family of hypersurfaces of degree $n+1$ is generically ordinary for every p (Mazur's conjecture). However, we do not know if the one-parameter family X_λ of hypersurfaces is generically ordinary for every p. Thus, we raise

QUESTION 7.1. *Is the one-parameter family X_λ of degree $n+1$ hypersurfaces in $\mathbb{P}^n$ generically ordinary for every prime number p not dividing $(n+1)$?*

The answer is yes if $p \equiv 1 \pmod{n+1}$ since the fibre for $\lambda = 0$ is already ordinary if $p \equiv 1 \pmod{n+1}$. It is also true if $n \leq 3$. The first unknown case is when $n = 4$, the quintic case.

Next, we recall what is known about $Z(Y_\lambda, T)$. Let $\lambda \in \mathbb{F}_q$ such that g is Δ-regular. This is equivalent to assuming that $(-\lambda)^{n+1} \neq (n+1)^{n+1}$. Then the zeta function of the affine toric hypersurface $g = 0$ over $\mathbb{F}_q$ in $\mathbb{G}_m^n$ has the following form (see [**15**]):

$$Z(g, T) = Q(\lambda, T)^{(-1)^n} \prod_{i=0}^{n-1} (1 - q^i T)^{(-1)^{n-i}\binom{n}{i+1}},$$

where $Q(\lambda, T) \in 1 + T\mathbb{Z}[T]$ is a polynomial of degree n, pure of weight $n-1$. The product of the trivial factors in $Z(g, T)$ is simply the zeta function of the sequence

$$\frac{(q^k - 1)^n + (-1)^{n+1}}{q^k}, \ k = 1, 2, \cdots.$$

¿From this and (11), one deduces that the zeta function of the projective toric hypersurface Y_λ has the form

$$Z(Y_\lambda, T) = \frac{Q(\lambda, T)^{(-1)^n}}{(1-T)(1-qT)\cdots(1-q^{n-1}T)}. \tag{14}$$

By the results in [**10**][**14**], this one-parameter family Y_λ of toric hypersurfaces is generically ordinary for every n and every prime number p.

Now we are ready to compare the two zeta functions $Z(X_\lambda, T)$ and $Z(Y_\lambda, T)$. Let now $\lambda \in \mathbb{F}_q$ such that X_λ is smooth and g is Δ-regular. The above description shows that

$$\frac{Z(X_\lambda, T)}{Z(Y_\lambda, T)} = (\frac{P(\lambda, T)}{Q(\lambda, T)})^{(-1)^n}.$$

To understand this quotient of zeta functions, it suffices to understand the quotient $P(\lambda, T)/Q(\lambda, T)$.

LEMMA 7.2. *Assume that $(n+1)|(q-1)$. Then the polynomial $Q(\lambda, T)$ divides $P(\lambda, T)$.*

Proof. Since $(n+1)|(q-1)$, the finite map $X_\lambda \to Y_\lambda$ is a Galois covering with Galois group G, where $G = (\mathbb{Z}/(n+1)\mathbb{Z})^{n-1}$ is an abelian group. For an ℓ-adic representation $\rho : G \to \mathrm{GL}(V_\rho)$, let $L(X_\lambda, \rho, T)$ denote the corresponding L-function of ρ associated to this Galois covering. Then we have the standard factorization

$$Z(X_\lambda, T) = \prod_\rho L(X_\lambda, \rho, T),$$

where ρ runs over all irreducible (necessarily one-dimensional) ℓ-adic representations of G. If $\rho = 1$ is the trivial representation, then

$$L(X_\lambda, 1, T) = Z(Y_\lambda, T).$$

For a prime number $\ell \neq p$, the ℓ-adic trace formula for $Z(X_\lambda, T)$ is

$$Z(X_\lambda, T) = \prod_{i=0}^{2(n-1)} \det(I - T\mathrm{Frob}_q | H^i(X_\lambda \otimes \bar{\mathbb{F}}_q, \mathbb{Q}_\ell))^{(-1)^{i-1}},$$

where Frob_q denotes the geometric Frobenius element over $\mathbb{F}_q$. Since X_λ is a smooth projective hypersurface of dimension $n-1$, one has the more precise form of the zeta function:

$$(15) \qquad Z(X_\lambda, T) = \frac{\det(I - T\mathrm{Frob}_q | H^{n-1}(X_\lambda \otimes \bar{\mathbb{F}}_q, \mathbb{Q}_\ell))^{(-1)^n}}{(1-T)(1-qT)\cdots(1-q^{n-1}T)}.$$

Similarly, the ℓ-adic trace formula for the L-function is

$$L(X_\lambda, \rho, T) = \prod_{i=0}^{2(n-1)} \det(I - T(\mathrm{Frob}_q \otimes 1) | (H^i(X_\lambda \otimes \bar{\mathbb{F}}_q, \mathbb{Q}_\ell) \otimes V_\rho)^G)^{(-1)^{i-1}}.$$

For odd $i \neq n-1$,

$$H^i(X_\lambda \otimes \bar{\mathbb{F}}_q, \mathbb{Q}_\ell) = 0, \ (H^i(X_\lambda \otimes \bar{\mathbb{F}}_q, \mathbb{Q}_\ell) \otimes V_\rho)^G = 0.$$

For even $i = 2k \neq n-1$ with $0 \leq k \leq n-1$,

$$H^{2k}(X_\lambda \otimes \bar{\mathbb{F}}_q, \mathbb{Q}_\ell) = \mathbb{Q}_\ell(-k), \ (H^{2k}(X_\lambda \otimes \bar{\mathbb{F}}_q, \mathbb{Q}_\ell) \otimes V_\rho)^G = 0$$

for non-trivial irreducible ρ. This proves that for irreducible $\rho \neq 1$, we have

$$L(X_\lambda, \rho, T) = \det(I - T(\mathrm{Frob}_q \otimes 1) | (H^{n-1}(X_\lambda \otimes \bar{\mathbb{F}}_q, \mathbb{Q}_\ell) \otimes V_\rho)^G)^{(-1)^n}.$$

Similarly, taking $\rho = 1$, one finds that

$$(16) \qquad Z(Y_\lambda, T) = \frac{\det(I - T\mathrm{Frob}_q | (H^{n-1}(X_\lambda \otimes \bar{\mathbb{F}}_q, \mathbb{Q}_\ell))^G)^{(-1)^n}}{(1-T)(1-qT)\cdots(1-q^{n-1}T)}.$$

Comparing (13)-(16), we conclude that

$$P(\lambda, T) = \det(I - T\mathrm{Frob}_q | H^{n-1}(X_\lambda \otimes \bar{\mathbb{F}}_q, \mathbb{Q}_\ell)),$$

$$Q(\lambda, T) = \det(I - T\mathrm{Frob}_q | (H^{n-1}(X_\lambda \otimes \bar{\mathbb{F}}_q, \mathbb{Q}_\ell))^G).$$

Furthermore, the quotient

$$\frac{P(\lambda,T)}{Q(\lambda,T)} = \prod_{\rho\neq 1} \det(I - T(\mathrm{Frob}_q \otimes 1) | (H^{n-1}(X_\lambda \otimes \bar{\mathbb{F}}_q, \mathbb{Q}_\ell) \otimes V_\rho)^G)$$

is a polynomial with integer coefficients of degree $\frac{n(n^n-(-1)^n)}{n+1} - n$, pure of weight $n-1$. The lemma is proved.

This lemma together with Theorem 6.1 gives the following result.

THEOREM 7.3. *Assume that* $(n+1)|(q-1)$. *There is a polynomial* $R_n(\lambda, T) \in 1 + T\mathbb{Z}[T]$ *which is pure of weight* $n-3$ *and of degree* $\frac{n(n^n-(-1)^n)}{n+1} - n$, *such that*

$$\frac{P(\lambda,T)}{Q(\lambda,T)} = R_n(\lambda, qT).$$

We conjecture that the condition $(n+1)|(q-1)$ is not necessary in the above lemma and theorem. In the appendix, D. Haessig proves this divisibility in the case $(n+1, q-1) = 1$ and $n+1$ is a prime. In particular, the divisibility conjecture is always true if $(n+1)$ is a prime.

The polynomial $R_n(\lambda, T)$ measures how far the zeta function of Y_λ differs from the zeta function of X_λ. Being of integral pure weight $n-3$, the polynomial $R_n(\lambda, T)$ should come from the zeta function of a variety (or motive $M_n(\lambda)$) of dimension $n-3$. It would be interesting to find this variety or motive $M_n(\lambda)$ parameterized by λ. In this direction, the following is known.

If $n = 2$, then $n - 3 < 0$, $M_2(\lambda)$ is empty and we have $R_2(\lambda, T) = 1$. If $n = 3$, then $n - 3 = 0$ and

$$R_3(\lambda, T) = \prod_{i=1}^{18} (1 - \alpha_i(\lambda)T)$$

is a polynomial of degree 18 with $\alpha_i(\lambda)$ being roots of unity. In fact, Dwork [**4**] proved that all $\alpha_i(\lambda) = \pm 1$ in this case. Thus, $R_3(\lambda, T)$ comes from the the zeta function of a zero-dimensional variety $M_3(\lambda)$ parameterized by λ. What is this zero-dimensional variety $M_3(\lambda)$? For every p and generic λ, the slope zeta function has the form $S_p(Y_\lambda, u, T) = 1$ and

$$S_p(X_\lambda, u, T) = \frac{1}{(1-T)^2(1-uT)^{20}(1-u^2T)^2}.$$

Note that Y_λ is singular and not a smooth mirror of X_λ yet. Thus, it is not surprising that the two slope zeta functions $S_p(X_\lambda, u, T)$ and $S_p(Y_\lambda, u, T)$ do not satisfy the expected slope mirror symmetry.

If $n = 4$, then $n - 3 = 1$ and

$$R_4(\lambda, T) = \prod_{i=1}^{200} (1 - \alpha_i(\lambda)T)$$

is a polynomial of degree 200 with $\alpha_i(\lambda) = \sqrt{q}$. Thus, $M_4(\lambda)$ should come from some curve parameterized by λ. This curve has been constructed explicitly in a recent paper by Candelas, de la Ossa and F. Rodriguez-Villegas [**3**]. For every p and generic λ, we know that $S_p(Y_\lambda, u, T) = 1$, but as indicated at the beginning of this section, we do not know if the slope zeta function of X_λ for a generic λ has the form

$$S_p(X_\lambda, u, T) = \frac{(1-T)(1-uT)^{101}(1-u^2T)^{101}(1-u^3T)}{(1-T)(1-uT)(1-u^2T)(1-u^3T)}.$$

For general n and $\lambda \in K$ for some field K, in terms of ℓ-adic Galois representations, the pure motive $M_n(\lambda)$ is simply given by

$$M_n(\lambda) = (\bigoplus_{\rho \neq 1} (H^{n-1}(X_\lambda \otimes \bar{K}, \mathbb{Q}_\ell) \otimes V_\rho)^G) \otimes \mathbb{Q}_\ell(-1),$$

where $\mathbb{Q}_\ell(-1)$ denotes the Tate twist. If λ is in a number field K, this implies that the Hasse-Weil zeta functions of X_λ and Y_λ are related by

$$\zeta(X_\lambda, s) = \zeta(Y_\lambda, s) L(M_n(\lambda), s-1).$$

8. Slope zeta functions

The slope zeta function satisfies a functional equation. This follows from the usual functional equation which in turn is a consequence of Poincaré duality for ℓ-adic cohomology.

PROPOSITION 8.1. *Let X be a connected smooth projective variety of dimension d over $\mathbb{F}_q$. Then the slope zeta function $S_p(X, u, T)$ satisfies the following functional equation:*

$$S_p(X, u, \frac{1}{u^d T}) = S_p(X, u, T)(-u^{d/2} T)^{e(X)}, \tag{17}$$

where $e(X)$ denotes the ℓ-adic Euler characteristic of X.

Proof. Let $P_i(T)$ denote the characteristic polynomial of the geometric Frobenius acting on the i-th ℓ-adic cohomology of $X \otimes \bar{\mathbb{F}}_q$. Then,

$$Z(X, T) = \prod_{i=0}^{2d} P_i(T)^{(-1)^{i+1}}.$$

Let s_{ij} $(j = 1, \cdots, b_i)$ denote the slopes of the polynomial $P_i(T)$, where b_i is the degree of $P_i(T)$, which is the i-th Betti number. Write

$$Q_i(T) = \prod_{j=1}^{b_i} (1 - u^{s_{ij}} T).$$

Then, by the definition of the slope zeta function, we have

$$S_p(X, u, T) = \prod_{i=0}^{2d} Q_i(T)^{(-1)^{i+1}}.$$

For each $0 \leq i \leq 2d$, the slopes of $P_i(T)$ satisfy the determinant relation

$$\sum_{j=1}^{b_i} s_{ij} = \frac{i}{2} b_i.$$

Using this, one computes that

$$Q_i(\frac{1}{T}) = (-1/T)^{b_i} u^{ib_i/2} \prod_{j=1}^{b_i} (1 - u^{-s_{ij}} T).$$

Replacing T by $u^d T$, we get

$$Q_i(\frac{1}{u^d T}) = (\frac{-1}{u^d T})^{b_i} u^{ib_i/2} \prod_{j=1}^{b_i} (1 - u^{d - s_{ij}} T).$$

The functional equation for the usual zeta function $Z(X,T)$ implies that $d - s_{ij}$ $(j = 1, \cdots, b_i)$ are exactly the slopes for $P_{2d-i}(T)$. Thus,

$$Q_i(\frac{1}{u^dT}) = (\frac{-1}{u^dT})^{b_i} u^{ib_i/2} Q_{2d-i}(T).$$

We deduce that

$$S_p(X, u, \frac{1}{u^dT}) = \prod_{i=0}^{2d} \left(Q_{2d-i}(T)(\frac{-1}{u^dT})^{b_i} u^{ib_i/2}\right)^{(-1)^{i+1}}.$$

Since $b_i = b_{2d-i}$, it is clear that

$$\sum_{i=0}^{2d} (-1)^i \frac{i}{2} b_i = \frac{d}{2} e(X).$$

We conclude that

$$S_p(X, u, \frac{1}{u^dT}) = S_p(X, u, T)(-T)^{e(X)} u^{\frac{d}{2}e(X)}.$$

The proposition is proved.

In the rest of this section, we assume that X is a smooth projective scheme over $W(\mathbb{F}_q)$. Assume that the reduction $X \otimes \mathbb{F}_q$ is ordinary, i.e., the p-adic Newton polygon coincides with the Hodge polygon [**9**]. This means that the slopes of $P_i(T)$ are exactly j $(0 \leq j \leq i)$ with multiplicity $h^{j,i-j}(X)$. In this case, one gets the explicit formula

$$S_p(X \otimes \mathbb{F}_q, u, T) = \prod_{j=0}^{d} (1 - u^j T)^{e_j(X)}, \tag{18}$$

where

$$e_j(X) = (-1)^j \sum_{i=0}^{d} (-1)^{i-1} h^{j,i}(X). \tag{19}$$

If X and Y form a mirror pair over the Witt ring $W(\mathbb{F}_q)$, the Hodge symmetry $h^{j,i}(X) = h^{j,d-i}(Y)$ implies for each j,

$$e_j(X) = (-1)^j \sum_{i=0}^{d} (-1)^{i-1} h^{j,d-i}(Y) = (-1)^d e_j(Y).$$

We obtain the following result.

PROPOSITION 8.2. *Let X and Y be a mirror pair of d-dimensional smooth projective Calabi-Yau schemes over $W(\mathbb{F}_q)$. Assume that both $X \otimes \mathbb{F}_q$ and $Y \otimes \mathbb{F}_q$ are ordinary. Then we have the following symmetry for the slope zeta function:*

$$S_p(X \otimes \mathbb{F}_q, u, T) = S_p(Y \otimes \mathbb{F}_q, u, T)^{(-1)^d}.$$

The converse of this proposition may not always be true. The slope mirror conjecture follows from the following slightly stronger

CONJECTURE 8.3 (Generic ordinarity conjecture). *Let $d \leq 3$. Suppose that $\{X, Y\}$ form a generic mirror pair of d-dimensional smooth projective Calabi-Yau schemes over $W(\mathbb{F}_q)$. Then both $X \otimes \mathbb{F}_q$ and $Y \otimes \mathbb{F}_q$ are generically ordinary.*

For $d \leq 3$, it should be possible to prove this conjecture in the toric hypersurface case using the results in [**10**][**14**]. For $d \geq 4$, we expect that the same conjecture holds if $p \equiv 1 \pmod{D}$ for some positive integer D. This should again be provable in the toric hypersurface case using the results in [**10**]. But we do not know if we can always take $D = 1$, even in the toric hypersurface case if $d \geq 4$.

9. Appendix (by C. Douglas Haessig)

The purpose of this appendix is twofold. First, we demonstrate that under certain conditions we may extend the Arithmetic Mirror Theorem (Theorems 1.1 and 6.1). Second, we apply this extension to the study of the quotient of the zeta functions of X_λ and Y_λ.

As in the introduction, with $\lambda \in \mathbb{C}$ we may define a family of complex projective hypersurfaces X_λ in $\mathbb{P}^n_{\mathbb{C}}$ by

$$x_1^{n+1} + \cdots + x_{n+1}^{n+1} + \lambda x_1 \cdots x_{n+1} = 0.$$

With the group

$$G := \{(\zeta_1, \ldots, \zeta_{n+1}) \mid \zeta_i \in \mathbb{C},\ \zeta_i^{n+1} = 1, \zeta_1 \cdots \zeta_{n+1} = 1\}$$

we may define the (singular) mirror variety Y_λ as the quotient X_λ / G where G acts by coordinate multiplication. It turns out that Y_λ is a toric hypersurface and may be explicitly described as the projective closure in $\mathbb{P}_\Delta$ of the affine toric hypersurface

$$g(x_1, \ldots, x_n) := x_1 + \cdots + x_n + \frac{1}{x_1 \cdots x_n} + \lambda = 0.$$

Note that $\mathbb{P}_\Delta$ is the toric variety obtained from the polytope in $\mathbb{R}^n$ with vertices $\{e_1, \ldots, e_n, -(e_1 + \cdots + e_n)\}$, where the e_i are the standard basis vectors of $\mathbb{R}^n$. From this description of Y_λ, if we let $\mathbb{F}_q$ denote the finite field with q elements of characteristic p, it makes sense to discuss $\mathbb{F}_{q^k}$-rational points of X_λ and its mirror Y_λ when the parameter λ lies in $\mathbb{F}_q$.

When the $\gcd(n+1, q^k - 1) = 1$, there are no $(n+1)$-st roots of unity in the field $\mathbb{F}_{q^k}$. Viewing G as a group scheme over $\mathbb{Z}$, this means there are no $\mathbb{F}_{q^k}$-rational points of G. This leads one to suspect a direct relation between the $\mathbb{F}_{q^k}$-rational points of X_λ and Y_λ:

THEOREM 9.1. *For every positive integer k such that $\gcd(n+1, q^k - 1) = 1$, we have the equality $\#X_\lambda(\mathbb{F}_{q^k}) = \#Y_\lambda(\mathbb{F}_{q^k})$.*

If W_λ is a smooth crepant resolution of Y_λ, then there is a rational map from W_λ to Y_λ which is injective on rational smooth points. Thus, if none of the $\mathbb{F}_{q^k}$-rational points on Y_λ are singular points, we see that $\#Y_\lambda(\mathbb{F}_{q^k}) = \#W_\lambda(\mathbb{F}_{q^k})$. Consequently, we have:

COROLLARY 9.2. *Suppose the singular locus of Y_λ contains no $\mathbb{F}_{q^k}$-rational points. If $\gcd(n+1, q^k-1) = 1$, then we have $\#X_\lambda(\mathbb{F}_{q^k}) = \#Y_\lambda(\mathbb{F}_{q^k}) = \#W_\lambda(\mathbb{F}_{q^k})$.*

Next, when $\gcd(n+1, q^k - 1) > 1$ we may prove:

THEOREM 9.3. *Let $d := \gcd(n+1, q^k - 1) > 1$. Then*

1. *$\#X_\lambda(\mathbb{F}_{q^k}) \equiv 0 \bmod d$,*
2. *if $n+1$ is a power of a prime ℓ, then, writing $\lambda = -(n+1)\psi$ in the new parameter ψ, we have*

$$\#X_\lambda(\mathbb{F}_{q^k}) \equiv 0 \bmod \ell d \quad \text{and} \quad \#Y_\lambda(\mathbb{F}_{q^k}) \equiv \begin{cases} 1 & \psi^{n+1} = 1 \\ 0 & \text{otherwise} \end{cases} \bmod \ell.$$

Thus, combining Theorems 1.1 and 9.3 with the Chinese Remainder Theorem yields:

COROLLARY 9.4. *Suppose $n+1$ is a power of a prime ℓ and $gcd(n+1,q)=1$. Set $\lambda=-(n+1)\psi$. If $\psi^{n+1}\neq 1$, then for every positive integer k, we have*

$$\#X_\lambda(\mathbb{F}_{q^k})\equiv \#Y_\lambda(\mathbb{F}_{q^k}) \mod \ell q^k.$$

Before discussing the proofs of Theorems 9.1 and 9.3, let us apply Theorem 9.1 to the quotient of the zeta functions of X_λ and Y_λ.

9.1. Application to zeta functions. From Theorem 7.3, when $(n+1)|q-1$ then the quotient of the zeta functions of X_λ and Y_λ, when raised to the $(-1)^n$ power, is a polynomial of specified degree. As mentioned in Section 7, we conjecture that the divisibility $(n+1)|q-1$ is unnecessary and may be removed without disturbing the conclusion. Evidence for this is the following:

THEOREM 9.5. *Let $n+1$ be a prime such that $gcd(n+1,q)=1$. Let k be the smallest positive integer such that $q^k\equiv 1 \mod n+1$. Assume X_λ is non-singular and $\lambda^{n+1}\neq(-(n+1))^{n+1}$. Then there are positive integers $\rho_1,\ldots,\rho_s$, each divisible by k, and polynomials $Q_1,\ldots,Q_s\in 1+T\mathbb{Z}[T]$ which are pure of weight $n-3$ and irreducible over $\mathbb{Z}$, such that*

$$\left(\frac{Z(X_\lambda/\mathbb{F}_q,T)}{Z(Y_\lambda/\mathbb{F}_q,T)}\right)^{(-1)^n}=Q_1(q^kT^k)^{\rho_1/k}\cdots Q_s(q^kT^k)^{\rho_s/k}$$

Furthermore, $\rho_1+\cdots+\rho_s=\frac{n(n^n-(-1)^n)}{n+1}-n$. (Note the polynomials Q_i depend on n and λ.)

PROOF. For every nonnegative integer s and $j=1,\ldots,k-1$, we have $\gcd(n+1,q^{sk+j}-1)=1$. So, by Theorem 9.1, we have $\#X_\lambda(\mathbb{F}_{q^{sk+j}})=\#Y_\lambda(\mathbb{F}_{q^{sk+j}})$ for every $s\geq 0$ and $j=1,\ldots,k-1$. This implies

$$\frac{Z(X_\lambda/\mathbb{F}_q,T)}{Z(Y_\lambda/\mathbb{F}_q)}=\frac{\exp\sum_{s\geq 1}\frac{\#X_\lambda(\mathbb{F}_{q^{ks}})}{ks}T^{ks}}{\exp\sum_{s\geq 1}\frac{\#Y_\lambda(\mathbb{F}_{q^{ks}})}{ks}T^{ks}}=\left(\frac{Z(X_\lambda/\mathbb{F}_{q^k},T^k)}{Z(Y_\lambda/\mathbb{F}_{q^k},T^k)}\right)^{1/k} \tag{20}$$

where the first equality uses the previous sentence and the second equality is simply a definition. By Theorem 7.3, there exists a polynomial $R_n(\lambda,T)\in 1+T\mathbb{Z}[T]$ of degree $\frac{n(n^n-(-1)^n)}{n+1}-n$, pure of weight $n-3$, such that

$$\left(\frac{Z(X_\lambda/\mathbb{F}_{q^k},T)}{Z(Y_\lambda/\mathbb{F}_{q^k},T)}\right)^{(-1)^n}=R_n(\lambda,q^kT).$$

Combining this with (20) shows us that

$$\left(\frac{Z(X_\lambda/\mathbb{F}_q,T)}{Z(Y_\lambda/\mathbb{F}_q,T)}\right)^{(-1)^n}=R_n(\lambda,q^kT^k)^{1/k}.$$

Therefore, factorizing $R_n(\lambda,T)=Q_1(T)^{\rho_1}\cdots Q_s(T)^{\rho_s}$ into irreducibles over $\mathbb{Z}$ proves the theorem. □

As a side remark, for $n+1=5$, Theorem 9.5 explains the form of the zeta function of $Z(X_\lambda/\mathbb{F}_q,T)$ found in [**3**].

We note that, for the quintic $(n+1=5)$, it follows from [**3**, Equation 10.3] and [**3**, Equation 10.7], in which they empirically compute the zeta functions of X_λ and W_λ, that for X_λ smooth,

$$\frac{Z(X_\lambda/\mathbb{F}_q,T)}{Z(W_\lambda/\mathbb{F}_q,T)}=R_{\mathcal{A}}(q^kT^k,\lambda)^{20/k}R_{\mathcal{B}}(q^kT^k,\lambda)^{30/k}$$

where k is the smallest positive integer such that $q^k \equiv 1$ modulo 5 and the R's are quartic polynomials over $\mathbb{Z}$ which are not necessarily irreducible. Note that $k = 1$, 2 or 4. Furthermore, they have constructed auxiliary curves $\mathcal{A}$ and $\mathcal{B}$, both of genus 4, whose zeta functions experimentally correspond to $R_{\mathcal{A}}$ and $R_{\mathcal{B}}$, respectively. It would be interesting to find these "auxiliary varieties" for general $n+1$ and see how they fit into the framework of mirror symmetry (if at all).

9.2. The proof of Theorem 9.1. Without loss, we will write q instead of q^k in the following proof.

9.2.1. *Formulas for* $X_\lambda(\mathbb{F}_q)$ *and* $Y_\lambda(\mathbb{F}_q)$ *in terms of Gauss sums.* Define the Gauss sums $G(k)$ as in Section 2. Also, let M be the $(n+2) \times (n+2)$-matrix defined in Section 3.

Define the set

$$E := \{k \in \mathbb{Z}^{n+2} | 0 \leq k_i \leq q-1 \text{ and } Mk \equiv 0 \bmod (q-1)\}.$$

For each $k \in \mathbb{Z}^{n+2}$, define $s(k)$ as the number of non-zero entries in $Mk \in \mathbb{Z}^{n+2}$. Next, define

$$E_1 := \{k \in E | \text{not all } k_1, \ldots, k_{n+1} \text{ are the same, but } 0 \leq k_{n+2} \leq q-1\}$$
$$E_2 := \{k \in E | k_1 = k_2 = \cdots = k_{n+1}, 0 \leq k_{n+2} \leq q-1\}$$
$$E_2^* := \{k \in E_2 | 0 < k_1 < q-1, s(k) = n+2\}$$
$$S_k := \frac{q^{n+1-s(k)}}{(q-1)^{n+3-s(k)}} \left(\prod_{j=1}^{n+2} G(k_j) \right) \chi(\lambda)^{k_{n+2}}.$$

Now, Section 3 demonstrated that

$$\#X_\lambda(\mathbb{F}_q) = \frac{-1}{q-1} + \sum_{k \in E_1} S_k + \sum_{k \in E_2} S_k.$$

Consider $k \in E$. Suppose $k_1 = \cdots = k_{n+1} = 0$. If $k_{n+2} = 0$, then $S_k = q^{n+1}/(q-1)$; otherwise, if $k_{n+2} = q-1$, then $S_k = -(q-1)^n$. Similarly, suppose $k_1 = \cdots = k_{n+1} = q-1$. If $k_{n+2} = 0$, then $S_k = (-1)^{n+1} q^n$; otherwise, if $k_{n+2} = q-1$, then $S_k = (-1)^n q^{n+1}/(q-1)$.

Next, notice

$$Mk = \begin{pmatrix} k_1 + \cdots + k_{n+2} \\ (n+1)k_1 + k_{n+2} \\ (n+1)k_2 + k_{n+2} \\ \vdots \\ (n+1)k_{n+1} + k_{n+2} \end{pmatrix} \in \mathbb{Z}^{n+2}. \tag{21}$$

If one of the rows equals zero, then we must have $k_i = 0$ for some $1 \leq i \leq n+1$. Thus, if $k \in E_2$ such that $0 < k_1 < q-1$, then all the rows of Mk must be non-zero; that is, $s(k) = n+2$. Putting this together with the last paragraph, we find that for $\lambda \neq 0$, then

$$\#X_\lambda(\mathbb{F}_q) = \frac{q^{n+1} + (-1)^n q^n - 1 - (q-1)^{n+1}}{q-1} + \sum_{k \in E_1} S_k + \sum_{k \in E_2^*} S_k. \tag{22}$$

If $\lambda = 0$, then Section 2 tells us that k_{n+2} is forced to equal zero. Thus, in the above calculation, we need to neglect all terms in which $k_{n+2} \neq 0$. Doing this, we obtain

$$\#X_0(\mathbb{F}_q) = \sum_{k\in E_1} S_k + N_0^* + \frac{q^{n+1}-1}{q-1} + (-1)^{n+1}q^n + \frac{(-1)^n - (q-1)^n}{q}.$$

Let N_λ^* denote the number of $\mathbb{F}_q$-rational points on the affine (toric) variety defined by

$$g(x_1,\ldots,x_n) := x_1 + \cdots + x_n + \frac{1}{x_1\cdots x_n} + \lambda = 0.$$

In the proof of Theorem 5.1, we saw that for $\lambda \neq 0$

$$N_\lambda^* = \frac{(q-1)^n}{q} + \frac{(-1)^n}{q(q-1)} + \sum_{k\in E_2^*} S_k. \tag{23}$$

Also, if $\lambda = 0$, we may calculate that

$$N_0^* = \frac{(q-1)^n}{q} + \frac{(-1)^{n+1}}{q} + \sum_{k\in E_2^*} S_k.$$

For ease of reference, recall from Section 2 that, for all $\lambda \in \mathbb{F}_q$), we have

$$\#Y_\lambda(\mathbb{F}_q) = N_\lambda^* - \frac{(q-1)^n}{q} + \frac{(-1)^n}{q} + \frac{q^n-1}{q-1}. \tag{24}$$

9.2.2. *Finishing the proof of Theorem 9.1.* For $\lambda \neq 0$, combining equations (22), (23), and (24) yields

$$\#X_\lambda(\mathbb{F}_q) - \#Y_\lambda(\mathbb{F}_q) = \sum_{k\in E_1} S_k - (q-1)^n + \frac{q^{n+1} + (-1)^n q^n + (-1)^{n+1} - q^n}{q-1}.$$

Similarly, if $\lambda = 0$, then

$$\#X_0(\mathbb{F}_q) - \#Y_0(\mathbb{F}_q) = q^n[(-1)^{n+1} + 1] + \sum_{k\in E_1} S_k.$$

We may now prove Theorem 9.1 by demonstrating that the right-hand sides of the above two formulas equal zero when $\gcd(n+1, q-1) = 1$.

LEMMA 9.6. *If $\gcd(n+1, q-1) = 1$, then the right-hand sides are zero.*

PROOF. Let $k \in E$. Suppose $k_i \neq 0$ for $1 \leq i \leq q-1$. Then, by (21), we see that $(n+1)k_i \equiv (n+1)k_j \mod q-1$ for every $1 \leq i, j \leq q-1$. By hypothesis, $n+1$ is invertible in $\mathbb{Z}/(q-1)$, and so $k_i = k_j$. This means that, if $k \in E_1$ then at least one of the first $n+1$ coordinates must be zero.

Let $1 \leq i \leq n$. Suppose $k \in E_1$ and its first i coordinates are zero. Then (21) tells us that k_{n+2} is either zero or $q-1$. In the *first case*, we have $k_{i+1} = \cdots = k_{n+1} = q-1$ and $s(k) = n+2$. In the *second case*, again we have $k_{i+1} = \cdots = k_{n+2} = q-1$, but $s(k) = (n+2) - i$. This leads to the following formulas:

When $k_{n+2} = 0$ (first case): $S_k = (-1)^{(n+1)-i}q^n$

When $k_{n+2} = q-1$ (second case): $S_k = (-1)^{n-i}(q-1)^{i-1}q^{(n+1)-i}$.

(Note, if $\lambda = 0$, then k_{n+2} must be zero, and so the second case never occurs.) Permuting these zeros around in $\binom{n+1}{i}$ ways among the first $n+1$ coordinates gives us all possible points in E_1. That is, if we set

$$A := \sum_{i=1}^{n} \binom{n+1}{i} (-1)^{(n+1)-i} q^n \qquad \text{(counts first case)}$$

and

$$B := \sum_{i=1}^{n} \binom{n+1}{i} (-1)^{n-i} (q-1)^{i-1} q^{(n+1)-i} \qquad \text{(counts second case)},$$

then we have $\sum_{k \in E_1} S_k = A + B$ for $\lambda \neq 0$; otherwise, $\sum_{k \in E_1} S_k = A$ for $\lambda = 0$. Now, by the binomial theorem, we see that

$$A = q^n[(1-1)^{n+1} - (-1)^{n+1} - 1] = q^n[(-1)^n - 1]$$

and

$$\begin{aligned} B &= (q-1)^{-1}(-1)^n[(-(q-1)+q)^{n+1} - q^{n+1} - (-1)^{n+1}(q-1)^{n+1}] \\ &= (q-1)^{-1}[(-1)^n + (-1)^{n+1}q^{n+1} + (q-1)^{n+1}]. \end{aligned}$$

Thus, for $\lambda \neq 0$, we have

$$\sum_{k \in E_1} S_k = q^n[(-1)^n - 1] + \frac{(-1)^n + (-1)^{n+1}q^{n+1} + (q-1)^{n+1}}{q-1}$$

which proves the lemma. □

9.3. The proof of Theorem 9.3. Let us recall what we will prove.

Theorem 9.3. *Let $d := gcd(n+1, q^k - 1) > 1$. Then*

(1) $\#X_\lambda(\mathbb{F}_{q^k}) \equiv 0 \mod d$.
(2) *Writing $\lambda = -(n+1)\psi$ in the new parameter ψ, if $n+1$ is a power of a prime ℓ, then*

$$\#X_\lambda(\mathbb{F}_{q^k}) \equiv 0 \mod \ell d \quad \textit{and} \quad \#Y_\lambda(\mathbb{F}_{q^k}) \equiv \begin{cases} 1 & \psi^{n+1} = 1 \\ 0 & \textit{otherwise} \end{cases} \mod \ell.$$

Proof. Without loss, we will write q instead of q^k in the following proof. First, let us prove the congruences on X_λ. We do this by gathering all the points of X_λ that have the same number of coordinates zero. For each $1 \leq i \leq n-1$, define M_i^* as the number of $\mathbb{F}_q$-rational points in $\mathbb{P}^{n-i}_{\mathbb{F}_q^*}$ which lie on the diagonal hypersurface

$$x_{i+1}^{n+1} + \cdots + x_{n+1}^{n+1} = 0.$$

Notice that the group

$$G_i(\mathbb{F}_q) := \{(\zeta_{i+1}, \ldots, \zeta_{n+1}) \mid \zeta_j \in \mathbb{F}_q, \zeta_j^{n+1} = 1\}/\{(\zeta, \ldots, \zeta) | \zeta^{n+1} = 1\}$$

acts freely on the set of points defining M_i^*. Since there are $d := \gcd(n+1, q-1)$ many $(n+1)$-st roots of unity in $\mathbb{F}_q$, we have $\#G_i(\mathbb{F}_q) = d^{n+1-i}/d = d^{n-i}$. Consequently, d^{n-i} divides M_i^*. Next, let M_0^* be the number of $\mathbb{F}_q$-rational points in $\mathbb{P}^n_{\mathbb{F}_q^*}$ which lie on X_λ. The group

$$G(\mathbb{F}_q) := \{(\zeta_1, \ldots, \zeta_{n+1}) | \zeta_i \in \mathbb{F}_q, \zeta_i^{n+1} = 1, \zeta_1 \cdots \zeta_{n+1} = 1\}/\{(\zeta, \ldots, \zeta) | \zeta^{n+1} = 1\}$$

acts freely on the points defining M_0^*, and so $\#G(\mathbb{F}_q) = d^n$ divides M_0^*. Putting this together, we have

$$\#X_\lambda(\mathbb{F}_q) = M_0^* + \sum_{i=1}^{n-1} \binom{n+1}{i} M_i^*.$$

This proves the first part of the theorem since each M_i^* are divisible by d. If $n+1$ is a power of a prime ℓ, then not only are the M_i^* divisible by d, but each of the binomial factors is divisible by ℓ; this proves the congruence on X_λ in the second part of the theorem.

We now assume $n+1$ is a power of a prime ℓ. Let us prove the congruence on Y_λ. With $\lambda = -(n+1)\psi$, recall from (24) that

$$\#Y_\lambda(\mathbb{F}_q) = N_\lambda^* - \frac{(q-1)^n}{q} + \frac{(-1)^n}{q} + \frac{q^n-1}{q-1}$$

where N_λ^* is the number of $\mathbb{F}_q$-rational points in $\mathbb{A}_{\mathbb{F}_q}^{n+1}$ that satisfy

$$\begin{cases} x_1 + \cdots + x_{n+1} - (n+1)\psi = 0 \\ x_1 \cdots x_{n+1} = 1 \end{cases} . \tag{25}$$

We claim that $\#Y_\lambda(\mathbb{F}_q) \equiv N_\lambda^* \mod \ell$. Since $\gcd(n+1, q-1) > 1$, $q \equiv 1 \mod \ell$. Using the fact that $\frac{q^n-1}{q-1} = q^{n-1} + \cdots + q + 1$, we have: if ℓ is an odd prime (the even case is similar), then

$$-\frac{(q-1)^n}{q} + \frac{(-1)^n}{q} + \frac{q^n-1}{q-1} \equiv -0 + 1 + n \equiv 0 \mod \ell.$$

This proves the claim.

Since we now have $\#Y_\lambda(\mathbb{F}_q) \equiv N_\lambda^* \mod \ell$, we will concentrate on computing N_λ^*. Consider counting the points on (25) as follows: suppose a point $x := (x_1, \ldots, x_{n+1}) \in \mathbb{A}_{\mathbb{F}_q}^{n+1}$ has two coordinates equal. Then we may permute these two around in $\binom{n+1}{2}$ ways without changing the *order* of the other coordinates. Thus, the orbit of the point x under this type of permutation contains $\binom{n+1}{2}$ points contained in the affine toric variety defined by (25). Note that we are not overcounting the points which have multiple pairs of coordinates being the same, like $(1,1,2,2,2)$. If all the coordinates of x are different then we may permute these around in $(n+1)!$ ways.

Putting this together, we find, modulo ℓ:

$$N_\lambda^*(\mathbb{F}_q) \equiv \#\{x \in \mathbb{A}_{\mathbb{F}_q}^{n+1} | \text{ all coordinates are equal and } x \text{ satisfies (25)}\}.$$

If all the coordinates are equal then we have the system $(n+1)x - (n+1)\psi = 0$ and $x^{n+1} = 1$. By hypothesis, $n+1$ is invertible in $\mathbb{F}_q$, thus, we have $x = \psi$ for the first equation, and so $\psi^{n+1} = 1$ for the second. Therefore,

$$N_\lambda^* \equiv \begin{cases} 1 & \psi^{n+1} = 1 \\ 0 & \text{otherwise} \end{cases} \mod \ell.$$

□

References

[1] V. Batyrev, Dual polyhedra and mirror symmetry for Calabi-Yau hypersurfaces in toric varieties, J. Algebraic Geom., 3(1994), no. 3, 493-535.

[2] V. Batyrev and D.I. Dais, Strong McKay correspondence, string-theoretic Hodge numbers and mirror symmetry, Topology, 35(1996), no. 4, 901-929.

[3] P. Candelas, X. de la Ossa and F. Rodriguez-Villegas, Calabi-Yau Manifolds over Finite Fields, II, Fields Institute Communiucations, Vol. 38, N. Yui and James D. Lewis (eds.), pp. 121–157, 2003. arXiv:hep-th/0402133 v1

[4] B. Dwork, p-Adic cycles, Publ. Math., IHES, 39(1969), 327-415.

[5] B. Dwork, Normalized period matrices II, Ann. Math., 98(1973), 1-57.

[6] L. Fu and D. Wan, Mirror congruence for rational points on Calabi-Yau varieties, arXiv:math.AG/0503703, 30 March 2005.

[7] G. Kempf, F. Knudsen, D. Mumford and B. Saint-Donat, Toroidal Embeddings, Lecture Notes in Math., 339, Spriner-Verlag, 1973.

[8] L. Illusie, Ordinarité des intersections complètes générales, in Grothendieck Festschrift, Volume II (1990), 375-405.

[9] B. Mazur, Frobenius and the Hodge filtration, Bull. Amer. Math. Soc., 78(1972), 653-667.
[10] D. Wan, Newton polygons of zeta functions and L-functions, Ann. Math., 137(1993), 247-293.
[11] D. Wan, Dwork's conjecture on unit root zeta functions, Ann. Math., 150(1999), 867-927.
[12] D. Wan, Higher rank case of Dwork's conjecture, J. Amer. Math. Soc., 13(2000), 807-852.
[13] D. Wan, Rank one case of Dwork's conjecture, J. Amer. Math. Soc., 13(2000), 853-908.
[14] D. Wan, Variation of p-adic Newton polygons for L-functions of exponential sums, Asian J. Math., Vol. 8, 3(2004), 427-474.
[15] D. Wan, Zeta functions of toric Calabi-Yau hypersurfaces, Course notes given at the Arizona Winter School 2004, 22 pages. "http://swc.math.arizona.edu/oldaws/04GenlInfo.html".

Institute of Mathematics, Chinese Academy of Sciences, Beijing, P.R. China

Department of Mathematics, University of California, Irvine, CA 92697-3875
E-mail address: dwan@math.uci.edu

AMS/IP Studies in Advanced Mathematics
Volume 38, 2006

The L-Series of Calabi–Yau Orbifolds of CM Type

Noriko Yui

ABSTRACT. We study the 147 Calabi–Yau orbifolds defined over $\mathbb{Q}$ in weighted projective 4-spaces which are obtained from hypersurfaces of Fermat type via the orbifolding procedure (which introduces abelian quotient singularities). These Calabi–Yau orbifolds are varieties of CM-type, and are of interest for mirror symmetry. We will determine the L–series of these Calabi–Yau orbifolds of CM type. This is done by decomposing the Calabi–Yau orbifolds into the direct sum of motives. We define the Hodge and Betti numbers of these motives. We determine the L-series of these motives. Then, gluing all the motivic L-series, and also determining the L-series arising from singular loci, we obtain the L-series of our Calabi–Yau orbifolds. Along the way we re-prove the orbifold Euler characteristic formula of Vafa for these orbifolds, using motives. Also we present a few new examples of Calabi–Yau orbifolds for which the generalized Hodge conjecture holds true. Finally, we discuss the modularity of our Calabi–Yau orbifolds, in particular, we establish the modularity of some weighted Fermat curves appearing in the singular loci.

Contents

2000 *Mathematics Subject Classification.* Primary 14J32, 14G10, 11G40, 11F80; Secondary 14J15, 11F23, 14K15, 11G15.

Key words and phrases. Calabi–Yau orbifolds, CM type varieties, L-series, Jacobi sums, motives.

This work is partially supported by a Discovery Grant from NSERC, Canada.

1. Introduction

In this article, we will consider a special class of smooth projective varieties defined over number fields, the so-called *Calabi–Yau* varieties. Here is the definition.

DEFINITION 1.1. A smooth projective threefold X (defined over $\mathbb{C}$) is a Calabi–Yau threefold if it satisfies the following two conditions:

(1) $H^1(X, \mathcal{O}_X) = H^2(X, \mathcal{O}_X) = 0$, and

(2) X has trivial canonical bundle.

If X is defined over $\mathbb{Q}$ or a number field $\mathbb{K}$, X is a Calabi–Yau threefold if its complexification $X_{\mathbb{C}} := X \otimes \mathbb{C}$ is a Calabi–Yau threefold.

The numerical invariants of Calabi–Yau threefolds $X_{\mathbb{C}}$ are computed as follows:

(a) The (i,j)-th *Hodge* number $h^{i,j}(X)$ of X is defined by

$$h^{i,j}(X) := \dim_{\mathbb{C}} H^j(X_{\mathbb{C}}, \Omega^i_{X_{\mathbb{C}}}).$$

It is immediate from the duality of the Hodge cohomology groups that $h^{i,j}(X) = h^{j,i}(X)$, and that $h^{i,j}(X) = 0$ for all pairs i, j except for those $0 \leq i, j \leq 3$. The condition (1) in (1.1) for Calabi–Yau threefolds implies that $h^{1,0}(X) = h^{2,0}(X) = 0$, and the condition (2) asserts that $h^{3,0}(X) = h^{0,3}(X) = 1$ (which is nothing but the geometric genus, $p_g(X)$, of X). $h^{2,1}(X)$ is the number of deformations of X, and $h^{1,1}(X)$ is the number of Hodge $(1,1)$-cycles (Kähler classes) on X. The Hodge numbers are concocted to define the *Hodge diamond*:

$$\begin{array}{ccccccc} & & & 1 & & & \\ & & 0 & & 0 & & \\ & 0 & & h^{1,1} & & 0 & \\ 1 & & h^{1,2} & & h^{1,2} & & 1 \\ & 0 & & h^{2,2} & & 0 & \\ & & 0 & & 0 & & \\ & & & 1 & & & \end{array}$$

(b) The n-th *Betti* number of X is defined by

$$B_n(X) := \dim_{\mathbb{Q}_\ell} H^n(X \otimes \bar{\mathbb{Q}}, \mathbb{Q}_\ell) = \dim_{\mathbb{C}} H^n(X_{\mathbb{C}}, \mathbb{C})$$

where the cohomology group is the ℓ-adic étale or the singular cohomology group. There is the Hodge decomposition:

$$H^n(X_{\mathbb{C}}, \mathbb{C}) = \oplus_{i+j=n} H^j(X_{\mathbb{C}}, \Omega^i_{X_{\mathbb{C}}}).$$

Taking the dimensions of both sides, we obtain:

$$B_n(X) = \sum_{i+j=n} h^{i,j}(X) = \sum_{i=0}^{n} h^{i,n-i}(X).$$

Using this, we can express the Betti numbers of Calabi–Yau threefolds in terms of Hodge numbers:

$$B_0(X) = B_6(X) = 1, \quad B_1(X) = B_5(X) = 0,$$
$$B_2(X) = h^{1,1}(X) = h^{2,2}(X) = B_4(X)$$

and

$$B_3(X) = 2(1 + h^{1,2}(X)).$$

(c) The *Euler characteristic* of a Calabi–Yau threefold X is

$$\chi(X) = \sum_{n=0}^{6}(-1)^n B_n(X) = 2(h^{1,1}(X) - h^{1,2}(X)).$$

REMARK 1.1. About the Euler characteristic of Calabi–Yau threefolds X, there is a conjecture which asserts the existence of an absolute constant C such that $|\chi(X)| < C$ for any Calabi–Yau threefold. The currently known upper bound for $|\chi(X)|$ is 960, which is by now about 15 years old, first observed for 7,555 Calabi–Yau threefolds in weighted projective 4-spaces in the paper of Candelas, Lynker and Schimmrigk [CLS90]. Since then nearly 500,000,000 families of Calabi–Yau threefolds have been constructed using Batyrev's method of reflexive polytopes, and 960 is still the upper bound for the Euler characteristic of all these Calabi–Yau threefolds. (Refer to the websites of Schimmrigk [S] and Kreuzer and Skarke [KS].) There is also the counter-conjecture by Miles Reid ([Re02]) that there is no such absolute bound on the Euler characteristic of Calabi–Yau threefolds.

The main goal of this paper is to study the L-series of Calabi–Yau orbifolds of dimension 3 of Fermat type defined over number fields $\mathbb{K}$. Here, by the L-series of Calabi–Yau threefolds we mean the L–series of the ℓ-adic étale Galois representations attached to the Calabi–Yau threefolds in question. So we will first recall the definition of the L-series.

Let X be a Calabi–Yau threefold defined over a number field $\mathbb{K}$. Suppose that X is a hypersurface defined by an equation

$$f(X_1, X_2, X_3, X_4, X_5) = 0 \in R[X_1, X_2, X_3, X_4, X_5]$$

where $R = \mathcal{O}_{\mathbb{K}}$. (Or we may take R to be a slightly larger ring than $\mathcal{O}_{\mathbb{K}}$ which contains the inverse of a finite number of prime ideals $\mathfrak{p}$ in $\mathcal{O}_{\mathbb{K}}$.)

We may consider X as a variety in the ordinary projective 4–space $\mathbb{P}^4$ with homogeneous coordinates $X_1, \cdots, X_5$, or as a variety in a weighted projective 4-space where the coordinates $X_1, \cdots, X_5$ carry some weights in the sense that $\deg(X_i)$ are not necessarily equal to 1.

Let $\Gamma = \mathrm{Gal}(\bar{\mathbb{K}}/\mathbb{K})$ be the absolute Galois group of $\mathbb{K}$. For a prime $\mathfrak{p} \in \operatorname{Spec} \mathcal{O}_{\mathbb{K}}$, let $X_{\mathfrak{p}}$ denote the reduction of X modulo $\mathfrak{p}$. We say that $\mathfrak{p}$ is *good* if the reduction $X_{\mathfrak{p}}$ (extended over $\bar{\mathbb{F}}_p$) defines a smooth projective Calabi–Yau threefold, otherwise *bad*. Put $q := \#(\mathcal{O}_{\mathbb{K}}/\mathfrak{p}) = |\mathrm{Norm}\,\mathfrak{p}|$. For a good prime ideal $\mathfrak{p}$, let $\mathrm{Frob}_{\mathfrak{p}} \in \mathrm{Gal}(\bar{\mathbb{K}}/\mathbb{K})$ denote the Frobenius morphism induced from the map $x \to x^q$. Then $\mathrm{Frob}_{\mathfrak{p}}$ acts on the ℓ-adic étale cohomology groups $H^i_{et}(\bar{X}_{\mathfrak{p}}, \mathbb{Q}_\ell)$, where ℓ is a rational prime such that $(\ell, q) = 1$, for $i = 1, 2, 3$, and $\bar{X}_{\mathfrak{p}}$ is the base change of $X_{\mathfrak{p}}$ to the algebraic closure $\bar{\mathbb{F}}_p$. By the Grothendieck specialization theorem, $H^i_{et}(\bar{X}_{\mathfrak{p}}, \mathbb{Q}_\ell)$ is isomorphic to $H^i_{et}(\bar{X}, \mathbb{Q}_\ell)$, where $\bar{X} = X \times_{\mathbb{K}} \bar{\mathbb{K}}$ for each i, $0 \le i \le 6$, and the Frobenius morphism $\mathrm{Frob}_{\mathfrak{p}}$ induces a $\mathbb{Q}_\ell$–linear map $\mathrm{Frob}^*_{\mathfrak{p}}$ on $H^i_{et}(\bar{X}, \mathbb{Q}_\ell)$. There is the Poincaré duality for the cohomology groups $H^i_{et}(\bar{X}, \mathbb{Q}_\ell)$:

$$H^i_{et}(\bar{X}, \mathbb{Q}_\ell) \times H^{6-i}_{et}(\bar{X}, \mathbb{Q}_\ell) \to \mathbb{Q}_\ell.$$

Let

$$P_{\mathfrak{p},i}(T) := \det(1 - \mathrm{Frob}^*_{\mathfrak{p}}\, T \,|\, H^i_{et}(\bar{X}, \mathbb{Q}_\ell))$$

be the characteristic polynomial of ${\mathrm{Frob}^*}_{\mathfrak{p}}$, where T is an indeterminate. Then $P_{\mathfrak{p},i}(T) \in 1 + \mathbb{Z}[T]$ with degree $B_i(X)$. Furthermore, the reciprocal roots $\{\alpha_{i,j} \,|\, j =$

$1, 2, \cdots, B_i(X)\}$ are algebraic integers satisfying the Riemann Hypothesis, i.e., they all have absolute value $|\alpha_{i,j}| = q^{i/2}$ (see Deligne [De74]).

If $\mathfrak{p}$ is a *bad* prime, the polynomial $P_{\mathfrak{p},i}(T)$ can also be defined, though it is degenerate with degree $\leq B_i(X)$. Let S denote the set of bad primes for X. It is known that S is a finite set. For a bad prime $\mathfrak{p} \in S$ let $D_\mathfrak{p}$ and $I_\mathfrak{p}$ be the decomposition group and inertia group of $\mathfrak{p}$, respectively. Then $\text{Frob}_\mathfrak{p}$ belongs to $D_\mathfrak{p}$, and we can define, for each i, $0 \leq i \leq 6$:

$$P_{\mathfrak{p},i}(T) := \det(1 - \text{Frob}_\mathfrak{p}^* T | H^i_{et}(\bar{X}, \mathbb{Q}_\ell)^{I_\mathfrak{p}}).$$

For each i, $0 \leq i \leq 6$, we define the i–th (cohomological) L–series of X, denoted by $L_i(X, s)$, by

$$L_i(X, s) := L(H^i_{et}(\bar{X}, \mathbb{Q}_\ell), s)$$

where s is a complex variable. In terms of the characteristic polynomial $P_{\mathfrak{p},i}(T)$, we have

$$L_i(X, s) = \prod_{\mathfrak{p} \in \text{Spec}\, \mathcal{O}_\mathbb{K}} \frac{1}{P_{\mathfrak{p},i}(N\mathfrak{p}^{-s})}.$$

DEFINITION 1.2. Let X be a Calabi–Yau threefold defined over a number field $\mathbb{K}$. The (global) L-series of X, $L(X, s)$, is then defined by taking the product:

$$L(X, s) = \frac{L_0(X, s)L_2(X, s)L_4(X, s)L_6(X, s)}{L_1(X, s)L_3(X, s)L_5(X, s)}$$

REMARK 1.2. For a Calabi–Yau threefold X, the only non-trivial cohomology groups are $H^2_{et}(\bar{X}, \mathbb{Q}_\ell)$, $H^3_{et}(\bar{X}, \mathbb{Q}_\ell)$, and $H^4_{et}(\bar{X}, \mathbb{Q}_\ell)$, and the "interesting" factors of the L-series are $L_2(X, s)$ and $L_3(X, s)$, as $L_4(X, s) = L_2(X, s - 1)$.

We know by Batyrev [Ba99] (and also by Tetsushi Ito [I03], and others) that if X and Y are birationally equivalent Calabi–Yau threefolds over $\mathbb{C}$, then X and Y have the same Betti and Hodge numbers. As a corollary of its proof, it was shown that birational smooth minimal models of Calabi–Yau threefolds defined over $\mathbb{Q}$ (or a number field $\mathbb{K}$) have the same zeta-function at almost all reductions mod p (or mod $\mathfrak{p}$). Consequently, the partial L-series $L_2(X, s)$ and $L_3(X, s)$ are, up to a finite number of factors, birational invariants of smooth minimal Calabi–Yau threefolds over $\mathbb{Q}$ (or a number field $\mathbb{K}$).

Since our Calabi–Yau orbifolds X are smooth minimal (toroidal) resolutions of the singular Calabi–Yau threefolds Y, they satisfy the hypothesis of Batyrev and Ito. Therefore, the L-series $L(X, s)$ of our Calabi–Yau orbifolds are birational invariants.

Now we will describe the contents of this paper. First we will construct examples of Calabi–Yau threefolds defined over $\mathbb{Q}$. This is done by the so-called *orbifolding process*, that is, by taking the quotient of a hypersurface of Fermat type defined over $\mathbb{Q}$ by a finite abelian group scheme (which, after passing to $\bar{\mathbb{Q}}$, is identified with a subgroup of the automorphism group of the hypersurface). This construction is described in Section 2.

The orbifolding process gives rise to singularities. The analysis of singularities and the crepant resolutions are discussed in Section 3. The minimal crepant resolutions thus obtained are called *the Calabi–Yau orbifolds of Fermat type*. We recall the formula of Vafa for the orbifold Euler characteristic.

In Section 4, we will recall the notion of motives and motivic decompositions of hypersurfaces of Fermat type, which was first introduced in Shioda [Sh87]. We will obtain the motivic decompositions of our Calabi–Yau orbifolds. We then introduce the Hodge and Betti numbers of motives via Hodge and Betti realizations of these motives, and present explicit formulas for them. This enables us to recover the formula of Vafa for the orbifold Euler characteristic with a completely different approach.

In Section 5, we will determine the L–series of these Calabi–Yau orbifolds explicitly, up to the factors corresponding to bad primes. This is done by determining the L-series of each motive, and the L-series associated to singularities. The L-series of these Calabi–Yau orbifolds are expressed as a product of Hecke L-series with Jacobi sum Grössencharaktere of some cyclotomic fields over $\mathbb{Q}$, and the Hasse–Weil zeta-functions of smooth irreducible curves in the singular loci.

In Section 6, we will recall the notion of varieties of CM type. A variety X is said to be of *CM type* if its Hodge group is commutative (see Borcea [Bo92], Mumford [Mu66], [Mu69]). We show that our Calabi–Yau orbifolds of Fermat type constructed in Section 2 are all of CM type. Our results on the L-series provide evidence for the conjecture of Shafarevich ([S-PS73]):
Every variety of CM type has a model defined over a number field, and its L-series are expressed by means of the L-series of one-dimensional characters of the field of definition of the variety.

We also discuss the generalized Hodge conjecture for our Calabi–Yau orbifolds. For Fermat threefolds of degree ≤ 10, Shioda ([Sh82], [Sh83]) proved the generalized Hodge conjecture, from which we can show that the six Calabi–Yau orbifolds of degree ≤ 10 in our list satisfy the generalized Hodge conjecture. Besides these, we will present fifteen new examples of Calabi–Yau threefolds for which the generalized Hodge conjecture holds true.

In Section 7, we will address the modularity questions of our Calabi–Yau orbifolds, and motives. For singular $K3$ surfaces (dimension 2 Calabi–Yau varieties), Shioda and Inose [SI87] showed that they are all of CM type, and established the conjecture of Shafarevich. Their L-series is expressed in terms of the Dedekind zeta-function of the number field (of definition) and the Hecke L-series of a suitable Grössencharakter. Furthermore, for any singular $K3$ surface defined over $\mathbb{Q}$, the modularity of the L-series of the transcendental part was established by Livné [Li95] that there is a cusp form of weight 3 on some congruence subgroup of $PSL_2(\mathbb{Z})$ (possibly twisted by a character) which determines the L-series.

For our Calabi–Yau orbifolds, we have described their L-series in terms of Jacobi sum Grössencharaktere of cyclotomic fields over $\mathbb{Q}$. Therefore, the next natural question is the existence of modular forms which hopefully determine the L-series. However, the situation seems more complicated, and it is expected that if modular forms will enter the scene, not just a single modular form but a nest of modular forms will be needed to determine the L-series. We establish the modularity for some weighted Fermat curves appearing in the singular loci and that the L-series of the Jacobian varieties of these curves come from modular (cusp) forms.

In Section 8, we give some remarks, and list open problems for further discussions. We give a preview on the construction of one- or multiparameter deformations of our Calabi–Yau orbifolds, and their mirror families, which is topic of the subsequent paper(s). Indeed in the sequels to this paper, we will construct mirror

Calabi–Yau threefolds of one- or multiparameter deformations of our Calabi–Yau orbifolds, compute their L-series and finally interpret the mirror symmetry phenomenon in terms of their L-series.

In Appendix A1, we tabulate 147 Calabi–Yau orbifolds, their Hodge numbers and Betti numbers. In Appendix A2, singularities of dimension 1 for each orbifold are listed. Some of the results in this paper have been announced in Yui [Yui03].

Appendix B by Yasuhiro Goto discusses the zeta-function and L-series of cubic hypersurface fourfolds defined over number fields. Though these cubic fourfolds are not Calabi–Yau manifolds, but Fano fourfolds, their arithmetic properties have some relevance to the theme of this paper. Also, there is a brief discussion on the supersingularity and unirationality of the fourfolds in connection to the zeta-functions.

2. Construction of Calabi–Yau orbifolds

Our starting point is a hypersurface threefold of Fermat type of degree $m \geq 5$. Let $V = V(m, 3, \mathbf{c})$ denote a hypersurface threefold of degree $m \geq 5$ defined in the usual projective 4-space $\mathbb{P}^4_{\mathbb{Q}}$ by the equation:

$$V: \quad c_1X_1^m + c_2X_2^m + c_3X_3^m + c_4X_4^m + c_5X_5^m = 0 \subset \mathbb{P}^4_{\mathbb{Q}}$$

where $\mathbf{c} = (c_1, \cdots, c_5) \in (\mathbb{Q}^\times)^5$ is a *twist*. If all the c_i are equal, i.e., $\mathbf{c} = \mathbf{1}$, V defines the Fermat threefold. (We remark that geometric properties of V are independent of the choice of $\mathbf{c}$. However, properties of arithmetic nature will depend in an essential ways on twists.) The geometric genus of V is $p_g(V) = \binom{m-1}{4}$ (the binomial coefficient), and $h^{i,0}(V) = 0$ for $i = 1, 2$. So V is a Calabi–Yau threefold if and only if $m = 5$.

Now we will discuss the orbifolding construction, which yields more examples of Calabi–Yau threefolds defined over $\mathbb{Q}$, in weighted projective 4-spaces. For the basic properties of weighted projective spaces, the reader is referred to the article of Dolgachev [Dol82].

DEFINITION 2.1. Fix a hypersurface threefold of Fermat type $V = V(m, 3, \mathbf{c})$ of degree m and twist $\mathbf{c}$ over $\mathbb{Q}$. Let $Q = (q_1, q_2, q_3, q_4, q_5) \in \mathbb{N}^5$ be a vector consisting of 5-tuples of positive rational integers. We call Q a *weight* if it satisfies the following two conditions:

(1) $\gcd(q_1, \cdots, \widehat{q_j}, \cdots, q_5) = 1$ for every j, $1 \leq j \leq 5$, and

(2) $q_j \mid m = \deg(V)$ for every j, $1 \leq j \leq 5$.

(Here $\hat{}$ means the exclusion of that component.)

For each j, $1 \leq j \leq 5$, let μ_{q_j} denote the finite group scheme of q_j-th roots of unity over $\mathbb{Q}$, namely, $\mu_{q_j} := \mathrm{Spec}\,(\mathbb{Q}[T]/(T^{q_j} - 1))$. Take the direct product of the μ_{q_j} and define

$$\mu_Q = \mu_{q_1} \times \cdots \times \mu_{q_5}.$$

Then μ_Q is a finite automorphism group scheme of $V_{\mathbb{Q}}$, whose action over $\bar{\mathbb{Q}}$ is described as follows:

$$(X_1, \cdots, X_5) \mapsto (\zeta_{q_1}^{e_1} X_1, \cdots, \zeta_{q_5}^{e_5} X_5), \quad e_j \in \mathbb{Z}/q_j\mathbb{Z} \text{ for each } j$$

where ζ_{q_j} denotes a primitive q_j–th root of unity in $\bar{\mathbb{Q}}$ for every j, $1 \leq j \leq 5$.

REMARK 2.1. If we pass to a sufficiently large extension $\mathbb{K}$ of $\mathbb{Q}$ which contains all the m–th roots of unity, then each μ_{q_j} is indeed a cyclic group $\mathbb{Z}/q_j\mathbb{Z}$ of order q_j generated by a primitive q_j–th root of unity ζ_{q_j} and μ_Q is in fact an automorphism group of V over $\mathbb{K}$.

LEMMA 2.1. *Take the quotient, $Y := V/\mu_Q$. Then Y is, in general, a singular variety in the weighed projective 4-space $\mathbb{P}^4_{\mathbb{Q}}(Q)$ with a finite number of singular loci. Moreover, all singularities are at most abelian (indeed, cyclic) quotient singularities.*

In order to resolve singularities on Y, we will make use of the following result proved for varieties over $\mathbb{C}$.

PROPOSITION 2.2. (Greene, Roan and S.-T.Yau [GRY91]) *Let W be a singular algebraic variety of dimension ≤ 3 with at most abelian quotient singularities and whose canonical sheaf ω_W is trivial. Then there exists a resolution, Z of W which is smooth with trivial canonical bundle ω_Z. Such a smooth resolution is called a crepant resolution.*

REMARK 2.2. The condition ω_W is trivial may be relaxed to ω_W being *torsion*. A crepant resolution may not be uniquely determined. However, all crepant resolutions have the same Euler characteristic and the same Betti and Hodge numbers.

We apply the above Proposition 2.2 to our quotient threefolds Y over $\mathbb{Q}$ to prove the existence of crepant Calabi–Yau resolutions.

THEOREM 2.3. *Let $V = V(m, 3, \mathbf{c})$ be a hypersurface threefold of Fermat type of degree $m \geq 5$ with twist $\mathbf{c}$ defined over $\mathbb{Q}$. Let $Q = (q_1, q_2, q_3, q_4, q_5)$ be a weight satisfying the conditions in Definition (2.1). Let $Y = V/\mu_Q$ be the quotient threefold. Then the following assertions hold:*

(a) In the weighted projective 4-space $\mathbb{P}^4_{\mathbb{Q}}(Q)$, Y is defined by the equation

$$c_1 Y_1^{m/q_1} + c_2 Y_2^{m/q_2} + c_3 Y_3^{m/q_3} + c_4 Y_4^{m/q_4} + c_5 Y_5^{m/q_5} = 0 \subset \mathbb{P}^4_{\mathbb{Q}}(Q)$$

with $\deg(Y_j) = q_j$ for each j, $1 \leq j \leq 5$. Y has at most abelian quotient singularities.

(b) Resolving singularities on Y, we obtain a crepant resolution $\pi : X \to Y$ of Y. There exist exactly 147 pairs $< m, Q >$ for which the crepant resolution X of Y is a Calabi–Yau threefold defined over $\mathbb{Q}$.

PROOF. (a) For the defining equation for Y in the weighted projective space, see, for instance, Reid [Re80] (cf. Fletcher [Fl89]), where they considered weighted complete intersections. Our hypersurface threefolds of Fermat type and their quotients are complete intersections. From the choice of finite groups acting on V, Y has only abelian (cyclic) quotient singularities.

(b) The proof for Proposition 2.2 by Greene, Roan and S.-T. Yau was given for complex varieties of dimension ≤ 3. For the threefolds considered here, their proof can be carried over to number fields with special care paid to the fields of definition of the varieties and singularities. (Detailed discussions on singularities and their fields of definition are carried out in Section 3 below). Therefore the minimal toroidal resolution $\pi : X \to Y$ satisfies

$$\pi^* K_Y = K_X,$$

where K_X (resp. K_Y) denotes the canonical bundle of X (resp. Y). This implies that it is a crepant resolution, and it is indeed defined over $\mathbb{Q}$. (For this use an embedded resolution.)

Now we will show that there are exactly 147 pairs of degree m and weight Q for which X captures the structure of Calabi–Yau threefold. A theorem of Dolgachev [Dol82] characterizes the pairs $< m, Q >$ for which singular threefolds $Y = V/\mu_Q$ have trivial canonical bundle ω_Y. In fact, in weighted projective space $\mathbb{P}^4_{\bar{\mathbb{Q}}}(Q)$, the canonical bundle $\omega_Y \simeq \mathcal{O}_Y(m - q_1 - q_2 - q_3 - q_4 - q_5)$. Then Y has the trivial canonical bundle $\omega_Y \simeq \mathcal{O}_Y$ only if $m = q_1 + q_2 + q_3 + q_4 + q_5$. Now we search for pairs of degree m and weight $Q = (q_1, q_2, q_3, q_4, q_5)$ subject to the conditions:

(1) $m = q_1 + q_2 + q_3 + q_4 + q_5$,
(2) $\gcd(q_1, \cdots, \widehat{q_j} \cdots, q_5) = 1$ for every j, $1 \le j \le 5$, and
(3) $q_j \,|\, m$ for every j, $1 \le j \le 5$.

Put $m_j = m/q_j$ for each j, $1 \le j \le 5$. Then $m_j \in \mathbb{Z}$ by condition (3). Now condition (1) is equivalent to the identity $1 = \sum_{j=1}^{5} \frac{1}{m_j}$. Then it is easy to see that there are only finitely many 4-tuples $(m_1, \cdots, m_5)$ satisfying this identity. Consequently, there are only finitely many weights $Q = (q_1, \cdots, q_5)$ satisfying the above three conditions. □

Examples: *The Calabi–Yau threefolds obtained in Theorem 2.3 are crepant resolutions of Calabi–Yau orbifolds of Fermat type. The smallest possible degree is* $m = 5$ *with weight* $Q = (1, 1, 1, 1, 1)$, *and the largest possible is* $m = 1806$ *with weight* $Q = (1, 42, 258, 602, 903)$. *All the 147 Calabi–Yau orbifolds of Fermat type obtained in Theorem 2.3 are tablulated in Appendix A1.*

It should be remarked that the Calabi–Yau orbifolds constructed in Theorem 2.3 are non-rigid, i.e., $h^{2,1} \neq 0$. *(See Appendix A1 where Hodge numbers are computed for these Calabi–Yau orbifolds.*

REMARK 2.3. Every Calabi–Yau orbifold constructed above is dominated by some product variety. That is, the function field of a Calabi–Yau orbifold is a subfield of the function field of the product variety. For Fermat hypersurfaces

$$V : X_1^m + X_2^m + \cdots + X_5^m = 0 \subset \mathbb{P}^4$$

this was shown by Shioda [Sh78]. A more precise exposition can be found in the article of Hunt and Schimmrigk [HS99], which can be formulated as follows: *V is birationally equivalent to a quotient of the product of the Fermat curve of degree m and the Fermat surface of degree m by the group $(\mathbb{Z}/m\mathbb{Z})$ acting only on the first coordinate of each constituent by multiplication by a primitive m-th root of unity.* Inductively, the Fermat surface of degree m is dominated by the product of two Fermat curves of degree m.

Now this statement can be generalized to weighted Fermat hypersurfaces. Indeed, in each case, the orbifolding map is compatible with the dominant map.

3. Singularities and their resolutions

We now determine explicitly the singularities on Y and their resolutions in order to study arithmetic properties of our Calabi–Yau orbifolds. Let $\pi : X \to Y$ be a crepant resolution, let $\Sigma(Y)$ denote the singular locus of Y and let $\hat{\Sigma}(Y) = \pi^{-1}(\Sigma(Y))$.

PROPOSITION 3.1. *Let Y be a singular Calabi–Yau orbifold over $\mathbb{Q}$ corresponding to a pair $< m, Q >$ with $Q = (q_1, q_2, q_3, q_4, q_5)$. Then Y is defined by the equation*

$$Y \,:\, c_1 Y_1^{m/q_1} + c_2 Y_2^{m/q_2} + c_3 Y_3^{m/q_3} + c_4 Y_4^{m/q_4} + c_5 Y_5^{m/q_5} = 0 \subset \mathbb{P}^4_{\mathbb{Q}}(Q)$$

with $deg(Y_j) = q_j$ for each j, $1 \le j \le 5$. Put $m_j := m/q_j$ (for $1 \le j \le 5$).

Let ℓ be an integer ranging over the set $\{0, 1, \cdots, m-1\}$. For each ℓ, we define the two sets of indices:

$$S_\ell := \{\, j \,|\, 1 \le j \le 5, \ell/m_j \in \mathbb{Z}\,\} \quad \text{and} \quad I_\ell = \{1,2,3,4,5\} \setminus S_\ell.$$

Further put $c_{I_\ell} := gcd\{q_j \,|\, j \notin I_\ell\} = gcd\{q_j \,|\, j \in S_\ell\}$. Let

$$Y_{I_\ell} := Y \cap \{Y_j = 0,\, j \in I_\ell\}.$$

Then $Y_{I_\ell} \in \Sigma(Y)$, and it is defined over $\mathbb{Q}$.

More concretely, the following assertions hold:

(1) $\#S_\ell \neq 4$ for any ℓ, and $\#S_\ell = 5$ for $\ell = 0$.

(2) If $\#S_\ell = 3$ for some ℓ, then there is a set I_ℓ with $\#I_\ell = 2$, $c_{I_\ell} \ge 2$. In this case $\Sigma(Y)$ contains a dimension 1 singularity.

Suppose that C is an irreducible component of $Y_{I_\ell} = Y \cap \{Y_j = 0 \,|\, j \in I_\ell\}$; then C is a smooth weighted curve of Fermat type defined over $\mathbb{Q}$ of degree $m' := m/c_{I_\ell}$ and reduced weight $Q' := (q'_i, q'_j, q'_k)$ where $i, j, k \notin I_\ell$ and $q'_\cdot = q_\cdot/c_{I_\ell}$ for $\cdot \in \{i, j, k\}$. The multiplicity of C in Y_{I_ℓ} is $m_C := c_{I_\ell} - 1$. The genus $g(C)$ of C is given by the coefficient of $t^{m'-(q'_i+q'_j+q'_k)}$ in the formal power series

$$\frac{1 - t^{m'}}{(1 - t^{q'_i})(1 - t^{q'_j})(1 - t^{q'_k})}.$$

Furthermore, $\pi^{-1}(C) \in \hat{\Sigma}(Y)$ is a smooth ruled surface birationally equivalent to $C \times \mathbb{P}^1$, and it is defined over $\mathbb{Q}$.

(3) If $\#S_\ell = 2$ for some ℓ, then there are dimension 0 singularities. Let $P \in \Sigma(Y)$ be a singular point. Then P is a cyclic quotient singularity, and $\pi^{-1}(P)$ is a smooth rational surface birationally equivalent to $\mathbb{P}^2$. It is defined over a finite Galois extension of $\mathbb{Q}$; its Galois orbit is defined over $\mathbb{Q}$.

(4) If $\#S_\ell = 1$, Y has no singularity.

PROOF. The geometric assertions are proved using the same line of argument as in Roan [Ro90, Ro91], and Dimca [Di86], though we need to pay special attention to the fields of definition of singularities. The field of definition of each singularity on Y may be larger than $\mathbb{Q}$. However, the singular locus is indeed defined over $\mathbb{Q}$ as it is invariant under the action of the Galois group $\Gamma = \mathrm{Gal}(\bar{\mathbb{Q}}/\mathbb{Q})$.

The assertion (1) that $\#S_\ell \neq 4$ follows from the condition on the weight that for every j, $1 \le j \le 5$, $\gcd(q_1, \cdots, \hat{q_j}, \cdots, q_5) = 1$. □

EXAMPLE 3.1. (a) There are altogether 41 pairs $< m, Q >$ for which the orbifolds have no singularities of dimension 1. These are indicated with • under Remark in Table 5 in Appendix A.

(b) Let $m = 8$ and $Q = (1,1,2,2,2)$. Then for $\ell = 4$, $s_4 = \{3,4,5\}$ and $I_4 = \{1,2\}$. So the singular locus contains the smooth genus 3 curve $C :\ c_3 Y_3^4 + c_4 Y_4^4 + c_5 Y_5^4 = 0$ with multiplicity $1 = \gcd(2,2,2) - 1$. For all other $\ell > 0$, $S_\ell = \emptyset$.

(c) Let $m = 18$ and $Q = (1,1,1,6,9)$. Then there are no ℓ for which $\#S_\ell = 3$ so no singularities of dimension 1. For $\ell = 6, 12$, $S_\ell = \{4,5\}$ ($I_\ell = \{1,2,3\}$) and $c_{I_\ell} = 3$. This yields a singularity of dimension 0, arising from $c_4Y_4 + c_5Y_5^2 = 0$.

(d) Let $m = 24$ and $Q = (2,3,3,4,12)$. Then for $\ell = 8, 16$ (resp. 12), $S_\ell = \{2,3,5\}$ with $c_{I_\ell} = 3$ (resp. $S_{12} = \{1,4,5\}$ with $c_{I_{12}} = 2$). Then the singular locus contains the smooth curve $c_2Y_2^8 + c_3Y_3^8 + c_5Y_5^2 = 0$ of degree 8 and reduced weight $Q' = (1,1,4)$ of genus 2, and with multiplicity 2 (resp. $c_1Y_1^{12} + c_4Y_4^6 + c_5Y_5^2 = 0$ of degree 12 and reduced weight $Q' = (1,2,6)$ of genus 2, and with multiplicity 1).

For $\ell = 6, 18$, $S_\ell = \{4,5\}$ with $c_{I_\ell} = 4$. Then the singular locus contains a singularity of dimension 0, namely a point P (a cyclic quotient singularity) with multiplicity 3, arising from $c_4Y_4^6 + c_5Y_5^2 = 0$. For all other $\ell > 0$, $S_\ell = \emptyset$.

(e) Let $m = 84$ and $Q = (1,6,14,21,42)$. Then for $\ell = 12k$ with $1 \leq k \leq 6$, $S_\ell = \{3,4,5\}$ with $c_{I_\ell} = 7$. Then the singular locus contains smooth curves $c_3Y_3^6 + c_4Y_4^4 + c_5Y_5^2 = 0$ of degree $m' = 12$ and reduced weight $Q' = (2,3,6)$ of genus 0, and with multiplicity 6. For $\ell = 28, 56$, $S_\ell = \{2,4,5\}$ with $c_{I_\ell} = 3$ and the singular locus contains the curve $c_2Y_2^{14} + c_4Y_4^4 + c_5Y_5^2 = 0$ of degree 28 and reduced weight $(2,7,14)$ of genus 0, and with multiplicity 2.

Now we compute the third Betti number $B_3(X)$ of our Calabi–Yau orbifolds.

PROPOSITION 3.2. *Let $\pi : X \to Y$ be a crepant resolution of a Calabi–Yau orbifold Y of Fermat type corresponding to a pair $< m, Q >$ with weight $Q = (q_1, q_2, q_3, q_4, q_5)$. Then*

$$B_3(X) = B_3(Y) + \sum_C m_C\, B_1(C)$$

where C runs over all **distinct** *smooth irreducible curves in $C \in \Sigma(Y)$ of genus $g(C)$ with multiplicity $m_C := c_{I_\ell} - 1$ and $B_1(C) = 2g(C)$.*

Consequently,

$$h^{2,1}(X) = \frac{1}{2}(B_3(Y) + \sum_{C\in\Sigma(Y)} m_C B_1(C)) - 1 = \frac{1}{2}B_3(Y) + \sum_{C\in\Sigma(Y)} m_C\, g(C) - 1$$

where the sum runs over all distinct irreducible curves $C \in \Sigma(Y)$.

REMARK 3.1. A note might be in order on how to count the number in the second sum of the formula for $B_3(X)$. By Proposition 3.1, we know that curve singularities $C \in \Sigma(Y)$ correspond to some integer $\ell \in \{0, 1, \cdots, m-1\}$ with $\#S_\ell = 3$. If there is more than one value for ℓ giving rise to the same singularity, we just count one such ℓ, and hence count just once such a curve C with multiplicity $m_C = c_{I_\ell}$. This may be illustrated by the following example.

Let $m = 20$ and $Q = (1,4,5,5,5)$. Then $\Sigma(Y)$ contains singularities corresponding to $\ell = 4k\,(1 \leq k \leq 4)$ giving rise to the same curve singularity $C : Y_3^4 + Y_4^4 + Y_5^4 = 0$ of genus $g(C) = 3$ with multiplicity $m_C = 4$. So $B_1(C) = 6$. Though four values of ℓ (that is, 4, 8, 12, 16) give rise to the same curve singularity C, we just count C once in the second sum, as the sum is taken over all *distinct* smooth curves in $\Sigma(Y)$. Thus we obtain

$$B_3(X) = B_3(Y) + \sum_{C\in\Sigma(Y)} m_c B_1(C) = 108 + 4 \times 6 = 132.$$

PROOF OF PROPOSITION 3.2. We follow the exposition of Roan [Ro90, Proof of Theorem 2] with special attention to the field of definition of the varieties involved. We have

$$H^3(X) = H^3(Y) \oplus \oplus_E H^1(E)$$

where the sum runs over all exceptional divisors E in X. Let

$$\{E_1, \cdots, E_s; E'_1, \cdots, E'_t\}$$

denote the set of all irreducible components of $\hat{\Sigma}(Y)$ with $\pi(E_i) = C$ a smooth curve, and $\pi(E'_j) = P$ a point in $\Sigma(Y)$. By the construction of a resolution, E_i is a smooth ruled surface over a smooth curve C and E'_j is a smooth rational surface. Define

$$\hat{\Sigma}_C = \cup\{E_i \,|\, \pi(E_i) = C\}, \quad \hat{\Sigma}' = \coprod_C \hat{\Sigma}_C$$

and

$$\hat{\Sigma}_P = \cup\{E'_j \,|\, \pi(E'_j) = P\}, \quad \hat{\Sigma}'' = \coprod_P \hat{\Sigma}_P.$$

Then by the Mayer-Vietoris homology sequence for $\hat{\Sigma}'$ and $\hat{\Sigma}''$, we have

$$0 \to H_3(\hat{\Sigma}') \oplus H_3(\hat{\Sigma}'') \to H_3(\hat{\Sigma}(Y)) \to H_2(\hat{\Sigma}' \cap \hat{\Sigma}'') \xrightarrow{j} H_2(\hat{\Sigma}') \oplus H_2(\hat{\Sigma}'') \to \cdots.$$

Roan [Ro90] has shown that the map j is injective. Then it follows that

$$H_3(\hat{\Sigma}(Y)) = H_3(\hat{\Sigma}') \oplus H_3(\hat{\Sigma}'') = H_3(\hat{\Sigma}').$$

Passing to the dimensions, we get:

$$B_3(\hat{\Sigma}(Y)) = B_3(\hat{\Sigma}') = \sum_C m_C B_1(C) = 2 \sum_C m_C g_C$$

where g_C denotes the genus of C. The point of the identity for $B_3(X)$ is that singularities in $\Sigma(Y)$ of dimension 0, i.e., the points corresponding to the sets $\#S_\ell = 2$ in Proposition 3.1), do not contribute to $B_3(\hat{\Sigma}(Y))$, but they do to $B_2(X)$. □

REMARK 3.2. As a consequence of Proposition 3.2, we obtain an algorithm to compute the Euler characteristic $\chi(X)$ and the Betti numbers $B_3(X), B_3(Y)$. The formula for the Euler characteristic of an orbifold was predicted by Vafa [V89] based solely on reasoning from physics (see also [DHVW85-86]). A mathematical proof of Vafa's Euler characteristic formula was supplied by Roan [Ro90] using techniques in topology and algebraic geometry (cf. [HH90]). With the Euler characteristic $\chi(X)$ and the third Betti number $B_3(X)$ at our disposal, the second Betti number $B_2(X)$ (or equivalently, the Hodge number $h^{1,1}(X)$) can be deduced from the formula $\chi(X) = 2(h^{1,1}(X) - h^{2,1}(X)) = 2h^{1,1}(X) + 2 - B_3(X)$.

PROPOSITION 3.3. *Let $\pi : X \to Y$ be a crepant resolution of a Calabi–Yau orbifold of Fermat type, corresponding to a pair $< m, Q >$ in Theorem 2.3. Then $Q = (q_1, q_2, q_3, q_4, q_5)$ with $q_i | m$ for each i. For each integer $\ell \in \{0, 1, 2, \cdots, m-1\}$, let $S_\ell = \{j \,|1 \le j \le 5, \ell\frac{q_i}{m} \in \mathbb{Z}\}$ be the set of indices defined in Proposition 3.1, and define the integer β_ℓ by*

$$\beta_\ell := \frac{1}{m} \sum_{r=0}^{m-1} \prod_{\ell\frac{q_i}{m} \in \mathbb{Z}, r\frac{q_i}{m} \in \mathbb{Z}} (1 - \frac{m}{q_i}).$$

Here we employ the convention that the product $\prod_{\ell\frac{q_i}{m}, r\frac{q_i}{m}}(1-\frac{m}{q_i}) = 1$ *if there is no* q_i *with* $\ell\frac{q_i}{m} \in \mathbb{Z}, r\frac{q_i}{m} \in \mathbb{Z}$. *Then*

$$B_3(Y) = -\beta_0, \quad \textit{and} \quad B_3(X) = B_3(Y) - \sum_{\#s_\ell = 3} \beta_\ell,$$

where the sum runs over the set S_ℓ *consisting of three elements. The Euler characteristic of* X *is given by*

$$\chi(X) = \sum_{\ell=0}^{m-1} \beta_\ell.$$

Consequently,

$$B_2(X) = h^{1,1}(X) = -1 + \frac{1}{2}\sum_{\ell=0}^{m-1}\sum_{\#s_\ell<3}\beta_\ell = -1 + \frac{1}{2}(\chi(X) - \sum_{\ell,\#s_\ell\in\{3,5\}}\beta_\ell).$$

In particular, all cycles generating $H^{1,1}(X)$ *arise from singular points on* Y.

REMARK 3.3. The fact that β_ℓ is an integer for any ℓ can be proved as follows. The proof is due to Hendrik Lenstra. In fact, a more general statement holds true.

PROPOSITION. *Let* $d_1, \cdots, d_t; D_1, \cdots, D_t \in \mathbb{Z}_{>0}$ *such that* $d_i | D_i$ *for each* i, $1 \le i \le t$. *Let* $m \in \mathbb{Z}_{>0}$ *and* $D_i | m$ *for each* i, $0 \le i \le t$. *Then*

$$\sum_{r=0}^{m-1} \prod_{1\le i\le t,\, d_i|r} (1-D_i) = m \sum_{J\in\{1,2,\cdots,t\}} (-1)^{\#J} \frac{\prod_{j\in J} D_j}{lcm(d_j,\, j\in J)} \equiv 0 \pmod{m}.$$

In particular, take $D_i = d_i = m/q_{i-1}$ *for* i, $1 \le i \le 5$. *Then*

$$\beta_\ell \in \mathbb{Z} \quad \textit{for each } \ell \in \{0, 1, \cdots, m-1\}$$

PROOF. By induction on t. We start with $t = 0$ (i.e., the empty product), in which case $\sum_{r=0}^{m-1} 1 = m$. Let $t = 1$. Then $\sum_{r=0}^{m-1}(1-D_1) = (1-D_1)\frac{m}{D_1} = m - m = 0$.

Now suppose that the assertion holds true for t, and want to establish the congruence for $t+1$. We have

$$\sum_{r=0}^{m-1} \prod_{1\le i\le t+1,\, D_i|r} (1-D_i) = \sum_{r=0}^{m-1}\Big(\prod_{1\le i\le t,\, D_i|r}(1-D_i)\Big)\begin{cases}1-D_{t+1} & \text{if } D_{t+1}|r\\ 1 & \text{otherwise}\end{cases}$$

$$\equiv -D_{t+1} \sum_{r=0,\, D_{t+1}|r}^{m-1} \prod_{1\le i\le t, D_i|r} (1-D_i) \pmod{m}$$

The divisibility condition is equivalently formulated as

$$\text{lcm}(D_i, D_{t+1})|r,$$

so we may write $\text{lcm}(D_i, D_{t+1}) = d_i D_{t+1}$ for i, $1 \le i \le t$. Then by the inductive hypothesis, the above sum is equal to

$$= -D_{t+1} \sum_{s=0}^{m/D_{t+1}-1} \prod_{1\le i\le t,\, d_i|r} (1-d_i) = -D_{t+1} \times k\frac{m}{D_{t+1}} \equiv 0 \pmod{m}.$$

This completes the proof. □

EXAMPLE 3.2. Let $m=8$ and $Q=(1,1,2,2,2)$. Then there are altogether 64 pairs (ℓ,r), out of which only the pairs $(0,0)$, $(0,4)$, $(4,0)$ and $(4,4)$ give rise to non-trivial (i.e., $\neq 1$) values in the double product. Indeed, β_ℓ for each $\ell=0,1,\cdots,7$ is calculated as follows: First $\beta_\ell=1$ for all $\ell\neq 0,4$, and for $\ell=0$ and 4, we have

$$\beta_0=\frac{1}{8}\{(1-4)^3(1-8)^2+(1-4)^3+6\}=-168,$$

and

$$\beta_4=\frac{1}{8}\{(1-4)^3+(1-4)^3+6\}=-6.$$

Then

$$B_3(Y)=-\beta_0=168,\ B_3(X)=168-(-6)=174.$$

and

$$\chi(X)=\sum_{\ell=0}^{7}\beta_\ell=-168-6+6=-168.$$

Hence

$$h^{1,1}(X)=B_2(X)=-1+\frac{1}{2}(\beta_1+\beta_2+\beta_3+\beta_5+\beta_6+\beta_7)=-1+\frac{6}{2}=2$$

and hence $h^{2,1}(X)=86$.

EXAMPLE 3.3. Let $m=14$ and $Q=(1,2,2,2,7)$. Then there are altogether 296 pairs of (ℓ,r). We compute that

$$\beta_0=\frac{1}{14}\{(1-14)(1-7)^3(1-2)+(1-7)^3+6(1-2)+6\}=(-6)^3=-216,$$

and

$$\beta_{2k}=\frac{1}{14}\{(1-2)7+7\}=0\quad\text{for every } k\in\{1,2,3,4,5,6\}.$$

Calculating β_{2k+1} we have

$$\beta_7=\frac{1}{14}\{(1-7)^3+(1-7)^3+12\}=-30$$

and

$$\beta_{2k+1}=1\quad\text{for all } k\in\{0,1,2,4,5,6\}.$$

Therefore, $\chi(X)=-216-30+6=-240$. Now we calculate $B_2(X)$. We have

$$\begin{aligned}h^{1,1}(X)=B_2(X) &= -1+\frac{1}{2}(\sum_{\ell=0}^{13}\beta_\ell-(\beta_0+\beta_7))\\ &= -1+\frac{1}{2}(6)=-1+3=2,\end{aligned}$$

and $h^{2,1}(X)=122$.

EXAMPLE 3.4. Let $m=30$ and $Q=(2,2,5,6,15)$. There are altogether 900 pairs of (ℓ,r). We compute that $\beta_0=-156$, $\beta_{15}=-52$, $\beta_{3k}=1$ for every $k\in\{1,2,3,4,6,7,8,9\}$, $\beta_1=1$ and $\beta_p=1$ for all primes $p, 7\leq p\leq 29$. While $\beta_{2k}=0$ for every $k\in\{1,2,4,5,7,8,10,11,13,14\}$ and $\beta_{5k}=0$ for every $k\in\{1,2,4,5\}$. Therefore, we have

$$B_3(Y)=-\beta_0=156,\ B_3(X)=156-\beta_{15}=156+52=208$$

and $\chi(X)=\sum_{\ell=0}^{29}\beta_\ell=-156-52+8+8=-192$. Now,

$$\begin{aligned} h^{1,1}(X)=B_2(X) &= -1+\frac{1}{2}(\sum_\ell \beta_\ell-(\beta_0+\beta_{15}))\\ &= -1+\frac{1}{2}(-192+156+52)\\ &= -1+\frac{16}{2}=7, \end{aligned}$$

and $h^{2,1}(X)=103$.

EXAMPLE 3.5. We consider yet another example. Let $m=60$ and $Q=(3,5,12,20,20)$. There are altogether 3600 pairs of (ℓ,r). We compute that $\beta_0=-56$, $\beta_{12k}=-2$ for each k, $1\le k\le 4$. Also $\beta_{5k}=0$ for every k, $1\le k\le 11, k\ne 20,40$, while $\beta_{20}=\beta_{40}=4$. For β_{3k} with $1\le k\le 19$ and $3k$ is not a multiple of 12 nor 15, we have $\beta_{3k}=2$ and for the remaining cases, $\beta_\ell=1$. Therefore we get

$$B_3(Y)=-\beta_0=56,\ B_3(X)=56-\sum_{k=1}^{4}\beta_{12k}=56+8=64$$

and

$$\chi(X)=\sum_{\ell=0}^{59}\beta_\ell=-56-8+2\times 12+4\times 2+32=-56-8+24+8+32=0.$$

We also compute that $h^{1,1}(X)=-1+\frac{1}{2}(0+8+56)=-1+32=31$.

REMARK 3.4. There is another method of calculating Betti numbers $B_3(Y)$ and hence $B_3(X)$ as well as the Hodge numbers $h^{3,0}(X), h^{2,1}(X)$. This method makes use of Fermat motives and will be discussed in the following Section 4.

4. The motivic decomposition of Calabi–Yau orbifolds

In this section, we define the motives and decompose our Calabi–Yau orbifolds into the direct sum of motives. The notion of Fermat motives was first introduced in Shioda [Sh87]. We give the ℓ-adic realization of the motivic decomposition in terms of the third ℓ-adic étale cohomology groups. Then we give an algorithm to compute the Hodge numbers $h^{3,0}, h^{2,1}$, and the Betti number B_3 of each motive. Finally, summing over all these motivic invariants, we obtain the global Hodge numbers, Betti numbers and the Euler characteristic of our Calabi–Yau orbifolds. This algorithm is different from the algorithm given by Roan [Ro90] described in Proposition 3.3 and Examples.

4.1. Motives. We now recall the definition and basic properties of Fermat motives. The Fermat motives in this form were first introduced by Shioda [Sh87]. For a thorough treatment of Fermat motives, the reader is referred to Shioda [Sh87], Gouvêa and Yui [GY95, Chapter 3]. (Cf. Manin [Ma70].)

We fix a Fermat hypersurface threefold of degree $m\ge 5$:

$$V\quad:\quad X_1^m+X_2^m+X_3^m+X_4^m+X_5^m=0\subset\mathbb{P}^4_{\mathbb{Q}}.$$

Let μ_m denote the group of m-th roots of unity, and let Δ be the image of the diagonal inclusion $\mu_m\hookrightarrow\mu_m^5$. Let

$$\mathfrak{G}:=\mu_m^5/\Delta=\{\mathbf{g}=(g_1,g_2,g_3,g_4,g_5)\in\mu_m^5\}/\Delta.$$

We may identify $\mathfrak{G}$ as a subgroup of the automorphism group $\mathrm{Aut}(V)$ of V. Let $\widehat{\mathfrak{G}}$ be the character group of $\mathfrak{G}$. $\mathbb{L} = \mathbb{Q}(\zeta_m)$ (with $\zeta_m = e^{2\pi i/m}$) denotes the m-th cyclotomic field over $\mathbb{Q}$. Then $\widehat{\mathfrak{G}}$ is identified with the set

$$\{\mathbf{a} = (a_1, a_2, a_3, a_4, a_5) \mid a_i \in \mathbb{Z}/m\mathbb{Z}, \sum_{i=1}^{5} a_i \equiv 0 \pmod{m}\}$$

under the pairing $\widehat{\mathfrak{G}} \times \mathfrak{G} \to \mathbb{L}\ :\ (\mathbf{a}, \mathbf{g}) \mapsto \mathbf{a}(\mathbf{g}) := \prod_{i=1}^{5} g_i^{a_i}$. The group $(\mathbb{Z}/m\mathbb{Z})^\times$ acts on $\widehat{\mathfrak{G}}$ by $t \cdot (a_1, a_2, a_3, a_4, a_5) = (ta_1, ta_2, ta_3, ta_4, ta_5)$.

DEFINITION 4.1. For any $\mathbf{a} \in \widehat{\mathfrak{G}}$, let

$$p_{\mathbf{a}} := \frac{1}{\#\mathfrak{G}} \sum_{\mathbf{g} \in \mathfrak{G}} \mathbf{a}(\mathbf{g})^{-1}\mathbf{g} = \frac{1}{m^4} \sum_{\mathbf{g} \in \mathfrak{G}} \mathbf{a}(\mathbf{g})^{-1}\mathbf{g} \in \mathbb{Z}[\frac{1}{m}, \zeta_m][\mathfrak{G}].$$

The group $(\mathbb{Z}/m\mathbb{Z})^\times$ acts on $\widehat{\mathfrak{G}}$ by multiplication by $t \in (\mathbb{Z}/m\mathbb{Z})^\times$ on each component of $\mathbf{a}$. Let $A = [\mathbf{a}]$ be the $(\mathbb{Z}/m\mathbb{Z})^\times$-orbit of $\mathbf{a}$, and let

$$p_A := \sum_{\mathbf{a} \in A} p_{\mathbf{a}} \in \mathbb{Z}[\frac{1}{m}][\widehat{\mathfrak{G}}].$$

Then $p_{\mathbf{a}}$ and p_A are idempotents, and

$$\sum_{\mathbf{a} \in \widehat{\mathfrak{G}}} p_{\mathbf{a}} = \sum_{A \in O(\widehat{\mathfrak{G}})} p_A = 1$$

where $O(\widehat{\mathfrak{G}})$ denotes the set of $(\mathbb{Z}/m\mathbb{Z})^\times$-orbits in $\widehat{\mathfrak{G}}$. $p_{\mathbf{a}}$ and p_A are projectors. Identifying $\mathbf{g} \in \mathfrak{G} \subset \mathrm{Aut}(V)$ with its graph $\Gamma_{\mathbf{g}}$, $p_A \in \mathrm{End}(V) \otimes \mathbb{Z}[\frac{1}{m}]$ is regarded as an algebraic cycle on $V \times V$ over $\mathbb{F}_p$ with coefficients in $\mathbb{Z}[\frac{1}{m}]$ (cf. Manin [Ma70]). Therefore, the pair (V, p_A) defines a motive over the prime field $\mathbb{F}_p$ or $\mathbb{Q}$ corresponding to the $(\mathbb{Z}/m\mathbb{Z})^\times$-orbit $A = [\mathbf{a}]$ in $\widehat{\mathfrak{G}}$. This is called a *Fermat motive* and denoted by $\mathcal{M}_A$. (The same projectors define the twisted Fermat motives if $\mathbf{c} \neq \mathbf{1}$.) The construction gives the decomposition of V into the direct sum of Fermat motives:

$$V = \oplus_{A \in O(\widehat{\mathfrak{G}})} \mathcal{M}_A$$

corresponding to $\sum_A p_A = 1$.

Further, the motivic decomposition of V gives rise to the composition of relevant (e.g., Betti, Hodge, and ℓ-adic étale) cohomology groups:

$$H^n(V) = \oplus_{A \in O(\widehat{\mathfrak{G}})} H^n(\mathcal{M}_A).$$

Here we define $H^n(\mathcal{M}_A) := \oplus_{\mathbf{a} \in A} H^n(V)(\mathbf{a})$ where $H^n(V)(\mathbf{a})$ denotes the part of the cohomology group of $H^n(V)$ fixed by the kernel of $\mathbf{a}$. That is, $H^n(V)(\mathbf{a})$ may be viewed as the eigenspace of the automorphism $p_{\mathbf{a}}$ and $H^n(\mathcal{M}_A)$ as the eigenspace $H^n(V)(A)$ of $p_A = \sum_{\mathbf{a} \in A} p_{\mathbf{a}}$.

LEMMA 4.1. *Let $\mathfrak{A}$ be the subset of $\widehat{\mathfrak{G}}$ defined by*

$$\mathfrak{A} := \{\mathbf{a} \in \widehat{\mathfrak{G}} \mid a_i \not\equiv 0 \pmod{m} \quad \textit{for every } i \quad \}.$$

Then the numerical invariants for V are computed as follows:

(a) For $\mathbf{a}=(a_1,a_2,a_3,a_4,a_5)\in A$*, let*

$$\|\mathbf{a}\|:=\left(\sum_{j=1}^{5}\langle\frac{a_j}{m}\rangle\right)-1$$

where $\langle x\rangle$ *denotes the fractional part of* $x\in\mathbb{Q}$*. Then for* $i\in\{0,1,2,3\}$*, the Hodge realization of* $\mathcal{M}_A$ *determines the motivic Hodge numbers:*

$$h^{i,3-i}(\mathcal{M}_A)=\#\{\mathbf{a}\in A\mid\|\mathbf{a}\|=i\ \}.$$

Summing over all the motivic Hodge numbers, we obtain the global Hodge numbers:

$$h^{i,3-i}(V)=\sum_{A\in O(\widehat{\mathfrak{G}})}h^{i,3-i}(\mathcal{M}_A).$$

(b) The Betti realization of $\mathcal{M}_A$ *determines the motivic Betti number:*

$$B_3(\mathcal{M}_A)=\#A$$

Summing over all motivic Betti numbers, we have

$$B_3(V)=\sum_{A\in O(\widehat{\mathfrak{G}})}B_3(\mathcal{M}_A)=\#\mathfrak{A}.$$

For a proof, see Gouvêa and Yui [GY95], Chapter 3.

4.2. Numerical invariants of Calabi–Yau orbifolds. Here is an algorithm to compute Hodge numbers and Betti numbers of (a) singular Calabi–Yau orbifolds and (b) singular loci. For (a), we decompose the varieties in question into Fermat motives. However, we are not going to consider all motives $\mathcal{M}_A$, but only those motives which are compatible with the weight $Q=(q_1,q_2,q_3,q_4,q_5)$. What do we mean by "compatible with weight"? This will be made precise in Theorem 4.2. Then summing over the corresponding motivic Hodge numbers and Betti numbers for those motives, we obtain the global Hodge numbers, and Betti numbers of singular Calabi–Yau orbifolds. For (b), we need to compute the genus of each irreducible curve in the singular locus.

THEOREM 4.2. *Let* Y *be the singular Calabi–Yau orbifold corresponding to a pair* $<m,Q>$ *where* $Q=(q_1,q_2,q_3,q_4,q_5)$*. Let* X *be a smooth Calabi–Yau crepant resolution of* Y*.*

Let $\mathfrak{A}(Q)$ *be the subset of* $\mathfrak{A}$ *defined by*

$$\mathfrak{A}(Q):=\{\mathbf{a}\in\mathfrak{A}\mid a_i\in q_i\mathbb{Z}/m\mathbb{Z}\quad\text{for every } i\}.$$

For each $\mathbf{a}\in\mathfrak{A}(Q)$*, let* $\mathcal{M}_A$ *be the Fermat motive corresponding to the* $(\mathbb{Z}/m\mathbb{Z})^\times$*-orbit of* $\mathbf{a}$*, which is called a motive compatible with the weight* Q*. Let* $O(\mathfrak{A}(Q))$ *be the set of* $(\mathbb{Z}/m\mathbb{Z})^\times$*-orbits in* $\mathfrak{A}(Q)$*.*

(a) The weight $Q=(q_1,q_2,q_3,q_4,q_5)$ *is an element of* $\mathfrak{A}(Q)$*. Let* $\mathcal{M}_Q$ *denote the Fermat motive determined by the weight* Q*. Then the Hodge and Betti realizations of the Fermat motives compatible with* Q *determine the global Hodge and Betti numbers of* Y*:*

$$h^{0,3}(\mathcal{M}_Q)=1=h^{0,3}(Y)$$

and

$$h^{2,1}(Y)=\sum_{A\in O(\mathfrak{A}(Q))}h^{2,1}(\mathcal{M}_A).$$

TABLE 1. Fermat motives for $m = 5$ and $Q = (1,1,1,1,1)$

A	dim.	mult.	$h^{3,0}$	$h^{2,1}$	$\sum B_3(\mathcal{M}_A)$
$[1,1,1,1,1]$	4	1	1	1	4
$[1,1,1,3,4]$	4	20	0	2	80
$[1,1,2,2,4]$	4	30	0	2	120

Furthermore,

$$B_3(Y) = \sum_{A \in O(\mathfrak{A}(Q))} B_3(\mathcal{M}_A).$$

(b) Let $\Sigma(Y)$ be the singular locus of Y. Let $C \in \Sigma(Y)$ be an irreducible curve of genus g_C in the singular locus $\Sigma(Y)$. Then

$$h^{2,1}(\hat{\Sigma}(Y)) = \sum_C m_C g_C$$

where the sum runs over all **distinct** *irreducible curves on $\Sigma(Y)$ of genus g_C and multiplicity m_C. Hence*

$$h^{3,0}(X) = 1,$$

$$h^{2,1}(X) = h^{2,1}(Y) + h^{2,1}(\hat{\Sigma}(Y)) = h^{2,1}(Y) + \sum_C m_C g_C,$$

and

$$B_3(X) = B_3(Y) + \sum_C m_C B_1(C) = B_3(Y) + 2\sum_C m_C g_C.$$

REMARK 4.1. (a) The curves in $\Sigma(Y)$ of genus 0 do not contribute to $B_3(X)$. An interpretation of these rational curves will be discussed subsequently.

(b) We do not yet know how to calculate $B_2(X) = h^{1,1}(X)$ using motivic data. At the moment, for the calculation of $h^{1,1}(X)$ we will combine Proposition 3.3 for computing the Euler characteristic $\chi(X)$, and Theorem 4.2 for $h^{2,1}(X)$. Then $h^{1,1}(X)$ can be obtained by the formula $\frac{1}{2}(\chi(X) + h^{2,1}(X))$.

EXAMPLE 4.1. Let X be the Fermat quintic Calabi–Yau threefold corresponding to $m = 5$ and $Q = (1,1,1,1,1)$. In this case $Y = V$, and there are three types of Fermat motives which are tabulated together with their dimension, multiplicity and Hodge numbers as follows:
Summing over the motivic Hodge numbers and Betti numbers, we obtain

$$h^{3,0}(Y) = 1,\ h^{2,1}(Y) = 1 + 2 \cdot 20 + 2 \cdot 30 = 101$$

and

$$B_3(Y) = 4 + 80 + 120 = 204.$$

This Calabi–Yau threefold is smooth, so there is no singularity and hence we get

$$h^{3,0}(X) = 1,\ h^{2,1}(X) = 101 \quad \text{and} \quad B_3(X) = 204.$$

We compute that $h^{1,1}(X) = 1$ so that the Euler characteristic is $\chi = 2(1 - 101) = -200$.

TABLE 2. Fermat motives for $m = 8$ and $Q = (1, 1, 2, 2, 2)$

A	dim.	mult.	$h^{0,3}$	$h^{2,1}$	$\sum B_3(\mathcal{M}_A)$
[1, 1, 2, 2, 2]	4	1	1	1	4
[7, 3, 2, 2, 2]	4	1	0	2	4
[6, 4, 2, 2, 2]	2	2	0	2	4
[7, 1, 2, 2, 4]	4	6	0	12	24
[6, 2, 2, 2, 4]	2	6	0	6	12
[4, 4, 2, 2, 4]	2	3	0	3	6
[5, 1, 2, 2, 6]	4	3	0	6	12
[4, 2, 2, 2, 6]	2	6	0	6	12
[3, 3, 2, 2, 6]	4	3	0	6	12
[5, 1, 2, 4, 4]	4	3	0	6	12
[4, 2, 2, 4, 4]	2	6	0	6	12
[3, 3, 2, 4, 4]	4	3	0	6	12
[3, 1, 2, 4, 6]	4	6	0	12	24
[2, 2, 2, 4, 6]	2	6	0	6	12
[3, 1, 4, 4, 4]	4	1	0	2	4
[2, 2, 4, 4, 4]	2	1	0	1	2

EXAMPLE 4.2. Let X be a Calabi–Yau orbifold corresponding to $m = 8$ and $Q = (1, 1, 2, 2, 2)$. The characters giving rise to Fermat motives and their Hodge numbers are listed in Table 2, and by summing over the motivic Hodge numbers, we get

$$h^{0,3}(\mathcal{M}_Q) = 1 = h^{0,3}(Y), \quad h^{2,1}(Y) = 83, \quad \text{and,} \quad B_3(Y) = 168.$$

There is one singularity of dimension 1, namely, the curve $Y_3^4 + Y_4^4 + Y_5^4 = 0$ of genus 3. Hence $h^{2,1}(\hat{\Sigma}(Y)) = 3$. Therefore, we obtain

$$h^{3,0}(X) = 1,\ h^{2,1}(X) = 86 \quad \text{and} \quad B_3(X) = 174.$$

We compute that $h^{1,1}(X) = 2$ so that the Euler characteristic of X is $\chi(X) = 2(2 - 86) = -168$.

EXAMPLE 4.3. Let $m = 10$ and $Q = (1, 1, 1, 2, 5)$. The quotient Y is smooth, so $\Sigma(Y) = \emptyset$, and hence the Hodge numbers and Betti numbers of X are entirely determined by the Fermat motives $\mathcal{M}_A$ with $A \in O(\mathfrak{A}(Q))$.
We compute that

$$h^{0,3}(Y) = 1,\ h^{2,1}(Y) = 145 \quad \text{and hence} \quad B_3(X) = B_3(Y) = 292.$$

Also we calculate that $h^{1,1}(X) = 1$ so that the Euler characteristic is $\chi(X) = -288$.

EXAMPLE 4.4. Let X be a Calabi–Yau orbifold corresponding to $m = 14$ and $Q = (1, 2, 2, 2, 7)$. There are ten non-isomorphic Fermat motives and their dimension, multiplicity, Hodge numbers and Betti numbers are listed in Table 4. By adding up the motive Hodge numbers we obtain

$$h^{0,3}(Y) = 1, \quad h^{2,1}(Y) = 107, \quad B_3(Y) = 216.$$

There is one singularity of dimension 1, namely a smooth curve $Y_I : c_2Y_2^7 + c_3Y_3^7 + c_4Y_4^7 = 0$ of degree 7 and the reduced weight $Q' = (1, 1, 1)$ with multiplicity 1.

TABLE 3. Fermat motives for $m = 10$ and $Q = (1, 1, 1, 2, 5)$

A	dim.	mult.	$h^{0,3}$	$h^{1,2}$	$\sum B_3(\mathcal{M}_A)$
[1, 1, 1, 2, 5]	4	1	1	1	4
[9, 3, 1, 2, 5]	4	6	0	12	24
[9, 2, 2, 2, 5]	4	3	0	6	12
[8, 4, 1, 2, 5]	4	6	0	12	24
[8, 3, 2, 2, 5]	4	6	0	12	24
[7, 5, 1, 2, 5]	4	6	0	12	24
[7, 4, 2, 2, 5]	4	6	0	12	24
[7, 3, 3, 2, 5]	4	3	0	6	12
[6, 6, 1, 2, 5]	4	3	0	6	24
[6, 5, 2, 2, 5]	4	6	0	12	24
[6, 4, 3, 2, 5]	4	6	0	12	24
[5, 5, 3, 2, 5]	4	3	0	6	12
[5, 4, 4, 2, 5]	4	3	0	6	12
[1, 1, 7, 6, 5]	4	3	0	6	12
[1, 4, 4, 6, 5]	4	3	0	6	12
[1, 1, 5, 8, 5]	4	3	0	6	12
[1, 2, 4, 8, 5]	4	6	0	12	24

TABLE 4. Fermat motives for $m = 14$ and $Q = (1, 2, 2, 2, 7)$

A	dim.	mult.	$h^{0,3}$	$h^{2,1}$	$\sum B_3(\mathcal{M}_A)$
[1, 2, 2, 2, 7]	6	1	1	2	6
[13, 2, 2, 4, 7]	6	3	0	9	18
[11, 2, 2, 6, 7]	6	3	0	9	18
[9, 2, 2, 8, 7]	6	3	0	9	18
[7, 2, 2, 10, 7]	6	3	0	9	18
[5, 2, 2, 12, 7]	6	3	0	9	18
[9, 2, 4, 6, 7]	6	6	0	18	36
[7, 2, 4, 8, 7]	6	2	0	6	12
[5, 2, 4, 10, 7]	6	6	0	18	36
[3, 2, 4, 12, 7]	6	6	0	18	36

This is an irreducible curve of genus $g = 15$, which is denoted by C. So $B_1(C) = 30$ and $h^{2,1}(X) = 107 + 15 = 122$, and $B_3(X) = 216 + 30 = 246$. We compute that $h^{1,1}(X) = 2$ so that the Euler characteristic is $\chi(X) = 2(2 - 122) = -240$.

Finally we consider one more example.

EXAMPLE 4.5. Let X be a Calabi–Yau orbifold corresponding to $m = 60$ and $Q = (3, 5, 12, 20, 20)$. There are five non-isomorphic Fermat motives and their dimension, multiplicity, Hodge numbers and Betti numbers are listed in Table 5. By adding up the motivic Hodge numbers we obtain

$$h^{0,3}(Y) = 1,\ h^{2,1}(Y) = 27 \quad \text{and} \quad B_3(Y) = 56.$$

TABLE 5. Fermat motives for $m = 60$ and $Q = (3, 5, 12, 20, 20)$

A	dim.	mult.	$h^{3,0}$	$h^{2,1}$	$\sum B_3(\mathcal{M}_A)$
$[3, 5, 12, 20, 20]$	16	1	1	7	16
$[3, 45, 12, 20, 40]$	16	1	0	8	16
$[6, 10, 24, 40, 40]$	8	1	0	4	8
$[6, 30, 24, 20, 40]$	8	1	0	4	8
$[12, 20, 48, 20, 20]$	8	1	0	4	8

There are singularities of dimension 1 for each $\ell = 12k$, $1 \leq k \leq 4$, namely, a smooth curve $C : Y_2^{12} + Y_4^3 + Y_5^3 = 0$ of degree 12 and reduced weight $Q' = (1, 4, 4)$ and multiplicity 4. It has genus 1. So $B_1(C) = 2$. (The curve C could further be reduced to the cubic Fermat curve $Y_2^3 + Y_3^3 + Y_5^3 = 0$ with reduced weight $Q'' = (1, 1, 1)$, which is a curve of genus 1.) Therefore $h^{2,1}(X) = 27 + 4 \times 1 = 31$. Further $B_3(X) = 56 + 4 \times 2 = 64$. We compute that $h^{1,1}(X) = 31$ so that the Euler characteristic $\chi(X) = 0$.

5. The L-series of Calabi–Yau orbifolds

In the previous section, we discussed the Hodge and Betti realizations of our Fermat motives. In this section, we will look into the étale ℓ-adic realizations of our Fermat motives.

First we recall the results of Weil [W52], and of Gouvêa and Yui [GY95], describing the L–series of hypersurfaces V of Fermat type of degree m and dimension n by means of Hecke L-series with Jacobi sum Grössencharaktere of the m–th cyclotomic field $\mathbb{L} = \mathbb{Q}(\zeta_m)$. The notations of Section 4 remain in force.

DEFINITION 5.1. Fix a rational integer n (which is 1 or 3 in our case) and let $m \geq n+2$. Let $\mathfrak{p}$ be a prime ideal of $\mathbb{L} = \mathbb{Q}(\zeta_m)$ prime to m, and put $q = N\mathfrak{p}$; then $q \equiv 1 \pmod{m}$. There is a unique multiplicative character $\chi_\mathfrak{p} : \mathbb{F}_q^\times \to \mu_m$ such that $\chi_\mathfrak{p}(x) \equiv x^{q-1/m} \pmod{\mathfrak{p}}$ for every integer $x \in \mathbb{Z}[\zeta]$ prime to $\mathfrak{p}$, and $\chi_\mathfrak{p}(x) = 0$ for $x \equiv 0 \pmod{\mathfrak{p}}$.

Let Δ, $\mathfrak{G}$, $\widehat{\mathfrak{G}}$ and $\mathfrak{A}$ be as defined in Section 4 (for $n = 3$). (For $n = 1$, the definitions are entirely similar with three components instead of five.) That is,

$$\mathfrak{A} = \{\mathbf{a} = (a_1, \cdots, a_{n+2}) \in \widehat{\mathfrak{G}} \mid a_i \in (\mathbb{Z}/m\mathbb{Z}) \setminus \{0\}, \sum a_i \equiv 0 \pmod{m}\}.$$

The *Jacobi sum of dimension n (relative to $\chi_\mathfrak{p}$ and $\mathbf{a} \in \mathfrak{A}$)* is defined by

$$j(\mathbf{a}) = j_\mathfrak{p}(\mathbf{a}) = (-1)^n \sum \chi_\mathfrak{p}(x_2)^{a_2} \chi_\mathfrak{p}(x_3)^{a_3} \cdots \chi_\mathfrak{p}(x_{n+2})^{a_{n+2}}$$

where the summation is taken over all $(n+1)$-tuples $(x_2, \cdots, x_{n+2})$ with $x_i \pmod{\mathfrak{p}}$ subject to the relation $x_2 + \cdots + x_{n+2} = -1$.

If $\mathbf{c} = (c_1, c_2, \cdots, c_{n+2})$ is a twist, then the twisted Jacobi sum is defined by

$$\mathbf{c}(\mathbf{a}) = \bar{\chi}_\mathfrak{p}(c_1^{a_1} c_2^{a_2} \cdots c_{n+2}^{a_{n+2}}) j(\mathbf{a}).$$

Here are some properties of Jacobi sums and twisted Jacobi sums (cf. Weil [W52]) which are relevant to our subsequent discussions.

LEMMA 5.1. *(a) Let $G(\chi, \psi)$ denote the Gauss sum associated to χ and an additive character ψ of $\mathbb{F}_q$. Then the Jacobi sum $j(\mathbf{a})$ is expressed as a product of*

Gauss sums:

$$j(\mathbf{a}) = \frac{1}{q} G(\chi^{a_1}) G(\chi^{a_2}) \cdots G(\chi^{a_{n+2}}).$$

The Gauss sums are algebraic integers in the field $\mathbb{Q}(\zeta_m, \zeta_q)$ *with complex absolute value* $\sqrt{q}$. *It then follows that a Jacobi sum is an algebraic integer in* $\mathbb{L} = \mathbb{Q}(\zeta_m)$ *with complex absolute value* $|j(\mathbf{a})| = (N\mathfrak{p})^{n/2} = q^{n/2}$.

(b) Jacobi sums of dimension n *are Grössencharaktere of* $\mathbb{L} = \mathbb{Q}(\zeta_m)$ *with conductor* $\mathfrak{f}$, *and* $\mathfrak{f}$ *is a divisor of* $m^2 \times \prod_{i=1}^{n+2} c_i$.

(c) Let $V = V(m, n, \mathbf{1})$ *be the Fermat hypersurface of degree* m *and of odd dimension* n *defined over* $\mathbb{Q}$. *Then Jacobi sums* $j(\mathbf{a})$ *are eigenvalues of the* $\mathbb{Q}_\ell$-*linear transformation* $Frob_{\mathfrak{p}}^*$ *on* $H^n_{et}(\bar{V}, \mathbb{Q}_\ell)$ *induced from the Frobenius morphism* $Frob_{\mathfrak{p}}$. *Globally, the* n*-th partial* L*-series* $L_n(V, s)$ *is expressed in terms of the Hecke* L*–series with Jacobi sum Grössencharaktere of* $\mathbb{L} = \mathbb{Q}(\zeta_m)$*:*

$$L_n(V, s) = L_{Hecke,n}(s) :\doteq \prod_{\mathfrak{p} \not| \mathfrak{f}} \prod_{\mathbf{a} \in \mathfrak{A}_n^m} (1 - j(\mathbf{a}) N\mathfrak{p}^{-s})^{-1},$$

where $\doteq$ *indicates that the equality holds up to a finite number of factors corresponding to* $\mathfrak{p} \mid \mathfrak{f}$. *For all other values of* $i, 0 \leq i \leq 2n, i \neq n$,

$$L_i(V, s) = \begin{cases} 1 & \text{if } i \text{ is odd} \\ \zeta_{\mathbb{Q}}(s - i) & \text{if } i \text{ is even} \end{cases}$$

where $\zeta_{\mathbb{Q}}(s)$ *denotes the Riemann zeta-function.*

If $V = V(m, n, \mathbf{c})$ *is a diagonal hypersurface with non-trivial coefficients* $\mathbf{c}$, *then the same assertions hold with* $\mathbf{c}(\mathbf{a})$ *in place of* $j(\mathbf{a})$.

Now consider Jacobi sums in connection with the given weight

$$Q = (q_1, q_2, \cdots, q_5).$$

DEFINITION 5.2. For a fixed weight $Q = (q_1, \cdots, q_{n+2})$, let $\mathfrak{A}(Q)$ be the subset of $\mathfrak{A}$ defined as in Section 3 (for $n = 3$), that is,

$$\mathfrak{A}(Q) = \{\mathbf{a} = (a_1, \cdots, a_{n+2}) \in \mathfrak{A} \mid q_i \text{ divides } a_i \text{ for every } i\}.$$

The Hecke L–series with *Jacobi sum Grössencharaktere of dimension* n *compatible with the weight* Q is defined by

$$L^Q_{\text{Hecke},n}(s) :\doteq \prod_{\mathfrak{p} \not| \mathfrak{f}} \prod_{\mathbf{a} \in \mathfrak{A}(Q)} (1 - j(\mathbf{a})\, N\mathfrak{p}^{-s})^{-1}.$$

If the coefficient vector $\mathbf{c} \neq \mathbf{1}$, we use $\mathbf{c}(\mathbf{a})$ in place of $j(\mathbf{a})$.

Now for the Calabi–Yau orbifolds of Fermat type constructed in Theorem 2.3, we can determine the partial L–series $L_3(X, s) = L(H^3_{et}(\bar{X}, \mathbb{Q}_\ell), s)$ explicitly. Again we pass to Fermat motives. Note that when $\mathbf{a}$ runs over the $(\mathbb{Z}/m\mathbb{Z})^\times$-orbit $A \in O(\mathfrak{A}(Q))$, the corresponding Jacobi sums $j(\mathbf{a})$ are all conjugate algebraic integers in $\mathbb{L} = \mathbb{Q}(\zeta_m)$.

LEMMA 5.2. *Let* Y *be the singular Calabi–Yau orbifold of degree* m *and weight* $Q = (q_1, q_2, q_3, q_4, q_5)$. *Then the étale* ℓ*-adic realization of the Fermat motives gives rise to the decomposition*

$$L(Y, s) \doteq \prod_{A \in O(\mathfrak{A}(Q))} L(\mathcal{M}_A, s)$$

where

$$L(\mathcal{M}_A, s) = \prod_{\mathfrak{p} \nmid \mathfrak{f}} \prod_{\mathbf{a} \in A} (1 - j(\mathbf{a}) N\mathfrak{p}^{-s})^{-1}.$$

We also have

$$L(\hat{\Sigma}(Y), s) = \prod_{C} L(C, s)^{m_C}$$

where the product runs over all distinct irreducible curves in $\Sigma(Y)$ *with multiplicity* m_C*, and each* $L(C, s)$ *is again a product of* L*-series of Fermat motives of dimension* 1.

PROOF. First notice that the Fermat motives of V which are left invariant under the quotienting action by μ_Q are exactly those corresponding to the characters $\mathbf{a} \in \mathfrak{A}(Q)$. This implies that Y has the motivic decomposition

$$(Y, \mathrm{Id}) = \oplus_{\mathrm{A} \in \mathrm{O}(\mathfrak{A}(\mathrm{Q}))} (\mathcal{M}_{\mathrm{A}}, \mathrm{p}_{\mathrm{A}})$$

where the direct sum runs over the $(\mathbb{Z}/m\mathbb{Z})^{\times}$-orbits in $\mathfrak{A}(Q)$. The étale ℓ-adic cohomological realization of the Fermat motives yields the decomposition of the L-series:

$$L(Y, s) = \prod_{A \in O(\mathfrak{A}(Q))} L(\mathcal{M}_A, s).$$

Similarly the same line of arguments is applied for $L(C, s)$. Since the C are curves, the Fermat motives are of dimension 1. □

REMARK 5.1. For the actual calculation of $L(\mathcal{M}_A, s)$, we first fix $\mathfrak{p}$ and calculate the local pieces $\prod_A (1 - j(\mathbf{a})T)$ where T is an indeterminate. Let p be a prime such that $p \nmid m$, and let f be the order of p modulo m, that is, $p^f \equiv 1 \pmod{m}$.

For instance, let p be a prime such that $p \equiv 1 \pmod{m}$ and consider the polynomial $P_{[Q]}(T) := \prod_{\mathbf{a} \in [Q]} (1 - j(\mathbf{a})T)$ corresponding to the weight motive $[Q]$. Then it is always irreducible over $\mathbb{Q}$ of degree $\dim(\mathcal{M}_Q)$ over $\mathbb{Q}$ and the coefficient of the highest degree term is $p^{3/2 \times \dim(\mathcal{M}_Q)}$. The Newton polygon of $P_{[Q]}(T)$ has the slope sequence $0, 1, 2, 3$ with multiplicity $1, (\dim(\mathcal{M}_Q) - 2)/2, (\dim(\mathcal{M}_Q) - 2)/2, 1$, respectively. This means that the coefficient of T is not divisible by p but those of higher powers of T^i, $i \geq 2$ are divisible by a power of p. $P_{[Q]}(T)$ is always irreducible over $\mathbb{Q}$.

The polynomials $P_{\mathcal{M}_A}(T)$ for the motives $A \neq [Q]$ have the slope sequence $1, 2$, which in turn implies that all coefficients apart from the constant term are divisible by powers of p. There are many examples of Fermat motives where $P_{\mathcal{M}_A}(T)$ is not irreducible over $\mathbb{Q}$ but whose irreducible factors are of degree 2. This implies that the Fermat motives decompose, in their étale ℓ-adic realization, further into "rigid" Fermat motives. Note that for a rigid Calabi–Yau threefold over $\mathbb{Q}$, if p is a good prime, then the characteristic polynomial of the Frobenius endomorphism is an irreducible degree 2 polynomial of the form $1 - t_3(p)T + p^3T^2$ with $t_3(p) \in \mathbb{Z}$, $|t_3(p)| \leq 2p^{3/2}$ (cf. Saito and Yui [SY01]). When the characteristic polynomial of a Fermat motive has such a factor, we may call it a "rigid" Fermat motive. (See Examples 5.2, 5.3 and 5.4 below.)

For other primes p such that $p^f \equiv 1 \pmod{m}$ with $f > 1$, the polynomial $P_{[Q]}(T)$ can be computed as well.

We now recall the notion of ordinary, supersingular motives and motives of Hodge–Witt type from Gouvêa and Yui [GY95].

DEFINITION 5.3. Fix m and a prime p and let f be the smallest integer such that $p^f \equiv 1 \pmod{m}$. Put

$$H := \{\, p^i \bmod m \,|\, 0 \leq i < f \,\}$$

and define an arithmetic function

$$A_H(\mathbf{a}) := \sum_{t \in H} ||t\mathbf{a}||.$$

Let $\mathcal{M}_A$ be the Fermat motive constructed above. Then the Newton slopes of $\mathcal{M}_A$ are given by $\{\, \frac{A_H(\mathbf{a})}{f} \,|\, \mathbf{a} \in A \,\}$.

(a) $\mathcal{M}_A$ is said to be *ordinary* if the Hodge polygon and the Newton polygon of $\mathcal{M}_A$ coincide.

(b) $\mathcal{M}_A$ is said to be *supersingular* if the Newton polygon has the pure slope 3/2.

(c) $\mathcal{M}_A$ is said to be *of Hodge–Witt type* if the formal group is of finite height.

The characteristic polynomials are arithmetic invariants sensitive to the coefficients $\mathbf{c}$. All examples discussed below are for the trivial coefficient $\mathbf{c} = \mathbf{1}$.

EXAMPLE 5.1. Let X be the Fermat quintic Calabi–Yau threefold corresponding to $m = 5$ and $Q = (1, 1, 1, 1, 1)$. There are three motives $\mathcal{M}_A$ for X as listed in Table 1. They correspond to $A = [1, 1, 1, 1, 1]$, $[1, 1, 1, 3, 4]$ and $[1, 1, 2, 2, 4]$, each of dimension 4.

(a) Let p be a prime such that $p \equiv 1 \pmod 5$, e.g., $p = 11$. Then for the trivial twist $\mathbf{c} = \mathbf{1}$, we have

$$P_{[Q]}(T) = \prod_{\mathbf{a} \in [1,1,1,1,1]} (1 - j(\mathbf{a})T) = 1 + 89T + 3^3 \cdot 11 \cdot 13T^2 + 89 \cdot 11^3 T^3 + 11^6 T^4$$

which is irreducible over $\mathbb{Q}$, and its Newton slopes are $0, 1, 2, 3$ each with multiplicity 1, and hence the Newton polygon coincides with the Hodge polygon. Hence the motive $\mathcal{M}_Q$ is ordinary.

On the other hand, we have

$$\begin{aligned} \prod_{\mathbf{a} \in [1,1,1,3,4]} (1 - j(\mathbf{a})T) &= 1 - 11T - 3^2 \cdot 11^2 T^2 - 11^4 T^3 + 11^6 T^4 \\ &= \prod_{\mathbf{a} \in [1,1,2,2,4]} (1 - j(\mathbf{a})T) \end{aligned}$$

which is irreducible over $\mathbb{Q}$, and its Newton slopes are $1, 2$ each with multiplicity 2. This implies that the two Fermat motives $\mathcal{M}_{[1,1,1,3,4]}$ and $\mathcal{M}_{[1,1,2,2,4]}$ have isomorphic étale ℓ-adic realizations and hence the same L-series. These motives are neither ordinary nor supersingular.

(b) Let p be a prime such that $p \not\equiv 1 \pmod 5$, i.e., $f > 1$. In this case, $f = 2$ or 4. In fact, if $p \equiv 2$ or $3 \pmod 5$, then $f = 2$ and $H = \{\, 1, p \pmod m \,\}$, and if $p \equiv 4 \pmod 5$, then $f = 4$ and $H = \{\, 1, p, p^2, p^3 \pmod m \,\}$. In this case, all three motives have Newton slopes 3/2 with multiplicity 4. The characteristic polynomials are either two copies of irreducible degree 2 polynomials over $\mathbb{Q}$, or $\prod(1 - \zeta\, p^{3/2} T)$ with some root of unity ζ. All these three motives are supersingular.

The global L-series of the Fermat motives with trivial twist are obtained by taking the product of all local p factors.

$$L(\mathcal{M}_{[1,1,1,1,1]},s)=\prod_{p\equiv 1}P_{[Q]}(p^{-s})^{-1}\prod_{p\equiv 2,3}P_{[Q]}(p^{-4s})^{-1}\prod_{p\equiv 4}P_{[Q]}(p^{-2s})^{-1}.$$

(Here, for a natural number k, $p\equiv k$ stands for $p\equiv k \pmod 5$.) Similarly for

$$L(\mathcal{M}_{[1,1,1,3,4]},s)=L(\mathcal{M}_{[1,1,2,2,4]},s).$$

EXAMPLE 5.2. Let X be a Calabi–Yau orbifold corresponding to $m=8$ and $Q=(1,1,2,2,2)$. The Fermat motive $\mathcal{M}_Q$ corresponding to the weight $Q=(1,1,2,2,2)$ gives rise to the class of transcendental cycles of dimension 4. There are four distinct classes of primes modulo 8. They are (a) $p\equiv 1 \pmod 8$ and $H=\{1\}$, (b) $p\equiv 3 \pmod 8$ and $H=\{1,3\}$, (c) $p\equiv 5 \pmod 8$ and $H=\{1,5\}$ and (d) $p\equiv 7 \pmod 8$ and $H=\{1,7\}$.

(a) Let $p\equiv 1 \pmod 8$, e.g., $p=17$ and take the trivial twist $\mathbf{c}=\mathbf{1}$. Then calculate that

$$P_{[Q]}(T)=1-2^2\cdot 3^2\cdot 5T+2\cdot 17\cdot 467T^2-2^2\cdot 3^2\cdot 5\cdot 17^3T^3+17^6T^4$$

which is irreducible over $\mathbb{Q}$, and its Newton slopes are $0,1,2,3$ with multiplicity 1 for each slope. So $\mathcal{M}_Q$ is ordinary.

For the Fermat motives corresponding to $A=[1,1,2,6,6]$, $[1,1,4,4,6]$ (resp. $A=[1,3,2,4,6]$, $[1,3,4,4,4]$), we compute that

$$\begin{aligned} P_{\mathcal{M}_A}(T) &= 1\pm 2^2\cdot 3\cdot 17T+2\cdot 5\cdot 7\cdot 17^2T^2\pm 2^2\cdot 3\cdot 17^4T^3+17^6T^4\\ &= (1\pm 3\cdot 17T+17^3T^2)^2\end{aligned}$$

where the equations with the minus signs correspond to the Fermat motives with the respective A's. For the Fermat motives corresponding to $A=[1,5,2,2,6]$, $[1,5,2,4,4]$, and $[1,5,6,6,6]$, we compute that

$$P_{\mathcal{M}_A}(T)=(1-2\cdot 17T+17^3T^2)^2.$$

Similarly, for the Fermat motives corresponding to the remaining characters $A=[1,7,2,2,4]$, $[2,2,2,4,6]$, $[2,2,4,4,4]$, $[2,4,2,2,6]$, $[2,4,2,4,4$ $[2,4,6,6,6]$, $[2,6,2,2,4]$, $[4,4,2,4,4]$, we have

$$P_{\mathcal{M}_A}(T)=(1+2\cdot 17T+17^3T^2)^2.$$

These calculations imply that these Fermat motives are decomposable and the basic building blocks are "rigid" motives corresponding to the irreducible degree 2 polynomials over $\mathbb{Q}$.

(b) Let $p\equiv 3 \pmod 8$. Then the motives corresponding to the characters $[1,1,2,2,2]$, $[1,1,2,6,6]$, $[1,1,4,4,6]$, $[1,3,2,4,6]$ and $[1,3,4,4,4]$ all have Newton slopes $1,2$ each with multiplicity 2, and the motives corresponding to the characters $[1,5,2,2,6]$, $[1,5,2,4,4]$, $[1,5,6,6,6]$, $[1,7,2,2,4]$ and $[2,2,2,4,6]$ are supersingular.

(c) and (d) Let $p\equiv 5,7 \pmod 8$. Then we can determine Newton slopes of each motive in these cases as well.

The motivic L–series is given as follows:

$$L(\mathcal{M}_{[1,1,2,2,2]},s)=\prod_{p\equiv 1 \bmod 8}P_{[Q]}(p^{-s})^{-1}\prod_{p\equiv 3,5,7 \bmod 8}P_{[Q]}(p^{-2s})^{-1}$$

which is a Hecke L-series with Jacobi sum Grössencharakter of $\mathbb{Q}(e^{2\pi i/8})$.

Similarly for other motives.

Now we consider the L–series corresponding to singularities of dimension 1. We know that there is only one singularity of dimension 1, namely, a smooth curve: $Y_I : Y_2^4 + Y_3^4 + Y_4^4 = 0$ of degree 4 and the reduced weight $Q' = (1, 1, 1)$ with multiplicity 1. This is irreducible and gives rise to a curve C of genus $g = 3$. So $B_1(C) = 6$ and we obtain $B_3(X) = 168 + 6 = 174$. The L-series of C is again a Hecke L–series with a Jacobi sum Grössencharakter of dimension 1, associated to the cyclotomic field $\mathbb{L}' = \mathbb{Q}(e^{2\pi i/4}) = \mathbb{Q}(\sqrt{-1})$. We will discuss details of these L-series associated to singular loci in Section 7.

EXAMPLE 5.3. Let X be a Calabi–Yau orbifold corresponding to $m = 14$ and $Q = (1, 2, 2, 2, 7)$. There are four classes of primes modulo 14: (a) $p \equiv 1 \pmod{14}$ and $H = \{1\}$, (b) $p \equiv 3, 5 \pmod{14}$ and $H = \{1, 3, 9, 13, 11, 5\}$, (c) $p \equiv 9, 11 \pmod{14}$ and $H = \{1, 9, 11\}$ and (d) $p \equiv 13 \pmod{14}$ and $H = \{1, 13\}$.

(a) Let $p \equiv 1 \pmod{14}$, say, $p = 29$ and take the trivial twist $\mathbf{c} = \mathbf{1}$. Then for the weight motive $[Q]$, the polynomial $P_{[Q]}(T)$ is computed:

$$P_{[Q]}(T) = 1+5^3T+3^3\cdot 29\cdot 41T^2+3\cdot 29^2\cdot 31\cdot 73T^3+3^3\cdot 29^4\cdot 41T^4+5^3\cdot 29^6T^5+29^9T^6.$$

It is irreducible over $\mathbb{Q}$ and its Newton slopes are $0, 1, 2, 3$ with multiplicity $1, 2, 2, 1$, respectively. The Fermat motives corresponding to $A = [1, 2, 6, 12, 7]$, $[1, 4, 6, 10, 7]$, $[1, 6, 6, 8, 7]$ have isomorphic étale ℓ-adic realizations and the polynomials are computed as follows.

$$P_{\mathcal{M}_A}(T) = 1+2\cdot 3\cdot 29T-13\cdot 29^2T^2-2^2\cdot 29^3\cdot 79T^3-13\cdot 29^5T^4+2\cdot 3\cdot 29^7T^5+29^9T^6,$$

which is irreducible over $\mathbb{Q}$ and its Newton slopes are $1, 2$ each with multiplicity 3. The Fermat motive corresponding to $A = [7, 2, 4, 8, 7]$ has the polynomial

$$P_{\mathcal{M}_A}(T) = (1 + 2\cdot 29T + 29^3T^2)^3.$$

This implies that $\mathcal{M}_A$ decomposes into a triple product of a "rigid" Fermat motive whose characteristic polynomial is the irreducible polynomial $1 + 2\cdot 29T + 29^3T^2$ with Newton slopes $1, 2$.

For primes in the classes (b),(c),(d), the polynomials are also computed similarly, but it is too tedious to list them here. As above, the global L–series of each motive is obtained by taking the product over all primes, which are Hecke L-series with Jacobi sum Grössencharaktere of $\mathbb{Q}(e^{2\pi i/14})$.

There is one singularity of dimension 1, namely a smooth curve $Y_I : Y_2^7 + Y_3^7 + Y_4^7 = 0$ of degree 7 and the reduced weight $Q' = (1, 1, 1)$ with multiplicity 1. This is an irreducible curve of genus $g = 15$, which is denoted by C. So $B_1(C) = 30$. Then the L-series of C is again expressed in terms Hecke L-series with Jacobi sum Grössencharaktere of $\mathbb{L}' = \mathbb{Q}(e^{2\pi i/7})$.

We will discuss one more example.

EXAMPLE 5.4. Let $m = 60$ and $Q = (3, 5, 12, 20, 20)$. Let $p = 61 \equiv 1 \pmod{60}$. Then there are five classes of Fermat motives: $[Q] = [3, 5, 12, 20, 20]$ and $A = [3, 45, 12, 20, 40]$, , $[6, 10, 24, 40, 40]$, $[6, 30, 24, 20, 40]$ and $[12, 20, 48, 20, 20]$. We compute the polynomials $P_{[Q]}(T)$ and $P_{\mathcal{M}_A}(T)$: $P_{[Q]}(T)$ is irreducible over $\mathbb{Q}$ of degree 16 with slope sequence $0, 1, 2, 3$ of multiplicity $1, 7, 7, 1$, respectively. For $A = [6, 10, 24, 40, 40]$, $P_{\mathcal{M}_A}(T)$ is irreducible over Q of degree 16 with slope sequence $1, 2$ of multiplicity 8 each. These calculations imply that these two Fermat motives are indecomposable over $\mathbb{Q}$.

For $A = [3, 45, 12, 20, 40]$, we compute

$$P_{\mathcal{M}_A}(T) = (1 - 2^3 \cdot 61T + 2^7 \cdot 61^2T^2 - 2^3 \cdot 61^3 \cdot 167T^3 + 2 \cdot 5 \cdot 61^4 \cdot 883T^4 \\ -2^3 \cdot 61^6 \cdot 167T^5 + 2^7 \cdot 61^8T^6 - 2^3 \cdot 61^{10}T^7 + 61^{12}T^8)^2$$

where the degree 8 polynomial is irreducible over $\mathbb{Q}$. This implies that the Fermat motive $\mathcal{M}_A$ is built from an indecomposable motives corresponding to the degree 8 polynomial.

For $A = [6, 30, 24, 20, 40]$, $P_{\mathcal{M}_A}(T)$ is highly factorizable over $\mathbb{Q}$:

$$P_{\mathcal{M}_A}(T) = (1 - 2^4 \cdot 61T + 2 \cdot 61^2 \cdot 83T^2 - 2^4 \cdot 61^4T^3 + 61^6T^4)^4$$

and the degree 4 polynomial is irreducible over $\mathbb{Q}$. So the Fermat motive $\mathcal{M}_A$ is built from an indecomposable motive corresponding to the degree 4 polynomial.

Finally for $A = [12, 20, 48, 20, 20]$, we have

$$P_{\mathcal{M}_A}(T) = (1 - 61T + 61^3T^2)^8$$

where the degree 2 polynomial is irreducible over $\mathbb{Q}$, and the Fermat motive $\mathcal{M}_A$ is built from a "rigid" Fermat motive corresponding to the degree 2 irreducible polynomial.

There are two singularities of dimension 1. One is a smooth curve $Y_I : Y_2^{12} + Y_4^3 + Y_5^3 = 0$ of degree 12 and the reduced weight $Q' = (1, 4, 4)$, which we denote by C. C is an irreducible curve of genus $g = 1$, so $B_1(C) = 2$. It occurs with multiplicity 4. Another curve singularity is given by $Y_I : Y_3^5 + Y_4^3 + Y_5^3 = 0$ of degree 15 and the reduced weight $Q' = (3, 5, 5)$ and with multiplicity 3. The curve has genus 0, and gives no contribution to the L-series $L(\hat{\Sigma}(Y), s)$.

Now we can state the main result of this paper, that the L-series of our Calabi–Yau orbifolds are described in terms of the Hecke L-series with Jacobi sum Grössencharaktere of cyclotomic fields over $\mathbb{Q}$.

THEOREM 5.3. *Let X be a Calabi–Yau orbifold defined over $\mathbb{Q}$, which is a crepant resolution of the singular Calabi–Yau orbifold Y of the Fermat hypersurface of degree m and weight $Q = (q_1, q_2, q_3, q_4, q_5)$ in Theorem 2.3. Let $L_i(X, s) = L(H^i_{et}(\bar{X}, \mathbb{Q}_\ell), s)$ denote the i-th L-series of X for $i = 0, 1, 2, \cdots, 6$. Then the following assertions hold.*

(a) The conjecture of Shafarevich (stated in the Introduction) holds true for X. More precisely,

$$L_3(X, s) \doteq L^Q_{Hecke,3}(s) \times \prod_C L(C,\, s-1)^{m_C}$$

where C runs over all distinct smooth irreducible curves in $\Sigma(Y)$ with multiplicity m_C.
Here $L^Q_{Hecke,3}(s)$ is the Hecke L-series with Jacobi sum Grössencharaktere of $\mathbb{L} = \mathbb{Q}(\zeta_m)$ of dimension 3 compatible with the weight Q. $L(C, s)$ is the L-series $L(H^1(\bar{C}, \mathbb{Q}_\ell), s)$ of a smooth irreducible curve $C \in \Sigma(Y)$. Furthermore, $L(C, s)$ is expressed as a Hecke L–series with Jacobi sum Grössencharacters of the cyclotomic field $\mathbb{L}' = \mathbb{Q}(\zeta_{m'})$ of dimension 1 compatible with the reduced weight $Q' = (q'_i, q'_j, q'_k)$. That is,

$$L(C, s) = L^{Q'}_{Hecke,1}(s).$$

The symbol $\doteq$ means that the equality is valid up to the factors corresponding to bad primes.

(b) All cycles generating $H^{1,1}(X)$ arise from the hyperplane section and singular points. If they are all defined over $\mathbb{Q}$, then they are birationally isomorphic to a rational surface $\mathbb{P}^2$ over $\mathbb{Q}$, and

$$L_2(X,s)=\zeta_{\mathbb{Q}}(s-2)^{h^{1,1}(X)}.$$

(c) In totality, the L-series of X is given by

$$L(X,s)=\frac{L_0(X,s)L_2(X,s)L_4(X,s)L_6(X,s)}{L_1(X,s)L_3(X,s)L_5(X,s)}$$

where $L_3(X,s)$ and $L_2(X,s)$ are given above, and

$$L_1(X,s)=L_5(X,s)=1,\ L_0(X,s)=\zeta_{\mathbb{Q}}(s),\ L_6(X,s)=\zeta_{\mathbb{Q}}(3-s)$$

and

$$L_4(X,s)=L_2(X,2-s).$$

(d) $L(X,s)$ satisfies the functional equation under $s\mapsto 4-s$.

PROOF. By Proposition 3.2, and its proof,

$$H^3(X)=H^3(Y)\oplus H^3(\hat{\Sigma}(Y)).$$

Furthermore, we have that $H^3(Y)=H^3(V)^{\mu_Q}$. Hence

$$B_3(Y)=B_3(V)^{\mu_Q}=2(h^{3,0}(Y)+h^{2,1}(Y))$$

and

$$B_3(\hat{\Sigma}(Y))=\sum_C m_C\,B_1(C)$$

where C runs over all the smooth irreducible curves in $\Sigma(Y)$.

By Lemma 5.1, the $L_3(V,s)$ of a hypersurface threefold V of Fermat type is described in terms of the Hecke L–series with Jacobi sum Grössencharakters of $\mathbb{L}=\mathbb{Q}(\zeta_m)$. Then the L–series corresponding to the cohomology group $H^3(Y)=H^3(V)^{\mu_Q}$ is given by Hecke L–series with Jacobi sum Grössencharaktere of $\mathbb{L}=\mathbb{Q}(\zeta_m)$ of dimension 3 compatible with the weight Q. For a good prime p, let $\mathfrak{p}$ be a prime ideal in the ring of integers $\mathcal{O}_L$ of $\mathbb{L}=\mathbb{Q}(\zeta_m)$ lying above p. So $N\,\mathfrak{p}=q=p^f\equiv 1 \pmod{m}$. Then

$$L(H^3(Y),s)\doteq L^Q_{\text{Hecke},3}(s)=\prod_{p:\text{good}}\ \prod_{\mathbf{a}\in\mathfrak{A}(Q)}(1-j(\mathbf{a})\,p^{-fs})^{-1}.$$

The L–series corresponding to the singular locus is given by

$$L(H^3(\hat{\Sigma}(Y)),s)=\prod_C\zeta(C,s-1)^{m_C}$$

where the product runs over all the smooth irreducible **distinct** curves C in $\Sigma(Y)$ with multiplicity m_C. By Proposition 3.1, $\pi^{-1}(C)\simeq C\times\mathbb{P}^1$, and the L–series of $C\times\mathbb{P}^1$ is given by the L-series of C shifted by $s\to s-1$.

Now C is again a weighted curve of Fermat type of degree m' and a reduced weight Q'. The notions of *good* primes, the congruence zeta-function and the global L–series, Jacobi sum Grössencharaktere, etc. are defined for these weighted curves of Fermat type with m' (resp. Q') in place of m (resp. Q). Then $L(C,s)$ is a Hecke

L-series with Jacobi sum Grössencharaktere of $\mathbb{L}' = \mathbb{Q}(\zeta_{m'}) \subset \mathbb{L}$ of dimension 1 compatible with the reduced weight Q',

$$L(C, s) \doteq L^{Q'}_{\text{Hecke},1}(s).$$

Finally, $H^{1,1}(X)$ is generated by a hyperplane section and rational surfaces arising from singular points. If they are all defined over $\mathbb{Q}$, then

$$L_2(s) = \zeta_{\mathbb{K}}(s-2)^{h^{1,1}(X)}.$$

□

6. Calabi–Yau orbifolds of CM type

In this section, we will introduce the notion of *CM type* varieties after Shafarevich [S-PS73]. We follow the exposition of Borcea [Bo92]. Let X be a Calabi–Yau threefold. We put $H_{\mathbb{Z}} := H^3(X, \mathbb{Z})$. Then there is the Hodge decomposition:

$$H_{\mathbb{C}} := H^3(X, \mathbb{C}) = H^{3,0} \oplus H^{2,1} \oplus H^{1,2} \oplus H^{0,3} \quad \text{with } H^{j,i} = \overline{H^{i,j}}.$$

There is a natural polarization

$$H^3(X, \mathbb{Q}) \times H^3(X, \mathbb{Q}) \to \mathbb{Q} \qquad \text{given by} \quad \langle x, y \rangle := -\int_X x \wedge y.$$

This gives rise to a rational polarized Hodge structure h of weight 3 and level 3 on $H^3(X, \mathbb{Q})$. There is the Hodge filtration

$$H^3(X, \mathbb{C}) = F^0 \supset F^1 \supset F^2 \supset F^3 \supset \{0\}$$

where

$$F^1 = H^{3,0} \oplus H^{2,1} \oplus H^{1,2}, \quad F^2 = H^{3,0} \oplus H^{2,1}, \quad F^3 = H^{3,0}.$$

The (Griffiths) intermediate Jacobian $J^2(X)$ of X is defined by

$$J^2(X) := \frac{H^3(X, \mathbb{C})}{F^2 H^3(X, \mathbb{C}) + H_3(X, \mathbb{Z})} = \frac{H^{1,2}(X) \oplus H^{0.3}(X)}{H_3(X, \mathbb{Z})}.$$

It is a complex torus of dimension $1 + h^{1,2}$ (which becomes an abelian variety if it is equipped with a polarization).

The notion of the Hodge group of a rational Hodge structure was first introduced by Mumford [Mu66] for abelian varieties, and it was proved in [Mu69] that an abelian variety is of CM type if and only if the Hodge group is commutative, i.e., an algebraic torus group (some power of $\mathbf{G}_m$. Here we recall the definitions of the Hodge group and varieties of CM type.

DEFINITION 6.1. (a) For a rational polarized Hodge structure h of weight 3, the *Hodge group* of h, $\text{Hdg} := \text{Hdg}(h)$, is the smallest algebraic subgroup of $GL(H^3(X, \mathbb{Q}))$, defined over $\mathbb{Q}$, such that the real points $\text{Hdg}(\mathbb{R})$ contain the image $h(U^1)$ where $U^1 = \{ z \in \mathbb{C}^* \mid z\bar{z} = 1\}$.

(b) A Calabi–Yau threefold X is said to be *of CM type* if the Hodge group Hdg is commutative (i.e., a torus algebraic group, copies of $\mathbf{G}_m = \mathbb{C}^*$).

(c) If X is a Calabi–Yau threefold of CM type, then the Hodge group Hdg is contained in the isotropy subgroup of h:

$$\{\gamma \in GL(H^3(X, \mathbb{Q}) \mid \langle \gamma x, \gamma y \rangle = \langle x, y \rangle \}.$$

We define the endomorphism algebra

$$\mathcal{E} := \mathrm{End}_{\mathrm{Hdg}}(H^3(X,\mathbb{Q})) := \{\gamma \in GL(H^3(X,\mathbb{Q})) \mid \gamma \text{ commutes with Hdg}\}.$$

PROPOSITION 6.1 (Borcea). *Let X be a Calabi–Yau threefold. Suppose that there is a simple rational Hodge structure h on $H^3(X,\mathbb{Q})$. Then the following conditions are equivalent.*

(i) X is of CM type.

(ii) The Hodge group Hdg is commutative, in fact, $\mathbf{G}_m^r$.

(iii) The intermediate Jacobian $J^2(X)$ is of CM type (as an abelian variety provided that $J^2(X)$ is projective).

(iv) $\mathcal{E}$ is a CM field with $[\mathcal{E}:\mathbb{Q}] = 2(1+h^{2,1})$, that is, $\mathcal{E}$ is a totally imaginary quadratic extension of a totally real algebraic number field of degree $1+h^{2,1}$.

PROOF. See Borcea [Bo92]. □

REMARK 6.1. In general, a rational Hodge structure associated to Calabi–Yau threefolds is not necessarily simple. Note that rigid Calabi–Yau threefolds have simple rational Hodge structures. However, all the Calabi–Yau threefolds constructed in Theorem 2.3 are non-rigid. Thus the above theorem of Borcea cannot be applied directly to our Calabi–Yau orbifolds.

By Deligne [De71], the category of rational Hodge structure is a semi-simple category. Therefore Borcea's theorem will be valid if we restrict ourselves to simple rational sub-Hodge structures. For this, we will now decompose the Hodge structure h over $\mathbb{Q}$ as follows:

Let $T(h)$ denote the smallest rational Hodge sub-structure of h that contains the line $H^{3,0}$. Clearly $T(h)$ is polarized by the restriction of the polarization on h. The orthogonal complement $T(h)^\perp$ of $T(h)$ is a rational Hodge sub-structure of level 1, and we have the splitting

$$h = T(h) \oplus T(h)^\perp \quad \text{over } \mathbb{Q}.$$

In order to discuss $T(h)$ and $T(h)^\perp$ in our case, we need to invoke the generalized Hodge conjectures.

The classical and generalized Hodge conjectures. We follow the exposition of Lewis [Le04] on this subject. Let X be a smooth projective variety of dimension n over $\mathbb{C}$. For d, $0 < d < n$, let $C^d(V,\mathbb{Z})$ denote the group of algebraic cycles of codimension d, with integer coefficients. Then $C^d(V,\mathbb{Z}) \subset H^{d,d}(V) \cap H^{2d}(V,\mathbb{Z})$. The original question of Hodge was whether this inclusion is an equality. However, there is a counter-example to this claim by Atiyah and Hirzebruch, and Hodge's problem needed to be modified to $\mathbb{Q}$-coefficients. The following is what is known as the classical Hodge conjecture:

The classical Hodge conjecture Hodge (V,d): *Let*

$$C^d(V,\mathbb{Q}) := C^d(V,\mathbb{Z}) \otimes \mathbb{Q} \subset H^{2d}(V,\mathbb{Q}).$$

Then the classical Hodge conjecture Hodge(V,d) asserts:

$$C^d(V,\mathbb{Q}) = H^{2d}(V,\mathbb{Q}) \cap H^{d,d}(V).$$

Incidentally, the classical Hodge conjecture is true for any threefold X as (i) $H^{1,1}(X,\mathbb{Q})$ is algebraic by the Lefschetz $(1,1)$ theorem, and (ii) $H^{2,2}(X,\mathbb{Q})$ is algebraic by (i) with the hard Lefschetz theorem.

A generalized Hodge conjecture Hodge (V, F^pH^i): *Let*

$$F'^pH^i(V,\mathbb{Q}) = \cup_{W\in Z^p(V)} Ker\{H^i(V,\mathbb{Q}) \to H^i(V-W,\mathbb{Q})\}$$

where the union ranges over all codimension $\geq p$ *subvarieties* $W \subset V$. *Then the generalized Hodge conjecture Hodge*(V, F^pH^i) *asserts*

$$F'^pH^i(V,\mathbb{Q}) = \textit{the largest sub-Hodge structure of} \quad H^i(V,\mathbb{Q}) \cap F^pH^i(V,\mathbb{C})$$

where F^p *is the usual Hodge filtration.*

REMARK 6.2. Grothendieck [Gr69] pointed out that $H^i(V,\mathbb{Q}) \cap F^pH^i(V,\mathbb{C})$ is not necessarily a Hodge structure, giving a counterexample of the triple product of an elliptic curve without complex multiplication. The generalized Hodge conjecture had to be amended by adding *the largest sub-Hodge structure* to $H^i(V,\mathbb{Q}) \cap F^pH^i(V,\mathbb{C})$.

However, if V is a variety of CM type or a Fermat variety, $H^i(V,\mathbb{Q}) \cap F^pH^i(V,\mathbb{C})$ *is* a Hodge structure, and hence the generalized Hodge conjecture takes the simpler form:

$$F'^pH^i(V,\mathbb{Q}) = H^i(V,\mathbb{Q}) \cap F^pH^i(V,\mathbb{C}).$$

The generalized Hodge conjecture has been established for Fermat varieties of various degrees and dimensions by Shioda [Sh82], [Sh83] and Ran [Ra80]. In particular, for Fermat curves, the conjecture is true, and for Fermat hypersurface threefolds V of degree m, the generalized Hodge conjecture Hodge(V, F^1H^3) is true for degree $m \leq 10$. This was established using the method of geometric structures called *inductive structure* by Shioda and Katsura [SK79]), and reducing the problem to questions on combinatorics. However, their method fell short of proving the generalized Hodge conjecture when $m = 11$, for instance.

We cannot apply Shioda's criterion for the validity of the generalized Hodge conjecture in the original form as described in [Sh82], [Sh83]. Here we reformulate Shioda's criterion for our singular Calabi–Yau orbifolds Y with degree m and weight $Q = (q_1, q_2, \cdots, q_5)$.

DEFINITION 6.2. Let Y be a singular Calabi–Yau orbifold in Theorem 2.3 corresponding to a pair $< m, Q >$. Let

$$\mathfrak{A}(Q) = \{\mathbf{a} = (a_1,a_2,a_3,a_4,a_5) \,|\, a_i \in q_i\mathbb{Z}/m\mathbb{Z}, \sum_{i=1}^{5} a_i \equiv 0 \pmod{m}\}.$$

An element $\mathbf{a} \in \mathfrak{A}(Q)$ is said to be *admissible* if it is either *decomposable* or *quasi-decomposable*. Here $\mathbf{a}$ is called decomposable if $a_i + a_j \equiv 0 \pmod{m}$ for some $i \neq j$, and quasi-decomposable if $a_4 + a_5 \not\equiv 0 \pmod{m}$ and $(a_1, a_2, a_3, \frac{a_4+a_5}{m})$ has length $||ta_1 + ta_2 + ta_3 + t < a_4 + a_5 > || = 1$ for any $t \in (\mathbb{Z}/m\mathbb{Z})^\times$. This should hold up to permutations of the a_i. Here $< a >$ stands for the least positive residue of a modulo m.

It is immediate from the definition that the motive $\mathcal{M}_{[Q]}$ corresponding to the weight Q is not admissible.

TABLE 6. The generalized Hodge conjecture is true

#	m	Q
#1	5	$(1, 1, 1, 1, 1)$
#2	6	$(1, 1, 1, 1, 2)$
#3	8	$(1, 1, 1, 1, 4)$
#4	8	$(1, 1, 2, 2, 2)$
#5	9	$(1, 1, 1, 3, 3)$
#6	10	$(1, 1, 1, 2, 5)$
#12	12	$(1, 2, 3, 3, 3)$
#13	12	$(2, 2, 3, 3, 3)$
#53	36	$(2, 4, 9, 9, 12)$
#56	40	$(2, 5, 5, 8, 20)$
#61	42	$(2, 6, 6, 7, 21)$
#79	60	$(3, 10, 12, 15, 20)$
#84	60	$(4, 6, 15, 15, 20)$
#86	70	$(1, 10, 10, 14, 35)$
#91	78	$(1, 6, 6, 26, 39)$
#96	84	$(2, 12, 21, 21, 28)$
#105	100	$(1, 4, 20, 25, 50)$
#132	294	$(1, 6, 42, 98, 147)$
#140	462	$(2, 33, 42, 154, 231)$
#141	546	$(6, 7, 78, 182, 273)$
#147	1806	$(1, 42, 258, 602, 903)$

Criterion (cf. Shioda [Sh82]). If every element $\mathbf{a} \in \mathfrak{A}(Q)$, $\mathbf{a} \neq Q$ is admissible, then the generalized Hodge conjecture holds for X.

Now we test the above criterion for our Calabi–Yau orbifolds.

EXAMPLE 6.1. The first two examples follows from the validity of the generalized Hodge conjecture for Fermat hypersurface threefolds proved by Shioda ([Sh82],[Sh83]).

(a) Let $m = 5$ and $Q = (1, 1, 1, 1, 1)$. In Table 1, both $\mathbf{a} = (1, 1, 1, 3, 4)$ and $\mathbf{a} = (1, 1, 2, 2, 4)$ are decomposable.

(b) Let $m = 8$ and $Q = (1, 1, 2, 2, 2)$. Then in Table 2, all $\mathbf{a}$ from the third column on are decomposable, while $\mathbf{a} = (7, 3, 2, 2, 2)$ is quasi-decomposable.

(c) Let $m = 14$ and $Q = (1, 2, 2, 2, 7)$. Then we see that $\mathbf{a} = (13, 2, 2, 4, 7)$ $(9, 2, 4, 6, 7)$ in the Table 4 are neither decomposable nor quasi-decomposable, so not admissible.

(d) Let $m = 60$ and $Q = (3, 5, 12, 20, 20)$. Then in Table 5, $\mathbf{a} = (6, 10, 24, 40, 40)$ is not admissible.

As the above examples (c) and (d) illustrate, the criterion is not always satisfied. In fact, we have the following result.

THEOREM 6.2. *Among the 147 Calabi–Yau orbifolds in Theorem 2.3, the generalized Hodge conjecture Hodge*(X, F^1H^3) *can be verified in the affirmative by the above method for the* 21 *pairs* $< m, Q >$ *listed in Table 6.*

PROOF. For any weighted Fermat curves appearing in the singular locus $\Sigma(Y)$, the generalized Hodge conjecture is true. For singular Calabi–Yau orbifolds Y corresponding to pairs $< m, Q >$, these are the only ones that satisfy the above criterion. The first six pairs are can be deduced from the results of Shioda [Sh82],[Sh83] of Fermat hypersurface threefolds. The remaining fifteen are new examples. □

For our Calabi–Yau orbifolds X, we can describe $T(h)$ and $T(h)^{\perp}$ explicitly.

THEOREM 6.3. *Let X be a Calabi–Yau orbifold over $\mathbb{Q}$ constructed in Theorem 2.3 corresponding to a pair $< m, Q >$. Let h be a rational polarized Hodge structure on X. Then the following assertions hold.*

(a) $T(h)$ is determined solely by the weight motive $\mathcal{M}_Q$, that is, it is "transcendental" of dimension $\varphi(m)$, and it is of CM type. More precisely, the endomorphism algebra of the rational Hodge structure is isomorphic to the m-th cyclotomic field $\mathbb{Q}(\zeta_m)$, which is a CM field of degree $\varphi(m)$ over $\mathbb{Q}$.

(b) The orthogonal complement $T(h)^{\perp}$ corresponds to $\oplus_{A\neq[Q]}\mathcal{M}_A$ where

$$\mathcal{M}_A = \{\mathbf{a} = (a_1, a_2, a_3, a_4, a_5) \in \mathfrak{A}(Q) \,|\, ||t\mathbf{a}|| = 1 \text{ or } 2 \quad \text{for all } t \in (\mathbb{Z}/m\mathbb{Z})^{\times}\}.$$

$T(h)^{\perp}$ decomposes into a product of factors corresponding to Fermat motives $\mathcal{M}_A$ with $A \in \mathfrak{A}(Q)$ not equal to $\mathcal{M}_Q$, and singularities of dimension 1. Furthermore, each of these factors is of CM type, i.e., the endomorphism algebra of rational Hodge structure is isomorphic to a subfield of the cyclotomic field $\mathbb{Q}(\zeta_m)$.

PROOF. First we need the following fact due to Borcea [Bo92]. The direct product of rational polarized Hodge structures is of CM type if and only if each rational polarized Hodge structure is of CM type. In particular, h is of CM type if and only if $T(h)$ and $T(h)^{\perp}$ are of CM type. Thus we ought to show that both $T(h)$ and $T(h)^{\perp}$ are of CM type. Now $T(h)$ corresponds to the weight motive $\mathcal{M}_Q$, while $T(h)^{\perp}$ corresponds to other Fermat motives, and singularities of dimension 1. As the étale ℓ-adic realizations of these motives indicate, all these motives are of CM type, that is, their endomorphism algebras are all CM fields (in fact, cyclotomic fields) over $\mathbb{Q}$. Therefore, the corresponding rational polarized Hodge structures are all of CM type. □

REMARK 6.3. If we assume the validity of the generalized Hodge conjecture for X, $T(h)^{\perp}$ should be the "algebraic" part of h, and should correspond to the image of the Abel–Jacobi map of the (Griffiths) intermediate Jacobian $J^2(X)$ on the Chow group of $CH^2(X)_{alg}$ consisting of codimension 2 algebraic cycles on X, algebraically equivalent to zero. See Borcea [Bo92] for more detail. Further, Shioda [Sh83] asserts that the $J^2(X)$ decomposes into abelian varieties of CM type by cyclotomic fields.

As a consequence of this theorem, we obtain the following result.

COROLLARY 6.4. *The Calabi–Yau orbifolds constructed in Theorem 2.3 are all of CM type.*

This result can be established without invoking Hodge structures, Hodge groups and Borcea's result. This can also be established by studying geometric structures of our Calabi–Yau orbifolds and making recourse to the classical notion of CM type varieties due to Shimura and Taniyama [ST61]. A simple abelian variety A is of CM type if the endomorphism algebra $\mathrm{End}(A)\otimes\mathbb{Q}$ is a CM field over $\mathbb{Q}$. The proof along this line was outlined in Yui [Yui03], and is done in several steps. In particular, this proof is independent of the Hodge and generalized Hodge conjectures.

An alternative proof of Corollary 6.4:

Step 1: *The Fermat hypersurface V is of CM type.*

This follows from the fact that the Fermat hypersurface V is dominated by a product of Fermat curves by a result of Schoen [Sch96] (cf. Remark 2.3), or by the result of Hunt and Schimmrigk [HS99]. Moreover, it is a well-known fact that the Jacobian varieties of the Fermat curves are isogenous to a product of abelian varieties of CM type (cf. Rohrlich [Roh78], Aoki [Ao91]). Therefore, V itself is of CM type.

Step 2: *The quotient $Y = V/\mu_Q$ is of CM type.*

This follows from the fact that the quotient map commutes with the dominant map.

Step 3: *The Calabi–Yau orbifold X is of CM type.*

This follows since the singular loci consist of weighted Fermat curves, and they are all of CM type as they are dominated by Fermat curves by Step 1. By Step 2, Y is of CM type. Therefore X is of CM type.

EXAMPLE 6.2. This example is well known and discussed in Shioda ([Sh83], [Sh82]); we reinterpret it in terms of Fermat motives and their endomorphism algebras. Let X be the Fermat quintic threefold: $\sum_{i=1}^5 X_i^5 = 0 \subset \mathbb{P}^4$ with weight $Q = (1, 1, 1, 1, 1)$. Then $T(h)$ corresponds to the weight motive $\mathcal{M}_{[Q]}$ of dimension 4 with the endomorphism algebra of rational Hodge structure isomorphic to the 5-th cyclotomic field $\mathbb{Q}(e^{2\pi i/5})$ over $\mathbb{Q}$, which is a CM field. The orthogonal complement $T(h)^\perp$ corresponds of the (isomorphic) Fermat motives $\mathcal{M}_{[1,1,1,3,4]}$ and $\mathcal{M}_{[1,1,2,2,4]}$ of dimension 4, whose endomorphism algebras are again $\mathbb{Q}(e^{2\pi i/5})$.

EXAMPLE 6.3. Consider the Calabi–Yau orbifold X corresponding to the pair $< m, Q >$ where $m = 8$ and $Q = (1, 1, 2, 2, 2)$. Then $T(h)$ corresponds to the weight motive $\mathcal{M}_{[1,1,2,2,2]}$ of dimension 4 and its endomorphism algebra is $\mathbb{Q}(e^{2\pi i/8})$, which is a CM field over $\mathbb{Q}$ of degree 4. Then $T(h)^\perp$ corresponds to two kinds of objects, one being the remaining Fermat motives (see Table 2) all of whose endomorphism algebras are again CM subfields of $\mathbb{Q}(e^{2\pi i/8})$, the other arising from the singularity of dimension 1 whose endomorphism algebra is $\mathbb{Q}(e^{2\pi i/4}) = \mathbb{Q}(\sqrt{-1})$ (cf. Example 5.2).

EXAMPLE 6.4. Consider the Calabi–Yau orbifold X of degree $m = 60$ and weight $Q = (3, 5, 12, 20, 20)$ (cf. Table 5 for the motives). Then $T(h)$ corresponds to the weight motive $\mathcal{M}_{[3,5,12,20,20]}$ whose endomorphism algebra is $\mathbb{Q}(e^{2\pi i/60})$ which is a CM field over $\mathbb{Q}$ of degree 16. Then $T(h)^\perp$ corresponds to the following Fermat motives: $\mathcal{M}_{[6,10,24,40,40]}$ with endomorphism algebra again $\mathbb{Q}(e^{2\pi i/60})$; $\mathcal{M}_{[3,45,12,20,40]}$ with endomorphism algebra a CM subfield of $\mathbb{Q}(e^{2\pi i/60})$ of degree 8 over $\mathbb{Q}$; $\mathcal{M}_{[6,30,24,20,40]}$ with endomorphism algebra a CM subfield of $\mathbb{Q}(e^{2\pi i/60})$ of degree 4 over $\mathbb{Q}$ and $\mathcal{M}_{12,20,48,20,20]}$ with endomorphism algebra a CM subfield of $\mathbb{Q}(e^{2\pi i/60})$ of degree 2 over $\mathbb{Q}$. Singularities of dimension 1 gives rise to a CM factor.

EXAMPLE 6.5. Consider the Calabi–Yau orbifold X of degree 1806 and weight $Q = (1, 42, 258, 602, 903)$. Then $T(h)$ has dimension 504 and $T(h)^{\perp} = \emptyset$. Therefore, the motive $\mathcal{M}_{[Q]}$ is of CM type with $\mathbb{Q}(e^{2\pi i/1806})$ as its endomorphism algebra.

7. Modularity questions

We have not yet discussed the modularity of our Calabi–Yau orbifolds, and in this section we will address exactly that.

We are in the similar situation as singular (extremal) $K3$ surfaces.

Shioda and Inose [SI87] proved that any singular K3 surface is of CM type and that it is defined over some number field, K, and furthermore, its L-series is expressed in terms of Hecke L-series and the Dedekind zeta-function of the field K.

The modularity of the transcendental part of a singular $K3$ surface defined over $\mathbb{Q}$ has been established by Livné [Li95] (cf. Yui [Yui03] for details).

For our Calabi–Yau orbifolds, they are all defined over $\mathbb{Q}$ and we have shown that they are all of CM type, and that L-series are expressed in terms of L-series with Jacobi sum Grössencharaktere of cyclotomic fields over $\mathbb{Q}$.

The modularity is an obvious next question. We will divide our discussions into two parts:

- Modularity of singular loci: Since curves appearing in singular loci are all smooth curves of Fermat type in weighted projective 2-spaces, we will discuss modular forms associated to Fermat curves.
- Modularity of singular Calabi–Yau orbifolds: Since they decompose into Fermat motives compatible with weights, we will investigate the modularity of each Fermat motive.

Modularity of singular loci: Genus 1 curves. Let Y be a singular Calabi–Yau orbifold defined over $\mathbb{Q}$ and let C be an irreducible curve contained in the singular locus $\Sigma(Y)$. Then we have seen that C is a smooth curve of Fermat type in some weighted projective 2-space with genus ≥ 0. We consider the Jacobian variety $J(C)$ of C. First we recall some classical results on the Jacobian varieties of Fermat curves in the usual projective 2-spaces. Fermat curves, in particular, simple factors in their Jacobians, have appeared in rational conformal field theory (RCFT) in connection with $SU(3)$ modular invariance. (See [BCIR97].)

Let C be a Fermat curve in $\mathbb{P}^2$. It is classically known that $J(C)$ is highly reducible and, up to isogeny, it splits into a product of simple abelian varieties over $\mathbb{Q}$, and each simple factor is further isogenous over $\bar{\mathbb{Q}}$ to a product of elliptic curves (Rohrlich [Roh78], Koblitz [Ko78], and Aoki [Ao91]).

LEMMA 7.1. *Let C be a Fermat curve of degree m in $\mathbb{P}^2$, and let $J(C)$ be its Jacobian variety. Then $J(C)$ contains a simple abelian variety which splits over $\bar{\mathbb{Q}}$ into a product of elliptic curves when m takes one of the following values:*

$$3, 4, 6, 7, 8, 12, 15, 16, 18, 20, 21, 22, 24, 30, 39, 40, 48, 60.$$

In particular, $J(C)$ itself splits over $\bar{\mathbb{Q}}$ into a product of elliptic curves if and only if m is at most 12 and divides 24, i.e., $m \in \{3, 4, 6, 8, 12\}$.

In Lemma 7.1, the decomposition occurs over $\bar{\mathbb{Q}}$. Koblitz [Ko78] addresses the question of determining the smallest field extension of $\mathbb{Q}$ for which this decomposition actually occurs. In particular, if each one-dimensional factor is defined over $\mathbb{Q}$, then its modularity follows from Wiles' theorem, which we formulate as a proposition.

PROPOSITION 7.2. *Every elliptic curve over $\mathbb{Q}$ in a weighted projective 2-space is modular, and there is a cusp form of weight 2 on some congruence subgroup of $PSL_2(\mathbb{Z})$.*

PROOF. The proof of Wiles et al. is valid for any elliptic curve defined over $\mathbb{Q}$ in any ordinary projective space. Since weighted projective spaces can be embedded into some ordinary projective space, the modularity of elliptic curves in any weighed projective space follows. □

In particular, this will apply to the weighted Fermat curves of genus 1 arising in the singular loci of our Calabi–Yau orbifolds.

COROLLARY 7.3. *Any weighted Fermat curve of genus 1 occurring in the singular locus $\Sigma(Y)$ of a Calabi–Yau orbifold Y is modular. That is, its L-series is determined by a cusp form of weight 2 on $\Gamma_0(N)$ for some N.*

Now we consider weighted Fermat curves of genus $g > 1$ appearing in singular loci whose Jacobian varieties split into a product of genus 1 curves.

COROLLARY 7.4. *If a weighted Fermat curve of genus $g > 1$ is such that its Jacobian variety splits into a product of genus 1 weighted Fermat curves over $\mathbb{Q}$, then it is modular, and there is a nest of cusp forms of weight 2 that determines its L-series.*

In particular, if a weighted Fermat curve has reduced degree $m' \leq 12$ dividing 24, then it is modular.

Now we will determine cusp forms associated to weighted Fermat curves and Fermat curves illustrating the above corollaries.

EXAMPLE 7.1. Let $E_3 : x^3+y^3+z^3 = 0$ be a Fermat curve defined over $\mathbb{Q}$. Then E_3 is an elliptic curve with CM by $\mathbb{Q}(e^{2\pi i/3})$. It is modular, that is, there is a cusp form $\eta(q^3)^2\eta(q^9)^2$ of level 27 on $\Gamma_0(27)$ such that $L(E_3, s) = L(\eta(q^3)^2\eta(q^9)^2, s)$. In fact, we have

$$\begin{aligned}\eta(q^3)^2\eta(q^9)^2 &= q - 2q^4 - q^7 + 5q^{13} + 4q^{16} - 7q^{19} - 5q^{25} + 2q^{28} \\ &\quad -4q^{31} + 11q^{37} + 8q^{43} - 6q^{49} - 10q^{52} + \cdots .\end{aligned}$$

(Schimmrigk and Underwood [SU03] have given a Kac–Moody algebra interpretation of this L-series, which is the first of such examples of modular forms arising from string theory.)

EXAMPLE 7.2. Consider the weighted Fermat curve $E_4 : x^4 + y^4 + z^2 = 0$ in $\mathbb{P}^2(Q)$ with $Q = (1,1,2)$. Then E_4 is an elliptic curve over $\mathbb{Q}$ with CM by $\mathbb{Q}(\sqrt{-1})$, and E_4 is modular. Note that 2 is obviously a bad prime. We also have $\mathfrak{A}(Q) = \{(1,1,2),(3,3,2)\}$ and $L(E_4, s) = L(f, s)$ where

$$f(q) = q+2q^5-3q^9-6q^{13}+2q^{17}-q^{25}+10q^{29}+2q^{37}+10q^{41}-6q^{45}-7q^{49}-14q^{53}+\cdots$$

is a cusp form of weight 2 on $\Gamma_0(64)$. The curve E_4 appears in the singular loci of several Calabi–Yau orbifolds, for instance, $\langle m, Q\rangle = \langle 12, (1,2,2,3,4)\rangle\langle 24, (3,3,4,6,8)\rangle$.

Now the Fermat curve $F_4 : x^4 + y^4 + z^4 = 0$ (in the usual projective 2-space) is a double cover of E_4 (indeed, there is a map from $F_4 \to E_4 : (x, y, z) \mapsto (x, y, -z)$).

The Fermat curve F_4 is of genus 3 and $\mathfrak{A} = \{ (1,1,2)(3), (3,3,2)(3) \}$ with $\#\mathfrak{A} = 6 = B_1(F_4)$. As $\mathfrak{A}$ has three copies of $\mathfrak{A}(Q)$, it follows that the Jacobian variety $J(F_4)$ is isogenous to the triple product of the elliptic curve E_4 in $\mathbb{P}^2(1,1,2)$. Therefore, $L(F_4, s) = L(J(F_4), s) = L(E_4, s)^3$, and since E_4 is modular, we may conclude that F_4 is modular. Note that F_4 appears in the singular locus of the Calabi–Yau orbifold of degree $m = 8$ and weight $Q = (1,1,2,2,2)$.

EXAMPLE 7.3. Consider the weighted Fermat curve $E_6 : x^6 + y^3 + z^2 = 0 \in \mathbb{P}^2(Q)$ with $Q = (1,2,3)$. Then E_6 is a genus 1 curve, and hence defines an elliptic curve over $\mathbb{Q}$. Note that 2 and 3 are obviously bad primes. We also have $\mathfrak{A}(Q) = \{ (1,2,3), (5,5,2) \}$ and $L(E_6, s) = L(f, s)$, where

$$f(q) = q + 4q^7 + 2q^{13} - 8q^{19} - 5q^{25} + 4q^{31} - 10q^{37} - 8q^{43} + 9q^{49} + 14q^{61} + 16q^{67} + \cdots$$

is a cusp form of weight 2 on $\Gamma_0(144)$. (Note that $f(q)$ is a twist of the cusp form of weight 2 on $\Gamma_0(36)$ by the unique Dirichlet character of conductor 4. This character was determined by William Stein using MAGMA. Independently, Schimmrigk computed $f(q)$.) We also see that E_6 has CM by $\mathbb{Q}(e^{2\pi i/3})$.

Now the Fermat curve $F_6 : x^6 + y^6 + z^6 = 0$ (in the usual projective 2-space) has genus 10 and

$$\mathfrak{A} = \{ (1,2,3)(6), (5,5,3)(6), (2,2,2), (4,4,4), (1,1,4)(3), (5,5,2)(3) \}$$

with $\#\mathfrak{A} = 20 = B_1(F_6)$. Therefore, the Jacobian variety $J(F_6)$ of F_6 is isogenous to the product of 6 copies of E_6, an elliptic curve corresponding to $(2,2,2)$ and 3 copies of an elliptic curve corresponding to $(1,1,4)$, and

$$L(J(F_6), s) = L(E_6, s)^6 L([2,2,2], s) L([1,1,4], s)^3.$$

EXAMPLE 7.4. Consider the Fermat curve $F_{12} : x^{12} + y^{12} + z^{12} = 0$. Then its Jacobian variety $J(F_{12})$ of dimension 55 splits, up to isogeny, into the product of elliptic curves. In fact, there are two isogeny classes of elliptic curves, one with CM by $\mathbb{Q}(e^{2\pi i/3})$ and the other by $\mathbb{Q}(\sqrt{-1})$. Therefore, F_{12} is modular. If a weighted Fermat curve has degree $m' = 12$, then it is modular.

We consider the Fermat curve $F_{24} : x^{24} + y^{24} + z^{24} = 0$; then its Jacobian variety $J(F_{24})$ has dimension 253, and it factors into 24 copies of a simple abelian variety of dimension 4 and 157 elliptic curves. If a weighted Fermat curve has reduced degree $m' = 24$, then its Jacobian has some of the factors of $J(F_{24})$. We have not checked which factors will survive the orbifolding process.

EXAMPLE 7.5. If a weighted Fermat curve in a singular locus $\Sigma(Y)$ has genus $g > 1$ and reduced degree m' not dividing 24, the Jacobian variety may have an elliptic curve factor if the reduced degree m' takes one of the values in Lemma 7.1, namely, $7, 15, 16, 18, 20, 21, 22, 30, 39, 40, 48$ and 60.

For instance, if $m' = 7$, then there are two non-isomorphic motives: $\mathcal{M}_{[1,2,4]}$ of dimension 6 and multiplicity 2, and $\mathcal{M}_{[1,3,3]}$ of dimension 6 and multiplicity 3. The motive $\mathcal{M}_{[1,2,4]}$ splits into a product of 3 copies of an elliptic curve over $\mathbb{Q}$, where the elliptic curve factor is indeed given by $y^2 = x^3 - \frac{5}{7}x + \frac{2}{7}$ with CM by $\mathbb{Z}[\frac{1+\sqrt{-7}}{2}]$ (see Koblitz [Ko78]).

For the other cases, it remains to check which motives would give rise to elliptic curve factors over $\mathbb{Q}$.

Modularity of singular loci: Curves of genus > 1. Now consider Fermat curves of genus > 1 appearing in singular loci which do not have split Jacobians into 1-dimensional pieces. For these cases, we cannot invoke Wiles' result. Yet we can associate modular forms to them. The idea is to use the cartography of a Riemann surface which is uniformized by $\mathfrak{H}/B$ for some Fuchsian group B (indeed a subgroup of $PSL_2(\mathbb{Z})$), to uniformize the Fermat curves. We first follow the exposition of Bauer, Coste, Itzykson and Ruelle [BCIR97], and then we will generalize their arguments to weighted Fermat curves.

Let Σ denote a compact Riemann surface equipped with a holomorphic map h to the Riemann sphere ramified only over the three points $0, 1, \infty$. The Riemann sphere has the standard triangulation. The inverse image of this triangulation by h gives rise to a triangulation on Σ. By a flag of a triangulation, we mean an ordered triple (v, e, f) where v is a vertex, e an edge containing (starting from) v and f a face containing e. The cartographic group of a triangulation, $\hat{\mathfrak{C}}$, is generated by three specific permutations on flags. (The cartographic group of the standard triangulation of the Riemann sphere is the symmetric group S_3.) It is a fact that the cartographic group is always realized as a quotient group of $PSL_2(\mathbb{Z})$, and the cartographic group S_3 of the standard triangulation of the Riemann sphere $\mathbb{P}^1$ is identified with the quotient group $PSL_2(\mathbb{Z})/\Gamma(2)$, where $\Gamma(2)$ is a genus zero principal congruence subgroup of $PSL_2(\mathbb{Z})$ of index 6. Let $\mathfrak{C}$ be the kernel of the homomorphism from $PSL_2(\mathbb{Z})$ to $\hat{\mathfrak{C}}$. Then $\mathfrak{C}$ is a subgroup of $\Gamma(2)$.

Each pair (Σ, h) gives rise to a subgroup $\mathfrak{C} \subset \Gamma(2)$, and an invariant subgroup G (the intersection of all conjugates of $\mathfrak{C}$ in $PSL_2(\mathbb{Z})$ so that G is a subgroup of $\mathfrak{C}$) such that the flags are parametrized by the set of left cosets $\mathfrak{C} \setminus PSL_2(\mathbb{Z})$ and the cartographic group is isomorphic to $PSL_2(\mathbb{Z})/G$ acting on $\mathfrak{C} \setminus PSL_2(\mathbb{Z})$ on the right.

For example, the following groups are subgroups of $\Gamma(2)$ of genus zero: $\Gamma(2)$, $\Gamma(4)$, $\Gamma_0(4) \cap \Gamma(2)$, $\Gamma_1(8:4,1,2) = \{\begin{pmatrix} 1+4a & 2b \\ 4c & 1+4d \end{pmatrix} \in \Gamma(2) \,|\, a \equiv c \pmod 2\}$. In fact, $\Gamma(2)$ and $\Gamma(4)$ are the only genus zero principal congruence subgroups contained in $\Gamma(2)$. For more examples of genus zero subgroups of $\Gamma(2)$, see Sebbar [Se01], e.g., Table 1.

Let $F_m : x^m + y^m + z^m = 0 \subset \mathbb{P}^2$ be the Fermat curve of degree m. Then it is known that a basis of holomorphic differential 1-forms on F_m is given by

$$\{\, w_{r,s,t} \,|\, 1 \le r, s, t \le n, r + s + t = m \,\}$$

where for $u \neq 0, w \neq 0$,

$$w_{r,s,t} = u^{r-1} v^{s-m} w^{t+1} d(-\frac{u}{w}).$$

Furthermore, if $m > 3$, the automorphism group of F_m is order $6m^2$ and it is isomorphic to the semi-direct product of the symmetric group S_3 of degree 3 and $(\mathbb{Z}/m\mathbb{Z}) \times (\mathbb{Z}/m\mathbb{Z})$.

To the Fermat curve F_m one can associate the cartographic group. Starting with $F_1 : x + y + z = 0$, $t = -y/z$ gives an isomorphism from F_1 to $\mathbb{P}^1$. The base points $x = 0, y = 0, z = 0$ of F_1 are respectively mapped to $0, 1, \infty$, and the inverse gives a triangulation of F_1. Now there is a canonical map $F_m \to F_1$, $(x, y, z) \mapsto (x^m, y^m, z^m)$ ramified only at these base points. The inverse of the standard triangulation of F_1 gives rise to a triangulation of F_m with $3m$ vertices, $3m^2$

edges and $2m^2$ faces. Then the cartographic group of this triangulation, denoted by $\hat{\mathfrak{C}}_m$, is shown to be the semi-direct product of S_3 and $(\mathbb{Z}/m\mathbb{Z}) \times (\mathbb{Z}/m\mathbb{Z})$, which is nothing but the automorphism group of F_m when $m > 3$. This yields a subgroup $\mathfrak{C}_m$ of $\Gamma(2)$ such that the Fermat curve F_m is isomorphic, as a Riemann surface, to the quotient $\mathfrak{H}/\mathfrak{C}_m$. The covering map $F_m \to F_1$, $(x, y, z) \mapsto (x^m, y^m, z^m)$ gives a map $\mathfrak{H}/\mathfrak{C}_m \to \mathfrak{H}/\Gamma(2)$.

Now we note that the Riemann sphere $\mathbb{P}^1$ has a unique holomorphic map λ : $\mathfrak{H}/\Gamma(2) \to \mathbb{P}^1$ given by

$$\lambda(q) = -U(q)/W(q)$$

where $U(q), V(q)$ and $W(q)$ are the functions uniformizing $\mathfrak{H}/\Gamma(2)$ and are given as follows:

$$U(q) = \prod_{n=1}^{\infty}(1 - q^{n-\frac{1}{2}})^8, \quad V(q) = -\prod_{n=1}^{\infty}(1 + q^{n+\frac{1}{2}})^8,$$

$$W(q) = 16q^{\frac{1}{2}}\prod_{n=1}^{\infty}(1 + q^n)^8$$

and $U(q)+V(q)+W(q) = 0$. (In fact, if $j(q)$ denotes the standard modular function for $PSL_2(\mathbb{Z})$, then j and λ are related by the equation $j = 256\frac{(1-\lambda+\lambda^2)^3}{\lambda^2(1-\lambda)^2}$.)

Now take the m-th roots of the functions $U(q), V(q), W(q)$, and put

$$u(q) = \prod_{n=1}^{\infty}(1 - q^{n-\frac{1}{2}})^{8/m}, \quad v(q) = \zeta\prod_{n=1}^{\infty}(1 + q^{n-\frac{1}{2}})^{8/m},$$

and

$$w(q) = (16q^{\frac{1}{2}})^{\frac{1}{m}}\prod_{n=1}^{\infty}(1 + q^n)^{\frac{8}{m}}$$

where ζ is a primitive 8-th root of unity. Then the modular form associated to $w_{r,s,t}$ is given by

$$q^{\frac{1}{2m}}\prod_{n=1}^{\infty}(1 - q^{\frac{n}{2}})^4(1 + q^{\frac{n}{2}})^{-4}(1 + q^n)^{\frac{8(s+2t)}{m}}(1 + q^{n-\frac{1}{2}})^{\frac{8(2s+t)}{m}}$$

This modular form has weight 2 for the group $\mathfrak{C}_m$ if m is a divisor of 24.

We summarize the above discussion.

PROPOSITION 7.5. (Bauer–Coste–Itzykson–Ruelle [BCIR97]) *Let $F_m : x^m + y^m + z^m = 0 \in \mathbb{P}^2$ be the Fermat curve of degree m, where m is a divisor of 24. Then, corresponding to F_m, there is a modular form of weight 2 for a subgroup $\mathfrak{C}_m \subset \Gamma(2)$.*

Now we ought to extend the above argument to weighted Fermat curves appearing in the singular loci $\Sigma(Y)$ of our Calabi–Yau orbifolds Y. Recall that for the Fermat curve $F_m : x^m + y^m + z^m = 0 \subset \mathbb{P}^2$ of degree m, there corresponds the set of characters:

$$\mathfrak{A} = \{\mathbf{a} = (a_1, a_2, a_3) \mid a_i \in (\mathbb{Z}/m\mathbb{Z}) \setminus \{0\},\ a_1 + a_2 + a_3 \equiv 0 \pmod{m}\}.$$

Now recall from Proposition 3.1 that a weighted Fermat curve of our interest is of the form

$$C : Y_i^{m_i} + Y_j^{m_j} + Y_k^{m_k} = 0 \quad \text{where } m_i := m/q_i,\ m_j = m/q_j,\ m_k = m/q_k$$

where $\gcd(q_i, q_j, q_k) = c_{I_\ell} > 1$. Put $m' = m/c_{I_\ell}$ and $q'_\bullet = q_\bullet/c_{I_\ell}$. It is easy to see that there is a map

$$F_m \longrightarrow C : (Y_i, Y_j, Y_k) \mapsto (Y_i^{q_i}, Y_j^{q_j}, Y_k^{q_k}).$$

The curve C is the weighted Fermat curve of degree m' with reduced weight $Q' = (q'_i, q'_j, q'_k)$ realizing C as a quotient of the usual Fermat curve F_m. Such weighted Fermat curves arise when we consider singular loci of our Calabi–Yau orbifolds. Here is a typical example. Consider the singular Calabi–Yau orbifold Y of degree 24 and $Q = (1, 3, 4, 4, 12)$. Then $m' = 24/4 = 6$ and the weighted Fermat curve $C : Y_3^6 + Y_4^6 + Y_5^2 = 0$ of degree $m' = 6$ and $Q' = (1, 1, 3)$ occurs in the singular locus $\Sigma(Y)$. There is a map $F_{24} \to C : (Y_3, Y_4, Y_5) \mapsto (Y_3^4, Y_4^4, Y_5^{12})$, realizing C as the quotient of F_{24} by this map. (In Appendix A2, we list all weighted Fermat curves occurring in this way.)

Now we try to associate to a weighted Fermat curve C some modular group and modular form using the method of cartographic groups described above.

COROLLARY 7.6. *Let $C : Y_i^{m_i} + Y_j^{m_j} + Y_k^{m_k} = 0 \in \mathbb{P}(q_i, q_j, q_k)$ be a weighted Fermat curve of degree m' and reduced weight Q' occurring the singular locus $\Sigma(Y)$ of a Calabi–Yau orbifold Y. Suppose that C has genus $g > 1$ and m' is a divisor of 24. Then we can associate to C, via the above procedure, a modular group $\mathfrak{C} \subset \Gamma(2) \subset PSL_2(\mathbb{Z})$ and a modular form of weight 2 for the group $\mathfrak{C}$.*

PROOF. Since a weighted Fermat curve is covered by a Fermat curve in the usual projective 2-space, its modularity follows from that of a Fermat curve in $\mathbb{P}^2$. □

EXAMPLE 7.6. Let $m = 24$ and $Q = (3, 3, 4, 6, 8)$. Then there is a genus 9 weighted Fermat curve $Y_1^8 + Y_2^8 + Y_4^4 = 0$ of reduced degree 8 and reduced weight $Q' = (1, 1, 2)$. A natural question which we have not yet settled is to find the cartographic group and a modular form corresponding to this curve.

Modularity of singular Calabi–Yau orbifolds. Now we turn our discussion to modularity of singular Calabi–Yau orbifolds. This will be done by passing to Fermat motives arising from our singular Calabi–Yau threefolds Y, and discussing the modularity of each Fermat motive.

As we observed in Section 5, there are many examples of Fermat motives whose characteristic polynomials are powers of irreducible degree 2 polynomials.

For a rigid Calabi–Yau threefold defined over $\mathbb{Q}$, for a good prime p, the characteristic polynomial is an irreducible polynomial of degree 2. We [SY01] have formulated the modularity conjecture that every rigid Calabi–Yau threefold over $\mathbb{Q}$ is modular in the sense that its L-series is determined by a cusp form of weight 4 on some $\Gamma_0(N)$. We now extend this modularity conjecture to "rigid" Fermat motives.

CONJECTURE 7.1. *Every Fermat motive whose characteristic polynomial is a power of a degree 2 irreducible polynomial over $\mathbb{Q}$ is modular. That is, there exists an imaginary quadratic field over $\mathbb{Q}$ associated to a Jacobi sum Grössencharakter, and it corresponds to a cusp form of weight 4 on some congruence subgroup of $PSL_2(\mathbb{Z})$.*

As for a Fermat motive whose characteristic polynomial is irreducible of degree ≥ 4, again there exists a CM field over $\mathbb{Q}$ associated to Jacobi sum Grössencharaktere.

We should expect that there would correspond modular forms, or more generally, automorphic forms associated to these Fermat motives. We will formulate this as a problem in the next section.

8. Open problems, further discussions

PROBLEM 8.1. Let Y be a singular Calabi–Yau orbifold and let $\mathcal{M}_A$ be a motivic factor of Y. Suppose that the characteristic polynomial of $\mathcal{M}_A$ is irreducible of degree ≥ 4. Find a modular group and a modular form that determine the L-series of $\mathcal{M}_A$.

PROBLEM 8.2. As we noted in Appendix A1, some of our Calabi–Yau orbifolds are equipped with $K3$ fibrations. What would be geometric structures of these Calabi–Yau orbifolds with $K3$ fibrations? How are zeta-functions and L-series dependent on these $K3$ fibrations? For an initial investigation on this problem, see Goto and Yui [GY04].

PROBLEM 8.3. We have not discussed constructions of mirror Calabi–Yau threefolds of our Calabi–Yau orbifolds. Since our Calabi–Yau orbifolds are non-rigid, there exist mirror Calabi–Yau threefolds. The construction of mirrors will be discussed in the paper [Yui], following the method of Greene and Plesser [GP90], and Roan [Ro91] (cf. Also Klemm and Theisen [KT93]). That is, first take (one-or-more parameter) deformation of our Calabi–Yau orbifold. Let G be a finite abelian group which is a discrete symmetry group of symmetries of the deformed Calabi–Yau orbifold leaving a unique holomorphic 3-form invariant. Taking the quotient of the deformed Calabi–Yau orbifold by G and resolve singularities. Then the resulting Calabi–Yau manifold is a mirror partner.

This construction imposes several challenging and hence interesting problems from arithmetic points of view.

(1) If we pass to deformations of our Calabi–Yau orbifolds, the resulting Calabi–Yau manifolds are no longer dominated by product varieties by a theorem of Scheon [Sch96], and consequently, they lose the property of CM type varieties.

(2) The computations of zeta-functions and L-series of deformed Calabi–Yau orbifolds will get much more challenging. For instance, the number of $\mathbb{F}_{p^r}$-rational points should be expressed in terms of polynomials involving deformation parameters. In fact, this approach has been explored by Candelas, de la Ossa, Villegas and Kadir, using Dwork's p-adic method.

(3) To recapitulate mirror symmetry phenomenon at the level of zeta-functions and L-series, we need a good definition of "quantum" version of these objects.

(4) The concept of "modularity" in one way or the other ought to be introduced in this setting.

PROBLEM 8.4. According to Greg Moore [Moo98], Calabi–Yau threefolds of CM type are related to attractor theory. There are "arithmetic attractor conjectures" which state that at an attractor point the periods and the algebraic complex structure coordinates are both arithmetic, and the relation between them generalizes the relation associating to imaginary quadratic argument τ the singular value $j(\tau)$ in a class field. For instance, for $N = 8$ supersymmetry, a product of three elliptic curves with CM, and for $N = 4$ supersymmetry, a product of a singular (extremal) $K3$ surface and an elliptic curve, appear as solutions to the attractor equations.

Note that any singular (extremal) $K3$ surface is of CM type as it corresponds to an even positive definite binary quadratic form.

The $N = 2$ supersymmetry attractor varieties are still not well understood; the attractor conjecture has been studied only for a very small number of examples. For instance, the attractor conjecture has been verified in the affirmative for the two-parameter deformation family of Calabi–Yau orbifold X of Fermat type of degree 8 in $\mathbb{P}^4(1,1,2,2,2)$ in Candelas et al. [COFKM94], where the CM field involved was imaginary quadratic over $\mathbb{Q}$.

Along the same line, Lynker–Periwal and Schimmrigk [LPS03] discussed black hole attractors of Calabi–Yau threefolds with CM, but without elliptic factors.

The arithmetic attractor conjectures should be studied for our Calabi–Yau orbifolds of CM type, as the CM fields involved here are not necessarily quadratic but have higher degree over $\mathbb{Q}$, i.e., cyclotomitic fields over $\mathbb{Q}$.

Moore [Moo98] has shown that the Calabi–Yau orbifolds corresponding to $\langle m, Q\rangle \in \{\langle 6,(1,1,1,1,2)\rangle, \langle 8,(1,1,2,2,2)\rangle\}$ are attractors, while those corresponding to $\langle m, Q\rangle \in \{\langle 5,(1,1,1,1,1)\rangle, \langle 8,(1,1,1,1,4)\rangle, \langle 10,(1,1,1,2,5)\rangle\}$ are not.

A questions is: What distinguishes Calabi–Yau orbifolds which are attractor varieties from those which are not?

Along the same line, Lynker, Periwal and Schimmrigk [LPS03] discussed black hole attractors of Calabi–Yau threefolds with CM, though, without elliptic factors. For a more recent discussion on this problem, see also Lynker et al. [LSS04].

PROBLEM 8.5. Calabi–Yau threefolds of CM type, especially, our Calabi–Yau orbifolds discussed in this paper are linked to rational conformal field theory, e.g., see for instance, Bauer–Coste–Itzykson–Ruelle [BCIR97], Schimmrigk [Sch03], Lynker–Schimmrigk–Stewart [LSS04] and Gukov–Vafa [GV02]. Any rational conformal field theories are linked to a cyclotomic field over $\mathbb{Q}$. In particular, there should be links between $su(3)$ modular invariants and simple factors of Jacobian varieties of Fermat curves, or more generally, Calabi–Yau orbifolds of CM type and weighted Fermat curves and their Jacobians discussed in this paper. This connection should be explored in fuller detail.

Acknowledgments. During the course of this work, the author held visiting positions at a number of institutions. This include University of Cambridge, Department of Pure Mathematics and Mathematical Statistics, and Newnham College (Spring 1997 and Spring 2002), The Newton Institute Cambridge (Spring 2002), Max-Planck-Institut-für-Mathematik Bonn (Spring 1999, 2001, 2004), MSRI Berkeley (Fall 2000), University of Leiden (Winter 2004), Hannover (Spring 2004), Mainz (Spring 2004), and Tsuda College (Summer 2004). She is grateful for the hospitality of these institutions.

This work was presented at seminars at several institutions, e.g., Cambridge, Oxford, Max-Planck Institut für Mathematik Bonn, Berkeley, MSRI, Kyoto, Tohoku, Leiden, Hannover, Mainz and also at the workshop on Calabi–Yau Varieties and Mirror Symmetry at the Fields Institute (2001). I thank the audience for their helpful comments and suggestions.

Special thanks go to Rolf Schimmrigk who has read through the manuscript at an earlier stage pointing out some inaccuracies (in particular, calculations of L-series in Examples 7.1 and 7.3) and suggestions for improvements. I also thank

James Lewis, Yasuhiro Goto and Stefan Müller-Stach for their comments and suggestions. Finally, I thank Lloyd Elliott who made computer programs for the calculations of motives, curve singularities and Jacobi sums in this paper, and William Stein for his help in determining the twist of the L-series in Example 7.3.

Appendix A. Tables

Appendix A1 contains the 147 Calabi–Yau orbifolds, while Appendix A2 lists the dimension 1 singularities of Calabi–Yau orbifolds in A1.

A.1. Calabi–Yau orbifolds. We tabulate the 147 Calabi–Yau orbifolds in Tables 7,8,9 and Table 10. The second and the third columns present degree and weight, respectively. The Hodge numbers $h^{1,1}$, $h^{2,1}$, the third Betti number B_3 and the Euler characteristic χ are listed in the fourth, fifth, sixth and the seventh columns, respectively. The eighth column exhibits a $K3$ fibration when it exists. To find a $K3$ fibration, we use the characterization of Hosono–Lian–Yau [HLY96]. A weight $Q = (q_1, q_2, q_3, q_4, q_5)$ is said to be *admissible* if the generic weighted degree m hypersurface in $\mathbb{P}^4(Q)$ is smooth away from the origin. Given an admissible weight $Q = (q_1, q_2, q_3, q_4, q_5)$, let Y be the generic Calabi–Yau threefold. We intersect this variety with the coordinate hyperplane $Y_i = 0$ for some i, $1 \leq i \leq 5$. A pencil of $K3$ surfaces is obtained when $Y \cap \{Y_i = 0\}$ is isomorphic to a weighted $K3$ hypersurface determined by Reid and Yonemura [Yo90] with some additional conditions on the weight of this $K3$ surface (e.g., the most crucial condition is (ii) in [HLY96]). In the final column, $\bullet$ indicates if the orbifold has no quotient singularities of dimension 1 (i.e., curve singularities). (Note that some of the orbifolds may still have singularities of dimension 0, namely, singular points, whose resolutions contribute to $(1,1)$ forms, and hence to $h^{1,1}$.)

TABLE 7. Calabi–Yau orbifolds

#	m	Q	$h^{1,1}$	$h^{2,1}$	B_3	χ	$K3$ fibr.	C.S.
1	5	(1, 1, 1, 1, 1)	1	101	204	−200		•
2	6	(1, 1, 1, 1, 2)	1	103	208	−204		•
3	8	(1, 1, 1, 1, 4)	1	149	300	−296		•
4	8	(1, 1, 2, 2, 2)	2	86	174	−168	(1, 1, 1, 1)	
5	9	(1, 1, 1, 3, 3)	4	112	226	−216		•
6	10	(1, 1, 1, 2, 5)	1	145	292	−288		•
7	12	(1, 1, 1, 3, 6)	3	165	332	−324		•
8	12	(1, 1, 2, 2, 6)	2	128	258	−252	(1, 1, 1, 3)	
9	12	(1, 1, 2, 4, 4)	5	101	204	−192	(1, 1, 2, 2)	
10	12	(1, 1, 3, 3, 4)	2	89	180	−168		•
11	12	(1, 2, 2, 3, 4)	2	74	150	−144		
12	12	(1, 2, 3, 3, 3)	3	69	140	−132	(1, 1, 1, 1)	
13	12	(2, 2, 2, 3, 3)	6	60	122	−108	(1, 1, 1, 3)	
14	14	(1, 2, 2, 2, 7)	2	122	246	−240		
15	15	(1, 1, 3, 5, 5)	7	103	208	−192		•
16	15	(1, 3, 3, 3, 5)	3	75	152	−144	(1, 1, 1, 1)?	
17	16	(1, 1, 2, 4, 8)	4	148	298	−288	(1, 1, 2, 4)	
18	18	(1, 1, 1, 6, 9)	2	272	546	−540		•
19	18	(1, 2, 3, 3, 9)	3	99	200	−192	(1, 1, 1, 3)	
20	18	(1, 2, 3, 6, 6)	7	79	160	−144	(1, 1, 2, 2)	
21	18	(2, 2, 2, 3, 9)	4	112	226	−216		
22	20	(1, 1, 4, 4, 10)	7	143	288	−272	(1, 2, 2, 5)	•
23	20	(1, 2, 2, 5, 10)	6	120	242	−228		
24	20	(1, 4, 5, 5, 5)	5	65	132	−120	(1, 1, 1, 1)	
25	20	(2, 4, 4, 5, 5)	15	39	80	−48	(1, 2, 2, 5)	
26	21	(1, 3, 3, 7, 7)	17	65	132	−96		
27	24	(1, 1, 2, 8, 12)	3	243	488	−480	(1, 1, 4, 6)	
28	24	(1, 1, 4, 6, 12)	8	164	330	−312	(1, 2, 3, 6)	•
29	24	(1, 1, 6, 8, 8)	11	131	264	−240	(1, 3, 4, 4)	•
30	24	(1, 2, 3, 6, 12)	6	114	230	−216	(1, 1, 2, 4)	
31	24	(1, 3, 4, 4, 12)	6	90	182	−168	(1, 1, 1, 3)	
32	24	(1, 3, 4, 8, 8)	10	70	142	−120	(1, 1, 2, 2)	
33	24	(1, 3, 6, 6, 8)	8	68	138	−120		
34	24	(2, 3, 3, 4, 12)	6	66	134	−120	(1, 2, 3, 6)	
35	24	(2, 3, 3, 8, 8)	19	43	88	−48	(1, 3, 4, 4)	
36	24	(3, 3, 4, 6, 8)	7	55	112	−96	(2, 3, 3, 4)	
37	28	(1, 2, 4, 7, 14)	8	116	234	−216		
38	30	(1, 1, 3, 10, 15)	5	251	504	−492		
39	30	(1, 2, 2, 10, 15)	4	208	418	−408		
40	30	(1, 2, 6, 6, 15)	9	117	236	−216	(1, 2, 2, 5)	
41	30	(1, 3, 5, 6, 15)	7	91	184	−168		

TABLE 8. Calabi–Yau orbifolds: continued

#	m	Q	$h^{1,1}$	$h^{2,1}$	B_3	χ	$K3$ fibr.	C.S.
42	30	$(1,3,6,10,10)$	19	67	136	-96		•
43	30	$(2,2,5,6,15)$	7	103	208	-192		
44	30	$(2,3,5,10,10)$	12	48	98	-72		
45	30	$(2,3,5,5,15)$	7	61	124	-108		
46	30	$(3,5,6,6,10)$	15	39	80	-48	$(1,2,2,5)$	
47	36	$(1,1,4,12,18)$	7	271	544	-528	$(1,2,6,9)$	•
48	36	$(1,2,3,12,18)$	5	185	372	-360	$(1,1,4,6)$	
49	36	$(1,2,6,9,18)$	12	126	254	-228	$(1,2,3,6)$	
50	36	$(1,2,9,12,12)$	16	100	202	-168	$(1,3,4,4)$	
51	36	$(1,4,4,9,18)$	19	91	184	-144		•
52	36	$(2,3,4,9,18)$	10	70	142	-120	$(1,2,3,6)$	
53	36	$(2,4,9,9,12)$	19	43	88	-48	$(1,2,6,9)$	
							$(2,3,3,4)$	
54	40	$(1,1,8,10,20)$	11	227	456	-432	$(1,4,5,10)$	•
55	40	$(1,4,5,10,20)$	12	96	194	-168	$(1,1,2,4)$	
56	40	$(2,5,5,8,20)$	7	63	128	-112	$(1,1,2,4)$	
							$(1,4,5,10)$	
57	42	$(1,3,3,14,21)$	6	180	362	-348		
58	42	$(1,6,7,14,14)$	17	65	132	-96	$(1,1,2,2)$	
59	42	$(1,6,7,7,21)$	8	86	174	-156	$(1,1,1,3)$	
60	42	$(2,2,3,14,21)$	7	151	304	-288		
61	42	$(2,6,6,7,21)$	17	65	132	-96		
62	45	$(1,5,9,15,15)$	19	67	136	-96		
63	48	$(1,1,6,16,24)$	9	321	644	-624	$(1,3,8,12)$	•
64	48	$(1,3,12,16,16)$	26	86	174	-120	$(1,3,4,4)$	•
65	48	$(1,3,4,16,24)$	8	164	330	-312	$(1,1,4,6)$	
66	48	$(1,3,8,12,24)$	16	112	226	-192	$(1,2,3,6)$	
67	48	$(2,3,3,16,24)$	7	127	256	-240	$(1,3,8,12)$	
68	54	$(1,2,6,18,27)$	10	214	430	-408	$(1,2,6,9)$	
69	60	$(1,12,12,15,20)$	23	47	96	-48		•
70	60	$(1,2,12,15,30)$	17	173	348	-312	$(1,4,5,10)$	
71	60	$(1,3,6,20,30)$	10	178	358	-336		
72	60	$(1,4,10,15,30)$	20	104	210	-168	$(1,2,3,6)$	
73	60	$(1,4,15,20,20)$	26	86	174	-120	$(1,3,4,4)$	
74	60	$(1,4,5,20,30)$	11	155	312	-288	$(1,1,4,6)$	
75	60	$(1,5,12,12,30)$	25	85	172	-120	$(1,2,2,5)$	•
76	60	$(2,3,10,15,30)$	20	74	150	-108		
77	60	$(2,3,15,20,20)$	31	55	112	-48		
78	60	$(2,3,5,20,30)$	10	106	214	-192		
79	60	$(3,10,12,15,20)$	15	39	80	-48	$(2,3,3,4)$	
80	60	$(3,3,4,20,30)$	11	107	216	-192	$(2,3,10,15)$	

TABLE 9. Calabi–Yau orbifolds: continued

#	m	Q	$h^{1,1}$	$h^{2,1}$	B_3	χ	$K3$ fibr.	C.S.
81	60	(3, 5, 10, 12, 30)	13	49	100	−72	(1, 4, 5, 10)	
							(1, 2, 3, 6)	
82	60	(3, 5, 12, 20, 20)	31	31	64	0	(1, 3, 4, 4)	
83	60	(4, 5, 6, 15, 30)	18	42	86	−48		
84	60	(4, 6, 15, 15, 20)	27	27	56	0	(2, 3, 10, 15)	
85	66	(2, 3, 6, 22, 33)	9	129	260	−240		
86	70	(1, 10, 10, 14, 35)	23	71	144	−96		•
87	70	(2, 5, 14, 14, 35)	27	59	120	−64		
88	72	(1, 2, 9, 24, 36)	14	242	486	−456	(1, 3, 8, 12)	
89	72	(1, 3, 8, 24, 36)	15	183	368	−336	(1, 2, 6, 9)	
90	72	(1, 8, 9, 18, 36)	19	91	184	−144	(1, 1, 2, 4)	
91	78	(1, 6, 6, 26, 39)	23	143	288	−240		•
92	84	(1, 1, 12, 28, 42)	11	491	984	−960	(1, 6, 14, 21)	•
93	84	(1, 6, 14, 21, 42)	30	96	194	−132	(1, 2, 3, 6)	
94	84	(1, 6, 21, 28, 28)	40	76	154	−72	(1, 3, 4, 4)	
95	84	(1, 6, 7, 28, 42)	15	147	296	−264	(1, 1, 4, 6)	
96	84	(2, 12, 21, 21, 28)	35	35	72	0	(2, 3, 3, 4)	
97	84	(2, 7, 12, 21, 42)	23	59	120	−72	(1, 2, 3, 6)	
98	84	(3, 4, 14, 21, 42)	28	52	106	−48		
99	84	(3, 4, 21, 28, 28)	41	41	84	0		
100	84	(3, 4, 7, 28, 42)	16	76	154	−120		
101	90	(1, 5, 9, 30, 45)	17	155	312	−276		
102	90	(2, 10, 15, 18, 45)	19	67	136	−96		
103	90	(2, 3, 10, 30, 45)	18	126	254	−216		
104	96	(1, 3, 12, 32, 48)	18	222	446	−408	(1, 3, 8, 12)	
105	100	(1, 4, 20, 25, 50)	33	141	284	−216	(1, 4, 5, 10)	•
106	110	(1, 10, 22, 22, 55)	39	79	160	−80	(1, 2, 2, 5)	•
107	120	(1, 15, 20, 24, 60)	33	69	140	−72		•
108	120	(1, 15, 24, 40, 40)	55	55	112	0		•
109	120	(1, 4, 15, 40, 60)	24	204	410	−360	(1, 3, 8, 12)	
110	120	(1, 5, 24, 30, 60)	33	141	284	−216	(1, 4, 5, 10)	
111	120	(2, 3, 15, 40, 60)	23	143	288	−240		
112	120	(3, 5, 12, 40, 60)	22	82	166	−120	(1, 3, 8, 12)	
113	126	(1, 2, 18, 42, 63)	17	377	756	−720	(1, 6, 14, 21)	
114	126	(1, 6, 14, 42, 63)	28	160	332	−264	(1, 2, 6, 9)	
115	140	(1, 14, 20, 35, 70)	44	80	162	−72		•
116	140	(2, 5, 28, 35, 70)	39	87	176	−96		
117	156	(1, 12, 13, 52, 78)	23	143	288	−240	(1, 1, 4, 6)	
118	156	(1, 12, 26, 39, 78)	53	89	180	−72	(1, 2, 3, 6)	
119	156	(1, 12, 39, 52, 52)	71	71	144	0	(1, 3, 4, 4)	•
120	168	(1, 3, 24, 56, 84)	23	335	672	−624	(1, 6, 14, 21)	

TABLE 10. Calabi–Yau orbifolds: continued

#	m	Q	$h^{1,1}$	$h^{2,1}$	B_3	χ	$K3$ fibr.	C.S.
121	168	$(1,6,21,56,84)$	34	190	382	-312	$(1,3,8,12)$	
122	168	$(3,4,21,56,84)$	33	105	212	-144		
123	180	$(1,9,20,60,90)$	42	150	302	-216	$(1,2,6,9)$	•
124	180	$(4,5,36,45,90)$	55	55	112	0		
125	198	$(2,9,22,66,99)$	45	93	188	-96		
126	210	$(2,3,30,70,105)$	29	221	444	-384		
127	210	$(6,14,15,70,105)$	43	43	88	0		
128	216	$(1,8,27,72,108)$	48	180	302	-264	$(1,3,8,12)$	•
129	220	$(1,10,44,55,110)$	65	125	252	-168	$(1,4,5,10)$?	•
130	240	$(1,15,24,80,120)$	50	134	270	-168		•
131	264	$(3,8,33,88,132)$	57	81	164	-48		
132	294	$(1,6,42,98,147)$	47	287	576	-480	$(1,6,14,21)$	•
133	312	$(1,12,39,104,156)$	66	174	350	-216	$(1,3,8,12)$	•
134	330	$(3,22,30,110,165)$	59	59	120	0	$(2,3,10,15)$	
135	336	$(1,7,48,112,168)$	47	287	576	-480	$(1,6,14,21)$	
136	342	$(1,18,38,114,171)$	71	143	288	-144	$(1,2,6,9)$	•
137	378	$(2,7,54,126,189)$	53	173	348	-240		
138	420	$(1,20,84,105,210)$	119	119	240	0	$(1,4,5,10)$	•
139	420	$(3,7,60,140,210)$	59	131	264	-144		
140	462	$(2,33,42,154,231)$	79	79	160	0		
141	546	$(6,7,78,182,273)$	83	83	168	0		
142	600	$(1,24,75,200,300)$	119	167	336	-96	$(1,3,8,12)$	
143	630	$(1,14,90,210,315)$	95	263	528	-336	$(1,6,14,21)$	•
144	714	$(3,14,102,238,357)$	107	107	216	0		
145	924	$(1,21,132,308,462)$	137	257	516	-240	$(1,6,14,21)$	•
146	966	$(2,21,138,322,483)$	143	143	288	0		
147	1806	$(1,42,258,602,903)$	251	251	504	0	$(1,6,14,21)$	•

A.2. Singular Loci. In this appendix, we list examples of singularities of Calabi–Yau orbifolds. Let Y be a singular Calabi–Yau threefold of Fermat type of degree m and weight $Q=(q_1,q_2,q_3,q_4,q_5)$. It is defined by the hypersurface in the weighted projective 4-space $\mathbb{P}^4(Q)$:

$$Y : Y_1^{m/q_1}+Y_2^{m/q_2}+Y_3^{m/q_3}+Y_4^{m/q_4}+Y_5^{m/q_5}=0.$$

Let $\Sigma(Y)$ denote the singular locus consisting of dimension 1 singularities. In the table below, we list irreducible components in $\Sigma(Y)$.

First we illustrate with examples the algorithm for determining singularities of dimension 0 and dimension 1.

We tabulate a complete list of singularities of dimension 1.

TABLE 11. $m=14$ and $Q=(1,2,2,2,7)$

ℓ	S_ℓ	$\#S_\ell$	I_ℓ	c_{I_ℓ}	C.S	Rmk	Mult
0	$\{1,2,3,4,5\}$	5	$\emptyset$				
1	$\emptyset$	0	$\{1,2,3,4,5\}$				
2	$\{5\}$	1	$\{1,2,3,4\}$	7			
3	$\emptyset$	0	$\{1,2,3,4,5\}$				
4	$\{5\}$	1	$\{1,2,3,4\}$	7			
5	$\emptyset$	0	$\{1,2,3,4,5\}$				
6	$\{5\}$	1	$\{1,2,3,4\}$	7			
7	$\{2,3,4\}$	3	$\{1,5\}$	2	$Y_2^7+Y_3^7+Y_4^7=0$	$g=15$	1
8	$\{5\}$	1	$\{1,2,3,4\}$	7			
9	$\emptyset$	0	$\{1,2,3,4,5\}$				
10	$\{5\}$	1	$\{1,2,3,4\}$	7			
11	$\emptyset$	0	$\{1,2,3,4,5\}$				
12	$\{5\}$	1	$\{1,2,3,4\}$				
13	$\emptyset$	0	$\{1,2,3,4,5\}$				

TABLE 12. $m=18$ and $Q=(1,2,3,6,6)$

ℓ	S_ℓ	$\#S_\ell$	I_ℓ	c_{I_ℓ}	C.S.	Rmk	Mult
0	$\{1,2,3,4,5\}$	5	$\emptyset$				
1	$\emptyset$	0	$\{1,2,3,4,5\}$				
2	$\emptyset$	0	$\{1,2,3,4,5\}$				
3	$\{4,5\}$	2	$\{1,2,3\}$	6			
4	$\emptyset$	0	$\{1,2,3,4,5\}$				
5	$\emptyset$	0	$\{1,2,3,4,5\}$				
6	$\{3,4,5\}$	3	$\{1,2\}$	3	$Y_3^6+Y_4^3+Y_5^3=0$	$g=1$	2
7	$\emptyset$	0	$\{1,2,3,4,5\}$				
8	$\emptyset$	0	$\{1,2,3,4,5\}$				

TABLE 13. $m = 18$ and $Q = (1, 2, 3, 6, 6)$: continued

ℓ	S_ℓ	$\#S_\ell$	I_ℓ	c_{I_ℓ}	C.S.	Rmk	Mult
9	$\{2,4,5\}$	3	$\{1,3\}$	2	$Y_2^9 + Y_4^3 + Y_5^3 = 0$	$g = 1$	1
10	$\emptyset$	0	$\{1,2,3,4,5\}$				
11	$\emptyset$	0	$\{1,2,3,4,5\}$				
12	$\{3,4,5\}$	3	$\{1,2\}$	3	$Y_3^6 + Y_4^3 + Y_5^3 = 0$	$g = 1$	2
13	$\emptyset$	0	$\{1,2,3,4,5\}$				
14	$\emptyset$	0	$\{1,2,3,4,5\}$				
15	$\{4,5\}$	2	$\{1,2,3\}$	6			
16	$\emptyset$	0	$\{1,2,3,4,5\}$				
17	$\emptyset$	0	$\{1,2,3,4,5\}$				

TABLE 14. $m = 24$ and $Q = (1, 2, 3, 6, 12)$

ℓ	s_ℓ	$\#S_\ell$	I_ℓ	c_{I_ℓ}	C.S.	Rmk	Mult
0	$\{1,2,3,4,5\}$	5	$\emptyset$				
1	$\emptyset$	0	$\{1,2,3,4,5\}$				
2	$\{5\}$	1	$\{1,2,3,4\}$	1			
3	$\emptyset$	0	$\{1,2,3,4,5\}$				
4	$\{4,5\}$	2	$\{1,2,3\}$	6	$A_{3,4}$		
5	$\emptyset$	0	$\{1,2,3,4,5\}$				
6	$\{5\}$	1	$\{1,2,3,4\}$	1			
7	$\emptyset$	0	$\{1,2,3,4,5\}$				
8	$\{3,4,5\}$	3	$\{1,2\}$	3	$Y_3^8 + Y_4^4 + Y_5^2 = 0$	$g = 1$	2
9	$\emptyset$	0	$\{1,2,3,4,5\}$				
10	$\{5\}$	1	$\{1,2,3,4\}$	1			
11	$\emptyset$	0	$\{1,2,3,4,5\}$				
12	$\{2,4,5\}$	3	$\{1,3\}$	2	$Y_2^{12} + Y_4^4 + Y_5^2 = 0$	$g = 1$	1
13	$\emptyset$	0	$\{1,2,3,4,5\}$				
14	$\{5\}$	1	$\{1,2,3,4\}$	1			
15	$\emptyset$	0	$\{1,2,3,4,5\}$				
16	$\{3,4,5\}$	3	$\{1,2\}$	3	$Y_3^8 + Y_4^4 + Y_5^2 = 0$	$g = 1$	2
17	$\emptyset$	0	$\{1,2,3,4,5\}$				
18	$\{5\}$	1	$\{1,2,3,4\}$	1			
19	$\emptyset$	0	$\{1,2,3,4,5\}$				
20	$\{4,5\}$	2	$\{1,2,3\}$	6	$A_{3,4}$		
21	$\emptyset$	0	$\{1,2,3,4,5\}$				
22	$\{5\}$	1	$\{1,2,3,4\}$	1			
23	$\emptyset$	0	$\{1,2,3,4,5\}$				

TABLE 15. Dimension 1 singularities

#	m	Q	ℓ	c_{I_ℓ}	m'	Q'	Σ	g
4	8	$(1,1,2,2,2)$	4	2	4	$(1,1,1)$	$Y_3^4+Y_4^4+Y_5^4$	3
8	12	$(1,1,2,2,6)$	6	2	6	$(1,1,3)$	$Y_3^6+Y_4^6+Y_5^2$	2
9	12	$(1,1,2,4,4)$	6	2	6	$(1,2,2)$	$Y_3^6+Y_4^3+Y_5^3$	1
11	12	$(1,2,2,3,4)$	6	2	6	$(1,1,2)$	$Y_2^6+Y_3^6+Y_5^3$	4
12	12	$(1,2,3,3,3)$	$4,8$	3	4	$(1,1,1)$	$Y_3^4+Y_4^4+Y_5^3$	3
13	12	$(2,2,2,3,3)$	6	2	6	$(1,1,1)$	$Y_1^6+Y_2^6+Y_3^6$	10
14	14	$(1,2,2,2,7)$	7	2	7	$(1,1,1)$	$Y_2^7+Y_3^7+Y_4^7$	15
16	15	$(1,3,3,3,5)$	5	3	5	$(1,1,1)$	$Y_2^5+Y_3^5+Y_4^5$	6
17	16	$(1,1,2,4,8)$	8	2	8	$(1,2,4)$	$Y_3^8+Y_4^4+Y_5^2$	1
19	18	$(1,2,3,3,9)$	6	3	6	$(1,1,3)$	$Y_3^6+Y_4^6+Y_5^2$	2
20	18	$(1,2,3,6,6)$	$6,12$	3	6	$(1,2,2)$	$Y_3^6+Y_4^3+Y_5^3$	1
			9	2	9	$(1,3,3)$	$Y_2^9+Y_4^3+Y_5^3$	1
21	18	$(2,2,2,3,9)$	9	2	9	$(1,1,1)$	$Y_1^9+Y_2^9+Y_3^9$	28
22	20	$(1,1,4,4,10)$	10	2	10	$(2,2,5)$	$Y_3^5+Y_4^5+Y_5^2$	0
23	20	$(1,2,2,5,10)$	10	2	10	$(1,1,5)$	$Y_2^{10}+Y_3^{10}+Y_5^2$	4
24	20	$(1,4,5,5,5)$	$4,8,12,16$	5	4	$(1,1,1)$	$Y_3^4+Y_4^4+Y_5^4$	3
25	20	$(2,4,4,5,5)$	10	2	10	$(1,2,2)$	$Y_1^{10}+Y_2^5+Y_3^5$	6
27	24	$(1,1,2,8,12)$	12	2	12	$(1,4,6)$	$Y_3^{12}+Y_4^3+Y_5^2$	1
28	24	$(1,1,4,6,12)$	12	2	12	$(2,3,6)$	$Y_3^6+Y_4^4+Y_5^2$	0
29	24	$(1,1,6,8,8)$	12	2	12	$(3,4,4)$	$Y_3^4+Y_4^3+Y_5^3$	0
30	24	$(1,2,3,6,12)$	$8,16$	3	8	$(1,2,4)$	$Y_3^8+Y_4^4+Y_5^2$	1
			12	2	12	$(1,3,6)$	$Y_2^{12}+Y_4^4+Y_5^2$	1
31	24	$(1,3,4,4,12)$	$6,12,18$	4	6	$(1,1,3)$	$Y_3^6+Y_4^6+Y_5^2$	2
32	24	$(1,3,4,8,8)$	$6,12,18$	4	6	$(1,2,2)$	$Y_3^6+Y_4^3+Y_5^3$	1
33	24	$(1,3,6,6,8)$	$8,16$	3	8	$(1,2,2)$	$Y_2^8+Y_3^4+Y_4^4$	3
			12	2	12	$(3,3,4)$	$Y_3^4+Y_4^4+Y_5^3$	0
34	24	$(2,3,3,4,12)$	$8,16$	3	8	$(1,1,4)$	$Y_1^8+Y_3^8+Y_5^2$	3
			12	2	12	$(1,2,6)$	$Y_2^{12}+Y_4^6+Y_5^2$	2
35	24	$(2,3,3,8,8)$	12	2	12	$(1,4,4)$	$Y_1^{12}+Y_4^3+Y_5^3$	1
36	24	$(3,3,4,6,8)$	$8,16$	3	8	$(1,1,2)$	$Y_1^8+Y_2^8+Y_4^4$	9
			12	2	12	$(2,3,4)$	$Y_3^6+Y_4^4+Y_5^3$	1
37	28	$(1,2,4,7,14)$	14	2	14	$(1,2,7)$	$Y_2^{14}+Y_4^7+Y_5^2$	3
39	30	$(1,2,2,10,15)$	15	2	15	$(1,1,5)$	$Y_2^{15}+Y_3^{15}+Y_4^3$	13
40	30	$(1,2,6,6,15)$	$10,20$	3	10	$(2,2,5)$	$Y_3^5+Y_4^5+Y_5^2$	0
			15	2	15	$(1,3,3)$	$Y_2^{15}+Y_3^5+Y_4^5$	6
41	30	$(1,3,5,6,15)$	10	2	10	$(1,2,5)$	$Y_2^{10}+Y_3^5+Y_5^2$	2
42	30	$(1,3,6,10,10)$	15	2	15	$(3,5,5)$	$Y_3^5+Y_4^3+Y_5^3$	0
43	30	$(2,2,5,6,15)$	15	2	15	$(1,1,3)$	$Y_1^{15}+Y_2^{15}+Y_4^5$	26
44	30	$(2,3,5,10,10)$	15	2	15	$(1,5,5)$	$Y_1^{15}+Y_4^3+Y_5^3$	1
			$6,12,18,24$	5	6	$(1,2,2)$	$Y_3^6+Y_4^6+Y_5^3$	1
45	30	$(2,3,5,5,15)$	$6k (1\leq k\leq 4)$	5	6	$(1,1,3)$	$Y_3^6+Y_4^6+Y_5^2$	2
46	30	$(3,5,6,6,10)$	15	2	15	$(3,3,5)$	$Y_3^5+Y_4^5+Y_5^3$	0
			$10,20$	3	10	$(1,2,2)$	$Y_1^{10}+Y_3^5+Y_4^5$	6
47	36	$(1,1,4,12,18)$	18	2	18	$(2,6,9)$	$Y_3^9+Y_4^3+Y_5^2$	0
48	36	$(1,2,3,12,18)$	$12,24$	3	12	$(1,4,6)$	$Y_3^{12}+Y_4^3+Y_5^2$	1
			18	2	18	$(1,6,9)$	$Y_2^{18}+Y_4^3+Y_5^2$	1
49	36	$(1,2,6,9,18)$	$12,24$	3	12	$(2,3,6)$	$Y_3^6+Y_4^4+Y_5^2$	0
			18	2	18	$(1,3,9)$	$Y_2^{18}+Y_3^6+Y_5^2$	2
50	36	$(1,2,9,12,12)$	$12,24$	3	12	$(3,4,4)$	$Y_3^4+Y_4^3+Y_5^3$	0
			18	2	18	$(1,6,6)$	$Y_2^{18}+Y_4^3+Y_5^3$	1
51	36	$(1,4,4,9,18)$	18	2	18	$(2,2,9)$	$Y_2^9+Y_3^9+Y_5^2$	0

TABLE 16. Dimension 1 singularities: continued

#	m	Q	ℓ	c_{I_ℓ}	m'	Q'	Σ	g
52	36	$(2,3,4,9,18)$	$12,24$	3	12	$(1,3,6)$	$Y_2^{12}+Y_4^4+Y_5^2$	1
			18	2	18	$(1,2,9)$	$Y_1^{18}+Y_3^9+Y_5^2$	4
53	36	$(2,4,9,9,12)$	$12,24$	3	12	$(3,3,4)$	$Y_3^4+Y_4^4+Y_5^3$	0
			18	2	18	$(1,2,6)$	$Y_1^{18}+Y_2^9+Y_5^2$	7
54	40	$(1,1,8,10,20)$	20	2	20	$(4,5,10)$	$Y_3^{20}+Y_4^4+Y_5^2$	0
55	40	$(1,4,5,10,20)$	20	2	20	$(2,5,10)$	$Y_3^{10}+Y_4^4+Y_5^2$	0
			$8k(1\le k\le 4)$	5	8	$(1,2,4)$	$Y_3^8+Y_4^4+Y_5^2$	1
56	40	$(2,5,5,8,20)$	$8k(1\le k\le 4)$	5	8	$(1,1,4)$	$Y_2^8+Y_3^8+Y_5^2$	3
			20	2	20	$(1,4,10)$	$Y_1^{20}+Y_4^5+Y_5^2$	2
57	42	$(1,3,3,14,21)$	$14,28$	3	14	$(1,1,7)$	$Y_2^{14}+Y_3^{14}+Y_5^2$	6
58	42	$(1,6,7,14,14)$	$6k(1\le k\le 6)$	7	6	$(1,2,2,)$	$Y_3^6+Y_4^3+Y_5^3$	1
			21	2	21	$(3,7,7)$	$Y_2^7+Y_4^3+Y_5^3$	0
59	42	$(1,6,7,7,21)$	$6k(1\le k\le 6)$	7	6	$(1,1,3)$	$Y_3^7+Y_4^7+Y_5^2$	2
60	42	$(2,2,3,14,21)$	21	2	21	$(1,1,7)$	$Y_1^{21}+Y_2^{21}+Y_4^3$	19
61	42	$(2,6,6,7,21)$	$14,28$	3	14	$(2,2,7)$	$Y_2^6+Y_3^7+Y_5^2$	0
			21	2	21	$(1,3,3)$	$Y_1^{21}+Y_2^7+Y_3^7$	15
62	45	$(1,5,9,15,15)$	$9k(1\le k\le 4)$	5	9	$(1,3,3)$	$Y_2^9+Y_4^3+Y_5^3$	1
			$15,30$	3	15	$(3,5,5)$	$Y_3^5+Y_4^3+Y_5^3$	0
63	48	$(1,1,6,16,24)$	24	2	24	$(3,8,12)$	$Y_3^8+Y_4^3+Y_5^2$	0
64	48	$(1,3,12,16,16)$	$12,24,36$	4	12	$(3,4,4)$	$Y_3^4+Y_4^3+Y_5^3$	0
65	48	$(1,3,4,16,24)$	$12,24,36$	4	12	$(1,4,6)$	$Y_3^{12}+Y_4^3+Y_5^2$	1
66	48	$(1,3,8,12,24)$	$12,24,36$	4	12	$(2,3,6)$	$Y_3^6+Y_4^4+Y_5^2$	0
			$16,32$	3	18	$(1,4,8)$	$Y_2^{18}+Y_4^4+Y_5^2$	1
67	48	$(2,3,3,16,24)$	$16,32$	3	16	$(1,1,8)$	$Y_2^{16}+Y_3^{16}+Y_5^2$	7
			24	2	24	$(1,8,12)$	$Y_1^{24}+Y_4^3+Y_5^2$	1
68	54	$(1,2,6,18,27)$	$18,36$	3	18	$(2,6,9)$	$Y_3^9+Y_4^3+Y_5^2$	0
			27	2	27	$(1,3,9)$	$Y_2^{27}+Y_3^9+Y_4^3$	7
69	60	$(1,12,12,15,20)$	$15,30,45$	4	15	$(3,3,5)$	$Y_2^5+Y_3^5+Y_5^3$	0
			$20,40$	3	20	$(4,4,5)$	$Y_2^5+Y_3^5+Y_4^4$	0
70	60	$(1,2,12,15,30)$	$20,40$	3	20	$(4,5,10)$	$Y_3^5+Y_4^4+Y_5^2$	0
			30	2	30	$(1,6,15)$	$Y_2^{30}+Y_3^5+Y_5^2$	2
71	60	$(1,3,6,20,30)$	$20,40$	3	20	$(1,2,10)$	$Y_2^{20}+Y_3^{10}+Y_5^2$	4
			30	2	30	$(3,10,15)$	$Y_3^{10}+Y_4^3+Y_5^2$	0
72	60	$(1,4,10,15,30)$	$12k(1\le k\le 4)$	5	12	$(2,3,6)$	$Y_3^6+Y_4^4+Y_5^2$	0
			30	2	30	$(2,5,15)$	$Y_2^5+Y_3^6+Y_5^2$	1
73	60	$(1,4,15,20,20)$	$12k(1\le k\le 4)$	5	12	$(1,5,5)$	$Y_3^4+Y_4^3+Y_5^3$	1
			$15,30,45$	4	15	$(1,5,5)$	$Y_2^{15}+Y_4^5+Y_5^5$	1
74	60	$(1,4,5,20,30)$	$12k(1\le k\le 4)$	5	12	$(1,4,6)$	$Y_3^{12}+Y_4^3+Y_5^2$	1
			30	2	30	$(2,10,15)$	$Y_2^{15}+Y_4^3+Y_5^2$	0
75	60	$(1,5,12,12,30)$	$10k(1\le k\le 5)$	6	10	$(2,2,5)$	$Y_3^5+Y_4^5+Y_5^2$	0
76	60	$(2,3,10,15,30)$	$12k(1\le k\le 4)$	5	12	$(2,3,6)$	$Y_3^6+Y_4^4+Y_5^2$	0
			$20,40$	3	20	$(1,5,10)$	$Y_2^{20}+Y_4^4+Y_5^2$	1
			30	2	30	$(1,5,15)$	$Y_1^{30}+Y_3^6+Y_5^2$	2
77	60	$(2,3,15,20,20)$	$12k(1\le k\le 4)$	5	12	$(3,4,4)$	$Y_3^4+Y_4^3+Y_5^3$	0
			30	2	30	$(1,10,10)$	$Y_1^{30}+Y_4^3+Y_5^3$	1
78	60	$(2,3,5,20,30)$	$12k(1\le k\le 4)$	5	12	$(1,4,6)$	$Y_3^{12}+Y_4^3+Y_5^2$	1
			30	2	30	$(1,10,15)$	$Y_1^{30}+Y_4^3+Y_5^2$	1
79	60	$(3,10,12,15,20)$	$12k(1\le k\le 4)$	5	12	$(2,3,4)$	$Y_2^6+Y_4^4+Y_5^3$	1
			$20,40$	3	20	$(1,4,5)$	$Y_1^{20}+Y_3^5+Y_5^4$	6
			30	2	30	$(5,6,10)$	$Y_2^6+Y_4^5+Y_5^3$	0
80	60	$(3,3,4,20,30)$	$20,40$	3	20	$(1,1,10)$	$Y_1^{20}+Y_2^{20}+Y_5^4$	9
			30	2	30	$(2,10,15)$	$Y_3^{15}+Y_4^3+Y_5^2$	0

TABLE 17. Dimension 1 singularities: continued

#	m	Q	ℓ	c_{I_ℓ}	m'	Q'	Σ	g
81	60	$(3,5,10,12,30)$	$12k (1 \le k \le 4)$	5	12	$(1,2,6)$	$Y_2^{12}+Y_3^6+Y_5^2$	2
			$20,40$	3	20	$(1,4,10)$	$Y_1^{20}+Y_4^4+Y_5^2$	2
			30	2	30	$(5,6,15)$	$Y_3^6+Y_4^6+Y_5^2$	0
82	60	$(3,5,12,20,20)$	$12k (1 \le k \le 4)$	5	12	$(1,4,4)$	$Y_2^{12}+Y_4^3+Y_5^3$	1
			$15,30,45$	4	15	$(3,5,5)$	$Y_3^5+Y_4^3+Y_5^3$	0
83	60	$(4,5,6,15,30)$	$12k (1 \le k \le 4)$	5	12	$(1,3,6)$	$Y_2^{12}+Y_4^4+Y_5^2$	1
			$20,40$	3	20	$(2,5,10)$	$Y_3^{10}+Y_4^4+Y_5^2$	0
			30	2	30	$(2,3,15)$	$Y_1^{15}+Y_3^{10}+Y_5^2$	2
84	60	$(4,6,15,15,20)$	$12k (1 \le k \le 4)$	5	12	$(3,3,4)$	$Y_3^4+Y_4^4+Y_5^3$	0
			$20,40$	3	20	$(2,5,5)$	$Y_2^{10}+Y_3^4+Y_4^4$	1
			30	2	30	$(2,3,10)$	$Y_1^{15}+Y_3^{10}+Y_5^3$	4
85	66	$(2,3,6,22,33)$	$22,44$	3	22	$(1,2,11)$	$Y_2^{22}+Y_3^7+Y_5^2$	5
			33	2	33	$(1,3,11)$	$Y_1^{33}+Y_3^{11}+Y_4^3$	10
86	70	$(1,10,10,14,35)$	$14,28,52$	5	14	$(2,2,7)$	$Y_2^7+Y_3^7+Y_5^2$	0
			35	2	35	$(5,5,7)$	$Y_2^7+Y_3^7+Y_4^5$	0
87	70	$(2,5,14,14,35)$	$10k (1 \le k \le 6)$	7	10	$(2,2,5)$	$Y_3^5+Y_4^5+Y_5^2$	0
			35	2	35	$(1,7,7)$	$Y_1^{35}+Y_3^5+Y_4^5$	6
88	72	$(1,2,9,24,36)$	$24,48$	3	24	$(3,8,12)$	$Y_3^8+Y_4^3+Y_5^2$	0
			36	2	36	$(1,12,18)$	$Y_1^{36}+Y_4^3+Y_5^2$	1
89	72	$(1,3,8,24,36)$	$24,48$	3	24	$(1,8,12)$	$Y_2^{24}+Y_4^3+Y_5^2$	1
			$18,36,54$	4	18	$(2,6,9)$	$Y_3^9+Y_4^3+Y_5^2$	0
90	72	$(1,8,9,18,36)$	$8k (1 \le k \le 8)$	9	8	$(1,2,4)$	$Y_3^8+Y_4^4+Y_5^2$	1
			36	2	36	$(4,9,18)$	$Y_2^9+Y_4^4+Y_5^2$	0
91	78	$(1,6,6,26,39)$	$26,52$	3	26	$(2,2,13)$	$Y_2^{13}+Y_3^{13}+Y_5^2$	0
			39	2	39	$(3,3,13)$	$Y_2^{13}+Y_3^{13}+Y_4^3$	0
92	84	$(1,1,12,28,42)$	42	2	42	$(6,7,21)$	$Y_3^7+Y_4^6+Y_5^2$	0
93	84	$(1,6,14,21,42)$	$12k (1 \le k \le 6)$	7	12	$(2,3,6)$	$Y_3^6+Y_4^4+Y_5^2$	0
			$28,56$	3	28	$(2,7,14)$	$Y_2^{14}+Y_4^4+Y_5^2$	0
			42	2	43	$(3,7,21)$	$Y_2^{14}+Y_3^6+Y_5^2$	0
94	84	$(1,6,21,28,28)$	$12k (1 \le k \le 6)$	7	12	$(3,4,4)$	$Y_3^4+Y_4^3+Y_5^3$	1
			42	2	42	$(3,14,14)$	$Y_2^{14}+Y_4^3+Y_5^3$	0
95	84	$(1,6,7,28,42)$	$12k (1 \le k \le 6)$	7	12	$(1,4,6)$	$Y_3^{12}+Y_4^3+Y_5^2$	1
			42	2	42	$(3,14,21)$	$Y_2^{14}+Y_4^3+Y_5^2$	0
96	84	$(2,12,21,21,28)$	$12k (1 \le k \le 6)$	7	12	$(3,3,4)$	$Y_3^4+Y_4^4+Y_5^3$	0
			$28,56$	3	28	$(4,7,7)$	$Y_2^7+Y_3^4+Y_4^4$	0
			42	2	42	$(1,6,14)$	$Y_1^{42}+Y_2^7+Y_5^3$	6
97	84	$(2,7,12,21,42)$	$12k (1 \le k \le 6)$	7	12	$(1,3,6)$	$Y_2^{12}+Y_4^4+Y_5^2$	1
			$28,56$	3	28	$(4,7,14)$	$Y_3^7+Y_4^4+Y_5^2$	0
			42	2	42	$(1,6,21)$	$Y_1^{42}+Y_3^7+Y_5^2$	3
98	84	$(3,4,14,21,42)$	$12k (1 \le k \le 6)$	7	12	$(2,3,6)$	$Y_3^6+Y_4^4+Y_5^2$	0
			$28,56$	3	28	$(1,7,14)$	$Y_1^{28}+Y_4^4+Y_5^2$	1
			42	2	42	$(2,7,21)$	$Y_2^{21}+Y_3^6+Y_5^2$	1
99	84	$(3,4,21,28,28)$	$12k (1 \le k \le 6)$	7	12	$(3,4,4)$	$Y_3^4+Y_4^3+Y_5^3$	0
			$21,42,63$	4	21	$(1,7,7)$	$Y_2^{21}+Y_4^3+Y_5^3$	1
100	84	$(3,4,7,28,42)$	$12k (1 \le k \le 6)$	7	12	$(1,4,6)$	$Y_3^{12}+Y_4^3+Y_5^2$	1
			42	2	42	$(2,14,21)$	$Y_2^{21}+Y_4^3+Y_5^2$	0
101	90	$(1,5,9,30,45)$	$30,60$	3	30	$(3,10,15)$	$Y_3^{10}+Y_4^3+Y_5^2$	0
			$18,36,54$	5	18	$(1,6,9)$	$Y_2^{13}+Y_4^3+Y_5^2$	1
102	90	$(2,10,15,18,45)$	$30,60$	3	30	$(5,6,15)$	$Y_3^6+Y_4^3+Y_5^2$	0
			$18,36,54$	5	18	$(2,3,9)$	$Y_2^9+Y_3^6+Y_5^2$	1
			45	2	45	$(1,5,9)$	$Y_1^{45}+Y_2^9+Y_5^5$	16

TABLE 18. Dimension 1 singularities: continued

#	m	Q	ℓ	c_{I_ℓ}	m'	Q'	Σ	g
103	90	$(2,3,10,30,45)$	$30,60$	3	30	$(1,10,15)$	$Y_2^{30}+Y_4^3+Y_5^2$	1
			$18,36,54$	5	18	$(2,6,9)$	$Y_3^9+Y_4^6+Y_5^2$	0
			45	2	45	$(1,5,15)$	$Y_1^{45}+Y_3^7+Y_4^3$	7
104	96	$(1,3,12,32,48)$	$24,48,72$	4	24	$(3,8,12)$	$Y_3^8+Y_4^3+Y_5^2$	0
			$32,64$	3	32	$(1,4,16)$	$Y_2^{32}+Y_3^8+Y_5^2$	3
105	100	$(1,4,20,25,50)$	$20k(1\le k\le 4)$	5	20	$(4,5,10)$	$Y_3^5+Y_4^4+Y_5^2$	0
			50	2	50	$(2,10,25)$	$Y_2^{25}+Y_3^5+Y_5^2$	0
106	110	$(1,10,22,22,55)$	$10k(1\le k\le 9)$	11	10	$(2,2,5)$	$Y_3^5+Y_4^5+Y_5^2$	0
			55	2	55	$(5,11,11)$	$Y_2^5+Y_3^5+Y_4^5$	0
107	120	$(1,15,20,24,60)$	$24k(1\le k\le 4)$	5	24	$(3,4,12)$	$Y_2^8+Y_3^6+Y_5^2$	0
			$30k(1\le k\le 3)$	4	30	$(5,6,10)$	$Y_3^6+Y_4^5+Y_5^3$	0
			$40,80$	3	40	$(5,8,20)$	$Y_2^8+Y_4^5+Y_5^2$	0
108	120	$(1,15,24,40,40)$	$15k(1\le k\le 7)$	4	30	$(6,10,10)$	$Y_3^5+Y_4^3+Y_5^3$	0
			$24k(1\le k\le 4)$	5	24	$(3,8,8)$	$Y_2^8+Y_4^3+Y_5^3$	0
109	120	$(1,4,15,40,60)$	$24k(1\le k\le 4)$	5	24	$(3,8,12)$	$Y_2^{30}+Y_4^3+Y_5^2$	0
			$30k(1\le k\le 3)$	4	30	$(1,10,15)$	$Y_2^{30}+Y_4^3+Y_5^2$	1
110	120	$(1,5,24,30,60)$	$20k(1\le k\le 5)$	6	20	$(4,5,10)$	$Y_3^5+Y_4^4+Y_5^2$	0
			$24k(1\le k\le 4)$	5	24	$(1,6,12)$	$Y_2^{24}+Y_3^4+Y_4^2$	1
111	120	$(2,3,15,40,60)$	$24k(1\le k\le 4)$	5	24	$(3,8,12)$	$Y_3^8+Y_4^3+Y_5^2$	0
			$40,80$	3	40	$(1,5,20)$	$Y_2^{40}+Y_3^8+Y_5^2$	3
			60	2	60	$(1,20,30)$	$Y_1^{60}+Y_3^3+Y_5^2$	1
112	120	$(3,5,12,40,60)$	$24k(1\le k\le 4)$	5	24	$(1,8,12)$	$Y_2^{24}+Y_4^2+Y_5^2$	1
			$30k(1\le k\le 3)$	4	30	$(3,10,15)$	$Y_3^{10}+Y_4^3+Y_5^2$	0
			$40,80$	3	40	$(1,4,20)$	$Y_1^{40}+Y_3^{10}+Y_5^2$	4
113	126	$(1,2,18,42,63)$	$42,84$	3	42	$(6,14,21)$	$Y_3^7+Y_4^6+Y_5^2$	0
			63	2	63	$(1,9,21)$	$Y_2^{63}+Y_3^7+Y_4^3$	6
114	126	$(1,6,14,42,63)$	$18k(1\le k\le 6)$	7	18	$(2,6,9)$	$Y_3^7+Y_4^3+Y_5^2$	0
			63	2	63	$(3,7,21)$	$Y_2^{21}+Y_3^9+Y_4^3$	1
115	140	$(1,14,20,35,70)$	$28k(1\le k\le 4)$	5	28	$(4,7,14)$	$Y_3^7+Y_4^4+Y_5^2$	0
			$20k(1\le k\le 6)$	7	20	$(2,5,10)$	$Y_2^{10}+Y_3^4+Y_5^2$	0
			70	2	70	$(7,14,35)$	$Y_1^{70}+Y_3^5+Y_5^2$	0
116	140	$(2,5,28,35,70)$	$20k(1\le k\le 6)$	7	20	$(4,5,10)$	$Y_3^5+Y_4^4+Y_5^2$	0
			$28k(1\le k\le 4)$	5	28	$(1,7,14)$	$Y_2^{28}+Y_4^4+Y_5^2$	1
			70	2	70	$(1,14,35)$	$Y_1^{70}+Y_3^5+Y_5^2$	2
117	156	$(1,12,13,52,78)$	$12k(1\le k\le 12)$	13	12	$(1,4,6)$	$Y_3^{12}+Y_4^3+Y_5^2$	1
			78	2	78	$(6,26,39)$	$Y_2^{13}+Y_4^3+Y_5^2$	0
118	156	$(1,12,26,39,78)$	$12k(1\le k\le 12)$	13	12	$(2,3,6)$	$Y_3^6+Y_4^4+Y_5^2$	0
			$52,104$	3	52	$(4,13,26)$	$Y_2^{13}+Y_4^4+Y_5^2$	0
			78	2	78	$(6,13,39)$	$Y_2^{13}+Y_3^6+Y_5^2$	0
119	156	$(1,12,39,52,52)$	$12k(1\le k\le 12)$	13	12	$(3,4,4)$	$Y_3^4+Y_4^3+Y_5^3$	0
			$39k(1\le k\le 3)$	4	39	$(3,13,13)$	$Y_2^{13}+Y_4^3+Y_5^3$	0
120	168	$(1,3,24,56,84)$	$42k(1\le k\le 3)$	4	42	$(6,14,21)$	$Y_3^7+Y_4^3+Y_5^2$	0
			$56,112$	3	56	$(1,8,28)$	$Y_2^{56}+Y_3^7+Y_5^2$	3
121	168	$(1,6,21,56,84)$	$24k(1\le k\le 6)$	7	24	$(3,8,12)$	$Y_3^8+Y_4^3+Y_5^2$	0
			84	2	84	$(3,28,42)$	$Y_2^{28}+Y_4^3+Y_5^2$	0
			$56,112$	3	56	$(2,7,28)$	$Y_2^{28}+Y_3^8+Y_5^2$	1
122	168	$(3,4,21,56,84)$	$24k(1\le k\le 6)$	7	24	$(3,8,12)$	$Y_3^8+Y_4^3+Y_5^2$	0
			$42k(1\le k\le 3)$	4	42	$(1,14,21)$	$Y_2^{42}+Y_4^3+Y_5^2$	1
			$56,112$	3	56	$(1,7,28)$	$Y_1^{56}+Y_3^8+Y_5^2$	3
123	180	$(1,9,20,60,90)$	$18k(1\le k\le 9)$	10	18	$(2,6,9)$	$Y_3^9+Y_4^3+Y_5^2$	0
			$60,120$	3	60	$(3,20,30)$	$Y_2^{20}+Y_4^3+Y_5^2$	0

TABLE 19. Dimension 1 singularities: contiuned

#	m	Q	ℓ	c_{I_ℓ}	m'	Q'	Σ	g
124	180	$(4,5,36,45,90)$	$20k(1 \le k \le 8)$	9	20	$(4,5,10)$	$Y_3^5+Y_4^4+Y_5^2$	0
			$36k(1 \le k \le 4)$	5	36	$(1,9,18)$	$Y_2^{36}+Y_4^4+Y_5^2$	1
			90	2	90	$(2,18,45)$	$Y_1^{45}+Y_2^5+Y_5^2$	0
125	198	$(2,9,22,66,99)$	$18k(1 \le k \le 10)$	11	18	$(2,6,9)$	$Y_3^9+Y_4^3+Y_5^2$	0
			$66, 132$	3	66	$(3,22,33)$	$Y_2^{33}+Y_4^3+Y_5^2$	0
			99	2	99	$(1,11,33)$	$Y_1^{99}+Y_3^9+Y_4^3$	7
126	210	$(2,3,30,70,105)$	$42k(1 \le k \le 4)$	5	42	$(6,14,21)$	$Y_3^7+Y_4^3+Y_5^2$	0
			$70, 140$	3	70	$(1,10,35)$	$Y_2^{70}+Y_3^7+Y_5^2$	3
			105	2	105	$(1,15,35)$	$Y_1^{105}+Y_3^7+Y_5^3$	6
127	210	$(6,14,15,70,105)$	$42k(1 \le k \le 4)$	5	42	$(3,14,21)$	$Y_3^{14}+Y_4^3+Y_5^2$	0
			$30k(1 \le k \le 6)$	7	30	$(2,10,15)$	$Y_2^{15}+Y_4^3+Y_5^2$	0
			$70, 140$	3	70	$(2,5,35)$	$Y_1^{35}+Y_3^4+Y_4^2$	3
			105	2	105	$(3,7,35)$	$Y_1^{35}+Y_2^{15}+Y_4^3$	4
128	216	$(1,8,27,72,108)$	$24k(1 \le k \le 8)$	9	24	$(3,8,12)$	$Y_2^8+Y_3^3+Y_4^2$	0
			$54k(1 \le k \le 3)$	3	72	$(9,24,36)$	$Y_2^8+Y_3^3+Y_5^2$	0
129	220	$(1,10,44,55,110)$	$20k(1 \le k \le 10)$	11	20	$(4,5,10)$	$Y_3^5+Y_4^4+Y_5^2$	0
			$44k(1 \le k \le 4)$	5	44	$(2,11,22)$	$Y_2^{22}+Y_4^4+Y_5^2$	0
			110	2	110	$(1,22,55)$	$Y_1^{110}+Y_3^5+Y_5^2$	0
130	240	$(1,15,24,80,120)$	$30k(1 \le k \le 7)$	8	30	$(3,10,15)$	$Y_3^{10}+Y_4^3+Y_5^2$	0
			$48k(1 \le k \le 4)$	5	48	$(3,16,24)$	$Y_2^{16}+Y_4^3+Y_5^2$	0
			$80, 160$	3	80	$(5,8,40)$	$Y_2^{16}+Y_3^{10}+Y_5^2$	0
131	264	$(3,8,33,88,132)$	$24k(1 \le k \le 10)$	11	24	$(3,8,12)$	$Y_3^8+Y_4^3+Y_5^2$	0
			$66k(1 \le k \le 3)$	4	66	$(2,22,33)$	$Y_2^{33}+Y_4^6+Y_5^2$	0
			$88, 176$	3	88	$(1,11,44)$	$Y_1^{88}+Y_3^8+Y_5^2$	3
132	294	$(1,6,42,98,147)$	$42k(1 \le k \le 6)$	7	42	$(6,14,21)$	$Y_3^7+Y_4^3+Y_5^2$	0
			$98k(1 \le k \le 2)$	3	98	$(2,14,45)$	$Y_2^{46}+Y_3^7+Y_5^2$	0
			147	2	147	$(3,21,49)$	$Y_2^{49}+Y_2^7+Y_4^3$	0
133	312	$(1,12,39,104,156)$	$24k(1 \le k \le 12)$	13	24	$(3,8,12)$	$Y_3^8+Y_4^3+Y_5^2$	0
			$78k(1 \le k \le 3)$	2	156	$(6,52,78)$	$Y_2^{26}+Y_4^6+Y_5^2$	0
			$104, 208$	3	104	$(4,13,52)$	$Y_2^{26}+Y_3^8+Y_5^2$	0
134	330	$(3,22,30,110,165)$	$66k(1 \le k \le 4)$	5	66	$(6,22,33)$	$Y_3^{11}+Y_4^3+Y_5^2$	0
			$30k(1 \le k \le 10)$	11	30	$(2,10,15)$	$Y_2^{15}+Y_4^3+Y_5^2$	0
			$110, 220$	3	110	$(1,10,55)$	$Y_1^{110}+Y_3^{11}+Y_5^2$	5
135	336	$(1,7,48,112,168)$	$42k(1 \le k \le 7)$	8	42	$(7,14,21)$	$Y_3^6+Y_4^3+Y_5^2$	0
			$48k(1 \le k \le 6)$	7	48	$(1,16,24)$	$Y_2^{48}+Y_4^3+Y_5^2$	1
136	342	$(1,18,38,114,171)$	$18k(1 \le k \le 18)$	19	18	$(2,6,9)$	$Y_3^9+Y_4^3+Y_5^2$	0
			$114, 228$	3	114	$(6,38,57)$	$Y_2^{19}+Y_4^3+Y_5^2$	0
			171	2	171	$(9,19,57)$	$Y_2^{19}+Y_3^9+Y_5^3$	0
137	378	$(2,7,54,126,189)$	$42k(1 \le k \le 8)$	9	42	$(6,14,21)$	$Y_3^7+Y_4^3+Y_5^2$	0
			$54k(1 \le k \le 6)$	7	54	$(1,18,27)$	$Y_2^{54}+Y_4^3+Y_5^2$	1
			189	2	189	$(1,27,63)$	$Y_1^{189}+Y_3^7+Y_4^3$	6
138	420	$(1,20,84,105,210)$	$20k(1 \le k \le 20)$	21	20	$(4,5,10)$	$Y_3^5+Y_4^4+Y_5^2$	0
			$84k(1 \le k \le 4)$	5	84	$(4,21,42)$	$Y_2^{21}+Y_4^4+Y_5^2$	0
			210	2	210	$(10,42,105)$	$Y_2^{21}+Y_3^5+Y_5^2$	0
139	420	$(3,7,60,140,210)$	$42k(1 \le k \le 9)$	10	42	$(6,14,21)$	$Y_3^7+Y_4^3+Y_5^2$	0
			$60k(1 \le k \le 6)$	7	60	$(1,20,30)$	$Y_2^{60}+Y_4^3+Y_5^2$	1
			$140, 280$	3	140	$(1,20,70)$	$Y_1^{140}+Y_3^7+Y_5^2$	3
140	462	$(2,33,42,154,231)$	$42k(1 \le k \le 10)$	11	42	$(3,14,21)$	$Y_2^{12}+Y_4^3+Y_5^2$	0
			$66k(1 \le k \le 6)$	7	66	$(6,22,33)$	$Y_3^{11}+Y_4^3+Y_5^2$	0
			$154, 308$	3	154	$(11,14,77)$	$Y_2^{14}+Y_3^{11}+Y_5^2$	0
			231	2	231	$(1,21,77)$	$Y_1^{231}+Y_3^{11}+Y_4^3$	10

TABLE 20. Dimension 1 singularities: continued

#	m	Q	ℓ	c_{I_ℓ}	m'	Q'	Σ	g
141	546	$(6,7,78,182,273)$	$42k(1\le k\le 12)$	13	42	$(6,14,21)$	$Y_3^7+Y_4^3+Y_5^2$	0
			$78k(1\le k\le 8)$	7	78	$(1,26,39)$	$Y_2^{78}+Y_4^3+Y_5^2$	1
			$156k(1\le k\le 3)$	7	78	$(1,26,39)$	$Y_2^{78}+Y_4^3+Y_5^2$	1
			$182,364$	3	182	$(2,26,91)$	$Y_1^{91}+Y_3^7+Y_5^2$	0
142	600	$(1,24,75,200,300)$	$24k(1\le k\le 24)$	25	24	$(3,8,12)$	$Y_3^8+Y_4^3+Y_5^2$	0
			$150k(1\le k\le 3)$	4	150	$(6,50,75)$	$Y_2^{25}+Y_4^3+Y_5^2$	0
			$200,400$	3	200	$(8,25,100)$	$Y_2^{15}+Y_3^8+Y_4^2$	0
143	630	$(1,14,90,210,315)$	$42k(1\le k\le 24)$	15	42	$(6,14,21)$	$Y_3^7+Y_4^3+Y_5^2$	0
			$90k(1\le k\le 6)$	7	90	$(2,30,45)$	$Y_2^{45}+Y_4^3+Y_5^2$	0
			315	2	315	$(7,45,105)$	$Y_2^{45}+Y_3^7+Y_4^3$	0
144	714	$(3,14,102,238,357)$	$42k(1\le k\le 16)$	17	42	$(6,14,21)$	$Y_3^7+Y_4^3+Y_5^2$	0
			$102k(1\le k\le 6)$	7	102	$(2,34,51)$	$Y_2^{51}+Y_4^3+Y_5^2$	0
			$238k(1\le k\le 2)$	3	238	$(1,34,119)$	$Y_1^{238}+Y_3^7+Y_5^2$	3
			357	2	357	$(7,51,119)$	$Y_2^{51}+Y_3^7+Y_4^3$	0
145	924	$(1,21,132,308,462)$	$42k(1\le k\le 21)$	22	42	$(6,14,21)$	$Y_3^7+Y_4^3+Y_5^2$	0
			$132k(1\le k\le 6)$	7	132	$(3,44,66)$	$Y_2^{44}+Y_4^3+Y_5^2$	0
			$308,616$	3	308	$(7,44,154)$	$Y_2^{44}+Y_3^7+Y_5^2$	0
146	966	$(2,21,138,322,483)$	$42k(1\le k\le 22)$	23	42	$(6,14,21)$	$Y_3^7+Y_4^3+Y_5^2$	0
			$138k(1\le k\le 6)$	7	138	$(3,46,69)$	$Y_2^{46}+Y_4^3+Y_5^2$	0
			483	2	483	$(1,69,161)$	$Y_1^{483}+Y_3^7+Y_4^3$	6
147	1806	$(1,42,258,602,903)$	$42k(1\le k\le 42)$	43	42	$(6,14,21)$	$Y_3^7+Y_4^3+Y_5^2$	0
			$258k(1\le k\le 6)$	7	258	$(6,86,149)$	$Y_2^{43}+Y_4^3+Y_5^2$	0
			$602,1204$	3	602	$(14,86,301)$	$Y_2^{43}+Y_3^7+Y_5^2$	0
			903	2	903	$(21,129,301)$	$Y_2^{43}+Y_3^7+Y_4^3$	0

Appendix B. The L-Series of Cubic Hypersurface Fourfolds

Yasuhiro Goto
Department of Mathematics
Hokkaido University of Education at Hakodate
1-2 Hachiman-cho, Hakodate, 040-8567 Japan
E-mail: ygoto@cc.hokkyodai.ac.jp

The purpose of this appendix is to compute the zeta-functions and L-series of cubic hypersurface fourfolds over a number field. We also discuss the L-series of some related fourfolds, especially of those related by explicit birational maps. The basic machinery used for these computations is classical and our results are mostly expository. Based on the results on zeta-functions, we give a brief discussion on supersingular fourfolds in positive characteristic.

B.1. Zeta-functions and L-series. As before, let $\mathbb{K}$ be a number field and $\mathcal{O}_{\mathbb{K}}$ the ring of integers of $\mathbb{K}$. Let V be a cubic hypersurface fourfold over $\mathbb{K}$ defined by the equation:

$$c_1x_1^3 + c_2x_2^3 + c_3x_3^3 + c_4x_4^3 + c_5x_5^3 + c_6x_6^3 = 0$$

in $\mathbb{P}^5_{\mathbb{K}}$ with $\mathbf{c} := (c_1, c_2, c_3, c_4, c_5, c_6) \in (\mathbb{K}^\times)^6$. It is smooth over $\mathbb{K}$ and is a Fano fourfold. For a prime $\mathfrak{p} \in \operatorname{Spec} \mathcal{O}_{\mathbb{K}}$, let $V_{\mathfrak{p}}$ denote the reduction modulo $\mathfrak{p}$; write $k := \mathcal{O}_{\mathbb{K}}/\mathfrak{p}$ and $q := \#k =\mid N\mathfrak{p} \mid$.

If $\mathfrak{p} \not| 3$ and $\mathbf{c} \in (k^\times)^6$, then $V_{\mathfrak{p}}$ is a smooth fourfold over k (i.e. $\mathfrak{p}$ is a good prime). Otherwise, $V_{\mathfrak{p}}$ is either reducible or singular, and hence $\mathfrak{p}$ is bad. When $V_{\mathfrak{p}}$ has a singularity, we can still calculate the zeta-functions and L-series of V, but the description of them becomes very complicated. Therefore for the sake of simplicity, we choose the trivial twist $\mathbf{c} = (1, 1, 1, 1, 1, 1)$ for the rest of the appendix, so that V is a cubic Fermat fourfold

$$x_1^3 + x_2^3 + x_3^3 + x_4^3 + x_5^3 + x_6^3 = 0.$$

Furthermore, we fix a primitive cube root of unity, ω, in $\mathbb{C}$ and choose

$$\mathbb{K} = \mathbb{Q}(\omega).$$

Then the primes satisfying $\mathfrak{p} \not| 3$ are good and satisfy $q \equiv 1 \pmod 3$. There is a unique bad prime with $\mathfrak{p} \mid 3$, namely $\mathfrak{p} = (1 - \omega)$, and in this case, $V_{\mathfrak{p}}$ can be reduced to a rational fourfold

$$x_1 + x_2 + x_3 + x_4 + x_5 + x_6 = 0.$$

Recall that the zeta-function of $V_{\mathfrak{p}}$ has the form $Z(V_{\mathfrak{p}}, T) = \prod_{i=0}^{8} P_{\mathfrak{p},i}(T)^{(-1)^{i+1}}$,

where

$$P_{\mathfrak{p},i}(T) = \det(1 - \mathrm{Frob}^*{}_{\mathfrak{p}}\, T \mid H^i_{et}(\bar{X}, \mathbb{Q}_\ell))$$

and the L-series of V is defined to be

$$L(V, s) = \frac{L_0(V, s)L_2(V, s) \cdots L_8(V, s)}{L_1(V, s) \cdots L_7(V, s)}$$

with

$$L_i(V, s) = \prod_{\mathfrak{p} \in \operatorname{Spec} \mathcal{O}_{\mathbb{K}}} \frac{1}{P_{\mathfrak{p},i}(N\mathfrak{p}^{-s})}.$$

Recall also that $\zeta_{\mathbb{K}}(s)$ denotes the Dedekind zeta function of $\mathbb{K}$, i.e.

$$\zeta_{\mathbb{K}}(s) = \prod_{\mathfrak{p}\in \operatorname{Spec} \mathcal{O}_{\mathbb{K}}} \frac{1}{1 - N\mathfrak{p}^{-s}}.$$

For $\mathfrak{p} \in \operatorname{Spec} \mathcal{O}_{\mathbb{K}}$, let $\chi_{\mathfrak{p}}$ be a cubic character of $\mathcal{O}_{\mathbb{K}}/\mathfrak{p}$. As in Definition 5.1, we write $G(\chi_{\mathfrak{p}})$ for the Gauss sum associated to $\chi_{\mathfrak{p}}$ and

$$\mathfrak{A} = \{\mathbf{a} \in \widehat{\mathfrak{G}} \mid a_i \in \mathbb{Z}/3\mathbb{Z},\ a_i \not\equiv 0 \pmod 3),\ \sum a_i \equiv 0 \pmod 3)\ \}.$$

Given $\mathbf{a} = (a_1, a_2, \cdots, a_6) \in \mathfrak{A}$ and $\mathfrak{p} \not| 3$, the associated Jacobi sum (of dimension 4) is defined by

$$j_{\mathfrak{p}}(\mathbf{a}) = \sum \chi_{\mathfrak{p}}(x_2)^{a_2}\chi_{\mathfrak{p}}(x_3)^{a_3}\cdots\chi_{\mathfrak{p}}(x_6)^{a_6}$$

where the sum runs over all $(x_1, x_2, \cdots, x_6)$ with $x_i \pmod{\mathfrak{p}}$ subject to the relation $1 + x_2 + \cdots + x_6 \equiv 0 \pmod{\mathfrak{p}}$.

PROPOSITION B.1. *Assume $\mathfrak{p} \not| 3$. Put $q = | N\mathfrak{p} |$. Then the zeta-function of $V_{\mathfrak{p}}$ is described as follows:*

$$Z(V_{\mathfrak{p}}, T) = \frac{1}{(1-T)(1-qT)P_4(V_{\mathfrak{p}}, T)(1-q^3T)(1-q^4T)}$$

where

$$P_4(V_{\mathfrak{p}}, T) = (1-q^2T)^{21}(1-j_{\mathfrak{p}}(\mathbf{a}_1)T)(1-j_{\mathfrak{p}}(\mathbf{a}_2)T)$$

with $\mathbf{a}_1 := (1,1,1,1,1,1)$ *and* $\mathbf{a}_2 := (2,2,2,2,2,2)$.

PROOF. Since $q \equiv 1 \pmod 3$, we can apply Weil's result [W49] to compute $Z(V_{\mathfrak{p}}, T)$. Then $P_4(V_{\mathfrak{p}}, T)$ is the only non-trivial factor and it is given as

$$P_4(V_{\mathfrak{p}}, T) = (1-q^2T)\prod_{\mathbf{a}\in\mathfrak{A}}(1 - j_{\mathfrak{p}}(\mathbf{a})T)$$

where

$$\mathfrak{A} = \left\{ \begin{array}{c} (1,1,1,1,1,1), \quad (2,2,2,2,2,2) \\ 20 \text{ copies of } (1,1,1,2,2,2) \text{ up to permutation} \end{array} \right\}.$$

Now noting $G(\chi_{\mathfrak{p}})G(\chi_{\mathfrak{p}}^2) = G(\chi_{\mathfrak{p}})G(\overline{\chi_{\mathfrak{p}}}) = \chi_{\mathfrak{p}}(-1)q = \chi_{\mathfrak{p}}((-1)^3)q = q$, we see

$$qj_{\mathfrak{p}}((1,1,1,2,2,2)) = \{G(\chi_{\mathfrak{p}})G(\chi_{\mathfrak{p}}^2)\}^3 = q^3.$$

Hence $j_{\mathfrak{p}}((1,1,1,2,2,2)) = q^2$. This gives our result. □

COROLLARY B.2. *The odd Betti numbers are all equal to 0. All even Betti numbers except for the fourth one are equal to 1. The fourth Betti number of V is equal to 23. The Hodge diamond of this variety is given as follows:*

$$\begin{array}{ccccccccc}
 & & & & 1 & & & & \\
 & & & 0 & & 0 & & & \\
 & & 0 & & 1 & & 0 & & \\
 & 0 & & 0 & & 0 & & 0 & \\
0 & & 1 & & 21 & & 1 & & 0 \\
 & 0 & & 0 & & 0 & & 0 & \\
 & & 0 & & 1 & & 0 & & \\
 & & & 0 & & 0 & & & \\
 & & & & 1 & & & &
\end{array}$$

PROOF. This follows from Proposition B.1 by counting the degree of $P_{\mathfrak{p},i}(T)$ for $0 \le i \le 8$. $\square$

REMARK B.1. The Hodge numbers of this variety were computed by Rapoport [Rap72]. The Hodge polygon of V in dimension 4 has slopes $1, 2, 3$ with multiplicities $1, 21, 1$, respectively. If $\mathfrak{p} \not| 3$, then V is ordinary, that is, the Newton polygon of $P_{\mathfrak{p},4}(T)$ coincides with the Hodge polygon of V in dimension 4, and $V_{\mathfrak{p}}$ is of $K3$ type in the sense of Zarhin [Z04]. (Note also that the Hodge diamond above is very similar to that of a $K3$ surface.) Furthermore, the Tate conjecture for algebraic cycles (see Subsection B.3 below) has been established for ordinary $V_{\mathfrak{p}}$ by Levin [Lev01].

PROPOSITION B.3. *Assume* $\mathfrak{p} \mid 3$. *Then the zeta-function of* $V_{\mathfrak{p}}$ *is given as follows:*

$$Z(V_{\mathfrak{p}}, T) = \frac{1}{(1-T)(1-qT)(1-q^2T)(1-q^3T)(1-q^4T)}.$$

PROOF. In this case, $V_{\mathfrak{p}}$ is defined by $(x_1+x_2+x_3+x_4+x_5+x_6)^3 = 0$ and has the same number of rational points as the rational fourfold $x_1+x_2+x_3+x_4+x_5+x_6 = 0$ (although they are different as schemes). Hence the zeta-function of $V_{\mathfrak{p}}$ is computed as above. $\square$

THEOREM B.4. *Let* $\mathbb{K} = \mathbb{Q}(\omega)$ *and put* $\mathfrak{p}_0 = (1-\omega)$. *Let* V *be the cubic Fermat fourfold in* $\mathbb{P}^5_{\mathbb{K}}$ *defined by the equation*

$$V: \quad x_1^3 + x_2^3 + x_3^3 + x_4^3 + x_5^3 + x_6^3 = 0.$$

Write $\mathcal{M}_{[Q]}$ *for the motive corresponding to* $[Q] = [1,1,1,1,1,1]$. *Then the* L*-series of* V *is computed as follows:*

$$L(V, s) = \zeta_{\mathbb{K}}(s)\zeta_{\mathbb{K}}(s-1)L_4(V, s)\zeta_{\mathbb{K}}(s-3)\zeta_{\mathbb{K}}(s-4)$$

with

$$\begin{aligned} L_4(V, s) &= (1-3^{2-s})^{20}\zeta_{\mathbb{K}}(s-2)^{21}L(\mathcal{M}_{[Q]}, s) \\ &= (1-3^{2-s})^{20}\zeta_{\mathbb{K}}(s-2)^{21} \prod_{\mathfrak{p}:split} P_{[Q]}(p^{-s})^{-1} \prod_{\mathfrak{p}:inert} P_{[Q]}(p^{-2s})^{-1}. \end{aligned}$$

PROOF. It follows from Propositions B.1 and B.3 that $L_i(V, s) = 1$ when i is odd and $L_{2r}(V, s) = \zeta_{\mathbb{K}}(s-r)$ for $r = 0, 1, 3, 4$. Noting that $\mid N\mathfrak{p}_0 \mid= 3$, we find

$$\begin{aligned} L_4(V, s) &= \frac{1}{1-3^{2-s}} \prod_{\mathfrak{p} \neq \mathfrak{p}_0} \frac{1}{(1-q^{2-s})^{21}(1-j_{\mathfrak{p}}(\mathbf{a}_1)q^{-s})(1-j_{\mathfrak{p}}(\mathbf{a}_2)q^{-s})} \\ &= (1-3^{2-s})^{20}\zeta_{\mathbb{K}}(s-2)^{21} \prod_{\mathfrak{p} \neq \mathfrak{p}_0} \frac{1}{(1-j_{\mathfrak{p}}(\mathbf{a}_1)q^{-s})(1-j_{\mathfrak{p}}(\mathbf{a}_2)q^{-s})} \\ &= (1-3^{2-s})^{20}\zeta_{\mathbb{K}}(s-2)^{21}L(\mathcal{M}_{[Q]}, s) \end{aligned}$$

Furthermore, when $\mathfrak{p}$ is split (resp. inert), $N\mathfrak{p} = p$ (resp. $N\mathfrak{p} = p^2$). This gives rise to the formula we are after (cf. Section 5). $\square$

We also compute the L-series of some other related varieties and compare them with $L(V, s)$.

LEMMA B.5. *Let S be a smooth cubic Fermat surface over $\mathbb{K}$ in $\mathbb{P}^3$ defined by the equation*

$$x_1^3 + x_2^3 + x_3^3 + x_4^3 = 0.$$

(*a*) *If $\mathfrak{p} \not| 3$, then $S_{\mathfrak{p}}$ is a smooth cubic surface over $k = \mathcal{O}_{\mathbb{K}}/\mathfrak{p}$ and its zeta-function is given as follows:*

$$Z(S_{\mathfrak{p}}, T) = \frac{1}{(1-T)P_2(S_{\mathfrak{p}}, T)(1-q^2T)}$$

where

$$P_2(S_{\mathfrak{p}}, T) = (1-qT) \prod_{\mathbf{a} \in \mathfrak{A}_2} (1 - j_{\mathfrak{p}}(\mathbf{a})T) = (1-qT)^7$$

and $\mathfrak{A}_2$ is the set of 6 *copies of* $(1,1,2,2)$ *up to permutation of entries.*

(*b*) *If $\mathfrak{p} \mid 3$, then $S_{\mathfrak{p}}$ is reducible and the reduced component is a rational surface $x_1 + x_2 + x_3 + x_4 = 0$. Its zeta-function is given as follows:*

$$Z(S_{\mathfrak{p}}, T) = \frac{1}{(1-T)(1-qT)(1-q^2T)}.$$

PROOF. The formula can be obtained in a similar way as in Proposition B.1. For (a), we just note that $j_{\mathfrak{p}}((1,1,2,2)) = \frac{1}{q}(G(\chi_{\mathfrak{p}})G(\chi_{\mathfrak{p}}^2))^2 = q$. □

When $\mathfrak{p} \not| 3$, it is known (e.g. [Sh82]) that the ℓ-adic étale cohomology group $H^2_{et}(\bar{S}_{\mathfrak{p}}, \mathbb{Q}_\ell)$ is decomposed into one-dimensional pieces as

$$H^2_{et}(\bar{S}_{\mathfrak{p}}, \mathbb{Q}_\ell) \cong V(0) \oplus \bigoplus_{\mathbf{a} \in \mathfrak{A}_2} V(\mathbf{a}) \tag{1}$$

where ℓ is a prime satisfying $\ell \equiv 1 \pmod 3$, $V(0)$ is the subspace corresponding to a hyperplane section and

$$V(\mathbf{a}) = \{v \in H^2_{prim}(\bar{S}_{\mathfrak{p}}, \mathbb{Q}_\ell) \mid \gamma^*(v) = \omega_1^{a_1}\omega_2^{a_2}\omega_3^{a_3}\omega_4^{a_4}v, \ \forall\gamma = (\omega_1, \omega_2, \omega_3, \omega_4) \in \Gamma\}$$

with $\Gamma := \mu_3 \times \mu_3 \times \mu_3 \times \mu_3/(\text{diagonal elements})$ and $\mu_3 =< \omega >$ the group of cube roots of unity in $\mathbb{K}$. We use this decomposition in the proof of Lemma B.6.

Let $S \times S$ be the self-product of S. As in the Fermat inductive structure of Shioda and Katsura [SK79], we can define an action of μ_3 on $S \times S$ by

$$\begin{array}{ccc} S \times S & \longrightarrow & S \times S \\ ((y_1 : y_2 : y_3 : y_4), (z_1 : z_2 : z_3 : z_4)) & \longmapsto & ((y_1 : y_2 : y_3 : \omega y_4), (z_1 : z_2 : z_3 : \omega z_4)). \end{array}$$

LEMMA B.6. *Let S be the cubic Fermat surface in* Lemma B.5. *Write $W := (S \times S)/\mu_3$ for the quotient fourfold with respect to the μ_3-action above. Then W is defined over $\mathbb{K}$ and the following assertions hold.*

(*a*) *If $\mathfrak{p} \not| 3$, then the zeta-function of $W_{\mathfrak{p}}$ is given as follows:*

$$Z(W_{\mathfrak{p}}, T) = \frac{1}{(1-T)P_2(W_{\mathfrak{p}}, T)P_4(W_{\mathfrak{p}}, T)P_6(W_{\mathfrak{p}}, T)(1-q^4T)}$$

where

$$P_2(W_{\mathfrak{p}}, T) = (1-qT)^2$$

$$P_6(W_{\mathfrak{p}}, T) = (1-q^3T)^2$$

$$P_4(W_{\mathfrak{p}}, T) = (1-q^2T)^3 \prod_{\mathbf{a}, \mathbf{b} \in \mathfrak{A}_2'} (1 - j_{\mathfrak{p}}(\mathbf{a})j_{\mathfrak{p}}(\mathbf{b})T) = (1-q^2T)^{21}.$$

Here $\mathfrak{A}_2'$ *in the description of* $P_4(W_\mathfrak{p}, T)$ *means that* $\mathbf{a} = (a_1, a_2, a_3, a_4)$ *and* $\mathbf{b} = (b_1, b_2, b_3, b_4)$ *are subject to the condition* $a_4 + b_4 \equiv 0 \pmod 3$.

(b) *If* $\mathfrak{p} \mid 3$, *then* $W = S \times S$ *and it has the zeta-function*

$$Z(W_\mathfrak{p}, T) = \frac{1}{(1-T)(1-qT)^2(1-q^2T)^3(1-q^3T)^2(1-q^4T)}.$$

PROOF. (a) We write $Y := S \times S$ for this proof. By the Künneth formula on the ℓ-adic étale cohomology, we find the zeta-function of $Y_\mathfrak{p}$ as

$$Z(Y_\mathfrak{p}, T) = \frac{1}{(1-T)P_2(Y_\mathfrak{p}, T)P_4(Y_\mathfrak{p}, T)P_6(Y_\mathfrak{p}, T)(1-q^4T)}$$

where

$$\begin{aligned}
P_2(Y_\mathfrak{p}, T) &= (1-qT)^2 \prod_{\mathbf{a}\in\mathfrak{A}_2}(1 - j_\mathfrak{p}(\mathbf{a})T) \prod_{\mathbf{b}\in\mathfrak{A}_2}(1 - j_\mathfrak{p}(\mathbf{b})T) \\
P_6(Y_\mathfrak{p}, T) &= P_2(Y_\mathfrak{p}, q^2T) \\
P_4(Y_\mathfrak{p}, T) &= (1-q^2T)^3 \prod_{\mathbf{a},\mathbf{b}\in\mathfrak{A}_2}(1 - j_\mathfrak{p}(\mathbf{a})j_\mathfrak{p}(\mathbf{b})T) \\
&\quad \times \prod_{\mathbf{a}\in\mathfrak{A}_2}(1 - qj_\mathfrak{p}(\mathbf{a})T) \prod_{\mathbf{b}\in\mathfrak{A}_2}(1 - qj_\mathfrak{p}(\mathbf{b})T).
\end{aligned}$$

Note that the ℓ-adic étale cohomology satisfies $H^{2i}_{et}(\bar{W}_\mathfrak{p}, \mathbb{Q}_\ell) \cong H^{2i}_{et}(\bar{Y}_\mathfrak{p}, \mathbb{Q}_\ell)^{\mu_3}$, where $\bar{W}_\mathfrak{p}$ and $\bar{Y}_\mathfrak{p}$ denote the extension of coefficients to an algebraic closure of $\mathcal{O}_\mathbb{K}/\mathfrak{p}$. Considering the μ_3-action on $Y_\mathfrak{p}$, we see that $H^{2i}_{et}(\bar{Y}_\mathfrak{p}, \mathbb{Q}_\ell)^{\mu_3}$ is the direct sum of the eigenspaces corresponding to eigenvalues q^i and $j_\mathfrak{p}(\mathbf{a})j_\mathfrak{p}(\mathbf{b})$ with $a_4 + b_4 \equiv 0 \pmod 3$. In fact, for instance, the product $\prod_{\mathbf{a},\mathbf{b}\in\mathfrak{A}_2}(1 - j_\mathfrak{p}(\mathbf{a})j_\mathfrak{p}(\mathbf{b})T)$ in $P_4(Y_\mathfrak{p}, T)$ corresponds to the subspace

$$H := \left(\oplus_{\mathbf{a}\in\mathfrak{A}_2} V(\mathbf{a})\right) \otimes \left(\oplus_{\mathbf{b}\in\mathfrak{A}_2} V(\mathbf{b})\right) = \bigoplus_{\mathbf{a},\mathbf{b}\in\mathfrak{A}_2} V(\mathbf{a}) \otimes V(\mathbf{b})$$

of $H^4_{et}(\bar{Y}_\mathfrak{p}, \mathbb{Q}_\ell)$ in the decomposition (1) and H is closed under the μ_3-action. As $V(\mathbf{a})$ and $V(\mathbf{b})$ are subspaces of dimension one, non-zero vectors $v_1 \in V(\mathbf{a})$ and $v_2 \in V(\mathbf{b})$ form a basis for them respectively. Then $v_1 \otimes v_2$ is a basis for $V(\mathbf{a}) \otimes V(\mathbf{b})$ and

$$\gamma^*(v_1) = \omega_1^{a_1}\omega_2^{a_2}\omega_3^{a_3}\omega_4^{a_4} v_1, \qquad \gamma^*(v_2) = \omega_1^{b_1}\omega_2^{b_2}\omega_3^{b_3}\omega_4^{b_4} v_2$$

for $\gamma \in \Gamma$. Since μ_3 is a subgroup of Γ and it acts on the last coordinate of S, $\omega \in \mu_3$ sends $v_1 \otimes v_2$ to

$$\omega^*(v_1 \otimes v_2) = \omega^{a_4} v_1 \otimes \omega^{b_4} v_2 = \omega^{a_4+b_4} v_1 \otimes v_2.$$

Hence $V(\mathbf{a}) \otimes V(\mathbf{b})$ is invariant under the μ_3-action if and only if $a_4 + b_4 \equiv 0 \pmod 3$. This yields

$$H^{\mu_3} = \bigoplus_{\mathbf{a},\mathbf{b}\in\mathfrak{A}_2'} V(\mathbf{a}) \otimes V(\mathbf{b})$$

and the product $\prod_{\mathbf{a},\mathbf{b}\in\mathfrak{A}_2'}(1 - j_\mathfrak{p}(\mathbf{a})j_\mathfrak{p}(\mathbf{b})T)$ in $P_4(W_\mathfrak{p}, T)$ follows from this. Other factors of $Z(W_\mathfrak{p}, T)$ can be obtained similarly.

(b) If $\mathfrak{p} \mid 3$, then S becomes $x_1 + x_2 + x_3 + x_4 = 0$ and the μ_3-action is trivial on S. Hence $W = S \times S$ and the zeta-function of it follows immediately. □

THEOREM B.7. *Let $\mathbb{K} = \mathbb{Q}(\omega)$ and put $\mathfrak{p}_0 = (1-\omega)$. Let V be the cubic Fermat fourfold in $\mathbb{P}^5_{\mathbb{K}}$. Let S be the cubic Fermat surface in $\mathbb{P}^3$. Write $W = (S \times S)/\mu_3$. Then the L-series of V and W are related as follows:*

$$L(V,s) = \frac{(1-3^{2-s})^2 L(\mathcal{M}_{[Q]}, s)}{\zeta_{\mathbb{K}}(s-1)\zeta_{\mathbb{K}}(s-3)} L(W,s).$$

PROOF. It follows from Theorem B.4 that

$$L(V,s) = \zeta_{\mathbb{K}}(s)\zeta_{\mathbb{K}}(s-1)L_4(V,s)\zeta_{\mathbb{K}}(s-3)\zeta_{\mathbb{K}}(s-4)$$

with

$$L_4(V,s) = (1-3^{2-s})^{20}\zeta_{\mathbb{K}}(s-2)^{21}L(\mathcal{M}_{[Q]}, s).$$

By Lemma B.6, we find

$$L(W,s) = \zeta_{\mathbb{K}}(s)\zeta_{\mathbb{K}}(s-1)^2 L_4(W,s)\zeta_{\mathbb{K}}(s-3)^2\zeta_{\mathbb{K}}(s-4)$$

with

$$L_4(W,s) = (1-3^{2-s})^{18}\zeta_{\mathbb{K}}(s-2)^{21}.$$

Combining these equations, we obtain the asserted formula. □

The theorem shows a resemblance in two L-series. In Subsection B.2, we comment on this resemblance and describe a birational map between W and V. Here we also consider the L-series of some abelian varieties.

LEMMA B.8. *Let E be an elliptic curve over $\mathbb{K}$ defined by the equation*

$$E: \quad x_1^3 + x_2^3 + x_3^3 = 0.$$

The following assertions hold.

(a) *If $\mathfrak{p} \not| 3$, then $E_{\mathfrak{p}}$ is an elliptic curve over $k = \mathcal{O}_{\mathbb{K}}/\mathfrak{p}$ and its zeta-function is given as follows:*

$$Z(E_{\mathfrak{p}}, T) = \frac{P_1(E_{\mathfrak{p}}, T)}{(1-T)(1-qT)}$$

where $P_1(E_{\mathfrak{p}}, T) = (1 - j_{\mathfrak{p}}(1,1,1)T)(1 - j_{\mathfrak{p}}(2,2,2)T)$.

(b) *If $\mathfrak{p} \mid 3$, then $E_{\mathfrak{p}}$ is reducible and the reduced component is a rational curve $x_1 + x_2 + x_3 = 0$. Its zeta-function is given as follows:*

$$Z(E_{\mathfrak{p}}, T) = \frac{1}{(1-T)(1-qT)}.$$

PROOF. The formula can be obtained in a similar way as in Proposition B.1 and is well-known. □

LEMMA B.9. *Let E be the elliptic curve in* Lemma B.8. *Let $A = E \times E$ be the self-product of E. Then A is an abelian surface over $\mathbb{K}$ and the following assertions hold.*

(a) *If $\mathfrak{p} \not| 3$, then $A_{\mathfrak{p}}$ is an abelian surface and its zeta-function is given as follows:*

$$Z(A_{\mathfrak{p}}, T) = \frac{P_1(A_{\mathfrak{p}}, T)P_3(A_{\mathfrak{p}}, T)}{(1-T)P_2(A_{\mathfrak{p}}, T)(1-qT)}$$

where

$$\begin{aligned} P_1(A_{\mathfrak p},T) &= (1-j_{\mathfrak p}(1,1,1)T)^2(1-j_{\mathfrak p}(2,2,2)T)^2 = P_1(E_{\mathfrak p},T)^2\\ P_2(A_{\mathfrak p},T) &= (1-qT)^4(1-j_{\mathfrak p}(1,1,1)^2T)(1-j_{\mathfrak p}(2,2,2)T)\\ &= (1-qT)^4\Big(1-\frac{1}{q}j_{\mathfrak p}(\mathbf{a}_1)T\Big)\Big(1-\frac{1}{q}j_{\mathfrak p}(\mathbf{a}_2)T\Big)\\ P_3(A_{\mathfrak p},T) &= P_1(A_{\mathfrak p},qT) \end{aligned}$$

with $\mathbf{a}_1$ *and* $\mathbf{a}_2$ *being the characters defined in* Proposition B.1.

(*b*) *If* $\mathfrak p \mid 3$, *then* $A_{\mathfrak p}$ *is reducible and the reduced component has the zeta-function:*

$$Z(A_{\mathfrak p},T) = \frac{1}{(1-T)(1-qT)^2(1-q^2T)}.$$

PROOF. The result is a straightforward calculation. The only thing we should mention is

$$j_{\mathfrak p}(\mathbf{a}_1) = \frac{1}{q}G(\chi_{\mathfrak p})^6 = q\Big(\frac{1}{q}G(\chi_{\mathfrak p})^3\Big)^2 = qj_{\mathfrak p}(1,1,1)^2.$$

□

THEOREM B.10. *Let* $\mathbb{K} = \mathbb{Q}(\omega)$ *and put* $\mathfrak p_0 = (1-\omega)$. *Let* V *be the cubic Fermat fourfold in* $\mathbb{P}^5_{\mathbb{K}}$. *Write* $A = E\times E$ *for the abelian surface defined in* Lemma B.9. *Then the* L*-series of* V *can be expressed as follows:*

$$L(V,s) = \zeta_{\mathbb K}(s)\zeta_{\mathbb K}(s-1)L_4(V,s)\zeta_{\mathbb K}(s-3)\zeta_{\mathbb K}(s-4)$$

with

$$L_4(V,s) = (1-3^{2-s})^{18}\zeta_{\mathbb K}(s-2)^{17}L_2(A,s-1).$$

PROOF. By Lemma B.9, we find

$$L_2(A,s) = (1-3^{1-s})^2\zeta_{\mathbb K}(s-1)^4\prod_{\mathfrak p\neq\mathfrak p_0}\frac{1}{(1-j_{\mathfrak p}(\mathbf{a}_1)q^{-(s+1)})(1-j_{\mathfrak p}(\mathbf{a}_2)q^{-(s+1)})}$$

Hence

$$\prod_{\mathfrak p\neq\mathfrak p_0}\frac{1}{(1-j_{\mathfrak p}(\mathbf{a}_1)q^{-s})(1-j_{\mathfrak p}(\mathbf{a}_2)q^{-s})} = \frac{L_2(A,s-1)}{(1-3^{1-s})^2\zeta_{\mathbb K}(s-2)^4}. \tag{2}$$

On the other hand, recall Theorem B.4 that

$$L_4(V,s) = (1-3^{2-s})^{20}\zeta_{\mathbb K}(s-2)^{21}\prod_{\mathfrak p\neq\mathfrak p_0}\frac{1}{(1-j_{\mathfrak p}(\mathbf{a}_1)q^{-s})(1-j_{\mathfrak p}(\mathbf{a}_2)q^{-s})}.$$

Hence the formula follows from equation (2). □

B.2. Birational maps. In this subsection, we describe explicitly a birational map between V and $(S\times S)/\mu_3$, where S is a smooth cubic Fermat surface over $\mathbb K$ in $\mathbb P^3$ and μ_3 is the group of cubic roots of unity in $\mathbb K$. The birational map gives rise to the equivalence of the two Galois representations associated to $(S\times S)/\mu_3$ and V, and hence the same L-series, at least between a non-singular part of $(S\times S)/\mu_3$ and some open subset of V. Our discussion is motivated by the following conjecture by Tate (note that there is another conjecture due to Tate in subsection B.3).

CONJECTURE B.1 (The Tate conjecture). *If two varieties have the same L-series (i.e., have isomorphic Galois representations), then there should be an algebraic correspondence between the two varieties compatible with the Galois action.*

There have been found only a few examples so far in support of this conjecture. Some remarkable pieces of evidences for this conjecture were obtained by Schoen [Sch87] and [Sch95], by Hulek, Spandaw, van Geemen and van Straten [HSGS] and by van Geeman and Top [vGT].

EXAMPLE B.1. Recently, Kimura [Ki04] has found an explicit rational map between two rigid Calabi–Yau threefolds giving rise to the equivalence of the two 2-dimensional Galois representations, and hence the same L-series. One variety is V_{33} which is defined by the two cubic equations $X_0^3 + X_1^3 + X_2^3 + X_3^3 = 0$ and $X_2^3 + X_3^3 + X_4^3 + X_5^3 = 0$ in $\mathbb{P}^5$ and the other one is the triple product E^3 of the elliptic curve $E : y_0^3 + y_1^3 + y_2^3 = 0 \subset \mathbb{P}^2$. They have the same L-series, thereby establishing the Tate conjecture for that case.

THEOREM B.11. *Let $\mathbb{K} = \mathbb{Q}(\omega)$ be the quadratic extension of $\mathbb{Q}$ obtained by adjoining ω. Let S be a smooth cubic Fermat surface over $\mathbb{K}$ in $\mathbb{P}^3$ defined by the equation*

$$x_1^3 + x_2^3 + x_3^3 + x_4^3 = 0.$$

Then there is a birational map over $\mathbb{K}$ between $(S \times S)/\mu_3$ and V in support of the Tate conjecture, i.e., giving rise to the equivalence of the two Galois representations attached to a non-singular part of $(S \times S)/\mu_3$ and an open subset of V.

PROOF. Write $\phi : S \times S \longrightarrow V$ for the rational map of the Fermat inductive structure given by Shioda and Katsura [SK79], namely

$$\begin{array}{ccc} S \times S & \xrightarrow{\phi} & V \\ ((y_1 : y_2 : y_3 : y_4), (z_1 : z_2 : z_3 : z_4)) & \longmapsto & (y_1z_4 : y_2z_4 : y_3z_4 : \zeta y_4z_1 : \zeta y_4z_2 : \zeta y_4z_3) \end{array}$$

where $\zeta := -\omega$ is a primitive 6-th root of unity in $\mathbb{C}$. This map is defined over $\mathbb{K}$ and is a morphism on the open subset

$$U_0 := S \times S \setminus \{y_4 = z_4 = 0\}.$$

If we choose affine open subsets

$$\begin{array}{ll} U_1 := S \times S \setminus \{y_1 = 0 \text{ or } z_4 = 0\} & \text{on } S \times S \\ V_1 := V \setminus \{x_1 = 0\} & \text{on } V \end{array}$$

and write

$$\begin{aligned} \mathcal{O}(V_1) &= \mathbb{K}[x_2, x_3, x_4, x_5, x_6]/(1 + x_2^3 + \cdots + x_6^3) \\ \mathcal{O}(U_1) &= \mathbb{K}[y_2, y_3, y_4, z_1, z_2, z_3]/(1 + y_2^3 + y_3^3 + y_4^3,\ z_1^3 + z_2^3 + z_3^3 + 1) \end{aligned}$$

then ϕ induces a homomorphism on the coordinate rings:

$$\begin{array}{ccc} \mathcal{O}(V_1) & \xrightarrow{\overline{\phi}} & \mathcal{O}(U_1) \\ h(x_2, x_3, x_4, x_5, x_6) & \longmapsto & h(y_2, y_3, \zeta y_4z_1, \zeta y_4z_2, \zeta y_4z_3) \end{array}$$

Further, we take a smaller open subset

$$U = U_1 \setminus \{y_4 = 0 \text{ or } z_1 = 0\}.$$

Then U is an open subset of $U_0 \cap U_1$ and its coordinate ring is

$$\mathcal{O}(U) = \mathbb{K}[y_2, y_3, y_4, y_4^{-1}, z_1, z_1^{-1}, z_2, z_3]/(1 + y_2^3 + y_3^3 + y_4^3,\ z_1^3 + z_2^3 + z_3^3 + 1)$$

Write $\mu_3 =< \omega >$ for the group of cube roots of unity. It acts on $S \times S$ by

$$\begin{array}{ccc} S \times S & \xrightarrow{\omega} & S \times S \\ ((y_1 : y_2 : y_3 : y_4), (z_1 : z_2 : z_3 : z_4)) & \longmapsto & ((y_1 : y_2 : y_3 : \omega y_4), (z_1 : z_2 : z_3 : \omega z_4)) \end{array}$$

where we let ω also denote the morphism. It induces an action on U_1 as

$$\begin{array}{ccc} U_1 & \xrightarrow{\omega|_{U_1}} & U_1 \\ ((y_2, y_3, y_4), (z_1, z_2, z_3)) & \longmapsto & ((y_2, y_3, \omega y_4), (\omega^{-1} z_1, \omega^{-1} z_2, \omega^{-1} z_3)) \end{array}$$

By rewriting y_1^{-1}, y_2/y_1 and y_3/y_1 as y_1, y_2 and y_3, respectively, on $\mathcal{O}(U)$, we can describe the action of μ_3 on U as

$$\begin{array}{ccc} U & \xrightarrow{\omega|_U} & U \\ ((y_2, y_3, y_4, y_4^{-1}), (z_1, z_1^{-1}, z_2, z_3)) & \longmapsto & ((y_2, y_3, \omega y_4, \omega y_4^{-1}), (\omega z_1, \omega z_1^{-1}, z_2, z_3)) \end{array}$$

Taking the μ_3-invariant subring of $\mathcal{O}(U)$ and setting

$$u := y_4^{-3}, \ v := z_1^3, \ w := \frac{y_4}{z_1}$$

we can express the coordinate ring of U/μ_3 as

$$\mathcal{O}(U/\mu_3) = \mathbb{K}[y_2, y_3, u, v, w, z_2, z_3]/(1 + y_2^3 + y_3^3 + vw^3, \ 1 + v + z_2^3 + z_3^3, \ uvw^3 - 1).$$

Combining these results, we obtain an isomorphism

$$\begin{array}{ccc} \mathcal{O}(V_1) & \longrightarrow & \mathcal{O}(U/\mu_3) \\ h(x_2, x_3, x_4, x_5, x_6) & \longmapsto & h(y_2, y_3, \zeta w, \zeta z_2 w, \zeta z_3 w). \end{array}$$

At the level of varieties, the isomorphism can be given as

$$\begin{array}{ccc} U/\mu_3 & \longrightarrow & V_1 \\ (y_2, y_3, u, v, w, z_2, z_3) & \longmapsto & (y_2, y_3, \zeta w, \zeta z_2 w, \zeta z_3 w). \end{array}$$

(Note that u and v are determined by $u = 1/vw^3$ and $v = -(1+z_2^3+z_3^3)$, respectively. Hence they do not appear in the isomorphism above). Therefore the isomorphism gives rise to a birational map between $(S \times S)/\mu_3$ and V as desired. □

B.3. Supersingular fourfolds and unirational fourfolds. Finally, we discuss supersingular fourfolds. We use our results on zeta-functions to find a unirational but not supersingular fourfold in positive characteristic.

Let k be a finite field of characteristic p and $\overline{k}$ an algebraic closure of k, fixed once and for all.

DEFINITION B.1. (cf. Shioda and Katsura [SK79]) Let V_k be a smooth variety of even dimension d over k. We say that V_k (or $V := V_{\overline{k}}$) is *supersingular* (in the sense of Shioda) if its ℓ-adic étale cohomology $H_{et}^{2i}(V, \mathbb{Q}_\ell(i))$ (twisted i times) is spanned by algebraic cycles for every i $(0 \leq i \leq d)$, where ℓ is a prime different from p.

To comment on the zeta-function of a supersingular variety, let

$$cl_i : \ A^i(V) \otimes \mathbb{Q}_\ell \longrightarrow H_{et}^{2i}(V, \mathbb{Q}_\ell(i))$$

be the cycle class map of codimension i, where $A^i(V)$ denotes the group of algebraic cycles of codimension i modulo ℓ-adically homological equivalence. Write $A^i(V_k)$ for the subgroup of $A^i(V)$ generated by the cycles defined over k. Then Tate [T65] conjectured

$$A^i(V_k) \otimes \mathbb{Q}_\ell \cong H_{et}^{2i}(V, \mathbb{Q}_\ell(i))^{Gal(\overline{k}/k)}. \tag{3}$$

The dimension of $H^{2i}_{et}(V, \mathbb{Q}_\ell(i))^{Gal(\overline{k}/k)}$ as a $\mathbb{Q}_\ell$-vector space can be calculated from the zeta-function of V_k as the multiplicity of $T = q^{-i}$ in the roots of $P_{2i}(V_k, T) = 0$. In particular, if V_k is supersingular, then

$$A^i(V_k) \otimes \mathbb{Q}_\ell \cong H^{2i}_{et}(V, \mathbb{Q}_\ell(i))^{Gal(\overline{k}/k)} \cong H^{2i}_{et}(V, \mathbb{Q}_\ell(i))$$

and hence all the reciprocal roots of $P_{2i}(V_k, T) = 0$ are equal to q^i by taking a sufficiently large k.

PROPOSITION B.12. *Let $\mathbb{K}$ be the quadratic extension of $\mathbb{Q}$ obtained by adjoining ω. Let $V_{\mathbb{K}}$ be a cubic hypersurface fourfold over $\mathbb{K}$. Assume that a prime $\mathfrak{p} \in \mathcal{O}_{\mathbb{K}}$ splits in $\mathbb{K}/\mathbb{Q}$ (i.e. $N\mathfrak{p} = \mathfrak{p}\bar{\mathfrak{p}} = p \equiv 1 \pmod 3$). Then $V_{\mathfrak{p}}$ is not supersingular.*

PROOF. $P_4(V_{\mathfrak{p}}, T)$ is the only non-trivial factor of $Z(V_{\mathfrak{p}}, T)$. To show that $V_{\mathfrak{p}}$ is not supersingular, it suffices to prove that some reciprocal roots of $P_4(V_{\mathfrak{p}}, T)$ are not of the form $q^2\epsilon$ with some root of unity ϵ, where q is the cardinality of k.

Indeed, from Proposition B.1, $P_4(V_{\mathfrak{p}}, T)$ has a reciprocal root $j_{\mathfrak{p}}(\mathbf{a}_1) = \frac{1}{q}G(\chi_{\mathfrak{p}})^6$. It is shown by Shioda and Katsura [SK79] that $j_{\mathfrak{p}}(\mathbf{a}_1)/q^2$ is a root of unity if and only if $p \equiv 2 \pmod 3$, where p is the characteristic of the residue field $\mathcal{O}_{\mathbb{K}}/\mathfrak{p}$. Hence V is not supersingular if $p \equiv 1 \pmod 3$, that is, if $\mathfrak{p} \in \mathcal{O}_{\mathbb{K}}$ splits in $\mathbb{K}/\mathbb{Q}$. □

PROPOSITION B.13 (cf. Manin [Ma72], Kollár [Ko02]). *Let $\mathbb{K} = \mathbb{Q}(\omega)$ be the quadratic extension of $\mathbb{Q}$ obtained by adjoining a primitive cube root of unity ω. Then V is unirational over $\mathbb{K}$.*

PROOF. As before, let S be a cubic surface in $\mathbb{P}^3$ defined by the equation $y_1^3 + y_2^3 + y_3^3 + y_4^3 = 0$. By evaluating the arithmetic and second plurigenera of S to be $p_a = p_2 = 0$, we can find that S is a rational surface over $\mathbb{C}$. Hence the product $S \times S$ is a rational fourfold over $\mathbb{C}$. To show that S and $S \times S$ are rational over $\mathbb{K}$, we may refer to a result of Manin [Ma72] or Kollár [Ko02]. For instance, in Example 12.6 of [Ma72], it is proved that there is a birational map $\mathbb{P}^3_{\mathbb{K}} \longrightarrow S$ over $\mathbb{K}$. This induces a birational map $\mathbb{P}^3_{\mathbb{K}} \times \mathbb{P}^3_{\mathbb{K}} \longrightarrow S \times S$ over $\mathbb{K}$.

Finally, we have a quotient map $S \times S \longrightarrow (S \times S)/\mu_3$ over $\mathbb{K}$ (see Theorem B.11) and $(S \times S)/\mu_3$ is birational to V over $\mathbb{K}$. Therefore V is unirational over $\mathbb{K}$. □

COROLLARY B.14. *There is a unirational fourfold that is not supersingular in positive characteristic.*

PROOF. By Proposition B.13, $V_{\mathfrak{p}}$ is unirational for almost all $\mathfrak{p} \in \operatorname{Spec} \mathcal{O}_{\mathbb{K}}$ (precisely, if the quotient map $S \times S \longrightarrow (S \times S)/\mu_3$ and a birational map between $(S \times S)/\mu_3$ and V are defined mod $\mathfrak{p}$). Hence $V_{\mathfrak{p}}$ with $N\mathfrak{p} = p$ gives an example of a unirational fourfold that is not supersingular. □

REMARK B.2. The main reason for $V_{\mathfrak{p}}$ not being supersingular is that the construction of V from $S \times S$ (or from $(S \times S)/\mu_3$) involves blowings-up at some non-supersingular elliptic curves. Such a phenomenon does not arise on surfaces since the Lefschetz number (= the Betti number minus the maximum Picard number) is a birational invariant.

A similar argument can also be found in [SK79], Remark 4.4.

REMARK B.3. Let V_1 be the open subset of V defined in the proof of Theorem B.11. Then V_1 is supersingular at any reduction mod $\mathfrak{p}$. Hence unirationality and supersingularity are the same notion on V_1.

REMARK B.4. Our results in this appendix can be generalized, for instance, to a weighted hypersurface fourfold V of degree m defined by the equation

$$c_1 x_1^{m/q_1} + c_2 x_2^{m/q_2} + c_3 x_3^{m/q_3} + c_4 x_4^{m/q_4} + c_5 x_5^{m/q_5} + c_6 x_6^{m/q_6} = 0$$

in $\mathbb{P}^5_{\mathbb{K}}(q_1, q_2, q_3, q_4, q_5, q_6)$. In fact, let S_1 and S_2 be the weighted diagonal surfaces of degree m defined by the equations

$$\begin{array}{lll} S_1: & y_1^{m/q_1} + y_2^{m/q_2} + y_3^{m/q_3} + y_4^m = 0 & \text{in } \mathbb{P}^3(q_1, q_2, q_3, 1) \\ S_2: & z_1^{m/q_4} + z_2^{m/q_5} + z_3^{m/q_6} + z_4^m = 0 & \text{in } \mathbb{P}^3(q_4, q_5, q_6, 1) \end{array}$$

respectively. Write ζ for a primitive $2m$-th root of unity in $\mathbb{C}$ and let $\mathbb{K} = \mathbb{Q}(\zeta)$. Then there exists a rational map over $\mathbb{K}$ from $S_1 \times S_2$ to V as

$$\begin{aligned} &((y_1 : y_2 : y_3 : y_4), (z_1 : z_2 : z_3 : z_4)) \\ &\qquad \longmapsto (y_1 z_4^{q_1} : y_2 z_4^{q_2} : y_3 z_4^{q_3} : (\zeta y_4)^{q_4} z_1 : (\zeta y_4)^{q_5} z_2 : (\zeta y_4)^{q_6} z_3) \end{aligned}$$

(a more general construction is given by Hunt and Schimmrigk [HS99]). Writing $\mu =< \zeta >$, we find that μ acts on $S_1 \times S_2$ by

$$\begin{array}{ccc} S_1 \times S_2 & \xrightarrow{\zeta} & S_1 \times S_2 \\ ((y_1 : y_2 : y_3 : y_4), (z_1 : z_2 : z_3 : z_4)) & \longmapsto & ((y_1 : y_2 : y_3 : \zeta y_4), (z_1 : z_2 : z_3 : \zeta z_4)) \end{array}$$

and the discussions in B.1 and B.2 can similarly be extended to V.

Furthermore, if we choose S_1 and S_2 to be rational, then the supersingularity and unirationality of $V_{\mathfrak{p}}$ can be discussed similarly as in B.3.

Acknowledgments. This work is partially supported by the JSPS Grant-in-Aid for Scientific Research (C)(2) 15540001. The final version of the paper was written at the Fields Institute, Toronto, Canada. I thank the Fields Institute for its hospitality. My special thanks are due to Noriko Yui for inviting me to the Fields Institute and providing me with excellent working conditions.

References

[Ao91] Aoki, N., *Simple factors of a Fermat curve and the Picard number of a product of Fermat curves*, Amer. J. Math. **113** (1991), pp. 779–833.

[Ba99] Batyrev, V., *Birational Calabi–Yau n-folds have equal Betti numbers*, in *New trends in algebraic geometry (Warwick, 1996)*, 1–11, London Math. Soc. Lecture Note Ser. **264**, Cambridge University Press, 1999.

[BCIR97] Bauer, M., Coste,A., Itzykson, C., and Ruelle, P., *Comments on the links between su(3) modular invariants, simple factors in the Jacobian of Fermat curves, and rational triangle billiards*, J. Geom. Phys. **22** (1997), no.2, pp. 134–189 (previously known as R.C.P. 25. Vol. **48** pp. 1–56. Prépubl. Inst. Rech. Math. Av. 1997/42. Univ. Louis Pasteur Strasbourg 1997).

[Bo92] Borcea, C., *Calabi–Yau threefolds and complex multiplication*, in "Mirror Symmetry I", AMS/IP Studies in Advanced Math. **9**, International Press 1998. (Previously, [Yau92], 1992, pp. 489–502.)

[CLS90] Candelas, P., Lynker, M. and Schimmrigk, R., *Calabi–Yau manifolds in weighted* $\mathbb{P}^4$, Nuclear Phys. B **341** (1990), pp. 383–402.

[COFKM94] Candelas, P., de la Ossa, X., Font, A., Katz, S. and Morrison, D., *Mirror symmetry for two parameter models – I*, Nuclear Phys. B **416** (1994), pp. 481–538. (Also in [GYau94] pp. 483–543.)

[De71] Deligne, P., *Théorie de Hodge II*, Pub. Math. IHES **40** (1971), pp. 5–58.

[De74] Deligne, P., *La conjecture de Weil I*, Pub. Math. I.H.E.S. **43** (1974), pp. 237–307.

[Di86] Dimca, A., *Singularities and coverings of weighted complete intersections*, J. Reine Angew Math. **366** (1986), pp. 184–193.

[DHVW85-86] Dixon, L., Harvey, J., Vafa, C., and Witten, E., *Strings on orbifolds*, Nuclear Phys. B **261** (1985), pp. 678–686, *Strings on orbifolds II*, Nuclear Phys. B **274** (1986), pp. 285–314.

[Dol82] Dolgachev, I., *Weighted projective varieties*, in "Group Actions and Vector Fields" Lecture Notes in Mathematics **956** (1982), pp. 24–71, Springer–Verlag.

[Fl89] Fletcher, A., *Working with weighted complete intersections*, Max–Planck–Institut für Mathematik, Preprint 1989.

[vGT] van Geemen, B., and Top, J., *An Isogeny of $K3$ surfaces*, mathAG/0309272 (2003).

[GP90] Greene, B., and Plesser, M.R., *Duality in Calabi–Yau moduli spaces*, Nuclear Physics B **338** (1990), no. 1, pp. 15–37.

[Gr69] Grothendieck, A., *Hodge's general conjecture is false for trivial reasons*, Topology **8** (1969), pp. 299–303.

[GV02] Gukov, S., and Vafa, C., *Rational conformal field theories and complex multiplication*, arXiv: htp-th/0203213 v2.

[GY04] Goto, Y., and Yui, N., *Zeta-functions and L-series of certain $K3$-fibered Calabi–Yau threefolds*, preprint 2004.

[GY95] Gouvêa, F.Q., and Yui, N., *Arithmetic of diagonal hypersurfaces over finite fields*, London Mathematical Society Lecture Notes **209**, Cambridge University Press 1995.

[GRY91] Greene, B. R., Roan, S.-S. and Yau, S.-T., *Geometric singularities and spectra of Landau–Ginzburg models*, Commun. Math. Phys. **142** (1991), pp. 245–259.

[GYau96] Greene, R., and Yau, S.-T., *Mirror Symmetry II*, AMS/IP Studies in Advanced Mathematics, **1**, American Mathematical Society and International Press 1996.

[HH90] Hirzebruch. F., and Höfer, T., *On the Euler number of an orbifold*, Math. Ann. **286** (1990), pp. 255–260.

[HLY96] Hosono, S., Lian, B.H., and Yau, S.-T., *Calabi–Yau varieties and pencils of $K3$ surfaces*, arXiv: alg-geom/9603020.

[HSGS] Hulek, K., Spandaw, J., van Geemen, B., and van Straten, D., *The modularity of the Barth-Nieto quintic and its relatives*, Adv. Geom. **1** (2001), pp. 263–289 (also available as mathAG/0010049).

[HS99] Hunt, B., and Schimmrigk, R., *K3 fibered Calabi–Yau threefolds*, Internat. J. Math. **18** (1999), No. 7, pp. 833–867.

[I03] Ito, Tetsushi, *Birational smooth minimal models have equal Hodge numbers in all dimensions*, in *Calabi–Yau Varieties and Mirror Symmetry* (Yui, N. and Lewis, J.D., (eds.)), Fields Inst. Commun. **38** (2003), pp. 183–194.

[Ki04] Kimura, K., *A rational map between two threefolds*, in this volume.

[KT93] Klemm, A., and Theisen, S., *Considerations of one-modulus Calabi–Yau compactifications: Picard-Fuchs equations, Kähler potentials and mirror maps*, Nuclear Phys. B **389** (1993), pp. 153–180.

[Ko78] Koblitz, N., *Simple factors in the Jacobian of a Fermat curve*, Canad. J. Math. **30** (1978), no. 6, pp. 1183–1205.

[Ko02] Kollár, J., *Unirationality of cubic hypersurfaces*, J. Inst. Math. Jussieu **1** (2002), no. 2, pp. 467–476.

[KS] Kreuzer, M., and Skarke, H., *Calabi–Yau data*, (can be downloaded from the website: tph16.tuwien.ac.at/ kreuzer/CY.

[Lev01] Levin, N., *The Tate conjecture for cubic fourfolds over finite fields*, Compositio Math. **127** (2001), pp. 1–21.

[Le04] Lewis, James D., *Three lectures on the Hodge conjecture*, *Transcendental aspects of algebraic cycles*, London Math. Soc. Lecture Note Ser. **313**, Cambridge Univ. Press, Cambridge 2004. pp. 199–234.

[Li95] Livné, R., *Motivic orthogonal two-dimensional representations of Gal*$(\bar{\mathbb{Q}}/\mathbb{Q})$, Israel. J. Math. **92**, no. 1-3 (1995), pp. 148–156.

[LPS03] Lynker, M., Periwal, V., and Schimmrigk, R., *Complex multiplication symmetry of black hole attractors*, Nuclear Phys. B **667** (2003), pp. 484–504.

[LSS04] Lynker, M., Schimmrigk, R., and Stewart, S., *Complex multiplication of exactly solvable Calabi–Yau varieties*, Nuclear Phys. B **700** (2004), pp. 463–489.

[Ma70] Manin, Yuri, *Correspondences, motives and monoidal transformations*, Math. USSR-Sb. **77** (1970), pp. 475–506.

[Ma72] Manin, Yu. I., *Cubic Forms*, Nauka, 1972 (in Russian), English translation published by North-Holland 1974, 2nd enlarged ed. 1986.

[Moo98] Moore, G., *Arithmetic and attractors*, arXiv:hep-th/9807087, *Attractors and arithmetic*, arXiv:hep-th/9807056.

[Mu66] Mumford, D., *Families of abelian varieties*, in "Algebraic Groups and Discontinuous Subgroups", Proc. Symp. Pure Math. **9** (1966), pp. 66–74.

[Mu69] Mumford, D., *A note of Shimura's paper "Discontinuous groups and abelian varieties"*, Math. Ann. **181** (1969), pp. 345–351.

[Ra80] Ran, Z., *Cycles on Fermat hypersurfaces*, Compositio Math. **42** (1980/81), no.1, pp. 121–142.

[Rap72] Rapoport, M., *Complement à l'article de P. Deligne " La conjecture de Weil pour les surfaces K3"*, Invent. Math. **15** (1972), pp. 227–236.

[Re80] Reid, M., *Canonical 3-folds*, in Algebraic Geometry Angers 1979, Sijthoff and Noordhoff, 1980, pp. 273–310.

[Re02] Reid, M., *Update on 3-folds*, ICM 2002 Bejing. Invited Lectures, Vol. 2 (2002), pp. 513–524.

[Ro90] Roan, S.-S., *On Calabi–Yau orbifolds in weighted projective spaces*, Internat. J. Math., **1**, No.2 (1990), pp. 211–232.

[Ro91] Roan, S.-S., *The mirror of Calabi–Yau orbifold*, Internat. J. Math., **2**, No.4 (1991), 439–455.

[Roh78] Rohrlich, D., Appendix to B. Gross, *On the periods of abelian integrals and a formula of Chowla and Selberg*, Invent. Math. **45** (1978), pp. 193–211.

[SY01] Saito, Masahiko and Yui, N., *The modularity conjecture for rigid Calabi–Yau threefolds over* $\mathbb{Q}$, Kyoto J. Math. **41**, no.2 (2001), pp. 403–419.

[Sch87] Schoen, C., *Zero cycles modulo rational equivalence for some varieties over fields of transcendence degree one*, Proc. Sympos. Pure Math. **46**, Part 2 (1987), pp. 463–473.

[Sch95] Schoen, C., *On the computation of the cycle class map for nullhomologous cycles over the algebraic closure of a finite field*, Ann. Sci. Ecole Norm. Sup. (4) **28** (1995), pp. 1–50.

[Sch96] Schoen, C., *Varieties dominated by product varieties*, Internat. J. Math. **7**, No. 4 (1996), pp. 541–571.

[S] Schimmrigk, R., http://www.th.physik.uni-bonn.de/th/People/netah/cy.html

[Schi03] Schimmrigk, R., *Arithmetic of Calabi–Yau manifolds and rational conformal field theory*, J. Geom. Phys. **44** (2003), no. 4, pp. 555–569.

[SU03] Schimmrigk, R., and Underwood, S., *The Shimura–Taniyama conjecture and conformal field theory*, J. Geom. Phys. **48** (2003), pp. 169–189.

[Se01] Sebbar, A., *Torsion-free genus zero congruence subgroups of* $PSL_2(\mathbb{R})$, Duke Math. J., **10** (2001), no.2, pp. 377–396.

[S-PS73] Shafarevich, I., and Piateskij–Shapiro, I., *The arithmetic of K3 surfaces*, Mat. Inst. Steklov **132** (1973), pp. 44–54.

[Sh78] Shioda, T., *The Hodge conjecture for Fermat varieties*, Math. Ann. **245** (1978), pp. 175–185.

[SK79] Shioda, T., and Katsura, T., *Fermat varieties*, Tôhoku Math. J. (2) **31** (1979), no. 1, pp. 97–115.

[Sh82] Shioda, T., *Geometry of Fermat varieties*, in "Number Theory Related to Fermat's Last Theorem", Progress in Math. **26** (1982), pp. 45–56.

[Sh83] Shioda, T., *What is known about the Hodge conjecture?*, in "Algebraic Varieties and Analytic Varieties", Adv. Studies Pure Math. **1** (1983), pp. 55–68.

[Sh87] Shioda, T., *Some observations on Jacobi sums*, Advanced Studies in Pure Math. **12**, North Holland–Kinokuniya Shoten 1987, pp. 119–135.

[SI87] Shioda, T., and Inose, K., *On singular K3 surfaces*, in "Complex Analysis and Algebraic Geometry", eds. W. Baily and T. Shioda (1987), Kinokuniya Shoten and Cambridge University Press, 119–136.

[ST61] Shimura, G., and Taniyama, Y., *Complex Multiplication of Abelian Varieties and its Applications to Number Theory*, Publ. Math. Soc. Japan **6**, 1961 Tokyo.

[T65] Tate, J., *Algebraic cycles and poles of zeta functions*, in "Arithmetic Algebraic Geometry," New York, Harper & Row, 1965, pp. 93–110.

[V89] Vafa, C., *String vacua and orbifoldized LG models*, Mod. Phys. Lett. A **4** (1989), pp. 1169–1185.

[W49] Weil, A., *Number of solutions of equations over finite fields*, Bull. Amer. Math. Soc. **55** (1949), pp. 497–508.

[W52] Weil, A., *Jacobi sums as "Grössencharaktere"*, Trans. A.M.S. **73** (1952), pp. 487–495.

[Yau92] Yau, S.-T., *Essays in Mirror Symmetry*, (eds.) (1992), International Press, Hong Kong.

[Yo90] Yonemura, T., *Hypersurface simple $K3$ singularities*, Tôhoku Math. J. **42** (1990), pp. 351–380.

[Yui03] Yui, N., *Update on the modularity of Calabi–Yau varieties*, in *Calabi–Yau Varieties and Mirror Symmetry* (Yui, N. and Lewis, J.D. (eds.)), Fields Inst. Commun. **38** (2003), pp. 307–362.

[Yui] Yui, N., *The L-series of mirror pairs of Calabi–Yau orbifolds.* In preparation.

[Z04] Zarhin, Y., *The Tate conjecture for powers of ordinary cubic fourfolds over finite fields*, J. Number Theory **108** (2004), pp. 44–59.

Department of Mathematics, Queen's University, Kingston. Ontario Canada K7L 3N6

E-mail address: `yui@mast.queensu.ca; yui@fields.utoronto.ca`

PART II

GEOMETRIC ASPECTS

AMS/IP Studies in Advanced Mathematics
Volume 38, 2006

Integral Cohomology and Mirror Symmetry for Calabi-Yau 3-Folds

Victor Batyrev and Maximilian Kreuzer

ABSTRACT. In this paper we compute the integral cohomology groups for all examples of Calabi-Yau 3-folds obtained from hypersurfaces in 4-dimensional Gorenstein toric Fano varieties. Among 473 800 776 families of Calabi-Yau 3-folds X corresponding to 4-dimensional reflexive polytopes there exist exactly 32 families having non-trivial torsion in $H^*(X,\mathbb{Z})$. We arrive at an interesting observation that the torsion subgroups in H^2 and H^3 are exchanged by the mirror symmetry involution.

Introduction

Let X be a smooth projective Calabi-Yau d-fold over $\mathbb{C}$. We shall always assume that $h^i(X,\mathcal{O}_X) = 0$ for $0 < i < d$. Our main interest is the torsion in the integral cohomology ring $H^*(X,\mathbb{Z})$. For a finite abelian group G we denote by G^* the dual abelian group $\mathrm{Hom}(G,\mathbb{Q}/\mathbb{Z})$. Using the isomorphism from the universal coefficient theorem

$$\mathrm{Tors}(H_i(X,\mathbb{Z})) \cong [\mathrm{Tors}(H^{i+1}(X,\mathbb{Z}))]^*$$

and the Poincaré duality

$$\mathrm{Tors}(H_i(X,\mathbb{Z})) \cong \mathrm{Tors}(H^{2d-i}(X,\mathbb{Z})),$$

we obtain for $d = 3$:

$$H^0(X,\mathbb{Z}) \cong H^6(X,\mathbb{Z}) \cong \mathbb{Z}, \quad H^1(X,\mathbb{Z}) = 0,$$
$$H^2(X,\mathbb{Z}) \cong A(X) \oplus \mathbb{Z}^a, \quad H^4(X,\mathbb{Z}) \cong B(X)^* \oplus \mathbb{Z}^a$$
$$H^3(X,\mathbb{Z}) \cong B(X) \oplus \mathbb{Z}^{2b+2}, \quad H^5(X,\mathbb{Z}) \cong A(X)^*,$$

where the finite abelian groups

$$A(X) := \mathrm{Tors}(H^2(X,\mathbb{Z})), \ B(X) := \mathrm{Tors}(H^3(X,\mathbb{Z}))$$

determine completely the torsion in $H^*(X,\mathbb{Z})$. In particular, we have the isomorphisms

$$A(X) \oplus B(X)^* \cong \mathrm{Tors}(H^{\mathrm{even}}(X,\mathbb{Z})) \cong [\mathrm{Tors}(H^{\mathrm{odd}}(X,\mathbb{Z}))]^*.$$

2000 *Mathematics Subject Classification.* Primary 14J32, 14M25.
Key words and phrases. Calabi–Yau manifolds, mirror symmetry, toric varieties, Fano varieties, reflexive polytopes, integral cohomology groups.

It is known that mirror symmetry exchanges the integers a, b so that for a mirror Calabi-Yau 3-fold X^* one has

$$\mathbb{Q}^{2a+2} \cong H^{\text{even}}(X, \mathbb{Q}) \cong H^{\text{odd}}(X^*, \mathbb{Q}),$$
$$\mathbb{Q}^{2b+2} \cong H^{\text{odd}}(X, \mathbb{Q}) \cong H^{\text{even}}(X^*, \mathbb{Q}).$$

An interesting mathematical question is to understand the behavior of the torsion in $H^*(X, \mathbb{Z})$, i.e. the groups $A(X)$ and $B(X)$, under mirror symmetry. The group $B(X)$ is isomorphic to the torsion in the étale cohomology group $H^2_{\text{et}}(X, \mathcal{O}^*_X)$ and it is called the *cohomological Brauer group of* X. By a recent result of Kresch and Vistoli [**16**], $B(X)$ is also isomorphic to the Brauer group of Azumaya algebras on X. On the other hand, there are canonical isomorphisms

$$A(X) \cong \text{Hom}(\pi_1(X), \mathbb{Q}/\mathbb{Z}) \cong \text{Tors}(\text{Pic}(X)).$$

From the physical point of view it is more natural to consider the topological K-groups $K^0(X)$ and $K^1(X)$ together with some natural filtrations [**7, 8**]. Using the Atiyah-Hirzebruch spectral sequence (see [**7**], 2.5), one obtains two short exact sequences

$$0 \to A(X)^* \to \text{Tors}(K^1(X)) \to B(X) \to 0,$$
$$0 \to B(X)^* \to \text{Tors}(K^0(X)) \to A(X) \to 0.$$

It is expected that mirror symmetry exchanges K^0 and K^1 so that for a mirror pair (X, X^*) one has the isomorphisms

$$\text{Tors}(K^1(X)) \cong \text{Tors}(K^0(X^*)), \quad \text{Tors}(K^0(X)) \cong \text{Tors}(K^1(X^*)). \tag{1}$$

These isomorphisms agree with predictions of the SYZ construction and topological calculations of M. Gross [**12**]. The compatibility of the isomorphisms (1) with the natural filtrations in K^i would imply more subtle isomorphisms

$$A(X) \cong B(X^*), \quad B(X) \cong A(X^*). \tag{2}$$

The main purpose of this paper is to verify the isomorphisms (2) for all examples of Calabi-Yau 3-folds obtained from hypersurfaces in 4-dimensional Gorenstein toric Fano varieties. Recently we were informed that Ch. Doran and J. Morgan have independently verified the K-theoretic isomorphisms (1) for the same class of Calabi-Yau 3-folds.

There exist 473 800 776 families of Calabi-Yau 3-folds X corresponding to 4-dimensional reflexive polytopes [**17, 18**]. These families give rise to 30 108 different pairs of numbers (a, b). We show that among the 473 800 776 families of Calabi-Yau 3-folds X there exist exactly 32 families having non-trivial torsion in $H^*(X, \mathbb{Z})$. More precisely, there are exactly 16 families of simply-connected Calabi-Yau 3-folds X having non-trivial Brauer group $B(X) \cong \mathbb{Z}/p\mathbb{Z}$ $(p = 2, 3, 5)$. They are mirror dual to another 16 families of Calabi-Yau 3-folds X^* having trivial Brauer group and a non-trivial cyclic fundamental group of order $p = 2, 3, 5$ [**15**]. Thus, we come to the observation that for all families of Calabi-Yau 3-folds one has the isomorphisms (2). Although the groups $A(X)$ and $B(X)$ can be computed in a purely combinatorial way using lattice points in faces of 4-dimensional reflexive polytopes Δ (or dual reflexive polytopes Δ^*) we do not see immediately the isomorphisms (2) from the combinatorial duality between Δ and Δ^*. Therefore, it would be interesting to find a general mathematical explanation for the mirror symmetry isomorphism

$$B(X) \cong \text{Hom}(\pi_1(X^*), \mathbb{Q}/\mathbb{Z}).$$

If a finite group G acts on a smooth Calabi-Yau manifold V, then there exists a conformal field theory of the orbifold V/G associated with an element $\alpha \in H^2(G, U(1))$ (discrete torsion). In [**28**] Vafa and Witten have suggested that there should be some connection between elements in the Brauer group of V/G (B-fields) and elements in $H^2(G, U(1))$ (discrete torsion 2-cocycles). In this paper we show that all 16 families of Calabi-Yau 3-folds with non-trivial cyclic Brauer group are birational to Calabi-Yau orbifolds V/G ($G \cong \mathbb{Z}/p\mathbb{Z} \oplus \mathbb{Z}/p\mathbb{Z}$, $p = 2, 3, 5$) and the corresponding elements in the cyclic Brauer group $B(X) \cong H_2(G) \cong \mathbb{Z}/p\mathbb{Z}$ can be identified with the discrete torsion 2-cocycles.

It is easy to see that there are no algebraic curves in X representing the torsion classes in $H_2(X, \mathbb{Z})$. Aspinwall and Morrison suggested in [**2**] that the instanton expansion of the Yukawa coupling for a Calabi-Yau 3-fold with the Brauer group $B(X) \cong \mathbb{Z}/p\mathbb{Z}$ should be weighted by powers of p-th roots of unity. It would be interesting to verify this prediction for the above 16 toric families of Calabi-Yau 3-folds with non-trivial Brauer group.

Acknowledgements: We would like to thank the Fields Institute for Research in Mathematical Sciences in Toronto for hospitality during the spring of 2005, and DFG-Schwerpunkt "Global methods in complex geometry" for the support of the first author. The second author was supported by FWF grant Nr. P15584-N08. We are grateful to Noriko Yui for her interest and encouragement in our work. We thank also Volker Braun, Charles Doran, Mark Gross, Kentaro Hori, Klaus Hulek, Askold Khovanskiy, Tony Pantev, Duco van Straten, Bernd Siebert and Balázs Szendrői for useful discussions.

1. The fundamental group of toric hypersurfaces

First we recall a combinatorial computation of the fundamental group of a toric variety [**9**]. Let $M \cong \mathbb{Z}^d$ be a lattice of rank d and $N := \mathrm{Hom}(M, \mathbb{Z})$ the dual lattice. We denote by $\langle *, * \rangle$ the canonical pairing $M \times N \to \mathbb{Z}$. For an algebraic torus $\mathbb{T}^d := \mathrm{Spec}\, \mathbb{C}[M] \cong (\mathbb{C}^*)^d$ there are canonical isomorphisms

$$\pi_1^{\mathrm{top}}(\mathbb{T}^d) \cong N \cong H_1(\mathbb{T}^d, \mathbb{Z}), \quad M \cong H^1(\mathbb{T}^d, \mathbb{Z}).$$

If $\mathbb{P}_\Sigma$ is a smooth partial toric compactification of $\mathbb{T}^d$ defined by a fan $\Sigma \subset N_{\mathbb{R}} = N \otimes \mathbb{R}$, then $\mathbb{P}_\Sigma \setminus \mathbb{T}^d$ is a union of divisors $D_1, \ldots, D_n$ which 1-to-1 correspond to primitive lattice vectors $e_1, \ldots, e_n \in N$ generating 1-dimensional cones in Σ. If D_i° is a dense torus orbit in D_i, then we have $\mathbb{P}_i := \mathbb{T}^d \cup D_i^\circ \cong \mathbb{C} \times \mathbb{T}^{d-1}$. The embedding $\mathbb{T}^d \hookrightarrow \mathbb{P}_i$ defines the homomorphism of the fundamental groups

$$\rho_i \,:\, N = \pi_1^{\mathrm{top}}(\mathbb{T}^d) \to \pi_1^{\mathrm{top}}(\mathbb{P}_i) \cong N/\mathbb{Z}e_i.$$

Combining all the homomorphisms ρ_i ($i = 1, \ldots, n$), we obtain the isomorphisms

$$\pi_1^{\mathrm{top}}(\mathbb{P}_\Sigma) \cong \pi_1^{\mathrm{top}}(\mathbb{P}_{\Sigma^{(1)}}) \cong N/\sum_{i=1}^{n} \mathbb{Z}e_i,$$

where $\mathbb{P}_{\Sigma^{(1)}}$ is a toric variety associated to the subfan $\Sigma^{(1)} \subset \Sigma$ of all cones of dimension ≤ 1 in Σ. In particular, we have

$$H_1(\mathbb{P}_\Sigma, \mathbb{Z}) \cong H_1(\mathbb{P}_{\Sigma^{(1)}}, \mathbb{Z}) \cong N/\sum_{i=1}^{n} \mathbb{Z}e_i.$$

In order to compute the fundamental group of hypersurfaces in toric varieties one applies the following result of Oka [**22**]:

THEOREM 1.1. *Let $\Delta \subset M_{\mathbb{R}}$ be a lattice polytope of dimension d,*

$$f(t) = \sum_{m \in \Delta \cap M} c_m t^m \in \mathbb{C}[M] \cong \mathbb{C}[t_1^{\pm 1}, \ldots, t_d^{\pm 1}]$$

be a Δ-nondegenerate Laurent polynomial, and $W \subset \mathbb{T}^d$ the nondegenerate affine hypersurface defined by the equation $f(t) = 0$. Then the embedding $j : W \hookrightarrow \mathbb{T}^d$ is a homotopy $(d-1)$-equivalence, i.e. the homomorphism of the homotopy groups

$$j_* \, : \, \pi_i(W) \to \pi_i(\mathbb{T}^d)$$

is bijective for $i < d-1$ and surjective for $i = d-1$.

By Whitehead's theorem, we get:

COROLLARY 1.2. *The homomorphism*

$$j_* \, : \, H_i(W, \mathbb{Z}) \to H_i(\mathbb{T}^d, \mathbb{Z})$$

is bijective for $i < d-1$ and surjective for $i = d-1$.

Since $H_i(\mathbb{T}^d, \mathbb{Z})$ is a free abelian group of rank $\binom{d}{i}$, using the universal coefficients theorem, we also obtain:

COROLLARY 1.3. *For any abelian group $\mathcal{A}$, the homomorphism*

$$j^* \, : \, H^i(\mathbb{T}^d, \mathcal{A}) \cong \Lambda^i M \otimes_{\mathbb{Z}} \mathcal{A} \to H^i(W, \mathcal{A})$$

is bijective for $i < d-1$ and injective for $i = d-1$.

Let $\Sigma := \Sigma(\Delta) \subset N_{\mathbb{R}}$ be the normal fan of the polytope $\Delta \subset M_{\mathbb{R}}$. The cones $\sigma = \sigma(\theta) \in \Sigma(\Delta)$ 1-to-1 correspond to faces $\theta \subset \Delta$ in the following way:

$$\sigma(\theta) = \{y \in N_{\mathbb{R}} \, : \, \min_{x \in \Delta} \langle x, y \rangle = \langle x, z \rangle \ \forall z \in \theta\},$$

i.e. $\sigma(\theta)$ consists of those $y \in N_{\mathbb{R}}$ for which the minimum of the linear function $\langle *, y \rangle$ on Δ is attained at any point of θ. Sometimes we denote the corresponding projective toric variety $\mathbb{P}_{\Sigma} = \mathbb{P}_{\Sigma(\Delta)}$ simply by $\mathbb{P}_{\Delta}$. Let $\overline{W} \subset \mathbb{P}_{\Sigma}$ be the Zariski closure of W in $\mathbb{P}_{\Sigma}$. If $\Sigma^{(1)} \subset \Sigma$ is the subfan of all cones of dimension ≤ 1 in Σ, then $\mathbb{P}_{\Sigma^{(1)}}$ and $\overline{W}^{(1)} := \mathbb{P}_{\Sigma^{(1)}} \cap \overline{W}_f$ are smooth quasi-projective varieties.

THEOREM 1.4. *Let $d \geq 3$. Then the embedding $j : \overline{W}^{(1)} \hookrightarrow \mathbb{P}_{\Sigma^{(1)}}$ induces the isomorphism*

$$j_* \, : \, \pi_1(\overline{W}^{(1)}) \cong \pi_1(\mathbb{P}_{\Sigma^{(1)}}) \cong N / \sum_{i=1}^{n} \mathbb{Z} e_i,$$

where $e_1, \ldots, e_n$ are primitive generators of 1-dimensional cones in $\Sigma^{(1)}$.

Proof. Let $\mathbb{P}_{\Sigma^{(1)}} \setminus \mathbb{T}^d = D_1^\circ \cup \cdots \cup D_n^\circ$ ($D_i^\circ \cong \mathbb{T}^{d-1}$). Using the isomorphisms $\pi_1(W) \cong \pi_1(\mathbb{T}^d) \cong N$ and the surjectivity of $\pi_1(W) \to \pi_1(\overline{W}^{(1)})$, we obtain that

$$j_* \, : \, \pi_1(\overline{W}^{(1)}) \to \pi_1(\mathbb{P}_{\Sigma^{(1)}})$$

is surjective. Therefore the universal covering $\overline{U}^{(1)}$ of $\overline{W}^{(1)}$ is induced by some unramified covering U of $\mathbb{T}^d$.

In order to prove that j_* is an isomorphism it remains to show that the image of every element $e_i \in \pi_1(W) \cong N$ $(i = 1, \dots, n)$ is zero in $\pi_1(\overline{W}^{(1)})$. Let $\mathbb{P}_i := \mathbb{T}^d \cup D_i^\circ$ and $\overline{W}_i^{(1)} := \overline{W}^{(1)} \cap \mathbb{P}_i$. Using open inclusions $W \subset \overline{W}_i^{(1)} \subset \overline{W}^{(1)}$, it is enough to prove that the image of $e_i \in \pi_1(W)$ is zero already in $\pi_1(\overline{W}_i^{(1)})$ so that one obtains $\pi_1(\overline{W}_i^{(1)}) \cong N/\mathbb{Z}e_i$. We can consider $\overline{W}_i^{(1)}$ as an affine hypersurface in $\mathbb{P}_i \cong \mathbb{T}^{d-1} \times \mathbb{C} \cong \operatorname{Spec} \mathbb{C}[t_1^{\pm 1}, \dots, t_{d-1}^{\pm 1}, t_d]$ where the divisor $D_i^\circ \subset \mathbb{P}_i$ is defined by $t_d = 0$. Moreover, the element $e_i \in \pi_1(\mathbb{T}^d) = N$ is represented by a small 1-cycle $\gamma_i \subset \mathbb{C}^*$ around $0 \in \mathbb{C}$ which can be contracted to 0 in $\mathbb{T}^{d-1} \times \mathbb{C}$. The transversality of $\overline{W}_i^{(1)}$ and D_i° implies that the 1-cycle γ_i can be chosen inside W_f so that it will be contractible in $\overline{W}_i^{(1)}$. Therefore all elements $e_1, \dots, e_n$ are in the kernel of the homomorphism $\pi_1(W) \to \pi_1(\overline{W}^{(1)})$. We remark that the isomorphism $\pi_1(\overline{W}_i^{(1)}) \cong \pi_1(\mathbb{P}_i)$ can also be interpreted as a Lefschetz-type statement for a nondegenerate hypersurface $\overline{W}_i^{(1)}$ in $\mathbb{T}^{d-1} \times \mathbb{C}$. □

Since $e_1, \dots, e_n$ generate N over $\mathbb{Q}$, we obtain:

COROLLARY 1.5. *The fundamental group of the quasi-projective hypersurface $\overline{W}^{(1)}$ is a finite abelian group. In particular, $\pi_1(\overline{W}^{(1)})$ coincides with the algebraic fundamental group of $\overline{W}^{(1)}$ and the universal covering of $\overline{W}^{(1)}$ is a toric quasi-projective hypersurface $\widetilde{W}^{(1)}$ in the finite universal covering of $\mathbb{P}'_{\Sigma^{(1)}} \to \mathbb{P}_{\Sigma^{(1)}}$ determined by the finite index sublattice $N' = \sum_{i=1}^n \mathbb{Z}e_i \subset N$.*

THEOREM 1.6. *Assume that $d \geq 3$. Let $\mathbb{P}_{\widehat{\Sigma}}$ be a smooth d-dimensional projective toric variety defined by a fan $\widehat{\Sigma} \subset N_\mathbb{R}$ which is a subdivision of $\Sigma = \Sigma(\Delta)$. We denote by Γ the set of primitive generators v of 1-dimensional cones in $\widehat{\Sigma}^{(1)}$ such that the minimum of $\langle *, v\rangle$ on Δ is attained on a face $\theta \subset \Delta$ of dimension ≥ 1. Define the sublattice $N_\Delta^{(1)} \subset N$ to be generated by all $v \in \Gamma$. If $\widehat{W}$ is the closure of W in $\mathbb{P}_{\widehat{\Sigma}}$, then the fundamental group $\pi_1(\widehat{W})$ is isomorphic to the cyclic group*

$$N/N_\Delta^{(1)}.$$

Proof. Let $\mathcal{E} := \{e_1, \dots, e_n\}$ be the set of primitive generators of 1-dimensional cones in $\Sigma^{(1)}$. Since every linear function $\langle *, e_i\rangle$ attains its minimum of Δ on a $(d-1)$-dimensional face of Δ, we have $\mathcal{E} \subset \Gamma$ and we can write $\Gamma = \{e_1, \dots, e_n, \dots, e_{n+k}\}$. Moreover, $\overline{W}^{(1)}$ can be considered as a dense open subset of $\widehat{W}$. Therefore, we have a surjective homomorphism

$$\psi \, : \, \pi_1(\overline{W}^{(1)}) \to \pi_1(\widehat{W})$$

and the dual injective homomorphism

$$\psi^* \, : \, \operatorname{Hom}(\pi_1(\widehat{W}, \mathbb{Q}/Z) \to \operatorname{Hom}(\pi_1(\overline{W}^{(1)}), \mathbb{Q}/Z).$$

Since $\widehat{W} \setminus \overline{W}^{(1)} = Z_1 \cup \dots \cup Z_k$, where

$$Z_i = D_{n+i} \cap \widehat{W}, \;\; i = 1, \dots, k,$$

we obtain that an element $g \in \operatorname{Hom}(\pi_1(\overline{W}^{(1)}), \mathbb{Q}/Z) \cong \operatorname{Hom}(N/\sum_{i=1}^n \mathbb{Z}e_i, \mathbb{Q}/\mathbb{Z})$ belongs to $\operatorname{Im} \psi^*$ if and only if the cyclic Galois covering of $\overline{W}^{(1)}$ corresponding to g is unramified along all divisors $Z_1, \cdots, Z_k$. Using the same arguments as in the proof of 1.4, we obtain that the latter holds exactly when $g(e_{n+1}) = \cdots = g(e_{n+k}) = 0$. Therefore,

$$\operatorname{Im}\psi \cong \operatorname{Hom}(N/\sum_{i=1}^{n+k} \mathbb{Z}e_i, \mathbb{Q}/\mathbb{Z}) = \operatorname{Hom}(N/N_\Delta^{(1)}, \mathbb{Q}/\mathbb{Z})$$

and by duality $\pi_1(\widehat{W}) \cong N/N_\Delta^{(1)}$. The group $N/N_\Delta^{(1)}$ is cyclic, because the smoothness of $\widehat{W}$ implies that for any $(d-1)$-dimensional cone $\sigma = \sigma(\theta) \in \Sigma$ the subset $\sigma \cap \Gamma$ generates a direct summand of the rank $d-1$ of the lattice N (all $(d-1)$-dimensional cones in $\widehat{\Sigma}$ which subdivide σ are generated by part of a $\mathbb{Z}$-basis of N). □

COROLLARY 1.7. *For any lattice point $v \in N$, we denote by*

$$M_v := \{x \in M \ : \ \langle x, v\rangle = 0\}$$

the orthogonal complement to v in M. Then we have the exact sequence

$$\bigoplus_{v\in\Gamma} \Lambda^{d-1}M_v \to \Lambda^{d-1}M \to H^{2d-3}_{\mathrm{tor}}(\widehat{W}, \mathbb{Z}) \to 0.$$

Proof. The exactness of

$$\bigoplus_{v\in\Gamma} \Lambda^{d-1}M_v \to \Lambda^{d-1}M \to H^{2d-3}_{\mathrm{tor}}(\widehat{W}, \mathbb{Z}) \to 0$$

follows from the exactness of

$$\bigoplus_{v\in\Gamma} \mathbb{Z}[v] \to N \to \pi_1(\widehat{W}) \to 0$$

together with the canonical isomorphisms

$$\pi_1(\widehat{W}) \cong H^{2d-3}_{\mathrm{tor}}(\widehat{W}, \mathbb{Z}), \ \ \Lambda^{d-1}M_v \cong \mathbb{Z}, \ \ \Lambda^{d-1}M \cong N$$

after choosing an isomorphism (orientation) $\Lambda^d \cong \mathbb{Z}$. □

REMARK 1.8. Theorem 1.6 can be considered as a special case of a more general result of Oka (see [**24**], Chapter V, §5).

COROLLARY 1.9. *Let $\overline{W}$ be a nondegenerate $(d-1)$-dimensional Calabi-Yau hypersurface in the Gorenstein toric Fano variety $\mathbb{P}_\Delta$ associated with a d-dimensional reflexive polytope $\Delta \subset M_{\mathbb{R}}$. Assume $d \geq 3$ and that there exists a smooth projective crepant resolution $\widehat{W}$ of singularities in $\overline{W}$ induced by a convex (coherent) triangulation of the dual reflexive polytope $\Delta^* \subset \mathbb{N}_{\mathbb{R}}$. We denote by N'_{Δ^*} the sublattice in N generated by all lattice points in $\Delta^* \cap N$ which belong to faces of codimension > 1 of Δ^*. Then the fundamental group of $\widehat{W}$ is isomorphic to the cyclic group*

$$N/N'_{\Delta^*}.$$

Proof. The statement follows immediately from Theorem 1.6, because for a crepant desingularization $\widehat{W} \to \overline{W}$ the set of primitive generators Γ is exactly the set of all lattice points in faces of codimension < 1 of the dual reflexive polytope Δ^* (see [**3**]). □

REMARK 1.10. Recently Haase and Nill proved that for a reflexive polytope Δ^* of arbitrary dimension $d > 2$ the sublattice in N generated by **all** lattice points in $\Delta^* \cap N$ coincides with N'_{Δ^*} [**14**]. The proof of this result uses the fact that the interior lattice points in codimension-1 faces of Δ^* are Demazure roots for the automorphism group of the corresponding Gorenstein toric Fano variety $\mathbb{P}_{\Delta^*}$ [**20**]. In particular, this result easily implies that $N = N'_{\Delta^*}$ for all reflexive polyhedra Δ^* of dimension $d = 3$. The last statement also follows from Corollary 1.9 and from the fact that every smooth $K3$-surface is simply connected.

REMARK 1.11. It is known that for $d = 4$ there always exists a smooth projective crepant resolution $\widehat{W}$ of singularities in $\overline{W}$ induced by a convex (coherent) triangulation of the dual reflexive polytope $\Delta^* \subset N_{\mathbb{R}}$. For $d \geq 5$ such a resolution does not exist in general. However, one can use N/N'_{Δ^*} as a candidate for a **stringy fundamental group** of a singular Calabi-Yau variety $\overline{W}$ or its maximal partial crepant desingularization $\widehat{W}$ (similar to stringy Hodge numbers [**4**]). It would be interesting to know whether for $d \geq 5$ the dual group $(N/N'_{\Delta^*})^*$ coincides with the torsion subgroup in the Picard group of the Deligne-Mumford Calabi-Yau stack associated with the V-manifold $\widehat{W}$ (see [**6, 25, 26**]).

2. The list of non-simply connected Calabi-Yau 3-folds

Using the Calabi-Yau data [**18**] and the program package PALP [**19**] one can check all 473 800 776 reflexive polytopes Δ and find that there exist exactly 16 examples of reflexive polytopes Δ_i $(1 \leq i \leq 16)$ such that $|N/N'_{\Delta_i^*}| > 1$. In the latter case, $|N/N'_{\Delta_i^*}|$ is always a prime number $p = 2, 3, 5$.

The most known example of a non-simply connected Calabi-Yau 3-fold obtained from a hypersurface in a Gorenstein toric variety $\mathbb{P}_\Delta$ is a free μ_5-quotient of a smooth quintic 3-fold in $\mathbb{P}^4$ obtained as follows. Consider the action of the cyclic group $\mu_5 = \langle \zeta \rangle$ by $(1, \zeta, \zeta^2, \zeta^3, \zeta^4)$ on $\mathbb{P}^4$ and take $\mathbb{P}_{\Delta_1} := \mathbb{P}^4/\mu_5$. The Gorenstein toric Fano variety $\mathbb{P}_{\Delta_1}$ contains exactly 5 isolated singular points. A non-degenerate hypersurface $\overline{W}_1 \subset \mathbb{P}_{\Delta_1}$ is smooth, i.e. $\overline{W}_1 = \widehat{W}_1$, it has the Hodge numbers $h^{1,1}(\overline{W}_1) = 1$, $h^{2,1}(\overline{W}_1) = 21$, and $\pi_1(\overline{W}_1) \cong \mathbb{Z}/5\mathbb{Z}$. This is the unique example in the case $|N/N'_{\Delta_i^*}| = 5$.

Similarly one obtains a free μ_3-quotient $\overline{W}_2$ of a smooth hypersurface of bidegree $(3, 3)$ in $\mathbb{P}^2 \times \mathbb{P}^2$ using the action of the cyclic group $\mu_3 = \langle \zeta \rangle$ on $\mathbb{P}^2 \times \mathbb{P}^2$ by $(1, \zeta, \zeta^2)$ on each $\mathbb{P}^2$. The toric variety $\mathbb{P}_{\Delta_2} := \mathbb{P}^2 \times \mathbb{P}^2/\mu_3$ has 9 isolated singularities. A non-degenerate hypersurface $\overline{W}_2 = \widehat{W}_2 \subset \mathbb{P}_{\Delta_2}$ is smooth, it has the Hodge numbers $h^{1,1}(\overline{W}_2) = 2$, $h^{2,1}(\overline{W}_2) = 29$, and $\pi_1(\overline{W}_2) \cong \mathbb{Z}/3\mathbb{Z}$.

We remark that the dual reflexive polytopes $\Delta_1^*, \ldots, \Delta_{16}^*$ have at most 8 vertices. Δ_{14}^* is the single dual reflexive polytope with 8 vertices. The corresponding Gorenstein toric Fano variety $\mathbb{P}_{\Delta_{14}}$ is $\mathbb{P}^1 \times \mathbb{P}^1 \times \mathbb{P}^1 \times \mathbb{P}^1/\mu_2$, where the cyclic group $\mu_2 = \langle \zeta \rangle$ acts by $(1, \zeta)$ on each $\mathbb{P}^1$.

We collected the information about all 16 reflexive polytopes $\Delta_1, \ldots, \Delta_{16}$ in the table below. In this table $\mathcal{P}_\Delta = |\Delta \cap M|$, $\mathcal{V}_\Delta = |\text{vert}(\Delta)|$, $\mathcal{P}_{\Delta^*} = |\Delta^* \cap N|$, and $\mathcal{V}_{\Delta^*} = |\text{vert}(\Delta^*)|$ denote numbers of points and vertices, respectively. Since any reflexive polytope $\Delta_i \subset M_{\mathbb{R}}$ is uniquely determined by its dual polytope $\Delta_i^* \subset N_{\mathbb{R}}$, we prefer to use the small number of vertices $v_1, \ldots, v_{\mathcal{V}_{\Delta^*}}$ in Δ_i^* in order to describe Δ_i^* by $\mathcal{V}_{\Delta^*} - 4$ independent linear relations among them.

The lattice N is generated by the vertices $v_1, \ldots, v_{\mathcal{V}_{\Delta^*}}$ and by an additional vector v expressed as a rational linear combination of the vertices $\{v_i\}$. The sublattice $N'_{\Delta_i^*}$ is generated by $v_1, \ldots, v_{\mathcal{V}_{\Delta^*}}$ except for 4 cases: $i = 4, 9, 15, 16$. In the latter cases $N'_{\Delta_i^*}$ is generated by $v_1, \ldots, v_{\mathcal{V}_{\Delta^*}}$ and by the vector $2v$.

n°	$\lvert\pi_1\rvert$	$\mathcal{P}_\Delta$ $\mathcal{V}_\Delta$	$\mathcal{P}_{\Delta^*}$ $\mathcal{V}_{\Delta^*}$	$h^{1,1}$ $h^{2,1}$	vertices $\{v_i\}$ of Δ^* and v
1	5	26 5	6 5	1 21	$v_1+v_2+v_3+v_4+v_5=0$ $v=\frac{1}{5}(v_2+2v_3+3v_4+4v_5)$
2	3	34 9	7 6	2 29	$v_1+v_2+v_3=v_4+v_5+v_6=0$ $v=\frac{1}{3}(v_2+2v_3+v_5+2v_6)$
3	3	49 5	7 5	2 38	$3v_1+3v_2+v_3+v_4+v_5=0$ $v=\frac{1}{3}(v_1+2v_2+v_3+2v_4)$
4	2	53 5	9 5	3 43	$4v_1+v_2+v_3+v_4+v_5=0$ $v=\frac{1}{4}(2v_1+v_2+2v_3+3v_4)$
5	2	77 7	9 6	3 59	$4v_1+2v_2+v_3+v_4=0$ $2v_1+v_5+v_6=0$ $v=\frac{1}{2}(v_1+v_2+v_3+v_5)$
6	2	77 9	9 7	3 59	$2v_1+v_2+v_3=2v_1+v_4+v_5=0$ $2v_1+v_6+v_7=0$ $v=\frac{1}{2}(v_1+v_2+v_4+v_6)$
7	2	101 5	9 5	3 75	$8v_1+4v_2+2v_3+v_4+v_5=0$ $v=\frac{1}{2}(v_1+v_2+v_3+v_4)$
8	2	101 6	9 6	3 75	$4v_1+2v_2+v_3+v_4=0$ $4v_1+2v_2+v_5+v_6=0$ $v=\frac{1}{2}(v_1+v_2+v_3+v_5)$
9	2	29 8	9 6	4 28	$v_1+v_2+v_3+v_4=v_5+v_6=0$ $v=\frac{1}{4}(v_2+2v_3+3v_4+2v_6)$
10	2	53 8	9 6	4 44	$4v_1+2v_2+v_3+v_4=v_5+v_6=0$ $v=\frac{1}{2}(v_1+v_2+v_3+v_5)$
11	2	53 10	9 7	4 44	$2v_1+v_2+v_3=0$ $2v_1+v_4+v_5=0$ $v_6+v_7=0$ $v=\frac{1}{2}(v_1+v_2+v_4+v_6)$
12	2	41 9	9 6	4 36	$2v_1+v_2+v_3=0$ $2v_4+v_5+v_6=0$ $v=\frac{1}{2}(v_1+v_2+v_4+v_5)$
13	2	41 12	9 7	4 36	$2v_1+v_2+v_3=0$ $v_4+v_5=v_6+v_7=0$ $v=\frac{1}{2}(v_1+v_2+v_4+v_6)$
14	2	41 16	9 8	4 36	$v_1+v_2=v_3+v_4=0$ $v_5+v_6=v_7+v_8=0$ $v=\frac{1}{2}(v_1+v_3+v_5+v_7)$
15	2	29 5	9 5	5 29	$2v_1+2v_2+2v_3+v_4+v_5=0$ $v=\frac{1}{4}(v_1+2v_2+3v_3+2v_5)$
16	2	29 6	9 6	5 29	$v_1+v_2+v_3+v_4=0$ $v_3+v_4+v_5+v_6=0$ $v=\frac{1}{4}(v_1+3v_2+2v_4+2v_6)$

3. The Brauer group of toric hypersurfaces

Let X be a smooth quasi-projective variety over $\mathbb{C}$. The cohomological Brauer group $B'(X)$ of X is defined as the torsion subgroup in $H^2_{\text{et}}(X, \mathcal{O}_X^*)$, where $\mathcal{O}_X^*$ denotes the sheaf of units in $\mathcal{O}_X$. If $B(X)$ is the Brauer group of sheaves of Azumaya algebras on X, then there exist a canonical injective homomorphism $B(X) \to B'(X)$. Recently Kresch and Vistoli proved that this homomorphism is in fact an isomorphism [**16**].

Let us summarize some well-known results about the Brauer group.

THEOREM 3.1. *The subgroup $B(X)_r \subset B(X)$ of elements $x \in B(X)$ with $rx = 0$ can be included into the short exact sequence*

$$0 \to \text{Pic}(X)/r\text{Pic}(X) \to H^2_{\text{et}}(X, \mu_r) \to B(X)_r \to 0,$$

where μ_r denotes the subsheaf of r-th roots of unity in $\mathcal{O}_X^*$. More generally, there is a short exact sequence

$$0 \to \text{Pic}(X) \otimes_{\mathbb{Z}} \mathbb{Q}/\mathbb{Z} \to H^2_{\text{et}}(X, \mathbb{Q}/\mathbb{Z}) \to B(X) \to 0,$$

where $H^2_{\text{et}}(X, \mathbb{Q}/\mathbb{Z})$ coincides with the singular cohomology group $H^2(X, \mathbb{Q}/\mathbb{Z})$ in the usual topology of X as a manifold over $\mathbb{C}$.

Proof. See [**13**] II, Theorem 3.1. □

COROLLARY 3.2. *If X is a smooth projective variety such that $Pic(X) \cong H^2(X, \mathbb{Z})$, then one has the isomorphism*

$$B(X) \cong \text{Tors}(H^3(X, \mathbb{Z})).$$

In particular, the last statement holds for smooth Calabi-Yau d-folds ($d \geq 3$).

Proof. Follows from the universal coefficient theorem. □

THEOREM 3.3. *Let X be a smooth quasi-projective variety and $Y \subset X$ a closed subvariety. Then the natural homomorphism $B(X) \to B(X \setminus Y)$ is always injective. This homomorphism is bijective if Y has codimension ≥ 2. Moreover, if Y is a union of closed irreducible subvarieties $Y_1, \ldots, Y_n$ of codimension 1, then one has the exact sequence*

$$0 \to B(X) \to B(X \setminus Y) \to \bigoplus_{i=1}^{n} H^1_{\text{et}}(Y_i, \mathbb{Q}/\mathbb{Z}),$$

where $H^1_{\text{et}}(Y_i, \mathbb{Q}/\mathbb{Z})$ coincides with the singular cohomology group $H^1(Y_i, \mathbb{Q}/\mathbb{Z})$ in the usual topology if Y_i is smooth.

Proof. See [**13**] III, Corollary 6.2. □

The Brauer group of a d-dimensional algebraic torus $\mathbb{T}^d$ over $\mathbb{C}$ was computed e.g. by Magid in [**21**]:

THEOREM 3.4. *Using the canonical isomorphism between the cohomology ring $H^*(\mathbb{T}^d, \mathbb{Z})$ and the exterior algebra $\Lambda^* M$ of the $\mathbb{Z}$-module M, one has the following canonical isomorphisms:*

$$B(\mathbb{T}^d) \cong H^2(\mathbb{T}^d, \mathbb{Q}/\mathbb{Z}) \cong \Lambda^2 M \otimes_{\mathbb{Z}} \mathbb{Q}/\mathbb{Z} \cong \text{Hom}(\Lambda^2 N, \mathbb{Q}/\mathbb{Z}).$$

The elements in the Brauer group $B(\mathbb{T}^d)$ can be explicitly described as follows. Take an integer $r > 1$ and two characters $\alpha, \beta \in M \subset \mathcal{O}^*_{\mathbb{T}^d}$ of the algebraic torus $\mathbb{T}^d$. The **symbol algebra** $(\alpha, \beta)_r$ is generated by two elements x, y with relations

$$x^r = \alpha, y^r = \beta, xy = \zeta yx,$$

where ζ is a primitive r-th root of unity. If $r = 2$, then $(\alpha, \beta)_2$ is the classical quaternionic algebra associated with $\alpha, \beta \in \mathcal{O}^*_{\mathbb{T}^d}$.

Let $B(\mathbb{T}^d)_r$ be r-torsion elements in $B(\mathbb{T}^d)$; then $B(\mathbb{T}^d)_r$ is a free $\mathbb{Z}/r\mathbb{Z}$ module with the basis

$$(m_i, m_j)_r, \quad 1 \le i < j \le d,$$

where $m_1, \dots, m_d$ is a $\mathbb{Z}$-basis of the lattice M, i.e.

$$B(\mathbb{T}^d)_r = \Lambda^2(M) \otimes \mathbb{Z}/r\mathbb{Z}.$$

THEOREM 3.5. *Let $\Delta \subset M_{\mathbb{R}}$ be a d-dimensional lattice polytope and let $W \subset \mathbb{T}^d$ be a nondegenerate affine hypersurface defined by the equation*

$$f(t) = \sum_{m \in \Delta \cap M} c_m t^m = 0.$$

Then for $d \ge 4$ the natural homomorphism

$$j^* \; : \; B(\mathbb{T}^d) \to B(W)$$

defined by the embedding $j : W \hookrightarrow \mathbb{T}^d$ is an isomorphism.

Proof. By Theorem 3.1, we obtain the commutative diagram

$$\begin{array}{ccccccccc} 0 & \longrightarrow & \mathrm{Pic}(\mathbb{T}^d) \otimes_{\mathbb{Z}} \mathbb{Q}/\mathbb{Z} & \longrightarrow & H^2(\mathbb{T}^d, \mathbb{Q}/\mathbb{Z}) & \longrightarrow & B(\mathbb{T}^d) & \longrightarrow & 0 \\ & & \downarrow{\scriptstyle j_P^*} & & \downarrow{\scriptstyle j_H^*} & & \downarrow{\scriptstyle j^*} & & \\ 0 & \longrightarrow & \mathrm{Pic}(W) \otimes_{\mathbb{Z}} \mathbb{Q}/\mathbb{Z} & \longrightarrow & H^2(W, \mathbb{Q}/\mathbb{Z}) & \longrightarrow & B(W) & \longrightarrow & 0 \end{array}$$

By 1.3, j_H^* is an isomorphism. Moreover, $\mathrm{Pic}(\mathbb{T}^d) = 0$ since $\mathbb{T}^d$ is a Zariski open subset of $\mathbb{C}^d$. Therefore it suffices to prove that $\mathrm{Pic}(W) \otimes_{\mathbb{Z}} \mathbb{Q} = 0$ (since the surjectivity of $\mathrm{Pic}(W) \otimes_{\mathbb{Z}} \mathbb{Q} \to \mathrm{Pic}(W) \otimes_{\mathbb{Z}} \mathbb{Q}/\mathbb{Z}$ would imply that $\mathrm{Pic}(W) \otimes_{\mathbb{Z}} \mathbb{Q}/\mathbb{Z} = 0$). Using toric resolution of singularities, one can compactify W to a smooth projective semiample hypersurface $\widehat{W}$ in some smooth projective toric variety $\widehat{\mathbb{P}}$ such that $Z := \widehat{W} \setminus W$ is a union of smooth irreducible divisors. By the vanishing theorem for semiample line bundles on toric varieties, we obtain that

$$h^1(\widehat{W}, \mathcal{O}_{\widehat{W}}) = \cdots = h^{d-2}(\widehat{W}, \mathcal{O}_{\widehat{W}}) = 0.$$

In particular, $\mathrm{Pic}(\widehat{W}) \cong H^2(\widehat{W}, \mathbb{Z})$. Therefore,

$$\dim_{\mathbb{Q}} Pic(W) \otimes_{\mathbb{Z}} \mathbb{Q} = \dim_{\mathbb{C}} H^2(\widehat{W}, \mathbb{C}) = b_2(\widehat{W}) = b_{2d-4}(\widehat{W}).$$

Let $Z_1, \dots, Z_k$ be irreducible components of $Z = \widehat{W} \setminus W$. Using the exact sequence

$$\bigoplus_{i=1}^{k} \mathbb{Q}[Z_i] \to \mathrm{Pic}(\widehat{W}) \otimes \mathbb{Q} \to \mathrm{Pic}(W) \otimes \mathbb{Q} \to 0,$$

it remains to show that the classes $[Z_1], \ldots, [Z_k] \in \mathrm{Pic}(\widehat{W})$ generate $\mathrm{Pic}(\widehat{W}) \otimes \mathbb{Q}$. The latter follows from Poincaré duality together with the long exact sequence for cohomology with compact supports

$$\cdots \to H_c^{2d-4}(W, \mathbb{Q}) \to H_c^{2d-4}(\widehat{W}, \mathbb{Q}) \to H_c^{2d-4}(Z, \mathbb{Q}) \to \cdots,$$

where $H_c^{2d-4}(Z, \mathbb{Q}) \cong \mathbb{Q}^k$, $H_c^{2d-4}(\widehat{W}, \mathbb{Q}) \cong \mathrm{Hom}(\mathrm{Pic}(\widehat{W}), \mathbb{Q})$, because the Hodge component of Hodge type $(d-2, d-2)$ in $H_c^{2d-4}(W, \mathbb{Q})$ is trivial (see [**10**]).

□

Let us recall the computation of the Brauer group of a smooth (quasi-projective) toric variety $\mathbb{P}_\Sigma$ obtained by De Meyer and Ford [**11**].

By Theorem 3.3, $B(\mathbb{P}_\Sigma) \cong B(\mathbb{P}_{\Sigma^{(1)}})$, where $\Sigma^{(1)}$ is the subfan of all cones of dimension ≤ 1 in Σ. So we can assume without loss of generality that $\Sigma = \Sigma^{(1)}$ and $\mathbb{P}_\Sigma$ is a toric compactification of $\mathbb{T}^d$ by divisors $D_1^\circ, \ldots, D_n^\circ$ ($D_i^\circ \cong \mathbb{T}^{d-1}$) corresponding to lattice vectors $e_1, \ldots, e_n \in N$. By Theorem 3.3, we have the exact sequence

$$0 \to B(\mathbb{P}_\Sigma) \to B(\mathbb{T}^d) \to \bigoplus_{i=1}^n H^1(D_i^\circ, \mathbb{Q}/\mathbb{Z}).$$

For any $i = 1, \ldots, n$ we have the canonical isomorphisms

$$H^1(D_i^\circ, \mathbb{Q}/\mathbb{Z}) \cong \mathrm{Hom}(N/\mathbb{Z}e_i, \mathbb{Q}/\mathbb{Z}) \cong M_{e_i} \otimes \mathbb{Q}/\mathbb{Z},$$

where

$$M_{e_i} := \{m \in M \, : \, \langle x, e_i \rangle = 0\}.$$

Therefore, one obtains

$$B(\mathbb{T}^d) \supset B(\mathbb{P}_\Sigma) = \bigcap_{i=1}^n \mathrm{Ker}\, \varphi_i,$$

where φ_i denotes the ramification map

$$\varphi_i \, : \, B(\mathbb{T}^d) \to H^1(D_i^\circ, \mathbb{Q}/\mathbb{Z})$$

which associates to a given symbol algebra $(\alpha, \beta)_r$ its ramification along the divisor $D_i^\circ \subset \mathbb{P}_\Sigma$, i.e. a cyclic Galois covering of D_i° of degree r.

THEOREM 3.6. *Using the canonical isomorphisms*

$$B(\mathbb{T}^d) \cong \mathrm{Hom}(\Lambda^2 N, \mathbb{Q}/\mathbb{Z}), \quad H^1(D_i^\circ, \mathbb{Q}/\mathbb{Z}) \cong \mathrm{Hom}(N/\mathbb{Z}e_i, \mathbb{Q}/\mathbb{Z}),$$

one obtains the ramification map φ_i *from the homomorphism*

$$\psi_i \, : \, N/\mathbb{Z}e_i \to \Lambda^2 N$$

$$y \mapsto y \wedge e_i, \quad \forall y \in N$$

by applying the functor $\mathrm{Hom}(*, \mathbb{Q}/\mathbb{Z})$. *In particular, the image of* ψ_i *is the sublattice*

$$\mathrm{Im}\, \psi_i = N \wedge e_i \subset \Lambda^2 N.$$

Proof. These statements are contained in [**11**] Lemmas 1.5-1.7. □

Let us define the sublattice $N' := \sum_{i=1}^n \mathbb{Z}e_i \subset N$ and write

$$N/N' \cong \mathbb{Z}/c_1\mathbb{Z} \oplus \cdots \oplus \mathbb{Z}/c_d\mathbb{Z}$$

where $c_1, \ldots, c_d$ are non-negative integers such that $c_1|c_2|\cdots|c_d$. We can choose a $\mathbb{Z}$-basis $y_1, \ldots, y_d$ of N such that $\{c_iy_i\}_{c_i\neq 0}$ is a $\mathbb{Z}$-basis of N'. Consider another sublattice

$$N \wedge N' = \sum_{i=1}^n N \wedge e_i = \sum_{i=1}^n \operatorname{Im} \psi_i \subset \Lambda^2 N.$$

Then the elements $\{c_iy_i \wedge y_j\}_{c_i\neq 0, i<j}$ form a $\mathbb{Z}$-basis of $N \wedge N'$. Therefore,

$$\Lambda^2 N/(N \wedge N') \cong \bigoplus_{i<j} \mathbb{Z}/c_i\mathbb{Z}.$$

Thus, one obtains (see [**11**], Theorem 1.1):

THEOREM 3.7. *In the above notations, the Brauer group of a nonsingular toric variety $\mathbb{P}_\Sigma$ is isomorphic to*

$$\bigoplus_{i<j} \operatorname{Hom}(\mathbb{Z}/c_i\mathbb{Z}, \mathbb{Q}/\mathbb{Z}).$$

In particular, if $N' \subset N$ is a sublattice of finite index, then $\mathbb{P}_\Sigma$ has a finite Brauer group:

$$B(\mathbb{P}_\Sigma) \cong \operatorname{Hom}(\Lambda^2 N/(N \wedge N'), \mathbb{Q}/\mathbb{Z}) \cong \bigoplus_{i<j} (\mathbb{Z}/c_i\mathbb{Z})^*.$$

Let $\Delta \subset M_\mathbb{R}$ be a d-dimensional lattice polytope and let $W \subset \mathbb{T}^d$ be a nondegenerate affine hypersurface defined by the equation

$$f(t) = \sum_{m\in\Delta\cap M} c_m t^m = 0.$$

Denote by $\overline{W}$ the closure of W in the projective toric variety $\mathbb{P}_\Delta$ and by $\widehat{W}$ a smooth projective desingularization of $\overline{W}$ defined by the fan $\Sigma \subset N_\mathbb{R}$ which is a subdivision of the normal fan to Δ. We denote by $N_\Delta^{(2)}$ the sublattice in N generated by all lattice points $v \in N$ such that the minimum of the linear function $\langle *, v\rangle$ on Δ is attained on a face $\theta \subset \Delta$ of dimension ≥ 2. Our main result in this section is the following:

THEOREM 3.8. *In the above notation, for $d \geq 4$ the Brauer group $B(\widehat{W})$ of the toric hypersurface $\widehat{W}$ is a cyclic group isomorphic to*

$$\operatorname{Hom}(\Lambda^2 N/(N \wedge N_\Delta^{(2)}), \mathbb{Q}/\mathbb{Z}).$$

Proof. Let $e_1, \ldots, e_n$ be primitive lattice generators of 1-dimensional cones in Σ such that the minimum of the linear function $\langle *, v\rangle$ on Δ is attained on a face $\theta \subset \Delta$ of dimension ≥ 1. Denote by $\Sigma' \subset \Sigma^{(1)}$ the subfan of Σ consisting of 0 and all 1-dimensional cones $\mathbb{R}_{\geq 0}e_1, \ldots, \mathbb{R}_{\geq 0}e_n$. We set $\widehat{W}^{(1)} = \widehat{W} \cap \mathbb{P}_{\Sigma'}$. Since the complement to $\widehat{W}^{(1)}$ in $\widehat{W}$ has codimension ≥ 2, by Theorem 3.3 it suffices to compute the Brauer group $B(\widehat{W}^{(1)})$. By Theorem 3.3, we have the commutative diagram:

$$\begin{array}{ccccccc} 0 & \longrightarrow & B(\mathbb{P}_{\Sigma'}) & \longrightarrow & B(\mathbb{T}^d) & \xrightarrow{\varphi} & \bigoplus_{i=1}^n H^1(D_i^\circ, \mathbb{Q}/\mathbb{Z}) \\ & & \downarrow \alpha & & \downarrow \beta & & \downarrow \gamma \\ 0 & \longrightarrow & B(\widehat{W}^{(1)}) & \longrightarrow & B(W) & \xrightarrow{\phi} & \bigoplus_{i=1}^n H^1(Z_i^\circ, \mathbb{Q}/\mathbb{Z}) \end{array}$$

By Theorem 3.5, β is an isomorphism. In order to understand the homomorphisms α and γ we divide the set of lattice vectors $\mathcal{E} = \{e_1, \ldots, e_n\}$ into a union of two disjoint subsets $\mathcal{E}_1 \cup \mathcal{E}_2$. The set $\mathcal{E}_2$ consists of all $e_i \in \mathcal{E}$ such that the minimum of $\langle *, e_i \rangle$ on Δ is attained on a face $\theta \subset \Delta$ of dimension ≥ 2 and we define $\mathcal{E}_1 := \mathcal{E} \setminus \mathcal{E}_2$.

For any $e_i \in \mathcal{E}$ we denote by θ_i the maximal subface of Δ on which the minimum of $\langle *, e_i \rangle$ on Δ is attained.

If $e_i \in \mathcal{E}_2$, then $d_i := \dim \theta_i \geq 2$ and the divisor Z_i° is isomorphic to the product $V_i \times \mathbb{T}^{d-d_i-1}$, where V_i is a θ_i- nondegenerate affine hypersurface in a torus $\mathbb{T}^{d_i}$. By Corollary 1.3, the ramification homomorphism

$$\phi_i \; : \; B(W) \to H^1(Z_i^\circ, \mathbb{Q}/\mathbb{Z})$$

can be identified with the composition of the ramification homomorphism

$$\varphi_i \; : \; B(\mathbb{T}^d) \to H^1(D_i^\circ, \mathbb{Q}/\mathbb{Z})$$

with the monomorphism

$$H^1(D_i^\circ, \mathbb{Q}/\mathbb{Z}) \to H^1(Z_i^\circ, \mathbb{Q}/\mathbb{Z}).$$

Thus, we obtain

$$B := \bigcap_{e_i \in \mathcal{E}_2} \operatorname{Ker} \varphi_i = \bigcap_{e_i \in \mathcal{E}_2} \operatorname{Ker} \phi_i.$$

Using Theorem 3.6, we can describe the dual group B^* as the cokernel of the homomorphisms

$$\psi_i \; : \; N/\mathbb{Z}e_i \to \Lambda^2 N, \;\; e_i \in \mathcal{E}_2,$$

i.e. we have

$$\sum_{e_i \in \mathcal{E}_2} \operatorname{Im} \psi_i = N \wedge N_\Delta^{(2)} \;\text{ and }\; B^* = N/(N \wedge N_\Delta^{(2)}).$$

Therefore, it remains to prove that

$$B = \bigcap_{e_i \in \mathcal{E}} \operatorname{Ker} \phi_i = B(\widehat{W}^{(1)}),$$

i.e., that for any $e_i \in \mathcal{E}_1$ the image $\operatorname{Im} \phi_i = \operatorname{Im} \gamma \circ \psi_i$ is contained in $N \wedge N_\Delta^{(2)}$. Now we consider two cases for $e_i \in \mathcal{E}_1$:

Case 1: Let $e_i \in N_\Delta^{(2)}$. Then the image of ϕ_i is contained in $N \wedge e_i \subset N \wedge N_\Delta^{(2)}$.
Case 2: Let $e_i \notin N_\Delta^{(2)}$. Then the image of ϕ_i is a sublattice $N_i \wedge e_i$, where N_i is a direct summand of N of rank $d-1$ containing e_i (N_i is a sublattice of rank $d-1$ which is orthogonal to the 1-dimensional face θ_i). The smoothness of $\widehat{W}$ implies that both lattices N_i and $N_\Delta^{(2)}$ contain a direct summand of N of rank $d-2$. Therefore, $N_i/N_i \cap N_\Delta^{(2)}$ is a cyclic group and we have

$$N_i = \mathbb{Z}v + N_i \cap N_\Delta^{(2)}$$

for some lattice vector $v \in N_i$ which is proportional to e_i. By $v \wedge e_i = 0$, we have

$$N_i \wedge e_i \subset (\mathbb{Z}v + N_i \cap N_\Delta^{(2)}) \wedge e_i \subset N_\Delta^{(2)} \wedge e_i \subset N \wedge N_\Delta^{(2)}.$$

Thus, all the divisors Z_i ($e_i \in \mathcal{E}_1$) do not define non-trivial ramification conditions on the subgroup B of the Brauer group $B(W)$ and

$$B \cong B(\widehat{W}^{(1)}) \cong B(\widehat{W}) \cong \mathrm{Hom}(\Lambda^2 N/(N \wedge N_\Delta^{(2)}), \mathbb{Q}/\mathbb{Z}).$$

□

COROLLARY 3.9. *Let $\widehat{W}$ be a smooth projective Calabi-Yau d-fold obtained as a crepant desingularization of a nondegenerate projective hypersurface $\overline{W} \subset \mathbb{P}_\Delta$ corresponding to a reflexive polytope Δ of dimension $d \geq 4$. Then the Brauer group $B(\widehat{W})$ is a cyclic group isomorphic to*

$$\mathrm{Hom}(\Lambda^2 N/(N \wedge N''_{\Delta^*}), \mathbb{Q}/\mathbb{Z}),$$

where $N''_{\Delta^*} \subset N$ is the sublattice generated by all lattice points in $N \cap \Delta^*$ which are contained in faces of Δ^* of codimension > 2.

Proof. The statement follows immediately from Theorem 3.8 and from the easy combinatorial equality $N_\Delta^{(2)} = N''_{\Delta^*}$ using the duality between faces of dual reflexive polytopes $\Delta \subset M_\mathbb{R}$ and $\Delta^* \subset N_\mathbb{R}$. □

4. The Brauer group of Calabi-Yau 3-folds

Let Δ be a 4-dimensional reflexive polytope. Then

$$N''_{\Delta^*} := \sum_{v \in \theta^* \cap N, \dim \theta^* = 1} \mathbb{Z}v \subset N$$

is a sublattice of finite index. We write

$$N/N''_{\Delta^*} = \mathbb{Z}/c_1\mathbb{Z} \oplus \mathbb{Z}/c_2\mathbb{Z} \oplus \mathbb{Z}/c_3\mathbb{Z} \oplus \mathbb{Z}/c_4\mathbb{Z},$$

where $c_1|c_2|c_3|c_4$. In this case $c_1 = c_2 = 1$ and

$$B(\widehat{W}) \cong \mathrm{Hom}(\Lambda^2 N/(N \wedge N''_{\Delta^*}), \mathbb{Q}/Z) \cong \mathbb{Z}/c_3\mathbb{Z}.$$

Using the Calabi-Yau data [**18**], one can check all 473 800 776 reflexive polytopes Δ and find exactly 16 reflexive polytopes of dimension 4 with $c_3 > 1$. For each of these 16 polytopes, one obtains

$$N/N''_{\Delta^*} \cong \mathbb{Z}/p\mathbb{Z} \oplus \mathbb{Z}/p\mathbb{Z}, \quad p = 2, 3, 5.$$

So we have again 16 families with the cyclic Brauer group $\mathbb{Z}/p\mathbb{Z}$. The main observation is that these 16 families are exactly the mirrors of 16 non-simply connected families of Calabi-Yau 3-folds with the cyclic fundamental group $\mathbb{Z}/p\mathbb{Z}$ ($c = 2, 3, 5$) (see Section 2). Therefore, we can exchange the lattices $N \leftrightarrow M$ and polytopes $\Delta^* \leftrightarrow \Delta$ so that for all 4-dimensional reflexive polytopes we have the isomorphisms

$$\Lambda^2 M/(M \wedge M''_\Delta) \cong N/N'_{\Delta^*} \quad \text{and} \quad \Lambda^2 N/(N \wedge N''_{\Delta^*}) \cong M/M'_\Delta \tag{3}$$

Unfortunately we do not have any natural combinatorial explanation of the isomorphisms (3).

The sublattice $N''_{\Delta^*} \subset N$ defines a (ramified) Galois covering $\widehat{W}'' \to \widehat{W}$ with Galois group $G = N/N''_{\Delta^*} \cong \mathbb{Z}/p\mathbb{Z} \oplus \mathbb{Z}/p\mathbb{Z}$. If we consider G as a 2-dimensional vector space over the finite field $\mathbb{F}_p$, then the Brauer group $B(\widehat{W})$ is isomorphic to

$\mathrm{Hom}(\Lambda^2 G, \mathbb{Q}/\mathbb{Z})$, i.e, to a 1-dimensional $\mathbb{F}_p$-space. This allows us to look at $B(\widehat{W})$ as the discrete torsion group $H^2(G, U(1))$ for the orbifold $\widehat{W}''/G$.

We finish this section by describing a typical example $\widehat{W}_1^*$ of a Calabi-Yau 3-fold from our list (the mirror of $\widehat{W}_1$ from Section 2) with Brauer group $B(\widehat{W}_1^*) \cong \mathbb{Z}/5\mathbb{Z}$. This example gives a 1-parameter family of Calabi-Yau 3-folds $\widehat{W}_1^*$ having a close connection to the counterexample to global Torelli considered by Aspinwall-Morrison and B. Szendrői [**1, 27**].

Consider the action of the group $G := \mu_5 \times \mu_5$ on $\mathbb{P}^4$ by

$$\zeta_1 \to (1, \zeta_1, \zeta_1^2, \zeta_1^3, \zeta_1^4)$$
$$\zeta_2 \to (1, \zeta_2, \zeta_2^3, \zeta_2, 1).$$

This action is not free and the quotient $\mathbb{P}_{\Delta_1^*} := \mathbb{P}^4/G$ has cyclic singularities of type $\frac{1}{5}(3,1,1)$ along 10 curves. Let $\widehat{W}_1^*$ be the crepant desingularization of a generic Calabi-Yau hypersurface $\overline{W}_1^* \subset \mathbb{P}_{\Delta_1^*}$. Then

$$\pi_1(\widehat{W}_1^*) = 0 = B(\widehat{W}_1)$$
$$B(\widehat{W}_1^*) = \Lambda^2 G \cong \mu_5 \cong \pi_1(\widehat{W}_1).$$

References

[1] P. S. Aspinwall, D. R. Morrison, *Chiral Rings Do Not Suffice: N=(2,2) Theories with Nonzero Fundamental Group,* Phys.Lett. B334 (1994), 79-86.

[2] P. S. Aspinwall, D. R. Morrison (with an appendix by M. Gross), *Stable Singularities in String Theory,* Commun.Math.Phys. 178 (1996), 115-134.

[3] V. V. Batyrev, *Dual polyhedra and mirror symmetry for Calabi-Yau hypersurfaces in toric varieties.* J. Algebraic Geom. 3 (1994), no. 3, 493–535.

[4] V. V. Batyrev, *Stringy Hodge numbers of varieties with Gorenstein canonical singularities.* Integrable systems and algebraic geometry (Kobe/Kyoto, 1997), 1-32, World Sci. Publishing, River Edge, NJ, 1998.

[5] V. V. Batyrev, L.A. Borisov, *Mirror duality and string-theoretic Hodge numbers.* Invent. Math. 126 (1996), no. 1, 183–203.

[6] L. A. Borisov, L. Chen, G.G. Smith, *The orbifold Chow ring of toric Deligne-Mumford stacks.* J. Amer. Math. Soc. 18 (2005), no. 1, 193–215.

[7] I. Brunner, J. Distler, *Torsion D-Branes in Nongeometrical Phases* Adv.Theor.Math.Phys. 5 (2002), 265-309.

[8] I. Brunner, J. Distler, R. Mahajan, *Return of the Torsion D-Branes,* Adv.Theor.Math.Phys. 5 (2002), 311-352.

[9] V.I. Danilov, *The geometry of toric varieties,* Uspekhi Mat. Nauk 33 (1978), no. 2(200), 85–134, 247.

[10] V.I. Danilov, A.G. Khovanskiĭ, *Newton polyhedra and an algorithm for calculating Hodge-Deligne numbers.* Izv. Akad. Nauk SSSR Ser. Mat. 50 (1986), no. 5, 925–945.

[11] F. De Meyer, T. Ford, *On the Brauer group of toric varieties.* Trans. Amer. Math. Soc. 335 (1993), 559-577.

[12] M. Gross, *Special Lagrangian fibrations. I.* Winter School on Mirror Symmetry, Vector Bundles and Lagrangian Submanifolds (Cambridge, MA, 1999), 65–93, Amer. Math. Soc., Providence, RI, 2001.

[13] A. Grothendieck *Le groupe de Brauer I,II,III.* In: J. Giraud (ed) et al.: Dix exposés sur la cohomologie des schémas, pp. 46-189. North-Holland, Amsterdam (1968).

[14] C. Haase, B. Nill, *Lattices generated by skeletons of reflexive polytopes,* in preparation.

[15] A. Klemm, M. Kreuzer, E. Riegler, E. Scheidegger, *Topological String Amplitudes, Complete Intersection Calabi-Yau Spaces and Threshold Corrections,* hep-th/0410018.

[16] Andrew Kresch, Angelo Vistoli, *On coverings of Deligne-Mumford stacks and surjectivity of the Brauer map,* math.AG/0301249.

[17] M. Kreuzer, H. Skarke, *Complete classification of reflexive polyhedra in four dimensions,* Adv.Theor.Math.Phys. 4 (2002) 1209-1230, hep-th/0002240.

[18] M. Kreuzer, H. Skarke, http://hep.itp.tuwien.ac.at/~kreuzer/CY.html

[19] M. Kreuzer, H. Skarke, *PALP: A package for analyzing lattice polytopes with applications to toric geometry,* Computer Physics Commun. 157 (2004) 87–106, math.NA/0204356.

[20] B. Nill, *Complete toric varieties with reductive automorphism group,* math.AG/0407491.

[21] A. R. Magid, *Brauer groups of linear algebraic groups with characters.* Proc. Amer. Math. Soc. 71 (1978), no. 2, 164–168.

[22] M. Oka, *On the topology of the Newton boundary II,* J. Math. Soc. Japan, Vol **32**, n.1 (1980), 65–92.

[23] M. Oka, *A Lefschetz type theorem in a torus.* Singularity theory (Trieste, 1991), 574–593, World Sci. Publishing, River Edge, NJ, 1995.

[24] M. Oka, *Non-degenerate complete intersection singularity.* Actualités Mathématiques. Hermann, Paris, 1997.

[25] T. Pantev, E. Sharpe, *String compactifications on Calabi-Yau stacks,* hep-th/0502044.

[26] T. Pantev, E. Sharpe, *GLSM's for gerbes (and other toric stacks),* hep-th/0502053.

[27] B. Szendrői, *On an example of Aspinwall and Morrison,* Proc. AMS, vol 132, n. 3 (2003), 621–632.

[28] C. Vafa, E. Witten *Orbifolds with discrete torsion,* J. Geom. Phys. **15** (1995)

Mathematisches Institut, Universität Tübingen, Auf der Morgenstelle 10, 72076 Tübingen, Germany

E-mail address: victor.batyrev@uni-tuebingen.de

Institut für Theoretische Physik, Technische Universität Wien, Wiedner Hauptstr. 8-10/136, A-1040 Vienna, Austria

E-mail address: maximilian.kreuzer@tuwien.ac.at

AMS/IP Studies in Advanced Mathematics
Volume 38, 2006

The Real Regulator for a Product of $K3$ Surfaces

Xi Chen and James D. Lewis

ABSTRACT. Let X be a general complex algebraic $K3$ surface, and $\mathrm{CH}^2(X,1)$ Bloch's higher Chow group ([**Blo1**]). In [**C-L2**] it is proven that the real regulator $r_{2,1}:\mathrm{CH}^2(X,1)\otimes\mathbb{R}\to H^{1,1}(X,\mathbb{R})$ is surjective (Hodge-$\mathcal{D}$-conjecture). The question addressed in this paper is whether a similar story holds for products of $K3$ surfaces. We prove a general regulator result for a product of two surfaces, and then deduce, under the assumptions of [a variant of] the Bloch-Beilinson conjecture regarding the injectivity of the Abel-Jacobi map for smooth quasiprojective varieties defined over number fields, that the (induced) regulator map $\underline{r}_{3,1}:\mathrm{CH}^3(X\times Y,1)\to\big\{H^2_{\mathrm{tr}}(X,\mathbb{R})\otimes H^2_{\mathrm{tr}}(Y,\mathbb{R})\big\}\cap H^{2,2}(X\times Y)$ is trivial for a general pair of $K3$ surfaces, where $H^2_{\mathrm{tr}}(X,\mathbb{R})$ is the space of transcendental classes in $H^2(X,\mathbb{R})$.

0. Introduction

Let $f:W\to\Gamma$ be a dominating morphism of smooth complex projective varieties, with $\dim W=3$ and $\dim\Gamma=1$. For $t\in\Gamma$, put $W_t=f^{-1}(t)$, and $U=\{t\in\Gamma\mid W_t \text{ is smooth}\}$. Let $\Sigma=\Gamma\backslash U$. For $t\in\Sigma$ we will assume that the singular set of W_t is a finite set of ordinary double points. With respect to a polarization of W, and for $t\in U$, let $\mathrm{Prim}^2(W_t)$ be the primitive cohomology ([**G-H**], p. 122). For fixed $t_0\in U$, we will assume that the monodromy group representation $\pi_1(U)\to\mathrm{Aut}\big(\mathrm{Prim}^2(W_{t_0})\big)$ is irreducible. Let us further assume that $\mathrm{Prim}^{2,0}(W_t)\neq 0$ for $t\in U$. Note that for general $t\in\Gamma$, $\mathrm{Prim}^2(W_t)=H^2_{\mathrm{tr}}(W_t)$, where the latter is the space of transcendental cycles (= orthogonal complement of the algebraic classes under the cup product). We prove the following:

THEOREM 0.1. *Given the above assumptions, let Y be a smooth surface with the property that $\mathrm{CH}^1(Y;\mathbb{Q})\simeq\mathbb{Q}$. If*

$$\mathrm{CH}^2(W_t\times Y;\mathbb{Q})=\sum_{i=0}^{2}\mathrm{CH}^i(W_t;\mathbb{Q})\otimes\mathrm{CH}^{2-i}(Y;\mathbb{Q}),\tag{0.2}$$

for every $t\in\Gamma$, then the (induced) regulator map

$$\underline{r}_{3,1}:\mathrm{CH}^3(W_t\times Y,1)\to\big\{\mathrm{Prim}^2(W_t,\mathbb{R})\otimes H^2_{\mathrm{tr}}(Y,\mathbb{R})\big\}\cap H^{2,2}(W_t\times Y),$$

2000 *Mathematics Subject Classification.* Primary 14C25, 14F43, 14J28, 14K30.

Key words and phrases. Higher Chow group, real regulator, algebraic cycle.

Both authors partially supported by a grant from the Natural Sciences and Engineering Research Council of Canada.

is zero for general $t \in \Gamma$.

Next, we prove the following:

PROPOSITION 0.3. *Suppose that the complex Abel-Jacobi map is injective for smooth quasiprojective varieties defined over a number field (see Conjecture 3.0 below). If* $\{W_t\}_{t\in\Gamma}$ *is a general choice of* 1*-parameter* $K3$ *surfaces, and if* Y *is likewise a general choice of* $K3$ *surface, then statement* (0.2) *in Theorem 0.1 holds.*

Combining Proposition 0.3 and Theorem 0.1, we arrive at:

COROLLARY 0.4. *With the assumption in Proposition 0.3, let* X *and* Y *be a general pair of* $K3$ *surfaces. Then the (induced) regulator map*

$$\underline{r}_{3,1} : \mathrm{CH}^3(X \times Y, 1) \to \{H^2_{\mathrm{tr}}(X, \mathbb{R}) \otimes H^2_{\mathrm{tr}}(Y, \mathbb{R})\} \cap H^{2,2}(X \times Y),$$

is trivial.

We leave it to the reader to generalize Proposition 0.3 and Corollary 0.4 to other classes of surfaces (for example general complete intersection surfaces). In light of our knowledge of the Hodge-$\mathcal{D}$-conjecture for general $K3$ surfaces ([**C-L2**]), the authors regard Corollary 0.4 as rather surprising.

1. Some definitions

Unless otherwise stated, all varieties in this paper are complex varieties.

1.0 *Higher Chow groups.* The seminal definition of higher Chow groups $\mathrm{CH}^k(X, m)$ can be found in [**Blo1**]. We put $\mathrm{CH}^k(X, m; \mathbb{Q}) = \mathrm{CH}^k(X, m) \otimes \mathbb{Q}$. Note that $\mathrm{CH}^k(X, 0) = \mathrm{CH}^k(X)$, where the latter is the standard definition given in [**F**]. We are mainly interested in the case $m = 1$. For a quasiprojective Z, the following abridged definition of $\mathrm{CH}^k(Z, 1)$ will suffice [**La**] (cf. [**MS2**]).

DEFINITION 1.1. $\mathrm{CH}^k(Z, 1)$ is the homology of the middle term in the complex

$$\coprod_{\mathrm{cd}_Z Y = k-2} K_2(\mathbb{C}(Y)) \xrightarrow{\text{Tame}} \coprod_{\mathrm{cd}_Z Y = k-1} K_1(\mathbb{C}(Y)) \xrightarrow{\text{div}} \coprod_{\mathrm{cd}_Z Y = k} K_0(\mathbb{C}(Y)),$$

where we recall that $K_1(\mathbb{F}) = \mathbb{F}^\times$ and $K_0(\mathbb{F}) = \mathbb{Z}$, for a field $\mathbb{F}$, and Tame, div are respectively the tame symbol and divisor maps.

1.2 *Real regulator.* For a projective algebraic manifold X of dimension n, the following abridged definition of the real regulator

$$r_{k,1} : \mathrm{CH}^k(X, 1) \to H^{k-1,k-1}(X, \mathbb{R}) := H^{2k-2}(X, \mathbb{R}) \cap H^{k-1,k-1}(X),$$

will suffice. First of all, Serre duality gives

$$H^{k-1,k-1}(X, \mathbb{R}) \simeq H^{n-k+1,n-k+1}(X, \mathbb{R})^\vee.$$

Let $\xi = \sum_\alpha (f_\alpha, Z_\alpha)$ represent a class in $\mathrm{CH}^k(X, 1)$. Thus $\mathrm{codim}_X Z_\alpha = k - 1$ and $\sum_\alpha \mathrm{div}_{Z_\alpha}(f_\alpha) = 0$ in X. For $\omega \in H^{n-k+1,n-k+1}(X, \mathbb{R})$ we define

$$r_{k,1}(\xi)(\omega) = \sum_\alpha \int_{Z_\alpha} \log |f_\alpha| \omega. \tag{1.3}$$

Up to a multiplicative factor of a Tate twist, this definition agrees with that given in [**Be**] and [**Ja1**].

Next, for a surface X, the subspaces $\mathrm{Prim}^2(X,\mathbb{R})$, $H^2_{\mathrm{tr}}(X,\mathbb{R})$ of $H^2(X,\mathbb{R})$ are self-dual under the cup product pairing on $H^2(X,\mathbb{R})$. Thus for smooth projective surfaces X and Y, there are induced regulator maps:

$$\underline{r}_{3,1} : \mathrm{CH}^3(X\times Y,1) \to H^{2,2}(X,\mathbb{R}) \to \begin{bmatrix}\mathrm{Prim}^2(X,\mathbb{R})\otimes H^2_{\mathrm{tr}}(Y,\mathbb{R})\\ H^2_{\mathrm{tr}}(X,\mathbb{R})\otimes H^2_{\mathrm{tr}}(Y,\mathbb{R})\end{bmatrix} \cap H^{2,2}(X\times Y).$$

1.4 *Horizontal displacement.* Let $h : W \to S$ be a proper smooth morphism of quasiprojective varieties over $\mathbb{C}$, where say for simplicity $\dim S = 1$, with smooth projective fiber $W_t := h^{-1}(t)$. Fix a reference point $t_0 \in S$ and consider a disk Δ centered at t_0. It is well known that there is a diffeomeomorphism $h^{-1}(\Delta) \approx \Delta \times W_{t_0}$. For $t\in\Delta$, let $j_t : W_t \hookrightarrow h^{-1}(\Delta)$ be the inclusion. Then $j_t^* : H^\bullet(h^{-1}(\Delta),\mathbb{C}) \xrightarrow{\sim} H^\bullet(W_t,\mathbb{C})$ is an isomorphism. Thus for a cohomology class $\gamma := \gamma_{t_0} \in H^\bullet(W_{t_0})$, one can talk about its horizontal displacement $\gamma_t := j_t^*\circ\left(j_{t_0}^*\right)^{-1}(\gamma) \in H^\bullet(W_t)$, for $t\in\Delta$ and more generally for $t \in S$. Consider the Hodge decomposition $H^\bullet(W_t,\mathbb{C}) = \bigoplus_{p+q=\bullet} H^{p,q}(W_t)$, $\gamma_t = \oplus_{p+q=\bullet}\gamma_t^{p,q}$. We say that the Hodge (p,q) components deform horizontally if $\gamma_t^{p,q} = (\gamma^{p,q})_t$ for all $t\in\Delta$. By analytic considerations of Hodge subbundles, this is equivalent to saying that $\gamma_t^{p,q} = (\gamma^{p,q})_t$ for all $t\in S$.

1.5 The word "general" in this paper will have the following meaning. Given a family of varieties $\{V_t\}_{t\in\mathcal{T}}$ parameterized by some variety $\mathcal{T}$, V_t is general if t is a general point of $\mathcal{T}$, i.e. t belongs to the complement of a *countable* union of proper subvarieties of $\mathcal{T}$, governed by certain properties.

2. Proof of Theorem 0.1

Suppose to the contrary that

$$\underline{r}_{3,1} : \mathrm{CH}^3(W_t\times Y,1) \to \left\{\mathrm{Prim}^2(W_t,\mathbb{R})\otimes H^2_{\mathrm{tr}}(Y,\mathbb{R})\right\}\cap H^{2,2}(W_t\times Y), \tag{2.0}$$

is nontrivial for a general $t\in\Gamma$. Then by a standard argument (see [**C-L1**]), after a base change $\Gamma'\to\Gamma$ with $Z = W\times_\Gamma \Gamma'$, there exists $\xi\in\mathrm{CH}^3(Z_U\times Y,1)$ such that $\underline{r}_{3,1}(\xi_t)\neq 0$ when ξ is restricted to general $t\in U$, where U is a nonempty Zariski open subset of Γ' and $Z_U = \cup_{t\in U} Z_t$. As in [**C-L1**], the key here is to extend ξ to all of Γ'; once that is done, we apply a monodromy argument. Let $\xi = \sum_\alpha (f_\alpha, D_\alpha)$ and $\overline{\xi} = \sum_\alpha(\overline{f}_\alpha,\overline{D}_\alpha)$ be the closure of ξ in $Z\times Y$, i.e., $\overline{D}_\alpha$ is the closure of D_α in $Z\times Y$ and $\overline{f}_\alpha$ is just f_α on $\overline{D}_\alpha$ via the identification $\mathbb{C}(D_\alpha)^\times = \mathbb{C}(\overline{D}_\alpha)^\times$. Then

$$\mathrm{div}(\overline{\xi}) = \sum_\alpha \mathrm{div}_{\overline{D}_\alpha}(\overline{f}_\alpha),$$

is supported on $Z_B\times Y$, where $B = \Gamma'\backslash U$. Thus for $t\in U$, $\overline{\xi}_t = \xi_t$, i.e. when $\overline{\xi}$, ξ are restricted to $Z_t\times Y$. By our assumption on W_t, for each $b\in B$,

$$\begin{aligned}\mathrm{CH}^2(Z_b\times Y;\mathbb{Q}) = \mathrm{CH}^0(Z_b)\otimes\mathrm{CH}^2(Y;\mathbb{Q}) \\ \oplus\,\mathrm{CH}^1(Z_b;\mathbb{Q})\otimes\mathrm{CH}^1(Y;\mathbb{Q}) \\ \oplus\,\mathrm{CH}^2(Z_b;\mathbb{Q})\otimes\mathrm{CH}^0(Y;\mathbb{Q})\end{aligned} \tag{2.1}$$

So after replacing $\overline{\xi}$ by $\overline{\xi} + \sum_{b\in B}(g_b, S_b)$ where $S_b\subset Z_b\times Y$, $g_b\in\mathbb{C}(S_b)^\times$, we may write

$$\mathrm{div}(\overline{\xi}) = R + S + T \tag{2.2}$$

where $[R] \in \mathrm{CH}^0(Z_B;\mathbb{Q}) \otimes \mathrm{CH}^2(Y;\mathbb{Q})$, $[S] \in \mathrm{CH}^1(Z_B;\mathbb{Q}) \otimes \mathrm{CH}^1(Y;\mathbb{Q})$ and $[T] \in \mathrm{CH}^2(Z_B;\mathbb{Q}) \otimes \mathrm{CH}^0(Y;\mathbb{Q})$. Since $\mathrm{div}(\overline{\xi}) \sim_{\mathrm{rat}} 0$ on $Z \times Y$, each of R, S and T is rationally equivalent to zero in $\mathrm{CH}^3(Z \times Y)$. (To see this, we need only pull back $\mathrm{div}(\overline{\xi})$ to $Z \times \{p\}$ for any $p \in Y$ to deduce that $T \sim_{\mathrm{rat}} 0$. To show that $S \sim_{\mathrm{rat}} 0$, use the fact that $\mathrm{CH}^1(Y;\mathbb{Q}) \simeq \mathbb{Q}$.) From the projection $\pi_{\Gamma'} : Z \times Y \to \Gamma' \times Y$, we see that $R = \pi_{\Gamma'}^* L$ for some $L \subset \Gamma' \times Y$ and $L = 0$ in $\mathrm{CH}^3(\Gamma' \times Y;\mathbb{Q})$. Therefore there exists $\sum_\beta (g_\beta, M_\beta)$ with $M_\beta \subset \Gamma' \times Y$, $\dim M_\beta = 1$ and $g_\beta \in \mathbb{C}(M_\beta)^\times$ such that

$$\sum_\beta \mathrm{div}_{M_\beta}(g_\beta) = L \tag{2.3}$$

Therefore, if we replace $\overline{\xi}$ by $\overline{\xi} - \sum_\beta (\pi_{\Gamma'}^* g_\beta, \pi_{\Gamma'}^* M_\beta)$, we have $\mathrm{div}(\overline{\xi}) = S + T$ and $\underline{r}_{3,1}(\overline{\xi}_t)$ unchanged for a general $t \in \Gamma'$. Similarly, from the projection $\pi_Z : Z \times Y \to Z$, we see that $T = \pi_Z^* L$ for some $L \subset Z$ and $L = 0$ in $\mathrm{CH}^3(Z;\mathbb{Q})$. So there exists $\sum_\beta (g_\beta, M_\beta)$ with $M_\beta \subset Z$ such that (2.3) holds. Again, if we replace $\overline{\xi}$ by $\overline{\xi} - \sum_\beta (\pi_Z^* g_\beta, \pi_Z^* M_\beta)$, we have $\mathrm{div}(\overline{\xi}) = S$ and $\underline{r}_{3,1}(\overline{\xi}_t)$ unchanged for a general $t \in \Gamma$. Finally, we have to get rid of the contribution S. Fixing a smooth irreducible curve $C \subset Y$, we may assume that S is supported on $Z_B \times C$ since $\mathrm{CH}^1(Z_B;\mathbb{Q}) \otimes \mathrm{CH}^0(C;\mathbb{Q}) = \mathrm{CH}^1(Z_B;\mathbb{Q}) \otimes \mathrm{CH}^1(Y;\mathbb{Q})$, as $\mathrm{CH}^1(Y;\mathbb{Q}) \simeq \mathbb{Q}$. Since $\mathrm{CH}^2(Z;\mathbb{Q}) \otimes \mathrm{CH}^0(C;\mathbb{Q}) \hookrightarrow \mathrm{CH}^3(Z \times Y;\mathbb{Q})$ is an injection, $S = 0$ in $\mathrm{CH}^2(Z \times C;\mathbb{Q})$. So there exists $\eta = \sum_\beta (g_\beta, M_\beta)$ with $M_\beta \subset Z \times C$ such that $\mathrm{div}(\eta) = S$. If we replace $\overline{\xi}$ by $\overline{\xi} - \eta$, we have $\mathrm{div}(\overline{\xi}) = 0$. Of course, we have to show that $\underline{r}_{3,1}(\eta_t) = 0$ for $t \in \Gamma$ general. Obviously, $\eta_t \in \mathrm{CH}^2(Z_t \times C, 1)$ for t general. Then $\underline{r}_{3,1}(\eta_t) = 0$ follows from the following diagram:

$$\begin{array}{ccc} \mathrm{CH}^2(Z_t \times C, 1) & \to & \mathrm{CH}^3(Z_t \times Y, 1) \\ \downarrow & & \downarrow \\ H^{1,1}(Z_t \times C, \mathbb{R}) & \to & H^{2,2}(Z_t \times Y, \mathbb{R}) \\ \downarrow & & \downarrow \\ H^{1,1}(Z_t, \mathbb{R}) \otimes H^0(C, \mathbb{R}) & \to & H^{1,1}(Z_t, \mathbb{R}) \otimes H^{1,1}(Y, \mathbb{R}) \end{array} \tag{2.4}$$

Here the bottom two vertical arrows are given by Hodge-Künneth projections. The final step in our construction is a mixed Hodge structure argument. Notice that $\overline{\xi}$ restricts to a class $\mathrm{CH}^3(U_Z \times Y, 1)$ where U_Z is the smooth set of Z. The singular set $\Sigma_Z := Z \backslash U_Z$ is easily seen to be a finite number of points, as a result of possible ramification of $\Gamma' \to \Gamma$ over the singular set $\Sigma \subset \Gamma$ of $f : W \to \Gamma$. By applying the existence of the cycle class map to Deligne cohomology (see [**Blo2**]), together with the appropriate exact sequence in [**E-V**] (Cor. 2.10), we arrive at the Betti image[1]

$$\mathrm{CH}^3(U_Z \times Y, 1) \to H^5(U_Z \times Y, \mathbb{R}),$$

which is the result of the composite

$$\mathrm{CH}^3(U_Z \times Y, 1) \to H^5_{\mathcal{D}}(U_Z \times Y, \mathbb{R}(3)) \to H^5(U_Z \times Y, \mathbb{R}),$$

where $H^\bullet_{\mathcal{D}}$ is Deligne cohomology. This map factors accordingly as follows:

$$\mathrm{CH}^3(U_Z \times Y, 1) \to H^5_{\mathcal{D}}(U_Z \times Y, \mathbb{Q}(3)) \to F^3 H^5(U_Z \times Y, \mathbb{Q}) \to H^5(U_Z \times Y, \mathbb{R}),$$

[1] We are purposely ignoring Tate twists here.

where

$$F^3H^5(U_Z \times Y, \mathbb{Q}) := F^3H^5(U_Z \times Y, \mathbb{C}) \cap H^5(U_Z \times Y, \mathbb{Q}).$$

Let

$$[\Delta_Y] = \oplus_{p+q=4}[\Delta_Y(p,q)] \in \oplus_{p+q=4}H^p(Y,\mathbb{Q}) \otimes H^q(Y,\mathbb{Q}) = H^4(Y \times Y, \mathbb{Q}),$$

be the Künneth decomposition of the diagonal class of Y. One can argue that the $\Delta_Y(p,q)$'s are algebraic cycles. We can modify $\overline{\xi} \in \mathrm{CH}^3(U_Z \times Y, 1)$ by

$$\{\mathrm{Id}_{U_Z} \otimes \Delta_Y(2,2)\}_* \overline{\xi},$$

so that now the Betti image of $\overline{\xi}$ in $H^5(U_Z \times Y, \mathbb{R})$ lies in $H^3(U_Z, \mathbb{Q}) \otimes H^2(Y, \mathbb{Q})$. Next, by the nature of the singular set $\Sigma_Z \subset Z$ (see the Appendix), observe that

$$F^3\big(H^3(U_Z,\mathbb{Q}) \otimes H^2(Y,\mathbb{Q})\big) \subset H^3(U_Z,\mathbb{Q}) \otimes \mathbb{Q}[H_Y],$$

where $[H_Y] \in H^2(Y,\mathbb{Q})$ is a hyperplane class. The primitive projection $H^2(Y,\mathbb{Q}) \to \mathrm{Prim}^2(Y,\mathbb{Q}) = \mathbb{Q}[H_Y]^{\perp}$ is induced by an algebraic cycle $w \in \mathrm{CH}^2(Y \times Y; \mathbb{Q})$. [Explicitly $w = \big\{\Delta_Y - \frac{1}{\langle H_Y^2\rangle_Y} H_Y \times H_Y\big\}$.] If we further modify $\overline{\xi}$ by $(\mathrm{Id}_{U_Z} \otimes \{w\})_* \overline{\xi}$, then the corresponding Betti image of $\overline{\xi}$ in $H^5(U_Z \times Y, \mathbb{R})$ is zero. We now summarize what we have arrived at thus far. We have constructed a class $\overline{\xi} \in \mathrm{CH}^2(U_Z \times Y, 1)$ with the property that for $t \in U$, $\underline{r}_{3,1}(\overline{\xi}_t)$ is given by the restriction $\underline{r}_{3,1}(\overline{\xi})\big|_{Z_t \times Y}$, i.e. induced by the restriction

$$\{H^2(U_Z,\mathbb{R}) \otimes H^2(Y,\mathbb{R})\} \bigcap F^2H^4(U_Z \times Y, \mathbb{C})$$

$$\downarrow$$

$$\{H^2(Z_t,\mathbb{R}) \otimes H^2(Y,\mathbb{R})\} \bigcap H^{2,2}(Z_t \times Y)$$

$$\downarrow$$

$$\{\mathrm{Prim}^2(Z_t,\mathbb{R}) \otimes H^2_{\mathrm{tr}}(Y,\mathbb{R})\} \bigcap H^{2,2}(Z_t \times Y).$$

This is a result of the exact sequence (see [**E-V**])

$$\frac{H^4(U_Z \times Y, \mathbb{R})}{\big\{v + \overline{v} \mid v \in F^3H^4(U_Z \times Y, \mathbb{C})\big\}} \to H^5_{\mathcal{D}}(U_Z \times Y, \mathbb{R}(3)) \to H^5(U_Z \times Y, \mathbb{R}).$$

Now recall the notion of horizontal displacement in 1.4 of section 1.

LEMMA 2.5. *Let $\tilde{Z} \supset U_Z$ be a smooth projective compactification of U_Z, and for $t \in U$, let $j_t : Z_t \times Y \hookrightarrow \tilde{Z} \times Y$ be the inclusion. Then there is a class $R \in \big\{H^2(\tilde{Z},\mathbb{R}) \otimes H^2(Y,\mathbb{R})\big\} \bigcap H^{2,2}(\tilde{Z} \times Y)$ such that $\underline{r}_{3,1}(\xi_t) = j_t^*(R)$ for all $t \in U$. In particular, as $t \in U$ varies, the class $\underline{r}_{3,1}(\overline{\xi}_t)$ varies by horizontal displacement with the property that its Hodge (p,q) components also deform horizontally.*

PROOF. The latter statement is a well known consequence of the former. For the former statement, we will use the fact that $H^2(U_Z,\mathbb{Q}) \to H^2(Z_t,\mathbb{Q})$ is a morphism of mixed Hodge structures. It is likewise well known that the lowest weight $W_2H^2(U_Z,\mathbb{Q})$ is given by the image $H^2(\tilde{Z},\mathbb{Q}) \to H^2(U_Z,\mathbb{Q})$. Now since $t \in U$, $H^2(Z_t,\mathbb{Q})$ has a pure Hodge structure of weight 2. Thus, by a weight argument, the image of the composite

$$H^2(\tilde{Z},\mathbb{Q}) \to H^2(U_Z,\mathbb{Q}) \to H^2(Z_t,\mathbb{Q}),$$

coincides with the image of

$$H^2(U_Z, \mathbb{Q}) \to H^2(Z_t, \mathbb{Q}).$$

Lemma 2.5 easily follows from this. □

In summary, as $t \in U$ varies, the class $\underline{r}_{3,1}(\overline{\xi}_t)$ varies by a *horizontal* displacement whose Hodge (p,q) components displace *horizontally*, i.e. preserving the given Hodge type. But over the set where $\Gamma' \to \Gamma$ is unramified, one can find open sets $\Delta_{\Gamma'} \subset U \subset \Gamma'$, $\Delta \subset \Gamma$, in the strong topology, such that $\Delta_{\Gamma'} \simeq \Delta$. Thus $\underline{r}_{3,1}(\overline{\xi}_t) = 0$, by virtue of the fact that we are dealing with a product of surfaces $Z_t \times Y$, with Y *fixed,* together with:

LEMMA 2.6. *Consider a 1-parameter family $\{W_t\}_{t\in\Gamma}$ of surfaces given by a morphism $f : W \to \Gamma$, with smooth set $U \subset \Gamma$. Assume that* $\mathrm{Prim}^{2,0}(W_t) \neq 0$ *for $t \in U$ and that for fixed $t_0 \in U$, the monodromy group action $\pi_1(U) \to$* $\mathrm{Aut}\big(\mathrm{Prim}^2(W_{t_0})\big)$ *is irreducible. Let $\Delta \subset U$ be a disk, and assume given $\gamma_t \in$* $\mathrm{Prim}^2(W_t, \mathbb{C})$*, a horizontal displacement of a class γ for $t \in \Delta$. If the (p,q) components of γ_t also horizontally displace, then $\gamma_t = 0$.*

PROOF. This follows from a standard monodromy argument, together with the analyticity of Hodge subbundles. □

Thus we have arrived at a contradiction of our assumption in (2.0), and hence the proof of Theorem 0.1 is complete.

3. Proof of Proposition 0.3

The following conjecture stated in [**L**], can be thought of as a variant of the Bloch-Beilinson conjecture:

CONJECTURE 3.0. *Let $V/\overline{\mathbb{Q}}$ be a smooth quasiprojective variety. Then the Abel-Jacobi map* $\Phi_{k,\mathbb{Q}} : \mathrm{CH}^k_{\mathrm{hom}}(V/\overline{\mathbb{Q}}; \mathbb{Q}) \to J^k(V(\mathbb{C})) \otimes \mathbb{Q}$ *is injective.*

Here the definition of the Abel-Jacobi map for smooth quasiprojective varieties, which is an extension of Griffiths' prescription, involves Carlson's extension class interpretation of intermediate jacobians ([**Ca**]). A detailed description of this map for example can be found in [**Ja2**] (§9). We now make use of the following result:

THEOREM 3.1. ([**L**]) *Assume given a smooth projective variety X. Then for all k, there is a filtration*

$$\begin{aligned}\mathrm{CH}^k(X;\mathbb{Q}) = F^0 \supset F^1 \supset \cdots \supset F^\ell \supset F^{\ell+1}\\ \supset \cdots \supset F^k \supset F^{k+1} = F^{k+2} = \cdots,\end{aligned}$$

which satisfies the following

(i) $F^1 = \mathrm{CH}^k_{\mathrm{hom}}(X;\mathbb{Q})$

(ii) $F^2 \subset \ker \Phi_{k,\mathbb{Q}} : \mathrm{CH}^k_{\mathrm{hom}}(X;\mathbb{Q}) \to J^k(X) \otimes \mathbb{Q}$.

(iii) $F^\ell \bullet F^r \subset F^{\ell+r}$, *where $\bullet$ is the intersection product.*

(iv) F^ℓ *is preserved under push-forwards f_* and pull-backs f^*, where $f : X \to Y$ is a morphism of smooth projective varieties. [In short, F^ℓ is preserved under the action of correspondences between smooth projective varieties.]*

(v) $Gr_F^\ell := F^\ell/F^{\ell+1}$ *factors through the Grothendieck motive. More specifically, let us assume that the Künneth components of the diagonal class* $[\Delta] = \bigoplus_{p+q=2n}[\Delta(p,q)] \in H^{2n}(X \times X, \mathbb{Q})$ *are algebraic. Then*

$$\Delta(2n-2k+r, 2k-r)_*\Big|_{Gr_F^\ell \mathrm{CH}^k(X;\mathbb{Q})} = \begin{cases} \text{Identity} & \text{if } r = \ell \\ 0 & \text{otherwise} \end{cases}$$

(vi) *Let* $D^k(X) := \bigcap_\ell F^\ell$. *If Conjecture 3.0 above holds, then* $D^k(X) = 0$.

The reader may wish to consult [**Ja2**] (§11) and [**Mu**] for the connection between the above theorem and the general conjectures of Bloch/Beilinson/ Murre on this subject. Now consider a general choice of 1-parameter family of $K3$ surfaces $\{W_t\}_{t\in\Gamma}$, as well as a general choice of $K3$ surface Y. We can assume that if W_t is singular for a given $t \in \Gamma$, then its singular set is a single ordinary double point. In this case let $\widetilde{W}_t$ be its minimal desingularization. One can easily argue that a Künneth decomposition as in (0.2) of Theorem 0.1 for $\mathrm{CH}^2(\widetilde{W}_t \times Y;\mathbb{Q})$ implies a corresponding Künneth decomposition for $\mathrm{CH}^2(W_t \times Y;\mathbb{Q})$. Next, one can argue that $H^1(\overset{(\sim)}{W}_t) = H^3(\overset{(\sim)}{W}_t) = 0$ and that for all $t \in \Gamma$:

$$\{H^2(\overset{(\sim)}{W}_t, \mathbb{Q}) \otimes H^2(Y,\mathbb{Q})\} \cap H^{2,2}(\overset{(\sim)}{W}_t \times Y) = H^2_{\mathrm{alg}}(\overset{(\sim)}{W}_t, \mathbb{Q}) \otimes H^2_{\mathrm{alg}}(Y,\mathbb{Q}).$$

Thus we are reduced to proving the following:

LEMMA 3.2. *Let X and Y be smooth projective surfaces with the property that* $H^j(X,\mathbb{Q}) = H^j(Y,\mathbb{Q}) = 0$ *for odd j, and that*

$$\{H^2(X,\mathbb{Q}) \otimes H^2(Y,\mathbb{Q})\} \cap H^{2,2}(X \times Y) = H^2_{\mathrm{alg}}(X,\mathbb{Q}) \otimes H^2_{\mathrm{alg}}(Y,\mathbb{Q}).$$

If Conjecture 3.0 holds, then

$$\mathrm{CH}^2(X \times Y;\mathbb{Q}) = \sum_{i=0}^{2} \mathrm{CH}^i(X;\mathbb{Q}) \otimes \mathrm{CH}^{2-i}(Y;\mathbb{Q}).$$

PROOF. According to Theorem 3.1, together with Conjecture 3.0, we have for any smooth projective variety V for which the diagonal Δ admits an algebraic Künneth decomposition:

$$\begin{aligned}\mathrm{CH}^k(V;\mathbb{Q}) &= \bigoplus_{\ell=0}^{k} \Delta(2\dim V - 2k+\ell, 2k-\ell)_* \mathrm{CH}^k(V;\mathbb{Q}) \\ &= \bigoplus_{\ell=0}^{k} Gr_F^\ell \mathrm{CH}^k(V;\mathbb{Q}).\end{aligned}$$

Since X and Y are surfaces, and that the respective diagonals Künneth decompose algebraically, the same holds true for the diagonal of $X \times Y$ (see [**Kl**]). Since the odd cohomologies vanish, we have that

$$\mathrm{CH}^2(X \times Y;\mathbb{Q}) = Gr_F^0 \mathrm{CH}^2(X \times Y;\mathbb{Q}) \oplus Gr_F^2 \mathrm{CH}^2(X \times Y;\mathbb{Q}). \tag{3.3}$$

Note that

$$\begin{aligned}Gr_F^0 \mathrm{CH}^2(X \times Y;\mathbb{Q}) &\subset H^4(X \times Y,\mathbb{Q}) \cap H^{2,2}(X \times Y) \\ &\subset H^4(X,\mathbb{Q}) \otimes H^0(Y,\mathbb{Q}) \oplus \\ &H^2_{\mathrm{alg}}(X,\mathbb{Q}) \otimes H^2_{\mathrm{alg}}(Y,\mathbb{Q}) \ \oplus\ H^0(X,\mathbb{Q}) \otimes H^4(Y,\mathbb{Q})\end{aligned}$$

Thus $Gr_F^0\mathrm{CH}^2(X \times Y; \mathbb{Q})$ Künneth decomposes. By (3.3) it suffices to show that $Gr_F^2\mathrm{CH}^2(X \times Y; \mathbb{Q})$ Künneth decomposes. But

$$Gr_F^2\mathrm{CH}^2(X \times Y; \mathbb{Q}) = \Delta(6,2)_*\mathrm{CH}^2(X \times Y; \mathbb{Q}),$$

which by Theorem 3.1(i), depends only on the cohomology class

$$[\Delta(6,2)] \in \{H^6(X \times Y, \mathbb{Q}) \otimes H^2(X \times Y, \mathbb{Q})\} \cap H^{4,4}(X \times Y \times X \times Y).$$

The components of this cohomology class lie in:

$$\begin{aligned}
&(\mathrm{I}) : H^4(X,\mathbb{Q}) \otimes H^2_{\mathrm{alg}}(Y,\mathbb{Q}) \otimes H^2_{\mathrm{alg}}(X,\mathbb{Q}) \otimes H^0(Y,\mathbb{Q})\\
&(\mathrm{II}) : H^2_{\mathrm{alg}}(X,\mathbb{Q}) \otimes H^4(Y,\mathbb{Q}) \otimes H^0(X,\mathbb{Q}) \otimes H^2_{\mathrm{alg}}(Y,\mathbb{Q})\\
&(\mathrm{III}) : \{H^4(X,\mathbb{Q}) \otimes H^2(Y,\mathbb{Q}) \otimes H^0(X,\mathbb{Q}) \otimes H^2(Y,\mathbb{Q})\}\\
&\qquad \cap H^{4,4}(X \times Y \times X \times Y)\\
&(\mathrm{IV}) : \{H^2(X,\mathbb{Q}) \otimes H^4(Y,\mathbb{Q}) \otimes H^2(X,\mathbb{Q}) \otimes H^0(Y,\mathbb{Q})\}\\
&\qquad \cap H^{4,4}(X \times Y \times X \times Y)
\end{aligned}$$

It is easy to check that all terms of $[\Delta(6,2)]$ in (I) - (IV) are algebraic. For (I) and (II), this is obvious. For (III), the cycle is given by $\mathrm{Pr}_{24}^*(\Delta_Y(2,2)) \cap \mathrm{Pr}_{13}^*(\Delta_X(4,0))$, and for (IV), the cycle is given by

$$\mathrm{Pr}_{13}^*(\Delta_X(2,2)) \cap \mathrm{Pr}_{24}^*(\Delta_Y(4,0)),$$

where Pr_{ij} is the projection of $X \times Y \times X \times Y$ onto the ij-th factor. It is now an elementary exercise to show that the description of these terms implies the Künneth decomposition for $Gr_F^2\mathrm{CH}^2(X \times Y; \mathbb{Q})$. □

4. Appendix

Our purpose here is to correct a mistake in [**C-L1**] and give an alternative approach to our (first) main result Theorem 0.1. In [**C-L1**], an alternative proof for the triviality of regulator maps on a very general surface in $\mathbb{P}^3$ of degree $d \geq 5$ (the content of Theorem 1.1, op. cit.) was given, among other things. The precise statement is this. For a smooth complex projective surface Z, let

$$\underline{cl}_{2,1} : \mathrm{CH}^2(Z,1) \to \frac{H^2(Z,\mathbb{C})}{F^2H^2(Z,\mathbb{C}) + \mathrm{Hg}^1(Z) \otimes \mathbb{C} + H^2(Z,\mathbb{Q}(2))},$$

be the reduced cycle class map (such as can be found in in [**MS1**]), where $\mathrm{Hg}^1(Z)$ is the Hodge group (and where we include twists). "Sufficiently general" will have the same meaning as in 1.5 of section 1.

THEOREM 4.1. *(Theorem 1.1 in* [**C-L1**]*) For a sufficiently general surface $Z \subset \mathbb{P}^3$ of degree $d \geq 5$, the map $\underline{cl}_{2,1}$ is trivial.*

For completeness and for the convenience of the reader, we also state:

THEOREM 4.2. *(Theorem 1.2 in* [**C-L1**]*) Let $X = C_1 \times C_2$ be a product of two general curves (resp. hyperelliptic curves), where the genus $g(C_1) \geq 1$ and $g(C_2) \geq 2$. Then $\underline{cl}_{2,1}$ is trivial.*

The arguments for Theorem 0.1 here and other results in that paper are modelled after that proof of Theorem 1.1 in [**C-L1**]. Before sketching that proof and correcting the mistake, it is essential to explain the "yoga" of our construction in [**C-L1**] and which is also apparent in this present work. Suppose we are given a smooth and proper morphism $\rho : W \to T$ of smooth quasiprojective varieties. For $t \in T$, put $W_t = \rho^{-1}(t)$. Now consider a higher Chow cycle $\xi \in \mathrm{CH}^k(W,1)$, which of course restricts to a class $\xi_t \in \mathrm{CH}^k(W_t,1)$. Note that the real regulator takes its values $r_{k,1}(\xi_t) \in H^{k-1,k-1}(W_t,\mathbb{R})$. This is precisely due to the fact that W_t is a smooth projective variety. More specifically $r_{k,1}(\xi_t) \in H^{2k-1}_{\mathcal{D}}(W_t,\mathbb{R}(k))$ (real Deligne cohomology), but by a weight argument, and ignoring twists, $H^{2k-1}_{\mathcal{D}}(W_t,\mathbb{R}(k)) \simeq H^{k-1,k-1}(W_t,\mathbb{R})$. A similar statement for the image of $r_{k,1}(\xi)$ cannot be made, since W is not assumed complete, albeit smooth (the smoothness of W is required to get a cycle class map to cohomology). Let us assume for the moment that W is projective (and still smooth). Note that $r_{k,1}(\xi_t)$, being the restriction of $r_{k,1}(\xi)$ to W_t, arises from the restriction map $H^{k-1,k-1}(W,\mathbb{R}) \to H^{k-1,k-1}(W_t,\mathbb{R})$. This latter restriction translates to Betti rigidity, which implies that the class $r_{k,1}(\xi_t)$ varying with $t \in T$ is given by horizontal displacement, for which the Hodge (p,q) components also displace horizontally (which follows from the restriction map being a morphism of Hodge structures). In some cases W need not be complete to arrive at Betti rigidity, this being the import of our discussion in section 2, leading up to Lemma 2.5. Once Betti rigidity is established, a monodromy argument is used to show the triviality of a reduced or modified version of the real regulator for a general member of a specific family of varieties in question. To summarize our approach, a higher Chow cycle on a general smooth projective variety is spread out and extended to a cycle $\xi \in \mathrm{CH}^k(W,1)$, where W_t is now our general member for a general choice of $t \in T$. In section 2, our choice of W is $W = U_Z \times Y$. For a suitably modified ξ we establish Betti rigidity, and then employ the monodromy argument in Lemma 2.6. Alternatively, we could extend ξ to a class $\overline{\xi} \in \mathrm{CH}^k(\overline{W},1)$, where $\overline{W}$ is a *smooth* compactification of W. Then $r_{k,1}(\overline{\xi}) \in H^{k-1,k-1}(\overline{W},\mathbb{R})$ restricts to $r_{k,1}(\xi_t) \in H^{k-1,k-1}(W_t,\mathbb{R})$, viz., Betti rigidity. It is this latter approach that we employed in the proof of Theorem 1.1 of [**C-L1**], and for which we will now explain below.

Here is a sketch of that proof. For starters, we choose a general pencil $X \subset \mathbb{P}^1 \times \mathbb{P}^3$ of surfaces of degree $d \geq 5$. After a base change $\Gamma \to \mathbb{P}^1$, there exists a class $\xi \in \mathrm{CH}^2(Y_U,1)$ where $Y = X \times_{\mathbb{P}^1} \Gamma$, $U \subset \Gamma$ is a nonempty Zariski open set in Γ and $Y_U = \cup_{b\in U} Y_b$. Using an argument similar to that of Theorem 0.1, we show that ξ can be extended to all of Y. The key here is that $\mathrm{CH}^1(X_t)_{\mathbb{Q}} = \mathbb{Q}$ for every $t \in \mathbb{P}^1$, which follows from a result of M. Green on Noether-Lefschetz loci.

Note that Y might be singular at a point p which is a rational double point (node) of the fiber Y_t. Locally at p, Y is given by $x^2+y^2+z^2+t^\mu = 0$, where μ is the ramification index of $Y \to X$ at p. The singularity of Y at p can be resolved by a sequence of blowups $\widetilde{Y} = Y^{(m)} \to Y^{(m-1)} \to \cdots \to Y^{(0)} = Y$ over p, where $m = \lfloor \mu/2 \rfloor$. It was proved that ξ can be extended to a class $\widetilde{\xi} \in \mathrm{CH}^2(\widetilde{Y},1)$. Then a standard monodromy argument, as used here, concluded that the image of $\widetilde{\xi}$, under the regulator map, is trivial. However, the argument there that ξ can be extended on $\widetilde{Y}$ was flawed. Using the localization sequence, it is easy to see that the obstruction to extending ξ is the map

$$\mathrm{CH}^1(Q) \to \mathrm{CH}^2(\widetilde{Y})$$

where Q is the exceptional divisor of $Y^{(m)} \to Y^{(m-1)}$. It can be extended if and only if the above map is injective. If μ is odd, Q is a cone over a smooth conic curve and hence $\mathrm{CH}^1(Q) = \mathbb{Q}$; and the injectivity of $\mathrm{CH}^1(Q) \to \mathrm{CH}^2(\widetilde{Y})$ follows. If μ is even, $Q \cong \mathbb{P}^1 \times \mathbb{P}^1$ and this where the trouble originates. A flawed argument was given in [**C-L1**], where it was wrongly assumed that $\widetilde{Y}$ can be blown down along each ruling of Q with the resulting variety (only analytic *a priori*) projective. Indeed, it is a classical example that the small resolutions of a quartic threefolds with nodes are not projective (not even Kähler). It is a subtle problem to determine whether $\mathrm{CH}^1(Q) \to \mathrm{CH}^2(\widetilde{Y})$ is injective for a singular threefold Y with nodes, where $\widetilde{Y}$ is the blowup of Y at the nodes with Q the exceptional divisor. Take e.g. Hironaka's example of a nonsingular complete threefold which is Moishezon but not projective. Let X be a nonsingular projective threefold and $C \subset X$ be an irreducible curve on X with a node. Let Y be the blowup of X along C. Then Y has a node p over the node of C. Let $\widetilde{Y}$ be the blowup of Y at p and Z be the blowdown of $\widetilde{Y}$ along one of the two rulings of the exceptional divisor $Q \cong \mathbb{P}^1 \times \mathbb{P}^1$ of $\widetilde{Y} \to Y$. Then Z is Moishezon but not projective. We also see in this example that the two rulings of Q are linearly independent in $\mathrm{CH}^2(\widetilde{Y})$, i.e., $\mathrm{CH}^1(Q) \to \mathrm{CH}^2(\widetilde{Y})$ is injective if and only if C is rational. Hence some extra efforts are needed to justify the injectivity of $\mathrm{CH}^1(Q) \to \mathrm{CH}^2(\widetilde{Y})$ for our particular Y. Basically, we will prove the following:

PROPOSITION 4.3. *Let $X \subset \mathbb{P}^1 \times \mathbb{P}^3$ be a very general pencil of surfaces of degree $d \geq 2$ and let $\Gamma \to \mathbb{P}^1$ be a finite map from a smooth curve Γ to $\mathbb{P}^1$ and $Y = X \times_{\mathbb{P}^1} \Gamma$. Then for any desingularization $\widetilde{Y} \to Y$ with exceptional locus Q, $\mathrm{CH}^1(Q) \to \mathrm{CH}^2(\widetilde{Y})$ is an injection.*

Let $|\mathcal{O}_{\mathbb{P}^3}(d)|$ be the complete linear series of surfaces of degree d in $\mathbb{P}^3$ and $B \cong \mathbb{P}^2 \subset |\mathcal{O}_{\mathbb{P}^3}(d)|$ be a two-dimensional sublinear series constructed as follows.

We choose $V \subset \mathbb{P}^1 \times \mathbb{P}^3$ to be a general pencil of quadrics in $\mathbb{P}^3$ and fix a general surface $S \subset \mathbb{P}^3$ of degree $d-2$. We let $W \subset \mathbb{P}^1 \times \mathbb{P}^3$ be a pencil of surfaces of degree d whose fiber $R_t = V_t \cup S$ for every $t \in \mathbb{P}^1$. Let $b \in \mathbb{P}^1$ be a point where V_t is singular, i.e., a cone over a smooth conic with vertex p. This family $W/\mathbb{P}^1$ has the following important property at p.

After an arbitrary base change $\Gamma \to \mathbb{P}^1$, we obtain $\mathcal{W} = W \times_{\mathbb{P}^1} \Gamma$. At a point $q \in \mathcal{W}$ that lies over p, $\mathcal{W}$ is locally given by $x^2 + y^2 + z^2 + t^\mu = 0$, where μ is the ramification index of the map $\mathcal{W} \to W$ at q. In the case that $\mu > 1$, $\mathcal{W}$ is singular at q, where the singularity can be resolved by a sequence of $m = \lfloor \mu/2 \rfloor$ blowups: $\widetilde{\mathcal{W}} = \mathcal{W}^{(m)} \to \mathcal{W}^{(m-1)} \to \cdots \to \mathcal{W}^{(0)} = \mathcal{W}$. Let $E^{(i)}$ be the exceptional divisor of $\mathcal{W}^{(i)} \to \mathcal{W}^{(i-1)}$. It is not hard to see that $E^{(i)}$ is a cone over a conic curve for $\mu > 2i$ and $E^{(i)} \cong \mathbb{P}^1 \times \mathbb{P}^1$ for $\mu = 2i$, i.e., $i = m$ and $\mu = 2m$ being even. Let $E = E_1 + E_2 + \cdots + E_m$ be the exceptional divisor of $\widetilde{\mathcal{W}} \to \mathcal{W}$, where E_i is the proper transform of $E^{(i)}$. So $E_i \cong \mathbb{F}_2$ is the blowup of $E^{(i)}$ at the vertex of the cone for $i < m$ and $Q = E_m \cong E^{(m)}$ is a cone over a conic curve if μ is odd and $Q \cong \mathbb{P}^1 \times \mathbb{P}^1$ if μ is even. Also note that $E_i \cap E_{i+1}$ and $E_i \cap E_{i-1}$ are two disjoint sections of $E_i \cong \mathbb{F}_2$ over $\mathbb{P}^1$. We claim that the map

$$\mathrm{CH}^1(E) \hookrightarrow \mathrm{CH}^2(\widetilde{\mathcal{W}})$$

is an injection. It is not hard to see that $\mathrm{CH}^1(E_1 \cup E_2 \cup \ldots \cup E_{m-1}) \hookrightarrow \mathrm{CH}^2(\widetilde{\mathcal{W}})$ is an injection. Hence $\mathrm{CH}^1(E) \hookrightarrow \mathrm{CH}^2(\widetilde{\mathcal{W}})$ is injective if and only if

$$\mathrm{CH}^1(Q) \hookrightarrow \mathrm{CH}^2(\widetilde{\mathcal{W}})$$

is. This is more or less obvious when Q is a cone over a conic curve, i.e., when μ is odd. If μ is even, $Q \cong \mathbb{P}^1 \times \mathbb{P}^1$ as mentioned above; the two rulings l_1 and l_2 of Q are linearly independent in $\mathrm{CH}^2(\widetilde{\mathcal{W}})$ since in this case $\widetilde{\mathcal{W}}$ can be blown down along l_i with the resulting (analytic) variety projective (hence Kähler).

Let $B = \mathbb{P}^2 \subset |\mathcal{O}_{\mathbb{P}^3}(d)|$ and $X \subset B \times \mathbb{P}^3$ be a two-dimensional linear system of degree d surfaces that contains the pencil W. We choose X to be general with this property. Let $C \cong \mathbb{P}^1 \subset B$ be the line such that $W = X \times_B C$. Let

$$B_\Delta = \overline{\{t \in B - C : X_t \text{ singular}\}}$$

Due to our general choice of X, C meets B_Δ transversely at some point $b \in C \cap B_\Delta$ where V_b is singular at p. For simplicity, we may blow up B at a general point on C and realize B as a pencil, which we still denote by B. So $B \cong \mathbb{F}_1$ is a $\mathbb{P}^1$ bundle over $\mathbb{P}^1$, where we parameterize the base by s and let $B_0 = C$.

We assume that there is a class $\xi \in \mathrm{CH}^2(Y_U, 1)$ after a suitable base change $\phi : D \to B$, where $Y = X \times_B D$, U is a nonempty open set of D and $Y_U = X \times_B U$. Our purpose is to show that for $s \in \mathbb{P}^1$ general, ξ can be extended over Y_s after we desingularize Y over B_Δ. However, this is hard to prove directly. Our strategy is to find another base change $F \to B$ and replace D by $D \times_B F$.

For simplicity, we assume that $\phi : D \to B$ is simply ramified along $B_\Delta - b$ with multiplicity μ locally at b. (By simple ramification, we mean ramification over a smooth divisor.) If the preimage $\phi^{-1}(b)$ of b is not unique, we just do the following for every point in $\phi^{-1}(b)$. We let $F \to B$ be a finite map simply ramified along B_Δ with multiplicity μ locally at b. It is helpful to write down the local equations of D and F.

Let $\Delta^2_{s,t}$ be an analytic neighborhood of $b \in B$, where $B_0 = C = \{s = 0\}$ and $B_\Delta = \{t = 0\}$ locally. Then D and F are locally given by

$$w^\mu + s^\alpha t + f(s,t)t^2 = 0$$

and

$$u^\mu + t = 0,$$

respectively. Also note that X is

$$x^2 + y^2 + z^2 = t$$

locally at p.

Note that $D \times_B F$ is not normal if $\mu > 1$. Let G be an irreducible component of $D \times_B F$ after normalization. So we have the diagram

$$\begin{array}{ccccc} & & G & & \\ & \swarrow & & \searrow & \\ D & & & & F \\ & \searrow & & \swarrow & \\ & & B & & \end{array}$$

The important thing here is that the map $G \to F$ is not ramified over the locus $\{u = 0\}$.

Let $Z = X \times_B F$. Obviously, Z has singularities of type $x^2 + y^2 + z^2 = u^\mu$ over B_Δ, which can be resolved by a sequence of blowups: $\widetilde{Z} = Z^{(m)} \to Z^{(m-1)} \to \cdots \to Z^{(0)} = Z$ with $m = \lfloor \mu/2 \rfloor$. Let $E = E_1 + E_2 + ... + E_m$ be the exceptional divisor of $\widetilde{Z} \to Z$, where E_i is the proper transform of the exceptional divisor of $Z^{(i)} \to Z^{(i-1)}$. Let $Q = E_m$. Then due to our choice of the pencil $C = B_0 \subset B$,

$\mathrm{CH}^1(Q_0) \hookrightarrow \mathrm{CH}^2(\widetilde{Z}_0)$ is an injection. Hence $\mathrm{CH}^1(Q_s) \hookrightarrow \mathrm{CH}^2(\widetilde{Z}_s)$ is injective for $s \in \mathbb{P}^1$ general.

Let $\Sigma = \widetilde{Z} \times_F G$. The class $\xi \in \mathrm{CH}^2(Y_U, 1)$ can be lifted to Σ such that it is defined over an open set of G. We still denote the class by ξ. Let $P = \varphi^{-1}(Q)$, where φ is the map $\Sigma \to \widetilde{Z}$. For $s \in \mathbb{P}^1$, we have the diagram

$$\begin{array}{ccc} \mathrm{CH}^1(P_s) & \to & \mathrm{CH}^2(\Sigma_s) \\ \downarrow & & \downarrow \\ \mathrm{CH}^1(Q_s) & \to & \mathrm{CH}^2(\widetilde{Z}_s) \end{array}$$

Since $G \to F$ is not ramified over B_Δ, P_s is a disjoint union of copies of Q_s for $s \in \mathbb{P}^1$ general. Consequently, $\mathrm{CH}^1(P_s) \hookrightarrow \mathrm{CH}^2(\Sigma_s)$ is an injection since $\mathrm{CH}^1(Q_s) \hookrightarrow \mathrm{CH}^2(\widetilde{Z}_s)$ is. So ξ can be extended on Σ_s over $\varphi^{-1}(E)$ for s general. And since $G \to F$ is not ramified over B_Δ, Σ_s is smooth along $\varphi^{-1}(E)$ for s general. This is all we need when we apply the regulator map.

Now let us go back to the proof of Theorem 0.1. After we extend ξ to all of Z, an alternative approach is to desingularize Z and use the argument here to extend ξ to $\widetilde{Z}$. There is no change to the rest of the proof.

References

[Be] A. Beilinson, Higher regulators and values of L-functions, *J. Soviet Math.* **30** (1985), 2036-2070.

[Blo1] S. Bloch, Algebraic cycles and higher K-theory, *Advances in Mathematics* **61** (1986), 65-79.

[Blo2] ———, Algebraic cycles and the Beilinson conjectures, *Contemp. Math* **58 (1)**, (1986), 267-304.

[Ca] J. Carlson, Extensions of mixed Hodge structures, *Jour. de Géométrie Algébrique d'Angers 1979, Sijhoff and Nordhoff*, (1980), 107-127.

[C-L1] X. Chen and J. D. Lewis, Noether-Lefschetz for K_1 of a certain class of surfaces, *Bol. Soc. Mat. Mexicana* **10(3)**, (2004), 29-41.

[C-L2] ———, The Hodge-$\mathcal{D}$-conjecture for $K3$ and Abelian surfaces, *Journal of Algebraic Geometry* (To appear) **10(3)**, (2004).

[E-V] H. Esnault and E. Viehweg, Deligne-Beilinson cohomology, *in Beilinson's conjectures on special values of L-functions, Academic Press, Boston*, (1988), 43-92.

[F] W. Fulton, Intersection Theory, *Ergebnisse der Mathematik und ihrer Grenzgebiete, Springer-Verlag* **3(2)**, (1984).

[G-H] P. A. Griffiths and J. Harris, Principles of Algebraic Geometry, *Wiley*, (1978).

[Ja1] U. Jannsen, Deligne cohomology, Hodge-$\mathcal{D}$-conjecture, and motives *in Beilinson's Conjectures on Special Values of L-Functions, (Rapoport, Schappacher, Schneider, eds.), Perspect. Math., Academic Press* **4** (1988), 305-372.

[Ja2] ———, Mixed Motives and Algebraic K-Theory, *Lec. Notes in Mathematics, Springer-Verlag* **1400**, (1988).

[Kl] S. Kleiman, Algebraic cycles and the Weil conjectures, *Dix Exposés sur la Cohomologie des Schémas, North-Holland, Amsterdam* **35** (1968), 359-386.

[La] S. Landsberg, Relative Chow groups, *Illinois Jour. of Mathematics* **35** (1991), 618-641.

[L] J. D. Lewis, A filtration on the Chow groups of a complex projective variety, *Compositio Mathematica* **128(3)**, (2001), 299-322.

[MS1] S. Müller-Stach, Constructing indecomposable motivic cohomology classes on algebraic surfaces, *J. Algebraic Geometry* **6**, (1997), 513-543.

[MS2] ———, Algebraic cycle complexes, *Proceedings of the NATO Advanced Study Institute on the Arithmetic and Geometry of Algebraic Cycles, (Lewis, Yui, Gordon, Müller-Stach, S. Saito, eds.), Kluwer Academic Publishers, Dordrecht, The Netherlands* **548** (2000), 285-305.

[Mu] J. P. Murre, On a conjectural filtration on the Chow groups of an algebraic variety I & II, *Indag. Math. (N.S.)* **4** (1993), no. 2, 177-201.

632 Central Academic Building, University of Alberta, Edmonton, Alberta T6G 2G1, CANADA

E-mail address: `xichen@math.ualberta.ca`

632 Central Academic Building, University of Alberta, Edmonton, Alberta T6G 2G1, CANADA

E-mail address: `lewisjd@gpu.srv.ualberta.ca`

AMS/IP Studies in Advanced Mathematics
Volume 38, 2006

Derived Equivalence for Stratified Mukai Flop on $G(2,4)$

Yujiro Kawamata

1. Introduction

We consider a conjecture on the derived equivalence for K-equivalent varieties in the case of a certain flop between 9-dimensional varieties. The following is a very general version of this conjecture on the fully faithful derived embedding for log K-related varieties ([**Kw3**] Conjecture 2.2):

CONJECTURE 1.1. *Let (X,B) and (Y,C) be pairs of quasi-projective varieties with $\mathbb{Q}$-divisors such that there exist quasi-finite and surjective morphisms $\pi : U \to X$ and $\sigma : V \to Y$ from smooth varieties, which may be reducible, such that $\pi^*(K_X + B) = K_U$ and $\sigma^*(K_Y + C) = K_V$. Let $\mathcal{X} \to X$ and $\mathcal{Y} \to Y$ be natural morphisms from the associated Deligne-Mumford stacks. Assume that there are proper birational morphisms $\mu : W \to X$ and $\nu : W \to Y$ from a third variety W such that $\mu^*(K_X + B) \le \nu^*(K_Y + C)$. Then there exists a fully faithful exact functor $D^b(Coh(\mathcal{X})) \to D^b(Coh(\mathcal{Y}))$.*

The conjecture has been proved to be true in some cases in [**BO**], [**BKR**], [**B2**], [**Kw1**], [**Chen**], [**N1**], [**Kw2**], [**Kw3**] and [**BK**].

On the other hand, Namikawa [**N2**] proved that a naturally defined functor between the derived categories for the stratified Mukai flop on $G(2,4)$ is not an equivalence. This is a flop between 9-dimensional varieties which is the total space of a 3-parameter degeneration of standard 3-dimensional flops of $(-1,-1)$-curves and the fiber over the most degenerate point is isomorphic to $G(2,4)$.

So it is worthwhile to check the conjecture in this special case. We shall prove that there is nevertheless another functor between the same categories which is an equivalence.

2. Stratified Mukai flop

We recall the construction of stratified Mukai flops due to Markman [**M**] (see also [**N2**]).

Let $G = G(r,n)$ be the Grassmann variety of r-dimensional subspaces in an n-dimensional vector space V. We assume that $2r \le n$. Let S (resp. Q) be the tautological subbundle (resp. quotient) bundle on G:

$$0 \to S \to V \otimes \mathcal{O}_G \to Q \to 0.$$

2000 *Mathematics Subject Classification.* Primary 14E30; Secondary 14B05, 14C20.

Key words and phrases. Derived categories, equivalence of categories, stratified Mukai flop, Grassman variety.

The polarization of G is given by

$$\det Q = \mathcal{O}_G(1).$$

We have $\Omega^1_G \cong \mathcal{H}om(Q, S)$. Since

$$\mathrm{Ext}^1(\mathcal{O}_G, \Omega^1_G) \cong \mathrm{Ext}^1(Q, S) \cong \mathbb{C}$$

there exists a non-trivial extension of vector bundles on G induced by the natural homomorphism $\mathcal{O}_G \to \mathcal{H}om(S, S)$:

$$\begin{array}{ccccccccc} 0 & \longrightarrow & \Omega^1_G & \longrightarrow & \tilde{\Omega}^1_G & \longrightarrow & \mathcal{O}_G & \longrightarrow & 0 \\ & & \Vert & & \downarrow & & \downarrow & & \\ 0 & \longrightarrow & \mathcal{H}om(Q,S) & \longrightarrow & \mathcal{H}om(V \otimes \mathcal{O}_G, S) & \longrightarrow & \mathcal{H}om(S,S) & \longrightarrow & 0. \end{array}$$

Let $X_0 = T^*G$ and X be the total spaces of Ω^1_G and $\tilde{\Omega}^1_G$, respectively, and $\pi : X \to G$ the projection. We write $\pi^*S = S_X$ and $\pi^*\mathcal{O}_G(1) = \mathcal{O}_X(1)$. A point of X is given by a pair consisting of a point $p \in G$ and a homomorphism $A : V \to S_p \subset V$ which induces a homothety $t : S_p \to S_p$ ($t \in \mathbb{C}$). We have $(p, A) \in X_0$ if and only if $t = 0$. Let

$$Z = \{A \in \mathrm{End}(V) \mid \exists t \in \mathbb{C} \text{ s.t. } A^2 = tA, \mathrm{rank}\ A \le r\}$$

with a morphism $t : Z \to \mathbb{C}$, and set $Z_0 = t^{-1}(0)$. Then there is a natural morphism $\phi : X \to Z$ given by $\phi(p, A) = A$, which induces a morphism $\phi_0 : X_0 \to Z_0$. We know that ϕ_0 and ϕ are projective birational morphisms. We write

$$W^{(k)} = \{(p, A) \in X \mid \mathrm{rank}\ A \le k\}.$$

Then the exceptional loci of both ϕ_0 and ϕ are equal to $W = W^{(r-1)}$. We have $\dim G = r(n-r)$, $\dim X_0 = 2r(n-r)$, $\dim X = 2r(n-r)+1$, and $\dim W = 2(r-1)(n-r+1)+n-2r+1$.

We have a dual construction starting from the dual vector space V^*. Let G^+ be the Grassmann variety of r-dimensional subspaces in V^*, S^+ (resp. Q^+) the tautological subbundle (resp. quotient bundle) on G^+, $\det Q^+ = \mathcal{O}_{G^+}(1)$, X_0^+ (resp. X^+) the total space of $\Omega^1_{G^+}$ (resp. $\tilde{\Omega}^1_{G^+}$), $\pi^+ : X^+ \to G^+$ the projection, $\pi^{+*}S^+ = S^+_{X^+}$, and $\pi^{+*}\mathcal{O}_{G^+}(1) = \mathcal{O}_{X^+}(1)$. There is a natural projective birational morphism $\phi^+ : X^+ \to Z$ given by $\phi^+(q, B) = {}^tB$ which induces a projective birational morphism $\phi_0^+ : X_0^+ \to Z_0$. We write

$$W^{+(k)} = \{(q, B) \in X^+ \mid \mathrm{rank}\ B \le k\}.$$

The exceptional locus of ϕ is $W^+ = W^{+(r-1)}$. The diagram

$$X \xrightarrow{\ \phi\ } Z \xleftarrow{\ \phi^+\ } X^+$$

thus obtained is called a *stratified Mukai flop*. By restricting to the subspaces defined by $t = 0$, we have a smaller diagram

$$X_0 \xrightarrow{\ \phi\ } Z_0 \xleftarrow{\ \phi^+\ } X_0^+$$

which is also called a stratified Mukai flop.

If $r = 1$, then the above diagrams are reduced to a *standard flop* and a *Mukai flop*.

The birational map $(\phi^+)^{-1} \circ \phi : X - \to X^+$ is decomposed into blow-ups and downs ([**M**]). We consider only the case $r = 2$ in the following. Let $f_1 : X_1 \to X$ be the blowing up along the center $G = W^{(0)}$, the 0-section of the projection π. Then the strict transform W' of $W = W^{(1)}$ is smooth. Indeed, the projection $W' \to G$

is smooth and its fibers are isomorphic to the cone over $\mathbb{P}^1 \times \mathbb{P}^{n-3} \subset \mathbb{P}^{2n-5}$. Let $f_2 : X_2 \to X_1$ be the blowing up along the center W'. Let E_i be the exceptional divisor of f_i for $i = 1, 2$, and $E'_1 = f_{2*}^{-1} E_1$, the strict transform of E_1. We set $f = f_2 \circ f_1$. If $r \geq 3$, then we have a similar construction with more blow-ups ([**M**]).

On the dual side, let $f_1^+ : X_1^+ \to X^+$ be the blowing up along the center $G^+ = W^{+(0)}$, the 0-section of the projection π^+. Then the strict transform $(W^+)'$ of $W^+ = W^{+(1)}$ is smooth. Let $f_2^+ : X_2^+ \to X_1^+$ be the blowing up along the center $(W^+)'$. Let E_i^+ be the exceptional divisor of f_i^+ for $i = 1, 2$, and $(E_1^+)' = (f_{2*}^+)^{-1} E_1^+$. We set $f^+ = f_2^+ \circ f_1^+$.

The birational map $(\phi^+)^{-1} \circ \phi : X - \to X^+$ induces an isomorphism $X_2 \to X_2^+$ ([**M**]). We write $Y = X_2 = X_2^+$ under this identification. Then we have $E'_1 = (E_1^+)'$, $E_2 = E_2^+$, and a commutative diagram

$$\begin{array}{ccccccc}
A \in Z & \xleftarrow{\phi} & X & \xleftarrow{f_1} & X_1 & \xleftarrow{f_2} & X_2 \\
\| & & & & & & \| \\
A^+ \in Z^+ & \xleftarrow{\phi^+} & X^+ & \xleftarrow{f_1^+} & X_1^+ & \xleftarrow{f_2^+} & X_2^+
\end{array}$$

Since $\dim X = 4n - 7$, $\dim G = 2n - 4$ and $\dim W = 3n - 5$, we have $K_Y = f^* K_X + (2n - 4)E'_1 + (n - 3)E_2$.

The isomorphism $X_2 \to X_2^+$ can be described set theoretically in the following way.

(1) For a point $(p, A) \in X \setminus W$, we have rank $A = 2$ and $S_p = \text{Im } A$. The corresponding point $(q, {}^tA) \in X^+ \setminus W^+$ is given by $S_q^+ = \text{Im } {}^tA = (\text{Ker } A)^\perp$.

(2) For a point $\lim_{\epsilon \to 0}(p, \epsilon A_1) \in E_1 \setminus W'$, with rank $A_1 = 2$ and $S_p = \text{Im } A_1$, the corresponding point is $\lim_{\epsilon \to 0}(q, \epsilon {}^tA_1) \in E_1^+ \setminus (W^+)'$, where $S_q^+ = \text{Im } {}^tA_1$.

(3) For a point $(p, A) \in W \setminus G$, we have rank $A = 1$ and $S_p \supset \text{Im } A$. We take A_2 such that $S_p = \text{Im } (A + A_2)$. Then a point $\lim_{\epsilon \to 0}(p, A + \epsilon A_2) \in E_2 \setminus E'_1$ corresponds to a point $\lim_{\epsilon \to 0}(q_\epsilon, {}^t(A + \epsilon A_2)) \in E_2^+ \setminus (E_1^+)'$ which is over a point $q \in G^+$ given by $S_q^+ = \lim_{\epsilon \to 0} \text{Im } {}^t(A + \epsilon A_2)$.

(4) For a point $\lim_{\epsilon \to 0}(p, \epsilon A_1) \in E_1 \cap W'$ with rank $A_1 = 1$ and $S_p \supset \text{Im } A_1$, we take A_2 such that $S_p = \text{Im}(A_1 + A_2)$. Then a point $\lim_{\epsilon \to 0}(p, \epsilon A_1 + \epsilon^2 A_2) \in E'_1 \cap E_2$ corresponds to a point $\lim_{\epsilon \to 0}(q_\epsilon, {}^t(\epsilon A_1 + \epsilon^2 A_2)) \in (E_1^+)' \cap E_2^+$ which is over a point $q \in G^+$ given by $S_q^+ = \lim_{\epsilon \to 0} \text{Im } ({}^t(A_1 + \epsilon A_2))$.

The above description shows that the birational map $X_2 - \to X_2^+$ induced by $(\phi^+)^{-1} \circ \phi$ is an isomorphism by the Zariski main theorem. As a consequence, we have the following lemma:

LEMMA 2.1.
$$S_X^*|_{X \setminus W} = S_{X^+}^+|_{X^+ \setminus W^+}.$$
Moreover, $f^* S_X^*$ is a locally free subsheaf of $f^{+*} S_{X^+}^+$ such that a local section of the latter belongs to the former if and only if its value in S_q^+ at any point in $(f^+)^{-1}(x^+)$ for $x^+ = (q, B)$ is contained in Im B.

PROOF. For $(p, A) \in X \setminus W$, we have Ker ${}^tA = (\text{Im } A)^\perp$ from (1) above. Therefore, ${}^tA : V^* \to V^*$ induces an isomorphism $S_p^* = V^*/\text{Ker } {}^tA \to S_q^+$, hence the first assertion.

Any element $v \in V^*$ determines a global section of $f^* S_X^*$, and the sheaf $f^* S_X^*$ is generated by such sections. The value at the point $(q, {}^tA) \in X^+ \setminus W^+$ of the corresponding section of $S_{X^+}^+$ is given by ${}^tAv \subset S_q^+$, hence the second assertion. $\square$

By taking the determinants, we obtain:

COROLLARY 2.2.

$$f^*\mathcal{O}_X(1) = f^{+*}\mathcal{O}_{X^+}(-1) \otimes \mathcal{O}_Y(-2E'_1 - E_2).$$

□

Let P be a fiber of f_1 above a point in G, $P_0 = P \cap f_{1*}^{-1}X_0$ and $P_1 = P \cap W'$. Then the sequence $P_1 \subset P_0 \subset P$ is isomorphic to $\mathbb{P}^1 \times \mathbb{P}^{n-3} \subset \mathbb{P}^{2n-5} \subset \mathbb{P}^{2n-4}$, where the first inclusion is the Segre embedding and the second is linear. Let $P' = f_{2*}^{-1}P$, $P'_0 = f_{2*}^{-1}P_0$, and $P'_1 = f_2^{-1}(P_1)$. Let l_1 be a line on P_0 which meets W' at 2 points, $l'_1 = f_{2*}^{-1}l_1$, and l_2 a fiber of f_2 over a point in W'.

LEMMA 2.3. *The cone of curves $\overline{NE}(Y/X)$ is generated by the classes of l'_1 and l_2. The intersection numbers are given by the following table:*

$$(E'_1 \cdot l'_1) = -1, (E'_1 \cdot l_2) = 0, (E_2 \cdot l'_1) = 2, (E_2 \cdot l_2) = -1.$$

PROOF. P_1 is a determinantal variety defined by quadratic equations. Therefore we obtain the formula for $\overline{NE}(Y/X)$. The intersection numbers are obvious. □

3. Derived equivalence

Let $X \to Z \leftarrow X^+$ and $X_0 \to Z_0 \leftarrow X_0^+$ be stratified flops defined in the previous section. We shall compare the bounded derived categories of coherent sheaves $D^b(Coh(X))$ with $D^b(Coh(X^+))$ and $D^b(Coh(X_0))$ with $D^b(Coh(X_0^+))$. By [**Kw2**] Lemma 5.6 and Corollary 5.7, the derived equivalence for the former pair implies the latter.

We begin with the easy case $r = 1$, i.e., the standard flop and the usual Mukai flop.

PROPOSITION 3.1. *Let l be an arbitrary integer. Then the functor*

$$\Phi_l : D^b(Coh(X)) \to D^b(Coh(X^+))$$

defined by

$$\Phi_l(a) = f_*^+(f^*a \otimes \mathcal{O}_Y(lE))$$

is an equivalence.

PROOF. The adjoint functor $\Psi_l : D^b(Coh(X^+)) \to D^b(Coh(X))$ is given by

$$\Psi_l(b) = f_*(f^{+*}a \otimes \mathcal{O}_Y((n-1-l)E)).$$

The category $D^b(Coh(X))$ is spanned by the set of sheaves $\mathcal{O}_X(k)$ for $l - n + 1 \le k \le l$. By the Kodaira vanishing theorem, we have

$$\Phi_l : \mathcal{O}_X(k) \mapsto \mathcal{O}_Y(k,0)(lE) = \mathcal{O}_Y(0,-k)((l-k)E) \mapsto \mathcal{O}_{X^+}(-k)$$

and

$$\begin{aligned}\Psi_l : &\mathcal{O}_{X^+}(-k) \mapsto \mathcal{O}_Y(0,-k)((n-1-l)E) = \mathcal{O}_Y(k,0)((n-1-l+k)E) \\ &\mapsto \mathcal{O}_X(k).\end{aligned}$$

There is an adjunction morphism of functors $F : \mathrm{Id}_{D^b(Coh(X))} \to \Psi_l\Phi_l$, which is reduced to the identity when restricted to the open subset $X \setminus W$. By the above argument, we have isomorphisms $\omega \cong \Psi\Phi(\omega)$ for a spanning class $\Omega = \{\omega\}$. Since the ω are invertible sheaves, it follows that the morphisms $F(\omega)$ are isomorphisms. Therefore, the natural homomorphisms

$$\Phi : \mathrm{Hom}^p(\omega_1, \omega^2) \to \mathrm{Hom}^p(\omega_1, \Psi\Phi(\omega^2)) \to \mathrm{Hom}^p(\Phi(\omega_1), \Phi(\omega^2))$$

for any $\omega_1, \omega_2 \in \Omega$ and $p \in \mathbb{Z}$ are isomorphisms. Then Φ is an equivalence by [**BO**] and [**B1**]. □

We assume that $r = 2$ in the rest of the paper. We consider exact functors between bounded derived categories

$$\begin{aligned} \Phi &: D^b(Coh(X)) \to D^b(Coh(Y)) \\ \Psi &: D^b(Coh(Y)) \to D^b(Coh(X)) \end{aligned} \tag{3.1}$$

defined by

$$\begin{aligned} \Phi(a) &= Rf^+_*(Lf^*(a) \otimes \mathcal{O}_Y((2n-5)E'_1 + (n-3)E_2)) \\ \Psi(b) &= Rf_*(Lf^{+*}(b) \otimes \mathcal{O}_Y(E'_1)). \end{aligned}$$

They are adjoints to each other because $K_Y = f^*K_X + (2n-4)E'_1 + (n-3)E_2$.

LEMMA 3.2. *$D^b(Coh(X))$ is spanned by the set of the following locally free sheaves:*

$$Sym^i S^*_X \otimes \mathcal{O}_X(j)$$

for $0 \le i$, $0 \le j$ and $i + j \le n-2$.

PROOF. By [**K**], any point sheaf $\mathcal{O}_p$ for $p \in G$ has a finite locally free resolution whose terms are direct sums of the sheaves

$$Sym^i S^* \otimes \mathcal{O}_G(j)$$

for $0 \le i$, $0 \le j$ and $i + j \le n-2$. Hence $\mathcal{O}_{\pi^{-1}(p)}$ is resolved by our set. Then so is any point sheaf $\mathcal{O}_x$ for $x \in X$ because $\pi^{-1}(p)$ is an affine space. □

Let $\mathcal{E}^+$ be a subsheaf of $S^+_{X^+}$ such that a local section of the latter belongs to the former if and only if its value in S^+_q at a point (q, B) is contained in $\operatorname{Im} B$. We denote by $\mathcal{E}^+_i$ for $i > 0$ the image of the natural homomorphism $Sym^{i-1}S^+_{X^+} \otimes \mathcal{E}^+ \to Sym^i S^+_{X^+}$.

LEMMA 3.3. *(1) For $0 \le j \le n-3$,*

$$\begin{aligned} \Phi(\mathcal{O}_X(j)) &= \mathcal{O}_{X^+}(-j) \\ \Psi(\mathcal{O}_{X^+}(-j)) &= \mathcal{O}_X(j). \end{aligned}$$

(2) Let I_{W^+} be the ideal sheaf of $W^+ \subset X^+$. Then

$$\Phi(\mathcal{O}_X(n-2)) = I_{W^+} \otimes \mathcal{O}_{X^+}(-n+2).$$

(3) For $0 < i$, $0 \le j$ and $i + j \le n-3$,

$$\begin{aligned} \Phi(Sym^i S^*_X \otimes \mathcal{O}_X(j)) &= Sym^i S^+_{X^+} \otimes \mathcal{O}_{X^+}(-j) \\ \Psi(Sym^i S^+_{X^+} \otimes \mathcal{O}_{X^+}(-j)) &= Sym^i S^*_X \otimes \mathcal{O}_X(j). \end{aligned}$$

(4) For $0 < i \le n-2$,

$$\Phi(Sym^i S^*_X \otimes \mathcal{O}_X(n-2-i)) = \mathcal{E}^+_i \otimes \mathcal{O}_{X^+}(-n+2+i).$$

PROOF. (1) By Corollary 2.2

$$\begin{aligned} &f^*(\mathcal{O}_X(j)) \otimes \mathcal{O}_Y((2n-5)E'_1 + (n-3)E_2) \\ &= f^{+*}\mathcal{O}_{X^+}(-j) \otimes \mathcal{O}_Y((2n-5-2j)E'_1 + (n-3-j)E_2). \end{aligned}$$

Since $K_Y = f^*K_X + (2n-4)E'_1 + (n-3)E_2$, the higher direct images for f^+ vanish because $-(2j+1)E'_1 - jE_2$ is nef. Since $(2n-5-2j)E'_1 + (n-3-j)E_2$ is effective, we have the first formula.

Similarly, we have

$$f^{+*}(\mathcal{O}_{X^+}(-j)) \otimes \mathcal{O}_Y(E_1') = f^*\mathcal{O}_X(j) \otimes \mathcal{O}_Y((2j+1)E_1' + jE_2).$$

The higher direct images for f vanish because $(2j-2n+5)E_1' + (j-n+3)E_2$ is nef. $(2j+1)E_1' + jE_2$ is effective, hence the second formula.

(2) We already proved in (1) that the higher direct images vanish. We obtain our formula from

$$f_*^+\mathcal{O}_Y(-E_1' - E_2) = I_{W^+}.$$

(3) We have an exact sequence

$$0 \to \mathcal{O}_{E_2}(-E_1') \to f^{+*}S_{X^+}^+ \otimes \mathcal{O}_{E_2}(-E_1') \to \mathcal{O}_{E_2}(-E_1') \otimes f^{+*}\mathcal{O}_{X^+}(-1) \to 0.$$

where the first term is the image of $f^*S_X^*$. We define a decreasing filtration of the sheaf $f^*Sym^iS_X^*$ by locally free subsheaves $\mathcal{F}^{k,l}$ for $0 \le l \le k \le i$ by

$$\begin{aligned}\mathcal{F}^{k,l} =& f^*Sym^iS_X^* \cap f^{+*}\, Sym^iS_{X^+}^+(-iE_1' - kE_2)\\ &\cap (f^*\, Sym^iS_X^*(-lE_2) + f^{+*}\, Sym^iS_{X^+}^+(-iE_1' - (k+1)E_2)).\end{aligned}$$

We have

$$\begin{aligned}&f^*\, Sym^iS_X^* = F^{0,0} \supset F^{1,0} \supset F^{1,1} \supset F^{2,0} \supset\\ &\cdots \supset F^{i-1,i-1} \supset F^{i,0} = f^{+*}\, Sym^iS_{X^+}^+(-iE_1' - iE_2).\end{aligned}$$

and

$$\mathcal{F}^{k,l}/\mathcal{F}^{k,l+1} \cong \mathcal{O}_{E_2}(-iE_1' - kE_2) \otimes f^{+*}\mathcal{O}_{X^+}(-k+l)$$

for $0 \le l \le k < i$, where we put $\mathcal{F}^{k,k+1} = \mathcal{F}^{k+1,0}$. Since

$$R(f_2^+)_*\mathcal{O}_{E_2}(tE_2) = 0$$

for $0 < t \le n-3$, we have

$$R(f_2^+)_*(\mathcal{F}^{k,l}/\mathcal{F}^{k,l+1} \otimes \mathcal{O}_Y((2n-5-2j)E_1' + (n-3-j)E_2)) = 0$$

because $i + j \le n-3$. Therefore,

$$\begin{aligned}&\Phi(Sym^iS_X^* \otimes \mathcal{O}_X(j))\\ =& Sym^iS_{X^+}^+ \otimes \mathcal{O}_{X^+}(-j)\\ &\otimes Rf_*^+\mathcal{O}_Y((2n-5-i-2j)E_1' + (n-3-i-j)E_2)\\ =& Sym^iS_{X^+}^+ \otimes \mathcal{O}_{X^+}(-j) \otimes Rf_{1*}^+\mathcal{O}_{X_1^+}((2n-5-i-2j)E_1^+)\\ =& Sym^iS_{X^+}^+ \otimes \mathcal{O}_{X^+}(-j).\end{aligned}$$

For the inverse direction, we climb back along the filtration $\mathcal{F}^{k,l}$. Since

$$Rf_{2*}(\mathcal{F}^{k,l}/\mathcal{F}^{k,l+1} \otimes \mathcal{O}_Y((i+2j+1)E_1' + (i+j)E_2)) = 0$$

for $0 \le l \le k < i$, we have

$$\begin{aligned}&\Psi(Sym^iS_{X^+}^+ \otimes \mathcal{O}_{X^+}(-j))\\ &= Sym^iS_X^* \otimes \mathcal{O}_X(j) \otimes Rf_*\mathcal{O}_Y((i+2j+1)E_1' + (i+j)E_2)\\ &= Sym^iS_X^* \otimes \mathcal{O}_X(j) \otimes Rf_{1*}\mathcal{O}_{X_1}((i+2j+1)E_1)\\ &= Sym^iS_X^* \otimes \mathcal{O}_X(j).\end{aligned}$$

(4) We already proved in (3) that

$$\begin{aligned}&Rf_*^+(f^*(Sym^iS_X^* \otimes \mathcal{O}_X(n-2-i)) \otimes \mathcal{O}_Y((2n-5)E_1' + (n-3)E_2))\\ &\cong Rf_*^+(F^{i-1,0} \otimes f^{+*}\mathcal{O}_X(-n+2+i) \otimes \mathcal{O}_Y((2i-1)E_1' + (i-1)E_2)).\end{aligned}$$

Moreover

$$R^p f_*^+(f^{+*}(Sym^i S_{X^+}^+ \otimes \mathcal{O}_{X^+}(-n+2+i)) \otimes \mathcal{O}_Y((i-1)E_1' - E_2))$$
$$\to R^p f_*^+(F^{i-1,0} \otimes f^{+*}\mathcal{O}_X(-n+2+i) \otimes \mathcal{O}_Y((2i-1)E_1' + (i-1)E_2))$$

is surjective for $p > 0$. Therefore, there are no higher direct images for f^+. Since

$$0 \to F^{i-1,0} \to f^{+*}\, Sym^i S_{X^+}^+ \otimes \mathcal{O}_Y(-iE_1' - (i-1)E_2)$$
$$\to \mathcal{O}_{E_2}(-iE_1' - (i-1)E_2) \otimes f^{+*}\mathcal{O}_{X^+}(-i) \to 0$$

we have

$$0 \to f_*^+(F^{i-1,0} \otimes f^{+*}\mathcal{O}_X(-n+2+i) \otimes \mathcal{O}_Y((2i-1)E_1' + (i-1)E_2))$$
$$\to Sym^i S_{X^+}^+ \otimes \mathcal{O}_X(-n+2+i) \to \mathcal{O}_{W^+} \otimes \mathcal{O}_{X^+}(-n+2) \to 0.$$

On the other hand, we have

$$f_*^+(F^{i-1,0} \otimes f^{+*}\mathcal{O}_X(-n+2+i) \otimes \mathcal{O}_Y((2i-1)E_1' + (i-1)E_2))|_{X^+ \setminus G^+}$$
$$\cong \mathcal{E}_i^+ \otimes \mathcal{O}_{X^+}(-n+2+i)|_{X^+ \setminus G^+}.$$

By taking the direct image sheaves of both sides under the inclusion morphism $X^+ \setminus G^+ \to X^+$, we obtain

$$f_*^+(F^{i-1,0} \otimes f^{+*}\mathcal{O}_X(-n+2+i) \otimes \mathcal{O}_Y((2i-1)E_1' + (i-1)E_2))$$
$$\cong \mathcal{E}_i^+ \otimes \mathcal{O}_{X^+}(-n+2+i).$$

□

We assume that $n = 4$ as well as $r = 2$ in the following. Then $\dim G = 4$, $\dim X_0 = 8$, $\dim X = 9$, and $\dim W = 7$. In particular, W is locally complete intersection. We have $K_Y = f^*K_X + 4E_1' + E_2$.

In this case, Namikawa [**N2**] proved that the functors

$$\Phi' : D^b(Coh(X)) \to D^b(Coh(Y)), \Psi' : D^b(Coh(Y)) \to D^b(Coh(X))$$

defined by

$$\Phi'(a) = Rf_*^+(Lf^*(a) \otimes \mathcal{O}_Y(4E_1' + E_2)), \Psi'(b) = Rf_*(Lf^{+*}(b) \otimes \mathcal{O}_Y).$$

are *not* equivalences.

LEMMA 3.4. *Assume that* $n = 4$. *Then*
(1) $\Psi(I_{W^+} \otimes \mathcal{O}_{X^+}(-2)) = \mathcal{O}_X(2)$.
(2) $\Psi(\mathcal{E}_i^+ \otimes \mathcal{O}_{X^+}(i-2)) = Sym^i S^* \otimes \mathcal{O}_X(2-i)$ *for* $0 < i \leq 2$.

PROOF. (1) W^+ is a divisor on X_0^+ and ϕ^+ is the contraction of (-2)-curves along the generic points of $\phi(W^+)$. Hence $\mathcal{O}_{X_0^+}(-W^+) = \mathcal{O}_{X_0^+}(2)$. We have an exact sequence

$$0 \to \mathcal{O}_{X^+}(-2 - X_0^+) \to I_{W^+} \otimes \mathcal{O}_{X^+}(-2) \to \mathcal{O}_{X_0^+} \to 0.$$

Since X_0^+ is a Cartier divisor, we have $L_k f^{+*}\mathcal{O}_{X_0^+} = 0$ for $k > 0$. Hence

$$0 \to f^{+*}\mathcal{O}_{X^+}(-2 - X_0^+) \to f^{+*}(I_{W^+} \otimes \mathcal{O}_{X^+}(-2)) \to f^{+*}\mathcal{O}_{X_0^+} \to 0.$$

On the other hand,

$$0 \to f^{+*}\mathcal{O}_{X^+}(-2 - X_0^+) \to I_{W^+}\mathcal{O}_Y \otimes f^{+*}\mathcal{O}_{X^+}(-2) \to \mathcal{O}_{(X_0^+)''} \to 0$$

where $(X_0^+)'' = (f_*^+)^{-1}X_0^+$. Thus

$$0 \to \mathcal{O}_{(E_1^+)'+E_2^+}(-(X_0^+)'') \to f^{+*}(I_{W^+} \otimes \mathcal{O}_{X^+}(-2))$$
$$\to I_{W^+}\mathcal{O}_Y \otimes f^{+*}\mathcal{O}_{X^+}(-2) \to 0.$$

We have

$$\mathcal{O}_{(E_1^+)'+E_2^+}(-(X_0^+)'') \cong \mathcal{O}_{E_1'+E_2}(E_1' + E_2).$$

Since $Rf_{2*}\mathcal{O}_{E_2}(E_2) = 0$ and $Rf_{1*}\mathcal{O}_{E_1}(2E_1) = 0$, we have

$$Rf_*\mathcal{O}_{E_1'+E_2}(2E_1' + E_2) \cong Rf_{1*}\mathcal{O}_{E_1}(2E_1) \cong 0.$$

Therefore, we have

$$Rf_*(f^{+*}(I_{W^+} \otimes \mathcal{O}_{X^+}(-2)) \otimes \mathcal{O}_Y(E_1'))$$
$$\cong Rf_*(I_{W^+}\mathcal{O}_Y \otimes f^{+*}\mathcal{O}_{X^+}(-2) \otimes \mathcal{O}_Y(E_1')).$$

Let $F^+ = (X_0^+)'' \cap (E_1^+)' \subset Y$. The projection $f^+ : F^+ \to G^+$ is a $\mathbb{P}^3$-bundle whose fiber $(P_0^+)'$ is contracted by f. We have

$$I_{W^+}\mathcal{O}_Y = I_{F^+}(-E_1' - E_2)$$

Thus

$$I_{W^+}\mathcal{O}_Y \otimes f^{+*}\mathcal{O}_{X^+}(-2) \otimes \mathcal{O}_Y(E_1') = I_{F^+}f^*\mathcal{O}_X(2) \otimes \mathcal{O}_Y(4E_1' + E_2).$$

On the other hand, we have $\mathcal{O}_{(P_0^+)'}(E_1') \cong \mathcal{O}_{\mathbb{P}^3}(-1)$ and $\mathcal{O}_{(P_0^+)'}(E_2) \cong \mathcal{O}_{\mathbb{P}^3}(2)$. Hence $\mathcal{O}_{(P_0^+)'}(4E_1' + E_2) \cong \mathcal{O}_{\mathbb{P}^3}(-2)$. It follows that $Rf_*\mathcal{O}_{F^+}(4E_1' + E_2) = 0$. Therefore,

$$Rf_*(I_{F^+}\mathcal{O}_Y \otimes f^{+*}\mathcal{O}_{X^+}(-2) \otimes \mathcal{O}_Y(E_1')) \cong \mathcal{O}_X(2).$$

(2) We have an exact sequence

$$0 \to \mathcal{E}_i^+ \otimes \mathcal{O}_{X^+}(i-2) \to Sym^i S_{X^+}^+ \otimes \mathcal{O}_{X^+}(i-2) \to \mathcal{O}_{W^+} \otimes \mathcal{O}_{X^+}(-2) \to 0.$$

Thus the torsion part of $f^{+*}\mathcal{E}_i^+ \otimes \mathcal{O}_{X^+}(i-2)$ is isomorphic to that of $f^{+*}I_{W^+} \otimes \mathcal{O}_{X^+}(-2)$. Therefore, by the argument in (1), we have

$$\Psi(\mathcal{E}_i^+ \otimes \mathcal{O}_{X^+}(i-2)) \cong Rf_*(f^{+*}\mathcal{E}_i^+/\text{torsion} \otimes f^{+*}\mathcal{O}_{X^+}(i-2) \otimes \mathcal{O}_Y(E_1')).$$

We have an exact sequence

$$0 \to f^{+*}\mathcal{E}_i^+/\text{torsion} \otimes f^{+*}\mathcal{O}_{X^+}(i-2) \otimes \mathcal{O}_Y(E_1')$$
$$\to f^{+*}(Sym^i S_{X^+}^+ \otimes \mathcal{O}_{X^+}(i-2)) \otimes \mathcal{O}_Y(E_1')$$
$$\to f^{+*}(\mathcal{O}_{W^+} \otimes \mathcal{O}_{X^+}(-2)) \otimes \mathcal{O}_Y(E_1') \to 0.$$

Therefore, we have an exact sequence

$$0 \to f^{+*}\mathcal{E}_i^+/\text{torsion} \otimes f^{+*}\mathcal{O}_{X^+}(i-2) \otimes \mathcal{O}_Y(E_1')$$
$$\to F^{i-1,0} \otimes f^{+*}\mathcal{O}_{X^+}(i-2) \otimes \mathcal{O}_Y((i+1)E_1' + (i-1)E_2) \to Q \to 0$$

where the cokernel Q has a decomposition as follows

$$0 \to \mathcal{O}_{F^+}(-E_2) \otimes f^{+*}\mathcal{O}_{X^+}(-2) \to Q \to \mathcal{O}_{E_1'}(E_1' - E_2) \otimes f^{+*}\mathcal{O}_{X^+}(-2) \to 0.$$

As in (1), $F^+ = F = X_0'' \cap E_1'$ is a $\mathbb{P}^3$-bundle over G with fibers P_0' such that $\mathcal{O}_{P_0'}(E_1') \cong \mathcal{O}_{\mathbb{P}^3}(-1)$ and $\mathcal{O}_{P_0'}(E_2) = \mathcal{O}_{\mathbb{P}^3}(2)$. Thus

$$\mathcal{O}_{F^+}(-E_2) \otimes f^{+*}\mathcal{O}_{X^+}(-2) \otimes \mathcal{O}_{P_0'} \cong \mathcal{O}_{P_0'}(4E_1' + E_2) \cong \mathcal{O}_{P_0'}(-2)$$

and we have $Rf_*\mathcal{O}_{F^+}(-E_2) = 0$. We have also

$$\begin{aligned} &Rf_*\mathcal{O}_{E_1'}(E_1' - E_2) \otimes f^{+*}\mathcal{O}_{X^+}(-2) \\ &\cong Rf_*\mathcal{O}_{E_1'}(3E_1' + E_2) \otimes \mathcal{O}_X(2) \\ &\cong Rf_{1*}\mathcal{O}_{E_1}(3E_1) \otimes \mathcal{O}_X(2) \cong 0. \end{aligned}$$

Therefore

$$\begin{aligned} &\Psi(\mathcal{E}_i^+ \otimes \mathcal{O}_{X^+}(i-2)) \\ &\cong Rf_*(F^{i-1,0} \otimes f^{+*}\mathcal{O}_{X^+}(i-2) \otimes \mathcal{O}_Y((i+1)E_1' + (i-1)E_2)) \\ &\cong Rf_*(F^{i-1,0} \otimes \mathcal{O}_Y((5-i)E_1' + E_2)) \otimes \mathcal{O}_X(2-i). \end{aligned}$$

If $i = 1$, then $F^{0,0} = f^* Sym^i S^*$ and $Rf_*\mathcal{O}_Y(4E_1' + E_2) = \mathcal{O}_X$, hence the result.

If $i = 2$, then $\mathcal{F}^{0,0}/\mathcal{F}^{1,0} \cong \mathcal{O}_{E_2}(-2E_1')$, hence

$$Rf_*(\mathcal{F}^{0,0}/\mathcal{F}^{1,0} \otimes \mathcal{O}_Y(3E_1' + E_2)) = 0$$

and we complete the proof. □

THEOREM 3.5. *If $n = 4$, then Φ and Ψ in (3.1) are equivalences.*

PROOF. There is an adjunction morphism of functors $F : \mathrm{Id}_{D^b(Coh(X))} \to \Psi\Phi$, which is reduced to the identity when restricted to the open subset $X \setminus W$. By Lemmas 3.2, 3.3 and 3.4, we have isomorphisms $\omega \cong \Psi\Phi(\omega)$ for the spanning class $\Omega = \{\omega\}$ given by Lemma 3.2. Since the ω are locally free sheaves, it follows that the morphisms $F(\omega)$ are isomorphisms. Therefore, the natural homomorphisms

$$\Phi : \mathrm{Hom}^p(\omega_1, \omega^2) \to \mathrm{Hom}^p(\omega_1, \Psi\Phi(\omega^2)) \to \mathrm{Hom}^p(\Phi(\omega_1), \Phi(\omega^2))$$

for any $\omega_1, \omega_2 \in \Omega$ and $p \in \mathbb{Z}$ are isomorphisms. Then Φ is an equivalence by [**BO**] and [**B1**]. □

REMARK 3.6. (1) Let $F = \bigoplus \omega$ be the sum of the spanning sheaves given in Lemma 3.2. Then F is *not* an almost exceptional object in the sense of [**BK**] in our case $n = 4$ and $r = 2$, because $R^1\phi_*\mathcal{O}_X(-2) \neq 0$ implies $\mathrm{Hom}^1(F,F) \neq 0$.

(2) In order to extend our argument to the case $n \geq 5$, the *Eagon-Northcott resolution* on X_0^+ would be useful ([**E**]):

$$\begin{aligned} &0 \to \pi_0^{+*}(Sym^{n-2}S^{+2})^* \otimes \bigwedge^n V^{*n} \xrightarrow{B_*} \cdots \to \pi_0^{+*}(Sym^k S^{+2})^* \otimes \bigwedge^{k+2} V^{*n} \\ &\xrightarrow{B_*} \pi_0^{+*}(Sym^{k-1}S^{+2})^* \otimes \bigwedge^{k+1} V^{*n} \xrightarrow{B_*} \cdots \xrightarrow{B_*} \pi_0^{+*}(S^{+2})^* \otimes \bigwedge^3 V^{*n} \\ &\xrightarrow{B_*} \mathcal{O}_{X_0^+} \otimes \bigwedge^2 V^{*n} \xrightarrow{\wedge^2 B} \pi_0^{+*} \bigwedge^2 S^{+2} \to \mathcal{O}_{W^+}(-1) \to 0. \end{aligned}$$

References

[BK] R. Bezrukavnikov and D. Kaledin. *McKay equivalence for symplectic resolutions of singularities.* math.AG/0401002.

[BO] A. Bondal, D. Orlov. *Semiorthogonal decomposition for algebraic varieties.* alg-geom/9506012.

[B1] Tom Bridgeland. *Equivalences of triangulated categories and Fourier-Mukai transforms.* math.AG/9809114. Bull. London Math. Soc. 31 (1999), no. 1, 25–34.

[B2] Tom Bridgeland. *Flops and derived categories.* math.AG/0009053. Invent. Math. 147 (2002), no. 3, 613–632.

[BKR] Tom Bridgeland, Alastair King and Miles Reid. *Mukai implies McKay: the McKay correspondence as an equivalence of derived categories.* math.AG/9908027. J. Amer. Math. Soc. 14 (2001), no. 3, 535–554.

[Chen] J. -C. Chen. *Flops and equivalences of derived categories for threefolds with only Gorenstein singularities.* math.AG/0202005.

[E] David Eisenbud. *Commutative Algebra with a View Toward Algebraic Geometry.* Springer-Verlag, 1995.

[K] M. M. Kapranov. *Derived category of coherent sheaves on Grassmann manifolds.* (Russian) Izv. Akad. Nauk SSSR Ser. Mat. 48 (1984), no. 1, 192–202.

[Kw1] Yujiro Kawamata. *Francia's flip and derived categories.* math.AG/0111041. Algebraic geometry, 197–215, de Gruyter, Berlin, 2002.

[Kw2] Yujiro Kawamata. *D-equivalence and K-equivalence.* math.AG/0205287. J. Differential Geom. 61 (2002), no. 1, 147–171.

[Kw3] Yujiro Kawamata. *Log Crepant Birational Maps and Derived Categories.* math.AG/0311139.

[M] Eyal Markman. *Brill-Noether duality for moduli spaces of sheaves on K3 surfaces.* math.AG/9901072. Journal of Algebraic Geometry **10** (2001), no. 4, 623-694.

[N1] Yoshinori Namikawa. *Mukai flops and derived categories.* math.AG/0203287. J. Reine Angew. Math. 560 (2003), 65–76.

[N2] Yoshinori Namikawa. *Mukai flops and derived categories II.* math.AG/0305086.

Department of Mathematical Sciences, University of Tokyo, Komaba, Meguro, Tokyo, 153-8914, Japan

E-mail address: kawamata@ms.u-tokyo.ac.jp

AMS/IP Studies in Advanced Mathematics
Volume 38, 2006

A Survey of Transcendental Methods in the Study of Chow Groups of Zero-cycles

Matt Kerr

ABSTRACT. We review an assortment of candidate Bloch-Beilinson filtrations on $CH_0(X)$ (for X smooth projective), together with Hodge-theoretic invariants defined on their graded pieces. A large array of applications is given, especially to detecting nontrivial 0-cycles and the behavior of said invariants with respect to families. Ample background is provided, including many of the theorems and conjectures that have shaped the passage from classical to modern invariants.

Contents

2000 *Mathematics Subject Classification.* Primary 14D07, 14C15, 14C25, 14C30, 14J32.
Key words and phrases. algebraic cycle, Chow group, Bloch-Beilinson filtration, Hodge theory, higher Abel-Jacobi map, Mumford invariants, spreads.

1. Introduction

A driving force behind modern algebraic geometry has been the notion that certain topological, analytic or arithmetic invariants on a projective *algebraic* variety X should be represented or explained by *algebraically* defined objects "on X." So for example the HC (Hodge Conjecture) says that the rational (p,p) classes of X should be generated (modulo torsion) by the fundamental classes of its codimension-p subvarieties. Or for X defined over a number field, the Beilinson conjectures predict that special values of its L-functions should be computable by (determinants of) regulator images of elements in algebraic K-groups of X. These images ultimately boil down to integrals of algebraic differential forms (on X) over subsets of X bounded or cut out by algebraic equations.

In both instances, one has to get a handle on invariants of certain algebraic cycles — whether algebraic K-theory elements or linear combinations of subvarieties of X — and the study of such cycles modulo a suitable equivalence relation becomes a natural pursuit.

Zero-cycles — $\mathbb{Q}$-linear combinations of (zero-dimensional) points on X — are the simplest kind of algebraic cycles, and their study modulo rational equivalence is a rich area with many beautiful results (and unresolved conjectures). Historically, the first example is Abel's theorem, which links an algebraic question (when is a degree-0 divisor on an algebraic curve the divisor of a rational function?) with a transcendental invariant (the image of the divisor in the Jacobian, which involves integration along paths connecting the points).

For 0-cycles on varieties of dimension > 1, on the other hand, recent work has concentrated on the influence of the field of definition on their rational equivalence-classes. This paper is about new techniques and Hodge-theoretic invariants — e.g. higher cycle-classes and higher Abel-Jacobi classes — that reflect this influence, and about interesting examples of cycles they detect. It draws on work of Asakura, Bloch, Green, Griffiths, Lewis, M. Saito, S. Saito, Voevodsky, Voisin and others (and some of our own).

We offer some theorems (with proofs) about 0-cycles on products of curves and complete intersections, and a discussion of infinitesimal/topological invariants for "higher normal functions" arising from the invariants. A brisk review of background material is included, as this is rapidly becoming basic algebraic geometry and we believe it should be accessible.

We warmly thank D. Arapura, S. Bloch, P. Brosnan, M. Green, R. de Jeu, J. Lewis, and C. Voisin for helpful conversations and correspondence, and again thank P. Griffiths for introducing us to the subject. We thank the referee for helpful comments which have led to improvements in the exposition. This paper was written at the University of Chicago and MPI-Bonn; we are grateful to both institutions for their support.

2. Preliminaries

2.1. Cycles and rational equivalence. Even for complex algebraic geometers, a smooth (complex) projective variety X is ultimately cut out by homogeneous polynomials with coefficients belonging to some field $L \subseteq \mathbb{C}$ finitely generated over $\mathbb{Q}$; hence we may consider the underlying variety $/L$ with structure map $X \to Spec(L)$. If $K \supseteq L$ we write $X_K := X_{(L)} \times_L Spec(K)$ for the base change, and throughout the paper $d = \dim(X)$.

All cycle groups and cohomology groups are taken modulo torsion; so the algebraic cycle group $Z^p(X_K) = Z_{d-p}(X_K)$ denotes *rational* linear combinations of irreducible subvarieties of X, defined $/K$ and of codimension p. Hence a 0-cycle $\mathcal{Z} \in Z_0(X_K)$ is (after base-change to $\bar{K}$) of the form $\sum q_i z_i$, with $q_i \in \mathbb{Q}$ and z_i having coordinates in $\bar{K}$, and such that $\sum q_i z_i$ is invariant under the action of $Gal(\bar{K}/K)$ on $X(\bar{K})$. For example, on the affine line $\{\sqrt{3}\} + \{\sqrt{-3}\} - 5\{1\}$ is [the base-change to $\bar{\mathbb{Q}}$ of] a 0-cycle defined $/\mathbb{Q}$, but $2\{\sqrt{3}\} - 2\{\sqrt{-3}\}$ is not.

We say two cycles $\mathcal{Z}_1, \mathcal{Z}_2 \in Z^p(X_K)$ are *rationally equivalent* if (roughly speaking) one can get from $\mathcal{Z}_1$ to $\mathcal{Z}_2$ in a rationally parametrized family $\mathcal{W}$:

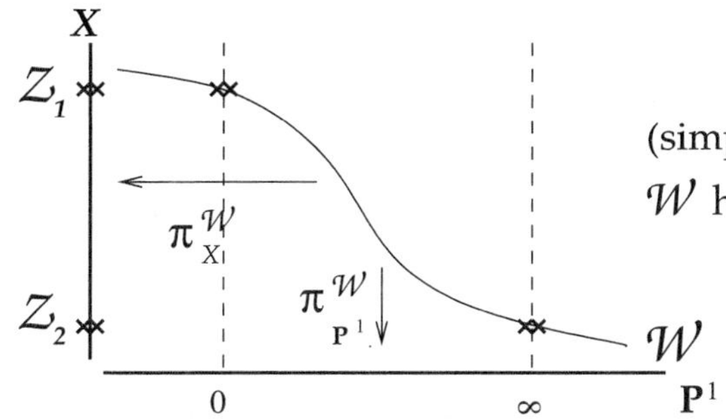

(simplest case:
$\mathcal{W}$ has one irreducible component.)

More precisely, let $\pi_X (\text{resp.}\, \pi_{\mathbb{P}^1}) : X \times \mathbb{P}^1 \to X\, (\text{resp.}\, \mathbb{P}^1)$ be the projections; then

$$\mathcal{Z}_1 \overset{\text{rat}}{\equiv} \mathcal{Z}_2 \quad (\text{or}\, \mathcal{Z}_1 - \mathcal{Z}_2 \overset{\text{rat}}{\equiv} 0) \quad \overset{\text{defn}}{\Longleftrightarrow}$$

$$\begin{array}{c} \exists\, \mathcal{W} \in Z^p(X \times \mathbb{P}^1_{/K}) \text{ such that} \\ \pi_{X_*}[\mathcal{W} \cdot (X \times \{0\}) - \mathcal{W} \cdot (X \times \{\infty\})] = \mathcal{Z}_1 - \mathcal{Z}_2 \end{array} \quad \Longleftrightarrow$$

$$\exists \left\{ \begin{array}{l} \widetilde{Y}_i \twoheadrightarrow Y_{i/K} \subseteq X \text{ irreducible codim. } p-1, \ (\text{composite } \iota_i) \\ f_i \in K(\widetilde{Y}_i)^* \text{ rational functions} \end{array} \right\} \begin{array}{c} \text{such that} \\ \sum \iota^i_*(f_i) = \mathcal{Z}_1 - \mathcal{Z}_2. \end{array}$$

(Here $\widetilde{Y}_i$ denotes a normalization of Y_i.) In the first definition, $\mathcal{W}$ may be reducible; so it changes nothing to replace $\mathbb{P}^1$ by a chain of $\mathbb{P}^1$'s. On the other hand, if one replaces $\mathbb{P}^1$ by a chain of curves of arbitrary genera, one has algebraic equivalence $\mathcal{Z}_1 \overset{\text{alg}}{\equiv} \mathcal{Z}_2$. Finally, a codimension-p cycle is homologous to zero iff on $X_{\mathbb{C}}$ it is the boundary of a topological chain of real dimension $2d - 2p + 1$,

$$\mathcal{Z}_1 \overset{\text{hom}}{\equiv} \mathcal{Z}_2 \quad \overset{\text{defn}}{\Longleftrightarrow} \quad \exists\, \Gamma \in C^{\text{top}}_{2d-2p+1}(X^{\text{an}}_{\mathbb{C}}, \mathbb{Q}) \ s.t.\ \partial\Gamma = \mathcal{Z}_1 - \mathcal{Z}_2,$$

or equivalently if its fundamental class (§2.2) vanishes.

Clearly $\overset{\text{rat}}{\equiv} 0 \implies \overset{\text{alg}}{\equiv} 0 \implies \overset{\text{hom}}{\equiv} 0$. For 0-cycles on a curve ($d = 1$), the second definition gives $\mathcal{Z} \overset{\text{rat}}{\equiv} 0 \iff \mathcal{Z} = (f)$ for $f \in K(X)^*$. If X is a surface ($d = 2$), a rational equivalence between 2 points may look as follows:

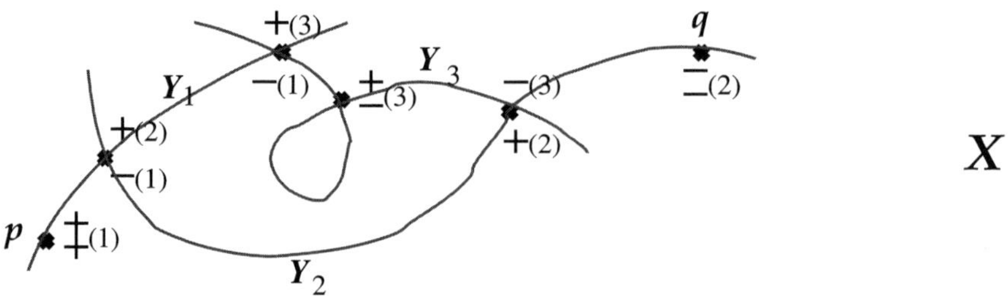

where e.g. $\overset{+}{+}$ (i) says f_i has a zero of order 2. (Since we work modulo torsion [i.e. $\otimes\mathbb{Q}$], $2p - 2q \overset{\text{rat}}{\equiv} 0 \implies p \overset{\text{rat}}{\equiv} q$.) Also, for 0-cycles in general, $\overset{\text{hom}}{\equiv}$ and $\overset{\text{alg}}{\equiv}$ are the same.

The object of study will be Chow groups of cycles $/K$ modulo rational equivalences defined $/K$,

$$CH^p(X_K) := \frac{Z^p(X_K)}{(\overset{\text{rat}}{\equiv})_K}.$$

(We also write $CH^p(X_{/K})$, especially when X is replaced by something like $Y \times \mathcal{S}$ so that it is clear that K applies to the whole product.) One consequence of working modulo torsion is that $CH^p(X_K) \hookrightarrow CH^p(X_{\mathbb{C}})$, so one does not have to worry about the field of definition of the $\overset{\text{rat}}{\equiv}$. When clarity is at stake (only), we have written $\langle \mathcal{Z} \rangle \in CH^*(X)$ for the $\overset{\text{rat}}{\equiv}$-class of a cycle $\mathcal{Z} \in Z^*(X)$; classes of cycles $\overset{\text{hom}}{\equiv}$ (resp. $\overset{\text{alg}}{\equiv}$)0 are written CH^*_{hom} (resp. CH^*_{alg}).

Our goal is to detect 0-cycles/ $\overset{\text{rat}}{\equiv}$ in the kernel of the Albanese map; since the invariants make heavy use of Hodge theory we review next what we need.

2.2. Hodge Structures. A HS (Hodge Structure) of weight m is a finite-dimensional $\mathbb{Q}$-vector space $\mathcal{H}$, together with a descending filtration $F^\bullet$ on $\mathcal{H}_{\mathbb{C}} := \mathcal{H} \otimes_{\mathbb{Q}} \mathbb{C}$ satisfying for each i: $F^i\mathcal{H}_{\mathbb{C}} \oplus \overline{F^{m-i+1}\mathcal{H}_{\mathbb{C}}} = \mathcal{H}_{\mathbb{C}} = F^0\mathcal{H}_{\mathbb{C}}$. Writing $\mathcal{H}^{i,m-i}_{(\mathbb{C})} := F^i\mathcal{H}_{\mathbb{C}} \cap \overline{F^{m-i}\mathcal{H}_{\mathbb{C}}}$, we get a Hodge decomposition $\mathcal{H}_{\mathbb{C}} = \oplus_{p+q=m} \mathcal{H}^{p,q}$. A $\mathbb{Q}$-subspace $\mathcal{G} \subseteq \mathcal{H}$ is a subHS iff $\mathcal{G}_{\mathbb{C}} = \oplus_{p+q=m}(\mathcal{G}_{\mathbb{C}} \cap \mathcal{H}^{p,q})$, and a $\mathbb{Q}$-linear transformation $\mathcal{H} \overset{\theta}{\to} {}'\mathcal{H}$ is a morphism of HS iff over $\mathbb{C}$ it takes the form $\oplus \mathcal{H}^{p,q} \overset{\oplus\theta^{p,q}}{\longrightarrow} \oplus {}'\mathcal{H}^{p,q}$ (relative to a pair of $\mathbb{C}$-bases subordinate to the resp. Hodge decompositions).

Intersections and sums of subHS are subHS, as are the image and kernel of a morphism of HS; quotients (e.g., $\mathcal{H}/\mathcal{G}$) have a natural HS ($\mathcal{H}_{\mathbb{C}}/\mathcal{G}_{\mathbb{C}} \cong \oplus_{p+q=m} \mathcal{H}^{p,q}/\mathcal{G}^{p,q}$), as do tensor products and duals (of HS). We define

$$F^j_h \mathcal{H}_{(\mathbb{Q})} := \text{largest subHS of } H_{(\mathbb{Q})} \text{ contained in } F^j\mathcal{H}_{\mathbb{C}} \cap \mathcal{H}_{(\mathbb{Q})},$$

and note that for $m > 2j$ equality does *not* in general hold. One also has the Tate HS $\mathbb{Q}(-d)$, of pure type (d,d) and weight $2d$.

From geometry comes a large assortment of examples. For X smooth projective, set $H^m(X) := H^m_{\text{sing}}(X^{(an)}_{\mathbb{C}}, \mathbb{Q})$, and identify $H^m(X,\mathbb{C}) \cong H^m\left\{\Gamma(X_{\mathbb{C}}, \Omega^\bullet_{X^\infty}), d\right\} \hookleftarrow H^m\left\{\Gamma(X_{\mathbb{C}}, F^j\Omega^\bullet_{X^\infty}), d\right\} =: F^jH^m(X,\mathbb{C})$ using C^∞-forms (de Rham cohomology). By the Hodge theorem $H^j(X)$ is a HS; we also see that $H^{p,q}(X,\mathbb{C})$ is represented by

C^∞-forms of type (p,q) (modulo coboundaries). Associated to an algebraic cycle $\mathcal{Z} \in Z^p(X)$ one has the fundamental class

$$[\mathcal{Z}] \in H^{2p}(X) \cap F^p H^{2p}(X,\mathbb{C}) =: Hg^p(X)$$

defined by the functional $\left\{\int_{\mathcal{Z}}(\cdot)\right\} \in \left\{H^{2d-2p}(X,\mathbb{C})\right\}^\vee$; and $\mathbb{Q}[\mathcal{Z}] \subseteq H^{2p}(X)$ is a (Tate) subHS. Write $H^{2p}_{\text{alg}}(X)$ for the span of all such classes. If $X = Y_1 \times Y_2$ then $H^m(X) \cong \oplus_{r+s=m} H^r(Y_1) \otimes H^s(Y_2)$ as HS, and $[\mathcal{Z}]$ has Künneth components $[\mathcal{Z}]_i \in H^{2p-i}(Y_1) \otimes H^i(Y_2)$ summing to $[\mathcal{Z}]$; again $\mathbb{Q}[\mathcal{Z}]_i$ is a subHS.

A polarization of a HS $\mathcal{H}$ is a choice of bilinear form $\mathcal{Q} : \mathcal{H}\times\mathcal{H} \to \mathbb{Q}$ (symmetric for m even, skew for m odd) which after tensoring with $\mathbb{C}$ obeys the Hodge-Riemann bilinear relations. Polarized HS are *semi-simple* in the sense that given $\mathcal{G} \subseteq \mathcal{H}$ subHS, $\mathcal{G}^\perp$ is a HS and $\mathcal{H} = \mathcal{G}\oplus\mathcal{G}^\perp$; so for example exact sequences (with polarized middle term) split. For X smooth projective, $H^m(X)$ has a natural polarization (using the Lefschetz decomposition and the hyperplane class); hence $H^m(X)$ and its subHS's are semisimple.

From above we have $H^{2p}_{\text{alg}}(X) \subseteq Hg^p(X)$; there is a large bank deposit waiting for whoever proves the equality [HC(p,X)] in general. The Generalized Hodge Conjecture GHC(j,m,X) predicts that $F^j_h H^m(X) = N^j H^m(X)$, where $N^\bullet$ is the descending filtration by coniveau (i.e., codimension of support); clearly HC(p,X) is GHC$(p,2p,X)$. (Note that $N^j \subseteq F^j_h$ always holds.) Now let $[\mathsf{H}] \in H^2(X)$ be the class of a hyperplane section; a consequence of the Hard Lefschetz theorem is that $\cup[\mathsf{H}]^i : H^{d-i}(X) \to H^{d+i}(X)$ is an $\cong$. This is clearly also induced by an algebraic correspondence (cycle) on $X \times X$; the Lefschetz standard conjecture (of Grothendieck) says the *inverse* is [algebraic-]cycle-induced (for all i). In this paper, we will write HLC(X) (Hard Lefschetz conjecture) for the apparently stronger statement that (for each i) there exists a cycle inducing $(\cup[\mathsf{H}]^i)^{-1} : H^{d+i}(X) \to H^{d-i}(X)$ and also inducing the zero-map in all other degrees. In fact, they are equivalent by [**Kl**]; and clearly HLC(X) is implied by HC$(d,\, X \times X)$.

2.3. Jacobians. Next assume $\mathcal{H}$ is of weight $2p-1$, and define the Jacobian

$$J^p(\mathcal{H}) := \frac{\mathcal{H}_{\mathbb{C}}}{F^p\mathcal{H}_{\mathbb{C}} + \mathcal{H}_{\mathbb{Q}}}.$$

Injective or surjective morphisms, direct sums, and quotients of HS (as well as commuting diagrams) induce the same behavior in the corresponding Jacobians. We write $J^p(H^{2p-1}(X)) =: J^p(X)$. (In the same way one can define for $'\mathcal{H}$ of weight $2p$ the Hodge group $Hg^p('\mathcal{H}) = {}'\mathcal{H} \cap F^p('\mathcal{H}_{\mathbb{C}})$, so that $Hg^p(H^{2p}(X)) = Hg^p(X)$.) Note that if $\mathcal{H} = \mathcal{K}^\vee \otimes \mathbb{Q}(-d)$ as HS then $J^p(\mathcal{H}) \cong \frac{(F^{d-p+1}\mathcal{K}_{\mathbb{C}})^\vee}{\text{im}(\mathcal{K}_{(\mathbb{Q})})^\vee}$.

Given $\mathcal{Z} \in Z^p_{\text{hom}}(X)$ and $\mathcal{H} \subseteq H^{2p-1}(X)$ (subHS) we may construct an element in $J^p(\mathcal{H})$ as follows. Let $\partial^{-1}\mathcal{Z} \in C^{\text{top}}_{2d-2p+1}(X,\mathbb{Q})$ be a fixed choice of chain bounding on $\mathcal{Z}$, and assume $\mathcal{H}$ and $\mathcal{K} \subseteq H^{2d-2p+1}(X)$ are dual under the perfect pairing $H^{2p-1}(X) \otimes H^{2d-2p+1}(X) \to \mathbb{Q}(-d)$. Since $\mathcal{Z}$ has complex dimension $d-p$, $\int_{\partial^{-1}\mathcal{Z}} d\alpha = \int_{\mathcal{Z}} \alpha = 0$ if $\alpha \in \Gamma(X, F^{d-p+1}\Omega^{2d-2p}_{X^\infty})$. Thus $\left\{\int_{\partial^{-1}\mathcal{Z}}(\cdot)\right\}$ gives a well-defined functional on[1]

$$F^{d-p+1}H^{2d-2p+1}(X,\mathbb{C}) \cong \frac{\ker(d) \subseteq \Gamma(X, F^{d-p+1}\Omega^{2d-2p+1}_{X^\infty})}{\text{im}\left\{d : \Gamma(X, F^{d-p+1}\Omega^{2d-2p}_{X^\infty}) \to \text{num}\right\}},$$

[1]here "num" denotes numerator.

which we may restrict to $F^{d-p+1}\mathcal{K}_{\mathbb{C}}$. Varying now the choice of $\partial^{-1}\mathcal{Z}$ (by topological cycles) changes the functional by rational functionals; hence we still get a well-defined result in $J^p(\mathcal{H})$, the Abel-Jacobi class $AJ_X(\mathcal{Z})$ if $\mathcal{H} = H^{2p-1}(X)$.

2.4. Deligne cohomology. For our cycle maps we will use Deligne (co)homology, which for X projective one may define (by **[J4]**) as cohomology of a complex:

$$H^i_{\mathcal{D}}(X, \mathbb{Q}(j)) :=$$

$$H^i\left\{C^{\rm top}_{2d-\bullet}\left(X, (2\pi\sqrt{-1})^j\mathbb{Q}\right) \oplus \Gamma\left(X, F^{j\,\prime}\mathcal{D}^\bullet_X\right) \oplus \Gamma\left(X, {}^{\prime}\mathcal{D}^{\bullet-1}_X\right), D\right\}$$

where[2] $D(\Gamma, \Omega, K) = (-\partial_{\rm top}\Gamma, -d[\Omega], d[K] - \Omega + \delta_\Gamma)$. In particular, one has the exact sequence

$$0 \to J^p(X) \to H^{2p}_{\mathcal{D}}(X, \mathbb{Q}(p)) \to Hg^p(X) \to 0.$$

A mixed Hodge structure (MHS) is a $\mathbb{Q}$-vector space $\mathcal{H}$ with an ascending (weight) filtration $W_\bullet$, plus a descending $F^\bullet$ on $\mathcal{H}_{\mathbb{C}}$, such that $Gr^W_m\mathcal{H}$ together with the induced $F^\bullet$ on $(Gr^W_m\mathcal{H})_{\mathbb{C}}$ is a weight m HS. If Y is smooth quasiprojective, $H^n(Y)$ has a natural MHS with $W_{n-1}H^n(Y) = \{0\}$ and $W_nH^n(Y) = \operatorname{im}\left\{H^n(\bar{Y}) \to H^n(Y)\right\} =: \underline{H^n}(Y)$ for *any* smooth compactification $\bar{Y}$ (see **[K2**, sec. 3**]**). For $\bar{Y}$ a *good* compactification ($\bar{Y} \setminus Y = D$ a normal-crossings divisor), one may interpret $W_{m+j}H^m(Y)$ in terms of forms with log poles along D locally no worse than $d\log z_{k_1} \wedge \cdots \wedge d\log z_{k_j}$ (for $D = \cup\{z_i = 0\}$ locally), see **[GS]**.

We will have extensive use for the following basic computations:

$$\begin{aligned} Hom_{\rm MHS}(\mathbb{Q}(-p), \mathcal{H}) &= W_{2p}\mathcal{H} \cap F^pW_{2p}\mathcal{H}_{\mathbb{C}}, \\ Ext^1_{\rm MHS}(\mathbb{Q}(-p), \mathcal{H}) &= \frac{W_{2p}\mathcal{H}_{\mathbb{C}}}{F^pW_{2p}\mathcal{H}_{\mathbb{C}} + W_{2p}\mathcal{H}_{(\mathbb{Q})}}. \end{aligned}$$

One notes that if $\mathcal{H}$ is *pure* of weight $2p - 1$ (i.e., a HS) then $Ext^1_{\rm MHS}(\mathbb{Q}(-p), \mathcal{H}) = J^p(\mathcal{H})$, and using this (together with the Ext long-exact sequence associated to $0 \to W_{2p-1} \to W_{2p} \to Gr^W_{2p} \to 0$) one shows that

$$\operatorname{im}\left\{Ext^1_{\rm MHS}(\mathbb{Q}(-p), H^{2p-1}(\bar{Y})) \to Ext^1_{\rm MHS}(\mathbb{Q}(-p), H^{2p-1}(Y))\right\} \cong$$
$$Ext^1_{\rm MHS}\left(\mathbb{Q}(-p), \underline{H^{2p-1}}(Y)\right)/\operatorname{im}\left\{Hom_{\rm MHS}\left(\mathbb{Q}(-p), Gr^W_{2p}H^{2p-1}(Y)\right)\right\}.$$

This quotient reflects the fact that in generalizing the construction of classes in Jacobians (§2.3) to the quasiprojective case, one needs to mod out essentially by all classes of cycles supported on $(\bar{Y} \setminus Y)$!

To appropriately generalize Deligne cohomology to Y, we need to take the weight filtration into account by using absolute Hodge cohomology $H^*_{\mathcal{H}}(Y, \mathbb{Q}(\cdot))$ (see (§5.5.2), or **[L1]** for a more conceptual explanation). While this reduces to Deligne if Y is projective, it isn't as obvious how to define it without passing to the derived category (owing to the difficulty of *controlling* poles of currents). Now set

$$\underline{H^{2p}_{\mathcal{D}}}(Y, \mathbb{Q}(p)) := \operatorname{im}\left\{H^{2p}_{\mathcal{D}}(\bar{Y}, \mathbb{Q}(p)) \to H^{2p}_{\mathcal{H}}(Y, \mathbb{Q}(p))\right\};$$

once more the image is independent of the choice of smooth compactification $\bar{Y}$ (see **[K2**, sec. 4**]**). One gets two exact sequences

[2] ${}^{\prime}\mathcal{D}^i_X$ denotes the sheaf of i-currents, a local section of which is a bounded linear functional on local sections of C^∞ $(2d-i)$-forms. Integration over topological $(2d-i)$-chains $\Gamma \in C^{\rm top}_{2d-i}(X, \mathbb{Q})$ includes them as "delta functions" $\delta_\Gamma \in \Gamma(X, {}^{\prime}\mathcal{D}^i_X)$.

$$
\begin{array}{ccccccc}
0 \to & Ext^1_{\mathrm{MHS}}\left(\mathbb{Q}(-p), H^{2p-1}(Y)\right) & \longrightarrow & H^{2p}_{\mathcal{H}}(Y, \mathbb{Q}(p)) & \to & Hom_{\mathrm{MHS}}\left(\mathbb{Q}(-p), H^{2p}(Y)\right) & \to 0 \\
& \uparrow & & \uparrow & & \| & \\
0 \to & \dfrac{Ext^1_{\mathrm{MHS}}\left(\mathbb{Q}(-p), \underline{H^{2p-1}}(Y)\right)}{Hom_{\mathrm{MHS}}\left(\mathbb{Q}(-p), Gr^W_{2p} H^{2p-1}(Y)\right)} & \to & \underline{H^{2p}_{\mathcal{D}}}(Y, \mathbb{Q}(p)) & \to & Hom_{\mathrm{MHS}}\left(\mathbb{Q}(-p), H^{2p}(Y)\right) & \to 0
\end{array}
$$

where we note that the right hand term is just $Hg^p(\underline{H^{2p}}(Y))$.

3. From classical invariants to the Bloch-Beilinson Conjectures

3.1. Cycle-class and Albanese. Take $\mathcal{Z} \in Z^p(X_K)$ as above (in this section K plays no role), and assume $\mathcal{Z} \overset{\text{rat}}{\equiv} 0$: then $\exists \mathcal{W} \in Z^p(X \times \mathbb{P}^1_{/\mathbb{C}})$ with irreducible components $\{\mathcal{W}_i\}$, such that

$$\mathcal{Z} = \sum_i \pi_X^{\widetilde{\mathcal{W}_i}}{}_* \left\{ \pi_{\mathbb{P}^1}^{\widetilde{\mathcal{W}_i}}{}^* \left(\{0\} - \{\infty\}\right) \right\}.$$

(Here e.g. $\pi_X^{\widetilde{\mathcal{W}_i}} = \pi_X \circ \iota^{\widetilde{\mathcal{W}_i}}$, with $\iota^{\widetilde{\mathcal{W}_i}} : \widetilde{\mathcal{W}_i} \twoheadrightarrow \mathcal{W}_i \subseteq X \times \mathbb{P}^1$ the normalization.) Setting $\Gamma = \sum_i \pi_X^{\widetilde{\mathcal{W}_i}}{}_* \left\{ \pi_{\mathbb{P}^1}^{\widetilde{\mathcal{W}_i}} \left(\overrightarrow{\infty.0}\right) \right\}$, for $\overrightarrow{\infty.0}$ a path on $\mathbb{P}^1$ (such as $\mathbb{R}^-$), gives $\partial\Gamma = \mathcal{Z}$ and $\mathcal{Z} \overset{\text{hom}}{\equiv} 0$. Thus (using Stokes's theorem) the fundamental (or cycle-) class induces a well-defined map

$$cl_X : CH^p(X_K) \overset{[\cdot]}{\to} Hg^p(X).$$

For 0-cycles this is $\deg : CH_0(X) \to H^{2d}(X) \cong \mathbb{Q}$ sending $\sum q_i z_i \mapsto \sum q_i$.

Above we showed how to obtain a map from $\{\ker([\cdot]) \subseteq Z^p(X_K)\} = Z^p_{\text{hom}}(X_K) \overset{AJ_X}{\to} J^p(X)$, by integration over an arbitrary chain $\partial^{-1}\mathcal{Z}$ bounding on $\mathcal{Z}$. Choosing now $\partial^{-1}\mathcal{Z} := \Gamma$ (still assuming $\mathcal{Z} \overset{\text{rat}}{\equiv} 0$) and $\omega \in \ker(d) \subseteq \Gamma\left(X, F^{d-p+1}\Omega^{2d-2p+1}_{X^\infty}\right)$, we have $\int_\Gamma \omega = \int_\infty^0 \sum_i \pi_{\mathbb{P}^1}^{\widetilde{\mathcal{W}_i}}{}_* \pi_X^{\widetilde{\mathcal{W}_i}}{}^* \omega$. Generically fibers of $\widetilde{\mathcal{W}_i} \to \mathbb{P}^1$ have dimension $d-p$; so intuitively $(\pi_{\mathbb{P}^1}^{\widetilde{\mathcal{W}_i}})_*$ (which integrates fiberwise) eats up $(d-p)$ each of dz's and $d\bar{z}$'s, and the integrand is d-closed of type $(1,0)$ on $\mathbb{P}^1$ (hence $\bar{\partial}$-closed). Now the push-forward *a priori* gives us a current, but a regularity lemma for $\bar{\partial}$ shows it is C^∞. So $\bar{\partial}$-closed + $C^\infty \implies$ holomorphic, which makes the integrand *zero* since $\Omega^1(\mathbb{P}^1) = \{0\}$, and thus $AJ_X(\mathcal{Z}) = 0 \in J^p(X)$. We have shown AJ_X is well-defined modulo $\overset{\text{rat}}{\equiv}$; this is Griffiths's Abel-Jacobi map

$$AJ_X : CH^p_{\text{hom}}(X_K) \to J^p(X)$$

(e.g., see **[Gr]**). For 0-cycles this is often called the Albanese, $Alb : CH_0^{\text{hom}}(X) \to \frac{\{\Omega^1(X)\}^\vee}{\operatorname{im} H_1(X,\mathbb{Q})}$, sending e.g. $\mathcal{Z} = p - q \longmapsto \left\{\omega \mapsto \int_q^p \omega\right\}$.

The Deligne cycle-class map

$$c_{\mathcal{D}} : CH^p(X_K) \to H^{2p}_{\mathcal{D}}(X^{\text{an}}_{\mathbb{C}}, \mathbb{Q}(p))$$

sends $\mathcal{Z}$ to the D-cohomology class of $\left((2\pi\sqrt{-1})^p \mathcal{Z}_{\text{top}},\ \delta_{\mathcal{Z}},\ 0\right)$ in the complex at the beginning of (§2.4). It allows one to write *cl* and *AJ* "in one piece" (in the sense of the short exact sequence). For Y smooth quasiprojective one has a corresponding $c_{\mathcal{H}}$ to $H^{2p}_{\mathcal{H}}$ and a diagram (with $\bar{Y}$ smooth)

$$\begin{array}{ccc} CH^p(\bar{Y}) & \xrightarrow{c_{\mathcal{D}}} & H^{2p}_{\mathcal{D}}(\bar{Y}_{\mathbb{C}}, \mathbb{Q}(p)) \\ \downarrow & & \downarrow \\ CH^p(Y) & \xrightarrow{c_{\mathcal{H}}} & H^{2p}_{\mathcal{H}}(Y_{\mathbb{C}}, \mathbb{Q}(p)), \end{array}$$

hence a (very useful) map

$$\underline{c_{\mathcal{D}}} : CH^p(Y) \to \underline{H^{2p}_{\mathcal{D}}}(Y_{\mathbb{C}}, \mathbb{Q}(p)).$$

3.2. Inadequacy of the classical invariants. Now we want to ask how good a job, in the case of 0-cycles, our two maps cl_X and Alb_X (or just $c_{\mathcal{D}}$) do in detecting $\overset{\text{rat}}{\equiv}$-classes.

Abel's Theorem. $X_{\mathbb{C}}$ a curve $\implies Alb_X$ is an isomorphism.

For $d = 1$, a great job; but for $d > 1$, miserable in general:

Theorem (Voisin, [V1]). Let $X_{\mathbb{C}} \subseteq \mathbb{P}^3$ be a smooth projective (hyper)surface, very general[3] of degree $D \geq 7$. Then no 2 points on X are rationally equivalent.

Now $H^1(X) = 0 \implies \Omega^1(X) = 0 \implies Alb_X = 0$, and thus cannot detect $[p - q](\neq 0) \in CH_0^{\text{hom}}(X)$. Fakhruddin (in [**Fa**]) has generalized this; the following is one case of his result:

Theorem (Fakhruddin). Given integers $d \geq 2$, $n \geq 2$; then for D sufficiently large, a generic degree D hypersurface $X_{\mathbb{C}} \subseteq \mathbb{P}^{d+1}$ has the following property: the classes of any n distinct points in $CH_0(X)$ are linearly independent. (Voisin: $d = 2$, $n = 2$.)

Here is another instance of nonvanishing of $\ker(Alb_X)$:

Theorem (Bloch, [B1]): Let $A^d_{\mathbb{C}}$ be a d-dimensional abelian variety. The group law induces a Pontryagin product "$*$" on cycles, and

$$CH_0^{\text{hom}}(A^d)^{*i} \begin{cases} = 0 & \text{for } i > d \\ \neq 0 & \text{for } i \leq d \end{cases}.$$

But already $CH_0^{\text{hom}}(A^d)^{*2} \subseteq \ker(Alb)$ (in fact $=$) by the "parallelogram law": say

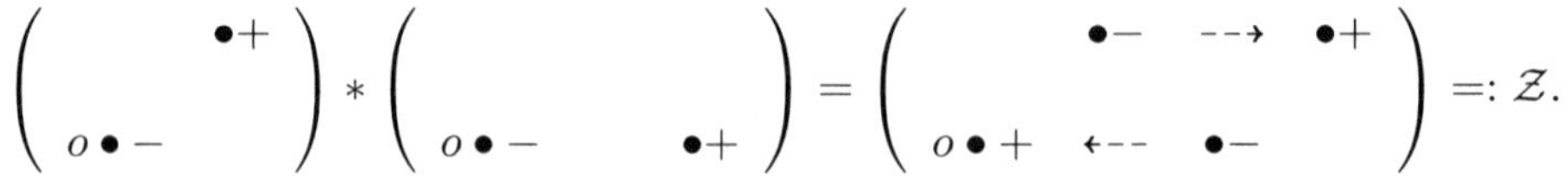

Then choosing $\partial^{-1}\mathcal{Z}$ as indicated (dotted arrows) and $\omega \in \Omega^1(A^d)$, $\int_{\partial^{-1}\mathcal{Z}} \omega = 0$ by cancellation.

The following result offers a partial "explanation" of the Albanese kernel: it's not just the 1-forms that control $\overset{\text{rat}}{\equiv}$ of cycles.

[3] i.e., of maximal transcendence degree in the parameter space

Mumford's Theorem ([Mu]). Let $X_{\mathbb{C}}$ be a smooth projective surface ($d = 2$) with $\Omega^2(X) \neq 0$. Then $\ker(Alb) \subseteq CH_0^{\text{hom}}(X)$ is "infinite-dimensional".

REMARK 3.1. (i) Roitman ([**R1**]) generalized this to $d \geq 2$: one has the same result if $\Omega^i(X) \neq 0$ for any $i > 1$. Also see [**L2**] for proofs of both.

(ii) "Infinite-dimensionality" means $\ker(Alb)$ "cannot be parametrized by a finite-dimensional algebraic variety" — concretely, this comes from increasing without bound the number of points z_i involved in 0-cycles $\mathcal{Z} = \sum q_i z_i$. This gives us as many parameters as we like, and the *key* is that rational equivalence (i.e., the dimension of the subspace of relations) doesn't keep up.

(iii) This theorem and Griffiths's result that $\overset{\text{hom}}{\equiv}$ and $\overset{\text{alg}}{\equiv}$ are different for higher-dimensional (not 0-)cycles have motivated vast quantities of research aimed at understanding structure / producing examples for well over 3 decades.

Together with Mumford, the following "converse" suggests that $\Omega^2(X)$ (or more accurately, $H^2(X)/F_h^1 H^2(X)$) "controls" the Albanese kernel.

Bloch Conjecture ([B2]). Let $X_{\mathbb{C}}$ be a smooth projective surface with $\Omega^2(X) = 0$. Then $\ker(Alb) = 0$.

More generally, for $d (= dim(X)) \geq 2$ this should hold if $\Omega^i(X) = 0$ for $1 < i \leq d$.

EXAMPLE 3.2. Bloch-Kas-Lieberman ([**BKL**]) proved BC for surfaces of Kodaira dimension $\kappa < 2$.

EXAMPLE 3.3. Consider a smooth hypersurface $X_{(\mathbb{C})} \subseteq \mathbb{P}^{d+1}$ of degree $D \leq d+1$. Then $\deg(K_X) = D - d - 2 < 0 \implies \Omega^d(X) = H^0(K_X) = 0$; also, $\Omega^i(X) = 0$ for $1 < i < d$ by the Lefschetz hyperplane theorem.

Theorem (Roitman, [R2]). For X as just described, $CH_0^{\text{hom}}(X) = 0$ (any two points are $\overset{\text{rat}}{\equiv}$).

That is, the BC holds for projective hypersurfaces (in fact, complete intersections) of low degree. We refer to [**L3**] (or [**R2**], [**L2**]) for a full proof, but explain things for the case $1 < D \leq d$ (for simplicity leaving out $D = d+1$). For any point $p \in X(\mathbb{C})$, the set of lines on X through p is described (in terms of local affine coordinates about p) as a complete intersection of multidegree $(1, 2, \ldots, D)$ in $\mathbb{P}^d$; since $D \leq d$ this is nonempty. So take individual lines L_p, $L_q \subseteq X$ through any given p, q; since $H^2(\mathbb{P}^{d+1}) \cong \mathbb{Q}$ and $CH_1^{\text{hom}}(\mathbb{P}^{d+1}) = \{0\}$, $L_p \overset{\text{rat}}{\equiv} L_q$ on $\mathbb{P}^{d+1}$. That is, $\exists$ surfaces $Y_i \subseteq \mathbb{P}^{d+1}$ and $F_i \in \mathbb{C}(\widetilde{Y}_i)^*$ with $\sum_i \iota_*^i(F_i) = L_p - L_q$. (We may also arrange that the Y_i and $Y_i \cap Y_j$ intersect X properly; e.g. a chain of 3 $\mathbb{P}^2$'s will suffice.) Restricting this rational equivalence to X we get curves $\mathcal{C}_i := X \cap Y_i \subseteq X$ and $f_i \in \mathbb{C}(\widetilde{\mathcal{C}}_i)^*$ with $[0 \overset{\text{rat}}{\equiv}] \sum' \iota_*^i(f_i) = \sum_{j=1}^D P_j - \sum_{j=1}^D Q_j$ where $P_j \in L_p$, $Q_j \in L_q$. Obviously $P_j \overset{\text{rat}}{\equiv} p$ and $Q_j \overset{\text{rat}}{\equiv} q$, so $D \cdot p \overset{\text{rat}}{\equiv} D \cdot q$. In fact, $CH_0^{\text{hom}}(X) [= CH_0^{\text{alg}}(X)]$ is divisible and so Roitman's result holds *without* going modulo torsion. Here is an oversimplified picture with only one Y_i:

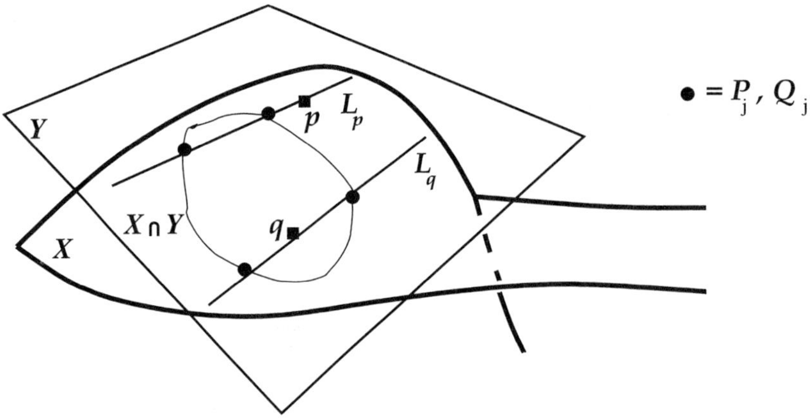

3.3. Albanese kernel and transcendence degree (Part I). It is instructive to consider the form of the cycles (in $\ker(Alb)$) guaranteed by (Roitman's generalization of) Mumford in case $X_{\mathbb{C}}$ has a model $/\bar{\mathbb{Q}}$. The first clue is given by:

Bloch-Beilinson Conjecture: Let X (smooth projective) be defined $/\bar{\mathbb{Q}}$. Then

$$c_{\mathcal{D}} : CH^p(X_{/\bar{\mathbb{Q}}}) \to H^{2p}_{\mathcal{D}}(X^{\mathrm{an}}_{\mathbb{C}}, \mathbb{Q}(p))$$

(or equivalently, $AJ : CH^p_{\mathrm{hom}}(X_{/\bar{\mathbb{Q}}}) \to J^p(X)$) is injective.

REMARK 3.4. Here one can replace $\bar{\mathbb{Q}}$ by a number field, bearing in mind that in this paper CH is taken modulo torsion. Otherwise the statement is *false*: for example, Murre and Ramakrishnan ([**MR**]) have shown that for any prime p there exists an elliptic curve E over a number field k and points $P, Q \in E(k)$ such that $(P,Q) - (0,Q) - (P,0) + (0,0) \in \ker(Alb) \subseteq CH^{\mathrm{hom}}_0(E \times E_{/k})$ is p-torsion.

For 0-cycles, $CH^{\mathrm{hom}}_0(X_{/\bar{\mathbb{Q}}})$ consists of $\sum q_i z_i$ with coordinates of all z_i in $\bar{\mathbb{Q}}$. If BBC holds, any such cycle in $\ker(Alb)$ is $\overset{\mathrm{rat}}{\equiv} 0$. So if $H_1(X) = 0$ (e.g. for $X_{(\mathbb{C})}$ simply connected), then $Alb_X = 0$ and BBC $\implies$ $CH^{\mathrm{hom}}_0(X_{/\bar{\mathbb{Q}}}) = 0$. Hence BBC implies that on a projective complete intersection (or K3, or CY) defined $/\bar{\mathbb{Q}}$, any 2 points $\in X(\bar{\mathbb{Q}})$ are rationally equivalent. This does not contradict Voisin's result because there X is not defined $/\bar{\mathbb{Q}}$ (hence doesn't even have $\bar{\mathbb{Q}}$-points).

The flip side of this is that according to BBC, $\ker(Alb) \subseteq CH^{\mathrm{hom}}_0(X_{\mathbb{C}})$ (for X defined $/\bar{\mathbb{Q}}$) is generated by cycles whose "minimal field of definition" $K \subseteq \mathbb{C}$ has $\mathrm{trdeg}(K/\bar{\mathbb{Q}}) \geq 1$. (We simply call this the transcendence degree of the cycle; here we are only considering X with a model $/\bar{\mathbb{Q}}$ because otherwise *all* cycles on X have $\mathrm{trdeg}_{/\bar{\mathbb{Q}}} \geq 1$.) To write down such cycles, we first describe the "very general" points we require in a broader context that will be used in our discussion of spreads below.

3.4. Interlude on points. Let $\mathcal{S}/\bar{\mathbb{Q}}$ be smooth projective, and choose some affine Zariski open subset $\mathcal{S}_0/\bar{\mathbb{Q}}$. The embedding $\bar{\mathbb{Q}}[\mathcal{S}_0] \hookrightarrow \bar{\mathbb{Q}}(\mathcal{S}_0) \cong \bar{\mathbb{Q}}(\mathcal{S})$ of the (affine) coordinate ring into its function field induces a morphism $\mathrm{Spec}\,\bar{\mathbb{Q}}(\mathcal{S}) \to \mathrm{Spec}\,\bar{\mathbb{Q}}[\mathcal{S}_0]$. Composing this with the inclusion $\mathcal{S}_0 \subseteq \mathcal{S}$ yields a morphism $\mathrm{Spec}\,\bar{\mathbb{Q}}(\mathcal{S}) \to \mathcal{S}$ which we call p_g, the "generic point" of $\mathcal{S}$. There are two totally different ways to get "points" on $\mathcal{S}_{\mathbb{C}}$ out of this.

(1). Set $p_g^{\mathbb{C}} := p_g \times_{\bar{\mathbb{Q}}} \operatorname{Spec}\mathbb{C}$ (i.e. $\operatorname{Spec}\mathbb{C}(\mathcal{S}) \to \mathcal{S}_{\mathbb{C}}$), the generic point of $\mathcal{S}_{\mathbb{C}}$. We approximate both generic points by shrinking $\mathcal{S}$ as follows: formally write $\eta_{\mathcal{S}} := \varprojlim \mathcal{U}$ (limit over $\mathcal{U}/\bar{\mathbb{Q}} \subseteq \mathcal{S}$ affine Zariski open) and $\eta_{\mathcal{S}}^{\mathbb{C}} := \varprojlim \mathcal{U}_{\mathbb{C}}$ (limit *still* over $\mathcal{U}$ defined $/\bar{\mathbb{Q}}$; $\mathcal{U}_{\mathbb{C}}$ is just $\mathcal{U} \times_{\bar{\mathbb{Q}}} \mathbb{C}$). The limit has practical meaning only under some functor $\mathfrak{F}$, usually contravariant (like H^*, $H^*_{\mathcal{H}}$, CH^*) so that $\mathfrak{F}(\eta_{\mathcal{S}}^{[\mathbb{C}]}) := \varinjlim \mathfrak{F}(\mathcal{U}_{[\mathbb{C}]})$. (We note in particular that direct limits are exact in the category of abelian groups.) The reason for introducing the η's is two-fold. On the one hand, CH^* cannot distinguish between p_g (i.e., pullback to $\operatorname{Spec}\bar{\mathbb{Q}}(\mathcal{S})$) and $\eta_{\mathcal{S}}$; e.g. for $X/\bar{\mathbb{Q}}$, $CH^*(X \times \eta_{\mathcal{S}/\bar{\mathbb{Q}}}) := \varinjlim CH^*(X \times \mathcal{U}_{/\bar{\mathbb{Q}}}) \cong CH^*(X \times \operatorname{Spec}\bar{\mathbb{Q}}(\mathcal{S})_{/\bar{\mathbb{Q}}}) \cong CH^*(X_{/\bar{\mathbb{Q}}(\mathcal{S})})$. But unlike $p_g^{\mathbb{C}}$, $\eta_{\mathcal{S}}^{\mathbb{C}}$ is a limit of analytic spaces (on which one can do Hodge theory); so one has singular and absolute Hodge cohomology groups, e.g. $H^*_{\mathcal{H}}(\eta_{\mathcal{S}}^{\mathbb{C}}, \mathbb{Q}(\cdot)) := \varinjlim H^*_{\mathcal{H}}(\mathcal{U}_{\mathbb{C}}^{\mathrm{an}}, \mathbb{Q}(\cdot))$. One can take the limit of the MHS's on the $H^m(\mathcal{U})$ to produce $W_\bullet$ and $F^\bullet$ on $H^m(\eta_{\mathcal{S}})$, and in particular

$$\underline{H^m}(\eta_{\mathcal{S}}) := W_m H^m(\eta_{\mathcal{S}}) = \operatorname{im}\left\{H^m(\mathcal{S}) \to \varinjlim H^m(\mathcal{U})\right\} \cong \frac{H^m(\mathcal{S})}{N^1 H^m(\mathcal{S})}$$

is a very important (finite-dimensional) HS.

(2). Given an embedding $\bar{\mathbb{Q}}(\mathcal{S}) \xrightarrow[\cong]{\mathrm{ev}} K \subseteq \mathbb{C}$ which restricts to the "identity" on $\bar{\mathbb{Q}}$, set $p := p_g \times_{\mathrm{ev}} \operatorname{Spec}\mathbb{C}$. What we mean by this slight abuse of notation is illustrated by the following pair of diagrams:

$$\begin{array}{ccc} \bar{\mathbb{Q}}(\mathcal{S}) & \hookleftarrow & \bar{\mathbb{Q}}[\mathcal{S}_0] \\ \downarrow{\scriptstyle \mathrm{ev}} & & \downarrow \\ \mathbb{C} & \xleftarrow{\widetilde{\mathrm{ev}}} & \mathbb{C}[\mathcal{S}_0] \end{array} \quad \xrightarrow{\text{induces}} \quad \begin{array}{ccc} \operatorname{Spec}\bar{\mathbb{Q}}(\mathcal{S}) & \xrightarrow{p_g} & \mathcal{S}_0 \quad \subseteq \mathcal{S} \\ \uparrow & & \uparrow \\ \operatorname{Spec}\mathbb{C} & \xrightarrow{p} & (\mathcal{S}_0)_{\mathbb{C}} \quad \subseteq \mathcal{S}_{\mathbb{C}} \end{array}$$

where $\widetilde{\mathrm{ev}}$ is (the unique lift of ev) chosen to make the first square commute. This is quite different from **(1)** in that p is 0-dimensional (over $\mathbb{C}$) and a geometric point. Now the restriction of $\widetilde{\mathrm{ev}}$ to $\bar{\mathbb{Q}}[\mathcal{S}_0] \subseteq \mathbb{C}[\mathcal{S}_0]$ factors through $\bar{\mathbb{Q}}(\mathcal{S})$, hence through $\bar{\mathbb{Q}}[\mathcal{U}]$ for any affine open $\mathcal{U} \subseteq \mathcal{S}_0$ defined $/\bar{\mathbb{Q}}$. It follows that for all such $\mathcal{U}$, p factors through all $\mathcal{U}_{\mathbb{C}}(= \mathcal{U} \times_{\bar{\mathbb{Q}}} \mathbb{C}) \subseteq \mathcal{S}_{\mathbb{C}}$. That is, p is in the complement of all (hence countably many) proper subvarieties defined $/\bar{\mathbb{Q}}$. We express this by saying $p \in \eta_{\mathcal{S}}(\mathbb{C})$, or that $p \in \mathcal{S}(\mathbb{C})$ is *very general*,[4] or of maximal transcendence degree (as its affine coordinates [in $\mathcal{S}_0$] generate over $\bar{\mathbb{Q}}$ a field of transcendence degree $\dim(\mathcal{S})$). Of course, in the above construction, p is really defined $/K$.

3.5. Albanese kernel and transcendence degree (Part II). Here now are some "canonical" examples of 0-cycles in $\ker(Alb_X)$ for X defined $/\bar{\mathbb{Q}}$. Let $\mathcal{C}_j/\bar{\mathbb{Q}}$ ($j = 1, \ldots, d$) be curves of positive genera, and select base points $o_j \in \mathcal{C}_j(\bar{\mathbb{Q}})$; set $X = \mathcal{C}_1 \times \cdots \times \mathcal{C}_d$. Take $p_j \in \mathcal{C}_j(\mathbb{C})$ very general and "algebraically independent" in the sense that $p_1 \times \cdots \times p_d \in X(\mathbb{C})$ is very general. Finally, let $\mathcal{W}_j \in CH_0^{\mathrm{hom}}(\mathcal{C}_{j/\bar{\mathbb{Q}}})$ (for each j) be such that $AJ_{\mathcal{C}_j}(\mathcal{W}_j) \neq 0$ in $J^1(\mathcal{C}_j)$. (We emphasize once more that for us torsion is zero in cycles and classes; in particular J is the rational Jacobian.)

[4]"general" is *not* the term we want here, as this denotes a point in the complement of only a *finite* union of Zariski closed subsets.

THEOREM 3.5. *For each $\ell \geq 2$, the 0-cycles*
$(A):\quad (p_1 - o_1) \times \cdots \times (p_\ell - o_\ell) \times o_{\ell+1} \times \cdots \times o_d,$
$(B):\quad (p_1 - o_1) \times \cdots \times (p_{\ell-1} - o_{\ell-1}) \times \mathcal{W}_\ell \times o_{\ell+1} \times \cdots \times o_d$
give nonzero classes in $\ker(Alb) \subseteq CH_0^{hom}(X_{\mathbb{C}})$.

REMARK 3.6. (i) For obvious reasons we call (A) "ℓ-box" (e.g., for $\ell = d = 2$ it is $(p_1,p_2)-(p_1,o_2)-(o_1,p_2)+(o_1,o_2))$; the respective transcendence degrees of (A) and (B) are clearly ℓ and $\ell-1$. (A) is well-known, while (B) follows for $n = \ell = 2$ from **[RS]** and in general from **[K1]** (see Theorem 6.1 below).

(ii) Any abelian variety A^d is covered by such a product of curves, and cycles of the above two types (A) and (B) project to nontrivial elements of $CH_0^{\hom}(A^d)^{*\ell}$ from the Bloch theorem.

(iii) A simple example of (B) with $d = \ell = 2$ is with the $\mathcal{C}_j$ both elliptic curves $E_i/\bar{\mathbb{Q}}$, $P \in E_1(\mathbb{C})$ very general, and $Q \in E_2(\bar{\mathbb{Q}})$ of infinite rank. Taking $\mathcal{W}_2 = \langle Q\rangle - \langle O\rangle$, we get $(\langle P\rangle - \langle O\rangle) \times (\langle Q\rangle - \langle O\rangle) \overset{\text{rat}}{\not\equiv} 0$ (in the sense of this paper, i.e. modulo torsion).

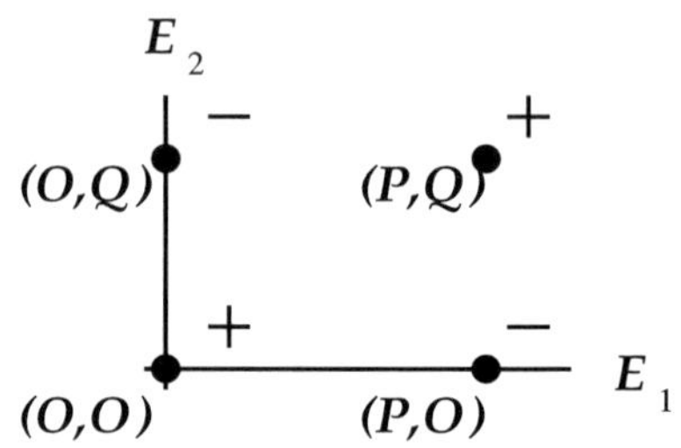

We will investigate cycles of the form e.g. $(p_1 - o_1) \times (p_1 - o_1)$ on $\mathcal{C}_1 \times \mathcal{C}_1$ in §6.2 below. On the other hand, $\mathcal{W}_1 \times \mathcal{W}_2$ "should" be $\overset{\text{rat}}{\equiv} 0$ since it is in $\ker(Alb) \cap CH_0^{\hom}(\mathcal{C}_1 \times \mathcal{C}_{2/\bar{\mathbb{Q}}})$, though this is only conjectural — even if $\mathcal{C}_1$ and $\mathcal{C}_2$ are elliptic curves.

Finally, Griffiths-Green-Paranjape **[GGP]** found an interesting example for any smooth projective surface $X/\bar{\mathbb{Q}}$ with $\Omega^2(X) \neq 0$. If $H_1(X) = 0$, here is how this works: take a Lefschetz pencil of X, which we write $\mathbb{P}^1 \overset{h}{\leftarrow} \mathfrak{H} \overset{\mathcal{P}}{\rightarrow} X$ ($\mathcal{P}$ is a blow-up) with $\mathcal{C}_t := \mathcal{P}\left(h^{-1}(t)\right) \subseteq X$ for $t \in \mathbb{P}^1$. For each t let $p_t \in X$ denote the inclusion of a very general point $\in \mathcal{C}_t(\mathbb{C})$, and $o \in X(\bar{\mathbb{Q}})$ lie in the base locus. Then for some $t_0 \in \mathbb{P}^1(\bar{\mathbb{Q}})$, $0 \overset{\text{rat}}{\not\equiv} p_{t_0} - o$ is a nontrivial cycle of transcendence degree 1 in the Albanese kernel. We will generalize this example in §7.3 below.

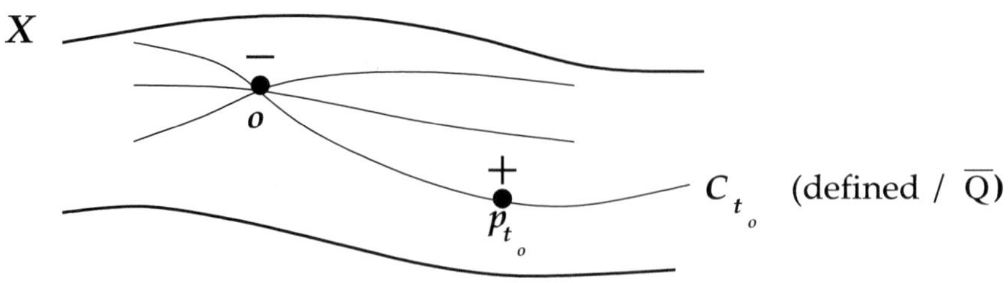

3.6. Bloch-Beilinson filtration and Chow-Künneth decomposition . For 0-cycles on a surface X (whether defined $/\bar{\mathbb{Q}}$ or not), cl (or deg) maps to $H^4(X)$, and AJ (or Alb) to $J^2(X)$ — which involves $H^3(X)$. Since $\ker(Alb)$ (which is conjecturally generated by cycles of $\mathrm{trdeg}_{/\bar{\mathbb{Q}}} \geq 1$) seems to depend on part of $H^2(X)$, we expect any invariant for it should involve $H^2(X)$.

More generally, for X smooth projective (defined over a subfield of $\mathbb{C}$), Bloch and Beilinson independently predicted a filtration $F^{\bullet}_{\mathrm{BB}}$ on $CH^p(X_{\mathbb{C}})$ with graded pieces $Gr^i_{F_{\mathrm{BB}}} CH^p(X_{\mathbb{C}})$ (a) generated by cycles of $\mathrm{trdeg}_{/\bar{\mathbb{Q}}} \geq i-1$ and (b) completely described by invariants involving $H^{2p-i}(X)$ (and no other $H^j(X)$). To be precise:

DEFINITION 3.7. A *Bloch-Beilinson filtration* (BBF) is a (conjectural) system of descending filtrations on all Chow groups of smooth projective varieties $/\mathbb{C}$. It must respect the intersection product (in the sense $F^i_{\mathrm{BB}} CH^p \cdot F^j_{\mathrm{BB}} CH^q \subseteq F^{i+j}_{\mathrm{BB}} CH^{p+q}$) and morphisms induced by correspondences ($\langle \Gamma \rangle \in CH^{p-q+d_X}(Y \times X) \implies \langle \Gamma \rangle_* F^i_{\mathrm{BB}} CH^q(Y) \subseteq F^i_{\mathrm{BB}} CH^p(X)$). One also demands $F^0_{\mathrm{BB}} CH^p = CH^p$, $F^{\ell} CH^p = \{0\}$ for ℓ sufficiently large, $F^1_{\mathrm{BB}} CH^p = CH^p_{\mathrm{hom}}$, and $F^2_{\mathrm{BB}} CH^p \subseteq \ker(AJ)$ (with $F^2_{\mathrm{BB}} CH^p \cap CH^p_{\mathrm{alg}} = \ker(AJ) \cap CH^p_{\mathrm{alg}}$).

But these last two (at least, the $\subseteq$'s) are really part of a larger demand corresponding to (b) above. For its statement, $[\Delta_X] \in H^{2d_X}(X \times X)$ must have "algebraic Künneth components" — that is, there must exist algebraic cycles $\tilde{\Delta}_{X,j}$ (or $\Delta(2d_X - j, j)$) on $X \times X$ with $[\tilde{\Delta}_{X,j}] = [\Delta_X]_j \in H^{2d_X - j}(X) \otimes H^j(X)$. The last requirement for $F^{\bullet}_{\mathrm{BB}}$ (essentially $\forall\, i, j \in \mathbb{Z}$) is then that the action of $\left\langle \tilde{\Delta}_{X,j} \right\rangle_*$ on $Gr^i_{F_{\mathrm{BB}}} CH^p(X)$ must be $\delta_{2p-i,\, j}$ times the identity, independent of the choice of $\{\tilde{\Delta}_{X,j}\}$. If we assume the HLC ($\Leftarrow$HC), not only do such $\{\tilde{\Delta}_{X,j}\}$ exist; they also may be chosen, for $j < p$, to be supported on $W \times X$ with $\mathrm{codim}_X(W) \geq d_X - p - 1$. For $i > p$, this yields $\left\langle \tilde{\Delta}_{X,2p-i} \right\rangle_* \Big|_{Gr^i_{F_{\mathrm{BB}}} CH^p(X)} = 0$, hence $Gr^i_{F_{\mathrm{BB}}} CH^p = 0$; and so $F^{p+1}_{\mathrm{BB}} CH^p(X) = 0$.

REMARK. The definition comes from **[J4]**.

Now, one can put multiplicative (hence ring) structures on $CH^{d_X}(X \times X)$, $H^{2d_X}_{\mathrm{alg}}(X \times X)$ by $A \circ B := \pi_{13_*}(\pi_{12}^*(A) \cdot \pi_{23}^*(B))$ (using the 3 obvious projections $X \times X \times X \to X \times X$); this corresponds to composing endomorphisms of $CH^*(X)$, $H^*(X)$ induced by correspondences. Under $\circ$, the Künneth components $\left\{[\tilde{\Delta}_{X,j}]\right\}$ are orthogonal idempotents in (still assuming HLC) $H^{2d_X}_{\mathrm{alg}}(X \times X)$ summing to $[\Delta_X]$; but the corresponding statements on the $CH^{d_X}(X \times X)$-level are much stronger, and false for the $\left\{\left\langle \tilde{\Delta}_{X,j} \right\rangle\right\}$ as we have described them. However, *if $F^{\bullet}_{\mathrm{BB}}$ exists as described above, then $CH^{d_X}_{hom}(X \times X)$ becomes a nilpotent ideal*[5] in $CH^{d_X}(X \times X)$. It follows (cf. **[J1]**, **[L2]**) that lifts $\{\Delta_{X,j}\}$ from the quotient $\frac{CH^{d_X}(X \times X)}{CH^{d_X}_{\mathrm{hom}}(X \times X)} \cong H^{2d_X}_{\mathrm{alg}}(X \times X)$ to $CH^{d_X}(X \times X)$ can be chosen so as to preserve the desired relations: (i) $\{\langle \Delta_{X,j} \rangle_*\}$ are orthogonal idempotents[6] under $\circ$, (ii) $\Delta_X \overset{\mathrm{rat}}{\equiv} \sum_j \Delta_{X,j}$, and (iii) $[\Delta_{X,j}] = [\Delta_X]_j$ as above.

[5]It is known that $CH^{d_X}_{\mathrm{alg}}(X \times X)$ is nilpotent; this clearly follows from §6.2 below.

[6]usually written $\{\pi_j\}$ in the literature, but we avoid this notation.

DEFINITION 3.8. A collection of cycles $\{\Delta_{X,j}\}_{j=0}^{2d_X} \subseteq Z^{d_X}(X \times X)$ with properties (i)-(iii), is called a Chow-Künneth decomposition (of Δ_X), terminology due to Murre. We shall call it also *good* (for our purposes with 0-cycles) if $\left\{\langle\Delta_{X,j}\rangle_*\right\}_{j<d_X}$ act as 0 on $CH_0(X)$ and $\left\{\ker\langle\Delta_{X,2d_X}\rangle_* \subseteq CH_0(X)\right\} = CH_0^{\hom}(X)$.

With a little more work the above discussion yields:

PROPOSITION 3.9. *HLC* $+ \exists$ *of* $F^\bullet_{BB} \implies \exists\ (\forall\ X)$ *of good C-K decomposition satisfying moreover* $\langle\Delta_{X,j}\rangle_*|_{Gr^i_{F_{BB}}CH^p(X)} = \delta_{2p-i,\,j}\cdot Id.$

However, one can construct good C-K decompositions *directly* (with no assumptions) for X a curve, surface, abelian variety, complete intersection, or arbitrary product of these.

4. Higher cycle- and AJ-classes

4.1. Spreads. We are about to study several actual filtrations on Chow groups, each of which gives a Bloch-Beilinson filtration under some conjectural conditions. For the construction of many of these, the following idea is crucial.

Let $K \subseteq \mathbb{C}$ be finitely generated $/\bar{\mathbb{Q}}$. Then using a transcendence basis, one constructs (e.g., see **[K1]** Lemma 1(a)) $\mathcal{S}/\bar{\mathbb{Q}}$ smooth projective and a very general point $s_0 \in \mathcal{S}(\mathbb{C})$ such that $\mathrm{ev}_{s_0} : \bar{\mathbb{Q}}(\mathcal{S}) \overset{\cong}{\to} K$ (in fact, $s_0 \in \mathcal{S}(K)$).

Assume first that X is defined $/\bar{\mathbb{Q}}$ and $\mathcal{Z} \in Z^p(X_K)$. Changing coefficients of $\mathcal{Z}$'s defining (homogeneous) equations from K to $\bar{\mathbb{Q}}(\mathcal{S})$, we have $\mathcal{Z}_g := \mathcal{Z} \times_{\mathrm{ev}_{s_0}^{-1}} \operatorname{Spec}\bar{\mathbb{Q}}(\mathcal{S}) \in Z^p(X_{/\bar{\mathbb{Q}}(\mathcal{S})})$. Now rational functions (in $\bar{\mathbb{Q}}(\mathcal{S})$) have denominators; we may clear these (nonuniquely; this introduces ambiguities) to yield bihomogeneous equations cutting out $\bar{\mathfrak{Z}} \in Z^p(X \times \mathcal{S}_{/\bar{\mathbb{Q}}})$. Owing to the ambiguities the $\overset{\text{rat}}{\equiv}$-class of $\bar{\mathfrak{Z}}$ is not well-defined, in contrast to that of its restriction $\mathfrak{Z} \in Z^p(X \times \eta_{\mathcal{S}/\bar{\mathbb{Q}}})$. (In particular, one may freely modify $\bar{\mathfrak{Z}}$ by cycles supported on $X \times D$, for $D \subseteq \mathcal{S}$ an arbitrary divisor defined $/\bar{\mathbb{Q}}$.) We refer to $\mathfrak{Z}$ and $\bar{\mathfrak{Z}}$ as the $\bar{\mathbb{Q}}$-spread and *complete* $\bar{\mathbb{Q}}$-spread of $\mathcal{Z}$; clever *choices* of the latter play a role in proofs.

Once one has spread out, going back is easy: tensor everything with K and pull back (restrict) along the inclusion $X = X \times \{s_0\} \hookrightarrow X \times \eta_{\mathcal{S}}$. As we can spread and restrict rational equivalences $\{(Y_i, f_i)\}$ too, $\mathcal{Z} \overset{\text{rat}}{\equiv} 0 \iff \mathfrak{Z} \overset{\text{rat}}{\equiv} 0$.

Now, different very general points $s \in \mathcal{S}(\mathbb{C})$ correspond to different embeddings $\mathrm{ev}_s : \bar{\mathbb{Q}}(\mathcal{S}) \hookrightarrow \mathbb{C}$ respecting $\bar{\mathbb{Q}}$. So if we tensor everything with $\mathbb{C}$ and set $\mathcal{Z}_s := \mathfrak{Z}|_{X\times\{s\}}$, letting s vary shows how the original cycle deforms in X as its defining equations' coefficients change along with $\mathrm{ev}_s(\bar{\mathbb{Q}}(\mathcal{S})) =: K_s$.

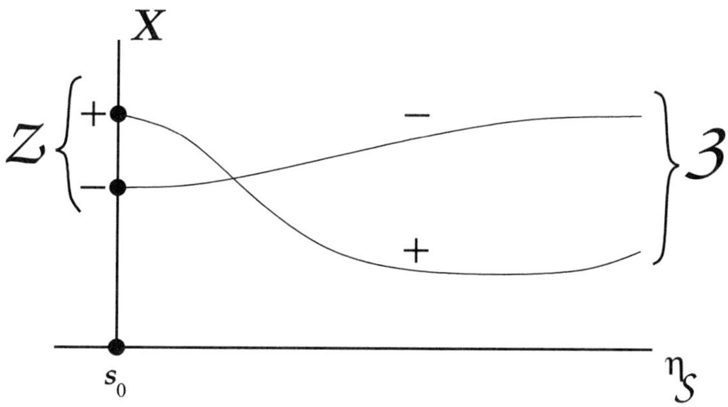

Clearly, spreading out exchanges field-theoretic information for additional geometric structure. That extra structure means that $c_{\mathcal{H}}(\mathfrak{Z})$ is going to capture more information than $c_{\mathcal{D}}(\mathcal{Z})$, and this is what the Lewis, M. Saito, and Griffiths-Green filtrations all capitalize on.

EXAMPLE 4.1. $X/\bar{\mathbb{Q}}$, $\mathcal{Z} = p \in X(\mathbb{C})$ very general $\implies$ $\mathcal{S} = X$ and (we can choose) $\bar{\mathfrak{Z}} = \Delta_X \in Z^d(X \times X_{/\bar{\mathbb{Q}}})$.

EXAMPLE 4.2. Consider a smooth curve $Y/\bar{\mathbb{Q}}$ on a surface $X/\bar{\mathbb{Q}}$, $o \in Y(\bar{\mathbb{Q}})$ and $p \in Y(\mathbb{C})$ very general; of course, p has a model $/K\,(\subseteq \mathbb{C})$ s.t. $K \cong \bar{\mathbb{Q}}(Y)$. We write $\iota : Y \hookrightarrow X$ and define $\mathcal{Z} := \iota_*(p - o) \in Z_0^{\text{hom}}(X_K)$. Taking $\mathcal{S} = Y$, we can choose $\bar{\mathfrak{Z}} = \mathsf{I}_*(\Delta_Y - Y \times \{o\}) \in Z^2(X \times Y)$ where $\mathsf{I} : Y \times Y \hookrightarrow X \times Y$.

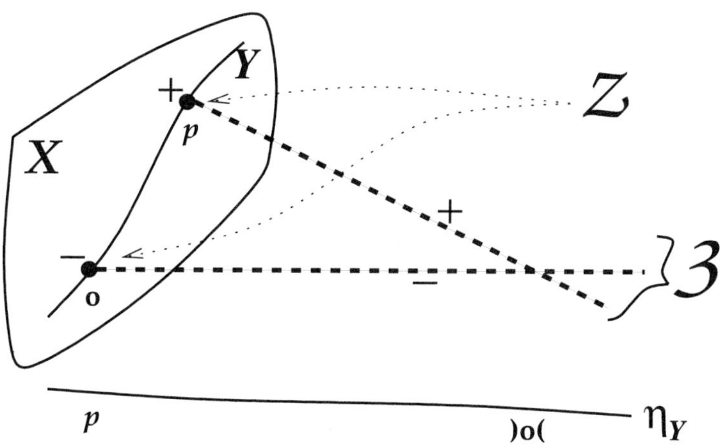

Now, so far we have only considered the case where "X does not spread". If X does *not* have a model $/\bar{\mathbb{Q}}$, it sits in something that does; and one constructs the $\bar{\mathbb{Q}}$-spread $\mathfrak{X}(\xrightarrow{\pi} \eta_S)$ of X_K just as one did $\mathfrak{Z}$. One can always choose $\bar{\mathfrak{X}}$ to be smooth, but not $\bar{\pi} : \bar{\mathfrak{X}} \to \mathcal{S}$ (which we note is of relative dimension d). So in general $\mathcal{Z}$ has $\bar{\mathbb{Q}}$-spreads $\mathfrak{Z} \in Z^p(\mathfrak{X}/\bar{\mathbb{Q}}) \leftarrow Z^p(\bar{\mathfrak{X}}/\bar{\mathbb{Q}}) \ni \bar{\mathfrak{Z}}$. (As above $\bar{\mathfrak{Z}}$ is defined modulo the addition of cycles supported on $\pi^{-1}(D)$, for any codim-1 $D_{/\bar{\mathbb{Q}}} \subseteq \mathcal{S}$.) We always refer to $\pi^{-1}(\mathcal{U} \subseteq \mathcal{S})$ as $\mathfrak{X}_{\mathcal{U}}$, so $\mathfrak{X}_{\mathcal{S}} = \bar{\mathfrak{X}}$ and $\mathfrak{X}_{\eta_{\mathcal{S}}} = \mathfrak{X}$; we write X_s for $\pi^{-1}(s)$. Of course, if on the other hand X is defined $/\bar{\mathbb{Q}}$, the ($\bar{\mathbb{Q}}$-)spread of X_K is $\mathfrak{X} = X \times \eta_{\mathcal{S}}$.

4.2. Lewis filtration. We will work with this for the remainder of the paper.

By identifying formally $\mathfrak{X}$ and $\varprojlim \pi^{-1}(\mathcal{U})$ (as we did above for $\eta_{\mathcal{S}}$ and $\varprojlim \mathcal{U}$), we obtain a composition

$$CH^p(X_{/K}) \underset{\text{spread}}{\overset{\cong}{\longrightarrow}} CH^p(\mathfrak{X}_{/\bar{\mathbb{Q}}}) \underset{c_{\mathcal{D}}}{\longrightarrow} \underline{H_{\mathcal{D}}^{2p}}(\mathfrak{X}, \mathbb{Q}(p))$$

which we denote by ψ. Lewis **[L1]** puts a Leray filtration $\mathcal{L}^\bullet$ (induced by π) on $\underline{H_{\mathcal{D}}^{2p}}(\mathfrak{X}_{\mathbb{C}}, \mathbb{Q}(p))$ with graded pieces

$$0 \to Gr_{\mathcal{L}}^{i-1} \underline{J^p}(\mathfrak{X}) \overset{\beta_{i-1}}{\to} Gr_{\mathcal{L}}^i \underline{H_{\mathcal{D}}^{2p}}(\mathfrak{X}_{\mathbb{C}}, \mathbb{Q}(p)) \overset{\alpha_i}{\to} Gr_{\mathcal{L}}^i \underline{Hg^p}(\mathfrak{X}) \to 0 \tag{4.1}$$

where

$$Gr_{\mathcal{L}}^i \underline{Hg^p}(\mathfrak{X}) := Hom_{\text{MHS}} \left(\mathbb{Q}(-p),\, [W_{2p}] H^i(\eta_{\mathcal{S}}, R^{2p-i}\pi_* \mathbb{Q}) \right)$$

and

$$Gr_{\mathcal{L}}^{i-1} \underline{J^p}(\mathfrak{X}) := \frac{Ext^1_{\text{MHS}} \left(\mathbb{Q}(-p),\, W_{2p-1} H^{i-1}(\eta_{\mathcal{S}}, R^{2p-i}\pi_* \mathbb{Q}) \right)}{\text{im}\, Hom_{\text{MHS}} \left(\mathbb{Q}(-p),\, Gr^W_{2p} H^{2p-1}(\mathfrak{X}, \mathbb{Q}) \right)}.$$

The notational abuse taking place here (e.g., $Gr^W_{2p} H^{2p-1}(\mathfrak{X}, \mathbb{Q})$ on its own is not finite-dimensional) refers as usual to the obvious limits. We define the Lewis filtration

$$\mathcal{L}^i CH^p(X_{/K}) := \psi^{-1} \left(\mathcal{L}^i \underline{H_{\mathcal{D}}^{2p}}(\mathfrak{X}_{\mathbb{C}}, \mathbb{Q}(p)) \right),$$

and for $\mathcal{Z} \in \mathcal{L}^i CH^p$ the ith higher cycle-class

$$cl_X^i(Z) := \alpha_i \left(Gr_{\mathcal{L}}^i \psi(\mathcal{Z}) \right) = \alpha_i \left(Gr_{\mathcal{L}}^i \underline{c_{\mathcal{D}}}(\mathfrak{Z}) \right) =: [\mathfrak{Z}]_i.$$

If the latter vanishes, we can define an $(i-1)^{\text{st}}$ higher AJ class

$$AJ_X^{i-1}(\mathcal{Z}) = [AJ(\mathfrak{Z})]_{i-1}$$

by demanding that β_{i-1} of it gives $Gr_{\mathcal{L}}^i \psi(\mathcal{Z}) = Gr_{\mathcal{L}}^i \underline{c_{\mathcal{D}}}(\mathfrak{Z})$.[7]

We may summarize our invariants in a chart

$$\begin{array}{c|c|c|c|c}
\mathcal{L}^0 \supseteq & \mathcal{L}^1 \supseteq & \mathcal{L}^2 \supseteq & \mathcal{L}^3 \supseteq & \cdots\cdots \\
[\mathfrak{Z}]_0 & [\mathfrak{Z}]_1 & [\mathfrak{Z}]_2 & [\mathfrak{Z}]_3 & \cdots\cdots \\
- & [AJ(\mathfrak{Z})]_0 & [AJ(\mathfrak{Z})]_1 & [AJ(\mathfrak{Z})]_2 &
\end{array}$$

(which mimics that in **[GG1]** although the invariants there are only verifiably the same if HC holds). Here

$$[\mathfrak{Z}]_0 \in Hom_{\text{MHS}} \left(\mathbb{Q}(-p),\, H^0(\eta_{\mathcal{S}}, R^{2p}\pi_* \mathbb{Q}) \right) \cong Hg^p(X)$$

[7]Note that by writing

$$\psi_i : \mathcal{L}^i CH^p(X_K) \to Gr_{\mathcal{L}}^i \underline{H_{\mathcal{D}}^{2p}}(\mathfrak{X}, \mathbb{Q}(p))$$

for the map $\mathcal{Z} \mapsto Gr_{\mathcal{L}}^i \psi(\mathcal{Z})$, one trivially expresses $\mathcal{L}^\bullet$ [$= \ker(\psi_{\bullet-1})$] as a filtration by kernels of successive maps.

identifies with $[\mathcal{Z}]$, while if $[\mathfrak{Z}]_0 = [\mathfrak{Z}]_1 = 0$ then $[AJ(\mathfrak{Z})]_0$ essentially identifies with $AJ(\mathcal{Z})$ (or more precisely, with a flat normal function); certainly $\mathcal{L}^2 \subseteq \ker(AJ)$. Writing $t = \mathrm{trdeg}(K/\bar{\mathbb{Q}}) = \dim(\mathcal{S})$, we have (mostly from **[L1]**) these facts:

PROPOSITION 4.3. *(a)* $Gr^k_{\mathcal{L}} CH^p(X_K) = 0$ *if* $k > p$ *or* $k > t+1$. *One may replace* $Gr^k_{\mathcal{L}}$ *by* $\mathcal{L}^k$ *provided* $\underline{c_{\mathcal{D}}}$ *is injective.*

(b) If X has algebraic Künneth components, then $\mathcal{L}^\bullet$ *satisfies the second paragraph of Defn.* 3.7.

Here (b) comes from the fact that in (4.1) only $R^{2p-i}\pi_*\mathbb{Q}$ appears, while (a) follows from the hard and weak Lefschetz theorems (respectively). Weak Lefschetz also implies $[\mathfrak{Z}]_{t+1} = 0$; however $[AJ(\mathfrak{Z})]_t$ (hence $Gr^{t+1}_{\mathcal{L}} CH^p$) need not vanish. In connection with (a) one makes the following Bloch-Beilinson-type conjecture for quasiprojective varieties:

Conjecture (BBC^q). For $Y/\bar{\mathbb{Q}}$ smooth quasiprojective, $\underline{c_{\mathcal{D}}} : CH^p(Y_{/\bar{\mathbb{Q}}}) \to \underline{H^{2p}_{\mathcal{D}}}(Y_{\mathbb{C}}, \mathbb{Q}(p))$ is injective.

Lewis (**[L1]**) deduces BBC^q from a conjecture of Jannsen (**[J2]**); one can also prove (with some work) that it follows from BBC+HC.

Now taking a limit over $K \subseteq \mathbb{C}$ f.g. $/\bar{\mathbb{Q}}$ (and using $CH^p(X_K) \hookrightarrow CH^p(X_{\mathbb{C}})$ and the corresponding limit of $\mathcal{S}$'s) we obtain filtrations and maps

$$\psi_i : \mathcal{L}^i CH^p(X_{\mathbb{C}}) \to \varinjlim_{\mathcal{S}} Gr^i_{\mathcal{L}} \underline{H^{2p}_{\mathcal{D}}}(\mathfrak{X}_{(\eta_S)}, \mathbb{Q}(p)).$$

While we prefer to work $/K$ (and without this extra limit) in what follows, this gives a candidate Bloch-Beilinson filtration.

Theorem (Lewis, [L1]). HLC+BBC^q $\implies \mathcal{L}^\bullet$ is a BBF.

If BBC^q holds then the above invariants (see chart) completely describe $CH^p(X_K)$, and in particular $[\mathfrak{Z}]_2$ and $[AJ(\mathfrak{Z})]_1$ describe $\ker(Alb)$ for 0-cycles on a surface$/\bar{\mathbb{Q}}$.[8] For computing these, we can use $\underline{c_{\mathcal{D}}}(\mathfrak{Z}) = \mathrm{im}(c_{\mathcal{D}}(\bar{\mathfrak{Z}}))$ for a convenient choice of $\bar{\mathfrak{Z}}$.

PROPOSITION 4.4. *If* $\mathcal{Z} \in \mathcal{L}^i CH^p(X_K)$ *then* $[\mathfrak{Z}]_i$ *is the image of* $[\bar{\mathfrak{Z}}]$ *under the projection*

$$Hg^p\left(H^{2p}(\bar{\mathfrak{X}})\right) \twoheadrightarrow Hg^p\left(Gr^i_{\mathcal{L}} \underline{H^{2p}}(\mathfrak{X})\right) = Gr^i_{\mathcal{L}} \underline{Hg^p}(\mathfrak{X}).$$

Here $Gr^i_{\mathcal{L}} \underline{H^{2p}}(\mathfrak{X}) = W_{2p} H^i(\eta_S, R^{2p-i}\pi_*\mathbb{Q})$ is justifiable by **[Ar]**.

According to Propositions 4.4 and 4.3(*a*), if $[\mathfrak{Z}]_0 = [\mathfrak{Z}]_1 = \cdots = [\mathfrak{Z}]_p = 0$, then $[\bar{\mathfrak{Z}}]$ projects to 0 in $Hg^p(\underline{H^{2p}}(\mathfrak{X}))$ (one says $[\mathfrak{Z}] = 0$). It follows (using **[De**, Cor. 8.2.8]) that for some divisor $D_{/\bar{\mathbb{Q}}} \subseteq \mathcal{S}$, $[\bar{\mathfrak{Z}}]$ comes from a class in $Hg^{p-1}(H^{2p-2}(\widetilde{\mathfrak{X}_D}))$. Assuming HC, this is the class of a cycle, by whose image (in $\bar{\mathfrak{X}}$, supported on $\mathfrak{X}_D$) we may modify our choice of $\bar{\mathfrak{Z}}$ to get $[\bar{\mathfrak{Z}}] = 0$. In specific cases (like those in §6), one may be able to explicitly construct (*without* HC) a complete spread $\bar{\mathfrak{Z}}$ which is $\overset{\text{hom}}{\equiv} 0$. Here is the reason we want this property:

[8] It is worth noting that $\mathcal{L}^2 = \ker(Alb)$ holds for 0-cycles if X (of any dimension) is defined $/\bar{\mathbb{Q}}$, see **[K2]**.

PROPOSITION 4.5. *Suppose $\mathcal{Z} \in \mathcal{L}^i CH^p(X_K)$ has a homologically trivial complete $\bar{\mathbb{Q}}$-spread $\bar{\mathfrak{Z}}$. Then $[AJ(\mathfrak{Z})]_i$ is the image of $AJ_{\bar{\mathfrak{X}}}(\bar{\mathfrak{Z}})$ under the projection*

$$J^p\left(H^{2p-1}(\bar{\mathfrak{X}})\right) \twoheadrightarrow J^p\left(Gr_{\mathcal{L}}^{i-1}\underline{H^{2p-1}}(\mathfrak{X})\right) \twoheadrightarrow Gr_{\mathcal{L}}^{i-1}\underline{J^p}(\mathfrak{X}).$$

If X is defined $/\bar{\mathbb{Q}}$ then we get simplifications: $\mathfrak{X} = X \times \eta_{\mathcal{S}}$ and (taking $p = d(=\dim(X))$ for 0-cycles)

$$\begin{aligned} Gr_{\mathcal{L}}^i \underline{Hg}^d(X \times \eta_{\mathcal{S}}) &= Hom_{\mathrm{MHS}}\left(\mathbb{Q}(-d),\, \underline{H^i}(\eta_{\mathcal{S}}) \otimes H^{2d-i}(X)\right) \\ &\cong Hom_{\mathrm{MHS}}\left(H^i(X),\, \underline{H^i}(\eta_{\mathcal{S}})\right) \twoheadrightarrow Hom_{\mathbb{C}}\left(\Omega^i(X),\, \Omega^i(\mathcal{S})\right); \end{aligned}$$

we write $[\mathfrak{Z}]_i^*$ for $[\mathfrak{Z}]_i$'s induced Hom's. Moreover,

$$Gr_{\mathcal{L}}^{i-1}\underline{J^d}(X \times \eta_{\mathcal{S}}) = \frac{Ext^1_{\mathrm{MHS}}\left(\mathbb{Q}(-d),\, \underline{H^{i-1}}(\eta_{\mathcal{S}}) \otimes H^{2d-i}(X)\right)}{\operatorname{im} Hom_{\mathrm{MHS}}\left(\mathbb{Q}(-d),\, Gr_i^W H^{i-1}(\eta_{\mathcal{S}}) \otimes H^{2d-i}(X)\right)}$$

$$\overset{\xi}{\twoheadrightarrow} J^d\left(\underline{H^{i-1}}(\eta_{\mathcal{S}}) \otimes \frac{H^{2d-i}(X)}{F_h^{d-i+1}H^{2d-i}(X)}\right),$$

where ξ is worked out in [**K2**, sec. 12]; we define $AJ_X^{i-1}(\mathcal{Z})^{tr} =$

$$[AJ(\mathfrak{Z})]_{i-1}^{tr} := \xi\left([AJ(\mathfrak{Z})]_{i-1}\right).$$

Here then is how to compute the invariants for 0-cycles:

COROLLARY 4.6. *Let $X/\bar{\mathbb{Q}}$ and $\mathcal{Z} \in \mathcal{L}^i CH_0(X_K)$; then*

(a) $[\mathfrak{Z}]_i$ is the image of the Künneth component $[\bar{\mathfrak{Z}}]_i \in H^{2d-i}(X) \otimes H^i(\mathcal{S})$, hence $[\mathfrak{Z}]_i^*$ is the composition of $[\bar{\mathfrak{Z}}]_i^*$ with e.g. $H^i(\mathcal{S}) \twoheadrightarrow \underline{H^i}(\eta_{\mathcal{S}})$.

(b) If in addition $\bar{\mathfrak{Z}} \overset{hom}{\equiv} 0$, then $[AJ(\mathfrak{Z})]_{i-1}^{tr}$ is the projected image of

$$\left\{\int_{\partial^{-1}\bar{\mathfrak{Z}}} (\cdot)\right\} \in \left(F^t\left\{H^{2t-i+1}(\mathcal{S},\mathbb{C}) \otimes H^i(X,\mathbb{C})\right\}\right)^{\vee} \Big/ \, periods$$

$\cong J^d\left(H^{i-1}(\mathcal{S}) \otimes H^{2d-i}(X)\right)$, *under the obvious projection of HS.*

EXAMPLE 4.7. Recall the case $d = \ell = 2$ of Thm. 3.5(A), the cycle $\mathcal{Z} = (p_1 - o_1) \times (p_2 - o_2) \in \ker(Alb) \subseteq CH_0^{\hom}(\mathcal{C}_1 \times \mathcal{C}_{2/K})$ where $\mathcal{C}_1$, $\mathcal{C}_2/\bar{\mathbb{Q}}$ and $K \cong \bar{\mathbb{Q}}(\mathcal{C}_1 \times \mathcal{C}_2)$. In fact this lives in $\mathcal{L}^2$, see §5.1. Noting that $X = \mathcal{C}_1 \times \mathcal{C}_2 = \mathcal{S}$, we need only check that $[\bar{\mathfrak{Z}}]_2^* : \Omega^2(X) \to \Omega^2(\mathcal{S})$ is nontrivial. This follows from the fact that the spread of (p_1, p_2) is $\Delta_{\mathcal{C}_1 \times \mathcal{C}_2}$ (which induces the identity on Ω^2), while the spreads of the other points of $\mathcal{Z}$ induce the zero map. Here the connection between $\Omega^2(X)$ and the Albanese kernel is quite plain! The other cases of Thm. 3.5(A) are dealt with similarly.

4.3. Definition of reduced higher AJ maps. Continue to assume X is defined $/\bar{\mathbb{Q}}$. At times it is too difficult to compute even $[AJ(\mathfrak{Z})]_{i-1}^{tr}$ fully; instead one can make use of the following quotient. Writing D for divisors$/\bar{\mathbb{Q}} \subseteq \mathcal{S}$, we can define

$$\overline{H^*}(\mathcal{S}_{\mathrm{rel}}) := (\mathrm{co})\mathrm{im}\left\{\varprojlim_D H^*(\mathcal{S}, D) \to H^*(\mathcal{S})\right\},$$

$$\underline{H_*}(\eta_{\mathcal{S}}) := (\mathrm{co})\mathrm{im}\left\{\varprojlim_D H_*(\mathcal{S} \setminus D) \to H_*(\mathcal{S})\right\};$$

the latter are homology classes on $\mathcal{S}$ which can be moved to avoid any divisor. The former is a (pure weight) HS.

By restricting functionals $\left\{F^{t+1}\left(\overline{H^{2t-i+1}}(\mathcal{S}_{\rm rel},\mathbb{C})\otimes\frac{H^i(X,\mathbb{C})}{F_h^1}\right)\right\}^\vee \twoheadrightarrow$

$$\left\{\overline{H^{2t-i+1}}(\mathcal{S}_{\rm rel},\mathbb{C})\otimes F^iH^i(X,\mathbb{C})\right\}^\vee \cong \left\{\underline{H_{i-1}}(\eta_{\mathcal{S}},\mathbb{Q})\otimes F^iH^i(X,\mathbb{C})\right\}^\vee,$$

we induce a map

$$J^d\left(\underline{H^{i-1}}(\eta_S)\otimes\frac{H^{2d-i}(X)}{F_h^{d-i+1}}\right)\twoheadrightarrow Hom_{\mathbb{Q}}\left(\underline{H_{i-1}}(\eta_S,\mathbb{Q}),\ \frac{\Omega^i(X)^\vee}{\operatorname{im}\{H_i(X,\mathbb{Q})\}}\right)$$

and the image of $[AJ(\mathfrak{Z})]_{i-1}^{tr}$ is denoted $\overline{[AJ(\mathfrak{Z})]_{i-1}}$. Assuming $\bar{\mathfrak{Z}}\overset{\rm hom}{\equiv}0$, recall that $[AJ(\mathfrak{Z})]_{i-1}^{tr}$ (as a functional) is induced by $\int_{\partial^{-1}\bar{\mathfrak{Z}}}$. Applying this to a class $[\hat{\gamma}]\otimes\omega$ for $\omega\in\Omega^i(X)$, $[\gamma]\in\underline{H_{i-1}}(\eta_S,\mathbb{Q})$ [with $[\hat{\gamma}]\in\overline{H^{2t-i+1}}(\mathcal{S}_{\rm rel},\mathbb{C})$ its Poincaré dual], gives

$$\int_{\partial^{-1}\bar{\mathfrak{Z}}}\hat{\gamma}\wedge\omega=\int_{\pi_X\left\{(\partial^{-1}\bar{\mathfrak{Z}})\cap(\gamma\times X)\right\}}\omega=:\int_\Gamma\omega,$$

where Γ is a topological i-chain bounding on $\bar{\mathfrak{Z}}_*(\gamma):=\pi_X\left\{\bar{\mathfrak{Z}}\cap(\gamma\times X)\right\}$.[9]

Now let $\{\omega\}=\{\omega^k\}_{k\in\mathsf{K}}\subseteq\Omega^i(X)$ denote a basis; evaluating functionals on it gives a map $\Omega^i(X)^\vee\underset{\mathrm{ev}_{\{\omega\}}}{\overset{\cong}{\longrightarrow}}\mathbb{C}^{|\mathsf{K}|}$, and we write $\Lambda_{\{\omega\}}$ for the $\mathrm{ev}_{\{\omega\}}$-image of $\operatorname{im}\{H_i(X,\mathbb{Q})\}$. Then the vector $\left(\int_\Gamma\omega^k\right)_{k\in\mathsf{K}}$ does not change modulo $\Lambda_{\{\omega\}}$ (= periods) if we choose a different Γ bounding on $\bar{\mathfrak{Z}}_*(\gamma)$ (provided we use the same Γ for all ω^k). We have proved the following:

PROPOSITION 4.8. *Given $\mathcal{Z}\in\mathcal{L}^iCH_0(X)$ with nullhomologous complete spread $\bar{\mathfrak{Z}}$, $\gamma\in\underline{H_{i-1}}(\eta_S,\mathbb{Q})$ and $\Gamma\in C_i^{top}(X,\mathbb{Q})$ with $\partial\Gamma=\bar{\mathfrak{Z}}_*(\gamma)$, $\overline{[AJ(\mathfrak{Z})]_{i-1}}(\gamma)$ evaluates to $\left(\int_\Gamma\omega^k\right)_{k\in\mathsf{K}}\in\mathbb{C}^{|\mathsf{K}|}/\Lambda_{\{\omega\}}$ on a basis $\{\omega\}$.*

The $\int_\Gamma\omega^k$ are called membrane integrals; for $i=2$ this once again makes the connection between $\overset{\rm rat}{\not\equiv}$ of cycles and holomorphic 2-forms quite clear. We also note that $\Lambda_{\{\omega\}}\subseteq\mathbb{C}^{|\mathsf{K}|}$ is a rational lattice of dimension *at least* $2|\mathsf{K}|$.

EXAMPLE 4.9. In the situation of Ex. 4.2 (assume $H^1(X)=0$), the membrane integrals are simply over chains $\subseteq X$ bounding loops $\subseteq Y$

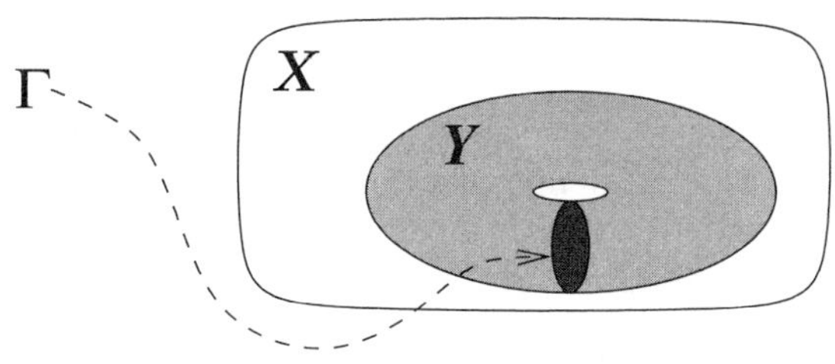

[9]One can ensure this intersection is proper by moving γ in its homology class on $\mathcal{S}$. The easiest way to do this is to choose a divisor $D\subseteq\mathcal{S}$ sufficiently "large" that the restriction of $\bar{\mathfrak{Z}}$ to $X\times(\mathcal{S}\setminus D)$ has relative dimension 0 over $\mathcal{S}\setminus D$, and move γ to avoid D. (This is possible because $[\gamma]\in\underline{H_{i-1}}(\eta_S,\mathbb{Q})$; of course, sometimes D can be chosen empty, as in example 4.2 above.)

These $\left\{\int_\Gamma \omega_X^{(2,0)}\right\}$ are related to extensions of MHS.

REMARK 4.10. (a) We can interpret $[\mathfrak{Z}]_{i-1}$ as corresponding to an element of $Hom_{\mathbb{Q}}\left(\underline{H_{i-1}}(\eta_S,\mathbb{Q}),\, H_{i-1}(X,\mathbb{Q})\right)$, and thus deduce existence of the membrane Γ (hence the membrane integrals) from vanishing of $[\mathfrak{Z}]_{i-1}$ (true for $\mathcal{Z}\in\mathcal{L}^i$) rather than $[\bar{\mathfrak{Z}}]$.

(b) With some work (see **[K2]**), one may furthermore relax the extra condition in Proposition 4.8 from $[\bar{\mathfrak{Z}}]=0$ to only $\underline{[\mathfrak{Z}]_i=0}$. This is a vast improvement, because it says the reduced invariants $\overline{[AJ(\mathfrak{Z})]_{i-1}}$ are computable by membrane integrals whenever they are defined — hence under weaker conditions than allow $[AJ(\mathfrak{Z})]_{i-1}^{(tr)}$ to be computed by Griffiths's prescription $\int_{\partial^{-1}\bar{\mathfrak{Z}}}(\cdot)$. This is needed to deal with the n-box cycles of §6.3 for $n>2$.

5. Additional filtrations and invariants

In this section we will study a few other "candidates" for $F^\bullet_{\mathrm{BB}}$. We are more interested in what one can do with them as is, than with the fact that under reasonable conjectural conditions they all coincide.

5.1. Murre filtration[10]**.** Assume X (defined / any subfield of $\mathbb{C}$) is such that Δ_X has a good Chow-Künneth decomposition. Writing $\Delta^X_{\leq a}:=\sum_{(0\leq)j\leq a}\Delta_{X,j}$ and $\Delta^X_{>b}:=\sum_{b<j(\leq 2d)}\Delta_{X,j}$, we set

$$F^i_M CH_0(X):=\mathrm{im}\left\langle\Delta^X_{\leq 2d-i}\right\rangle_*=\ker\left\langle\Delta^X_{>2d-i}\right\rangle_*.$$

Note that $F^1_M CH_0 = CH_0^{\mathrm{hom}}$, $F^2_M CH_0\subseteq\ker(Alb)$, $F^{d+1}_M CH_0=\{0\}$, and $\langle\Delta_{X,j}\rangle_*|_{Gr^i_{F_M}CH_0(X)}=\delta_{2d-i,\,j}\cdot Id$. This latter property does not necessarily hold for arbitrary algebraic Künneth components $\tilde{\Delta}_{X,j}$ (required of a BBF), but if $F^\bullet_{\mathrm{BB}}$ exists then one easily shows $F^\bullet_M$ must coincide with it. Under the assumptions of HLC+BBC^q, $\mathcal{L}^\bullet$ gives such a filtration, so then $\mathcal{L}^\bullet=F^\bullet_M$. Without such assumptions, always $F^\bullet_M\subseteq\mathcal{L}^\bullet$, and conveniently $\left\langle\Delta^X_{\leq 2d-i}\right\rangle_*$ gives a projection to F^i_M.

Example: products of curves. For one curve $\mathcal{C}$ with a base point o, set $\Delta_{\mathcal{C},2}=\mathcal{C}\times\{o\}$, $\Delta_{\mathcal{C},0}=\{o\}\times\mathcal{C}$, $\Delta_{\mathcal{C},1}=\Delta_{\mathcal{C}}-\Delta_{\mathcal{C},0}-\Delta_{\mathcal{C},2}$. If $X=\mathcal{C}_1\times\cdots\times\mathcal{C}_n$ (so $d=n$), write $\sigma:(\mathcal{C}_1\times\mathcal{C}_1)\times\cdots\times(\mathcal{C}_n\times\mathcal{C}_n)\to X\times X$ for the obvious permutation of factors; then

$$\sigma^*\Delta_{X,j}:=\sum_{j_1+\cdots+j_n=j}\Delta_{\mathcal{C}_1,j_1}\times\cdots\times\Delta_{\mathcal{C}_n,j_n}$$

yields a good C-K decomposition of Δ_X. We can obtain an alternate description of the resulting $F^\bullet_M$ as follows. Let Ξ^n_i denote the set of strictly increasing functions $\varepsilon:\{1,\ldots,i\}\hookrightarrow\{1,\ldots,n\}$ and (for each $\varepsilon\in\Xi^n_i$) write $\pi_\varepsilon: X\twoheadrightarrow X_\varepsilon:=\mathcal{C}_{\varepsilon(1)}\times\cdots\times\mathcal{C}_{\varepsilon(i)}$. Set $F^0_\times CH_0(X)=CH_0(X)$ and for $1\leq i\leq n+1$ define

$$F^i_\times CH_0(X):=\bigcap_{\varepsilon\in\Xi^n_{i-1}}\ker(\pi_{\varepsilon_*})\subseteq CH_0(X).$$

[10]We caution that this is not quite Murre's conjectural filtration (unless conjectures or further hypotheses on the $\{\Delta_{X,j}\}$ are assumed), see **[M]**.

Clearly $F_{\times}^{n+1}CH_0 = \{0\}$, and in fact $F_{\times}^{\bullet} = F_M^{\bullet}(\subseteq \mathcal{L}^{\bullet})$.

Choosing base points $o_j \in \mathcal{C}_j$ (implicit already in the above), we may also define inclusions $\iota_\varepsilon : X_\varepsilon \hookrightarrow X$. [For $\varepsilon \in \Xi_0^n$ the trivial map $\emptyset \hookrightarrow \{1, \ldots, n\}$, π_ε is the structure map and ι_ε incudes $\operatorname{Spec}\mathbb{C} \hookrightarrow X$ as $(o_1, \ldots, o_n)$.] Then we can write the projections $\langle \Delta^X_{\leq 2d-i} \rangle_*$ to the $F_{\times}^i$ as follows:

$$P_{\times}^{[i]}(\mathcal{Z}) := \mathcal{Z} + \sum_{j=0}^{i-1} (-1)^{i-j} \left(\sum_{\varepsilon \in \Xi_j^n} (\iota_\varepsilon \circ \pi_\varepsilon)_* \mathcal{Z} \right)$$

for $\mathcal{Z} \in Z_0(X)$. To write it in a more attractive way, define $Box^n(\mathcal{Z}) := P_{\times}^{[n]}(\mathcal{Z})$ so that e.g. for $\mathcal{Z}$ one point we have $Box^n\{(p_1, \ldots, p_n)\} = (p_1 - o_1) \times \cdots \times (p_n - o_n)$. (For $n = 2$ this is just $Box^2(p_1, p_2) = (p_1, p_2) - (p_1, o_2) - (o_1, p_2) + (o_1, o_2)$.) Then

$$P_{\times}^{[i]}(\mathcal{Z}) = \mathcal{Z} - \sum_{j=0}^{i-1} \sum_{\varepsilon \in \Xi_j^n} \iota_*^\varepsilon Box^j(\pi_*^\varepsilon \mathcal{Z}).$$

To be clear: we shall only take "Box^j" of a cycle defined on a product of j curves.

5.2. S. Saito filtration. The form of this filtration (we use the version in [**LS**]) precludes discussion exclusively for 0-cycles. Let Sm =category of smooth projective varieties $/\mathbb{C}$, and define inductively $\forall\, X \in Sm$, $p \in \mathbb{N}$: $F_S^0 CH^p(X) := CH^p(X)$,

$$F_S^{i+1}CH^p(X) := \sum_{\substack{Y \in Sm \\ q \in \mathbb{N}}} \sum_{\Gamma} \operatorname{im}\left\{ \langle \Gamma \rangle_* : F_S^i CH^q(Y) \to CH^p(X) \right\},$$

where the sum is over all $\Gamma \in Z^{p-q+d_X}(Y \times X)$ for which $[\Gamma]_* : H^{2q-i}(Y) \to H^{2p-i}(X)$ is the zero map.

PROPOSITION 5.1. *(i)* $F_S^\bullet \subseteq \mathcal{L}^\bullet$.

(ii) If there exists a C-K decomposition of Δ_X *then* $im \langle \Delta_{X,j} \rangle_* \subseteq F_S^i CH^p(X)$ $\forall\, j \leq 2p - i$.

(iii) In addition to the decomposition, if one assumes BBC^q *then* $F_S^\bullet = \mathcal{L}^\bullet$.

PROOF. [arbitrary codimension]

(i) Inductively assume $F_S^i \subseteq \mathcal{L}^i$ for all X, p; then fix X, p and let $\langle \mathcal{Z} \rangle \in F_S^{i+1}CH^p(X)$. That is, $\mathcal{Z} = \Gamma_* \mathcal{W}$ for $\Gamma \in Z^{p-q+d_X}(Y \times X)$, $\mathcal{W} \in Z^q(Y)$ all defined / some K f.g. $/\bar{\mathbb{Q}}$, and $\langle \mathcal{W} \rangle \in F_S^i CH^q(Y)$. Spreading out X, Y, $\mathcal{W}$, Γ, $\mathcal{Z}$ over the same $\mathcal{S}$, the spread $\tilde{\Gamma} \in Z^{p-q+d_X}(\mathfrak{Y} \times_{\mathcal{S}} \mathfrak{X})$ induces[11] a map $[\tilde{\Gamma}]_*^{\mathcal{D}} : \underline{H_{\mathcal{D}}^{2q}}(\mathfrak{Y}, \mathbb{Q}(q)) \to \underline{H_{\mathcal{D}}^{2p}}(\mathfrak{X}, \mathbb{Q}(p))$ respecting $\mathcal{L}^\bullet$. So $\psi(\mathcal{W}) \in \mathcal{L}^i \underline{H_{\mathcal{D}}^{2q}}(\mathfrak{Y}, \mathbb{Q}(q)) \implies \psi(\mathcal{Z}) = [\tilde{\Gamma}]_*^{\mathcal{D}} \psi(\mathcal{W}) \in \mathcal{L}^i$ and (using Lewis's description of the $Gr_{\mathcal{L}}^i \underline{H_{\mathcal{D}}^*}$) $Gr_{\mathcal{L}}^i \psi(\mathcal{Z}) = \left(Gr_{\mathcal{L}}^i [\tilde{\Gamma}]_*^{\mathcal{D}} \right) \psi(\mathcal{W}) = 0$.

(ii) For $\mathcal{Z} \in CH^p(X)$, one inductively assumes for $i < 2p - j$ that $\langle \Delta_{X,j} \rangle_* \mathcal{Z} \in F_S^i CH^p(X)$ and uses the definition of $F_S^\bullet$ to show $\langle \Delta_{X,j} \rangle_* \mathcal{Z} \in F_S^{i+1}$. Namely, taking $Y = X$, $\Gamma = \Delta_{X,j}$ we have $0 = [\Delta_{X,j}]_{2p-i} : H^{2p-i}(X) \to H^{2p-i}(X)$ (since $j \neq 2p - i$) and $\langle \Delta_{X,j} \rangle_* \langle \Delta_{X,j} \rangle_* \mathcal{Z} = \langle \Delta_{X,j} \rangle_* \mathcal{Z}$, done.

[11] using cup product with $\underline{c_{\mathcal{D}}}(\tilde{\Gamma})$ together with the (essentially fiberwise) pullbacks and pushforwards along $\mathfrak{X} \times_{\mathcal{S}} \mathfrak{Y} \twoheadrightarrow \mathfrak{X}$, $\mathfrak{Y}$ resp. In particular, pushforward is OK because fibers are compact.

(iii) If $\mathcal{Z} \in \mathcal{L}^i CH^p(X)$ then by part (ii), $\sum_{p\leq j\leq 2p-i} \langle \Delta_{X,j}\rangle_* \mathcal{Z} \in F_S^i\,[\subseteq \mathcal{L}^i]$; but restricting $\mathcal{Z} - \sum_{p\leq j\leq 2p-i} \langle \Delta_{X,j}\rangle_* \mathcal{Z}$ (iteratively for $k = i, i+1, \ldots, p$) to $Gr^k_{\mathcal{L}}$ yields $Gr^k_{\mathcal{L}}\left(\mathcal{Z} - \langle \Delta_{X,2p-k}\rangle_* \mathcal{Z}\right)$ which is zero (by property of $\mathcal{L}^\bullet$). Hence, $\mathcal{Z} \equiv \sum_{p\leq j\leq 2p-i} \langle \Delta_{X,j}\rangle_* \mathcal{Z} \mod \mathcal{L}^{p+1}$, which by BBC^q is zero; and so $\mathcal{Z} \in F_S^i$. □

Having set the conjectural merry-go-round into motion, we give it one more push:

COROLLARY 5.2. *(i) If a good C-K decomposition (hence $F_M^\bullet$) exists for Δ_X, then $F_M^\bullet \subseteq F_S^\bullet$ (on $CH_0(X)$).*
(ii) BBC^q+HLC $\Longrightarrow$ ($\exists\, F_M^\bullet$ and) $F_M^\bullet = F_S^\bullet = \mathcal{L}^\bullet$.

5.3. Mumford-Griffiths invariants. We now show how to define invariants on the $Gr^i_{F_S}$ which *together* carry the same information as the cl^i on the $Gr^i_{\mathcal{L}}$. This will involve putting a Leray filtration on $\mathbb{H}^{2d}(\mathfrak{X}, \Omega^{\bullet\geq d}_{\mathfrak{X}})$. (These invariants are really the only established way for computing when X is not defined $/\bar{\mathbb{Q}}$ and hence spreads along with $\mathcal{Z}$.)

As in §4 we let $\mathfrak{X}$, $\mathcal{S}$ be smooth projective $/\bar{\mathbb{Q}}$ with $\pi : \mathfrak{X}_{\mathcal{S}} \to \mathcal{S}$ proper, and let $\mathcal{U} = (\mathcal{S} \setminus D) \subseteq \mathcal{S}$ (affine Zariski open) be such that $\pi|_{\mathfrak{X}_{\mathcal{U}}}$ is smooth (the singular fibers are contained in $\mathfrak{X}_D$). Define a Leray filtration for forms on $\mathfrak{X}_{\mathcal{U}}$

$$\mathcal{L}^k \Omega^{\bullet\geq d}_{\mathfrak{X}_{\mathcal{U}}} := \mathrm{im}\left\{\pi^*\Omega^k_{\mathcal{U}} \otimes \Omega^{\bullet\geq d-k}_{\mathfrak{X}_{\mathcal{U}}}[-k] \to \Omega^{\bullet\geq d}_{\mathfrak{X}_{\mathcal{U}}}\right\}$$

so that $Gr^k_{\mathcal{L}}\Omega^{\bullet\geq d}_{\mathfrak{X}_{\mathcal{U}}} = \pi^*\Omega^k_{\mathcal{U}} \otimes \Omega^{\bullet\geq d-k}_{\mathfrak{X}_{\mathcal{U}}/\mathcal{U}}[-k]$; also set

$$\mathcal{L}^k\mathbb{H}^*(\mathfrak{X}_{\mathcal{U}}, \Omega^{\bullet\geq d}_{\mathfrak{X}_{\mathcal{U}}}) := \mathrm{im}\left\{\mathbb{H}^*(\mathfrak{X}, \mathcal{L}^k\Omega^{\bullet\geq d}_{\mathfrak{X}_{\mathcal{U}}}) \to \mathbb{H}^*(\mathfrak{X}_{\mathcal{U}}, \Omega^{\bullet\geq d}_{\mathfrak{X}_{\mathcal{U}}})\right\}.$$

Standard formal nonsense using the $\mathbb{H}^*$-long-exact sequences associated to $0 \to \mathcal{L}^{k+1} \to \mathcal{L}^k \to Gr^k_{\mathcal{L}} \to 0$ (for each k) leads to a spectral sequence computing $Gr^k_{\mathcal{L}}\mathbb{H}^{j+k}(\mathfrak{X}_{\mathcal{U}}, \Omega^{\bullet\geq d}_{\mathfrak{X}_{\mathcal{U}}}) = E^{j,k}_\infty$. This is

$$\begin{aligned} E_1^{j,k} &:= \mathbb{H}^{j+k}\left(\mathfrak{X}_{\mathcal{U}},\, Gr^k_{\mathcal{L}}\Omega^{\bullet\geq d}_{\mathfrak{X}_{\mathcal{U}}}\right) = \mathbb{H}^j\left(\mathfrak{X}_{\mathcal{U}},\, \pi^*\Omega^k \otimes \Omega^{\bullet\geq d-k}_{\mathfrak{X}_{\mathcal{U}}/\mathcal{U}}\right) \\ &= H^0\left(\mathcal{U},\, \Omega^k_{\mathcal{U}} \otimes \mathbb{R}^j\pi_*\Omega^{\bullet\geq d-k}_{\mathfrak{X}_{\mathcal{U}}/\mathcal{U}}\right) \end{aligned}$$

where $d_1 : E^{j,k} \to E^{j,k+1}$ is induced by ∇ (the Gauss-Manin connection). (The last equality comes from the Leray s.s. for $\mathfrak{X}_{\mathcal{U}} \xrightarrow{\pi} \mathcal{U}$ and observing that since $\mathcal{U}$ is affine and $\Omega^k_{\mathcal{U}} \otimes \mathbb{R}^j\pi_*\Omega^{\bullet\geq d-k}_{\mathfrak{X}_{\mathcal{U}}/\mathcal{U}}$ are coherent, all $H^{i>0}$'s vanish.) Writing[12] $\mathcal{F}^a\mathcal{H}^b_{\mathfrak{X}_{\mathcal{U}}/\mathcal{U}}$ for $\mathbb{R}^b\pi_*\Omega^{\bullet\geq a}_{\mathfrak{X}_{\mathcal{U}}/\mathcal{U}}$, we conclude that

$$Gr^k_{\mathcal{L}}\mathbb{H}^{2d}\left(\mathfrak{X}_{\mathcal{U}},\, \Omega^{\bullet\geq d}_{\mathfrak{X}_{\mathcal{U}}}\right) \cong H^k_\nabla\left\{H^0\left(\mathcal{U},\, \Omega^*_{\mathcal{U}} \otimes \mathcal{F}^{d-*}\mathcal{H}^{2d-k}_{\mathfrak{X}_{\mathcal{U}}/\mathcal{U}}\right)\right\}. \tag{5.1}$$

Next, consider the composition $\phi_{\mathcal{U}}$:

$$\begin{aligned} Hom_{\text{MHS}}\left(\mathbb{Q}(-d),\, H^{2d}(\mathfrak{X}_{\mathcal{U}}, \mathbb{Q})\right) \hookrightarrow F^dH^{2d}(\mathfrak{X}_{\mathcal{U}}, \mathbb{C}) \cong \mathbb{H}^{2d}\left(\bar{\mathfrak{X}},\, \Omega^{\bullet\geq d}_{\bar{\mathfrak{X}}}\langle\log \mathfrak{X}_D\rangle\right) \\ \to \mathbb{H}^{2d}\left(\mathfrak{X}_{\mathcal{U}},\, \Omega^{\bullet\geq d}_{\mathfrak{X}_{\mathcal{U}}}\right). \end{aligned}$$

[12]More generally, one writes $\mathcal{H}^b_{\mathfrak{X}/\mathcal{U}}$ for $\mathbb{R}^b\pi_*\Omega^\bullet_{\mathfrak{X}/\mathcal{U}} \cong (R^b\pi_*\mathbb{C}) \otimes \Omega_{\mathcal{U}}$, $\mathcal{H}^b_{\mathfrak{X}/\mathcal{U},\mathbb{Q}}$ for $\mathbb{R}^b\pi_*\mathbb{Q}$. (Also note: $\mathcal{H}^b_{X_s}$ means the same thing as $\mathcal{H}^b_{\mathfrak{X}/\mathcal{U}}$.)

One can show that this is independent of the choice of good compactification $\bar{\mathfrak{X}}$ of $\mathfrak{X}_{\mathcal{U}}$, and functorial for restrictions from $\mathcal{U}$ to $\mathcal{U}_0 \subseteq \mathcal{U}$. Hence it makes sense in the limit over $\mathcal{U} \subseteq \mathcal{S}$, and we may now consider Φ:

$$CH_0(X_K) \overset{\cong}{\to} CH^d(\mathfrak{X}_{(\bar{\mathbb{Q}})}) \overset{[\cdot]}{\to} Hom_{\text{MHS}}\left(\mathbb{Q}(-d),\, H^{2d}(\mathfrak{X},\mathbb{Q})\right) \overset{\phi_\eta}{\to} \mathbb{H}^{2d}(\mathfrak{X}, \Omega^{\bullet \geq d}_{\mathfrak{X}})$$

(where as before $\mathfrak{X} = \varprojlim_{\mathcal{U}} \pi^{-1}(\mathcal{U}) = \mathfrak{X}_\eta$). An argument exactly like that for Prop. 5.1(ii) above, shows

THEOREM 5.3. $\Xi\left(F^i_{\mathcal{S}} CH_0\right) \subseteq \mathcal{L}^i \mathbb{H}^{2d}$, *hence* $\exists$ *a map*

$$\Phi_i : F^i_{\mathcal{S}} CH_0(X_K) \to H^i_{\nabla}\left\{\Gamma\left(\eta_{\mathcal{S}},\, \Omega^*_{\mathcal{S}} \otimes \mathcal{F}^{d-*}\mathcal{H}^{2d-i}_{X_s}\right)\right\} =: \nabla J^{d,i}(\mathfrak{X}/\eta_{\mathcal{S}}).$$

REMARK 5.4. The notation $\nabla J^{\cdots}$ (due to S. Saito) suggests that we think of these Φ_i as infinitesimal invariants of higher normal functions of some kind (an analogous point of view will be considered below in §7). Accordingly, $\Phi_i(\mathcal{Z})$ will be denoted $\delta_i \mathfrak{Z}$.

The question arises as to whether $\delta_i \mathfrak{Z}$ and $[\mathfrak{Z}]_i$ record the same information *for each* i, or equivalently (see **[LS]**) whether Φ_i may be defined on $\mathcal{L}^i CH_0$. The problem is that it is not clear that ϕ_η maps $\mathcal{L}^i$ on $F^d H^{2d}(\mathfrak{X}_\eta, \mathbb{C})$ to $\mathcal{L}^i$ on $\mathbb{H}^{2d}(\mathfrak{X}_{\mathcal{U}}, \Omega^{\bullet \geq d}_{\mathfrak{X}_{\mathcal{U}}})$, owing to the lack of a de Rham-theoretic description[13] of the first $\mathcal{L}^i$. However, from the work of Lewis-Saito (*op. cit.*) [(i)] and the author (unpublished) [(ii)] one has the following partial results:

PROPOSITION 5.5. $\Phi\left(\mathcal{L}^i CH_0(X)\right) \subseteq \mathcal{L}^i \mathbb{H}^{2d}\left(\mathfrak{X},\, \Omega^{\bullet \geq d}_{\bar{\mathfrak{X}}}\right)$, *hence one can define* Φ_i *on* $\mathcal{L}^i$, *in case* either

(i) $\bar{\mathfrak{X}} \overset{\pi}{\to} \mathcal{S}$ *can be chosen to be smooth (no singular fibers in the compactified family),* or

(ii) $H^0(\Omega^i_X) = 0$ *for* $0 < i < d$.

REMARK 5.6. Lewis has reduced the general case to HC, a conjecture of Zucker-Brylinski, and a conjecture about morphisms of MHS involving intersection cohomology; this is still a vast improvement over HC+BBC^q.

EXAMPLE 5.7. *for* (i): $X/\bar{\mathbb{Q}}$ so that $\bar{\mathfrak{X}} = X \times \mathcal{S}$, though applicability is not limited to this.

for (ii): X a smooth complete intersection, $K3$ surface, CY 3-fold, etc. (defined / any field).

It is instructive to work out the target of Φ_i for $\bar{\mathfrak{X}} = X \times \mathcal{S}$: the connection ∇ is just differentiation along $\mathcal{S}$ and since $H^i_{d\mathcal{S}}\{\Gamma(\eta_{\mathcal{S}}, \Omega^*_{\mathcal{S}})\} = H^i(\eta_{\mathcal{S}}, \mathbb{C})$, $\nabla J^{d,i}(X \times \eta_{\mathcal{S}}/\eta_{\mathcal{S}}) = H^i(\eta_{\mathcal{S}}) \otimes F^{d-i}H^{2d-i}(X) = H^i(\eta_{\mathcal{S}}) \otimes H^{2d-i}(X)$. So in this case $\delta_i \mathfrak{Z}$ and $[\mathfrak{Z}]_i$ are precisely the same thing.

---

[13]Without something of this sort, $\mathcal{L}^i F^d H^{2d}(\mathfrak{X}_{\mathcal{U}}, \mathbb{C})$ (the group to which $[\cdot]$ takes the spread of a cycle in $\mathcal{L}^i CH_0$) is just $\mathcal{L}^i \mathbb{H}^{2d}(\mathfrak{X}_{\mathcal{U}}, \Omega^\bullet_{\mathfrak{X}_{\mathcal{U}}}) \cap \text{im}\left\{\mathbb{H}^{2d}(\bar{\mathfrak{X}}, \Omega^{\bullet \geq d}_{\bar{\mathfrak{X}}}\langle \log \mathfrak{X}_D \rangle) \to \mathbb{H}^{2d}(\mathfrak{X}_{\mathcal{U}}, \Omega^\bullet_{\mathfrak{X}_{\mathcal{U}}})\right\}$. Assuming $\bar{\mathfrak{X}} \overset{\pi}{\to} \mathcal{S}$ is log-D-regular, one can prove (using techniques from **[P]**) that the $E_2^{j,k}$-term of a spectral sequence computing $\mathbb{H}^*\left(\bar{\mathfrak{X}},\, \Omega^{\bullet \geq d}_{\bar{\mathfrak{X}}}\langle \log \mathfrak{X}_D \rangle\right)$ $[\cong F^d H^{2d}(\mathfrak{X}_{\mathcal{U}}, \mathbb{C})$ for $* = 2d]$ is $\mathbb{H}^k\left(\mathcal{S},\, \Omega^*_{\mathcal{S}}\langle \log D \rangle \otimes \mathbb{R}^j \pi_* \Omega^{\bullet \geq d-*}_{\bar{\mathfrak{X}}/\mathcal{S}}\langle \log(\mathfrak{X}_D/D) \rangle\right)$ but proving $E_2 = E_\infty$ or that E_∞ yields $Gr^\bullet_{\mathcal{L}}$ seems difficult.

Now we refine the maps Φ_i in a couple of different directions (the second in the next section).

The first "refinement" requires a momentary switching of gears, since up to this point all sheaf (hyper)cohomologies have been computed with respect to the analytic topology. Working in the $\bar{\mathbb{Q}}$-Zariski topology, one has the "dlog" (or El Zein) cycle-class map on $CH^d(\mathfrak{X}_{/\bar{\mathbb{Q}}})$ which we compose with the spread to get $\Phi^{\text{Zar}/\bar{\mathbb{Q}}} : CH_0(X_K) \to \mathbb{H}^{2d}_{\text{Zar}/\bar{\mathbb{Q}}}(\mathfrak{X}, \Omega^{\bullet \geq d}_{\mathfrak{X}/\bar{\mathbb{Q}}})$. (See **[LS]**, **[G1]**.) An argument practically identical to the one above, yields a Leray filtration on $\mathbb{H}^{2d}_{\text{Zar}/\bar{\mathbb{Q}}}$ and maps from $F^k_{\mathcal{S}} CH_0(X_K)$ to $Gr^k_{\mathcal{L}} \mathbb{H}^{2d}_{\text{Zar}/\bar{\mathbb{Q}}}$, which is

$$(5.2) \qquad H^k_{\nabla} \left\{ H^0_{\text{Zar}/\bar{\mathbb{Q}}} \left(\eta_{\mathcal{S}}, \, \Omega^*_{\eta_{\mathcal{S}}/\bar{\mathbb{Q}}} \otimes \mathbb{R}^{2d-k} \pi_* \Omega^{\bullet \geq d-*}_{\mathfrak{X}/\eta_{\mathcal{S}}} \right) \right\}$$
$$\cong H^0 \left(\eta_{\mathcal{S}}, \, \mathcal{H}^k_{\nabla, \text{Zar}/\bar{\mathbb{Q}}} \left\{ \Omega^*_{\eta_{\mathcal{S}}/\bar{\mathbb{Q}}} \otimes \mathcal{F}^{p-*} \mathcal{H}^{2d-k}_{\text{DR}}(\mathfrak{X}_{(\eta)}/\eta_{\mathcal{S}}) \right\} \right)$$

Here the first $\cong$ comes from Γ-acyclicity of $\eta_{\mathcal{S}}$ in the $\bar{\mathbb{Q}}$-Zariski topology. Now $H^0(\eta_{\mathcal{S}}, \Omega^*_{\eta_{\mathcal{S}}/\bar{\mathbb{Q}}}) \cong \Omega^*_{\bar{\mathbb{Q}}(\eta_{\mathcal{S}})/\bar{\mathbb{Q}}} \cong \Omega^*_{K/\bar{\mathbb{Q}}}$ (Kähler differentials where $\bar{\mathbb{Q}}(\eta_{\mathcal{S}}) \underset{\cong}{\overset{\text{ev}}{\to}} K$), while in the algebraic context $\mathfrak{X}_\eta = X_K \times_{\text{ev}^{-1}} \operatorname{Spec} \bar{\mathbb{Q}}(\eta_{\mathcal{S}})$, $\eta_{\mathcal{S}} = \operatorname{Spec} \bar{\mathbb{Q}}(\eta_{\mathcal{S}}) \cong \operatorname{Spec} K$. Hence we have maps

$$\Phi^{\text{Zar}}_i : F^i_{\mathcal{S}} CH_0(X_K) \to H^i_{\nabla} \left(\Omega^{\bullet}_{K/\bar{\mathbb{Q}}} \otimes F^{d-\bullet} H^{2d-i}_{\text{DR}}(X_K/K) \right)$$
$$=: \nabla J^{d,i}(X_K/K)$$

(here ∇ = "arithmetic" Gauss-Manin connection), which factor the Φ_i by mapping $(5.2) \to \varinjlim_{\mathcal{U}} (5.1)$. This is just the restriction from the Zariski to the analytic site; one conjectures it is injective. The Φ^{Zar}_i are called *Mumford-Griffiths invariants*, since (in the limit over $K \subseteq \mathbb{C}$) they pick up the information used in the proof of Mumford/Roitman, while also being a form of generalization of Griffiths's infinitesimal invariant.

5.4. Green-Voisin invariants and applications. On the other hand, the Γ-acyclic objects in the analytic topology are balls. To make a ball avoid any $D/\bar{\mathbb{Q}} \subseteq \mathcal{S}$ (hence live "in" $\eta_{\mathcal{S}}$), we take a limit over balls containing a very general point $s_0 \in \mathcal{S}(\mathbb{C})$, and obtain local invariants[14]

$$\Phi^{\text{an}}_i : F^i_{\mathcal{S}} CH_0(X_K) \to \varinjlim_{B \ni s_0} \Gamma \left(B, \mathcal{H}^i_{\nabla} \left\{ \Omega^*_{\mathcal{S}} \otimes \mathcal{F}^{d-\bullet} \mathcal{H}^{2d-i}_{X_s} \right\} \right).$$

The important point is that the cohomology sheaves $\mathcal{H}^i_{\nabla}$ may be filtered and graded by using the fiberwise Hodge filtration (on $\mathcal{H}^{2d-i}_{X_s}$). Specifically, one has a spectral sequence (*defined* to be 0 for $m < 0$) with

$$\mathcal{E}^{\ell,m}_1(i) := \mathcal{H}^{\ell+m}_{\bar{\nabla}} \left\{ \Omega^{\bullet}_{\mathcal{S}} \otimes \mathcal{H}^{d+m-\bullet, d-m-i+\bullet}_{X_s} \right\}$$

and $\mathcal{E}^{\ell,m}_{\infty}(i) =: Gr^m_{\mathsf{f}} \mathcal{H}^{\ell+m}_{\nabla} \left\{ \Omega^{\bullet}_{\mathcal{S}} \otimes \mathcal{F}^{d-\bullet} \mathcal{H}^{2d-i}_{X_s} \right\}$; but it does *not* always (for $\dim(X) > 1$, X not defined $/\bar{\mathbb{Q}}$) degenerate at $\mathcal{E}_1$ or even $\mathcal{E}_2$. (However, noting that d_j sends $\mathcal{E}^{\ell,m}_j \to \mathcal{E}^{\ell-j+1,m+j}_j$, we always have $\mathcal{E}^{\ell,0}_{\infty}(i) \subseteq \mathcal{E}^{\ell,0}_1(i)$.) Likewise, we warn that

[14] one can actually replace the r.h.s. by $\Gamma(\eta_{\mathcal{S}}, \cdots)$ but this will no longer be true for the $\Phi^{(m)}_i$ that follows.

while $\bar{\nabla}(=d_0)$ may be computed fiberwise (by cupping with a Kodaira-Spencer class), the higher $d_{j\geq 1}$'s cannot.

Since each B is acyclic and the (co)limit is exact, we denote the limit by $\hat{B}$ and obtain iteratively (for $m=0, 1, \ldots, i$)

$$\left\{ \varinjlim_{B\ni s_0} \Gamma\left(B, \mathcal{H}^i_{\nabla}\left\{\Omega^*_{\mathcal{S}}\otimes\mathcal{F}^{d-\bullet}\mathcal{H}^{2d-i}_{X_s}\right\}\right)\right\} \supset \ker\left(\Xi_i^{(m-1)}\right) \overset{\Xi_i^{(m)}}{\longrightarrow}$$

$$\varinjlim_{B\ni s_0} \Gamma\left(B, \mathcal{E}_\infty^{i-m,\,m}(i)\right) =: Gr^m_{\mathsf{f}}\nabla J^{d,i}\left(\mathfrak{X}_{\hat{B}}/\hat{B}\right),$$

and define $\Phi_i^{(m)}$ to be the composition

$$\left(\Phi_i^{\mathrm{an}}\right)^{-1}\left(\ker\,\Xi_i^{(m-1)}\right) =: \mathsf{f}^m F^i_{\mathcal{S}} CH_0(X_K) \overset{\Xi_i^{(m)}\circ\Phi_i^{\mathrm{an}}}{\longrightarrow} Gr^m_{\mathsf{f}}\nabla J^{d,i}\left(\mathfrak{X}_{\hat{B}}/\hat{B}\right);$$

write $\Phi_i^{(m)}(\mathcal{Z})=\delta_i^{(m)}\mathfrak{Z}$ for the corresponding *Green-Voisin invariants*. (These are, for example, essentially defined in **[G2]** from the VHS point of view.)

The reason for further "chopping up" the $\delta_i\mathfrak{Z}$ like this is for greater computability, esp. of $\delta_i^{(0)}\mathfrak{Z}$, by e.g. polynomial algebra. For example, suppose $X=\{\mathsf{F}(\{z_i\})=0\}$ is a very general degree-D hypersurface in $\mathbb{P}^{d+1}$ and $\mathcal{Z}\in\mathcal{L}^i=F^i_{\mathcal{S}}CH_0(X)$. Then $\delta_d^{(0)}\mathfrak{Z}\in\Gamma(\hat{B},\mathcal{E}_1^{d,0}(d))$, where $\mathcal{E}_1^{d,0}(d)=\frac{\Omega^d_{\mathcal{S}}\otimes\mathcal{H}^{0,d}_{X_s}}{\bar{\nabla}\left(\Omega^{d-1}_{\mathcal{S}}\otimes\mathcal{H}^{1,\,d-1}_{X_s}\right)}$ dualizes to $\ker\{\bigwedge^d\theta^1_{\mathcal{S}}\otimes\mathcal{H}^{d,0}_{X_s}\to\bigwedge^{d-1}\theta^1_{\mathcal{S}}\otimes\mathcal{H}^{d-1,\,1}_{X_s,\mathrm{prim}}\}\cong\ker\{\bigwedge^d S^D\otimes R^{D-d-2}_{\mathsf{F}}\overset{\mu}{\to}\bigwedge^{d-1}S^D\otimes R^{2D-d-2}_{\mathsf{F}}\}$; here $S^\bullet$ and $R^\bullet_{\mathsf{F}}=S^\bullet\Big/\left(\left\{\frac{\partial\mathsf{F}}{\partial z_i}\right\}\right)$ are polynomial and Jacobian rings (in the $\{z_i\}$) graded by homogeneous degree, and μ involves polynomial multiplication.

Voisin's result (§3.2) on rational inequivalence of points p, q on a very general surface $X_{(s_0)}\subseteq\mathbb{P}^3$ $(d=2)$ comes from showing $\delta_2^{(0)}\mathfrak{Z}\neq 0$ for $\mathcal{Z}=p-q$. Roughly speaking, the image of $\mathfrak{Z}$ in $\left.\frac{\Omega^2_{\mathcal{S}}\otimes\mathcal{H}^{0,2}_{X_s}}{\bar{\nabla}\left(\Omega^1_{\mathcal{S}}\otimes\mathcal{H}^{1,1}_{X_s}\right)}\right|_{s_0}$ is identified with the difference of the functionals $\tilde{p}_*,\,\tilde{q}_*:\,H^0\left(X_{s_0},\,\left.\Omega^{\dim(\mathcal{S})}_{\mathfrak{X}}\right|_{X_{s_0}}\right)\to\left.K_{\mathcal{S}}\right|_{s_0}\cong\mathbb{C}$ induced by the spreads of p and q. As a vector bundle, $\left.\Omega^{\dim(\mathcal{S})}_{\mathfrak{X}}\right|_{X_{s_0}}$ $(=V\to X_{s_0})$ is very ample for $D\geq 7$ (this is the technical part and involves the sort of polynomial algebra just described); hence the natural map $\mathbb{P}(V^\vee)\to\mathbb{P}\left(H^0(X_{s_0},\,\Omega^{\dim(\mathcal{S})}_{\mathfrak{X}}|_{X_{s_0}})^\vee\right)$ is injective. Since $\tilde{p}$, $\tilde{q}$ plainly yield elements of $V^\vee$ lying over different points of X_{s_0} (hence different in $\mathbb{P}V^\vee$) and mapping resp. to $\tilde{p}_*$, $\tilde{q}_*$, these latter are independent as desired. Fakhruddin elegantly generalized this approach to $d>2$ and essentially shows $\delta_d^{(0)}\mathfrak{Z}\neq 0$ (for $\mathcal{Z}$ any $\mathbb{Q}$-linear degree-0 combination of the n points in question).

Below in §6.2 we will study "general" cycles in $\ker(Alb)\subseteq CH_0(\mathcal{C}\times\mathcal{C}_{/\mathbb{C}})$ where $\mathcal{C}$ is defined $/\bar{\mathbb{Q}}$ (i.e., "special"), but there is the following result for a "special" cycle on the self-product of a general curve. Write $K_{\mathcal{C}}$ for the canonical divisor and $\iota^\Delta:\mathcal{C}=\Delta_{\mathcal{C}}\hookrightarrow\mathcal{C}\times\mathcal{C}$.

Theorem (Griffiths-Green, [GG2]). Let $\mathcal{C}$ be general of genus $g\geq 4$; then

$$\mathcal{Z}_{K_{\mathcal{C}}} := K_{\mathcal{C}}\times K_{\mathcal{C}}-(2g-2)\iota^\Delta_* K_{\mathcal{C}} \overset{\mathrm{rat}}{\not\equiv} 0$$

in $CH^2(\mathcal{C}\times\mathcal{C})$.

Once again the proof shows $\delta_2^{(0)}\mathfrak{Z}\neq 0$. Since Griffiths and Green use Schiffer variations to spread out $\mathcal{C}\times\mathcal{C}$, an explicit local computation replaces polynomial algebra.

We conclude with an example to demonstrate the entry of the higher differentials in the spectral sequence $\mathcal{E}_r^{\ell,m}(i)$. As a reference point, first consider curves $\mathcal{C}_1$, $\mathcal{C}_2\,/\bar{\mathbb{Q}}$ with $o_i\in\mathcal{C}_i(\bar{\mathbb{Q}})$, $p_i\in\mathcal{C}_i(\mathbb{C})$ very general (and such that $p_1\times p_2\in\mathcal{C}_1\times\mathcal{C}_2(\mathbb{C})$ is too). Then $p_i-o_i=:\mathcal{Z}_i$ each have $0\neq\delta_1^{(0)}\mathfrak{Z}_i\in\Gamma(\hat{B},\,\Omega^1_{\mathcal{C}_i}\otimes H^{0,1}(\mathcal{C}_i))$, while $(p_1-o_1)\times(p_2-o_2)=:\mathcal{Z}$ has spread $\mathfrak{Z}_1\times\mathfrak{Z}_2=:\mathfrak{Z}$ with essentially $[0\neq]\,\delta_1^{(0)}\mathfrak{Z}_1\otimes\delta_1^{(0)}\mathfrak{Z}_2=\delta_2^{(0)}\mathfrak{Z}\in\Gamma(\hat{B},\,\Omega^2_{\mathcal{C}_1\times\mathcal{C}_2}\otimes H^{0,2}(\mathcal{C}_1\times\mathcal{C}_2))$. (Note that the $\bar{\nabla}$'s are 0 as the $\mathcal{C}_i$ do not spread. The d_1's actually are not zero — $\mathcal{E}_2^{1,1}(2)$ and $\mathcal{E}_2^{0,2}(2)$ are 0 — but this does not affect $\mathcal{E}_2^{2,0}(2)$.)

Now we replace the $\mathcal{C}_i$ by two general elliptic curves. Write $E_\lambda=\overline{\{y^2=x(x-1)(x-\lambda)\}}\overset{\mathsf{X}_\lambda}{\to}\mathbb{P}^1_{[x]}$ for the fibers of the Legendre family $\mathcal{E}\overset{\pi}{\to}\mathcal{U}\subseteq\mathbb{A}^1_{[\lambda]}$ (omit singular fibers). Fix $x_0\in\mathbb{P}^1(\bar{\mathbb{Q}})\setminus\{0,1,\infty\}$ and choose continuously in λ a lift $q_\lambda\in\mathsf{X}_\lambda^{-1}(x_0)$; this will force us to pass to a double cover of Legendre, $\widetilde{\mathcal{E}}\to\widetilde{\mathcal{U}}$. Write $\mathcal{Q}=\bigcup_{\lambda\in\widetilde{\mathcal{U}}}\{q_\lambda\}$ and $o_\lambda=(0,0)\in E_\lambda$, $\mathcal{O}=\bigcup_{\lambda\in\widetilde{\mathcal{U}}}\{o_\lambda\}$. Take λ_1, λ_2 to be algebraically independent transcendental numbers.

Then $\mathcal{Z}_{\lambda_i}=q_{\lambda_i}-o_{\lambda_i}\in CH_0(E_{\lambda_i/\mathbb{C}})$ $(i=1,2)$ have the same spread $\mathfrak{Z}_1=\mathfrak{Z}_2=\mathcal{Q}-\mathcal{O}$ on $\widetilde{\mathcal{E}}$, which yields a normal function $\nu\in\Gamma(\widetilde{\mathcal{U}},\,\mathcal{J}^1_{\widetilde{\mathcal{E}}/\widetilde{\mathcal{U}}})$ defined by $\nu(\lambda)=AJ_{E_\lambda}(q_\lambda-o_\lambda)$. As this is 2-torsion for $\lambda=x_0$ but not elsewhere, it is nontrivial; by a monodromy argument with Picard-Lefschetz, it cannot be flat. Hence its infinitesimal invariant, which is $\delta_1\mathfrak{Z}_i\in\Gamma\left(\widetilde{\mathcal{U}},\,\frac{\Omega^1_{\widetilde{\mathcal{U}}}\otimes\mathcal{H}^1_{\widetilde{\mathcal{E}}/\widetilde{\mathcal{U}}}}{\nabla\left(\mathcal{O}_{\widetilde{\mathcal{U}}}\otimes\mathcal{F}^1\mathcal{H}^1_{\widetilde{\mathcal{E}}/\widetilde{\mathcal{U}}}\right)}\right)$, must be nonzero. However $\delta_1^{(0)}\mathfrak{Z}_i=0$ because $\bar{\nabla}:\mathcal{H}^{1,0}_{\widetilde{\mathcal{E}}/\widetilde{\mathcal{U}}}\to\Omega^1_{\widetilde{\mathcal{U}}}\otimes\mathcal{H}^{0,1}_{\widetilde{\mathcal{E}}/\widetilde{\mathcal{U}}}$ is surjective, hence $0\neq\delta_1^{(1)}\mathfrak{Z}_i\in\Gamma\left(B,\,\Omega^1_{\widetilde{\mathcal{U}}}\otimes\mathcal{H}^{1,0}_{\widetilde{\mathcal{E}}/\widetilde{\mathcal{U}}}=\mathcal{E}_\infty^{0,1}(1)\right)$.

Now the spread of $\mathcal{Z}=\mathcal{Z}_{\lambda_1}\times\mathcal{Z}_{\lambda_2}$ is $\mathfrak{Z}=\mathfrak{Z}_1\times\mathfrak{Z}_2$ (self-product) on $\widetilde{\mathcal{U}}\times\widetilde{\mathcal{U}}\times\widetilde{\mathcal{E}}\times\widetilde{\mathcal{E}}$, and one expects $0=\delta_1^{(0)}\mathfrak{Z}=\delta_1^{(1)}\mathfrak{Z}$, $0\neq\delta_1^{(1)}\mathfrak{Z}_1\otimes\delta_1^{(1)}\mathfrak{Z}_2=\delta_2^{(2)}\mathfrak{Z}\in\Gamma(B,\mathcal{E}_\infty^{0,2}(2))$. However, $\mathcal{E}_\infty^{0,2}(2)$ is not $\Omega^2_{\widetilde{\mathcal{U}}\times\widetilde{\mathcal{U}}}\otimes\mathcal{H}^{2,0}_{\widetilde{\mathcal{E}}\times\widetilde{\mathcal{E}}/\widetilde{\mathcal{U}}\times\widetilde{\mathcal{U}}}$ but a quotient; moreover, $\mathcal{Z}\overset{\text{rat}}{\equiv}0$, hence $\delta_2^{(2)}\mathfrak{Z}=0$. The rational equivalence is direct: let B (the bielliptic curve) denote the normalization of $E_{\lambda_1}\times_{\mathbb{P}^1_{[x]}}E_{\lambda_2}\subseteq E_{\lambda_1}\times E_{\lambda_2}$ and $\iota:B\to E_{\lambda_1}\times E_{\lambda_2}$ be the composition. Then clearly $\iota(B)\ni(q_1,q_2)$ and (o_1,o_2); let $\iota(o)=(o_1,o_2)$ and $\iota(q)=(q_1,q_2)$. Define $\Pi:B\times B\to E_{\lambda_1}\times E_{\lambda_2}$ by $\Pi(b,b')=(\pi_1(\iota(b)),\,\pi_2(\iota(b')))$, and note that $\Pi_*\{(q,q)-(0,q)-(q,0)+(0,0)\}=\mathcal{Z}$. But B is genus 2, hence hyperelliptic, and so by Beauville-Voisin (see §6.2 below) the cycle in brackets is $\overset{\text{rat}}{\equiv}0$.

Write $\mathcal{X}=\widetilde{\mathcal{E}}\times\widetilde{\mathcal{E}}$, $S=\widetilde{\mathcal{U}}\times\widetilde{\mathcal{U}}$. We conclude that the higher differentials — maps from $\ker\left\{\Omega^1_S\otimes\mathcal{H}^{2,0}_{\mathcal{X}/S}\overset{\bar{\nabla}}{\to}\Omega^2_S\otimes\mathcal{H}^{1,1}_{\mathcal{X}/S}\right\}$ and

$$\ker\left\{\ker\left\{\Omega^1_S\otimes\mathcal{H}^{1,1}_{\mathcal{X}/S}\to\Omega^2_S\otimes\mathcal{H}^{0,2}_{\mathcal{X}/S}\right\}\to\Omega^2_S\otimes\mathcal{H}^{1,1}_{\mathcal{X}/S}\right\}$$

to $\Omega^2_S \otimes \mathcal{H}^{2,0}_{\mathcal{X}/S}$ (or a quotient of this in the latter case) — must hit the class of $\delta_2^{(2)}\mathfrak{Z}$ locally, though this looks like a hard computation to do explicitly.

5.5. Other filtrations. We finish with a brief description of 3 additional filtrations obtained by taking the inverse image of a filtration under a cycle map of some kind.

(i) Raskind ([Ra]). This served as the original motivation for [**L1**]. Assume X/K, K finitely generated $/\mathbb{Q}$ (not $\bar{\mathbb{Q}}$). By work of Jannsen [**J3**], the Hochschild-Serre spectral sequence for continuous étale cohomology

$$E_2^{p,q}(d) = H^p_{\text{cont}}\left(Gal(\bar{K}/K),\, H^q_{et}(X_{\bar{K}}, \mathbb{Q}_\ell(d))\right) \implies H^{p+q}_{\text{cont}}(X, \mathbb{Q}_\ell(d))$$

degenerates at E_2. Pulling back the resulting filtration by his cycle map $CH_0(X) \to H^{2d}_{\text{cont}}(X, \mathbb{Q}_\ell(d))$ gives $F^\bullet_R$ and one has "higher ℓ-adic AJ" maps from $Gr^i_{F_R}$ to $E_2^{i,\,2d-i}(d)$. Clearly $F^\bullet_R \supseteq F^\bullet_S$; this is equality if HLC holds and Jannsen's cycle map is injective.

For the remaining filtrations, assume for simplicity that X is defined over a number field k, with cycles $/K$ f.g. $/k$.

(ii) M. Saito ([mS]). By means of various categories of realizations $\mathcal{M}$, one can put a variety of [hopefully not!] different filtrations $F^\bullet_{\mathcal{M}}$ on $CH^*(X_K)$. The idea is once again to pull back a suitable Leray filtration along the composition of spreading out $(/k)$ and taking some kind of Deligne (or absolute Hodge) class. For example, $F^\bullet_{\text{MHS}}$ would just be $\mathcal{L}^\bullet$. One always has $F^\bullet_S \subseteq F^\bullet_{\mathcal{M}}$.

We are interested (for §7.3 below) in $\mathcal{M} = \mathcal{M}_\ell$, whose objects consist roughly of MHS (w./ polarizable $Gr^W_\bullet$'s) and $\mathbb{Q}_\ell$-vector spaces with continuous $Gal(\bar{k}/k)$-action, together with comparison isomorphisms (under which polarizations and Galois action must be compatible). Cohomology $H^*(X/k, \mathbb{Q}) \in \mathcal{M}_\ell$ is then defined using $H^*(X_{\mathbb{C}})$ for the MHS and $H^*_{et}(X_{\bar{k}}, \mathbb{Q}_\ell)$ for the $\mathbb{Q}_\ell$-v.s.

Saito defines Deligne cohomology[15] for $\mathcal{M}$, e.g. for our purposes[16]

$$H^{2d}_{\mathcal{D},\mathcal{M}}(X \times \mathcal{U}/k, \mathbb{Q}(d)) := Ext^{2d}_{D^b\mathcal{M}}(\mathbb{Q}(-d),\, \mathcal{K}^\bullet_{\mathcal{H}}(X \times \mathcal{U}/k))$$

where $D^b\mathcal{M}$ is the bounded derived category and $\mathcal{K}^\bullet_{\mathcal{H}}$ is a complex of $\mathcal{M}$-objects with $H^i(\mathcal{K}^\bullet_{\mathcal{H}}) \cong H^i(X \times \mathcal{U}/k, \mathbb{Q})$. This is computed by a spectral sequence $E_2^{p,q} = Ext^p_{\mathcal{M}}(\mathbb{Q}(-d),\, H^q(X \times \mathcal{U}/k))$, in the sense that there is an increasing filtration $\mathcal{E}^\bullet$ on $H^{2d}_{\mathcal{D},\mathcal{M}}$ with $Gr^j_{\mathcal{E}} \cong E_\infty^{j,\,2d-j}$. (For $j \geq 2$ this is known to be zero for $\mathcal{M} = $ MHS but not $\mathcal{M}_\ell$.)

Moreover, one has a Leray filtration $\mathcal{L}^\bullet_{\mathcal{M}}$ (with respect to $X \times \mathcal{U} \xrightarrow{\pi} \mathcal{U}$) on whose graded pieces one can put a similar filtration $\mathcal{E}^\bullet$ so that [writing $H^*(\cdot)$ for $H^*(\cdot/k,\, \mathbb{Q})$]

$$Gr^0_{\mathcal{E}} Gr^i_{\mathcal{L}_{\mathcal{M}}} H^{2d}_{\mathcal{D},\mathcal{M}}\left(X \times \mathcal{U}/k,\, \mathbb{Q}(d)\right) \hookrightarrow Hom_{\mathcal{M}}\left(\mathbb{Q}(-d),\, \underline{H^i}(\mathcal{U}) \otimes H^{2d-i}(X)\right)$$

$$Gr^1_{\mathcal{E}} Gr^i_{\mathcal{L}_{\mathcal{M}}} H^{2d}_{\mathcal{D},\mathcal{M}}\left(X \times \mathcal{U}/k,\, \mathbb{Q}(d)\right) \hookrightarrow Ext^1_{\mathcal{M}}\left(\mathbb{Q}(-d),\, [W_i]H^{i-1}(\mathcal{U}) \otimes H^{2d-i}(X)\right)$$

$$\to Ext^1_{\mathcal{M}}\left(\mathbb{Q}(-d),\, H^{i-1}(\mathcal{U}) \otimes \frac{H^{2d-i}(X)}{F_h^{d-i+1}}\right)$$

---

[15]for $\mathcal{M} = MHS$ this is really absolute Hodge cohomology.

[16]as usual $\mathcal{U}[\subseteq \mathcal{S}]$ affine Zar. op.

and for $j \geq 2$, $Gr^j_{\mathcal{E}} Gr^i_{\mathcal{L}_{\mathcal{M}}}$ is a subquotient of $Ext^j_{\mathcal{M}}\left(\mathbb{Q}(-d), H^{i-j}(\mathcal{U}) \otimes H^{2d-i}(X)\right)$ (again, zero for $\mathcal{M} = \text{MHS}$ but not $\mathcal{M}_\ell$).

Finally, we take the limit over $\mathcal{U} \subseteq \mathcal{S}$ and pull $\mathcal{L}^\bullet_{\mathcal{M}}$ back by the cycle map $CH_0(X_K) \overset{\cong}{\to} CH^d(X \times \eta_{\mathcal{S}/k}) \to \varinjlim_{\mathcal{U}} H^{2d}_{\mathcal{D},\mathcal{M}}(X \times \mathcal{U}/k, \mathbb{Q}(d))$ to get $F^\bullet_{\mathcal{M}}$. As there are "forgetful" maps[17] from the graded pieces of $H^{2d}_{\mathcal{D},\mathcal{M}}$ to those of $H^{2d}_{\mathcal{D}(,\text{MHS})}$, $F^\bullet_{\mathcal{M}} \subseteq \mathcal{L}^\bullet$ on CH_0; if HC+BBC^q hold then they are equal.

(iii) Griffiths-Green ([GG1]). This filtration is closest in spirit to $\mathcal{L}^\bullet$. (Indeed, our presentation of the latter, including some notation, is influenced by [**GG1**].) In the construction of §4.2, instead of filtering $\underline{H^{2d}_{\mathcal{D}}}(X \times \eta_{\mathcal{S}}, \mathbb{Q}(d))$, filter $H^{2d}_{\mathcal{D}}(X \times \mathcal{S}, \mathbb{Q}(d))$ by $\mathcal{L}^\bullet_{\mathcal{S}}$ (Leray for $X \times \mathcal{S} \to \mathcal{S}$), and pull this back to $CH^d(X \times \mathcal{S}_{/\bar{\mathbb{Q}}})$. We say $\langle \mathcal{Z} \rangle \in CH^d(X_K)$ belings to F^i_{GG} iff it has a choice of complete spread $\bar{\mathfrak{Z}}$ with $\langle \bar{\mathfrak{Z}} \rangle \in \mathcal{L}^i_{\mathcal{S}} CH^d(X \times \mathcal{S}_{/\bar{\mathbb{Q}}})$ $[\Leftrightarrow c_{\mathcal{D}}(\bar{\mathfrak{Z}}) \in \mathcal{L}^i_{\mathcal{S}} H^{2d}_{\mathcal{D}}]$.

A priori this is a stronger condition than $\mathcal{L}^i$ on the cycle $\mathcal{Z}$. Under

$$H^{2d}_{\mathcal{D}}(X \times \mathcal{S}, \mathbb{Q}(d)) \to \underline{H^{2d}_{\mathcal{D}}}(X \times \eta_{\mathcal{S}}, \mathbb{Q}(d)),$$

one easily shows $\mathcal{L}^\bullet$ (on $\underline{H^{2d}_{\mathcal{D}}}$) is the *image* of $\mathcal{L}^\bullet_{\mathcal{S}}$ (on $H^{2d}_{\mathcal{D}}$), but $\mathcal{L}^\bullet_{\mathcal{S}}$ is *not* the preimage of $\mathcal{L}^\bullet$. Hence (for cycles) $\bar{\mathfrak{Z}} \in \mathcal{L}^i_{\mathcal{S}} \implies \mathfrak{Z} \in \mathcal{L}^i$ but not the other way around. However, if we assume the HC (only!), $F^\bullet_{GG} = \mathcal{L}^\bullet$ (on cycles $Z^d(X_K)$) by the argument in [**mS**, sec. 1.6].

6. Zero-cycles on products of varieties

We want to investigate the behavior of our invariants with respect to *products* of 0-cycles, considered on the *product* of the varieties on which they lie individually. The resulting cycles are called *exterior products*; this is clearly a well-defined operation on the level of Chow groups.

In the three subsections that follow, we consider the following cases:

(1) cycles $\mathcal{Z} = \mathcal{Z}_1 \times \mathcal{Z}_2$ such that $\mathfrak{Z} = \mathfrak{Z}_1 \times \mathfrak{Z}_2$ — the spread respects the product structure. In this case we can essentially take "products" of higher invariants of $\mathcal{Z}_1$ and $\mathcal{Z}_2$ to get higher invariants of $\mathcal{Z}$, and we have some quite general nonvanishing results.

(2) self-products $\mathcal{Z} \times \mathcal{Z}$ (and higher powers) of 0-cycles, especially for $\mathcal{Z}$ a divisor on a curve.

(3) 0-cycles on a product of curves whose spreads wildly disrespect the product structure.

The first section contains no proofs, as they are like more complicated versions of the proof of Asakura's result in §6.2 and appear in [**K1**]; stating the results is complicated enough anyway. A fourth subsection explains a situation similar to (3) for certain relative cycles.

[17]One caveat here is that $F^j_h(H^*(X))$ in the category of $\mathcal{M}_\ell$-objects (F^j_h always means the maximal subobject contained in F^j) may not map (under the forgetful functor $\mathcal{M}_\ell \to \text{MHS}$) to $F^j_h(H^*(X))$ in the category of MHS, but to something between this and $N^j(H^*(X))$ (which is the same for both categories). Of course, this is no problem if you assume GHC, but we don't. However, this issue can be safely ignored in §7.3.

6.1. Products of algebraically independent cycles. Let Y_1, Y_2 be smooth projective $/\bar{\mathbb{Q}}$ of respective dimensions d_1, d_2 and let $K \subseteq \mathbb{C}$ be finitely generated $/\bar{\mathbb{Q}}$ of trdeg. j. Take

(a) $\mathcal{V} \in \mathcal{L}^j CH_0(Y_{1/K})$ such that a complete spread $\overline{\mathfrak{V}} \in Z^{d_1}(Y_1 \times \mathcal{S})$ induces a nonzero map $\Omega^j(Y_1) \to \Omega^j(\mathcal{S})$,

(b) $\mathcal{W} \in CH_0^{\text{hom}}(Y_{2/\bar{\mathbb{Q}}})$ such that $0 \neq AJ_{Y_2}(\mathcal{W}) \in J^{d_2}(Y_2)$.

Then from [**K1**] one has the following.

THEOREM 6.1. *$\mathcal{Z} := \mathcal{V}\times\mathcal{W} \overset{rat}{\not\equiv} 0$ in $CH_0(Y_1\times Y_{2/K})$; in particular, $[AJ(3)]_j^{(tr)} \neq 0$.*

EXAMPLE 6.2. This immediately justifies Thm. 3.5(B), by taking

$$\mathcal{V} = (p_1 - o_1) \times \cdots \times (p_{\ell-1} - o_{\ell-1}) \qquad [d_1 = j = \ell - 1],$$

$$\text{and} \quad \mathcal{W} = \mathcal{W}_\ell \times o_{\ell+1} \times \cdots \times o_d \qquad [d_2 = d - \ell + 1].$$

REMARK 6.3. (i) For 0-cycles, the Theorem generalizes a result of Rosenschon and M. Saito ([**RS**]) which gives the special case $d_1 = 1 (= j)$. (This includes the $(\langle P\rangle - \langle O\rangle) \times (\langle Q\rangle - \langle 0\rangle)$ example of Remark 3.6(iii).) Note that in this case, one can replace (a) by the simpler statement that $\langle \mathcal{V}\rangle$ belongs to $CH_0^{\text{hom}}(Y_{1/\mathbb{C}})$ and is *not* defined $/\bar{\mathbb{Q}}$. (That is, no cycle rationally equivalent to $\mathcal{V}$ is defined $/\bar{\mathbb{Q}}$.)

(ii) The $(\langle P\rangle - \langle O\rangle)\times(\langle Q\rangle - \langle O\rangle)$ cycle can be used ([**K2**]) to construct explicit 0-cycles of trdeg. 1 in $\ker(Alb)$ on any special Kummer (K3) surface. In §8.2 we will show how to use the case $d_1 = 2$, $d_2 = 1$ to construct 0-cycles in $\mathcal{L}^3 CH_0$ on CY 3-folds arising from the Borcea-Voisin construction.

To generalize this even further, we define (for all $1 \leq j_2 \leq d_2$)

$$SF_h^{(1,\, d_2-j_2+1)}\left\{H^{j_2-1}(\mathcal{S}_2) \otimes H^{2d_2-j_2}(Y_2)\right\} :=$$

$$\text{largest subHS of } H^{j_2-1}(\mathcal{S}_2) \otimes H^{2d_2-j_2}(Y_2) \text{ lying in}$$

$$\ker\left(H^{j_2-1}(\mathcal{S}_2,\mathbb{C}) \otimes H^{2d_2-j_2}(Y_2,\mathbb{C}) \to H^{j_2-1,0}(\mathcal{S}_2,\mathbb{C}) \otimes H^{d_2-j_2,\, d_2}(Y_2,\mathcal{C})\right).$$

This contains

$$N^1H^{j_2-1}(\mathcal{S}_2) \otimes H^{2d_2-j_2}(Y_2) \; + \; H^{j_2-1}(\mathcal{S}_2) \otimes F_h^{d_2-j_2+1}H^{2d_2-j_2}(Y_2),$$

so one has a projection

$$J^{d_2}\left(\underline{H^{j_2-1}}(\eta_{\mathcal{S}_2}) \otimes \frac{H^{2d_2-j_2}(Y_2)}{F_h^{d_2-j_2+1}}\right) \overset{\Xi}{\twoheadrightarrow} J^{d_2}\left(\frac{H^{j_2-1}(\mathcal{S}_2) \otimes H^{2d_2-j_2}(Y_2)}{SF_h^{(1,\, d_2-j_2+1)}}\right)$$

and we define $\overline{[AJ(3)]_{j_2-1}^{tr}} := \Xi\left([AJ(3)]_{j_2-1}^{tr}\right)$. One can show that the cycles constructed by Theorem 6.1 have nontrivial $\overline{[AJ(3)]_j^{tr}}$.

Now take (subfields of $\mathbb{C}$) K_1, K_2 f.g. $/\bar{\mathbb{Q}}$ and

(a) $\mathcal{V} \in \mathcal{L}^{j_1}CH_0(Y_{1/K_1})$ with $cl_{Y_1}^{j_1}(\mathcal{V}) \neq 0$,

(b) $\mathcal{W} \in \mathcal{L}^{j_2}CH_0(Y_{2/K_2})$ with either
 (i) $cl_{Y_2}^{j_2}(\mathcal{W}) \neq 0$ OR
 (ii) $\overline{AJ_{Y_2}^{j_2-1}(\mathcal{W})^{tr}} \neq 0$ and $cl_{Y_2}^{j_2}(\mathcal{W}) = \cdots = cl_{Y_2}^{d_2}(\mathcal{W}) = 0$.

Set $K = \bar{\mathbb{Q}}(K_1, K_2)$, the field generated by K_1 and K_2 over $\bar{\mathbb{Q}}$ — this is the natural field of definition of the product $\mathcal{V} \times \mathcal{W}$. For the "algebraic independence" requirement (to force the spread of $\mathcal{V} \times \mathcal{W}$ to be $\mathfrak{V} \times \mathfrak{W}$), we demand $\mathrm{trdeg}_{/\bar{\mathbb{Q}}}(K) = \mathrm{trdeg}_{/\bar{\mathbb{Q}}}(K_1) + \mathrm{trdeg}_{/\bar{\mathbb{Q}}}(K_2)$. Then from **[K1]** we have:

THEOREM 6.4. *Assume the GHC; then* $0 \not\overset{rat}{\equiv} \mathcal{V} \times \mathcal{W} \in \mathcal{L}^{j_1+j_2} CH_0(Y_1 \times Y_{2/K})$. *In particular, when (i) holds,* $cl^{j_1+j_2}_{Y_1 \times Y_2}(\mathcal{V} \times \mathcal{W}) \neq 0$; *while if (ii) holds,* $AJ^{j_1+j_2-1}_{Y_1 \times Y_2}(\mathcal{V} \times \mathcal{W})^{(tr)} \neq 0$.

This, then, is the sense in which invariants "multiply" nontrivially. The theorem is obviously false without the independence requirement, e.g. for self-products ($\mathcal{V} = \mathcal{W}$, $K = K_1 = K_2$) — see the Beauville-Voisin theorem below.

6.2. Self-products of cycles. In this subsection we prove two results (approximate converses of one another) on the "symmetric square" of a divisor $p - o$ on a curve, concluding with a brief treatment of its higher "powers". As above, denote by $K_{\mathcal{C}}$ the canonical divisor of $\mathcal{C}$.

Theorem (Beauville-Voisin, [BV]). Let o, p be two points on a smooth hyperelliptic curve $\mathcal{C}/\mathbb{C}$ of genus $g \geq 1$, with $(2g-2)o \overset{\mathrm{rat}}{\equiv} K_{\mathcal{C}}$ in $CH_0(\mathcal{C})$. Then $\mathcal{Z} := (p - o) \times (p - o) \overset{\mathrm{rat}}{\equiv} 0$ in $CH_0(\mathcal{C} \times \mathcal{C})$.

PROOF. The assumption implies $o \in \mathcal{C}(\mathbb{C})$ is Weierstrass, hence a ramification point of the (intrinsic) canonical map $\varphi : \mathcal{C} \overset{\deg 2}{\to} \mathbb{P}^1 [\subseteq \mathbb{P}^{g-1}]$. Writing $\sigma : \mathcal{C} \to \mathcal{C}$ for the corresponding involution, we set $f(z) := \frac{z - \varphi(p)}{z - \varphi(o)}$ and $F := f \circ \varphi$, and note $(F) = p + \sigma(p) - 2 \cdot o$. Hence (and this is worth drawing for oneself)

$$(\iota_{\Delta_{\mathcal{C}}})_* (F) + (\iota_{\{o\} \times \mathcal{C}})_* \left(\frac{1}{F^2} \right) + (\iota_{\{p\} \times \mathcal{C}})_* (F) + (\iota_{\mathcal{C} \times \{\sigma(p)\}})_* \left(\frac{1}{F} \right) =$$

$$2 \cdot \{(p,p) - (o,p) - (p,o) + (o,o)\} \;=\; 2 \cdot \mathcal{Z}$$

shows that (always modulo torsion) $\mathcal{Z} \overset{\mathrm{rat}}{\equiv} 0$. □

In addition, this result holds *vacuously* if $\mathcal{C}$ is a general curve of genus $g \geq 3$; that is, there cannot exist $\omega \in \Omega^1(\mathcal{C})$ with $(\omega) = (2g-2) \cdot o$. Indeed, for such a curve all Weierstrass points are normal, hence cannot have $2g - 1$ as a gap value. (This is false for arbitrary nonhyperelliptic curves.)

Before giving the "converse", a beautiful theorem due to Asakura, we digress briefly on AJ maps. Consider Y_1, Y_2 smooth projective of dimensions d_1, d_2, $d_1 + d_2 =: d$, and let $\Pi_i : Y_1 \times Y_2 \to Y_i$ be the natural projections ($i = 1, 2$). Take $\mathcal{V} \in CH^{d-p}(Y_1)$ and $\mathfrak{Z} \in CH^p_{\mathrm{hom}}(Y_1 \times Y_2)$. The restriction of functionals

$$\left\{F^{d-p+1} H^{2d-2p+1}(Y_1 \times Y_2, \mathbb{C})\right\}^\vee \twoheadrightarrow \left\{F^{d-p+1}\left(\mathbb{Q}[\mathcal{V}] \otimes H^1(Y_2, \mathbb{C})\right)\right\}^\vee$$

$$\overset{\cong}{\to} \left\{F^1 H^1(Y_2, \mathbb{C})\right\}^\vee$$

is well-defined modulo periods, hence induces

$$\mathrm{pr}_{\mathcal{V}} : J^p(Y_1 \times Y_2) \to J^{d_2}(Y_2).$$

To describe this map on $\bar{\mathfrak{Z}}$, let $\partial^{-1}\bar{\mathfrak{Z}}$ be a bounding chain and restrict the functional $\int_{\partial^{-1}\bar{\mathfrak{Z}}}$ to $\mathbb{Q}[\mathcal{V}] \otimes F^1H^1(Y_2, \mathbb{C})$. If $\xi_{\mathcal{V}} \in \Gamma(\Omega^{d-p,\, d-p}_{Y_1^{\infty}})$ represents $[\mathcal{V}]$ and $\omega \in \Omega^1(Y_2)$, then

$$\int_{\partial^{-1}\bar{\mathfrak{Z}}} \Pi_1^*\xi_{\mathcal{V}} \wedge \Pi_2^*\omega = \int_{(\partial^{-1}\bar{\mathfrak{Z}})\cap(\mathcal{V}\times Y_2)} \Pi_2^*\omega = \int_{\Pi_2^*\{(\partial^{-1}\bar{\mathfrak{Z}})\cap(\mathcal{V}\times Y_2)\}} \omega.$$

Noting that $\partial\left(\Pi_2\left\{(\partial^{-1}\bar{\mathfrak{Z}})\cap(\mathcal{V}\times Y_2)\right\}\right) = \Pi_2\left\{\bar{\mathfrak{Z}}\cap(\mathcal{V}\times Y_2)\right\} =: \bar{\mathfrak{Z}}_*\mathcal{V}$, we may write this $\int_{\partial^{-1}(\bar{\mathfrak{Z}}_*\mathcal{V})} \omega$; it follows that

$$\mathrm{pr}_{\mathcal{V}}\left(AJ_{Y_1\times Y_2}(\bar{\mathfrak{Z}})\right) = Alb_{Y_2}\left(\bar{\mathfrak{Z}}_*\mathcal{V}\right).$$

Using this we now prove:

Theorem (Asakura, [As]). Let $\mathcal{C}$ be a smooth projective curve $/\bar{\mathbb{Q}}$, with $p \in \mathcal{C}(\mathbb{C})$ very general and $o \in \mathcal{C}(\bar{\mathbb{Q}})$. Then $(2g-2)o \overset{\text{rat}}{\not\equiv} K_{\mathcal{C}}$ in $CH_0(\mathcal{C}_{/\bar{\mathbb{Q}}}) \implies \mathcal{Z} := (p-o)\times(p-o) \overset{\text{rat}}{\not\equiv} 0$ in $CH_0(\mathcal{C}\times\mathcal{C}_{/\mathbb{C}})$; more precisely, $[AJ(\mathfrak{Z})]_1^{tr} \neq 0$.

Proof. We first construct a choice of complete $\bar{\mathbb{Q}}$-spread $\bar{\mathfrak{Z}} \in Z^2(\mathcal{C}\times\mathcal{C}\times\mathcal{C}_{/\bar{\mathbb{Q}}})$ for $\mathcal{Z}$. Write $\mathcal{P}_i$, $\mathcal{P}_{ij}$ and ι_i, ι_{ij} for projections and inclusions, e.g. $\iota_{13} : \mathcal{C}\times\mathcal{C} \hookrightarrow \mathcal{C}\times\mathcal{C}\times\mathcal{C}$ sends $(q_1, q_2) \mapsto (q_1, o, q_2)$ and $\mathcal{P}_{23} : \mathcal{C}\times\mathcal{C}\times\mathcal{C} \twoheadrightarrow \mathcal{C}\times\mathcal{C}$ sends $(q_1, q_2, q_3) \mapsto (q_2, q_3)$. Set $\Delta_{\mathcal{C}}^{[3]} := \{(Q,Q,Q) \mid Q \in \mathcal{C}\}$ and

$$\bar{\mathfrak{Z}} := \Delta_{\mathcal{C}}^{[3]} - \iota_*^{12}\Delta_{\mathcal{C}} - \iota_*^{13}\Delta_{\mathcal{C}} - \iota_*^{23}\Delta_{\mathcal{C}} + \iota_*^1\mathcal{C} + \iota_*^2\mathcal{C} + \iota_*^3\mathcal{C};$$

and note that $\bar{\mathfrak{Z}} \cap (\{p\}\times\mathcal{C}\times\mathcal{C}) = (p,p) - (o,p) - (p,o) + (o,o) = \mathcal{Z}$. Moreover $(\mathcal{P}_{23})_* : H^4(\mathcal{C}\times\mathcal{C}\times\mathcal{C})^{\mathfrak{S}_3} \to H^2(\mathcal{C}\times\mathcal{C})^{\mathfrak{S}_2}$ is an isomorphism[18] while $(\mathcal{P}_{23})_*\bar{\mathfrak{Z}} = 0$, so $\bar{\mathfrak{Z}} \overset{\text{hom}}{\equiv} 0$.

Consider the commuting diagram of Jacobians

$$\begin{array}{ccccc}
J^2\left(H^3(\mathcal{C}\times\mathcal{C}\times\mathcal{C})\right) & \twoheadrightarrow & J^2\left(H^1(\mathcal{C})\otimes H^2(\mathcal{C}\times\mathcal{C})\right) & \twoheadrightarrow & J^2\left(H^1(\mathcal{C})\otimes \frac{H^2(\mathcal{C}\times\mathcal{C})}{F_h^1H^2(\mathcal{C}\times\mathcal{C})}\right) \\
 & \searrow^{\mathrm{pr}_{\otimes 3}} & \downarrow & & \downarrow \\
J^2\left(\{H^1(\mathcal{C})^{\otimes 3}\}^{\mathfrak{S}_3}\right) & \overset{\mathcal{I}}{\hookrightarrow} & J^2\left(H^1(\mathcal{C})^{\otimes 3}\right) & \overset{\mathrm{pr}_F}{\twoheadrightarrow} & J^2\left(H^1(\mathcal{C})\otimes \frac{H^1(\mathcal{C})\otimes H^1(\mathcal{C})}{F_h^1\{H^1(\mathcal{C})^{\otimes 2}\}}\right) \\
 & & & & \| \\
 & & & & \frac{J^2\left(H^1(\mathcal{C})^{\otimes 3}\right)}{J^2\left(H^1(\mathcal{C})\otimes F_h^1\{H^1(\mathcal{C})^{\otimes 2}\}\right)}
\end{array}$$

where the composition in the top row sends $AJ_{\mathcal{C}\times\mathcal{C}\times\mathcal{C}}(\bar{\mathfrak{Z}})$ to $[AJ(\mathfrak{Z})]_1^{tr}$, which we must show nonzero. By the diagram, this is accomplished if we have $(\mathrm{pr}_F \circ \mathrm{pr}_{\otimes 3})(AJ(\bar{\mathfrak{Z}})) \neq 0$. The remainder of the proof therefore proceeds in 2 steps: we show first (i) that $0 \neq \mathrm{pr}_{\otimes 3}(AJ(\bar{\mathfrak{Z}})) \in \mathrm{im}(\mathcal{I})$, then (ii) that $\mathrm{pr}_F \circ \mathcal{I}$ is injective.

We apply the discussion preceding this theorem with $Y_1 = \mathcal{C}\times\mathcal{C}$, $Y_2 = \mathcal{C}$, and $\mathcal{V} = \Delta_{\mathcal{C}} - \{0\}\times\mathcal{C} - \mathcal{C}\times\{0\} \in CH^1(\mathcal{C}\times\mathcal{C})$. As the Künneth components $[\mathcal{V}]_0 \in H^0(\mathcal{C})\otimes H^2(\mathcal{C})$ and $[\mathcal{V}]_2 \in H^2(\mathcal{C})\otimes H^0(\mathcal{C})$ are zero, $[\mathcal{V}] = [\mathcal{V}]_1$ and $\mathbb{Q}[\mathcal{V}]\otimes H^1(\mathcal{C})$ is

[18] $\mathfrak{S}_n$ denotes the n^{th}symmetric group

a subHS of $H^1(\mathcal{C})^{\otimes 3}$. Hence $\mathrm{pr}_{\mathcal{V}} : J^2(\mathcal{C}\times\mathcal{C}\times\mathcal{C}) \to J^1(\mathcal{C})$ factors through $\mathrm{pr}_{\otimes 3}$; so if $Alb_{\mathcal{C}}(\bar{\mathfrak{Z}}^*\mathcal{V}) \neq 0$ then $\mathrm{pr}_{\otimes 3}(AJ(\bar{\mathfrak{Z}})) \neq 0$. Indeed, we have[19]

$$\begin{aligned}\bar{\mathfrak{Z}}\cdot\mathcal{P}_{12}^*\mathcal{V} &= \Delta_{\mathcal{C}}^{[3]}\cdot(\Delta_{\mathcal{C}}\times\mathcal{C}) - (\Delta_{\mathcal{C}}\times\{0\})\cdot(\Delta_{\mathcal{C}}\times\mathcal{C})\\ &= -(\iota^{\Delta_{\mathcal{C}}^{[3]}})_*K_{\mathcal{C}} + (\iota^{\Delta_{\mathcal{C}}\times\{0\}})_*K_{\mathcal{C}},\end{aligned}$$

and applying $(\mathcal{P}_3)_*$ gives $\bar{\mathfrak{Z}}^*\mathcal{V} = -K_{\mathcal{C}}+(2g-2)o$. Since $K_{\mathcal{C}} \overset{\mathrm{rat}}{\not\equiv} (2g-2)o$, $AJ_{\mathcal{C}}(\bar{\mathfrak{Z}}^*\mathcal{V}) \neq 0$.

Moreover, $\mathfrak{S}_3$ operates on $CH^2(\mathcal{C}\times\mathcal{C}\times\mathcal{C})$ and $J^2\left((H^1(\mathcal{C})^{\otimes 3})\right)$; and since $\bar{\mathfrak{Z}} \in \left(CH^2(\mathcal{C}\times\mathcal{C}\times\mathcal{C})\right)^{\mathfrak{S}_3}$, we have $[0 \neq]\, \mathrm{pr}_{\otimes 3}(AJ(\bar{\mathfrak{Z}})) \in \left\{J^2\left(H^1(\mathcal{C})^{\otimes 3}\right)\right\}^{\mathfrak{S}_3} = J^2\left(\left\{H^1(\mathcal{C})^{\otimes 3}\right\}^{\mathfrak{S}_3}\right)$ and (i) is verified.

Injectivity of the bottom composition will follow from the fact that

$$\left\{H^1(\mathcal{C})^{\otimes 3}\right\}^{\mathfrak{S}_3} \cap H^1(\mathcal{C})\otimes F_h^1\left\{H^1(\mathcal{C})^{\otimes 2}\right\} = \{0\}.$$

These are $\mathbb{Q}$-vector spaces. Given a basis $\{\omega_i\} \subseteq \Omega^1(\mathcal{C})$, an element of $F_h^1\left\{H^1(\mathcal{C})^{\otimes 2}\right\}$ may be written uniquely as $\sum_j \omega_j\otimes a_j + \sum_j \bar{\omega}_j\otimes\alpha_j$ for $\alpha_j \in H^{1,0}(\mathcal{C},\mathbb{C})$, $a_j \in H^{0,1}(\mathcal{C},\mathbb{C})$. For coefficients (hence periods) in $\mathbb{Q}$, the element must be fixed under conjugation, hence $a_j = \bar{\alpha}_j$. An element of $H^1(\mathcal{C})\otimes F_h^1\{H^1(\mathcal{C})^{\otimes 2}\}$ thus takes the form[20]

$$\sum_{i,j}\omega_i\otimes\omega_j\otimes\bar{\alpha}_j + \sum_{i,j}\omega_i\otimes\bar{\omega}_j\otimes\alpha_j + \sum_{i,j}\bar{\omega}_i\otimes\omega_j\otimes\bar{\beta}_j + \sum_{i,j}\bar{\omega}_i\otimes\bar{\omega}_j\otimes\beta_j.$$

If the last line is invariant under $\mathfrak{S}_3$ then, applying (12) and switching i and j, it must equal

$$\sum_{i,j}\omega_i\otimes\omega_j\otimes\bar{\alpha}_i + \sum_{i,j}\omega_i\otimes\bar{\omega}_j\otimes\bar{\beta}_i + \sum_{i,j}\bar{\omega}_i\otimes\omega_j\otimes\alpha_i + \sum_{i,j}\bar{\omega}_i\otimes\bar{\omega}_j\otimes\beta_i.$$

But if $\alpha_j = \bar{\beta}_i$ then they lie in $H^{1,0}\cap H^{0,1} = \{0\}$. □

REMARK 6.5. It would be interesting to see whether any similar results can be proved (using intersection theory) if we replace $\mathcal{C}$ by X of higher dimension. The generalization would not be straightforward, however. If X is a smooth projective surface with $H^1(X) = H^3(X) = 0$, the cycle analogous to $\mathcal{Z}$ in Asakura's theorem is $\overset{\mathrm{rat}}{\equiv} 0$ if BBC^q holds. This is because an exactly analogous choice of $\bar{\mathfrak{Z}}$ on $X\times X\times X$ gives $\bar{\mathfrak{Z}} \overset{\mathrm{hom}}{\equiv} 0$, and then necessarily $AJ(\bar{\mathfrak{Z}}) = 0$ since $H^7(X\times X\times X) = 0$. For X a $K3$, Beauville-Voisin ([**BV**]) have actually *proved* $\mathcal{Z} \overset{\mathrm{rat}}{\equiv} 0$ if $24\,o \overset{\mathrm{rat}}{\equiv} c_2(T_X)$, but it should be true for *any* choice of $o \in X(\bar{\mathbb{Q}})$.

Next we ask what happens if we take higher self-products $(p-o)\times\cdots\times(p-o)$ on $\mathcal{C}\times\cdots\times\mathcal{C}$? Here is a vanishing result (like the B-V theorem above).

---

[19]This uses the following self-intersection formula adapted from [**Fu**, pp. 59, 103]: let ι denote the composition $\mathcal{C} \underset{\sim}{\overset{\jmath}{\to}} \Delta_{\mathcal{C}} \overset{\iota^{\Delta}}{\hookrightarrow} \mathcal{C}\times\mathcal{C}$. Then $\jmath^*\left(N_{(\Delta_{\mathcal{C}})/(\mathcal{C}\times\mathcal{C})}\right) \cong (\iota^* T_{\mathcal{C}\times\mathcal{C}})/T_{\mathcal{C}} \cong T_{\mathcal{C}}$, and so $\iota^*\iota_*\langle\mathcal{C}\rangle = c_1(T_{\mathcal{C}})\cap\langle\mathcal{C}\rangle = -c_1(K_{\mathcal{C}})\cap\langle\mathcal{C}\rangle$, which is to say the negative of the canonical divisor (which we write $-K_{\mathcal{C}}$).

[20]in fact (applying complex conjugation again) $\alpha_j = \beta_j$, but we don't need this.

Theorem ([Vo][21]). Let $\mathcal{C}$ be a smooth projective curve $/\mathbb{C}$ with $\mathcal{Z} \in CH_0^{\text{hom}}(\mathcal{C})$. Then the n-fold self-product $\mathcal{Z}^{\times n} \overset{\text{rat}}{\equiv} 0$ in $CH_0(\mathcal{C}^{\times n})$ for $n \geq g+1$.

REMARK. One should note immediately that $\mathcal{Z}$ here is *not* limited to cycles of transcendence degree 1 (like $p - o$ above) on curves defined $/\bar{\mathbb{Q}}$. For that situation one expects a much stronger result, namely $(p-o)^{\times n} \overset{\text{rat}}{\equiv} 0$ for $n \geq 3$ (regardless of g). Indeed, if BBC^q holds then cycles of trdeg. 1 in $\mathcal{L}^3 CH_0(\mathcal{C}^{\times n})$ are $\overset{\text{rat}}{\equiv} 0$ (see Prop. 4.3(a)); while from §5.1 $(p-o)^{\times n} \in F_\times^n CH_0 \subseteq \mathcal{L}^n CH_0$.

PROOF. First let n be arbitrary, and write $[p_1, \ldots, p_n]$ for the image of $(p_1, \ldots, p_n)$ under the standard map $\tilde{\alpha} : \mathcal{C}^{\times n} \to S^n\mathcal{C}$ onto the nth symmetric product. Let $o \in \mathcal{C}$ be a choice of base point and define

$$\tilde{\xi}_k : S^k\mathcal{C} \to J^1(\mathcal{C}) \text{ by } [p_1, \ldots, p_k] \mapsto AJ_{\mathcal{C}}\left(\sum p_i - k \cdot o\right),$$

$$\tilde{\gamma} : S^{n-1}\mathcal{C} \to S^n\mathcal{C} \text{ by } [p_1, \ldots, p_{n-1}] \mapsto [p_1, \ldots, p_{n-1}, o];$$

denote the induced maps on CH_0 by α, ξ_k, γ. Next, writing $\tilde{\mathcal{W}}$ for a lift of $\mathcal{W} \in Z_0(S^k\mathcal{C})$ to $Z_0(\mathcal{C}^{\times k})$ and $Sym_k := \frac{1}{k!}\sum_{\sigma \in \mathfrak{S}_k} \sigma_* \in End\left(Z_0(\mathcal{C}^{\times k})\right)$ (and referring to §5.1 for $P_\times^{[k]}$, Box^k) we define

$$\beta : CH_0(S^n\mathcal{C}) \to CH_0(\mathcal{C}^{\times n}) \text{ by } Sym_n\left\{P_\times^{[n-1]}\tilde{\mathcal{W}}\right\},$$

$$b : CH_0(S^{n-1}\mathcal{C}) \to CH_0(\mathcal{C}^{\times n-1}) \text{ by}$$

$$b(\mathcal{W}) := Sym_{n-1}\left\{P_\times^{[n-1]}\tilde{\mathcal{W}}\right\} = Sym_{n-1}\left\{Box^{n-1}(\tilde{\mathcal{W}})\right\}.$$

Finally, let $\iota_j : \mathcal{C}^{\times n-1} \hookrightarrow \mathcal{C}^{\times n}$ be the inclusion putting $\{0\}$ in the jth coordinate and $\pi_j : \mathcal{C}^{\times n} \twoheadrightarrow \mathcal{C}^{\times n-1}$ be the projection forgetting that coordinate; and define

$$\iota_*^{(1)} : CH_0(\mathcal{C}^{\times n-1}) \to CH_0(\mathcal{C}^{\times n}) \text{ by } \iota_*^{(1)}(\mathcal{W}) := \frac{1}{n}\sum_{j=1}^{n}(\iota_j)_*\mathcal{W},$$

$$\pi_*^{(1)} : CH_0(\mathcal{C}^{\times n}) \to CH_0(\mathcal{C}^{\times n-1}) \text{ by } \pi_*^{(1)}(\mathcal{W}) := \sum_{j-1}^{n}(\pi_j)_*\mathcal{W}.$$

We summarize these maps in the diagram

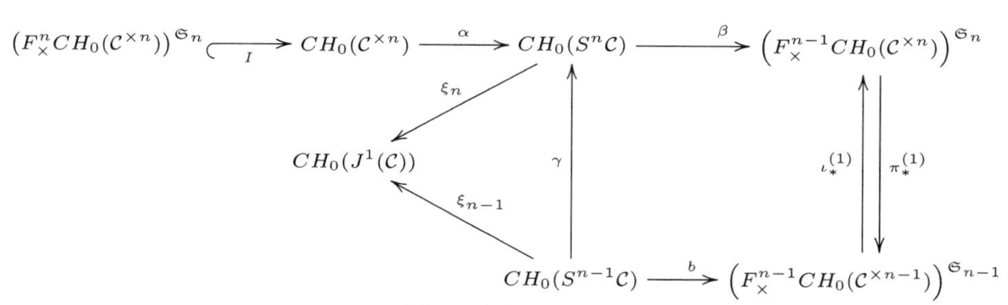

Clearly $\xi_{n-1} = \xi_n \circ \gamma$. While $\iota_*^{(1)} \circ \pi_*^{(1)}$ is not the identity, $(\pi_i)_* \circ (\iota_j)_* = \delta_{ij}$ on $\left(F_\times^{n-1} CH_0(\mathcal{C}^{\times n-1})\right)^{\mathfrak{S}_{n-1}} \implies \pi_*^{(1)} \circ \iota_*^{(1)} = $ identity there. It is also easy to see

[21] C. Voisin has pointed out that this is really due to Bloch [**B1**]. Indeed, it follows easily from the Theorem quoted in §3.2 above (take $A = J^1(\mathcal{C})$). Note also that Voevodsky arrives at the weaker lower bound $n \geq 2g$.

that $\beta \circ \gamma = \iota_*^{(1)} \circ b$, and applying $\pi_*^{(1)}$ on the left yields $\pi_*^{(1)} \circ \beta \circ \gamma = b$. Moreover, the top composition is just the obvious inclusion, and $\pi_*^{(1)} \circ \beta \circ \alpha \circ I = 0$. Since the $\overset{\text{rat}}{\equiv}$-class of $\mathcal{Z}^{\times n}$ belongs to $\left(F_\times^n CH_0(\mathcal{C}^{\times n})\right)^{\mathfrak{S}_n}$, it is obviously trivial if $\beta \circ \alpha \circ I$ is itself zero.

This is precisely what happens when we take $n \geq g+1$. Indeed, we claim in this case that ξ_n and ξ_{n-1}, hence γ, are isomorphisms.[22] Assuming this, we can apply $\gamma^{-1} \circ \alpha \circ I$ on the right to $\beta \circ \gamma = \iota_*^{(1)} \circ b$, obtaining

$$\beta \circ \alpha \circ I \;=\; \iota_*^{(1)} \circ b \circ \gamma^{-1} \circ \alpha \circ I.$$

On the other hand, applying $\gamma^{-1} \circ \alpha \circ I$ on the right and $\iota_*^{(1)}$ on the left to $b = \pi_*^{(1)} \circ \beta \circ \gamma$ gives

$$\iota_*^{(1)} \circ b \circ \gamma^{-1} \circ \alpha \circ I \;=\; \iota_*^{(1)} \circ \pi_*^{(1)} \circ \beta \circ \alpha \circ I \;=\; \iota_*^{(1)} \circ 0 \;=\; 0,$$

and we are done modulo our claim.

Recall that $\tilde{\xi}_k$ is surjective for $k \geq g$, with generic fiber $\mathbb{P}^{k-g}$. Indeed, all its fibers are rational: given $x \in J^1(\mathcal{C})$, pick any $[q_1, \ldots, q_k] \in \tilde{\xi}_k^{-1}(x)$; then $\tilde{\xi}_k^{-1}(x) \cong \mathbb{P}H^0(\mathcal{C}, \mathcal{O}(q_1 + \cdots + q_k)) \cong \mathbb{P}^M$ for some $M \geq k-g$. (These facts follow from Jacobi inversion and Riemann-Roch.) For $k = g$, one obtains that $\tilde{\xi}_g$ is birational; hence ξ_g is an isomorphism.[23] For $k = g+1$, one notes that $\operatorname{im}\{\tilde{\gamma} : S^g\mathcal{C} \to S^{g+1}\mathcal{C}\}$ meets every fiber of $\tilde{\xi}_{g+1}$, so any point of $S^{g+1}\mathcal{C}$ is $\overset{\text{rat}}{\equiv}$ to a point of $\operatorname{im}(\tilde{\gamma})$, hence γ is surjective; and since $\xi_{g+1} \circ \gamma = \xi_g$, γ is injective. Thus ξ_{g+1} is an isomorphism; repeating this argument shows ξ_k is an isomorphism for $k \geq g$ as claimed. □

This theorem has the exciting corollary that all cycles algebraically equivalent to 0 are "nilpotent" modulo $\overset{\text{rat}}{\equiv}$. To see this, let $Y \in CH_m^{\text{alg}}(X)$. Then there exist $\mathcal{Y}_i \in CH_{m+1}(X \times \mathcal{C}_i)$ and $p_i, q_i \in \mathcal{C}_i$ [=curve of genus g_i] s.t. $Y = \sum_i (\mathcal{Y}_i(p_i) - \mathcal{Y}_i(q_i))$, where $\mathcal{Y}_i(c) := \pi_X(\mathcal{Y}_i \cap (\{c\} \times X))$ for $c \in \mathcal{C}_i$. For each i one considers the map

$$\begin{aligned} \phi_n : CH_0(\mathcal{C}^{\times n}) &\to CH_{m \cdot n}(X^{\times n}) \\ c_1 \times \cdots \times c_n &\mapsto \mathcal{Y}(c_1) \times \cdots \times \mathcal{Y}(c_n) \end{aligned}$$

induced by $\mathcal{Y}^{\times n}$. Since $\phi_n\left\{(p-q)^{\times n}\right\} = \left(\mathcal{Y}(p) - \mathcal{Y}(q)\right)^{\times n}$ and $(p-q)^{\times n} \overset{\text{rat}}{\equiv} 0$ for $n \geq g+1$, the terms of the $\sum_i$ are nilpotent [i.e. $\left(\mathcal{Y}_i(p_i) - \mathcal{Y}_i(q_i)\right)^{\times(g_i+1)} \overset{\text{rat}}{\equiv} 0$]. It follows easily that $Y^{\times\{\sum_i (g_i+1)\}} \overset{\text{rat}}{\equiv} 0$.

Corollary 6.6. *All degree-0 zero-cycles are nilpotent.*

Voevodsky [**Vo**] also outlines an argument to the effect that extending the nilpotence statement from cycles $\overset{\text{alg}}{\equiv} 0$ to cycles $\overset{\text{num}}{\equiv} 0$ would amount to proving the Bloch conjecture. The same argument also appeared in Voisin's paper [**V3**]. Kimura ([**Ki**, sec. 7]) picks up this theme and proves nilpotence of cycles $\overset{\text{hom}}{\equiv} 0$ (and BC) for varieties of "finite dimensional" Chow motives. This includes (smooth) varieties covered by a product of curves (and the proof of this utilizes once again the classical fact regarding $S^{n \geq 2g-1}\mathcal{C} \to J^1(\mathcal{C})$ used in [**Vo**]).

---

[22] Voevodsky uses the classical fact that (essentially by Riemann-Roch) $S^k\mathcal{C} \overset{\tilde{\xi}_k}{\to} J^1(\mathcal{C})$ is a projective bundle for $k \geq 2g-1$, to see this for $n \geq 2g$.

[23] For this implication see the argument in [**V4**], §11.1.1 (p. 310), which is stated for surfaces but valid also for higher-dimensional varieties (e.g., see also [**L2**], pf. of 13.3).

6.3. Correspondence cycles. If $X = \mathcal{C}_1 \times \cdots \times \mathcal{C}_n \,/\bar{\mathbb{Q}}$ (here $n = d$) then we may consider 0-cycles of the form $\mathcal{Z} = Box^n(p_1, \ldots, p_n)$ for $p_\ell \in \mathcal{C}_\ell(\mathbb{C})$; we have automatically $\mathcal{Z} \in F^n_X \subseteq \mathcal{L}^n CH_0(\mathcal{C}_1 \times \cdots \times \mathcal{C}_{n/\mathbb{C}})$. Thus, to be *detectably* nontrivial, $\mathcal{Z}$ must *not* be defined over a field of transcendence degree less than $(n-1)$ over $\bar{\mathbb{Q}}$. The interesting situation is therefore where $\mathcal{S}_{0/\bar{\mathbb{Q}}} \subseteq \mathcal{C}_1 \times \cdots \times \mathcal{C}_n$ is a (possibly singular) $(n-1)$-dimensional subvariety, and $P = (p_1, \ldots, p_n)$ is the inclusion of a very general point $P_0 \in \mathcal{S}_0(\mathbb{C})$, so that the spread $\mathfrak{P}$ of P is the graph of a desingularization $\mathcal{S} \twoheadrightarrow \mathcal{S}_0 \subseteq X$ as a subvariety of $\mathcal{S} \times X$. We write in an obvious notation $\mathfrak{Z} := Box^n_{\mathcal{S}}(\mathfrak{P})$.

A technique for detecting such 0-cycles $\mathcal{Z}$ is developed in [**K2**, sec. 14], by writing an explicit $\Omega^n(X)^\vee$-valued $(n-1)$-current $R_{\mathfrak{Z}} = \sum_{k \in \mathsf{K}} [\omega^k]^\vee \otimes R^k_{\mathfrak{Z}}$ on $\mathcal{S}$. Integrating $R_{\mathfrak{Z}}$ on topological cycles $\gamma \in \operatorname{im}\{\varprojlim_D H_{n-1}(\mathcal{S} \setminus D) \to H_{n-1}(\mathcal{S})\}$ then computes $\overline{[AJ(\mathfrak{Z})]_{n-1}} \in Hom\left(\underline{H_{n-1}}(\eta_{\mathcal{S}}, \mathbb{Q}), \frac{\Omega^n(X)^\vee}{\operatorname{im} H_n(X,\mathbb{Q})}\right)$; i.e. as vectors $\in \mathbb{C}^{|\mathsf{K}|}$, $\left(\int_\gamma R^k_{\mathfrak{Z}}\right)_{k \in \mathsf{K}} \equiv \left(\int_\Gamma \omega^k\right)_{k \in \mathsf{K}}$ modulo $\Lambda_{\{\omega\}}$.[24] Remarkably, in case $n = 2$ (only) $\int_\gamma R^k_{\mathfrak{Z}}$ reduces to an iterated integral, so it is in this case that we review the construction.

Henceforth $\mathcal{S}_0 \subseteq \mathcal{C}_1 \times \mathcal{C}_2 = X$ are defined $/\bar{\mathbb{Q}}$ with $o_\ell \in \mathcal{C}_\ell(\bar{\mathbb{Q}})$, $P_0 \in \mathcal{S}_0(\mathbb{C})$ is very general with image $(p_1, p_2) \in X(\mathbb{C})$, and $\mathcal{Z} = (p_1 - o_1) \times (p_2 - o_2) \in \mathcal{L}^2 CH_0(X_K)$ where $K \cong \bar{\mathbb{Q}}(\mathcal{S})$.

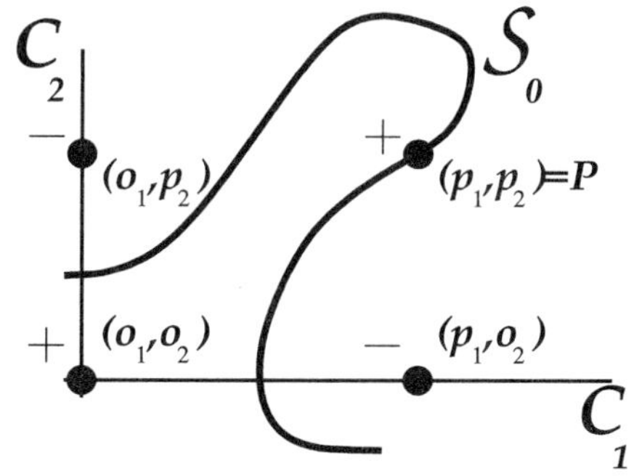

$\mathcal{Z}$ is called an "$\mathcal{S}_0$-" or "correspondence-" 0-cycle on $\mathcal{C}_1 \times \mathcal{C}_2$; in fact, we have already encountered the diagonal 0-cycle on $\mathcal{C} \times \mathcal{C}$ ($\mathcal{S}_0 = \Delta_{\mathcal{C}}$) and the bielliptic 0-cycle on $E_1 \times E_2$ ($\mathcal{S} = \mathcal{B}$). Note that since $[\mathfrak{Z}]_0 = [\mathfrak{Z}]_1 = [\mathfrak{Z}]_2 = 0$ (i.e. $[\mathfrak{Z}] = 0$ in $\underline{H^4}(\eta_{\mathcal{S}} \times \mathcal{C}_1 \times \mathcal{C}_2, \mathbb{Q})$), $[\bar{\mathfrak{Z}}] \in H^4(\mathcal{S} \times \mathcal{C}_1 \times \mathcal{C}_2)$ comes from a class in $H^2(\{Q\} \times \mathcal{C}_1 \times \mathcal{C}_2)$. As this must also be a rational $(1,1)$ class, Lefschetz $(1,1)$ says that it is the class of a cycle on $\{Q\} \times \mathcal{C}_1 \times \mathcal{C}_2$. If we modify $\bar{\mathfrak{Z}} = Box^2_{\mathcal{S}}(\mathfrak{P})$ by subtracting off this cycle, then the new $\bar{\mathfrak{Z}} \overset{\text{hom}}{\equiv} 0$.

Let $\{\alpha^j_\ell\}_{j=1,\ldots,2g_\ell}$, $\{\hat{\alpha}^j_\ell\}$ be dual bases of $\overline{H_1(\mathcal{C}_\ell, \mathbb{Q})}$ based at o_ℓ resp. o'_ℓ, while avoiding o'_ℓ resp. o_ℓ; and take $\mathbb{D}_\ell := \mathcal{C}_\ell \setminus \bigcup_j |\hat{\alpha}^j_\ell|$ as fundamental domain for $\mathcal{C}_\ell$. Writing $\{\omega^k_\ell\}_{k=1,\ldots,g_\ell} \subseteq \Omega^1(\mathcal{C}_\ell)$ for a basis, we obtain functions $z^k_\ell := \int^*_{o_\ell} \omega^k_\ell$ on $\mathbb{D}_\ell$ (which are 0 at o_ℓ), or equivalently 0-currents on $\mathcal{C}_\ell$ with "branch cuts" at the $\{\alpha^j_\ell\}$. One has in particular $d[z^k_\ell] = \omega^k_\ell - \sum_{j=1}^{2g_\ell} \left(\int_{\alpha^j_\ell} \omega^k_\ell\right) \cdot \delta_{\hat{\alpha}^j_\ell}$.

---

[24] We note that $\dim(\mathcal{S}) = n-1 \implies [\mathfrak{Z}]_n = 0 \implies \overline{[AJ(\mathfrak{Z})]_{n-1}}$ is defined and computed by the membrane integrals of Proposition 4.8. For $n > 2$ it is not clear (without HC) how to modify $\bar{\mathfrak{Z}}$ over $D \subseteq \mathcal{S}$ to get $\bar{\mathfrak{Z}} \overset{\text{hom}}{\equiv} 0$, so this uses Remark 4.10(b); for $n = 2$, see below.

Write $\underline{k} = (k_1, k_2)$, $\mathsf{K} = \{1, \ldots, g_1\} \times \{1, \ldots, g_2\}$ so that the basis $\{\omega^{\underline{k}}\} \subseteq \Omega^1(X)$ takes the form $\{\omega_1^{k_1} \wedge \omega_2^{k_2}\}_{(k_1,k_2)\in\mathsf{K}}$; also denote the pullbacks of z, ω, α, $\hat{\alpha}$ to $\mathcal{S}$ no differently. Then from **[K2]**

$$R_3^{(k_1,k_2)} = z_1^{k_1}\omega_2^{k_2} - \sum_{j=1}^{2g_1} \left(\int_{\alpha_1^j} \omega_1^{k_1} \right) z_2^{k_2} \cdot \delta_{\hat{\alpha}_1^j} \in {}'\mathcal{D}^1(\mathcal{S}).$$

Now assume $\mathcal{S}_0 \ni (o_1, o_2) =: o$ and also write o for a point on $\mathcal{S}$ mapping to it. Choose γ to pass through o (and such that $\gamma \cap \hat{\alpha}_1^{j_1} \cap \hat{\alpha}_2^{j_2} = \emptyset \ \forall\, j_1, j_2$) and let $\varphi\colon [0,1] \to \gamma \subseteq \mathcal{S}$ be a parametrization with $\varphi(0) = o$. Define the iterated integral

$$\int_{\gamma;\, o} \alpha \circ \beta := \int_{x=0}^{1} \left(\int_{t=0}^{x} \varphi_t^* \alpha \right) \varphi_x^* \beta$$

for any $\alpha, \beta \in \Omega^1(\mathcal{S})$. (*A priori* this is only an invariant of the homotopy class of $\gamma \in \pi_1(\mathcal{S}; o)$.) We claim that $\int_\gamma R_3^{(k_1,k_2)} = \int_{\gamma;\, o} \omega_1^{k_1} \circ \omega_2^{k_2}$ on the nose; to check this we drop the k_1,k_2 and write $R = z_1\omega_2 - \sum_j \left(\int_{\alpha_j} \omega_1 \right) z_2 \cdot \delta_{\hat{\alpha}_1^j}$. Pulling this back to γ gives

$$z_1\omega_2 - \sum_j \sum_{q \in \hat{\alpha}_1^j \cap \gamma} \left(\int_{\alpha_1^j} \omega_1 \right) z_2(q) \cdot \delta_{\{q\}}. \tag{6.1}$$

Denote the path along γ from q to o by $\overrightarrow{q.o}$; on this path let $\tilde{z}_2$ be the continuous function with $d\tilde{z}_2 = \omega_2$ and $\tilde{z}_2(o) = 0$. Then we can take d of the 0-current

$$-\sum_j \sum_{q \in \hat{\alpha}_1^j \cap \gamma} \left(\int_{\alpha_1^j} \omega_1 \right) \tilde{z}_2 \cdot \delta_{\overrightarrow{q.o}}$$

to obtain

$$+\sum_j \sum_{q \in \hat{\alpha}_1^j \cap \gamma} \left(\int_{\alpha_1^j} \omega_1 \right) \tilde{z}_2(q) \delta_q - \left\{ \sum_j \sum_{p \in \hat{\alpha}_1^j \cap \gamma} \left(\int_{\alpha_1^j} \omega_1 \right) \delta_{\overrightarrow{q.o}} \right\} \omega_2\, ;$$

adding this to (6.1) yields

$$\left\{ z_1 - \sum_j \sum_{p \in \hat{\alpha}_1^j \cap \gamma} (\omega_1)\, \delta_{\overrightarrow{q.o}} \right\} \omega_2 =: \tilde{z}_1 \cdot \omega_2$$

where $\tilde{z}_1$ is continuous except at o [essentially $(\tilde{z}_1 \circ \varphi)(x) = \int_0^x \varphi^* \omega_1$]. Along γ this is cohomologous to (6.1) by construction, so

$$\int_\gamma R = \int_\gamma \tilde{z}_1 \cdot \omega_2 = \int_{\gamma;\, o} \omega_1 \circ \omega_2$$

as desired.

THEOREM 6.7. *Let $\mathcal{S}_{0/\bar{\mathbb{Q}}} \subseteq \mathcal{C}_1 \times \mathcal{C}_2$ be a curve passing through $o = (o_1, o_2) \in (\mathcal{C}_1 \times \mathcal{C}_2)(\bar{\mathbb{Q}})$ with very general point (p_1, p_2), and set $\mathcal{Z} = (p_1, p_2) - (p_1, o_2) -$*

$\overline{(o_1,p_2)+(o_1,o_2)} \in \mathcal{L}^2CH_0(\mathcal{C}_1\times\mathcal{C}_{2/\mathbb{C}})$. *Then for* $\gamma\in H_1(\mathcal{S},\mathbb{Q})$, $\overline{[AJ(\mathfrak{Z})]_1}(\gamma)$ *evaluates to* $\left(\int_{\tilde{\gamma};\,o}\omega_1^{k_1}\circ\omega_2^{k_2}\right)_{(k_1,k_2)\in\mathsf{K}} \in \mathbb{C}^{|\mathsf{K}|}/\Lambda_{\{\omega\}}$ *on the basis* $\{\omega\}=\{\omega_1^{k_1}\wedge\omega_2^{k_2}\}$, *where* $\tilde{\gamma}$ *is any closed path* $\ni o$ *and* $\overset{hom}{\equiv}\gamma$.

Example: the Fermat 0-cycle. Specializing to $\mathcal{C}_1=\mathcal{C}_2=E$ an elliptic curve, we write $\Omega^1(E)=\mathbb{C}\langle dz\rangle$ so that $\omega=dz_1\wedge dz_2$ generates $\Omega^2(E\times E)$. If we take E to have CM, then $H^2(E\times E)=H^2_{\text{alg}}\oplus H^2_{\text{tr}}$ as HS where $H^2_{tr}(E\times E,\mathbb{C})=H^{2,0}\oplus H^{0,2}$; this has the effect of making ω's periods $\Lambda_\omega\subseteq\mathbb{C}$ yield an honest 2-dimensional rational lattice. Setting $E=\overline{\{v^2=1-u^4\}}$ with $o=(0,1)\,[=(u,v)]$ and $dz:=\frac{1-i}{4b}\frac{du}{v}$, where $b=\int_0^1\frac{dt}{\sqrt{1-t^4}}$, we have in fact $\Lambda_{\{dz\}}=\mathbb{Q}\langle 1,i\rangle$. Hence if $\mathcal{S}_0$, $\mathcal{Z}$, γ are as in Thm. 6.7, $\overline{[AJ(\mathfrak{Z})]_1}(\gamma)$ is computed by the number $\int_{\tilde{\gamma};\,o}\pi_1^*dz\circ\pi_2^*dz\in\mathbb{C}/\mathbb{Q}\langle 1,i\rangle$ (integration on $\mathcal{S}$). (It seems remarkable that the iterated integral is well-defined modulo $\mathbb{Q}\langle 1,i\rangle$ on *homology classes.*)

For $\mathcal{S}$ we shall take the Fermat curve $\mathfrak{F}=\overline{\{x^4+y^4=1\}}$, embedded in $E\times E$ by (π_1,π_2) where $\pi_1(x,y)=\left(-\frac{1+i}{\sqrt{2}}\frac{y}{x},\,\frac{1}{x^2}\right)$, $\pi_2(x,y)=(-y,\,x^2)$; note $o=(1,0)\in\mathcal{S}$ maps to $o_1\times o_2\in E\times E$. We shall also make use of a third map $\pi_3(x,y)=(x,\,y^2)$. Writing $\nu_i=\pi_i^*dz$ we have $\nu_1=\frac{1}{2b'}\frac{dx}{y^3}$ where $b'=\sqrt{2}b=\int_0^1\frac{dt}{(1-t^4)^{3/4}}$, $\nu_2=\frac{1-i}{4b}\frac{x\,dx}{y^3}$, $\nu_3=\frac{1-i}{4b}\frac{dx}{y^2}$. These span $\Omega^1(\mathfrak{F})$, and so $q\mapsto\left\langle\int_o^q\nu_i\right\rangle_{i=1,2,3}$ induces a map $\mathfrak{F}\overset{AJ_o}{\to}J^1(\mathfrak{F})$ factoring $(\pi_1,\pi_2,\pi_3):\mathfrak{F}\to E\times E\times E$, hence an isogeny $J^1(\mathfrak{F})\overset{\mu}{\to}E\times E\times E$.

The computation that follows is almost exactly the same as the one done in [**H**]: taking $\gamma=$path (on $\mathfrak{F}$) from $o=(1,0)\to(0,i)\to(i,0)\to(0,1)\to(1,0)$, we use the automorphism $\sigma(x,y)=(\frac{y}{x\sqrt{i}},\frac{1}{xi})$ to get

$$\int_{\gamma;\,o}\nu_1\circ\nu_2=\int_{\sigma(\gamma);\,(o,-i)}\nu_3\circ\nu_1=2(1+i)\int_{\overrightarrow{(0,1).o}}\nu_3\circ\nu_1$$
$$=2(1+i)\int_0^1\frac{(1-i)}{4b}\frac{dx}{\sqrt{1-x^4}}\circ\frac{1}{2b'}\frac{dx}{(1-x^4)^{3/4}}$$
$$=\frac{\int_0^1\frac{dt}{\sqrt{1-t^4}}\circ\frac{dt}{(1-t^4)^{3/4}}}{2\int_0^1\frac{dt}{\sqrt{1-t^4}}\cdot\int_0^1\frac{dt}{(1-t^4)^{3/4}}}=:\kappa\in\mathbb{R}/\mathbb{Q}\subseteq\mathbb{C}/\mathbb{Q}\langle 1,i\rangle\,.$$

Unfortunately, it appears to be unknown whether κ is irrational; but we can verify the Fermat 0-cycle $\mathcal{Z}\overset{\text{rat}}{\not\equiv}0$ by other means and so $\kappa\notin\mathbb{Q}$ would be implied by BBC^q.

Bloch has proved (in [**B3**], using the ℓ-adic Abel-Jacobi map) the algebraic inequivalence of $AJ_o(\mathfrak{F})$ and its mirror image $AJ_o(\mathfrak{F})^-$ as 1-cycles in $J^1(\mathfrak{F})$. Using the isogeny μ and exploiting symmetries of the image of $\mathfrak{F}$ in $E_{(1)}\times E_{(2)}\times E_{(3)}$, one first deduces that $Box^2_{E_3}(\text{im}(\mathfrak{F}))\overset{\text{alg}}{\not\equiv}0$ there, then that it cannot be $\overset{\text{alg}}{\equiv}$ (*a fortiori* $\overset{\text{rat}}{\equiv}$) to a cycle supported on $E_1\times E_2\times\{\text{pts.}\in E_3\}$. Consequently the spread $\bar{\mathfrak{Z}}$ of the Fermat 0-cycle $\mathcal{Z}$, which maps to $Box^2_{E_3}(\text{im}(\mathfrak{F}))$ via $E_1\times E_2\times\mathfrak{F}\overset{(\text{id},\text{id},\pi_3)}{\longrightarrow}E_1\times E_2\times E_3$, cannot be $\overset{\text{rat}}{\equiv}$ to a cycle supported over a divisor/$\bar{\mathbb{Q}}$ in $\mathfrak{F}$; and so $\mathcal{Z}\overset{\text{rat}}{\not\equiv}0$.

6.4. Regulators and relative 0-cycles. There is one product of "curves" for which one can completely describe the Chow group of 0-cycles and the higher AJ maps. This is the n-fold self-product of the *relative* curve $(\mathbb{P}^1,\{0,\infty\})$, which one may think of *intuitively* (only) as $\mathbb{P}^1$ with $\{0\}$ and $\{\infty\}$ attached — a degenerate elliptic curve. We give just the barest summary of the most striking result and refer the reader to [**K2**] and [**K3**] for full details (minus relative $\overset{\text{rat}}{\equiv}$, a discussion of which for these problems is in [**K4**]).

Set $X^n := (\mathbb{P}^1, \{0,\infty\})^{\times n}$, or equivalently for our purposes, $(\mathbb{P}^n, \bigcup_{i=0}^n \{z_i = 0\})$ (which one can think of as a degenerate CY n-fold). On X^n_L ($L \subseteq \mathbb{C}$) let $\{a_1, \ldots, a_n\}$ represent the relative $\overset{\text{rat}}{\equiv}$-class of $B(a_1, \ldots, a_n) := (\langle a_1 \rangle - \langle 1 \rangle) \times \cdots \times (\langle a_n \rangle - \langle 1 \rangle)$. We make the *ad hoc* definition

$$\mathcal{L}^n CH_0(X^n_L) := \{\{a_1, \ldots, a_n\} \mid a_i \in L^* \ (\forall i)\},$$

and observe that this gives Milnor K-theory $K^M_n(L)$ by a result of Totaro ([**To**]).

Now assume L is finitely generated $/\bar{\mathbb{Q}}$ of trdeg $< n$, so that one has $\bar{\mathbb{Q}}(\mathcal{S}) \underset{\text{ev}}{\overset{\cong}{\to}} L$ for some smooth projective $\mathcal{S}/\bar{\mathbb{Q}}$ of dimension $< n$. There is a map

$$\mathcal{R} : K^M_n(\bar{\mathbb{Q}}(\mathcal{S})) \to H^{n-1}(\eta_{\mathcal{S}}, \mathbb{C}/\mathbb{Q}(n))$$

called the Milnor regulator; one defines $\mathcal{R}\{f_1, \ldots, f_n\}$ to be the cohomology class of the $(n-1)$-current

$$\sum_{j=1}^n (-1)^{nj-1} (2\pi\sqrt{-1})^{j-1} \log f_j \, \mathrm{dlog} f_{j+1} \wedge \cdots \wedge \mathrm{dlog} f_n \cdot \delta_{f_1^{-1}(\mathbb{R}^-) \cap \cdots \cap f_{j-1}^{-1}(\mathbb{R}^-)}$$

on $\eta_{\mathcal{S}}$. Obviously one can consider the composition

$$\mathcal{L}^n CH_0(X^n_L) \cong K^M_n(L) \xrightarrow{\text{ev}^{-1}} K^M_n(\bar{\mathbb{Q}}(\mathcal{S})) \xrightarrow{\mathcal{R}} H^{n-1}(\eta_{\mathcal{S}}, \mathbb{C}/\mathbb{Q}(n)).$$

On the other hand, to try computing a higher AJ class for $\mathcal{Z} = B(a_1, \ldots, a_n)$, one would spread this out to $\mathfrak{Z}_{(\mathcal{U})} = B(f_1, \ldots, f_n)$ on $(\mathbb{P}^1, \{0,\infty\})^{\times n} \times \{\mathcal{U} \subseteq \mathcal{S}\}$ by taking $f_i := \text{ev}^{-1}(a_i)$. (Each component of $\mathfrak{Z}$ looks like the graph of n functions over $\mathcal{U}$.) One can then show that $\mathfrak{Z} \overset{\text{hom}}{\equiv} 0$ and obtain a generalized AJ class for $\mathfrak{Z}$, by integrating forms $\mathrm{dlog} z_1 \wedge \cdots \wedge \mathrm{dlog} z_n \wedge \omega_{\mathcal{U}}$ over a limit of chains bounding (in the limit) on $\mathfrak{Z}$. This boils down, by explicit computation and Hodge theory, to integration of the above current over topological cycles on $\mathcal{S}$. That is, the above composition gives the right generalization[25] of AJ^{n-1}_X, and one would conjecture it is injective.

As a sanity check, one can try out $n = 1$: $\mathcal{R}\{f\} = \log f \in H^0(\eta_{\mathcal{S}}, \mathbb{C}/\mathbb{Q}(1)) \cong \mathbb{C}/(2\pi\sqrt{-1})\mathbb{Q}$, where (since $\dim(\mathcal{S}) = 0$) really $f = \alpha \in \bar{\mathbb{Q}}^*$. The composition sends $(\langle \alpha \rangle - \langle 1 \rangle) \mapsto \log \alpha$; so we expect that for $\sum q_i = 0$, $\sum q_i \langle \alpha_i \rangle \overset{\text{rat}}{\equiv} 0$ (mod torsion) $\iff (\Pi \alpha_i^{q_i})^N = 1$ for some $N \in \mathbb{Z}^+$. Since on $(\mathbb{P}^1, \{0,\infty\})_{\bar{\mathbb{Q}}}$, rational equivalence is generated by divisors of functions $F \in \bar{\mathbb{Q}}(\mathbb{P}^1)^*$ with $F(0) = F(\infty) = 1$, this is indeed the case.

7. Behavior of higher Abel-Jacobi classes in families

7.1. Topological invariants for higher normal functions. Given a smooth family of d-dimensional projective varieties $\mathfrak{X}_{\mathcal{U}} \xrightarrow{\pi} \mathcal{U}\,[\subseteq \mathcal{S}]$ over an affine base,

[25] This result was the original motivation for introducing the reduced higher AJ maps of §4.3 and the currents R_3 in §6.3 computing them.

with $\mathfrak{Z}_{(\mathcal{U})} \in Z^d(\mathfrak{X}_{\mathcal{U}})$ such that $\mathcal{Z}_s := \mathfrak{Z}|_{X_s} \overset{\text{hom}}{\equiv} 0$ ($\forall\, s \in \mathcal{U}$), one can define the *normal function* $\nu_{\mathfrak{Z}} \in \Gamma(\mathcal{U}, \mathcal{J}^d_{\mathfrak{X}_{\mathcal{U}}/\mathcal{U}})$ by $\nu_{\mathfrak{Z}}(s) := Alb_{X_s}(\mathcal{Z}_s) \in J^d(X_s)$. Using the connecting homomorphism δ associated to the exact sequence $0 \to \mathcal{H}^{2d-1}_{\mathfrak{X}_{\mathcal{U}}/\mathcal{U},\mathbb{Q}} \to \mathcal{H}^{2d-1}_{\mathfrak{X}_{\mathcal{U}}/\mathcal{U}} \big/ \mathcal{F}^d\mathcal{H}^{2d-1}_{\mathfrak{X}_{\mathcal{U}}/\mathcal{U}} \to \mathcal{J}^d_{\mathfrak{X}_{\mathcal{U}}/\mathcal{U}} \to 0$, one obtains the topological invariant $\delta\nu_{\mathfrak{Z}} \in H^1(\mathcal{U}, \mathcal{H}^{2d-1}_{\mathfrak{X}_{\mathcal{U}}/\mathcal{U},\mathbb{Q}})$ of the normal function. In terms of the notation of §4.2, $\nu_{\mathfrak{Z}}$ is the family $\{[AJ(\mathfrak{Z}|_{X_s})]_0\}$ while $\delta\nu_{\mathfrak{Z}}$ is essentially $[\mathfrak{Z}]_1$; so $[\mathfrak{Z}]_1 \neq 0 \implies \{[AJ(\mathfrak{Z}|_{X_s})]_0\} \neq 0$ and $[\mathfrak{Z}]_1$ is an invariant of the family.

Below we will generalize this to show (for $i \geq 2$) $[\mathfrak{Z}]_i$ may be viewed (in some cases) as an invariant of a family of $(i-1)^{\text{st}}$-higher-AJ classes. First we explain (partly following [**K1**]) how such a family appears in terms of spreads and field extensions. Let K/L be a transcendental extension of subfields of $\mathbb{C}$ finitely generated $/\bar{\mathbb{Q}}$. Then there exist $\mathcal{S}/\bar{\mathbb{Q}}$ with $s_0 \in \mathcal{S}(\mathbb{C})$ such that $\text{ev}_{s_0} : \bar{\mathbb{Q}}(\mathcal{S}) \overset{\cong}{\to} K$, and $\mathcal{M}/\bar{\mathbb{Q}}$ with $\bar{\rho} : \mathcal{S} \to \mathcal{M}$ such that $\text{ev}_{\rho(s_0)} : \bar{\mathbb{Q}}(\mathcal{M}) \overset{\cong}{\to} L$; write $\mu_0 := \rho(s_0) \in \mathcal{M}(\mathbb{C})$. We get a foliation of $\mathcal{S}$ by subvarieties $S_\mu := \rho^{-1}(\mu) \overset{\iota_\mu}{\hookrightarrow} \mathcal{S}$ of dimension $\text{trdeg}(K/L) =:$ $t \geq 1$, and note that S_{μ_0} is defined $/L$ with $L(S_{\mu_0}) \underset{\text{ev}_{s_0}}{\overset{\cong}{\to}} K$. Finally, it is clear that $\bar{\rho}$ makes sense in the limit as a map $\eta_{\mathcal{S}} \to \eta_{\mathcal{M}}$; we shall denote this by ρ. Note that since this is a limit of affines over an affine base with t-dimensional affine fibers, $R^{t+1}\rho_*\mathbb{Q} = 0$. If $m = \dim \mathcal{M}$ then Leray for ρ yields $H^{t+m}(\eta_{\mathcal{S}}) \cong H^m(\eta_{\mathcal{M}}, R^t\rho_*\mathbb{Q})$.

Next consider X/K and $\mathcal{Z} \in Z^p(X/K)$; then one has the $\bar{\mathbb{Q}}$-spread $\mathfrak{Z} \in Z^p(\mathfrak{X}/\bar{\mathbb{Q}})$ with $\mathfrak{X} \overset{\pi}{\to} \eta_{\mathcal{S}}$ as above. But we don't have to spread clear on down to $\bar{\mathbb{Q}}$. Write $\eta_{S_\mu} := \varprojlim \mathcal{V}$ [$\mathcal{V} \subseteq S_\mu$ affine Zar. op. $/\bar{\mathbb{Q}}(\mu)$]$=\eta_{\mathcal{S}} \cap S_\mu$ and $\mathfrak{X}_\mu := \pi^{-1}(\eta_{S_\mu}) \overset{\mathcal{I}_\mu}{\hookrightarrow} \mathfrak{X}$. Then we can consider instead the (partial) L-spread $\mathfrak{Z}_{\mu_0} \in Z^p(\mathfrak{X}_{\mu_0/L})$, which is nothing other than $\mathcal{I}^*_{\mu_0}(\mathfrak{Z})$ (i.e., $\mathfrak{Z} \cap \mathfrak{X}_{\mu_0}$). We write $\mathcal{L}^\bullet_{K/\bar{\mathbb{Q}}}$, $\psi_i^{K/\bar{\mathbb{Q}}}$ for the $\mathcal{L}^\bullet$ and ψ_i defined on $Z^p(X_K)$ in §4.2 above; one can set up an analogous picture for the partial spread with invariants

$$\psi_{[i]}^{K/L} : [\mathcal{L}^i_{K/L}]\, CH^p(X_K) \to [Gr^i_{\mathcal{L}}]\, \underline{H^{2p}_{\mathcal{D}}}(\mathfrak{X}_{\mu_0}, \mathbb{Q}(p)).$$

(Here the $\mathcal{L}$ on $\underline{H^{2p}_{\mathcal{D}}}$ comes from Leray for $\mathfrak{X}_{\mu_0} \to \eta_{S_{\mu_0}}$.) One has $\psi^{K/L} = \mathcal{I}^*_{\mu_0} \circ \psi^{K/\bar{\mathbb{Q}}}$, $\mathcal{I}^*_{\mu_0}(\mathcal{L}^\bullet) \subseteq \mathcal{L}^\bullet$, hence $\mathcal{L}^\bullet_{K/\bar{\mathbb{Q}}} \subseteq \mathcal{L}^\bullet_{K/L}$ on $CH^p(X_K)$. Thus if $\mathcal{Z} \in \mathcal{L}^i_{K/\bar{\mathbb{Q}}} CH^p(X_K)$ has $\psi_i^{K/L}(\mathcal{Z}) \neq 0$, then trivially $\psi_i^{K/\bar{\mathbb{Q}}}(\mathcal{Z}) \neq 0$. It is a version of the converse that we shall be discussing.

Now the points $\mu \in \eta_{\mathcal{M}}(\mathbb{C})$ correspond to embeddings $\text{ev}_\mu : \bar{\mathbb{Q}}(\mathcal{M}) \hookrightarrow \mathbb{C}$ (with image $=: \bar{\mathbb{Q}}(\mu)$), while the points $s \in \eta_{S_\mu}(\mathbb{C})$ correspond to embeddings $\bar{\mathbb{Q}}(\mathcal{S}) \hookrightarrow \mathbb{C}$ which are ev_μ on $[\rho^*]\bar{\mathbb{Q}}(\mathcal{M})$. Having said this, for $\mu \neq \mu_0$, $\mathfrak{Z}_\mu$ is the partial $\bar{\mathbb{Q}}(\mu)$-spread of $\mathcal{Z}_s \in CH^p(X_{s/\bar{\mathbb{Q}}(s)})$ for any $s \in \eta_{S_\mu}(\mathbb{C})$. Thus we may speak of $\{\mathfrak{Z}_\mu\}_{\mu \in \mathcal{M}(\mathbb{C})}$ as a family of partial spreads, even if they are not all partial spreads of $\mathcal{Z} = \mathcal{Z}_{s_0}$.

So suppose that $\mathcal{Z} \in \mathcal{L}^i_{(K/\bar{\mathbb{Q}})} CH_0(X_K)$ and choose L (or equivalently $\mathcal{M}$) so that $t\,[= \text{trdeg}(K/L)] = i-1$. Clearly for each μ, $[\mathfrak{Z}_\mu]_i = 0$ as $\dim(S_\mu) = t < i$, and so the "higher normal function" $\nu_{\mathfrak{Z}}^{[i-1]}(\mu) := [AJ(\mathfrak{Z}_\mu)]_{i-1}$ is defined. (Note that for $\mu = \mu_0$ this is essentially $\psi_i^{K/L}(\mathcal{Z})$; for general μ, it is equivalent to $\psi_i^{\bar{\mathbb{Q}}(s)/\bar{\mathbb{Q}}(\mu)}(\mathcal{Z}_s)$

where $s \in \eta_{S_\mu}(\mathbb{C})$.)

$$\textbf{Expectation:}$$

$$(*)_{i+1}: \qquad [\mathfrak{Z}]_i \neq 0 \implies \text{the function } \nu_{\mathfrak{Z}}^{[i-1]} \text{ is nonzero (on } \eta_{\mathcal{M}}(\mathbb{C})) \text{ for some choice of } L.$$

That is, $[\mathfrak{Z}]_i$ should be an invariant of $\nu_{\mathfrak{Z}}^{[i-1]}$, its "topological invariant"; this gives a hint for the proof — that a connecting homomorphism δ is involved.

PROPOSITION 7.1. *Expectation (for all i) is (a) true if X is a surface with a model $/L$, and (b) true modulo GHC if X is of higher dimension with model $/\bar{\mathbb{Q}}$.*

REMARK 7.2. In the first instance, the $\bar{\mathbb{Q}}$-spread $\bar{\mathfrak{X}} \to \mathcal{S}$ is a fiber product $\bar{\mathfrak{Y}} \times_{\mathcal{M}} \mathcal{S}$ for some family $\bar{\mathfrak{Y}} \to \mathcal{M}$; hence X_s is constant along fibers of $\mathcal{S} \to \mathcal{M}$.

We prove the second statement (b); the first follows from a similar argument together with the fact that

$$H^2(\mathcal{U}, \mathcal{F}_h^1 \mathcal{H}^2_{\mathfrak{X}_{\mathcal{U}}/\mathcal{U},\mathbb{Q}}) \hookrightarrow H^2(\mathcal{U}, \mathcal{H}^2_{\mathfrak{X}_{\mathcal{U}}/\mathcal{U},\mathbb{Q}}) \cap F^1 H^2(\mathcal{U}, \mathcal{H}^2_{\mathfrak{X}_{\mathcal{U}}/\mathcal{U}})$$

for all $\mathcal{U} \subseteq \mathcal{S}$ s.t. $\mathfrak{X}_{\mathcal{U}} \to \mathcal{U}$ is smooth.

PROOF. Observe first that we may essentially reduce to the case where $(s :=)$ $\operatorname{trdeg}(K/\bar{\mathbb{Q}}) = i$. Choose a very general $(s-i)$-fold hyperplane section $\mathcal{S}_0 \subseteq \mathcal{S}$; this will be defined over $L_0[\subseteq K]$ such that $\operatorname{trdeg}(K/L_0) = i$. By affine weak Lefschetz one deduces $\underline{H}^i(\eta_{\mathcal{S}}) \hookrightarrow \underline{H}^i(\eta_{\mathcal{S}_0})$, hence

$$Hom_{\text{MHS}}\Big(\mathbb{Q}(-d),\, \underline{H}^i(\eta_{\mathcal{S}}) \otimes H^{2d-i}(X)\Big) \hookrightarrow Hom_{\text{MHS}}\Big(\mathbb{Q}(-d),\, \underline{H}^i(\eta_{\mathcal{S}_0}) \otimes H^{2d-i}(X)\Big)$$

and so $[\mathfrak{Z}]_i \neq 0 \implies [\mathfrak{Z}_0]_i \neq 0$ (where $\mathfrak{Z}_0 := \mathfrak{Z} \cdot (X \times \eta_{\mathcal{S}_0})$). A trivial modification of the argument below (with $\mathfrak{Z}_0$, $\mathcal{S}_0$, L_0 replacing $\mathfrak{Z}$, $\mathcal{S}$, $\bar{\mathbb{Q}}$ and $\mathcal{M}$ of dimension $s-i+1$) then gives the desired conclusion for any L such that $L_0 \overset{\text{trdeg } 1}{\subseteq} L \subseteq K$.

So assume $\dim \mathcal{S} = i$, $[\mathfrak{Z}]_i \neq 0$, and choose *any* $L \subseteq K$ with $\operatorname{trdeg}(L/\bar{\mathbb{Q}}) = 1$ $(= \dim \mathcal{M})$; we prove $\{[AJ(\mathfrak{Z}_\mu)]^{tr}_{i-1}\} \neq 0$. We will require the following notation for sheaves of MHS over $\eta_{\mathcal{M}}$: $\mathcal{H}^{i-1}_{\eta_{\mathcal{S}}/\eta_{\mathcal{M}},\mathbb{Q}} = R^{i-1}\rho_*\mathbb{Q}$, $\mathcal{H}^{i-1}_{\eta_{\mathcal{S}}/\eta_{\mathcal{M}}} = R^{i-1}\rho_*\mathbb{C} \otimes \mathcal{O}_{\eta_{\mathcal{M}}}$, with $\underline{\mathcal{H}}^{i-1}$ denoting lowest weight; and $\overline{\mathcal{H}}^{2d-i}_{X,\mathbb{Q}}$ (resp. $\overline{\mathcal{H}}^{2d-i}_X$) denotes the *constant* sheaf $\frac{H^{2d-i}(X)}{F_h^{d-i+1}\{\text{num}\}}$ (resp. $\frac{H^{2d-i}(X,\mathbb{C})}{F_h^{d-i+1}\{\text{num}\}}$). Consider the diagram

$$
\begin{array}{ccc}
\mathcal{L}^i_{(K/\bar{\mathbb{Q}})} CH_0(X_K) & & \\
\Big\downarrow \cong & & \\
\mathcal{L}^i CH^d(X \times \eta_{\mathcal{S}}{}_{/\bar{\mathbb{Q}}}) & \xrightarrow{\{[AJ(\,\cdot\,\mu)]^{tr}_{i-1}\}} & \\
\Big\downarrow [\,\cdot\,]_i & & \\
Hom_{\mathrm{MHS}}\left(\mathbb{Q}(-d), \underline{H}^i(\eta_{\mathcal{S}}) \otimes H^{2d-i}(X)\right) & & H^0\left(\eta_{\mathcal{M}}, Ext^1_{\mathrm{MHS}(\eta_{\mathcal{M}})}\left(\mathbb{Q}(-d), \underline{\mathcal{H}}^{i-1}_{\eta_{\mathcal{S}}/\eta_{\mathcal{M}},\mathbb{Q}} \otimes \overline{\mathcal{H}}^{2d-i}_{X,\mathbb{Q}}\right)\right) \\
\Big\downarrow (\dagger) & & \Big\downarrow \delta \\
\underline{H}^i(\eta_{\mathcal{S}}) \otimes \dfrac{H^{2d-i}(X)}{F_h^{d-i+1} H^{2d-i}} & & H^1\left(\eta_{\mathcal{M}}, \underline{H}^{i-1}_{\eta_{\mathcal{S}}/\eta_{\mathcal{M}},\mathbb{Q}} \otimes \overline{H}^{2d-i}_{X,\mathbb{Q}}\right) \\
\Big\downarrow \text{Leray for } \rho & & \Big\downarrow \\
H^1\left(\eta_{\mathcal{M}}, R^{i-1}\rho_*\mathbb{Q}\right) \otimes \dfrac{H^{2d-i}(X)}{F_h^{d-i+1} H^{2d-i}} & \xleftrightarrow{\cong} & H^1\left(\eta_{\mathcal{M}}, \mathcal{H}^{i-1}_{\eta_{\mathcal{S}}/\eta_{\mathcal{M}},\mathbb{Q}} \otimes \overline{\mathcal{H}}^{2d-i}_{X,\mathbb{Q}}\right)
\end{array}
$$

where GHC$(1, i, \mathcal{S}) \implies$ injectivity of $(\dagger)$ by [**K1**, Lemma II.2] and δ is induced by the short exact sequence

$$0 \to \underline{\mathcal{H}}^{i-1}_{\eta_{\mathcal{S}}/\eta_{\mathcal{M}},\mathbb{Q}} \otimes \overline{\mathcal{H}}^{2d-i}_{X,\mathbb{Q}} \to \frac{\underline{\mathcal{H}}^{i-1}_{\eta_{\mathcal{S}}/\eta_{\mathcal{M}}} \otimes \overline{\mathcal{H}}^{2d-i}_X}{\mathcal{F}^d(\mathrm{num})} \to Ext^1_{\mathrm{MHS}(\eta_{\mathcal{M}})}\left(\mathbb{Q}(-d), \underline{\mathcal{H}}^{i-1}_{\eta_{\mathcal{S}}/\eta_{\mathcal{M}},\mathbb{Q}} \otimes \overline{\mathcal{H}^{2d-i}_{X,\mathbb{Q}}}\right) \to 0$$

on $\eta_{\mathcal{M}}$.

To complete the argument, one needs the $\{[AJ(\mathfrak{Z}_\mu)]^{tr}_{i-1}\}$ to be obtained as quotients of $AJ(\overline{\mathfrak{Z}_\mu})$ as in §4.2 (corollary 4.6). This holds if $\overline{\mathfrak{Z}_\mu}$ can be made $\overset{\text{hom}}{\equiv} 0$, which follows from $[[\mathfrak{Z}_\mu]_0 = \cdots = [\mathfrak{Z}_\mu]_{i-1} = 0 \implies]\ \mathfrak{Z}_\mu \overset{\text{hom}}{\equiv} 0$ provided one assumes [GHC $\implies$] HC for X. In this case, the diagram then commutes because (quite generally) the topological invariant of a (1-D) family of AJ classes (of cycles on a family of projective varieties over $\mathcal{M}$ affine) gives the fundamental class of the family of cycles; we use this combined with the fiberwise Künneth decomposition (for $X \times S_\mu$). The desired conclusion follows at once. □

We note that in case X is a surface (hence $i = 2$) defined $/\bar{\mathbb{Q}}$, no conjectures need to be assumed: GHC$(1, 2, \mathcal{S})$ is just Lefschetz$(1, 1)$ for $\mathcal{S}$, and $[\mathfrak{Z}_\mu]_0 = [\mathfrak{Z}_\mu]_1 = [\mathfrak{Z}_\mu]_2 = 0 \implies \exists\ \overline{\mathfrak{Z}_\mu} \overset{\text{hom}}{\equiv} 0$ can be deduced from Lefschetz $(1, 1)$ for X; see argument in §6.3.

7.2. Idle speculations on the Leray filtration. There is an interesting relationship between the "expectation" above and a conjectural description of the Leray filtration on $\mathcal{H}^{2d}_{\mathcal{D}}(\mathfrak{X}, \mathbb{Q}(d))$ (and so ultimately on $CH^d(X_{/K})$). Always Leray is for $\pi : \mathfrak{X}_{\mathcal{U}} \to \mathcal{U}$ (or, in the limit, $\mathfrak{X} \to \eta_{\mathcal{S}}$) unless otherwise indicated.

For ordinary cohomology a coarse description of Leray uses the filtration on forms from §5.3,

$$\mathcal{L}^i H^{2d}(\mathfrak{X}_{\mathcal{U}}, \mathbb{C}) = \mathrm{im}\left\{\mathbb{H}^{2d}(\mathfrak{X}_{\mathcal{U}}, \mathcal{L}^i\Omega^\bullet_{\mathfrak{X}_{\mathcal{U}}}) \to \mathbb{H}^{2d}(\mathfrak{X}_{\mathcal{U}}, \Omega^\bullet_{\mathfrak{X}_{\mathcal{U}}}) \cong H^{2d}(\mathfrak{X}_{\mathcal{U}}, \mathbb{C})\right\}.$$

A more invariant version defined for $\mathbb{Q}$-coefficients uses the canonical filtration $\tau_{\leq \bullet}$, defined on a complex $K^\bullet$ by

$$\tau_{\leq \ell} K^j := \left\{ \begin{array}{ll} K^j\,, & j < \ell \\ (\ker d \subseteq K^\ell), & j = \ell \\ 0\,, & j > \ell \end{array} \right\}.$$

Writing $\mathcal{K}^\bullet$ for a π_*-acyclic resolution of $\mathbb{Q}$ (by sheaves on $\mathfrak{X}_\mathcal{U}$),

$$\mathcal{L}^i H^{2d}(\mathfrak{X}_\mathcal{U},\mathbb{Q}) = \mathrm{im}\left\{\mathbb{H}^{2d}(\mathcal{U},\tau_{\le 2d-i}\pi_*\mathcal{K}^\bullet) \to \mathbb{H}^{2d}(\mathcal{U},\pi_*\mathcal{K}^\bullet) \cong H^{2d}(\mathfrak{X}_\mathcal{U},\mathbb{Q})\right\}.$$

Finally, there is the following description due to Arapura [**Ar**] (worked out with Nori):

$$\mathcal{L}^i H^{2d}(\mathfrak{X}_\mathcal{U},\mathbb{Q}) = \bigcap_{\mathcal{V}\subseteq\mathcal{U}} \ker\left\{H^{2d}(\mathfrak{X}_\mathcal{U},\mathbb{Q}) \to H^{2d}(\mathfrak{X}_\mathcal{V},\mathbb{Q})\right\},$$

where the intersection is over $(i-2)$-dimensional smooth subvarieties $\mathcal{V}\subseteq\mathcal{U}$ defined $/\mathbb{C}$. In the limit this becomes

$$\mathcal{L}^i H^{2d}(\mathfrak{X},\mathbb{Q}) = \bigcap_{\mathcal{T}\subseteq\mathcal{S}} \ker\left\{H^{2d}(\mathfrak{X},\mathbb{Q}) \to H^{2d}(\mathfrak{X}_{\eta_\mathcal{T}},\mathbb{Q})\right\}$$

where $\mathcal{T}$ must be *very general* (of dim. $i-2$) in $\mathcal{S}$ for the restriction from $\eta_\mathcal{S}$ to $\eta_\mathcal{T}$ to make sense. (No rational function $\in \bar{\mathbb{Q}}(\mathcal{S})^*$ restricts to 0 on $\mathcal{T}$.) In terms of fields, if $\bar{\mathbb{Q}}(\mathcal{S}) \overset{\cong}{\to} K \subseteq \mathbb{C}$ is fixed, the choice of $\eta_\mathcal{T}\subseteq\eta_\mathcal{S}$ is essentially equivalent to a choice of $L\subseteq K$ s.t. $\mathrm{trdeg}(K/L) = i-2$ (which gives $\mathcal{M}$ with $\mathcal{S}\overset{\rho}{\to}\mathcal{M}$ of rel. dim. $i-2$) together with another independent embedding $L\hookrightarrow\mathbb{C}$ (which gives a choice of point $\mu\in\eta_\mathcal{M}(\mathbb{C})$, hence fiber of ρ). Alternatively, one could choose freely another complex embedding $\tilde{K}$ of $\bar{\mathbb{Q}}(\mathcal{S})$, then just choose $L\subseteq\tilde{K}$.

Does a similar description hold for lowest-weight absolute Hodge cohomology, viz.

$$(\sharp)_i:\quad \mathcal{L}^i \underline{H^{2d}_\mathcal{D}}(\mathfrak{X},\mathbb{Q}(d)) \overset{(?)}{=} \bigcap_{\substack{\mathcal{T}\subseteq\mathcal{S}\text{ v.g.}\\ (i-2)\text{-dim'l.}}} \ker\left\{\underline{H^{2d}_\mathcal{D}}(\mathfrak{X},\mathbb{Q}(d)) \overset{\mathcal{I}^*_{\eta_\mathcal{T}}}{\hookrightarrow} \underline{H^{2d}_\mathcal{D}}(\mathfrak{X}_{\eta_\mathcal{T}},\mathbb{Q}(d))\right\},$$

at least for the image $\left(\underline{H^{2d}_\mathcal{D}}(\mathfrak{X},\mathbb{Q}(d))\right)_{\mathrm{alg}}$ of the cycle map $\underline{c_\mathcal{D}}$? Modulo the following Christmas Wish we can show it is at least equivalent to the Expectation above:

Conjecture $(CW)_i$**:** If $\mathcal{Z}\in\mathcal{L}^{i-1}_{(K/\bar{\mathbb{Q}})}CH^d(X_{/K})$ with $[\mathfrak{Z}]_{i-1}=0$ and $[AJ(\mathfrak{Z}|_\mathcal{T})]_{i-2} = 0 \in Gr^{i-2}_\mathcal{L}\underline{J^d}(\mathfrak{X}_{\eta_\mathcal{T}})$ for all $\mathcal{T}\subseteq\mathcal{S}$ v.g. of dim. $(i-2)$, then $[AJ(\mathfrak{Z})]_{i-2} = 0 \in Gr^{i-2}_\mathcal{L}\underline{J^d}(\mathfrak{X})$.

REMARK 7.3. It is easy to see that under these circumstances $[AJ(\mathfrak{Z})]^{tr}_{i-2}=0$, but the extra little bit seems nontrivial.

PROPOSITION 7.4. *$(\sharp)^{alg}_i \implies (*)_i$; and assuming $(CW)_{\le i}$, $(*)_{\le i} \implies (\sharp)^{alg}_{\le i}$.*

PROOF. Consider the following five conditions on a cycle $\mathcal{Z}\in CH^d(X_K)$.

$(A)_i$: $\mathcal{Z}\in\mathcal{L}^i_{(K/\bar{\mathbb{Q}})}CH^d(X_K)$;

$(B)_i$: $\psi(\mathcal{Z})\in\mathcal{L}^i\underline{H^{2d}_\mathcal{D}}(\mathfrak{X},\mathbb{Q}(d))$;

$(C)_i$: $\forall\{\mathcal{T}\subseteq\mathcal{S}$ v.g. $(i-2)$-dim'l.$\}$, $\mathcal{I}^*_{\eta_\mathcal{T}}(\psi(\mathcal{Z})) = 0 \in \underline{H^{2d}_\mathcal{D}}(\mathfrak{X}_{\eta_\mathcal{T}},\mathbb{Q}(d))$;

$(D)_i$: $\forall\{\mathcal{M}/\bar{\mathbb{Q}},\ \rho:\mathcal{S}\to\mathcal{M}$ rel. dim. $i-2\}$, $0 = \{[AJ(\mathfrak{Z}_\mu)]_{i-2}\}_{\mu\in\eta_\mathcal{M}(\mathbb{C})}$;

$(E)_i$: $[\mathfrak{Z}]_{i-1}=0$ and $[AJ(\mathfrak{Z})]_{i-2}=0$.

(Note that $(D)_i$ and $(E)_i$ require an auxiliary condition, for example $(A)_{i-1}$, in order to have meaning.) Clearly $(A)_i$, $(B)_i$, and $(E)_{\le i}$ are equivalent; while

$(C)_i \implies (D)_i$ and $(A)_{i-1} + (D)_i \implies (C)_i$ from the above discussion. Since $\mathcal{L}^i \underline{H^{2d}_{\mathcal{D}}}(\mathfrak{X}_{\eta_\mathcal{T}}, \mathbb{Q}(d)) = 0$ and $\mathcal{I}^*_{\eta_\mathcal{T}}(\mathcal{L}^\bullet) \subseteq \mathcal{L}^\bullet$, $(B)_i \implies (C)_i$; the converse $(C)_i \implies (B)_i$ is the nontrivial content of $(\sharp)^{\text{alg}}_i$. Trivially $(A)_{i-1} + (E)_i \implies (D)_i$; the reverse $(A)_{i-1} + (D)_i \implies (E)_i$ is equivalent to $(*)_i + (CW)_i$.

So for the first statement let $(\sharp)^{\text{alg}}_i$ be given; then $(A)_{i-1} + (D)_i \implies (C)_i \overset{(\sharp)^{\text{alg}}_i}{\implies} (B)_i \implies (E)_i$ so $(*)_i$ holds.

For the "converse", inductively assume $(*)_{\leq i-1} + (CW)_{\leq i-1} \implies \{(C)_{\leq i-1} \implies (A)_{\leq i-1}\} + (\sharp)^{\text{alg}}_{\leq i-1}$. Then if $(*)_{\leq i} + (CW)_{\leq i}$, we have $(C)_i \implies (C)_{\leq i} \overset{\text{ind. hyp.}}{\implies} (D)_{\leq i} + (A)_{\leq i-1} \overset{(*)_i + (CW)_i}{\implies} (E)_{\leq i} \implies (A)_i + (B)_i$. □

If it held, $(\sharp)^{\text{alg}}$ would offer an interesting interpretation of $\mathcal{L}^\bullet$ on $CH^d(X_K)$. Let $\iota : K \overset{\cong}{\to} \tilde{K} \subseteq \mathbb{C}$ be *any* embedding of K (even if K comes already equipped with one) that is the identity on $\bar{\mathbb{Q}}$, and $L \subseteq K$ be any subfield $[\supset \bar{\mathbb{Q}}]$ with $t = \text{trdeg}(\tilde{K}/L) = i - 2$. Write $\psi_K^{\tilde{K}/L}$ for the composition $\psi^{\tilde{K}/L} \circ \iota^*$:

$$CH^d(X_K) \underset{\iota^*}{\overset{\cong}{\to}} CH^d(X_{\tilde{K}}) \underset{\text{part. sprd.}}{\overset{\cong}{\longrightarrow}} CH^d(\mathfrak{X}_{\eta_\mathcal{T}/L}) \underset{c_\mathcal{D}}{\to} \underline{H^{2d}_{\mathcal{D}}}(\mathfrak{X}_{\eta_\mathcal{T}}, \mathbb{Q}(d));$$

then $(\sharp)^{\text{alg}}_i \implies$

$$\mathcal{L}^i CH^d(X_K) = \bigcap_{\substack{\iota, \tilde{K}, L \\ [t=i-2]}} \ker(\psi_K^{\tilde{K}/L}).$$

Note that the $c_\mathcal{D}$ in the composition is *not* supposed to be injective (unless $L = \bar{\mathbb{Q}}$).

7.3. Specialization to lower transcendence degree. We now derive a powerful technique for producing 0-cycles with nontrivial higher AJ invariant (due to [**GGP**] for surfaces and M. Saito [**mS**] in the general case). Here is the idea: given $\mathcal{Z} \in \mathcal{L}^i CH_0(X_{/K})$ we can lift its $\bar{\mathbb{Q}}$-spread $\mathfrak{Z}$ to $\bar{\mathfrak{Z}} \in CH^d(\bar{\mathfrak{X}}_{/\bar{\mathbb{Q}}})$, and restrict along the inclusion $\mathcal{I}_\mathcal{T} : \mathfrak{X}_\mathcal{T} \hookrightarrow \mathfrak{X}_\mathcal{S} = \bar{\mathfrak{X}}$ (for $\mathcal{T} \subseteq \mathcal{S}$ a smooth $\bar{\mathbb{Q}}$-subvariety) to obtain $\overline{\mathfrak{Z}_\mathcal{T}} \in CH^d(\mathfrak{X}_{\mathcal{T}/\bar{\mathbb{Q}}}) \mapsto \mathfrak{Z}_\mathcal{T} \in CH^d(\mathfrak{X}_{\eta_\mathcal{T}/\bar{\mathbb{Q}}}) \overset{(\mathrm{I}^{-1})^*}{\longmapsto} \mathcal{Z}_\mathcal{T} \in CH_0(X_{/K_\mathcal{T}})$ where $\mathrm{I} : \bar{\mathbb{Q}}(\mathcal{T}) \overset{\cong}{\to} K_\mathcal{T} [\subseteq \mathbb{C}]$. We call this a $K_\mathcal{T}$-specialization of $\mathcal{Z}$, and it is not a well-defined operation (nor is $\mathfrak{X}_{\eta_\mathcal{T}}$ a subvariety of $\mathfrak{X}_{(\eta_\mathcal{S})}$, nor is there directly a map of function fields $\bar{\mathbb{Q}}(\mathcal{S}) \to \bar{\mathbb{Q}}(\mathcal{T})$). This is in contrast to the much nicer situation (§§7.1 − 2) when $\mathcal{T} \subseteq \mathcal{S}$ is very general, because here $\mathfrak{Z} \cap \mathfrak{X}_{\eta_\mathcal{T}}$ makes no sense. Hence the lift from $\mathfrak{Z}$ to $\bar{\mathfrak{Z}}$, which is not well-defined, plays an essential role in the passage from $\mathfrak{Z}$ to $\mathfrak{Z}_\mathcal{T}$.

However, the map $\underline{H^*}(\eta_\mathcal{S}) \to \underline{H^*}(\eta_\mathcal{T})$ *is* well-defined; consequently so are the maps of invariants $[\,\cdot\,]_j$ and $[AJ(\,\cdot\,)]_j^{tr}$ for the specialization of spreads from $\mathfrak{X}$ to $\mathfrak{X}_{\eta_\mathcal{T}}$. In fact, using a version of affine weak Lefschetz, one shows these are injective for $j \leq \dim(\mathcal{T})$ (cf. [**K1**, Lemma 3]). This of course does nothing for us unless e.g. both $[\mathfrak{Z}]_i$ *and* $[\mathfrak{Z}_\mathcal{T}]_i$ are defined, i.e. $\mathfrak{Z}$ and $\mathfrak{Z}_\mathcal{T}$ belong to $\mathcal{L}^i$ of $CH_0(X_{/K})$ and $CH_0(X_{/K_\mathcal{T}})$ (resp.). Even when $X/\bar{\mathbb{Q}}$, $\mathcal{Z} \in \mathcal{L}^i$ does not necessarily imply[26] $\mathcal{Z}_\mathcal{T} \in \mathcal{L}^i$

[26]The difficulty here is with the full invariants $[AJ(\cdot)]_j$ for $j < i - 1$ (as opposed to the quotient indicated by "tr") — it does *not* seem clear that $[AJ(\mathfrak{Z})]_j = 0 \implies [AJ(\mathfrak{Z}_\mathcal{T})]_j = 0$.

— one needs the stronger condition $c_{\mathcal{D}}(\bar{\mathfrak{Z}}) \in \mathcal{L}^i_{\mathcal{S}} H^{2d}_{\mathcal{D}}(X \times \mathcal{S}, \mathbb{Q}(d))$,[27] which may be obtained from $\underline{c_{\mathcal{D}}(\mathfrak{Z}) \in \mathcal{L}^i H^{2d}_{\mathcal{D}}}(X \times \eta_{\mathcal{S}}, \mathbb{Q}(d))$ [$\Leftarrow \mathcal{Z} \in \mathcal{L}^i$] only after assuming HC (see [**mS**, sec. 1.6]). One "exception" would be if $\mathcal{Z} = \langle \Delta_{X,\, 2d-i} \rangle_* \mathcal{Z}$ is the image of a Chow-Künneth component; then choosing $\bar{\mathfrak{Z}}$ with $c_{\mathcal{D}}(\bar{\mathfrak{Z}}) \in \mathcal{L}^i_{\mathcal{S}}$ is trivial.

If one does assume HC, it is easy to specialize $\mathcal{Z} \in \mathcal{L}^i CH_0(X_{/K})$ with $[\mathfrak{Z}]_i \neq 0$ to a cycle $\mathcal{Z}_{\mathcal{T}}$ of $\mathrm{trdeg}_{/\bar{\mathbb{Q}}}(K_{\mathcal{T}}) = \dim(\mathcal{T}) = i$ with $[\mathfrak{Z}_{\mathcal{T}}]_i \neq 0$. One has to go much deeper to find a transcendence-degree-$(i-1)$ specialization (necessarily $[\mathfrak{Z}_{\mathcal{T}}]_i = 0$) such that $[AJ(\mathfrak{Z}_{\mathcal{T}})]^{tr}_{i-1} \neq 0$. This is now our objective. Intuitively, we want a version of $(\sharp)^{\mathrm{alg}}_i$ with $\mathcal{T}/\bar{\mathbb{Q}}$ replacing $\mathcal{T}$ very general; unfortunately, in this case $\mathcal{I}^*_{\eta_{\mathcal{T}}}$ is not well-defined.

A better analogy is $(*)_{i+1}$, but with $\nu^{[i-1]}_{\mathfrak{Z}}(\mu) \neq 0$ *for some* $\mu \in \mathcal{M}(\bar{\mathbb{Q}})$, instead of $\mu \in \mathcal{M}(\mathbb{C})$ very general. Here now is a precise statement of what holds (where we take $\mathrm{trdeg}(K/k) = i$ since we have some idea of how to specialize *to* this situation):

PROPOSITION 7.5. *Let X be defined over a number field k, $\mathcal{Z} \in \mathcal{L}^i CH_0(X_K)$ with $K \cong k(\mathcal{S})$ f.g.$_{/\bar{\mathbb{Q}}}$ of trdeg. i and $[\mathfrak{Z}]_i \neq 0$. Take a lift $\bar{\mathfrak{Z}} \in CH^d(X \times \mathcal{S}_{/k})$ of $\mathfrak{Z}$ with $c_{\mathcal{D}}(\bar{\mathfrak{Z}}) \in \mathcal{L}^i_{\mathcal{S}}$ (assuming HC), and a Lefschetz pencil[28] $\mathcal{S} \xrightarrow{\bar{\rho}} \mathbb{P}^1_k$ with fibers $S_\mu := \bar{\rho}^{-1}(\mu)$. Assume $\mathcal{L}^\bullet = F^\bullet_{\mathcal{M}_\ell}$ (or BBC^q) and the GHC. Then $\exists\ \mu \in \mathbb{P}^1(\bar{\mathbb{Q}})$ s.t. $[AJ(\mathfrak{Z}_\mu)]_{i-1} \neq 0$, where $\mathfrak{Z}_\mu$ is the restriction to $X \times \eta_{S_\mu}$ of $\bar{\mathfrak{Z}} \cdot (X \times S_\mu)$. So we get a specialized cycle $\mathcal{Z}_\mu \in \mathcal{L}^i CH_0(X_{/\bar{\mathbb{Q}}(S_\mu)}) \setminus \mathcal{L}^{i+1}$ of trdeg. $i-1$.*

PROOF. We work in the category $(\mathcal{M} =)\mathcal{M}_\ell$ unless otherwise stated, see §5.5(ii) (and also [**mS**] for additional details in what follows).

Pick $\mathsf{U} \subseteq \mathbb{P}^1_k$ affine Zariski open and sufficiently small that $\mathcal{S}_{\mathsf{U}} \xrightarrow{\rho} \mathsf{U}$ has smooth fibers; let $\mu \in \mathsf{U}(k)$. Set $\mathcal{H} := H^i(X) \cong H^{2d-i}(X)^\vee$, $\underline{\mathcal{H}} := \frac{H^i(X)}{F^1_h} \cong \left(\frac{H^{2d-i}(X)}{F^{d-i+1}_h}\right)^\vee$; pulling back along the structure morphism $\mathsf{U} \xrightarrow{a} \operatorname{Spec} k$ yields $\mathcal{H}_{\mathsf{U}} := a^*\mathcal{H}$, $\underline{\mathcal{H}}_{\mathsf{U}} := a^*\underline{\mathcal{H}}$.

In $\mathcal{M}(\mathsf{U})$ we have by semisimplicity a decomposition

$$R^{i-1}\rho_*\mathbb{Q} \;=\; L' \oplus L'' \;=\; \text{variable} \oplus \text{constant}$$

where $H^0(\mathsf{U}, L') = 0$ and $W_i H^1(\mathsf{U}, L'') = W_1 H^1(\mathsf{U}) \otimes L''_\mu = 0$ (since $\mathsf{U} \subseteq \mathbb{P}^1$). Hence one has noncanonically

$$H^i(\mathcal{S}_{\mathsf{U}}) \cong H^0(\mathsf{U}, R^i\rho_*\mathbb{Q}) \oplus H^1(\mathsf{U}, L') \oplus H^1(\mathsf{U}, L'')$$

and (using in addition that the fibers of ρ are $(i-1)$-dimensional)

$$W_i H^i(\mathcal{S}_{\mathsf{U}}) \cong W_i H^1(\mathsf{U}, L').$$

Moreover, the short exact sequence $\mathcal{N}^1(R^{i-1}\rho_*\mathbb{Q}) \cap L' =: \mathcal{N}^1 L' \to L' \to L'/\mathcal{N}^1 L'$ splits and $H^1(\mathsf{U}, \mathcal{N}^1 L') \subseteq F^1_h H^1(\mathsf{U}, L') \implies$

$$\left(W_i H^i(\mathcal{S}_{\mathsf{U}})\right) / \, F^1_h \;\cong\; \left(W_i H^1(\mathsf{U},\, L'/\mathcal{N}^1 L')\right) / \, F^1_h \,.$$

Now $W_i\underline{\mathcal{H}}/F^1_h \;=\; \underline{\mathcal{H}}$ and so

$$Hom_{\mathcal{M}}\left(\underline{\mathcal{H}},\, H^i(\mathcal{S}_{\mathsf{U}})\right) \cong Hom_{\mathcal{M}}\left(\underline{\mathcal{H}},\, H^1(\mathsf{U}, L'/\mathcal{N}^1 L')\right). \tag{7.1}$$

[27] $\mathcal{L}^i_{\mathcal{S}}$ denotes Leray for $X \times \mathcal{S} \to \mathcal{S}$

[28] Here either $\mathcal{S}$ itself is suitably blown up or this is really $\mathcal{S} \xleftarrow{b} \mathfrak{S} \xrightarrow{\bar{\rho}} \mathbb{P}^1_k$. It makes no difference since we restrict immediately to open $\mathsf{U} \subseteq \mathbb{P}^1_k$; in fact, the pencil need not even be Lefschetz since all we use is that $\mathcal{S}_{\mathsf{U}} \xrightarrow{\rho} \mathsf{U}$ is smooth projective.

For any object $\mathcal{G} \in \mathcal{M}(\mathsf{U})$ one has a spectral sequence[29]

$$E_2^{p,q} = Ext^p_{\mathcal{M}}\left(\underline{\mathcal{H}},\, H^q(\mathsf{U},\mathcal{G})\right) \implies Ext^{p+q}_{\mathcal{M}(\mathsf{U})}\left(\underline{\mathcal{H}}_{\mathsf{U}},\, \mathcal{G}\right),$$

and in particular an exact sequence

$$0 \to Ext^1_{\mathcal{M}}\left(\underline{\mathcal{H}},\, H^0(\mathsf{U},\mathcal{G})\right) \to Ext^1_{\mathcal{M}(\mathsf{U})}\left(\underline{\mathcal{H}}_{\mathsf{U}},\, \mathcal{G}\right) \to$$
$$Hom_{\mathcal{M}}\left(\underline{\mathcal{H}},\, H^1(\mathsf{U},\mathcal{G})\right) \to Ext^2_{\mathcal{M}}\left(\underline{\mathcal{H}},\, H^0(\mathsf{U},\mathcal{G})\right).$$

Applying this to $\mathcal{G} = L'/\mathcal{N}^1L'$, with the observation $H^0(\mathsf{U}, L'/\mathcal{N}^1L') = 0$, gives

$$(7.2) \qquad Hom_{\mathcal{M}}\left(\underline{\mathcal{H}},\, H^1(\mathsf{U}, L'/\mathcal{N}^1L')\right) \cong Ext^1_{\mathcal{M}(\mathsf{U})}\left(\underline{\mathcal{H}}_{\mathsf{U}},\, L'/\mathcal{N}^1L'\right).$$

Referring to §5.5(*ii*), consider the following diagram

(7.3)

$$\begin{array}{ccc}
 & & Hom_{\mathcal{M}}\left(\underline{\mathcal{H}},\, H^i(\mathcal{S}_{\mathsf{U}})\right) \\
 & & \updownarrow \cong \\
 & & Hom_{\mathcal{M}}\left(\mathbb{Q}(-d),\, \underline{H}^i(\mathcal{S}_{\mathsf{U}}) \otimes \frac{H^{2d-i}(X)}{F_h^{d-i+1}}\right) \\
 & & \downarrow \\
Hom_{\mathcal{M}}\left(\mathbb{Q}(-d),\, \underline{H}^i(\eta_S) \otimes H^{2d-i}(X)\right) & \longrightarrow & Hom_{\mathcal{M}}\left(\mathbb{Q}(-d),\, \underline{H}^i(\eta_S) \otimes \frac{H^{2d-i}(X)}{F_h^{d-i+1}}\right) \\
\downarrow & & \downarrow \\
Hom_{\mathrm{MHS}}\left(\mathbb{Q}(-d),\, \underline{H}^i(\eta_S) \otimes H^{2d-i}(X)\right) & \longrightarrow & Hom_{\mathrm{MHS}}\left(\mathbb{Q}(-d),\, \underline{H}^i(\eta_S) \otimes \frac{H^{2d-i}(X)}{F_h^{d-i+1}}\right)
\end{array}$$

in the lower left corner of which lies $[\mathfrak{Z}]_i$. Since we are assuming $\mathcal{L}^i = F^i_{\mathcal{M}}$ on $CH_0(X_K)$, $\mathcal{Z}$ has a class $[\mathfrak{Z}]_i^{\mathcal{M}} \mapsto [\mathfrak{Z}]_i$ (in the diagram); and since we are assuming GHC, the horizontal arrows are injective. Since $[\mathfrak{Z}]_i \neq 0$ by hypothesis, we get a nonzero element $\widetilde{[\mathfrak{Z}]_i^{\mathcal{M}}} \in Hom_{\mathcal{M}}(\underline{\mathcal{H}}, H^i(\mathcal{S}_{\mathsf{U}}))$, hence one in $Ext^1_{\mathcal{M}_\ell(\mathsf{U})}(\underline{\mathcal{H}}_{\mathsf{U}}, L'/\mathcal{N}^1L')$ by (7.1) and (7.2).

The crucial step is that $\mu \in \mathsf{U}(k)$ can be chosen so that the specialization of this nontrivial extension class to $Ext^1_{\mathcal{M}_\ell}(\underline{\mathcal{H}}, L'_\mu/N^1L'_\mu)$ is still nonzero.[30] (The fiber S_μ over μ is a $\bar{\mathbb{Q}}$-member of the pencil.) This is really a statement about the underlying

[29]Grothendieck spectral sequence for derived functors of $Hom_{\mathcal{M}_\ell(\mathsf{U})}(\underline{\mathcal{H}}_{\mathsf{U}}, —) = Hom_{\mathcal{M}_\ell}(\underline{\mathcal{H}}, \Gamma(\mathsf{U}, —))$; $\mathcal{M}_\ell(\mathsf{U})$ is sheaves of $\mathcal{M}_\ell$-objects.

[30]Technical note: *a priori* one gets a nonzero specialized class only in $Ext^1_{\mathcal{M}_\ell}(\underline{\mathcal{H}}, L'_\mu/(\mathcal{N}^1L')_\mu)$, where $(\mathcal{N}^1L')_\mu$ may be a proper submodule of $N^1(L'_\mu)$. The choice of μ (see [**mS**, sec. 2.1]) allows $N^1(L'_\mu)$ to be extended to a subsheaf $L'_1 \subseteq L'$ on a finite cover of U. Repeating the arguments leading to (7.1), (7.2) with L'_1 replacing $\mathcal{N}^1L'$ (as L'_1 is likewise $\subseteq F^1_hL'$) shows that $Ext^1_{\mathcal{M}_\ell(\mathsf{U})}(\underline{\mathcal{H}}_{\mathsf{U}}, L'/\mathcal{N}^1L') \overset{\cong}{\to} Ext^1_{\mathcal{M}_\ell(\mathsf{U})}(\underline{\mathcal{H}}_{\mathsf{U}}, L'/L'_1)$. One then argues that the image of our extension class in the latter has nonzero specialization (at the same μ) in $Ext^1_{\mathcal{M}_\ell}(\underline{\mathcal{H}}, L'_\mu/L'_{1,\mu} = L'_\mu/N^1L'_\mu)$. See the last 2 paragraphs of [**mS**, sec. 2.3] for details.

ℓ-adic structure of the $\mathcal{M}_\ell$-objects, *not the MHS*: i.e., nontrivial extensions of ℓ-adic sheaves on U may be specialized to nontrivial extensions of ℓ-adic $Gal(\bar{k}/k)$ representations. This follows from the Hilbert irreducibility theorem, see **[mS]**, **[Te]**, **[GGP]** for details.

Now $\underline{H^{i-1}}(\eta_{S_\mu})$ is just the stalk of $(R^{i-1}\rho_*\mathbb{Q})/\mathcal{N}^1$ at μ, so $L'_\mu/N^1L'_\mu$ is a sub-$\mathcal{M}_\ell$-object, and

$$Ext^1_{\mathcal{M}_\ell}\left(\underline{\mathcal{H}},\, L'_\mu/N^1L'_\mu\right) \hookrightarrow Ext^1_{\mathcal{M}_\ell}\left(\underline{\mathcal{H}},\, \underline{H^{i-1}}(\eta_{S_\mu})\right)$$

$$\cong Ext^1_{\mathcal{M}_\ell}\left(\mathbb{Q}(-d),\, \underline{H^{i-1}}(\eta_{S_\mu}) \otimes \frac{H^{2d-i}(X)}{F_h^{d-i+1}}\right)$$

$$\hookrightarrow Ext^1_{\mathcal{M}_\ell}\left(\mathbb{Q}(-d),\, H^{i-1}(\eta_{S_\mu}) \otimes \frac{H^{2d-i}(X)}{F_h^{d-i+1}}\right).$$

Let ξ_μ denote the ultimate (nonzero) image of our class in the last term. Since we are assuming (G)HC, the specialized cycle $\mathcal{Z}_\mu$ lies in $\mathcal{L}^i$, hence (by the other assumption) $F^i_{\mathcal{M}}$; obviously $[\mathfrak{Z}_\mu]_i$ and $[\mathfrak{Z}_\mu]_i^{\mathcal{M}}$ are 0. Therefore $\mathcal{Z}_\mu$ has an invariant "$[AJ(\mathfrak{Z}_\mu)]_i^{\mathcal{M}}$" in $Gr^1_{\mathcal{E}}Gr^i_{\mathcal{L}_{\mathcal{M}_\ell}}H^{2d}_{\mathcal{D},\mathcal{M}_\ell}\left(X\times\eta_{S_\mu}/k,\,\mathbb{Q}(d)\right)$ *also* mapping to ξ_μ. Hence $[AJ(\mathfrak{Z}_\mu)]_{i-1}^{\mathcal{M}} \neq 0$ and $\mathcal{Z}_\mu \notin F^{i+1}_{\mathcal{M}}$. Using our assumption on the filtrations once more ($F^{i+1}_{\mathcal{M}} = \mathcal{L}^{i+1}$), we get $\mathcal{Z}_\mu \notin \mathcal{L}^{i+1}$; *therefore* $[AJ(\mathfrak{Z}_\mu)]_{i-1} \neq 0$. Without the assumption there is no reason why this should be so! □

Now suppose $X/\bar{\mathbb{Q}}$ is a smooth (d-dimensional) complete intersection in $\mathbb{P}^N$ with a nontrivial holomorphic d-form. The diagonal Δ_X has a Chow-Künneth decomposition $/\bar{\mathbb{Q}}$ and the 0-cycle $\mathcal{Z} = p - o$, for $p \in X(\mathbb{C})$ very general and $o/\bar{\mathbb{Q}}$, satisfies $\mathcal{Z} = \langle\Delta_{X,d}\rangle_*\,\mathcal{Z}$. Therefore $\mathcal{Z}$ and any specializations (*without* using HC) lie in $F^d_S CH_0(X_{/\mathbb{C}})$, *a fortiori* in $F^d_{\mathcal{M}}$ and $\mathcal{L}^d$. Furthermore, taking $\mathcal{S} = X$, $\mathfrak{Z}$ induces a nontrivial map $\Omega^i(X) \to \Omega^i(\mathcal{S})$ and hence one need not use GHC to get the nonzero classes on the right-hand side of the diagram (7.3). So by the proof of Prop. (7.5) above, assuming *no* conjectures, we have $\mathcal{Z}_\mu \notin F^{d+1}_{\mathcal{M}}$. Taking $N = d+r$, we have proved the following

THEOREM 7.6. [with no conjectural assumptions] *Let $X \subseteq \mathbb{P}^{d+r}$ be a smooth complete intersection defined $/\bar{\mathbb{Q}}$, of multidegree $(D_1, \ldots, D_r)$ with $\sum D_j \geq d+r+1$, and let $o \in X(\bar{\mathbb{Q}})$. Then there exists a $\bar{\mathbb{Q}}$-hyperplane section $Y \overset{\iota_Y}{\hookrightarrow} X$ containing o, such that if $q \in Y(\mathbb{C})$ is very general then $\iota_Y(q) - o \overset{rat}{\not\equiv} 0$ in $\mathcal{L}^d CH_0(X_{/\mathbb{C}})$.*

We emphasize that under the identification $\mathcal{S} = X$, $Y = S_\mu$ and $\mathcal{Z}_\mu = \iota_Y(q) - o$. (Moreover, $\mathcal{S}[\leftarrow \mathfrak{S}] \overset{\bar{\rho}}{\to} \mathbb{P}^1$ is taken to be a pencil of hyperplane sections with base locus containing o.) Incidentally, it is obvious that $q - o \overset{rat}{\not\equiv} 0$ on Y, since $\Omega^{d-1}(Y) \neq \{0\}$ and the spread of q induces the identity $\Omega^{d-1}(Y) \to \Omega^{d-1}(S_\mu)$. So $q - o$ gives a nonzero element of $Gr^{d-1}_{\mathcal{L}} CH_0(Y_{/\mathbb{C}})$. What make things hard on X is that fact that $\iota_Y(q) - o$ is of trdeg. $d-1$ and $\Omega^{d-1}(X) = \{0\}$.

In the Proposition and Theorem the spread base $\mathcal{S}$ has been fibered over a *rational* curve (unlike the $\mathcal{M}$ in §7.1); what if the curve (say, $\mathcal{C}$) is of positive genus? The only result in this direction takes $\mathcal{S} = S \times \mathcal{C}$, and is adapted from **[LS]**:

Theorem (Lewis): Let $\mathcal{C}$, S, X be smooth projective $/\bar{\mathbb{Q}}$ of resp. dimensions 1, $i-1$, d. Take $\mathcal{Z} \in \mathcal{L}^i CH_0(X_K)$ to be a cycle (of trdeg. i) with complete $\bar{\mathbb{Q}}$-spread $\bar{\mathfrak{Z}} \in CH^d(\mathcal{C} \times S \times X_{/\bar{\mathbb{Q}}})$ inducing a nontrivial map $\Omega^i(X) \to \Omega^i(\mathcal{C} \times S)$. Then there exists $\gamma \in CH_0^{\text{hom}}(\mathcal{C}_{/\bar{\mathbb{Q}}})$ s.t. writing $\bar{\mathfrak{Z}}_\gamma := \pi_{S\times X}\left\{\pi_{\mathcal{C}}^*(\gamma)\cdot \bar{\mathfrak{Z}}\right\}$, we have $[AJ(\mathfrak{Z}_\gamma)]_{i-1}^{tr} \neq 0$. Pulling $\mathfrak{Z}_\gamma$ back along the inclusion $\{s_0\}\times X \hookrightarrow S\times X$ for $s_0 \in S(\mathbb{C})$ v.g., we get the specialized cycle $\mathcal{Z}_\gamma \in \mathcal{L}^i CH_0(X_{/K_0}) \setminus \mathcal{L}^{i+1}$ of trdeg. $i-1$.

The crucial step in the proof (which plays the role of the Hilbert-Terasoma argument) is the existence of a $\bar{\mathbb{Q}}$-point of infinite rank on any abelian variety $/\bar{\mathbb{Q}}$ of positive dimension, applied to a certain sub-abelian-variety of $CH_0^{\text{hom}}(\mathcal{C}_{/\bar{\mathbb{Q}}})$. A similar result for symmetric squares of curves, together with the proof of this fact, can be found in [**GGP**].

8. Applications to Calabi-Yau varieties

In this final section we further illustrate the use of theorems from §§4, 6, and 7, on terrain probably quite familiar to readers of this volume. The results here are in the nature of straightforward corollaries rather than profound new theorems.

For us a "CY n-fold" shall mean a smooth projective variety X (of dimension n, defined over a subfield of $\mathbb{C}$), such that $X_{\mathbb{C}}^{(an)}$ has trivial canonical bundle and $H^0(\Omega_X^k)$ vanishes for $0<k<n$. So all CY's (and also K3 surfaces) in this section are algebraic (in fact, they will all be defined $/\bar{\mathbb{Q}}$). Our main purpose will be to construct 0-cycles in $\ker(Alb_X)$ which are not rationally equivalent to zero, hence (as we work $\otimes\mathbb{Q}$) of infinite order in $CH_0(X_{\mathbb{C}})$.

8.1. Projective complete intersections. Let $X_{/\bar{\mathbb{Q}}} \subseteq \mathbb{P}^{n+r}$ be smooth of multidegree $(D_1,\ldots,D_r)$, $\sum D_i =: D = n+r+1$. This is a CY n-fold, a basic example being the Fermat quintic 3-fold in $\mathbb{P}^4$. The diagonal class $[\Delta_X] \in [H^{2n}(X\times X,\mathbb{Q})$ has algebraic Künneth components

$$\tilde{\Delta}_{X,j} = \begin{cases} 0 & j \text{ odd}, \neq n \\ \frac{1}{D} H_X^{n-i}\times H_X^i & j \text{ even} = 2i, \neq n \\ \Delta_X - \sum_{k\neq n}\tilde{\Delta}_{X,k} & j = n \end{cases} \quad \text{in}\, CH_0(X\times X_{/\bar{\mathbb{Q}}}),$$

where (say) $H_X^i := X\cap\{z_1=0\}\cap\cdots\cap\{z_i=0\}$. Now let K be finitely generated $/\bar{\mathbb{Q}}$ and $\mathcal{Z} \in CH_0^{\text{hom}}(X_K)\,[=\mathcal{L}^1 CH_0]$; then Proposition 4.3($b$) implies that $\mathcal{Z} - \sum_{2n>j>n}\left\langle\tilde{\Delta}_{X,j}\right\rangle_* \mathcal{Z} \in \mathcal{L}^n CH_0(X_K)$. But $H_X^{n-i}\cdot\mathcal{Z} = 0$ for $n>i$, and so $\mathcal{Z} \in \mathcal{L}^n CH_0$. Consequently, any difference of two points (say, of $X(\mathbb{C})$) already belongs to $\mathcal{L}^n CH^n(X_{\mathbb{C}})$.

For a cycle of transcendence degree n, take $\mathcal{Z} = p - o$, where $p\in X(\mathbb{C})$ is very general and $o \in X(\bar{\mathbb{Q}})$. We can take as complete spread $\bar{\mathfrak{Z}} = \Delta_X - \{o\}\times X$ in $X\times(\mathcal{S}=X)$, and clearly $[\mathfrak{Z}]_n^* : \Omega^n(X) \to \Omega^n(\mathcal{S})$ is nontrivial; hence $[\mathfrak{Z}]_n \neq 0 \implies \mathcal{Z} \overset{\text{rat}}{\not\equiv} 0$.

To obtain a cycle of transcendence degree $n-1$, one considers a pencil

$$\mathbb{P}^1 \overset{f}{\leftarrow} \tilde{X} \overset{b}{\rightarrow} X$$

of hyperplane sections with $o\in$ base-locus; write $Y_\mu := b(f^{-1}(\mu)) \overset{\mathcal{I}^\mu}{\hookrightarrow} X$, $q\in Y_\mu(\mathbb{C})$ very general, $'\mathcal{Z}_\mu := (\mathcal{I}_*^\mu(q) - o) \in Z_0(X_{\mathbb{C}})$ for any $\mu \in \mathbb{P}^1(\bar{\mathbb{Q}})$. One may think of these as "general" cycles on "special" hyperplane sections. By theorem 7.6, there exists $\mu_0 \in \mathbb{P}^1(\bar{\mathbb{Q}})$ such that $'\mathcal{Z}_{\mu_0} \overset{\text{rat}}{\not\equiv} 0$ (and under conjectural conditions, also $[AJ('\mathfrak{Z}_{\mu_0})]_{n-1} \neq 0$); the theorem does not tell *which* μ_0 will suffice.

Cycles of transcendence degree $n-2$ in $CH_0^{\text{hom}} = \mathcal{L}^n$ should, assuming BBC^q, be $\overset{\text{rat}}{\equiv} 0$ (modulo torsion!).

Clearly transcendence degree $n-1$ is the interesting case; this is what we seek now to extend to a different set of examples, specifically for $n=3$.

8.2. Borcea-Voisin threefolds. We first recall the construction (due to [**Bo1, V2**]), which will begin with two ingredients:

(I). an elliptic curve $E/\bar{\mathbb{Q}}$ with $o \in E(\bar{\mathbb{Q}})$, $dz \in \Omega^1(E)$, and involution ι fixing o. One has $\iota^*(dz) = -dz$, fixed-point set $FP(\iota) = \{4 \text{ 2-torsion points}\} =: E^{[2]}$, and $E \overset{\pi^\iota}{\to} (E/\iota) \cong \mathbb{P}^1$. Write $\mathbb{T}$ for the π^ι-image of the torsion points of E relative to o (which all belong to $E(\bar{\mathbb{Q}})$), and $\mathbb{T}_{[2]}$ for $\pi^\iota(E^{[2]})$.

(II). a (smooth) K3 surface $S/\bar{\mathbb{Q}}$ with $\omega \in \Omega^2(S)$ and involution $\jmath$ with $\jmath^*(\omega) = -\omega$. One has $\mathcal{C} := FP(\jmath)$ either empty or a disjoint union of smooth curves (essentially clear from the fact that ω is nowhere-vanishing; see [**Bo1**] for requirements on their genera). Quotienting by the action of $\jmath$ yields a (branched) double cover $S \overset{\pi^\jmath}{\to} (S/\jmath) =: \mathcal{R}$. Plenty of exampes were tabulated by Nikulin (again see [**Bo1**]); we just mention a few:

(a). If $\mathcal{C} = \emptyset$ then $S \overset{\pi^\jmath}{\to} \mathcal{R}$ is an étale cover of an Enriques surface. [Otherwise $\mathcal{R}$ is rational (birational to $\mathbb{P}^2$).]

(b). $S \overset{\pi^\jmath}{\to} \mathbb{P}^2 = \mathcal{R}$ the double cover branched along a smooth sextic (genus 10) curve ($= \mathcal{C}$).

(c). Let $\hat{S}$ be the double cover of $\mathbb{P}^2$ branched over 6 lines in general position; this is singular over the lines' 15 intersection points. Blowing $\hat{S}$ up at these (and lifting the involution associated to $\hat{S} \overset{2}{\to} \mathbb{P}^2$) yields $S \overset{\pi^\jmath}{\to} \mathcal{R}$ ($S =$K3, $\mathcal{R}$ rational with $h^{1,1}(\mathcal{R}) = 16$) branched over 6 *disjoint* rational curves. If the 6 lines in $\mathbb{P}^2$ were common tangents to a conic Q, then S is the Kummer surface[31] associated to the Jacobian of the (genus 2) double cover of Q branched over the 6 points of tangency ([**Be**]).

(d). If E_2, E_3 are elliptic curves then $(E_2 \times E_3/\{1, (\iota_2, \iota_3)\}) =: \hat{S}$ is the (singular) 2-cover of $\mathbb{P}^1 \times \mathbb{P}^1$ branched over 8 $\mathbb{P}^1$'s intersecting in 16 points; blowing these points up yields the Kummer surface S of $E_2 \times E_3$, with $S \overset{\pi^\jmath}{\to} \mathcal{R}$ branched over 8 disjoint rational curves (and $h^{1,1}(\mathcal{R}) = 18$). Here $\jmath$ is induced from $(\iota_2, 1) \sim (1, \iota_3)$ on $\hat{S}$.

Now the involution $(\iota, \jmath)$ on $E \times S$ has FP-set $E^{[2]} \times \mathcal{C}$, which is a disjoint union of curves (e.g., $4 \times 6 = 24$ rational curves if S comes from case (c) above). Under the quotient, $FP(\iota, \jmath)$ becomes the singular set of $\hat{X} = E \times S/(\iota, \jmath)$. All we use from the Borcea-Voisin construction is that the blow-up of $\hat{X}$ along these curves gives a

[31]Recall that given an abelian variety A of dimension 2 with involution $\sigma = -id$, the Kummer surface associated to A is the resolution of $A/\{1, \sigma\}$ obtained by blowing up at the 16 singular points arising from the 2-torsion points of A.

(smooth) CY 3-fold $X/\bar{\mathbb{Q}}$. (In particular, $dz \wedge \omega$ is $(\iota, \jmath)^*$-invariant, so descends to $\Omega \in \Omega^3(X)$.)

For any $\xi \in \mathbb{P}^1(\mathbb{C}) \setminus \mathbb{T}_{[2]}$, one has a diagram of analytic varieties $/\mathbb{C}$:[32]

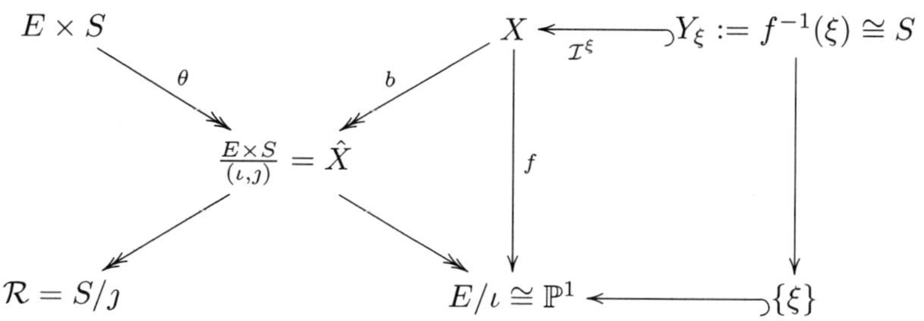

In analogy to §8.1, we want to show the following

Claim. There exists $\mu \in \mathbb{P}^1(\bar{\mathbb{Q}})$ such that (with $q \in Y_\mu(\mathbb{C})$ very general, $q_0 \in Y_\mu(\bar{\mathbb{Q}})$ any algebraic point) $Z_\mu := \mathcal{I}^\mu_*(q - q_0) \overset{\text{rat}}{\not\equiv} 0$ on X.

Here Proposition 7.5 (i.e., in the form of a corollary like Theorem 7.6) is no longer appropriate, and we shall use the Lewis result.

Since b^{-1} is well-defined on $\theta\left((E \setminus E^{[2]}) \times S\right)$, we have a map $\theta^{\setminus 2} : (E\setminus E^{[2]}) \times S \to X$. Let $q_0 \in S(\bar{\mathbb{Q}})$ be any algebraic point (which need not avoid $\mathcal{C}$), and write $\theta^{\setminus 2}_{q_0}$ for the composite of $\theta^{\setminus 2}$ with the internal projection $\pi^{q_0} : E \times S \to E \times \{q_0\} \hookrightarrow E \times S$ (restricted to $(E \setminus E^{[2]}) \times S$). The graphs of $\theta^{\setminus 2}$, $\theta^{\setminus 2}_{q_0}$ embed in $(E\setminus E^{[2]}) \times S \times X$, and taking their closures in $E \times S \times X$ gives correspondences Γ, $\Gamma_{q_0} \in Z^3(E \times S \times X)$. Next take $p \in E(\mathbb{C})$, $q \in S(\mathbb{C})$ very general and algebraically independent; and consider $\mathcal{Z}_0 := (p, q) - (o, q) - (p, q_0) + (o, q_0) \in \mathcal{L}^3 CH_0(E \times S_{/\mathbb{C}})$. Obviously $(p) + (\iota(p)) - 2(o) \overset{\text{rat}}{\equiv} 0$; we also need the following:

LEMMA. $\jmath^*\jmath_*((q) - (q_0)) = (q) - (q_0) + (\jmath(q)) - (\jmath(q_0)) \overset{rat}{\equiv} 0$.

PROOF. $\mathcal{R}$ is either an Enriques surface or a blow-up of $\mathbb{P}^2$ (which replaces points by rational curves). Clearly in the latter case any 2 points are $\overset{\text{rat}}{\equiv}$ on $\mathcal{R}$; this holds also for Enriques by [**BKL**]. Apply this to $\jmath(q)$ and $\jmath(q_0)$. □

It clearly follows that $\mathcal{Z}_0 \overset{\text{rat}}{\equiv} (\iota, \jmath)_* \mathcal{Z}_0$, hence that

$$\begin{aligned} \mathcal{Z}_0 &\overset{\text{rat}}{\equiv} \frac{1}{2}\left(\mathcal{Z}_0 + (\iota, \jmath)_*\mathcal{Z}_0\right) \\ &\overset{\text{rat}}{\equiv} \frac{1}{2}\left\{(p, q) - (p, q_0) + (\iota(p), \jmath(q)) - (\iota(p), \jmath(q_0))\right\} =: \mathcal{Z}_0'. \end{aligned}$$

[32]In particular $f^{-1}(\xi)$ is smooth. If $\xi \in \mathbb{P}^1(\bar{\mathbb{Q}})$ then it is also a diagram of $\bar{\mathbb{Q}}$-varieties. However if $\xi[\in \mathbb{P}^1(\mathbb{C})]$ is *not* algebraic, the isomorphism $S \cong f^{-1}(\xi)$ is *not* defined over $\bar{\mathbb{Q}}(\xi)$ but over a degree-2 extension isomorphic to $\bar{\mathbb{Q}}(E)$. In this case one must specify whether $\mathcal{I}^\xi$ means the map $S \hookrightarrow X$ (with the degree-2 extension) or $f^{-1}(\xi) \hookrightarrow X$, and for us it will be the former. So for $\xi \in \mathbb{P}^1(\mathbb{C})$, $q \in S(\mathbb{C})$ very general (and algebraically independent in the sense that $\xi \times q \in \mathbb{P}^1 \times S$ is very general), the natural spread base $\mathcal{S}$ for $\mathcal{I}^\xi_*\{q\}$[=general point on special fiber] is not X but $E \times S$ (or a birationally equivalent variety). This is used implicitly below.

By [**L1**, Thm. 1.2(iv)] the action of Γ preserves $\mathcal{L}^\bullet$ and so

$$\mathcal{Z} := \Gamma_* \mathcal{Z}_0' := \pi_*^X \left\{ \pi_{E\times S}^*(\mathcal{Z}_0') \cdot \Gamma \right\} = \mathcal{I}_*^{\pi^\iota(p)}(q - q_0)$$

belongs to $\mathcal{L}^3 CH_0(X_\mathbb{C})$. (Note $\pi^\iota(p) \in \mathbb{P}^1(\mathbb{C}) \setminus \mathbb{P}^1(\bar{\mathbb{Q}})$.) A choice of complete spread (for $\mathcal{Z}$) is $\Gamma - \Gamma_{q_0} =: \bar{\mathfrak{Z}}$, and its action sends Ω to $dz \wedge \omega$. This shows $\mathcal{Z} \overset{\text{rat}}{\not\equiv} 0$ (a trdeg.-3 cycle).

To get our desired transcendence-degree-2 cycle, observe that Lewis's theorem (§7.3) now guarantees the existence of $\gamma \in CH_0^{\text{hom}}(E_{/\bar{\mathbb{Q}}})$ such that $\mathcal{Z}_\gamma := \bar{\mathfrak{Z}}_\gamma\big|_{\{q\}\times X}$ is $\overset{\text{rat}}{\not\equiv} 0$ (with $[AJ(\mathfrak{Z}_\gamma)]_2 \neq 0$). Up to $\overset{\text{rat}}{\equiv}$, γ is of the form $(p_0) - (o)$ for some $p_0 \in \{E(\bar{\mathbb{Q}}) \setminus \text{torsion}\}$, and noting that $(\Gamma_{q_0})_* = \Gamma_* \circ \pi_*^{q_0}$,

$$\begin{aligned}\mathcal{Z}_{p_0 - o} &= \pi_*^X \left\{ \pi_{\mathcal{C}}^*(p_0 - o) \cdot \pi_S^*(q) \cdot \bar{\mathfrak{Z}} \right\} \\ &= \Gamma_* \{(p_0, q) - (o, q)\} - (\Gamma_{q_0})_* \{(p_0, q) - (o, q)\} \\ &= \Gamma_* \{(p_0, q) - (o, q) - (p_0, q_0) + (o, q_0)\}.\end{aligned}$$

Using the above computation for $\mathcal{Z}_0$ with p_0 replacing p,

$$\mathcal{Z}_{p_0 - o} \overset{\text{rat}}{\equiv} \mathcal{I}_*^\mu (q - q_0), \quad \text{where } \mu := \pi^\iota(p_0) \in \mathbb{P}^1(\bar{\mathbb{Q}}).$$

The claim is proved, and moreover $AJ_X^2(\mathcal{I}_*^\mu(q - q_0)) \neq 0$. As in §8.1, the proof doesn't say which values of μ suffice; this time we can do better.

THEOREM 8.1. *For $X/\bar{\mathbb{Q}}$ a Borcea-Voisin CY 3-fold (as described above), take any algebraic point $q_0 \in S(\bar{\mathbb{Q}})$ and* ***any*** *$\mu \in \mathbb{P}(\bar{\mathbb{Q}}) \setminus \mathbb{T}$; let $q \in S(\mathbb{C})$ be very general. Then writing $\mathcal{I}^\mu : S \cong f^{-1}(\mu) \hookrightarrow X$, $Z := \mathcal{I}_*^\mu(q - q_0)$ gives a nontrivial (nontorsion) class in $\mathcal{L}^3 CH_0(X_\mathbb{C})$.*

PROOF. Let p_0 be one of the two points in $(\pi^\iota)^{-1}(\mu) \subseteq E(\bar{\mathbb{Q}})$; $(p_0) - (o)$ is a nontorsion class. As above, $\mathcal{Z} \in \mathcal{L}^3$ follows by writing $\mathcal{Z}$ as the Γ_*-image of the cycle

$$\mathcal{Z}_0' := \frac{1}{2}\left\{(p_0, q) - (p_0, q_0) + (\iota(p_0), \jmath(q)) - (\iota(p_0), \jmath(q_0))\right\},$$

since this is

$$\overset{\text{rat}}{\equiv} \mathcal{Z}_0 := (p_0, q) - (o, q) - (p_0, q_0) + (o, q_0) \in \mathcal{L}^3 CH_0(E \times S).$$

Moreover, by Theorem 6.1 (with $j = 2$, $Y_1 = E$, $Y_2 = S$, $\mathcal{W} = p_0 - o$, $\mathcal{V} = q - q_0$), one has $\mathcal{Z}_0 = \mathcal{W} \times \mathcal{V} \overset{\text{rat}}{\not\equiv} 0$; hence $\mathcal{Z}_0' \overset{\text{rat}}{\not\equiv} 0$. But then

$$\theta^*(b_* \mathcal{Z}) = \theta^* \left(\theta_* \{(p_0, q) - (p_0, q_0)\}\right) = 2\mathcal{Z}_0'$$

implies $b_* \mathcal{Z} \overset{\text{rat}}{\not\equiv} 0$ on $\hat{X}$; so $\mathcal{Z} \overset{\text{rat}}{\not\equiv} 0$ on X, done. □

Now let E_1, E_2, and E_3 be elliptic curves defined $/\bar{\mathbb{Q}}$ (with origins o); take S to be the Kummer of $E_2 \times E_3$, and X to be the Borcea-Voisin CY of $E_1 \times S$. An alternate construction of X due to [**Bo2**] proceeds as follows. Write $\Theta : E_1 \times E_2 \times E_3 \to \frac{E_1 \times E_2 \times E_3}{V_4} =: \hat{X}'$ for the quotient by the action of

$$V_4 = \{(1,1,1), (\iota_1, \iota_2, 1), (\iota_1, 1, \iota_3), (1, \iota_2, \iota_3)\}.$$

Now $\hat{X}'$ is singular along the 48 rational curves $\Theta(E_1 \times E_2^{[2]} \times E_3^{[2]})$, $\Theta(E_1^{[2]} \times E_2 \times E_3^{[2]})$, and $\Theta(E_1^{[2]} \times E_2^{[2]} \times E_3)$; blowing up *first* at the $\Theta(E_1 \times E_2^{[2]} \times E_3^{[2]})$ then at the rest, yields $X \overset{B}{\to} \hat{X}'$.

Next we take points p, Q_2, Q_3 on E_1, E_2, E_3 (resp.); assume either

(i) p is algebraic and Q_2, Q_3 are (alg. indep.) very general, OR

(ii) Q_2 or Q_3 is algebraic and the other two are (alg. indep.) very general.

Write q, q_0 for points mapping (under $S \twoheadrightarrow \hat{S}$) to the images of (Q_2, Q_3), (Q_2, o) (resp.) under $E_2 \times E_3 \twoheadrightarrow \hat{S}$. Then by Example 6.2 and explicit rational equivalences one has on $E_1 \times E_2 \times E_3$:

$$\begin{aligned} 0 \overset{\text{rat}}{\not\equiv} 4\,Box^3(p,\,Q_2,\,Q_3) &\overset{\text{rat}}{\equiv} \sum_{\sigma\in V_4} \sigma_* Box^3(p,\,Q_2,\,Q_3) \\ &\overset{\text{rat}}{\equiv} \sum_{\sigma\in V_4} \sigma_*\{(p,\,Q_2,\,Q_3) - (p,\,Q_2,\,o)\} \\ &= \Theta^*\left(\Theta_*\{(p,\,Q_2,\,Q_3) - (p,\,Q_2,\,o)\}\right). \end{aligned}$$

This implies on $\hat{X}'$:

$$0 \overset{\text{rat}}{\not\equiv} \Theta_*\{(p,\,Q_2,\,Q_3) - (p,\,Q_2,\,o)\} = B_*\left\{\mathcal{I}_*^{\pi^\iota(p)}(q - q_0)\right\},$$

hence on X:

$$\mathcal{Z} := \mathcal{I}_*^{\pi^\iota(p)}(q - q_0) \overset{\text{rat}}{\not\equiv} 0.$$

In case (i) this is the cycle $\mathcal{Z}_{\pi^\iota(p)}$ of Theorem 8.1 — a transcendence-degree-2 cycle on an algebraic fiber of f, included into X. In case (ii), $\mathcal{Z}$ is a transcendence-degree-1 cycle[33] on a general fiber of f, included into X.

REMARK 8.2. A similar approach also yields nontrivial 0-cycles on the (smooth) CY 3-folds arising from the recent construction in **[PR]**.

8.3. 0-cycles and differential characters. To conclude the section we revisit the reduced higher AJ maps of §4.3 for X a CY n-fold defined $/\bar{\mathbb{Q}}$, as there are some nice simplifications in this case. For one thing, the basis $\{\omega\}$ (for $i = n$) is replaced by a single form $\Omega \in \Omega^n(X)$; and so the period lattice is

$$\Lambda = \text{im}\left\{\int_{(\cdot)} \Omega : H_n(X, \mathbb{Q}) \to \mathbb{C}\right\}.$$

Now assume $K \cong \bar{\mathbb{Q}}(\mathcal{S})$ is finitely generated $/\bar{\mathbb{Q}}$ of transcendence degree $n-1$, and note that $Gr^n_{\mathcal{L}}\underline{Hg}^n(X \times \eta_{\mathcal{S}}) = 0$ (as weak Lefschetz $\implies H^n(\mathcal{S}) = N^1H^n(\mathcal{S}) \implies \underline{H^n}(\eta_{\mathcal{S}}) = 0$). This means that AJ_X^{n-1} is defined directly on $\mathcal{L}^nCH_0(X_K)$, and composing this with the projections of §4.3 gives

$$\overline{AJ_X^{n-1}} : \mathcal{L}^nCH_0(X_{/K}) \to Hom_{\mathbb{Q}}\left(\underline{H_{n-1}}(\eta_{\mathcal{S}}, \mathbb{Q}), \mathbb{C}/\Lambda\right). \tag{8.1}$$

Using Proposition 4.8 and Remark 4.10(b) we may describe this as follows, for $\langle\mathcal{Z}\rangle \in \mathcal{L}^nCH_0$ and $[\gamma] \in \underline{H_{n-1}}(\eta_{\mathcal{S}}, \mathbb{Q})\,[\subseteq H_{n-1}(\mathcal{S}, \mathbb{Q})]$. Choosing a complete spread $\bar{\mathfrak{Z}} \in Z^n(X \times \mathcal{S}_{/\bar{\mathbb{Q}}})$ for $\mathcal{Z}$, take $\mathcal{U}_{\bar{\mathfrak{Z}}} \subseteq \mathcal{S}$ Zariski open such that $\bar{\mathfrak{Z}}|_{X\times\mathcal{U}_{\bar{\mathfrak{Z}}}}$ has relative dimension 0 over $\mathcal{U}_{\bar{\mathfrak{Z}}}$. We may also select a representative γ_0 of the class $[\gamma] \in H_{n-1}(\mathcal{S}, \mathbb{Q})$ which is supported in $\mathcal{U}_{\bar{\mathfrak{Z}}}$. Then $\pi_X\{\bar{\mathfrak{Z}} \cap (\gamma_0 \times X)\} \in Z^{\text{top}}_{n-1}(X, \mathbb{Q})$ is well-defined and (since $[\mathfrak{Z}]_{n-1} = 0$) a boundary, say $= \partial\Gamma$ (Γ a topological n-chain). One then sets

[33] i.e., the cycle has trdeg. 1 over the *field of definition* of the general fiber.

$$\overline{AJ_X^{n-1}}(\mathcal{Z})[\gamma] := \int_\Gamma \Omega \in \mathbb{C}/\Lambda, \tag{8.2}$$

which is independent of all choices.

A generalization of the Bloch conjecture says that the vanishing of $H^0(\Omega^i_X)$ for $0 < i < n$ should imply $Gr^i_{\mathcal{L}} CH_0(X_{/K}) = 0$ in that range. So $\overline{AJ_X^{n-1}}$ should be defined already on $CH_0^{\mathrm{hom}}(X_{/K})$. Moreover, under the duality $H^{n-1}(\mathcal{S},\mathbb{Q}) \cong H_{n-1}(\mathcal{S},\mathbb{Q})^\vee$ induced by integration, the dual of $\underline{H_{n-1}}(\eta_\mathcal{S},\mathbb{Q})$ (considered as subgroup of $H_{n-1}(\mathcal{S},\mathbb{Q})$) is $H^{n-1}(\mathcal{S},\mathbb{Q})/N^1H^{n-1}(\mathcal{S},\mathbb{Q})$. So assuming the "GBC", $\overline{AJ_X^{n-1}}$ may be written as a map

$$CH_0^{\mathrm{hom}}(X_{/K}) \longrightarrow \frac{H^{n-1}(\mathcal{S},\mathbb{C})}{N^1H^{n-1}(\mathcal{S},\mathbb{C}) + H^{n-1}(\mathcal{S},\Lambda)}. \tag{8.3}$$

EXAMPLE. For the complete intersection CY's considered in §8.1, we *know* (without conjecture) that $\mathcal{L}^nCH_0 = CH_0^{\mathrm{hom}}$. Consider a smooth $\bar{\mathbb{Q}}$-hyperplane section $Y \overset{\iota_Y}{\hookrightarrow} X$ with $q \in Y(\mathbb{C})$ very general, $o \in Y(\bar{\mathbb{Q}})$; take $\mathcal{Z} = \iota_*^Y(q-o)$ (defined $/K \cong \bar{\mathbb{Q}}(Y)$), $\mathcal{S} = Y$, and $\bar{\mathfrak{Z}}$ essentially as in Example 4.2. Then (as in Example 4.9) $\pi_X\{\bar{\mathfrak{Z}} \cap (\gamma \times X)\} = \iota_Y(\gamma)$ for $\gamma \in Z^{\mathrm{top}}_{n-1}(Y)$ with class $[\gamma] \in \underline{H_{n-1}}(\eta_Y,\mathbb{Q})$, and $\overline{AJ_X^{n-1}}(\mathcal{Z})[\gamma] = \int_\Gamma \Omega$ where $\partial\Gamma = \iota_Y(\gamma)$. If any membrane integral of this form gives a value not equal to a *period* of Ω (an integral over a topological n-cycle), then $\mathcal{Z} \overset{\mathrm{rat}}{\not\equiv} 0$ on X. [Note that since Y has a holomorphic form, $(H^{n-1}(Y)/N^1) \neq 0$; hence $\underline{H_{n-1}}(\eta_Y,\mathbb{Q}) \neq 0$.] However, nonvanishing of such integrals may be hard to establish, as one typically has $\dim_\mathbb{Q}(\Lambda) > 2$ (the exception being the case in which the subgroup $H^{n,0}(X) \oplus H^{0,n}(X) \subseteq H^n(X,\mathbb{C})$ is [the complexification of] a subHS).

We have (so far) restricted to the case $\mathrm{trdeg}(K/\bar{\mathbb{Q}}) = n-1$; obviously trdeg $<$ $n-1$ makes $\overline{AJ_X^{n-1}} = 0$. On the other hand, a more complicated picture arises in the event that trdeg $> n-1$. It is worked out in **[K2]** and we summarize pertinent details for the CY case. Since $Gr^n_{\mathcal{L}}\underline{Hg}^n(X \times \eta_\mathcal{S})$ no longer vanishes, $\overline{AJ_X^{n-1}}$ is not defined. Consequently, the $\mathbb{C}/\Lambda$-valued functional $\gamma \mapsto \int_{\partial_X^{-1}\{\bar{\mathfrak{Z}}_*(\cdot)\}} \Omega$ on (admissible) topological cycles (this is essentially the prescription from (8.2)) no longer descends to (admissible) homology classes. Instead, on a boundary $\gamma = \partial\mathcal{G}$ on $\mathcal{S}$, the functional evaluates (mod Λ) to $\int_{\bar{\mathfrak{Z}}_*\mathcal{G}} \Omega = \int_\mathcal{G} \{\bar{\mathfrak{Z}}^*\Omega =: \Omega_\mathcal{Z}\}$.[34]

This may be formalized as follows: replacing (8.1) is a map

$$\chi_X^n : \mathcal{L}^nCH_0(X_{/K}) \to \varinjlim \underline{H^n_\mathcal{D}}(\mathcal{S}\setminus D,\ \Lambda\{n\}) \tag{8.4}$$

(limit over divisors $D \subseteq \mathcal{S}$ defined $/\bar{\mathbb{Q}}$) also factoring through the map $Gr^n_{\mathcal{L}}\psi$ to $Gr^n_{\mathcal{L}}\underline{H^{2n}_\mathcal{D}}(X \times \eta_\mathcal{S},\mathbb{Q}(n))$; it sends $\mathcal{Z}$ to the above-described functional. We do not

[34]Here $\bar{\mathfrak{Z}}^* : \Omega^n(X) \to \Omega^n(\mathcal{S})$ is the induced map of holomorphic forms (see **[K2**, sec. 2**]**); this is the same as $[\mathfrak{Z}]^*_n$ in §4.

define the right-hand side of (8.4), but simply say that it sits in a short exact sequence

$$(8.5)\qquad \underline{H^{n-1}}(\eta_{\mathcal{S}}, \mathbb{C}/\Lambda) \xrightarrow{\overline{\beta_{n-1}}} \underline{H^n_{\mathcal{D}}}(\mathcal{S}\setminus D, \Lambda\{n\}) \xrightarrow{\overline{\alpha_n}} F^n\underline{H^n}(\eta_{\mathcal{S}}, \mathbb{C}) \cap \underline{H^n}(\eta_{\mathcal{S}}, \Lambda)$$

and informally explain the terms. In the limit, the left-hand term (of (8.5)) is just the right-hand side of (8.1) (or (8.3)). The right-hand term (also in the limit) identifies with the subset

$$\left\{ \omega \in \Omega^n(\mathcal{S}_{\{\mathbb{C}\}}) \;\middle|\; \int_\gamma \omega \in \Lambda \;\; \forall [\gamma] \in \underline{H_n}(\eta_{\mathcal{S}}, \mathbb{Q}) \right\} \subseteq \Omega^n(\mathcal{S}_{\mathbb{C}}).$$

The middle term may be interpreted as the space of "$\mathbb{C}/\Lambda$-valued differential $(n-1)$-characters on $\mathcal{S}\setminus D$", or functionals $\mathfrak{F}$ on $Z^{\text{top}}_{n-1}(\mathcal{S}\setminus D)$ with an associated holomorphic form $\overline{\alpha_n}(\mathfrak{F})$ such that $\mathfrak{F}(\partial\mathcal{G}) = \int_{\mathcal{G}} \overline{\alpha_n}(\mathfrak{F})$. Hence, $\overline{\alpha_n}\left(\chi^n_X(\mathcal{Z})\right)$ is precisely $\Omega_{\mathcal{Z}}$.

We hope that such differential-geometric descriptions of higher cycle maps (extended also beyond $CH_0(X)$ to all higher Chow groups $CH^p(X,m)$) will ultimately shed some light on the role played by algebraic K-theory and motivic cohomology in the context of mirror symmetry.

References

[Ar] D. Arapura, *The Leray spectral sequence is motivic*, to appear in Invent. Math.

[As] M. Asakura, *Arithmetic Hodge structures and nonvanishing of the cycle-class of* 0*-cycles*, K-Theory 27 (2002), no. 3, pp. 273-280.

[Be] A. Beauville, "Surfaces algébriques complexes", Asterisque 54, Soc. Math France, 1978.

[BV] A. Beauville and C. Voisin, *On the Chow ring of a* $K3$ *surface*, J. Alg. Geom. 13 (2004), no. 3, pp. 417-426.

[B1] S. Bloch, *Some elementary theorems about algebraic cycles on Abelian varieties*, Invent. Math. 37 (1976), pp. 215-228.

[B2] ———, "Lectures on algebraic cycles," Duke Univ. Math. Series IV, 1980.

[B3] ———, *Algebraic cycles and values of L-functions*, J. Reine Angew. Math. 350 (1984), pp. 94-108.

[BKL] S. Bloch, A. Kas and D. Lieberman, *Zero cycles on surfaces with* $p_g = 0$, Compositio Math. 33 (1976), no.2, 135-145.

[Bo1] C. Borcea, *K3 surfaces with involution and mirror pairs of Calabi-Yau manifolds*, AMS/IP Studies in Advanced Math. 1, 1997, pp. 717-743.

[Bo2] ———, *Calabi-Yau threefolds and complex multiplication*, Essays on mirror manifolds, Internat. Press, Hong Kong, 1992, pp. 489-502.

[De] P. Deligne, *Théorie de Hodge III*, Publ. Math IHES 44 (1974), pp. 5-77.

[Fa] N. Fakhruddin, *Zero-cycles on generic hypersurfaces of large degree*, preprint math.AG/0208180.

[Fu] W. Fulton, "Intersection theory" (2nd edition), Springer-Verlag, Berlin, 1998.

[G1] M. Green, Notes from the Banff conference on algebraic cycles, June 1998.

[G2] ———, *Griffiths' infinitesimal invariant and the Abel-Jacobi map*, J. Differential Geom. 29 (1989), no. 3, 545-555.

[Gr] P. Griffiths, *On the periods of certain rational integrals II*, Ann. of Math. (2) 90 (1969), pp. 496-541.

[GG1] M. Green and P. Griffiths, *Hodge-theoretic invariants for algebraic cycles*, Intl. Math. Res. Not. 2003, no. 9, pp. 477-510.

[GG2] ———, *An interesting* 0*-cycle*, Duke Math. J. 119 (2003), no. 2, 261-313.

[GGP] M. Green, P. Griffiths and K. Paranjape, *Cycles over fields of transcendence degree one*, Michigan Math. J. 52 (2004), no. 1, pp. 181-187.

[GS] P. Griffiths and W. Schmid, *Recent developments in Hodge theory: a discussion of techniques and results*, Discrete subgroups of Lie groups and applications to moduli (Bombay 1973), Oxford Univ. Press, 1975, pp. 31-127.

[H] B. Harris, *Homological versus algebraic equivalence in a Jacobian*, Proc. Nat. Acad. Sci. U.S.A. 80 (1983), no. 4 i., 1157-1158.

[J1] U. Jannsen, *Motivic sheaves and filtrations on Chow groups*, Motives (Seattle, WA, 1991), Proc. Sympos. Pure Math. 55(I), AMS, Providence, RI, 1994.

[J2] ———, *Mixed motives and algebraic K-theory*, Lecture Notes in Math. 1400, Springer-Verlag, Berlin, 1990.

[J3] ———, *Continuous étale cohomology*, Math. Ann. 280 (1988), pp. 207-245.

[J4] ———, *Deligne homology, Hodge-$\mathcal{D}$-conjecture, and motives*, Beilinson's Conjectures on Special Values of L-functions, pp. 305-372, Perspect. Math. 4, Academic Press, Boston, MA, 1988.

[K1] M. Kerr, *Exterior products of* 0*-cycles*, preprint, 2005.

[K2] ———, *Higher Abel-Jacobi maps for* 0*-cycles*, preprint, 2003.

[K3] ———, *A regulator formula for Milnor K-groups*, K-Theory 29 (2003), pp. 175-210.

[K4] ———, *Geometric construction of regulator currents with applications to algebraic cycles*, Princeton thesis, 2003.

[Ki] S-I. Kimura, *Chow groups can be finite dimensional in some sense*, preprint.

[Kl] S. Kleiman, *Algebraic cycles and the Weil conjectures*, in "Dix exposés sur la cohomologie des schémas", North-Holland, Amsterdam, 1968, pp. 359-386.

[L1] J. Lewis, *A filtration on the Chow groups of a complex projective variety*, Compositio Math. 128 (2001), no. 3, pp. 299-322.

[L2] ———, "A survey of the Hodge conjecture" (2nd edition), AMS, Providence, 1999.

[L3] ———, *Lectures on algebraic cycles*, Bol. Soc. Mat. Mexicana (3) 7 (2001), no. 2, 137-192.

[LS] J. Lewis and S. Saito, *Algebraic cycles and Mumford-Griffiths invariants*, preprint, 2004.

[Mu] D. Mumford, *Rational equivalence of* 0*-cycles on surfaces*, J. Math. Kyoto Univ. 9 (1968), pp. 195-204.

[M] J. Murre, *On a conjectural filtration on the Chow group of an algebraic variety (I)*, Indag. Math. (N.S.) 4 (1993), no. 2, 177-188.

[MR] J. Murre and D. Ramakrishnan, *Galois symbols on the square of an elliptic curve*, preprint.

[P] K. Paranjape, *Some spectral sequences for filtered complexes with applications*, J. Algebra 186 (1996), no. 3, 793-806.

[PR] K. Paranjape and D. Ramakrishnan, *Quotients of E^n by $\mathfrak{a}_{n+1}$ and Calabi-Yau manifolds*, to appear.

[Ra] W. Raskind, *Higher ℓ-adic Abel-Jacobi mappings and filtrations on Chow groups*, Duke Math. J. 78 (1995), no.1, pp. 33-57.

[R1] A. Roitman, *Rational equivalence of zero-dimensional cycles*, Math. USSR Sbornik 18 (1972), no. 4, pp. 571-588.

[R2] ———, *The torsion group of* 0*-cycles modulo rational equivalence*, Ann. Math. 111 (1980), pp. 553-569.

[RS] A. Rosenschon and M. Saito, *Cycle map for strictly decomposable cycles*, Amer. J. Math. 125 (2003), no. 4, pp. 773-790.

[mS] M. Saito, *Filtrations on Chow groups and transcendence degree*, Publ. Res. Inst. Math. Sci. 40 (2004), no. 3, pp. 933-948.

[sS] S. Saito, *Higher normal functions and Griffiths groups*, J. Alg. Geom. 11 (2002), no. 1, pp. 161-201.

[Te] T. Terasoma, *Complete intersections with middle Picard number* 1 *defined over* $\mathbb{Q}$, Math. Z. 189 (1985), no. 2, pp. 289-296.

[To] B. Totaro, *Milnor K-theory is the simplest part of algebraic K-theory*, K-Theory 6 (1992), no.2, pp. 177-189.

[V1] C. Voisin, *Variations de struture de Hodge et zéro-cycles sur les surfaces générales.* (French), Math. Ann. 299 (1994), no. 1, pp. 77-103.

[V2] ———, *Miroirs et involutions sur les surfaces K3*, in "Journeés de Géométrie Algébrique d'Orsay", Astérisque 218, Soc. Math. France, 1993, pp. 273-323.

[V3] ———, *Remarks on zero-cycles of self-products of varieties*, in "Moduli of vector bundles" (Proceedings of the Taniguchi symposium on vector bundles), Maruyama Ed., Dekker, 1996, pp. 265-285.

[V4] ———, "Hodge theory and complex algebraic geometry. II", Cambridge Studies in Advenced Mathematics 77, Cambridge University Press, Cambridge, 2003.

[Vo] V. Voevodsky, *A nilpotence theorem for cycles algebraically equivalent to zero*, Intl. Math. Res. Notices 1995, no. 4, pp. 187-198 (electronic).

Department of Mathematics, University of Chicago, Chicago, IL 60637 USA
E-mail address: matkerr@math.uchicago.edu

AMS/IP Studies in Advanced Mathematics
Volume 38, 2006

Geometry and Arithmetic of Non-rigid Families of Calabi–Yau 3-Folds; Questions and Examples

Eckart Viehweg and Kang Zuo

ABSTRACT. We speculate about the structure of maximal product subvarieties of moduli stacks of Calabi–Yau manifolds. We discuss an example of a family of quintic hypersurfaces in $\mathbb{P}^4$, parameterized by the product of two ball quotients, one of dimension two, the second one of dimension 12.

Let $\mathcal{M}_h(\mathbb{C})$ denote the set of isomorphism classes of minimal polarized manifolds F with fixed Hilbert polynomial h, and let $\mathcal{M}_h$ be the corresponding moduli functor, i.e.

$$\mathcal{M}_h(U) = \left\{ \begin{array}{c} (f : V \to U, \mathcal{L}); f \text{ smooth and} \\ (f^{-1}(u), \mathcal{L}|_{f^{-1}}(u)) \in \mathcal{M}_h(\mathbb{C}), \text{ for all } u \in U \end{array} \right\}$$

There exists a quasi-projective coarse moduli scheme M_h for $\mathcal{M}_h$. Fixing a projective manifold $\bar{U}$ and the complement U of a normal crossings divisor, we want to consider

$$\mathbf{H} = \left\{ \begin{array}{c} \varphi : (\bar{U}, U) \to (\overline{M}_h, M_h) \quad \text{induced} \\ \text{by polarized families } f : X \to U \end{array} \right\}.$$

Since M_h is just a coarse moduli scheme, it is not clear whether $\mathbf{H}$ has a scheme structure. However, by [**6**], if all $F \in \mathcal{M}(\mathbb{C})$ admit a locally injective Torelli map, there exists a fine moduli scheme M_h^N with a level structure N and étale over M_h. By abuse of notation, we will replace $\mathcal{M}_h$ by the moduli functor of polarized manifolds with a level N structure, and fix some compactification $\overline{M}_h$. Then $\mathbf{H}$ parameterizes all morphisms from

$$\varphi : (\bar{U}, U) \longrightarrow (\overline{M}_h, M_h),$$

hence it is a scheme. Moreover there exists a universal family $f : X \to \mathbf{H} \times U$.

As Kovács, Bedulev-Viehweg, Oguiso-Viehweg, and Viehweg-Zuo have shown, $\mathbf{H}$ is of finite type.

2000 *Mathematics Subject Classification.* Primary 14D07; Secondary 14D05, 14J32, 32Q25.

Key words and phrases. Variation of Hodge structures, Structure of families, Calabi–Yau manifolds, mirror symmetry.

This work has been supported by the "DFG-Schwerpunktprogramm Globale Methoden in der Komplexen Geometrie", and by the DFG-Leibniz program.

DEFINITION 1. $\varphi : U \to M_h$ called rigid if the component of $\mathbf{H}$ containing φ is zero-dimensional.

PROBLEM 2. Study the geometry of $\mathbf{H}$ and the arithmetic properties (for example the Mumford-Tate group) of the universal family $f : X \longrightarrow \mathbf{H} \times U$.

Acknowledgment. We thank the referee for pointing out several misprints and ambiguities.

This note was finished when the second named author visited the Institute of Mathematical Sciences at the Chinese University of Hong Kong. He would like to thank the members of the Institute for their hospitality.

1. Splitting of variations of Hodge structures

Let us start by recalling some of the properties of complex polarized variations of Hodge structures, and of families of Calabi–Yau manifolds.

PROPOSITION 3. *If $\mathbb{V}$ is an irreducible complex polarized variation of Hodge structures over $U_1 \times \cdots \times U_\ell$ then*

$$\mathbb{V} = p_1^*(\mathbb{V}_1) \otimes \cdots \otimes p_\ell^*(\mathbb{V}_\ell),$$

for complex polarized variations of Hodge structures $\mathbb{V}_i$ over U_i.

PROOF. The proof (see [**8**], Section 3, for the details) uses Schur's Lemma and Deligne's semi-simplicity of complex polarized variations of Hodge structures. □

2. Products in moduli stacks of Calabi–Yau manifolds

Since Calabi–Yau manifolds are unobstructed, the fine moduli scheme M_h is smooth, and we choose a smooth projective compactification $\overline{M}_h$ such that $\overline{M}_h \setminus M_h$ is a normal crossings divisor. Let $g : \mathcal{X} \to M_h$ be the universal family. We will assume moreover, that the local monodromies of $R^m g_* \mathbb{C}_{\mathcal{X}}$ around the components of $\overline{M}_h \setminus M_h$ are unipotent, where $m = \deg(h)$ is the dimension of the fibres.

Consider a smooth family

$$f : X \longrightarrow U_1 \times \cdots \times U_\ell = U$$

of Calabi–Yau m-folds, such that $\varphi : U \to M_h$ is generically finite. We assume that the factors U_i are nonsingular, and that $\dim(U_i) > 0$. Let $\mathbb{V} \subset R^m f_*(\mathbb{C}_X)$ be the irreducible sub-variation of Hodge structures with system of Hodge bundles

$$\bigoplus_{p+q=m} E^{p,q}$$

such that $E^{m,0} = f_* \Omega^m_{X/U}$.

FACT 4. The Kodaira–Spencer map is injective and factors through

$$d\varphi : T_U \longrightarrow E^{m-1,1} \otimes E^{m,0^{-1}} \subset \varphi^* T_{M_h}.$$

By Proposition 3 one has a decomposition $\mathbb{V} = \mathbb{V}_1 \otimes \cdots \otimes \mathbb{V}_\ell$. Let us write

$$\bigoplus_{p+q=m} F_i^{p,q}$$

for the system of Hodge bundles of $\mathbb{V}_i$, and $\varphi_i : U \to U_i \to \mathcal{D}_i$ for the corresponding period map. Then

$$d\varphi_i : T_{U_i} \longrightarrow F_i^{m_i-1,1} \otimes F_i^{m_i,0^{-1}} \subset \varphi_i^* T_{\mathcal{D}_i},$$

for $1 \leq i \leq \ell$. As in [**8**], 3.5, a comparison of Hodge bundles on both sides gives rise to

PROPOSITION 5.

i. *The cup-product*

$$\bigoplus_{1 \leq i_1 < \cdots < i_k \leq \ell} T_{U_{i_1}} \otimes \cdots \otimes T_{U_{i_k}} \longrightarrow R^k f_* T^k_{X/U}$$

is injective for $1 \leq k \leq \ell$.

ii. *If* $\varphi : U_1 \times \cdots \times U_\ell \to M_h$ *is an embedding and if* $\ell = m$ *is the dimension of the fibres of* f *then* $U_1 \times \cdots \times U_\ell$ *is a product of curves, and uniformized by* $\mathbb{V}_1 \oplus \cdots \oplus \mathbb{V}_\ell$ *over an algebraic number field.*

QUESTION 6. When will $U_1 \times \cdots \times U_m$ be a product of Shimura curves?

REMARK 7. A similar argument shows that part (i) of Proposition 5 also holds true for moduli stacks of hypersurfaces in $\mathbb{P}^n$ (see [**8**], 3.5 (b)).

QUESTION 8. Does Proposition 5, (1) hold true for moduli stacks of minimal polarized manifolds?

If $U_1 \times \cdots \times U_\ell$ maps generically finitely to a moduli stack $\mathcal{M}_h$ of minimal polarized manifolds, then it has been shown in [**7**], Corollary 6.4, that

$$\ell \leq m = \deg(h).$$

QUESTION 9. Can one improve this bound for certain moduli stacks and, fixing ℓ, what are optimal bounds for the dimensions of the U_i?

Since we assumed M_h to be a fine moduli space, deformations of the morphism $\varphi : U \to M_h$ correspond to deformations of the family $f : X \to U$. If one assumes that U has a compactification $\bar{U}$ such that φ extends to $\varphi : \bar{U} \to \overline{M}_h$, in such a way that the pre-image of $S = \overline{M}_h \setminus M_h$ remains a reduced normal crossings divisor, the first order deformations of the first type are classified by

$$H^0(\bar{U}, \varphi^* T_{\overline{M}_h}(-\log S)).$$

PROPOSITION 10. *Assume in addition that* f *extends to a proper morphism* $f : \bar{X} \to \bar{U}$, *semi-stable in codimension one, and that* $f^* f_* \omega_{\bar{X}/\bar{U}} \to \omega_{\bar{X}/\bar{U}}$ *is an isomorphism outside of* $f^{-1}(Z)$ *for some* $Z \subset \bar{U}$ *closed and of codimension at least two. Then*

$$\dim H^0(\bar{U}, \varphi^* T_{\overline{M}_h}(-\log S))$$

is invariant under infinitesimal deformations.

In particular, by Ran's T^1*-lifting property, deformations of those families* $f : X \to U$ *of Calabi–Yau manifolds with* U *fixed are unobstructed.*

REMARK 11. We expect that Proposition 10 holds true under weaker and more natural conditions on the boundary.

PROOF. Since we are only interested in global sections, taking complete intersection we may assume that $\dim \bar{U} = 1$, that all fibres are semi-stable and that

$$f^* f_* \omega_{\bar{X}/\bar{U}} \longrightarrow \omega_{\bar{X}/\bar{U}}$$

is an isomorphism.

Recall that (choosing a level N structure) we assumed the existence of a universal family $f : \mathcal{X} \to M_h$. The pullback of the logarithmic Higgs field

$$\theta : E \longrightarrow E \otimes \Omega^1_{\overline{M}_h}(\log S)$$

of the variation of Hodge structures $R^m f_* \mathbb{Q}_{\mathcal{X}}$ to $\bar{U}$ corresponds to a subsheaf

$$\varphi^* T_{\overline{M}_h}(-\log S) \longrightarrow (\mathcal{E}nd(\varphi^* E), \theta^{\mathcal{E}nd}).$$

By ([**10**], Prop. 2.1) $\theta^{\mathcal{E}nd}(\varphi^* T_{\overline{M}_h}(-\log S)) = 0$. This means that the above subsheaf is a Higgs subsheaf.

We need the following theorem on intersection cohomology and Higgs cohomology of a complex polarized variation of Hodge structures $\mathbb{W}$ with unipotent local monodromy around S. Let (F, θ) denote the logarithmic Higgs bundle of $\mathbb{W}$. We consider the complex of sheaves defined by the Higgs field

$$F \xrightarrow{\theta} F \otimes \Omega^1_{\bar{U}}(\log S) \xrightarrow{\theta} F \otimes \Omega^2_{\bar{U}}(\log S) \longrightarrow \cdots.$$

In [**9**] (for $\dim \bar{U} = 1$ in an implicit way) and in [**4**] (in general) one finds the definition of an algebraic L_2- subcomplex of sheaves

$$\begin{array}{ccccccc}
F & \xrightarrow{\theta} & F \otimes \Omega^1_{\bar{U}}(\log S) & \xrightarrow{\theta} & F \otimes \Omega^2_{\bar{U}}(\log S) & \longrightarrow & \cdots \\
\cup & & \cup & & \cup & & \\
F_{(2)} & \xrightarrow{\theta} & (F \otimes \Omega^1_{\bar{U}}(\log S))_{(2)} & \xrightarrow{\theta} & (F \otimes \Omega^2_{\bar{U}}(\log S))_{(2)} & \longrightarrow & \cdots
\end{array}$$

determined by an algebraic condition on $F|_S$ imposed by the weight filtration of

$$\mathrm{res}(\theta) : F|_S \longrightarrow F|_S.$$

Note that for a subsheaf $F' \subset \mathrm{Ker}(\theta)$, one has $F' \subset F_{(2)}$.

THEOREM 12 ([**9**] for $\dim \bar{U} = 1$, [**4**]).

$$\mathbb{H}^i\big(F_{(2)} \xrightarrow{\theta} (F \otimes \Omega^1_{\bar{U}}(\log S))_{(2)} \xrightarrow{\theta} \cdots\big) \simeq H^i_{\text{intersection}}(\mathbb{W}).$$

Going back to our situation, the exact sequence of complexes of sheaves

$$0 \longrightarrow (\varphi^* T_{\overline{M}_h}(-\log S), 0) \longrightarrow (\mathcal{E}nd(\varphi^* E), \theta^{\mathcal{E}nd}) \longrightarrow (Q, \theta) \longrightarrow 0$$

gives rise to a long exact sequence

$$\begin{array}{llll}
\cdots \to & \mathbb{H}^{i-1}(Q,\theta) \to & H^i(\varphi^* T_{\overline{M}_h}(-\log S)) & \to \mathbb{H}^i(\mathcal{E}nd(\varphi^* E)_{(2)}, \theta^{\mathcal{E}nd}) \\
\to & \mathbb{H}^i(Q,\theta) \to & H^{i+1}(\varphi^* T_{\overline{M}_h}(-\log S)) & \to \mathbb{H}^{i+1}(\mathcal{E}nd(\varphi^* E)_{(2)}, \theta^{\mathcal{E}nd}) \\
\to & \cdots & &
\end{array}$$

Since we assumed the fibres $f^{-1}(p)$ of f to be semi-stable and minimal, [**5**] implies that $f^{-1}(p)$ has no obstruction to deformations in any direction. This means that the pullback of the Kodaira–Spencer map of the moduli space to $\bar{U}$

$$(\varphi^* T_{\overline{M}_h}(-\log S), 0) \longrightarrow (\mathcal{E}nd(\varphi^* E)_{(2)}, \theta^{\mathcal{E}nd}) \longrightarrow (\varphi^* E^{m-1,1} \otimes \varphi^* E^{0,m}, 0)$$

is an isomorphism. Taking into account that these are maps between complexes of sheaves, we find

$$H^i(\varphi^* T_{\overline{M}_h}(-\log S)) \longrightarrow \mathbb{H}^i(\mathcal{E}nd(\varphi^* E)_{(2)}, \theta^{\mathcal{E}nd})$$

to be injective for all i. Hence there is a splitting

$$\mathbb{H}^i(\mathcal{E}nd(\varphi^*E)_{(2)}, \theta^{\mathcal{E}nd}) = H^i((\varphi^* T_{\overline{M}_h}(-\log S)) \oplus \mathbb{H}^i(Q, \theta).$$

By Theorem 12 $\mathbb{H}^i(\mathcal{E}nd(\varphi^*E)_{(2)}, \theta^{\mathcal{E}nd})$ is isomorphic to the intersection cohomology, hence is invariant under small deformations. Using the semi-continuity of the hypercohomology of complexes of sheaves one shows that both

$$H^i((\varphi^* T_{\overline{M}_h}(-\log S)) \qquad \text{and} \qquad \mathbb{H}^i(Q, \theta)$$

are invariant under small deformations. □

COROLLARY 13. *Under the assumptions made in 10 the scheme* $\mathbf{H}$ *is smooth.*

3. Applications

Again $f : X \to U$ denotes a smooth family of Calabi–Yau 3-folds, such that $\varphi : U \to M_h$ is generically finite. We keep the assumption, that M_h has a universal family. Moreover, we choose a compactification $\overline{M}_h$ with $\overline{M}_h \setminus M_h$ a normal crossings divisor, such that $U \to M_h$ extends to $\bar{U} \to \overline{M}_h$.

Starting with

$$\mathbf{H}_1 = \mathrm{Hom}((\overline{U}, U), (\overline{M}_h, M_h)),$$

consider

$$\mathbf{H}_2 = \mathrm{Hom}((\overline{\mathbf{H}}_1 \times \{0\}, \mathbf{H}_1 \times \{0\}), (\overline{M}_h, M_h)), \quad \{0\} \in U,$$

together with the induced family $f : X \to \mathbf{H}_1 \times \mathbf{H}_2 = \mathbf{H}$.

Let $\mathbb{V} \subset R^3 f_*(\mathbb{C}_X)$ be the irreducible sub-variation of Hodge structures with Hodge decomposition

$$\bigoplus_{p+q=3} F^{p,q} \quad \text{with} \quad F^{3,0} = f_*\Omega^3_{X/\mathbf{H}}.$$

Recall that by Proposition 3 one has a decomposition $\mathbb{V} = \mathbb{V}_1 \otimes \mathbb{V}_2$, where $\mathbb{V}_i$ is the pullback of a $\mathbb{C}$ variation of Hodge structures on $\mathbf{H}_i$. Comparing the possible Hodge numbers, one finds:

PROPOSITION 14. $\mathbb{V}_i$ *has one of the following Hodge types:*

a. $F_i^{1,0} \oplus F_i^{0,1}$, $\quad \mathrm{rk}\, F_i^{1,0} = 1$.
b. $F_i^{2,0} \oplus F_i^{1,1} \oplus F_i^{0,2}$, $\quad \mathrm{rk} F_i^{2,0} = \mathrm{rk}\, F_i^{0,2} = 1$, *and* $\mathbb{V}_i$ *is real.*
c. *Moreover, if* $\mathbb{V}_1$ *is of type (b), then* $\mathrm{rk}\, \mathbb{V}_2 = 2$.

It is well known that the period domains $\mathcal{D}_i$ of Hodge structures of types (a) and (b) are the bounded symmetric domains of the algebraic group $U(1, \mathrm{rk}\, \mathbb{V}_i^{0,1})$, and $SO(2, \mathrm{rk}\, \mathbb{V}_i^{1,1})$, respectively.

The unobstructedness for deformations of families implies that the generically finite period map $\tilde{\mathbf{H}}_i \to \mathcal{D}_i$ has to be dominant. Let us assume that $U \to M_h$ is injective.

QUESTION 15.

1. Is $\mathbf{H}_i^s \simeq \mathcal{D}_i/\Gamma_i$ for some Γ_i a partial compactification of $\mathbf{H}_i$?
2. What is the moduli-interpretation of points in $\mathbf{H}_i^s \setminus \mathbf{H}_i$?

4. An example of a non-rigid family of Calabi–Yau quintic threefolds

Let $f_5(x_2, x_1, x_0) \in \mathbb{C}[x_2, x_1, x_0]$ be the polynomial of a quintic plane curve in $\mathbb{P}^2$. Then

$$x_3^5 + f_5(x_2,\, x_1,\, x_0)$$

defines a quintic hypersurface in $\mathbb{P}^3$, and

$$x_4^5 + x_3^5 + f_5(x_2,\, x_1,\, x_0)$$

a Calabi–Yau quintic 3-fold in $\mathbb{P}^4$.

Obviously this construction can also be done locally over the moduli stack $M_{5,2}$ of quintic plane curves in $\mathbb{P}^2$, starting with the universal family $f : X \to M_{5,2}$ of curves. Replacing $M_{5,2}$ by some covering, one can glue those families as family of subvarieties in some projective bundle (see [**8**]). The resulting family of surfaces will be denoted by $g_1 : Z_1 \to M_{5,2}$, and that of threefolds by $g_2 : Z_2 \to M_{5,2}$.

REMARK 16. As pointed out by S.T. Yau, this family has been studied by S. Ferrara and J. Louis [**3**]. They have shown that the Yukawa coupling is zero and that the monodromy lies in $SU(2,1)$. In [**8**] the exact length of the Yukawa coupling is calculated for such families.

One can play a similar game, starting with 5 points in $\mathbb{P}^1$. say with equation $h_5(x_1,\, x_0) \in \mathbb{C}[x_1,\, x_0]$. Then $x_2^5 + h_5(x_1,\, x_0)$ defines a quintic plane curve. Again, one can do such a construction starting with the universal family $P \to M_{5,1}$ of 5 points in $\mathbb{P}^1$, and one obtains a family $g_0 : Z_0 \to M_{5,1}$ of quintic plane curves.

Finally Σ_5 denotes the Fermat curve $x_2^5 + x_1^5 + x_0^5 = 0$ of degree 5.

PROPOSITION 17. *The fibre product $Z_1 \times \Sigma_5 \to M_{5,2}$ admits a $\mathbb{Z}_5$-action over $M_{5,2}$, given fibrewise by*

$$(x_3, x_2, x_1, x_0), (y_2, y_1, y_0) \mapsto (e^{2\pi i/5}x_3, x_2, x_1, x_0), (e^{2\pi i/5}y_2, y_1, y_0).$$

1 *The family of Calabi–Yau quintics $g_2 : Z_2 \to M_{5,2}$ can be reconstructed as:*

$$\begin{array}{ccccc} (Z_1 \times \Sigma_5)/\mathbb{Z}_5 & \xleftarrow{\text{blowup}} & (\widehat{Z_1 \times \Sigma_5})/\mathbb{Z}_5 & \xrightarrow{\text{blowdown}} & Z_2 \\ & \searrow^{g_1} & \downarrow & \swarrow^{g_2} & \\ & & M_{5,2}. & & \end{array}$$

2. *The construction in (1) extends to the product family*

$$\begin{array}{ccccc} (Z_1 \times Z_0)/\mathbb{Z}_5 & \xleftarrow{\text{blow up}} & (\widehat{Z_1 \times Z_0})/\mathbb{Z}_5 & \xrightarrow{\text{blow down}} & \mathcal{Z}_2 \\ & \searrow^{(g_1,g_0)} & \downarrow & \swarrow^{h_2} & \\ & & M_{5,2} \times M_{5,1}. & & \end{array}$$

3. *The family $h_2 : \mathcal{Z}_2 \to M_{5,2} \times M_{5,1}$ of Calabii–Yau quintics is a universal family of the form*

$$h_2 : \mathcal{Z}_2 \longrightarrow \mathbf{H}_1 \times \mathbf{H}_2,$$

i.e. for suitable compactifications $\overline{M}_h$, $\overline{M}_{5,2}$ *and* $\overline{M}_{5,1}$ *and for some base point* $u \in M_{5,2}$ *and* $u' \in M_{5,1}$

$$M_{5,2} = \mathbf{H}_1 = \mathrm{Hom}((\{u\} \times \overline{M}_{5,1}, \{u\} \times M_{5,1}), (\overline{M}_h, M_h)),$$

and

$$M_{5,1} = \mathbf{H}_2 = \mathrm{Hom}((\overline{M}_{5,2} \times \{u'\}, M_{5,2} \times \{u'\}), (\overline{M}_h, M_h)).$$

Moreover, a partial compactification $\mathbf{H}_1^s$ *of* $\mathbf{H}_1$ *is a 2-dimensional complex arithmetic ball quotient, and a partial compactification* $\mathbf{H}_2^s$ *of* $\mathbf{H}_2$ *is a 12-dimensional complex non-arithmetic ball quotient.*

SKETCH OF THE PROOF. (1) and (2) have been shown in ([**8**], Proposition 6.4). For (3) consider the eigenspace decompositions

$$R^1 g_{0*}(\bar{\mathbb{Q}}_{Z_0}) = \bigoplus_{i=1}^{4} R^1 g_{0*}(\bar{\mathbb{Q}}_{Z_0})_i, \quad \text{and} \quad R^2 g_{1*}(\bar{\mathbb{Q}}_{Z_1}) = \bigoplus_{i=1}^{4} R^2 g_{1*}(\bar{\mathbb{Q}}_{Z_1})_i$$

for the $\mathbb{Z}_5$-action. Recall that the restriction of $R^1 g_{0*}(\bar{\mathbb{Q}}_{Z_0})_i$ to a point in $M_{5,1}$ is a Hodge structure with $H^0(\mathbb{P}^1, \Omega^1_{\mathbb{P}^1}(5-i))$ in degree $(1,0)$ and $H^1(\mathbb{P}^1, \mathcal{O}_{\mathbb{P}^1}(-i))$ in degree $(0,1)$. Hence $R^1 g_{0*}(\bar{\mathbb{Q}}_{Z_0})_i$ is unitary for $i = 1$ and $i = 4$.

$R^1 g_{0*}(\bar{\mathbb{Q}}_{Z_0})_3$ and $R^1 g_{0*}(\bar{\mathbb{Q}}_{Z_0})_2$ are not unitary, and are dual to each other. By Deligne-Mostow [**1**] $M^s_{5,1}$ is uniformized by $R^1 g_{0*}(\bar{\mathbb{Q}}_{Z_0})_3$ as a 2-dimensional arithmetic ball quotient, which is a component of the moduli space, parameterizing Abelian varieties of dimension 6 with complex multiplication $\mathbb{Q}(\zeta)$ for $\zeta = e^{2\pi i/5}$.

For g_1 the situation is more complicated. One easily computes that the Higgs bundle of $R^2 g_{1*}(\bar{\mathbb{Q}}_{Z_1})_2$ is trivial in degree $(0,2)$, of a rank one in degree $(2,0)$ and of a rank 12 in degree $(1,1)$. A similar argument as the one used by Deligne–Mostow allows one to show that $M^s_{5,2}$ is uniformized by $R^2 g_{1*}(\bar{\mathbb{Q}}_{Z_1})_2$ as a 12-dimensional complex ball quotient.

There is a Galois conjugate $R^2 g_{1*}(\bar{\mathbb{Q}}_{Z_1})_2^\sigma$, which is neither the dual of $R^2 g_{1*}(\bar{\mathbb{Q}}_{Z_1})_2$, nor unitary. As in Deligne-Mostow this implies that the ball quotient is not arithmetic.

The quotient by $\mathbb{Z}_5$, together with blowing up and blowing down, gives rise to a $\mathbb{Q}$-Hodge isometry (see [**8**], 7.4) $(R^3 h_{2*}\mathbb{Q}_{\mathcal{Z}_2}) \simeq \mathbb{V}' \oplus \mathbb{T}$ with

$$\mathbb{V}' \otimes \mathbb{Q}(\zeta) = \bigoplus_{i=1}^{4} \mathbb{V}_{i,5-i} \quad \text{for} \quad \mathbb{V}_{i,5-i} = R^2 g_{1*}(\mathbb{Q}(\zeta)_{Z_1})_i \otimes R^1 g_{0*}(\mathbb{Q}(\zeta)_{Z_0})_{5-i}$$

$$\text{and with} \quad \mathbb{T} = \bigoplus^{4} R^1 f_* \mathbb{Q}_X(1),$$

where (1) denotes the Tate twist. Of course, we should write

$$\mathbb{V}_{i,5-i} = \mathrm{pr}_1^* R^2 g_{1*}(\mathbb{Q}(\zeta)_{Z_1})_i \otimes \mathrm{pr}_2^* R^1 g_{0*}(\mathbb{Q}(\zeta)_{Z_0})_{5-i},$$

but we suppress the pullback under the projections in our notation. Note that $\mathbb{T}$ is the part of the variation of Hodge structures, coming from the blow-ups. $\mathbb{V}_{i,5-i}$ is an irreducible subvariation of Hodge structures in $(R^3 h_{2*}\mathbb{Q}(\zeta)_{\mathcal{Z}_2})$, and for the corresponding $\mathbb{C}$ variation of Hodge structures, the fibre over a point y has

$$H^0(\mathbb{P}^2, \Omega^2_{\mathbb{P}^2}(5-i)) \otimes H^0(\mathbb{P}^1, \Omega^1_{\mathbb{P}^1}(i))$$

in degree $(3,0)$. This is zero for $i=1$ and $i=4$, and the $\mathbb{C}$ variation of Hodge structures given by

$$\mathbb{V} = \mathbb{V}_{2,3} \oplus \bar{\mathbb{V}}_{2,3} = \mathbb{V}_{2,3} \oplus \mathbb{V}_{3,2}$$

contains the Hodge bundle $h_{2*}\Omega^3_{\mathcal{Z}_2/M_{5,2}\times M_{5,1}}$.

By abuse of notation we will regard $\mathbb{V}$ and the $\mathbb{V}_{i,5-i}$ as $\bar{\mathbb{Q}}$ variations of Hodge structures. Write $(R^3h_{2*}\bar{\mathbb{Q}}_{\mathcal{Z}_2}) = \mathbb{V}\oplus\mathbb{W}$, and let $(F^{2,1}\oplus F^{1,2}\oplus F^{0,3},\theta)$ denote the system of Hodge bundles corresponding to $\mathbb{W}$. The missing part of 17, (3), follows from the next two claims. □

CLAIM 18. There is no nontrivial extension

$$\begin{array}{ccc} \mathcal{Z}_2 & \xrightarrow{h_2} & M_{5,2}\times M_{5,1} \\ \cap\downarrow & & \downarrow\cap \\ \mathcal{Z}_2' & \xrightarrow{h_2'} & N\times M_{5,1}, \end{array}$$

such that the induced morphism $\varphi: N\times M_{5,1}\to M_h$ is generically finite over its image.

PROOF. A deformation $N\times M_{5,1}$ of $M_{5,1}=\{u\}\times M_{5,1}$ which does not lie in $M_{5,2}\times M_{5,1}$ induces a non-zero flat section τ of $\mathrm{End}(\mathbb{V}\oplus\mathbb{W})|_{M_{5,1}}$ of type $(-1,1)$, which does not respect the direct sum decomposition $\mathbb{V}\oplus\mathbb{W}$.

In fact, if it does, one has $\tau(\mathbb{V}|_{M_{5,1}})\subset\mathbb{V}|_{M_{5,1}}$. The restrictions of $\mathbb{V}_{2,3}$ and $\mathbb{V}_{3,2}$ to $M_{5,1}$ are direct sums of local systems isomorphic to $R^1g_{0*}(\bar{\mathbb{Q}}_{Z_0})_3$ or $R^1g_{0*}(\bar{\mathbb{Q}}_{Z_0})_2$, respectively. As uniformizing variations of Hodge structures of a ball quotient, both are irreducible and, since the Hodge numbers are different, $R^1g_{0*}(\bar{\mathbb{Q}}_{Z_0})_3$ is not isomorphic to $R^1g_{0*}(\bar{\mathbb{Q}}_{Z_0})_2$. So τ respects the decomposition $\mathbb{V}=\mathbb{V}_{2,3}\oplus\mathbb{V}_{3,2}$ and it is induced by an endomorphism of $R^2g_{1*}(\bar{\mathbb{Q}}_{Z_1})_2|_{\{u\}}$ of type $(-1,1)$. The calculation of Hodge numbers, indicated above, shows that they lie in a 12-dimensional vector space. So they correspond to all the deformations along $M_{5,2}$.

On the other hand, if τ does not respect the direct sum decomposition, one obtains a non-trivial morphism $\mathbb{V}|_{M_{5,1}}\to\mathbb{W}|_{M_{5,1}}$. Since $\mathbb{W}|_{M_{5,1}}$ is a direct sum of unitary local systems, and since for $i=2$ or $i=3$ there exists no non-trivial morphism

$$R^1g_{0*}(\bar{\mathbb{Q}}_{Z_0})_i \longrightarrow \text{unitary local system},$$

this leads to a contradiction. □

CLAIM 19. There is no nontrivial extension

$$\begin{array}{ccc} \mathcal{Z}_2 & \xrightarrow{h_2} & M_{5,2}\times M_{5,1} \\ \cap\downarrow & & \downarrow\cap \\ \mathcal{Z}_2' & \xrightarrow{h_2'} & M_{5,2}\times N, \end{array}$$

such that the induced morphism $\varphi: M_{5,2}\times N\to M_h$ is generically finite over its image.

PROOF. Again, the deformations of $M_{5,2}\times\{u\}$ correspond to flat sections τ of $\mathrm{End}(\mathbb{V}\oplus\mathbb{W})|_{M_{5,2}}$ of type $(-1,1)$. If τ respects the direct sum decomposition, the irreducibility of $R^2g_{1*}(\bar{\mathbb{Q}}_{Z_1})_2$ and $R^2g_{1*}(\bar{\mathbb{Q}}_{Z_1})_3$ implies that τ is induced by an endomorphism of $R^1 g_{0*}(\bar{\mathbb{Q}}_{Z_0})_3|_{\{u\}}$ of type $(-1,1)$. Those form a 2-dimensional vector space, corresponding to the deformations along $M_{5,1}$.

So a deformation $M_{5,2} \times N$ which does not lie in $M_{5,2} \times M_{5,1}$ induces a non-zero flat section τ of $\mathrm{End}(\mathbb{V} \oplus \mathbb{W})|_{M_{5,2}}$ of type $(-1,1)$. which does not respect the direct sum decomposition. So one finds a non-trivial morphism

$$\tau : \mathbb{V}|_{M_{5,2}} = \bigoplus R^2 g_{1*}(\bar{\mathbb{Q}}_{Z_1})_3 \oplus \bigoplus R^2 g_{1*}(\bar{\mathbb{Q}}_{Z_1})_2 \longrightarrow \mathbb{W}|_{M_{5,2}}.$$

On the other hand, $\mathbb{W}|_{M_{5,2}}$ is a direct sum of several copies of the local systems

$$R^2 g_{1*}(\bar{\mathbb{Q}}_{Z_1})_1,\ R^2 g_{1*}(\bar{\mathbb{Q}}_{Z_1})_4, \quad \text{and} \quad R^1 f_* \bar{\mathbb{Q}}_X(1).$$

Note that the uniformization local system $R^2 g_{1*}(\bar{\mathbb{Q}}_{Z_1})_2$ for $M_{5,2}$ is irreducible, as well as its dual $R^2 g_{1*}(\bar{\mathbb{Q}}_{Z_1})_3$.

The local system $R^1 f_* \bar{\mathbb{Q}}_X(1)$ is the variation of Hodge structures attached to the universal family of plane curves of degree 5, hence it is irreducible by Deligne's irreducibility theorem [**2**]. $R^2 g_{1*}(\bar{\mathbb{Q}}_{Z_1})_1$ and $R^2 g_{1*}(\bar{\mathbb{Q}}_{Z_1})_4$ are both irreducible, by a generalization of Deligne's irreducibility theorem proved in [**8**], Lemma 4.1.

On the other hand, all the irreducible local systems considered above have different Hodge types. So there exists no non-zero morphism between them, a contradiction. □

REMARK 20. In [**8**] we consider the subscheme

$$M_{5,1} \times M_{5,1} \subset M_{5,2} \times M_{5,1},$$

and the restriction of

$$h_2 : \mathcal{Z}_2 \longrightarrow M_{5,2} \times M_{5,1}$$

to this subscheme. It is shown there that the set of CM-points $y \in M_{5,1} \times M_{5,1}$ is dense in $M_{5,1} \times M_{5,1}$, i.e. the set of points y for which the Hodge structure $H^3(h_2^{-1}(y), \mathbb{Q})$ has complex multiplication.

Since $M^s_{5,2}$ is a non-arithmetic ball quotient one should expect, according to the André-Oort conjecture, that the only positive dimensional component of the Zariski closure of the set of CM-points in $M_{5,2} \times M_{5,1}$ is $M_{5,1} \times M_{5,1}$.

References

[1] Deligne, P., Mostow, G.:Monodromy of hypergeometric functions and non-lattice integral monodromy. IHES **63** (1986) 5–90

[2] Deligne, P., Katz, N.: Groupes de Monodromie en Géométrie Algébrique (SGA VII, 2) Lecture Notes in Math. **340** (1973) Springer, Berlin Heidelberg New York

[3] Ferrara, S., Louis, J.: Picard-Fuchs equations and flat holomorphic connections from $N = 2$ supergravity. Essays on Mirror Manifolds, edited by Shing-Tung Yau, International Press (1992) 301–315

[4] Jost, J., Yang,Y-H., Zuo, K.: The cohomology of a variation of polarized Hodge structures over a quasi-compact Kähler manifold. preprint, AG/0312145

[5] Kawamata, Y., Namikawa, Y.: Logarithmic deformations of normal crossings varieties and smoothing degenerated Calabi–Yau varieties. Invent. Math. **118** (1994) 395–409

[6] Popp, H.: Moduli theory and classification theory of algebraic varieties. Lect. Notes in Math. **620** Springer-Verlag, Berlin-New York, 1977

[7] Viehweg, E., Zuo K.: Base spaces of non-isotrivial families of smooth minimal models. In: Complex Geometry (Collection of Papers dedicated to Hans Grauert), Springer Verlag (2002) 279–328

[8] Viehweg, E., Zuo, K.: Complex multiplication, Griffiths-Yukawa couplings, and rigidity for families of hypersurfaces. preprint, AG/0307398. To appear in J. of Alg. Geom.

[9] Zucker, S.: Hodge theory with degenerating coefficients: L^2–cohomology in the Poincaré metric. Ann. of Math. **109** (1979) 415–475

[10] Zuo, K.: On the negativity of kernels of Kodaira–Spencer maps on Hodge bundles and applications. The Asian J. of Math. **4** (2000) 279–302

Universität Duisburg-Essen, Mathematik, 45117 Essen, Germany
E-mail address: viehweg@uni-essen.de

Universität Mainz, Fachbereich 17, Mathematik, 55099 Mainz, Germany
E-mail address: kzuo@mathematik.uni-mainz.de

AMS/IP Studies in Advanced Mathematics
Volume 38, 2006

Some Results on Families of Calabi–Yau Varieties

Yi Zhang

Abstract

In section one, we study the analogue of the Shafarevich conjecture for Calabi–Yau varieties and we present some results on the rigidity of families of Calabi–Yau varieties. In section two, we study the algebraic monodromy of those families and Calabi–Yau varieties endowed with complex multiplication.

CONTENTS

1. Geometry of moduli spaces of polarized Calabi–Yau varieties

In this section, we study the analogue of the Shafarevich conjecture for Calabi–Yau varieties in order to understand the global geometry of their moduli spaces.

• Moduli spaces of polarized Calabi–Yau varieties.

Calabi–Yau varieties.

THEOREM 1.1 (Yau's solution to the Calabi conjecture cf. [**49**]). *For any compact Kähler n-fold X with* $K_X = \wedge^n \Omega_X^1 = \mathcal{O}_X$, *there exists a unique Ricci flat metric on* M, *i.e. a Kähler metric* g_{ij} *such that* $R(g)_{ij} = 0$.

By the Berger decomposition theorem, one has

2000 *Mathematics Subject Classification.* Primary 14D05; Secondary 14J32, 14G35.

Key words and phrases. Polarized Calabi–Yau varieties, moduli spaces, rigidity, algebraic monodromy, complex multiplication, Shafarevich's conjecture.

COROLLARY 1.2 (cf. [**4**]). *Let X be a simply connected compact complex n-fold. The following two conditions are equivalent :*

I. *X admits a Riemannian metric with nonzero holonomy group $H \subseteq \mathrm{SU}(n)$.*
II. *X is a compact Kähler manifold with trivial canonical line bundle.*

Moreover, if the holonomy group $H = \mathrm{SU}(n)$, X is projective and

$$h^i(X, \mathcal{O}_X) = h^{0,i}(X) = h^0(X, \Omega_X^i) = h^{n,n-i}(X) = 0 \,\forall i \text{ with } 0 < i < n.$$

DEFINITION 1.3. A *Calabi–Yau* manifold is a projective n-fold X satisfying that

(a) the canonical line bundle K_X is trivial;
(b) $h^i(X, \mathcal{O}_X) = 0$, $\forall i$ with $0 < i < n$.

Remark. Classically, a Calabi–Yau manifold is a Kähler n-fold X with an $\mathrm{SU}(n)$-holonomy group and a finite fundamental group $\pi_1(X)$. These two definitions are equivalent by the theorem 1.1 of Yau.

EXAMPLE 1.4 (Complete intersections). Let $X \subset \mathbb{P}^{n+k}$ be a variety defined by $F_1 = F_2 = \cdots = F_k = 0$ with $\deg F_i = d_i$. For a generic choice of F_i, X is a smooth n-fold of dimension n, and by the Lefschetz hyperplane theorem $H^0(\Omega_X^i) = 0 \,\forall i$ with $0 < i < n$. If $\sum d_i = n + k + 1$, the canonical line bundle K_X is trivial.

EXAMPLE 1.5 (Double covering). Taking a double cover of $\mathbb{P}^3$ branched over 8 planes in general position and blowing up along the 28 singular lines of the covering space, we will obtain a Calabi–Yau threefold.

Coarse moduli spaces and algebraic families.

THEOREM 1.6 (Viehweg cf. [**41**]). *The coarse quasi-projective moduli space $\mathfrak{M}$ exists for minimal polarized manifolds. In particular it exists for polarized K3 surfaces, Calabi–Yau manifolds and Abelian varieties.*

Let $\mathfrak{M}_h$ be the coarse moduli space of polarized Calabi–Yau n-folds with a fixed Hilbert polynomial h of degree n. The definition of 'Coarse Moduli' implies that each closed point of the scheme $\mathfrak{M}_h$ represents a polarized variety, i.e.

$$\mathfrak{M}_h(\mathbb{C}) \longleftrightarrow \{\text{CY } n\text{-folds with the fixed Hilbert polynomial } h\}.$$

We say that the morphism $\phi : Y_0 \to \mathfrak{M}_h$ from a quasi-projective manifold factorizes through the moduli stack if ϕ is induced by a family $f : X_0 \to Y_0$ (so we denote such a morphism by ϕ_f). Therefore, the study of moduli stacks can be reduced to the study of families of polarized manifolds. A family $g : \mathcal{X} \to Z$ is called *isotrivial* if $\mathcal{X}_a \simeq \mathcal{X}_b$ for a, b in an open dense subscheme of Z. Thus, for a family $f : X \to Y$ of polarized Calabi–Yau varieties smooth over an open dense subscheme Y_0, the moduli map $\phi_f : Y_0 \to \mathfrak{M}_h$ is generically finite if and only if there is no isotrivial subfamily containing a general fiber.

The compactification of moduli spaces is a key question in the moduli programme. In order to compactify a moduli space of polarized smooth n-folds, one has to add points at the boundary, corresponding to limits of smooth varieties. One way to construct these limits is to take a one-parameter semistable family of n-folds.

For any smooth family over a smooth quasi-projective manifold, we always assume that the moduli morphism extends to a projective manifold such that the boundary is a normal crossings divisor. However, the extended family will not be flat (nor have a modular interpretation). Flatness holds of course in codimension one in the base.

For higher-dimensional base, one can always consider the following (good partial compactification cf. [**43**]): Let U be a manifold and Y be a smooth projective compactification of U with a reduced normal crossings divisor boundary $D_\infty = Y \setminus U$. Starting with a smooth family $f : V \to U$, we first choose a smooth projective compactification X of V, such that $V \to U$ extends to $f : X \to Y$. Then one leaves out codimension 2 subschemes of Y, such that the restricted morphism is flat. Replacing Y by some finite covering, one can also assume that $f^{-1}(D_\infty)$ is reduced.

Analytic theory of moduli spaces. Consider families in the category of complex analytic spaces. Let (X, L) be a polarized Calabi–Yau manifold, $\pi : (\mathfrak{X}, (X, L)) \to (\mathfrak{M}_{c_1(L)}, 0)$ be the polarized Kuranishi family and D be the classifying space of polarized $\mathbb{Q}$-Hodge structures with Hodge numbers $\{h^{p,q}\}$ given by $H^n_{prim}(X, \mathbb{C})$.

THEOREM 1.7 (Bogomolov–Todorov–Tian). *Let X be a Calabi–Yau manifold. Then the Kuranishi family of (X, L) is universal, and the base is an open smooth subset in $\mathbb{C}^l$, where*

$$l = \dim_{\mathbb{C}} H^1(X, \Theta)_{c_1(L)} :=$$
$$\dim_{\mathbb{C}}(\operatorname{Ker}(H^1(X, \Theta_X) \xrightarrow{\wedge c_1(L)} H^2(X, \mathcal{O}_X))).$$

Remark. The unobstructedness of the moduli space of polarized Calabi–Yau manifolds is also proven by Ran and Kawamata with purely algebraic methods. It implies that the moduli stack is nonsingular.

Let $f : (\mathcal{X}, (X, L)) \to (Z, 0)$ be a family of polarized Calabi–Yau varieties. In a neighborhood of $0 \in Z$, one has a commutative diagram of period maps

$$\begin{array}{ccc} \mathfrak{M}_{c_1(L)} & \xleftarrow{\iota} & Z \\ & \Phi \searrow \quad \swarrow \Phi_Z & \\ & D & \end{array}$$

and $\rho = (d\iota)_0$ is just the Kodaira–Spencer map, $T_{\mathfrak{M}_{c_1(L)},0} = H^1(X, \Theta_X)_{c_1(L)}$. Denote $\mu_0 = (d\Phi)_0$, $\lambda_0 = (d\Phi_Z)_0$.

THEOREM 1.8 (Infinitesimal Torelli Theorem). $\lambda_0 = \mu_0 \circ \rho$, *and if the canonical line bundle K_X of X is trivial, then*

$$\mu_0 : T_{\mathfrak{M}_{c_1(L)},0} \to \bigoplus_p \operatorname{Hom}(H^{n-p,p}_{prim}, H^{n-p-1,p+1}_{prim})$$

is injective.

By Yau's solution to the Calabi conjecture, there is a unique Ricci flat metric $g(t)$ on $\mathfrak{X}_t$ in the given polarization $[\omega(t)]$. The *Weil–Petersson metric* G_{WP} is defined by

$$G_{WP}(v, w) := \int_{\mathfrak{X}_t} < \rho_t(v), \rho_t(w) >_{g(t)} \quad \text{for } v, w \in T_{\mathfrak{M}_{c_1(L)},t}.$$

By the *infinitesimal Torelli theorem*, G_{WP} is a positive definite Kähler metric on $\mathfrak{M}_{c_1(L)}$, Tian and Todorov showed that its Kähler form is

$$\omega_{WP} = -\frac{\sqrt{-1}}{2}\partial\bar{\partial}\log h = c_1(\pi_* K_{\mathfrak{X}/\mathfrak{M}}, h)$$

where h is the induced Hodge metric on $\pi_* K_{\mathfrak{X}/\mathfrak{M}}$.

- **History of the Shafarevich problem.**

Deformation of algebraic families. Let $f : \mathcal{X} \to M$ be a algebraic family of projective manifolds. A *deformation of the family* f over a quasi-projective varieties T^0 with base point $0 \in T^0$ is a smooth projective morphism $g : \mathfrak{X} \to M \times T^0$ with a commutative diagram

$$\begin{array}{ccccc}
\mathcal{X} & \xrightarrow{\simeq} & g^{-1}(M \times \{0\}) & \xrightarrow{\subset} & \mathfrak{X} \\
{\scriptstyle f}\downarrow & & \downarrow & & \downarrow{\scriptstyle g} \\
M & \xrightarrow{\simeq} & M \times \{0\} & \xrightarrow{\subset} & M \times T^0
\end{array} .$$

Two families $\mathcal{X}_1 \to M$ and $\mathcal{X}_2 \to M$ are said to have the same *deformation type* if they can be deformed into each other, i.e. if there exists a family $\mathfrak{X} \to M \times T^0$ such that there are $t_1, t_2 \in T^0$ with $(\mathfrak{X}_{t_i} \to M \times t_i) \simeq (\mathcal{X}_i \to M)$ for $i = 1, 2$.

In the category of complex analytic spaces, we replace T^0 by a small disk. Then we should say *infinitesimal rigidity* instead of *rigidity*. It is obvious that an algebraic family which is *infinitesimally rigid* in the complex topology will automatically be *rigid*. If there is no ambiguity, we often do not distinguish between the two notions.

Analogues of the Shafarevich Conjecture:

(a) To prove that there are only finitely many deformation types (*Boundedness*).
(b) To find conditions implying that the family does not admit nontrivial deformations (*Rigidity*).

Typical Examples.

EXAMPLE 1.9 (Paršin and Arakelov in the 1970s cf. [**3**],[**33**]). The Shafarevich conjectures are true for families of curves.

EXAMPLE 1.10 (Faltings in the 1980s cf. [**17**]). Faltings dealt with the case of Abelian varieties, and he formulated a Hodge theoretic condition (Deligne–Faltings $(*)$-condition) for the fiber space to be rigid: *A smooth family* $f : \mathcal{X}_0 \to C \setminus S$ *of Abelian varieties is said to satisfy the* $(*)$*-condition if any anti-symmetric endomorphism* σ *of* $\mathbb{V}_{\mathbb{Z}} = R^1 f_* \mathbb{Z}$ *defines an endomorphism of* $\mathcal{X}_0$ *(so* σ *is of type* $(0,0)$*).* By the global Torelli theorem, a polarization of the Abelian scheme induces a symplectic bilinear form Q on $R^1 f_* \mathbb{Z}$. Faltings showed that the $(*)$-condition is equivalent to

$$\mathrm{End}^Q(\mathbb{V}_{\mathbb{Q}}) \otimes \mathbb{C} = (\mathrm{End}^Q(\mathbb{V}_{\mathbb{Q}}) \otimes \mathbb{C})^{0,0}.$$

On the other hand, the Zariski tangent space of the moduli space of Abelian schemes over $C - S$ with a fixed polarization is isomorphic to

$$(\mathrm{End}^Q(\mathbb{V}_{\mathbb{Q}}) \otimes \mathbb{C})^{-1,1}.$$

EXAMPLE 1.11 (Jost–Yau in the 1990s cf. [**20**]). As studying deformations of a family can be reduced to studying deformations of the corresponding period map, Jost and Yau analyzed the set $Sh(C, S, Z)$ introduced below for a large class of varieties by using techniques from harmonic maps. They provided analytic methods to solve the *rigidity* and gave differential geometric proofs of the Shafarevich conjectures. They started to study Higgs bundles with singular Hermitian metrics and their applications to Shafarevich problems.

Mok and Eyssidieux also have many results on deformations of the period map in case that the period domain is Hermitian symmetric (cf. [**16**]). They showed *gap rigidity* (which is stronger than *rigidity* in our case) for locally bounded symmetric domains of certain type (including all tube domains).

Shafarevich problems for families with high dimensional fiber. Through studying the analogous Shafarevich conjecture for families of high dimensional polarized varieties, we can understand part of the structure of moduli spaces.

Let $Sh(C, S, K)$ be the set of all equivalence classes of non-isotrivial families over C such that a general fiber $\mathcal{X}_b$ is a smooth projective variety with 'type' K. Two such families are equivalent if they are isomorphic over $C - S$ as families with polarized fiber.

Frame of the Shafarevich Problem : Fix (C, S) and a Hilbert polynomial h.

- (**B**) The elements of $Sh(C, S, K)$ are parameterized by points of a scheme $\mathbb{T}$ of finite type over $\mathbb{C}$.
- (**R**) When does one have $\dim \mathbb{T} = 0$?
- (**H**) (*Algebraic hyperbolicity*) $\mathbb{T} \neq \emptyset \Rightarrow 2g(C) - 2 + \#S > 0$.
- (**WB**) (*Weak Boundedness*) For a family $(f : X \to C) \in Sh(C, S, K)$, $\deg f_*\omega^m_{X/C}$ is bounded above in terms of $g(C), \#S, h, m$.

Remark. (**WB**) always implies (**H**). In the case of curves, (**WB**) always implies (**B**). The *rigidity* (**R**) fails in general.

EXAMPLE 1.12. Migliorini, Kovács and Zhang (cf. [**26**],[**22**],[**53**]) proved that any family of minimal algebraic surfaces of general type over a curve of genus g and m singular points such that $2g(C) - 2 + \#S \leq 0$ is isotrivial. Oguiso and Viehweg proved it for families of elliptic surfaces (cf. [**32**]).

EXAMPLE 1.13. Consider an algebraic family over a fixed smooth curve, let F be a general fiber. Bedulev–Viehweg proved (**B**) if F is an algebraic surface of general type. They also proved (**WB**) if F is a canonically polarized variety (cf. [**5**]). As explained in [**43**] one actually obtains (**B**). Precisely, Viehweg–Zuo obtained that (**WB**) $\Rightarrow$ (**B**) if F is a minimal model of general type or if the family is nondegenerate and F is a minimal model of Kodaira dimension zero (cf. [**43, 45**]), and they also showed that (**WB**) holds if F is a minimal model of general type or F has semi-ample ω_F.

EXAMPLE 1.14. For families of curves with genus $g \geq 2$, Arakelov proved that

$$\deg f_* \omega_{\mathcal{X}/Y} \leq \frac{g}{2}(2g(Y) - 2 + \#S).$$

S-L.Tan showed that $\leq$ must become $<$ for semi-stable families (cf. [**40**]). Recently Jost–Zuo and very recently Viehweg–Zuo obtained Arakelov–Yau type inequalities (sharp forms) in the case of higher-dimensional fibers (cf. [**19**],[**47**]). Inequalities of such type can be regarded as effective Schwarz–Yau inequalities in algebraic geometry.

• **Shafarevich problems for families of Calabi–Yau varieties.**

Rigidity for Families of Calabi–Yau varieties. There are recent important results of Viehweg–Zuo.

A complex analytic space $\mathcal{N}$ is called *Brody hyperbolic* if every holomorphic map $\mathbb{C} \to \mathcal{N}$ is constant.

THEOREM 1.15 (cf. [**42, 43, 44**]). *Brody hyperbolicity holds for moduli spaces of canonically polarized complex manifolds, thus the boundedness of $Sh(C, S, Z)$ for arbitrary Z with ω_Z semi-ample holds. Moreover, the automorphism group of moduli stacks of polarized manifolds is finite and rigidity holds for a general family.*

The "hyperbolicity" implies that one can use the *Schwarz–Yau lemma* (cf. [**24**]) to get the *boundedness*. Liu–Todorov–Yau–Zuo originally used the *Schwarz–Yau lemma* (cf. [**50**]) and *Bishop compactness* to give another proof of the *boundedness*; they also constructed a nonrigid family of Calabi–Yau manifolds.

EXAMPLE 1.16 (Key Observation cf. [**48**]). Very recently, Viehweg–Zuo constructed a nontrivial family of Calabi–Yau manifolds such that the closed fibers are Calabi–Yau manifolds endowed with *complex multiplication* over a dense set. Precisely, they obtained a family $g : \mathcal{Z} \to S$ of quintic hypersurfaces in $\mathbb{CP}^4$ such that S is finite dominant over a ball quotient (which is a Shimura variety) and S has a dense set of *CM* points (see section 2). Furthermore, they got an important counterexample for the *rigidity* part of the Shafarevich problem by showing that there exists a product of moduli spaces of hypersurfaces of degree d in $\mathbb{P}^n$ and that this product can be embedded into the moduli space of hypersurfaces of degree d in $\mathbb{P}^N$ for some $N > n$.

Altogether, the remaining step for the Shafarevich problem is to find nice conditions for rigidity.

A criterion for rigidity and its applications. It is well-known that a polarized VHS (V, H, ∇) over an algebraic manifold M induces a Higgs bundles (E, θ) with $\theta \wedge \theta = 0$ (cf. [**36, 37**]). By analyzing the Higgs bundle of the families over the product varieties and using generalized *Donaldson–Simpson–Uhlenbeck–Yau correspondence* (cf. [**58**],[**56**]), we obtain a general criterion for rigidity. The starting point is :

THEOREM 1.17 (Faltings, Jost–Yau and Peters cf. [**17**],[**57**],[**56**]). *Let $f : \mathcal{X}_0 \to Y_0$ be a non-isotrivial smooth family of polarized n-folds. If f is nonrigid, there is a non-zero flat section σ of $\mathrm{End}(R^n_{prim} f_*(\mathbb{C}))^{-1,1}$. Moreover, the Zariski tangent space of the deformation space of f is into $\mathrm{End}(R^n_{prim} f_*(\mathbb{C}))^{-1,1}$.*

We are going to show applications of the criterion and prove the *rigidity* of some families of Calabi–Yau varieties.

THEOREM 1.18 (cf. [**56**]). *Let $f : \mathfrak{X} \to \mathbb{P}^1$ be a non-isotrivial Lefschetz pencil. Assume that $f^0 : \mathfrak{X}_0 = f^{-1}(\mathbb{P}^1 \setminus S) \to \mathbb{P}^1 \setminus S$ is smooth and smooth fibers are Calabi–Yau n-folds; then n must be odd and*

$$R^n_{prim} f^0_*(\mathbb{Q}) = \mathbb{V}_{\mathbb{Q}} \oplus (R^n_{prim} f^0_*(\mathbb{Q}))^{\pi_1(\mathbb{P}^1 \setminus S)},$$

where $\mathbb{V}_{\mathbb{Q}}$ is the local system of vanishing cycles. Moreover, $\mathbb{V}_{\mathbb{Q}}$ is absolutely irreducible and

$$(R^n_{prim} f^0_*(\mathbb{C}))^{n,0}, (R^n_{prim} f^0_*(\mathbb{C}))^{0,n} \subset \mathbb{V}_{\mathbb{C}}.$$

Remarks. The proof depends on the *infinitesimal Torelli theorem* and the Kazhdan–Margulis theorem which can be found in [**11**] :

For any Lefschetz pencil $f : \mathcal{X} \to \mathbb{P}^1$, the local system of vanishing cycles is absolutely irreducible under the action of the fundamental group $\pi_1(\mathbb{P}^1 \setminus S)$, where S is the set of singular values, i.e. the local system of vanishing cycles has big monodromy.

THEOREM 1.19 (cf. [**56**]). *Any non-isotrivial Lefschetz pencil of odd-dimensional Calabi–Yau varieties is rigid.*

SKETCH OF THE PROOF. Otherwise, let $f : \mathcal{X} \to \mathbb{P}^1$ be a nonrigid Lefschetz pencil of Calabi–Yau varieties smooth over $C_0 = \mathbb{P}^1 \setminus S$. We have a nontrivial deformation

$$\begin{array}{ccc} \mathcal{X} \setminus f^{-1}(S) & \xrightarrow{\subset} & \mathfrak{X} \\ {\scriptstyle f^0}\downarrow & & \downarrow{\scriptstyle g} \\ C_0 \times \{0\} & \xrightarrow{\subset} & C_0 \times T^0 \end{array} .$$

where T^0 is a smooth quasi-projective curve. Let (E, θ) be the Higgs bundle associated to $R^n_{prim} f^0_*(\mathbb{C})$. By the criterion 1.17, we have :

LEMMA 1.20. *There is a splitting of the Higgs bundle*

$$(E, \theta) = \mathrm{Ker}(\sigma) \bigoplus (\mathrm{Ker}(\sigma))^{\perp}$$

by a nonzero flat section $\sigma \in \mathrm{End}(R^n_{prim} f^0_*(\mathbb{C}))^{-1,1}$.

The morphism $\sigma : E^{n,0}|_{C_0} \to E^{n-1,1}|_{C_0}$ must be injective at one point $s_0 \in \mathbb{P}^1 \setminus S$ because of the nontriviality of the deformation of the family. Hence at s_0,

$$E^{n,0} \not\subset \mathrm{Ker}(\sigma) \quad \text{and} \quad E^{0,n} \subset \mathrm{Ker}(\sigma)$$

as $E^{0,n} \subset \mathrm{Ker}(\sigma)$ along C_0. On the other hand, by 1.18 both $E^{n,0}$ and $E^{0,n}$ are in the irreducible local system $\mathbb{V}_{\mathbb{C}}$. We have a contradiction. □

DEFINITION 1.21 (Strongly degenerate family). Let $f : \mathcal{X} \to C$ be a family over a smooth projective curve C with singular values $\{c_0, \cdots, c_k\}$. f is called *strongly degenerate* at c_0 if f satisfies the following conditions :

(a) $X_i = f^{-1}(c_i) = X_{i1} + \cdots + X_{ir_i}$ is a reduced fiber of a union of transversally crossing smooth divisors for all i, i.e. f is a semi-stable family;
(b) the cohomology of each component of $X_0 = f^{-1}(c_0)$ has pure type (p, p).

EXAMPLE 1.22. Any one-parameter family in $\mathbb{P}^n$ with type

$$F(X_0, \cdots, X_n) + t\prod_{i=1}^{d} X_{\tau(i)} = 0$$

(F is a smooth homogeneous polynomial of degree d) is *strongly degenerate* after a semi-stable reduction, where $\tau : \{1, \cdots, d\} \to \{1, \cdots, n\}$ is an injective map.

THEOREM 1.23 (cf. [**56**]). *Any non-isotrivial strongly degenerate family (not only for families of Calabi–Yau manifolds) is rigid.*

SKETCH OF THE PROOF. Let $f : \mathcal{X} \to C$ be a non-isotrivial *strongly degenerated* family smooth over C_0. Assume $f^0 : \mathcal{X}_0 = f^{-1}(C_0) \to C_0$ is not rigid. Using the criterion 1.17 we have a nonzero $(-1, 1)$ morphism

$$\sigma : R^n_{prim} f^0_*(\mathbb{C}) \to R^n_{prim} f^0_*(\mathbb{C})$$

which is flat under the Gauss–Manin connection. The family $\pi : \mathcal{Y} = \mathcal{X} \times_C \mathcal{X} \to C$ is also *strongly degenerate*, and it has a smooth subfamily $\pi = f^0 \times_{C_0} f^0 : \mathcal{Y}_0 = \mathcal{X}_0 \times_{C_0} \mathcal{X}_0 \to C_0$. Fix $t \in C_0$; σ is actually in $(R^{n+1}\pi_*(\Omega^{n-1}_{\mathcal{Y}_0/C_0}))^{\pi_1(C_0,t)}$.

On the other hand, the $(n-1, n+1)$ component of $R^{n+1}\pi_*(\Omega^{n-1}_{\mathcal{Y}_0/C_0}))^{\pi_1(C_0,t)}$ should be zero by Deligne's Hodge theory: there is a commutative diagram

$$\begin{array}{ccc} H^n(\mathcal{Y}, \mathbb{C}) & \xrightarrow{\quad i^* \quad} & H^n(\mathcal{Y}_0, \mathbb{C}) \\ & {}_{\bar{i}_t^*}\searrow \quad \swarrow_{i_t^*} & \\ & H^n(\mathcal{Y}_t, \mathbb{C})^{\pi_1(C_0,t)} & \end{array}$$

where $i_t : \mathcal{Y}_t \hookrightarrow \mathcal{Y}_0$, $\bar{i}_t : \mathcal{Y}_t \hookrightarrow \mathcal{Y}$ are the natural embeddings, and there is an isomorphism of Hodge Structures $H^n(\mathcal{Y}_t, \mathbb{C})^{\pi_1(C_0,t)} \cong H^0(C_0, R^n\pi_*(\mathbb{C}))$. □

COROLLARY 1.24 (Weakly Arakelov Theorem cf. [**56**]). *Let $f : \mathcal{X} \to C$ be a semi-stable family over a smooth projective curve C with at least one singular fiber, which is a normal crossings divisor and such that each of its components is dominated by a projective space. Then f is rigid.*

Remark. In fact, we have proven a stronger property: *A non-trivial deformation of a family gives rise to a non-trivial Betti-cohomology class in the cohomology group of the total space of the product with itself of the original family.* This class is very special. It is not of Hodge type and does not vanish along fibers. Thus, there are relations between the geometry of total spaces of families and the *rigidity*.

Rigidity and Yukawa coupling. We are going to study some families which are important in string theory. Explored by Viehweg–Zuo and Liu–Todorov–Yau-Zuo, there are relations between the *rigidity* and the *Yukawa coupling.* Here is a general criterion of Viehweg–Zuo for rigidity of families of algebraic manifolds.

CRITERION 1.25 (cf. [**44, 45**]). Let h be a fixed Hilbert polynomial of degree n and $\mathfrak{M}_h$ be a coarse moduli space of polarized n-folds with semi-ample canonical line bundle (including Calabi–Yau manifolds) or of general type. Assume $\mathfrak{M}_h$ has a nice compactification and carries a universal family $\pi : \mathfrak{X} \to \mathfrak{M}_h$. (In the real situation, one needs to work on stacks). Let $k_{\mathfrak{M}}$ be the largest integer such that the $k_{\mathfrak{M}_h}$-times iterated Kodaira–Spencer map of this universal family is not zero (obviously, $1 \leq k_{\mathfrak{M}_h} \leq n$). *If the $k_{\mathfrak{M}_h}$-times iterated Kodaira–Spencer map for a family $f : X \to Y$ is not zero, then the family f must be rigid.*

DEFINITION 1.26 (Yukawa coupling). Let $f^0 : X_0 \to Y_0$ be a smooth family of CY n-folds over a quasi-projective manifold Y_0 and let (E, θ) be the Higgs bundle corresponding to $R^n_{prim} f^0_*(\mathbb{C})$. The *Yukawa coupling* is just the n-iterated Higgs field

$$\theta^n : E \to E \otimes S^n \Omega^1_{Y_0}.$$

The criterion 1.25 implies a result of Viehweg–Zuo and Liu–Todorov–Yau–Zuo.

COROLLARY 1.27 (cf. [**44, 45**],[**24**]). *Let $f : \mathcal{X} \to C$ be a family of Calabi–Yau n-folds over a projective curve, and $f^0 : \mathcal{X}_0 = f^{-1}(C_0) \to C_0$ be the maximal smooth family. If the Yukawa coupling is nonzero at some point, then the family f is rigid.*

SKETCH OF THE PROOF. The VHS $R^n_{prim} f^0_*(\mathbb{Q})$ induce a Higgs bundle

$$(E, \theta) = \Big(\bigoplus_{p+q=n} E^{p,q}, \bigoplus \theta^{p,q}\Big),$$

with $\theta^{p,q} : E^{p,q} \to E^{p-1,q+1} \otimes \Omega^1_{C_0}$. The *Yukawa coupling* is then

$$\theta^n : S^n T_{C_0} \to \mathcal{H}om(E^{n,0}, E^{0,n}) = ((R^0 f^0_* \Omega^n_{\mathcal{X}_0/C_0})^*)^{\otimes 2}.$$

Consider Deligne's quasi-canonical extension $(\overline{E}, \overline{\theta})$ of the Higgs bundle (E, θ) and the log deformation theory; we have a natural commutative diagram of morphisms :

$$\begin{array}{ccc} & S^n(T_C(-\log S)) & \\ {}^{\tau^n(f)}\swarrow & & \searrow^{\overline{\theta}^n(f)} \\ R^n f_*(\mathcal{L}^{-1}) & \longrightarrow & R^n f_* \mathcal{O}_{\mathcal{X}} \otimes (f_* \mathcal{L})^{-1} \end{array}$$

where $S = C \backslash C_0$, $\Delta = f^* S$, $\mathcal{L} = \Omega^n_{\mathcal{X}/C}(\log \Delta) = \omega_{\mathcal{X}/C}(\Delta_{\text{red}} - \Delta)$, $\tau^n(f)$ is the n-iterated log Kodaira–Spencer map and $\overline{\theta}^n(f)$ is the n-iterated map of the extended Higgs field. Hence f is rigid by the criterion of Viehweg–Zuo. $\square$

Remark. Unless $\operatorname{rank} \overline{E}^{n,0} = 1$, it is difficult to prove a similar result only by studying variations of Hodge structures. So one should study geometric deformations more precisely.

Let $f : \mathcal{X} \to C$ be a family of n-folds smooth over $C_0 = C \setminus S$ and $(\mathcal{V}, \nabla)$ be the locally free sheaf with the Gauss–Manin connection given by the VHS $R^n_{prim} f^0_* \mathbb{Q}$. $(\mathcal{V}, \nabla)$ always admits quasi-unipotent monodromy and one has the quasi-canonical extension $\overline{\mathcal{V}}$ with $\overline{\nabla} : \overline{\mathcal{V}} \to \overline{\mathcal{V}} \otimes \Omega^1_C(\log S)$. The VHS and its extension bundle correspond to the Higgs bundle (E, θ) and the quasi-canonical extension $(\overline{E}, \overline{\theta})$ with $\overline{\theta} : \overline{E} \to \overline{E} \otimes \Omega^1_C(\log S)$. Restricting $f : \mathcal{X} \to C$ to a small disk Δ, we consider the local family $f_{loc} : \mathcal{X}_{loc} \to (\Delta, t)$ which is degenerate only at 0. The local monodromy T is called *unipotent* if $(T-1)^{k+1} = 0$ and $(T-1)^k \neq 0$ for an integer k with $0 \leq k \leq n$. Moreover, if $k = n$, T is called *maximal unipotent monodromy*.

THEOREM 1.28 (cf. [**56**]). *Let $f : \mathcal{X} \to C$ be a family of Calabi–Yau varieties. If f admits a degeneration with maximal unipotent monodromy, then f must be rigid.*

The theorem depends on the following result of Schmid and Simpson.

PROPOSITION 1.29 (cf. [**34**],[**37**]). *The nilpotent part of* $\mathrm{Res}_0(\overline{\nabla})$ *is isomorphic to the nilpotent part of* $\mathrm{Res}_0(\overline{\theta})$ *under* $\overline{E}|_0 \cong \overline{\mathcal{V}}|_0$. *If* $\mathbb{V}$ *is a polarized VHS and the local monodromy* T *around* 0 *is unipotent, then* $\mathrm{Res}_0(\overline{\theta}) \cong \mathrm{Res}_0(\overline{\nabla}) \cong \frac{-1}{2\pi\sqrt{-1}} \log T$.

EXAMPLE 1.30 (cf. [**23**]). Lian–Todorov–Yau recently studied a family of Calabi–Yau varieties in $\mathbb{P}^{n+k}$ for $n \geq 4$ and $k \geq 1$ defined by $G_{1,t} = tF_1 - \prod_{i=0}^{n_1} x_i = 0, ..., G_{k,t} = tF_k - \prod_{j=n_1+..+n_{k-1}}^{n_k} x_j = 0$ *with* $t \in \mathbb{P}^1$ *and* $[x_0, \cdots, x_{n+k}] \in \mathbb{P}^{n+k}$ *where the system* $F_1 = ... = F_k = 0$ *defines a smooth Calabi–Yau manifold by* $n_i = \deg F_i \geq 2$ *and* $\sum n_i = n + k + 1$. They showed that this family admits a degeneration with *maximal unipotent monodromy*, hence it has to be rigid.

EXAMPLE 1.31. As a special case, Morrison showed that the *Yukawa coupling* is nonzero at 0 for the family defined by

$$X_0^5 + X_1^5 + X_2^5 + X_3^5 + X_4^5 - 5\lambda X_0 X_1 X_2 X_3 X_4 = 0$$

where $[X_0, \cdots, X_4] \in \mathbb{P}^4$ and $\lambda \in \mathbb{P}^1$ is a parameter. This family admits a degeneration with maximal unipotent monodromy at 0.

Remark. We should point out that the two examples above are both *strongly degenerate* families (after a semi-stable reduction), so that their rigidity also follows from 1.23.

2. Arithmetic of moduli spaces of polarized Calabi–Yau varieties

• The Mumford–Tate group and algebraic monodromy of polarized families. Let $\mathbb{G}_m$ be the multiplicative group; the scalar group

$$\mathbb{S} = R_{\mathbb{C}/\mathbb{R}}\mathbb{G}_m := \prod_{\sigma \in \mathrm{Gal}(\mathbb{C}/\mathbb{R})} \mathbb{G}_m^\sigma$$

is invariant under complex conjugation and it is a real group with $\mathbb{S}(\mathbb{R}) = \mathbb{C}^*$.

LEMMA 2.1 (Definition-Lemma). *A real Hodge structure is equivalent to a pair* (V, h) *consisting of a finite dimensional* $\mathbb{R}$*-vector space* V *and a homomorphism of real algebraic groups* $h : \mathbb{S} \to \mathbf{GL}(V)_\mathbb{R}$. *A rational Hodge structure is a* $\mathbb{Q}$*-vector space* V *endowed with a real Hodge structure on* $V_\mathbb{R}$.

LEMMA 2.2. *Let* (V, h) *be a rational polarized Hodge structure. The Mumford–Tate group* $\mathrm{MT}(V, h)$ *of* (V, h) *is the smallest* $\mathbb{Q}$*-algebraic subgroup* $\mathrm{G} \subset \mathbf{GL}(V)$ *such that* $h(\mathbb{S}(\mathbb{R})) \subset \mathrm{G}(\mathbb{R})$, *and the Hodge–Tate group (or special Mumford–Tate group)* $\mathrm{Hdg}(V, h)$ *is the smallest* $\mathbb{Q}$*-algebraic group* G *of* $\mathbf{GL}(V)$ *with* $h(U^1(\mathbb{R})) \subset \mathrm{G}(\mathbb{R})$ *where* $U^1(\mathbb{R}) = \{z \in \mathbb{C}^* \mid z\overline{z} = 1\}$.

PROPOSITION 2.3 (Mumford cf. [**29**]). *Let* (V, h, Q) *be a polarized rational Hodge structure.*

(1) *Both* $\mathrm{MT}(V, h)$ *and* $\mathrm{Hdg}(V, h)$ *are connected and reductive* $\mathbb{Q}$*-algebraic groups.*
(2) $\mathrm{MT}(V, h)$ *is the largest subgroup of* $\mathbf{GL}(V)$ *fixing all Hodge tensors, i.e.* $(0, 0)$*-type tensors in* $\bigoplus_{(n_1,n_2)\in\mathbb{N}^2} V^{\otimes n_1} \otimes (V^\vee)^{\otimes n_2}$.

(3) $\mathrm{Hdg}(V)$ *is the largest* $\mathbb{Q}$*-subgroup of* $\mathbf{GL}(V)$ *fixing all tensors* $\eta \in V^{\otimes n_1} \otimes (V^\vee)^{\otimes n_2}$ *of Hodge type* $(\frac{1}{2}r(n_1 - n_2), \frac{1}{2}r(n_1 - n_2)) \forall (n_1, n_2) \in \mathbb{N}^2$ *where* r *is the weight of* (V, h).

(4) *If* V *is pure of weight 1,* $\mathrm{Hdg}(V)$ *is the largest* $\mathbb{Q}$*-subgroup of* $\mathbf{GL}(V)$ *fixing all Hodge tensors* $\eta \in (V^{\otimes 2n})^{n,n} \forall n \in \mathbb{N}$.

Adding the condition that the Hodge tensors η have flat deformations, the definitions of $\mathrm{MT}(V, h)$ and $\mathrm{Hdg}(V)$ extend to variations of Hodge structures. Let $(\mathbb{V}, \nabla)$ be a polarized $\mathbb{Q}$-VHS of weight n over a connected algebraic manifold M, and $V := \mathbb{V}_{s_0}$ be a $\mathbb{Q}$-vector bundle where s_0 is a base point in M.

PROPOSITION 2.4 (Deligne cf. [**12**]). *For each* $s \in M(\mathbb{C})$*, the Mumford–Tate group* $\mathrm{MT}_s = \mathrm{MT}(V, h_s)$ *for the Hodge structure* (V, h_s) *is a subgroup in* $\mathbf{GL}(V)$*. Then there exists a nowhere dense set* Σ *of* $M(\mathbb{C})$ *(complex topology) such that:*

(1) MT_s *is a constant group for* $s \in U = M(\mathbb{C}) \setminus \Sigma$.

(2) MT_s *is a subgroup of this constant group for* $\forall s \in \Sigma$.

This constant group coincides with the *Mumford–Tate group* $\mathrm{MT}(\mathbb{V})$ of the VHS $\mathbb{V}$. One has a similar result for the *Hodge–Tate group* $\mathrm{Hdg}(\mathbb{V})$ of the VHS $\mathbb{V}$. The subset $U \subset M(\mathbb{C})$ in 2.4 is the *Hodge generic locus*, and $\Sigma = M(\mathbb{C}) \setminus U$ is the *Hodge exceptional locus*. Let $\rho : \pi_1(M, s_0) \to \mathrm{Aut}(V, Q)$ be the representation corresponding to $\mathbb{V}$ where Q is the polarization. *The algebraic monodromy group* G^{Mon} for $\mathbb{V}$ is defined to be the Zariski closure of the image of ρ, thus G^{Mon} is a reductive $\mathbb{Q}$-subgroup of $A(V) = \mathrm{Aut}(V, Q)$ (by Deligne's complete reducibility). The connected component of identity G_0^{Mon} is defined to be the *connected algebraic monodromy group* for $\mathbb{V}$ and it is also a reductive $\mathbb{Q}$-group.

THEOREM 2.5 (Deligne cf. [**12**]). *Let* MT *be the Mumford–Tate group of the VHS* $\mathbb{V}$.

(1) *There exists a subgroup* $\mathcal{G} \subset G^{\mathrm{Mon}}(\mathbb{Q}) \cap \mathrm{MT}(\mathbb{Q})$ *with* $[G^{\mathrm{Mon}}(\mathbb{Q}) : \mathcal{G}] < \infty$.

(2) $G_0^{\mathrm{Mon}}(\mathbb{Q}) \subset \mathcal{G}$ *and* G_0^{Mon} *is a* $\mathbb{Q}$*-subgroup of* $\mathrm{MT}(\mathbb{V})$*. Moreover,* G_0^{Mon} *is a* $\mathbb{Q}$*-subgroup of* $\mathrm{Hdg}(\mathbb{V})$.

QUESTION 2.6. We always have $G_0^{\mathrm{Mon}} \subset \mathrm{Hdg}(\mathbb{V}) \subset A(V)$.

(A) For which VHS $\mathbb{V}$, is $G_0^{\mathrm{Mon}} = \mathrm{Hdg}(\mathbb{V})$?

(B) For which VHS $\mathbb{V}$, is $\mathrm{Hdg}(\mathbb{V}) = A(V)$?

EXAMPLE 2.7. Here are examples for the above questions:

(1) Let $\mathbb{V}$ be a polarized weight one VHS from a family of Abelian varieties over a curve. If the Arakelov–Yau inequality of that family attains the bound, then $G_0^{\mathrm{Mon}} = \mathrm{Hdg}(\mathbb{V})$ (cf. [**46**]).

(2) For a non-isotrivial Lefschetz pencil of odd-dimensional varieties with at least one singular fiber, there is an absolutely irreducible VHS $(\mathbb{V}, Q)$ corresponding to the vanishing cycles space. By the Kazhdan–Margulis theorem, one has $\mathrm{Hdg}(\mathbb{V}) = A(V) = \mathrm{Sp}(V, Q)$ (cf. [**11**]).

QUESTION 2.8. If one of $\subset$ becomes $=$ in 2.6, what type should the base manifold be? Moreover, if the VHS comes from a family, is the family rigid? Recently Viehweg–Zuo have obtained a characterization of certain Shimura curves in the moduli stack of Abelian varieties, and some rigid Shimura curves in the moduli stack of even-dimensional polarized Calabi–Yau manifolds. Therefore, the relation between the Mumford–Tate group and the *rigidity* is worth studying.

Let $\mathbb{V}$ be a rational polarized VHS on a quasi-projective manifold Y with monodromy

$$\rho : \pi_1(Y, y_0) \to \mathrm{GL}(V),\ V := \mathbb{V}_{y_0}.$$

The monodromy is called *big* if the connected algebraic monodromy group G_0^{mon} acts irreducibly on $\mathbb{V}_{\mathbb{C}}$. Recall a theorem : *"Non-isotrivial Lefschetz pencils of Calabi–Yau manifolds of odd dimension must be rigid"*; we have that the theorem is based on the *bigness* of the monodromy representation of the local system of vanishing cycles of a Lefschetz pencil. Inspired by this theorem, we find that there is a strong relation between the monodromy representation of the base curve and the *rigidity* :

THEOREM 2.9 (cf. [**56**]). *Let $f : X \to Y$ be a non-isotrivial smooth family of Calabi–Yau n-folds. If $R^n_{prim} f_*(\mathbb{Q})$ has a sub $\mathbb{Q}$-VHS $\mathbb{V}$ with big monodromy and the bottom Hodge filtration of $R^n_{prim} f_*(\mathbb{Q})$ is in $\mathbb{V}$, then the family f must be rigid.*

• **Calabi–Yau varieties with complex multiplication.** Let X be an Abelian variety of dimension g. If X is simple, the endomorphism algebra $\mathrm{End}(X)$ can be embedded into $M_{2g}(\mathbb{Z})$. Denote $\mathfrak{D} = \mathrm{End}^0(X) = \mathrm{End}(X) \otimes_{\mathbb{Z}} \mathbb{Q}$. $\mathfrak{D}$ is a division algebra over $\mathbb{Q}$ with a positive definite anti-involution $'$ (*Rosati involution*), and it is a finite-dimensional simple algebra over $\mathbb{Q}$. Let $\mathbb{K}$ be the center of $\mathfrak{D}$; it is a subfield of $\mathfrak{D}$. By Albert's classification theorem for simple central algebras (cf. [**28**]),

$$d^2 = [\mathfrak{D} : \mathbb{K}],\ e = [\mathbb{K} : \mathbb{Q}] \text{ and } ed|2g. \tag{2.9.1}$$

DEFINITION 2.10. (Abelian variety with complex multiplication).

(a) A simple Abelian variety X has *complex multiplication* (*CM*) if $ed = 2g$.
(b) If X is not simple and isogenous to $X_1^{n_1} \times \ldots \times X_k^{n_k}$ where $n_1 g_1 + \cdots + n_k g_k = g$ and X_i is simple, then X is called *CM* if all X_i are *CM*.

PROPOSITION 2.11 (cf. [**28**]). Albert's classifying theorem

(1) *If X is a CM simple Abelian variety, $\mathbb{K}$ is equal to $\mathfrak{D}$ and $d = 1$. Moreover $\mathbb{K}$ must be a totally imaginary quadratic extension of a totally real field $\mathbb{K}^0$ of degree g over $\mathbb{Q}$, i.e.* $\mathrm{End}^0(X) \cong \mathbb{K}^0(\sqrt{-m})$.
(2) *If an Abelian variety X has CM but is not simple, then*

$$\mathrm{End}^0(X) \cong M_{n_1}(\mathbb{K}_1^0(\sqrt{-m_1})) \oplus \cdots \oplus M_{n_k}(\mathbb{K}_k^0(\sqrt{-m_k}))$$

where each $\mathbb{K}_i^0$ is a totally real field of degree g_i over $\mathbb{Q}$ and each $m_i \in \mathbb{K}_i^0$ is a positive integer. $\mathrm{End}^0(X)$ *is a semi-simple algebra over $\mathbb{Q}$. Denote $\mathcal{R} = (Id_{n_1} \otimes \mathbb{K}_1^0(\sqrt{-m_1})) \oplus \cdots \oplus (Id_{n_k} \otimes \mathbb{K}_k^0(\sqrt{-m_k}))$. Then $\mathcal{R}$ is a semi-simple and commutative $\mathbb{Q}$-subalgebra of* $\mathrm{End}^0(X)$ *and $[\mathcal{R} : \mathbb{Q}] = 2g$.*
(3) *Altogether, an Abelian variety X is CM if and only if there is a semi-simple commutative $\mathbb{Q}$-subalgebra $\mathcal{R}$ of* $\mathrm{End}^0(X)$ *such that $[\mathcal{R} : \mathbb{Q}] \geqslant 2g$ (hence $[\mathcal{R} : \mathbb{Q}] = 2g$).*

THEOREM 2.12 (Mumford cf. [**29**]). *An Abelian variety X is CM if and only if* $\mathrm{Hdg}(X)$ *is a torus algebraic group.*

The proof is purely Hodge-theoretic, hence we can define *CM-type* Hodge structures.

DEFINITION 2.13. A polarized $\mathbb{Q}$-Hodge structure (V, h) is called *CM* if $\mathrm{Hdg}(V, h)$ is commutative.

DEFINITION 2.14. A Calabi–Yau n-fold X is called *CM* if its period point $h(X)$ is *CM*, i.e. the Hodge–Tate group of the Hodge structure $H^n_{prim}(X, \mathbb{Q})$ is commutative.

¿From mathematical physics, we know that a Calabi–Yau sigma model is completely characterized by its complex and Kähler moduli, i.e. determined by a mirror pair of Calabi–Yau manifolds (M, W) where M is a Calabi–Yau manifold and W is its mirror. Gukov–Vafa recently showed an interesting phenomena in mathematical physics, thus we begin to study *CM* Calabi–Yau manifolds from this motivation.

THEOREM 2.15 (cf. [**18**]). *The sigma model on a Calabi–Yau manifold corresponding to the pair (M, W) is an RCFT (Rational Conformal Field Theory) if and only if M and W admit complex multiplication over the same number field.*

EXAMPLE 2.16 (cf. [**6**]). Let h_1 and h_2 be rational polarized Hodge structures on V with weights w_1 and w_2. The Hodge structure $h_3 = h_1 \otimes h_2$ is *CM* if and only if h_1 and h_2 are *CM*. Let E_1, E_2, E_3 be three elliptic curves. The group $G_4 = \mathbb{Z}_2 \times \mathbb{Z}_2$ acts on $E_1 \times E_2 \times E_3$ by the involution on $E_1 \times E_2$ and by the identity on the third factor so that the action preserves the holomorphic 3-form on the product variety. The quotient variety $E_1 \times E_2 \times E_3/G_4$ has a Calabi–Yau threefold resolution X with Euler number 96 and $h^{2,1}(X) = 3$. Let h_i be the Hodge structure of $H^1(E_i, \mathbb{C})\, \forall i$. The Hodge structure of $H^3(X, \mathbb{C})$ is then $h_1 \otimes h_2 \otimes h_3$, i.e. the G_4- invariant part of $H^3(E_1 \times E_2 \times E_3, \mathbb{C})$. Hence X is *CM* if and only if all E_i are *CM*.

Let us indicate some open questions.

Let D be a Hermitian symmetric domain of type $G(\mathbb{R})/K_\infty$ where G is a connected semisimple algebraic group defined over $\mathbb{Q}$ and K_∞ is the maximal compact $\mathbb{R}$-subgroup of $G(\mathbb{R})$ (certainly, $G(\mathbb{R})/K_\infty = G(\mathbb{R})^+/K_\infty^+$). Choose a $\mathbb{Z}$-structure $G_\mathbb{Z}$ on G, let Γ be a arithmetic lattice, i.e. a subgroup of $G(\mathbb{Q}) \cap G(\mathbb{R})^+$ and commensurable to $G_\mathbb{Z}(\mathbb{Z})$. The Baily–Borel theorem says that $\Gamma\backslash D$ has a unique algebraic structure of quasi-projective variety over $\mathbb{C}$ ($\Gamma\backslash D$ is smooth if Γ is torsion-free. $\Gamma\backslash D$ is a projective variety if and only if $\mathrm{Hom}(G, \mathbb{G}_m) = 0$, i.e. $\Gamma\backslash D$ is of finite volume, and $G(\mathbb{Q})$ contains no unipotent element other than the identity). Assume G has no simple factor G_i over $\mathbb{Q}$ which is compact (i.e. such that $G_i(\mathbb{R})$ is compact) and Γ contains a congruence group. Then $\Gamma\backslash D$ is a *connected Shimura variety* (cf. [**25**]).

THEOREM 2.17 (Shimura–Mumford–Deligne cf. [**29**],[**13**]). *Let D be a Hermitian symmetric domain as above. If D is a period domain of a polarized VHS, then the subset of CM points in $\Gamma\backslash D$ is Zariski dense.*

Remark. As shown by Deligne any irreducible Hermitian symmetric domain can be regarded as a connected component of a period domain of a suitable polarized VHS; hence a *connected Shimura variety* has a Zariski dense set of *CM* points.

It is believed that the set of *CM* points is dense in a moduli space only if the moduli space itself is of the form G/K. For example, the moduli spaces of elliptic curves, polarized Abelian varieties and polarized $K3$ surfaces are of this type, so that we can predict such a moduli space will have a dense set of *CM* (*RCFT*) points. Unfortunately, moduli spaces of polarized Calabi–Yau manifolds may not be homogeneous spaces and the deformation space of a Calabi–Yau manifold is into the horizonal tangent space of the corresponding period domain; thus we now can say little about the distribution of *CM* points in the moduli space of Calabi–Yau manifolds.

For sufficiently large genus g ($g \geq 4$), it is conjectured by Coleman that there should be only a finite number of Riemann surfaces C of genus g admitting *CM* (cf. [**8**]), and in order that a (sub)family of varieties contain a dense set of *CM* points, the base should be a Shimura (sub)variety due to the André–Oort conjecture.

CONJECTURE 2.18. [André–Oort cf. [**1, 2**],[**31**]] Assume S is a *connected Shimura variety.* Let $Z \subset S$ be an irreducible algebraic subvariety. Then Z contains a Zariski dense set of *CM* points if and only if Z is a *Shimura subvariety* of S.

Remark. Moonen showed the following: Let Z be an irreducible subvariety of a *Shimura variety.* Z is a *Shimura subvariety* if and only if it contains a *CM* point and the completion of the local ring of any smooth point of Z is formally linear (cf. [**27**]).

It is a motivation for us to find necessary conditions for a smooth family of Calabi–Yau manifolds with an open dense set of *CM* points in the base. However, moduli spaces of polarized Calabi–Yau manifolds are like homogeneous spaces. Hence, it is reasonable to consider the analogue of the André–Oort conjecture for moduli spaces of Calabi–Yau manifolds. We believe this is a meaningful question. Suppose that the conjecture holds and the set of *CM* points is open dense in a quasi-projective curve C in a moduli space of Calabi–Yau manifolds $\mathfrak{M}_h$. Then, C should be a Shimura curve and the induced family $f : \mathcal{X} \to C$ should be rigid, because any Shimura curve has a model defined over a number field and so a Shimura curve in the moduli space $\mathfrak{M}_h$ cannot move freely in $\mathfrak{M}_h$. We have interest in studying the relation between the distribution of *CM* points in moduli spaces and the *rigidity.* Recall in 1.16 that Viehweg–Zuo recently constructed a rigid family of Calabi–Yau varieties over a *Shimura variety* with a dense set of *CM* points in the base.

QUESTION 2.19 (cf. [**55**]). Let X be a *CM* Calabi–Yau n-fold. Assume that X is fibered by $(n-1)$-dimensional Calabi–Yau varieties over $\mathbb{P}^1$. Is the set of *CM* points dense in $\mathbb{P}^1$? Even in the case $n = 3$, i.e. for $K3$ fibrations, this does not seem to be known.

QUESTION 2.20 (cf. [**55**]). Fix a pair (C, S) where C is a smooth projective curve and $S \subset C$ is a finite subset. Let $f : \mathcal{X} \to C$ be a non-isotrivial family of Calabi–Yau varieties smooth over $C - S$ and $\mathfrak{C}$ be the set of *CM* points in $C \setminus S$. Does there exist a number N independent of f such that the family f would be rigid if $\#\mathfrak{C} > N$?

There is a new question related to those questions.

QUESTION 2.21 (Viehweg–Zuo cf. [**47**]). Does there exist a smooth family of Calabi–Yau $2k$-folds ($k \geq 2$) over a smooth projective curve with a strictly maximal Higgs field? (The strict maximality of the Higgs field implies that the base curve is a *Shimura curve* or a *modular curve*). No such family seems to be known.

• **A geometric introduction to Shimura curves.** Denote

$$\mathbf{PSL}(2,\mathbb{R}) = \{z \mapsto T(z) = \frac{az+b}{cz+d} | \ ad - bc = 1\}.$$

$T \in \mathbf{PSL}(2,\mathbb{R})$ is elliptic if $\mathrm{Tr}(T) := |a+d| < 2$, parabolic if $\mathrm{Tr}(T) = 2$ and hyperbolic otherwise. A *Fuchsian group* is defined to be a discrete subgroup of $\mathbf{PSL}(2,\mathbb{R})$. Let $\mathcal{H}$ be the upper half plane, or Poincaré disk.

THEOREM 2.22. *A subgroup Γ of $\mathbf{PSL}(2,\mathbb{R})$ is a Fuchsian group if and only if it acts properly discontinuously on $\mathcal{H}$.*

Let $\mathbb{F}$ be a totally real number field of degree n and φ_i $(1 \le i \le n)$ be the n distinct embeddings of $\mathbb{F}$ into $\mathbb{R}$. We assume that $\varphi_1 = \mathrm{id}$.

DEFINITION 2.23 (Archimedean ramification). Let $A = (\frac{l,q}{\mathbb{F}})$ be a quaternion algebra over a totally real number field $\mathbb{F}$. A possesses *Archimedean ramification* if A is unramified at φ_1 and ramified at the other infinite places $\varphi_i (2 \le i \le n)$, i.e. there exist $\mathbb{R}$-isomorphisms ρ_i with

$$\rho_1 : A^{\varphi_1} \otimes \mathbb{R} \to M(2,\mathbb{R}),\ \rho_i : A^{\varphi_i} \otimes \mathbb{R} \to \mathbb{H}\ (2 \le i \le n).$$

Remark. $A = (\frac{l,q}{\mathbb{F}})$ is a subalgebra of $M(2,\mathbb{F}(\sqrt{l}))$, i.e. an $\mathbb{F}$-linear injective homomorphism $\phi :\ A \hookrightarrow M(2,\mathbb{F}(\sqrt{l}))$, such that

$$\phi(x) = g_x = \begin{bmatrix} x_0 + x_1\sqrt{l} & x_2 + x_3\sqrt{l} \\ q(x_2 + x_3\sqrt{l}) & x_0 - x_1\sqrt{l} \end{bmatrix}.$$

THEOREM 2.24. *Let A be a quaternion algebra over a totally real number field $\mathbb{F} \neq \mathbb{Q}$. Assume $\mathbb{F}$ possesses Archimedean ramification. Then A is a division algebra.*

EXAMPLE 2.25. (Criteria for division algebra). Let A be a quaternion algebra and Nrd be the norm function on A.

(i) A is a division algebra $\Longleftrightarrow$ $\mathrm{Nrd}(x) = 0$ only at $x = 0$.
(ii) $A = (\frac{a,b}{\mathbb{F}})$ is a division algebra if A is not isomorphic to $M(2,\mathbb{F})$.

Let $\mathcal{O}$ be an order of A and $\mathcal{O}^1$ be the group of units with reduced norm 1 in $\mathcal{O}$; then $\rho_1(\mathcal{O}^1)$ is a subgroup of $\mathbf{SL}(2,\mathbb{R})$ and $\Gamma(A,\mathcal{O}) = \rho_1(\mathcal{O}^1)/\{+1,-1\}$ is a subgroup of $\mathbf{PSL}(2,\mathbb{R})$. One has

$$\mathrm{Nrd}(x) = \det(\rho_1(x)),\ \mathrm{Trd}(x) = \mathrm{Tr}(\rho_1(x)),$$
$$\varphi_i(\mathrm{Nrd}(x)) = \mathrm{Nrd}_{\mathbb{H}}(\rho_i(x)),\ \varphi_i(\mathrm{Trd}(x)) = \mathrm{Trd}_{\mathbb{H}}(\rho_i(x)).$$

PROPOSITION 2.26. *$\Gamma(A,\mathcal{O})$ is a Fuchsian group of the first kind, i.e. the volume $\mu(\mathcal{H}/\Gamma(A,\mathcal{O}))$ is finite.*

DEFINITION 2.27. One says that a Fuchsian group Γ is *derived from a quaternion algebra A* if Γ is a subgroup of $\Gamma(A,\mathcal{O})$ of finite index. Moreover, the Fuchsian group Γ is called *arithmetic* if Γ is commensurable with $\Gamma(A,\mathcal{O})$.

THEOREM 2.28 (Takeuchi cf. [**39**]). *Let Γ be a Fuchsian group. Assume Γ is of the first kind. Then Γ is* derived from a quaternion algebra A *over a totally real number field $\mathbb{F}$ if and only if γ satisfies the following conditions :*

(1) *Let k_1 be the field $\mathbb{Q}(\mathrm{Tr}(T)|T \in \Gamma)$ generated by the set $\mathrm{Tr}(T)$ over $\mathbb{Q}$. Then k_1 is an algebraic number field of finite degree, and $\mathrm{Tr}(T)$ is contained in $\mathcal{O}_{k_1}$ the ring of integers of k_1,*

(2) *let* φ *be any embedding of* k_1 *into* $\mathbb{C}$ *such that* $\varphi \neq$ *the identity; then* $\varphi(\text{Tr}(\Gamma))$ *is bounded in* $\mathbb{C}$.

Remarks. (Results related to the arithmeticity of Fuchsian groups).

1. Γ is *arithmetic* if and only if $\{T^2 | T \in \Gamma\}$ is *derived from a quaternion algebra.*
2. Let $A(\Gamma)$ be a vector space spanned by Γ over k_1 in $M(2, \mathbb{R})$; then $A(\Gamma)$ is a quaternion algebra over k_1. Moreover, Γ satisfies the condition (I) and the submodule $\mathcal{O}(\Gamma)$ of $A(\Gamma)$ spanned by Γ over $\mathcal{O}_{k_1}$ is an order of $A(\Gamma)$.
3. Let Γ be a Fuchsian group with $\mu(\mathcal{H}/\Gamma) < \infty$. Assume Γ satisfies (I) and (II). Then
   a) k_1 is a totally real number field. Moreover, if φ is not the identity, then $\varphi(\text{Tr}(T))$ is contained in $[-2, 2]$.
   b) $A(\Gamma)$ satisfies the Archimedean ramified condition.

EXAMPLE 2.29. (Quaternion algebras over $\mathbb{Q}$).

(i) Let Γ be a Fuchsian group with $\mu(\mathcal{H}/\Gamma) < \infty$. Then Γ is *derived from a quaternion algebra* A over $\mathbb{Q}$ if and only if $\text{Tr}(T) \in \mathbb{Z}\ \forall T \in \Gamma$.
(ii) If $T \in \mathbf{PSL}(2, \mathbb{R})$ is elliptic and $\text{Tr}(T) \in \mathbb{Z}$, T is then of order 2 or 3.
(iii) By (i) and (ii), a Fuchsian group *derived from a quaternion algebra* over $\mathbb{Q}$ only has order 2 or 3 elliptic elements.

THEOREM 2.30 (Shimura curve cf. [**14**]). *Suppose a Fuchsian group* Γ *is derived from a division quaternion algebra; then the quotient* $\mathcal{H}/\Gamma$ *is a projective curve, called Shimura curve.*

EXAMPLE 2.31. Let A be a quaternion algebra over a totally real number field $\mathbb{F} \neq \mathbb{Q}$. Assume that A satisfies the Archimedean ramified condition. Then A is a division algebra, and thus $\mathcal{H}/\Gamma(A, \mathcal{O})$ is compact.

ACKNOWLEDGEMENTS

The research was partially supported by The Institute of Mathematical Sciences and the Mathematics Department at The Chinese University of Hong Kong. I am grateful to Shing-Tung Yau and Kang Zuo for encouragement and instruction. I also wish to thank Eckart Viehweg for valuable suggestions, Noriko Yui and others of BIRS for organizing the Workshop on Calabi–Yau Varieties and Mirror Symmetry.

References

[1] André, Y. : Mumford–Tate groups of mixed Hodge structures and the theorem of the fixed part. Compositio Math. **82** (1992), no. 1, 1–24.

[2] André, Y. : Distribution des points CM sur les sous-variétés de modules de variétés abéliennes, 1997.

[3] Arakelov, S.Ju. : Families of algebraic curves with fixed degeneracies. (Russian) Izv. Akad. Nauk SSSR Ser. Mat. **35** (1971), 1269–1293

[4] Beauville, A. : Variétés Kähleriennes dont la première class de Chern est nulle, J.D.G. **18** (1983),755-782

[5] Bedulev, E. ; Viehweg, E. : On the Shafarevich conjecture for surfaces of general type over function fields, Inv. Math. **139** (2000), 603-615

[6] Borcea, C. : Calabi–Yau threefolds and complex multiplication. Essays on mirror manifolds. Internat. Press, Hong Kong (1992) 489–502

[7] Borel, A. : Linear algebraic groups. Second edition. Graduate Texts in Mathematics, **126**. Springer-Verlag, New York, 1991.

[8] Coleman, R. : Torsion points on curves. In Galois representations and arithmetic algebraic geometry (Ihara, Y. ed.), Adv. Studies Pure Math. **12** (1987) 235.
[9] De Jong, J. ; Noot, R. : Jacobians with Complex Multiplication. In *Arithmetic Algebraic Geometry* Birkhäuser, 1991.
[10] Deligne, P. : Théorie de Hodge II. I.H.É.S. Publ. Math. **40** (1971) 5–57
[11] Deligne, P. : La conjecture de Weil I. I.H.E.S Publ. Math. **43** (1974),273-307
[12] Deligne, P. : La conjecture de Weil pour les surfaces $K3$. Invent. math. **15** (1972) 206–226
[13] Deligne, P. : Hodge cycles on Abelian varieties (Notes by J. S. Milne). Springer Lecture Notes in Math. **900** (1982) 9–100
[14] Deligne, P. : Travaux de Shimura (French). Séminaire Bourbaki, **23** éme année (1970/71) Exp. No. 389, pp. 123–165. Lecture Notes in Math., Vol. 244, Springer, Berlin, 1971.
[15] Deligne, P. : Local behavior of Hodge structures at infinity. Mirror symmetry, II, 683–699, AMS/IP Stud. Adv. Math., **1**, Amer. Math. Soc., Providence, RI, 1997.
[16] Eyssidieux, P. ; Mok, N.M. : Characterization of certain holomorphic geodesic cycles on Hermitian locally symmetric manifolds of the noncompact type. Modern methods in complex analysis, 79-117, Ann. of Math. Stud., **137**, Princeton Univ. Press, Princeton, NJ, 1995.
[17] Faltings, G. : Arakelov's theorem for Abelian varieties. Inv. Math. **73** (1983) 337-347
[18] Gukov, S. ; Vafa, C. : Rational conformal field theories and complex multiplication. Preprint (2002), **hep-th/0203213**
[19] Jost, J ; Zuo, K. : Arakelov type inequalities for Hodge bundles over algebraic varieties, Part 1: Hodge bundles over algebraic curves. J. Alg. Geom. **11** (2002) 535-546
[20] Jost, J ; Yau, S.T. : Harmonic mappings and algebraic varieties over function fields. Amer. J. Math. **115** (1993), No. 6, 1197–1227
[21] Kempf, G. ; Knudsen, F. ; Mumford, D.; Saint-Donat, B. : Toroidal embeddings I. Lecture Notes in Math. **339** (1973) Springer, Berlin, Heidelberg, New York.
[22] Kovács, S.J. : Families Over a Base with Birationally Nef Tangent Bundle, J.Reine Angew. Math. **487** (1997), 171-177
[23] Lian, B.H ; Todorov, A ; Yau, S.T. : Maximal unipotent monodromy for complete intersection CY manifolds. Preprint (2000), **Math.AG/0008061**
[24] Liu, K. ; Todorov, A. ; Yau, S-T. ; Zuo, K. : Shafarevich's Conjecture for CY Manifolds I. Preprint (2003), **Math.AG/0308209**
[25] Milne, J.S. : Introduction to Shimura Varieties. 2004 http://www.jmilne.org/math/
[26] Migliorini, L. : A smooth family of minimal surfaces of general type over a curve of genus at most one is trivial. J. Alg. Geom. **4**(1995) 353-361
[27] Moonen, B. : Linearity properties of Shimura varieties. I. J. Alg. Geom. **7** (1998), no. 3, 539–567. II. Compositio Math. **114** (1998), no. 1, 3–35.
[28] Mumford, D. : Abelian Varieties. (1970) Oxford Univ. Press, Oxford
[29] Mumford, D. : Families of Abelian varieties. Proc. Sympos. Pure Math. **9** (1966) 347–351
[30] Mumford, D. : A note of Shimura's paper: Discontinuous groups and Abelian varieties. Math. Ann. **181** (1969) 345–351
[31] Oort, F. : Canonical lifts and dense sets of CM-points. Arithmetic Geometry, Proc. Cortona symposium 1994
[32] Oguiso, K. ; Viehweg, E.: On the isotriviality of families of elliptic surfaces. J. Alg. Geom. **10** (2001), No. 3, 569–598.
[33] Paršin, A.N. : Algebraic curves over function fields. I. (Russian) Izv. Akad. Nauk SSSR Ser. Mat. **32** 1968 1191–1219
[34] Schmid, W. : Variation of Hodge structure: The singularities of the period mapping. Inv. math. **22** (1973) 211–319
[35] Schoen, C. : Varieties dominated by product varieties. Int. J. Math. **7** (1996) 541–571
[36] Simpson, C. : Higgs bundles and local systems. Publ. Math. IHES. **75** (1992) 5-95
[37] Simpson, C. : Harmonic bundles on noncompact curves. Journal of the AMS **3** (1990) 713–770
[38] Sun, X.-T. ; Tan, S.L. ; Zuo, K. : Families of $K3$ surfaces over curves reaching upper bounds of Arakelov–Yau's type and modularity. Math. Res. Lett. **10** (2003), no. 2-3, 323–342
[39] Takeuchi, K. : A characterization of arithmetic Fuchsian groups. J. Math. Soc. Japan **27** (1975), no. 4, 600–612.

[40] Tan,S-L. : The minimal number of singular fibers of a semistable curve over $\mathbb{P}^1$. J. Alg. Geom. **4**(1995) 591-596.

[41] Viehweg, E. : Quasi-projective Moduli for Polarized Manifolds. Ergebnisse der Mathematik, 3. Folge **30** (1995), Springer Verlag, Berlin-Heidelberg-New York

[42] Viehweg, E. ; Zuo, K. : On the isotriviality of families of projective manifolds over curves. J. Alg. Geom. **10** (2001) 781-799

[43] Viehweg, E. ; Zuo, K. : Base spaces of non-Isotrivial families of smooth minimal models, Complex geometry (Göttingen, 2000), 279–328, Springer, Berlin, 2002

[44] Viehweg, E. ; Zuo, K. : On the Brody hyperbolicity of moduli spaces for canonically polarized manifolds, Duke Math Journal **118** No.1(2003), 103-150

[45] Viehweg, E. ; Zuo, K. : Discreteness of minimal models of Kodaira dimension zero and subvarieties of moduli stacks. Surveys in differential geometry, Vol. VIII (Boston, MA, 2002), 337–356, Surv. Differ. Geom., VIII, Int. Press, Somerville, MA, 2003.

[46] Viehweg, E. ; Zuo, K. : A characterization of certain Shimura curves in the moduli stack of abelian varieties. Preprint (2002) **Math.AG/0207228**

[47] Viehweg, E. ; Zuo, K. : Families over curves with a strictly maximal Higgs field. Asian Journal of Mathematics, **7**(2003), No.4, 575-598

[48] Viehweg, E. ; Zuo, K. : Complex multiplication, Griffiths-Yukawa couplings, and rigidity for families of hypersurfaces. Preprint (2003) **Math.AG/0307398.**

[49] Yau, S-T. : Calabi's conjecture and some new results in algebraic geometry. Proc. Nat. Acad. Sci. U.S.A. **74**(1977), no. 5, 1798–1799.

[50] Yau, S-T. : A generalized Schwarz lemma for Kähler manifolds. A. J. Mathematics, vol. **100**, Issue 1(1978), 197-203.

[51] Yui, N. : Update on the modularity of the Calabi–Yau varieties. Fields institute Communications 2003.

[52] Zarhin, Y. : Hodge group of $K3$ surfaces. J.Reine Angew. Math. **341**(1983),193-220

[53] Zhang, Q. : Global holomorphic one-forms on projective manifolds with ample canonical bundles. J. Alg. Geom. **6** (1997), 777-787

[54] Zhang, Y. : On families of Calabi–Yau manifolds, Ph.D. Thesis, CUHK 2003

[55] Zhang, Y. : The Mumford–Tate group of families of Calabi–Yau varieties I. Preprint 2003

[56] Zhang, Y. : Rigidity for families of polarized Calabi–Yau varieties. **Math.AG/0308034**. to apper in J. Diff. Geom.

[57] Zuo, K. : On the negativity of kernels of Kodaira–Spencer maps on Hodge bundles and applications. Asian Journal of Mathematics, **4**, (2000), No.1, 279-302

[58] Zuo, K. : Representations of fundamental groups of algebraic varieties. Lecture Notes in Mathematics, **1708**, Springer-Verlag, Berlin, 1999.

Center of Mathematical Sciences, Zhejiang University (Mailbox 1511), Hangzhou 310027, P. R. China,

E-mail address: `yzhang@cms.zju.edu.cn`

PART III

DIFFERENTIAL GEOMETRIC AND MATHEMATICAL PHYSICS ASPECTS

AMS/IP Studies in Advanced Mathematics
Volume 38, 2006

Boundary RG Flows of N=2 Minimal Models

Kentaro Hori

ABSTRACT. We study boundary renormalization group flows of $\mathcal{N}=2$ minimal models using the Landau-Ginzburg description of B-type. A simple algebraic relation of matrices is relevant. We determine the pattern of the flows and identify the operators that generate them. As an application, we show that the charge lattice of B-branes in the level k minimal model is torsion, $\mathbb{Z}_{k+2}$. We also reproduce the fact that the charge lattice for the A-branes is $\mathbb{Z}^{k+1}$, applying the B-brane analysis on the mirror LG orbifold.

1. Introduction

Many systems in statistical mechanics and quantum field theory have effective descriptions of Landau-Ginzburg (LG) type. In particular, in $(2,2)$ supersymmetric field theories in $1+1$ dimensions, LG models provide effective descriptions of a large class of theories, both conformal and massive, from which one can extract intuitive pictures as well as exact results, such as the set and character of vacua, dimensions of operators and chiral rings. The simplest example is the single variable model with superpotential

$$W = X^{k+2} \tag{1}$$

labeled by a positive integer k. It flows in the infrared limit to the $\mathcal{N}=2$ minimal model at level k — a $(2,2)$ superconformal field theory with central charge $c=\frac{3k}{k+2}$ [**1, 2, 3, 4, 5**], as argued in [**6, 7, 8**]. The LG description gives us a clear picture of renormalization group (RG) flows between conformal field theories. The system with superpotential (1) has left and right $U(1)$ R-symmetries which become a part of the superconformal algebra. A vector $U(1)$ is lost if we add a lower order term

$$W = X^{k+2} + \varepsilon X^{\ell}. \tag{2}$$

This can be regarded as a supersymmetric perturbation of the minimal model by a relevant operator, and one can immediately tell by looking at the superpotential that it flows to the model of lower level $\ell-2$.

In this paper, we study the boundary RG flows of Landau-Ginzburg models with an unbroken $\mathcal{N}=2$ supersymmetry. Boundary RG flows are being studied from two viewpoints — statistical mechanics of two-dimensional critical systems

2000 *Mathematics Subject Classification.* Primary 13C14, 19L99, 32S30, 81R10.
Key words and phrases. D-branes, Landau-Ginzburg models, matrix factorizations.

and string theory. In the latter, boundary RG flows describe, from the worldsheet perspective, the tachyon condensation on the worldvolume of unstable D-brane systems [**9**]. The subject of unstable D-brane systems [**10**] has proved to be extremely rich: it motivated the development of string field theory [**11, 12**], led to the K-theory characterization of D-brane charge [**13**] and its refinement [**14**], gave us a physical interpretation of matrix models [**15**], and provided workable models of time-dependent string theory [**16**]. In most of them, an important role is played by Chan-Paton factors, which are simple matrix factors that live on the worldsheet boundary.

A useful LG description of boundary RG flow already exists. This is in the context of *A-branes* which are wrapped on Lagrangian submanifolds and support flat gauge fields. In the minimal model and its deformations, the branes are D1-branes at the wedge-shaped lines that reside in the pre-image of $W \in \mathbb{R}$ [**17**]. (The coset construction provides a similar and sometimes useful picture where the branes are straight segments in a disc stretched between special points on the boundary circle [**18**].) This LG description provides a useful and geometrical picture of RR-charge, the Witten index, as well as the appearance of the Verlinde algebra [**17**]. In this description, the boundary RG flow is simply the annihilation of the brane and antibrane (see Figure 1), or a recombination of the branes at the intersection

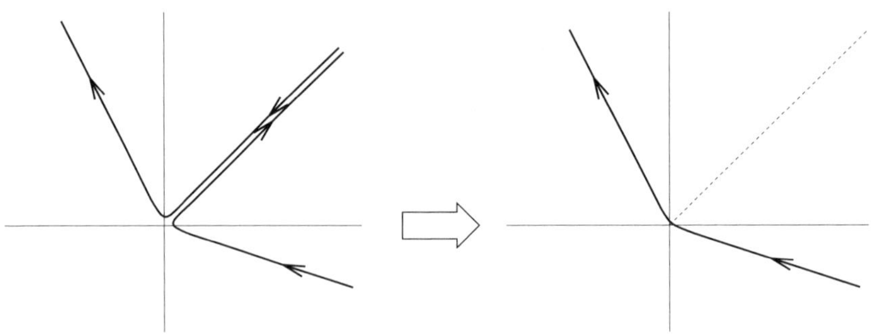

FIGURE 1. Flow of A-type boundary conditions in the minimal model

points. One can find the pattern of RG flows at a glance. This has been remarked, for example, in [**19**] (and also in the disc picture in [**18, 20**]). However, it is not easy to identify the operator that generates a given flow.

There is another class of branes, *B-branes*, which are wrapped on complex submanifolds and support holomorphic bundles. The purpose of this work is to find a useful picture of boundary RG flows using B-branes in Landau-Ginzburg models, hopefully to the same extent as the A-branes or even to complement what is missing in the A-type picture. B-branes in LG models have been studied in [**21, 22, 23, 24, 25, 26, 27**]. In particular, we use the recent description by Kontsevich; see [**24, 25, 26, 27**] (an independent and alternative description is in [**23**]) that uses the factorization of the superpotential on the Chan-Paton factor. What is relevant here, it turns out, is the continuous deformation of the matrices[1]

$$M_t = \begin{pmatrix} A & 0 \\ 0 & 1 \end{pmatrix} \begin{pmatrix} \cos t & -\sin t \\ \sin t & \cos t \end{pmatrix} \begin{pmatrix} 1 & 0 \\ 0 & B \end{pmatrix} \begin{pmatrix} \cos t & \sin t \\ -\sin t & \cos t \end{pmatrix}$$

[1]The author learned this in [**28**] where it is used in the proof of Bott periodicity.

that yields the flow

$$\begin{pmatrix} A & 0 \\ 0 & B \end{pmatrix} \Rightarrow \begin{pmatrix} AB & 0 \\ 0 & 1 \end{pmatrix}. \tag{3}$$

This simple algebraic relation provides the B-type counterpart of the A-type flow as in Figure 1. Also, by a basis change, $M_t \to M_t\left(\begin{smallmatrix} \cos t & -\sin t \\ \sin t & \cos t \end{smallmatrix}\right)$, the matrix for small t can be written as

$$M_\varepsilon \cong \begin{pmatrix} A & 0 \\ 0 & B \end{pmatrix} + \varepsilon \begin{pmatrix} 0 & -AB \\ 1 & 0 \end{pmatrix}. \tag{4}$$

This leads to the boundary analog of (2) where the perturbing term breaks the $U(1)$ R-symmetry and generates a flow of the boundary condition. In this way, one can identify the operator that generates a given flow, as well as identify the IR limit of a given perturbation.

As an application, we determine the charge lattice of the D-branes in the minimal model. For B-branes, it turns out that the lattice is torsion,

$$\Lambda_B \cong \mathbb{Z}_{k+2}.$$

This is obtained through the relation (3) or more explicitly

$$\begin{pmatrix} x & & & \\ & \ddots & & \\ & & \ddots & \\ & & & x \end{pmatrix} \Rightarrow \begin{pmatrix} x^{k+2} & & & \\ & 1 & & \\ & & \ddots & \\ & & & 1 \end{pmatrix}.$$

We also reproduce the charge lattice of A-branes, which was known by the (A-type) LG picture to be

$$\Lambda_A \cong \mathbb{Z}^{k+1}.$$

This is done by using the mirror symmetry between the minimal model and its $\mathbb{Z}_{k+2}$-orbifold, where A and B are exchanged, and applying (3) to the latter.

Note: While this work has being written, we noticed a paper [**29**] which studies boundary RG flows of non-supersymmetric minimal models in LG description. Also, a paper [**30**] has appeared, which has some overlaps in the discussion of B-branes in LG orbifolds.

2. B-branes in Landau-Ginzburg models

Let X be a non-compact Calabi-Yau manifold with a (local) Kähler potential $K(\phi, \overline{\phi})$ and a global holomorphic function $W(\phi)$. We consider the $(2,2)$ supersymmetric LG model with the action

$$S = S_K + S_W = \int_\Sigma \mathrm{d}^2x \left[\int \mathrm{d}^4\theta\, K(\Phi, \overline{\Phi}) + \mathrm{Re} \int \mathrm{d}^2\theta\, W(\Phi) \right]. \tag{5}$$

For superspace and superfields we use the convention of [**22**]. We are interested in B-branes of this system. Namely, the boundary conditions and interactions preserving the B-type $\mathcal{N} = 2$ supersymmetry $Q = Q_+ + Q_-$, $\overline{Q} = \overline{Q}_+ + \overline{Q}_-$ [**31**].

Under the transformation $\delta = \epsilon Q - \overline{\epsilon}\overline{Q}$, the Kähler potential term (D-term) S_K is invariant with the ordinary supersymmetric boundary condition for D-branes wrapped on a complex submanifold of X. On the other hand, the superpotential term (F-term) varies as [**33, 21**]

$$\delta S_W = \mathrm{Re}\int_{\partial\Sigma,B} \mathrm{d}t\,\mathrm{d}\theta(-i\overline{\epsilon})\,W(\Phi)\,, \tag{6}$$

where $\int_{\partial\Sigma,B}$ is the integration on the B-boundary in which $\theta^+=\theta^-=:\theta$ (see [**21**] for conventions on "boundary superspace"). This vanishes if the D-brane lies in a level set of the superpotential W [**32, 17**]. There is actually an alternative way to preserve the B-type supersymmetry [**34**]. Suppose the superpotential can be written as the product

$$W(\Phi)=if(\Phi)g(\Phi). \tag{7}$$

Then, we add a boundary term

$$S_{\mathrm{bdry}}=\int_{\partial\Sigma}\mathrm{d}t\left[\frac{1}{2}\int_B \mathrm{d}\theta\mathrm{d}\overline{\theta}\,\overline{\Gamma}\Gamma+\mathrm{Re}\int_B\mathrm{d}\theta\,\Gamma f(\Phi)\right]. \tag{8}$$

where Γ is a fermionic superfield on the B-boundary which fails to be chiral,

$$\overline{D}\Gamma=g(\Phi). \tag{9}$$

Under the B-type supersymmetry the boundary term varies as

$$\begin{aligned}\delta S_{\mathrm{bdry}} &= \mathrm{Re}\int_{\partial\Sigma,B}\mathrm{d}t\mathrm{d}\theta\,\left(\left[\epsilon\mathcal{Q}-\overline{\epsilon}\overline{\mathcal{Q}}\right]\Gamma f(\Phi)\right)\Big|_{\overline{\theta}=0}\\ &= \mathrm{Re}\int_{\partial\Sigma,B}\mathrm{d}t\mathrm{d}\theta\,(-\overline{\epsilon}\overline{D}\Gamma f(\Phi))\Big|_{\overline{\theta}=0}=\mathrm{Re}\int_{\partial\Sigma,B}\mathrm{d}t\mathrm{d}\theta\,i\overline{\epsilon}W(\Phi)\,,\end{aligned}$$

where $\mathcal{Q}$ and $\overline{\mathcal{Q}}$ are the odd differential operator as defined in [**21**]. We see that δS_{bdry} indeed cancels δS_W. Thus we find a B-brane for each factorization of the superpotential (7). More generally, if the superpotential is expressed as $W=i\sum_{\alpha=1}^r a_\alpha b_\alpha$, one can do the same by introducing Γ_α obeying $\overline{D}\Gamma_\alpha=b_\alpha(\Phi)$ with the boundary superpotential $\sum_{\alpha=1}^r\Gamma_\alpha a_\alpha(\Phi)$. It is straightforward to generalize this to the case of gauged linear or non-linear sigma models with superpotential.

The component expression of a superfield with constraint $\overline{D}\Gamma=g(\Phi)$ is $\Gamma=\eta+\theta G-i\theta\overline{\theta}\dot{\eta}-\overline{\theta}g(\Phi)$ and the boundary term reads as

$$\begin{aligned}S_{\mathrm{bdry}} &= \int_{\partial\Sigma}\mathrm{d}t\left\{i\overline{\eta}\dot{\eta}-\frac{1}{2}|g(\phi)|^2-\mathrm{Re}\Big(\overline{\eta}\partial_i g(\phi)\psi^i\Big)\right.\\ &\qquad\left.+\frac{1}{2}|G|^2+\mathrm{Re}\Big(-\eta\partial_i f(\phi)\psi^i+Gf(\phi)\Big)\right\}\\ &= \int_{\partial\Sigma}\mathrm{d}t\left\{i\overline{\eta}\dot{\eta}-\frac{1}{2}|g(\phi)|^2-\frac{1}{2}|f(\phi)|^2-\mathrm{Re}\Big(\overline{\eta}\partial_i g(\phi)\psi^i+\eta\partial_i f(\phi)\psi^i\Big)\right\}\end{aligned}$$

where $\psi^i:=\psi^i_++\psi^i_-$. The auxiliary field G is eliminated by solving $G=-\overline{f(\phi)}$. The supersymmetry variation of the fermion η is $\delta\eta=\epsilon G-\overline{\epsilon}g(\phi)=-\epsilon\overline{f(\phi)}-\overline{\epsilon}g(\phi)$. Let us formulate the system on the segment $0\le\sigma\le\pi$ where we put the fermions η_0 and η_π at the two boundaries with the boundary terms corresponding to the factorizations

$$W=if_0g_0=if_\pi g_\pi.$$

The boundary at $\sigma = 0$ is oriented toward the past $t \to -\infty$ while the $\sigma = \pi$ boundary is oriented toward the future $t \to +\infty$. By the Noether procedure we find the supercharge $Q = Q_{\text{bulk}} + Q_{\text{bdry}}$, $\overline{Q} = \overline{Q}_{\text{bulk}} + \overline{Q}_{\text{bdry}}$ where

$$\overline{Q}_{\text{bulk}} = \int_0^\pi \mathrm{d}\sigma \Big\{ (\overline{\psi}_+ + \overline{\psi}_-) \cdot \partial_t \phi + (\overline{\psi}_+ - \overline{\psi}_-) \cdot \partial_\sigma \phi + i(\psi_- - \psi_+)^i \partial_i W(\phi) \Big\}$$

$$\overline{Q}_{\text{bdry}} = -i \Big[\eta f(\phi) + \overline{\eta} g(\phi) \Big]_0^\pi.$$

Using the canonical (anti)commutation relation, we find

$$\overline{Q}^2_{\text{bulk}} = -iW\Big|_0^\pi, \quad \{\overline{Q}_{\text{bulk}}, \overline{Q}_{\text{bdry}}\} = 0, \quad \overline{Q}^2_{\text{bdry}} = iW\Big|_0^\pi, \tag{10}$$

and indeed the total supercharge $\overline{Q}$ is nilpotent. The fermion at $\sigma = \pi$ is represented on a two-dimensional vector space $\mathbb{C}^2_\pi$ spanned by $|0\rangle_\pi, \overline{\eta_\pi}|0\rangle_\pi$. This is the Chan-Paton factor of the brane. The Chan-Paton factor for the brane at $\sigma = 0$ is likewise $\mathbb{C}^2_0$ spanned by $|0\rangle_0, \overline{\eta_0}|0\rangle_0$. Both spaces are $\mathbb{Z}_2$ graded by the fermion number — $|0\rangle$ is bosonic and $\overline{\eta}|0\rangle$ is fermionic. Open string states take values in $\mathrm{Hom}(\mathbb{C}^2_0, \mathbb{C}^2_\pi)$ on which the boundary part of the supercharge is represented as

(11)

$$i\overline{Q}_{\text{bdry}} \begin{pmatrix} a & b \\ c & d \end{pmatrix} = \begin{pmatrix} 0 & f_\pi(\phi) \\ g_\pi(\phi) & 0 \end{pmatrix} \begin{pmatrix} a & b \\ c & d \end{pmatrix} - \begin{pmatrix} a & -b \\ -c & d \end{pmatrix} \begin{pmatrix} 0 & f_0(\phi) \\ g_0(\phi) & 0 \end{pmatrix}.$$

Since X is assumed to be Calabi-Yau, the supersymmetric ground states are determined by the zero mode analysis. The zero mode Hilbert space is represented as the space of $\mathrm{Hom}(\mathbb{C}^2_0, \mathbb{C}^2_\pi)$-valued antiholomorphic forms on X on which the supercharge acts as the Dolbeault operator plus (11):

$$i\overline{Q} = \overline{\partial} + i\overline{Q}_{\text{bdry}}$$

(In the zero mode sector, the left and the right boundaries are mapped to the same point and thus $\overline{Q}^2_{\text{bulk}} = \overline{Q}^2_{\text{bdry}} = 0$.) Note that this forms a $\mathbb{Z}_2$-graded complex, where the grading comes from the mod 2 reduction of form degree and the fermion number in $\mathrm{Hom}(\mathbb{C}^2_0, \mathbb{C}^2_\pi)$. The space of supersymmetric ground states can be identified as the $\overline{Q}$-cohomology group. If X is a Stein space, this reduces to the cohomology of $\overline{Q}_{\text{bdry}}$ acting on the space of $\mathrm{Hom}(\mathbb{C}^2_0, \mathbb{C}^2_\pi)$-valued holomorphic functions on X:

$$\mathcal{H}_{\text{SUSY}} \cong H(\Gamma_{hol}(X, \mathrm{Hom}(\mathbb{C}^2_0, \mathbb{C}^2_\pi)); \overline{Q}_{\text{bdry}}). \tag{12}$$

If the brane corresponds to a more general "factorization", $W = i \sum_{\alpha=1}^r a_\alpha b_\alpha$, the Chan-Paton factor is $\mathbb{C}^{2^r}$ spanned by the exterior powers of $\overline{\eta}_\alpha$ multiplied to the state $|0\rangle$ annihilated by η_α. The operator $i\overline{Q}_{\text{bdry}}$ acts as $\sum_{\alpha=1}^r (\eta_\alpha a_\alpha + \overline{\eta}_\alpha b_\alpha)$, exchanging the bosonic and fermionic subspaces of $\mathbb{C}^{2^r}$. Let f (resp. g) be the restriction of this operator to the fermionic (resp. bosonic) subspace. Then fg and gf are both proportional to $-iW$. One can further generalize the Chan-Paton factor to an arbitrary $\mathbb{Z}_2$-graded vector space $V = V_+ \oplus V_-$ on which there are operators $g : V_+ \to V_-$ and $f : V_- \to V_+$ such that

$$fg = -iW\mathbf{1}_{V_+}, \qquad gf = -iW\mathbf{1}_{V_-}. \tag{13}$$

The boundary action is given by the super Wilson line for the superconnection

$$\mathcal{A} = \frac{1}{2}\begin{pmatrix} ff^\dagger + g^\dagger g & \psi^i\partial_i f + \overline{\psi}^{\bar{i}}\partial_{\bar{i}} g^\dagger \\ \psi^i\partial_i g + \overline{\psi}^{\bar{i}}\partial_{\bar{i}} f^\dagger & f^\dagger f + gg^\dagger \end{pmatrix}.$$

This is the same as the standard super Wilson line factor for the brane-antibrane system [**35, 36**] corresponding to the tachyon field $T = f + g^\dagger : V_- \to V_+$ (up to the shift by $\mathrm{Im}(W)$). Let $(V_- \overset{f}{\underset{g}{\rightleftarrows}} V_+)_0$ and $(V_- \overset{f}{\underset{g}{\rightleftarrows}} V_+)_\pi$ be two such branes and consider the open string stretched between them. The space of supersymmetric ground states is isomorphic to the cohomology group

$$\mathcal{H}_{\mathrm{SUSY}} \cong H(\Gamma_{hol}(X, \mathrm{Hom}(V_0, V_\pi)); \overline{Q}_{\mathrm{bdry}}), \tag{14}$$

where $\overline{Q}_{\mathrm{bdry}}$ is defined as in (11).

This realization of the space of supersymmetric ground states was obtained by Kontsevich, whose work is interpreted in the above form in [**24**]. An independent work on the same subject was done by the author [**21, 22, 23**]. In [**23**] a different form of the cohomological realization is obtained (see also [**22**] for a preliminary version which gives the derivation). We also note that some mathematical predictions of these results combined with Mirror Symmetry [**37**] were confirmed in [**38, 39**].

3. B-branes in $\mathcal{N} = 2$ minimal models

Let us consider the Landau-Ginzburg model of a single variable X with the Kähler potential $K = |X|^2$ and superpotential

$$W = iX^{k+2}.$$

Since the superpotential is homogeneous, this model has a vector R-symmetry $U(1)_V : X(\theta^\pm) \to \mathrm{e}^{\frac{2i\alpha}{k+2}} X(\mathrm{e}^{-i\alpha}\theta^\pm)$, in addition to the axial R-symmetry $U(1)_A : X(\theta^\pm) \to X(\mathrm{e}^{\mp i\beta}\theta^\pm)$. There is also a discrete $\mathbb{Z}_{k+2}$ symmetry generated by

$$\gamma : X \to \omega X, \qquad \omega := \mathrm{e}^{\frac{2\pi i}{k+2}}. \tag{15}$$

The system flows in the infrared limit to a $(2,2)$ sueprconformal field theory of central charge $c = \frac{3k}{k+2}$, called the $\mathcal{N} = 2$ minimal model of level k. The two R-symmetries of the LG model define the $U(1)$ currents of the $(2,2)$ superconformal algebra.

3.1. The B-branes $\mathscr{B}_L$. Let us apply the result of the previous section to this Landau-Ginzburg model. The superpotential can be factorized as $W = ix^{L+1} \cdot x^{k+1-L}$. Thus, for each $L = -1, 0, ..., k, k+1$, we find a B-brane $\mathscr{B}_L$ given by the boundary action

$$\int_{\partial\Sigma} \mathrm{d}t \left[\frac{1}{2}\int_B \mathrm{d}\theta \mathrm{d}\overline{\theta}\, \overline{\Gamma}\Gamma + \mathrm{Re}\int_B \mathrm{d}\theta\, \Gamma X^{L+1}\right]$$

where Γ obeys the constraint $\overline{D}\Gamma = X^{k+1-L}$. The boundary term is invariant under the $\mathbb{Z}_{k+2}$ discrete symmetry if we let γ act on the superfield Γ by

$$\gamma : \Gamma \to \omega^{-L-1}\Gamma. \tag{16}$$

Also, the vector R-symmetry is preserved under the transformation

$$U(1)_V : \Gamma(\theta) \to \mathrm{e}^{i\frac{k-2L}{k+2}\alpha}\Gamma(\mathrm{e}^{-i\alpha}\theta). \tag{17}$$

If the brane $\mathscr{B}_L$ defines a conformal boundary condition in the IR limit, we expect that $U(1)_V$ becomes a part of the $\mathcal{N}=2$ superconformal algebra.

3.2. Supersymmetric ground states. Here we will find the supersymmetric ground states of the open strings. We first consider the string with both ends at $\mathscr{B}_L$. Using (11), we find

$$i\overline{Q}_{\rm bdry}\begin{pmatrix} a(x) & b(x) \\ c(x) & d(x)\end{pmatrix} = \begin{pmatrix} x^{L+1}c(x)+b(x)x^{k+1-L} & x^{L+1}(d(x)-a(x)) \\ x^{k+1-L}(a(x)-d(x)) & x^{k+1-L}b(x)+c(x)x^{L+1}\end{pmatrix}$$

It vanishes if and only if $a(x)=d(x)$ and $x^{k+1-L}b(x)+c(x)x^{L+1}=0$. Then, it is $\overline{Q}_{\rm bdry}$-exact when $a(x)$ is divisible by x^{L+1} or x^{k+1-L}, $b(x)$ is divisible by x^{L+1} and $c(x)$ is divisible by x^{k+1-L}. If $L=-1$ or $L=k+1$, any $\overline{Q}_{\rm bdry}$-closed state is $\overline{Q}_{\rm bdry}$-exact and thus there is no supersymmetric ground state. For $0\le L\le k$, there are non-zero $\overline{Q}_{\rm bdry}$-cohomology classes represented by

$$|j\rangle^b_{LL} = \begin{pmatrix} x^j & 0 \\ 0 & x^j\end{pmatrix},\quad |j\rangle^f_{LL} = \begin{pmatrix} 0 & x^{L-j} \\ -x^{k-L-j} & 0\end{pmatrix} \tag{18}$$
$$j=0,1,...,\min\{L,k-L\}.$$

Similarly, for the string stretched from $\mathscr{B}_{L_1}$ to $\mathscr{B}_{L_2}$, the $\overline{Q}_{\rm bdry}$-cohomology group is non-trivial for $0\le L_1,L_2\le k$ and the elements of the basis are represented by

(19)
$$|j\rangle^b_{L_1L_2} = \begin{pmatrix} x^{j-\frac{L_1-L_2}{2}} & 0 \\ 0 & x^{j+\frac{L_1-L_2}{2}}\end{pmatrix},\quad |j\rangle^f_{L_1L_2} = \begin{pmatrix} 0 & x^{\frac{L_1+L_2}{2}-j} \\ -x^{k-\frac{L_1+L_2}{2}-j} & 0\end{pmatrix}$$

$$j=\tfrac{|L_1-L_2|}{2},\tfrac{|L_1-L_2|}{2}+1,...,\min\left\{\tfrac{L_1+L_2}{2},k-\tfrac{L_1+L_2}{2}\right\}. \tag{20}$$

In all cases, there is an equal number of bosonic and fermionic supersymmetric ground states. This in particular means that the Witten index vanishes:

$$\operatorname*{Tr}_{\mathscr{B}_{L_1}\mathscr{B}_{L_2}}(-1)^F=0. \tag{21}$$

$\mathbb{Z}_{k+2}$ *action.* The $\mathbb{Z}_{k+2}$ symmetry $X\to\omega X$ acts on the boundary fermion for the brane $\mathscr{B}_L$ as $\eta\to\omega^{-L-1}\eta$ according to (16). The action on the Chan-Paton factor is determined by the action on the ground state $|0\rangle_L$, which depends on a choice

$$\rho_M(\gamma):|0\rangle_L\to\omega^{-\frac{M+L+1}{2}}|0\rangle_L \tag{22}$$

parametrized by a mod $2(k+2)$ integer M such that $L+M+1$ is even. The action on the ground states for the open string stretched from $(\mathscr{B}_{L_1},\rho_{M_1})$ to $(\mathscr{B}_{L_2},\rho_{M_2})$ is

$$\gamma:|j\rangle^b_{L_1L_2}\mapsto\omega^{j+\frac{M_1-M_2}{2}}|j\rangle^b_{L_1L_2},\quad |j\rangle^f_{L_1L_2}\mapsto\omega^{-j-1+\frac{M_1-M_2}{2}}|j\rangle^f_{L_1L_2}. \tag{23}$$

The open string Witten index twisted by the symmetry γ^m is given by

$$\operatorname*{Tr}_{(\mathscr{B}_{L_1},\rho_{M_1})(\mathscr{B}_{L_2},\rho_{M_2})}(-1)^F\gamma^m=\omega^{m\frac{M_1-M_2}{2}}\sum_{j\in(20)}\left(\omega^{mj}-\omega^{-m(j+1)}\right), \tag{24}$$

where the sum is over j in Eq (20) above.

R-charges. Let us next analyze the R-charges of these supersymmetric ground states. By (17), the R-symmetry acts on the boundary fermion for the brane $\mathscr{B}_L$ as $\eta \to \lambda^{k-2L}\eta$, where $\lambda = \mathrm{e}^{\frac{i\alpha}{k+2}}$. We let the Chan-Paton state $|0\rangle_L$ transform as $|0\rangle_L \to \lambda^{\frac{k}{2}-L}|0\rangle_L$ so that the two states $|0\rangle_L$ and $\overline{\eta}|0\rangle_L$ have opposite charges. This yields the following R-transformation of the L_1-L_2 open string states:

$$U(1)_V : \begin{pmatrix} a(x) & b(x) \\ c(x) & d(x) \end{pmatrix} \longrightarrow \lambda^{\bullet} \begin{pmatrix} \lambda^{-L_2+L_1}a(\lambda^2 x) & \lambda^{-L_2+k-L_1}b(\lambda^2 x) \\ \lambda^{L_2-k+L_1}c(\lambda^2 x) & \lambda^{L_2-L_1}d(\lambda^2 x) \end{pmatrix}$$

where $\lambda^{\bullet}$ is a factor to be determined. The ground states transform as $|j\rangle^b_{L_1L_2} \mapsto \lambda^{2j+\bullet}|j\rangle^b_{L_1L_2}$ and $|j\rangle^f_{L_1L_2} \mapsto \lambda^{k-2j+\bullet}|j\rangle^f_{L_1L_2}$. We now require that the charge spectrum be symmetric under the sign flip $q \to -q$. This fixes $\lambda^{\bullet}$ to be $\lambda^{-\frac{k}{2}}$. Then, we find that the R-charges of $|j\rangle^b_{L_1L_2}$ and $|j\rangle^f_{L_1L_2}$ are

$$q_{|j\rangle^b_{L_1L_2}} = -q_{|j\rangle^f_{L_1L_2}} = \frac{2j}{k+2} - \frac{k}{2(k+2)}. \tag{25}$$

3.3. Boundary chiral primaries. Just as for the closed string, there is a one-to-one correspondence with the open string supersymmetric ground states and the boundary chiral ring elements:

$$\mathscr{O}^b_{j,L_1L_2} \longleftrightarrow |j\rangle^b_{L_1L_2}, \qquad \mathscr{O}^f_{j,L_1L_2} \longleftrightarrow |j\rangle^f_{L_1L_2}, \tag{26}$$

Their R-charges are obtained from (25) by the spectral flow $q \to q + \frac{c}{6}$:

$$q_{\mathscr{O}^b_{j,L_1L_2}} = \frac{2j}{k+2}, \qquad q_{\mathscr{O}^f_{j,L_1L_2}} = \frac{k-2j}{k+2} \tag{27}$$

Under the assumption that $\mathscr{B}_L$ defines conformal boundary conditions, the boundary chiral ring elements define boundary chiral primary fields of the boundary CFT. As usual [**40**], the R-charge q determines the conformal dimension of the operator $h = \frac{q}{2}$. Thus, the operators $\mathscr{O}^{b,f}_{j,L_1L_2}$ have dimension $h^b_{j,L_1L_2} = \frac{j}{k+2}$ and $h^f_{j,L_1L_2} = \frac{\frac{k}{2}-j}{k+2}$.

For the boundary preserving operators, $L_1 = L_2 = L$, they can be expressed in terms of the elementary fields as

$$\mathscr{O}^b_{j,LL} = x^j, \qquad \mathscr{O}^f_{j,LL} = \eta x^{L-j} - \overline{\eta} x^{k-L-j}. \tag{28}$$

They indeed represent the non-trivial cohomology classes of $\overline{\delta}x = 0, \overline{\delta}\eta = -\overline{\epsilon}x^{k+1-L}$ and $\overline{\delta}\overline{\eta} = -\overline{\epsilon}x^{L+1}$, and have the right R-charge (27) under $x \to \mathrm{e}^{\frac{2i\alpha}{k+2}}x$, $\eta \to \mathrm{e}^{\frac{k-2L}{k+2}i\alpha}\eta$, $\overline{\eta} \to \mathrm{e}^{-\frac{k-2L}{k+2}i\alpha}\overline{\eta}$. The fermionic operator $\mathscr{O}^f_{j,LL}$ is the lowest component of the superfield $\Gamma X^{L-j} - \overline{\Gamma}X^{k-L-j}$, which is indeed chiral since $\overline{D}\Gamma = X^{k+1-L}$ and $\overline{D\Gamma} = X^{L+1}$ (on shell). Thus, the corresponding deformation is given by the boundary F-term

$$\Delta S_{\mathrm{bdry}} = \mathrm{Re}\int_{\partial\Sigma,B} \mathrm{d}t\mathrm{d}\theta \left(\Gamma X^{L-j} - \overline{\Gamma}X^{k-L-j}\right)$$

The t integrand is an operator of dimension

$$\frac{\frac{k}{2}-j}{k+2} + \frac{1}{2} = \frac{k+1-j}{k+2} < 1,$$

and therefore is relevant.

3.4. The branes $\mathscr{B}_{-1}$ and $\mathscr{B}_{k+1}$. The branes $\mathscr{B}_{-1}$ and $\mathscr{B}_{k+1}$ are special in the sense that there is no supersymmetric ground state on the open strings stretched between themselves as well as between them and any other brane. Also, either f or g of the factorization of W is 1 (identity) and the term $|f|^2+|g|^2$ of the boundary action is larger than or equal to 1 everywhere on the field space. This is analogous to the situation of "constant tachyon" where we expect that the brane decays to nothing. From these facts, we claim that the branes $\mathscr{B}_{-1}$ and $\mathscr{B}_{k+1}$ can be regarded as "nothing" or "zero". Namely, if they comprise a summand of the brane, like $\mathscr{B}_{-1}\oplus\mathscr{B}'$, the boundary condition in the infrared limit is identified as $\mathscr{B}'$ only.

3.5. Comparison with RCFT results. A detailed study of the boundary states of $\mathcal{N}=2$ minimal models is done in a part of [**41**], based on an earlier work [**18**] which studies the D-branes in the coset model $\frac{SU(2)_k\times U(1)_2}{U(1)_{k+2}}$ using the standard RCFT technique [**42**]. ([**43**] considers a free field realization.) The coset model is obtained from the minimal model by a particular non-chiral GSO projection, and what is done in [**41**] is to carefully identify the boundary state before that GSO projection. It is found that there is a boundary condition $\mathscr{B}_{L,M,S}$ labeled by $L\in\{0,1,...,k\}$, $M\in\mathbb{Z}_{2(k+2)}$ and $S\in\mathbb{Z}_4$ with $L+M+S$ even, which preserves the supersymmetry $Q_+-(-1)^SQ_-$ (thus S odd ones are relevant for us). It is invariant under the $\mathbb{Z}_{k+2}$ discrete symmetry and the boundary state on the circle twisted by γ^m is[2]

(29)
$$|\mathscr{B}_{L,M,S}\rangle_{\gamma^m}=(2k+4)^{\frac{1}{4}}\,\mathrm{e}^{-\pi i\frac{Mm}{k+2}+\pi i\frac{S}{2}}\sum_{\substack{l\in\{0,1,\dots,k\}\\ \nu\in\{0,1\}}}\frac{S_{L\,l}}{\sqrt{S_{0\,l}}}(-1)^{S\nu}|l,m,2\nu+1\rangle\!\rangle_B,$$

(30)
$$|\mathscr{B}_{L,M,S}\rangle_{(-1)^F\gamma^m}=(2k+4)^{\frac{1}{4}}\,\mathrm{e}^{-\pi i\frac{Mm}{k+2}}\sum_{\substack{l\in\{0,1,\dots,k\}\\ \nu\in\{0,1\}}}\frac{S_{L\,l}}{\sqrt{S_{0\,l}}}(-1)^{S\nu}|l,m,2\nu\rangle\!\rangle_B,$$

where $|j,m,s\rangle\!\rangle_B$ is the B-type Ishibashi state. From this one can read off the full spectrum of open strings, including the supersymmetric ground states with their $\mathbb{Z}_{k+2}$-charges. In particular, the $\mathbb{Z}_{k+2}$-twisted Witten index can be computed as the overlap (see Eqn (5.43) of [**41**]):

$$\begin{aligned}&{}_{\gamma^m}\langle\mathscr{B}_{L_1,M_1,S_1}|\,\mathrm{e}^{\pi iJ_0}q_t^H\,|\mathscr{B}_{L_2,M_2,S_2}\rangle_{\gamma^m}\\ &=\mathrm{e}^{\pi i\left(\frac{m(M_1-M_2)}{k+2}-\frac{S_1-S_2}{2}\right)}\sum_{l\in\{0,1,\dots,k\}}N_{L_1L_2}^{l}\left(\mathrm{e}^{\pi i\frac{ml}{k+2}}-\mathrm{e}^{-\pi i\frac{m(l+2)}{k+2}}\right),\end{aligned}\tag{31}$$

where $N_{l\,l'}^{l''}$ are the $SU(2)_k$ fusion coefficients. This agrees with (24) under the identification of $(\mathscr{B}_L,\rho_M)$ and $\mathscr{B}_{L,M,1}$. The NSNS part of the boundary state ($m=0$ in (30)) shows that the boundary entropy [**44**] of the brane $\mathscr{B}_L$ is given by

$$g_L:=\langle 0|\mathscr{B}_L\rangle_{\mathrm{NSNS}}=(2k+4)^{\frac{1}{4}}\frac{S_{L\,0}}{\sqrt{S_{0\,0}}}=\sqrt{2}\frac{\sin(\pi\frac{L+1}{k+2})}{\sqrt{\sin(\frac{\pi}{k+2})}}.\tag{32}$$

[2]This differs slightly from the boundary state $|\mathscr{B}_{[j,s]}\rangle$ used in [**41**] by a phase as follows: $|\mathscr{B}_{[j,s]}\rangle_{a^{2m}}=\mathrm{e}^{\frac{\pi inm}{k+2}-\frac{\pi is}{2}}|\mathscr{B}_{2j,n,s}\rangle_{\gamma^m}$ and $|\mathscr{B}_{[j,s]}\rangle_{(-1)^Fa^{2m}}=\mathrm{e}^{\frac{\pi inm}{k+2}}|\mathscr{B}_{2j,n,s}\rangle_{(-1)^F\gamma^m}$.

We also need to stress that we do not have the so-called "short-orbit branes" $\widehat{\mathscr{B}}$ in our system. They are the oriented branes in the GSO projected model (coset model) [**18**], where nothing is wrong with the coexistence of $\widehat{\mathscr{B}}$ and our branes. In the model with the other GSO projection, opposite in the RR sector, our branes become oriented and short-orbit branes are unoriented, but again there is no problem with the coexistence. However, in the model before the GSO projection, there is an odd number of real fermions between our branes $\mathscr{B}_L$ and the short-orbit branes $\widehat{\mathscr{B}}$, which is problematic in quantization [**13**]. Thus our branes and the short-orbit branes cannot coexist before GSO projection. This fact is very important in the construction of rational B-branes in Gepner models directly from the B-branes in the minimal models [**45**]. The LG description of the short-orbit branes is given in the model with superpotential $W = X^{k+2} - Y^2$. A related discussion is given in the third paper of [**24**].

4. Brane charges and RG flows

We have seen that the Witten index of the open string for any pair of B-branes vanishes (21). By factorization, this implies that the overlap of the boundary state and the RR ground states all vanish ${}_{\rm RR}\langle i|\mathscr{B}_L\rangle_{\rm RR} = 0$, which is indeed the case: (29) has no overlap with the supersymmetric ground states $|\ell, \ell+1, 1\rangle \otimes |\ell, -\ell-l, -1\rangle$ on the untwisted circle. However, this does not mean that the D-brane charge vanishes. There could be a torsion charge that cannot be measured by the overlaps, ${}_{\rm RR}\langle i|\mathscr{B}\rangle_{\rm RR}$. In this section, we show that our B-branes $\mathscr{B}_L$ with $L = 0, 1, ..., k$ indeed carry such torsion charges. What we use extensively is the homotopy relation between matrices of the type (3),

$$\begin{pmatrix} A & 0 \\ 0 & B \end{pmatrix} \overset{0\leftarrow t}{\longleftarrow} R_t \begin{pmatrix} A & 0 \\ 0 & B \end{pmatrix} R_t^{-1} \overset{t\to\frac{\pi}{2}}{\longrightarrow} \begin{pmatrix} B & 0 \\ 0 & A \end{pmatrix}$$

where

$$R_t := \begin{pmatrix} \cos t & -\sin t \\ \sin t & \cos t \end{pmatrix}. \tag{33}$$

Furthermore, we determine the pattern of boundary RG flows and identify the operators that generate them.

4.1. The charges of the B-branes. Let us consider the one-parameter family of 2×2 matrix pairs

$$f_t(x) = R_t \begin{pmatrix} 1 & 0 \\ 0 & x \end{pmatrix} R_t^{-1} \begin{pmatrix} x & 0 \\ 0 & 1 \end{pmatrix} \tag{34}$$

$$g_t(x) = \begin{pmatrix} 1 & 0 \\ 0 & x \end{pmatrix} R_t \begin{pmatrix} x^{k+1} & 0 \\ 0 & x^k \end{pmatrix} R_t^{-1}, \tag{35}$$

where R_t is the matrix given by (33). One can readily see that the condition of B-type supersymmetry is preserved at each t, $f_t g_t = x^{k+2}\mathbf{1}_2$, $g_t f_t = x^{k+2}\mathbf{1}_2$. At $t = 0$, the linear maps are

$$f_0(x) = \begin{pmatrix} x & 0 \\ 0 & x \end{pmatrix}, \quad g_0(x) = \begin{pmatrix} x^{k+1} & 0 \\ 0 & x^{k+1} \end{pmatrix},$$

representing the sum of two copies of the $L = 0$ brane, $\mathscr{B}_0 \oplus \mathscr{B}_0$. At $t = \frac{\pi}{2}$, the linear maps are

$$f_{\frac{\pi}{2}}(x) = \begin{pmatrix} x^2 & 0 \\ 0 & 1 \end{pmatrix}, \quad g_{\frac{\pi}{2}}(x) = \begin{pmatrix} x^k & 0 \\ 0 & x^{k+2} \end{pmatrix},$$

representing the sum of $L = 1$ and $L = -1$ branes, $\mathscr{B}_1 \oplus \mathscr{B}_{-1}$. Since the $L = -1$ brane is empty, we find that the two brane configurations are continuously connected with each other.

$$\mathscr{B}_0 \oplus \mathscr{B}_0 \simeq \mathscr{B}_1.$$

Similarly, by the following family of configurations:

$$f_t(x) = R_t \begin{pmatrix} 1 & 0 \\ 0 & x^{L_2+1} \end{pmatrix} R_t^{-1} \begin{pmatrix} x^{L_1+1} & 0 \\ 0 & 1 \end{pmatrix}, \tag{36}$$

$$g_t(x) = \begin{pmatrix} x^{k-L_1-L_2} & 0 \\ 0 & x^{k+1-L_2} \end{pmatrix} R_t \begin{pmatrix} x^{L_2+1} & 0 \\ 0 & 1 \end{pmatrix} R_t^{-1} \tag{37}$$

we find the homotopy relation of the branes

$$\mathscr{B}_{L_1} \oplus \mathscr{B}_{L_2} \simeq \mathscr{B}_{L_1+L_2+1}. \tag{38}$$

Here we have assumed that $L_1 + L_2 \leq k$. If $L_1 + L_2 \geq k$, we find $\mathscr{B}_{L_1} \oplus \mathscr{B}_{L_2} \simeq \mathscr{B}_{L_1+L_2-k-1}$ by a suitable modification of the homotopy. Using (38) repeatedly, we find

$$\mathscr{B}_L \simeq \underbrace{\mathscr{B}_0 \oplus \cdots \oplus \mathscr{B}_0}_{L+1}. \tag{39}$$

As the special case $L = k + 1$, we find

$$\underbrace{\mathscr{B}_0 \oplus \cdots \oplus \mathscr{B}_0}_{k+2} \simeq 0. \tag{40}$$

The homotopy relations (39) and (40) imply that the RR-charge of the B-branes is

$$\Lambda_B \cong \mathbb{Z}_{k+2} \tag{41}$$

generated by $\mathscr{B}_0$.

4.2. Mirror picture: A-branes in the $\mathbb{Z}_{k+2}$ LG orbifold. The LG model with superpotential $W = X^{k+2}$ is mirror to the LG orbifold $\widetilde{W} = \widetilde{X}^{k+2}$ with respect to the group $\mathbb{Z}_{k+2}$ [**46**] which acts on the fields as $\widetilde{X} \to \omega\widetilde{X}$. The B-branes in the model $W = X^{k+2}$ are mapped to the A-branes in the LG orbifold. A-branes in the model before orbifolding are the D1-branes at the wedge-shaped lines with apex at $\widetilde{X} = 0$, which are mapped to the (positive) real line of the $\widetilde{W}$-plane. For each $L = 0, 1, ..., k$ and $M \in \mathbb{Z} \pmod{2(k+2)}$ such that $L + M + 1$ is even, there is an A-brane $\mathscr{A}_{L,M}$ with the wedge coming in from the direction $\arg(\widetilde{x}) = \pi\frac{M-L-1}{k+2}$ and going out to the direction $\arg(\widetilde{x}) = \pi\frac{M+L+1}{k+2}$. See Figure 2. The opening angle and the mean-direction of the wedge are determined by L and M respectively as $2\pi\frac{L+1}{k+2}$ and $\frac{\pi M}{k+2}$.

The orbifold group $\mathbb{Z}_{k+2}$ acts on the branes as $\mathscr{A}_{L,M} \to \mathscr{A}_{L,M+2}$, which is free of fixed points and stabilizers. Thus A-branes in the orbifold model are just the sums over images. We denote by $\mathscr{A}_L$ the brane obtained from $\mathscr{A}_{L,M}$ which have opening angle $2\pi\frac{L+1}{k+1}$. Let us consider the sum of two such branes $\mathscr{A}_{L_1} \oplus \mathscr{A}_{L_2}$ which are obtained from the sums over images of $\mathscr{A}_{L_1,M_1}$ and $\mathscr{A}_{L_2,M_2}$. One can find a pair (M_1, M_2) such that the outgoing direction of $\mathscr{A}_{L_1,M_1}$ agrees with the incoming

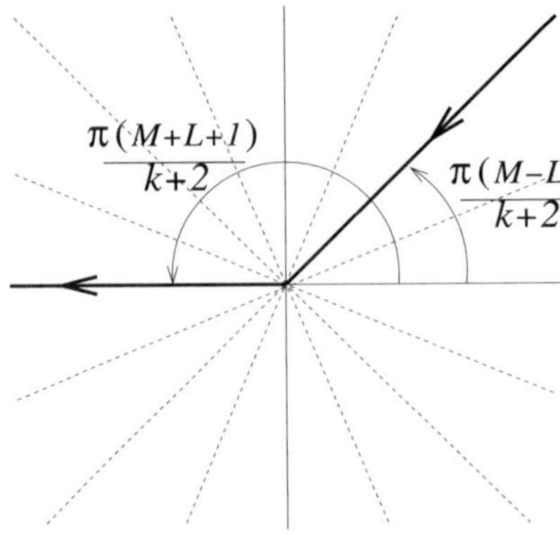

FIGURE 2. The A-brane $\mathscr{A}_{L,M}$. This is the example $L = 2$, $M = 5$ for $k = 6$.

direction of $\mathscr{A}_{L_2,M_2}$. Under such an arrangement, the charge of the sum $\mathscr{A}_{L_1,M_1} \oplus \mathscr{A}_{L_2,M_2}$ is the same as the charge of the brane $\mathscr{A}_{L_1+L_2+1,M'}$ for some M'. (We will

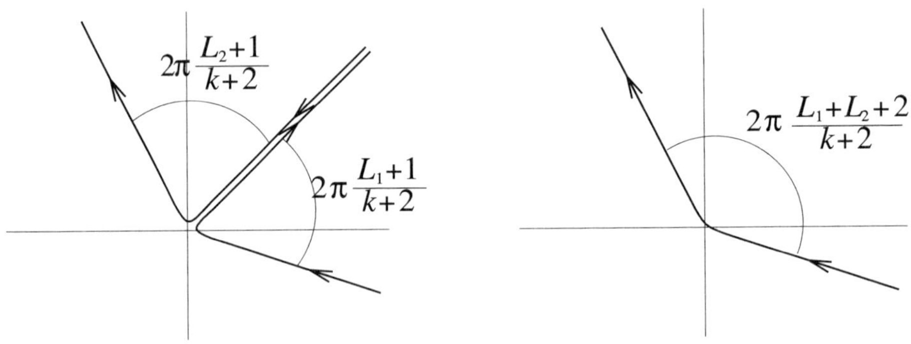

FIGURE 3. Cancellation of out-going and in-coming rays of two A-branes.

discuss the A-brane charge in Section 5.1 in more detail.) This is understood as the cancellation of the two parallel rays with the opposite orientations. See Figure 3. Thus, we find the homotopy relation

$$\mathscr{A}_{L_1} \oplus \mathscr{A}_{L_2} \simeq \mathscr{A}_{L_1+L_2+1}. \tag{42}$$

This is the mirror counterpart of the relation (38) under the identification

$$\mathscr{B}_L = \mathscr{A}_L. \tag{43}$$

In particular, the relation (40) is mirror to the process that the sum of $(k+2)$ copies of the brane $\mathscr{A}_0$ of opening angle $\frac{2\pi}{k+2}$ annihilates to nothing, by cancellation of the outgoing ray of one brane and the incoming ray of the next brane. See Figure 6 in Section 5 for the corresponding annihilation in the model before orbifolding.

4.3. Boundary RG flows. The cancellation of the parallel rays of opposite orientations we have seen is nothing but annihilation of brane and antibrane, which can be regarded as a tachyon condensation on the worldvolume [**47**]. In the worldsheet perspective, open string tachyon condensations can be described as the boundary renormalization group flows generated by boundary relevant operators [**9, 48, 49, 50**]. We would like to view the homotopy relation $\mathscr{B}_{L_1} \oplus \mathscr{B}_{L_2} \to \mathscr{B}_{L_1+L_2+1}$ as such a boundary RG flow, directly for B-branes in the model without orbifolding, and identify the relevant operator that generates it.

To this end, we make a change of basis of the Chan-Paton factors that simplifies the expression of the intermediate configurations (36)-(37). (Recall that we are assuming $L_1 + L_2 \leq k$. Other cases can be treated with an obvious modification.) We change the basis of $\mathbb{C}^2_+$ by R_t, so that the matrix expression changes as $f_t(x) \to R_t^{-1} f_t(x)$, $g_t(x) \to g_t(x) R_t$. More explicitly we have

$$\begin{aligned} f_t(x) &= \begin{pmatrix} 1 & 0 \\ 0 & x^{L_2+1} \end{pmatrix} R_t^{-1} \begin{pmatrix} x^{L_1+1} & 0 \\ 0 & 1 \end{pmatrix} \\ &= \begin{pmatrix} x^{L_1+1}\cos t & \sin t \\ -x^{L_1+L_2+2}\sin t & x^{L_2+1}\cos t \end{pmatrix}, \end{aligned} \tag{44}$$

$$\begin{aligned} g_t(x) &= \begin{pmatrix} x^{k-L_1-L_2} & 0 \\ 0 & x^{k+1-L_2} \end{pmatrix} R_t \begin{pmatrix} x^{L_2+1} & 0 \\ 0 & 1 \end{pmatrix} \\ &= \begin{pmatrix} x^{k+1-L_1}\cos t & -x^{k-L_1-L_2}\sin t \\ x^{k+2}\sin t & x^{k+1-L_2}\cos t \end{pmatrix}. \end{aligned} \tag{45}$$

In this expression, we see that the configuration for $t = \varepsilon \ll 1$ can be expanded as

$$f_\varepsilon(x) = f_0(x) + \varepsilon \begin{pmatrix} 0 & 1 \\ -x^{L_1+L_2+2} & 0 \end{pmatrix} + \cdots, \tag{46}$$

$$g_\varepsilon(x) = g_0(x) + \varepsilon \begin{pmatrix} 0 & -x^{k-L_1-L_2} \\ x^{k+2} & 0 \end{pmatrix} + \cdots. \tag{47}$$

To identify the operator that gives this perturbation, we note that the matrix entries of $f(x)$ and $g(x)$ have the following invariant meaning:

$$f(x) = \begin{pmatrix} |0\rangle_{11}\langle 0|\eta_1 & |0\rangle_{12}\langle 0|\eta_2 \\ |0\rangle_{21}\langle 0|\eta_1 & |0\rangle_{22}\langle 0|\eta_2 \end{pmatrix}, \quad g(x) = \begin{pmatrix} \overline{\eta}_1|0\rangle_{11}\langle 0| & \overline{\eta}_1|0\rangle_{12}\langle 0| \\ \overline{\eta}_2|0\rangle_{21}\langle 0| & \overline{\eta}_2|0\rangle_{22}\langle 0| \end{pmatrix}.$$

This is enough to find that the perturbation corresponds to the following fermionic states in the $\mathscr{B}_{L_1}$-$\mathscr{B}_{L_2}$ sector and the $\mathscr{B}_{L_2}$-$\mathscr{B}_{L_1}$ sector:

$$|\mathrm{pert}\rangle^f_{L_1L_2} = \begin{pmatrix} 0 & -x^{L_1+L_2+2} \\ x^{k+2} & 0 \end{pmatrix}, \quad |\mathrm{pert}\rangle^f_{L_2L_1} = \begin{pmatrix} 0 & 1 \\ -x^{k-L_1-L_2} & 0 \end{pmatrix} \tag{48}$$

Comparing with (19), these states can be formally identified as the "supersymmetric ground states" $-|j_{12}\rangle^f_{L_1L_2}$ and $|j_{21}\rangle^f_{L_2L_1}$ respectively, where $j_{12} = -\frac{L_1+L_2}{2} - 2$ and $j_{21} = \frac{L_1+L_2}{2}$. Note that j_{21} is in the range (20) but j_{12} is not. Thus, $|\mathrm{pert}\rangle^f_{L_2L_1} = |j_{21}\rangle^f_{L_2L_1}$ is a true ground state but $|\mathrm{pert}\rangle^f_{L_1L_2} = -|j_{12}\rangle^f_{L_1L_2}$ is not. In fact, $|\mathrm{pert}\rangle^f_{L_1L_2}$ is fine in the sense that it is annihilated by the supercharge $\overline{Q}_{\mathrm{bdry}}$ but it is a $\overline{Q}_{\mathrm{bdry}}$-exact state. We can interpret what we have seen as follows: *The perturbation (46)-(47) is generated by the boundary F-term corresponding to the fermionic state*

$$|j_{21}\rangle^f_{L_2L_1}; \qquad j_{21} = \frac{L_1+L_2}{2}. \tag{49}$$

In order to define a deformation which obeys the condition of supersymmetry, $f(x)g(x) = -iW(x)$ and $g(x)f(x) = -iW(x)$, it needs to be accompanied by a $\overline{Q}$-exact part, which is $-|j_{12}\rangle^f_{L_1L_2}$, $j_{12} = -\frac{L_1+L_2}{2} - 2$.

It is conjectured that the boundary entropy $g = \langle 0|\mathscr{B}\rangle_{\rm NSNS}$ must decrease under the boundary RG flows — "g-theorem" [**44, 51**]. Let us check this in the present case. The boundary entropy of the brane $\mathscr{B}_L$ is given by (32) or

$$g_L = c_k \cdot \sin\left(\pi \frac{L+1}{k+2}\right),$$

where c_k is an L-independent constant. A nice picture to help in understanding it is the one regarding the $\mathcal{N}=2$ minimal model or its orbifold as the dilatonic sigma model on the disc where the A-branes are given by the straight segments connecting $k+2$ special points on the boundary [**18**].[3] The boundary entropy of a brane is proportional to the length of the segment. In this picture it is clear that the boundary entropy decreases under the flow $\mathscr{B}_{L_1} \oplus \mathscr{B}_{L_2} \to \mathscr{B}_{L_1+L_2+1}$. It is simply the triangle inequality (see Figure 4).

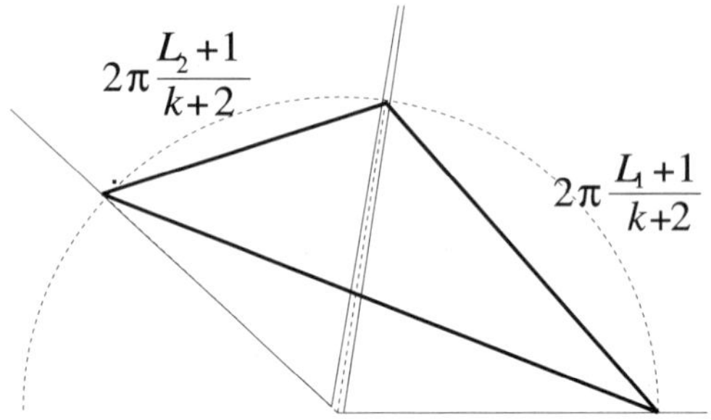

FIGURE 4. The disc picture of the mirror A-branes. The boundary entropy is given by the length of the segment. It decreases under the RG flow $\mathscr{B}_{L_1} \oplus \mathscr{B}_{L_2} \to \mathscr{B}_{L_1+L_2+1}$ by the triangle inequality.

Remarks.
(i) Although $g_{L_1+L_2+1}$ is less than the sum $g_{L_1} + g_{L_2}$, it is larger than the individual entropies g_{L_1}, g_{L_2}. This is related to the fact that we need to add the $\overline{Q}$-trivial piece $|\text{pert}\rangle^f_{L_1L_2}$ whose corresponding operator formally has dimension $\frac{k+1-j_{12}}{k+2} = 1 + \frac{L_1+L_2+2}{2(k+2)}$ larger than 1.
(ii) It would be a very interesting problem to define the boundary entropy throughout the RG flow, not just the UV and IR limits, and see if it continuously decreases. In fact, the disc partition function has such a property [**52, 53, 54**] for any $\mathcal{N}=1$ preserving boundary flow. See also [**55**] for a proposal in the context of supersymmetric field theories in (bulk) four dimensions.

4.3.1. *More general perturbations.* The perturbation operator corresponding to (49) has dimension

$$\Delta_{\rm pert} = \frac{k+1-j_{21}}{k+2} = \frac{k+1-\frac{L_1+L_2}{2}}{k+2}.$$

[3] However, one can also compute it in the LG picture using the proportionality to the "RR-charge" $\langle 0|\mathscr{B}\rangle_{\rm RR}$ which is identified as the weighted integral $\int_L e^{-iW}d\widetilde{X}$ [**17**]. The disc picture may be regarded as focusing on the origin $\widetilde{X}=0$ of the LG model in the IR limit. The sharp apex of the wedge may be modified to be the straight segment of the disc (private communication with J. Maldacena, 2002).

It is the most relevant operator since the maximum value of j for $|j\rangle^f_{L_1L_2}$ or $|j\rangle^f_{L_2L_1}$ is $j_{\rm max}=\min\{\frac{L_1+L_2}{2},k-\frac{L_1+L_2}{2}\}$ (which is $\frac{L_1+L_2}{2}$ in the present case where $L_1+L_2\leq k$ is assumed). We would now like to study the perturbation generated by other relevant operators, $j=\frac{|L_1-L_2|}{2},\ldots,j_{\rm max}-1$.

We consider the following generalization of the family of configurations:

$$f_t(x) = \begin{pmatrix}1&0\\0&x^a\end{pmatrix}R_t^{-1}\begin{pmatrix}x^{L_1+1}&0\\0&x^{L_2+1-a}\end{pmatrix} = \begin{pmatrix}x^{L_1+1}\cos t & x^{L_2+1-a}\sin t\\ -x^{L_1+1+a}\sin t & x^{L_2+1}\cos t\end{pmatrix}, \tag{50}$$

$$g_t(x) = \begin{pmatrix}x^{k+1-L_1-a}&0\\0&x^{k+1-L_2}\end{pmatrix}R_t\begin{pmatrix}x^a&0\\0&1\end{pmatrix} = \begin{pmatrix}x^{k+1-L_1}\cos t & -x^{k+1-L_1-a}\sin t\\ x^{k+1-L_2+a}\sin t & x^{k+1-L_2}\cos t\end{pmatrix} \tag{51}$$

for some integer a such that the matrix entries of the right hand sides are all monomials in x:

$$\max\{-L_1-1,\,L_2-k-1\}\leq a\leq\min\{L_2+1,\,k+1-L_1\}. \tag{52}$$

As before we can identify this as the perturbation generated by the operator corresponding to the fermionic states $-|j_{12}\rangle^f_{L_1L_2}$ and $|j_{21}\rangle^f_{L_2L_1}$, where

$$j_{12}=\frac{L_2-L_1}{2}-a-1,\qquad j_{21}=\frac{L_1-L_2}{2}+a-1.$$

Note that $j_{12}+j_{21}=-2$ and hence it is impossible for both to satisfy the condition (20) — at most only one of them can satisfy it. If we assume $L_1\geq L_2$ and $L_1+L_2\leq k$ in addition to (52), the condition is

$$|j_{12}\rangle^f_{L_1L_2}\in\mathcal{H}_{\rm SUSY}\iff a\leq L_2-L_1-1, \tag{53}$$

$$|j_{21}\rangle^f_{L_2L_1}\in\mathcal{H}_{\rm SUSY}\iff a\geq 1. \tag{54}$$

Only if either one of these conditions is met can one consider $f_\varepsilon(x),g_\varepsilon(x)$ as a relevant perturbation of $\mathscr{B}_{L_1}\oplus\mathscr{B}_{L_2}$. The final configuration is the one with $t=\frac{\pi}{2}$. After the change of basis of $\mathbb{C}^2_+$ by $R_{\frac{\pi}{2}}=\left(\begin{smallmatrix}0&-1\\1&0\end{smallmatrix}\right)$, it is expressed as

$$\widetilde{f}(x)=\begin{pmatrix}x^{L_1+1+a}&0\\0&x^{L_2+1-a}\end{pmatrix},\quad \widetilde{g}(x)=\begin{pmatrix}x^{k+1-L_1-a}&0\\0&x^{k+1-L_2+a}\end{pmatrix}, \tag{55}$$

which is the configuration of the sum $\mathscr{B}_{L_1+a}\oplus\mathscr{B}_{L_2-a}$. Note that $\{L_1+a,L_2-a\}=\{\frac{L_1+L_2}{2}+j+1,\frac{L_1+L_2}{2}-j-1\}$ for $j=j_{12}$ or $j=j_{21}$. Thus we find that the relevant operator corresponding to $|j\rangle^f_{L_1L_2}\in\mathcal{H}_{\rm SUSY}$ or $|j\rangle^f_{L_2L_1}\in\mathcal{H}_{\rm SUSY}$ generates an RG flow

$$\mathscr{B}_{L_1}\oplus\mathscr{B}_{L_2}\longrightarrow\mathscr{B}_{\frac{L_1+L_2}{2}+j+1}\oplus\mathscr{B}_{\frac{L_1+L_2}{2}-j-1}. \tag{56}$$

Note that the difference of the two opening angles $|L_1-L_2|$ increases after the flow $|(\frac{L_1+L_2}{2}+j+1)-(\frac{L_1+L_2}{2}-j-1)|=2j+2>|L_1-L_2|$. In the disc picture of the mirror minimal model, we indeed see that the total boundary entropy decreases (see Figure 5):

$$g_{L_1}+g_{L_2}>g_{\frac{L_1+L_2}{2}+j+1}+g_{\frac{L_1+L_2}{2}-j-1}. \tag{57}$$

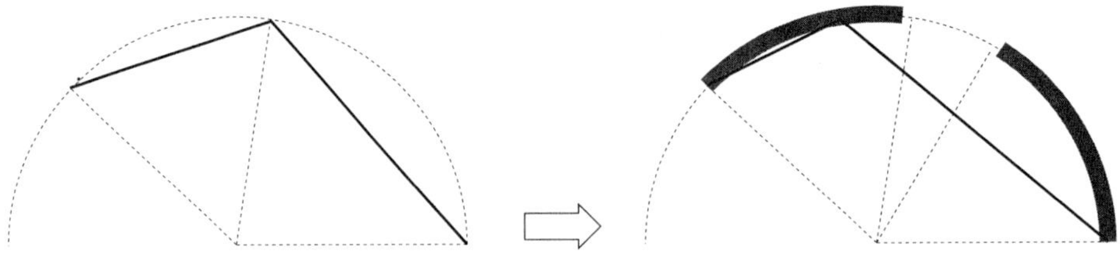

FIGURE 5. The RG flow $\mathscr{B}_{L_1}\oplus\mathscr{B}_{L_2}\to\mathscr{B}_{\frac{L_1+L_2}{2}+j+1}\oplus\mathscr{B}_{\frac{L_1+L_2}{2}-j-1}$ generated by the operator corresponding to $|j\rangle^f_{L_1L_2}$ or $|j\rangle^f_{L_2L_1}$. The mid-point after the flow can only be in the shaded regions. The two components correspond to whether the flow is generated by $|j\rangle^f_{L_1L_2}$ or $|j\rangle^f_{L_2L_1}$. It is evident that the sum of lengths decreases under the flow.

Finally, we would like to comment on operators in the $\mathscr{B}_{L_i}$-$\mathscr{B}_{L_i}$ sectors. As we have seen, there are indeed fermionic chiral primary operators that may correspond to relevant deformations. However, one cannot find a deformation that obeys the supersymmetry condition $f(x)g(x)=-iW$ and $g(x)f(x)=-iW$. Thus, it seems that these operators cannot induce a finite supersymmetric deformation of the system. This guarantees the conservation of the torsion D-brane charge $\Lambda_B\cong\mathbb{Z}_{k+2}$ we have claimed.

5. B-branes in LG orbifold $W=X^{k+2}/\mathbb{Z}_{k+2}$

In this section, we study the charges and the boundary RG flows for B-branes in the LG orbifold of $W=X^{k+2}$ with respect to the group $\mathbb{Z}_{k+2}$ generated by $\gamma: X\to\omega X$. This is mirror to A-branes in the LG models without orbifolding which are studied in [**17**]. We first describe the known cases and then see how the result is reproduced and extended using B-branes of the orbifold.

5.1. Mirror picture: A-branes in $W=\widetilde{X}^{k+2}$. We recall that we have an A-brane $\mathscr{A}_{L,M}$ in the LG model $W=\widetilde{X}^{k+2}$ for each (L,M) with $L+M+1$ even. It is the wedge coming from $\arg(\widetilde{x})=\pi\frac{M-L-1}{k+2}$ and going to $\arg(\widetilde{x})=\pi\frac{M+L+1}{k+2}$ (see Figure 2). The Witten index of the open string stretched between two such branes is

$$I(\mathscr{A}_1,\mathscr{A}_2)=\#(\mathscr{A}_1^-\cap\mathscr{A}_2^+) \tag{58}$$

where $\mathscr{A}^\pm$ is the rotation of $\mathscr{A}$ by a small positive/negative angle (see [**22**]). The space of supersymmetric ground states of the string is

$$\mathcal{H}^b_{\rm SUSY}\oplus\mathcal{H}^f_{\rm SUSY}=\begin{cases}\mathbb{C}\oplus 0 & \text{if } I=1\\ 0\oplus\mathbb{C} & \text{if } I=-1\\ 0\oplus 0 & \text{if } I=0.\end{cases} \tag{59}$$

If $L_1\geq L_2$, $k\geq L_1+L_2$, the index is given by

$$I(\mathscr{A}_{L_1,M_1},\mathscr{A}_{L_2,M_2})=\begin{cases}1 & \text{if } \frac{M_2-L_2-1}{2}+1\leq\frac{M_1+L_1+1}{2}\leq\frac{M_2+L_2+1}{2}\\ -1 & \text{if } \frac{M_2-L_2-1}{2}+1\leq\frac{M_1-L_1-1}{2}\leq\frac{M_2+L_2+1}{2}\\ 0 & \text{otherwise.}\end{cases} \tag{60}$$

The charge lattice of the A-branes is $H_1(\mathbb{C}, B^+)$ where B^+ is the region in which $\mathrm{Im}(W)$ is large and positive. It is

$$\Lambda_A \cong \mathbb{Z}^{k+1} \tag{61}$$

generated by $[\mathscr{A}^+_{0,M}]$, $M = 0, 2, ..., 2(k+1)$, which obey the linear relation

$$[\mathscr{A}^+_{0,0}] + [\mathscr{A}^+_{0,2}] + \cdots + [\mathscr{A}^+_{0,2k+2}] = 0 \qquad \text{in } H_1(\mathbb{C}, B^+).$$

Figure 6 describes this relation in the example of $k = 6$.

FIGURE 6. The branes $\mathscr{A}_{0,M}$, $M = 0, 2, ..., 2(k+1)$ (for the case $k = 6$). The shaded regions are B^+. With a small positive rotation, each represent a class $[\mathscr{A}^+_{0,M}] \in H_1(\mathbb{C}, B^+)$. The sum annihilates in $H_1(\mathbb{C}, B^+)$ because of the cancellation of the outgoing ray of one brane and the incoming ray of the next brane.

5.1.1. *RG flow as brane recombination.* Let us study the supersymmetric boundary RG flows. We first consider a single brane $\mathscr{A}_{L,M}$. This represents a non-trivial charge $[\mathscr{A}^+_{L,M}] \in H_1(\mathbb{C}, B^+)$ and must be stable. Indeed, since $I(\mathscr{A}_{L,M}, \mathscr{A}_{L,M}) = 1$, there is only one supersymmetric ground state and it is bosonic. Therefore there is no fermionic chiral primary field and hence no supersymmetric deformation operator. We next consider the sum of two branes $\mathscr{A}_{L_1,M_1} \oplus \mathscr{A}_{L_2,M_2}$. Whether there is another brane configuration to which it can decay depends on the intersection numbers of the two branes. There are four cases to consider.

(i) No intersection

If the two do not intersect, there is no supersymmetric ground state, and hence no supersymmetric deformation of the brane configuration, both in the 1-2 and 2-1 string sectors. Since there is neither in the 1-1 and 2-2 sectors (as we have seen for the single brane case), there is no supersymmetric deformation of the brane. Thus the brane is stable.

(ii) "Transverse" intersection

This is the case where the two intersect transversely as in Figure 7(Left). In this case, one of the intersection numbers is +1 and other is −1. In the Figure we have $I(1,2) = 1$ and $I(2,1) = -1$. In this case, there is one fermionic supersymmetric ground state in the 2-1 string sector. Depending on the dimension of the corresponding deformation operator, the brane can decay into another brane configuration. The wedge-picture suggests that the brane recombination occurs, and it ends up as

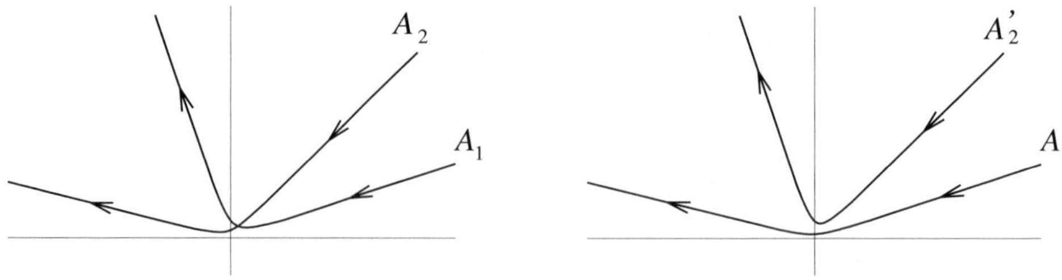

FIGURE 7. Two branes intersecting "transversely" (Left). It is expected to recombine to two other branes without intersection (Right).

a configuration of two other branes. The end point is as in Figure 7 (Right) which is the sum $\mathscr{A}_{L_1',M_1'} \oplus \mathscr{A}_{L_2',M_2'}$, where $L_1' = \frac{L_1+L_2-M_1+M_2}{2}$, $M_1' = \frac{-L_1+L_2+M_1+M_2}{2}$, $L_2' = \frac{L_1+L_2+M_1-M_2}{2}$, $M_2' = \frac{L_1-L_2+M_1+M_2}{2}$.

(iii) "Non-transverse" intersection, -1

If the incoming ray of one wedge is the same as the outgoing ray of the other, one of the intersections is 0 but the other is ± 1 depending on the orientations of the two branes. Here we consider the -1 case, as in Figure 8(Left) where $I(1,2) = 0$ and $I(2,1) = -1$. In this case, there is one fermionic supersymmetric ground state from

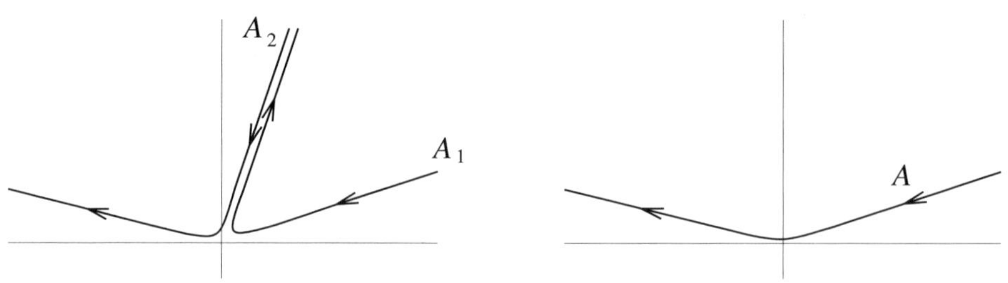

FIGURE 8. Two branes intersecting "non-transversely" (Left). They are expected to combine into one brane (Right) by canceling the parallel rays of opposite orientation.

the 2-1 string sector. If the dimension of the corresponding deformation operator is less than 1, it can decay to a new brane configuration. The wedge-picture suggests it is the brane obtained by deleting the overlapping and oppositely oriented rays of the two branes. For the case as in Figure 8 where $\pi\frac{M_1+L_1+1}{k+2} = \pi\frac{M_2-L_2-1}{k+2}$ (mod 2π) and $L_1+L_2 \leq k$, we have $\mathscr{A}_{L,M}$ with $L = L_1+L_2+1$ and $M = \frac{-L_1+L_2+M_1+M_2}{2}$.

(iv) "Non-transverse" intersection, $+1$

Next, we consider the $+1$ case, as in Figure 9 where $I(1,2) = 0$ and $I(2,1) = 1$, which is obtained from case (iii) by flipping the orientation of one of the two branes. In this case, there is one bosonic ground state but no fermionic ground state. Thus, there is no supersymmetric deformation and the brane is stable.

5.2. B-branes in the orbifold. We now describe the same system in the mirror LG orbifold.

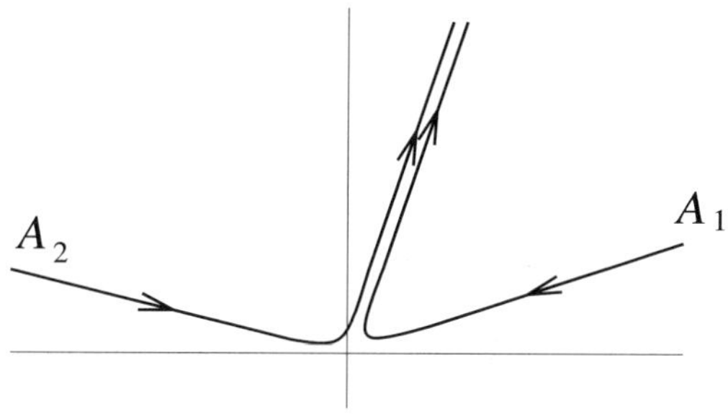

FIGURE 9. Two branes intersecting "non-transversely". They cannot combine into one brane because the orientation of the parallel rays are the same.

Let us denote by $\mathscr{B}_{L,M}$ the B-brane in the orbifold obtained from the B-brane $(\mathscr{B}_L, \rho_M)$ of the original model, where ρ_M is the Chan-Paton representation of the orbifold group $\mathbb{Z}_{k+2}$ given by (22),

$$\gamma : |0\rangle_L \mapsto \omega^{-\frac{M+L+1}{2}}|0\rangle_L, \quad \overline{\eta}|0\rangle_L \mapsto \omega^{-\frac{M-L-1}{2}}\overline{\eta}|0\rangle_L.$$

Note that the "brane configuration" $(\mathbb{C}\overline{\eta}|0\rangle_L \overset{f}{\underset{g}{\rightleftharpoons}} \mathbb{C}|0\rangle_L)$ is invariant under the orbifold group:

$$f(x) = x^{L+1} \to \omega^{-\frac{M+L+1}{2}} f(\omega x)\omega^{\frac{M-L-1}{2}} = f(x),$$
$$g(x) = x^{k+1-L} \to \omega^{-\frac{M-L-1}{2}} g(\omega x)\omega^{\frac{M+L+1}{2}} = g(x).$$

5.2.1. *Open string ground states.* Let us first analyze the supersymmetric ground states of the open string stretched between two such branes, $\mathscr{B}_{L_1,M_2}$ and $\mathscr{B}_{L_2,M_2}$. The first thing to note is that the $\mathbb{Z}_{k+2}$-equivariant $\overline{Q}$-cohomology is the same as the $\mathbb{Z}_{k+2}$-invariant states of the ordinary $\overline{Q}$-cohomology. Let us explain this. Let (C, Q) be a complex and let Γ be a symmetry group. Namely, Γ acts on C and each $\gamma \in \Gamma$ preserves the degree and commutes with Q. Then the Γ-invariant elements of C form a complex C^Γ. We assume Γ is a compact group. Then we have the
Theorem. *The Γ-invariant part of the cohomology group is the same as the cohomology of the Γ-invariant part of the complex:*

$$H(C,Q)^\Gamma = H(C^\Gamma, Q). \tag{62}$$

Proof: The proof assumes that Γ is finite but it is clear that this also applies to a group with an invariant measure with volume 1. We have a map

$$Z^\Gamma \longrightarrow (Z/QC)^\Gamma. \tag{63}$$

We want to show (i) it is surjective, and (ii) the kernel is $Q(C^\Gamma)$. To show (i) let $f \in C$ represent a Γ-invariant cohomology class. Namely, $Qf = 0$ and, for any $\gamma \in \Gamma$, $\gamma f = f + Qf'_\gamma$ for some $f'_\gamma \in C$. Then

$$f_1 := \frac{1}{|\Gamma|}\sum_{\gamma\in\Gamma} \gamma f = f + Qf'$$

for some $f' \in C$. Namely, $f \equiv f_1 \in Z^\Gamma$. This shows (i). To show (ii), let $f \in Z^\Gamma$ be mapped to 0 by (63). Namely, $\gamma f = f (\forall \gamma \in \Gamma)$, $Qf = 0$, and $f = Qf'$ for some

$f' \in C$. Then,

$$f = \frac{1}{|\Gamma|}\sum_{\gamma\in\Gamma}\gamma f = \frac{1}{|\Gamma|}\sum_{\gamma\in\Gamma}\gamma Q f' = Q\left(\frac{1}{|\Gamma|}\sum_{\gamma\in\Gamma}\gamma f'\right).$$

This means $f \in Q(C^\Gamma)$. Thus (ii) is shown.

The orbifold group action on the supersymmetric ground states of the original theory has been found in (23). There is at most one invariant state. It is $|j_*\rangle^b_{L_1L_2}$ or $|j_*'\rangle^f_{L_1L_2}$ if j_* or j_*' defined by

$$j_* \equiv \frac{M_2 - M_1}{2}, \qquad j_*' \equiv \frac{M_1 - M_2}{2} - 1 \qquad \mathrm{mod}\ (k+2),$$

is in the range (20), and there is none if both of them are outside that range. Thus, we find

(64)

$$\mathcal{H}^b_{\rm SUSY} \oplus \mathcal{H}^f_{\rm SUSY} = \begin{cases} \mathbb{C}\oplus 0 & \text{if } \frac{|L_1-L_2|}{2} \leq \left[\frac{M_2-M_1}{2}\right]_{k+2} \leq \min\left\{\frac{L_1+L_2}{2}, k - \frac{L_1+L_2}{2}\right\} \\ 0\oplus\mathbb{C} & \text{if } \frac{|L_1-L_2|}{2} \leq \left[\frac{M_1-M_2}{2} - 1\right]_{k+2} \leq \min\left\{\frac{L_1+L_2}{2}, k - \frac{L_1+L_2}{2}\right\} \\ 0\oplus 0 & \text{otherwise,} \end{cases}$$

where $[c]_{k+2}$ is the smallest non-negative number in $c + (k+2)\mathbb{Z}$. This agrees with the ground state spectrum for the pair $\mathscr{A}_{L_1,M_1}$-$\mathscr{A}_{L_2,M_2}$. For example, compare with (60) for the case $L_1 \geq L_2$ and $k \geq L_1 + L_2$. Thus we find the following identification:

$$\mathscr{A}_{L,M} = \mathscr{B}_{L,M}. \tag{65}$$

What we learn from this is the dimension of the operator $\mathscr{O}$ for the supersymmetric deformations. The identification given in Section 3 of the R-charge of the ground states and corresponding operators goes through without modification also in the orbifold theory. In particular, the dimension of the operator that corresponds to the fermionic supersymmetric ground state (present in the middle case of (64)) is

$$\Delta = \frac{k+1-j_*'}{k+2} \equiv 1 - \frac{M_1 - M_2}{2(k+2)} \quad \mathrm{mod}\ (k+2). \tag{66}$$

It is indeed a relevant operator, $\Delta < 1$.

5.2.2. *RG flows.* Let us consider the sum of two branes $\mathscr{B}_{L_1,M_1} \oplus \mathscr{B}_{L_2,M_2}$. The Chan-Paton factor is given by $\mathbb{C}^2_+ \oplus \mathbb{C}^2_-$, where $\mathbb{C}^2_+$ is spanned by $\{|0\rangle_{L_1}, |0\rangle_{L_2}\}$ and $\mathbb{C}^2_-$ is spanned by $\{\overline{\eta}_1||0\rangle_{L_1}, \overline{\eta}_2|0\rangle_{L_2}\}$ on which the orbifold group acts as

$$\gamma_+ = \begin{pmatrix} \omega^{-\frac{M_1+L_1+1}{2}} & 0 \\ 0 & \omega^{-\frac{M_2+L_2+1}{2}} \end{pmatrix}, \quad \gamma_- = \begin{pmatrix} \omega^{-\frac{M_1-L_1-1}{2}} & 0 \\ 0 & \omega^{-\frac{M_2-L_2-1}{2}} \end{pmatrix},$$

respectively. The maps $f : \mathbb{C}^2_- \to \mathbb{C}^2_+$ and $g : \mathbb{C}^2_+ \to \mathbb{C}^2_-$ are represented by the matrices

$$f(x) = \begin{pmatrix} x^{L_1+1} & 0 \\ 0 & x^{L_2+1} \end{pmatrix}, \quad g(x) = \begin{pmatrix} x^{k+1-L_1} & 0 \\ 0 & x^{k+1-L_2} \end{pmatrix},$$

which are indeed invariant under the orbifold group action, $f(x) \to \gamma_+ f(\omega x)\gamma_-^{-1}$, $g(x) \to \gamma_- g(\omega x)\gamma_+^{-1}$. We would like to identify the supersymmetric deformation given by the relevant operator corresponding to $|j'_*\rangle^f_{L_1L_2}$ when it is present. Let us try the family of configurations given by (50) and (51):

$$f_t(x) = \begin{pmatrix} x^{L_1+1}\cos t & x^{L_2+1-a}\sin t \\ -x^{L_1+1+a}\sin t & x^{L_2+1}\cos t \end{pmatrix}, \tag{67}$$

$$g_t(x) = \begin{pmatrix} x^{k+1-L_1}\cos t & -x^{k+1-L_1-a}\sin t \\ x^{k+1-L_2+a}\sin t & x^{k+1-L_2}\cos t \end{pmatrix} \tag{68}$$

where a is an integer such that

$$\max\{-L_1-1,\, L_2-k-1\} \le a \le \min\{L_2+1,\, k+1-L_1\}. \tag{69}$$

In the present case, we need to make sure that the deformation is invariant under the orbifold group $\mathbb{Z}_{k+2}$ which acts as $f_t(x) \to \gamma_+ f_t(\omega x)\gamma_-^{-1}$, $g_t(x) \to \gamma_- g_t(\omega x)\gamma_+^{-1}$. This requires the condition

$$\omega^a = \omega^{\frac{M_2+L_2-M_1-L_1}{2}}. \tag{70}$$

If j'_* is in the allowed range (20), one can solve (70) for a within (69). Then it is clear that $f_\varepsilon(x), g_\varepsilon(x)$ can be regarded as the perturbation by the operator corresponding to the state $|j'_*\rangle_{L_1L_2}$ (cf. Section 4.3). The perturbed theory flows in the IR limit to the configuration with $t = \frac{\pi}{2}$. After the basis change of $\mathbb{C}^2_+$ by $R_{\frac{\pi}{2}} = \left(\begin{smallmatrix} 0 & -1 \\ 1 & 0 \end{smallmatrix}\right)$, it is given by

$$f'(x) = \begin{pmatrix} x^{L_1+a+1} & 0 \\ 0 & x^{L_2-a+1} \end{pmatrix}, \quad g'(x) = \begin{pmatrix} x^{k-L_1-a+1} & 0 \\ 0 & x^{k-L_2+a+1} \end{pmatrix}$$

$$\gamma'_+ = \begin{pmatrix} \omega^{-\frac{M_2+L_2+1}{2}} & 0 \\ 0 & \omega^{-\frac{M_1+L_1+1}{2}} \end{pmatrix}, \quad \gamma'_- = \gamma_-.$$

This is the configuration for $\mathscr{B}_{L'_1,M'_1} \oplus \mathscr{B}_{L'_2,M'_2}$ where

$$L'_1 = L_1 + a \equiv \tfrac{L_1+L_2-M_1+M_2}{2} \pmod{k+2}, \quad M'_1 = \tfrac{-L_1+L_2+M_1+M_2}{2}, \tag{71}$$

$$L'_2 = L_2 - a \equiv \tfrac{L_1+L_2+M_1-M_2}{2} \pmod{k+2}, \quad M'_2 = \tfrac{L_1-L_2+M_1+M_2}{2}. \tag{72}$$

Thus, we found that the deformation by the operator corresponding to $|j'_*\rangle^f_{L_1L_2}$ generates the RG flow

$$\mathscr{B}_{L_1,M_1} \oplus \mathscr{B}_{L_2,M_2} \longrightarrow \mathscr{B}_{L'_1,M'_1} \oplus \mathscr{B}_{L'_2,M'_2} \tag{73}$$

where the two sets of labels are related by (71) and (72). The decrease of the boundary entropy can be shown in the same way as in Section 4.3. This process is the mirror of the brane-recombination of the A-branes as described in Case (ii) of Section 5.1.1, see Figure 7. This is enough to see that the brane charge is generated by $\mathscr{B}_{0,M}$ with $M = 0, 2, ..., 2(k+1)$ which are related by

$$[\mathscr{B}_{0,0}] + [\mathscr{B}_{0,2}] + \cdots + [\mathscr{B}_{0,2k+2}] = 0.$$

The mirror of Case (iii) where two branes $\mathscr{A}_{L_1,M_1} \oplus \mathscr{A}_{L_2,M_2}$ combine into one $\mathscr{A}_{L,M}$ is described by the flow with $a = L_2 + 1$. The $\mathbb{Z}_{k+2}$-invariance condition (70) in this case is nothing but $\pi\frac{M_1+L_1+1}{k+2} \equiv \pi\frac{M_2-L_2-1}{k+2} \pmod{2\pi}$, which is the condition that the outgoing ray of $\mathscr{A}_{L_1,M_1}$ agrees with the incoming ray of $\mathscr{A}_{L_2,M_2}$.

Acknowledgement

The author thanks J. Walcher for many useful discussions including motivating ones at the initial stage. He also thanks A. Kapustin, C. Lazaroiu and W. Lerche for discussions, and D. Kutasov and S. Schäfer-Nameki for useful information. He thanks the KITP, Santa Barbara for hospitality where some part of this work is done. This work is supported in part by Connaught Foundation, NSERC and the Alfred P. Sloan Foundation.

Note added: K-theory for the D-brane charges in the coset model $\frac{SU(2)_k \times U(1)_2}{U(1)_{k+2}}$ has been computed in [**56**]. The result shows the A-brane charges $\mathbb{Z}^{k+1}$ but there is no sign of the torsion $\mathbb{Z}_{k+2}$ charges. This is not a contradiction - the coset model is different from the model we study, as discussed in Section 3.5. (See [**57**] for a more recent discussion about this point.)

References

[1] A. B. Zamolodchikov and V. A. Fateev, "Disorder fields in two-dimensional conformal quantum field theory and N=2 extended supersymmetry", Sov. Phys. JETP **63** (1986) 913 [Zh. Eksp. Teor. Fiz. **90** (1986) 1553].

[2] Z. Qiu, "Nonlocal current algebra and N=2 superconformal field theory in two dimensions", Phys. Lett. B **188** (1987) 207.

[3] M. Ademollo *et al.*, "Supersymmetric strings and color confinement", Phys. Lett. B **62** (1976) 105; "Dual string with U(1) color symmetry", Nucl. Phys. B **111** (1976) 77.

[4] F. Ravanini and S. K. Yang, "Modular invariance in N=2 superconformal field theories", Phys. Lett. B **195** (1987) 202.

[5] Z. Qiu, "Modular invariant partition functions for N=2 superconformal field theories", Phys. Lett. B **198** (1987) 497.

[6] E. J. Martinec, "Algebraic Geometry And Effective Lagrangians," Phys. Lett. B **217** (1989) 431; "Criticality, Catastrophes And Compactifications," In Brink, L. (ed.) et al.: *Physics and mathematics of strings, V.G. Knizhnik memorial vol.* 389-433.

[7] C. Vafa and N. P. Warner, "Catastrophes And The Classification Of Conformal Theories," Phys. Lett. B **218** (1989) 51.

[8] E. Witten, "On the Landau-Ginzburg description of N=2 minimal models," Int. J. Mod. Phys. A **9** (1994) 4783 [hep-th/9304026].

[9] J. A. Harvey, D. Kutasov and E. J. Martinec, "On the relevance of tachyons," hep-th/0003101.

[10] A. Sen, "Non-BPS states and branes in string theory," hep-th/9904207.

[11] A. Sen and B. Zwiebach, "Tachyon condensation in string field theory," JHEP **0003** (2000) 002 [hep-th/9912249].

[12] A. A. Gerasimov and S. L. Shatashvili, "On exact tachyon potential in open string field theory," JHEP **0010** (2000) 034 [hep-th/0009103];
D. Kutasov, M. Marino and G. W. Moore, "Some exact results on tachyon condensation in string field theory," JHEP **0010** (2000) 045 [hep-th/0009148].

[13] E. Witten, "D-branes and K-theory," JHEP **9812** (1998) 019 [hep-th/9810188].

[14] M. R. Douglas, "D-branes, categories and N = 1 supersymmetry," J. Math. Phys. **42** (2001) 2818 [hep-th/0011017].

[15] J. McGreevy and H. Verlinde, "Strings from tachyons: The c = 1 matrix reloaded," JHEP **0312** (2003) 054 [hep-th/0304224];
I. R. Klebanov, J. Maldacena and N. Seiberg, "D-brane decay in two-dimensional string theory," JHEP **0307** (2003) 045 [hep-th/0305159].

[16] A. Sen, "Rolling tachyon," JHEP **0204** (2002) 048 [hep-th/0203211];
M. Gutperle and A. Strominger, "Spacelike branes," JHEP **0204** (2002) 018 [hep-th/0202210];
N. Lambert, H. Liu and J. Maldacena, "Closed strings from decaying D-branes," hep-th/0303139.

[17] K. Hori, A. Iqbal and C. Vafa, "D-branes and mirror symmetry," hep-th/0005247.

[18] J. M. Maldacena, G. W. Moore and N. Seiberg, "Geometrical interpretation of D-branes in gauged WZW models," JHEP **0107** (2001) 046 [hep-th/0105038].

[19] K. D. Kennaway, "A geometrical construction of Recknagel-Schomerus boundary states in linear sigma models," Nucl. Phys. B **647** (2002) 471 [hep-th/0203266].

[20] S. Fredenhagen and V. Schomerus, "D-branes in coset models," JHEP **0202** (2002) 005 [hep-th/0111189]; "On boundary RG-flows in coset conformal field theories," Phys. Rev. D **67** (2003) 085001 [hep-th/0205011];
S. Fredenhagen, "Organizing boundary RG flows," Nucl. Phys. B **660** (2003) 436 [hep-th/0301229].

[21] K. Hori, "Linear models of supersymmetric D-branes," hep-th/0012179.

[22] K. Hori, S. Katz, A. Klemm, R. Pandharipande, R. Thomas, C. Vafa, R. Vakil and E. Zaslow, *Mirror Symmetry*, (AMS-CMI, 2003).

[23] K. Hori, "Mirror symmetry and quantum geometry," Proceedings ICM, Vol. III (Beijing 2002), 431–443 [hep-th/0207068].

[24] A. Kapustin and Y. Li, "D-branes in Landau-Ginzburg models and algebraic geometry," JHEP **0312** (2003) 005 [hep-th/0210296];
"Topological correlators in Landau-Ginzburg models with boundaries," hep-th/0305136;
"D-branes in topological minimal models: The Landau-Ginzburg approach," hep-th/0306001.

[25] D. Orlov, "Triangulated categories of singularities and D-branes in Landau-Ginzburg models," Tr. Mat. Inst. Steklov **246** (2004), 240–262 [math.ag/0302304].

[26] I. Brunner, M. Herbst, W. Lerche and B. Scheuner, "Landau-Ginzburg realization of open string TFT," hep-th/0305133.

[27] C. I. Lazaroiu, "On the boundary coupling of topological Landau-Ginzburg models," hep-th/0312286.

[28] M. F. Atiyah, *K-theory*, Benjamin, New York, 1967.

[29] A. Cappelli, G. D'Appollonio and M. Zabzine, "Landau-Ginzburg description of boundary critical phenomena in two dimensions," JHEP 2004, no. 4, 010 [hep-th/0312296].

[30] S. A. Ashok, E. Dell'Aquila and D. E. Diaconescu, "Fractional branes in Landau-Ginzburg orbifolds," Adv. Theor. Math. Phys **8** (2004),no. 3, 461–513 [hep-th/0401135].

[31] H. Ooguri, Y. Oz and Z. Yin, "D-branes on Calabi-Yau spaces and their mirrors," Nucl. Phys. B **477** (1996) 407 [hep-th/9606112].

[32] S. Govindarajan, T. Jayaraman and T. Sarkar, "Worldsheet approaches to D-branes on supersymmetric cycles," Nucl. Phys. B **580** (2000) 519 [hep-th/9907131].

[33] N. P. Warner, "Supersymmetry in boundary integrable models," Nucl. Phys. B **450** (1995) 663 [hep-th/9506064].

[34] S. Hellerman, S. Kachru, A. E. Lawrence and J. McGreevy, "Linear sigma models for open strings," JHEP **0207** (2002) 002 [hep-th/0109069].

[35] T. Takayanagi, S. Terashima and T. Uesugi, "Brane-antibrane action from boundary string field theory," JHEP **0103** (2001) 019 [hep-th/0012210].

[36] P. Kraus and F. Larsen, "Boundary string field theory of the $D\bar{D}$ system," Phys. Rev. D **63** (2001) 106004 [hep-th/0012198].

[37] K. Hori and C. Vafa, "Mirror Symmetry," hep-th/0002222.

[38] C.-H. Cho and Y.-G. Oh, "Floer cohomology and discs instantons of Lagrangian torus fibers in Fano toric manifolds," Int. Math. Res. Not. 2004, no. 35, 1803–1843 [math.SG/0308225].

[39] C.-H. Cho, "Holomorphic disc, spin structures and Floer cohomology of the Clifford torus," math.SG/0308224.

[40] W. Lerche, C. Vafa and N. P. Warner, "Chiral Rings In N=2 Superconformal Theories," Nucl. Phys. B **324** (1989) 427.

[41] I. Brunner and K. Hori, "Orientifolds and mirror symmetry," JHEP 2004, no. 11, 005 [hep-th/0303135].

[42] J. L. Cardy, "Boundary Conditions, Fusion Rules And The Verlinde Formula," Nucl. Phys. B **324** (1989) 581.

[43] S. E. Parkhomenko, "Free field construction of D-branes in N = 2 superconformal minimal models," Nucl. Phys. B **671** (2003) 325 [hep-th/0301070].

[44] I. Affleck and A. W. W. Ludwig, "Universal Noninteger 'Ground State Degeneracy' In Critical Quantum Systems," Phys. Rev. Lett. **67** (1991) 161.

[45] I. Brunner, K. Hori, K. Hosomichi and J. Walcher, "Orientifolds of Gepner Models", hep-th/0401137.

[46] C. Vafa, "String Vacua And Orbifoldized L-G-Models," Mod. Phys. Lett. A **4** (1989) 1169.

[47] A. Sen, "Tachyon condensation on the brane antibrane system," JHEP **9808** (1998) 012 [hep-th/9805170].
[48] C. G. Callan, I. R. Klebanov, A. W. W. Ludwig and J. M. Maldacena, "Exact solution of a boundary conformal field theory," Nucl. Phys. B **422** (1994) 417 [hep-th/9402113].
[49] J. Polchinski and L. Thorlacius, "Free fermion representation of a boundary conformal field theory," Phys. Rev. D **50** (1994) 622 [hep-th/9404008].
[50] P. Fendley, H. Saleur and N. P. Warner, "Exact solution of a massless scalar field with a relevant boundary interaction," Nucl. Phys. B **430** (1994) 577 [hep-th/9406125].
[51] I. Affleck and A. W. W. Ludwig, "Exact conformal-field-theory results on the multichannel Kondo effect: Single-fermion Green's function, self-energy, and resisitivity," Phys. Rev. B **48** (1993) 7297.
[52] D. Kutasov, M. Marino and G. W. Moore, "Remarks on tachyon condensation in superstring field theory," hep-th/0010108.
[53] M. Marino, "On the BV formulation of boundary superstring field theory," JHEP **0106** (2001) 059 [hep-th/0103089].
[54] V. Niarchos and N. Prezas, "Boundary superstring field theory," Nucl. Phys. B **619** (2001) 51 [hep-th/0103102].
[55] D. Kutasov, "New results on the 'a-theorem' in four dimensional supersymmetric field theory," hep-th/0312098.
[56] S. Schäfer-Nameki, "D-branes in N = 2 coset models and twisted equivariant K-theory," hep-th/0308058.
[57] S. Schäfer-Nameki, "K-theoretical boundary rings in N = 2 coset models," Nucl. Phys. B **706** (2005) 531 [hep-th/0408060].

Department of Physics, University of Toronto, 60 St. George Street, Toronto, Ontario, M5S 1A7, Canada
E-mail address: hori@physics.utoronto.ca

AMS/IP Studies in Advanced Mathematics
Volume 38, 2006

Central Charges, Symplectic Forms, and Hypergeometric Series in Local Mirror Symmetry

Shinobu Hosono

ABSTRACT. We study a cohomology-valued hypergeometric series which naturally arises in the description of (local) mirror symmetry. We identify it as a central charge formula for BPS states and study its property from the viewpoint of Kontsevich's homological mirror symmetry. In the case of local mirror symmetry, we will identify a symplectic form, and will conjecture an integral and symplectic monodromy property of a relevant hypergeometric series of Gel'fand–Kapranov–Zelevinski type.

Table of Contents

2000 *Mathematics Subject Classification.* Primary 14J32; Secondary 14D05, 14D07, 32Q20, 32Q25.

Key words and phrases. Calabi–Yau manifolds, mirror symmetry, local mirror symmetry, homological mirror symmetry, monodromy, variation of Hodge structures, GKZ hypergeometric systems, BPS states.

1. Introduction

Let us consider a (famous) hypergeometric series of one variable $[1,x] \in \mathbf{P}^1$

$$w(x) = \sum_{n\geq 0} \frac{(5n)!}{(n!)^5} x^n \tag{1.1}$$

This is a hypergeometric series of type ${}_4F_3(\frac{1}{5},\frac{2}{5},\frac{3}{5},\frac{4}{5};1,1,1;x)$ which arises in the mirror symmetry of the quintic hypersurface $X_5 \subset \mathbf{P}^4$. This hypergeometric series (1.1) represents one of the period integrals of the mirror quintic $X_5^\vee$ and satisfies the following differential equation (Picard–Fuchs) equation:

$$\{\theta_x^4 - 5^5 x(\theta_x + \frac{4}{5})(\theta_x+\frac{3}{5})(\theta_x+\frac{2}{5})(\theta_x+\frac{1}{5})\}w(x) = 0\ , \tag{1.2}$$

where $\theta_x := x\frac{d}{dx}$. See the original work by Candelas et al [**CdOGP**] for the description of the mirror family and its period integrals.

As is clear in the form of differential equation, the regular singularity at $x=0$ has a distinguished property, i.e., the monodromy around this point is *maximally unipotent* [**Mor**]. In physics, the point $x=0$ is called a *large complex structure limit* and plays important roles, e.g., near this point, we evaluate the quantum corrections to the classical geometry of the σ-model on the quintic X_5. Here we focus on the construction of local solutions about $x=0$ by the classical Frobenius method

$$w_0(x) := w(x)\ ,\ w_1(x) := \frac{\partial}{\partial\rho} w(x;\rho)|_{\rho=0}\ ,$$
$$w_2(x) := \frac{\partial^2}{\partial\rho^2} w(x;\rho)|_{\rho=0}\ ,\ w_3(x) := \frac{\partial^3}{\partial\rho^3} w(x;\rho)|_{\rho=0}\ ,$$

where $w(x;\rho) := \sum_{n\geq 0}\frac{\Gamma(1+5(n+\rho))}{\Gamma(1+(n+\rho))^5}x^{n+\rho}$. With the mirror symmetry of X_5 in $\mathbf{P}^4$ and $X_5^\vee$ in mind, we introduce the following cohomology-valued hypergeometric series;

$$w(x;\frac{J}{2\pi i}) := w(x) + w_1(x)(\frac{J}{2\pi i}) + w_2(x)(\frac{J}{2\pi i})^2 + w_3(x)(\frac{J}{2\pi i})^3\ , \tag{1.3}$$

where J is the ample, integral generator of $\mathrm{Pic}(X_5) = H^{1,1}(X_5)\cap H^2(X_5,\mathbf{Z})$. In this form, we note that the classical Frobenius method is concisely summarized as the Taylor expansion $w(x;\rho)|_{\frac{J}{2\pi i}}$ with respect to the nilpotent element J. Although this seems just an advantage in bookkeeping, the following observation reported in [**Hos**] indicates that we have *more than* that in (1.3):

Observation: *Arrange the Taylor series expansion of the cohomology-valued hypergeometric series* $w(x;\frac{J}{2\pi i})$ *as*

(1.4)

$$w(x;\frac{J}{2\pi i}) = w^{(0)}(x) + w^{(1)}(x)(J - \frac{c_2(X_5)J}{12} - \frac{11}{2}\frac{J^2}{5}) + w^{(2)}(x)\frac{J^2}{5} + w^{(3)}(x)(-\frac{J^3}{5}).$$

Then the monodromy matrices of the coefficient hypergeometric series $w^{(0)}(x)$, $w^{(1)}(x)$, $w^{(2)}(x)$, $w^{(3)}(x)$ *are integral and symplectic (with respect to the symplectic form of the mirror, see below).*

The integral and symplectic properties of the solutions $w^{(k)}(x)$ $(k=0,1,2,3)$ stem from those of the middle homology group of the mirror $H_3(X_5^\vee,\mathbf{Z})$. The point here is that we can recover these integral, symplectic properties from the series $w(x;\frac{J}{2\pi i})$ through suitably arranging a basis of $H^{even}(X,\mathbf{Q})$ near the large complex structure limit. Since there is a symplectic, integral structure on $H^{even}(X, \mathbf{Q})$ which comes from $K(X)$, the Grothendieck group of algebraic vector bundles on X, it is natural to conjecture (Conjecture 2.2) that the cohomology-valued hypergeometric series describes a "pairing" between $K(X)$ and $H_3(X_5^\vee, \mathbf{Z})$, as proposed in [**Hos**].

The aims (and main results) of this paper are : (1) to make an interpretation of the cohomology-valued hypergeometric series from the viewpoint of homological mirror symmetry, and to give a general definition of the so-called central charge of BPS states (**Conjecture 2.2**, **Conjecture 6.3**), and (2) to present supporting evidence for the interpretation in the cases arising from local mirror symmetry in dimensions two and three (Section 5, Section 6).

The idea of local mirror symmetry is simply to focus on the Calabi–Yau geometry near rigid curves or (Fano) surfaces in a Calabi–Yau manifold [**CKYZ**]. Recently this idea has been providing us useful testing grounds for several duality symmetries proposed in string theory. In particular, when the relevant Calabi–Yau geometry is toric, remarkable progresses have been made (see [**AKMV**] and references therein). In this paper, for the local Calabi–Yau geometries, we will consider the crepant resolutions $\widehat{\mathbf{C}^2/\mathbf{Z}_{\mu+1}}$ of the two-dimensional canonical singularity and also $\widehat{\mathbf{C}^3/G}$ with finite abelian group $G \subset SU(3, \mathbf{C})$. Although we restrict our attention to these cases, our arguments should extend to more general local Calabi–Yau geometries.

When studying the relevant Gel'fand–Kapranov–Zelevinski (GKZ) hypergeometric series, we will find its connection to the primitive forms introduced by K. Saito in the deformation theory of singularities (**Proposition 4.2**, **Proposition 4.4**). This seems to be interesting in its own right, since GKZ hypergeometric series may provide a simple way to express the 'period integrals' of the primitive forms (or oscillating integrals) in the theory of singularities.

The construction of this paper is as follows: In section 2, we give a definition of the central charge assuming Kontsevich's homological mirror symmetry conjecture. Following the observations made in [**Hos**], we interpret the cohomology-valued hypergeometric series as the central charge for the cases of X compact. In section 3, we consider the local mirror symmetry of $X = \widehat{\mathbf{C}^2/\mathbf{Z}_{\mu+1}}$ and study in detail the GKZ hypergeometric series in this case. In section 4, we connect the GKZ system to K. Saito's system of differential equations which describes the deformation of singularities. In section 5, we extract a (homological) mirror map $mir : K^c(X) \tilde{\rightarrow} H_2(Y, \mathbf{Z})$ from our interpretation of the cohomology-valued hypergeometric series, and find a consistency with the mirror symmetry in terms of the hyperkähler rotation. In section 6, we formulate our conjecture on cohomology-valued hypergeometric series for three-dimensional cases $X = \widehat{\mathbf{C}^3/G}$. We also find a closed formula for the prepotential. Two explicit examples are presented to support our conjecture. In Appendices A and B, we evaluate period integrals over vanishing cycles for the mirror geometry of $X = \widehat{\mathbf{C}^3/\mathbf{Z}_3}$. There, we see that a certain non-compact vanishing cycle arises in the description of the isomorphism $mir : K^c(X) \tilde{\rightarrow} H_3(Y, \mathbf{Z})$. The relation to the period integral of a primitive form is also elucidated.

Acknowledgments: The author would like to thank T. Terasoma for discussions on the GKZ hypergeometric system. He also would like to thank K. Oguiso and S.-T.Yau for valuable comments on this work.

2. Central charge formula in terms of $w(x; \frac{J}{2\pi i})$

Here, we will interpret the cohomology-valued hypergeometric series from the viewpoint of Kontsevich's homological mirror symmetry [**Ko**].

Let X be a Calabi–Yau 3 fold and Y be a mirror of X. Following Kontsevich, we consider the bounded derived category $D^b(Coh(X))$ of coherent sheaves (D-branes of B type) on X. On the other hand, for the mirror side, we consider the derived Fukaya category $DFuk(Y, \beta)$ with the Kähler form viewed as a symplectic form β. The objects of the latter category consist of (graded) Lagrangian submanifolds with a flat $U(1)$ bundle on each of them (D-branes of type A) and morphisms are given by the Floer homology for Lagrangian submanifolds, and this category forms a triangulated category (see [**FO3**] for a more precise definition). Kontsevich proposed that these two different categories are equivalent (as triangulated categories) when X and Y are mirror symmetric, and also that this should be a mathematical definition of mirror symmetry. This conjecture itself is of great interest; however let us consider this conjecture at a more tractable level, i.e. at the level of cohomology or K-groups as shown in the second line below;

$$\begin{array}{ccc} D^b(Coh(X)) & \xrightarrow{Mir} & DFuk(Y,\beta) \\ \downarrow & & \downarrow \\ K(X) \text{ or } H^{even}(X,\mathbf{Q}) & \xrightarrow{mir} & H_3(Y,\mathbf{Z}) \end{array} \tag{2.1}$$

where the left vertical arrow represents the natural map from $D^b(Coh(X))$ to the K-group of algebraic vector bundles $K(X)$, and its composition with the Chern character homomorphism $ch(\cdot)$ if we further map to $H^{even}(X, \mathbf{Q}) = \oplus_{p=0}^{3} H^{2p}(X, \mathbf{Q})$. The right vertical arrow is given simply by taking the homology classes of the graded Lagrangian cycles. In the second line, the equivalence, Mir, of the two categories becomes simply an isomorphism, mir, between the K-group and $H_3(Y, \mathbf{Z})$. We should note that this is not simply an isomorphism but an isomorphism with the symplectic structures, i.e.

$$mir: \ (K(X), \chi(E,F)) \ \xrightarrow{\sim} \ (H_3(Y,\mathbf{Z}), \#(L_E \cap L_F)),$$

where $\chi(E,F) = \int_X ch(E^{\vee}) ch(F) Todd_X$ and $\#(L_E \cap L_F) := \int_Y \mu_{L_E} \cup \mu_{L_F}$ with the Poincaré duals $\mu_{L_E}, \mu_{L_F} \in H^3(Y, \mathbf{Z})$ of the mirror homology cycles $L_E := mir(E), L_F := mir(F)$. Here we remark that the Euler number $\chi(E,F)$ is anti-symmetric due to Serre duality and $K_X = 0$, and also non-degenerate. Thus $\chi(E,F)$ introduces a symplectic structure on $K(X)$, which is the mirror of the symplectic structure on $H_3(Y, \mathbf{Z})$.

In the diagram (2.1), we assume that a complex structure is fixed to define $D^b(Coh(X))$. Correspondingly the symplectic form (Kähler form) β, determined by the mirror map, is fixed in the right hand side. On the other hand, we may change the (complexified) Kähler class of X which corresponds to the complex structure moduli of Y under the mirror map. Changing the (complexified) Kähler structure amounts to changing the polarization and thus results in varying the stability condition on the sheaves on X. This change of the stability (Π-stability) condition has been studied in [**Do**] as a stability of BPS D-branes and its mathematical aspects are elaborated in [**Br**]. Here, we will not go into the detailed definition of Π-stability, but we will propose a closed formula for the *central charge* which is indispensable for an explicit description of Π-stability.

DEFINITION 2.1. **(Central charge formula.)** Assume $K(X)$ is torsion-free, and let $E_1, \cdots, E_r$ be a $\mathbf{Z}$-basis of $K(X)$. Let $\Omega(Y_x)$ be a holomorphic 3-form of the mirror family $\{Y_x\}_{x \in \mathcal{B}}$ of X with $x = (x_1, \cdots, x_r)$ being the local parameter near a large complex structure limit. Denote by $\mathbf{C}\{x\}[\log x]$ the polynomial ring of convergent power series in the variables $\log x_1, \cdots, \log x_r$ over the ring $\mathbf{C}\{x\}$. Under the mirror symmetry (2.1), we define the following $\mathcal{Z}_x$ as an element in $K(X) \otimes \mathbf{C}\{x\}[\log x]$:

$$\mathcal{Z}_x := \sum_{i,j} \int_{mir(E_i)} \Omega(Y_x)\, \chi^{ij}\, E_j^{\vee} \tag{2.2}$$

with $(\chi^{ij}) := (\chi(E_i, E_j))^{-1}$. Then the central charge of $F \in K(X)$ is defined by

$$Z_t(F) = \int_X ch(F)\, ch(\mathcal{Z}_x)\, Todd_X \quad , \tag{2.3}$$

where $t = t(x)$ (or $x = x(t)$) is the mirror map connecting the local parameters $x_1, \cdots, x_r$ to those $t_1, \cdots, t_r$ of the (complexified) Kähler moduli space of X.

In the above definition, it should be noted that $\mathcal{Z}_x$ does not depend on the choice of a basis $E_1, \cdots, E_r$. Also the central charge $Z_t(F)$ contains full 'quantum corrections' as a function of $t_1, \cdots, t_r$ (cf. the asymptotic formula given in [**Do**]).

Let us connect our hypergeometric series $w(x; \frac{J}{2\pi i})$ to the central charge above. Before doing this, we remark that, in the mirror symmetry of Calabi–Yau hypersurfaces due to Batyrev[**Ba1**], the hypergeometric series (1.1) is naturally generalized to the Gel'fand–Kapranov–Zelevinski (GKZ) hypergeometric series of multi-variables $x_1, \cdots, x_r$ [**GKZ1**],[**Ba2**]. Using the GKZ hypergeometric series, and also suitable integral, (semi-)ample generators $J_1, \cdots, J_r$ of $H^{1,1}(X) \cap H^2(X, \mathbf{Z})$, we have a cohomology-valued hypergeometric series $w(x; \frac{J}{2\pi i})$ as a generalization of (1.3) (see Sect.2 of [**Hos**] for more details).

CONJECTURE 2.2. *The cohomology-valued hypergeometric series (1.3) gives rise to an expression for* $ch(\mathcal{Z}_x) \in H^{even}(X) \otimes \mathbf{C}\{x\}[\log x]$*;*

$$w(x_1, \cdots, x_r; \frac{J_1}{2\pi i}, \cdots, \frac{J_r}{2\pi i}) = \sum_{i,j} \int_{mir(E_i)} \Omega(Y_x)\, \chi^{ij}\, ch(E_j^{\vee})\,. \tag{2.4}$$

Using this, and also the mirror map $t = t(x)$*, we can write the central charge* $Z_t(F)$ *of* $F \in K(X)$ *as*

$$Z_t(F) = \int_X ch(F)\, w(x; \frac{J}{2\pi i})\, Todd_X \quad . \tag{2.5}$$

Here we note that the hypergeometric series has a finite radius of convergence and shows a monodromy property when it is analytically continued around its (regular) singularities. As noticed by Kontsevich, this monodromy property should be mirrored to some linear (symplectic) transformations on $ch(E_i)$ which come from Fourier–Mukai transforms on $D^b(Coh(X))$. If we postulate that the cohomology-valued hypergeometric series has an invariant meaning under these monodromy actions, our cohomology-valued hypergeometric series $w(x; \frac{J}{2\pi i})$ provides a connection between these two different 'monodromy' transforms on the two sides. The conjectural formula (2.4) has been tested in the case X is an elliptic curve, (lattice polarized) K3 surfaces, and several Calabi–Yau hypersurfaces in [**Hos**]. Cohomology-valued hypergeometric series are utilized also in [**Gi**],[**LLY**],[**Sti**] in a slightly different form. We remark that, for our Conjecture 2.2 to work, the definition given in [**Hos**] is crucial.

As studied in [**Mu**] for the cases of K3 surfaces and abelian varieties, and in [**Or**] for general, the Fourier–Mukai transform is a self-equivalence of the category $D^b(Coh(X))$ which takes the form

$$\Phi^{\mathcal{P}}(\cdot) = \mathbf{R}p_{2*}(p_1^*(\cdot) \overset{\mathbf{L}}{\otimes} \mathcal{P})$$

where $\mathcal{P}$ is an object in $D^b(Coh(X \times X))$, called the *kernel*, and p_1 and p_2 are, respectively, the natural projections to the first and the second factor from $X \times X$. Due to a result in [**Or**], we may always assume the above form, i.e., there exists a suitable kernel $\mathcal{P}$, for any equivalence $\Phi : D^b(Coh(X)) \simeq D^b(Coh(X))$ as triangulated category. It is rather easy to see that the monodromy transforms around the large complex structure limit are given by tensoring invertible sheaves, which may be expressed by the kernels;

$$\mathcal{P} : \cdots \to 0 \to \mathcal{O}_\Delta \times p_2^*(\mathcal{O}_X(D)) \to 0 \cdots ,$$

with $D \in Pic(X)$ and Δ representing the diagonal in $X \times X$. Kontsevich predicted that a monodromy transform associated to a vanishing cycle, the Picard–Lefschetz transform, has its mirror FM transform with its kernel,

$$\mathcal{P} : \cdots \to 0 \to \mathcal{O}_{X \times X} \to \mathcal{O}_X \to 0 \cdots .$$

Seidel and Thomas [**ST**] (and Horja [**Hor**]) generalized the above kernel, associating it to so-called spherical objects $\mathcal{E} \in D^b(Coh(X))$ with the defining property: $Ext^i(\mathcal{E}, \mathcal{E}) = 0$ $(i \neq 0, n)$, $= \mathbf{C}$ $(i = 0, n)$ where $n = \dim X$. For each spherical object, we have a kernel given by the mapping cone;

$$\mathcal{P} = Cone(\mathcal{E}^\vee \otimes^{\mathbf{L}} \mathcal{E} \to \mathcal{O}_\Delta) .$$

The equivalence $\Phi^{\mathcal{P}}$ is called a Seidel-Thomas twist. We will see these equivalences in the corresponding monodromy property of our hypergeometric series.

3. Local mirror symmetry I — $X = \widehat{\mathbf{C}^2/\mathbf{Z}_{\mu+1}}$

In this section and the subsequent sections, we will test our Conjecture 2.2 for the case of mirror symmetry of non-compact toric Calabi–Yau manifolds (local mirror symmetry). Batyrev's idea of mirror symmetry [**Ba1**] still makes sense for such non-compact toric Calabi–Yau manifolds as well as the attractive proposal by Strominger–Yau–Zaslow (SYZ) [**SYZ**], which is closely related to the homological mirror symmetry (2.1). Mirror symmetry of non-compact toric Calabi–Yau manifolds and also Fano varieties are formulated in the language of Landau–Ginzburg theory in [**HIV**].

(3-1) Mirror symmetry and hyperkähler rotation. Let us consider the minimal resolution of a two-dimensional simple singularity; $X = \widehat{\mathbf{C}^2/\mathbf{Z}_{\mu+1}}$. This is an example of a two-dimensional, non-compact, toric Calabi–Yau manifold. Two-dimensional Calabi–Yau manifolds are hyperkähler, and it is known that their mirror symmetry is well-described by the hyperkähler rotation, see e.g. [**GW**],[**Huy**].

Our minimal resolution X has a natural hyperkähler structure, and therefore its mirror is X itself with a different complex structure after a suitable rotation. To describe the mirror symmetry, let us first write the quotient $\mathbf{C}/\mathbf{Z}_{\mu+1}$ by a hypersurface $UV = W^{\mu+1}$ in $\mathbf{C}^3$. Blowing up the singularities at the origin μ times results in the minimal resolution X, and thereby we introduce exceptional curves $C_i \cong \mathbf{P}^1$ $(i = 1, \cdots, \mu)$. On the other hand, we may deform the defining equation $UV = W^{\mu+1}$ to $UV = a_1 + a_2 W + \cdots + a_{\mu+2} W^{\mu+1}$ with the introduction of finite sizes to vanishing cycles $L_i \approx S^2$ $(\mu = 1, \cdots, \mu)$. Note that the number of vanishing cycles is given by the Milnor number $\mu = \dim R_J$, where R_J is the Jacobian ring of the singularity $UV = W^{\mu+1}$. The vanishing cycles are Lagrangians, and become holomorphic cycles under a suitable hyperkähler rotation. The holomorphic geometry after the rotation is bi-holomorphic to the blown-up geometry of X. If we forget about the role of the B-fields, this describes the mirror symmetry of X. (See e.g. [**Huy**] and references therein for full details of the mirror symmetry via the hyperkähler rotation.) Here we note that the intersection forms of these vanishing cycles are given in both holomorphic and Lagrangian geometry by

$$(C_i \cdot C_j) = (\# L_i \cap L_j) = -\mathcal{C}_{ij} \ ,$$

where $\mathcal{C}_{ij}$ $(1 \le i, j \le \mu)$ is the Cartan matrix for the root system of A_μ.

(3-2) GKZ hypergeometric series. The minimal resolution $X = \widehat{\mathbf{C}^2/\mathbf{Z}_{\mu+1}}$ is a (non-compact) toric variety whose resolution is described by a two-dimensional fan Σ with its integral generators ν_i for one-dimensional cones (see Fig.1). We set

$$A = \{ \nu_1, \nu_2, \cdots, \nu_{\mu+2} \} = \{ (1,0), (1,1), \cdots, (1, \mu+1) \} \ .$$

The half-lines $\overline{o\nu_i}$ $(i = 1, \cdots, \mu+2)$ from the origin $o = (0,0)$ constitute the one-dimensional cones of the resolution diagram Σ. In Batyrev's mirror symmetry, the resolution diagram of X, up to flop operations, is identified with the Newton polytope of the defining equation of its mirror Y. This construction has been extended to the cases of non-compact toric Calabi–Yau manifolds in [**CKYZ**],[**HIV**], and for $X = \widehat{\mathbf{C}^2/\mathbf{Z}_{\mu+1}}$, the mirror Y is proposed to be a hypersurface $U^2 + V^2 + f_\Sigma(W) = 0 \subset \mathbf{C}^2 \times \mathbf{C}^*$ with

$$f_\Sigma(W) = a_1 + a_2 W^1 + a_3 W^2 + \cdots + a_{\mu+2} W^{\mu+1} \ ,$$

where $(U, V, W) \in \mathbf{C}^2 \times \mathbf{C}^*$. In the case of local mirror symmetry, the meaning of the period integrals of the holomorphic two-form becomes less clear than the compact case. Here, motivated by [**CKYZ**], we consider the following integral

$$\Pi_\gamma(a) := \frac{i}{2\pi^2} \int_\gamma \frac{1}{U^2 + V^2 + f_\Sigma(a; W)} dU dV \frac{dW}{W} \ , \tag{3.1}$$

for each cycle γ in the complement of the hypersurface

$$\gamma \in H_3(\mathbf{C}^2 \times \mathbf{C}^* \setminus Y, \mathbf{Z}).$$

Note that we have an isomorphism $H_3(\mathbf{C}^2 \times \mathbf{C}^* \setminus Y, \mathbf{Z}) \simeq H_2(Y, \mathbf{Z})$ for the hypersurface $Y \subset \mathbf{C}^2 \times \mathbf{C}^*$ (see [**Di**], page 46, for example).

In the next section, we will relate this 'period integral' to K. Saito's primitive form (or Gel'fand–Leray form) in the deformation theory of singularities [**Sa**],[**AGV**]. Here we set up a system of hypergeometric differential equations (GKZ system) satisfied by $\Pi_\gamma(a)$. This GKZ hypergeometric system is also referred to as the A-hypergeometric system since it is described by the set A through the *lattice of relations*,

$$\mathcal{L} = \{ (l_1, l_2, \cdots, l_{\mu+2}) \in \mathbf{Z}^{\mu+2} \,|\, l_1\nu_1 + l_2\nu_2 + \cdots + l_{\mu+2}\nu_{\mu+2} = (0,0) \}. \tag{3.2}$$

Using this lattice of relations, the system is written as

$$\Box_l \Pi_\gamma(a) = 0 \ (l \in \mathcal{L}) \ , \ \ \mathcal{Z}_i \Pi_\gamma(a) = 0 \ (i = 1, 2) \ , \tag{3.3}$$

where

$$\Box_l = \left(\frac{\partial}{\partial a}\right)^{l^+} - \left(\frac{\partial}{\partial a}\right)^{l^-} \ , \ \ \begin{pmatrix} \mathcal{Z}_1 \\ \mathcal{Z}_2 \end{pmatrix} = \begin{pmatrix} \theta_1 + \theta_2 + \cdots + \theta_{\mu+2} \\ \theta_2 + 2\theta_3 \cdots + (\mu+1)\theta_{\mu+2} \end{pmatrix}$$

with $l = l^+ - l^-$ ($l_i^\pm := \frac{1}{2}(l_i \pm |l_i|)$ and $\theta_k = a_k \frac{\partial}{\partial a_k}$. From the formal solutions of this system in [**GKZ1**], it is straightforward to write down our $w(x; \frac{J}{2\pi i})$ which generates the period integrals $\Pi_\gamma(a)$ with suitable cycles (see (3-3) below). An important aspect of this system is that there is a natural toric compactification $\mathcal{M}_{Sec(\Sigma)}$ of the parameter space $\{(a_1, \cdots, a_{\mu+2}) \in (\mathbf{C}^*)^{\mu+2}/(\mathbf{C}^*)^2\}$ in terms of the *secondary fan* $Sec(\Sigma)$ (see [**GKZ2**] and references therein), where the quotient by $(\mathbf{C}^*)^2$ is represented by the (scaling) operators $\mathcal{Z}_1$ and $\mathcal{Z}_2$. This compactification plays an important role in the applications of mirror symmetry to Gromov–Witten invariants, since the large radius limit appears as an intersection point of the boundary toric divisors. Connecting the GKZ system to K. Saito's differential equations in singularity theory, we will see in section 4 that this compactification will also provide a natural way to compactify the deformation space of singularities which is local in nature (see (4-1) for a brief description of the deformation space).

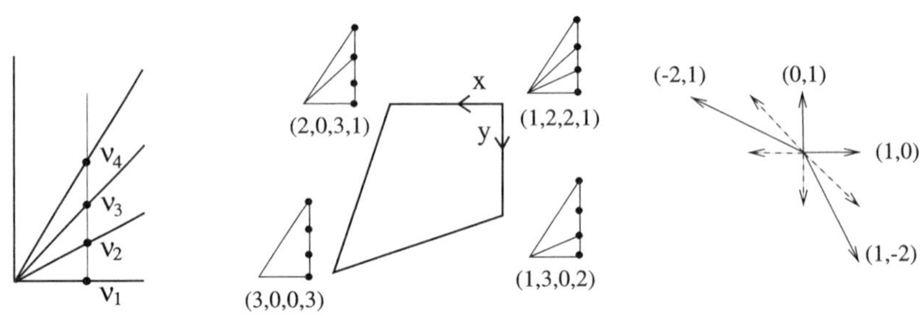

Fig.1. *The resolution diagram (left), the secondary polytope (middle), and the secondary fan $Sec(\Sigma)$ (right) for $\mu = 2$. The secondary polytope has its vertices parametrized by (regular) triangulations of the polytope as shown. For each triangulation T, the corresponding vertex is determined by a vector $v_T = (\varphi_T(\nu_1), \varphi_T(\nu_2), \varphi_T(\nu_3), \varphi_T(\nu_4))$ with $\varphi_T(\nu_i) = \sum_{\nu_i \prec \sigma} Vol(\sigma)$. As we see in this example ($\mu = 2$), the convex hull of these vertices lies on $\mathcal{L}_{\mathbf{R}} + v_T$ with a vector v_T, where $\mathcal{L}_{\mathbf{R}} = \mathcal{L} \otimes \mathbf{R}$. Normal cones of the secondary polytope determine the secondary fan.*

(3-3) Example ($\mu = 2$). Here we present some explicit calculations of cohomology-valued hypergeometric series for the case $\mu = 2$, instead of giving general formulas valid for any μ. In this case, the cohomology-valued hypergeometric series is simple and takes the following form

$$w(x; y; \frac{J_1}{2\pi i}, \frac{J_2}{2\pi i}) = w(x; y; \rho_1, \rho_2)\big|_{\rho_1 = \frac{J_1}{2\pi i}, \rho_2 = \frac{J_2}{2\pi i}} \ ,$$

where $w(x;y;\rho_1,\rho_2)=\sum_{n,m\geq 0} c(n+\rho_1,m+\rho_2)x^{n+\rho_1}y^{m+\rho_2}$ with

$$c(n,m)=1/\big(\Gamma(1+n)\Gamma(1-2n+m)\Gamma(1+n-2m)\Gamma(1+m)\big)\ .$$

J_1 and J_2 are semi-ample classes which are dual to the exceptional curves C_1 and C_2 (i.e. the toric divisors D_{ν_2} and D_{ν_3}), respectively. The local parameters $x:=\frac{a_1a_3}{a_2^2}, y:=\frac{a_2a_4}{a_3^2}$ are depicted in Fig.1. Here we remark that the secondary polytope in Fig.1 sits on a translation of $\mathcal{L}_{\mathbf{R}}$, i.e. $\mathcal{L}_{\mathbf{R}}+v_T$, and the summation in $\sum_{n,m\geq 0} c(m,m)x^m y^n$ is in fact that over the integral points inside (the dual of) the normal cone from the vertex $v_T=(1,2,2,1)$. We may recognize this fact in the relation $x^ny^m=a^{nl^{(1)}+ml^{(2)}}$, where by definition $l^{(1)}=(1,-2,1,0)$ and $l^{(2)}=(0,1,-2,1)$ generate (the dual of) the normal cone form the vertex v_T. We remark also that the coefficients $c(n,m)$ are determined from the entries of $nl^{(1)}+ml^{(2)}$;

$$(n\ m)\begin{pmatrix}1 & -2 & 1 & 0\\ 0 & 1 & -2 & 1\end{pmatrix}=\begin{pmatrix}n & -2n & n & 0\\ 0 & m & -2m & m\end{pmatrix}\ ,$$

where the column vectors determine the arguments of the gamma functions in the denominator of $c(n,m)$. In this form, it is rather clear how the formula generalizes to arbitrary μ (see [**CKYZ**],[**dOFS**] for example).

Since the ring $H^{even}(X,\mathbf{Z})$ is generated by $1,J_1,J_2$, we have the expansion

$$w(\vec{x},\frac{\vec{J}}{2\pi i})=1+w_1(x,y)J_1+w_2(x,y)J_2\ ,$$

with $w_1(x,y)=\frac{1}{2\pi i}\log x+\cdots$, $w_2(x,y)=\frac{1}{2\pi i}\log y+\cdots$, from which we read off the period integral (3.1) as a solution of the GKZ system. The mirror map is defined from the following relations:

$$q_1:=e^{2\pi i w_1(x,y)}=x(1+g_1(x,y))\ ,\quad q_2:=e^{2\pi i w_2(x,y)}=y(1+g_2(x,y))\ , \tag{3.4}$$

where $g_1(x,y),g_2(x,y)$ represent power series in x and y. Then

$$t_1:=\frac{1}{2\pi i}\log q_1(=w_1(x,y))\quad\text{and}\quad t_2:=\frac{1}{2\pi i}\log q_2(=w_2(x,y))$$

are the complexified Kähler moduli and measure the volumes of the exceptional curves C_1, and C_2, respectively. The inverse relation $x=x(q_1,q_2),y=y(q_1,q_2)$ of (3.4) is often referred to as the mirror map, and has the following properties (see Proposition 4.4 for a proof):

PROPOSITION 3.1.

(1) *The mirror map* $x=x(q_1,q_2),y=y(q_1,q_2)$ *is rational and is expressed by* $x=\frac{a_1a_3}{a_2^2},y=\frac{a_2a_4}{a_3^2}$ *with* a_i*'s determined through*

$$a_1+a_2W+a_3W^2+a_4W^3=(1+W)(1+q_1W)(1+q_1q_2W)\ . \tag{3.5}$$

Concretely, it has the form

$$x=\frac{q_1(1+q_2+q_1q_2)}{(1+q_1+q_1q_2)^2}\ ,\ y=\frac{q_2(1+q_1+q_1q_2)}{(1+q_1+q_1q_2)^2}\ . \tag{3.6}$$

(2) *The discriminant of the GKZ system (3.3) consists of three components:* $x=0,y=0$ *and* $dis(x,y)=0$ *with*

$$dis(x,y)=1-4x-4y+18xy-27x^2y^2=\frac{(1-q_1)^2(1-q_2)^2(1-q_1q_2)^2}{(1+q_1+q_1q_2)^2(1+q_2+q_1q_2)^2}\ .$$

It is easy see that $x = 0, y = 0$ are the toric boundary divisors whose intersection point defines the large complex structure. Over the zeros of the discriminant $dis(x, y)$, we see vanishing cycles in $f_\Sigma(a, W) + U^2 + V^2 = 0 \subset \mathbf{C}^2 \times \mathbf{C}^*$. In fact, in the holomorphic picture, $q_1 = 1 (q_2 = 1)$ represents a vanishing volume limit of the exceptional curve $C_1(C_2)$. After a hyperkähler rotation, these vanishing volumes are identified as the corresponding vanishing of the Lagrangian cycles L_1, L_2.

We remark that the above Proposition 3.1, and also the interpretation made above, generalize to arbitrary μ in a straightforward way. For example, the relation (3.5) should read

$$\textstyle\sum_{k=0}^{k=\mu+1} a_{k+1} W^k \quad = (1 + W)(1 + q_1 W)(1 + q_1 q_2 W) \cdots (1 + q_1 \cdots q_\mu W) \ ,$$

and the discriminant is essentially given by the product of pair-wise differences of the roots $-1, -1/q_1, \cdots, -1/(q_1 \cdots q_\mu)$ (see Propositions 4.1 and 4.4).

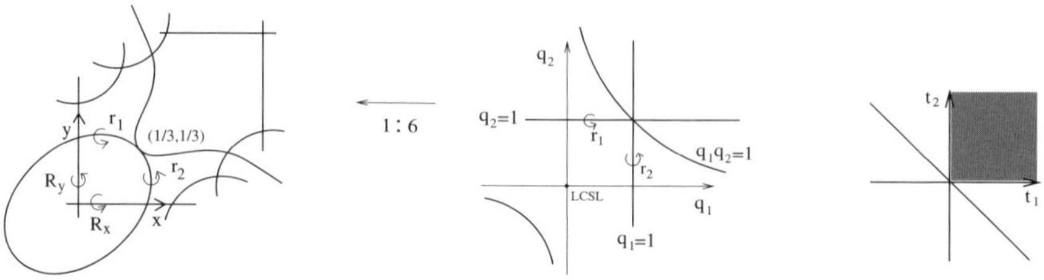

Fig.2. *The discriminant $dis(x, y) = 0$ in $\hat{\mathcal{M}}_{Sec(\Sigma)}$ (left), the mirror map (3.6) from the $q_1 q_2$-plane to the xy-plane (middle), and the complexified Kähler moduli t_1, t_2 with the complexified Kähler cone (right). The discriminant is an elliptic curve with a node at $(x, y) = (\frac{1}{3}, \frac{1}{3})$, which corresponds to $(q_1, q_2) = (1, 1)$. The mirror map is $6 : 1$ at generic (q_1, q_2). Over the discriminant, it is $3 : 1$ and the inverse image of $dis(x, y) = 0$ consists of the three lines $q_1 = 1, q_2 = 1, q_1 q_2 = 1$ in the q-plane.*

In the rest of this subsection, we look at the mirror map (3.6), and summarize its monodromy property. Let us first note that the moduli space $\mathcal{M}_{Sec(\Sigma)}$ is a two-dimensional singular toric variety, and may be desingularized to $\hat{\mathcal{M}}_{Sec(\Sigma)} = Bl_4(\mathbf{P}^1 \times \mathbf{P}^1)$ after blowing up four points (see the dashed lines in Fig.1). Then the discriminant $dis(x, y) = 0$ describes a nodal elliptic curve depicted in Fig.2. Note that the mirror map from (q_1, q_2) to (x, y) is six to one at generic points because (3.6) is invariant under the reflections;

$$(3.7) \qquad r_1 : q_1 \mapsto 1/q_1 \ , \ q_2 \mapsto q_1 q_2 \quad ; \quad r_2 : q_2 \mapsto q_1 q_2 \ , \ q_1 \mapsto 1/q_2 \ ,$$

which satisfy $r_1^2 = r_2^2 = (r_1 r_2)^3 = 1$ and thus generate the symmetric group of degree 3. By making an analytic continuation of the hypergeometric series, we can derive the above invariance group actions (3.7) as the monodromy actions for the loops r_1 and r_2 depicted in Fig.2. Together with the monodromy matrices R_x, R_y about the large complex structure limit, we summarize the monodromy generators with respect to a basis ${}^t(1, w_1(x, y), w_2(x, y)) = {}^t(1, \frac{1}{2\pi i} \log q_1, \frac{1}{2\pi i} \log q_2)$;

$$r_1 = \begin{pmatrix} 1 & 0 & 0 \\ 0 & -1 & 0 \\ 0 & 1 & 1 \end{pmatrix} , \ r_2 = \begin{pmatrix} 1 & 0 & 0 \\ 0 & 1 & 1 \\ 0 & 0 & -1 \end{pmatrix} ; \ R_x = \begin{pmatrix} 1 & 0 & 0 \\ 1 & 1 & 0 \\ 0 & 0 & 1 \end{pmatrix} , \ R_y = \begin{pmatrix} 1 & 0 & 0 \\ 0 & 1 & 0 \\ 1 & 0 & 1 \end{pmatrix} .$$

In general, we may state the monodromy property as follows;

PROPOSITION 3.2. *The mirror map (3.6) uniformizes the solutions of the GKZ system (3.3) with the symmetric group $\mathcal{S}_{\mu+1}$ of order $\mu+1$, up to the shifts*

$$w_i(x_1,\cdots,x_\mu)\mapsto w_i(x_1,\cdots,x_\mu)+1 \quad (i=1,\cdots,\mu).$$

In other words, the mirror coordinates $q_1,\cdots,q_\mu$ uniformize the hypergeometric series with the symmetric group $\mathcal{S}_{\mu+1}$ of order $\mu+1$ [1].

The above result seems to be interesting from the viewpoint of the uniformization of hypergeometric series [**Yo**], since our GKZ system provides an infinite number of examples for which multi-valued hypergeometric series are uniformizable. However the result itself is not surprising, as we will show in the next section that our GKZ system has a close relation to K. Saito's differential equations for which the monodromy property is well-studied.

4. K. Saito's differential equations

In the deformation theory of a singularity, we have a notion of so-called *primitive form* which satisfies a set of differential equations, K. Saito's differential equations [**Sa**]. The primitive form is also referred to as the Gel'fand–Leray form [**AGV**]. Here we briefly introduce the primitive form and K. Saito's differential equations, and then connect them to the 'period integral' (3.1) and the GKZ system (3.3).

(4-1) K. Saito's system for a primitive form $\mathcal{U}_0(a)$. Let us first note that, in the deformation of the singularity $U^2+V^2+W^{\mu+1}$, the polynomial equation $f_\Sigma(a,W)+U^2+V^2=0$ will be considered in $\mathbf{C}^3$ and the deformation parameters are set to $a_{\mu+2}=1, a_{\mu+1}=0$ by a coordinate change of W. Namely we take the defining equation of the form $f_\Sigma(a,W)+U^2+V^2=a_1+f_1(a,U,V,W)$ with

$$f_1(a,U,V,W)=a_2W+\cdots+a_\mu W^{\mu-1}+W^{\mu+1}+U^2+V^2\ ,$$

and regard the parameters $a_1,a_2,\cdots,a_\mu$ as giving a deformation of the singularity $W^{\mu+1}+U^2+V^2=0\subset\mathbf{C}^2$ at the origin. Since the parameter a_1 plays a distinguished role from the others, we set the local parameters $(a_2,\ldots,a_\mu)$ as coordinates of $T:=\mathbf{C}^{\mu-1}$. The full parameters $(a_1,a_2,\cdots,a_\mu)$ will be regarded as coordinates of $S:=\mathbf{C}\times T$. We consider the product space $\mathfrak{X}:=\mathbf{C}^3\times T$ with coordinates $(U,V,W,a_2,\cdots,a_\mu)$. Then we have a natural map $\varphi:\mathfrak{X}\to S$ by

$$(U,V,W,a_2,\cdots,a_\mu)\mapsto(-f_1(a,U,V,W),a_2,\cdots,a_\mu)\ .$$

This map plays an important role in describing the deformation of the singularity. Consider the sheaf $\Omega^p_{\mathfrak{X}/T}$ of germs of relative holomorphic p-forms for the natural projection $\pi:\mathfrak{X}=\mathbf{C}^3\times T\to T$. We may consider the following sheaves on S:

$$\mathcal{H}^{(0)}=\varphi_*\Omega^3_{\mathfrak{X}/T}/df_1\wedge d(\varphi_*\Omega^1_{\mathfrak{X}/T})\ ,\ \ \mathcal{H}^{(-1)}=\varphi_*\Omega^2_{\mathfrak{X}/T}/(df_1\wedge\varphi_*\Omega^1_{\mathfrak{X}/T}+d(\varphi_*\Omega^1_{\mathfrak{X}/T}))\ .$$

A primitive form ζ is an element in $H^0(S,\mathcal{H}^{(0)})$ satisfying certain conditions (see [**Sa**] for details). Instead of ζ, hereafter we consider its image $\mathcal{U}_0$ in $H^0(S,\mathcal{H}^{(-1)})$

---

[1]It was pointed out to the author by Prof. K. Saito that the monodromy group of this GKZ system should be the affine Weyl group of the root system A_μ. This is the mirror dual to the affine Weyl group which appears in the recent study of $\mathrm{Auteq}\,D(\widehat{\mathbf{C}^2/\mathbf{Z}_{\mu+1}})$ in [**IU**].

under an isomorphisms $\mathcal{H}^{(0)} \cong \mathcal{H}^{(-1)}$ (see [**Sa**]), which we may write explicitly as

$$\mathcal{U}_0(a) = Res_{\{a_1+f_1=0\}} \left(\frac{dU \wedge dV \wedge dW}{a_1 + f_1(a, U, V, W)} \right) .$$

In this from, the primitive form is also called as the Gel'fand–Leray form in the study of oscillatory integrals (see e.g. [**AGV**]). We note the similarity of $\mathcal{U}_0(a)$ to our 'period integral' (3.1), although, in (3.1), we do not set $a_{\mu+2} = 1, a_{\mu+1} = 0$ but consider torus actions $(\mathbf{C}^*)^2$ instead.

K. Saito's system is defined as a set of differential equations satisfied by the primitive form $\mathcal{U}_0(a) = \mathcal{U}_0(a_1, \cdots, a_\mu)$:

$$\begin{aligned} P_{ij}\mathcal{U}_0(a) &= \left\{ \frac{\partial^2}{\partial a_i \partial a_j} - \nabla_{\frac{\partial}{\partial a_i}} \frac{\partial}{\partial a_j} - (\frac{\partial}{\partial a_i} * \frac{\partial}{\partial a_j}) \frac{\partial}{\partial a_1} \right\} \mathcal{U}_0(a) = 0 \ , \\ Q(\tfrac{\partial}{\partial a_i})\mathcal{U}_0(a) &= \left\{ w(\frac{\partial}{\partial a_i}) \frac{\partial}{\partial a_1} - N(\frac{\partial}{\partial a_i}) + \frac{3}{2}\frac{\partial}{\partial a_i} \right\} \mathcal{U}_0(a) = 0 \ , \end{aligned}$$

see [**Sa**] for detailed definitions. This system is defined in a general setting in the deformation theory of singularities, and is also known to define a holonomic system. A proof of the following proposition may be found in the Appendix by Ambai in [**Oda1**], for example. Although, we restrict our attention to the A_μ case, similar results are also known for other (D or E) types of singularities.

PROPOSITION 4.1. *Let $\beta_0(a), \cdots, \beta_\mu(a)$ be the roots of $\psi_0(W) := a_1 + a_2 W + \cdots + a_\mu W^{\mu-1} + W^{\mu+1}$ which satisfy $\beta_0(a) + \cdots + \beta_\mu(a) = 0$. Then the space of solutions of K. Saito's system is generated by*

$$1 \ , \ \beta_0(a) - \beta_1(a) \ , \ \cdots \ , \ \beta_{\mu-1}(a) - \beta_\mu(a) \ . \tag{4.1}$$

The system has a regular singularity at the discriminant locus

$$dis\psi_0(a) = \Pi_{1 \le i,j \le \mu} (\beta_i(a) - \beta_j(a))^2 = 0 \ ,$$

and the monodromy group about the discriminant coincides with the symmetric group $S_{\mu+1}$ of order $\mu+1$ acting as permutations among the roots $\beta_i(a)$.

As we have noted in (3-1), μ vanishing cycles appears in the deformation of the singularity. Now it is easy to find that the above solutions $\alpha_i := \beta_i(a) - \beta_{i-1}(a)$ represent the integrals $\int_C \mathcal{U}_0(a)$ over the corresponding vanishing cycles (see Appendix A). Also we note that there is a residue pairing $I(\alpha_i, \alpha_j)$ among the solutions which reproduces the intersection pairing $\# L_i \cap L_j$ among the vanishing cycles [**Sa**].

(4-2) GKZ system for a primitive form $\mathcal{U}(a)$. As defined above, the primitive form $\mathcal{U}_0(a)$ is parametrized by $(a_1, \cdots, a_\mu) \in S$ by setting $a_{\mu+1} = 0, a_{\mu+2} = 1$ in the defining equation $f_\Sigma(a, W) + U^2 + V^2$. Instead of this, one may consider the natural torus action of $(\mathbf{C}^*)^2$ on $(a_1, \cdots, a_{\mu+2})$ by $a_i \mapsto \lambda a_i$ and $a_i \mapsto \lambda^{i-1} a_i$ ($\lambda \in \mathbf{C}^*$). With this slight change of the parameters ('gauge'), we may connect K. Saito's system to a GKZ system. Let us define a period integral of the primitive form

$$\int_C \mathcal{U}(a) = \int_C Res_{f_\Sigma(W)+U^2+V^2=0} \left(\frac{dW \wedge dU \wedge dV}{f_\Sigma(a, W) + U^2 + V^2} \right) , \tag{4.2}$$

where C is a two-cycle on $f_\Sigma(a,W)+U^2+V^2=0\subset \mathbf{C}^3$ and $f_\Sigma(a,W)=a_1+a_2W+\cdots+a_{\mu+2}W^{\mu+1}$. The period integral above is similar to (3.1), and thus satisfies a similar GKZ system to (3.3). The only difference appears in the scaling properties expressed by the linear operators $\mathcal{Z}_i$.

PROPOSITION 4.2. *(1) The period integral (4.2) satisfies*

$$\square_l \int_C \mathcal{U}(a)=0\ (l\in\mathcal{L})\ ,\quad \mathcal{Z}'_i \int_C \mathcal{U}(a)=0\ (i=1,2)\ , \tag{4.3}$$

where the operators $\square_l$ and the lattice $\mathcal{L}$ are the same as in (3.3), and $\mathcal{Z}'_i\,(i=1,2)$ are given by

$$\begin{pmatrix}\mathcal{Z}'_1\\ \mathcal{Z}'_2\end{pmatrix}=\begin{pmatrix}\theta_1+\theta_2+\cdots+\theta_{\mu+2}\\ \theta_2+2\theta_3\cdots+(\mu+1)\theta_{\mu+2}+1\end{pmatrix}$$

(2) The system (4.3) above is reducible of rank $\mu+1$ with its irreducible part of rank μ. The μ independent solutions of the irreducible part are given by

$$\beta_0(a)-\beta_1(a),\cdots,\beta_{\mu-1}(a)-\beta_\mu(a)\ ,$$

where $\beta_i(a)$ are roots of $\psi(W)=a_1+a_2W+\cdots+a_{\mu+2}W^{\mu+1}=0$.

Derivation of (1) above is straightforward, but it should be noted that K. Saito's system is replaced by a different but simpler GKZ system. Also, it is known that the number of independent solutions of the GKZ system is given by the volume of a relevant polytope [**GKZ1**], which in our case is $Vol(\Sigma)=\mu+1$. One may verify that the system is reducible by observing the factorization of a differential operator in our GKZ system when expressing $\square_l$ operators in the affine coordinates of $\mathcal{M}_{Sec(\Sigma)}$. After the factorization, the reduced GKZ system has μ independent solutions as claimed. (It has been observed in [**HKTY1**] that the GKZ systems in mirror symmetry are often reducible in this way.) One can derive the μ independent solutions in (2) by evaluating the period integral (4.2) over the vanishing cycles (see Appendix A). Or, more directly, one may verify that the roots β_k are solutions of (4.3) with one relation $\beta_0+\cdots+\beta_\mu=-\frac{a_{\mu+1}}{a_{\mu+2}}$ from the following property:

PROPOSITION 4.3. *The root $\beta(a)\,(=\beta_0(a),\cdots,\beta_\mu(a))$ of $\psi(W)=0$ satisfies*

$$\frac{\partial\beta}{\partial a_i}=-\frac{\beta^{i-1}}{\psi'(\beta)}\ ,\quad \frac{\partial^2\beta}{\partial a_i\partial a_j}=\frac{1}{\psi'(\beta)}\frac{d}{dx}\left(\frac{x^{i+j-2}}{\psi'(x)}\right)\Big|_{x=\beta}\ . \tag{4.4}$$

PROOF. The first relation follows from the differentiation with respect to a_i of $\psi(\beta(a))=a_1+a_2\beta(a)+\cdot\cdot+a_{\mu+2}\beta(a)^{\mu+1}=0$. For the second relation, we note the following:

$$\frac{\partial\psi'(\beta)}{\partial a_j}=(j-1)\beta^{j-2}+\psi''(\beta)\frac{\partial\beta}{\partial a_j}=(j-1)\beta^{j-2}-\frac{\psi''(\beta)}{\psi'(\beta)}\beta^{j-1}\ .$$

Then the second relation follows by differentiating the first equation with respect to a_i. ∎

Now a proof of Proposition 4.2 is straightforward by the relations above. Here, since the arguments are similar, we present the solutions of the GKZ system (3.3) for local mirror symmetry, which is the same as the GKZ system (4.3) for the primitive form except for the scalings operators.

PROPOSITION 4.4. *(1) The independent solutions of the GKZ system (3.3) are given by*

$$1 \ , \ \log\beta_0(a)-\log\beta_1(a) \ , \ \cdots \ , \ \log\beta_{\mu-1}(a)-\log\beta_\mu(a) \ .$$

(2) These solutions, up to monodromy transformations, are related to the expansion $w(x;\frac{J}{2\pi i})=1+\sum_{k=1}^{\mu}w_k(a)J_k$ *near the large complex structure limit by*

$$2\pi i\, w_k(x)=\log\beta_{k-1}(a)-\log\beta_k(a) \ \ (k=1,\cdots,\mu).$$

PROOF. (1) By the relations (4.4), it it straightforward to evaluate $\mathcal{Z}_1\log\beta$ and $\mathcal{Z}_2\log\beta$ as follows:

$$\mathcal{Z}_1\log\beta=\frac{1}{\beta}\sum_{i=1}^{\mu+2}\theta_{a_i}\beta=-\frac{1}{\beta}\sum_{i=1}^{\mu+2}\frac{1}{\psi'(\beta)}a_i\beta^{i-1}=-\frac{1}{\beta\psi'(\beta)}\psi(\beta)=0 \ ,$$

$$\mathcal{Z}_2\log\beta=\frac{1}{\beta}\sum_{i=1}^{\mu+2}(i-1)\theta_{a_i}\beta=-\frac{1}{\psi'(\beta)}\sum_{i=1}^{\mu+2}(i-1)a_i\beta^{i-2}=-1 \ .$$

Therefore the differences $\log\beta_k-\log\beta_{k-1}$ $(k=1,\cdots,\mu)$ are annihilated by the linear operators $\mathcal{Z}_1,\mathcal{Z}_2$. As for the operators $\Box_l$ $(l\in\mathcal{L})$, we first note

$$\frac{\partial^2}{\partial a_i\partial a_j}\log\beta=\frac{\partial^2}{\partial a_k\partial a_l}\log\beta \quad \text{if} \ \ i+j=k+l \ , \tag{4.5}$$

which follows from

$$\frac{\partial^2}{\partial a_i\partial a_j}\log\beta=-\frac{1}{\beta^2}\frac{\beta^{i+j-2}}{(\psi'(\beta))^2}+\frac{1}{\beta}\frac{1}{\psi'(\beta)}\frac{d}{dx}\left(\frac{x^{i+j-2}}{\psi'(x)}\right)\Big|_{x=\beta} \ .$$

Since the entries of $l=(l_1,l_2,\cdots,l_{\mu+2})$ satisfies $\sum_i l_i=0$ and $\sum_i(i-1)l_i=0$, we have $\sum_i i\,l_i^+=\sum_i i\,l_i^-$ for $l_i^\pm:=\frac{1}{2}(l_i\pm|l_i|)$. Now, by repeated use of (4.5), one may deduce

$$\prod_i\left(\frac{\partial}{\partial a_i}\right)^{l_i^+}\log\beta=\prod_i\left(\frac{\partial}{\partial a_i}\right)^{l_i^-}\log\beta \ ,$$

which means $\Box_l\log\beta=0$. Since the linear independence of $\log\beta_k-\log\beta_{k-1}$ $(k=1,\cdots,\mu)$ is clear, we obtain the claim including the trivial solution 1.

(2) By definition, we have $2\pi i\,w_k(x)\sim\log x_k$, where $x_k=\frac{a_ka_{k+2}}{a_{k+1}^2}$. Since, by extending the classical arguments due to Frobenius, it is straightforward to show that the $w_k(x)$ are solutions of the GKZ hypergeometric system (3.3), we only need to connect these solutions to those in terms of the roots β_k. For this, let us assume that the roots are given by $\beta_k=-\frac{1}{\lambda_k}$ and also the following relation:

$$\psi(W)=a_1+a_2W+\cdots+a_{\mu+2}W^{\mu+1}=\prod_k(\lambda_kW+1) \ .$$

Assume also that $|\lambda_\mu|<<|\lambda_{\mu-1}|<<\cdots<<|\lambda_0|$. Then from the relations between the roots and the coefficients, it is easy to see

$$\log a_1=0 \ , \ \log a_2=\log\lambda_0+A_2(\lambda) \ , \ \log a_3=\log(\lambda_0\lambda_1)+A_3(\lambda) \ , \ \cdots \ ,$$
$$\log a_{\mu+2}=\log(\lambda_0\lambda_1\cdots\lambda_\mu)+A_{\mu+2}(\lambda) \ ,$$

where $A_k(\lambda) \in \mathbf{C}\{\frac{\lambda_1}{\lambda_0}, \frac{\lambda_2}{\lambda_1}, \cdots, \frac{\lambda_\mu}{\lambda_{\mu-1}}\}$. Now we have

$$\log x_k = \log \frac{a_k a_{k+2}}{a_{k+1}^2} = \log \lambda_k - \log \lambda_{k-1} + B_k(\lambda) ,$$

with $B_k(\lambda) := A_k(\lambda) + A_{k+2}(\lambda) - 2A_{k+1}(\lambda)$, which entails the claimed relation since $2\pi i w_k(x) \sim \log x_k$. (Note that if we invert the series relations $x_k = \frac{\lambda_k}{\lambda_{k-1}} e^{B_k(\lambda)}$ in $C\{\frac{\lambda_1}{\lambda_0}, \frac{\lambda_2}{\lambda_1}, \cdots, \frac{\lambda_\mu}{\lambda_{\mu-1}}\}$, then we obtain the $\frac{\lambda_k}{\lambda_{k-1}}$ in terms of the x_k, and thus recover the series expansion of $2\pi i w_k(x)$ by $2\pi i w_k(x) = \log \lambda_k - \log \lambda_{k-1} = \log x_k - B_k(\lambda)$.) □

The above form of the solutions may be connected to the piecewise linear functions on the fan Σ [**Ba3**]. Since we have established a relation of the GKZ system (3.3) of $\widehat{\mathbf{C}^2/\mathbf{Z}_{\mu+1}}$ to K. Saito's system of the primitive form, for which the integral monodromy property is well studied (see Proposition 4.1), it is clear that we have the uniformization property of the mirror map stated in Proposition 3.2.

5. Central charge formula and G-Hilb

In the last two sections, we have looked at the local mirror symmetry of $\widehat{\mathbf{C}^2/\mathbf{Z}_{\mu+1}}$ paying attention to the monodromy property of the associated GKZ system. Here we come back to our claim for the central charge formula (2.4) in this case.

Let us first recall that the non-compact Calabi–Yau manifold $X = \widehat{\mathbf{C}^2/\mathbf{Z}_{\mu+1}}$ is given as the Hilbert scheme of points on $\mathbf{C}^2$, G-Hilb $\mathbf{C}^2$. Here G-Hilb $\mathbf{C}^n$ is defined for a finite subgroup $G \subset SU(n, \mathbf{C})$ and consists of zero-dimensional subschemes Z in $\mathbf{C}^n$ of length equal to $|G|$ such that G acts on Z and $H^0(\mathcal{O}_Z)$ is the regular representation of G. The following results are due to Gonzalez–Sprinberg and Verdier [**GV**] for $n = 2$ and their generalizations to $n = 3$ have been done in [**Na**],[**IN**],[**BKR**]. Here, we summarize the relevant results for our cases where $G \subset SU(n, \mathbf{C})$ is a finite abelian subgroup:

(i) The K-group $K(X)$ of algebraic vector bundles on X is generated by the so-called tautological bundles $\mathcal{F}_0, \mathcal{F}_1, \cdots, \mathcal{F}_d$ $(d = |G| - 1)$, where the subscripts refer to the one-dimensional representations of G with 0 for the trivial representation.

(ii) Let $K^c(X)$ be the K-group of complexes of algebraic vector bundles which are exact off $\pi^{-1}(0)$ where $\pi : X \to \mathbf{C}^n/G$. Then there exists a perfect pairing $\langle \ , \ \rangle : K^c(X) \times K(X) \to \mathbf{Z}$, and also there exists the dual basis $S_0, S_1, \cdots, S_d$ of $K^c(X)$ which satisfies

$$\langle S_i, \mathcal{F}_j \rangle := \int_X ch(S_i) ch(\mathcal{F}_j) Todd_X = \delta_{ij} \quad . \tag{5.1}$$

(iii) The dual basis $S_k (k = 0, \cdots, d)$ defines a symmetric form for $n = 2$ (a symplectic form for $n = 3$) on $K^c(X)$ by

$$\chi(S_i, S_j) := \int_X ch(S_i^\vee) ch(S_j) Todd_X \ \ (0 \le i, j \le d). \tag{5.2}$$

In two-dimensional case with $G = \mathbf{Z}_{\mu+1}$, the S_k for $k \ne 0$ have a simple form. They are given by $S_k = \mathcal{O}_{C_k}(-1)$ with $C_k \cong \mathbf{P}^1$ being the exceptional curves.

Including S_0, they satisfy $\chi(S_i, S_j) = -\hat{C}_{ij}$ $(0 \le i, j \le d)$ where $\hat{C}_{ij}$ is the extended Cartan matrix of the root system of type A_μ.

Now with these properties about the geometry of $X = \widehat{\mathbf{C}^2/\mathbf{Z}_{\mu+1}}$, let us recall our Conjecture 2.2 encoded in the following formula,

$$w(x_1, \cdots, x_r; \frac{J_1}{2\pi i}, \cdots, \frac{J_r}{2\pi i}) = \sum_{i,j} \int_{mir(E_i)} \Omega(Y_x)\, \chi^{ij}\, ch(E_j^\vee)\,.$$

One should note that this is a conjecture for X a compact Calabi–Yau manifold. It is rather clear how to modify this relation for our non-compact X if we notice that the expression $\sum_j \chi^{ij} ch(E_j^\vee)$ satisfies $\int_X ch(E_k) \sum_j \chi^{ij} ch(E_j^\vee) Todd_X = \delta^i{}_k$, i.e., provides a dual basis to $ch(E_k)$. From this, we claim that the central charge formula for $X = \widehat{\mathbf{C}^2/\mathbf{Z}_{\mu+1}}$ is given by

$$w(x_1, \cdots, x_\mu; \frac{J_1}{2\pi i}, \cdots, \frac{J_\mu}{2\pi i}) = \sum_k \int_{mir(S_k)} \Omega(Y_x)\, ch(\mathcal{F}_k)\,, \tag{5.3}$$

where the holomorphic two-form $\Omega(Y_x)$ should be understood as a holomorphic two-form associated to the integrand of (3.1), i.e.,

$$\Omega(Y_x) = \frac{i}{2\pi^2} Res\left(\frac{1}{f(W) + U^2 + V^2} dU dV \frac{dW}{W}\right)\,.$$

We will state the corresponding claim in Conjecture 6.3 for the three-dimensional case.

Now, since the tautological bundles satisfy

$$\int_{C_j} c_1(\mathcal{F}_k) = \delta_{jk} \ \ (1 \le j, k \le \mu),$$

for the exceptional curves C_j $(1 \le j \le \mu)$, we have $c_1(\mathcal{F}_k) = J_k$ $(k \ge 1)$ for our generators J_k of $H^2(X, \mathbf{Z})$. Therefore we have

$$w(x; \frac{J}{2\pi i}) = 1 + \sum_k w_k(x) J_k = (1 - \sum_{k=1}^{\mu} w_k(x)) ch(\mathcal{F}_0) + \sum_{k=1}^{\mu} w_k(x) ch(\mathcal{F}_k)\,,$$

where we use $ch(\mathcal{F}_k) = 1 + c_1(\mathcal{F}_k) = 1 + J_k$. From this, one may read off the central charge of the sheaf $S_k = \mathcal{O}_{C_k}(-1)$ $(k \ne 0)$ as

$$Z_t(S_k) = \int_{mir(S_k)} \Omega(Y_x) = w_k(x) = \frac{1}{2\pi i}\left(\log \beta_{k-1}(a) - \log \beta_k(a)\right)\,. \tag{5.4}$$

In Appendix A, we obtain the above central charge as a period integral over a vanishing cycle. Therefore, we have obtained a refinement of the mirror symmetry sketched in (3-1) that the mirror image $mir(S_k)$ of the sheaf $S_k = \mathcal{O}_{C_k}(-1)$ is a vanishing cycle whose period is given by $Z_t(S_k)$ above.

A similar relation holds also for S_0 with its central charge $1 - \sum_k w_k(x)$. However, since S_0 is not a sheaf but a complex of sheaves (see [**IN**]), we replace it with the skyscraper sheaf $\mathcal{O}_p$ supported at a point p. Then the sheaves $\mathcal{O}_p, \mathcal{O}_{C_k}(-1)$ $(k = 1, \cdots, \mu)$ form another basis of $K^c(X)$ with relations

$$\mathcal{O}_p = S_0 + S_1 + \cdots + S_\mu\,,\ \ \mathcal{O}_{C_k}(-1) = S_k \ (k = 1, \cdots, \mu)\,,$$

where we use $\langle \mathcal{O}_p, \mathcal{F}_j \rangle = 1$ $(j = 0, \cdots, \mu)$ to derive the first equality. In (6-2), this basis of $K^c(X)$ will be generalized as a *symplectic D-brane basis* for the three-dimensional cases. Now, for the sheaf $\mathcal{O}_p$, we have

$$Z_t(\mathcal{O}_p) = \int_{mir(\mathcal{O}_p)} \Omega(Y_x) \equiv 1 \ .$$

This is identified with the period integral over a cycle T_0 whose topology is $S^1 \times S^1$, see Appendix A.

Finally, we note that the sheaves S_k $(k \neq 0)$ are spherical and thus define self-equivalences of $D^b(Coh(X))$, the Seidel–Thomas twists summarized in section 2. In [ST, Proposition 3.19], it is shown that these spherical objects, which form the so-called (A_μ)-configuration, generate a weak braid group action on $D^b(Coh(X))$. This braid group action should be mirror to the corresponding Dehn twists (Picard–Lefschetz transformations) on the symplectic side. Our formula (5.4) explicitly shows this mirror correspondence as the linear transformations on the central charges.

6. Local mirror symmetry II — $X = \widehat{\mathbf{C}^3/G}$

In the two-dimensional case, the local mirror symmetry is rather clear from the hyperkähler rotation as summarized in (3-1), and our results in the previous sections are consistent with those arguments from the hyperkähler rotation. In this section, we will argue that our central charge formula for a crepant resolution of the three dimensional singularity $\mathbf{C}^3/G$ with a finite abelian group $G \subset SU(3, \mathbf{C})$. The crepant resolution we use is the so-called G-Hilb $\mathbf{C}^3$ whose definition has been given in the previous section. Examples of GKZ hypergeometric systems in this situation have been also studied in [**dOFS**].

(6-1) Period integrals and GKZ systems. As above, let us consider a non-compact Calabi–Yau manifold $X = G\text{-Hilb}\ \mathbf{C}^3$ with a finite, abelian group $G \subset SU(3, \mathbf{C})$. For G abelian, X is given as a toric resolution of a singularity $\mathbf{C}^3/G$ as follows: Let us first denote by M_G the lattice corresponding to the invariants $\mathbf{C}[x^{\pm}, y^{\pm}, z^{\pm}]^G$, and its dual lattice by N_G. Then the invariant monomials in $\mathbf{C}[x, y, z]^G$ define a cone in $M_G \otimes \mathbf{R}$, with its dual cone Σ_G^0 in $N_G \otimes \mathbf{R}$. The toric (Calabi–Yau) resolution which corresponds to G-Hilb $\mathbf{C}^3$ is given by a fan obtained by subdividing the cone Σ_G^0, which we will denote by Σ_G hereafter. (See [**Re**] for detailed construction of the fan Σ_G.) We will denote by $\nu_1, \nu_2, \cdots, \nu_{r+3}$ the integral generators of the one-dimensional cones in the fan Σ_G. Note that, since the resolution is crepant, these generators satisfy

$$(m, \nu_k) = 1 \ \ (k = 1, \cdots, r+3) \ ,$$

with some $m \in M_G$, where $(\ ,\ ) : M_G \times N_G \to \mathbf{Z}$ is the dual pairing.

To describe the mirror configuration of X, let us fix an isomorphism $\varphi : N_G \simeq \mathbf{Z}^3$ satisfying $\varphi(\nu_k) = (1, \bar{\nu}_k)$. Consider the defining equation

$$F(a; U, V, W_1, W_2) = U^2 + V^2 + \sum_k a_k W^{\bar{\nu}_k} \ , \tag{6.1}$$

then the mirror configuration of X claimed in [**CKYZ**],[**HIV**] is the hypersurface $Y = (F(a; U, V, W_1, W_2) = 0) \subset \mathbf{C}^2 \times (\mathbf{C}^*)^2$. Note that Y does not depend on the choice of the isomorphism $\varphi : N_G \tilde{\rightarrow} \mathbf{Z}^3$ with the described property, since W_1, W_2 are considered in $\mathbf{C}^*$ and negative powers of them are allowed. One important criterion for the mirror configuration Y is the fact that we can extract the right Gromov–Witten invariants for X from the "period integrals" of Y (cf. [**CKYZ**]).

DEFINITION 6.1. For the mirror configuration Y given above, we define the period integral of a cycle $L \in H_3(Y, \mathbf{Z})$ by

$$\Pi_L(a) = \frac{1}{4\pi^3} \int_L Res_{F=0} \left(\frac{1}{U^2 + V^2 + \sum_k a_k W^{\bar{\nu}_k}} dU dV \frac{dW_1}{W_1} \frac{dW_2}{W_2} \right) , \tag{6.2}$$

where the residue is taken around $F(a, U, V, W) = 0$ in $\mathbf{C}^2 \times (\mathbf{C}^*)^2$.

In (6-3), we will describe how one can determine the Gromov–Witten invariants from our cohomology-valued hypergeometric series. Here, as in the two-dimensional cases, we see that the following GKZ system is satisfied by the period integral:

PROPOSITION 6.2. *Let $\mathcal{L}$ be the lattice of relations defined by*

$$\mathcal{L} = \{(l_1, \cdots, l_{r+3}) \in \mathbf{Z}^{r+3} \mid \sum_k l_k \nu_k = \vec{0}\ \}.$$

Then the period integral (6.2) satisfies the following set of differential equations:

$$\Box_l \Pi_L(a) = 0 \ (l \in \mathcal{L}) \ , \ \ \mathcal{Z}_i \Pi_L(a) = 0 \ (i = 1, 2, 3) \ , \tag{6.3}$$

where

$$\Box_l = \left(\frac{\partial}{\partial a}\right)^{l^+} - \left(\frac{\partial}{\partial a}\right)^{l^-} , \quad \begin{pmatrix} \mathcal{Z}_1 \\ \mathcal{Z}_2 \\ \mathcal{Z}_3 \end{pmatrix} = \begin{pmatrix} \theta_1 + \theta_2 + \cdots + \theta_{r+3} \\ \sum_k \bar{\nu}_{k,2} \theta_k \\ \sum_k \bar{\nu}_{k,3} \theta_k \end{pmatrix}$$

with $l = l^+ - l^-$ ($l_i^\pm := \frac{1}{2}(l_i \pm |l_i|)$) and $\theta_k = a_k \frac{\partial}{\partial a_k}$. $\bar{\nu}_{k,i}$ represents the i-th component of the vector $\varphi(\nu_k) = (1, \bar{\nu}_k)$ with the isomorphism $\varphi : N_G \tilde{\rightarrow} \mathbf{Z}^3$ fixed above.

Remark. As we have shown in section 4, by a slight modification of the period integral, we can connect (6.2) to K. Saito's primitive form of a three-dimensional singularity. The primitive form satisfies the same GKZ hypergeometric system but with a different scaling property, i.e., $\mathcal{Z}_i (i = 2, 3)$ being replaced by $\mathcal{Z}_i' := \mathcal{Z}_i + 1 (i = 2, 3)$. One may observe for several examples that, when the singularity of $\mathbf{C}^3/G$ is isolated, the resulting GKZ system for the primitive form is reducible and its irreducible part shows a slightly different degeneration at 'infinity', i.e. at the large complex structure limit (see Appendix B). □

(6-2) Central charge formula and symplectic forms. The cohomology-valued hypergeometric series for the GKZ system (6.3) are defined by constructing the secondary fan $Sec(\Sigma_G)$. As described for an example in (3-3) and Fig.1, the secondary fan is constructed combinatorially from the secondary polytope in (a translate of) $\mathcal{L}_{\mathbf{R}}$. We refer to [**HLY**],[**Hos**],[**Sti**] and references therein for details of the construction in our context of mirror symmetry.

Let us denote by $l^{(1)}, \cdots, l^{(r)}$ the integral generators of $\mathcal{L}$ which represents the large complex structure limit. We denote the affine coordinate by $x_k = a^{l^{(k)}}$ $(k = 1, \cdots, r)$. Then, we have cohomology-valued hypergeometric series

$$w(x_1, \cdot\cdot, x_r; \frac{J_1}{2\pi i}, \cdot\cdot, \frac{J_r}{2\pi i}) = \sum_{m \in \mathbf{Z}^r_{\geq 0}} c(m+\rho) x^{m+\rho}|_{\rho = \frac{J}{2\pi i}}, \tag{6.4}$$

where $c(m) := c(m_1, \cdots, m_r)$ is defined by

$$c(m_1, \cdots, m_r) := \frac{1}{\prod_{i=1,\ldots,r+3} \Gamma(1 + \sum_k m_k l_i^{(k)})},$$

and $J_1, \cdots, J_r \in H^2(X, \mathbf{Z})$ are dual (semi-ample) generators to $l^{(1)}, \cdots, l^{(r)}$ (see [Sect.1, Hos] for a quick review of the construction). It is found in [**HKTY2**] for X compact Calabi–Yau hypersurfaces, and used in [**CKYZ**] for non-compact cases that the above cohomology-valued hypergeometric series generate the solutions of the GKZ system (6.3) when expanded in $H^{even}(X, \mathbf{Q})$. If we recall the integral structure of $K(X)$ and also the natural symplectic structure in $K^c(X)$ which are summarized in section 4, we come to the following conjecture:

CONJECTURE 6.3. *(1) Consider the expansion of the cohomology-valued hypergeometric series (6.4) with respect to the basis* $ch(\mathcal{F}_0), \cdots, ch(\mathcal{F}_d)$ *of* $H^{even}(X, \mathbf{Q})$; $w(x; \frac{J}{2\pi i}) = \sum_k f_k(x) ch(\mathcal{F}_k)$. *Then the coefficient hypergeometric series* $f_k(x)$ *may be identified with the period integrals over the cycles* $mir(S_k)$, *i.e., we have*

$$w(x_1, \cdots, x_r; \frac{J_1}{2\pi i}, \cdots, \frac{J_r}{2\pi i}) = \sum_k \int_{mir(S_k)} \Omega(Y_x)\, ch(\mathcal{F}_k), \tag{6.5}$$

where $mir : K^c(X) \tilde{\rightarrow} H_3(Y, \mathbf{Z})$ *is the (homological) mirror map and* $\Omega(Y_x) = \frac{1}{4\pi^3} Res_{F=0}(\frac{1}{F(a,U,V,W)} dU dV \frac{dW_1}{W_1} \frac{dW_2}{W_2})$.
(2) The monodromy of the coefficient hypergeometric series $f_k(x) (= \int_{mir(S_k)} \Omega(Y_x))$ *is integral and symplectic with respect to the symplectic form:*

$$\chi(S_i, S_j) = \int_X ch(S_i^\vee) ch(S_j) Todd_X \quad (0 \leq i, j \leq d = |G| - 1).$$

(3) Using the cohomology-valued hypergeometric series, the central charge of $F \in K^c(X)$ *is given by* $Z_t(F) = \int_X ch(F) w(x; \frac{J}{2\pi i}) Todd_X$.

In section 4 and section 5, we have verified explicitly the corresponding conjecture for $X = \widehat{\mathbf{C}^2/\mathbf{Z}_{\mu+1}}$ by connecting our period integral to that of the primitive form in singularity theory. For three dimensions, however, little is known about the explicit form of the period integral of the primitive form. In the following sections, we will test our conjecture for the cases, $G = \mathbf{Z}_3, \mathbf{Z}_5$. In particular, for the former case, we will evaluate the period integrals of the primitive form, and connect them to the periods in mirror symmetry (see Appendix A). In this case, we will be able to see the mirror correspondence $mir : K^c(X) \tilde{\rightarrow} H_3(Y, \mathbf{Z})$ from our Conjecture 6.3, which generalizes the correspondence between the spherical objects and the vanishing cycles presented in section 5.

As another aspect of our conjecture, we will define the so-called prepotential which gives predictions of Gromov–Witten invariants on the geometry X. When X is compact, the existence of the prepotential (special Kähler geometry) is ensured by a canonical symplectic basis of $H_3(Y, \mathbf{Z})$ and also by the property called

Griffiths transversality [**St**]. Correspondingly, in our non-compact cases, we have the symplectic structure on $H_3(Y,\mathbf{Z})$ given by $(\chi(S_i,S_j))$ under the homological mirror map, $mir : K^c(X) \tilde{\to} H_3(Y,\mathbf{Z})$. We will find that our cohomology-valued hypergeometric series combined with the symplectic structure suffices to present a closed formula for the prepotential.

(6-3) Prepotential and a symplectic D-brane basis. In string theory, the so-called D-branes play important roles for duality symmetries such as mirror symmetry. In our non-compact Calabi–Yau manifolds, roughly speaking, the D-branes are elements in $K^c(X)$ or more precisely elements in the category $D^b_c(Coh(X))$ of compact support. In what follows, we will work on $K^c(X)$(or $H^{even}_c(X,\mathbf{Q})$ taking the Chern character). In our case, it is easy to see that there are natural D-branes like $\mathcal{O}_p$, the skyscraper sheaf, $\mathcal{O}_C$, the torsion sheaf supported on a curve C, and $\mathcal{O}_D$, that supported on an exceptional divisor D. Depending on the dimensions of the support, these sheaves are called $D0$-, $D2$-, $D4$-branes, respectively, see [**Do**] and references therein. We will connect the bases S_i's to these D-branes.

(1) *Symplectic D-brane basis and its dual*: First, let us recall that we have introduced the (semi-ample) generators $J_1,\cdots,J_r$ of $H^2(X,\mathbf{Z})$ when writing down our cohomology-valued hypergeometric series. Assume that these generators measure the volume of the respective curves $C_1,\cdots,C_r$, i.e.,

$$\int_{C_i} J_k = \delta_{ik} \quad .$$

Assume also that the exceptional divisors $D_1,\cdots,D_s$ provide a basis of $H_4(X,\mathbf{Z})$. Then in such cases, we seek a basis of $K^c(X)$ of the form,

$$\{\mathcal{B}_I\} := \{\ \mathcal{O}_p\,;\,\mathcal{O}_{C_1}(-J_1),\cdots,\mathcal{O}_{C_r}(-J_r)\,;\,\mathcal{O}_{D_1}(\mathcal{L}_1),\cdots,\mathcal{O}_{D_s}(\mathcal{L}_s)\ \} \tag{6.6}$$

where $\mathcal{O}_C(-J)=\mathcal{O}_C\otimes\mathcal{O}_X(-J)$ and $\mathcal{O}_{D_k}(\mathcal{L}_k) := \mathcal{O}_D\otimes\mathcal{L}_k$ with $\mathcal{L}_k = \mathcal{O}_X(-a_{k,1}J_1 - \cdots - a_{k,r}J_r)$. Here we fix the integers $a_{k,1},\dots,a_{k,r}$ requiring

$$\chi(\mathcal{O}_{D_j}(\mathcal{L}_j),\mathcal{O}_{D_k}(\mathcal{L}_k)) = 0 \tag{6.7}$$

for $j,k=1,\cdots,s$. The integers $a_{k,1},\dots,a_{k,r}$ satisfying the above requirement are not unique in general. However, the reason for our requirement (6.7) will soon become clear when we present a closed formula of the prepotential. Hereafter, we will call a basis $\{\mathcal{B}_I\}$ satisfying the requirement as a *symplectic D-brane basis* of $K^c(X)$.

Now let us consider the dual basis of $K(X)$ to the basis $\{\mathcal{B}_I\}$ (6.6) under the pairing (5.1). Taking $ch : K(X) \to H^{even}(X,\mathbf{Q})$, we construct (charges) $\{Q_J\} := \{Q_0; Q_2^1,..,Q_2^r; Q_4^1,..,Q_4^s\}$ which satisfy

$$\int_X ch(\mathcal{B}_I)\, Q_J\, Todd_X = \delta_{IJ} \quad .$$

We note that, due to the support property of the D-brane basis (6.6) and the above relation, the base Q_0 starts from the degree zero element 1 to higher degree terms. Similarly Q_2^a starts from J_a to higher degree terms, and Q_4^b is written by a linear combination of the degree four elements of $H^{even}(X,\mathbf{Q})$. The set of charges $\{Q_J\}$ provides a basis of $H^{even}(X,\mathbf{Q})$.

(2) *Expansion of* $w(x; \frac{J}{2\pi i})$ *and the prepotential*: Now we expand our cohomology-valued hypergeometric series using the basis $\{Q_J\}$ of $H^{even}(X, \mathbf{Q})$;

$$w(x; \frac{J}{2\pi i}) = w_0(x) Q_0 + \sum_{a=1}^{r} w_a^{(1)}(x) Q_2^a + \sum_{b=1}^{s} w_b^{(2)}(x) Q_4^b .$$

Then our conjecture says that the coefficient hypergeometric series $\{w_0(x), w_a^{(1)}(x), w_b^{(2)}(x)\}$ represents the central charges of the sheaves $\{\mathcal{B}_I\}$ in (6.6), whose monodromy matrices are integral and symplectic with respect to the symplectic form

$$(\chi(\mathcal{B}_I, \mathcal{B}_J)) = \begin{pmatrix} 0 & 0 & 0 \\ 0 & 0 & C_{ab} \\ 0 & -C_{ab} & 0 \end{pmatrix}, \tag{6.8}$$

where we set $C_{ab} = \chi(\mathcal{O}_{C_a}(-J_a), \mathcal{O}_{D_b}(\mathcal{L}_b))$ and used the fact $\chi(\mathcal{O}_p, \mathcal{B}_I) = 0$ for sheaves $\mathcal{B}_I$ of compact support. The above form should be contrasted to the canonical symplectic form on $H_3(Y, \mathbf{Z})$ for the compact (mirror) Calabi–Yau manifolds, by which we define the special Kähler geometry on the moduli spaces (see e.g. [**St**]). In fact, we may define the prepotential for local mirror symmetry given in [**CKYZ**] in terms of the symplectic form (6.8).

For simplicity, let us assume $\dim H_c^2(X, \mathbf{Q}) = \dim H_c^4(X, \mathbf{Q})$ (i.e. $r = s$), and thus (C_{ab}) is a square matrix. Also we assume $\det(C_{ab}) \neq 0$. In this case, inverting the matrix (C_{ab}), we may define the 'symplectic duals' of $\mathcal{O}_{C_1}(-J_1), \cdots, \mathcal{O}_{C_r}(-J_r)$ by $\sum_b C^{ab} \mathcal{O}_{D_b}(\mathcal{L}_b)$ $(a = 1, \cdots, r)$ where $(C^{ab}) := (C_{ab})^{-1}$. We should note that these symplectic duals are not in $K^c(X)$ in general, but are elements in $K^c(X) \otimes \mathbf{Q}$. Correspondingly we consider the symplectic duals of the mirror cycles $mir(\mathcal{O}_{C_a}(-J_a))$ by $\sum_b C^{ab} mir(\mathcal{O}_{D_b}(\mathcal{L}_b))$. From the period integrals of these cycles, we define the prepotential $F(t)$ by the special Kähler geometry relation

$$(1, t_a, \frac{\partial F}{\partial t_a}) = (1, w_a^{(1)}(x), \sum_b C^{ab} w_b^{(2)}(x)), \tag{6.9}$$

where we use the fact $w^{(0)}(x) = Z(\mathcal{O}_p) \equiv 1$ for local mirror symmetry of $X = \widehat{\mathbf{C}^3/G}$ (cf. section 5). Integrating this special Kähler geometry relation, we obtain the prepotential, which gives the right predictions for Gromov–Witten invariants (see [**CKYZ**] for examples).

For general G, the matrix (C_{ab}) may not be square. Even if it is square it might be singular. In such cases, some of the sheaves $\mathcal{O}_{C_a}(-J_a)$ do not have symplectic duals. However the special Kähler relation (6.9) still makes sense under suitable modifications.

(6-4) Examples. Here we will present two examples to verify Conjecture 6.3. Some detailed physical analysis of D-branes and Π-stability on the first example may be found in [**DFR**] (see also [**Ma**],[**Do**] and references therein). The hypergeometric series (with a different definition) of the second example are also studied in [**dOFS**] (see also [**MR**]).

Example (1): As the first example, we consider the case $X = \widehat{\mathbf{C}^3/\mathbf{Z}_3}$ with the group action specified by the weights $\frac{1}{3}(1, 1, 1)$. The toric resolution is the same as the G-Hilb and may be identified with the total space of the canonical bundle $K_{\mathbf{P}^2}$. The cohomology $H^{even}(X, \mathbf{Q})$ is generated by $1, J, J^2$ where J is the class dual to a line C in $\mathbf{P}^2$. As shown in the toric diagram in Fig.3, the toric divisor D represents the section $\mathbf{P}^2$. The construction of the secondary polytope is easy, and we find $l^{(1)} = (1,1,1,-3)$ for the local parameter $x = a^{l^{(1)}}$ of the GKZ hypergeometric system. With these data, we may write down the cohomology-valued hypergeometric series,

$$w(x;\frac{J}{2\pi i}) = \sum_{n=0}^{\infty} \frac{1}{\Gamma(1+n+\rho)^3\Gamma(1-3(n+\rho))} x^{n+\rho}|_{\rho=\frac{J}{2\pi i}} . \tag{6.10}$$

The Taylor series expansion with respect to ρ and the ring structure of $H^{even}(X,\mathbf{Q})$ define the hypergeometric series of our interest. As conjectured in Conjecture 6.3, the integral and symplectic structure of the hypergeometric series come from $K(X)$ and $K^c(X)$, respectively.

The tautological line bundles form an integral basis of $K(X)$ and, following [**Re**],[**Cr**], they are easily determined as

$$\mathcal{F}_0 = \mathcal{O}_X \ , \ \mathcal{F}_1 = \mathcal{O}_X(J) \ , \ \mathcal{F}_2 = \mathcal{O}_X(2J) \ .$$

We can determine the dual basis S_k following [**IN**]. Here, instead, we fix the D-brane basis $\{\mathcal{B}_I\}$ to $\{\mathcal{O}_p, \mathcal{O}_C(-1), \mathcal{O}_D(-2)\}$, where $\mathcal{O}_C(-1) := \mathcal{O}_C \otimes \mathcal{O}_X(-J)$ and $\mathcal{O}_D(-2) := \mathcal{O}_D \otimes \mathcal{O}_X(-2J)$. The relations to the basis $\{ S_k \}$ are easily worked out to be

$$\mathcal{O}_p = S_0 + S_1 + S_2 \ , \ \ \mathcal{O}_C(-1) = S_1 + 2S_2 \ , \ \ \mathcal{O}_D(-2) = S_2, \tag{6.11}$$

and also the symplectic form (6.8) becomes

$$(\chi(\mathcal{B}_I, \mathcal{B}_J)) = \begin{pmatrix} 0 & 0 & 0 \\ 0 & 0 & -3 \\ 0 & 3 & 0 \end{pmatrix} . \tag{6.12}$$

From the relations in (6.11), we find a basis $\{\mathcal{F}_0, \mathcal{F}_1 - \mathcal{F}_0, \mathcal{F}_0 - 2\mathcal{F}_1 + \mathcal{F}_2\}$ of $K(X)$, which is dual to $\{\mathcal{O}_p, \mathcal{O}_C(-1), \mathcal{O}_D(-2)\}$. Taking $ch : K(X) \to H^{even}(X,\mathbf{Q})$, we then finally arrive at a basis $\{Q_0, Q_2, Q_4\}$ of $H^{even}(X,\mathbf{Q})$,

$$Q_0 = ch(\mathcal{F}_0) \ \ , \ \ Q_2 = ch(\mathcal{F}_1) - ch(\mathcal{F}_0) \ \ , \ \ Q_0 = ch(\mathcal{F}_2) - 2ch(\mathcal{F}_1) + ch(\mathcal{F}_0) \ \ ,$$

in which we expand our cohomology-valued hypergeometric series,

$$w(x;\frac{J}{2\pi i}) = w^{(0)}(x)Q_0 + w^{(1)}(x)Q_2 + w^{(2)}(x)Q_4 \ .$$

For convenience, we write explicitly the coefficient hypergeometric series,

$$w^{(0)}(x) = 1 \ , \ w^{(1)}(x) = \partial_{\tilde\rho} w(x) \ , \ w^{(2)}(x) = \frac{1}{2}\partial_{\tilde\rho}^2 w(x) - \frac{1}{2}\partial_{\tilde\rho} w(x) \ , \tag{6.13}$$

where $\partial_{\tilde\rho} := \frac{1}{2\pi i}\frac{\partial}{\partial\rho}$.

Our Conjecture 6.3 predicts that the coefficient hypergeometric series $w^{(0)}(x)$, $w^{(1)}(x)$, $w^{(2)}(x)$ thus obtained have integral and symplectic monodromy properties with respect to the symplectic form (6.12). We verify these properties in Appendix A. Moreover our Conjecture 6.3 predicts that these hypergeometric series represent period integrals over the respective mirror cycles of $\mathcal{O}_p, \mathcal{O}_C(-1), \mathcal{O}_D(-2)$ under $mir : K^c(X) \xrightarrow{\sim} H_3(Y,\mathbf{Z})$, i.e.,

$$\begin{aligned} w^{(0)}(x) &= \int_{mir(\mathcal{O}_p)} \Omega(Y_x) \ , \\ w^{(1)}(x) &= \int_{mir(\mathcal{O}_C(-1))} \Omega(Y_x) \ , \ \ w^{(2)}(x) = \int_{mir(\mathcal{O}_D(-2))} \Omega(Y_x) \ , \end{aligned} \tag{6.14}$$

where Y_x is the mirror hypersurface $F(a; U, V, W_1, W_2) = 0$ in $\mathbf{C}^2 \times (\mathbf{C}^*)^2$ defined in (6.1). In Appendix A, we construct explicitly the cycles which represent $mir(\mathcal{O}_p), mir(\mathcal{O}_C(-1))$ and $mir(\mathcal{O}_D(-2))$.

We remark that, in the case of the two-dimensional singularity, the hyperkähler rotation helped us identify the mirror cycles in $H_2(Y, \mathbf{Z})$. In the present case, however, it is not so obvious to see that there is a canonical (geometric) way to construct the isomorphism $mir : K^c(X) \xrightarrow{\sim} H_3(Y, \mathbf{Z})$. Nevertheless, our Conjecture 6.3 predicts that there is in fact a canonical isomorphism encoded indirectly in our cohomology-valued hypergeometric series. This situation is parallel to the mirror symmetry of compact Calabi–Yau manifolds, where the Strominger–Yau–Zaslow construction should provide a recipe for the canonical mirror correspondence.

Finally we determine the prepotential. For this, we construct the symplectic dual to $\mathcal{O}_C(-1)$ by $-\frac{1}{3}\mathcal{O}_D(-2)$. Then from (6.14) the special Kähler relation becomes

$$(1, t, \frac{\partial F}{\partial t}) = (1, w^{(1)}(x), -\frac{1}{3}w^{(2)}).$$

Integrating this relation, we obtain the Gromov–Witten invariants of the geometry $X = K_{\mathbf{P}^2}$ (the first column of Table 1 below). It should be noted that the symplectic dual $\frac{1}{3}\mathcal{O}_D(-2)$ is no longer in $K^c(X)$ but in $K^c(X) \otimes \mathbf{Q}$.

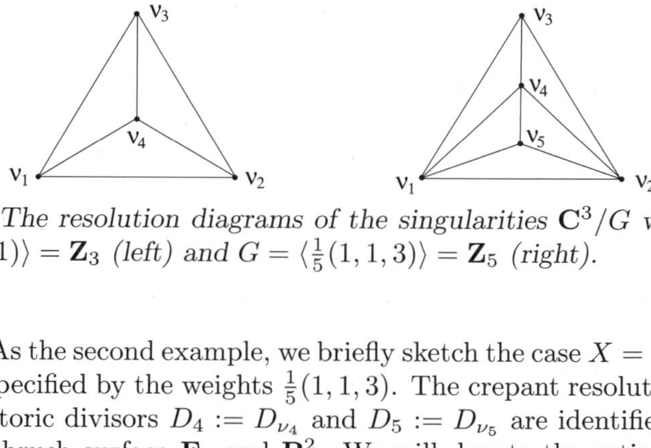

Fig.3. *The resolution diagrams of the singularities* $\mathbf{C}^3/G$ *with* $G = \langle \frac{1}{3}(1,1,1) \rangle = \mathbf{Z}_3$ *(left) and* $G = \langle \frac{1}{5}(1,1,3) \rangle = \mathbf{Z}_5$ *(right).*

Example (2): As the second example, we briefly sketch the case $X = \widehat{\mathbf{C}^3/\mathbf{Z}_5}$ with the group action specified by the weights $\frac{1}{5}(1,1,3)$. The crepant resolution is described in Fig.3. The toric divisors $D_4 := D_{\nu_4}$ and $D_5 := D_{\nu_5}$ are identified, respectively, with the Hirzebruch surface $\mathbf{F}_3$ and $\mathbf{P}^2$. We will denote the rational curves that appear as the zero section of $\mathbf{F}_3$ by $b \equiv C_1$, and also the fiber by $f \equiv C_2$. Thus the intersection numbers on $\mathbf{F}_3$ are $b \cdot b = -3$, $b \cdot f = 1$, $f \cdot f = 0$. $H^2(X, \mathbf{Z})$ is generated by J_1 and J_2, which measure the volumes $C_1 = b$ and $C_2 = f$, respectively (i.e. $\int_{C_i} J_k = \delta_{ik}$). By the standard toric method (see, e.g., [**Oda2**],[**Fu**]), we can determine the ring structure of the cohomology as

$$H^{even}(X, \mathbf{Q}) = \mathbf{Q}[J_1, J_2]/(J_1^2, J_2^2 - 3J_1J_2, J_2^3).$$

The construction of the secondary fan $Sec(\Sigma_G)$ is straightforward and we find $l^{(1)} = (1,1,0,1,-3)$ and $l^{(2)} = (0,0,1,-2,1)$ for the local parameters $x_1 = x := a^{l^{(1)}}$ and $x_2 = y := a^{l^{(2)}}$ of the hypergeometric series. With these data we can write down the cohomology-valued hypergeometric series (6.4) with

$$c(n, m) = \frac{1}{\Gamma(1+n)^2\Gamma(1+m)\Gamma(1+n-2m)\Gamma(1-3n+m)}.$$

For the expansion of the cohomology-valued hypergeometric series, we first fix a symplectic D-brane basis of $K^c(X)$ by $\{\mathcal{B}_J\} = \{\mathcal{O}_p, \mathcal{O}_b(-J_1), \mathcal{O}_f(-J_2), \mathcal{O}_{\mathbf{P}^2}(-2J_1), \mathcal{O}_{\mathbf{F}_3}(-2J_2)\}$. Then it is straightforward to express these bases in terms of the S_k by evaluating the pairings of $\mathcal{B}_J$'s with the tautological bundles [**Re**],[**Cr**];

$$\mathcal{F}_0 = \mathcal{O}_X \ , \ \mathcal{F}_1 = \mathcal{O}_X(J_1) \ , \ \mathcal{F}_2 = \mathcal{O}_X(2J_1) \ , \ \mathcal{F}_3 = \mathcal{O}_X(J_2) \ , \ \mathcal{F}_4 = \mathcal{O}_X(J_1 + J_2) \ .$$

Here we summarize the results together with their duals under the pairing $K^c(X) \times K(X) \to \mathbf{Z}$:

$$\begin{array}{rll} & K^c(X) & K(X) \\ \mathcal{O}_p & = S_0 + S_1 + S_2 + S_3 + S_4 & -\mathcal{F}_0 + 2\mathcal{F}_1 + 2\mathcal{F}_3 - 2\mathcal{F}_4 \\ \mathcal{O}_b(-J_1) & = S_1 + 2S_2 + S_4 & -\mathcal{F}_3 + \mathcal{F}_4 \\ \mathcal{O}_f(-J_2) & = S_3 + S_4 & \mathcal{F}_0 - 2\mathcal{F}_1 - \mathcal{F}_3 + 2\mathcal{F}_4 \\ \mathcal{O}_{\mathbf{P}^2}(-2J_1) & = S_2 & \mathcal{F}_0 - 2\mathcal{F}_1 + \mathcal{F}_2 \\ \mathcal{O}_{\mathbf{F}_3}(-2J_2) & = 2S_0 + S_1 & \mathcal{F}_0 - \mathcal{F}_1 - \mathcal{F}_3 + \mathcal{F}_4 \end{array} \tag{6.15}$$

The symplectic form on $K^c(X)$ may be evaluated using [Corollary 5.3. in IN] as

$$(\chi(\mathcal{B}_I, \mathcal{B}_J)) = \begin{pmatrix} 0 & 0 & 0 & 0 & 0 \\ 0 & 0 & 0 & -3 & 1 \\ 0 & 0 & 0 & 1 & -2 \\ 0 & 3 & -1 & 0 & 0 \\ 0 & -1 & 2 & 0 & 0 \end{pmatrix} . \tag{6.16}$$

Evaluating the Chern characters of the tautological bundles in (6.15), we have for the basis $\{Q_J\}$ of $H^{even}(X, \mathbf{Q})$,

$$Q_0 = 1 - 2J_1J_2 \ ; \ Q_2^1 = J_1 + \frac{1}{2}J_1^2 + J_1J_2 \ , \ Q_2^2 = J_2 + \frac{7}{2}J_1J_2 \ ; \ Q_4^1 = J_1^2 \ , \ Q_4^2 = J_1J_2 \ ,$$

by which we arrange the cohomology-valued hypergeometric series as

$$w(x; \frac{J}{2\pi i}) = w^{(0)}(x)Q_0 + \sum_{a=1}^{2} w_a^{(1)}(x)Q_2^a + \sum_{b=1}^{2} w_b^{(2)}(x)Q_4^b \ ,$$

(where $w^{(0)}(x) \equiv 1$ from the form $c(n,m)$ above). Our Conjecture 6.3 predicts that the coefficient hypergeometric series $\{w^{(0)}(x), w_a^{(1)}(x), w_b^{(2)}(x)\}$ should be integral and symplectic with respect to the symplectic form (6.16).

As a check of the mirror symmetry, we determine the prepotential. First, we write the special Kähler relation (6.9),

$$(1 \, , \, t_a \, , \, \frac{\partial F}{\partial t_a}) = (1 \, , \, w_a^{(1)}(x) \, , \, \sum_b C^{ab} w_b^{(2)}(x)) \ ,$$

with the matrix $(C^{ab}) = \begin{pmatrix} -3 & 1 \\ 1 & -2 \end{pmatrix}^{-1}$. From this relation, it is straightforward to obtain

$$\begin{aligned} \frac{\partial F}{\partial t_1} &= -\frac{1}{5}(\partial_{\tilde{\rho}_1}^2 + \partial_{\tilde{\rho}_1}\partial_{\tilde{\rho}_2} + \frac{3}{2}\partial_{\tilde{\rho}_2}^2) w(x) + \frac{2}{5}t_1 + \frac{7}{10}t_2 - \frac{2}{5} \ , \\ \frac{\partial F}{\partial t_2} &= -\frac{1}{10}(\partial_{\tilde{\rho}_1}^2 + 6\partial_{\tilde{\rho}_1}\partial_{\tilde{\rho}_2} + 9\partial_{\tilde{\rho}_2}^2) w(x) + \frac{7}{10}t_1 + \frac{27}{10}t_2 - \frac{6}{5} \ , \end{aligned}$$

where $\partial_{\tilde{\rho}_k} = \frac{1}{2\pi i}\partial_{\rho_k}$. Just as in the case for compact Calabi–Yau manifolds, we can integrate the special Kähler geometry relation with the prepotential

$$F(t) = \frac{1}{6}(-\frac{2}{5}t_1^3 - \frac{3}{5}t_1^2t_2 - \frac{9}{5}t_1t_2^2 - \frac{9}{5}t_2^3) + \frac{7}{10}t_1t_2 + \frac{1}{5}t_1^2 + \frac{27}{20}t_2^2 - \frac{2}{5}t_1 - \frac{6}{5}t_2 + F_{inst}(t) \ ,$$

where $F_{inst}(t) := \sum_{n,m\geq 0} N(n,m)q_1^n q_2^m$ represents the quantum corrections. Here $N(n,m)$ are the genus zero Gromov–Witten invariants of the homology class $\beta = nb + mf \in H_2(X,\mathbf{Z})$, i.e. $N(n,m) = N_0(nb+mf)$ in the standard notation $N_g(\beta)$ of Gromov–Witten invariants of genus g in literatures. In Table 1, we have listed the integral 'invariants' called Gopakumar–Vafa invariants $\tilde{N}(n,m)$ [**GoV**]. One can verify that the Gromov–Witten invariants coincide with the results from the topological vertex technique (see [**AKMV**]).

We may observe the similar integrability of the special Kähler relation for many other examples. In compact situations, this integrability (or more specifically the existence of the so-called Griffiths–Yukawa coupling) is a consequence of Griffiths transversality (see [**St**]) for which we use integration over (mirror) manifolds. However it should be noted that the existence of the prepotential for non-compact cases seems non-trivial, since integrating over a non-compact manifold needs some special care.

n\m	0	1	2	3	4	5	6
0	0	-2	0	0	0	0	0
1	3	4	3	5	7	9	11
2	-6	-10	-12	-12	-24	-56	-140
3	27	64	91	108	150	294	675
4	-192	-572	-980	-1332	-1808	-2982	-5992
5	1695	6076	12259	18912	26983	42005	76608
6	-17064	-71740	-166720	-289440	-443394	-689520	-1192644
7	188454	909760	2394779	4632120	7665776	12254816	20764870

Table 1. Gopakumar–Vafa numbers $\tilde{N}_0(n,m) = \tilde{N}_0(nb+mf)$ for $X = \widehat{\mathbf{C}^3/\mathbf{Z}_5}$. $\tilde{N}_0(n,0)$ coincides with the Gopakumar–Vafa numbers for $\widehat{\mathbf{C}^3/\mathbf{Z}_3}$.

7. Conclusion and discussions

We have studied in detail the cohomology-valued hypergeometric series for the case of non-compact Calabi–Yau manifolds. Giving an interpretation of $w(x; \frac{J}{2\pi i})$ as the central charge formula (Conjecture 6.3), we have provided supporting evidence for the conjecture in the cases of $X = \widehat{\mathbf{C}^2/\mathbf{Z}_{\mu+1}}$, and also $X = G$-Hilb $\mathbf{C}^3$ with finite abelian subgroups $G = \mathbf{Z}_3, \mathbf{Z}_5 \subset SU(3,\mathbf{C})$. We have also clarified the structure of the prepotential for the non-compact Calabi–Yau geometries $X = G$-Hilb $\mathbf{C}^3$.

As a byproduct of our study, we have found that K. Saito's system of differential equations satisfied by the primitive forms may be replaced by a suitable (resonant and reducible) GKZ system, whose solutions we may set up easily.

As addressed after Conjecture 2.2, our cohomology-valued hypergeometric series (or the central charge formula) connects two different 'monodromy' properties, Fourier–Mukai transforms and the monodromy transforms of hypergeometric series (Dehn twists in the symplectic mapping class group). We see that the latter monodromy property arises associated with the discriminant locus in $\mathcal{M}_{Sec(\Sigma)}$.

As shown in section 3 for $X = \widehat{\mathbf{C}^2/\mathbf{Z}_{\mu+1}}$, the discriminant splits into several irreducible components in the q-coordinate and the monodromy transform around each irreducible component may be identified with a suitable twist functor. From these facts, it is conceivable that the group of self-equivalences of $D^b(Coh(X))$, i.e. Auteq $D^b(X)$, is generated by these 'monodromy' transformations up to the shift functors,

$$\mathrm{Auteq} D^b(\widehat{\mathbf{C}^2/\mathbf{Z}_{\mu+1}})/\{\,[k]\,|\,k\in\mathbf{Z}\,\} = \left\langle R_1,\cdots,R_\mu, T_{\mathcal{O}_{C_1}(-1)},\cdots,T_{\mathcal{O}_{C_\mu}(-1)}\right\rangle\,,$$

where R_k represents the functor tensoring the tautological line bundle $\mathcal{F}_k$ and $T_{\mathcal{O}_{C_k}(-1)}$ is the Seidel-Thomas twist. (This has been proved affirmatively in a recent paper by Ishii and Uehara [**IU**].) In Proposition 4.4, we have seen explicitly that these functors are represented by the corresponding actions on the roots $\beta_k(a)$, i.e., changing the phases of the roots and permuting (braiding) of the roots $\beta_k(a)$. This simple picture should provide us an intuition about the possible forms of the self-equivalences in $DFuk(Y,\beta)$.

A similar correspondence between monodromy transforms and Fourier–Mukai transforms, and also the canonical mirror isomorphism $mir: K^c(X)\tilde{\to}H_3(Y,\mathbf{Z})$, are contained in our Conjecture 6.3 for three-dimensional cases $X = \widehat{\mathbf{C}^3/G} = G$-Hilb $\mathbf{C}^3$ [**IN**],[**BKR**],[**CI**]. In three dimensions, there can be several Calabi–Yau resolutions of the singularity $\mathbf{C}^3/G$, and correspondingly there are many different types of (Fourier–Mukai) transforms connecting these different resolutions [**BKR**],[**CI**]. It is interesting to study the details of the canonical mirror isomorphism mir from our cohomology-valued hypergeometric series in such situations.

As we have done for $X = \widehat{\mathbf{C}^3/\mathbf{Z}_3}$, in some cases with lower $\dim H^2(X,\mathbf{Q})$, we can determine explicitly the integral cycles and their monodromy properties to provide a consistency check for our Conjecture 6.3. We can also read off the isomorphism $mir: K^c(X)\tilde{\to}H_3(Y,\mathbf{Z})$ from our cohomology-valued hypergeometric series. This isomorphism mir should also follow from the geometric framework like the Strominger–Yau–Zaslow construction, which provides a canonical equivalence $Mir: D^b_c(Coh(X))\tilde{\to}DFuk(Y,\beta)$.

Finally, as for compact Calabi–Yau (hypersurfaces), much supporting evidence for our Conjecture 2.2 is reported in [**Hos**], especially for dimensions one and two. However, for example, the observation (1.4) made for the quintic hypersurface X_5 still needs justification. Namely we have to work out a symplectic basis of the K-group $K(X_5)$ which yields the specific forms of the 'charges'

$$1\,,\; J-\frac{c_1(X_5)J}{12}-\frac{11}{2}\frac{J^2}{5}\,,\;\frac{J^2}{5}\,,\;-\frac{J^3}{5}\,,$$

and also their mirror cycles in $H_3(X_5^\vee,\mathbf{Z})$ with their period integrals

$$w^{(0)}(x)\,,\; w^{(1)}(x)\,,\; w^{(2)}(x)\,,\; w^{(3)}(x).$$

From the SYZ construction [**SYZ**],[**LYZ**] we may expect that $K(X_5)$ is generated by the structure sheaf $\mathcal{O}_X$, a skyscraper sheaf $\mathcal{O}_p$, and additional sheaves $\mathcal{E}$ and $\mathcal{F}$ which, respectively, have their support on a divisor and a curve (,i.e. $D4$- and $D2$-branes). The cycles which give the period integrals $w^{(0)}(x)$ and $w^{(3)}(x)$ are known from the original work [**CdOGP**]. In fact, claiming these cycles to be the mirror images of $\mathcal{O}_p$ and $\mathcal{O}_X$, i.e. the Lagrangian torus cycle T^3 and the section S^3 of the fibration, respectively, was the starting point of the SYZ construction. However we still lack knowledge for the remaining cycles and also the corresponding sheaves $\mathcal{E}$ and $\mathcal{F}$.

Appendix A. Integrating over vanishing cycles

(A-1) $\mathbf{C}^2/\mathbf{Z}_{\mu+1}$. As a warm-up, let us first evaluate the period integral (4.2) of the primitive form. We will obtain the central charge $Z_t(S_k)$ in (5.4) of the sheaf $S_k = \mathcal{O}_{C_k}(-1)$ as the integral over the vanishing cycle $L_k = mir(\mathcal{O}_{C_k}(-1))$.

As in Proposition 4.2, we denote the roots of the polynomial $\psi(W) = a_0 + a_1 W + \cdots + a_{\mu+1} W^{\mu+1}$ by $\beta_0, \cdots, \beta_\mu$. Then, for the defining equation of the singularity, we have

$$f(a,W) + U^2 + V^2 = (W - \beta_0)\cdots(W - \beta_\mu) + U^2 + V^2 \,,$$

where we set $a_{\mu+1} = 1$ by scaling the variable W. Assume that all the roots β_i are real and $\beta_0 < \beta_1 < \cdots < \beta_\mu$. When $f(a,W) \le 0$ holds for $\beta_{k-1} \le W \le \beta_k$, we may construct a vanishing cycle L_k as $L_k = L_k^+ \cup L_k^- \approx S^2$ with

$$L_k^{\pm} = \{(\pm\sqrt{|f(a,W)| - V^2}, V, W) \in \mathbf{R}^3 \mid f_\Sigma(a,W) + V^2 \le 0,\ \beta_{k-1} \le W \le \beta_k\ \}\,.$$

Similarly, when $f(a,W) \ge 0$ for $\beta_{k-1} \le W \le \beta_k$, we construct a vanishing cycle by changing the variables: $U \to \sqrt{-1}U, V \to \sqrt{-1}V$. For the first case, the evaluation of the period integral over L_k^+ proceeds as

$$\text{(A.1)}\quad \begin{aligned} \int_{L_k^+} Res_{F_\Sigma=0}\left(\frac{dW dU dV}{f(W) + U^2 + V^2}\right) &= \int_{\beta_{k-1}}^{\beta_k} dW \int_{-\sqrt{|f(W)|}}^{\sqrt{|f(W)|}} \frac{(\frac{1}{2})\,dV}{\sqrt{|f(W)| - V^2}} \\ = \frac{1}{2}\int_{\beta_{k-1}}^{\beta_k} dW\ \mathrm{Sin}^{-1}\left(\frac{V}{\sqrt{|f(W)|}}\right)\Bigg|_{-\sqrt{|f(W)|}}^{\sqrt{|f(W)|}} &= \frac{\pi}{2}\int_{\beta_{k-1}}^{\beta_k} dW \,, \end{aligned}$$

where $F_\Sigma := f(W) + U^2 + V^2$. We have a similar result for L_k^-, and in total $\pi(\beta_k - \beta_{k-1})$ for the period integral. Since we have μ-independent vanishing cycles in this way, we come to the well-known results summarized in Proposition 4.1, and also Proposition 4.2. In the above evaluation, one should note that the period integrals of the primitive form measure the volumes of the vanishing cycles.

The period integral $\Pi_\gamma(a)$ (3.1) has a slightly different shape in the integration measure. First let us take $\gamma = L_k \times S^1$ for the cycle $\gamma \in H_3(\mathbf{C}^2 \times \mathbf{C}^* \setminus (F_\Sigma = 0), \mathbf{Z})$ with a loop S^1 which encircles the hypersurface. Since the integration over S^1 produces the holomorphic two-form $\Omega(Y_x) = \frac{i}{2\pi^2} Res(\frac{1}{f(W)+U^2+V^2} dU dV \frac{dW}{W})$, we have

$$\text{(A.2)}\quad \int_{L_k^+} \Omega(Y_x) = \frac{i}{4\pi^2}\int_{\beta_{k-1}}^{\beta_k} \frac{dW}{W}\int_{-\sqrt{|f(W)|}}^{\sqrt{|f(W)|}} \frac{dV}{\sqrt{|f(W)| - V^2}} = \frac{i}{4\pi}\int_{\beta_{k-1}}^{\beta_k} \frac{dW}{W} \,,$$

and a similar result for L_k^-. Thus we arrive at the central charge of the sheaves $S_k = \mathcal{O}_{C_k}(-1)$ in (5.4) and justify our claim that $mir(\mathcal{O}_{C_k}(-1)) = L_k$ which follows from the cohomology-valued hypergeometric series (5.3). One should note that the logarithmic behaviors of the solutions in Proposition 4.4 come from the torus invariant measure $\frac{dW}{W}$ in (3.1). The claimed relation $mir(\mathcal{O}_p) = T_0$ for a cycle $T_0 \approx S^1 \times S^1$ can also be justified by evaluating the period integral $\int_{T_0} \Omega(Y_x) = 1$

(see the corresponding cycle T_0 and the evaluation of the period integral over it in three dimensions below).

(A-2) $\mathbf{C}^3/\mathbf{Z}_3$. The evaluations of the period integrals in (A.1),(A.2) extend to three dimensions. In three dimensions, however, we see that some vanishing cycles require careful treatment and in fact become non-compact cycles with non-trivial fundamental groups. This non-compactness of certain (vanishing) cycles comes from the tori $(\mathbf{C}^*)^2$ of the ambient space $\mathbf{C}^2 \times (\mathbf{C}^*)^2$, and is detected only by the invariant measure of the period integral of the local mirror symmetry (A.2). To illustrate this, we first construct all the cycles which reproduce the period integrals (6.14). We will then contrast the results with the corresponding period integrals of the primitive form.

• *(Non-compact) Vanishing cycles and period integrals.*

Let us start with the toric data for the crepant resolution $\widehat{\mathbf{C}^3/\mathbf{Z}^3}$, i.e., $\nu_1 = (1,0,0), \nu_2 = (0,1,0), \nu_3 = (0,0,1), \nu_4 = (1/3,1/3,1/3)$ in N_G. We fix an isomorphism $\varphi : N_G \cong \mathbf{Z}^3$ such that $\varphi(\nu_1) = (1,0,0), \varphi(\nu_2) = (1,2,1), \varphi(\nu_3) = (1,1,2)$, $\varphi(\nu_4) = (1,1,1)$, and consider the following deformation family of hypersurfaces in $\mathbf{C}^2 \times (\mathbf{C}^*)^2$:

$$F_\Sigma(a,U,V,W) := a_1 + a_2 W_1^2 W_2 + a_3 W_1 W_2^2 + a_4 W_1 W_2 + U^2 + V^2 = 0 \,,$$

with $(a_1, \cdots, a_4) \in (\mathbf{C}^*)^4$. The period integral of the local mirror symmetry is defined in general in (6.2) and satisfies the GKZ hypergeometric system (6.3) in Proposition 6.2. This GKZ hypergeometric system introduces the toric compactification $\mathcal{M}_{Sec(\Sigma)}$ in terms of the secondary fan $Sec(\Sigma)$ for the deformation parameters $(a_1, \cdots, a_4)$. In the present case, we have $\mathcal{M}_{Sec(\Sigma)} \cong \mathbf{P}^1$ with the local parameter $x = \frac{a_1 a_2 a_3}{a_4^3}$ near the large complex structure limit. By scaling the variables, we may set $(a_1, a_2, a_3, a_4) = (a, 1, 1, 1)$, and hence we have the mirror family $\{Y_x\}$ with the period integral parametrized by $x = a$;

$$\text{(A.3)} \quad \int_L \Omega(Y_x) = \frac{1}{4\pi^3} \int_L Res_{F_\Sigma=0} \left(\frac{1}{a + f_1(W) + U^2 + V^2} dU dV \frac{dW_1}{W_1} \frac{dW_2}{W_2} \right) ,$$

where $a + f_1(W) = a + W_1 W_2 (1 + W_1 + W_2)$ and L is a cycle in $H_3(Y_x, \mathbf{Z})$. In what follows, we construct integral cycles L and evaluate the above period integral.

First, we construct a torus cycle $T_0 \approx S^1 \times S^1 \times S^1$ which fixes the normalization of the period integral. To do this, we introduce new variables $\tilde{U} = U + iV, \tilde{V} = U - iV$ and write the defining equation as

$$F_\Sigma(a,U,V,W) = a + f_1(W) + U^2 + V^2 = a + f_1(W) + \tilde{U}\tilde{V} \ .$$

We define the cycle to be

$$T_0 := \{ (\tilde{U}, \tilde{V}, W_1, W_2) \in \mathbf{C}^2 \times (\mathbf{C}^*)^2 \mid \tilde{U} = \frac{-a - f_1(W)}{\tilde{V}}, \ |\tilde{V}| = |W_1| = |W_2| = \varepsilon \} \,.$$

Then it is straightforward to evaluate $\int_{T_0} \Omega(Y_x) = 1$ since

$$\frac{1}{4\pi^3} Res_{F_\Sigma=0} \left(\frac{(-2i)^{-1} d\tilde{U} d\tilde{V}}{a + f_1(W) + \tilde{U}\tilde{V}} \frac{dW_1}{W_1} \frac{dW_2}{W_2} \right) = \frac{1}{(2\pi i)^3} \frac{d\tilde{V}}{\tilde{V}} \frac{dW_1}{W_1} \frac{dW_2}{W_2} \,.$$

This normalizes the period integral and verifies the first relation in (6.14) with $mir(\mathcal{O}_p) = T_0$.

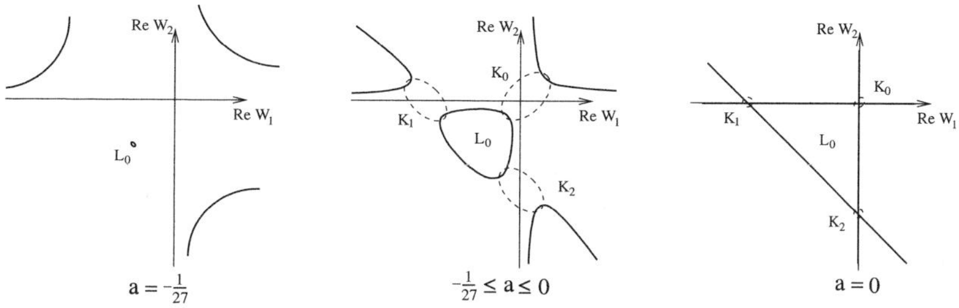

Fig.4. *Vanishing cycles near the critical points* $(0,0), (-1,0), (0,-1), (-\frac{1}{3}, -\frac{1}{3})$ *of the hypersurface* $a + W_1W_2(1+W_1+W_2) + U^2 + V^2 = 0$ *in* $\mathbf{C}^2 \times (\mathbf{C}^*)^2$ $(-\frac{1}{27} \le a \le 0)$. *They are termed* K_0, K_1, K_2 *and* L_0, *respectively, and have topologies* $K_i \approx S^3 \setminus S^1$ $(i = 0,1,2)$ *and* $L_0 \approx S^3$.

Secondly, we describe a vanishing cycle which verifies the third relation in (6.14). Following general arguments on singularities, e.g. in [**AGV**], we may obtain vanishing cycles by looking at the critical points of $f_1(W)$ (more precisely $f_1(W) + U^2 + V^2$). We see that $f_1(W) = W_1W_2(1+W_1+W_2)$ has four critical points,

$$(W_1, W_2)_{crit} = (0,0), (-1,0), (0,-1), (-1/3, -1/3),$$

with critical values $0, 0, 0, \frac{1}{27}$, respectively. The vanishing cycle at $(-1/3, -1/3)$ is easy to describe in a way similar to (A-1). The result is $L_0 := L_0^+ \cup L_0^- \approx S^3$ with

$$L_0^\pm = \left\{ (\sqrt{-1}u, \sqrt{-1}v, W_1, W_2) \,\middle|\, \begin{matrix} v = \pm\sqrt{a + f_1(W) - u^2}, \;\; a + f_1(W) \ge 0, \\ -\sqrt{a+f_1(W)} \le u \le \sqrt{a + f_1(W)} \end{matrix} \right\},$$

where u, v, W_1, W_2 above are assumed to be real, and also $-\frac{1}{27} \le a \le 0$ (see Fig.4). The evaluation of the period integral is quite parallel to (A.2), and we have

$$\text{(A.4)} \qquad \int_{L_0} \Omega(Y_x) = \frac{1}{4\pi^2} \int_{a+W_1W_2(1+W_1+W_2)\ge 0} \frac{dW_1}{W_1} \frac{dW_2}{W_2},$$

where $\int_{a+W_1W_2(1+W_1+W_2)\ge 0}$ means the integral over the closed domain which appears in the third quadrant of the real W_1, W_2 plane for $-\frac{1}{27} \le a \le 0$ (see Fig.4).

We may evaluate the integral (A.4) near $a = 0$ in a straightforward way, by setting $X = W_1, Y = W_2$, as

$$\frac{1}{4\pi^2} \int_{Y_{min}(a)}^{Y_{max}(a)} \frac{dY}{Y} \int_{X_{min}(a,Y)}^{X_{max}(a,Y)} \frac{dX}{X} = \frac{1}{8\pi^2} (\log(-a))^2 + \cdots,$$

where $X_{min}(a,Y) < X_{max}(a,Y)$ are roots of $a + XY(1+X+Y) = 0$, and $Y_{min}(a)$, $Y_{max}(a)$ are determined from the conditions

$$a + XY(1+X+Y) = \frac{\partial}{\partial X}(a + XY(1+X+Y)) = 0$$

and

$$Y_{min}(-\frac{1}{27}) = Y_{max}(-\frac{1}{27}) = -1/3.$$

The behavior $\frac{1}{8\pi^2}(\log(-a))^2 + \cdots$ is easily seen by observing

$$\int_{X_{min}(a,Y)}^{X_{max}(a,Y)} \frac{dX}{X} = \log(-a) - \log(-Y(1+Y)^2) + O(a)$$

and

$$Y_{min}(a) = -1 - 2\sqrt{-a} - 2a + \cdots, \qquad Y_{max}(a) = 4a - 32a^2$$

for small $|a|$. Clearly this period integral vanishes at $a = -\frac{1}{27}$. This vanishing property and the $(\log(-a))^2$ behavior near $a = 0$ suffice to identify the period integral with $w^{(2)}(x) = \frac{1}{2}\partial_{\rho}^2 w(x) - \frac{1}{2}\partial_{\rho} w(x)$ in (6.13), which has the same vanishing property at $x = -\frac{1}{27}$ and also the expansion near $x = 0$,

$$w^{(2)}(x) = -\frac{1}{8\pi^2}(\log(-x))^2 + \frac{1}{8} + \frac{3}{2\pi^2} x \log(-x) + \frac{9}{4\pi^2} x - \frac{423}{16\pi^2} x^2 + \cdots .$$

The difference in sign is simply a matter of the orientation of the vanishing cycle L_0. Thus we verify the third relation in (6.14) with $mir(\mathcal{O}_D(-2)) = L_0 \approx S^3$ up to the orientation of the cycle.

Finally, let us look carefully at the vanishing cycles at the critical points $(0,0)$, $(-1,0)$, $(0,-1)$, which we call K_0, K_1, K_2, respectively. Usually vanishing cycles have the topology of S^3. However this is not the case for the cycles K_i since we consider the mirror geometry in $\mathbf{C}^2 \times (\mathbf{C}^*)^2$, (see Fig 4). It will turn out that they have topology $S^3 \setminus S^1$ and hence are non-compact.

Let us first describe the vanishing cycle at the critical point $(0,0)$. Similarly to L_0, we define $K_0 = K_0^+ \cup K_0^-$ using real variables A, B, U, V by

$$K_0^{\pm} = \left\{ (U, V, A+iB, A-iB) \,\middle|\, \begin{array}{l} U = \pm\sqrt{|a + f_1(A,B)| - V^2}, \;\; a + f_1(A,B) \le 0, \\ -\sqrt{|a + f_1(A,B)|} \le V \le \sqrt{|a + f_1(A,B)|} \end{array} \right\},$$

where $a + f_1(A,B) = a + (A^2+B^2)(1+2A)$. Note that the locus $A = B = 0$ is not contained in $K_0^{\pm}$ because $W_1 = A+iB, W_2 = A-iB$ live in $\mathbf{C}^*$. Since the excluded loci $A = B = 0$ in K_0^+ and K_0^- make up S^1, we see that $K_0 = K_0^+ \cup K_0^- \approx S^3 \setminus S^1$. After some calculations similar to (A.2), we obtain

$$\int_{K_0} \Omega(Y_x) = \frac{1}{4\pi^3}\pi \int_{a+(A^2+B^2)(1+2A)\le 0} \frac{-2i dA dB}{(A^2+B^2)},$$

where we mean by $a + (A^2+B^2)(1+2A) \le 0$ the closed domain which appears near the origin in the real A, B plane for $-\frac{1}{27} \le a \le 0$. Note that the integral diverges if the locus $A = B = 0$ is contained in K_0. We may isolate this divergent contribution from the origin by introducing the polar coordinate $r\cos\theta = \sqrt{1+2A}\,A$, $r\sin\theta = \sqrt{1+2A}\,B$ satisfying $r^2\cos^2\theta = (1+2A)A^2$. The calculation proceeds as follows:

$$\text{(A.5)} \qquad \begin{aligned} \int_{K_0} \Omega(Y_x) &= \frac{-i}{2\pi^2}\int_{\sqrt{\epsilon}}^{\sqrt{-a}} \frac{dr}{r} \int_0^{2\pi} d\theta \frac{1+2A(r,\theta)}{1+3A(r,\theta)} \\ &= \frac{1}{2\pi i}\frac{1}{3}(-6a + 45a - 560a^3 + \cdots) + \frac{1}{2\pi i}(\log(-a) - \log\epsilon), \end{aligned}$$

where $\epsilon > 0$ is a small constant to handle the non-compact cycle $K_0 \approx S^3 \setminus S^1$, and the divergent term $\log\epsilon$ when $\epsilon \to 0$ will be ignored in the following. (Here is an ambiguity about the cut of the function log. Precisely, we should say that we drop the divergent term $\log(-\epsilon) = \log\epsilon - \pi i$.) This series expansion should be compared with the hypergeometric series $w^{(1)}(x) = \partial_{\tilde\rho} w(x)$ in (6.13), which has the form near $x = 0\,(x = a)$

$$w^{(1)}(x) = \frac{1}{2\pi i}\log(x) + \frac{1}{2\pi i}(-6x + 45x^2 - 560x^3 + \frac{17325}{2}x^4 + \cdots)\ .$$

The slight difference in the factor $\frac{1}{3}$ can be explained by the contribution from the similar non-compact vanishing cycles K_1, K_2 at the critical points $(-1, 0)$ and $(0, -1)$, respectively. Let us describe the cycle $K_1 = K_1^+ \cup K_1^-$ by

$$K_1^{\pm} = \left\{ (U, V, -1-2A, A+iB) \,\middle|\, \begin{array}{l} U = \pm\sqrt{|a + f_1(A,B) - V^2|},\ a + f_1(A,B) \le 0, \\ -\sqrt{|a + f_1(A,B)|} \le V \le \sqrt{|a + f_1(A,B)|} \end{array} \right\}.$$

Here, similarly to K_0, $a + f_1(A,B) = a + (A^2+B^2)(1+2A)$ and we mean for $a + f_1(A,B) \le 0$ to represent the closed domain near the origin in the real A, B plane which appears for $-\frac{1}{27} \le a \le 0$. As is the case for K_0, the locus $A = B = 0$ is excluded and hence we have $K_1 \approx S^3 \setminus S^1$. The integration over this non-compact cycle K_1 may be done by changing variables to polar coordinates. After some manipulations, we get a similar result to (A.5);

$$\text{(A.6)}\qquad \begin{aligned} \int_{K_0} \Omega(Y_x) &= \frac{-i}{2\pi^2}\int_{\sqrt{\epsilon}}^{\sqrt{-a}} \frac{dr}{r}\int_0^{2\pi} d\theta \frac{-A(r,\theta)}{1 + 3A(r,\theta)} \\ &= \frac{1}{2\pi i}\frac{1}{3}(-6a + 45a - 560a^3 + \cdots)\ , \end{aligned}$$

which does not contain the divergent term $\log(-a) - \log\epsilon$. We should note in the difference between (A.5) and (A.6) that K_0 is *not* homologous to K_1 in $H_3(Y_x, \mathbf{Z})$ although they have the same topology $S^3 \setminus S^1$.

By symmetry under $W_1 \leftrightarrow W_2$, we have a similar construction of the cycle $K_2 = K_2^+ \cup K_2^- \approx S^3 \setminus S^1$ and obtain the same result as above for $\int_{K_2}\Omega(Y_x)$. Putting all the above results together, we obtain $\int_{K_0+K_1+K_2}\Omega(Y_x) = w^{(1)}(x)$, which entails $mir(\mathcal{O}_C(-1)) = K_0 + K_1 + K_2$ from (6.14).

To summarize, we read the mirror isomorphism $mir : K^c(X) \xrightarrow{\sim} H_3(Y, \mathbf{Z})$ from the first part of our Conjecture 6.3 (and the equation (6.14)), up to orientations of cycles, as

$$\text{(A.7)}\quad mir(\mathcal{O}_p) = T_0\ ,\quad mir(\mathcal{O}_C(-1)) = K_0 + K_1 + K_2\ ,\quad mir(\mathcal{O}_D(-2)) = L_0\ ,$$

where the topologies of the cycles are $T_0 \approx S^1 \times S^1 \times S^1$, $K_i \approx S^3 \setminus S^1 (i = 0, 1, 2)$ and $L_0 \approx S^3$.

- *Monodromy and the symplectic form.*

We have identified the period integrals over the cycles with the hypergeometric series by $\int_{T_0}\Omega(Y_x) = w^{(0)}(x)$, $\int_{K_0+K_1+K_2}\Omega(Y_x) = w^{(1)}(x)$, $\int_{L_0}\Omega(Y_x) = w^{(2)}(x)$, where $(w^{(0)}(x), w^{(1)}(x), w^{(2)}(x)) = (1, \partial_{\tilde\rho} w(x), \frac{1}{2}\partial_{\tilde\rho}^2 w(x) - \frac{1}{2}\partial_{\tilde\rho} w(x))$. The calculation of the monodromy of the hypergeometric series ${}^t(w^{(0)}(x), w^{(1)}(x), w^{(2)}(x))$ is straightforward (see e.g. [**Hos**]), and we obtain the monodromy matrices about $x = 0, x = -\frac{1}{27}, x = \infty$, respectively, as

$$M_0 = \begin{pmatrix} 1 & 0 & 0 \\ 1 & 1 & 0 \\ 0 & 1 & 1 \end{pmatrix}\ ,\quad M_1 = \begin{pmatrix} 1 & 0 & 0 \\ 0 & 1 & -3 \\ 0 & 0 & 1 \end{pmatrix}\ ,\quad M_\infty = \begin{pmatrix} 1 & 0 & 0 \\ -1 & 1 & 3 \\ 1 & -1 & -2 \end{pmatrix}\ ,$$

where $M_\infty := (M_1 M_0)^{-1}$. Now we may verify the second part of Conjecture 6.3 by observing the equality of the two symplectic forms

$$(\chi(\mathcal{B}_I, \mathcal{B}_J)) = (\# L_I \cap L_J) = \begin{pmatrix} 0 & 0 & 0 \\ 0 & 0 & -3 \\ 0 & 3 & 0 \end{pmatrix} ,$$

for the basis $\{\mathcal{B}_I\} = \{\mathcal{O}_p, \mathcal{O}_C(-1), \mathcal{O}_D(-2)\}$ of $K^c(X)$ and its mirror image $\{L_I\} = \{T_0, K_0 + K_1 + K_2, L_0\}$ in $H_3(Y, \mathbf{Z})$. The intersection numbers among the cycles L_I's are easily determined from their definitions (see also Fig.4,) under suitable orientations. Now the second part of Conjecture 6.3 follows from the Picard–Lefschetz theory of the cycles L_I. By the mirror isomorphism $mir : K^c(X) \overset{\sim}{\to} H_3(Y, \mathbf{Z})$, it is conjectured that the cycles $T_0, K_0 + K_1 + K_2, L_0$ constitute an integral basis of $H_3(Y, \mathbf{Z})$.

- *Period integral of the primitive form.*

Here to have things to be contrasted, let us evaluate the period integrals of the primitive form,

$$\text{(A.8)} \qquad \int_L \mathcal{U}(a) = \int_L Res_{F_\Sigma = 0} \left(\frac{dU dV dW_1 dW_2}{a + f_1(W) + U^2 + V^2} \right)$$

for the cycles $L = T_0, K_0, K_1, K_2, L_0$ which are considered in the hypersurface $a + f_1(W) + U^2 + V^2 = 0$ in $\mathbf{C}^4$. Firstly, it is rather clear that the homology class T_0 is trivial and we have $\int_{T_0} \mathcal{U}(a) = 0$. Also, all the cycles K_0, K_1, K_2 and L_0 are simply vanishing cycles having the topology of S^3. The evaluations of the period integral are similar to the previous case of the local mirror symmetry, and it is straightforward to obtain

$$\int_{L_0} \mathcal{U}(a) = \pi \int_{a + f_1(W) \geq 0} dW_1 dW_2 = \frac{2\pi^2}{9\sqrt{3}} (1 + 27a)\, {}_2F_1(\frac{1}{3}, \frac{2}{3}, 2, 1 + 27a),$$

where $f_1(W) = W_1 W_2 (1 + W_1 + W_2)$ and, as before, $a + f_1(W) \geq 0$ represents the closed domain in the third quadrant in the real W_1, W_2 plane for $-\frac{1}{27} \leq a \leq 0$. For the cycles K_i $(i = 0, 1, 2)$, we have the same results,

$$\int_{K_i} \mathcal{U}(a) = (-2i)\pi \int_{a + f_1(A,B) \leq 0} dA dB = 2\pi^2 i\, a\, {}_2F_1(\frac{1}{3}, \frac{2}{3}, 2, -27a) \ ,$$

where $f_1(A, B) = (A^2 + B^2)(1 + 2A)$ and $a + f_1(A, B) \leq 0$ represents a closed domain near the origin of the real A, B plane which exists for $-\frac{1}{27} \leq a \leq 0$. The above result indicates that the cycles K_0, K_1, K_2 are homologically equivalent. In fact, we may verify the following Picard–Fuchs equation of second order satisfied by the period integral (A.8);

$$\text{(A.9)} \qquad \{\theta_a(\theta_a - 1) + 3a(3\theta_a - 2)(3\theta_a - 1)\} \int_L \mathcal{U}(a) = 0 \ ,$$

where $\theta_a = a\frac{\partial}{\partial a}$. This equation indicates that there exist only two homologically independent cycles. In appendix B, we will derive this Picard–Fuchs equation starting from the GKZ system appearing in the Remark after Proposition 6.2. This equation should be compared with the differential equation

$$\text{(A.10)} \qquad \{\theta_x^2 + 3x(3\theta_x + 2)(3\theta_x + 1)\}\theta_x \int_L \Omega(Y_x) = 0 \ ,$$

where $\theta_x = x\frac{\partial}{\partial x}$. The latter third order differential equation, which also follows from the GKZ system in Proposition 6.2, characterizes the period integrals over the cycles (A.7).

Appendix B. Differential equation (A.9)

Here we derive the second order differential equation (A.9) from the GKZ system for the primitive form (cf.(6.3)) by observing the factorization of a first order differential operator. The same second order differential equation can be derived from the K. Saito's system.

As in (A-2), we fix an isomorphism $\varphi : N_G \xrightarrow{\sim} \mathbf{Z}^3$ so that

$$(\varphi(\nu_1)\ \varphi(\nu_2)\ \varphi(\nu_3)\ \varphi(\nu_4)) = \begin{pmatrix} 1 & 1 & 1 & 1 \\ 0 & 2 & 1 & 1 \\ 0 & 1 & 2 & 1 \end{pmatrix} .$$

From this, we may write the linear operators $\mathcal{Z}'_i$;

$$\mathcal{Z}'_1 = \theta_1 + \theta_2 + \theta_3 + \theta_4 \ , \ \mathcal{Z}'_2 = 2\theta_2 + \theta_3 + \theta_4 + 1 \ , \ \mathcal{Z}'_3 = 3\theta_3 + \theta_4 + 1 .$$

Since the period integral $\int_L \mathcal{U}(a)$ is annihilated by these operators, we have the following relations:

(B.1) $$\theta_2 = \theta_1 - 1 \ , \ \theta_3 = \theta_1 - 1 \ , \ \theta_4 = -3\theta_1 + 2 \ ,$$

when acting on $\int_L \mathcal{U}(a)$. Now consider the operator $\Box_l$ for $l = (1,1,1,-3)$. We have

$$a_1a_2a_3\Box_l \int_L \mathcal{U}(a) = \left(\theta_1\theta_2\theta_3 - \frac{a_1a_2a_3}{a_4^3}(\theta_4-2)(\theta_4-1)\theta_4\right) \int_L \mathcal{U}(a) = 0 .$$

Then it is easy to observe the following factorization of the operator when we eliminate $\theta_{a_2}, \theta_{a_3}, \theta_{a_4}$ by (B.1):

$$(\theta_1 - 1)\left\{\theta_1(\theta_1-1) + 3\frac{a_1a_2a_3}{a_4^3}(3\theta_1-2)(3\theta_1-1)\right\} \int_C \mathcal{U}(a) = 0 .$$

The inhomogeneous terms (i.e. constants) in (B.1) may be removed if we define $\Pi_L(a) := \frac{a_2a_3}{a_4^2}\int_L \mathcal{U}(a)$. Even in that case, the shape of the above factorized differential operator does not change. The resulting homogeneous linear operators imply that $\Pi_L(a) = \Pi_L(x)$ with $x = \frac{a_1a_2a_3}{a_4^3}$, and enable us to set $a_1 = a, a_2 = a_3 = a_4 = 1$ in which we have $\Pi_L(x) = \int_L \mathcal{U}(a)$ with $x = a$. Now we arrive at the second order differential equation

$$\{\theta_a(\theta_a - 1) + 3a(3\theta_a - 2)(3\theta_a - 1)\}\, \Pi_L(a) = 0 \ ,$$

by taking the irreducible part. This completes our derivation of (A.9).

A similar factorization property of the GKZ systems for primitive forms may be observed for G such that $\mathbf{C}^3/G$ has an isolated singularity. This factorization property is reminiscent of the factorizations observed in [**HKTY1**] for the (extended) GKZ systems which appear in applications of mirror symmetry, although the irreducible part shows slightly different degenerations, as we have seen above (cf. (A.10)).

References

[AKMV] M. Aganagic, A. Klemm, M. Marino and Cumrun Vafa, *The Topological Vertex* , Commu. Math. Phys. 254 (2000), 425–478.

[AGV] V.I. Arnold, S.M. Gusein-Zade and A.N. Varchenko, *Singularities of Differential Maps, Vol.II*, Birkhäuser (1988).

[Ba1] V.V. Batyrev, *Dual polyhedra and mirror symmetry for Calabi–Yau hypersurfaces in toric varieties*, J. Alg. Geom. 3 (1994) 493–535.

[Ba2] V.V. Batyrev, *Variations of the mixed Hodge structure of affine hypersurfaces in algebraic tori*, Duke Math. J. 69(1993), 349 – 409.

[Ba3] V.V. Batyrev, Talk given at Department of Mathematics, Tokyo Institute of Technology (September, 2003).

[Br] T. Bridgeland, *Stability conditions on triangulated categories*, math.AG/0212237.

[BKR] T. Bridgeland, A. King and M. Reid, *The McKay correspondence as an equivalence of derived categories*, J. Amer. Math. Soc. 14(2001),no.3,535–554.

[CdOGP] P. Candelas, X.C. de la Ossa, P.S. Green, and L.Parkes, *A pair of Calabi–Yau manifolds as an exactly soluble superconformal theory*, Nucl.Phys. B356(1991), 21-74.

[CKYZ] T.-T. Chiang, A. Klemm, S.-T. Yau and E. Zaslow, *Local Mirror Symmetry: Calculations and Interpretations*, Adv.Theor.Math.Phys. 3(1999),495–565.

[Cr] A. Craw, *An explicit construction of the McKay correspondence for A-Hilb* $\mathbf{C}^3$, Duke Math. J., 124 (2004), 259–307.

[CI] A. Craw and A. Ishii, *Flops of G-Hilb and equivalences of derived categories by variation of GIT quotient*, math.AG/0211360.

[dOFS] X. de la Ossa, B. Florea and H. Skarke, *D-Branes on Noncompact Calabi–Yau Manifolds: K-Theory and Monodromy*, Nucl.Phys. B644 (2002) 170–200.

[Di] A. Dimca, *Singularities and Topology of Hypersurfaces*, Universitext, Springer-Verlag, New York, 1992.

[Do] M.R. Douglas, *Dirichlet branes, homological mirror symmetry, and stability*, math.AG/0207021, in the 2002 ICM proceedings.

[DFR] M. Douglas, B. Fiol and C. Römelsberger, *The spectrum of BPS branes on a noncompact Calabi–Yau*, hep-th/0003263.

[FO3] K. Fukaya, Y.G. Oh, H. Ohta and K. Ono, *Lagrangian intersection Floer theory – anomaly and obstruction*, preprint (2000) available at http://www.kusm.kyoto-u.ac.jp/~fukaya .

[Fu] W. Fulton, *Introduction to Toric Varieties*, Ann. of Math. Studies 131, Princeton University Press, Princeton, New Jersey, 1993.

[Gi] A.B. Givental, *Equivariant Gromov–Witten invariants*, Internat. Math. Res. Notices 13 (1996) 613-663.

[GKZ1] I.M. Gel'fand, A. V. Zelevinski, and M.M. Kapranov, *Equations of hypergeometric type and toric varieties*, Funktsional Anal. i. Prilozhen. 23(1989), 12 – 26; English transl. Functional Anal. Appl. 23(1989), 94-106.

[GKZ2] I.M. Gel'fand, A. V. Zelevinski, and M.M. Kapranov, *Discriminants, Resultants and Multidimensional Determinants*, Birkhäuser, Boston, 1994.

[GV] G. Gonzalez-Sprinberg and J.-L. Verdier, *Construction géométrique de la correspondance de McKay*, Ann. Sci. École Norm. Sup. 16(1983) 409–449.

[GoV] R. Gopakumar and C. Vafa, *M-Theory and Topological Strings–II*, hep-th/9812127.

[GW] M. Gross and P.M.H. Wilson, *Mirror symmetry via 3-tori for a class of Calabi–Yau threefolds*, Math. Ann. 309 (1997) 505-531.

[HIV] K. Hori, A. Iqbal and C. Vafa, *D-branes and mirror symmetry*, hep-th/0005247.

[Hor] P. Horja, *Hypergeometric functions and mirror symmetry in toric varieties*, math.AG/9912109.

[Hos] S. Hosono, *Local Mirror Symmetry and Type IIA Monodromy of Calabi–Yau Manifolds*, Adv. Theor. Math. Phys. 4(2000), 335–376.

[HKTY1] S. Hosono, A. Klemm, S. Theisen, and S.-T. Yau, *Mirror symmetry, mirror map and applications to Calabi–Yau hypersurfaces*, Commun. Math. Phys. 167(1995), 301-350, hep-th/9308122.

[HKTY2] S. Hosono, A. Klemm, S. Theisen, and S.-T. Yau, *Mirror symmetry, mirror map and applications to complete intersection Calabi–Yau spaces*, Nucl. Phys. B433 (1995) 501-554, hep-th/9406055.

[HLY] S. Hosono, B.H. Lian, and S.-T. Yau, *GKZ-Generalized hypergeometric systems in mirror symmetry of Calabi–Yau hypersurfaces*, Commun. Math. Phys. 182 (1996) 535–577.

[Huy] D. Huybrechts, *Moduli spaces of hyperkäher manifolds and mirror symmetry*, math.AG/0210219.

[IU] A. Ishii and H. Uehara, *Autoequivalences of derived categories on the minimal resolutions of A_n-singularities on surfaces* , math.AG/0409151.

[IN] Y. Ito and H. Nakajima, *The McKay correspondence and Hilbert schemes in dimension three*, Topology 39 (2000), 1155–1191.

[Ko] M. Kontsevich, *Homological algebra of mirror symmetry*, Proceedings of the International Congress of Mathematicians (Zürich, 1994) Birkhäuser (1995) pp. 120 –139.

[LYZ] C.N. Leung, E. Zaslow and S.-T. Yau, *From Special Lagrangian to Hermitian-Yang-Mills via Fourier–Mukai Transform*, Adv.Theor.Math.Phys. 4 (2002) 1319-1341.

[LLY] B.H. Lian, K. Liu and S.-T. Yau, *Mirror principle I*, Asian J. Math. 1 (1997), no. 4, 729–763, alg-geom/9712011.

[Mor] D. R. Morrison, *Picard–Fuchs equations and mirror maps for hypersurfaces*, Essays on Mirror Manifolds (S.-T. Yau, ed.), Internal Press, Hong Kong, (1992), 241 – 264.

[Ma] P. Mayr, *Phases of Supersymmetric D-branes on Kaehler Manifolds and the McKay correspondence*,JHEP 0101 (2001) 018, hep-th/0010223.

[Mu] S. Mukai, *On the moduli space of bundles on K3 surfaces I*, in: Vector bundles on algebraic varieties Oxford Univ. Press (1987) 341–413.

[MR] S. Mukhopadhyay and K. Ray, *Fractional branes on a non-compact orbifolds*, JHEP 0107(2001)007.

[Na] I. Nakamura, *Hilbert schemes of Abelian group orbits*, J.Alg.Geom.10(2000)757–779.

[Oda1] T. Oda, *Introduction to Algebraic Singularities (with an Appendix by T. Ambai)* , preprint (1989).

[Oda2] T. Oda, *Convex bodies and Algebraic Geometry, An Introduction to the Theory of Toric Varieties*, A Series of Modern Surveys in Mathematics, Springer-Verlarg, New York, 1985.

[Or] D. Orlov, *Equivalences of derived categories and K3 surfaces*, math.AG/9606006, Algebraic geometry, 7. J. Math. Sci. (New York)84, no. 5 (1997) 1361–1381.

[Re] M. Reid, *McKay correspondence* , alg-geom/9702016.

[Sa] K. Saito, *Period Mapping Associated to a Primitive Form*, Publ. RIMS, Kyoto Univ. 19(1983)1231–1264.

[ST] P. Seidel and R. Thomas, *Braid group and actions on derived categories of coherent sheaves*, math.AG/0001043, Duke Math. J. 108, no. 1 (2001) 37–108.

[Sti] J. Stienstra, *Resonant hypergeometric systems and mirror symmetry*, in *Integrable systems and algebraic geometry (Kobe/Kyoto 1997)*, World Sci. Publishing, River Edge, NJ, 1998, 412-452.

[St] A. Strominger, *Special Geometry*, Commun. Math. Phys. 133 (1990) 163-180.

[SYZ] A. Strominger, S.-T. Yau and E. Zaslow *Mirror symmetry is T-Duality*, Nucl. Phys. B479 (1996) 243-259.

[Yo] M. Yoshida, *Hypergeometric Functions, My Love*, Vieweg (1997).

Graduate School of Mathematical Sciences, University of Tokyo, Komaba Meguro-ku, Tokyo 153-8914, Japan

E-mail address: hosono@ms.u-tokyo.ac.jp

AMS/IP Studies in Advanced Mathematics
Volume 38, 2006

Extracting Gromov–Witten Invariants of a Conifold from Semi-stable Reduction and Relative GW-invariants of Pairs

Chien-Hao Liu and Shing-Tung Yau

From C.-H.L.: In memory of Barbara M. Willman

ABSTRACT. The study of open/closed string duality and large N duality suggests a Gromov–Witten theory for conifolds that sits on the border of both a closed Gromov–Witten theory and an open Gromov–Witten theory. The work of Jun Li on Gromov–Witten theory for a projective singular variety of the gluing form $Y_1 \cup_D Y_2$, where D is a smooth divisor on smooth Y_1 and Y_2, suggests two methods to study Gromov–Witten invariants for a projective conifold: one by a direct generalization of his construction to the conifold singularity and the other by an appropriate semi-stable reduction of a degeneration to a conifold and then applying his results on this new degeneration to extract Gromov–Witten invariants of the original conifold. In this work we carry out the second method. Suggested by the semi-stable reduction, we associate to a conifold Y with singular locus $\{p_1, \ldots\}$ a set of smooth variety-divisor pairs $(\widetilde{Y}, E)$, (Y_i, D_i), $i = 1, \ldots$, and a canonical morphism $\widetilde{Y} \cup_{\coprod_i D_i} \coprod_i Y_i \to Y$, where $\widetilde{Y}$ is the blow-up of Y at the conifold singularities, $E \simeq \coprod_i D_i$ is the exceptional divisor, and Y_i is a smooth quadric hypersurface in $\mathbb{P}^4$ with a smooth hyperplane section $D_i \simeq \mathbb{P}^1 \times \mathbb{P}^1$, and i runs through the labels of the conifold singularities p_i. The existence of a $\mathbb{Z}/2\mathbb{Z}$-action on the quadric hypersurface Y_i that restricts to an exchange of the two product factor $\mathbb{P}^1$'s of D_i implies that the variety $\widetilde{Y} \cup_{\coprod_i D_i} \coprod_i Y_i$ obtained from gluing these pairs is uniquely determined by Y up to isomorphism. Jun Li's relative Gromov–Witten theory and invariants for smooth variety-divisor pairs $(Y_0, \coprod_i D_i)$ and (Y_i, D_i) and a refinement of his degeneration formula worked out here for the current situation are then employed to extract Gromov–Witten invariants of Y.

2000 *Mathematics Subject Classification.* Primary 14N35; Secondary 81T30.

Key words and phrases. String world-sheet instanton, open/closed string duality, conifold, stable morphism, Gromov–Witten invariant, semi-stable reduction, quadric hypersurface, degeneration formula.

The work is supported by NSF grants DMS-9803347 and DMS-0074329.

0. Introduction and outline

Introduction: Gromov–Witten theory for conifolds in an open/closed string duality and a large N duality.

Given the 3-sphere S^3, let X_0 be the conifold from the degeneration of $X := T^*S^3$, as a complex 3-fold that pinches the zero-section 3-cycle S^3 of T^*S^3, and X' be the complex 3-fold from the small resolution of X_0 with exceptional locus $\simeq \mathbb{P}^1$. (X' is isomorphic to the total space of the bundle $\mathcal{O}_{\mathbb{P}^1}(-1) \oplus \mathcal{O}_{\mathbb{P}^1}(-1)$ and is a non-compact Calabi–Yau 3-fold.) Gopakumar and Vafa [G-V] conjecture the following correspondence that relates the $U(N)$ or $SU(N)$ Chern-Simons gauge theory on S^3 and an A-model topological closed string theory on X' via the mechanisms indicated below:

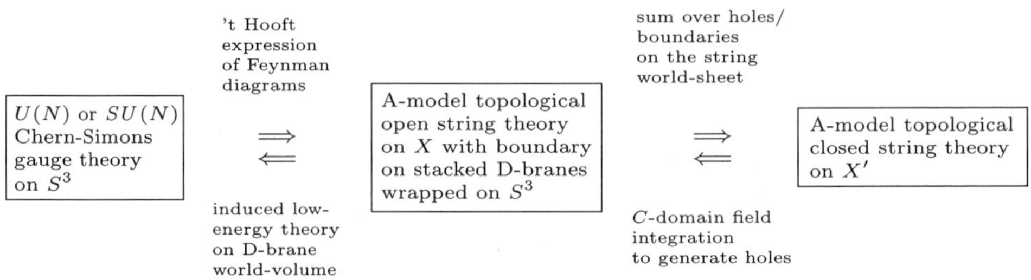

Under this correspondence,

$U(N)$ or $SU(N)$ Chern-Simons gauge theory on S^3	$\Longleftrightarrow$	A-model topological closed string theory on X'
· 't Hooft coupling $\lambda := Ng_s$		· B-field magnitude on $\mathbb{P}^1$
· Wilson loop observable		· quantity in the effective theory of associated brane-probe in X'
· large N limit of Chern-Simons gauge theory on S^3		· A-model topological string theory on conifold X_0

This diagram involves a large N duality and an open/closed string duality, and an A-model topological string theory with a conifold as the target space serves as a geometric mediator and transition-point for these dualities. The dualities were tested/examined from *five* different aspects of stringy dualities: the *string world-sheet* aspect, the *target space-time* aspect, the *low dimensional effective field theory* aspect, the *brane-probe* aspect, and the *Wilson theory space* aspect. See [G-P], [G-V], [O-V1], [O-V2], [Va], [Wi1], [Wi2] for string-theoretical details and insights, [G-R] for an introductory mathematical review and more references, and e.g. [A-M-V] and [D-F-G] for generalizations to more general non-compact Calabi–Yau 3-folds involving toric geometry.

Here a *conifold* X_0 (e.g. [B-L], [C-dlO-G-P], and [St]) is by definition the singular variety obtained from a degeneration of Calabi–Yau 3-fold X via a deformation of complex structures that pinches isolated smoothly-embedded 3-spheres S^3 in a smooth 3-fold X. The partition function of an A-model topological string theory with target X_0 is supposed to compute the string world-sheet instanton numbers that can be interpreted either as a counting of holomorphic maps from (complex) curves to X_0 or from bordered Riemann surfaces to X_0 with boundary components, if non-empty, mapped to the isolated singularities of X_0. In other words, this is a *Gromov–Witten theory for conifolds* on the mathematical side. Also, the open/closed string duality from string theory reveals a very distinguished feature of it as a theory on the border of *both* a closed Gromov–Witten theory on one geometry and an *open* Gromov–Witten theory on another geometry related to the previous geometry via the conifold transition.

To develop a Gromov–Witten theory for conifolds, there are two paths one may attempt to follow. The first one is to generalize the techniques in [Li1] and [Li2] directly to a conifold singularity. This is technical. The second one is to try to replace a conifold as the degenerate fiber of a smooth $\mathbb{A}^1$-family $W/\mathbb{A}^1$ by the degenerate fiber of a semi-stable reduction of $W/\mathbb{A}^1$ and see if one can reduce the problem to the case already dealt with in [Li1] and [Li2] and use it to extract the Gromov–Witten invariants of the conifold. Surely, for general singularities one will not expect the second path would immediately work either, since one still lacks the understanding of Gromov–Witten theory for a pair (Y, D), where Y is smooth and D is a divisor on Y with simple normal crossings singularities, and a related degeneration/gluing formula for gluing Gromov–Witten invariants from a collection of such pairs. As if given by God, it turns out that such potentially existing difficulties along the second path are not really there for a conifold singularity, (Sec. 1). This accidental simplicity for conifolds together with similar discussions and results in [L-Y] that refine Jun Li's degeneration formula to one for curve classes in $H_2(\,\cdot\,;\mathbb{Z})$ or $A_1(\,\cdot\,)$, (Sec. 2), enable us to extract some Gromov–Witten invariants of conifolds, (Sec. 3). We explain the details of this second path in this work.

Convention. This work is parallel to [L-Y] and follows the notations and the terminology of [Li1], [Li2], and [L-Y] closely, except where confusions may occur. Other notations follow [Hart], [Fu], [De], and [K-M]. All schemes are over $\mathbb{C}$ and all points referred to are closed points. All conifolds are assumed to be projective. Though appearing a few times in the Introduction for easy comparisons with the literature in string theory, the Calabi–Yau condition is not relevant in this work but will be relevant in an application.

Preliminary. Readers are referred to [Li1: Sec. 0], [Li2: Sec. 0], and [L-Y: Sec. 1, Sec. 2] for definitions and an outline of Jun Li's work that are needed for the current work.

Outline.

1. A semi-stable reduction of a conifold degeneration.

Semi-stable reduction and the associated smooth pairs of a conifold.

Let $\pi : W \longrightarrow \mathbb{A}^1$ be a flat family of (complex) 3-dimensional projective varieties with smooth general fibers W_t, $t \neq \mathbf{0}$, and smooth total space W such that W_t degenerates to a conifold $W_0 = Y$ over $\mathbf{0} \in \mathbb{A}^1$. For simplicity of notation and presentation, we assume that Y has only one conifold singularity. An example of such a family can be obtained from a degeneration of quintic 3-folds in $\mathbb{P}^4$, cf. [C-dlO-G-P]. In a local analytic germ, the degeneration to a conifold singularity is modeled on the morphism

$$\begin{array}{ccc} \mathit{Spec}\,\mathbb{C}[[\,x,y,z,w\,]] & \longrightarrow & \mathit{Spec}\,\mathbb{C}[[t]] \\ xy - zw & \longleftarrow & t\,. \end{array}$$

This gives a 3-dimensional isolated hypersurface singularity of multiplicity 2 at the fiber over $\mathbf{0} := (t)$. The family $W/\mathbb{A}^1$ can be semi-stabilized by the following sequence of blow-ups and finite base change obtained from a straightforward computation:

$$\begin{array}{ccccccc} W^{(3)} & \xrightarrow{\varphi_3} & W^{(2)} & \xrightarrow{\varphi_2} & W^{(1)} & \xrightarrow{\varphi_1} & W^{(0)} := W \\ \pi_3 \downarrow & & \pi_2 \downarrow & & \downarrow \pi_1 & & \downarrow \pi_0 = \pi \\ \mathbb{A}^1 & = & \mathbb{A}^1 & \xrightarrow{\alpha} & \mathbb{A}^1 & = & \mathbb{A}^1 \end{array}\,,$$

where

- $\varphi_1 : W^{(1)} \rightarrow W^{(0)}$ is the blow-up of W at the conifold singularity of W_0;
- recall that W is smooth and hence the exceptional locus of φ_1 is a $\mathbb{P}^3$; the degenerate fiber $\pi_1^{-1}(\mathbf{0})$ of π_1 contains a nonreduced irreducible component of multiplicity 2 supported on this $\mathbb{P}^3$; as divisors in $W^{(1)}$, $\pi_1^{-1}(\mathbf{0}) = \widetilde{Y} + 2\,\mathbb{P}^3$, where $\widetilde{Y}$ is the resolution of Y by a blow-up at the conifold singularity and $\widetilde{Y} \cap \mathbb{P}^3$ is a smooth quadric surface in $\mathbb{P}^3$;
- $\alpha : (\mathbb{A}^1, \mathbf{0}) \rightarrow (\mathbb{A}^1, \mathbf{0})$ is a finite morphism of degree 2 branched over $\mathbf{0}$, φ_2 and π_2 on $W^{(2)}$ are from the fiber product of α and π_1;
- $W^{(2)}$ is now a singular scheme whose singularities are modelled on the Whitney umbrella; the singular locus $\mathit{Sing}\,(W^{(2)})$ of $W^{(2)}$ with the reduced subscheme structure has multiplicity 2, lies over $\mathbf{0}$, and is isomorphic to $\mathbb{P}^3$;
- normalization/blow-up of $W^{(2)}$ along $\mathit{Sing}\,(W^{(2)}) \simeq \mathbb{P}^3$ gives $W^{(3)}$ which is smooth with $\pi_3^{-1}(\mathbf{0}) = Y_0 \cup Y_1$, where $Y_0 := \widetilde{Y}$ and Y_1 is naturally realized as a double cover of $\mathbb{P}^3$ branched over a smooth quadric surface in $\mathbb{P}^3$.

In local germs or local formal schemes, the morphisms in the diagram above are given by

$$\begin{array}{ccccccc} & \mathit{Spec}\,\mathbb{C}[[x_1,y_1,z_1,w_1]] & & \xrightarrow{\varphi_1} & & \mathit{Spec}\,\mathbb{C}[[x_0,y_0,z_0,w_0]] & \\ w_1^2(x_1y_1 - z_1) & & x_1w_1\,;\; y_1w_1\,;\; z_1w_1\,;\; w_1 & \xleftarrow{\varphi_1^\sharp} & x_0\,;\; y_0\,;\; z_0\,;\; w_0 & & x_0y_0 - z_0w_0 \\ \pi_1^\sharp \uparrow & \pi_1 \downarrow & & & & \downarrow \pi_0 & \uparrow \pi_0^\sharp \\ t_1 & & t_1 & \longleftarrow & t_0 & & t_0 \\ & \mathit{Spec}\,\mathbb{C}[[t_1]] & & \longrightarrow & & \mathit{Spec}\,\mathbb{C}[[t_0]] & \end{array}\,,$$

$$
\begin{array}{cccccccc}
 & \multicolumn{2}{c}{Spec\,\mathbb{C}[[x_1,y_1,z_1,w_1,t_2]]/(t_2^2-w_1^2(x_1y_1-z_1))} & \xrightarrow{\varphi_2} & \multicolumn{3}{c}{Spec\,\mathbb{C}[[x_1,y_1,z_1,w_1]]} & \\
t_2 & & x_1\,;\,y_1\,;\,z_1\,;\,w_1 & \xleftarrow{\varphi_2^\sharp} & x_1\,;\,y_1\,;\,z_1\,;\,w_1 & & w_1^2(x_1y_1-z_1) & \\
\pi_2^\sharp\uparrow & \pi_2\downarrow & & & & \downarrow\pi_1 & \uparrow\pi_1^\sharp & \\
t_2 & & t_2^2 & \longleftarrow & t_1 & & t_1 & \\
 & Spec\,\mathbb{C}[[t_2]] & & \longrightarrow & & Spec\,\mathbb{C}[[t_1]] & & ;
\end{array}
$$

consider the ring isomorphism generated by $x_1 \mapsto x_2$, $y_1 \mapsto y_2$, $z_1 \mapsto x_2y_2 - z_2$, $w_1 \mapsto w_2$, and $t_2 \mapsto t_2$ and rewrite

$$Spec\,\mathbb{C}[[x_1,y_1,z_1,w_1,t_2]]/(t_2^2 - w_1^2(x_1y_1 - z_1)) \quad \text{as} \quad Spec\,\mathbb{C}[[x_2,y_2,z_2,w_2,t_2]]/(t_2^2 - w_2^2z_2)\,,$$

which reveals the Whitney umbrella transverse surface singularity of $W^{(2)}$; then

$$
\begin{array}{cccccccc}
 & \multicolumn{2}{c}{Spec\,\mathbb{C}[[x_3,y_3,z_3,w_3]]} & \xrightarrow{\varphi_3} & \multicolumn{3}{c}{Spec\,\mathbb{C}[[x_2,y_2,z_2,w_2,t_2]]/(t_2^2-w_2^2z_2)} & \\
z_3w_3 & & x_3\,;\,y_3\,;\,z_3^2\,;\,w_3\,;\,z_3w_3 & \xleftarrow{\varphi_3^\sharp} & x_2\,;\,y_2\,;\,z_2\,;\,w_2\,;t_2 & & t_2 & \\
\pi_3^\sharp\uparrow & \pi_3\downarrow & & & & \downarrow\pi_2 & \uparrow\pi_2^\sharp & \\
t_3 & & t_3 & \longleftarrow & t_2 & & t_2 & \\
 & Spec\,\mathbb{C}[[t_3]] & & \longrightarrow & & Spec\,\mathbb{C}[[t_2]] & & .
\end{array}
$$

(Cf. FIGURE 1-1.)

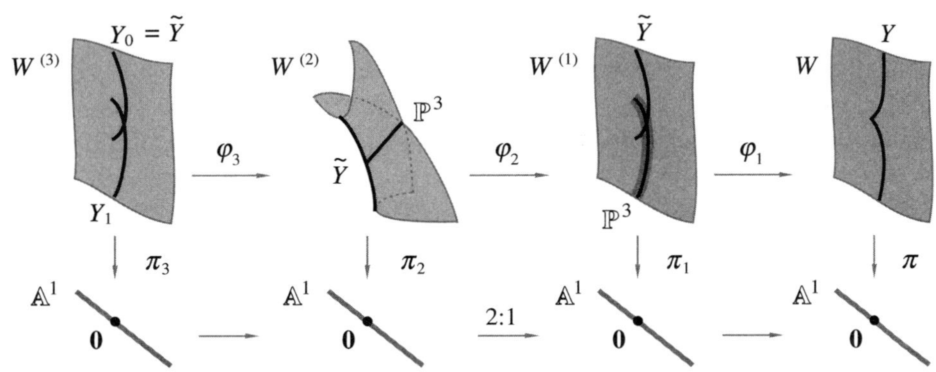

FIGURE 1-1. A semi-stable reduction of a conifold degeneration. The degenerate fiber over $\mathbf{0} \in \mathbb{A}^1$ in each family is indicated.

Definition 1.1 [canonical semi-stable reduction]. We shall call $W^{(3)}/\mathbb{A}^1$ obtained above the *canonical semi-stable reduction* of the conifold degeneration $W/\mathbb{A}^1$.

After this semi-stable reduction of the conifold degeneration $\pi : W \to \mathbb{A}^1$, one obtains a new family $\pi_3 : W^{(3)} \to \mathbb{A}^1$ of 3-folds with $W^{(3)}$ smooth, $W_t^{(3)} = W_{t^2}$ for $t \neq \mathbf{0}$, and where $W_0^{(3)} := \pi_3^{-1}(\mathbf{0}) = Y_0 \cup_D Y_1$ is a variety with simple normal crossings singularity from gluing smooth Y_0 and Y_1 along the isomorphic smooth divisor D. This happens to be exactly the type of degeneration whose Gromov–Witten theory is studied in [Li1] and [Li2].

Definition 1.2 [associated canonical pairs to conifold]. We shall call the set of pairs (Y_0, D) and (Y_1, D) the *set of canonical smooth pairs associated to the conifold Y*.

By construction, there is a canonical morphism $p = p_0 \cup_D p_1 : Y_0 \cup_D Y_1 \rightarrow Y$, where $p_0 : Y_0 \rightarrow Y$ is the resolution of Y whose exception locus is the smooth divisor D and $p_1 : Y_1 \rightarrow Y$ pinches Y_1 to the conifold singularity of Y.

Special properties of Y_1.

The 3-fold Y_1 is naturally realized as a quadric hypersurface in $\mathbb{P}^4$ with $D \simeq \mathbb{P}^1 \times \mathbb{P}^1$ realized as the intersection of Y_1 with a hyperplane H in $\mathbb{P}^4$. It follows from the Bruhat cellular decomposition of a smooth quadric hypersurface in a projective space and the Lefschetz hyperplane theorem that the only non-vanishing Chow groups and (singular) homology groups for Y_1 coincide and are given by

$$A_k(Y_1) \simeq H_{2k}(Y_1;\mathbb{Z}) \simeq \mathbb{Z}, \quad \text{for } k = 0,\, 1,\, 2,\, 3\,.$$

In particular, $A_1(Y_1) \simeq H_2(Y_1;\mathbb{Z})$ is generated by a complex line γ_2 from either of the rulings of D, $A_2(Y_1) \simeq H_4(Y_1;\mathbb{Z})$ is generated by the complex surface $\gamma_4 := D$, and $\gamma_2 \cap \gamma_4 = +1$ on Y_1. (See [Fu], [G-H], [Harr], and [P-S] for details.)

Fix an embedding $D = Y_1 \cap H \subset Y_1 \hookrightarrow \mathbb{P}^4$ and an isomorphism $D \simeq \mathbb{P}^1 \times \mathbb{P}^1$ from the rulings of D. Up to a $PGL\,(5,\mathbb{C})$-action on $\mathbb{P}^4$, one can write this explicitly in terms of homogeneous coordinates $[u : x : y : z : w]$ of $\mathbb{P}^4$ as

$$D = V(u\,,\, u^2 + xy - zw) \subset Y_1 = V(u^2 + xy - zw) \subset \mathbb{P}^4 \quad \text{with} \quad H = V(u)\,.$$

The $\mathbb{Z}/2\mathbb{Z}$-action on D that interchanges the two $\mathbb{P}^1$-factors extends to a linear $\mathbb{Z}/2\mathbb{Z}$-action on $\mathbb{P}^4$ that leaves Y_1 invariant. Explicitly, $D \simeq \mathbb{P}^1 \times \mathbb{P}^1$ via $\{\, xy - zw = 0\,\} \rightarrow ([x:z],[x:w])$ and one can choose this $\mathbb{Z}/2\mathbb{Z}$-action to be the exchange $z \leftrightarrow w$. It follows that:

Lemma 1.3 [uniqueness of gluing]. *Any variety obtained from gluing the smooth pairs (Y_0, D) and (Y_1, D) associated to Y via an automorphism of D is isomorphic to any other so obtained.*

The monodromy of $\pi : W \rightarrow \mathbb{A}^1$.

To be able to refine Jun Li's degeneration formula from one with respect to a π_3-ample line bundle on $W^{(3)}/\mathbb{A}^1$ to one with respect to a curve class of a general fiber W_t of $W/\mathbb{A}^1$, the monodromy around the degenerate fiber $W_0 = Y$ of $W/\mathbb{A}^1$ is required to be well-behaved. Let us thus take a look at this. In this part, we will leave the algebro-geometric category over $\mathbb{C}$ and enter the topological category over $\mathbb{R}$. In particular, '$\simeq$' means 'topologically homeomorphic'.

Topologically, the degeneration $W/\mathbb{A}^1$ pinches a real 3-cycle realized as a smoothly embedded 3-sphere S^3 in W_t, $t \neq \mathbf{0}$, to the conifold singularity of $W_0 = Y$. Over a loop $S^1 \hookrightarrow \mathbb{A}^1$ around $\mathbf{0} \in \mathbb{A}^1$ (here we will think of $\mathbb{A}^1$ as the usual complex plane $\mathbb{C}$ with the analytic topology and S^1 is an oriented circle therein), $\pi^{-1}(S^1)$ is the mapping torus associated to a smooth automorphism $\phi : W_t \xrightarrow{\sim} W_t$. ϕ can be homotoped to an automorphism, still denoted by ϕ, on W_t that is not the identity map only in a small tubular neighborhood $N_\varepsilon(S^3/W_t) \simeq S^3 \times B^3$ of S^3 in W_t, where B^3 is the unit (real) 3-ball and there is an orientation on (S^3, B^3) that is compatible with the complex structure on $N_\varepsilon(S^3/W_t)$. Indeed, ϕ is a *generalized Dehn twist* along S^3 and, up to homotopy, can be described explicitly as follows. Recall that S^3 is the underlying topology of the Lie group $SU(2)$. Let e be its identity element and fix a degree 1 map $\phi_e : (B^3, \partial B^3) \to (S^3, e)$ that is a homeomorphism from $B^3 - \partial B^3$ to $S^3 - \{e\}$. Define $\phi : N_\varepsilon(S^3/W_t) \to N_\varepsilon(S^3/W_t)$ by

$$\begin{array}{cccc}\phi: & S^3 \times B^3 & \longrightarrow & S^3 \times B^3 \\ & (g, x) & \longmapsto & (g \cdot \phi_e(x),\, x)\end{array} ,$$

where the $\cdot$ in the definition is the group multiplication of $SU(2)$. By construction, ϕ restricts to the identity map on $\partial(N_\varepsilon(S^3/W_t))$ and extends by the identity map to the automorphism ϕ on W_t.

Since $H_i(S^3 \times B^3, \partial(S^3 \times B^3); \mathbb{Z}) = 0$ for $i = 0,\ 1,\ 2$, the monodromy of ϕ on $H_i(W_t; \mathbb{Z})$ is the identity map for $i = 0,\ 1,\ 2$. (The monodromy on $H_3(W_t; \mathbb{Z})$ is the much studied Picard-Lefschetz operation $\gamma \mapsto \gamma + \langle \gamma, [S^3] \rangle\, [S^3]$, where $\langle\, ,\, \rangle$ is the intersection pairing on $H_3(W_t; \mathbb{Z})$. But we do not need this for this work.) Since the conifold $W_0 = Y$ is topologically the quotient space of W_t, $t \neq \mathbf{0}$, by identifying the vanishing 3-cycle S^3 to a point, Y is homotopically equivalent to $W_t \cup_{S^3 \simeq \partial B^4} B^4$, where $t \neq \mathbf{0}$ and the 4-ball B^4 is attached to W_t along an isomorphism of ∂B^4 to the vanishing 3-cycle S^3. This implies that $H_i(W_0; \mathbb{Z}) \simeq H_i(W_t; \mathbb{Z})$ canonically for $i = 0,\ 1,\ 2$. In particular, $R^\bullet \pi_* \mathbb{Z}_W$ has a summand that is the constant sheaf on $\mathbb{A}^1$ associated to the group $H_2(W_t; \mathbb{Z}) = H_2(Y; \mathbb{Z})$.

To summarize,

Lemma 1.4 [canonical identification and trivial monodromy on H_2]. *The monodromy of $W/\mathbb{A}^1$ around the conifold $W_0 = Y$ is trivial on $H_2(W_t; \mathbb{Z})$. Each $H_2(W_t; \mathbb{Z})$, $t \neq \mathbf{0}$, is canonically isomorphic to $H_2(W_0; \mathbb{Z}) = H_2(Y; \mathbb{Z})$.*

We now resume to proceed in the domain of algebraic geometry.

2. A degeneration formula of Gromov–Witten invariants with respect to a curve class

Once the semi-stable degeneration $W^{(3)}/\mathbb{A}^1$ of the conifold degeneration $W/\mathbb{A}^1$ is understood, the second ingredient toward a Gromov–Witten theory for Y via extracting information from gluing pairs (Y_0, D) and (Y_1, D) is the degeneration/gluing formula of Gromov–Witten invariants for the family $W^{(3)}/\mathbb{A}^1$ that is intrinsic to curve classes on Y (note: *not* on $W_0^{(3)}$). The goal of this section is to derive such a formula from [Li2]. The reasoning and the discussion of this section are parallel to [L-Y: Sec. 2 and Sec. 3]. Readers are referred to [L-Y] for more explanations in a simpler situation and [Li1] and [Li2] for the thorough related details.

A refinement of Jun Li's degeneration formula.

Recall first the definitions of the following: *admissible weighted graph* ([Li1: Definition 4.6], [L-Y: Definition 1.1]; *admissible triple* ([Li1: Definition 4.11], [L-Y: Definition 1.2]), the integer triple $|\eta|$ for an admissible triple $\eta = (\Gamma_0, \Gamma_1, I)$, and Jun Li's *degeneration formula* ([Li2: Sec. 0, Theorem 3.15, and Corollary 3.16]).

Fix a π_3-ample line bundle $\mathcal{L}$ on $W^{(3)}/\mathbb{A}^1$ and an $\mathcal{L}$-degree d; then the stack of stable morphisms from prestable curves of genus g, with n marked points, into fibers of the universal family of expanded degenerations associated to $W^{(3)}/\mathbb{A}^1$ of $\mathcal{L}$-degree d is a Deligne-Mumford stack $\mathfrak{M}(\mathfrak{W}^{(3)}, (g, n; d))$ with a perfect obstruction theory. By construction, $\mathfrak{M}(\mathfrak{W}^{(3)}, (g, n; d))$ fibers naturally over $\mathbb{A}^1$ with the fiber over $\mathbf{0}$

denoted by $\mathfrak{M}(\mathfrak{W}_0^{(3)}, (g,n;d))$. The perfect obstruction theory on $\mathfrak{M}(\mathfrak{W}^{(3)}, (g,n;d))$ restricts to a perfect obstruction theory on $\mathfrak{M}(\mathfrak{W}_0^{(3)}, (g,n;d))$. The Gromov–Witten theory and invariants of the singular variety $Y_0 \cup_D Y_1$ from gluing are defined via the virtual fundamental class $[\mathfrak{M}(\mathfrak{W}_0^{(3)}, (g,n;d))]^{virt}$ on $\mathfrak{M}(\mathfrak{W}_0^{(3)}, (g,n;d))$. On the other hand, for each of the smooth variety-divisor pairs (Y_i, D), Jun Li constructed a general theory of relative Gromov–Witten invariants, which consists of the moduli stack $\mathfrak{M}(\mathfrak{Y}_i^{rel}, \Gamma_i)$ of stable morphisms of topological type Γ_i from prestable curves to the fibers of the universal family of the stack $\mathfrak{Y}_i^{rel}$ of expanded relative pairs associated to (Y_i, D) and a perfect obstruction theory thereon. For each admissible triple η with $|\eta| = (g,n;d)$, there is a morphism (the r below is the total root weight of Γ_0 in η),

$$\Phi_\eta \;:\; \mathfrak{M}(\mathfrak{Y}_0^{rel}, \Gamma_0) \times_{D^r} \mathfrak{M}(\mathfrak{Y}_1^{rel}, \Gamma_1) \;\longrightarrow\; \mathfrak{M}(\mathfrak{W}_0^{(3)}, (g,n;d))$$

that is finite étale to its image $\mathfrak{M}(\mathfrak{Y}_0^{rel} \sqcup \mathfrak{Y}_1^{rel}, \eta)$ and has degree a combinatorial factor $|\mathrm{Eq}(\eta)|$ from η, cf. [Li1: Sec. 4]. Up to the difference from a nonreduced structure that has to be taken care of ([Li2: Sec. 4.4]), $\mathfrak{M}(\mathfrak{W}_0^{(3)}, (g,n;d))$ is the union of the substacks $\mathfrak{M}(\mathfrak{Y}_0^{rel} \sqcup \mathfrak{Y}_1^{rel}, \eta)$ in $\mathfrak{M}(\mathfrak{W}_0^{(3)}, (g,n;d))$. Going from $\mathfrak{M}(\mathfrak{W}_t^{(3)}, (g,n;d))$, $t \neq \mathbf{0}$, which gives the usual Gromov–Witten theory of the smooth projective W_t to $\mathfrak{M}(\mathfrak{W}_0^{(3)}, (g,n;d))$, which gives the Gromov–Witten theory on $Y_0 \cup_D Y_1$, and recasting it in a form from gluing relative Gromov–Witten theory of the pairs (Y_0, D) and (Y_1, D), gives a degeneration formula that relates Gromov–Witten invariants of W_t to a combination of relative Gromov–Witten invariants of (Y_0, D) and (Y_1, D) in the degenerate fiber $W_0^{(3)}$. See [Li1] and [Li2] for the complete technical details.

Since the monodromy on $H_2(W_t;\mathbb{Z})$ around $\mathbf{0} \in \mathbb{A}^1$ is trivial, so is the monodromy on $H_2(W_t^{(3)};\mathbb{Z})$ around $\mathbf{0}$. Recall also the canonical isomorphism

$$H_2(W_t;\mathbb{Z}) \simeq H_2(Y;\mathbb{Z}).$$

Thus, a $\beta' \in H_2(Y;\mathbb{Z})$ determines a unique class in $H_2(W_t;\mathbb{Z})$, $t \neq \mathbf{0}$, and hence a unique class, still denoted by β', in $H_2(W_t^{(3)};\mathbb{Z})$, $t \neq \mathbf{0}$. Define the $\mathcal{L}$-degree $\mathcal{L}\cdot\beta'$ of $\beta' \in H_2(Y;\mathbb{Z})$ via this identification. Let d be the $\mathcal{L}$-degree for a given curve class $\beta \in H_2(Y;\mathbb{Z})$ and define $C_{(\mathcal{L},d)}(Y) := \{\beta' \in H_2(Y;\mathbb{Z}) \,:\, \mathcal{L}\cdot\beta' = d\}$. Then

$$\mathfrak{M}(\mathfrak{W}^{(3)}, (g,n;d)) \;=\; \coprod_{\beta' \in C_{(\mathcal{L},\,d)}(Y)} \mathfrak{M}(\mathfrak{W}^{(3)}, (g,n;\beta'))\,,$$

where $\mathfrak{M}(\mathfrak{W}^{(3)}, (g,n;\beta'))$ is the stack of stable morphisms of $\mathcal{L}$-degree d from prestable curves of genus g with n marked points to the fibers of the universal family of the stack $\mathfrak{W}^{(3)}$ of expanded degenerations associated to $W^{(3)}/\mathbb{A}^1$ such that after the post-composition with the morphisms

$$\mathfrak{W}^{(3)}/\mathbb{A}^1 \;\longrightarrow\; W^{(3)}/\mathbb{A}^1 \;\stackrel{\varphi_1\circ\varphi_2\circ\varphi_3}{\longrightarrow}\; W/\mathbb{A}^1\,,$$

the images of the stable morphisms lie in the curve class $\beta' \in H_2(Y;\mathbb{Z})$. From its definition, $\mathfrak{M}(\mathfrak{W}^{(3)}, (g,n;\beta'))$ depends not just on (g,n,β') but also on $\mathcal{L}$. (Indeed, since Y_1 is mapped to the conifold singularity of Y, the choice of curve classes from $H_2(Y_0 \cup_D Y_1;\mathbb{Z})$ does depend on $\mathcal{L}$ in general.)

Recall the $\mathcal{L}$-dependent set $\Omega_{(g,k;d)}$ of admissible triples η such that $|\eta| = (g,k;d)$ and the quotient set $\overline{\Omega}_{(g,k;d)}$ defined in [Li2: Sec. 0]. Recall also the morphism $p = p_0 \cup_D p_1 \,:\, Y_0 \cup_D Y_1 \to Y$. For an admissible weighted graph Γ for a relative pair, let $b(\Gamma) := \sum_{v\in V(\Gamma)} b(v)$. Define the *$\beta$-compatible subset* of $\Omega_{(g,n;d)}$ by

$$\Omega^{\mathcal{L}}_{(g,k;\beta)} := \left\{ \eta = (\Gamma_0, \Gamma_1, I) \in \Omega_{(g,k;d)} \mid p_{0*}b(\Gamma_0) = \beta \right\} .$$

The same discussions as in [L-Y: Sec. 2] imply the existence of a perfect obstruction theory on the moduli stack $\mathfrak{M}(\mathfrak{W}^{(3)}, (g,n;\beta))$ over $\mathbb{A}^1$ and its fiber $\mathfrak{M}(\mathfrak{W}^{(3)}_0, (g,n;\beta))$ over $\mathbf{0}$, inherited from those constructed in [Li2]. This gives a well-defined Gromov–Witten theory and Gromov–Witten invariants on $Y_0 \cup_D Y_1$ associated to $\beta \in H_2(Y;\mathbb{Z})$, Jun Li's degeneration formula [Li2] then implies the following degeneration formula:

Lemma 2.1 [Jun Li's degeneration formula]. (Cf. [L-Y: Corollary 2.2] and explanations of notations there and in [Li1] and [Li2].) *Let $\alpha \in H^0_c(R^*\pi_{3*}\mathbb{Q}_{W^{(3)}})^{\times n}$, whose restriction to $W^{(3)}_t$ will be denoted by $\alpha(t)$, and $\zeta \in A_*(\mathfrak{M}_{g,n})$. Denote by $\Psi^{W^{(3)}_t}_{(g,n;\beta)}(\alpha(t),\zeta)$ the usual Gromov–Witten invariant of $W^{(3)}_t$ associated to these data. For $\eta \in \overline{\Omega}^{\mathcal{L}}_{(g,n;\beta)}$, assume that $G^*_\eta(\zeta) = \sum_{j\in K_\eta} \zeta_{\eta,0,j} \boxtimes \zeta_{\eta,1,j}$, where $G_\eta : \mathfrak{M}_{\Gamma^\circ_0} \times \mathfrak{M}_{\Gamma^\circ_1} \to \mathfrak{M}_{g,n}$ is the natural morphism between the related moduli stacks of nodal curves. Then*

$$\Psi^{W^{(3)}_t}_{(g,n;\beta)}(\alpha(t),\zeta) = \sum_{\eta\in\overline{\Omega}^{\mathcal{L}}_{(g,n;\beta)}} \frac{\mathbf{m}(\eta)}{|\mathrm{Eq}(\eta)|} \sum_{j\in K_\eta} \left[\Psi^{Y^{rel}_0}_{\Gamma_0}(j^*_0\alpha(0), \zeta_{\eta,0,j}) \bullet \Psi^{Y^{rel}_1}_{\Gamma_1}(j^*_1\alpha(0), \zeta_{\eta,1,j}) \right]_0 ,$$

where $j_i : Y_i \to W^{(3)}_0$ is the inclusion map, and for $i = 0, 1$,

$$\Psi^{Y^{rel}_i}_{\Gamma_i}(j^\alpha_i(0), \zeta_{\eta,i,j}) = \mathbf{q}_{i\,*}\left(ev^*(j^*_i\alpha(0)) \cdot \pi^*_{\Gamma_i}(\zeta_{\eta,i,j}) \cdot [\mathfrak{M}(\mathfrak{Y}^{rel}_i, \Gamma_i)]^{virt} \right) \in H_*(D^r)$$

defines the relative Gromov–Witten invariants of the pair (Y_i, D) with values in $H_*(D^r)$. Here r is the number of roots in η, $\bullet$ stands for the intersection product on $H_*(D^r)$, and $[\,\cdot\,]_0$ denotes the degree-0 component of an element in $H_*(D^r)$. And $\mathbf{m}(\eta)$ is the product of weights of the roots in η.

In cycle form,

$$[\mathfrak{M}(\mathfrak{W}^{(3)}_0, (g,n;\beta))]^{virt} = \sum_{\eta\in\overline{\Omega}^{\mathcal{L}}_{(g,n;\beta)}} \frac{\mathbf{m}(\eta)}{|\mathrm{Eq}(\eta)|} \Phi_{\eta*}\Delta^! \left([\mathfrak{M}(\mathfrak{Y}^{rel}_0, \Gamma_0)]^{virt} \times [\mathfrak{M}(\mathfrak{Y}^{rel}_1, \Gamma_1)]^{virt} \right) ,$$

where $\mathfrak{M}(\mathfrak{W}^{(3)}_0, (g,n;\beta))$ is the fiber of $\mathfrak{M}(\mathfrak{W}^{(3)}, (g,n;\beta))$ over $\mathbf{0} \in \mathbb{A}^1$, and $\Delta^!$ is the Gysin map associated to the diagonal map $\Delta : D^r \to D^r \times D^r$ for the relevant D^r in each summand.

We want to make things as intrinsic to Y as possible so that we can appropriately combine the Gromov–Witten invariants of $Y_0 \cup_D Y_1$ defined from [Li2] to a quantity that is justifiable to be called a Gromov–Witten invariant of Y. In particular, we want to remove the possible $\mathcal{L}$-dependence (so far on *both* sides of the equation) in the above gluing formula. Let us now deal with this issue.

The $\mathcal{L}$-(in)dependence of $\Omega^{\mathcal{L}}_{(g,k;\beta)}$ and $\mathfrak{M}(\mathfrak{W}_0^{(3)},(g,n;\beta))$.

Recall from Sec. 1 the intermediate families over $\mathbb{A}^1$ that occur in the semi-stable reduction of $W/\mathbb{A}^1$:

$$\begin{array}{ccccccc} W^{(3)} & \xrightarrow{\varphi_3} & W^{(2)} & \xrightarrow{\varphi_2} & W^{(1)} & \xrightarrow{\varphi_1} & W^{(0)} := W \\ \pi_3\downarrow & & \pi_2\downarrow & & \downarrow\pi_1 & & \downarrow\pi_0=\pi \\ \mathbb{A}^1 & = & \mathbb{A}^1 & \xrightarrow{\alpha} & \mathbb{A}^1 & = & \mathbb{A}^1 \end{array} .$$

Denote the exceptional divisor of φ_1 on $W^{(1)}$ by E_1 ($\simeq \mathbb{P}^3$) and recall that the exceptional divisor of φ_3 on $W^{(3)}$ is Y_1. Recall [C-H] or [L-Y: Remark 3.1]. Let $\mathcal{L}_0$ be a sufficiently very ample line bundle on $W^{(0)} = W$. Then $\mathcal{L}_1 := (\varphi_1^*\mathcal{L}_0)(-E_1)$ is very ample on $W^{(1)}$. The pull-back $\mathcal{L}_2 := \varphi_2^*\mathcal{L}_1$ on $W^{(2)}$ is π_2-ample. Thus there is an open subset U of $\mathbb{A}^1$, containing $\mathbf{0}$, such that $\mathcal{L}_2$ is ample on $\pi_2^{-1}(U)$. By removing finitely many fibers of the families/$\mathbb{A}^1$ in the above diagram and with the abuse of notation that we denote $\pi_2^{-1}(U)$ over U still by $W^{(2)}/\mathbb{A}^1$ (and since it is only a neighborhood of Y in W that matters), we will say that $\mathcal{L}_2$ is ample on $W^{(2)}$. By taking $k >> 0$, $\mathcal{L}_2^{\otimes k}$ becomes sufficiently very ample on $W^{(2)}$ and its pull-back with a twist $\mathcal{L} := (\varphi_3^*\mathcal{L}_2^{\otimes k})(-Y_1)$ becomes very ample on $W^{(3)}$.

Lemma 2.2 [**$\mathcal{L}$-independence of $\Omega^{\mathcal{L}}_{(g,n;\beta)}$**]. *Let $\mathcal{L}$ be a very ample line bundle on $W^{(3)}$ as constructed above and d be the $\mathcal{L}$-degree of $\beta \in H_2(Y;\mathbb{Z})$. Then $\Omega^{\mathcal{L}}_{(g,n;\beta)}$ that appears in the degeneration formula in Lemma 2.1 is independent of $\mathcal{L}$.*

Proof. The proof follows the same reasoning as in the proof of [L-Y: Lemma 3.2]. Recall the blow-up resolution $p_0 : Y_0 \to Y$ with exceptional divisor identical to D ($= Y_0 \cap Y_1$) and p_1 that sends Y_1 to the conifold singularity of Y. It follows from a careful chasing through the sequence of pull-backs and twists in the construction of $\mathcal{L}$ that

$$\mathcal{L}^{(0)} := \mathcal{L}|_{Y_0} \simeq (p_0^*\mathcal{L}_0|_Y)^{\otimes k}(-(k+1)D) \quad \text{and} \quad \mathcal{L}^{(1)} := \mathcal{L}|_{Y_1} \simeq \mathcal{O}_{Y_1}(-(k+1)D)\,.$$

Recall the generator γ_2 of $H_2(Y_1;\mathbb{Z})$. Let $\gamma_{2,1}$ and $\gamma_{2,2}$ be the curve classes on Y_0 from the two rulings of D. Then they generate the semi-group of curve classes in the relative Mori cone $NE(p_0)$. Observe that $D\cdot\gamma_{2,i} = -1$ on Y_0 for $i = 1,2$ while $D\cdot\gamma_2 = +1$ on Y_1. Let $NE(Y)_\beta$ be the set of all the curve classes in $NE(Y)_{\mathbb{Z}}$ that represent β in $H_2(Y;\mathbb{Z})$. Then, for each $\gamma \in NE(Y)_\beta$, $p_{0*}^{-1}(\gamma)$ in $NE(Y_0)_{\mathbb{Z}}$ is of the form $\widetilde{\gamma} + \mathbb{Z}_{\geq 0}\gamma_{2,1} + \mathbb{Z}_{\geq 0}\gamma_{2,2}$ for a unique $\widetilde{\gamma} \in NE(Y_0)_{\mathbb{Z}}$ determined by γ.

Now let $\eta = (\Gamma_0, \Gamma_1, I) \in \Omega^{\mathcal{L}}_{(g,n;\beta)}$. Then the pairs $(b(\Gamma_0), b(\Gamma_1))$ are characterized by the conditions:

$$p_{0*}b(\Gamma_0) = \beta \quad \text{and} \quad \mathcal{L}^{(0)}\cdot b(\Gamma_0) + \mathcal{L}^{(1)}\cdot b(\Gamma_1) = \mathcal{L}\cdot\beta = d\,.$$

Solving it explicitly as in [L-Y: proof of Lemma 3.2], one concludes that

$$\Omega^{\mathcal{L}}_{(g,n;\beta)} = \coprod_{\gamma\in NE(Y)_\beta} \left\{ \begin{array}{l} \eta = (\Gamma_0,\Gamma_1,I) \\ \text{admissible} \\ \text{triple} \\ \text{for } Y_0 \cup_D Y_1 \end{array} \;\middle|\; \begin{array}{l} \bullet\ b(\Gamma_0) = \widetilde{\gamma} + l_{0,1}\gamma_{2,1} + l_{0,2}\gamma_{2,2},\ b(\Gamma_1) = l_1\gamma_2\,, \\ \quad (l_{0,1}+l_{0,2}) + l_1 = D\cdot\widetilde{\gamma},\ l_{0,1}, l_{0,2}, l_1 \in \mathbb{Z}_{\geq 0}\,; \\ \bullet\ g(\eta) = g\,,\ k_1 + k_2 = n\,; \\ \bullet\ \sum_i \mu_{0,i} = l_1\,; \\ \bullet\ I \subset \{1,\ldots,n\}\,,\ |I| = k_1\,. \end{array} \right\}$$

This set is indeed independent of $\mathcal{L}$. □

We shall denote the $\mathcal{L}$-independent set of admissible triples worked out explicitly at the end of the proof above by $\Omega_{(g,n;\beta)}$.

The proof of the above lemma also implies that, with this choice of $\mathcal{L}$ on $W^{(3)}/\mathbb{A}^1$, the potentially $\mathcal{L}$-dependent stack $\mathfrak{M}(\mathfrak{W}^{(3)},(g,n;\beta))$ can be re-defined without referring to $\mathcal{L}$ at all and, hence, depends only on W and $(g,k;\beta)$. Its fiber $\mathfrak{M}(\mathfrak{W}_0^{(3)},(g,n;\beta))$ over $\mathbf{0}\in\mathbb{A}^1$ is thus also $\mathcal{L}$-independent. Adding this $\mathcal{L}$-independence into the statements in Lemma 2.1, we obtain:

Theorem 2.3 [intrinsic to $W/\mathbb{A}^1$ and β]. (Cf. [L-Y: Theorem 3.3].) *The degeneration/gluing formulas, one in the numerical form and the other in the equivalent cycle form, in Lemma 2.1 are independent of $\mathcal{L}$ and are intrinsic to W and $(g,n;\beta)$ when $\mathcal{L}$ is chosen as above: (the cycle form omitted)*

$$\Psi^{W_t^{(3)}}_{(g,n;\beta)}(\alpha(t),\zeta) \;=\; \sum_{\eta\in\overline{\Omega}_{(g,n;\beta)}} \frac{\mathbf{m}(\eta)}{|\mathrm{Eq}(\eta)|}\sum_{j\in K_\eta}\left[\Psi^{Y_0^{rel}}_{\Gamma_0}(j_0^*\alpha(0),\zeta_{\eta,0,j})\bullet\Psi^{Y_1^{rel}}_{\Gamma_1}(j_1^*\alpha(0),\zeta_{\eta,1,j})\right]_0 .$$

Remark 2.4 [*W-dependence of $\mathfrak{M}(\mathfrak{W}_0^{(3)},(g,n;\beta))$*]. From the details in Section 4.4 of [Li2], the dependence of the stack $\mathfrak{M}(\mathfrak{W}_0^{(3)},(g,n;\beta))$ on the neighborhood of Y in W, if any, can be only mild. The fact that the morphism

$$\coprod_{\eta\in\Omega_{(g,n;\beta)}}\Phi_\eta \;:\; \coprod_{\eta\in\Omega_{(g,n;\beta)}}\mathfrak{M}(\mathfrak{Y}_0^{rel},\Gamma_0)\times_{D^r}\mathfrak{M}(\mathfrak{Y}_1^{rel},\Gamma_1)\;\longrightarrow\;\mathfrak{M}(\mathfrak{W}_0^{(3)},(g,n;\beta))$$

is natural and surjective ([Li1: Sec. 4] and [Li2: Sec. 4.4]) on the underlying topological space of points with the Zariski topology implies that different choices of W for a given Y give stacks $\mathfrak{M}(\mathfrak{W}_0^{(3)},(g,n;\beta))$ of homeomorphic underlying topologies, (cf. [L-MB: Chapter 5]). Since each Φ_η is finite étale of degree a W-independent combinatorial factor $|\mathrm{Eq}(\eta)|$ to its image $\mathfrak{M}(\mathfrak{Y}_0^{rel}\sqcup\mathfrak{Y}_1^{rel},\eta)$ ([Li1: Sec. 4]), as long as these stacks have the same relative multiplicity to the corresponding image of Φ_η - which is proved to be true and this relative multiplicity of the component of $\mathfrak{M}(\mathfrak{W}_0^{(3)},(g,n;\beta))$ labelled by η is given by $\mathbf{m}(\eta)$ irrespective of W ([Li2: Sec. 4.4]) - any detailed difference will not affect the resulting value of Gromov–Witten invariants of Y to be extracted from these stacks for a given $(g,n;\beta)$. (Such indifference is implicit in [Li2] in order to define Gromov–Witten invariants of the degenerate fiber of a family that is family-independent, and indeed the degeneration formula implies this indifference as well.)

3. Extracting GW-invariants of a conifold from pairs

The results in Sec. 1 and Sec. 2 together with the constant-under-deformation requirement for any good definition of Gromov–Witten invariants of conifolds implies the following route to extract Gromov–Witten invariants of the conifold Y from the pairs (Y_0,D) and (Y_1,D) (in this section, by "Gromov–Witten invariants of Y", we mean the invariants of Y that would be defined via the intersection theory on the moduli stack $\overline{\mathcal{M}}_{g,n}(Y,\beta)$ of stable maps to Y, should $\overline{\mathcal{M}}_{g,n}(Y,\beta)$ exists and admits a perfect obstruction theory):

$$(g,n;\beta) \Rightarrow \left\{ \left(\mathfrak{M}(\mathfrak{Y}_0^{rel},\Gamma_0)\,,\, \mathfrak{M}(\mathfrak{Y}_1^{rel},\Gamma_1) \right) \right\}_{\eta=(\Gamma_0,\Gamma_1,I)\in\overline{\Omega}_{(g,n;\beta)}} \overset{\substack{\text{gluing formula}\\ \text{from}\\ \text{Theorem 2.3}}}{\Rightarrow} \begin{array}{l}\text{Gromov–Witten}\\ \text{invariants of } Y\\ \text{as would be}\\ \text{defined via}\\ \overline{\mathcal{M}}_{g,n}(Y,\beta)\,.\end{array}$$

There is one last ingredient, though, that we have not yet discussed: when the number n of marked points on the prestable curve is non-zero, the definition of Gromov–Witten invariants of Y involves choices of cycles on Y. In this case, we need to know whether we can choose canonically (perhaps up to some equivalence relation) a set of cycles on $Y_0 \cup_D Y_1$ to feed into the expression in the definition of relative Gromov–Witten invariants for (Y_0, D) and (Y_1, D) in the gluing formula. Let us now turn to this last issue.

Suppose that $Y = W_0$ is in a degeneration $W/\mathbb{A}^1$. Let $\tau_i : A_*(Y_i) \to A_{*-1}(D)$, $i = 0, 1$, be the group homomorphism defined by taking intersection with D. Let Z be a cycle in W flat over $\mathbb{A}^1$; then $(Z \cdot Y_0) \cdot D = -(Z \cdot Y_1) \cdot D$. Since $Y_0 \cdot D = -Y_1 \cdot D$ on W, the equality $\tau_0(Z \cdot Y_0) = \tau_1(Z \cdot Y_1)$ must hold. This motivates the following definition:

Definition 3.1 [pre-deformable class on $Y_0 \cup_D Y_1$]. Let $\xi_i \in A_*(Y_i)$, $i = 0, 1$. We say that (ξ_0, ξ_1) gives a *pre-deformable class* (i.e. $\xi_0 + \xi_1$) on $Y_0 \cup_D Y_1$ if $\tau_0(\xi_0) = \tau_1(\xi_1)$ in $A_*(D)$. Such (ξ_0, ξ_1) will be called an *admissible pair* of cycle classes with respect to the gluing $Y_0 \cup_D Y_1$.

Readers may note that the image of a pre-deformable map from a (not necessarily connected) prestable curve to $Y_0 \cup_D Y_1$ as defined in [Li1: Sec. 2; in particular, Definition 2.9] gives an example of a pre-deformable class on $Y_0 \cup_D Y_1$.

Recall ([Fu]) that the rational map $\varsigma_0 : Y \dashrightarrow Y_0$ (the inverse of $p_0 : Y_0 \to Y$) defines a group homomorphism $\varsigma_{0*} : A_*(Y) \to A_*(Y_0)$ by taking intersection product with the closure of the graph of ς_0 in $Y \times Y_0$.

Definition 3.2 [admissible class on Y]. A class $\xi \in A_*(Y)$ will be called *admissible* if $\varsigma_{0*}(\xi)$ in $A_*(Y_0)$ admits a unique extension to an admissible pair $(\varsigma_{0*}(\xi), \xi')$ with respect to $Y_0 \cup_D Y_1$. Denote the set of all such classes in $A_*(Y)$ by $A_*^{ad}(Y)$.

Note that $\tau_1 : A_*(Y_1) \to A_*(D)$ is injective on $A_{\geq 1}(Y_1)$ with $\tau_1(A_{\geq 1}(Y_1))$ spanned by $[pt]$, $[(1,1)\text{-curve}]$, and $[D]$. Consequently, effective classes of constant dimension ξ_1 in $A_{\geq 1}(Y_0)$ whose intersection with D lies in this subspace have a unique extension to an admissible pair (ξ_1, ξ_2) with respect to $Y_0 \cup_D Y_1$. The image of such ξ_1 on Y will span $A_*^{ad}(Y)$. The constant-under-deformation requirement and multi-linearity in the axioms of Gromov–Witten invariants for any variety imply the following:

Corollary 3.3 [extraction of GW-invariants of Y from pairs]. *The Gromov–Witten invariants of Y that involve only classes in $A_*^{ad}(Y)$ can be extracted from the relative Gromov–Witten invariants of the pairs (Y_0, D) and (Y_1, D) by the gluing formula.*

In other words, let $\alpha \in (A_*^{ad}(Y)_{\mathbb{Q}})^{\times n}$ and $\zeta \in A_*(\mathfrak{M}_{g,n})$. Denote by $\Psi^Y_{(g,n;\beta)}(\alpha,\zeta)$ the Gromov–Witten invariant of Y associated to the topological type $(g,n;\beta)$ and classes α and ζ that is defined via *any* standard procedure, i.e. a construction of $\overline{\mathcal{M}}_{g,n}(Y,\beta)$ with a perfect obstruction theory, ..., etc. in such a way that the Gromov–Witten axioms and the constant-under-deformation property are satisfied and that, when the construction is applied to a smooth variety, it recovers the Gromov–Witten invariants from the equivalent construction of [B-M] and [L-T]. Let $(\alpha_0,\alpha_1) \in A_*(Y_0)^{\times n} \times A_*(Y_1)^{\times n}$ be the tuple from the admissible pairs with respect to $Y_0 \cup_D Y_1$ of the class in the entries of α. For $\eta = (\Gamma_0,\,\Gamma_1,\,I) \in \overline{\Omega}_{(g,n;\beta)}$, assume that $G_\eta^*(\zeta) = \sum_{j\in K_\eta} \zeta_{\eta,0,j} \boxtimes \zeta_{\eta,1,j}$, where $G_\eta : \mathfrak{M}_{\Gamma_0^\circ} \times \mathfrak{M}_{\Gamma_1^\circ} \to \mathfrak{M}_{g,n}$ is the natural morphism between the related moduli stack of nodal curves, cf. [Li2: Sec. 0]. Then

$$\Psi^Y_{(g,n;\beta)}(\alpha,\zeta) \;=\; \sum_{\eta\in\overline{\Omega}_{(g,n;\beta)}} \frac{\mathbf{m}(\eta)}{|\mathrm{Eq}(\eta)|} \sum_{j\in K_\eta} \left[\Psi^{Y_0^{rel}}_{\Gamma_0}(\alpha_0,\zeta_{\eta,0,j}) \,\bullet\, \Psi^{Y_1^{rel}}_{\Gamma_1}(\alpha_1,\zeta_{\eta,1,j}) \right]_0 .$$

While the left-hand side of the identity remains to be truly constructed (though one may turn this identity into a *definition* of Gromov–Witten invariants of Y for α admissible if one wishes), the right-hand side is completely determined by and canonically/intrinsically associated to the data: Y, $(g,n;\beta)$, α, and ζ.

Note that when Y is a Calabi–Yau conifold (i.e. it arises from a degeneration of Calabi–Yau 3-folds), the expected dimension of the would-be moduli stack $\overline{\mathcal{M}}_{g,0}(Y,\beta)$ is zero. In this case, no issue of choice of cycles is involved.

Corollary 3.4 [Calabi–Yau conifold]. *When Y is a Calabi–Yau conifold, all the Gromov–Witten invariants of Y associated to the topological type $(g,0\,;\beta)$ can be extracted from the relative Gromov–Witten invariants of pairs (Y_0,D) and (Y_1,D) by the gluing formula.*

Remark 3.5 [*general conifold*]. For a general conifold Y with $Sing(Y) = \{p_1,\cdots\}$, the discussions, expressions, and formulas in Sec. 2 and Sec. 3 generalize immediately by appropriately doing the following replacement: D in Y_0 by $\coprod_i D_i$ in Y_0, (Y_1,D) by isomorphic copies of smooth quadric pairs $(Y_i,D_i) \simeq (Y_1,D)$, summands involving the index 1 of the pair (Y_1,D) by the summation $\sum_i$, ..., etc., in the expressions involved, where $i = 1,\ldots.$ corresponds to each conifold singularity p_i of Y.

Remark 3.6 [*a general picture*]. For a general irreducible singular variety Y, the result in the current work suggests the following picture whose details remain to be studied: an assignment to Y of a collection of relative pairs: a principal relative pair $(Y_0, D_0 = \cup_i D_{0,i})$, where $Y_0 \to Y$ is a resolution of Y whose exceptional locus D_0 is a divisor with only simple normal crossings (snc) singularities, and a set of auxiliary (smooth variety, snc divisor) pairs (Y_j,D_j) that are determined by the germ/formal scheme $\hat{Y}_{sing}$ of the singular locus Y_{sing} in Y and the resolution $Y_0 \to Y$ only, and a gluing diagram Γ that encodes the gluing of all the these relative pairs so that

(1) the relative Gromov–Witten invariants of (Y_0,D_0) are obtained from a combination of those for the smooth pairs

$$(Y_0,D_{0,i}),\;\; (D_{0,i},D_{0,i}\cap\cup_{i'\neq i}D_{0,i'}),\;\; \cdots\,,\text{etc},$$

and similarly for (Y_j,D_j) (cf. a "quantum inclusion-exclusion principle");

(2) the Gromov–Witten invariants of Y are then obtained from a combination of the relative Gromov–Witten invariants of (Y_0, D_0) and (Y_j, D_j)'s;
(3) the combination rules in Items (1) and (2) above are functorial/universal in the same sense that Jun Li's degeneration formula is functorial/universal (i.e. independent of details of Y away from the singularities, curve classes chosen, cycle classes chosen, ... etc.).

(Cf. FIGURE 3-1.)

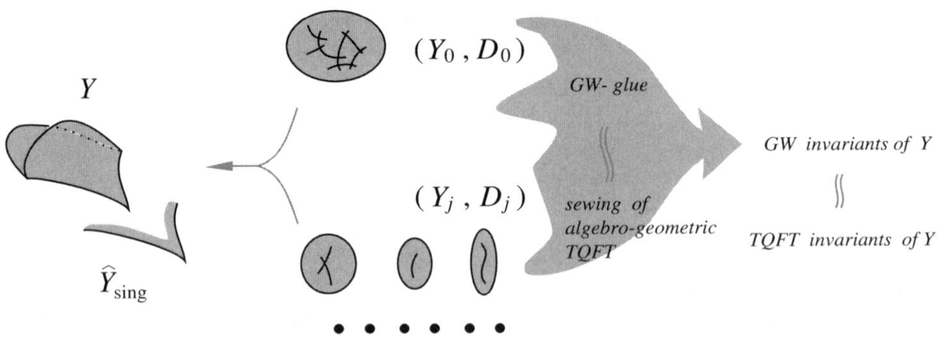

FIGURE 3-1. A gluing construction of Gromov–Witten invariants of a singular variety Y. Since Gromov–Witten invariants arise from A-model topological string theory on the physics side [Wi3], such a gluing construction should resonate closely with the Atiyah-Segal formulation of topological quantum field theory ([At] and [Se]; see also [E-G-H] and [B-P]).

Acknowledgements

We thank Joe Harris, Jun Li, Kefeng Liu, and Cumrun Vafa for many influences on this project via lectures, communications, and/or very profound works. C.-H.L. would like to thank in addition K.L. for serving as a sounding board to many of his raw ideas; Lisa Randall, Itay Yavin for valuable lectures; Andrew Strominger, Xiaowei Wang for discussions; Department of Mathematics of UCLA for hospitality; Rev. Campbell Willman and Ann Willman for a conversation; and Ling-Miao Chou for tremendous moral support.

References

[At] M. Atiyah, *Topological quantum field theories*, Publ. Math. I.H.E.S. **68** (1989), pp. 175 - 186.

[A-GZ-V] V.I. Arnold, S.M. Gusein-Zade, and A.N. Varchenko, *Singularities of differentiable maps*, vol. I. *The classification of critical points, caustics and wave fronts*; vol. II, *Monodromy and asymptotics of integrals*, Birkhäuser, 1988.

[A-M-V] M. Aganagic, M. Mariño, and C. Vafa, *All loop topological string amplitudes from Chern-Simons theory*, Commun. Math. Phys. **247** (2004), pp. 467 - 512.

[Be] K. Behrend, *Gromov–Witten invariants in algebraic geometry*, Invent. Math. **127** (1997), pp. 601 - 617.

[B-L] V. Braun and C.-H. Liu, *On the isolated singularity of a 7-space obtained by rolling Calabi–Yau 3-folds through extremal transitions*, `hep-th/9801175`.

[B-M] K. Behrend and Yu. Manin, *Stacks of stable maps and Gromov–Witten invariants*, *Duke Math. J.* **85** (1996), pp. 1 - 60.

[B-P] J. Bryan and R. Pandharipande, *The local Gromov–Witten theory of curves*, math.AG/0411037.

[C-dlO-G-P] P. Candelas, X.C. de la Ossa, P.S. Green, and L. Parkes, *A pair of Calabi–Yau manifolds as an exactly soluble superconformal theory*, *Nucl. Phys.* **B359** (1991), pp. 21 - 74.

[C-H] S.D. Cutkosky and J. Herzog, *Cohen-Macaulay coordinate rings of blowup schemes*, *Comment. Math. Helv.* **72** (1997), pp. 605 - 617.

[De] O. Debarre, *Higher-dimensional algebraic geometry*, Springer-Verlag, 2001.

[D-F-G] D.-E. Diaconescu, B. Florea, and A. Grassi, *Geometric transitions and open string instantons*, *Adv. Theor. Math. Phys.* **6** (2002), pp. 619 - 642.

[E-G-H] Y. Eliashberg, A. Givental, and H. Hofer, *Introduction to symplectic field theory*, *Geom. Funct. Anal.* **2000**, *Special Volume, Part II*, pp. 560- 673.

[Fu] W. Fulton, *Intersection theory*, Ser. Mod. Surv. Math. 2, Springer-Verlag, 1984.

[Gr] A. Grothendieck, *Éléments de géométrie algébrique II: Étude globale élémentaire de quelques classes de morphismes*, *I.H.E.S. Publ. Math.* **8** (1961), pp. 1 - 222.

[G-H] P. Griffiths and J. Harris, *Principles of algebraic geometry*, John-Wiley & Sons, 1978.

[G-P] M.B. Green and J. Polchinski, *Summing over world-sheet boundaries*, *Phys. Lett.* **B335** (1994), pp. 377 - 382.

[G-R] A. Grassi and M. Rossi, *Large N dualities and transitions in geometry*, math.AG/0209044.

[G-V] R. Gopakumar and C. Vafa, *On the gauge theory/geometry correspondence*, *Adv. Theor. Math. Phys.* **3** (1999), pp. 1415 - 1443.

[Harr] J. Harris, *Algebraic geometry, a first course*, GTM 133, Springer-Verlag, 1992.

[Hart] R. Hartshorne, *Algebraic geometry*, GTM 52, Springer-Verlag, 1977.

[Hi] M.W. Hirsch, *Differential topology*, GTM 33, Springer-Verlag, 1976.

[Ko] J. Kollár, *Flips, flops, minimal models, etc.*, in *Surveys in differential geometry*, **1** (1991), S.-T. Yau and B.H. Lawson, Jr. eds., pp. 113 - 199.

[K-K-M-SD] G. Kempf, F. Knudsen, D. Mumford, and B. Saint-Donat, *Toroidal embeddings, I*, Lect. Notes Math. 339, Springer-Verlag, 1973.

[K-M] J. Kollár and S. Mori, *Birational geometry of algebraic varieties*, Cambridge Tracts Math. 134, Cambridge Univ. Press, 1998.

[Li1] J. Li, *Stable morphisms to singular schemes and relative stable morphisms*, *J. Diff. Geom*, **57** (2001), pp. 509 - 578.

[Li2] ——, *A degeneration formula of GW-invariants*, *J. Diff. Geom.* **60** (2002), pp. 199 - 293.

[L-T] J. Li and G. Tian, *Virtual moduli cycles and Gromov–Witten invariants of algebraic varieties*, *J. Amer. Math. Soc.* **11** (1998), pp. 119 - 174.

[L-Y] C.-H. Liu and S.-T. Yau, *A degeneration formula of Gromov–Witten invariants with respect to a curve class for degenerations from blow-ups*, math.AG/0408147.

[L-MB] G. Laumon and L. Moret-Bailly, *Champs algébriques*, Springer-Verlag, 2000.

[O-V1] H. Ooguri and C. Vafa, *Knot invariants and topological strings*, *Nucl. Phys.* **B577** (2000), pp. 419 - 438.

[O-V2] ——, *World-sheet derivation of a large N duality*, *Nucl. Phys.* **B641** (2002), pp. 3 - 34.

[P-S] K.H. Paranjape and V. Srinivas, *Continuous self-maps of quadric hypersurfaces*, in *Proceedings of the Indo-French conference on geometry, Bombay, 1989*, S. Ramanan and A. Beauville eds., pp. 135 - 148, Hindustan Book, Delhi, 1993.

[Se] G. Segal, *The definition of conformal field theory*, in *Topology, geometry and quantum field theory*, U. Tillmann ed., pp. 421 - 577, (draft widely circulated since 1989), London Math. Soc. Lect. Notes Ser. 308, r Cambridge Univ. Press, 2004.

[Sp] E.H. Spanier, *Algebraic topology*, Springer-Verlag, 1966.

[St] A. Strominger, *Massless black holes and conifolds in string theory*, *Nucl. Phys.* **B451** (1995), pp. 96 - 108.

[Va] C. Vafa, *Superstrings and topological strings at large N*, *J. Math. Phys.* **42** (2001), pp. 2798 - 2817.

[Vi] J.W. Vick, *Homology theory - an introduction to algebraic topology*, Academic Press, 1973.

[Wi1] E. Witten, *Phases of $N = 2$ theories in two dimensions*, Nucl. Phys. **B403** (1993), pp. 159 - 222.

[Wi2] ——, *Chern-Simons gauge theory as a string theory*, in The Floer memorial volume, H. Hofer, C.H. Taubes, A. Weinstein, and Z. Zehnder eds., Prog. Math. 133, pp. 637 - 678, Birkhäuser, 1995.

[Wi3] ——, *Mirror manifolds and topological field theory*, in Mirror symmetry I, S.-T. Yau ed., pp. 121 - 160, Studies Adv. Math. 9, Amer. Math. Soc. & International Press, 1998.

Department of Mathematics, Harvard University, Cambridge, MA 02138
E-mail address: `chienliu@math.harvard.edu`

Department of Mathematics, Harvard University, Cambridge, MA 02138
E-mail address: `yau@math.harvard.edu`

AMS/IP Studies in Advanced Mathematics
Volume 38, 2006

Generalized Special Lagrangian Torus Fibrations for Calabi–Yau Hypersurfaces in Toric Varieties II

Wei-Dong Ruan

ABSTRACT. In this paper we construct monodromy representing generalized special Lagrangian torus fibrations for Calabi–Yau hypersurfaces in toric varieties near the large complex structure limit.

1. Introduction

This paper is a sequel to [**11**]. The aim of this series of papers is to construct generalized special Lagrangian torus fibrations for Calabi–Yau hypersurfaces in toric varieties near the large complex structure limit. In this paper we will use the same notations as in [**11**] unless specified otherwise.

Let (P_Δ, ω) be a toric variety whose moment map image (with respect to the toric Kähler form ω) is the real convex polyhedron $\Delta \subset M_{\mathbb{R}}$. Also assume that the anti-canonical class of P_Δ is represented by an integral reflexive convex polyhedron $\Delta_0 \subset M$ and the unique interior point of Δ_0 is the origin of M. Integral points $m \in \Delta_0$ correspond to holomorphic toric sections s_m of the anti-canonical bundle. For the unique interior point m_o of Δ_0, s_{m_o} is the section of the anti-canonical bundle that vanishes to order 1 along each toric divisor of P_Δ.

Let $\{w_m\}_{m\in\Delta_0}$ be a strictly convex function on Δ_0 such that $w_m > 0$ for $m \in \Delta_0 \setminus \{m_o\}$ and $w_{m_o} \ll 0$. Define

$$\tilde{s}_t = s_{m_o} + ts, \quad s = \sum_{m\in\Delta_0\setminus\{m_o\}} a_m s_m, \text{ where } |a_m| = \tau^{w_m}, \text{ for } m \in \Delta_0 \setminus \{m_o\}.$$

(As in the case of the Fermat type quintic, in general Δ_0 need not contain all the integral m in the real polyhedron spanned by Δ_0.) Let $X_t = \{\tilde{s}_t^{-1}(0)\}$. Then $\{X_t\}$ is a 1-parameter family of Calabi–Yau hypersurfaces in P_Δ. $X_0 = \{s_{m_o}^{-1}(0)\}$ is the so-called large complex structure limit. X_t is said to be near the large complex

2000 *Mathematics Subject Classification.* Primary 32Q25; Secondary 14J32, 14M25, 53C38.
Key words and phrases. Special Lagrangian fibration, torus fibration, toric variety, Calabi–Yau hypersurface, large complex structure limit.
The author was partially supported by NSF Grant DMS-0104150.

structure limit if τ and t are small and $t \leq \tau^{-w_{m_o}}$.

Since X_0 is toric, the moment map induces the standard generalized special Lagrangian fibration $\hat{\pi}_0 : X_0 \rightarrow \partial\Delta$ with respect to the toric holomorphic volume form. In [**7, 8, 9, 10**], we constructed Lagrangian torus fibrations for X_t when X_t is near the large complex structure limit, using the Hamiltonian-gradient flow to deform this fibration $\hat{\pi}_0$ for X_0 symplectically to the desired Lagrangian fibration $\hat{\pi}_t : X_t \rightarrow \partial\Delta$ for such X_t. The (topological) singular set of the fibration map $\hat{\pi}_t$ is $C = X_t \cap \mathrm{Sing}(X_0)$, which is independent of t. The corresponding singular locus $\tilde{\Gamma} = \hat{\pi}_0(C)$ is also independent of t. When X_t is near the large complex structure limit, $\tilde{\Gamma} \subset \partial\Delta$ exhibits an amoeba structure that is a fattening of a graph $\Gamma \subset \partial\Delta$. It was conjectured in [**8**] that the singular locus for the (generalized) special Lagrangian torus fibration should resemble the singular locus $\tilde{\Gamma}$ of the Lagrangian torus fibration.

[**11**] initiated the program of using a similar idea to construct generalized special Lagrangian fibrations for X_t by deformation from the standard fibration $\hat{\pi}_0$ of the large complex structure limit X_0. As pointed out in [**11**], due to the canonical nature of the generalized special Lagrangian fibration on a Calabi–Yau when we fix the Kähler metric, one can construct the generalized special Lagrangian fibration over different parts of the base $\partial\Delta$ separately and they will automatically match on the overlaps.

Recall from [**11**] that a (generalized) special Lagrangian fibration of smooth tori over an open set $U \subset \partial\Delta$ is said to represent the monodromy if $\partial\Delta \setminus U$ is a fattening of $\tilde{\Gamma}$ that retracts to $\tilde{\Gamma}$.

There is a natural toric map from P_Δ to its anti-canonical model P_{Δ_0} that induces a map $\pi : \partial\Delta \rightarrow \partial\Delta_0$. Recall that Δ_0 is an integral reflexive polyhedron. We will use $\Delta_0^{(k)}$ to denote the set of integral points in the k-skeleton of Δ_0, where the coefficients of the polynomial defining the family of Calabi–Yau hypersurfaces are non-zero. Let $U \subset \partial\Delta$ be the union of the (slightly shrunk) interiors of top dimensional faces of $\partial\Delta$ and a tubular neighborhood of $\pi^{-1}(\Delta_0^{(n-1)})$ in $\partial\Delta$. (Here $n = \dim \partial\Delta$.) According to our discussion of the singular locus $\tilde{\Gamma}$ in [**9, 10**], for such U, $\partial\Delta \setminus U$ is a fattening of $\tilde{\Gamma}$ that retracts to $\tilde{\Gamma}$. Consequently, a fibration over such U represents monodromy.

In [**11**], we constructed generalized special Lagrangian fibrations for X_t over two types of regions in $\partial\Delta$. The type one regions are the interiors of top dimensional faces in $\partial\Delta$ (also see [**2**]). The type two regions are small neighborhoods of vertices of $\partial\Delta$. Notice from [**9, 10**] that these two types of regions are exactly dual to each other under the so-called symplectic SYZ mirror symmetry proved in [**10**] for Calabi–Yau hypersurfaces in toric varieties. This construction will be enough to a produce monodromy representing generalized special Lagrangian fibration if $\Delta_0^{(n-1)} = \Delta_0^{(0)}$ and $\dim \pi^{-1}(\Delta_0^{(0)}) = 0$, which is the case for Fermat type Calabi–Yau hypersurfaces. The difficulty in generalizing the construction for these two types of regions to the rest of $\pi^{-1}(\Delta_0^{(n-1)})$ in general is the need to deal with "thin torus" fibres that have small circles in some directions and big circles in other directions.

In this paper, we will construct generalized special Lagrangian fibrations for X_t over two other types of regions in $\partial\Delta$. The first type is in the neighborhood of $\pi^{-1}(\Delta_0^{(0)})$. (The case when $\dim \pi^{-1}(\Delta_0^{(0)}) = 0$ was already dealt with in [**11**].) The second type is in the neighborhood of $\pi^{-1}(\Delta_0^{(n-1)} \setminus \Delta_0^{(0)})$. Our construction for the thin torus in regions of the first type is rather complete. Notice that the mirror of the quintic Calabi–Yau satisfies $\Delta_0^{(n-1)} = \Delta_0^{(0)}$. Therefore, this construction will be enough to produce monodromy representing generalized special Lagrangian fibrations for the mirror family of the quintic Calabi–Yau. Our construction for the thin torus in regions of the second type is less satisfactory, since it will require stronger sense of being near the large complex structure limit for the hypersurfaces and toric metric. Nevertheless, this construction will produce monodromy representing generalized special Lagrangian fibrations for such Calabi–Yau hypersurfaces with a suitable toric metric.

In the sequels of this paper, we will give a more satisfactory treatment of the second type of thin torus and discuss some singular fibres.

This paper is organized as follows. In section 2 we lay out the local setting that each of our constructions in section 5 and 6 will reduce to and clarify the estimates for the thin torus needed to apply the implicit function theorem. These estimates are worked out in section 3. In section 4, applying the estimates of the thin torus from section 3, we work out a precise quantitative version of theorem 5.1 in [**11**], which will be needed for the construction in section 6. Sections 5 and 6 are devoted to the construction of generalized special Lagrangian fibrations over the two types of regions based on the estimates from section 3. As a consequence, we are able to construct monodromy representing generalized special Lagrangian torus fibrations for a large class of Calabi–Yau hypersurfaces in toric varieties including the quintic Calabi–Yau 3-fold (corollary 6.2) and its mirror (corollary 5.1). We also indicated how to generalize our construction to general Calabi–Yau hypersurfaces in toric variety in a remark in section 6.

2. Basic setting

Toric terminology, though straightforward, can be daunting or confusing for people not familiar with it. Fortunately, in each of the regions we will consider, using suitable toric coordinates around the region we are interested, we can locally identify the hypersurface X_t in P_Δ with a hypersurface in $\mathbb{C}^{n+1}$:

$$X_t = \{z \in \mathbb{C}^{n+1} : \tilde{p}_t(z) = 0\}, \quad \text{where } \tilde{p}_t(z) = z^m + tp(z),\ z^m = \prod_{k=0}^{n} z_k^{m_k}, m_k \leq 1.$$

Under such an identification one can forget about toric terminology and work strictly inside $\mathbb{C}^{n+1}$.

Let $I' = \{k : m_k \leq 0\}$ and $I'' = \{k : m_k = 1\}$. Then we have the natural decomposition $\{0, \cdots, n\} = I' \cup I''$. Correspondingly we have decompositions

$m = (m_0, \cdots, m_n) = (m', m'')$ and $z = (z_0, \cdots, z_n) = (z', z'')$.

In the case of the construction of fibrations over top-dimensional faces of $\partial\Delta$ discussed in section 4, one only needs to consider $m = (1, 0, \cdots, 0)$, in which case $I'' = \{0\}$. The generalized special Lagrangian torus fibration of $X_t = \{z \in \mathbb{C}^{n+1} | z_0 + tp(z) = 0\}$ will be constructed through deformation from the standard generalized special Lagrangian torus fibration of $X_0 = \{z \in \mathbb{C}^{n+1} | z_0 = 0\}$.

In the case of the construction of fibrations near $\pi^{-1}(\tilde{m})$ for $\tilde{m} \in \Delta_0^{(n-1)}$ discussed in sections 5 and 6, $z^m = s_{m_o}/s_{\tilde{m}}$ and $p(z)$ is a small perturbation of 1 in the region of interest. The generalized special Lagrangian torus fibration of X_t in this case will be constructed through a different deformation starting with the following local model.

Local model: In $\mathbb{C}^{n+1}$, consider the family of hypersurfaces $\{X_{t,0}\}$ defined as

$$X_{t,0} = \{z^m = t\}.$$

Clearly $X_{t,0}$ are all toric varieties. For a toric metric on $\mathbb{C}^{n+1}$ with Kähler form ω, $\omega_{t,0} = \omega|_{X_{t,0}}$ is a toric Kähler form on $X_{t,0}$. Therefore the family of natural real tori

$$L_{t,0} = \{z \in X_{t,0} : |z_k| = r_k(\text{constant}), \text{ for } 0 \leq k \leq n\}$$

forms the generalized special Lagrangian torus fibration of $X_{t,0}$ with respect to $\omega_{t,0}$. The fibration can also be defined using the fibration map $\hat{\pi}_{t,0} : X_{t,0} \to \partial\Delta$ defined as $\hat{\pi}_{t,0}(z) = \{|z_k|^2 - r_{\min}^2\}_{k=0}^n$, where Δ denotes the first quadrant in $\mathbb{R}^{n+1}$ and $r_{\min} = \min\limits_{0 \leq k \leq n} |z_k|$. □

Let $\check{p}(z) = \log p(z)$, and

$$X_{t,s} = \{\tilde{p}_{t,s}(z) = 0\}, \quad \text{where } \tilde{p}_{t,s}(z) = z^m + te^{s\check{p}(z)}.$$

We will use the family $\{X_{t,s}\}_{s\in[0,1]}$ to connect the local model $X_{t,0}$ and $X_t = X_{t,1}$. Our idea here is to deform the generalized special Lagrangian torus fibration of the local model $X_{t,0}$ to the generalized special Lagrangian torus fibration of $X_t = X_{t,1}$.

Hamiltonian-gradient vector fields were discussed in section 4 of [**11**]. Here we will need explicit expressions of them. Generally, consider complex hypersurfaces Y_t defined by $t = u/v$, where u and v are holomorphic functions. t is meromorphic when the family $\{Y_t\}$ has non-empty base locus. Let V be the Hamiltonian-gradient vector field with respect to the family $\{Y_t\}$.

LEMMA 2.1. *When restricted to the complex hypersurface Y_t, where t is a constant, we have*

$$V = \frac{\nabla f}{|\nabla f|^2} = \frac{\mathrm{Re}(\bar{v}\nabla(u - tv))}{|d(u - tv)|^2}, \quad \text{where } f = \mathrm{Re}(t).$$

In particular, when Y_t are all smooth, V will be a smooth vector field on the total space $\mathcal{Y}$.

Proof: For $t = u/v$, we have

$$dt = \frac{du - tdv}{v}.$$

When restricted to the complex hypersurface Y_t, where t is a constant, we have

$$\nabla t = \frac{1}{|v|^2}\bar{v}\nabla(u - tv).$$

$$\nabla f = \frac{1}{|v|^2}\mathrm{Re}(\bar{v}\nabla(u - tv)).$$

$$|\nabla f|^2 = |df|^2 = \frac{1}{2}|dt|^2 = \frac{1}{2}\frac{|du - tdv|^2}{|v|^2}.$$

$$V = \frac{\nabla f}{|\nabla f|} = \frac{2\mathrm{Re}(\bar{v}\nabla(u - tv))}{|d(u - tv)|^2}. \qquad \square$$

In lemma 2.1, we can take $u = \log(-z^m/t)$, $v = \check{p}$ and replace t by s, Y_t by $X_{t,s}$. Then along $X_{t,s}$,

$$d(u - sv) = \sum_{k=0}^{n} q_k \frac{dz_k}{z_k}, \quad \text{where } q_k = m_k - s z_k \check{p}_k.$$

The normalized Hamiltonian-gradient vector field for the parameter s is

$$V = 2\mathrm{Re}\left(\check{p}\sum_{i=0}^{n} \lambda_i z_i \frac{\partial}{\partial z_i}\right), \quad \text{where}$$

$$\lambda_i = \left(\sum_{j=0}^{n} \frac{g^{i\bar{j}}\bar{q}_j}{z_i \bar{z}_j}\right)\left(\sum_{k,j=0}^{n} \frac{g^{k\bar{j}} q_k \bar{q}_j}{z_k \bar{z}_j}\right)^{-1} \tag{2.1}$$

The normalized Hamiltonian-gradient vector field for the parameter t is

$$V_t = 2\mathrm{Re}\left(\frac{1}{t}\sum_{i=0}^{n} \lambda_i z_i \frac{\partial}{\partial z_i}\right).$$

In our construction of generalized special Lagrangian torus fibrations in section 5 and 6 (resp. in section 4), we will use the normalized Hamiltonian-gradient vector field V (resp. V_t) to deform the standard generalized special Lagrangian torus fibration for $X_{t,0}$ (resp. X_0) to a Lagrangian torus fibration for $X_{t,1} = X_t$ (resp. X_t) that is almost generalized special up to a small error that only depends on t. Using the quantified implicit function theorem from [**11**], we would be able to construct and control the generalized special Lagrangian torus fibration for X_t when t is small. For this purpose, we would need estimates of V (resp. V_t) that will be worked out in the next section. In the rest of this section, we will introduce the Banach spaces, define the functional and prove the uniform invertibility of the derivative of the functional in proposition 2.1. Let

$$\Omega = \frac{1}{p}\bigwedge_{i=0}^{n} dz_i, \quad \Omega_t = \imath(V_t)\Omega|_{X_{t,s}}, \quad \text{where } p = e^{s\check{p}}.$$

(Here Ω, Ω_t, V_t and V also depend on s. We are omitting the subscript for simplicity of notation.) Notice that according to lemma 4.1 of [**11**], Ω_t so defined when $s = 1$ coincides with the holomorphic volume form Ω_t on $X_t = X_{t,1}$. Let ϕ_s denote the flow for V. For fixed t, the family $(X_{t,s}, \omega_{t,s}, \Omega_t)$ of Calabi–Yau manifolds can be reduced to the equivalent family of Calabi–Yau structures $\{\Omega_{t,s} = \phi_s^*\Omega_t\}$ on the fixed symplectic manifold $(X_{t,0}, \omega_{t,0})$.

For the generalized special Lagrangian submanifold $L_{t,0}$ in $(X_{t,0}, \omega_{t,0}, \Omega_{t,0})$, there is an identification $X_{t,0} \cong T^*L_{t,0}$ near $L_{t,0}$ such that $\omega_{t,0}$ is identified with the canonical symplectic form on T^*L_0. (See [**13**] for the general case. In our toric situation, such an identification is actually explicit and does not need the general result from [**13**].) Any Lagrangian submanifold L near $L_{t,0}$ can be identified with the graph of dh for some smooth function h on $L_{t,0}$. In the standard coordinates (x, y) on $T^*L_{t,0}$, $\omega_{t,0} = dx \wedge dy$ and L is locally the graph $(x, \frac{\partial h}{\partial x})$.

Locally, (as in [**11**], with a slight adjustment of notation) we can write

$$\Omega_{t,s} = \eta_s \bigwedge_{k=1}^{n} (dx_k + u_{s,k} dy),$$

$$\Omega_{t,s}|_L = \eta_s \left(x, \tfrac{\partial h}{\partial x}\right) \det\left(I + U_s \frac{\partial^2 h}{\partial x^2}\right) \bigwedge_{k=1}^{n} dx_k,$$

where $U_s = (u_{s,1}, \cdots, u_{s,n}) = (u_{s,jk})$ and $u_{s,k}dy = \sum_{j=1}^{n} u_{s,jk} dy_j$.

$$F(h, s) = \operatorname{Im}(\log \Omega_{t,s}|_L) = \operatorname{Im}\left(\log \eta_s \left(x, \tfrac{\partial h}{\partial x}\right) + \log\det\left(I + U_s \frac{\partial^2 h}{\partial x^2}\right)\right)$$

defines a map $F : \mathcal{B}_1 \times \mathbb{R} \to \mathcal{B}_2$, where $\mathcal{B}_1 = C^{2,\alpha}(L_{t,0})$ and $\mathcal{B}_2 = C^{\alpha}(L_{t,0})$ are Banach spaces. We intend to apply the implicit function theorem to F to construct the family of generalized special Lagrangians $L_{t,s}$ with respect to $(X_{t,0}, \omega_{t,0}, \Omega_{t,s})$.

Straightforward computations give us

$$\delta_h F(h, s) = \frac{\partial F}{\partial h}(h, s)\delta h = a_s^{ij} \frac{\partial^2 \delta h}{\partial x_i \partial x_j} + b_s^i \frac{\partial \delta h}{\partial x_i},$$

where

$$a_s^{ij}(x) = \operatorname{Im}\left(\left(I + U_s \frac{\partial^2 h}{\partial x^2}\right)^{-1} U_s\right)\left(x, \tfrac{\partial h}{\partial x}\right), \tag{2.2}$$

$$b_s^i(x) = \operatorname{Im}\left(\frac{1}{\eta_s}\frac{\partial \eta_s}{\partial y_i} + \operatorname{Tr}\left(\left(I + U_s \frac{\partial^2 h}{\partial x^2}\right)^{-1} \frac{\partial U_s}{\partial y_i}\frac{\partial^2 h}{\partial x^2}\right)\right)\left(x, \tfrac{\partial h}{\partial x}\right).$$

In particular,

$$\delta_h F(0, 0) = \frac{\partial F}{\partial h}(0, 0)\delta h = a^{ij} \frac{\partial^2 \delta h}{\partial x_i \partial x_j} + b^i \frac{\partial \delta h}{\partial x_i},$$

where

$$a^{ij}(x) = \operatorname{Im}((U_0)_{ij})(x, 0), \quad b^i(x) = \operatorname{Im}\left(\frac{1}{\eta_0}\frac{\partial \eta_0}{\partial y_i}\right)(x, 0).$$

In our toric situation, in the coordinate $x_k = \theta_k$ for $1 \leq k \leq n$, $g_{t,0}|_{L_{t,0}}$ is a flat metric on $L_{t,0}$. According to (3.1) in [**11**], we have $a^{ij}(x) = (g_{t,0}|_{L_{t,0}})^{ij}$ being a constant positive definite matrix and $b^i(x) = 0$. Therefore

$$\delta_h F(0,0) = \frac{\partial F}{\partial h}(0,0)\delta h = (g_{t,0}|_{L_{t,0}})^{ij}\frac{\partial^2 \delta h}{\partial x_i \partial x_j}$$

is just the standard Laplace operator on $L_{t,0}$.

PROPOSITION 2.1. *Assume that the torus $(L_{t,0}, g_{t,0}|_{L_{t,0}})$ has bounded diameter. Then there exists a constant C (only depending on the dimension) such that*

$$\|\delta h\|_{\mathcal{B}_1} \leq C \left\| \left(\frac{\partial F}{\partial h}(0,0)\right)\delta h \right\|_{\mathcal{B}_2}.$$

Proof: One only needs to show that C is independent of the thinness of the torus. It is easy to see that there exists a finite covering map $\pi : \tilde{L} \to L_{t,0}$ such that $(\tilde{L}, \tilde{g} = \pi^* g_{t,0}|_{L_{t,0}})$ is of bounded geometry. We will use those symbols with "~" to denote the corresponding pullback objects by π. By standard Schauder estimates, we have

$$\|\tilde{\delta h}\|_{\tilde{\mathcal{B}}_1} \leq C \left\| \left(\frac{\partial \tilde{F}}{\partial \tilde{h}}(0,0)\right)\tilde{\delta h} \right\|_{\tilde{\mathcal{B}}_2},$$

where the constant C only depends on the dimension. Since $\tilde{\delta h}$ and $\left(\frac{\partial \tilde{F}}{\partial \tilde{h}}(0,0)\right)\tilde{\delta h}$ on $\tilde{L}$ are invariant under deck transformations, we have

$$\|\delta h\|_{\mathcal{B}_1} = \|\tilde{\delta h}\|_{\tilde{\mathcal{B}}_1} \leq C \left\| \left(\frac{\partial \tilde{F}}{\partial \tilde{h}}(0,0)\right)\tilde{\delta h} \right\|_{\tilde{\mathcal{B}}_2} = C \left\| \left(\frac{\partial F}{\partial h}(0,0)\right)\delta h \right\|_{\mathcal{B}_2}. \square$$

According to the implicit function theorem (theorem 3.2 in [**11**]), proposition 2.1 has reduced the construction of the family of generalized special Lagrangians $L_{t,s}$ with respect to $(X_{t,0}, \omega_{t,0}, \Omega_{t,s})$ to the estimates of

$$\left\| \frac{\partial F}{\partial h}(h,s) - \frac{\partial F}{\partial h}(0,0) \right\|, \text{ and } \|F(0,s)\|_{\mathcal{B}_2},$$

which will be discussed in the following section. In the following sections, we will always assume that $\|h\|_{\mathcal{B}_1} = \|h\|_{C^{2,\alpha}}$ is bounded (independent of t), which will also be part of the condition to apply the implicit function theorem.

3. Basic estimates

In this section, we will derive the basic estimates along the flow of V. The parallel estimates along the flow of V_t will also be mentioned in the remarks following the corresponding results for V. The estimate for V_t is only needed for the construction in section 4 of the fibration over the top-dimensional faces of $\partial\Delta$. The estimate for V is needed for the construction in section 5 and 6 of the fibration near $\pi^{-1}(\Delta_0^{(n-1)})$.

Let $Z = \mathrm{Diag}(z_1, \cdots, z_n)$, $G = G^* = (g_{i\bar{j}})_{n\times n}$. G is called **T-bounded** if ZGZ^{-1} is bounded. Notice that, as a consequence of T-boundedness, we also have $\bar{Z}^{-1}G\bar{Z}$ and G being bounded.

LEMMA 3.1. *If G is T-bounded and* $\det(G) \geq c > 0$, *then* G^{-1} *is T-bounded.*

Proof: G being bounded and $\det(G) \geq c > 0$ imply that G^{-1} is bounded. Since $\det(ZGZ^{-1}) = \det(G)$, we also have $(ZGZ^{-1})^{-1} = ZG^{-1}Z^{-1}$ being bounded. Consequently, G^{-1} is T-bounded. □

Remark: When applying this lemma, it is often convenient to verify the stronger conditions $G \geq cI$ $(c > 0)$ or G^{-1} being bounded in place of $\det(G) \geq c > 0$.

Example: Assume that ω_g is a toric Kähler metric on $\mathbb{C}^{n+1}$ with the toric Kähler potential ρ (as a function of $\{|z_k|^2\}_{k=0}^n$). Then

$$g_{j\bar{k}} = \rho_k\delta_{jk} + \rho_{jk}\bar{z}_j z_k, \quad \text{where } \rho_k = \frac{\partial\rho}{\partial|z_k|^2}, \ \rho_{jk} = \frac{\partial^2\rho}{\partial|z_j|^2\partial|z_k|^2}.$$

Let $G = (g_{j\bar{k}})$. Then $\dfrac{g_{j\bar{k}}z_j}{z_k} = \rho_k\delta_{jk} + \rho_{jk}|z_j|^2$ (which are in fact smooth functions of $\{|z_k|^2\}_{k=0}^n$) are clearly bounded. Therefore, G is T-bounded. □

Recall that ϕ_s denotes the flow for V. Let $(z_0, \cdots, z_n) = \phi_s(z_0^0, \cdots, z_n^0)$ and

$$\|\check{p}\| = |\check{p}| + \sum_{k=0}^n \left| z_k \frac{\partial\check{p}}{\partial z_k} \right|.$$

In the following, we will always assume that $|z_0^0|$ is the smallest among $|z_k^0|$ for $0 \leq k \leq n$, and use $(z_1, \cdots, z_n)$ $((z_1^0, \cdots, z_n^0))$ as the local coordinate of $X_{t,s}$ $(X_{t,0})$.

LEMMA 3.2.
$$\lambda_k = O\left(\left|\frac{z_0^0}{z_k^0}\right|^2\right), \quad \frac{\partial\log z_k}{\partial\log z_j^0} = \delta_{jk} + O\left(\|\check{p}\|\left|\frac{z_0^0}{z_k^0}\right|^2\right), \quad \frac{\partial\log z_k}{\partial\log \bar{z}_j^0} = O\left(\|\check{p}\|\left|\frac{z_0^0}{z_k^0}\right|^2\right).$$

More generally, all the multi-derivatives of $\log z_k - \log z_k^0$ *or* λ_k *with respect to* $(\log z^0, \log \bar{z}^0)$ *are of order* $O\left(\|\check{p}\|\left|\frac{z_0^0}{z_k^0}\right|^2\right)$ *or* $O\left(\left|\frac{z_0^0}{z_k^0}\right|^2\right)$.

Proof: According to (2.1),

$$\lambda_i = \left(\frac{1}{|z_i|^2}\sum_{j=0}^n \bar{q}_j \frac{g^{i\bar{j}}\bar{z}_i}{\bar{z}_j}\right)\left(\sum_{k,j=0}^n \frac{g^{k\bar{j}}q_k\bar{q}_j}{z_k\bar{z}_j}\right)^{-1}$$

Since the matrix $(g_{j\bar{k}})$ is T-bounded and $(g^{j\bar{k}})$ is bounded, by lemma 3.1, $(g^{j\bar{k}})$ is also T-bounded. Consequently, $\dfrac{g^{i\bar{j}}\bar{z}_i}{\bar{z}_j}$ is bounded. Hence $\lambda_k = O\left(\left|\frac{z_0}{z_k}\right|^2\right)$. Since λ_k are bounded smooth functions of the bounded terms $\left\{z_k, \bar{z}_k, \left|\frac{z_0}{z_k}\right|^2\right\}_{k=1}^n$, it is straightforward to verify that all of the multi-derivatives of λ_k with respect to $(\log z, \log \bar{z})$ are also of order $O\left(\left|\frac{z_0}{z_k}\right|^2\right)$.

$$\frac{d}{ds}\log z_k = V(\log z_k) = \check{p}\lambda_k.$$

$$\begin{aligned}\frac{d}{ds}\frac{\partial \log z_k}{\partial \log z_j^0} &= \frac{\partial \check{p}\lambda_k}{\partial \log z_i}\frac{\partial \log z_i}{\partial \log z_j^0} + \frac{\partial \check{p}\lambda_k}{\partial \log z_0}\frac{\partial \log z_0}{\partial \log z_j^0} + \frac{\partial \check{p}\lambda_k}{\partial \log \bar{z}_i}\frac{\partial \log \bar{z}_i}{\partial \log z_j^0} + \frac{\partial \check{p}\lambda_k}{\partial \log \bar{z}_0}\frac{\partial \log \bar{z}_0}{\partial \log z_j^0} \\ &= \left(\frac{\partial \check{p}\lambda_k}{\partial \log z_i} - \frac{q_i}{q_0}\frac{\partial \check{p}\lambda_k}{\partial \log z_0}\right)\frac{\partial \log z_i}{\partial \log z_j^0} + \left(\frac{\partial \check{p}\lambda_k}{\partial \log \bar{z}_i} - \frac{\bar{q}_i}{\bar{q}_0}\frac{\partial \check{p}\lambda_k}{\partial \log \bar{z}_0}\right)\frac{\partial \log \bar{z}_i}{\partial \log z_j^0}.\end{aligned}$$

$$\frac{d}{ds}\frac{\partial \log z_k}{\partial \log \bar{z}_j^0} = \left(\frac{\partial \check{p}\lambda_k}{\partial \log z_i} - \frac{q_i}{q_0}\frac{\partial \check{p}\lambda_k}{\partial \log z_0}\right)\frac{\partial \log z_i}{\partial \log \bar{z}_j^0} + \left(\frac{\partial \check{p}\lambda_k}{\partial \log \bar{z}_i} - \frac{\bar{q}_i}{\bar{q}_0}\frac{\partial \check{p}\lambda_k}{\partial \log \bar{z}_0}\right)\frac{\partial \log \bar{z}_i}{\partial \log \bar{z}_j^0}.$$

Notice that

$$\frac{\partial \check{p}\lambda_k}{\partial \log z_i} - \frac{q_i}{q_0}\frac{\partial \check{p}\lambda_k}{\partial \log z_0} = O\left(\|\check{p}\|\,|\lambda_k|\right) = O\left(\|\check{p}\|\left|\frac{z_0}{z_k}\right|^2\right).$$

Similarly

$$\left(\frac{\partial \check{p}\lambda_k}{\partial \log \bar{z}_i} - \frac{\bar{q}_i}{\bar{q}_0}\frac{\partial \check{p}\lambda_k}{\partial \log \bar{z}_0}\right) = O\left(\|\check{p}\|\left|\frac{z_0}{z_k}\right|^2\right).$$

By standard estimates of linear ODE, we have

$$\log z_k = \log z_k^0 + O\left(\|\check{p}\|\left|\frac{z_0}{z_k}\right|^2\right),\ \frac{z_k}{z_k^0} = 1 + O\left(\|\check{p}\|\left|\frac{z_0}{z_k}\right|^2\right),$$

$$\frac{\partial \log z_k}{\partial \log z_j^0} = \delta_{jk} + O\left(\|\check{p}\|\left|\frac{z_0}{z_k}\right|^2\right),\ \frac{\partial \log z_k}{\partial \log \bar{z}_j^0} = O\left(\|\check{p}\|\left|\frac{z_0}{z_k}\right|^2\right).$$

Consequently

$$\frac{z_0^0}{z_k^0} = \frac{z_0}{z_k} + O\left(\|\check{p}\|\left|\frac{z_0}{z_k}\right|\right),\ O\left(\left|\frac{z_0}{z_k}\right|^2\right) = O\left(\left|\frac{z_0^0}{z_k^0}\right|^2\right).$$

Since all the multi-derivatives of λ_k with respect to $(\log z, \log \bar{z})$ are of order $O\left(|\frac{z_0}{z_k}|^2\right)$, by induction and similar ODE estimates, all the multi-derivatives of $\log z_k - \log z_k^0$ or λ_k with respect to $(\log z^0, \log \bar{z}^0)$ are of order $O\left(\|\check{p}\|\left|\frac{z_0^0}{z_k^0}\right|^2\right)$ or $O\left(\left|\frac{z_0^0}{z_k^0}\right|^2\right)$. □

Remark: Along the flow of V_t, we have similar estimates as in lemma 3.2. In the proof, one needs to replace $\|\check{p}\|$ by $\frac{1}{|t|}$ and the parameter s by t. When $\left|\frac{z_0}{z_k}\right|^2 = O(t^{\alpha'})$ for a constant $\alpha' > 0$, we have the conclusion

$$\lambda_k = O\left(\left|\frac{z_0}{z_k}\right|^2\right),\ \frac{\partial \log z_k}{\partial \log z_j^0} = \delta_{jk} + O\left(\left|\frac{z_0}{z_k}\right|^2\right),\ \frac{\partial \log z_k}{\partial \log \bar{z}_j^0} = O\left(\left|\frac{z_0}{z_k}\right|^2\right).$$

More generally, all the multi-derivatives of $\log z_k - \log z_k^0$ or λ_k with respect to $(\log z^0, \log \bar{z}^0)$ are of order $O\left(\left|\frac{z_0}{z_k}\right|^2\right)$.

Remark on terminology: In this section, a bounded term or $O(1)$ quite often means a bounded smooth function of the bounded terms $\left\{z_k^0, \bar{z}_k^0, \left|\frac{z_0^0}{z_k^0}\right|^2\right\}_{k=1}^n$. It is easy to see that the derivatives of such a function with respect to $\left\{\log|z_k^0|^2\right\}_{k=1}^n$ will be functions of the same type, and therefore are also bounded or $O(1)$. □

Let $\omega_{t,0} = \omega|_{X_{t,0}} = \partial\bar{\partial}\tilde{\rho} = \sum_{i,j=1}^n \tilde{g}_{i\bar{j}} dz_i^0 d\bar{z}_j^0$ and $\omega_{t,0}^0 = \sum_{k=1}^n dz_k^0 d\bar{z}_k^0$, where $\tilde{\rho} = \rho|_{X_{t,0}}$.

Lemma 3.3. *There exists a constant $c_1 > 0$ such that*

$$c_1\omega_{t,0}^0 \leq \omega_{t,0} \leq c_1^{-1}\omega_{t,0}^0.$$

Both $(\tilde{g}_{i\bar{j}})_{n\times n}$ and its inverse matrix $(\tilde{g}^{i\bar{j}})_{n\times n}$ are T-bounded.

Proof: Clearly $\omega^0 = \sum_{k=0}^n dz_k^0 d\bar{z}_k^0$ is quasi-isometric to ω. When restricted to $X_{t,0}$,

$$dz_0^0 = -\sum_{k=1}^n m_k \frac{z_0^0}{z_k^0} dz_k^0.$$

Since we are considering the region where $|z_0^0|$ is the smallest among all $|z_k^0|$, $\omega_{t,0}^0$ is quasi-isometric to $\omega^0|_{X_{t,0}}$, which is quasi-isometric to $\omega_{t,0} = \omega|_{X_{t,0}}$.

$$\tilde{g}_{i\bar{j}} = g_{i\bar{j}} - g_{0\bar{j}}\frac{m_i z_0^0}{z_i^0} - g_{i\bar{0}}\frac{m_j \bar{z}_0^0}{\bar{z}_j^0} + g_{0\bar{0}}\frac{m_i m_j |z_0^0|^2}{z_i^0 \bar{z}_j^0}.$$

It is straightforward to check that $(\tilde{g}_{i\bar{j}})_{n\times n}$ is T-bounded based on the T-boundedness of $(g_{i\bar{j}})_{(n+1)\times(n+1)}$. Then by lemmas 3.1 and the first part of this lemma, $(\tilde{g}^{i\bar{j}})_{n\times n}$ is also T-bounded. □

Notice that

$$\omega_{t,0} = \omega|_{X_{t,0}} = \sum_{k=0}^n d\theta_k^0 d(|z_k^0|^2\rho_k) = \sum_{k=1}^n d\theta_k^0 d(|z_k^0|^2\rho_k - m_k|z_0^0|^2\rho_0).$$

When we choose $x_k = \theta_k^0$, we may take $y_k = |z_k^0|^2\tilde{\rho}_k - C_k = |z_k^0|^2\rho_k - m_k|z_0^0|^2\rho_0 - C_k$ for suitable constants C_k, $1 \leq k \leq n$, so that $\omega_{t,0} = \sum_{k=1}^n dx_k dy_k$ and $y|_{L_{t,0}} = 0$. We have

Lemma 3.4.

$$\frac{\partial \log|z_j^0|^2}{\partial y_k} = \min\left(\frac{1}{|z_j^0|^2}, \frac{1}{|z_k^0|^2}\right) O(1).$$

Consequently

$$\frac{\partial f}{\partial y_k} = \frac{1}{|z_k^0|^2} O\left(\frac{\partial f}{\partial \log|z^0|^2}\right).$$

Proof:

$$\frac{\partial y_j}{\partial \log |z_k^0|^2} = \tilde{g}_{j\bar{k}} z_j^0 \bar{z}_k^0.$$

$$\frac{\partial \log |z_j^0|^2}{\partial y_k} = \frac{1}{z_j^0 \bar{z}_k^0} \tilde{g}^{j\bar{k}} = \frac{1}{|z_j^0|^2} \frac{\tilde{g}^{j\bar{k}} \bar{z}_j^0}{\bar{z}_k^0} = \frac{1}{|z_k^0|^2} \frac{\tilde{g}^{j\bar{k}} z_k^0}{z_j^0}$$

By lemma 3.3, $(\tilde{g}^{j\bar{k}})_{n\times n}$ is T-bounded. Consequently, $\frac{\tilde{g}^{j\bar{k}} \bar{z}_j^0}{\bar{z}_k^0}$ and $\frac{\tilde{g}^{j\bar{k}} z_k^0}{z_j^0}$ are bounded, which implies the first claim of the lemma. Then

$$\frac{\partial f}{\partial y_k} = \frac{\partial f}{\partial \log |z_j^0|^2} \frac{\partial \log |z_j^0|^2}{\partial y_k} = \frac{1}{|z_k^0|^2} O\left(\frac{\partial f}{\partial \log |z^0|^2}\right). \qquad \square$$

For a function f on $\mathbb{R}^n$, we will use $[f]_{C^{k,\alpha}}$ to denote the (k,α)-portion of the $C^{k,\alpha}$-norm of f. For example,

$$[f]_{C^{1,\alpha}} = \sup_{\substack{1 \le i \le n \\ x,\, y \in \mathbb{R}^n}} \left(\frac{|\partial_i f(x) - \partial_i f(y)|}{|x-y|^\alpha}\right), \quad [f]_{C^1} = \sup_{\substack{1 \le i \le n \\ x \in \mathbb{R}^n}} \left(|\partial_i f(x)|\right).$$

Let h be a function on $L_{t,0}$; we have the following estimates.

LEMMA 3.5.

$$\left[\frac{\partial h}{\partial x_k}\right]_{C^0} \le C(r_k^0)^{2+\alpha}[h]_{C^{2,\alpha}}, \quad \left[\frac{\partial^2 h}{\partial x_j \partial x_k}\right]_{C^0} \le C r_j^0 r_k^0 \min((r_j^0)^\alpha, (r_k^0)^\alpha)[h]_{C^{2,\alpha}}.$$

$$\left[\frac{\partial h}{\partial x_k}\right]_{C^\alpha} \le C(r_k^0)^2[h]_{C^{2,\alpha}}, \quad \left[\frac{\partial^2 h}{\partial x_j \partial x_k}\right]_{C^\alpha} \le C r_j^0 r_k^0 [h]_{C^{2,\alpha}}.$$

Proof: We will prove the most challenging estimate

$$\left[\frac{\partial h}{\partial x_k}\right]_{C^\alpha} \le C(r_k^0)^2[h]_{C^{2,\alpha}}.$$

When $|x - x'| \ge 10 r_k^0$, we have

$$\frac{1}{|x-x'|^\alpha}\left|\frac{\partial h}{\partial x_k}(x) - \frac{\partial h}{\partial x_k}(x')\right| \le \frac{1}{|x-x'|^\alpha}\left(\left|\frac{\partial h}{\partial x_k}(x)\right| + \left|\frac{\partial h}{\partial x_k}(x')\right|\right)$$

$$\le \frac{C}{(r_k^0)^\alpha}\left[\frac{\partial h}{\partial x_k}\right]_{C^0} \le \frac{C}{(r_k^0)^\alpha}(r_k^0)^{2+\alpha}[h]_{C^{2,\alpha}} \le C(r_k^0)^2[h]_{C^{2,\alpha}}.$$

Rescale the coordinate and metric as $\hat{x} = x/r_k^0$, $\hat{g} = g/(r_k^0)^2$. Then the estimate is reduced to

$$\left[\frac{\partial h}{\partial \hat{x}_k}\right]_{C^\alpha} \le C[h]_{C^{2,\alpha}}.$$

It is easy to check that this estimate is always true in a region with bounded diameter. In our remaining case, $|x - x'| \le 10 r_k^0$ or equivalently $|\hat{x} - \hat{x}'| \le 10$ ensures that the region of discussion is bounded with respect to the rescaled metric. The other estimates are similar and easier to prove. $\square$

$$\frac{dz_i}{z_i} = \frac{dr_i}{r_i} + i d\theta_i = \left(\frac{\partial \log r_i}{\partial \log r_k^0} + i\frac{\partial \theta_i}{\partial \log r_k^0}\right)\frac{dr_k^0}{r_k^0} + \left(\frac{\partial \log r_i}{\partial \theta_k^0} + i\frac{\partial \theta_i}{\partial \theta_k^0}\right) d\theta_k^0.$$

$$\Omega_t = \imath(V_t)\Omega|_{X_t} = -\frac{1}{q_0}\prod_{i=1}^n \frac{dz_i}{z_i}$$

$$= \eta_s \prod_{i=1}^n \left(d\theta_i^0 + \left(\frac{\partial \log r_j}{\partial \theta_i^0} + i\frac{\partial \theta_j}{\partial \theta_i^0}\right)^{-1} \left(\frac{\partial \log r_j}{\partial \log r_k^0} + i\frac{\partial \theta_j}{\partial \log r_k^0}\right) \frac{dr_k^0}{r_k^0} \right).$$

$$\eta_s = -\frac{1}{q_0}\det\left(\frac{\partial \log r_i}{\partial \theta_k^0} + i\frac{\partial \theta_i}{\partial \theta_k^0}\right).$$

$$(U_s)_{ij} = \left(\frac{\partial \log r_l}{\partial \theta_i^0} + i\frac{\partial \theta_l}{\partial \theta_i^0}\right)^{-1} \left(\frac{\partial \log r_l}{\partial \log r_k^0} + i\frac{\partial \theta_l}{\partial \log r_k^0}\right) \frac{\partial \log r_k^0}{\partial y_j}.$$

LEMMA 3.6.

$$\left|\log |z_k^0|^2 \left(x, \tfrac{\partial h}{\partial x}\right) - \log |z_k^0|^2 (x, 0)\right| \leq C[h]_{C^2}.$$

Proof: With the help of lemma 3.4 and 3.5, we have

$$\left|\log |z_k^0|^2 \left(x, \tfrac{\partial h}{\partial x}\right) - \log |z_k^0|^2 (x, 0)\right| \leq \sum_{j=1}^n \left|\frac{\partial \log |z_k^0|^2}{\partial y_j}\right| \left|\frac{\partial h}{\partial x_j}\right|$$

$$\leq C \max_{1\leq j\leq n}\left(\frac{1}{|z_j^0|^2}\left|\frac{\partial h}{\partial x_j}\right|\right) \leq C[h]_{C^2}. \qquad \square$$

Remark: According to lemma 3.6, $O\left(|z_k^0|^2 \left(x, \frac{\partial h}{\partial x}\right)\right) = O\left(|z_k^0|^2 (x, 0)\right)$ when $[h]_{C^2}$ is bounded. For this reason, in the following arguments we will often omit the mentioning of such implicit dependence of $|z_k^0|^2$ on h.

Let $\xi_\beta = \max\limits_{1\leq l\leq n}\left(\frac{|z_0^0|^2}{|z_l^0|^{2+\beta}}\right)$ for $\beta = 0$ or α (for the flow of V_t, replace z^0 by z in ξ_β). We have

LEMMA 3.7. *For $\beta = 0$ or α,*

$$\left[\eta_s\left(x, \tfrac{\partial h}{\partial x}\right) - \eta_0(x,0)\right]_{C^\beta} \leq C(\|\check{p}\|\xi_\beta + (r_0^0)^{1-\beta}).$$

$$\left[\frac{\partial \eta_s}{\partial y_k}\left(x, \tfrac{\partial h}{\partial x}\right)\right]_{C^\beta} \leq \frac{C}{(r_k^0)^2}(\|\check{p}\|\xi_\beta + (r_0^0)^{1-\beta}).$$

$$\left[(U_s)_{ij}\left(x, \tfrac{\partial h}{\partial x}\right) - (U_0)_{ij}(x,0)\right]_{C^\beta} \leq \frac{C}{r_i^0 r_j^0}\left(\|\check{p}\|\xi_\beta + [h]_{C^{2,\beta}}\right).$$

$$\left[\left(\frac{\partial U_s}{\partial y_k}\right)_{ij}\left(x, \tfrac{\partial h}{\partial x}\right) - \left(\frac{\partial U_0}{\partial y_k}\right)_{ij}(x,0)\right]_{C^\beta} \leq \frac{C}{(r_k^0)^2 r_i^0 r_j^0}\left(\|\check{p}\|\xi_\beta + [h]_{C^{2,\beta}}\right).$$

(When $\beta = 0$, $C^{2,\beta}$ will mean C^2. Notice that the second term on the left hand side of each of the estimates is constant.)

Proof: The proofs for the four estimates are straightforward and somewhat tedious applications of lemmas 3.2 and 3.4. We will carefully prove the third estimate here; the proofs of the other three are very similar.

We may write $U_s = U_s^1 U_s^2$, where

$$(U_s^1)_{ik} = \left(\frac{\partial \log r_l}{\partial \theta_i^0} + i\frac{\partial \theta_l}{\partial \theta_i^0}\right)^{-1}\left(\frac{\partial \log r_l}{\partial \log r_k^0} + i\frac{\partial \theta_l}{\partial \log r_k^0}\right),\ (U_s^2)_{kj} = \frac{\partial \log r_k^0}{\partial y_j}.$$

Estimates in lemma 3.2 imply that

$$\left[(U_s^1)_{ij}\left(x, \tfrac{\partial h}{\partial x}\right) - (U_0^1)_{ij}(x,0)\right]_{C^0} \leq C\|\check{p}\|\frac{(r_0^0)^2}{(r_i^0)^2} \leq C\|\check{p}\|\xi_0.$$

$$\left[(\frac{\partial U_s^1}{\partial x})_{ij}\left(x, \tfrac{\partial h}{\partial x}\right)\right]_{C^0} \leq C\|\check{p}\|\frac{(r_0^0)^2}{(r_i^0)^2} \leq C\|\check{p}\|\xi_0.$$

$$\left[(\frac{\partial U_s^1}{\partial y_k})_{ij}\left(x, \tfrac{\partial h}{\partial x}\right)\right]_{C^0} \leq \frac{C}{(r_k^0)^2}\|\check{p}\|\frac{(r_0^0)^2}{(r_i^0)^2} \leq \frac{C}{(r_k^0)^2}\|\check{p}\|\xi_0.$$

(The last estimate also needs lemma 3.4.) For $\alpha > 0$, we have

$$\left[(U_s^1)_{ij}\left(x, \tfrac{\partial h}{\partial x}\right) - (U_0^1)_{ij}(x,0)\right]_{C^\alpha} = \left[(U_s^1)_{ij}\left(x, \tfrac{\partial h}{\partial x}\right)\right]_{C^\alpha}$$

$$\leq C\left[(\frac{\partial U_s^1}{\partial x})_{ij}\left(x, \tfrac{\partial h}{\partial x}\right)\right]_{C^0}\max_{1\leq l\leq n}\left(|z_l^0|^{-\alpha}\right)$$

$$+C\max_{1\leq k\leq n}\left(\left[(\frac{\partial U_s^1}{\partial y_k})_{ij}\left(x, \tfrac{\partial h}{\partial x}\right)\right]_{C^0}\left[\frac{\partial h}{\partial x_k}\right]_{C^\alpha}\right)$$

$$\leq C\left(\|\check{p}\|\frac{(r_0^0)^2}{(r_i^0)^2}\max_{1\leq l\leq n}\left(|z_l^0|^{-\alpha}\right) + \frac{(r_0^0)^2}{(r_i^0)^2}[h]_{C^{2,\alpha}}\right) \leq C\left(\|\check{p}\|\xi_\alpha + [h]_{C^{2,\alpha}}\right).$$

Estimates in lemma 3.2 imply that

$$\left[(U_s^2)_{ij}\left(x, \tfrac{\partial h}{\partial x}\right)\right]_{C^0} \leq \frac{C}{(r_j^0)^2}.$$

Since U_s^2 is independent of s and x, we have

$$\left[(U_s^2)_{ij}\left(x, \tfrac{\partial h}{\partial x}\right) - (U_0^2)_{ij}(x,0)\right]_{C^\beta}$$

$$\leq C\max_{1\leq k\leq n}\left(\left[(\frac{\partial U_s^2}{\partial y_k})_{ij}\left(x, \tfrac{\partial h}{\partial x}\right)\right]_{C^0}\left[\frac{\partial h}{\partial x_k}\right]_{C^\beta}\right) \leq \frac{C}{r_i^0 r_j^0}[h]_{C^{2,\beta}}$$

for $\beta = 0$ or α. Combining the estimates for U_s^1 and U_s^2 gives the estimate for U_s. One subtle point is that the estimate of $(U_s^1)_{ik}$ involves the factor $\frac{(r_0^0)^2}{(r_i^0)^2}$ except that one term in $[(U_s^1)_{ik}]_{C^0}$ involves δ_{ik}. The estimate of $(U_s^2)_{kj}$ involves the factor $\frac{1}{r_k^0 r_j^0}$ (see lemma 3.4). To get the factor $\frac{1}{r_i^0 r_j^0}$ in both cases for the final estimate of $(U_s)_{ij}$, we use the following:

$$\frac{(r_0^0)^2}{(r_i^0)^2}\frac{1}{r_k^0 r_j^0} = \frac{1}{r_i^0 r_j^0}\frac{(r_0^0)^2}{r_i^0 r_k^0} \leq \frac{1}{r_i^0 r_j^0}\xi_0 \text{ and } \delta_{ik}\frac{1}{r_k^0 r_j^0} = \frac{1}{r_i^0 r_j^0}.$$

For the estimates of η_s, other than those partial derivative terms that can be similarly dealt with as in the third estimate, there is also an additional factor $q_0 = 1 - sz_0\check{p}_0$ that needs the following estimate

$$[q_0 - 1]_{C^\beta} \leq C(r_0^0)^{1-\beta},$$

which is straightforward to verify. □

Recall the expressions of $a_s^{ij}(x)$ and $b_s^i(x)$ from (2.2). Applying lemma 3.7, it is easy to derive the following

LEMMA 3.8. *For $\beta = 0$ or α,*

$$[a_s^{ij} - a^{ij}]_{C^\beta} \le \frac{C}{r_i^0 r_j^0} \left(\|\check{p}\|\xi_\beta + [h]_{C^{2,\beta}}\right).$$

$$[b_s^i - b^i]_{C^\beta} \le \frac{C}{(r_i^0)^2} \left(\|\check{p}\|\xi_\beta + [h]_{C^{2,\beta}} + (r_0^0)^{1-\beta}\right).$$

Proof: Let $\Lambda = \mathrm{Diag}(r_1^0, \cdots, r_n^0)$, $(a_s^{ij})^\Lambda = \Lambda(a_s^{ij})\Lambda$, $U_s^\Lambda = \Lambda U_s \Lambda$, $\left(\frac{\partial^2 h}{\partial x^2}\right)^\Lambda = \Lambda^{-1}\frac{\partial^2 h}{\partial x^2}\Lambda^{-1}$. Then

$$(a_s^{ij})^\Lambda = \mathrm{Im}\left(\left(I + U_s^\Lambda \left(\frac{\partial^2 h}{\partial x^2}\right)^\Lambda\right)^{-1} U_s^\Lambda\right)\left(x, \tfrac{\partial h}{\partial x}\right),$$

$$b_s^i(x) = \mathrm{Im}\left(\frac{1}{\eta_s}\frac{\partial \eta_s}{\partial y_i} + \mathrm{Tr}\left(\left(I + U_s^\Lambda \left(\frac{\partial^2 h}{\partial x^2}\right)^\Lambda\right)^{-1} \frac{\partial U_s^\Lambda}{\partial y_i}\left(\frac{\partial^2 h}{\partial x^2}\right)^\Lambda\right)\right)\left(x, \tfrac{\partial h}{\partial x}\right).$$

Estimates in lemma 3.7 imply that

$$\left[U_s^\Lambda\left(x, \tfrac{\partial h}{\partial x}\right) - U_0^\Lambda(x, 0)\right]_{C^\beta} \le C\left(\|\check{p}\|\xi_\beta + [h]_{C^{2,\beta}}\right).$$

Also notice that $U_0^\Lambda(x,0)$ is a bounded constant matrix and

$$\left[\left(\frac{\partial^2 h}{\partial x^2}\right)^\Lambda\right]_{C^\beta} = [h]_{C^{2,\beta}}.$$

Consequently

$$\left[(a_s^{ij})^\Lambda - (a^{ij})^\Lambda\right]_{C^\beta} \le C\left(\|\check{p}\|\xi_\beta + [h]_{C^{2,\beta}}\right).$$

The estimate for b_s^i is similar by applying lemma 3.7. □

Remark: The corresponding estimates along the flow of V_t are

$$[a_t^{ij} - a^{ij}]_{C^\beta} \le \frac{C}{r_i r_j}\left(\xi_\beta + [h]_{C^{2,\beta}}\right).$$

$$[b_t^i - b^i]_{C^\beta} \le \frac{C}{(r_i)^2}\left(\xi_\beta + [h]_{C^{2,\beta}} + |z_0|^{1-\beta}\right). \quad \square$$

PROPOSITION 3.1.

$$\left\|\frac{\partial F}{\partial h}(h,s) - \frac{\partial F}{\partial h}(0,0)\right\| \le C\left(\|\check{p}\|\xi_\alpha + [h]_{C^{2,\alpha}} + (r_0^0)^{1-\alpha}\right).$$

$$\|F(0,s)\|_{\mathcal{B}_2} \le C(\|\check{p}\|\xi_\alpha + (r_0^0)^{1-\alpha}).$$

Proof: Recall that

$$\left(\frac{\partial F}{\partial h}(h,s) - \frac{\partial F}{\partial h}(0,0)\right)\delta h = (a_s^{ij} - a^{ij})\frac{\partial^2 \delta h}{\partial x_i \partial x_j} + (b_s^i - b^i)\frac{\partial \delta h}{\partial x_i}$$

and

$$F(0,s) = \mathrm{Im}(\log \eta_s(x,0)).$$

Lemmas 3.8, 3.5 will imply the first estimate. The second estimate is implied by lemma 3.7. □

Remark: We again have the corresponding estimate along the flow of V_t.

$$\left\|\frac{\partial F}{\partial h}(h,t)-\frac{\partial F}{\partial h}(0,0)\right\|\leq C\left(\xi_\alpha+[h]_{C^{2,\alpha}}+|z_0|^{1-\alpha}\right).$$

$$\|F(0,t)\|_{\mathcal{B}_2}\leq C(\xi_\alpha+|z_0|^{1-\alpha}).$$ □

4. Fibration over top-dimensional faces of $\partial\Delta$

In [**11**] section 5, we constructed the generalized special Lagrangian fibration over $U_t^{\rm top}$, which is in the interior of the top-dimensional faces of $\partial\Delta$ (also see [**2**]). In this section, using our estimates in section 3 for the thin torus, we will give a more quantitative characterization of $U_t^{\rm top}$, which is crucial for our argument in section 6.

Locally in the face $r_0^0=0$ of $\partial\Delta$, we may use $(r_1^0,\cdots,r_n^0)$ as local coordinate. Take $X_0\subset P_\Delta$ to be the large complex structure limit, with the natural generalized special Lagrangian fibration $X_0\to\partial\Delta$. The smooth part of X_0 is a union of top-dimensional complex tori fibred over top-dimensional faces in $\partial\Delta$.

THEOREM 4.1. *There exists a constant $\beta>0$ such that when $\min_{1\leq k\leq n}((r_k^0)^2)\geq Ct^\beta$, the corresponding L_0 in theorem 5.1 of* [**11**] *will be able to deform to $L_t\subset X_t$ that form the generalized special Lagrangian fibration over the open set $U_t^{\rm top}$ in the top-dimensional faces of $\partial\Delta$.*

Proof: According to the remark after lemma 3.2

$$\min_{1\leq k\leq n}((r_k)^2)\geq C\min_{1\leq k\leq n}((r_k^0)^2)\geq Ct^\beta.$$

When $\beta>0$ is chosen suitably (for example $\beta<\frac{2}{n+2}$), there exists $\alpha'>0$ such that $(\xi_\alpha+|z_0|^{1-\alpha})\leq Ct^{\alpha'}$. Start with $\|h\|_{\mathcal{B}_1}\leq C_1$, where C_1 is small depending on the constant in proposition 2.1. The estimates in proposition 2.1 and the remark after proposition 3.1 will enable us to apply the implicit function theorem (theorem 3.2 in [**11**]) to prove the theorem except for the fact that the L_t form a fibration. In particular, h for the actual solution will satisfy $\|h\|_{\mathcal{B}_1}\leq C_2t^{\alpha'}$.

For fixed small t, to show the generalized special Lagrangian tori L_t form a fibration when the fibre L_0 varies, it is sufficient to show that non-trivial deformation 1-forms of L_t have no zeroes on L_t. It is straightforward to check that the deformation 1-forms of L_t are spanned by $\{dx_k-df_k\}_{k=1}^n$, where $\{f_k\}_{k=1}^n$ are functions on L_t satisfying

$$a_t^{ij}\frac{\partial^2 f_k}{\partial x_i\partial x_j}+b_t^i\frac{\partial f_k}{\partial x_i}=b_t^k,$$

which can be rewritten as

$$a^{ij}\frac{\partial^2 f_k}{\partial x_i\partial x_j}=(a^{ij}-a_t^{ij})\frac{\partial^2 f_k}{\partial x_i\partial x_j}+(b^i-b_t^i)\frac{\partial f_k}{\partial x_i}+(b_t^k-b^k).$$

Propositions 2.1, lemma 3.5 and the remark after lemma 3.8 imply that

$$|f_k|_{C^{2,\alpha}} \leq C\left(\xi_\alpha + [h]_{C^{2,\alpha}} + |z_0|^{1-\alpha}\right)(|f_k|_{C^{2,\alpha}} + \frac{1}{r_k^2}).$$

Since $[h]_{C^{2,\alpha}} \leq Ct^{\alpha'}$, we have

$$|f_k|_{C^{2,\alpha}} \leq \frac{C}{r_k^2}\left(\xi_\alpha + [h]_{C^{2,\alpha}} + |z_0|^{1-\alpha}\right) \leq \frac{Ct^{\alpha'}}{r_k^2}.$$

Applying lemma 3.5, we have

$$\left[\frac{\partial f_k}{\partial x_k}\right]_{C^0} \leq Cr_k^{2+\alpha}|f_k|_{C^{2,\alpha}} \leq Ct^{\alpha'}r_k^{\alpha}.$$

Hence the dx_k-coefficient of $dx_k - df_k$ is of order $1 + O(t^{\alpha'})$, which is non-zero when t is small. Consequently $dx_k - df_k$ does not vanish anywhere. □

5. Fibration near $\pi^{-1}(\Delta_0^{(0)})$

Consider the local situation

$$X_{t,s} = \{z \in \mathbb{C}^{n+1} | \tilde{p}_{t,s}(z) = 0\}, \quad \text{where } \tilde{p}_{t,s}(z) = \prod_{k=0}^{n} z_k + tp(z), \quad p(z) = e^{s\check{p}(z)}.$$

Let $z = (z_0, \cdots, z_n) = (z', z'')$. Correspondingly we have $\{0, \cdots, n\} = I' \cup I''$. Assume that $\check{p}(z', 0) = 0$. Recalling the notations from section 2, we have

PROPOSITION 5.1. *When t is small enough, for the generalized special Lagrangian thin tori $L_{t,0}$ satisfying*

$$\max_{k \in I''} r_k^0 \leq C \min_{1 \leq k \leq n} ((r_k^0)^\alpha),$$

there exists the deformation family of generalized special Lagrangian thin torus $\{L_{t,s}\}_{s \in [0,1]}$ with respect to $(X_{t,0}, \omega_{t,0}, \Omega_{t,s})$. Furthermore, $L_{t,s}$ is part of a generalized special Lagrangian fibration in $(X_{t,0}, \omega_{t,0}, \Omega_{t,s})$.

Proof: Since $\check{p}(z', 0) = 0$,

$$\|\check{p}\| = O\left(\max_{k \in I''} r_k^0\right).$$

Taking $\alpha > 0$ smaller if necessary, there exists a constant $\alpha' > 0$ such that

$$\|\check{p}\|\xi_\alpha + (r_0^0)^{1-\alpha} \leq Ct^{\alpha'}.$$

Start with $\|h\|_{\mathcal{B}_1} \leq C_1$, where C_1 is small depending on the constant in proposition 2.1. The estimates in proposition 2.1 and 3.1 will enable us to apply the implicit function theorem (theorem 3.2 in [**11**]) to prove the claim. In particular, h for the actual solution will satisfy $\|h\|_{\mathcal{B}_1} \leq C_2 t^{\alpha'}$. Using similar arguments as in theorem 4.1, one can show that the non-trivial deformation 1-forms of $L_{t,s}$ have no zeroes, therefore $L_{t,s}$ is part of a generalized special Lagrangian fibration in $(X_{t,0}, \omega_{t,0}, \Omega_{t,s})$. □

THEOREM 5.1. *When t is small enough, there exists an open neighborhood $U_t^{(0)}$ of $\pi^{-1}(\Delta_0^{(0)}) \subset \partial\Delta$ such that for any generalized special Lagrangian fibre $L_{t,0}$ in $X_{t,0}$ over $U_t^{(0)}$, there exists a smooth family $\{L_{t,s}\}_{s\in[0,1]}$, where $L_{t,s}$ is a generalized special Lagrangian torus in $(X_{t,0}, \omega_{t,0}, \Omega_{t,s})$ that is Hamiltonian equivalent to $L_{t,0}$. When the fibre $L_{t,0}$ varies over $U_t^{(0)}$, $\phi_1(L_{t,1}) \subset X_{t,1} = X_t$ will form a generalized special Lagrangian fibration over $U_t^{(0)}$. This fibration will coincide with the fibration over $U_t^{\rm top}$ (in theorems 5.1 and 5.2 of* **[11]** *and theorem 4.1) on the overlaps.*

Proof: Generically, a connected component of $\pi^{-1}(\Delta_0^{(0)})$ is either a point or a contractible union of top-dimensional faces of $\partial\Delta$. The first case was considered in **[11]**. We will deal with the second case here. Near a vertex of $\pi^{-1}(\Delta_0^{(0)})$, $\pi^{-1}(\Delta_0^{(0)})$ can locally be identified with a union of several codimension 1 faces of the first quadrant in $\mathbb{R}^{n+1}$. Assume that $\sigma_0 = \{(r_0^0)^2 = |z_0^0|^2 = 0\}$ is one of these faces. Then all the subfaces of σ_0 are $\sigma_I = \{(r_i^0)^2 = 0 | i \in I\}$ for $I \ni 0$. The fibration over a tubular neighborhood of σ_I can be discussed using proposition 5.1 with $I'' = I$.

Without loss of generality, we assume that r_k^0 is in ascending order with respect to k. When $|I| = 2$, $I'' = I = \{0, 1\}$, we always have

$$\min_{1\leq k\leq n} r_k^0 = r_1^0 = \max_{k\in I''} r_k^0,$$

and proposition 5.1 will always apply. Therefore the fibration over the interior of σ_0 will extend over each codimension 1 edge of σ_0 to adjacent codimension 1 faces of $\partial\Delta$.

In general, let

$$X_{t,0}^I = \{z^0 \in X_{t,0} | |z_k^0|^2 = \mu \leq |z_j^0|^2, \quad \text{for any } k \in I,\ j \notin I.\}$$

$$U_{t,0}^I = \{z^0 \in X_{t,0} | \min_{1\leq k\leq n}((r_k^0)^\alpha) \geq \max_{k\in I} r_k^0 \leq |z_j^0|^2, \quad \text{for any } j \notin I.\}$$

Notice that $X_{t,0}^I \subset U_{t,0}^I$ and proposition 5.1 applies to $U_{t,0}^I$. $X_{t,0}^I$ fibres over σ_I, and $U_{t,0}^I$ fibres over a neighborhood U_{σ_I} of σ_I in $\partial\Delta$. $U_{t,0}^{\sigma_0} = \bigcup_{I\ni 0} U_{t,0}^I$ fibres over $U^{\sigma_0} = \bigcup_{I\ni 0} U_{\sigma_I}$, which is an open neighborhood of σ_0. We may take $U_{t,0}^{(0)}$ ($U_t^{(0)}$) to be the union of such neighborhoods $U_{t,0}^{\sigma_0}$ (U^{σ_0}). $U_t^{(0)}$ is an open neighborhood of $\pi^{-1}(\Delta_0^{(0)})$. Applying proposition 5.1, we have an open set $U_{t,1}^{(0)} \subset X_{t,1}$ and a fibration $\hat{\pi}_{t,1} : U_{t,1}^{(0)} \to U_t^{(0)}$.

Let σ be a top-dimensional face of $\partial\Delta$ that is adjacent to $\pi^{-1}(\Delta_0^{(0)})$. What we have just proved is enough to make the open neighborhood $U_t^{(0)}$ of $\pi^{-1}(\Delta_0^{(0)})$ and $U_t^{\rm top} \cap \sigma$ touch. We need to show that the two fibrations coincide on the overlap. From the definitions, it is easy to check that for suitable $\{r_k\}_{k=1}^n$ satisfying that $\log r_k$ are bounded for $1 \leq k \leq n$, in the local models, $L_{t,0}$ (L_0) is a fibre over $U_t^{(0)}$ ($U_t^{\rm top} \cap \sigma$), where

$$L_{t,0} = \{z \in X_{t,0} | |z_k| = r_k(\text{constant}), \ \text{for } 0 \leq k \leq n\},$$

$$L_0 = \{z_0 = 0, |z_k| = r_k(\text{constant}),\ \text{for } 1 \leq k \leq n\}.$$

Notice that $L_{t,0}$ and L_0 are tori of bounded geometry. This is exactly the same situation as in the proof of theorem 7.1 of [**11**]. The proof of theorem 7.1 of [**11**] implies that after suitable adjustment of $\{r_k\}_{k=1}^n$ in L_0, the deformed torus $L_{t,1}$ coincide with L_1. □

As we mentioned in the introduction, the mirror of the quintic Calabi–Yau satisfies $\Delta_0^{(n-1)} = \Delta_0^{(0)}$. Applying theorem 5.1, we have

COROLLARY 5.1. *For the mirror of the quintic Calabi–Yau or more generally for the Calabi-Yau family $\{X_t\}$ in a toric variety P_Δ satisfying $\Delta_0^{(n-1)} = \Delta_0^{(0)}$ with the restriction of any toric Kähler metric from P_Δ, when t is small enough there exists a smooth monodromy representing a generalized special Lagrangian torus (partial) fibration for X_t.* □

6. Fibration near $\pi^{-1}(\Delta_0^{(n-1)} \setminus \Delta_0^{(0)})$

Starting from [**11**], our discussion so far is valid for any toric Kähler metric on P_Δ. The construction in this section is of a somewhat different nature, where the precise meaning of near the large complex structure limit discussed in the introduction will be very essential and the toric Kähler metric on P_Δ also has to depend on the Calabi–Yau hypersurface $X_t \subset P_\Delta$.

The concept of "near the large complex structure limit" in this section is slightly different from the one in the introduction. In this section, X_t is said to be near the large complex structure limit if τ and $t = \tau^{-w_{m_o}}$ are small. Namely, t is not allowed to get small independently of τ once w is determined.

Define $\tilde{w}_m = w_m$ for $m \in \Delta_0 \setminus \{m_o\}$ and $\tilde{w}_{m_o} = 0$. $\{\tilde{w}_m\}_{m\in\Delta_0}$ is clearly still strictly convex on Δ_0 and determines the same simplicial decomposition of Δ_0 as $\{w_m\}_{m\in\Delta_0}$. Let $\tilde{\Delta}_0$ denote those integral m in the real polyhedron spanned by Δ_0. Let $\tilde{\Delta}_0^{(k)}$ denote the k-skeleton of $\tilde{\Delta}_0$. (Δ_0 ($\Delta_0^{(k)}$) can be reinterpreted as containing those integral m in $\tilde{\Delta}_0$ ($\tilde{\Delta}_0^{(k)}$) such that $a_m \neq 0$.) It is easy to see that one can extend $\{\tilde{w}_m\}_{m\in\Delta_0}$ to $\{\tilde{w}_m(\geq 0)\}_{m\in\tilde{\Delta}_0}$, which is also strictly convex.

Assumption*: For any $\tilde{m} \in \Delta_0^{(l)} \setminus \Delta_0^{(l-1)}$ ($l \leq n-1$), we assume that there exists a primitive top-dimensional simplex $S(\ni \tilde{m})$ in the simplicial decomposition of $\tilde{\Delta}_0$ determined by $\{\tilde{w}_m(\geq 0)\}_{m\in\tilde{\Delta}_0}$ such that $\dim S \cap \tilde{\Delta}_0^{(l)} = l$. We also assume that P_Δ is equivalent to its anti-canonical model P_{Δ_0}.

The following argument will apply to the situation when assumption* is satisfied. It will include most notably the situation of Calabi–Yau quintics in $\mathbb{CP}^4$ with a primitive or a standard simplicial decomposition of Δ_0. (By standard simplicial decomposition we mean any simplicial subdivision of the standard polyhedron decomposition of Δ_0 resulting from cutting by integral hyperplanes that are parallel to the faces of Δ_0.) We will indicate in a remark at the end of this section what is needed to generalize the discussion to the general case of Calabi-Yau hypersurfaces in toric varieties.

By adjustment of an affine function, we can make $\tilde{w}_m = 0$ for $m \in S$, $\tilde{w}_m > 0$ for $m \in \tilde{\Delta}_0 \setminus S$. By further adjustment of a suitable affine function, we can make $\tilde{w}_{\tilde{m}} = 0$, $\tilde{w}_m = \delta$ for $m \in S \setminus \{\tilde{m}\}$, $\tilde{w}_m > C\delta$ for $m \in \tilde{\Delta}_0 \setminus S$, where $\delta > 0$ is a small constant and $C > 0$ is a constant depending on $\tilde{\Delta}_0$. Consequently, $w_{\tilde{m}} = 0$, $w_m > 0$ for $m \in \Delta_0 \setminus \{m_o, \tilde{m}\}$. In other words, we have a toric coordinate z, whose components are made up of $\{\tau^{-\delta} s_m/s_{\tilde{m}}\}_{m \in S\setminus\{\tilde{m}\}}$, such that

$$\partial\bar{\partial}\log\left(\sum_{m\in\tilde{\Delta}_0} \tau^{2\tilde{w}_m}|s_m|^2\right) = \partial\bar{\partial}\log\left(1 + \tau^{2\delta}|z|^2 + \sum_{m\in\tilde{\Delta}_0\setminus S} \tau^{2\tilde{w}_m}|s_m/s_{\tilde{m}}|^2\right)$$

and

$$\tilde{s}_t = s_{m_o} + t\left(s_{\tilde{m}} + \sum_{m\in\Delta_0\setminus\{m_o,\tilde{m}\}} a_m s_m\right), \quad \text{where } |a_m| = \tau^{w_m}.$$

For X_t, we will use the restriction of the toric Kähler metric on P_Δ (used in [**9**]) with the Kähler potential

$$\rho_t = \tau^{-2\delta}\log\left(\sum_{m\in\tilde{\Delta}_0} \tau^{2\tilde{w}_m}|s_m|^2\right). \tag{6.1}$$

One can observe that locally the situation can be reduced to the basic setting of section 2, with

$$\omega_{t,s} = \omega_t|_{X_{t,s}}, \quad \omega_t = \partial\bar{\partial}\rho_t = \sum_{k=0}^{n} dz_k d\bar{z}_k + O(\tau^+).$$

$$\tilde{p}_t(z) = \tilde{s}_t/s_{\tilde{m}} = z^m + tp(z), \quad p(z) = 1 + \sum_{m\in\Delta_0\setminus\{m_o,\tilde{m}\}} a_m s_m/s_{\tilde{m}}.$$

$$\check{p}(z) = \log p(z) = O(\tau^{\alpha'}), \quad \alpha' = \min_{m\in\Delta_0\setminus\{m_o,\tilde{m}\}} w_m.$$

$$\tilde{p}_{t,s}(z) = z^m + te^{s\check{p}(z)}.$$

THEOREM 6.1. *When assumption* is satisfied and t is small enough, there exists an open neighborhood $U_t^{(n-1)}$ of $\pi^{-1}(\Delta_0^{(n-1)}) \subset \partial\Delta$ such that for any generalized special Lagrangian fibre $L_{t,0}$ in $X_{t,0}$ over $U_t^{(n-1)}$, there exists a smooth family $\{L_{t,s}\}_{s\in[0,1]}$, where $L_{t,s}$ is a generalized special Lagrangian torus in $(X_{t,0}, \omega_{t,0}, \Omega_{t,s})$ that is Hamiltonian equivalent to $L_{t,0}$. When the fibre $L_{t,0}$ varies over $U_t^{(n-1)}$, $\phi_1(L_{t,1}) \subset X_{t,1} = X_t$ will form a generalized special Lagrangian fibration over $U_t^{(n-1)}$. This fibration will coincide with the fibration over $U_t^{\rm top}$ (in theorems 5.1 and 5.2 of* [**11**] *and theorem 4.1) on the overlaps.*

Proof: For $\tilde{m} \in \Delta_0^{(l)} \setminus \Delta_0^{(l-1)}$ $(l \le n-1)$, our situation locally is reduced to the basic setting of section 2, with $I' = (S \setminus \{\tilde{m}\}) \cap \tilde{\Delta}_0^{(l)}$ and $I'' = S \setminus (I' \cup \{\tilde{m}\})$. For a bounded constant $c_1 > 1$, consider the region $\tilde{U}_{\tilde{m}}$, where $|z_k| \le c_1$ for all k and $|z_k| \ge c_1^{-1}$ for $k \in I'$. Locally $X_{t,s}$ is defined by

$$\prod_{k\in I''} z_k = -te^{s\check{p}(z)} \prod_{k\in I'} z_k^{-m_k},$$

where $m = m_o - \tilde{m} = (m_0, \cdots, m_n) = (m', m'')$. Therefore in $\tilde{U}_{\tilde{m}}$, we have $|z_k| \geq Ct$ for $k \in I''$. Taking a suitable α satisfying $\gamma = \alpha' + w_{m_o}\alpha > 0$, we have

$$\|\check{p}\|\xi_\alpha + (r_0^0)^{1-\alpha} \leq C\tau^\gamma, \quad \text{in } \tilde{U}_{\tilde{m}}.$$

Let $U_{\tilde{m}}$ be the region in $\partial\Delta$ over which $\tilde{U}_{\tilde{m}} \cap X_{t,0}$ is fibred. Let $U_t^{(n-1)} = \bigcup_{\tilde{m}\in\Delta_0^{(n-1)}} U_{\tilde{m}}$. Then for any generalized special Lagrangian fibre $L_{t,0}$ in $X_{t,0}$ over $U_t^{(n-1)}$, when τ is small, according to the discussion in section 2, the estimates in propositions 2.1 and 3.1 together with the implicit function theorem (theorem 3.2 in [**11**]) imply the existence of the family of generalized special Lagrangians $L_{t,s}$ with respect to $(X_{t,0}, \omega_{t,0}, \Omega_{t,s})$.

Since the toric Kähler form depends on τ, the deformation (section 4) over a top-dimensional face σ of $\partial\Delta$ needs some additional comment. When doing the deformation, we use the fixed Kähler form for τ, we modify the meaning of t by fixing τ and take $t \in [0, \tau^{-w_{m_o}}]$. Then we carry out the deformation as in section 4 for the parameter t. Such a toric Kähler metric is not of bounded geometry. It can be turned into a metric with bounded geometry in the area of discussion via a toric transformation of order $O(\tau^{|\tilde{w}|})$. It is easy to see that the conclusion of theorem 4.1 is still true for such a Kähler metric with perhaps a slightly smaller $\beta > 0$. (β in theorem 4.1 may need to be adjusted by $O(\frac{|\tilde{w}|}{|w_{m_o}|})$, which is small because $|w_{m_o}| \gg |\tilde{w}|$.)

Let σ be a top-dimensional face of Δ that is adjacent to $\tilde{m}$. In the local toric coordinate on $\tilde{U}_{\tilde{m}}$, assume that σ corresponds to the toric divisor $\{z_0 = 0\}$. It is easy to see that there exists a torus $L_{t,0}$ in $\tilde{U}_{\tilde{m}}$ satisfying $\log r_k$ being bounded for $1 \leq k \leq n$. Such a $L_{t,0}$ is of bounded geometry. Based on this, the same argument as in the proof of theorem 5.1 will imply that $U_{\tilde{m}}$ touch $U_t^{\text{top}} \cap \sigma$ in theorem 4.1 and the two fibrations coincide on the overlap. □

Remark: The result in this section can be extended to Calabi–Yau hypersurfaces in general toric varieties. The assumption* in theorem 6.1 can be removed with a suitable construction of the toric metric ω_t similar to the construction in [**10**]. Such a construction is straightforward using the Legendre transformation but is somewhat tedious. We will omit such a construction here and defer the general case to a sequel of this paper, where, using a different method, we will give a stronger construction of a thin torus near $\pi^{-1}(\Delta_0^{(n-1)} \setminus \Delta_0^{(0)})$ under any fixed toric metric that does not need to depend on t.

Combining theorems 4.1 and 6.1, we have

COROLLARY 6.1. *For the Calabi-Yau family $\{X_t\}$ in P_Δ satisfying assumption* with the restriction of the family of toric Kähler metrics ω_t from P_Δ, when t is small enough there exists a smooth monodromy representing generalized special Lagrangian torus (partial) fibration for X_t.* □

COROLLARY 6.2. *For the quintic Calabi–Yau family $\{X_t\}$ in $\mathbb{CP}^4$ that determines a primitive or a standard simplicial decomposition for Δ_0, with the restriction* *of the family of toric Kähler metrics* ω_t *from* $\mathbb{CP}^4$, *when* t *is small enough, there exists a smooth monodromy representing generalized special Lagrangian torus (partial) fibration for* X_t. □

References

[1] Gilbarg, D. and Trudinger, N.S., *Elliptic partial differential equations of second order*, Second edition, Springer-Verlag, 1983.

[2] Goldstein, E., *Calibrated fibrations on noncompact manifolds via group action*, Duke Math. J. 110, No. 2, (2001), 309-343.

[3] Harvey, R. and Lawson, H.B., *Calibrated Geometries*, Acta Math. 148 (1982), 47-157.

[4] Joyce, D., *Singularities of special Lagrangian fibrations and the SYZ Conjecture*, Math.DG/0011179.

[5] Joyce, D., *Lectures on Calabi–Yau and special Lagrangian geometry*, Math.DG/0108088.

[6] Mclean, R., *Deformations of calibrated submanifolds*, Comm. Anal. Geom. 6 (1998), 705-747.

[7] Ruan, W.-D., *Lagrangian torus fibration of quintic Calabi–Yau hypersurfaces I: Fermat type quintic case*, math.DG/9904012, in *Proceedings of Winter School in Mirror Symmetry*, edited by Yau and Vafa, AMS and International Press, 2001.

[8] ———, *Lagrangian torus fibration of quintic Calabi–Yau hypersurfaces II: Technical results on gradient flow construction*, Journal of Symplectic Geometry, Volume 1 (2002), 435-521.

[9] ———, *Lagrangian torus fibration of quintic Calabi–Yau hypersurfaces III: general quintic case*, J. Differential Geometry, Volume 63 (2003), 171–229.

[10] ———, *Lagrangian torus fibration and mirror symmetry of Calabi–Yau hypersurfaces in toric variety*, Math.DG/0007028.

[11] ———, *Generalized special Lagrangian torus fibration for Calabi–Yau hypersurfaces in toric variety I*, math.DG/0303114.

[12] Strominger, A.,Yau, S.-T. and Zaslow, E, *Mirror Symmetry is T-duality*, Nuclear Physics B 479 (1996),243-259.

[13] Weinstein, A., *Symplectic manifolds and their Lagrangian submanifolds*, Advances in Math. 6 (1971),329-346.

Department of Mathematics, University of Illinois at Chicago, Chicago, IL 60607
E-mail address: ruan@math.uic.edu

PART IV

GEOMETRIC ANALYTIC ASPECTS:

PICARD–FUCHS EQUATIONS

AMS/IP Studies in Advanced Mathematics
Volume 38, 2006

Differential Equations, Mirror Maps and Zeta Values

Gert Almkvist and Wadim Zudilin

ABSTRACT. The aim of this work is an analytic investigation of differential equations producing mirror maps as well as presentation of new examples of mirror maps; one of these examples is related to (rational approximations to) $\zeta(4)$. We also indicate certain observations that might become a subject of further research.

The existence of this paper is due to the following observations. From Apéry's proof of the irrationality of $\zeta(2)$ and $\zeta(3)$ followed certain 2nd- and 3rd-order linear differential equations (see [**7**]–[**9**], [**12**]). Doing a similar construction for $\zeta(4)$ resulted in a 5th-order differential equation. This equation was similar to the linear differential equations occuring in Calabi–Yau theory (except the order was 5 instead of 4). Computing the Lambert series of an analogue of the Yukawa coupling we got integer coefficients divisible by the square of the degree (not by the cube as in the Calabi–Yau case). Then we managed to pull back the 5th-order differential equation to one of order 4, which had all the properties of a Calabi–Yau equation. This part is the main objective of Sections 1–4. Quite recently we found two more examples relating fourth order differential equations to simultaneous rational approximations to di- and trilogarithms (see the end of Section 4).

Collecting known cases of the Calabi–Yau equations, of which 14 were classical hypergeometric and the 15 remaining were discovered by V. V. Batyrev, D. van Straten et al in [**5**], [**6**], [**35**], we tried to extend the list by application of the algorithm of creative telescoping due to Gosper and Zeilberger. This seems to be pure machinery (cf. Section 5 below), for which we cannot predict any end. We also tried to figure out possible algebraic transformations between them and our new example related to $\zeta(4)$. We did not succeed in this, but doing that we found two other Calabi–Yau equations emanating from quadratic transformations of hypergeometric Calabi–Yau solutions of type ${}_4F_3$. This is the subject of Section 6. Section 7 describes other (rather rich!) possibilities for deriving fake 4th-order

2000 *Mathematics Subject Classification.* Primary 14J32; Secondary 11M06, 11B83, 32S40, 33C20, 34M15, 34M25.

Key words and phrases. Calabi–Yau equation, maximal unipotent monodromy, mirror map, instanton number, zeta value, binomial sum, recurrence relation, Hadamard product, Lambert series, supercongruence.

The work of the second author is supported by an Alexander von Humboldt research fellowship and partially supported by grant no. 03-01-00359 of the Russian Foundation for Basic Research.

linear differential equations. Summarizing the methods of Sections 6 and 7 resulted in new classes of transformations which we present in Section 8.

The divisibility properties of the coefficients of the Lambert series for the Yukawa coupling $K(q)$ are equivalent to high-order Kummer congruences between the coefficients of the power series for $K(q)$. The latter coefficients can be realized as the number of fixed points of iterates of a set map $T\colon X \to X$, where the number of orbits of order n is divisible by n^2. We figure out these very natural relations in Section 9.

Our final table (mostly represented in the joint contribution [**3**], Table A) contains more than 200 cases of Calabi–Yau equations (see also van Enckevort's electronic database [**14**], which provides more advanced knowledge, like instanton numbers up to 20, in some cases also the number of elliptic curves, monodromy matrices, etc.). In almost all cases, the mirror map $z(q)$ divided by q seems to be a high power of a series with integer coefficients (cf. [**23**]). For this we have no explanation at all, but we collect this experimental knowledge in another table [**3**], Table B. The last table [**3**], Table C, presents a brief systematic guide to the main table from [**3**], Table A, with entries ordered according to instanton numbers[1].

ACKNOWLEDGEMENTS. First of all we want to thank Duco van Straten, who has shown great interest in our work and suggested the use of Hadamard products. He and Christian van Enckevort contributed by finding several new Calabi–Yau equations, usually coming from geometry. Combining our examples and methods with those from [**15**] resulted in the joint project [**3**]. Frits Beukers gave us details of computations on [**10**] and inspiration to start this work; his comments helped us a lot in improving the first version of our manuscript. Helena Verrill communicated to us a couple of new equations and Arne Meurman showed us an algorithm to find the Hadamard product. Special gratitude is due to Noriko Yui who kindly proposed to us to contribute to the BIRS workshop volume on Calabi–Yau varieties and mirror symmetry.

CONTENTS

[1]With the kind permission of our coauthors D. van Straten and C. van Enckevort, we include the last table from [**3**] in the version of the present paper appearing in the BIRS workshop volume on Calabi–Yau varieties and mirror symmetry. The E-print version of this article at `http://arXiv.org/math.NT/0402386` also contains Section 10 and proofs of Propositions 1 and 3.

1. Maximal unipotent monodromy

Consider a linear differential equation

$$y^{(s)}+a_{s-1}(z)y^{(s-1)}+\cdots+a_1(z)y'+a_0(z)y=0, \tag{1.1}$$

where the prime stands for the z-derivative and $a_0,a_1,\dots,a_{s-1}$ are meromorphic functions in the variable z. Assuming that $z=0$ is a regular singularity of the differential equation (1.1), we may write its coefficients in the form

$$a_{s-j}(z)=z^{-j}\tilde{a}_{s-j}(z), \qquad j=1,\dots,s,$$

where the functions $\tilde{a}_{s-j}(z)$ are analytic at $z=0$. The roots of the *indicial equation*

$$\lambda(\lambda-1)\cdots(\lambda-s+1)+\tilde{a}_{s-1}(0)\lambda(\lambda-1)\cdots(\lambda-s+2)+\cdots+\tilde{a}_1(0)\lambda+\tilde{a}_0(0)=0$$

determine the *exponents* of the differential equation (1.1) at the point $z=0$. Following [**26**], we will say that our differential equation (1.1) is of *maximal unipotent monodromy* (briefly, MUM) if its exponents at $z=0$ are all zero. The Frobenius method (see, e.g., [**18**], Section 16.1) gives one a constructive way for writing a basis of solutions to (1.1) in the form

$$\begin{gathered} y(z,\rho)=\sum_{n=0}^{\infty}A(n,\rho)z^{n+\rho} \pmod{\rho^s}=y_0(z)+y_1(z)\rho+\cdots+y_{s-1}(z)\rho^{s-1}, \\ A(0,\rho)=1 \pmod{\rho^s} \end{gathered} \tag{1.2}$$

(in particular, $A(0,0)=y_0(0)=1$), where by definition we set

$$z^{\rho} \pmod{\rho^j}=1+\log z\cdot\rho+\frac{\log^2 z}{2}\cdot\rho^2+\cdots+\frac{\log^{j-1}z}{(j-1)!}\cdot\rho^{j-1}.$$

The components $y_0,y_1,\dots,y_{s-1}$ in (1.2) define the *Frobenius basis* of the MUM differential equation (1.1) at $z=0$.

2. 4th-order linear differential equations

Consider a linear homogeneous differential equation of order 4,

$$y^{(4)}+a_3(z)y^{(3)}+a_2(z)y''+a_1(z)y'+a_0(z)y=0, \tag{2.1}$$

where a_0,a_1,a_2,a_3 are rational functions of the variable z. Suppose that (2.1) is MUM and let y_0,y_1,y_2,y_3 be the Frobenius basis of solutions to (2.1). Set $t=y_1/y_0$.

PROPOSITION 1 ([**2**]). *In the above notation, the condition*

$$a_1=\frac{1}{2}a_2a_3-\frac{1}{8}a_3^3+a_2'-\frac{3}{4}a_3a_3'-\frac{1}{2}a_3'' \tag{2.2}$$

is equivalent to either of the following two possibilities:

$$\frac{\mathrm{d}^2}{\mathrm{d}t^2}(y_2/y_0)=\frac{\exp\left(-\frac{1}{2}\int a_3(z)\,\mathrm{d}z\right)}{y_0^2\left(\frac{\mathrm{d}t}{\mathrm{d}z}\right)^3} \quad \textit{and} \quad \frac{\mathrm{d}^2}{\mathrm{d}t^2}(y_3/y_0)=t\frac{\mathrm{d}^2}{\mathrm{d}t^2}(y_2/y_0). \tag{2.3}$$

If the differential equation (2.1) is MUM and assumption (2.2) holds, we define the *mirror map* $z=z(q)$ as the inverse of $q(z)=e^t=\exp(y_1/y_0)$ and the *Yukawa coupling* as

$$K(q)=N_0\cdot\frac{\mathrm{d}^2}{\mathrm{d}t^2}(y_2/y_0),$$

where N_0 is some integer different from 0. The interesting (*Calabi–Yau*) cases are those when $z(q) \in \mathbb{Z}[[q]]$ and

$$K(q) = \sum_{n=0}^{\infty} C_n q^n = N_0 + \sum_{l=1}^{\infty} \frac{N_l l^3 q^l}{1-q^l} \tag{2.4}$$

with

$$N_0 = C_0 \in \mathbb{Z} \qquad \text{and} \qquad N_l = \frac{1}{l^3}\sum_{d|l} \mu\Big(\frac{l}{d}\Big) C_d \in \mathbb{Z} \quad \text{for } l = 1, 2, \dots$$

and μ is the Möbius function (and, usually, $N_l > 0$).

PROPOSITION 2. *Under hypothesis* (2.2) *we have the identity of Wronskian determinants:*

$$\begin{vmatrix} y_0 & y_3 \\ y_0' & y_3' \end{vmatrix} = \begin{vmatrix} y_1 & y_2 \\ y_1' & y_2' \end{vmatrix}.$$

PROOF. The required identity is equivalent to $y_0^2(y_3/y_0)' = y_1^2(y_2/y_1)'$, i.e.

$$\left(\frac{y_3}{y_0}\right)' = \left(\frac{y_1}{y_0}\right)^2 \left(\frac{y_0}{y_1} \cdot \frac{y_2}{y_0}\right)' = t^2 \left(\frac{1}{t}\frac{y_2}{y_0}\right)'.$$

Since $\frac{\mathrm{d}}{\mathrm{d}z} = \frac{\mathrm{d}t}{\mathrm{d}z} \cdot \frac{\mathrm{d}}{\mathrm{d}t}$, verification of the last identity is equivalent to showing that $Y = 0$, where

$$Y = \frac{\mathrm{d}}{\mathrm{d}t}\left(\frac{y_3}{y_0}\right) - t^2 \frac{\mathrm{d}}{\mathrm{d}t}\left(\frac{1}{t}\frac{y_2}{y_0}\right) = \frac{\mathrm{d}}{\mathrm{d}t}\left(\frac{y_3}{y_0}\right) + \frac{y_2}{y_0} - t\frac{\mathrm{d}}{\mathrm{d}t}\left(\frac{y_2}{y_0}\right).$$

But

$$\frac{\mathrm{d}Y}{\mathrm{d}t} = \frac{\mathrm{d}^2}{\mathrm{d}t^2}\left(\frac{y_3}{y_0}\right) - t\frac{\mathrm{d}^2}{\mathrm{d}t^2}\left(\frac{y_2}{y_0}\right) = 0$$

by Proposition 1. Therefore $Y = \text{const}$, and verifying near $z = 0$ we get $Y = 0$ (all log-terms disappear). This completes the proof. □

3. Wronskian formalism

Consider an arbitrary pair $y(z), \tilde{y}(z)$ of linearly independent solutions of a 4th-order differential equation (2.1) (not necessarily MUM). Define the Wronskian determinant

$$Y = \begin{vmatrix} y & \tilde{y} \\ y' & \tilde{y}' \end{vmatrix}.$$

PROPOSITION 3. *The function $Y = Y(z)$ satisfies a 6th-order linear differential equation with coefficients depending only on the coefficients of* (2.1) (*i.e., independent of the choice of the pair $y, \tilde{y}$*), *called the exterior square of* (2.1). *Moreover, this 6th-order differential equation reduces to a 5th-order differential equation* (*i.e., the coefficient of $Y^{(6)}$ is zero*) *if and only if condition* (2.2) *holds.*

REMARK 1. In other words, condition (2.2) is equivalent to the exterior square having order 5. As pointed out by Beukers, by this equivalence properties (2.3) are derived properties which are handy in calculating the Yukawa coupling.

REMARK 2. The claim of Proposition 3 may be proved in `Maple` by the `DEtools` command `exterior_power` (see also the E-print version of this paper).

From now on, we assume condition (2.2) for a given 4th-order MUM differential equation (2.1). To construct a 5th-order MUM differential equation we modify the above construction by taking

$$w(z) = zY = z\begin{vmatrix} y & \tilde{y} \\ y' & \tilde{y}' \end{vmatrix}.$$

Denote the resulting 5th-order differential equation for the function w by

$$w^{(5)} + b_4(z)w^{(4)} + b_3(z)w^{(3)} + b_2(z)w'' + b_1(z)w' + b_0(z)w = 0. \tag{3.1}$$

REMARK. Clearly, condition (2.2) induces a certain condition on the coefficients of the differential equation (3.1). We leave the corresponding (rather boring) relation to the reader as a `Maple` exercise.

PROPOSITION 4. *The differential equation* (3.1) *is MUM; its Frobenius basis is given by the following formulas:*

$$\begin{gathered} w_0 = z\begin{vmatrix} y_0 & y_1 \\ y_0' & y_1' \end{vmatrix}, \quad w_1 = z\begin{vmatrix} y_0 & y_2 \\ y_0' & y_2' \end{vmatrix}, \quad w_2 = z\begin{vmatrix} y_0 & y_3 \\ y_0' & y_3' \end{vmatrix} = z\begin{vmatrix} y_1 & y_2 \\ y_1' & y_2' \end{vmatrix}, \\ w_3 = \frac{z}{2}\begin{vmatrix} y_1 & y_3 \\ y_1' & y_3' \end{vmatrix}, \quad w_4 = \frac{z}{2}\begin{vmatrix} y_2 & y_3 \\ y_2' & y_3' \end{vmatrix}. \end{gathered} \tag{3.2}$$

REMARK. The converse is not true in the following sense. If we take a MUM linear differential equation of order 5 with monodromy group O_5, then its 4th-order differential pullback is not necessarily MUM. An example of this is the 5th-order differential operator

$$D = \theta^5 - 6^6 z(\theta + \tfrac{1}{6})(\theta + \tfrac{2}{6})(\theta + \tfrac{3}{6})(\theta + \tfrac{4}{6})(\theta + \tfrac{5}{6})$$

(which is a natural generalization of the 4th-order operator

$$D = \theta^4 - 5^5 z(\theta + \tfrac{1}{5})(\theta + \tfrac{2}{5})(\theta + \tfrac{3}{5})(\theta + \tfrac{4}{5})$$

corresponding to the Calabi–Yau differential equation).

PROOF. The five functions in (3.2) are linearly independent solutions to (3.1). Developing their $(\log z)$-expansions (with the known structure of the Frobenius basis y_0, y_1, y_2, y_3 to (2.1)) shows that they form the Frobenius basis of (3.1), hence the latter differential equation is MUM. □

COROLLARY (Beukers's relations). *We have*

$$2w_0w_4 - 2w_1w_3 + w_2^2 = 0, \qquad 2w_0'w_4' - 2w_1'w_3' + {w_2'}^2 = 0. \tag{3.3}$$

PROOF. Write

$$w_2^2 = z\begin{vmatrix} y_0 & y_3 \\ y_0' & y_3' \end{vmatrix} \cdot z\begin{vmatrix} y_1 & y_2 \\ y_1' & y_2' \end{vmatrix}.$$

Then the first identity is trivial by expanding. In the same vein we obtain the identity for the functions $\tilde{w}_j$, in which the second line of the determinant contains y_i's replaced by y_i''s. Since $w_j' = w_j/z + \tilde{w}_j$, we finally arrive at the second identity in (3.3). □

REMARK. Beukers proved the above relations for a concrete example of a 5th-order differential equation, which we discuss in the next section.

In the case of the 5th-order MUM differential equation (3.1) we can define analogues of a mirror map and a Yukawa coupling. Namely, we can define $\tilde{z}(q)$ as the inverse of $q(z) = \exp(w_1/w_0)$ and

$$\tilde{K}(q) = \tilde{N}_0 \cdot \frac{\mathrm{d}^2}{\mathrm{d}t^2}(w_2/w_0), \qquad t = w_1/w_0.$$

Developing the last q-series as the Lambert series,

$$\tilde{K}(q) = \tilde{N}_0 + \sum_{l=1}^{\infty} \frac{l^2 \tilde{N}_l q^l}{1-q^l} \tag{3.4}$$

(we put l^2 instead of l^3 here), we have the following experimental observation: $\tilde{z}(q) \in K[[q]]$ and $\tilde{N}_l \in K$, where K is either $\mathbb{Z}$ or $\mathbb{Z}[1/p]$ for a certain prime p (the possibilities $p = 3, 5, 7, 23$ occur).

4. Strange integrality related to $\zeta(4)$

In the work [**39**], the following 3-term polynomial recursion is given:

$$\begin{aligned} &(n+1)^5 A_{n+1} - 3(2n+1)(3n^2+3n+1)(15n^2+15n+4)A_n \\ &\quad - 3n^3(3n-1)(3n+1)A_{n-1} = 0 \qquad \text{for} \quad n \geqslant 1. \end{aligned}$$

If we take two linearly independent solutions $\{A_n\}$ and $\{B_n\}$ to the above recursion given by the initial data

$$A_0 = 1, \quad A_1 = 12, \qquad \text{and} \qquad B_0 = 0, \quad B_1 = 13,$$

then

$$\lim_{n\to\infty} \frac{B_n}{A_n} = \zeta(4) = \frac{\pi^4}{90}$$

(see [**39**] for details). This recursion was previously proved in [**13**] and [**33**] but without any indication of arithmetic properties of the sequences $\{A_n\}$ and $\{B_n\}$. In [**39**], it is proved that

$$D_n A_n, D_n^5 B_n \in \mathbb{Z}, \qquad \text{where} \quad D_n = \text{the least common multiple of } 1, 2, \dots, n,$$

and some stronger inclusions were conjectured that were finally proved in [**21**]. In particular, from [**21**] we have the inclusions

$$A_n, D_n^4 B_n \in \mathbb{Z},$$

and even the following explicit formula (see also [**42**]):

$$A_n = \sum_{j,k} \binom{n}{j}^2 \binom{n}{k}^2 \binom{n+j}{n} \binom{n+k}{n} \binom{j+k}{n},$$

where the binomial coefficients $\binom{a}{b}$ are zero if $b < 0$ or $a < b$.

A natural object related to the above recursion is the 5th-order MUM differential equation

$$\begin{aligned} &z^5(27z^2+270z-1)w^{(5)} + z^4(405z^2+3375z-10)w^{(4)} \\ &\quad + z^3(1752z^2+11502z-25)w^{(3)} + z^2(2412z^2+11259z-15)w'' \\ &\quad + z(816z^2+2130z-1)w' + 12z(2z+1)w = 0. \end{aligned} \tag{4.1}$$

Denote by w_0, w_1, w_2, w_3, w_4 the Frobenius basis of (4.1) near $z = 0$, so that

$$w_0 = w_0(z) = \sum_{n=0}^{\infty} A_n z^n = 1 + 12z + 804z^2 + 88680z^3 + 12386340z^4 + \cdots.$$

It was pointed out by F. Beukers [**10**] that the (Zariski closure of the) monodromy group of the differential equation (4.1) turns out to be O_5. This result (as well as identities (3.3) for the latter functions w_j) are consequences of the following statement.

PROPOSITION 5. *The 5th-order MUM differential equation* (4.1) *is constructed from the 4th-order MUM differential equation* (2.1) *with*

$$a_3 = \frac{6(36z^2+315z-1)}{z(27z^2+270z-1)},$$

$$a_2 = \frac{10530z^4+180387z^3+759417z^2-4671z+7}{z^2(27z^2+270z-1)^2},$$

$$a_1 = \frac{96228z^6+2421009z^5+18416565z^4+49339854z^3-450054z^2+1107z-1}{z^3(27z^2+270z-1)^3},$$

$$a_0 = \frac{\begin{array}{l}3(8748z^7+253692z^6+10875303z^5-37601010z^4\\ \qquad +13643328z^3+432135z^2+13224z+11)\end{array}}{z^3(27z^2+270z-1)^4}$$

by the algorithm of Section 3.

PROOF. By direct computation (using `Maple`). □

REMARK. The original proof by Beukers [**11**] of relations (3.3) for the differential equation (4.1) follows the principle 'the proof is trivial once one has found the explicit formula'. We reproduce the arguments from [**11**] here. The symmetric square of the 5th-order equation (4.1) has order 14, one order less than the expected 15. This means that there should be a homogeneous quadratic relation between the solutions $w_0, \dots, w_4$ of (4.1). It suffices to find a relation which vanishes at $z=0$ of order one higher than the local exponents given by the symmetric square of order 14. (Clearly, any non-trivial relation satisfies the 14th-order equation; if we find one with one vanishing order higher than predicted by this 14th-order equation, it must be the trivial solution.) If we replace the w_js by their derivatives, we get quadratic forms that form a one-dimensional monodromy representation. Looking at the local exponents we can infer that they should be rational functions with poles only at infinity. Hence the expressions are polynomial of degree again given by the local exponents at infinity of the symmetric square.

Here are some observations related to the 5th-order differential equation (4.1) and its 4th-order pullback given in Proposition 5 (we denote by y_0, y_1, y_2, y_3 the Frobenius basis of the latter differential equation).

OBSERVATION 1. We have $y_0 \in \mathbb{Z}[1/2][[z]]$ but not $y_0 \in \mathbb{Z}[[z]]$, while $z(q), K(q) \in \mathbb{Z}[[q]]$ and for $K(q)$ we have the Lambert expansion (2.4) with $N_l \in \mathbb{Z}$ but $N_l/N_0 < 0$.

OBSERVATION 2. We have $w_0 = u_0^2$, where

$$u_0 = 1 + 6z + 384z^2 + 42036z^3 + 5867226z^4 + \cdots \in \mathbb{Z}[[z]].$$

OBSERVATION 3. One has the following generalized relation of Beukers: for $k \geqslant 2$

$$2w_0^{(k)}w_4^{(k)} - 2w_1^{(k)}w_3^{(k)} + (w_2^{(k)})^2 = \sum_\nu c_{k,\nu}(9z)^\nu,$$

where $c_{k,\nu} \in \mathbb{Z}$.

OBSERVATION 4. Put $t = w_1/w_0$, $q(z) = e^t = z + O(z^2)$ and denote by $\tilde{z}(q)$ its inverse. Then $q(z) \in \mathbb{Z}[[z]]$ and, hence, $\tilde{z}(q) \in \mathbb{Z}[[q]]$. Moreover,

$$\begin{aligned} q(z) &= z(1 + 15z + 1145z^2 + \cdots)^3, \\ \tilde{z}(q) &= q(1 - 15q - 245q^2 - 20138q^3 - 2043703q^4 + \cdots)^3. \end{aligned}$$

OBSERVATION 5. Expand w_0 in Lambert series:

$$2w_0(z) = 2 + \sum_{n=1}^{\infty} \frac{n^3 L_n z^n}{1 - z^n}.$$

Then $L_n \in \mathbb{Z}$ and $L_n > 0$. What is the enumerative meaning of the numbers L_n (i.e., do they count some geometric objects)?

OBSERVATION 6. Write

$$\tilde{K}(q) = \frac{\mathrm{d}^2}{\mathrm{d}t^2}(w_2/w_0) = 1 + \sum_{l=1}^{\infty} \frac{l^2 \tilde{N}_l q^l}{1 - q^l}.$$

Then $\tilde{N}_l \in \mathbb{Z}$ and $\tilde{N}_l > 0$. What is the enumerative meaning of the numbers $\tilde{N}_l$ (i.e., do they count some geometric objects)?

OBSERVATION 7 (Kummer supercongruences). If we write w_0 and $\tilde{K}$ as functions of z,

$$w_0(z) = \sum_{n=0}^{\infty} A(n) z^n, \qquad \tilde{K}(q(z)) = \sum_{n=0}^{\infty} C(n) z^n,$$

then

$$A(np^r) \equiv A(np^{r-1}) \pmod{p^{3r}}, \qquad C(np^r) \equiv C(np^{r-1}) \pmod{p^{2r}},$$

for primes p and $(n,p) = 1$; $r = 1, 2, \ldots$.

Recently the second author [**43**] (see also Theorem 4 in [**40**]), when approximating $\mathrm{Li}_3(z)$ (see definition (9.3) below), found two more *fourth order* examples related to simultaneous rational approximations to $\zeta(2)$ and $\zeta(3)$. About the same time the first author discovered compact formulas

$$A_n = \sum_{k,l} \binom{n}{k}^2 \binom{n}{l}^2 \binom{k+l}{l} \binom{n+k}{n} \qquad n = 0, 1, 2, \ldots, \tag{4.2}$$

and

$$A_n = \binom{2n}{n} \sum_{k} \binom{n}{k}^2 \binom{n+k}{n} \binom{n+2k}{n}, \qquad n = 0, 1, 2, \ldots,$$

for generating series $y_0(z) = \sum_{n=0}^{\infty} A_n z^n$ satisfying exactly the same differential equations. These examples are cases #195 and #209 in [**3**], Table A. The coefficients (4.2) are of the form similar to those in [**6**] (cf. case #26), and van Straten has confirmed that equation #195 indeed 'comes from geometry'. Thus we have a direct link between number theory and geometry.

5. Strategy of finding new Calabi–Yau equations

How much do we know about Calabi–Yau differential equations? We discovered a list of 29 interesting cases taken from different papers on the subject; these cases initiated our systematization of Calabi–Yau equations and started the table in [**3**][2]. The first 14 cases in the table correspond to all known hypergeometric Calabi–Yau equations, while the origin of the remaining 15 known cases is due to certain variations of multiple hypergeometric series (or multiple binomial sums). The latter circumstance prompted us to look at those linear MUM differential equations, whose analytic solutions $y_0(z) \in \mathbb{Z}[[z]]$ are single or multiple sums of certain hypergeometric terms. There is an algorithmic way, which is due to Gosper and Zeilberger, known as *creative telescoping*, to find polynomial recursions for such sums [**28**]. Unfortunately, the algorithm for multiple summations (being a part of the general *WZ-theory* developed by H. Wilf and D. Zeilberger [**37**]) is not implemented in `Maple`, but solving auxiliary systems of linear equations is. The recursions may be easily transferred to linear differential equations satisfied by $y_0(z)$. That was our general recipe for verifying known cases #15–29, and then we continued our experiments by constructing several new cases, also in combination with other hypergeometric methods—quadratic transformations (Section 6), Hadamard products (Section 7) and other transformations (Section 8). In several cases the algorithm of creative telescoping led to reducible differential equations, but then we used `DEtools` implemented in `Maple` to factorize the corresponding differential operators.

The way *à la Apéry* of producing new examples of the mirror maps and Yukawa couplings decribed in the previous section seems to be very sporadic: we do not really have other examples of rational approximations to 'interesting' constants (but see Section 10 in the E-print version). Nevertheless, the method of finding Calabi–Yau equations by pulling back some 'nice' 5th-order linear differential equations works in several other situations as well. We stress that the 5th- to 4th-order reduction did not appear before in connection with problems of Calabi–Yau manifolds.

Our table in [**3**] extends the list of 29 known cases and contains a sporadic list of more than two hundred Calabi–Yau (conjecturally) differential equations. The table is long but we have found many more equations that we decided not to include (not to stretch the reader's patience too far!). In particular, we have many examples of type $A_n = \sum_k (-1)^k 2^{n-2k} \binom{n}{2k}\binom{2k}{k}(\,\cdot\,)$. The most unfortunate thing seems to be the fact that we cannot see any reasonable end of our list and, therefore, we do not expect to be much criticized by the reader for missing a hundred or more other interesting cases.

The analysis of Table A in [**3**] shows that, besides the 14 hypergeometric functions (cases #1–14), we have plenty of other 'simple' Calabi–Yau differential operators of the type

$$\theta^4 - c_1 z(a\theta^2 + a\theta + b)P(\theta) + c_2 z^2 Q(\theta),$$

where $P(\theta)$ is of the form $(2\theta+1)^2$, $(3\theta+1)(3\theta+2)$, $(4\theta+1)(4\theta+3)$ or $(6\theta+1)(6\theta+5)$, and $Q(\theta)$ splits into linear factors of similar shapes. Here $a \in \{2, 3, 5, 7, 8, 9, 11, 13, 17\}$ and c_1, c_2 are usually powers of 2 and (or) 3. We also have 5th-order differential equations similar to the 'simple' 4th-order ones. Their 4th-order pullbacks (derived by the method of Section 3) are much more complicated. Case #32 seems to be even more complicated:

$$\begin{aligned}\theta^5 &- 3z(2\theta+1)(3\theta^2+3\theta+1)(15\theta^2+15\theta+4)\\ &- 3z^2(\theta+1)^3(3\theta+2)(3\theta+4)\end{aligned}$$

[2]Unfortunately, we are not quite sure that the nice arithmetic properties of mirror maps and Yukawa couplings are really proved in all presented cases! In this sense, Table A in [**3**] involves cases that are only expected to be 'arithmetically nice'.

(this is the differential operator that started this paper; it corresponds to equation (4.1)). Then there are the cases when both the 4th-order differential equation and the 5th-order equation (obtained by the Wronskian construction) are very complicated. The corresponding linear differential operators usually appear as factors of higher order differential operators, for which we have no pattern at all.

One can suspect that the equations in the list from Table A in [**3**] are Picard–Fuchs equations of one-dimensional families of Calabi–Yau threefolds. In many cases, such as the hypergeometric ones and the Hadamard products, this should not be hard to see. For other cases, such as the one related to $\zeta(4)$, this fact will be quite non-trivial, but perhaps all the more interesting. This is a reason why the equations are called Calabi–Yau equations.

REMARK. We are quite surprised by the enomous amount of 4th-order linear differential equations with (expectedly) nice arithmetic properties. After discovering equation #9, the first author for a long time thought that there were no more examples but #1–14. Then after finding the papers [**5**], [**6**] and examples #15–28 there, it was still hard to find any new cases. The use of the Gosper–Zeilberger algorithm changed this drastically. For a while we thought that whenever A_n was a binomial sum (even multiple) and it gave a 4th-order MUM differential equation, then we automatically got an integer mirror map $z(q)$ and the Yukawa coupling $K(q)$ as in (2.4) with integers N_l. There is, however, the following example: the generating series $y_0 = \sum_{n=0}^{\infty} A_n z^n$, where

$$A_n = \sum_k (-1)^k 4^{n-4k} \binom{n}{4k} \frac{(5k)!}{k!^5},$$

satisfies a 4th-order MUM differential equation with polynomial coefficients of degree 8; the equation produces the non-integer mirror map $z(q)$ (actually, $z(q) \in \mathbb{Z}[1/2][[q]]$) but the Yukawa coupling has integers N_l. There are more examples, also for the cases of non-integers N_l; we indicate one in Section 7 below.

6. Quadratic transformations and mirror maps

A purely analytic, hypergeometric machinery—quadratic and higher-order transformations—may be put forward for constructing new examples from the old ones. The method works powerfully in the case of 2nd- and 3rd-order differential equations, when we have to substitute a suitable modular function in place of the variable z; see, for example, [**17**].

Our starting point is the existence of quadratic transformations for ${}_2F_1$- and ${}_3F_2$-series, like

$$ {}_2F_1\left(\begin{matrix} a, \quad b \\ 1+a-b \end{matrix}\,\middle|\, z\right) = (1-z)^{-a} \cdot {}_2F_1\left(\begin{matrix} \frac12 a, \; \frac12 + \frac12 a - b \\ 1+a-b \end{matrix}\,\middle|\, -\frac{4z}{(1-z)^2}\right) \tag{6.1}$$

due to Gauß and

$$\begin{aligned} &{}_3F_2\left(\begin{matrix} a, \quad b, \quad c \\ 1+a-b, \, 1+a-c \end{matrix}\,\middle|\, z\right) \\ &\quad = (1-z)^{-a} \cdot {}_3F_2\left(\begin{matrix} \frac12 a, \; \frac12 + \frac12 a, \; 1+a-b-c \\ 1+a-b, \; 1+a-c \end{matrix}\,\middle|\, -\frac{4z}{(1-z)^2}\right) \end{aligned} \tag{6.2}$$

due to Whipple. Substituting $a=b=c=\frac12$ in them, we get algebraic relations between different mirrors that are necessarily *modular* maps in these cases. These quadratic transformations have a natural generalization to the case of higher dimensional hypergeometric series (see [**41**], Sections 6 and 7). Similar but specialized results are given in the following statement.

PROPOSITION 6. *The following quadratic transformations are available for* ${}_4F_3$-*series:*

$$
\begin{aligned}
&{}_4F_3\left(\begin{matrix} a,\ b,\ c,\ d \\ 1+a-b,\ 1+a-c,\ 1+a-d \end{matrix}\,\middle|\, z\right) \\
&\qquad = \frac{1}{(1+z)^a}\sum_{n=0}^{\infty}\frac{(\frac12 a)_n\,(\frac12+\frac12 a)_n}{(1+a-b)_n\,(1+a-c)_n}\left(\frac{4z}{(1+z)^2}\right)^n \\
&\qquad\quad \times\sum_{\nu=0}^{n}\frac{(b)_\nu\,(c)_\nu\,(1+a-b-c)_{n-\nu}}{\nu!\,(n-\nu)!\,(1+a-d)_\nu}
\end{aligned}
\tag{6.3}
$$

$$
\begin{aligned}
&\qquad = \frac{1}{(1-z)^a}\sum_{n=0}^{\infty}\frac{(\frac12 a)_n}{(1+a-b)_n}\left(-\frac{4z}{(1-z)^2}\right)^n \\
&\qquad\quad \times\sum_{\mu=0}^{n}\frac{(b)_\mu\,(\frac12+\frac12 a)_\mu\,(\frac12+\frac12 a-b)_{n-\mu}}{(n-\mu)!\,(1+a-c)_\mu} \\
&\qquad\quad \times\sum_{\nu=0}^{\mu}\frac{(c)_\nu}{\nu!\,(\mu-\nu)!\,(1+a-d)_\nu}(-1)^\nu.
\end{aligned}
\tag{6.4}
$$

PROOF. Writing

$$
\begin{aligned}
&{}_4F_3\left(\begin{matrix} a,\ b,\ c,\ d \\ 1+a-b,\ 1+a-c,\ 1+a-d \end{matrix}\,\middle|\, z\right) \\
&\quad = \sum_{n=0}^{\infty}\frac{(a)_n\,(b)_n\,(c)_n}{n!\,(1+a-b)_n\,(1+a-c)_n}(-z)^n\cdot{}_2F_1\left(\begin{matrix} -n,\ a+n \\ 1+a-d \end{matrix}\,\middle|\, 1\right) \\
&\quad = \sum_{\nu=0}^{\infty}\frac{(-1)^\nu}{\nu!\,(1+a-d)_\nu}\sum_{n=\nu}^{\infty}\frac{(a)_{\nu+n}\,(b)_n\,(c)_n}{(n-\nu)!\,(1+a-b)_n\,(1+a-c)_n}(-z)^n \\
&\quad = \sum_{\nu=0}^{\infty}\frac{(a)_{2\nu}(b)_\nu(c)_\nu}{\nu!\,(1+a-b)_\nu\,(1+a-c)_\nu\,(1+a-d)_\nu}z^\nu \\
&\quad\quad \times{}_3F_2\left(\begin{matrix} a+2\nu,\ b+\nu,\ c+\nu \\ 1+a-b+\nu,\ 1+a-c+\nu \end{matrix}\,\middle|\, -z\right).
\end{aligned}
\tag{6.5}
$$

To the latter ${}_3F_2$-series we apply the quadratic transformation (6.2) and reorder summations to get (6.3).

For the proof of (6.4), we proceed as in (6.5) to deduce

$$
\begin{aligned}
&{}_3F_2\left(\begin{matrix} a,\ b,\ c \\ 1+a-b,\ 1+a-c \end{matrix}\,\middle|\, z\right) \\
&\quad = \sum_{\lambda=0}^{\infty}\frac{(a)_{2\lambda}(b)_\lambda}{\lambda!\,(1+a-b)_\lambda\,(1+a-c)_\lambda}z^\lambda\cdot{}_2F_1\left(\begin{matrix} a+2\lambda,\ b+\lambda \\ 1+a-b+\lambda \end{matrix}\,\middle|\, -z\right).
\end{aligned}
\tag{6.6}
$$

To the inner ${}_2F_1$-series we now apply the transformation (6.1) to get, as before, the double series for the left-hand side of (6.6). Then substituting the resulting formula for the ${}_3F_2$-series into (6.5) we arrive at the desired identity (6.4). □

Plugging in $a=b=c=d=\frac{1}{2}$ we obtain

$$\sum_{n=0}^{\infty}\binom{2n}{n}^4\left(\frac{z}{2^8}\right)^n = (1+z)^{-1/2}\sum_{n=0}^{\infty}A_n^{(+)}\left(\frac{z}{2^8(1+z)^2}\right)^n$$
$$= (1-z)^{-1/2}\sum_{n=0}^{\infty}A_n^{(-)}\left(-\frac{z}{2^8(1-z)^2}\right)^n,$$

where

$$A_n^{(+)} = \frac{(4n)!}{(2n)!\,n!^2}\sum_{\nu=0}^{n}2^{2(n-\nu)}\binom{2\nu}{\nu}^2\binom{2(n-\nu)}{n-\nu}, \tag{6.7}$$

$$A_n^{(-)} = 2^{2n}\frac{\prod_{j=0}^{n-1}(1+4j)}{n!}\sum_{\mu=0}^{n}2^{4(n-\mu)}\binom{2\mu}{\mu}\frac{\prod_{j=0}^{n-\mu-1}(1+4j)}{(n-\mu)!}\,\frac{\prod_{j=0}^{\mu-1}(3+4j)}{\mu!} \times\sum_{\nu=0}^{\mu}2^{2(\mu-\nu)}\binom{2\nu}{\nu}\binom{\mu}{\nu}(-1)^{\nu}. \tag{6.8}$$

The resulting series $\sum_{n=0}^{\infty}A_n^{(+)}z^n$ and $\sum_{n=0}^{\infty}A_n^{(-)}z^n$ are solutions of 4th-order linear differential equations. They gave two new examples (cases #30 and #31) of differential equations producing mirror maps and Yukawa couplings with desired integrality properties. It seems to be an interesting problem to find quadratic and high-order transformations for other cases from [**3**], Table A.

Besides hypergeometric transformations, there is also an 'obvious' way to produce infinitely many linear MUM differential equations starting with the sole such example (1.1).

PROPOSITION 7. *Let $y=y(z)$ be a generic solution of equation (1.1) of order s with coefficients in $\mathbb{C}(z)$ and let $u=u(z)$ be any function satisfying $u'/u\in\mathbb{C}(z)$. Then the function $w=y/u$ is a solution of a linear differential equation of order s with rational coefficients, and this equation depends only on (1.1) but not on its solution y.*

PROOF. Substituting $y=uw$ in (1.1) we get the equation

$$(uw)^{(s)}+a_{s-1}(z)(uw)^{(s-1)}+\dots+a_1(z)(uw)'+a_0(z)uw=0 \tag{6.9}$$

in w unknown. Developing the derivatives by the formulas

$$(uw)^{(k)}=\sum_{j=0}^{k}\binom{k}{j}u^{(j)}w^{(k-j)}, \qquad k=1,2,\dots,s,$$

and dividing the left-hand side in (6.9) by u, we obtain the sth-order linear differential equation for w with rational coefficients since $u^{(j)}/u\in\mathbb{C}(z)$ for any $j=0,1,2,\dots$. □

By Proposition 7, taking a known example of a Calabi–Yau equation with the corresponding series $y_0=\sum_{n=0}^{\infty}A_nz^n\in\mathbb{Z}[[z]]$ and choosing any $u\in 1+z\mathbb{Z}[[z]]$ with the property $u'/u\in\mathbb{Q}(z)$, we see that $w_0=y_0/u$ is in $\mathbb{Z}[[z]]$ and $w=y/u$ also satisfies a Calabi–Yau equation, with the same mirror map $z(q)$ and Yukawa coupling $K(q)$, since $w_j/w_0=y_j/y_0$ for $j=1,2,3$. For instance, the choice $u=$

$\sqrt{1-4pz}$, where p is an arbitrary integer, leads to an infinite family of Calabi–Yau equations for the sequences

$$\widehat{A}_n = \sum_{k=0}^{n} p^{n-k}\binom{2n-2k}{n-k}A_k.$$

Clearly, examples of such type are not presented in [**3**], Appendix A.

We also found experimentally another (doubly) infinite classes of Calabi–Yau differential equations depending on positive integer parameters p and r.

PROPOSITION 8. *Given A_n, one of the cases presented in* [**3**], *Table* A, *with the corresponding Yukawa coupling $K(q)$, define*

$$\widehat{A}_n = \sum_{k} p^{n-rk}\binom{n}{rk}A_k.$$

Then $\widehat{y}_0 = \sum_{n=0}^{\infty}\widehat{A}_n z^n$ satisfies a 4th-order MUM differential equation with Yukawa coupling $\widehat{K}(q) = K(q^r)$.

REMARK. It can occur that the mirror map $\widehat{z}(q)$ does not lie in $\mathbb{Z}[[q]]$ (but the denominators of its coefficients are necessarily divisors of powers of r).

PROOF. We only give a sketch of the proof, without indicating computational details. Suppose that D is the linear 4th-order differential operator annihilating the given series $y_0(z) = \sum_{n=0}^{\infty} A_n z^n$ and y_0, y_1, y_2, y_3 is the Frobenius basis of the differential equation $Dy = 0$. Change the variable $z \mapsto Z(z) = \bigl(z/(1-pz)\bigr)^r$; then substituting

$$\delta = z\frac{\mathrm{d}}{\mathrm{d}z} = \frac{r}{1-pz}\cdot Z\frac{\mathrm{d}}{\mathrm{d}Z}$$

into the differential equation $Dy = 0$ leads to a new 4th-order MUM differential equation that annihilates $Y_0(z) = y_0(Z(z))$. There is no difficulty in computing the Frobenius basis of the new differential equation (cf. (1.2)): it is

$$Y_j(z) = r^{-j}y_j(Z(z)), \qquad j = 0,1,2,3.$$

Therefore, $T(z) = Y_1(z)/Y_0(z) = \frac{1}{r}t(Z(z))$ and the new mirror map $Z(q)$ is related to the old one $z(q)$ in accordance with the formula

$$z(q^r) = \left(\frac{Z(q)}{1-pZ(q)}\right)^r.$$

Finally,

$$\frac{\mathrm{d}^2}{\mathrm{d}T^2}\left(\frac{Y_2}{Y_0}\right) = K(q^r).$$

Applying now the composition

$$y(z) \mapsto \widetilde{y}(z) = \frac{1}{1-pz}\cdot y(Z(z)) = \frac{1}{1-pz}\cdot y\left(\left(\frac{z}{1-pz}\right)^r\right)$$

with the help of Proposition 7 we conclude that the differential equation for $\widetilde{y}(z)$ is of order 4 and MUM, and the corresponding Yukawa coupling is $K(q^r)$. It remains to verify that

$$\widetilde{y}_0(z) = \sum_{n=0}^{\infty}\widetilde{A}_n z^n = \frac{1}{1-pz}\sum_{n=0}^{\infty}A_n\left(\frac{z}{1-pz}\right)^{rn} = \frac{1}{1-pz}\cdot y_0\left(\left(\frac{z}{1-pz}\right)^r\right).$$

□

REMARK. As pointed out to us by Beukers, Propositions 7 and 8 might be superfluous by choosing a normalization for Calabi–Yau differential equations. Natural requirements are as follows:

(a) the smallest local exponent at the finite singularities is zero;
(b) the point $z = \infty$ is a singularity.

However, the propositions remain useful if these conditions are abandoned.

7. Hadamard products

Using the algorithm of creative telescoping we found almost a hundred new Calabi–Yau differential equations. Then D. van Straten suggested that using Hadamard products of solutions to 'nice' 2nd-order equations could give the required 4th-order equations. Surprisingly, this was also the case for many of the examples we had found just by accident.

Let

$$u = \sum_{n=0}^{\infty} b_n z^n, \qquad v = \sum_{n=0}^{\infty} c_n z^n$$

be two D-finite (i.e., satisfying a linear differential equation of finite order with polynomial coefficients) power series. Then the Hadamard product

$$y = \sum_{n=0}^{\infty} a_n z^n = u * v = \sum_{n=0}^{\infty} b_n c_n z^n \tag{7.1}$$

is also D-finite (see [**34**], p. 194). If D_u and D_v are linear differential operators annihilating u and v, respectively, by $D_u * D_v$ we denote the differential operator annihilating (7.1). We do not know a general algorithm for computing $D_u * D_v$, but for a given pair of operators D_u, D_v the problem is easily solved by linear algebra arguments. We have found about 30 second order MUM differential equations coming from binomial coefficients; Zagier's manuscript [**38**] provides us with 36 second order examples with 'nice' arithmetic properties, although a closed binomial-sum formula is not known in all cases. Consider the following examples for which we were successful in finding the differential equations $Dy = 0$ for their Hadamard products:

(a) $A_n = \sum_{k=0}^{n} \binom{n}{k}^3$, $\qquad D = \theta^2 - z(7\theta^2 + 7\theta + 2) - 8z^2(\theta + 1)^2$;

(b) $A_n = \sum_{k=0}^{n} \binom{n}{k}^2 \binom{n+k}{k}$, $\qquad D = \theta^2 - z(11\theta^2 + 11\theta + 3) - z^2(\theta + 1)^2$;

(c) $A_n = \sum_{k=0}^{n} \binom{n}{k}^2 \binom{2k}{k}$, $\qquad D = \theta^2 - z(10\theta^2 + 10\theta + 3) + 9z^2(\theta + 1)^2$;

(d) $A_n = \sum_{k=0}^{n} \binom{n}{k} \binom{2k}{k} \binom{2n-2k}{n-k}$, $\qquad D = \theta^2 - 4z(3\theta^2 + 3\theta + 1) + 32z^2(\theta + 1)^2$;

(e) $A_n = \sum_{k=0}^{n} 4^{n-k} \binom{2k}{k}^2 \binom{2n-2k}{n-k}$, $\qquad D = \theta^2 - 4z(8\theta^2 + 8\theta + 3) + 256(\theta + 1)^2$.

Taking all possible Hadamard products of these examples (including squares) we get 15 new differential equations. Squares will be of degree 5 and the others of degree 8. The products are listed in #100–107, 113–115 and 120–123.

We may also extend the above list by the following examples (f)–(h) due to D. Zagier [**38**] and (i), (j) due to C. van Enckevort and D. van Straten [**15**]:

(f) $A_n = \sum_{k=0}^{\lfloor n/3 \rfloor} (-1)^k 3^{n-3k} \binom{n}{3k} \frac{(3k)!}{k!^3}$, $D = \theta^2 - 3z(3\theta^2 + 3\theta + 1) + 27z^2(\theta+1)^2$;

(g) $D = \theta^2 - z(17\theta^2 + 17\theta + 6) + 72z^2(\theta+1)^2$;

(h) $A_n = 27^n \sum_k (-1)^k \binom{-2/3}{k} \binom{-1/3}{n-k}^2$,

$D = \theta^2 - 3z(18\theta^2 + 18\theta + 7) + 729z^2(\theta+1)^2$;

(i) $A_n = 64^n \sum_k (-1)^k \binom{-3/4}{k} \binom{-1/4}{n-k}^2$,

$D = \theta^2 - 4z(32\theta^2 + 32\theta + 13) + 4096z^2(\theta+1)^2$;

(j) $A_n = 432^n \sum_k (-1)^k \binom{-5/6}{k} \binom{-1/6}{n-k}^2$,

$D = \theta^2 - 12z(72\theta^2 + 72\theta + 31) + 186624z^2(\theta+1)^2$.

In case (g), an explicit formula for A_n may be extracted from the generating series

$$\sum_{n=0}^{\infty} A_n z^n = (1-12z)^{-1/4}(1-252z+3888z^2-15552z^3)^{-1/4}$$
$$\times {}_2F_1\left(\begin{matrix}1/12,\, 5/12\\ 1\end{matrix}\,\middle|\, \frac{-1728z(1-8z)^2(1-9z)^3}{(1-12z)^3(1-252z+3888z^2-15552z^3)^3}\right),$$

which is borrowed from [**25**]. In other cases the explicit formulas for A_n were found in the following way.

The Legendre function

$$P_a(z) = {}_2F_1\left(\begin{matrix}-a,\, a+1\\ 1\end{matrix}\,\middle|\, \frac{1-z}{2}\right)$$

satisfies the differential equation

$$(1-z^2)\frac{\mathrm{d}^2y}{\mathrm{d}z^2} - 2z\frac{\mathrm{d}y}{\mathrm{d}z} + a(a+1)y = 0.$$

Therefore, the function

$$y_0(z) = \frac{1}{1-z} P_a\left(\frac{1+z}{1-z}\right)$$

is annihilated by the differential operator

$$\theta^2 - z(2\theta^2 + 2\theta + a^2 + a + 1) + z^2(\theta+1)^2$$

(see [**38**]). We have

$$(1-z)^a P_a\left(\frac{1+z}{1-z}\right) = \sum_{n=0}^{\infty} \binom{a}{n}^2 z^n$$

which gives

$$\frac{1}{1-cz} P_a\left(\frac{1+cz}{1-cz}\right) = \sum_{n=0}^{\infty} A_n z^n,$$

where

$$A_n = c^n \sum_k (-1)^k \binom{-1-a}{k} \binom{a}{n-k}^2.$$

Putting $a = -1/3, -1/4, -1/6$ for a suitable choice of c we get cases (h)–(j).

Let further

$$\text{(A)} \quad A_n = \binom{2n}{n}^2, \qquad D = \theta^2 - 4z(2\theta+1)^2;$$

$$\text{(B)} \quad A_n = \frac{(3n)!}{n!^3}, \qquad D = \theta^2 - 3z(3\theta+1)(3\theta+2);$$

$$\text{(C)} \quad A_n = \frac{(4n)!}{n!^2(2n)!}, \qquad D = \theta^2 - 4z(4\theta+1)(4\theta+3);$$

$$\text{(D)} \quad A_n = \frac{(6n)!}{n!(2n)!(3n)!}, \qquad D = \theta^2 - 12z(6\theta+1)(6\theta+5).$$

One can prove the following formula for the Hadamard product by the method presented below (P and Q are quadratic polynomials and c is an arbitrary constant):

$$\big(\theta^2 - zP(\theta) - cz^2(\theta+1)^2\big) * \big(\theta^2 - zQ(\theta)\big) = \theta^4 - zP(\theta)Q(\theta) - cz^2Q(\theta+1)Q(\theta).$$

Then we get the following table of cases[3] from [**3**], Table A, corresponding to Hadamard products:

	(A)	(B)	(C)	(D)
(a)	45	15	68	62
(b)	25	24	51	63
(c)	58	70	69	64
(d)	36	48	38	65
(e)	111	110	30	112
(f)	133	134	135	136
(g)	137	138	139	140
(h)	141	142	∅	143

The empty set at place (C)$*$(h) means that the Yukawa coupling $K(q)$ is constant.

In our computations we applied the following method communicated to us by A. Meurman. The method is, in a sense, a discrete analogue of the Wronskian formalism.

METHOD TO FIND HADAMARD PRODUCTS. Assume

$$A_{n+2} = P_1(n)A_{n+1} + Q_1(n)A_n, \qquad B_{n+2} = P_2(n)B_{n+1} + Q_2(n)B_n,$$

and we want to find a recursion formula for $C_n = A_nB_n$. For this, define

$$R_j(n) = P_j(n+1)P_j(n) + Q_j(n+1), \quad S_j(n) = P_j(n+1)Q_j(n), \qquad j = 1,2,$$

$$\begin{aligned} U_j(n) &= P_j(n+2)R_j(n) + Q_j(n+2)P_j(n), \\ V_j(n) &= P_j(n+2)S_j(n) + Q_j(n+2)Q_j(n), \end{aligned} \qquad j = 1,2,$$

[3] We will not write out in [**3**], Table A, the equations in cases #133–143, since the reader can easily derive them himself by the formula above. We have, however, checked that one gets integers everywhere.

and take

$$T_2(n) = R_2S_1U_1V_2 - R_1S_2U_2V_1, \quad T_3(n) = P_1Q_2U_2V_1 - P_2Q_1U_1V_2,$$
$$T_4(n) = R_1S_2P_2Q_1 - R_2S_1P_1Q_2,$$

and

$$W_0(n) = Q_1Q_2T_2 + S_1S_2T_3 + V_1V_2T_4, \quad W_1(n) = P_1P_2T_2 + R_1R_2T_3 + U_1U_2T_4.$$

Then

$$T_4(n)C_{n+4} + T_3(n)C_{n+3} + T_2(n)C_{n+2} - W_1(n)C_{n+1} - W_0(n)C_n = 0.$$

When $A_n = B_n$, the method has to be modified. Assume

$$A_{n+2} = P(n)A_{n+1} + Q(n)A_n,$$

where $P(n)$ and $Q(n)$ are rational. Then

$$A_{n+3} = R(n)A_{n+1} + S(n)A_n,$$

where

$$R(n) = P(n+1)P(n) + Q(n+1), \quad S(n) = P(n+1)Q(n).$$

Letting $C_n = A_n^2$ and $T(n) = R(n)Q(n) - P(n)S(n)$, we see that

$$P(n)Q(n)C_{n+3} - R(n)S(n)C_{n+2} - P(n)R(n)T(n)C_{n+1} + Q(n)S(n)T(n)C_n = 0.$$

The procedure usually leads (in cases (a)–(h)) to a 8th-order differential operator (6th-order if $A_n = B_n$) having a 4th-order factor. Cases #167–179 were computed by this way.

Actually we found four 2nd-order differential equations $Dy = 0$ somewhat suitable for Hadamard products:

(k) $$A_n = 9^n \sum_k (-1)^k \binom{-1/3}{k}\binom{-2/3}{n-k}\binom{n}{k},$$
$$D = \theta^2 - 3z(2\theta+1) - 81z^2(\theta+1)^2;$$

(l) $$A_n = 8^n \sum_k (-1)^k \binom{-1/4}{k}\binom{-3/4}{n-k}\binom{n}{k},$$
$$D = \theta^2 - 4z(2\theta+1) - 64z^2(\theta+1)^2;$$

(m) $$A_n = 36^n \sum_k (-1)^k \binom{-1/6}{k}\binom{-5/6}{n-k}\binom{n}{k},$$
$$D = \theta^2 - 24z(2\theta+1) - 1296z^2(\theta+1)^2;$$

(n) $$A_n = 4^n \sum_k (-1)^k \binom{-1/2}{k}\binom{-1/2}{n-k}\binom{n}{k}, \quad D = \theta^2 - 16z^2(\theta+1)^2.$$

Only (n) gives Calabi-Yau differential equations when we form (a) $*$ (n), (b) $*$ (n), $\dots$, (n) $*$ (n) but this is just the substitution $z \mapsto z^{1/2}$. The Hadamard squares of (k), (l), (m) correspond to Calabi–Yau differential operators found from the formula

$$\begin{aligned}
&(\theta^2 - Az(2\theta+1) - Bz^2(\theta+1)^2)^{*2} \\
&= \theta^4 - z(2B\theta^4 + A^2(2\theta+1)^2) - Bz^2(2\theta+1)(2B\theta^2 + 2B\theta + 4A^2 + B) \\
&\quad + B^2z^3(2B\theta^4 + 8B\theta^3 + (4A^2+12B)\theta^2 + (4A^2+B)\theta + A^2 + 2B) \\
&\quad - B^4z^4(\theta+1)^4.
\end{aligned}$$

As a curiosity we mention that in the case (m) $*$ (m) we have

$$q = z(1 + 21z + 3840z^2 + \cdots)^{12960},$$

where the series in the brackets is expected to lie in $\mathbb{Z}[[z]]$.

Of all non-square Hadamard products only (k)$*$(m) gives a Calabi–Yau operator

$$\begin{aligned}&\theta^4 - 4\cdot 18z(2\theta+1)^2 - 2\cdot 18^3 z^2(27\theta^4 + 36\theta^3 + 74\theta^2 + 76\theta + 24)\\ &\quad - 32\cdot 18^5 z^3(2\theta+1) + 2\cdot 18^7 z^4(27\theta^4 + 72\theta^3 + 128\theta^2 + 72\theta + 13)\\ &\quad + 4\cdot 18^9 z^5(2\theta+1)^2 - 18^{12} z^6(\theta+1)^4.\end{aligned}$$

To get integer instanton numbers we need to make the substitution

$$\widetilde{K}(q) = K(q/3) = 1 + \sum_{k=1}^{\infty} \frac{k^3 \widetilde{N}_k q^k}{1-q^k};$$

then

$$\widetilde{N}_1 = -48, \quad \widetilde{N}_2 = -126, \quad \widetilde{N}_3 = -2864, \quad \widetilde{N}_4 = 77958, \quad \widetilde{N}_5 = 4942032, \quad \dots .$$

We can also get 5th-order equations by taking the Hadamard product of (A)–(D) with 3rd-order equations of the following type:

(α) $$A_n = \sum_{k=0}^{n} \binom{n}{k}^2 \binom{2k}{k}\binom{2n-2k}{n-k},$$
$$D = \theta^3 - 2z(2\theta+1)(5\theta^2+5\theta+2) + 64z^2(\theta+1)^3;$$

(β) $$A_n = \sum_{k=0}^{n} \binom{2k}{k}^2 \binom{2n-2k}{n-k}^2,$$
$$D = \theta^3 - 8z(2\theta+1)(2\theta^2+2\theta+1) + 256z^2(\theta+1)^3;$$

(γ) $$A_n = \sum_{k=0}^{n} \binom{n}{k}^2 \binom{n+k}{k}^2,$$
$$D = \theta^3 - z(2\theta+1)(17\theta^2+17\theta+5) + z^2(\theta+1)^3;$$

(δ) $$A_n = \sum_{k=0}^{\lfloor n/3\rfloor} (-1)^k 3^{n-3k} \binom{n}{3k}\binom{n+k}{k}\frac{(3k)!}{k!^3},$$
$$D = \theta^3 - z(2\theta+1)(7\theta^2+7\theta+3) + 81z^2(\theta+1)^3;$$

(ϵ) $$A_n = \sum_{k} \binom{n}{k}^2 \binom{2k}{n}^2,$$
$$D = \theta^3 - 4z(2\theta+1)(3\theta^2+3\theta+1) + 16z^2(\theta+1)^3;$$

(ζ) $$A_n = \sum_{k,l} \binom{n}{k}^2 \binom{n}{l}\binom{k}{l}\binom{k+l}{n},$$
$$D = \theta^3 - 3z(2\theta+1)(3\theta^2+3\theta+1) - 27z^2(\theta+1)^3;$$

(η) $$D = \theta^3 - z(2\theta+1)(11\theta^2+11\theta+5) + 125z^2(\theta+1)^3;$$

(ϑ) $$A_n = (-64)^n \sum_{k} \binom{-1/2}{k}\binom{-1/2}{n-k}^3 = 64^n \sum_{k} \binom{-1/4}{k}^2 \binom{-3/4}{n-k}^2,$$
$$D = \theta^3 - 8z(2\theta+1)(8\theta^2+8\theta+5) + 4096z^2(\theta+1)^3;$$

(ι) $A_n = 27^n \sum_k \binom{-1/3}{k}^2 \binom{-2/3}{n-k}^2,$

$D = \theta^3 - 3z(2\theta+1)(9\theta^2+9\theta+5) + 729z^2(\theta+1)^3;$

(κ) $A_n = 432^n \sum_k \binom{-1/6}{k}^2 \binom{-5/6}{n-k}^2,$

$D = \theta^3 - 24z(2\theta+1)(18\theta^2+18\theta+13) + 186624z^2(\theta+1)^3.$

The last 6 examples were found in [**15**] (without formulas for A_n).

The general formula for the Hadamard product in this case is as follows:

$$\begin{aligned}&(\theta^2 - z(2\theta+1)P(\theta) - cz^2(\theta+1)^3) * (\theta^2 - zQ(\theta))\\ &\quad= \theta^5 - z(2\theta+1)P(\theta)Q(\theta) - cz^2(\theta+1)Q(\theta+1)Q(\theta)\end{aligned}$$

and we obtain the following table of corresponding cases[4]:

	(A)	(B)	(C)	(D)
(α)	39	61	37	66
(β)	40	49	43	67
(γ)	44	53	52	149
(δ)	150	151	152	153

In addition, we have another type of Hadamard products. If $\{A_n\}$ is given by any of the cases (α)–(ϵ) with operator

$$\theta^3 - z(2\theta+1)P(\theta) + cz^2(\theta+1)^3,$$

then the generating series for the sequence $\{\binom{2n}{n}A_n\}$ is annihilated by the 4th-order differential operator

$$\theta^4 - 2z(2\theta+1)^2P(\theta) + 4cz^2(\theta+1)^2(2\theta+1)(2\theta+3).$$

Thus we get cases #16, 35, 29, 41, and 42, respectively.

PROPOSITION 9. *Any differential equation of type $D'y = 0$, where*

$$D' = \theta^3 - z(2\theta+1)(a\theta^2+a\theta+b) + cz^2(\theta+1)^3,$$

is the symmetric square of $Dy = 0$, where

$$D = \theta^2 - z(2a\theta^2 + a\theta + \tfrac{b}{2}) + cz^2(\theta+\tfrac{1}{2})^2. \tag{7.2}$$

PROOF. In [**2**] it is proved that the differential equation

$$y^{(3)} + s_2(z)y'' + s_1(z)y' + s_0(z)y = 0 \tag{7.3}$$

is a symmetric square of

$$u'' + p_1(z)u' + p_0(z)u = 0 \tag{7.4}$$

if and only if

$$\frac{s_1s_2}{3} - \frac{2s_2^3}{27} + \frac{s_1'}{2} - \frac{s_2''}{6} - \frac{s_2s_2'}{3} - s_0 = 0,$$

and then

$$p_1 = \frac{s_2}{3}, \qquad p_0 = \frac{s_1}{4} - \frac{s_2^2}{18} - \frac{s_2'}{12}.$$

[4]We computed all the other pullbacks of the Hadamard products (one of those has trivial $K(q)$) and found no new instanton numbers: they are all identical with earlier known Hadamard products of type $(a) * (b)$ etc. The pullbacks are contained in the database [**14**].

Here we get

$$s_0 = \frac{-b + cz}{1 - 2az + cz^2}, \qquad s_1 = \frac{1 - (6a + 2b)z + 7cz^2}{1 - 2az + cz^2}, \qquad s_2 = \frac{3(1 - 3az + 2cz^2)}{1 - 2az + cz^2},$$

which satisfy the identity above. We also get

$$p_1 = \frac{1 - 3az + 2cz^2}{z(1 - 2az + cz^2)}, \qquad p_0 = \frac{-2b + cz}{4z(1 - 2az + cz^2)},$$

and the operator $\mathrm{d}^2/\mathrm{d}z^2 + p_1(z)\,\mathrm{d}/\mathrm{d}z + p_0(z)$ is easily converted to (7.2). □

REMARK. The Frobenius basis of the 3rd-order differential equation (7.3) is

$$y_0 = u_0^2, \quad y_1 = u_0 u_1, \quad y_2 = \frac{1}{2}u_1^2,$$

where u_0, u_1 form the Frobenius basis of (7.4). Hence

$$\frac{y_2}{y_0} = \frac{1}{2}\left(\frac{u_1}{u_0}\right)^2 = \frac{1}{2}\left(\frac{y_1}{y_0}\right)^2 = \frac{t^2}{2}$$

and

$$K(q) = \frac{\mathrm{d}^2}{\mathrm{d}t^2}\left(\frac{y_2}{y_0}\right) = 1 = \text{constant}.$$

The substitution $z \mapsto 4z$ makes also $u_0 = y_0^{1/2}$ have integer coefficients. Since θ is invariant under this substitution we get

$$D \mapsto \theta^2 - z(8a\theta^2 + 4a\theta + 2b) + 4cz^2(2\theta + 1)^2.$$

Finding such differential operators is equivalent to finding integer solutions to the recursion

$$(n+1)^2 u_{n+1} = 2(a \cdot 2n(2n+1) + 2b)u_n - 4c(2n-1)^2 u_{n-1}$$

which is similar to Zagier's computation in [**38**].

On the other hand, we found that only 2nd-order differential operators of type $\theta^2 - zP(\theta) + cz^2(\theta+1)^2$, where $P(\theta)$ is a polynomial of degree 2, were useful for producing 4th-order Calabi–Yau by taking the Hadamard product of two of them. This, however, coincides with Zagier's list [**38**] of integer solutions to

$$(n+1)^2 A_{n+1} - an(n+1)A_n + bn^2 A_{n-1} = \lambda A_n.$$

Zagier's list has 36 entries. It contains our cases (a)–(h), (A)–(C) (but not (D) since the coefficients are too large). If A_n is a polynomial in n (that corresponds to 8 entries in the list), then the corresponding 2nd-order differential operator factors. Excluding these polynomial cases, hypergeometric cases (A)–(C) and terminating cases, we found that certain remaining cases contain examples that we do not really like. For instance, the recurrence

$$(n+1)^2 A_{n+1} - 32n(n+1)A_n + 256A_{n-1} = 28A_n$$

admits the following integer-valued solution:

$$A_n = 2^{4n}\sum_{k=0}^{n}\binom{1/2}{n-k}^2\binom{k+1/2}{k} = \sum_{k=0}^{n} 2^{2k}\frac{2k+1}{(2n-2k+1)^2}\binom{2n-2k}{n-k}^2\binom{2k}{k};$$

the 4th-order differential operator annihilating the series $\sum_{n=0}^{\infty}\binom{2n}{n}^2 A_n z^n$ is as follows:

$$\theta^4 - 16z(2\theta+1)^2(8\theta^2 + 8\theta + 7) + 4096z^2(2\theta+1)^2(2\theta+3)^2$$

and the corresponding differential equation satisfies relation (2.2); then computations show that $z(q) \notin \mathbb{Z}[[q]]$ (and N_l in (2.4) are not integers). This is an example promised in Remark of Section 5.

8. More transformations

8.1. Transformations of the first 14 hypergeometric cases. The success with negative rational numbers in binomial coefficients caused us to make the following experiment. Taking the hypergeometric operator in case #9,

$$\theta^4 - 12^6 z\Big(\theta+\frac{1}{12}\Big)\Big(\theta+\frac{5}{12}\Big)\Big(\theta+\frac{7}{12}\Big)\Big(\theta+\frac{11}{12}\Big),$$

consider

$$A_n = 1728^n\binom{2n}{n}\sum_k\binom{-1/12}{k}\binom{-5/12}{k}\binom{-7/12}{n-k}\binom{-11/12}{n-k}, \qquad n=0,1,2,\dots.$$

Then the series $y_0(z) = \sum_{n=0}^{\infty} A_n z^n$ satisfies the differential equation $Dy = 0$, where

$$D = \theta^4 - 48z(2\theta+1)^2(72\theta^2+72\theta+41) + 2^{14}\cdot 3^4 z^2(2\theta+1)(2\theta+3)(3\theta+2)(3\theta+4),$$

an equation found in [**15**] and mentioned as case #9* in [**3**], Table A. We get the instanton numbers

$$N_1 = -480, \quad N_2 = -226848, \quad N_3 = 16034720,$$
$$N_4 = 1094330202744, \quad N_5 = 4352645747040, \quad \dots.$$

We also note that

$$y_0 = (1+246z+768132z^2+\cdots)^8, \qquad q = z(1+12z+33333z^2+\cdots)^{288},$$

where the series in brackets are expectively in $\mathbb{Z}[[z]]$. Interchanging $5/12$ and $7/12$, we obtain case 9**:

$$\begin{aligned} A_n &= 1728^n\binom{2n}{n}\sum_k\binom{-1/12}{k}\binom{-7/12}{k}\binom{-5/12}{n-k}\binom{-11/12}{n-k} \\ &= 432^n\binom{2n}{n}^2\sum_k(-1)^k\binom{-5/6}{k}\binom{-1/6}{n-k}^2, \end{aligned}$$

with the corresponding differential operator

$$D = \theta^4 - 48z(2\theta+1)^2(72\theta^2+72\theta+31) + 2^{12}\cdot 3^6 z^2(2\theta+1)^2(2\theta+3)^2;$$

the coupling $K(q)$ is the same as in #9*.

Trying the above experiment for all equations #1–#14 we find that cases #1, #2, #11, #12 and #14 give integer mirror map and Yukawa coupling but *there seems to be no way to get integer instanton numbers* (by doing the substitution $z \to cz$). Case #5* has trivial $K(q)$. We have, however, a couple of 'good' cases #3*, #4*, #4**, #6*, #7*, #7**, #8*, #8**, #10*, #10**, #13* and #13** (see Appendix A). This (together with cases #9*, #9** given above) ends the list of the 14 new equations.

By some unusual substitions we can also consider the following additional cases. In case #14*, we take

$$A_n = 144^n\binom{2n}{n}\sum_k\binom{-1/6}{k}\binom{-1/2}{k}\binom{-5/6}{n-k}\binom{-1/2}{n-k}$$

and obtain the corresponding differential operator

$$D=\theta^4-2^4\cdot 3^2z(2\theta+1)^2(2\theta^2+2\theta+1)+2^{10}\cdot 3^2z^2(2\theta+1)(2\theta+3)(3\theta+2)(3\theta+4).$$

Then, choosing

$$\widetilde{K}(q)=K\big((q/3)^{1/2}\big)=1+\sum_{k=1}^{\infty}\frac{k^3\widetilde{N}_kq^k}{1-q^k},$$

we have

$$\widetilde{N}_1=-2592,\quad \widetilde{N}_2=-307800,\quad \widetilde{N}_3=81451104,$$
$$\widetilde{N}_4=144135316512,\quad \widetilde{N}_5=98667659422368,\quad \dots.$$

In case #2*, when

$$A_n=2000^n\binom{2n}{n}\sum_k\binom{-1/10}{k}\binom{-3/10}{k}\binom{-7/10}{n-k}\binom{-9/10}{n-k},$$
$$D=\theta^4-2^4\cdot 5z(2\theta+1)^2(50\theta^2+50\theta+33)$$
$$+2^{10}\cdot 5^4z^2(2\theta+1)(2\theta+3)(5\theta+4)(5\theta+6),$$

we need even more drastic methods; the 'instanton numbers' land in $\mathbb{Z}[\sqrt{5}]$. Taking

$$\widetilde{K}(q)=K(q/\sqrt{5})=1+\sum_{k=1}^{\infty}\frac{k^3\widetilde{N}_kq^k}{1-q^k},$$

we obtain

$$\widetilde{N}_1=-256\sqrt{5},\quad \widetilde{N}_2=32\sqrt{5}-35260,\quad \widetilde{N}_3=1004288\sqrt{5},$$
$$\widetilde{N}_4=835297220,\quad \widetilde{N}_5=102454248704\sqrt{5},\quad \dots.$$

Case #2** corresponds to the interchange of 3/10 and 7/10, hence the change of the factor $50\theta^2+50\theta+33$ by $50\theta^2+50\theta+17$ in the differential operator and conjugate $\widetilde{K}(q)$. Is there a geometric interpretation of the numbers $\widetilde{N}_k$?

8.2. Differential equations inspired by Guillera's formulas. The formulas discovered in [**16**] caused us to try the following (again we mimic case #9). Let

$$A'_n=3456^{2n}\sum_{k=0}^{n}(-1)^k\binom{n}{k}\frac{(1/2)_k(1/12)_k(5/12)_k(7/12)_k(11/12)_k}{k!^5}.$$

Then $w_0=\sum_{n=0}^{\infty}A'_nz^n$ satisfies $D'w=0$, where

$$\begin{aligned}D'=\theta^5&-288z(2\theta+1)(103680\theta^4+207360\theta^3+262944\theta^2+159264\theta+41087)\\&+2^{20}3^8z^2(\theta+1)(207360\theta^4+829440\theta^3+1514592\theta^2+1370304\theta+498143)\\&-2^{38}3^{17}z^3(\theta+1)(\theta+2)(2\theta+3)(240\theta^2+720\theta+793)\\&+2^{53}3^{22}z^4(\theta+1)(\theta+2)(\theta+3)(360\theta^2+1440\theta+1633)\\&-2^{69}3^{30}z^5(\theta+1)(\theta+2)(\theta+3)(\theta+4)(2\theta+5).\end{aligned}$$

The 4th-order pullback of the latter differential operator is

$$\begin{aligned}D=\theta^4&-2^43^2z(248832\theta^4+414720\theta^3+318528\theta^2+111168\theta+14497)\\&+2^{22}3^{10}z^2(4\theta+3)(432\theta^3+1116\theta^2+886\theta+165)\\&-2^{34}3^{18}z^3(4\theta+1)(4\theta+3)(4\theta+7)(4\theta+9)\end{aligned}$$

and, remarkably, it has lower degree than the 5th-order equation (cf. case #129). We get the instanton numbers

$$N_1 = 2710944,\ N_2 = -717640978896,\ N_3 = 302270555492914464,$$
$$N_4 = -171507700573958028578832,\ N_5 = 113303073680022744870130144224,\ \dots,$$

and note that

$$y_0 = (1 + 260946z + 1405445560884z^2 + \cdots)^8,$$
$$q = x(1 + 13295z + 67507583411z^2 + \cdots)^{576}, \quad K(q) = (1 + 56478q - \cdots)^{48},$$

where the parentheses are supposed to have integer coefficients.

Doing the same thing with case #3, i.e. taking

$$A'_n = 1024^n \sum_{k=0}^{n} (-1)^k \binom{n}{k} \frac{(1/2)_k^5}{k!^5},$$

we get the same Yukawa coupling as for case #115 which is an Hadamard square with

$$A_n = \left\{ \sum_k 4^{n-k} \binom{2k}{k}^2 \binom{2n-2k}{n-k}^2 \right\}^2$$

But the differential equations are quite different as are the solutions y_0.

We list all possible variations on the theme as cases $\#\widehat{1}$–$\widehat{14}$ in [**3**], Table A.

There is an even simpler way to arrive at essentially the same Yukawa coupling $K(q)$ as above. Consider the 5th-order hypergeometric operator

$$\theta^5 - 4 \cdot 12^6 z\Big(\theta + \frac{1}{2}\Big)\Big(\theta + \frac{1}{12}\Big)\Big(\theta + \frac{5}{12}\Big)\Big(\theta + \frac{7}{12}\Big)\Big(\theta + \frac{11}{12}\Big),$$

with the solution

$$w_0 = \sum_{n=0}^{\infty} \frac{(12n)!}{(6n)!n!^6} \frac{z^n}{\binom{4n}{2n}}$$

of the corresponding differential equation $D'w = 0$, and with the 4th-order pullback

$$\begin{aligned} D = \theta^4 &- 2^4 3^2 z(331776\theta^4 + 82944\theta^3 + 13248\theta^2 - 28224\theta - 14497) \\ &+ 2^{19} 3^8 z^2 (248832\theta^4 + 124416\theta^3 + 25056\theta^2 - 4176\theta + 21143) \\ &- 2^{32} 3^{14} z^3 (331776\theta^4 + 248832\theta^3 + 60480\theta^2 + 21312\theta - 4453) \\ &+ 2^{51} 3^{22} z^4 \theta(2\theta + 1)(12\theta + 1)(12\theta + 5) \end{aligned}$$

Observe that here the pullback increased the degree from 1 to 4. The instanton numbers are

$$N_1 = -2710944, \quad N_2 = -717640301160, \quad \dots,$$

i.e. the same as above up to a sign. We also note

$$y_0 = (1 - 86982z - 577991455848z^2 - \cdots)^{24},$$
$$q = x(1 + 7441z + 32486988467z^2 + \cdots)^{576}.$$

A third variation of the theme is the following. Consider

$$A'_n = (4 \cdot 12^6)^n \binom{-1/2}{n} \sum_{k=0}^{n} \binom{n}{k} \binom{-1/12}{k} \binom{-5/12}{k} \binom{-7/12}{k} \binom{-11/12}{k}$$

with corresponding 5th-order differential operator

$$\begin{aligned} D' = \theta^5 &+ 2\cdot 12^2 z(2\theta+1)(1244160\theta^4+2488320\theta^3+2347200\theta^2+1103040\theta+21121)\\ &+ 112\cdot 12^{10} z^2(\theta+1)(2\theta+1)(2\theta+3)(72\theta^2+144\theta+101)\\ &+ 16\cdot 12^{16} z^3(2\theta+1)(2\theta+3)(2+5)(1152\theta^2+3456\theta+2831)\\ &+ 12^{26} z^4(\theta+2)(2\theta+1)(2\theta+3)(2\theta+5)(2\theta+7)\\ &+ 64\cdot 12^{30} z^5(2\theta+1)(2\theta+3)(2\theta+5)(2+7)(2\theta+9) \end{aligned}$$

with pullback

$$\begin{aligned} D = \theta^4 &+ 2^4 3^2 z(995328\theta^4+497664\theta^3+220608\theta^2-28224\theta-35233)\\ &+ 2^{18}3^8 z^2(5142528\theta^4+5308416\theta^3+2946816\theta^2+9792\theta-292949)\\ &+ 2^{35}3^{14} z^3(18662400\theta^4+29859840\theta^3+20113200\theta^2+1429200\theta-208501)\\ &+ 2^{48}3^{20} z^4(66562560\theta^4+146810880\theta^3+117328320\theta^2+13269600\theta-1309273)\\ &+ 2^{64}3^{28} z^5(4\theta-1)(518400\theta^3+1607040\theta^2+1779080\theta+64537)\\ &+ 2^{79}3^{34} z^6(4\theta-1)(4\theta+3)(44640\theta^2+135360\theta+10985)\\ &+ 2^{97}3^{43} z^7(\theta+2)(4\theta-1)(4\theta+3)(4\theta+7)\\ &+ 2^{108}3^{48} z^8(4\theta-1)(4\theta+3)(4\theta+7)(4\theta+11). \end{aligned}$$

We have

$$y_0 = (1+211398z-2341345644648z^2+\cdots)^{24}$$

and

$$q = z(1-28177z+245565832115z^2-\cdots)^{576}.$$

The instanton numbers are the same as above. Thus we have three 4th-order differential equations of degree 3, 4, and 8, respectively. The instanton numbers, however, are invariants (possibly, with some geometric interpretation). It would be interesting to find the transformations between the three different solutions.

8.3. The mirror at infinity. In [**31**] E. Rødland studied case #27 at infinity. We choose instead #124 from [**15**]:

$$\begin{aligned} D = 61^2\theta^4 &- 61z(3029\theta^4+5572\theta^3+4677\theta^2+1891\theta+305)\\ &+ z^2(1215215\theta^4+3428132\theta^3+4267228\theta^2+2572675\theta+611586)\\ &- 3^4 z^3(39370\theta^4+140178\theta^3+206807\theta^2+142191\theta+37332)\\ &+ 3^8 z^4(566\theta^4+2230\theta^3+3356\theta^2+2241\theta+558) - 3^{13} z^5(\theta+1)^4, \end{aligned}$$

where the explicit formulas for A_n are known and one should take $N_0 = 61$ in order to get integer instanton numbers (cf. (2.4)). The substitution $z \mapsto 3^{-5}z^{-1}$, $y \mapsto zy$ results in $\theta \to -\theta - 1$ (this works only if the highest degree term has the form $(\theta+1)^4$) and we obtain the 'dual' differential operator

$$\begin{aligned} D^* = \theta^4 &- z(566\theta^4+34\theta^3+62\theta^2+45\theta+9)\\ &+ 3z^2(39370\theta^4+17302\theta^3+22493\theta^2+8369\theta+1140)\\ &- 3^2 z^3(1215215\theta^4+1432728\theta^3+1274122\theta^2+538245\theta+93222)\\ &+ 3^7 61 z^4(3029\theta^4+6544\theta^3+6135\theta^2+2863\theta^2+548) - 3^{12}61^2 z^5(\theta+1)^4. \end{aligned}$$

We do not know a formula for the corresponding A_n here.

Many differential equations with highest degree term $(\theta+1)^4$ (in particular, all Hadamard products) are self-dual. There is, however, the class with

$$A_n = \sum_k \binom{n}{k}^{5-2r} \binom{2k}{k}^r \binom{2n-2k}{n-k}^r$$

for $r = 0, 1, 2, 3, 4, 5$ (cases #22, #21, #23, #56, #71, #118, respectively). Then the duality maps $r \leftrightarrow 5-r$.

Also, differential equations with highest degree term $(2\theta+1)^4$ can be reflected in ∞. Let us take, e.g., case #55 with

$$\begin{aligned} D = 9\theta^4 &- 12z(208\theta^4 + 224\theta^3 + 163\theta^2 + 51\theta + 6) \\ &+ 2^9 z^2 (32\theta^4 - 928\theta^3 - 1606\theta^2 - 837\theta - 141) \\ &+ 2^{16} z^3 (144\theta^2 + 576\theta^3 + 467\theta^2 + 144\theta + 15) - 2^{24} z^4 (2\theta+1)^4. \end{aligned}$$

Then the transformation $z \to 2^{-18} z^{-1}$ maps θ to $-\theta - 1/2$ and gives the dual

$$\begin{aligned} D^* = \theta^4 &- 2^4 z(576\theta^4 - 1152\theta^3 - 724\theta^2 - 148\theta - 13) \\ &- 2^{17} z^2 (32\theta^4 + 992\theta^3 - 166\theta^2 - 57\theta - 6) \\ &+ 2^{26} 3 z^3 (832\theta^4 + 768\theta^3 + 556\theta^2 + 192\theta + 25) - 2^{40} 3^2 z^4 (2\theta+1)^4 \end{aligned}$$

Also, cases #33, #99, #154 can be treated in the same way (case #154 is self-dual).

9. Supercongruences and k-realizable series

It is time to stop and discuss the phenomenon of integer coefficients of Yukawa couplings number-theoretically.

Let $T\colon X \to X$ be a map of a set. Define the number of points x with period n:

$$f_n(T) = \#\{x \in X : T^n x = x\},$$

and the number of points x with the smallest period n:

$$f_n^*(T) = \#\{x \in X : T^n x = x,\ \#\{T^l x : l \in \mathbb{Z}\} = n\}.$$

Then $f_n^0(T) = \frac{1}{n} f_n^*(T)$ determines the number of orbits of length n. Applying the Möbius inversion formula to $f_n(T) = \sum_{d|n} f_d^*(T)$ we obtain

$$f_n^0(T) = \frac{1}{n} \sum_{d|n} \mu\Big(\frac{n}{d}\Big) f_d(T). \tag{9.1}$$

Proposition 10 ([**29**]). *Let $\{A_n\}_{n=1}^\infty$ be a sequence with all $a_n \geqslant 0$. Then there exists a map $T\colon X \to X$ such that $A_n = f_n(T)$ if and only if the numbers*

$$B_n = \frac{1}{n} \sum_{d|n} \mu\Big(\frac{n}{d}\Big) A_d, \qquad n = 1, 2, \dots,$$

are all non-negative integers.

Proof. The '$\Longrightarrow$'-part was already done in (9.1).

In order to prove the '$\Longleftarrow$'-part, we take $X = \{0, 1, 2, \dots\}$ and let $T\colon X \to X$ be a permutation consisting of B_n cycles of length n for all $n = 1, 2, \dots$. Then $A_n = f_n(T)$ as required. □

To illustrate the above construction of T in the proof, we consider the following

EXAMPLE. Let $\{A_n\}_{n=1}^{\infty} = \{1, 5, 7, 17, 31, 65, \dots\}$ be given by the recurrence $A_{n+1} = A_n + 2A_{n-1}$ for $n = 1, 2, \dots$. Then $\{B_n\}_{n=1}^{\infty} = \{1, 2, 2, 3, 6, 9, \dots\}$ and a choice of T is as follows:

$$T = \underbrace{(0)}_{B_1=1} \ \underbrace{(1\ 2)(3\ 4)}_{B_2=2} \ \underbrace{(5\ 6\ 7)(8\ 9\ 10)}_{B_3=2} \ \underbrace{(11\ 12\ 13\ 14)(15\ 16\ 17\ 18)(19\ 20\ 21\ 22)}_{B_4=3} \dots .$$

The points of order 4 are

$$\{11, 12, \dots, 22,\ 1, 2, 3, 4,\ 0\},$$

hence $f_4(T) = 4 \cdot 3 + 2 \cdot 2 + 1 = 17 = A_4$.

Assume now that for some $k \geqslant 1$ we have

$$\frac{1}{n^k} \sum_{d|n} \mu\Big(\frac{n}{d}\Big) A_d \in \mathbb{N} \qquad \text{for all } n = 1, 2, \dots .$$

This means that the number of orbits of length n is divisible by n^{k-1}. The sequence $\{A_n\}_{n=1,2,\dots}$ is said to be *k-realizable* if

$$B_n = \frac{1}{n^k} \sum_{d|n} \mu\Big(\frac{n}{d}\Big) A_d \in \mathbb{Z} \qquad \text{for all } n = 1, 2, \dots \tag{9.2}$$

holds. Call a power series $y = \sum_{n=0}^{\infty} A_n z^n \in \mathbb{Q}[[z]]$ *k-realizable* if there exists an integer C such that $\{CA_n\}_{n=1}^{\infty}$ is k-realizable.

The most remarkable property of the examples of the Calabi–Yau differential equations is that the Yukawa coupling (2.4) is (expected to be) 3-realizable; the pseudo-coupling (3.4) constructed in Section 3 seems to be 2-realizable (where in cases #15–29 we have to replace $\mathbb{Z}$ in (9.2) by $\mathbb{Z}[1/p]$ for a suitable choice of the prime p). Another striking property of the first 14 (hypergeometric) cases from Table A in [**3**] is that $y_0(z)$ is also 3-realizable.

The property of being k-realizable is related to Kummer supercongruences by the following result.

PROPOSITION 11. *The following are equivalent:*

(i) $\{A_n\}$ *is k-realizable;*

(ii) $A_{mp^r} \equiv A_{mp^{r-1}} \pmod{p^{rk}}$ *for p prime and* $(m, p) = 1$*;*

(iii) $\sum_{n=1}^{\infty} A_n z^n / n^k = \sum_{n=1}^{\infty} B_n \operatorname{Li}_k(z^n)$, *where B_n are the integers defined in* (9.2) *and*

$$\operatorname{Li}_k(x) = \sum_{n=1}^{\infty} \frac{x^n}{n^k} \tag{9.3}$$

denotes the k-th polylogarithm.

PROOF. The proofs on pp. 3–6 in [**30**] for $k = 1$ are easily generalized; note that $\operatorname{Li}_1(x) = -\log(1 - x)$. □

REMARK. For the first n numbers $A_1, A_2, \dots, A_n$, the case $k = 1$ is well described in Exercise 5.2 of [**34**]. Then there is also an $(n \times n)$-matrix M with integer entries such that $A_k = \operatorname{Tr}(M^k)$ for $k = 1, 2, \dots, n$ (see [**1**]).

Superseeker of Calabi–Yau differential equations

	N_0	$\lvert N_1\rvert$	$\lvert N_3\rvert$	Hadamard product	# in [**3**], Table A
1	1	2	8		184
2	1	2	104		41
3	3	3	28		28
4	1	4	44		253
5	3	4	44		23
6	1	4	84		84
7	5	4	108		246
8	1	4	284		206
9	1	4	3252	(p)$*$(p)	3*
10	1	5	100	(a)$*$(a)	100
11	1	6	104	(k)$*$(k)	4*
12	1	6	170		245
13	1	6	325	(a)$*$(f)	160
14	1	9	748	(f)$*$(f)	165
15	17	10	170		279
16	7	10	295		22
17	7	10	508		235
18	1	10	664	(c)$*$(c)	103
19	1	10	870		60
20	6	12	140		130
21	2	12	208		46
22	10	12	236		17
23	3	12	644		16
24	1	12	3204	(A)$*$(a)	45
25	1	13	2650	(b)$*$(b)	101
26	2	16	208	(a)$*$(d)	105
27	10	16	304		21
28	1	16	1232	(A)$*$(d)	36
29	1	16	1744	(d)$*$(d)	107
30	2	16	2000		42
31	2	16	2106	(a)$*$(b)	102
32	1	16	3280		56
33	6	18	490		20
34	9	18	3820		199
35	7	18	5676		234
36	20	20	100		205
37	4	20	1680		281
38	5	20	1820		18

	N_0	$\lvert N_1\rvert$	$\lvert N_3\rvert$	Hadamard product	# in [**3**], Table A
39	2	20	2036		244
40	1	20	5924	(a)∗(e)	114, 150
41	1	20	8220		25
42	1	21	15894		15
43	3	24	1552		188
44	3	27	14201		216
45	1	27	18089	(B)∗(ι)	70
46	1	28	1036	(b)∗(e)	121
47	5	28	1268		262
48	6	28	1820		27
49	3	28	3892		119
50	1	32	1440	(l)∗(l), (A)∗(β)	10**, 40
51	1	32	7584		201
52	1	32	26016	(C)∗(d), (A)∗(θ)	3, 30, 31, 72
53	3	32	38880	(A)∗(d)	111
54	3	33	3600	(b)∗(c)	113
55	7	35	2184		28
56	9	36	556		183
57	10	36	1286		266
58	3	36	3020	(d)∗(f)	163
59	3	36	3284	(A)∗(f)	133
60	6	36	3648		185
61	1	36	8076	(B)∗(e)	110
62	1	36	41421	(B)∗(b)	24
63	1	36	128217204	(B)∗(j)	
64	1	37	15270	(g)∗(g)	144
65	13	39	1621		197
66	6	42	2542	(a)∗(c)	104
67	44	44	308		182
68	5	44	2480		249
69	1	44	22500	(C)∗(g)	139
70	4	48	1424	(b)∗(d)	106
71	1	48	2864	(B)∗(β), (A)∗(h)	8**, 49, 141
72	3	48	11056	(A)∗(c)	58
73	2	48	11664	(B)∗(d)	48
74	1	48	32368	(d)∗(e), (A)∗(ε)	122
75	1	48	73328	(C)∗(d)	38
76	19	49	1761		186
77	21	51	5095		217
78	20	52	1356		203

	N_0	$\|N_1\|$	$\|N_3\|$	Hadamard product	# in [3], Table A
79	3	52	52284		117
80	1	52	220220		68
81	1	54	40552		50
82	3	54	64744		223
83	1	55	11655		118
84	24	56	3552		248
85	5	59	22503		224
86	3	60	1684	(A)∗(g)	137
87	3	60	28820		255
88	1	60	134292		5, 90, 91, 93, 157
89	1	60	307860	(B)∗(i)	
90	1	63	96866	(f)∗(h)	172
91	1	64	23360		116
92	1	64	246848	(C)∗(c)	69
93	2	66	59386	(B)∗(δ)	151
94	4	68	95246	(c)∗(g)	175
95	24	72	3496		286
96	12	72	3900	(a)∗(f)	160
97	4	72	20708	(B)∗(f)	134
98	5	76	10500		270
99	5	76	24836		261
100	3	76	144196		55
101	1	76	415420	(a)∗(i), (C)∗(δ)	152
102	28	80	2912		212
103	23	80	4655		19
104	1	80	104976		233
105	1	80	174096		83
106	1	80	249872		236
107	11	84	9052		198
108	6	84	20848		29
109	2	84	83412		243
110	3	84	113304		124^{∞}
111	1	90	151648		73
112	1	92	585396	(C)∗(b)	51
113	1	96	12064	(C)∗(β)	7^*, 43
114	3	96	26208	(c)∗(e), (A)∗(α)	39, 120
115	1	100	126580	(b)∗(i), (C)∗(η)	
116	6	108	910		242
117	12	108	968	(c)∗(f)	162
118	6	108	3136		259

	N_0	$\|N_1\|$	$\|N_3\|$	Hadamard product	# in [**3**], Table A
119	12	108	4916	(b)∗(f)	161
120	4	108	10472	(B)∗(g)	138
121	12	108	12580		251
122	6	108	19598	(f)∗(g)	178
123	3	108	62596	(e)∗(f), (A)∗(ζ)	164
124	2	108	81104	(c)∗(h)	169
125	3	108	206716	(C)∗(f)	135
126	1	108	49457556	(D)∗(g)	140
127	1	112	378800		71
128	19	113	8515		202
129	5	116	186172		275
130	1	117	713814		4
131	1	117	844872		280
132	17	126	11700		194
133	1	128	382592		220
134	1	128	800384		256
135	1	128	5282176	(C)∗(α)	37
136	7	129	41441		193
137	12	132	9736	(a)∗(g)	173
138	4	132	52204		32
139	4	132	118772	(a)∗(h)	167
140	7	138	42984		218
141	14	140	24136		26
142	12	144	7312	(c)∗(d)	123
143	6	144	30896	(d)∗(g)	176
144	15	147	6032		274
145	1	160	9310		19
146	2	160	539680	(e)∗(e), (c)∗(i), (C)∗(θ)	$\widehat{3}$, 6^*, 115
147	1	160	1956896		6, 75, 96, 146
148	1	160	5870688		76
149	61	163	4795		124
150	4	180	28320	(b)∗(h)	168
151	4	180	110940	(B)∗(h)	142
152	1	180	21847076	(D)∗(f)	136
153	15	186	20300		226
154	5	188	450516		260
155	47	189	9277		196
156	12	192	156	(b)∗(g)	174
157	1	196	2993772		33
158	3	204	18628		228

	N_0	$\lvert N_1\rvert$	$\lvert N_3\rvert$	Hadamard product	# in [3], Table A
159	1	208	1218192		237
160	1	208	1863312	(d)∗(i), (C)∗(ε)	
161	3	220	267636		215
162	3	228	278988	(e)∗(g), (A)∗(γ)	44, 177
163	13	231	38037		240
164	5	232	122168		252
165	9	234	103520		214
166	34	236	22848		213
167	56	240	6944		59
168	8	240	117056		74
169	1	240	19105840	(D)∗(d)	65
170	38	241	17458		284
171	1	243	513936		278
172	20	252	56064		273
173	4	252	387464	(B)∗(ζ)	
174	1	252	1162036		154
175	3	252	1522388		35
176	2	270	835370	(g)∗(h)	53, 179
177	7	274	281388		208
178	20	276	116324		222
179	29	285	40626		195
180	1	288	2339616	(e)∗(h)	$\widehat{5}$, 98, 171
181	1	288	96055968	(D)∗(e), (A)∗(κ)	112
182	1	291	7935104		230
183	23	308	70799		250
184	1	320	19748928		241
185	2	324	1502052		17^{∞}
186	1	324	10792428		11, 95
187	4	336	595280	(d)∗(h), (B)∗(ε)	170
188	5	364	6324580		282
189	1	372	71562236	(D)∗(a)	62
190	5	379	1364199		232
191	1	384	164736	(m)∗(m)	13^{*}, 47
192	3	384	1546624	(c)∗(i), (C)∗(α)	37
193	14	420	159040		189
194	4	428	485244		187
195	1	432	78259376	(D)∗(c)	64
196	5	444	1501908		210
197	1	444	19050964		238
198	3	460	894404		231

	N_0	$\lvert N_1\rvert$	$\lvert N_3\rvert$	Hadamard product	# in [**3**], Table A
199	3	468	3687996	(f)$*$(i), (C)$*$(ζ)	
200	5	468	59427420		272
201	17	478	285760		209
202	1	480	4215904		258
203	1	480	16034720	(D)$*$(β)	9^*, 9^{**}, 67
204	5	492	872164		221
205	1	492	136094428	(a)$*$(j), (D)$*$(δ)	153
206	15	498	360988		219
207	1	522	9879192	(h)$*$(h)	$\widehat{4}$, 145
208	3	564	2422620		283
209	1	575	63441275		1, 79, 87, 128
210	1	608	22293216		247
211	1	612	51318900	(b)$*$(j), (D)$*$(η)	
212	1	624	43406256		180
213	13	647	942613		99
214	72	684	398428		200
215	1	684	195638820	(D)$*$(b)	63
216	1	736	26911072	(e)$*$(i)	$\widehat{6}$, 77, 78, 97
217	5	828	4270932		268
218	1	864	147560800	(c)$*$(j), (D)$*$(α)	66
219	1	900	8364884		227
220	1	928	170809536		10, 54
221	1	992	63721056		257
222	1	1008	607849200	(D)$*$(h), (B)$*$(κ)	143
223	3	1020	15174100	(g)$*$(i), (C)$*$(γ)	52
224	3	1056	15001120		229
225	1	1056	138459552		265
226	1	1116	349462868	(f)$*$(j), (D)$*$(ζ)	
227	1	1248	683015008		14, 85, 86, 156
228	1	1312	58156708		263
229	1	1344	109320512	(h)$*$(i)	$\widehat{11}$, 94
230	7	1434	18676572		109
231	1	1488	517984144	(d)$*$(j), (D)$*$(ε)	
232	1	1584	171534960		239
233	1	1818	467810538		267
234	4	2300	253765100		148
235	1	2400	2956977632		211
236	1	2450	623291900		$\widehat{1}$, 131
237	1	2484	1327731388	(g)$*$(j), (D)$*$(γ)	149
238	1	2592	81451104	(C)$*$(j)	

	N_0	$\lvert N_1 \rvert$	$\lvert N_3 \rvert$	Hadamard product	# in [**3**], Table A
239	1	2628	3966805740		8, 92, 125, 158
240	44	3180	6378752		285
241	1	3488	1142687008	(i)∗(i)	$\widehat{10}$, 155
242	1	3936	10892932064	(D)∗(i)	
243	1	4192	2124587232		277
244	2	4900	1246583800		80, 81
245	1	5408	22147077792		254
246	1	5472	6444589536	(e)∗(j)	$\widehat{14}$, 88, 89
247	1	7776	66942277344		12
248	1	10080	24400330080	(h)∗(j)	$\widehat{8}$, 82, 126, 127, 129
249	1	10912	71557619232		271
250	1	14752	711860273440		7, 147
251	1	26400	230398034080	(i)∗(j)	$\widehat{12}$
252	1	37216	464865119712		264
253	1	41184	5124430612320	(D)∗(j), (D)∗(κ)	61
254	1	57760	3869123234080		$\widehat{7}$
255	1	67104	28583248229280		13, 57, 108
256	1	70944	3707752060576		207
257	1	93984	25265152551072		225
258	1	188832	101990911789344		276
259	1	201888	40177844666400	(j)∗(j)	$\widehat{13}$, 166
260	1	231200	1700894366474400		2, 159
261	1	549216	5134247872650720		269
262	1	678816	69080128815414048		9
263	1	791200	4288711075194400		$\widehat{2}$
264	1	2710944	302270555492914464		$\widehat{9}$

COMMENTS. We list $|N_1|$ and $|N_3|$. The reason for not using N_2 is that it is not invariant under the transformation $q \mapsto -q$. There are many differential equations in the table that are just transformations (by Proposition 8)

$$y_0(z) \mapsto \frac{1}{1-pz} y_0\left(\left(\frac{z}{1-pz}\right)^r\right)$$

which transforms the Yukawa coupling $K(q) \mapsto K(q^r)$ These differential equations should not be in the table but we did not know this transformation when we found them. In the Superseeker table we identify $K(q)$ and $K(q^r)$ as the instanton numbers are just thinned out.

References

[1] G. Almkvist, *Integrity of ghosts*, Preprint, Lund, 1983.

[2] G. Almkvist, *Strängar i månsken I*, Normat **51** (2003), no. 1, 22–33; *II*, Normat **51** (2003), no. 2, 63–79.

[3] G. Almkvist, C. van Enckevort, D. van Straten and W. Zudilin, *Tables of Calabi–Yau equations*, in preparation (2005).

[4] G. E. Andrews, R. Askey and R. Roy, Special functions, Encyclopedia of Mathematics and its Applications **71**, Cambridge Univ. Press, Cambridge, 1999. MR 1688958

[5] V. V. Batyrev and D. van Straten, *Generalized hypergeometric functions and rational curves on Calabi–Yau complete intersections in toric varieties*, Commun. Math. Phys. **168** (1995), 493–533; `alg-geom/9310001` (1993). MR 1328251

[6] V. V. Batyrev, I. Ciocan-Fontanine, B. Kim, and D. van Straten, *Conifold transitions and mirror symmetry for Calabi–Yau complete intersections in Grassmannians*, Nuclear Phys. B **514** (1998), no. 3, 640–666; `alg-geom/9710022` (1997). MR 1619529

[7] F. Beukers, *Irrationality proofs using modular forms*, Journées arithmétiques (Besançon, 1985), Astérisque **147–148** (1987), 271–283. MR 0891433

[8] F. Beukers, *Consequences of Apéry's work on* $\zeta(3)$, Preprint, Utrecht, 2003.

[9] F. Beukers, *On Dwork's accessory parameter problem*, Math. Z. **241** (2002), no. 2, 425–444. MR 1935494

[10] F. Beukers, *Some Galois theory associated to Zudilin's recurrences*, Tagunsbericht 12/2003, Elementare und Analytische Zahlentheorie (09.03.—15.03.2003), Mathematisches Forschungsinstitut Oberwolfach, Germany, 2003, p. 3.

[11] F. Beukers, *Personal communication* (March 2004).

[12] F. Beukers and C. A. M. Peters, *A family of K3 surfaces and* $\zeta(3)$, J. Reine Angew. Math. **351** (1984), 42–54. MR 0749676

[13] H. Cohen, *Accélération de la convergence de certaines récurrences linéaires*, Séminaire de Théorie des nombres de Bordeaux (Année 1980–81), exposé 16, 2 pp.

[14] C. van Enckevort, *Calabi–Yau equations*, A database, `http://enriques.mathematik.uni-mainz.de/enckevort/db/` (2004).

[15] C. van Enckevort and D. van Straten, *Monodromy calculations for fourth order equations of Calabi–Yau type*, Calabi–Yau Varieties and Mirror Symmetry, The BIRS workshop volume (06–11 December 2003) (J. Lewis, S.-T. Yau and N. Yui, eds.), Amer. Math. Soc. & International Press, Providence, RI, 2005; `math.AG/0412539` (2004).

[16] J. Guillera, *About a new kind of Ramanujan-type series*, Experiment. Math. **12** (2003), no. 4, 507–510. MR 2044000

[17] J. Harnad and J. McKay, *Modular solutions to equations of generalized Halphen type*, Royal Soc. Lond. Proc. Ser. A Math. Phys. Eng. Sci. **456** (2000), no. 1994, 261–294; `solv-int/9804006` (1998). MR 1811320

[18] E. L. Ince, Ordinary differential equations, Longman, Green & Co., London, 1927. MR 0010757

[19] A. Klemm and S. Theisen, *Considerations of one-modulus Calabi–Yau compactifications: Picard–Fuchs equations, Kähler potentials and mirror maps*, Nuclear Phys. B **389** (1993), no. 1, 153–180; `hep-th/9205041` (1992). MR 1202211

[20] A. Klemm and S. Theisen, *Mirror maps and instanton sums for complete intersections in weighted projective space*, Modern Phys. Lett. A **9** (1994), no. 20, 1807–1817; `hep-th/9304034` (1993). MR 1287096

[21] C. Krattenthaler and T. Rivoal, *Hypergéométrie et fonction zêta de Riemann*, `math.NT/0311114v3` (2004) (French).

[22] B. H. Lian, K. Liu and S.-T. Yau, *Mirror principle I*, Asian J. Math. **1** (1997), no. 4, 729–763; *II*, Asian J. Math. **3** (1999), no. 1, 109–146. MR 1621573, 1701925

[23] B. H. Lian and S.-T. Yau, *The nth root of the mirror map*, Calabi–Yau varieties and mirror symmetry (Toronto, ON, 2001), Fields Inst. Commun. **38**, Amer. Math. Soc., Providence, RI, 2003, pp. 195–199. MR 2019153

[24] A. Libgober and J. Teitelbaum, *Lines in Calabi–Yau complete intersections, mirror symmetry, and Picard–Fuchs equations*, Intern. Math. Research Notices **1** (1993), no. 1, 29–39; `alg-geom/9301001` (1993). MR 1201748

[25] R. S. Maier, *Algebraic hypergeometric transformations of modular origin*, `math.NT/0501425` (2005).

[26] D. R. Morrison, *Picard–Fuchs equations and mirror maps for hypersurfaces*, Essays on mirror manifolds, Hong Kong, International Press, 1992, pp. 241–264; hep-th/9111025 (1991). MR 1191426

[27] M. A. Perlstadt, *Some recurrences for sums of powers of binomial coefficients*, J. Number Theory **27** (1987), 304–309. MR 0915502

[28] M. Petkovšek, H. S. Wilf and D. Zeilberger, $A = B$, A. K. Peters, Ltd., Wellesley, MA, 1996. MR 1379802

[29] Y. Puri and T. Ward, *Arithmetic and growth of periodic orbits*, J. Integer Seq. **4** (2001), no. 2, #01.2.1, 18 pp. MR 1873399

[30] J. A. de Reyna, *Dynamical zeta functions and Kummer congruences*, math.NT/0309190 (2003).

[31] E. A. Rødland, *The Pfaffian Calabi-Yau, its mirror, and their link to the Grassmannian* $G(2,7)$, Compositio Math. **122** (2000), no. 2, 135–149; math.AG/9801092 (1998). MR 1775415

[32] A. L. Schmidt and Y. Jin, *On recurrences for sums of powers of binomial coefficients*, Preprint, Copenhagen, 1996.

[33] V. N. Sorokin, *One algorithm for fast calculation of* π^4, Preprint, Russian Academy of Sciences, M. V. Keldysh Institute for Applied Mathematics, Moscow, 2002, 59 pp. (Russian); http://www.wis.kuleuven.ac.be/applied/intas/Art5.pdf.

[34] R. Stanley, Enumerative combinatorics, vol. 2, Cambridge Studies in Advanced Mathematics **62**, Cambridge Univ. Press, Cambridge, 1999. MR 1676282

[35] D. van Straten, *Personal communications* (October 2003–April 2004).

[36] H. A. Verrill, *Sums of squares of binomial coefficients, with applications to Picard–Fuchs equations*, math.CO/0407327 (2004).

[37] H. S. Wilf and D. Zeilberger, *An algorithmic proof theory for hypergeometric (ordinary and "q") multisum/integral identities*, Invent. Math. **108** (1992), no. 3, 575–633. MR 1163239

[38] D. Zagier, *Integral solutions of Apéry-like recurrence equations*, Manuscript (2003).

[39] W. Zudilin, *An Apéry-like difference equation for Catalan's constant*, Electron. J. Combin. **10** (2003), no. 1, #R14, 10 pp.; math.NT/0201024 (2002). MR 1975764

[40] W. Zudilin, *A third-order Apéry-like recursion for* $\zeta(5)$, Mat. Zametki **72** (2002), no. 5, 796–800; English transl., Math. Notes **72** (2002), 733–737; math.NT/0206178 (2002). MR 1963141

[41] W. Zudilin, *Well-poised hypergeometric transformations of Euler-type multiple integrals*, J. London Math. Soc. **70** (2004), no. 1, 215–230. MR 2064759

[42] W. Zudilin, *Binomial sums related to rational approximations to* $\zeta(4)$, Mat. Zametki **75** (2004), no. 4, 637–640; English transl., Math. Notes **75** (2004), 594–597; math.CA/0311196 (2003). MR 2068773

[43] W. Zudilin, *Approximations to -, di- and tri- logarithms*, math.NT/0409023 (2004).

Matematikcentrum, Lunds Universitet, Matematik MNF, Box 118, SE-221 00 Lund, Sweden

E-mail address: gert@maths.lth.se

Department of Mechanics and Mathematics, Moscow Lomonosov State University, Vorobiovy Gory, GSP-2, 119992 Moscow, Russia

E-mail address: wadim@ips.ras.ru

AMS/IP Studies in Advanced Mathematics
Volume 38, 2006

Mirror Symmetry and Integral Variations of Hodge Structure Underlying One-parameter Families of Calabi-Yau Threefolds

Charles F. Doran and John W. Morgan

ABSTRACT. This proceedings note introduces aspects of the authors' work relating mirror symmetry and integral variations of Hodge structure. The emphasis is on their classification of the integral variations of Hodge structure which can underlie families of Calabi-Yau threefolds over $\mathbb{P}^1 \setminus \{0, 1, \infty\}$ with $b^3 = 4$, or equivalently $h^{2,1} = 1$, and the related issues of geometric realization of these variations. The presentation parallels that of the first author's talk at the BIRS workshop.

1. Integral structures in mirror symmetry

1.1. The first examples. Since its introduction to the mathematical community through the seminal papers of Greene-Plesser [**GP**] and Candelas-de la Ossa-Green-Parkes [**CdOGP**], Mirror Symmetry has been the source of the most persistently rich and subtle novel mathematics yet to emerge from the study of string dualities. One of its key features is its resistance to rigorous mathematical definition and even more to being described within any single traditional mathematical setting. Research in mathematics related to mirror symmetry is thus driven by the goal of discerning deeper, more complete, and purely mathematical avatars of this very physical duality.

For example, one of the earliest mirror symmetric mathematical predictions was that a "mirror pair" of Calabi-Yau threefolds X and $\widetilde{X}$ should have the property that their Hodge numbers "mirror" one another, i.e., that $h^{1,1}(X) = h^{2,1}(\widetilde{X})$ and $h^{2,1}(X) = h^{1,1}(\widetilde{X})$.

The first proposal for such a pair was introduced in [**GP**] where it was applied to Calabi-Yau threefolds presented as hypersurfaces in a weighted projective space. The construction involves the operation of quotienting by a finite group (or "orbifolding").

2000 *Mathematics Subject Classification.* Primary 14D07, 14J32; Secondary 14M25, 19L64, 14D05.

Key words and phrases. Variation of Hodge structures, Calabi–Yau manifolds, mirror symmetry, toric varieties, Newton polyhedra, computations, geometric applications.

The simplest of Calabi-Yau threefold hypersurfaces, the generic quintic hypersurface in $\mathbb{P}^4$, often denoted by $\mathbb{P}^4[5]$, has $h^{1,1} = 1$ (coming from the polarization class) and $h^{2,1} = 101$ (corresponding to 101 complex structure moduli). This last follows from Kodaira-Spencer theory [**Kod**] and the theorem of Bogomolov-Tian-Todorov [**Tian**] that any Calabi-Yau manifold admits a locally universal deformation over a smooth base. In this case all of the complex structure deformations arise from varying the coefficients of the defining equations for quintic hypersurfaces in $\mathbb{P}^4$. Let G denote the finite group $(\mathbb{Z}/5\mathbb{Z})^3$ presented as

$$\left\{ (a_1, a_2, a_3, a_4, a_5) \in (\mathbb{Z}/5\mathbb{Z})^5 \;\middle|\; \sum_{i=1}^{5} a_i = 0 \right\} / (\mathbb{Z}/5\mathbb{Z}) \,,$$

where the action of $\mathbb{Z}/5\mathbb{Z}$ is the diagonal one. Upon quotienting the projective space $\mathbb{P}^4$ by G, acting as

$$(x_1 : x_2 : x_3 : x_4 : x_5) \mapsto (\mu^{a_1} x_1 : \mu^{a_2} x_2 : \mu^{a_3} x_3 : \mu^{a_4} x_4 : \mu^{a_5} x_5)$$

(where $\mu = \exp 2\pi\imath/5$), resolving the resulting orbifold singularities, and keeping track of the hypersurface throughout the process, another family of Calabi-Yau threefolds is constructed — no longer as hypersurfaces in projective space, but now with the property that generically $h^{1,1} = 101$ and $h^{2,1} = 1$. It follows that these threefolds sit in a one-parameter family of complex structures. The specific presentation of the mirror family $\widetilde{X}_z$ in $\mathbb{P}^4/G$ is given by

$$x_1^5 + x_2^5 + x_3^5 + x_4^5 + x_5^5 - 5z^{-1/5} x_1 x_2 x_3 x_4 x_5 = 0 \,.$$

Here the complex deformation parameter is identified with a coordinate on the base of the family $\mathbb{P}^1 \setminus \{0, 1, \infty\}$.

REMARK 1.1. In [**GP**, Table 1] this construction is extended to include intermediate quotients by subgroups of G. In particular, instead of the quintic itself one can consider the quotient of the quintic by the free $\mathbb{Z}/5\mathbb{Z}$-action given by

$$(x_1 : x_2 : x_3 : x_4 : x_5) \mapsto (x_1 : \mu x_2 : \mu^2 x_3 : \mu^3 x_4 : \mu^4 x_5) \,,$$

resulting in a family of Calabi-Yau threefolds with $h^{1,1} = 1$ and $h^{2,1} = 21$. These "quintic twin" Calabi-Yau threefolds have fundamental group $\mathbb{Z}/5\mathbb{Z}$. The mirror family is constructed in [**GP**] as another intermediate quotient, and consists (as expected) of simply connected Calabi-Yau threefolds with $h^{1,1} = 21$ and $h^{2,1} = 1$. Moreover the parameter space of this quintic twin mirror family $\widetilde{X}_z^{twin}$ is again $\mathbb{P}^1 \setminus \{0, 1, \infty\}$ as for the original quintic mirror family.

Since the mirror quintic family $\widetilde{\mathfrak{X}} = \{\widetilde{X}_z\}$ depends on just one parameter, an old method of Griffiths-Dwork [**CoKa**, §5.3] can be readily applied to determine the ordinary differential equation satisfied by the periods of the (unique up to complex scaling) holomorphic three-form on the Calabi-Yau threefold. More formally, consider the local system denoted $R^3\varphi_*\mathbb{C}$ arising from the third cohomology (with complex coefficients) of the fibers $\widetilde{X}_z$ of the algebraic family $\varphi\colon \widetilde{\mathfrak{X}} \to B := \mathbb{P}^1 \setminus \{0, 1, \infty\}$. Integration over 3-cycles defines a pairing, making this local system dual to the local system whose fiber over z is $H_3(\widetilde{X}_z, \mathbb{C})$. Every class $\gamma \in H_3(\widetilde{X}_z, \mathbb{C})$ extends uniquely as a multi-valued flat section of this latter local system. Consider the holomorphic bundle $R^3\varphi_*\mathbb{C} \otimes \mathcal{O}_B$; the local system $R^3\varphi_*\mathbb{C}$ determines a flat structure on this vector bundle and hence a flat connection, ∇_z. The sub-line bundle $\mathcal{H}^{3,0}$ of classes of relative holomorphic 3-forms is of special interest. A holomorphic section $\varpi(z)$

of $\mathcal{H}^{3,0}$ amounts to a holomorphic family of holomorphic 3-forms. Generically $\varpi(z)$ and its first three covariant derivatives, $\nabla_z^i(\varpi)$ $(1 \leq i \leq 3)$ are linearly independent, and $\nabla_z^4(\varpi)$ is expressible as a linear combination of these with coefficients meromorphic functions on the base B. This relationship is the differential equation associated to the family of cohomology classes $[\varpi(z)]$ — the *Picard-Fuchs equation.* There is also the *Picard-Fuchs ODE* obtained by replacing Θ by $z\frac{d}{dz}$. An integral lattice in the space of all solutions of this ODE is given by the periods $\int_\gamma \varpi(z)$ as γ varies over all (integral) three-cycles. There is a holomorphic map of the dual of the vector bundle of solutions of the Picard-Fuchs ODE to $R^3\varphi_*\mathbb{C}\otimes\mathcal{O}_B$. This map is an isomorphism near any regular singular point of the ODE.

For the quintic mirror family $\widetilde{\mathfrak{X}}$, letting $\Theta = z\nabla_z$, for an appropriate holomorphic family $\varpi(z)$ of holomorphic 3-forms the Picard-Fuchs equation is the generalized hypergeometric differential equation

$$\left[\Theta^4 - z\left(\Theta+\frac{1}{5}\right)\left(\Theta+\frac{2}{5}\right)\left(\Theta+\frac{3}{5}\right)\left(\Theta+\frac{4}{5}\right)\right]\varpi(z) = 0\ .$$

The Picard-Fuchs ODE has 3 singular points $0, 1, \infty$, all of which are regular singular points. Let γ_0 be an indivisible integral cycle invariant under monodromy about $z = 0$. A holomorphic solution to the Picard-Fuchs ODE is given by the hypergeometric-type series

$$\int_{\gamma_0}\varpi(z) = \sum_{n\geq 0}\frac{(5n)!}{(n!)^5}\left(\frac{z}{5^5}\right)^n$$

defined on the punctured unit disk. This function extends across the origin, reflecting the fact that it is integration of a family of holomorphic 3-forms over a flat family of 3-cycles invariant under monodromy around $z = 0$. This power series is absolutely convergent inside the unit disk. Of course, it also admits a multi-valued analytic continuation to the whole thrice-punctured sphere given explicitly in terms of Meijer functions (Mellin-Barnes type integrals) on $\mathbb{P}^1 \setminus \{0, 1, \infty\}$ [**HTF, GL**].

REMARK 1.2. The story for the quintic twin mirror family is similar, with everything working just as above. In fact, for appropriate choices of the family of holomorphic three-forms, the Picard-Fuchs ODE defined as above and satisfied by the periods of the two families of holomorphic forms on $\widetilde{X}_z$ and $\widetilde{X}_z^{twin}$ are exactly the same. Thus, the difference between the two families is not reflected in the Picard-Fuchs differential equations. It turns out that it is reflected in the lattices spanned by the periods of the holomorphic three-form over a basis of integral cycles. Of course, the difference is also seen from the fact that the ranks of the even-dimensional cohomologies of the two families are different.

Mirror symmetry predicts that the periods over a suitable integral basis of three-cycles on the mirror quintic contain extremely subtle information about the geometry of the original quintic hypersurface in $\mathbb{P}^4$. In particular, mirror symmetry predicts the number of rational curves of a given degree lying on the hypersurface [**CdOGP, CoKa**]. These curve counts can be read off from an appropriate generating function built from the "mirror map" q-series — the single-valued local inverse to the projectivized period mapping about a regular singular point of maximal unipotent monodromy (or, in terms of the geometry of the family of Calabi-Yau threefolds, of maximal degeneracy). In many cases these curve counts, and their descriptions in terms of hypergeometric functions, have now been rigorously established [**LLY1, LLY2, Giv**]. We will not be pursuing this approach here; our focus is the integral lattice structure and monodromy rather than enumerative geometry.

1.2. Classical Hodge theory. In order to study more general families of Calabi-Yau threefolds than just the mirror quintic or mirror quintic twin hypersurfaces we introduce here Hodge structures and their variations. Consider a Calabi-Yau threefold $\widetilde{X}$ with $h^{2,1} = 1$. The cohomology $H^3(\widetilde{X}, \mathbb{R})$ is polarized by the intersection form, denoted $\langle \cdot, \cdot \rangle$, which is unimodular and skew, and has a Hodge decomposition

$$H^3(\widetilde{X}, \mathbb{R}) \otimes_{\mathbb{R}} \mathbb{C} = H^3(\widetilde{X}, \mathbb{C}) = H^{3,0} \oplus H^{2,1} \oplus H^{1,2} \oplus H^{0,3}$$

with associated Hodge filtration

$$F^3 = H^{3,0} \;, \; F^2 = H^{3,0} \oplus H^{2,1} \;, \; F^1 = H^{3,0} \oplus H^{2,1} \oplus H^{1,2} \; .$$

Furthermore, for all i, $0 \leq i \leq 3$ these filtrants satisfy the conditions

$$(F^i)^{\perp} = F^{4-i} \tag{1.1}$$

and

$$F^i \oplus \overline{F^{4-i}} = H^3(\widetilde{X}, \mathbb{C}) \; . \tag{1.2}$$

Notice that we recover the Hodge decomposition from the filtration since $F^i \cap \overline{F^{3-i}} = H^{i,3-i}$.

For us a Calabi-Yau threefold has holonomy all of $SU(3)$ and hence $H^1(\widetilde{X}, \mathbb{C}) = 0$, implying that all of H^3 is primitive. Thus, the Hermitian form

$$h(\omega_1, \omega_2) = \imath \int_M \omega_1 \wedge \overline{\omega}_2 = \imath \langle \omega_1, \overline{\omega}_2 \rangle$$

is positive definite on $H^{3,0}$ and negative definite on $H^{2,1}$, i.e., for $\omega \neq 0$,

$$h(\omega, \omega) \;\text{ is }\; \begin{cases} > 0 & \text{when } \omega \in H^{3,0} \\ < 0 & \text{when } \omega \in H^{2,1} \; . \end{cases} \tag{1.3}$$

Fix a rank 4 real vector space V with a nondegenerate skew form. Weight-three Hodge structures with $h^{2,1} = 1$ on V (and by this we shall always mean primitive and polarized weight-three Hodge structures) are classified by the period domain

$$\mathcal{D} = \mathrm{Sp}(4, \mathbb{R})/\mathrm{U}(1) \times \mathrm{U}(1) \; .$$

$\mathcal{D}$ is embedded as a domain inside $\mathrm{Sp}(4, \mathbb{C})/P$, where P is the maximal parabolic subgroup. A holomorphic family of such structures on V parameterized by a complex curve U consists of a holomorphic filtration by subbundles

$$\mathcal{F}^3 \subset \mathcal{F}^2 \subset \mathcal{F}^1 \subset \mathcal{F}^0 = V \otimes_{\mathbb{R}} \mathcal{O}_U,$$

with Conditions (1.1), (1.2), and (1.3) holding on every fiber, resulting in a family of (polarized) Hodge structures. Given such a holomorphic family, the *period map* is the (holomorphic) classifying map $\Pi \colon U \to \mathcal{D}$.

There is an additional geometric condition on the period map $\Pi \colon U \to \mathcal{D}$ if the family of Hodge structures is geometric, i.e., arises from a family of smooth varieties. This is the *Griffiths transversality condition*, also called horizontality, [**Grif**, Condition (2.8) on page 143] which says that $(\mathcal{F}^i)' \subset \mathcal{F}^{i-1}$ (where the derivative is taken with respect to a local coordinate on U). Horizontal families of Hodge structures are called *variations of Hodge structure (VHS)*, or $\mathbb{Z}$-VHS, to emphasize the underlying flat integral structure. Horizontality can be reformulated in terms of a (non-integrable) distribution, called the *horizontal distribution*, on $\mathcal{D}$. Any period map $\Pi\colon U \to \mathcal{D}$ from a horizontal family is tangent to this distribution.

For the Hodge structures we are considering, there is the natural map

$$p_3\colon \mathcal{D} \to \mathbb{P}(V_{\mathbb{C}})$$

associating to each Hodge filtration the line F^3 in $V_{\mathbb{C}} = V \otimes_{\mathbb{R}} \mathbb{C}$. The horizontality condition for a curve $C \subset \mathcal{D}$ is equivalent to

$$T(p_3(C))_p \subset p^{\perp}/\langle p\rangle \subset (V_{\mathbb{C}})/\langle p\rangle = T\left(\mathbb{P}(V_{\mathbb{C}})\right)_p .$$

Conversely, given an analytic curve $C \subset \mathbb{P}(V_{\mathbb{C}})$ satisfying this condition, there is a unique lifting $\mathcal{C} \subset \mathcal{D}$ of C that is horizontal: Given $p \in C$, the Hodge filtration is given as follows: $F^3 = \langle p\rangle$, F^2 is the preimage in $V_{\mathbb{C}}$ of $T(C)_p \subset (V_{\mathbb{C}})/\langle p\rangle$, and $F^1 = (F^3)^{\perp}$. For a fuller explanation of this see [**BG**].

Now let $V = H^3(\widetilde{X}, \mathbb{R})$ and let U be a local simply connected open subset in the (one-dimensional) moduli space of complex structures on $\widetilde{X}$. Over U, the family of holomorphic 3-forms $\varpi(z)$ is a holomorphic section of $V_{\mathbb{C}}$. It satisfies a differential equation of the form

$$\nabla_z^4(\varpi(z)) + \sum_{i=0}^{3} P_{4-i}(z)\nabla^i(\varpi(z)) = 0 .$$

Using a local C^∞-trivialization of the family, we extend any 3-cycle $\gamma \in H_3(\widetilde{X}_{z_0}, \mathbb{C})$ to a family of 3-cycles in all neighboring fibers. The periods $\int_\gamma \varpi(z)$ form a (local) holomorphic function on the base. Since the family of cycles γ is a parallel section, the function

$$\varphi_\gamma(z) = \int_\gamma \varpi(z)$$

satisfies the Picard-Fuchs ODE (i.e., the ODE associated to the original Picard-Fuchs equation for ϖ):

$$\left(\frac{d}{dz}\right)^4 \varphi_\gamma + \sum_{i=1}^{4} P_{4-i}(z)\left(\frac{d}{dz}\right)^i \varphi_\gamma = 0.$$

The 4-dimensional space of periods as γ ranges over all of $H_3(\widetilde{X}_b; \mathbb{C})$ is the space of all solutions of the Picard-Fuchs ODE. This gives an identification (over a contractible neighborhood of $b \in B$) between the space of solutions to the Picard-Fuchs ODE and the vector bundle $(R^3\varphi_*\mathbb{C})^* \otimes \mathcal{O}_B$. Fixing a basis $\gamma_1^*, \ldots, \gamma_4^*$ of the lattice $H^3(\widetilde{X}, \mathbb{Z})/\{\text{Torsion}\}$, we have

$$\varpi(z) = \sum_{i=1}^{4} f^i(z) \cdot \gamma_i^* .$$

Of course the $f^i(z)$ are simply the periods

$$\int_{\gamma_i} \varpi(z)$$

where $\gamma_1, \ldots, \gamma_4$ is the basis of $H_3(\widetilde{X}_z, \mathbb{Z})/\{\text{Torsion}\}$ dual to $\gamma_i^*, \ldots, \gamma_4^*$. The periods $f^i(z)$ are a basis of solutions of the Picard-Fuchs ODE.

Let $\varphi\colon \widetilde{\mathfrak{X}} \to B$ be a family of complex structures on $\widetilde{X}$ with a not necessarily simply connected base. The fiberwise sublattice $R^3\varphi_*\mathbb{Z}$ produces a flat connection ∇, the Gauss-Manin connection, on $R^3\varphi_*\mathbb{C}\otimes\mathcal{O}_B$ which cannot (usually) be globally trivialized. A $\mathcal{C}^\infty$ trivialization of the family along paths in B, with basepoint b, defines a monodromy representation

$$\rho : \pi_1(B,b) \to \mathrm{Aut}\left(H^3(\widetilde{X}_b,\mathbb{Z}), \langle\cdot,\cdot\rangle\right) =: \Gamma\ .$$

Of course Γ acts naturally on $\mathcal{D}$, and the distribution associated to the transversality condition is Γ-invariant. The local period maps globalize to a horizontal holomorphic map

$$\Pi\colon B \to \mathcal{D}/\Gamma\ .$$

These maps are classifying Integral Variations of Hodge Structure ($\mathbb{Z}$-VHS), i.e., locally liftable horizontal variations of Hodge structure together with the integral local system, see [**Grif**, page 165]. The image of such maps is contained in the smooth locus of $\mathcal{D}/\Gamma$. These objects were abstracted and axiomatized by Deligne [**Del1**]. Deligne even established a (non-effective) finiteness theorem for $\mathbb{Z}$-VHS over a base with a fixed discriminant (e.g., over $\mathbb{P}^1\setminus\{0,1,\infty\}$) [**Del3**]. In particular Deligne's result implies that there are finitely many complete horizonal locally liftable integral curves isomorphic to $\mathbb{P}^1\setminus\{0,1,\infty\}$ in the smooth locus of the period domain. (It is elementary to see [**Grif**, Theorem (9.8) on page 167] that there are no such curves isomorphic to $\mathbb{P}^1\setminus A$ where A has cardinality at most two.) Given a holomorphic section $\varpi(z)$ of the line bundle $\mathcal{F}^3$ one still has the Picard-Fuchs equation satisfied by $\varpi(z)$ and the $\nabla^i\varpi(z), 1\le i\le 4$. There is a meromorphic map from the trivial bundle of solutions of the corresponding Picard-Fuchs ODE to the vector bundle given by the dual local system. This map is a holomorphic isomorphism at every regular point for the Picard-Fuchs ODE, and it identifies the monodromy of the Picard-Fuchs ODE with the monodromy of the Gauss-Manin connection on the dual local system.

1.3. Toric examples. The proposals for just what constitutes a mirror pair of Calabi-Yau threefolds have evolved significantly since [**GP**]. The most commonly used is the Batyrev-Borisov construction [**Bat1, Bat2, Bor, BB1, BB2**] in the toric setting, which reduces the description of mirror pairs of Calabi-Yau hypersurfaces and complete-intersections in Gorenstein Fano toric varieties to the classical "polar duality" of reflexive polytopes (with nef partitions in the complete intersection case). In this setting Batyrev-Borisov [**BB2**] establish the basic mirror prediction for Hodge numbers of the proposed mirror pairs, and as a consequence these pairs are widely believed to be examples of mirror pairs [**MSCMM**].

Let N be a lattice and M the dual lattice. As is well known [**Ful**], convex lattice polytopes $\Delta\subset N\otimes\mathbb{R}$ define toric varieties $\mathbb{P}_\Delta = \mathrm{Proj}(S_\Delta)$, where S_Δ is the polytope ring of monomials indexed by $n\in k\Delta\cap N$ and graded by total degree k, and where the torus $T_N = N\otimes_{\mathbb{Z}}\mathbb{C}^*$ acts. Recall that a lattice polytope $\Delta\subset N\otimes\mathbb{R}$ is *reflexive* if the *polar* polytope

$$\Delta^\circ = \{x\in M\otimes\mathbb{R} \big| \langle x,\Delta\rangle \ge -1\}$$

is a lattice polytope with respect to M. When the polytope Δ is reflexive [**CoKa**, §3.5], the toric variety is Fano, and the "divisor at infinity" D_∞, i.e., the union of the lower-dimensional T_N-orbits, is an anti-canonical divisor. Generic sections of $\mathcal{O}(D_\infty)$ are then Calabi-Yau threefolds. In general $\mathbb{P}_\Delta$ is sufficiently singular so that all sections of $\mathcal{O}(D_\infty)$ produce singular threefolds. To remedy this defect one takes a simplicial decomposition Σ of $\partial\Delta$ with vertices the set of lattice points. Then the fan consisting of the cones on the simplices of Σ determines a toric variety with isolated singularities[1], providing a crepant resolution of $\mathbb{P}_\Delta$. Because the singularities are isolated and the resolution is crepant, the generic section of the pullback of $\mathcal{O}(D_\infty)$ is a smooth Calabi-Yau 3-fold. The proposal of Batyrev [**Bat2**] is that the Calabi-Yau X obtained from a maximal triangulation of a reflexive polytope Δ is mirror to the Calabi-Yau $\widetilde{X}$ obtained from a maximal triangulation of the polar Δ°.

EXAMPLE 1.3. The Fano toric variety $\mathbb{P}_\Delta = \mathbb{P}^4$ is specified by the reflexive polytope Δ with vertices

$$\{(1,0,0,0),(0,1,0,0),(0,0,1,0),(0,0,0,1),(-1,-1,-1,-1)\}\ .$$

Here the "divisor at infinity" is a union of the five coordinate hyperplanes in $\mathbb{P}^4$. The quintic family $\mathbb{P}^4[5]$ is interpreted in this context as the family of hypersurfaces in this (ample) anticanonical divisor class. The ambient toric variety $\mathbb{P}_{\Delta^\circ}$ in which the quintic mirror family of hypersurfaces is constructed is given by the polar reflexive polytope Δ° with vertices

$$\{(4,-1,-1,-1),(-1,4,-1,-1),(-1,-1,4,-1),(-1,-1,-1,4),(-1,-1,-1,-1)\}.$$

This time, however, there are many more integral points in the polytope besides the vertices and origin and hence one must take a simplicial decomposition to obtain smooth hypersurfaces. A combinatorial formula of Batyrev allows one to determine the Hodge numbers of both the quintic and quintic mirror from the arrangement of these integral points on the facets of Δ and Δ°.

The quintic twin family also has such a toric hypersurface representation, this time with reflexive polytope the simplex with vertices

$$\{(1,0,0,0),(0,1,0,0),(0,0,1,0),(1,2,3,5),(-2,-3,-4,-5)\}$$

and its polar polytope is

$$\{(-1,-1,4,-2),(4,-1,-1,0),(-1,4,-1,-1),(-1,-1,-1,2),(-1,-1,-1,1)\}.$$

As before, the families of both the quintic twin and quintic twin mirror hypersurfaces are the families of generic sections of the anti-canonical line bundle over the toric varieties (appropriately resolved). Notice that the sublattice spanned by the integral points of the quintic twin lattice polytope has index 5 in the integral lattice.

What of the associated Picard-Fuchs equations for the family of holomorphic 3-forms? In this toric setting the machinery of Gel'fand-Graev-Zelevinsky [**GGZ**] and Gel'fand-Kapranov-Zelevinsky [**GKZ**] applies to construct, from the reflexive polytope corresponding to a given Calabi-Yau family, a differential equation, or more generally a system of such equations in the case of multiparameter families, satisfied by the periods of the mirror family (GKZ-system). This is a system of generalized hypergeometric equations. This system may possess more solutions than the genuine Picard-Fuchs system, but at least in the hypersurface case there are methods for describing the reduction to the actual Picard-Fuchs system [**HLY1, HLY2**].

[1]In higher dimension these varieties are smooth through codimension 3.

The situation is easier to describe for one-dimensional families coming from toric geometry. When the family $\widetilde{\mathfrak{X}}$ of Calabi-Yau threefolds that are complete intersections in a Gorenstein Fano toric variety have $h^{2,1}(\widetilde{X}_z) = 1$, then the corresponding Picard-Fuchs ODE is always a generalized hypergeometric ordinary differential equation [**CoKa**, §5.5]. These equations have base $\mathbb{P}^1$ and have exactly three regular singular points. The parameter space B of the variation can always be taken to be $\mathbb{P}^1 \setminus \{0, 1, \infty\}$. The local monodromy about the point at infinity in B (conventionally taken to be $z = 0$) corresponding to the toric divisor at infinity D_∞ is maximally unipotent. Let γ_0 be an invariant cycle for this monodromy. The period on γ_0 of a family of holomorphic 3-forms (defined over all of B) is always represented in the unit disk as a GKZ-generalized hypergeometric series. A second regular singular point (which we locate at $z = 1$) has local monodromy unipotent of rank one and is called a *conifold singularity* in the physics literature. Of course, by general principles [**BN**], the latter has quasi-unipotent monodromy.

For the quintic mirror and the quintic twin mirror this generalized hypergeometric series is the one given in Section 1.1.

1.4. Homological Mirror Symmetry. Much more than a correspondence of Hodge numbers of mirror pairs is predicted by mirror symmetry. Naïvely, for a mirror pair $(X, \widetilde{X})$ deformation of the complex structure on $\widetilde{X}$ is mirror to deformation of the symplectic (i.e., Kähler) structure on X. So far there has been no general construction of a moduli space of symplectic structure deformations mirror to the usual moduli space of complex structure deformations. (For such a proposed construction in the case of $K3$ surfaces, see [**Bri**].) In particular, there is no known analogue for X of the $\mathbb{Z}$-VHS for $\widetilde{X}$. Nevertheless, there is a proposal due to Kontsevich which significantly refines the statement that $b^{\mathrm{even}}(X) = b^{\mathrm{odd}}(\widetilde{X})$. This is his Homological Mirror Symmetry (HMS) Conjecture. It posits that mirror pairs $(X, \widetilde{X})$ should have the property that the derived category of bounded complexes of coherent sheaves on X (clearly a generalization of the even cohomology) should be equivalent to a category built from the Fukaya symplectic category of the mirror $\widetilde{X}$, taking into account certain flat line bundles on the special Lagrangian cycles (which is a generalization of the third cohomology of $\widetilde{X}$). It is not yet a precise conjecture, and much work in algebraic and symplectic geometry is underway to make it so. The conjecture was inspired by the earliest inklings of what eventually became D-brane physics, and, through the introduction of the derived category, has itself provided a language for physicists to express that theory [**MSCMM**]. In the context of the Batyrev-Borisov mirror pairs, illustrative examples have been studied by many (see for example [**Asp1, Asp2, BD, BDM, DJP, Doug, Hor, Hos1, Hos2, Mor**] and references therein).

As we noted before, there is more structure underlying the family $\widetilde{X}$ than just the Gauss-Manin connection, or equivalently the Picard-Fuchs differential equation satisfied by the family of holomorphic 3-forms. There is the $\mathbb{Z}$-VHS where the integral local system arises by taking periods over integral 3-cycles. This is one of the main objects of our interest in this investigation. There is a general result, due to Griffiths, Deligne, and Schmid (see the end of this section) that tells us in this case that the integral monodromy representation determines the $\mathbb{Z}$-VHS. Now by considering "D-brane charges", i.e., by taking cohomology in the categories on both sides of the HMS correspondence, and retaining the information about categorical automorphisms on both sides, we obtain a mathematical prediction from mirror symmetry. By this reasoning, mirror symmetry predicts that there is an equivalence between the monodromy representation of the $\mathbb{Z}$-VHS for $\widetilde{\mathfrak{X}}$ and automorphisms of integral even topological K-theory $K^0(X)$ obtained from the categorical autoequivalences. There is a precise conjecture [**Kon**] for the K-theory automorphism mirror to the maximally unipotent monodromy for any Calabi-Yau family. There is also a conjecture [**BD, BDM, DJP**] (generalizing a proposal of Kontsevich in the case of the quintic) for the K-theory automorphism mirror to the monodromy about $z = 1$ for one-parameter families, or more generally for the monodromy about the conifold locus in the compactification of the moduli space.

Let us give more details about these proposed K-theory automorphisms. Let $(X, \widetilde{X})$ be a mirror pair with $h^{1,1}(X) = h^{2,1}(\widetilde{X}) = 1$. The maximally unipotent monodromy for the one-parameter $\mathbb{Z}$-VHS for $\widetilde{\mathfrak{X}}$ is proposed to be mirror to the K-theory automorphism $K(X) \to K(X)$ given by

$$\xi \mapsto \xi \otimes \mathcal{L} \tag{1.4}$$

where $c_1(\mathcal{L})$ is the positive generator of $H^{1,1}(X, \mathbb{Z})$. The conifold monodromy (conventionally taken about $z = 1$) is proposed to be mirror to the Fourier-Mukai "push-pull" transform

$$\xi \mapsto \sum_{\lambda} (p_1)_* \left(\{(\lambda^{-1} \boxtimes \lambda) \to \mathcal{O}_\Delta\} \otimes p_2^*(\xi)\right) \tag{1.5}$$

where $\Delta \subset X \times X$ is the diagonal, p_i are the projections onto the factors, and λ runs over flat line bundles on X. This proposal implies that the monodromy automorphism of $H^3(\widetilde{X}, \mathbb{Z})$ around the conifold locus is then a sum of terms each of which is a Picard-Lefschetz transformation on a vanishing cycle. The vanishing cycles are all equivalent modulo torsion, so that this monodromy is divisible by the number of terms in the sum, which is $|\pi_1(X)|$. The remaining monodromy (conventionally taken at ∞), whether of finite order (as in the case of the quintic family) or not, is determined as the product of the monodromies at 0 and 1.

2. Classification of $\mathbb{Z}$-VHS

For the purpose of investigating this proposed Hodge-theoretic/K-theoretic mirror relationship it is very useful to have a complete description of the possible integral variations of Hodge structure that can underlie a family of Calabi-Yau threefolds over $\mathbb{P}^1 \setminus \{0, 1, \infty\}$ with $h^{2,1} = 1$, subject to the assumption of the existence of a point of maximally unipotent monodromy (at $z = 0$), a point of unipotent monodromy of rank one (at $z = 1$), and quasi-unipotency of their product (necessary at $z = \infty$ by the Monodromy Theorem [**BN**]. In this case the monodromy representation of the $\mathbb{Z}$-VHS determines that variation up to isomorphism. To see this let $\mathcal{H}$ and $\mathcal{H}'$ be $\mathbb{Z}$-VHS over the same connected algebraic base B with $h^{3,0} = h^{2,1} = 1$ and with isomorphic irreducible monodromy ((2) of Theorem 2.1 establishes irreducibility for our specific case). The bundle $\mathcal{H}om(\mathcal{H}, \mathcal{H}')$ with its natural induced flat connection, integral structure and Hodge filtration is a $\mathbb{Z}$-VHS of weight zero over B. By [**Sch**, Theorem (7.22)], the $\pi_1(B)$-invariant subbundle is a sub $\mathbb{Z}$-VHS. Of course, the invariant subbundle is the flat bundle whose fiber over any point $x \in B$ is the $\pi_1(B, x)$-invariant subspace of $\mathrm{Hom}_{\mathbb{C}}(\mathcal{H}_x, \mathcal{H}'_x)$. Since the monodromies of $\mathcal{H}$ and $\mathcal{H}'$ are irreducible, by Schur's lemma, the invariant subbundle is one-dimensional. But a weight zero Hodge structure on a one-dimensional space is of type $(0,0)$. This means that any non-trivial $\pi_1(B,x)$-invariant homomorphism $\mathcal{H}_x \to \mathcal{H}'_x$ preserves the Hodge filtration and hence is an isomorphism of Hodge structures, showing that $\mathcal{H}$ and $\mathcal{H}'$ are isomorphic as $\mathbb{Z}$-VHS over B.

In this section we classify $\mathbb{Z}$-VHS over $\mathbb{P}^1\setminus\{0,1,\infty\}$ with $h^{3,0}=h^{2,1}=1$ subject to the conditions that the local monodromy about $z=0$ is maximally unipotent and the local monodromy about $z=1$ is unipotent of rank one. As noted above, the key to this classification is the classification of possible monodromy representations underlying such variations.

2.1. Classification results for $\rho\colon \pi_1(\mathbb{P}^1\setminus\{0,1,\infty\}) \to Sp(4,\mathbb{Z})$. We begin by classifying real representations. Let V be a four-dimensional real vector space. Let $T\colon V\to V$ be a unipotent automorphism of V and let N be the nilpotent endomorphism of V defined by $N = T - \mathrm{Id}$. Then $N^{\dim V}=0$. T is said to be *maximally unipotent* if $N^{\dim V-1}\neq 0$. Also, the *rank* of T is defined to be the dimension of the image of N, so that T is maximally unipotent if and only if its rank is one less than the dimension of V.

Throughout we fix $x\in\mathbb{P}^1\setminus\{0,1,\infty\}$ and a representation

$$\rho\colon \pi_1(\mathbb{P}^1\setminus\{0,1,\infty\},x)\to GL(V).$$

We have geometric generators γ_0,γ_1, and γ_∞ for $\pi_1(X,x)$ which are represented by loops in $\mathbb{P}^1\setminus\{0,1,\infty\}$ encircling $0,1,\infty$ respectively. These are oriented positively so that they satisfy the relation $\gamma_0\gamma_1\gamma_\infty=1$. We denote the images of these generators under ρ by T_0,T_1,T_∞, respectively, and in view of the eventual application to flat bundles over $\mathbb{P}^1\setminus\{0,1,\infty\}$ call them the *monodromy* around $0,1,\infty$.

THEOREM 2.1. *Suppose that the monodromy T_0 is maximally unipotent and that the monodromy T_1 is unipotent of rank 1. Set $N_i = T_i - Id$ for $i=0,1$. Define an invariant $m\in\mathbb{R}$ by choosing a non-zero vector $v\in\operatorname{Ker} N_0$ and setting $N_0^3(N_1(v)) = -mv$. Then:*

(1) *m is independent of the choice of v.*
(2) *The representation ρ is irreducible if and only if $m\neq 0$.*
(3) *If the monodromy at infinity has an invariant vector, then $m=0$.*
(4) *In the case when $m\neq 0$, the representation ρ is determined, up to conjugation by an element of $GL(V)$, by the characteristic polynomial of the monodromy transformation around infinity. Direct computation shows that the characteristic polynomial of $T_\infty^{-1}=T_0T_1$ is*

$$x^4+(a-4)\,x^3+(6-2a+bm)\,x^2+(a-4+m-bm)\,x+1\,. \tag{2.1}$$

Notice that a and b are determined by this polynomial. The real numbers a,b,m are complete invariants of the conjugacy class of ρ as a representation into $GL(V)$.

(5) *For $m\neq 0$ take as basis for V the vectors*

$$N_1(-v),\, N_0(N_1(-v)),\, \frac{N_0^2(N_1(-v))}{m},\, v\,. \tag{2.2}$$

Then in this basis matrices for T_0 and T_1 are

$$T_0 = \begin{pmatrix} 1 & 0 & 0 & 0 \\ 1 & 1 & 0 & 0 \\ 0 & m & 1 & 0 \\ 0 & 0 & 1 & 1 \end{pmatrix} \tag{2.3}$$

$$T_1 = \begin{pmatrix} 1 & -a & -b & -1 \\ 0 & 1 & 0 & 0 \\ 0 & 0 & 1 & 0 \\ 0 & 0 & 0 & 1 \end{pmatrix} \tag{2.4}$$

For the rest of this section we assume that T_0 is maximally unipotent, that T_1 is a unipotent of rank one, and that ρ is irreducible, i.e., $m \neq 0$, and we let m, a, b be as in the statement of the previous lemma.

COROLLARY 2.2. *There is a non-degenerate symplectic form on V invariant under ρ if and only if $b = 1$. In this case, this form is unique up to non-zero scalar multiplication and the matrix for this form in the basis (2.2) is (up to a positive scalar)*

$$\langle \cdot, \cdot \rangle = \pm \begin{pmatrix} 0 & -a & -1 & -1 \\ a & 0 & 1 & 0 \\ 1 & -1 & 0 & 0 \\ 1 & 0 & 0 & 0 \end{pmatrix} \tag{2.5}$$

We now fix a non-degenerate skew pairing $\langle \cdot, \cdot \rangle$ on V.

COROLLARY 2.3. *There are exactly 14 conjugacy classes of representations $\rho\colon \pi_1(\mathbb{P}^1 \setminus \{0, 1, \infty\}) \to Sp(V)$ with T_0 and T_1 as given in Theorem 2.1 and with T_∞ being quasi-unipotent. These are classified by the pairs of nonzero integers (m, a) (with $b = 1$) and a complete set of such pairs is given in the first two columns of Table 1.*

We are actually interested in representations to $Sp(4, \mathbb{Z})$, or equivalently in lattices $L \subset V$ invariant under ρ and on which the symplectic form is a perfect integral pairing. Here is a first result along these lines.

COROLLARY 2.4. *Suppose that $L \subset V$ is an integral lattice invariant under ρ such that the restriction of $\langle \cdot, \cdot \rangle$ to L is a perfect integral pairing. Then $b = 1$ and $a, m \in \mathbb{Z}$. Conversely, if $b = 1$ and $a, m \in \mathbb{Z}$, then there is an integral lattice L_0 in V invariant under ρ on which the pairing is a perfect integral pairing. Furthermore, there is a basis for L_0 in which T_0, T_1 and the pairing are given by the matrices in Equations (2.3),and (2.4), respectively, and the symplectic form is given, up to sign, by Equation (2.5).*

Notice that we are not claiming that, given a lattice $L \subset V$ invariant under ρ and on which the symplectic form is a perfect integral pairing, the lattice L_0 as in Corollary 2.4 is related in any way to L. In fact, they are closely related (though not necessarily equal), but that relationship is somewhat delicate and depends on the arithmetic properties of m and a.

Now suppose that $b = 1$ and that a and $m \neq 0$ are integers. Let e_1, e_2, e_3, e_4 be a basis of V in which the symplectic pairing $\langle \cdot, \cdot \rangle$ is given by Equation (2.5), and T_0 and T_1 are given by Equations (2.3) and (2.4), respectively. Our goal is to study all possible integral lattices invariant under ρ and on which the symplectic pairing is a perfect integral pairing. By Corollary 2.4 the lattice L_0 spanned by $\{e_1, \ldots, e_4\}$ is such a lattice.

We introduce the (increasing) weight filtration $W_*(V)$ associated with T_0 by setting $W_{2i} = W_{2i+1} = \mathrm{Ker}(N_0)^{i+1}$ for $i = 0, \ldots, 3$. This is the monodromy weight filtration in the limiting mixed Hodge structure as defined by Deligne [**Del1**]. This filtration has the property that $N_0\colon W_j \to W_{j-2}$ and N_0 induces an isomorphism on the associated gradeds: $N_0\colon W_j/W_{j-1} \to W_{j-2}/W_{j-3}$. Fix an integral lattice L on which the pairing is a perfect integral pairing and which is invariant under ρ. Then $W_{2i}(L) = L \cap W_{2i}$ has rank equal to $i+1$, and hence the quotients $W_{2i}(L)/W_{2i-2}(L) \cong \mathbb{Z}$ for $i = 0, \ldots, 3$. Notice in particular that for the lattice L_0

$$N_1\colon W_0(L_0) \to L_0/W_4(L_0)$$

and

$$N_0\colon L_0/W_4(L_0) \to W_4(L_0)/W_2(L_0)$$

are isomorphisms.

Lemma 2.5. *Suppose that $\ell_1, \ell_2, \ell_3, \ell_4$ is a basis for L such that for each i the subset $\ell_{4-i}, \ldots, \ell_4$ is a basis for $W_{2i}(L)$. Then the matrix for the symplectic form is*

$$\begin{pmatrix} 0 & \alpha & \beta & \pm 1 \\ -\alpha & 0 & \pm 1 & 0 \\ -\beta & \pm 1 & 0 & 0 \\ \pm 1 & 0 & 0 & 0 \end{pmatrix}$$

for appropriate integers α and β.

Remark 2.6. We now fix our conventions regarding integral bases for L pertaining to the above isomorphisms on associated gradeds. From now on we shall consider bases for L satisfying:

(1) For every $i \le 4$ the set $\{\ell_{4-i}, \ldots, \ell_4\}$ forms a basis for $W_{2i}(L)$;
(2) $\langle \ell_1, \ell_4 \rangle = -1$;
(3) $\langle \ell_2, \ell_3 \rangle = 1$.
(4) Duality tells us that there is a non-zero integer r such that $N_0(\ell_1) \equiv r\ell_2 \pmod{\ell_3, \ell_4}$ and $N_0(\ell_3) = r\ell_4$. By changing the signs of ℓ_2 and ℓ_3 if necessary, we arrange that $r > 0$.
(5) We define s by requiring that $N_0(\ell_2) \equiv s\ell_3 \pmod{\ell_4}$.
(6) We define t such that $N_1(\ell_4) = -t\ell_1 \pmod{\ell_2, \ell_3, \ell_4}$.

Of course, r is the divisibility of N_0 as a map $W_6/W_4 \to W_4/W_2$ and also $W_2/W_0 \to W_0$ while s is the divisibility of $N_0\colon W_4/W_2 \to W_2/W_0$ and t is the divisibility of $N_1\colon W_0 \to W_6/W_4$.

In particular, we have:

Theorem 2.7. *The integers r, s and t are invariants of the isomorphism class of $(L, \rho, \langle \cdot, \cdot \rangle)$.*

Notice that these integral invariants are related to the integer invariant m of the real conjugacy class of ρ by

$$m = r^2 st \quad \text{and} \quad t | a. \tag{2.6}$$

For each i, $1 \leq i \leq 4$, the real subspace of V spanned by $\{\ell_i, \dots \ell_4\}$ is equal to that spanned by $\{e_i, \dots, e_4\}$, which means the matrix whose columns express the ℓ_i in terms of the e_j is weakly lower triangular. Since the image of the representation ρ is Zariski dense in $Sp(V)$ general considerations imply that the coefficients in this matrix are contained in $\mathbb{Q}[\sqrt{(1/d)}]$ for some integer d. In fact, a much more explicit result holds.

As an indication of how the basis $\{\ell_i\}$ of the new lattice L is expressed in terms of the original basis $\{e_i\}$ of the original lattice L_0, consider the following lemma.

LEMMA 2.8. *There exists a basis $\ell_1, \ell_2, \ell_3, \ell_4$ for L as described above such that*

$$\begin{aligned}
\ell_1 &= \frac{e_1}{\sqrt{t}} + \frac{\alpha e_2}{rt\sqrt{t}} + \frac{\beta e_3}{\sqrt{t}} + \frac{\gamma e_4}{rt\sqrt{t}} \\
\ell_2 &= \frac{e_2}{r\sqrt{t}} + \frac{\delta e_3}{\sqrt{t}} + \frac{\mu e_4}{r\sqrt{t}} \\
\ell_3 &= r\sqrt{t}e_3 + \frac{\alpha e_4}{\sqrt{t}} \\
\ell_4 &= \sqrt{t}e_4
\end{aligned}$$

for appropriate $\alpha, \beta, \gamma, \delta, \mu \in \mathbb{Z}$.

Once again let us impose the condition that T_∞ is quasi-unipotent, i.e., that the pair (m, a) is one of the 14 pairs listed in Table 1. Case-by-case analysis results in a complete classification of the possible $\alpha, \beta, \gamma, \delta, \mu$ corresponding to each allowable r, s, t. Adding the numbers in Column 4 of Table 1 one sees that there are 112 possibilities.

In Equations (1.4) and (1.5) we gave explicit formulas for the proposed automorphisms of K-theory. Translating these by mirror symmetry gives the following conditions on T_0 and T_1, which we express in terms of the invariants r and t:

CONJECTURE 2.9. *For any $\mathbb{Z}$-VHS arising from a family of Calabi-Yau threefolds $\widetilde{\mathfrak{X}}$ with $h^{2,1} = 1$ over $\mathbb{P}^1 \setminus \{0, 1, \infty\}$ we have:*

(2.7) $r = 1$ (mirror to the fact that $c_1(\mathcal{L})$ is a generator of $H^2(X, \mathbb{Z})$)

(2.8) $N_1 = t \cdot \overline{N}_1$ for some indivisible integral transformation $\overline{N}_1$.

Furthermore, t is equal to the order of the fundamental group of the mirror Calabi-Yau threefold X.

REMARK 2.10. A stronger version of the first equation was conjectured by Morrison to hold for the local maximally unipotent monodromy in arbitrary dimensional families of Calabi-Yau threefolds, [**CoKa**, p. 150]. Condition (2.8) and the last statement are conjectured to hold for the local monodromy around the principal discriminant for arbitrary dimensional families of Calabi-Yau threefolds.

The invariants m and a encode natural geometric information about the Calabi-Yau threefolds X in the last column of Table 1. As observed by Borcea [**Borc**] for the 13 examples with $t = 1$, the integers we identified as invariants (m, a) above have a mirror interpretation as $m = \mathcal{L}^3$ and $a = \dim(H^0(X, \mathcal{O}(\mathcal{L})))$ respectively. In fact, this interpretation applies to all the geometric examples listed in Table 1.

The classification of representations into $Sp(4, \mathbb{Z})$ satisfying the two conditions in Conjecture 2.9 is simpler to state than the general case:

PROPOSITION 2.11. *An integral lattice $L \subset V$ on which the symplectic pairing is a perfect integral pairing and which is invariant under ρ and which satisfies Conditions (2.7) and (2.8) is determined by t.*

Columns 5 and 6 of Table 1 list the 23 of the 112 representations that satisfy these conditions. Notice that without assuming Conditions (2.7) and (2.8) the invariants m, a, r, s, t do not in general determine the isomorphism class of the lattice.

2.2. From representations to $\mathbb{Z}$-VHS. As we argued at the end of Section 1.4, for a $\mathbb{Z}$-VHS over a quasi-projective base with completely irreducible monodromy representation, the isomorphism class of the monodromy representation determines the isomorphism class of the $\mathbb{Z}$-VHS. According to a theorem of Borel and Narasimhan [**BN**] the local monodromy of any $\mathbb{Z}$-VHS is quasi-unipotent.

THEOREM 2.12. *Suppose we have a primitive weight-three $\mathbb{Z}$-VHS $\mathcal{H}$ over $\mathbb{P}^1 \setminus \{0, 1, \infty\}$ satisfying:*

(1) $h^{3,0} = h^{2,1} = 1$,
(2) *the underlying monodromy representation ρ into $Sp(4, \mathbb{Z})$ is irreducible,*
(3) *T_0 is maximally unipotent, and*
(4) *T_1 is unipotent of rank one.*

Then ρ is conjugate in $Sp(4, \mathbb{Z})$ to one of the 112 representations enumerated in Table 1 and the isomorphism class of $\mathcal{H}$ as a $\mathbb{Z}$-VHS is determined by the $Sp(4, \mathbb{Z})$-conjugacy class of ρ.

The remaining question is one of existence: Which of the 112 representations underlie a $\mathbb{Z}$-VHS?

Each of the 14 real representations comes from a generalized hypergeometric ODE. That is to say, for each of the fourteen pairs (m, a) in Table 1 we have the following data:

(1) a flat bundle with a real structure,
(2) a real symplectic form parallel under the flat connection,
(3) a holomorphic section ϖ of the complexification of the flat bundle,

such that the Picard-Fuchs equation satisfied by ϖ is the generalized hypergeometric equation

$$\left[\Theta^4 - z(\Theta + a_1)(\Theta + a_2)(\Theta + a_3)(\Theta + a_4)\right]\varpi(z) = 0\ , \tag{2.9}$$

where the a_i are the entries in Column 3 of Table 1 corresponding to (m, a) and $\Theta = z\nabla_z$. The putative Hodge filtration on this flat bundle is given by: $\mathcal{F}^{3-i}$ is the span of the holomorphic section and its first i covariant derivatives. This filtration is automatically horizontal and satisfies the first Hodge condition (1.1). It is not clear that the filtration satisfies the second Hodge condition (1.2), but if it does, then the polarization condition (1.3) follows.

By direct geometric constructions we know that all 14 of these representations arise as local systems (or subquotients of local systems) underlying families of Calabi-Yau threefolds. Of these, 13 are local systems of third cohomology for one-parameter families of complete intersection Calabi-Yau threefolds with $h^{2,1} = 1$ in toric varieties (see Section 3). Thus, these thirteen are in fact real representations underlying $\mathbb{R}$-VHS. Consequently, for these 13 real types there is a Hodge filtration making this an $\mathbb{R}$-VHS. In the remaining case for which $(m, a) = (1, 4)$ we do not know whether Conditions (1.2) and (1.3) hold.

Since Conditions (1.1), (1.2), and (1.3) make reference only to the symplectic form and the real structure, a parallel lattice in a $\mathbb{R}$-VHS on which the symplectic form is a perfect integral pairing determines a $\mathbb{Z}$-VHS. Hence, all integral representations whose real types satisfy $(m, a) \neq (1, 4)$ underlie (unique isomorphism classes of) $\mathbb{Z}$-VHS. Since the real representation corresponding to $(m, a) = (1, 4)$ has a unique integral lattice, there is only one of the 112 integral representations enumerated in Table 1 which might not come from a $\mathbb{Z}$-VHS. Thus, up to this one ambiguity we have completely classified $\mathbb{Z}$-VHS satisfying the conditions given in Theorem 2.12. This means that, up to the same ambiguity, we have classified the complete curves in the smooth locus of the period space $\mathcal{D}/\Gamma$ isomorphic to $\mathbb{P}^1 \setminus \{0, 1, \infty\}$ with the given local monodromy conditions.

3. Geometric realization of abstract variations by one-parameter families of Calabi-Yau threefolds

The first two columns of Table 1 give the invariants m and a for the 14 $\mathbb{R}$-VHS in the classification. The third column gives the coefficients of the hypergeometric ODE, see Equation (2.9), or equivalently $(1/2\pi\imath)$ times the logs of the eigenvalues of T_∞. The fourth column enumerates the integral lattices within each real representation. The fifth column indicates the number of integral lattices that are "mirror-consistent" in the sense that they satisfy both conditions (2.7) and (2.8). The sixth column shows the t values for each of the mirror consistent representations enumerated in the fifth. For each of these, the final column indicates known geometric examples with integral variations of Hodge structure of this type. Since the examples listed all have $h^{1,1} = 1$, to read each entry preface with "The $\mathbb{Z}$-VHS of H^3 of the Batyrev-Borisov mirror of ...".

The two examples with $(m, a) = (2, 3)$ and $(6, 5)$ have an asterisk to indicate that, even though these families of Calabi-Yau complete intersections in weighted projective space are of the correct type, the Newton polytopes of the weighted projective spaces are not reflexive and so these examples don't quite fit into the Batyrev-Borisov description of mirror pairs (though the Greene-Plesser construction does apply [**KT**]). Denoting the vertices of the $(2, 3)$ polytope by $\{e_1, \ldots, e_5, -e_1 - e_2 - 2e_3 - 2e_4 - 3e_5\}$, and those of the $(6, 5)$ polytope by $\{e_1, \ldots, e_5, -e_1 - e_2 - e_3 - e_4 - 2e_5\}$, suitable reflexive polytopes are obtained by adding in each case the new vertex $-e_5$. These have the property that they possess nef partitions of bidegrees $[4, 6]$ and $[3, 4]$ respectively, with Hodge numbers matching those of the original Calabi-Yau hypersurfaces in the two weighted projective spaces.

The example denoted A is the quintic twin, which we already recognized as the free quotient of the quintic by $\mathbb{Z}/5\mathbb{Z}$. The rest of the examples labeled B through H arise in a quite similar fashion, by finding a suitable free action by a group of order $t \neq 1$ on a simply-connected Calabi-Yau of that class (specifically the one described in the corresponding $t = 1$ entry of the fifth column) and taking the quotient. When that action is compatible with the toric structure of the original example, the Batyrev-Borisov mirror construction applies. In the classification of 4D reflexive polytopes by Kreuzer-Skarke [KrSkCY] there are precisely five polytopes which correspond to toric hypersurfaces with $h^{1,1} = 1$, and four of these are among the weighted projective spaces listed in the seventh column of Table 1 — the quintic,

m	a	a_1, a_2, a_3, a_4	$\# L_{\mathbb{Z}}$	$\# L_{\mathrm{MC}}$	t	Geometric examples
1	4	$\frac{1}{12}, \frac{5}{12}, \frac{7}{12}, \frac{11}{12}$	1	1	1	I
1	3	$\frac{1}{10}, \frac{3}{10}, \frac{7}{10}, \frac{9}{10}$	1	1	1	$\mathbb{WP}^4_{1,1,1,2,5}[10]$
2	4	$\frac{1}{8}, \frac{3}{8}, \frac{5}{8}, \frac{7}{8}$	2	2	1	$\mathbb{WP}^4_{1,1,1,1,4}[8]$
					2	II
5	5	$\frac{1}{5}, \frac{2}{5}, \frac{3}{5}, \frac{4}{5}$	2	2	1	$\mathbb{P}^4[5]$
					5	A
1	2	$\frac{1}{6}, \frac{1}{6}, \frac{5}{6}, \frac{5}{6}$	1	1	1	$\mathbb{WP}^5_{1,1,2,2,3,3}[6,6]$
2	3	$\frac{1}{6}, \frac{1}{4}, \frac{3}{4}, \frac{5}{6}$	1	1	1	$\mathbb{WP}^5_{1,1,1,2,2,3}[4,6]$ *
3	4	$\frac{1}{6}, \frac{1}{3}, \frac{2}{3}, \frac{5}{6}$	1	1	1	$\mathbb{WP}^4_{1,1,1,1,2}[6]$
4	5	$\frac{1}{6}, \frac{1}{2}, \frac{1}{2}, \frac{5}{6}$	11	1	1	$\mathbb{WP}^5_{1,1,1,1,1,3}[2,6]$
4	4	$\frac{1}{4}, \frac{1}{4}, \frac{3}{4}, \frac{3}{4}$	8	3	1	$\mathbb{WP}^5_{1,1,1,1,2,2}[4,4]$
					2	B
					4	C
6	5	$\frac{1}{4}, \frac{1}{3}, \frac{2}{3}, \frac{3}{4}$	1	1	1	$\mathbb{WP}^5_{1,1,1,1,1,2}[3,4]$ *
8	6	$\frac{1}{4}, \frac{1}{2}, \frac{1}{2}, \frac{3}{4}$	14	2	1	$\mathbb{P}^5[2,4]$
					2	D
9	6	$\frac{1}{3}, \frac{1}{3}, \frac{2}{3}, \frac{2}{3}$	8	2	1	$\mathbb{P}^5[3,3]$
					3	E
12	7	$\frac{1}{3}, \frac{1}{2}, \frac{1}{2}, \frac{2}{3}$	11	1	1	$\mathbb{P}^6[2,2,3]$
16	8	$\frac{1}{2}, \frac{1}{2}, \frac{1}{2}, \frac{1}{2}$	50	4	1	$\mathbb{P}^7[2,2,2,2]$
					2	F
					4	G
					8	H

TABLE 1. Table of "Mirror-Consistent" $\rho \colon \pi_1(\mathbb{P}^1 \setminus \{0,1,\infty\}) \to Sp(4)$

the sextic, the octic, and the dectic. The remaining example is the alternative derivation of the quintic twin from Example 1.3 for which the Batyrev mirror pair construction applies. Note that the fact that the "octic twin" does not arise in this way means that if there does exist a free quotient of the octic hypersurface in $\mathbb{WP}^4_{1,1,1,1,4}[8]$ by the group $\mathbb{Z}/2\mathbb{Z}$, then the action cannot be compatible with the toric structure (i.e., arise from a natural involution on the weighted projective ambient space).

The only known construction of example H, a Calabi-Yau whose mirror defines the $t = 8$ integral structure in the same real class as the mirror of the complete intersection of four quadrics in $\mathbb{P}^7$, requires a manifestly non-toric group action. It comes from Jae Park's recent construction [**Park**] of the mirror of Beauville's Calabi-Yau threefold with nonabelian fundamental group. Example H is the Beauville manifold, the quotient of $\mathbb{P}^7[2,2,2,2]$ by the group Q_8 of unit quaternions. Examples F and G are then the intermediate quotients obtained through the same action, but restricted to the subgroups $\mathbb{Z}_2$ and $\mathbb{Z}_4$ of Q_8 respectively.

For Example II the authors don't know of a family of Calabi-Yau threefolds with $h^{2,1} = 1$ over $\mathbb{P}^1 \setminus \{0, 1, \infty\}$ realizing this $\mathbb{Z}$-VHS, and for Example I the authors don't know whether there is a geometric realization of this representation or indeed whether there is a $\mathbb{Z}$-VHS with this representation.

REMARK 3.1. All of these toric examples satisfy Conjecture 2.9. Notice that geometric families realizing a given (m, a) but different t values give examples of geometric $\mathbb{Z}$-VHS that are isomorphic as $\mathbb{R}$-VHS. The quintic mirror and the quintic twin mirror families comprise the simplest example of this phenomenon. More generally, to the best of the authors' knowledge, all complete intersection Calabi-Yau threefolds arising from a nef partition of the anti-canonical divisor at infinity in Gorenstein Fano toric varieties satisfy the multiparameter generalization of this conjecture.

Some positive results along these lines are:

THEOREM 3.2. *For any Calabi-Yau threefold that is an anti-canonical hypersurface in a Gorenstein Fano toric variety the local monodromy* $T_1 = \mathrm{Id} + N_1$ *around the prinicpal discriminant satisfies Condition (2.8) with* t *equal to the order of the fundamental group of the Batyrev mirror Calabi-Yau.*

For any Calabi-Yau threefold $\widetilde{X}$ *that is a complete intersection arising from a nef partition of the anti-canonical divisor at infinity in a Gorenstein Fano toric variety and with* $h^{2,1}(\widetilde{X}) = 1$ *the parameter space of complex moduli can be identified with* $\mathbb{P}^1 \setminus \{0, 1, \infty\}$ *in such a way that the monodromies* T_0 *and* T_1 *satisfy the assumptions of Theorem 2.12. Furthermore,* T_0 *satisfies Condition (2.7).*

There are five one-parameter families of Calabi-Yau threefold hypersurfaces in a Gorenstein Fano toric variety with $h^{2,1} = 1$. By this theorem each of their $\mathbb{Z}$-VHS is one of the 23 listed in column 5 of Table 1, where (m, a) are determined by the local monodromy matrices T_0 and T_1 and t is the order of the fundamental group of the mirror. We conjecture that the same is true for Calabi-Yau threefold complete intersections in Gorenstein Fano toric varieties with $h^{2,1} = 1$.

4. Extensions and final comments

4.1. Extensions. The classification obtained above for $\mathbb{Z}$-VHS underlying $h^{2,1} = 1$ Calabi-Yau threefold moduli over $\mathbb{P}^1 \setminus \{0, 1, \infty\}$ should be extended both to $h^{2,1} > 1$ multiparameter hypergeometric variations and to cases of $h^{2,1} = 1$ Calabi-Yau threefold moduli over curves of higher genus and with more punctures.

A small but growing number of the latter sorts of Calabi-Yau threefold families are known to exist, mostly via complete intersection constructions in partial flag varieties [**BCFKvS**], though there is a great deal of indirect evidence for their existence. In particular, recent work of Almkvist-Zudilin [**AZ**] on ordinary differential equations whose associated formal "Yukawa coupling" series possess integrality and positivity properties that suggest they could be Yukawa couplings for actual Calabi-Yau threefolds — so-called "fake Picard-Fuchs equations" — highlights the importance of classifying compatible integral structures, both as a means of eliminating some cases from consideration and in order to obtain more information about compatible Calabi-Yau manifolds if they do exist (e.g., possible nontrivial fundamental groups).

For a large number of the Almkvist-Zudilin examples (with $t = 1$) Christian van Enckevort and Duco van Straten [**vEvS**] have used a numerical approach, motivated by HMS and the integrality of formal Gopakumar-Vafa invariants, to determine the integral monodromy representations. One very interesting question is whether their methods can be modified to also detect alternate integral structures with $t > 1$. These would correspond conjecturally to more families of mirrors of non-simply connected Calabi-Yau threefolds.

For each integral structure with $t > 1$ in a class for which a geometric construction of the $t = 1$ representative is known, one can ask whether there is a free action of a finite group of order t on the known example such that the mirror family realizes the $t > 1$ variation. In particular, for the examples in Table 1 the only entry for which this remains an open question is case II. Thus we ask: Is there a free, holomorphic 3-form-preserving action of $\mathbb{Z}/2\mathbb{Z}$ on the "double octic"?

4.2. Final comments on Table 1. As noted already, the authors do not know of any geometric constructions of one-parameter families of Calabi-Yau threefolds with $h^{2,1} = 1$ whose integral monodromy representations, underlying their weight three $\mathbb{Z}$-VHS, correspond to the representations we found with $(m, a, t) = (1, 4, 1)$ or $(2, 4, 2)$. In fact, we do not even know of candidate mirror families with $h^{1,1} = 1$ and the correct geometric invariants. Nevertheless, these integral monodromy representations do "come from geometry" through restriction from multiparameter, higher rank variations of the same weight.

The task of finding the 4D polytope (in case II) and 5D polytope with nef partition (in case I) with appropriate subloci is not simple, but knowledge of the corresponding integral structures helps to point the way.

The $(m, a) = (1, 4)$ integral structure is unique, so it suffices to merely find a sub-local system of the appropriate hypergeometric type and then to prove that the restricted monodromy is actually integral. The structure of the generalized hypergeometric series ${}_4F_3(\frac{1}{12}, \frac{5}{12}, \frac{7}{12}, \frac{11}{12})$ suggests looking for a complete intersection of bidegree $[2, 12]$ in $\mathbb{WP}^5_{1,1,1,1,4,6}$. Unfortunately, there is not a well-defined mirror of such a complete intersection (difficulties with the singular locus). This is reflected in the fact that the the Newton polytope for the weighted projective space is not reflexive. We can correct this by considering instead the reflexive polytope (provided by Kreuzer and Scheidegger) with vertices given by $\{[e_1, e_3], [e_2, e_4, e_5, -e_1 - e_2 - e_3 - 4e_4 - 6e_5, -2e_4 - 3e_5]\}$, where the brackets denote the two pieces of the nef partition. The complete intersection Calabi-Yau threefold of bidegree $[2, 12]$ in this Gorenstein Fano toric variety has $h^{1,1} = 3$, but $h^{1,1}_{toric} = 2$. In terms of the natural (torically defined) complex structure coordinates z_1, z_2 for the polynomial deformations, the restriction locus is just the locus $z_1 = 0$. Following Batyrev, the generalized hypergeometric series for the holomorphic solution in these coordinates is

$$\sum_{k,m} \frac{(2m)!(6k+12m)!}{(3k+6m)!(m!)^4 k!(2k+4m)!} z_1^k z_2^m ,$$

which upon restriction yields the desired series. See [**KKRS**] for a discussion of the singular geometry and physics of this subfamily.

The 4D polytope for the $(m, a) = (2, 4)$ example is found by a similar method. Here one knows that the desired $\mathbb{Z}$-VHS has $t = 2$, so one expects if it arises by restriction that it will do so from a family of Calabi-Yau threefolds with fundamental group $\mathbb{Z}/2\mathbb{Z}$. Also, the form of the hypergeometric series ${}_4F_3(\frac{1}{8}, \frac{3}{8}, \frac{5}{8}, \frac{7}{8})$ suggests looking for a hypersurface of degree 8 in $\mathbb{WP}^4_{1,1,1,1,4}$. Instead of taking this weighted projective space as the starting point, since Kreuzer-Skarke have completely classified the 473,800,776 equivalence classes of 4D reflexive polytopes, one can imagine searching through these for those hypersurfaces with nontrivial fundamental group. In fact the authors have done this, and a unique polytope was found with the property that the hypersurface has fundamental group $\mathbb{Z}/2\mathbb{Z}$ and there is a sublocus of moduli on which the hypergeometric series restricts to the desired form.

References

[AZ] Almkvist, Gert and Zudilin, Wadim. Differential equations, mirror maps and zeta values, these proceedings.

[Asp1] Aspinwall, Paul S. String theory and duality. Proceedings of the International Congress of Mathematicians, Vol. II (Berlin, 1998). Doc. Math. 1998, Extra Vol. II, 229–238 (electronic).

[Asp2] Aspinwall, Paul S. Some navigation rules for D-brane monodromy. J. Math. Phys. 42 (2001), no. 12, 5534–5552.

[Bat1] Batyrev, Victor V. Variations of the mixed Hodge structure of affine hypersurfaces in algebraic tori. Duke Math. J. 69 (1993), no. 2, 349–409.

[Bat2] Batyrev, Victor V. Dual polyhedra and mirror symmetry for Calabi-Yau hypersurfaces in toric varieties. J. Algebraic Geom. 3 (1994), no. 3, 493–535.

[BB1] Batyrev, Victor V. and Borisov, Lev A. On Calabi-Yau complete intersections in toric varieties. Higher-dimensional complex varieties (Trento, 1994), 39–65, de Gruyter, Berlin, 1996.

[BB2] Batyrev, Victor V. and Borisov, Lev A. Mirror duality and string-theoretic Hodge numbers. Invent. Math. 126 (1996), no. 1, 183–203.

[BCFKvS] Batyrev, Victor V., Ciocan-Fontanine, Ionuţ, Kim, Bumsig and van Straten, Duco. Mirror symmetry and toric degenerations of partial flag manifolds. Acta Math. 184 (2000), no. 1, 1–39.

[BH] Beukers, F. and Heckman, G. Monodromy for the hypergeometric function ${}_nF_{n-1}$. Invent. Math. 95 (1989), no. 2, 325–354.

[Borc] Borcea, C. $K3$ surfaces with involution and mirror pairs of Calabi-Yau manifolds. In B. Greene and S.-T. Yau, editors, Mirror Symmetry II, Volume 1 of Studies in Advanced Mathematics, pages 717–739. Am. Math. Soc., 1991.

[BN] Borel, A. and Narasimhan, R. Uniqueness conditions for certain holomorphic mappings. Invent. Math. 2 (1967), 247–255.

[Bor] Borisov, Lev A. Towards the Mirror Symmetry for Calabi-Yau Complete intersections in Gorenstein Fano Toric Varieties, alg-geom/9310001.

[Bri] Bridgeland, T. Stability conditions on $K3$ surfaces. math.AG/0307164.

[BD] Brunner, Ilka and Jacques Distler. Torsion D-branes in nongeometrical phases. Adv. Theor. Math. Phys. 5 (2001), no. 2, 265–309.

[BDM] Brunner, Ilka, Distler, Jacques and Mahajan, Rahul. Return of the torsion D-branes. Adv. Theor. Math. Phys. 5 (2001), no. 2, 311–352.

[BG] Bryant, Robert L. and Griffiths, Phillip A. Some observations on the infinitesimal period relations for regular threefolds with trivial canonical bundle. Arithmetic and geometry, Vol. II, 77–102, Progr. Math., 36, Birkhäuser Boston, Boston, MA, 1983.

[CdOGP] Candelas, Philip, de la Ossa, Xenia C., Green, Paul S. and Parkes, Linda. A pair of Calabi-Yau manifolds as an exactly soluble superconformal theory. Essays on mirror manifolds, 31–95, Internat. Press, Hong Kong, 1992.

[CoKa] Cox, David A. and Katz, Sheldon. Mirror symmetry and algebraic geometry. Mathematical Surveys and Monographs, 68. American Mathematical Society, Providence, RI, 1999.

[Del1] Deligne, Pierre. Théorie de Hodge. I. Actes du Congrès International des Mathématiciens (Nice, 1970), Tome 1, pp. 425–430. Gauthier-Villars, Paris, 1971.

[Del2] Deligne, Pierre. Théorie de Hodge. II. Inst. Hautes tudes Sci. Publ. Math. No. 40, (1971), 5–57.

[Del3] Deligne, P. Un théorème de finitude pour la monodromie. Discrete groups in geometry and analysis (New Haven, Conn., 1984), 1–19, Progr. Math., 67, Birkhuser Boston, Boston, MA, 1987.

[DJP] Distler, Jacques, Jockers, Hans, and Park, Hyukjae. D-Brane Monodromies, Derived Categories and Boundary Linear Sigma Models, hep-th/0206242.

[Doug] Douglas, Michael R. Lectures on D-branes on Calabi-Yau manifolds. 2001 Spring School on Superstrings and Related Matters (Trieste), 211–271 (electronic), ICTP Lect. Notes, VII, Abdus Salam Int. Cent. Theoret. Phys., Trieste, 2002.

[vEvS] van Enckevort, Christian and van Straten, Duco. Monodromy calculatons of fourth order equations of Calabi-Yau type, these proceedings.

[Ful] Fulton, William. Introduction to toric varieties. Annals of Mathematics Studies, 131. The William H. Roever Lectures in Geometry. Princeton University Press, Princeton, NJ, 1993.

[GGZ] Gel'fand, I. M., Graev, M. I. and Zelevinsky, A. V. Holonomic systems of equations and series of hypergeometric type. (Russian) Dokl. Akad. Nauk SSSR 295 (1987), no. 1, 14–19; translation in Soviet Math. Dokl. 36 (1988), no. 1, 5–10.

[GKZ] Gel'fand, I. M., Kapranov, M. M. and Zelevinsky, A. V. Discriminants, resultants, and multidimensional determinants. Mathematics: Theory & Applications. Birkhuser Boston, Inc., Boston, MA, 1994.

[Giv] Givental, Alexander B. Equivariant Gromov-Witten invariants. Internat. Math. Res. Notices 1996, no. 13, 613–663.

[GL] Greene, B. R.; Lazaroiu, C. I. Collapsing D-branes in Calabi-Yau moduli space. I. Nuclear Phys. B 604 (2001), no. 1-2, 181–255.

[GP] Greene, B. R. and Plesser, M. R. Duality in Calabi-Yau moduli space. Nuclear Phys. B 338 (1990), no. 1, 15–37.

[Grif] Griffiths, Phillip A. Periods of integrals on algebraic manifolds, III (Some global differential-geometric properties of the period mapping). In Selected works of Phillip A. Griffiths with commentary. Part 3. Variations of Hodge structure. Edited by C. Herbert Clemens and David R. Morrison. American Mathematical Society, Providence, RI; International Press, Somerville, MA, 2003.

[Hor] R.P. Horja. Hypergeometric functions and mirror symmetry in toric varieties, math.AG/9912109.

[Hos1] Hosono, S. Local mirror symmetry and type IIA monodromy of Calabi-Yau manifolds. Adv. Theor. Math. Phys. 4 (2000), no. 2, 335–376.

[Hos2] Hosono, S. Central charges, symplectic forms, and hypergeometric series in local mirror symmetry, hep-th/0404043.

[HLY1] Hosono, S., Lian, B. H. and Yau, S.-T. GKZ-generalized hypergeometric systems in mirror symmetry of Calabi-Yau hypersurfaces. Comm. Math. Phys. 182 (1996), no. 3, 535–577.

[HLY2] Hosono, S., Lian, B. H. and Yau, S.-T. Maximal degeneracy points of GKZ systems. J. Amer. Math. Soc. 10 (1997), no. 2, 427–443.

[HTF] Erdélyi, Arthur, Magnus, Wilhelm. Oberhettinger, Fritz, Tricomi, Francesco G. Higher transcendental functions. Vols. I, II. Based, in part, on notes left by Harry Bateman. McGraw-Hill Book Company, Inc., New York-Toronto-London, 1953.

[KKRS] Klemm, A., Kreuzer, M., Riegler, E. and Scheidegger, E. Topological string amplitudes, complete intersection Calabi-Yau spaces, and threshold corrections, hep-th/0410018.

[KT] Klemm, Albrecht and Theisen, Stefan. Mirror maps and instanton sums for complete intersections in weighted projective space. Modern Phys. Lett. A 9 (1994), no. 20, 1807–1817.

[Kod] Kodaira, Kunihiko. Complex manifolds and deformation of complex structures. Translated from the Japanese by Kazuo Akao. With an appendix by Daisuke Fujiwara. Grundlehren der Mathematischen Wissenschaften [Fundamental Principles of Mathematical Sciences], 283. Springer-Verlag, New York, 1986.

[Kon] Kontsevich, Maxim. Homological algebra of mirror symmetry. Proceedings of the International Congress of Mathematicians, Vol. 1, 2 (Zürich, 1994), 120–139, Birkhäuser, Basel, 1995.

[LLY1] Lian, Bong H., Liu, Kefeng, Yau, Shing-Tung. Mirror principle. I and II. Surveys in differential geometry: differential geometry inspired by string theory, 405–454 and 455–509, Surv. Differ. Geom., 5, Int. Press, Boston, MA, 1999.

[LLY2] Lian, Bong H., Liu, Kefeng, Yau, Shing-Tung. Mirror principle. III and IV. Surveys in differential geometry, 433–474 and 475–496, Surv. Differ. Geom., VII, Int. Press, Somerville, MA, 2000.

[Mor] Morrison, David R. Geometric aspects of mirror symmetry. Mathematics unlimited—2001 and beyond, 899–918, Springer, Berlin, 2001.

[MSCMM] Hori, Kentaro, Katz, Sheldon, Klemm, Albrecht, Pandharipande, Rahul, Thomas, Richard, Vafa, Cumrun. Vakil, Ravi. Zaslow, Eric. Mirror symmetry. With a preface by Vafa. Clay Mathematics Monographs, 1. American Mathematical Society, Providence, RI; Clay Mathematics Institute, Cambridge, MA, 2003.

[Park] Park, Hyukjae. Finding the Mirror of the Beauville Manifold, hep-th/0312056.

[Sch] Schmid, Wilfried. Variation of Hodge structure: the singularities of the period mapping, Inventiones Math. 22, 211–319 (1973).

[Tian] Tian, Gang. Smoothness of the universal deformation space of compact Calabi-Yau manifolds and its Petersson-Weil metric. Mathematical aspects of string theory (San Diego, Calif., 1986), 629–646, Adv. Ser. Math. Phys., 1, World Sci. Publishing, Singapore, 1987.

Department of Mathematics, University of Washington, Seattle, Washington 98195
E-mail address: `doran@math.washington.edu`

Department of Mathematics, Columbia University, New York, New York 10027
E-mail address: `jm@math.columbia.edu`

AMS/IP Studies in Advanced Mathematics
Volume 38, 2006

Monodromy Calculations of Fourth Order Equations of Calabi-Yau Type

Christian van Enckevort and Duco van Straten

ABSTRACT. This paper contains a preliminary study of the monodromy of certain fourth order differential equations, that were called of Calabi-Yau type in [**3**]. Some of these equations can be interpreted as the Picard-Fuchs equations of a Calabi-Yau manifold with one complex modulus, which links up the observed integrality to the conjectured integrality of the Gopakumar-Vafa invariants. A natural question is if in the other cases such a geometrical interpretation is also possible. Our investigations of the monodromies are intended as a first step in answering this question. We use a numerical approach combined with some ideas from homological mirror symmetry to determine the monodromy for some further one-parameter models. Furthermore, we present a conjectural identification of the Picard-Fuchs equation for 5 new examples from Borcea's list and conjecture the existence of some new Calabi-Yau threefolds. The paper does not contain any theorems or proofs but is, we think, nevertheless of interest.

1. Introduction

A differential operator of order n on $\mathbb{P}^1$ has the form

$$L := a_n(z)\frac{d^n}{dz^n} + a_{n-1}(z)\frac{d^{n-1}}{dz^{n-1}} + \cdots + a_0(z), \tag{1}$$

where the $a_i(z)$ are polynomials. The set $\Sigma \subset \mathbb{P}^1$ of singular points is given by the zeros of $a_n(z)$ and possibly $z = \infty$. The solutions to the equation $Ly = 0$ can be considered as a $\mathbb{C}$-local system $\mathbb{L}$ of rank n on $S := \mathbb{P}^1 \setminus \Sigma$. After the choice of a base point $s \in \mathbb{P}^1 \setminus \Sigma$, the information of $\mathbb{L}$ is given by the monodromy representation

$$\pi_1(S, s) \longrightarrow \mathrm{Aut}(\mathbb{L}_s) = \mathrm{GL}_n(\mathbb{C})$$

A power series $y_0(x) \in \mathbb{Z}[[x]]$ that satisfies a homogeneous linear differential equation as above is a G-function and a folklore conjecture that goes back to Bombieri and Dwork states that all such power series and differential operators have a *geometrical origin* (see [**29**]). This means that the operator should occur as a *factor* of a Picard-Fuchs operator describing the variation of a cohomology of a family $\rho : \mathcal{Y} \longrightarrow \mathbb{P}^1$, with singular fibres over Σ and defined over a number field. The local system $\mathbb{L}$ should then be a summand of a local system $\mathbb{L}_{\mathbb{C}} := R^d\rho_*(\mathbb{C}_{\mathcal{Y}})_{|S}$, where

2000 *Mathematics Subject Classification.* Primary 14J32; Secondary 14D045, 32S40, 81T30.
Key words and phrases. Calabi–Yau manifolds, mirror symmetry, monodromy, relations with differential equations and D-modules, structure of families, string and superstring theories.

d is the complex dimension of the fibres of ρ. It follows among other things that the equation has regular singularities with all exponents rational. It can be shown that the set of power series of geometric origin in this sense is closed under the ordinary product of power series and under the coefficientwise Hadamard product of series. On the level of local systems, the Cauchy product corresponds to the tensor product, whereas the Hadamard product corresponds to the convolution of local systems. We refer to the books [**5**] and [**22**] for details.

The fourth order equations in [**3**] were collected with a stricter notion of geometrical origin in mind: by requiring that the operator admits an invariant symplectic form and gives rise to integral instanton numbers, it starts making sense asking for the existence of a one-parameter family $\mathcal{Y}\longrightarrow\mathbb{P}^1$ of *Calabi-Yau threefolds*, whose associated Picard-Fuchs operator for $\mathrm{H}^3(Y_s)$ is the given one. The instanton numbers then should have the interpretation of counting curves on a mirror manifold X with Picard number one. The first 14 equations in the list are in fact the much studied hypergeometric cases (see [**14**], [**33**], [**26**], [**9**], [**35**], [**15**]). Mirror pairs of Calabi-Yau threefolds obtained from Batyrev's polar duality of reflexive polytopes [**7**] yield a plethora of examples but usually with high Picard number (see [**25**]). By taking restrictions to carefully chosen one-dimensional sub-loci these examples sometimes give rise to equations of Calabi-Yau type, but the instanton numbers computed in this way represent *sums over different homology classes* and there will not exist a Calabi-Yau threefold X with Picard number one with the given instanton numbers. Case 15 is an example of this phenomenon: it is the equation belonging to the diagonal restriction of Calabi-Yau family in $\mathbb{P}^3\times\mathbb{P}^3$ (see [**9**]). The list contains many more such examples. The question is how can one see this from the differential equation alone.

In order to find the cases that are potentially of strictly geometric origin, we remark that a geometrical local system $\mathbb{L}_{\mathbb{C}}$ carries an integral lattice $\mathbb{L}_{\mathbb{Z}} = R^d\rho_*(\mathbb{Z}_{\mathcal{Y}})|_S$ and that Poincaré duality provides it with a *unimodular* pairing $\langle\cdot,\cdot\rangle$, which in our case is alternating. Hence the monodromy representation is in the symplectic group $\mathrm{Sp}(4,\mathbb{Z})$. For differential equations of hypergeometric type, the monodromy representation is explicitly known, essentially because the associated local system is rigid (Levelt's theorem, [**12**],[**22**]). This leads to the 14 hypergeometric cases mentioned above. For equations with three singular points which are not of hypergeometric type or for equations with more than three singular points, the monodromy representation is in general not determined by local data alone and we have the problem of accessory parameters. We do not know of any general method to determine the monodromy representation in such cases. We use a brute force numerical approach combined with ideas from homological mirror symmetry to conjecturally determine the monodromy for some further one-parameter models.

Acknowledgements: Our thanks go to G. Almkvist, C. Doran, A. Klemm, and W. Zudilin for their interest in the project. In particular we thank C. Doran for his explanation of the integral basis and A. Klemm for the suggestion of using the genus one instanton numbers as an extra integrality check and for explanations on higher genus computations.

2. Sketch of homological mirror symmetry

According to Kontsevich [**27**], the phenomenon of mirror symmetry between Calabi-Yau spaces X and Y should be formulated in terms of equivalence of categories. To a Calabi-Yau space X one can associate two triangulated categories, namely the derived category of coherent sheaves $D^b(X)$ and a derived Fukaya-category $D\mathcal{F}(X)$ of Lagrangian cycles (graded, with local systems on them) in X (see [**17**]). The first category depends only on the holomorphic moduli, the second only on the symplectic (or Kähler) moduli. Mirror symmetry between Calabi-Yau spaces X and Y is then expressed as equivalences of categories.

$$\mathrm{Mir} : D^b(X) \xrightarrow{\approx} D\mathcal{F}(Y), \quad D^b(Y) \xrightarrow{\approx} D\mathcal{F}(X)$$

These equivalences induce isomorphisms between the corresponding K-groups. Via the Chern character they descend to cohomology:

$$\mathrm{mir} : \mathrm{H}^{\mathrm{ev}}(X, \mathbb{Q}) \xrightarrow{\approx} \mathrm{H}^d(Y, \mathbb{Q}), \quad \mathrm{H}^{\mathrm{ev}}(Y, \mathbb{Q}) \xrightarrow{\approx} \mathrm{H}^d(X, \mathbb{Q}),$$

where $d = \dim_C X = \dim_C Y$. This also induces an isomorphism between the Kähler moduli $\mathrm{H}^{1,1}(X)$ of X and the complex moduli of $\mathrm{H}^{d-1,1}(Y)$ of Y. In the Strominger-Yau-Zaslow picture of mirror symmetry (see [**37**], [**20**]) X and Y are represented as (real) singular torus fibrations over a common base B. The fibres are dual tori and mirror symmetry should correspond to fibrewise T-duality. From this one can get some intuitive understanding of the mirror transformation on objects. In particular, the structure sheaf $\mathcal{O}_p$ of a point $p \in X$ gets mapped to a SYZ-fibre $\mathbf{T}$ (with a local system on it) in Y and the structure sheaf $\mathcal{O}_X$ should map to the image $\mathbf{S}$ of a section $\sigma : B \longrightarrow Y$ of the fibration.

For any pair $(\mathcal{E}, \mathcal{F})$ of objects of $D^b(X)$ the Euler bilinear form is defined by

$$\langle \mathcal{E}, \mathcal{F} \rangle := \chi(\mathcal{E}, \mathcal{F}) = \sum_i (-1)^i \dim \mathrm{Hom}(\mathcal{E}, \mathcal{F}[i]).$$

which by Serre duality and triviality of the canonical bundle is $(-1)^d$ symmetric. It descends via the Chern character to a bilinear form $\langle \cdot, \cdot \rangle$ on the cohomology $\mathrm{H}^{\mathrm{ev}}(X, \mathbb{Q})$ of X, which by Riemann-Roch is given by

$$\langle \alpha, \beta \rangle = \int_X \tilde{\alpha} \cup \beta \cup \mathrm{td}(X),$$

where $\tilde{\alpha} = (-1)^k \alpha$ for $\alpha \in \mathrm{H}^{2k}(X, \mathbb{Q})$.

Under the mirror transformation the form $\langle \cdot, \cdot \rangle$ should correspond to the intersection form $\langle \cdot, \cdot \rangle$ of the corresponding Lagrangians. One instance of this can easily be checked:

$$\langle \mathcal{O}_p, \mathcal{O}_X \rangle = 1 = \langle \mathbf{T}, \mathbf{S} \rangle.$$

3. Monodromy in one-parameter models

From now on we assume that X and Y are strict Calabi-Yau threefolds and furthermore that they satisfy $h^{2,1}(Y) = 1 = h^{1,1}(X)$. This is the case of so called one-parameter models: Y varies in a one-dimensional moduli space and X has one Kähler modulus, i.e., $\mathrm{Pic}(X) = \mathbb{Z}$. In such a case one has $\dim \mathrm{H}^3(Y) = 4 = \dim \mathrm{H}^{\mathrm{ev}}(X)$.

To be specific, we assume that we have a proper map $\rho : \mathcal{Y} \longrightarrow \mathbb{P}^1$, smooth outside singular fibres that sit over points from $\Sigma \subset \mathbb{P}^1$ and furthermore that Y is the fibre over a base-point $s \in \mathbb{P}^1 \setminus \Sigma =: S$. As the geometrical monodromy along a path $\gamma \in \pi_1(S, s)$ can be realised as a symplectic map $M(\gamma) : Y_s \longrightarrow Y_s$, $M(\gamma)$ induces an autoequivalence of its symplectic invariant $D\mathcal{F}(Y)$, thus setting up a homomorphism

$$\pi_1(S, s) \longrightarrow \mathrm{Auteq}(D\mathcal{F}(Y))$$

which is a refined version of the ordinary monodromy representation of $\pi_1(S,s)$ on $\mathrm{H}^{\mathrm{odd}}(Y)$. The group $\pi_1(S,s)$ is generated by paths that encircle one of the singular fibres of the family. The induced transformation is determined by the specific properties of the singular fibre. If the fibre acquires the simplest type of singularity, namely an A_1-singularity ('conifold'), there is a vanishing Lagrangian 3-sphere. The geometrical monodromy is then a Dehn-twist along this sphere and its effect on homology is given by the classical Picard-Lefschetz transformation [**30**], [**6**], [**32**]:

$$\alpha \mapsto S_\delta(\alpha) := \alpha - \langle \delta, \alpha \rangle \delta$$

where δ is the homology class of the vanishing cycle. In the situation of mirror symmetry there also will be a point of degeneration with maximal unipotent monodromy. The fibre will typically have normal crossing singularities and there will be a 'vanishing n-torus', invariant under the monodromy.

Using the mirror equivalence Mir we get a representation

$$\pi_1(S,s) \longrightarrow \mathrm{Auteq}(D^b(X))$$

and one may ask what sort of autoequivalences correspond to specific types of degenerations of Y.

In [**36**] Seidel and Thomas described a type of autoequivalence in $D^b(X)$ to mirror a symplectic Dehn-twist. It is the Seidel-Thomas twist $T_{\mathcal{E}}$ by a so called spherical object $\mathcal{E}$ of $D^b(X)$, which has the property that $\dim(\mathrm{Ext}^*(\mathcal{E},\mathcal{E})) = \dim \mathrm{H}^*(\mathbf{S})$ and is given by the triangle

$$\longrightarrow (\mathcal{E},\mathcal{F}) \otimes \mathcal{E} \longrightarrow \mathcal{F} \longrightarrow T_{\mathcal{E}}(\mathcal{F}) \xrightarrow{+1}$$

The structure sheaf $\mathcal{O}_X$ is the basic spherical object in $D^b(X)$, but also each line bundle $L \in \mathrm{Pic}(X)$ is spherical. Another particularly simple type of autoequivalence is the operation $\otimes L$ of tensoring with a line bundle L. Note that $\mathcal{O}_p \otimes L = \mathcal{O}_p$. This fits on the mirror side with the monodromy tranformation around a point of maximal unipotent monodromy, with invariant vanishing torus $\mathbf{T}$.

Let us write out these transformations on the level of cohomology. Let $L = \mathcal{O}(H)$ be the ample generator of $\mathrm{Pic}(X)$. The powers $1, H, H^2, H^3$ form a basis for $\mathrm{H}^{\mathrm{ev}}(X,\mathbb{Q})$. With respect to this basis, the matrix T of tensoring with L is given by

$$T = \begin{pmatrix} 1 & 0 & 0 & 0 \\ 1 & 1 & 0 & 0 \\ \frac{1}{2} & 1 & 1 & 0 \\ \frac{1}{6} & \frac{1}{2} & 1 & 1 \end{pmatrix}, \tag{2}$$

as easily follows from $\mathrm{ch}(L \otimes \mathcal{E}) = \mathrm{ch}(L) \cup \mathrm{ch}(\mathcal{E}) = e^H \cup \mathrm{ch}(\mathcal{E})$.

The twist $T_{\mathcal{O}_X}$ on the level of cohomology is given by $\gamma \mapsto \gamma - \int_X \gamma \cup \mathrm{td}(X) \cdot 1$ and hence its matrix is given by

$$S = \begin{pmatrix} 1 & -c & 0 & -d \\ 0 & 1 & 0 & 0 \\ 0 & 0 & 1 & 0 \\ 0 & 0 & 0 & 1 \end{pmatrix} \tag{3}$$

where

$$d := H^3, \qquad c := c_2 \cdot H/12.$$

The matrix Q representing the bilinear form $\langle \cdot, \cdot \rangle$ in this basis is given by

$$Q = \begin{pmatrix} 0 & c & 0 & d \\ -c & 0 & -d & 0 \\ 0 & d & 0 & 0 \\ -d & 0 & 0 & 0 \end{pmatrix}$$

Now Kontsevich [**28**] observed the miracle that for the quintic and its mirror the matrices T and S indeed correspond to monodromy matrices of the Picard-Fuchs operator

$$\theta^4 - 5^5 z(\theta + \tfrac{1}{5})(\theta + \tfrac{2}{5})(\theta + \tfrac{3}{5})(\theta + \tfrac{4}{5})$$

where $\theta = z\frac{d}{dz}$. It has 0, $1/5^5$ and ∞ as singular points. In an appropriate basis, the monodromy around 0 is given by T and around $1/5^5$ by S.

We see that apparently the following happens: there is a point of maximal unipotent monodromy, corresponding to $\otimes\mathcal{O}(H)$ in $\mathrm{Auteq}(D^b(X))$ and there is a conifold point, corresponding to the twist along $\mathcal{O}_X$.

Similar things occur in all the 14 hypergeometric cases. As there are only three singular points in these cases, these two monodromies generate the monodromy group. We refer to [**21**] for a generalisation to Calabi-Yaus in more general toric manifolds.

Calabi-Yau spaces with Picard number one seem to be rather scarce. Apart from the 13 hypergeometric families there is a list (not claiming completeness in any sense) by Borcea [**13**] containing 11 further cases. The examples are ramified covers and complete intersections in Fano varieties with Picard number one. We know of a few other cases. Basic invariants for such X are the *degree* $d := H^3$, the *second Chern number* $c_2 \cdot H$ and the Euler number $c_3 = \chi_{top}$, of which the first two can be read off from the matrix S.

It is sometimes more convenient to work with a different representation based on the one used by C. Doran and J. Morgan (see [**15**]). That basis can be obtained from the one above using the coordinate transformation given by the matrix

$$W = \begin{pmatrix} 0 & 0 & 0 & 1 \\ 0 & 0 & 1 & -1 \\ 0 & \frac{1}{d} & -\frac{1}{2} & \frac{1}{3} - \frac{c}{d} \\ \frac{1}{d} & 0 & -\frac{c}{d} & \frac{c}{d} \end{pmatrix}$$

where c and d are as above. This yields the following representation:

$$T_{\mathrm{DM}} = W^{-1}TW = \begin{pmatrix} 1 & 1 & 0 & 0 \\ 0 & 1 & d & 0 \\ 0 & 0 & 1 & 1 \\ 0 & 0 & 0 & 1 \end{pmatrix}, \quad S_{\mathrm{DM}} = W^{-1}SW = \begin{pmatrix} 1 & 0 & 0 & 0 \\ -k & 1 & 0 & 0 \\ -1 & 0 & 1 & 0 \\ -1 & 0 & 0 & 1 \end{pmatrix}$$

$$Q_{\mathrm{DM}} = W^tQW = \begin{pmatrix} 0 & 0 & 0 & 1 \\ 0 & 0 & -1 & 1 \\ 0 & 1 & 0 & -k \\ -1 & -1 & k & 0 \end{pmatrix}$$

Here $d = H^3$ and $k = \frac{c_2 \cdot H}{12} + \frac{H^3}{6}$. This last number has a simple interpretation as the dimension $\dim(H^0(X, \mathcal{O}(H))$ of the linear system $|H|$.

4. Computation of the monodromy

Our starting point for the computation of the monodromy is the following working hypothesis

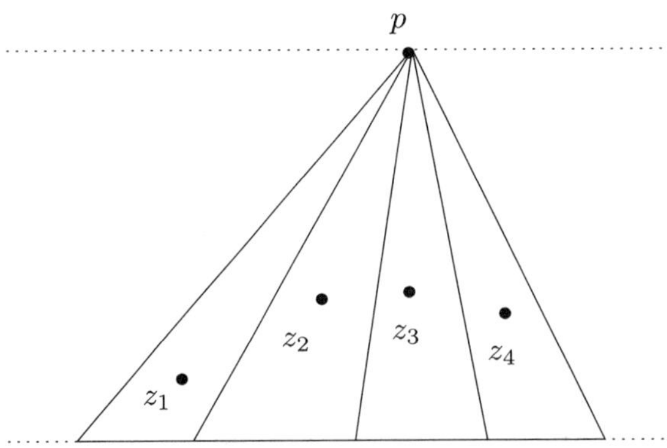

FIGURE 1. Piecewise linear loops around the critical points z_i

HYPOTHESIS 1. *For any differential equation of Calabi-Yau type which is strictly geometrical and for which the instanton numbers have an interpretation as the numbers of curves on a mirror manifold, the monodromy should satisfy the following conditions:*

(H1) *There is a point of maximal unipotent monodromy, correspond-ing to* $\otimes\mathcal{O}(H)$ *in* $\mathrm{Auteq}(D^b(X))$.

(H2) *There is a conifold point, corresponding to the twist along* $\mathcal{O}_X$.

By construction all the equations in the list from [**2**] have a point of maximal unipotent monodromy at $z = 0$. The non-obvious part is to find a conifold point. We observed that in the cases where we know the conifold point the *spectrum*, i.e., the set of zeros of the indicial equation at that point, was $\{0, 1, 1, 2\}$. This is also suggested by Hodge theory. Therefore as a practical selection criterion, we computed the indicial equations at the singular points of all equations and found the equations with at least one singular point with spectrum $\{0, 1, 1, 2\}$. As of the time of writing of this article there were 178 such equations in our database. In many cases there are several such points, but there are also some notable exceptions, where no such singular point exists. An example is equation 32, which is related to $\zeta(4)$ (see [**3**]). For the moment, we are unable to find integral or even just rational lattices for these cases.

For all the 178 equations that do have at least one singular point with spectrum $\{0, 1, 1, 2\}$ we computed high precision numerical approximations for a set of generators of the monodromy group. These computations were done in `Maple`. The first step was to determine the critical points $z_1, \ldots, z_\ell$ and to choose a reference point p. Next, for each of the critical points z_i except the point $z = \infty$ we choose a piecewise linear loop starting and ending at the reference point p and enclosing only one critical point, namely z_i (see Figure 1).

Using the `Maple`-function `dsolve` we can numerically integrate the differential equation along these paths. It turns out to be a bit tricky to obtain the precision needed for the next steps. We used the following options: method=gear, relerr=10^{-15}, abserr=10^{-15} and also increased `Digits` to 100. This yielded the monodromy matrices with respect to an arbitrary basis and produces fully filled 4×4-matrices with seemingly random complex entries.

At this point there is a simple consistency check that we can do. If there exists an integral lattice, the characteristic polynomial of each of the monodromies should be a polynomial with integral coefficients. As a further check, the roots of the indicial equations at the corresponding singular points should be logarithms of the roots of the characteristic polynomial. For the MUM-point and the points with spectrum $\{0,1,1,2\}$ the characteristic polynomial should be $(1-\lambda)^4$. This provides an indication of the precision we have achieved.

The next step is to try and find a simultaneous base change that makes all matrices integral, i.e., to find a monodromy invariant lattice Λ. The crucial observation is the following. The monodromy S around an A_1-singularity has the property that $\mathrm{rk}(S - \mathrm{Id}) = 1$. The one-dimensional image of $S - \mathrm{Id}$ is the span of the vanishing cycle. Now choose one of the singular points with spectrum $\{0,1,1,2\}$ and call the monodromy around the loop enclosing this singular point S. As we are working with numerical approximations we cannot expect $S - \mathrm{Id}$ to have rank 1, but we can hope that the columns of the matrix $S - \mathrm{Id}$ are nearly proportional. In that case we can pick an arbitrary vector and apply $S - \mathrm{Id}$ to it. In this way we find a vector v_0 that should be a good approximation to a lattice vector.

Further lattice vectors $v_1, \ldots, v_k$ can be obtained by applying words in the numerically computed monodromy matrices to v_0. By picking n independent vectors among the ones found in this way, we should find a basis for $\Lambda \otimes \mathbb{Q}$. When we transform the monodromy matrices to this basis, the resulting matrices should have rational entries. Of course this will not be exact, but we can try to find rational matrices close to the matrices that we do find. For this we used continued fractions. It may happen that we get very large denominators or that the rational approximation is not very accurate. In that case we can try another set of n independent vectors among the v_i. If that is not successful, we can try another point with spectrum $\{0,1,1,2\}$, if there is any. As a consistency check, we can compute the characteristic polynomials of these rational matrices and check that they have integral coefficients. As noticed above, at the MUM-point and the conifold point the characteristic polynomial should be $(1-\lambda)^4$, which we can also check. If any of these checks fails, we have to try again with a different basis or a different singular point with spectrum $\{0,1,1,2\}$. However, it can and does happen that we try all potential conifold points and several choices of a basis in each case, but do not find a rational basis. We did find a rational basis in 143 of the 178 investigated cases.

The rational basis found in this way is still rather arbitrary. However, a major advantage is that at this point we expect to be working with the *exact* monodromy matrices. This allows us to do linear algebra without worrying about the extra complications of working with non-exact numerical approximations. Provided that the monodromy matrices around the MUM-point and the conifold point have the right Jordan structure, we can find a new basis such that with respect to this basis they have the standard form (2) and (3). In a geometrical situation we expect the transformed matrices to be integral. This happens in 64 cases. When we have the monodromies around the MUM-point and the conifold point in the standard form, we can read off the invariants H^3 and $c_2 \cdot H$ and try to match the invariants with those of known Calabi-Yau spaces.

Despite our efforts to identify equivalent Calabi-Yau equations our list probably still includes some Calabi-Yau equations that correspond to the same geometrical situation. Transformations in the parameter z are a way of constructing seemingly different equations that actually describe the same geometrical situation. In a geometrical language this corresponds to pullback under a map $f : \mathbb{P}^1 \to \mathbb{P}^1$. If the map f is not injective, this may increase the number of singular points. As long as the map f is unramified around the MUM-point and the conifold point it does not change the monodromies and we ought to find the same H^3 and $c_2 \cdot H$. So as a practical way of trying to group together the equations that correspond to the same geometry, we sort the 64 integral equations we found according to H^3 and $c_2 \cdot H$. If we find several equations with the same H^3 and $c_2 \cdot H$, it turns out that the genus zero instanton numbers also coincide. This is a strong indication that these equations are equivalent.

5. Conifold-period and Euler characteristic

If Ω is a family of holomorphic three-forms on Y_s (that is, a section of $\mathcal{L} := \rho_*(\omega_{\mathcal{Y}}/S)$) and Γ is a horizontal family of cycles, then the *periods*

$$\int_\Gamma \Omega$$

are the solutions of the associated Picard-Fuchs equation. In our situation we identified two cycles, namely the torus $\mathbf{T}$ near the MUM-point, and the vanishing sphere $\mathbf{S}$ near the conifold point z_c. Correspondingly we have the fundamental period $\int_{\mathbf{T}} \Omega$ which is the unique holomorphic solution near the MUM-point. Equally important is the period $\int_{\mathbf{S}} \Omega$ which we call the *conifold-period* and which was called $z_2(t)$ in the paper [**14**]. As the local monodromy around the conifold point is supposed to be a symplectic reflection in S, this special period can be determined directly from the differential equation as follows. At such a conifold point there exists a basis of solutions to the Calabi-Yau equation around this point that consists of three power series solutions and one solution of the form

$$y(z) = f(z)\log(z - z_c) + g(z).$$

Going around z_c once y is replaced by $y(z) + 2\pi i f(z)$. The power series $f(z)$ represents a special solution around $z = z_c$ that is determined up to a multiplicative scalar and which we call the *conifold-period.* This function can be continued analytically around an arbitrary path which avoids the singularities of the differential equation. It is a remarkable fact that in all (but one, namely nr. 224) of the examples we know, the point z_c is the singular point that is *closest to the origin.* So there is a preferred path from z_c to 0 by going along a straight line and we consider the analytic continuation along this path. In [**14**] the expansion of the conifold-period around 0 is derived for the quintic. It has the form

$$z_2(t) = \frac{H^3}{6}t^3 + \frac{c_2 \cdot H}{24}t + \frac{c_3}{(2\pi i)^3}\zeta(3) + O(q). \tag{4}$$

The term $O(q)$ stands for any terms containing $q = e^{2\pi i t}$ and $t = \frac{1}{2\pi i}\frac{y_1(z)}{y_0(z)} = \frac{1}{2\pi i}\log z + \frac{1}{2\pi i}\frac{f_1(z)}{f_0(z)}$ instead of (6). Remarkable here is the 'constant term' $\frac{c_3}{(2\pi i)^3}\zeta(3)$. This term is related to the four-loop correction to the free energy F_0 introduced in [**14**].[1]

One can conjecture this expansion to hold in all cases, which leads to the following algorithm to determine c_3. One can easily compute an expansion of $y(z)$ to an arbitrary number of terms, e.g., using the `Maple`-function `formal_sol` from the `DEtools`-package, as we did. This allows us to find $f(z)$ as the coefficient of

---

[1] To be precise, one has an expansion (see [**23**]):

$$F_0 = H^3\frac{t^3}{3!} + (c_2 \cdot H)t + \frac{\chi}{2}\zeta(3) + \sum_{d=1}^{\infty} n_d^0 \mathrm{Li}_3(q^d)$$

where $\mathrm{Li}_3(x) := \sum_{k=1}^{\infty} k^{-3}x^k$ is the classical trilogarithm.

$\log(z - z_c)$. Around the MUM-point $z = 0$ we can compute expansions of the elements $y_i(z)$ of the Frobenius basis (see Appendix B). We suppose that the domains of convergence of the solutions around $z = 0$ and those around $z = z_c$ overlap. That enables us to pick some point z_* where both expansions converge. Computing numerically $f^{(k)}(z_*)$ $(k = 0, \ldots, 3)$ and $y_i^{(k)}(z_*)$ $(k, i = 0, \ldots, 3)$, we can consider the equations

$$f^{(k)}(z_*) = \sum_{i=0}^{3} c_i y_i^{(k)}(z_*).$$

These equations can be solved for the c_i and determine the analytic continuation z_2 around 0 of $f(z)$ as a linear combination of the $y_i(z)$,

$$z_2 = \sum_{i=0}^{3} c_i y_i(z).$$

From this we can readily read of the expansion of z_2 in t. At this point we can already check that the coefficient of t^2 vanishes. As the conifold-period $f(z)$ was only determined up to a constant, of course $z_2(t)$ is determined up to a constant. If we suppose that H^3 is known, then one can multiply the expansion for $z_2(t)$ by a constant such that the coefficient of t^3 is $\frac{H^3}{6}$. We can then read off $c_2 \cdot H$ and c_3. In practice we find H^3 as discussed above from the monodromy generators. This also yields $c_2 \cdot H$, so we have one more consistency check. It is remarkable that in all cases we indeed find an integral value of c_3!

6. Comments on the table of Calabi-Yau equations

In Table 1 the heading Sings denotes the number of singular point of the (first mentioned) differential equation. An additional $*$ indicates that an apparent singularity is present, around which there is no monodromy. The notation $X(\ldots)$ denotes a complete intersection of the indicated degrees in the indicated manifold. Apart from the familiar 13 hypergeometric cases and the cases from complete intersections in Grassmannians that were studied in in [**8**], one finds a few notable further cases. First there is the elusive 14th hypergeometric case, observed in [**4**] and [**15**]. Any complete intersection $X(2, 12)$ inside $\mathbb{P}(1, 1, 1, 1, 4, 6)$ has a singular point of type $A_1/(\mathbb{Z}/2)$, which does not admit a crepant resolution. However, the inferred value $h^{2,1} = 243$ fits with the proposal in [**24**].

The case $X \xrightarrow{2:1} B_5$ is the Calabi-Yau double cover of the Fano threefold B_5, which is nothing but the three-dimensional section of Grass$(2, 5)$, which is no. 14 in the list of Borcea. We found a fit with the equation 51 from [**2**]. A mirror for this Calabi-Yau is not known, but we conjecture the Picard-Fuchs equation to be the indicated one. We find similar fits for

- **$X(1, 1, 1, 1, 1, 1, 2) \subset X_{10}$:** Here $X_{10} \subset \mathbb{P}^{15}$ is the 10-dimensional spinor variety of isotropic 4-planes in the 8-dimensional quadric.
- **$X(1, 1, 2) \subset$ LGrass$(3, 6)$:** LGrass$(3, 6)$ $=$ Sp$(3, \mathbb{C})/P(\alpha_3) \subset$ Grass$(3, 6)$ is the Lagrangian Grassmannian.
- **$X(1, 2) \subset X_5$:** Here $X_5 = G_2/P(\alpha_{\text{long}}) \subset$ Grass$(5, 7)$ is the space of 5-dimensional subspaces isotropic for a 4-form on a 7-dimensional space.

These are complete intersections inside homogeneous spaces. In principle one can calculate the Picard-Fuchs equation for the instanton numbers for these cases and verify our conjecture. The first method consists in computing the quantum cohomology of these homogeneous examples (for example by fixed point localisation)

TABLE 1. **Calabi-Yau equations with integral monodromy**

H^3	$c_2 \cdot H$	c_3	$\|H\|$	Sings	Database	Description	Reference
1	10	48?	1	4*	225		
1	22	−120	2	3	13	$X(6,6) \subset \mathbb{P}^5(1,1,2,2,3,3)$	[**26**]
1	34	−288	3	3	2	$X(10) \subset \mathbb{P}^4(1,1,1,2,5)$	[**33**]
1	46	−484	4	3	9	$X(2,12) \subset \mathbb{P}^5(1,1,1,1,4,6)$(?)	[4],[**15**]
2	20	−44	2	4*	271		
2	32	−156	3	3	12	$X(3,4) \subset \mathbb{P}^5(1,1,1,1,1,2)$	[**26**]
2	44	−296	4	3	7	$X(8) \subset \mathbb{P}^5(1,1,1,1,4)$	[**33**]
3	42	−204	4	3	8, 125	$X(6) \subset \mathbb{P}^4(1,1,1,1,2)$	[**33**]
4	40	−144	4	3	10	$X(4,4) \subset \mathbb{P}^5(1,1,1,1,2,2)$	[**26**]
4	52	−256	5	3	14, 85, 86	$X(2,6) \subset \mathbb{P}^5(1,1,1,1,1,3)$	[**26**]
5	38	−100*	4	4*	302		
5	50	−200	5	3	1, 79, 87, 128	$X(5) \subset \mathbb{P}^4$	[**14**]
5	62	−310	6	4	63		
6	36	−72	4	4*	33		
6	48	−156	5	3	11, 95	$X(4,6) \subset \mathbb{P}^5(1,1,1,2,2,3)$	[**26**]
7	46	−120*	5	4*	109		
8	32	−8	4	4	291		
8	56	−176	6	3	6, 75, 76, 96	$X(2,4) \subset \mathbb{P}^6$	[**31**]
9	30	12?	4	4	73		
9	54	−144	6	3	4	$X(3,3) \subset \mathbb{P}^5$	[**31**]
10	40	−50	5	5*	118		
10	40	−32	5	4*	292		
10	52	−116*	6	4*	263		
10	64	−200	7	4	51	**Conj:** $X \xrightarrow{2:1} B_5$	[**13**, nr. 14]
12	36	−32	5	5*	117		
12	48	−60	6	5*	267		
12	60	−144	7	3	5, 90, 91, 93	$X(2,2,3) \subset \mathbb{P}^6$	[**31**]
13	58	−120	7	4*	99	**Conj:** 5×5-Pfaffian $\subset \mathbb{P}^6$	[**40**]
14	56	−98	7	5*	222	7×7-Pfaffian $\subset \mathbb{P}^6$	[**34**]
14	56	−100	7	5*	289		
15	54	−78	7	?	?	$\mathrm{To}_{15} \subset \mathbb{P}^6$	[**40**]
15	66	−150	8	4	24	$X(1,1,3) \subset \mathrm{Grass}(2,5)$	[**8**]
16	52	−60	7	?	?	$\mathrm{To}_{16} \subset \mathbb{P}^6$	[**40**]
16	64	−128	8	3	3, 72, 224	$X(2,2,2,2) \subset \mathbb{P}^7$	[**31**]
17	50	−44	7	?	?	$\mathrm{To}_{17} \subset \mathbb{P}^6$	[**40**]
18	60	−88	8	4	266		
20	68	−120	9	4	25	$X(1,2,2) \subset \mathrm{Grass}(2,5)$	[**8**]
21	66	−102	9	5*	254		
21	66	−100	9	5*	270		
24	72	−116	10	4	29	**Conj:** $X(1,1,1,1,1,1,2) \subset X_{10}$	[**13**, nr. 6]
25	70	−100*	10	5*	101		
28	76	−116	11	4	26	$X(1,1,1,1,2) \subset \mathrm{Grass}(2,6)$	
29	74	−100*	11	5*	256		
32	80	−116	12	4	42	**Conj:** $X(1,1,2) \subset \mathrm{LGrass}(3,6)$	[**13**, nr. 8]
33	78	−102*	12	5*	259		
34	76	−88	12	4*	255		
36	72	−72	12	5*	100		
36	84	−120	13	4	184	**Conj:** $X(1,2) \subset X_5$	[**13**, nr. 9]
42	84	−98	14	5*	27	$X(1,1,1,1,1,1,1) \subset \mathrm{Grass}(2,7)$	[**8**]
42	84	−96	14	4	28	$X(1,1,1,1,1,1) \subset \mathrm{Grass}(3,6)$	[**8**]
44	92	−128	15	?	?	$X \xrightarrow{2:1} A_{22}$ or A'_{22}	[**13**, nr. 10]
47	86	−90*	15	6**	257		
56	92	−92	17	?	?	$X(1,1,1,1) \subset F_1(Q_5)$	[**13**, nr. 24]
57	90	−84	17	5*	247	Tjøtta's example	[**38**]

and then using the quantum Lefschetz hyperplane principle. A second method consists of finding a toric degeneration and then using polar duality. Such toric degenerations have been constructed for all spherical varieties in [**1**]. Both methods were used in [**8**] for the case of complete intersections in Grassmannians.

In his thesis [**40**], F. Tonoli considers Calabi-Yau varieties in $\mathbb{P}^6$ from degree 12 up to 17. The first one is the complete intersection $X(2,2,3)$, the second one the 5×5-Pfaffian, for which we found a fit with the data from equation 99. The 7×7-Pfaffian was considered in [**34**]. The remaining three cases are new Calabi-Yau threefolds for which we have not yet found corresponding Picard-Fuchs equations.

The column for the Euler characteristic c_3 was determined using the expansion of the conifold-period around the MUM-point. It is a miracle that we found integral values in all cases (except 224). This checked with the known Euler number in those cases where a geometrical interpretation was known. However, there are two notable cases where we get a *positive* value for c_3, which excludes an interpretation as a Calabi-Yau space with Picard number one. Furthermore, the conjectural integrality of elliptic intanton numbers implies some congruence property on c_3. In most cases this was satisfied, giving a strong indication that a Calabi-Yau threefold with the indicated invariants should exist. In some cases however, we found non-integral n^1_d in this way. This is indicated with a $*$ after the value for c_3.

In the database column we indicate the number of the equation in the electronic database of Calabi-Yau equations that can be found at the web address

http://enriques.mathematik.uni-mainz.de/enckevort/db

Up to 180 these numbers coincide with the ones used in [**2**]. For higher numbers one should check the source field in the electronic database. If it contains Almkvist[n] the corresponding number in [**2**] is n.

7. Some examples

Let us now discuss a few typical examples from Table 1 in more detail. For full information on the other cases, we refer to the database mentioned above.

Example 1. The first equation we want to study is equation 28 from [**2**], which is given by the following operator:

$$L = \theta^4 - z(65\,\theta^4 + 130\,\theta^3 + 105\,\theta^2 + 40\,\theta + 6) + 4z^2(4\,\theta + 3)(\theta + 1)^2(4\,\theta + 5),$$

where $\theta = z\frac{d}{dz}$. This differential operator has four singular points, namely 0, 1/64, 1, and ∞. The Riemann scheme is

$$P\left\{\begin{array}{cccc} 0 & 1/64 & 1 & \infty \\ \hline 0 & 0 & 0 & 3/4 \\ 0 & 1 & 1 & 1 \\ 0 & 1 & 1 & 1 \\ 0 & 2 & 2 & 5/4 \end{array}\right\}$$

Here the columns are the spectra, i.e., the set of solutions to the indicial equation at the singular point indicated above the line. The points 1/64 and 1 have spectrum $\{0, 1, 1, 2\}$, so they are potential conifold points. Using the algorithm discussed above we computed the monodromies around the critical points and found an integral lattice. With respect to this lattice the monodromy matrices are as follows:

$$T = T_0 = \begin{pmatrix} 1 & 1 & 0 & 0 \\ 0 & 1 & 42 & 0 \\ 0 & 0 & 1 & 1 \\ 0 & 0 & 0 & 1 \end{pmatrix}, \quad S = T_{\frac{1}{64}} = \begin{pmatrix} 1 & 0 & 0 & 0 \\ -14 & 1 & 0 & 0 \\ -1 & 0 & 1 & 0 \\ -1 & 0 & 0 & 1 \end{pmatrix},$$

$$T_1 = \begin{pmatrix} 37 & 12 & -252 & 156 \\ -126 & -41 & 882 & -546 \\ -12 & -4 & 85 & -52 \\ -18 & -6 & 126 & -77 \end{pmatrix},$$

$$T_\infty = (T_1 T_{\frac{1}{64}} T_0)^{-1} = \begin{pmatrix} 77 & 29 & -588 & 348 \\ -112 & -41 & 840 & -504 \\ -6 & -2 & 43 & -27 \\ -17 & -6 & 126 & -77 \end{pmatrix}.$$

From this, one can read off the invariants

$$H^3 = 42, \quad c_2 \cdot H = 84.$$

In this case we know that the equation is the Picard-Fuchs equation of the complete intersection $X(1,1,1,1,1,1)$ in Grass$(3,6)$ and we can easily check that these numbers coincide with the ones computed from the geometry. Of course, the value c_3 computed from the expansion of the conifold-period gives the right value -96.

One can easily check that $T_{\frac{1}{64}}$ and T_1 are of the Picard-Lefschetz form $S_{\lambda,v}$, with the vector v given by

$$v_{\frac{1}{64}} = \begin{pmatrix} 0 \\ 14 \\ 1 \\ 1 \end{pmatrix}, \quad v_1 = \begin{pmatrix} 6 \\ -21 \\ -2 \\ -3 \end{pmatrix}.$$

For $T_{\frac{1}{64}}$ we have $\lambda = 1$, but for T_1 we have $\lambda = 2$. So the critical point $z = 1$ is *not* an ordinary conifold point. This $\lambda = 2$ is exactly what is needed to get integral genus one instanton numbers with the recipe from Appendix B. The first few elliptic instanton numbers that we find in this way are

$$n_1^1 = n_2^1 = n_3^1 = n_4^1 = 0, \quad n_5^1 = 84, \quad n_6^1 = 74382, \quad n_7^1 = 8161452.$$

So it appears that there is an $\mathbb{R}P^3$ vanishing at the point 1. The derived category of coherent sheaves on a Grassmannian is reasonably well understood (see [**22**]) and so one can hope to study in detail what happens in $D^b(X)$. This will be pursued at another place [**41**].

Example 2. Our second example has been discussed in [**34, 39**]. It is interesting because there are two points with maximal unipotent monodromy, both of which have a geometrical interpretation. Because our convention is to have the point of maximal unipotent monodromy that we are considering at $z = 0$ this example occurs twice in our list: once as 27 and once as 222.

In the former case the differential operator is given by

$$\begin{aligned} L = {} & 3^2\,\theta^4 - 3z(173\,\theta^4 + 340\,\theta^3 + 272\,\theta^2 + 102\,\theta + 15) \\ & - 2z^2(1129\,\theta^4 + 5032\,\theta^3 + 7597\,\theta^2 + 4773\,\theta + 1083) \\ & + 2z^3(843\,\theta^4 + 2628\,\theta^3 + 2353\,\theta^2 + 675\,\theta + 6) \\ & - z^4(295\,\theta^4 + 608\,\theta^3 + 478\,\theta^2 + 174\,\theta + 26) + z^5(\theta+1)^4 \end{aligned}$$

The Riemann scheme of equation 27 is

$$P\left\{\begin{array}{cccccc} \zeta_1 & 0 & \zeta_2 & 3 & \zeta_3 & \infty \\ \hline 0 & 0 & 0 & 0 & 0 & 1 \\ 1 & 0 & 1 & 1 & 1 & 1 \\ 1 & 0 & 1 & 3 & 1 & 1 \\ 2 & 0 & 2 & 4 & 2 & 1 \end{array}\right\},$$

where $\zeta_1 < \zeta_2 < \zeta_3$ are the (real) roots of $z^3 - 289z^2 - 57z + 1$. The monodromies can be determined with our usual recipe:

$$T_{\zeta_1} = \begin{pmatrix} 15 & 7 & -98 & 49 \\ -28 & -13 & 196 & -98 \\ -2 & -1 & 15 & -7 \\ -4 & -2 & 28 & -13 \end{pmatrix}, \quad T = T_0 = \begin{pmatrix} 1 & 1 & 0 & 0 \\ 0 & 1 & 42 & 0 \\ 0 & 0 & 1 & 1 \\ 0 & 0 & 0 & 1 \end{pmatrix},$$

$$S = T_{\zeta_2} = \begin{pmatrix} 1 & 0 & 0 & 0 \\ -14 & 1 & 0 & 0 \\ -1 & 0 & 1 & 0 \\ -1 & 0 & 0 & 1 \end{pmatrix}, \quad T_3 = \mathrm{Id}, \quad T_{\zeta_3} = \begin{pmatrix} 1 & 0 & 0 & 0 \\ -84 & 1 & 392 & -392 \\ -9 & 0 & 43 & -42 \\ -9 & 0 & 42 & -41 \end{pmatrix},$$

$$T_\infty = (T_{\zeta_3} T_3 T_{\zeta_2} T_0 T_{\zeta_1})^{-1} = \begin{pmatrix} 85 & 6 & -448 & 399 \\ -266 & -13 & 1330 & -1232 \\ -26 & -1 & 127 & -120 \\ -42 & -2 & 210 & -195 \end{pmatrix}.$$

Here the monodromy operators T_{ζ_1}, T_{ζ_2}, and T_{ζ_3} can be written in the Picard-Lefschetz form $S_{1,v}$ with the vector v given by

$$v_{\zeta_1} = \begin{pmatrix} 7 \\ -14 \\ -2 \\ -2 \end{pmatrix}, \quad v_{\zeta_2} = \begin{pmatrix} 0 \\ 14 \\ 1 \\ 1 \end{pmatrix}, \quad v_{\zeta_3} = \begin{pmatrix} 0 \\ 28 \\ 3 \\ 3 \end{pmatrix}.$$

The operator for equation 222 can be obtained by replacing $y(z)$ by $w^{-1}y(w^{-1})$ where $w = 1/z$. In this way one finds the operator

$$\begin{aligned} L = \theta^4 &- z(295\,\theta^4 + 572\,\theta^3 + 424\,\theta^2 + 138\,\theta + 17) \\ &+ 2z^2(843\,\theta^4 + 744\,\theta^3 - 473\,\theta^2 - 481\,\theta - 101) \\ &- 2z^3(1129\,\theta^4 - 516\,\theta^3 - 725\,\theta^2 - 159\,\theta + 4) \\ &- 3z^4(173\,\theta^4 + 352\,\theta^3 + 290\,\theta^2 + 114\,\theta + 18) + 3^2 z^5(\theta + 1)^4. \end{aligned}$$

The Riemann scheme of 222 also follows from that of 27:

$$P\left\{\begin{array}{cccccc} 1/\zeta_1 & 0 & 1/\zeta_3 & 1/3 & 1/\zeta_2 & \infty \\ \hline 0 & 0 & 0 & 0 & 0 & 1 \\ 1 & 0 & 1 & 1 & 1 & 1 \\ 1 & 0 & 1 & 3 & 1 & 1 \\ 2 & 0 & 2 & 4 & 2 & 1 \end{array}\right\}.$$

For the monodromies the relation is not so obvious. Doing the standard computation we find

$$T_{\zeta_1^{-1}} = \begin{pmatrix} 29 & 14 & -98 & 49 \\ -56 & -27 & 196 & -98 \\ -8 & -4 & 29 & -14 \\ -16 & -8 & 56 & -27 \end{pmatrix}, \quad T = T_0 = \begin{pmatrix} 1 & 1 & 0 & 0 \\ 0 & 1 & 14 & 0 \\ 0 & 0 & 1 & 1 \\ 0 & 0 & 0 & 1 \end{pmatrix},$$

$$S = T_{\zeta_3^{-1}} = \begin{pmatrix} 1 & 0 & 0 & 0 \\ -7 & 1 & 0 & 0 \\ -1 & 0 & 1 & 0 \\ -1 & 0 & 0 & 1 \end{pmatrix}, \quad T_{1/3} = \mathrm{Id}, \quad T_{\zeta_2^{-1}} = \begin{pmatrix} 1 & 0 & 0 & 0 \\ -105 & 1 & 294 & -294 \\ -25 & 0 & 71 & -70 \\ -25 & 0 & 70 & -69 \end{pmatrix},$$

$$T_\infty = (T_{\zeta_2^{-1}} T_{1/3} T_{\zeta_3^{-1}} T_0 T_{\zeta_1^{-1}})^{-1} = \begin{pmatrix} 155 & 13 & -476 & 427 \\ -420 & -27 & 1260 & -1162 \\ -76 & -4 & 225 & -211 \\ -126 & -8 & 378 & -349 \end{pmatrix}.$$

Again the monodromies $T_{\zeta_1^{-1}}$, $T_{\zeta_3^{-1}}$, and $T_{\zeta_2^{-1}}$ can be written in the Picard-Lefschetz form $S_{1,v}$ with the vector v given by

$$v_{\zeta_1^{-1}} = \begin{pmatrix} 7 \\ -14 \\ -2 \\ -4 \end{pmatrix}, \quad v_{\zeta_3^{-1}} = \begin{pmatrix} 0 \\ 7 \\ 1 \\ 1 \end{pmatrix}, \quad v_{\zeta_2^{-1}} = \begin{pmatrix} 0 \\ 21 \\ 5 \\ 5 \end{pmatrix}.$$

Despite the fact that we are really dealing with the same equation in a different formulation, the monodromies look rather different. Of course the monodromy groups generated by these matrices are isomorphic, but this is not so easy to see.

Example 3. The next example is equation 29 from [**2**]. The operator is

$$L = \theta^4 - 2z(2\theta+1)^2(17\theta^2+17\theta+5) + 2^2z^2(2\theta+1)(\theta+1)^2(2\theta+3).$$

In this case the Riemann scheme is

$$P\left\{\begin{array}{cccc} 0 & \zeta_1 & \zeta_2 & \infty \\ \hline 0 & 0 & 0 & 1/2 \\ 0 & 1 & 1 & 1 \\ 0 & 1 & 1 & 1 \\ 0 & 2 & 2 & 3/2 \end{array}\right\},$$

where $\zeta_1 < \zeta_2$ are the (real) roots of $1 - 136z + 16z^2$. The monodromy matrices are

$$T = T_0 = \begin{pmatrix} 1 & 1 & 0 & 0 \\ 0 & 1 & 24 & 0 \\ 0 & 0 & 1 & 1 \\ 0 & 0 & 0 & 1 \end{pmatrix}, \quad S = T_{\zeta_1} = \begin{pmatrix} 1 & 0 & 0 & 0 \\ -10 & 1 & 0 & 0 \\ -1 & 0 & 1 & 0 \\ -1 & 0 & 0 & 1 \end{pmatrix},$$

$$T_{\zeta_2} = \begin{pmatrix} 51 & 20 & -240 & 140 \\ -130 & -51 & 624 & -364 \\ -15 & -6 & 73 & -42 \\ -25 & -10 & 120 & -69 \end{pmatrix},$$

$$T_\infty = (T_{\zeta_2} T_{\zeta_1} T_0)^{-1} = \begin{pmatrix} 71 & 31 & -360 & 200 \\ -120 & -51 & 600 & -340 \\ -10 & -4 & 49 & -29 \\ -24 & -10 & 120 & -69 \end{pmatrix}.$$

One can check that the T_{ζ_i} can be written as $S_{1,v}$ with v given by

$$v_{\zeta_1} = \begin{pmatrix} 0 \\ 10 \\ 1 \\ 1 \end{pmatrix}, \quad v_{\zeta_2} = \begin{pmatrix} 10 \\ -26 \\ -3 \\ -5 \end{pmatrix}.$$

From the expressions for T and S we find $H^3 = 24$, $c_2 \cdot H = 72$. We also have enough information to compute the elliptic instanton numbers as a function of c_3. By equating $n_1^1 = 0$ we find $c_3 = -116$ and all the n_i^1 we computed are integral. Thes same value for c_3 is obtained from the expansion of the conifold-period. It turns out that we are *lucky* and that there is exactly one 1-parameter Calabi-Yau known with these invariants, namely $X(1,1,1,1,1,1,2) \subset X_{10}$ (see [**13**]). So we conjecture that equation 29 is the Picard-Fuchs equation corresponding to this Calabi-Yau. In the same way we conjecturally identified the Picard-Fuchs equations of four more 1-parameter Calabi-Yau spaces. We labelled these equations in Table 1 by writing **Conj:** in front of the conjectured Calabi-Yau.

Example 4. As our final example we will use equation 270 (218 in the numbering from [**2**]) which is given by the differential operator

$$\begin{aligned} L = {} & 7^2\,\theta^4 - 42z(192\,\theta^4 + 396\,\theta^3 + 303\,\theta^2 + 105\,\theta + 14) \\ & + 2^2 \cdot 3z^2(1188\,\theta^4 + 11736\,\theta^3 + 20431\,\theta^2 + 12152\,\theta + 2436) \\ & + 2^2 \cdot 3^3 z^3(532\,\theta^4 + 504\,\theta^3 - 3455\,\theta^2 - 3829\,\theta - 1036) \\ & - 6^4 z^4(2\,\theta + 1)(36\,\theta^3 + 306\,\theta^2 + 421\,\theta + 156) \\ & - 2^6 \cdot 3^4 z^5(2\,\theta + 1)(3\,\theta + 2)(3\,\theta + 4)(2\,\theta + 3). \end{aligned}$$

The Riemann scheme is

$$P\left\{\begin{array}{cccccc} -7/12 & 0 & \zeta_1 & \zeta_2 & \zeta_3 & \infty \\ \hline 0 & 0 & 0 & 0 & 0 & 1/2 \\ 1 & 0 & 1 & 1 & 1 & 2/3 \\ 3 & 0 & 1 & 1 & 1 & 4/3 \\ 4 & 0 & 2 & 2 & 2 & 3/2 \end{array}\right\},$$

where ζ_1 is the real root of $1296z^3 - 864z^2 + 168z - 1$ and ζ_2, ζ_3 are its complex roots with $\operatorname{im}\zeta_2 < 0$ and $\zeta_3 = \bar{\zeta}_2$. The monodromies can be computed and turn out to be integral:

$$T_{-\frac{7}{12}} = \mathrm{Id}, \quad T = T_0 = \begin{pmatrix} 1 & 1 & 0 & 0 \\ 0 & 1 & 21 & 0 \\ 0 & 0 & 1 & 1 \\ 0 & 0 & 0 & 1 \end{pmatrix}, \quad T_{\zeta_1} = \begin{pmatrix} 1 & 0 & 0 & 0 \\ -9 & 1 & 0 & 0 \\ -1 & 0 & 1 & 0 \\ -1 & 0 & 0 & 1 \end{pmatrix},$$

$$T_{\zeta_2} = \begin{pmatrix} 16 & 5 & -60 & 40 \\ -45 & -14 & 180 & -120 \\ -6 & -2 & 25 & -16 \\ -9 & -3 & 36 & -23 \end{pmatrix}, \quad T_{\zeta_3} = \begin{pmatrix} 11 & 5 & -45 & 25 \\ -18 & -8 & 81 & -45 \\ -2 & -1 & 10 & -5 \\ -4 & -2 & 18 & -9 \end{pmatrix},$$

$$T_\infty = (T_{\zeta_3} T_{\zeta_2} T_{\zeta_1} T_0 T_{-\frac{7}{12}})^{-1} = \begin{pmatrix} 8 & 5 & -45 & 21 \\ -12 & -5 & 60 & -36 \\ -1 & 0 & 4 & -4 \\ -3 & -1 & 15 & -10 \end{pmatrix}.$$

So we find $H^3 = 21$ and $c_2 \cdot H = 66$. The T_{ζ_i} can be written in Picard-Lefschetz form with $\lambda = 1$ and v given by

$$v_{\zeta_1} = \begin{pmatrix} 0 \\ 9 \\ 1 \\ 1 \end{pmatrix}, \quad v_{\zeta_2} = \begin{pmatrix} 5 \\ -15 \\ -2 \\ -3 \end{pmatrix}, \quad v_{\zeta_3} = \begin{pmatrix} 5 \\ -9 \\ -1 \\ -2 \end{pmatrix}.$$

We can also compute the elliptic instanton numbers. Setting $n_1^1 = 0$ we find $c_3 = -100$ and with this value of c_3 all computed n_d^1 turn out to be integers. The same value of c_3 was obtained from the expansion of the conifold-period. So we have a Calabi-Yau equation that as far as we can check looks like the Picard-Fuchs equation of a Calabi-Yau manifold. However, we do not know any 1-parameter Calabi-Yau with the geometric invariants that we computed. In Table 1 there are some more equations which look geometrical in every respect, but for which we have not yet found any Calabi-Yau.

8. Open problems

The work described in this paper is no more than a start and there are many open problems left. We have found quite a few Calabi-Yau equations that look in every respect like the Picard-Fuchs equation of a Calabi-Yau manifold, but for which we do not know if a Calabi-Yau manifold exists. We know the degree, the second Chern number, the Euler characteristic and the instanton numbers.

To determine an integral lattice we need to single out two singular points, where we bring the monodromies into the the standard forms T_{DM} and S_{DM}. When there are no singular points with spectrum $\{0, 1, 1, 2\}$ we do not have a good candidate for S_{DM} and cannot even start our procedure for determining an integral lattice. It would be interesting to see what can be done in such cases. We also did not look for other integral lattices as in [**15**].

The conjectural appearance of the constant term $c_3\zeta(3)/(2\pi i)^3$ in the expansion of the conifold-period (and the free energy) is very intriguing. Is this a mathematical theorem?

The key obstacle to computing the elliptic instanton numbers is finding the holomorphic function $f(z)$ in (11). Our ansatz in combination with our recipe for determining the exponents works in many cases, but it is no more than an educated guess. A better understanding of the genus one computation in terms of the BCOV-torsion as in [**16**] will probably be helpful.

Many of the equations from the list in [**2**] come from Hadamard products. The singular points of a Hadamard product are given by products of the singular points of the factors. Maybe it is also possible to determine the monodromies of the Hadamard product in terms of the monodromies of the factors.

Appendix A. Orbifolds of A_1

In many examples one encounters monodromy tranformations that are not described by the usual Picard-Lefschetz formula, but rather are *powers* of such operations. We offer a possible explanation of this phenomena, which is only visible on the integral level.

Consider a lattice Λ with bilinear form $\langle \cdot, \cdot \rangle$. For $\beta \in \Lambda$ and $\lambda \in \mathbb{Z}$ consider the transformation

$$S_{\lambda,\beta} : \Lambda \longrightarrow \Lambda, \quad S_{\lambda,\beta}(\alpha) = \alpha - \lambda\langle \beta, \alpha \rangle \beta. \tag{5}$$

The transformation $S_{\lambda,\beta}$ preserves $\langle\cdot,\cdot\rangle$ in the symmetric case only when $\lambda = 2/Q(\beta,\beta)$ (or $\lambda = 0$). In that case $S_{\lambda,\beta}$ has order two and is a reflection. When $\langle\cdot,\cdot\rangle$ is antisymmetric, there is no restriction on λ and $S_{\lambda,\beta} \circ S_{\lambda',\beta} = S_{\lambda+\lambda',\beta}$. So in that case $S_{\lambda,\beta}$ does not have finite order.

Such transformations occur as monodromy transformations where not a sphere, but rather a quotient S^3/G by a finite group G is vanishing, as we will explain now. Consider the function defining the three-dimensional A_1-singularity:

$$f : \mathbb{C}^4 \longrightarrow \mathbb{C}, \quad f(x,y,z,t) = x^2 + y^2 + z^2 + t^2$$

The fibre F_s of f over $s \in \mathbb{C} \setminus \{0\}$ is called the Milnor fibre and can be identified with the cotangent bundle to the sphere $\{(x,y,z,t) \in \mathbb{R}^4 \mid x^2+y^2+z^2+t^2 = s\}$, which is vanishing when $s \to 0$. We choose an orientation and let δ be the homology class of this sphere. There is also a covanishing cycle ϵ in the dual group $\mathrm{H}_3^{\mathrm{cl}}(F_s,\mathbb{Z})$ (homology with closed supports). One has

$$\mathrm{H}_3(F_s,\mathbb{Z}) = \mathbb{Z}\delta, \quad \mathrm{H}_3^{\mathrm{cl}}(F_s,\mathbb{Z}) = \mathbb{Z}\epsilon, \quad \langle \delta, \epsilon\rangle = 1$$

Let $G \subset \mathrm{SU}(2) = S^3$ be a finite subgroup. G then acts linearly on $\mathbb{R}^4$ and by complexification on $\mathbb{C}^4$, leaving invariant the function f defining the A_1-singularity. Consider the quotient map $\pi : \mathbb{C}^4 \longrightarrow X := \mathbb{C}^4/G$. The space X will be singular, but f descends to a function $g : X \longrightarrow \mathbb{C}$, such that $f = \pi \circ g$. So the fibre $G_s := g^{-1}(s)$ is the quotient of F_s by G. In the fibre G_s there is a cycle S^3/G vanishing when $s \to 0$, with homology class $d \in \mathrm{H}_3(G_s,\mathbb{Z})$. As above there also exists a covanishing cycle $e \in \mathrm{H}_3^{\mathrm{cl}}(G_s,\mathbb{Z})$ such that

$$\mathrm{H}_3(G_s,\mathbb{Z}) = \mathbb{Z}d, \quad \mathrm{H}_3^{\mathrm{cl}}(G_s,\mathbb{Z}) = \mathbb{Z}e, \quad \langle d, e\rangle = 1.$$

The map π induces maps π^* and π_* between the homology groups of F_s and G_s and one easily sees that

$$\pi_*(\delta) = |G|d, \quad \pi_*(\epsilon) = e, \quad \pi^*(d) = \delta, \quad \pi^*(e) = |G|\epsilon$$

The Picard-Lefschetz formula tells us that under the monodromy of f the cycle δ remains fixed, whereas the cycle ϵ gets mapped to $\epsilon - \delta$. From the fact that the monodromy commutes with the group action we obtain, by taking π_*, that d remains fixed, whereas e gets mapped to $e - |G|d$. From this one deduces in the usual way that the occurrence of a singularity of type A_1/G will lead to the monodromy transformation described by the modified Picard-Lefschetz formula (see [**6**])

$$\gamma \mapsto \gamma - |G|\langle d, \gamma\rangle d$$

The cycle d should should give rise to a spherical object in the derived category of the mirror, but only the $|G|$th power of the Seidel-Thomas twist would arise from a monodromy transformation.

Appendix B. Computation of instanton numbers

According to [**18, 19**] (see also [**23**]) we have the following expansion for the partition function F of the topological string

$$F = \sum_{g=0}^{\infty} \lambda^{2g-2} F_g = \sum_{g=0}^{\infty} \sum_{d} \sum_{m=0}^{\infty} n_d^g \frac{1}{m} \left(2 \sin \frac{m\lambda}{2}\right)^{2g-2} q^{dm}.$$

The partion function F can be defined physically or mathematically using Gromov-Witten invariants. The above formula can then be considered to define the Gopakumar-Vafa invariants n_d^g. In contrast to e.g., the Gromov-Witten invariants, the Gopakumar-Vafa invariants are conjectured to be always integral.

We will restrict to the genus zero and genus one invariants. Furthermore, we will only consider the 1-parameter case. In that case we have the following formulas (see [**23**])

$$\partial_t^3 F_0 = n_0^0 + \sum_{\ell=1}^{\infty} \frac{n_\ell^0 \ell^3 q^\ell}{1-q^\ell}$$

with $n_0^0 = H^3$ and

$$\partial_t F_1 = \frac{c_2 \cdot H}{24} + \sum_{d=1}^{\infty}\sum_{k=1}^{\infty} (\tfrac{1}{12} n_d^0 + n_d^1) d q^{kd}.$$

Here we define the coordinate t by $q = e^{-t}$. So if we can compute the left hand sides of these equations the invariants n_d^0 and n_d^1 can easily be determined.

To do so, we first introduce a special basis of solutions for the equation (1) around a point of maximal unipotent monodromy, i.e., a singular point where $\lambda = 0$ is the only solution to the indicial equation. In physical terms such a point (also called a MUM-point) corresponds to a large radius limit point.

Suppose $z = 0$ is a MUM-point. Then we can use the Frobenius method. The idea is to consider a solution with values in the ring $\mathbb{C}[\rho]/(\rho^n)$. We make the following ansatz for such a solution:

$$\tilde{y}(z) = \sum_{n=0}^{\infty} A(n,\rho) z^{n+\rho} = y_0(z) + y_1(z)\rho + \cdots + y_{n-1}(z)\rho^{n-1},$$

where we define

$$z^\rho = e^{\log z \cdot \rho} = 1 + \log z \cdot \rho + \frac{\log^2 z}{\rho} \cdot \rho^2 + \cdots + \frac{\log^{n-1} z}{(n-1)!} \cdot \rho^{n-1}.$$

Using $\theta z^{n+\rho} = (n+\rho) z^{n+\rho}$, where $\theta = z\frac{d}{dz}$, we can translate the equation $L\tilde{y} = 0$ into a recursion relation for the $A(n,\rho)$. As initial condition for the recursion we use $A(0,\rho) = 1$. The y_i we find in this way are called the *Frobenius basis*.

Define power series f_i by the following expression:

$$\sum_{n=0}^{\infty} A(n,\rho) z^n = f_0(z) + f_1(z)\rho + \cdots + f_{n-1}(z)\rho^{n-1}.$$

Because $z^\rho \sum_{i=0}^{n-1} f_i(z)\rho^i = \sum_{i=0}^{n-1} y_i(z)\rho^i$, we find

$$y_i(z) = \sum_{j=0}^{i} \frac{\log^i z}{i!} f_{j-i}(z).$$

Using the Frobenius basis we can define a new coordinate

$$t = y_1(z)/y_0(z) = \log z + \frac{f_1(z)}{f_0(z)}. \tag{6}$$

There are basically two ways to compute $\partial_t^3 F_0$. The starting point of the first one is the *Yukawa coupling* in the z coordinate

$$K_{zzz} = \exp(-\tfrac{1}{2}\textstyle\int a_3(z) \mathrm{d}z),$$

where $a_3(z)$ is one of the coefficients from (1). The claim is that $\partial_t^3 F$ is the following transformation of this function to the t-coordinate defined in (6):

$$\partial_t^3 F(t) = \frac{K_{zzz}(z(t))}{y_0^2(z(t))\left(\frac{dt}{dz}\right)^3} \tag{7}$$

Now recall that a Calabi-Yau equation has to satisfy a list of conditions (see [**3, 2**]). One of these can be written as

$$a_1 = \frac{1}{2}a_2a_3 - \frac{1}{8}a_3^3 + a_2' - \frac{3}{4}a_3a_3' - \frac{1}{2}a_3'' \tag{8}$$

According to Proposition 1 from [**3**] this condition is equivalent to the two conditions

$$\frac{d^2}{dt^2}\frac{y_2}{y_0} = \frac{\exp(-\frac{1}{2}\int a_3(z)dz)}{y_0^2\left(\frac{dt}{dz}\right)^3}, \tag{9}$$

$$\frac{d^2}{dt^2}\frac{y_3}{y_0} = t\frac{d^2}{dt^2}\frac{y_2}{y_0}. \tag{10}$$

The second condition is equivalent to the existence of a function G and a constant c such that

$$\Pi(t) := \frac{1}{y_0}\begin{pmatrix} y_0 \\ y_1 \\ y_2 \\ y_3 \end{pmatrix} = \begin{pmatrix} 1 \\ t \\ \partial_t G - c \\ t\partial_t G - 2G \end{pmatrix}.$$

The vector $\Pi(t)$ is called the *normalized period vector.* Using the second condition (7) translates to

$$\partial_t^3 F_0 = \partial_t^3 G(t) = \partial_t^2 \frac{y_2}{y_0}.$$

This yields a different way of computing $\partial_t^3 F_0$ and therefore also the instanton numbers n_d^0.

To compute $\partial_t F_1$ we use the recipe from [**10, 11**] based on an analysis of the so called holomorphic anomaly. We will use the following formula from [**10**] (using our notation and adapted slightly for the case we are studying):

$$\partial_t F_1 = \partial_t \log\left(\frac{z^{1+\frac{c_2\cdot H}{12}} f(z)}{y_0^{4-\frac{c_3}{12}}\frac{\partial t}{\partial z}}\right). \tag{11}$$

In this formula one needs the geometrical data $c_2 \cdot H$ and c_3 which can usually be determined from the monodromy calculation and/or conifold-period. However, the main problem with this formula is the function f, which is a holomorphic function of z that still has to be determined. We will use an ansatz for f to reduce this problem to the determination of a finite number of parameters. To describe this ansatz note that because of the special form of a Calabi-Yau equation we can write

$$a_4(z) = z^4\Delta(z) = z^4\prod_i(\Delta_i(z))^{k_i},$$

for some polynomial $\Delta(z)$, which we call the *discriminant.* The $\Delta_i(z)$ are the irreducible factors (over $\mathbb{R}$) of $\Delta(z)$. Our ansatz is then the following

$$f(z) = \prod_i(\Delta_i(z))^{s_i},$$

where the exponents $s_i \in \mathbb{Q}$ still have to be determined. The (apparent) singular points of the operator are the zeros of the discriminant $\Delta(z)$ (and 0 and ∞). So each of the factors $\Delta_i(z)$ corresponds via its zeros to one or more (apparent) singular points. To determine the exponents we look at the monodromies around the corresponding singular points. When the singular point is a conifold, i.e., the monodromy is of the form $S_{1,v}$, then the exponent is generally assumed to be $-\frac{1}{6}$. We generalize this to $-\frac{\lambda}{6}$ for monodromies of the form $S_{\lambda,v}$ for arbitrary λ. When the monodromy is the identity, we let the exponent to zero. These rules already allow us to deal with many equations. However, monodromies of other types for which we do not know of a sensible guess do occur.

References

[1] V. Alexeev and M. Brion. Toric degenerations of spherical varieties. math.AG/0403379.

[2] Gert Almkvist, Christian van Enckevort, Duco van Straten, and Wadim Zudilin. Tables of Calabi-Yau equations. 2005. math.AG/0503xxx.

[3] Gert Almkvist and Wadim Zudilin. Differential equations, mirror maps and zeta values. In N. Yui, editor, *Mirror Symmetry V*, Studies in Advanced Mathematics. AMS/International Press, 2005. math.NT/0402386.

[4] G. Almqvist. Strängar i månsken I. *Normat*, no.1:22–33, 2003.

[5] Y. André. *G-Functions and Geometry*, volume 13 of *Aspects of Mathematics*. Vieweg, 1989.

[6] V. Arnold, S. Gusein, and A. Varchenko. *Singularities of Differentiable Maps, Volume II.* Monographs in Mathematics. Birkhäuser, 1986.

[7] V. Batyrev. Dual polyhedra and mirror symmetry for Calabi-Yau hypersurfaces in toric varieties. *J. Algebraic Geom*, 3:493–535, 1994. alg-geom/9310003.

[8] V. Batyrev, I. Ciocan-Fontanine, B. Kim, and D. van Straten. Conifold transitions and mirror symmetry for Calabi-Yau complete intersections in Grassmannians. *Nucl. Phys.*, B 514:640–666, 1998. alg-geom/9710022.

[9] V. Batyrev and D. van Straten. Generalised hypergeometric functions and rational curves on Calabi-Yau complete intersections in toric varities. *Commun. Math. Phys.*, 168:493–533, 1995. alg-geom/9307010.

[10] M. Bershadsky, S. Cecotti, H. Ooguri, and C. Vafa. Holomorphic anomalies in topological field theories. *Nucl. Phys.*, B 405:279–304, 1993. hep-th/9302103.

[11] M. Bershadsky, S. Cecotti, H. Ooguri, and C. Vafa. Kodaira-Spencer theory of gravity and exact results for quantum string amplitudes. *Commun. Math. Phys.*, 165:311–428, 1994. hep-th/9309140.

[12] F. Beukers and G. Heckman. Monodromy for the hypergeometric function ${}_nF_{n-1}$. *Inv. Math.*, 95:325–354, 1989.

[13] C. Borcea. K3 surfaces with involution and mirror pairs of Calabi-Yau manifolds. In B. Greene and S.-T. Yau, editors, *Mirror Symmetry II*, volume 1 of *Studies in Advanced Mathematics*, pages 717–739. Am. Math. Soc., 1991.

[14] P. Candelas, X. de la Ossa, P. Green, and L. Parkes. A pair of Calabi-Yau manifolds as an exactly solvable superconformal theory. *Nucl. Phys*, B 359:21–74, 1991.

[15] C. Doran and J. Morgan. Mirror Symmetry and Integral Variations of Hodge Structure Underlying One Parameter Families of Calabi-Yau Threefolds. In N. Yui, editor, *Mirror Symmetry V*, Studies in Advanced Mathematics. AMS/International Press, 2005.

[16] Hao Fang and Zhiqin Lu. Generalized Hodge metrics and BCOV torsion on Calabi-Yau moduli. 2003. math.DG/0310007v1.

[17] K. Fukaya, Y. Oh, H. Ohta, and K. Ono. Lagrangian intersection Floer theory, anomaly and obstruction. Available from http://www.math.kyoto-u.ac.jp/~fukaya/fukaya.html, 2000.

[18] R. Gopakumar and C. Vafa. M-theory and topological strings—I. 1998. hep-th/9809187.

[19] R. Gopakumar and C. Vafa. M-theory and topological strings—II. 1998. hep-th/9812127.

[20] M. Gross. Topological mirror symmetry. *Inv. Math.*, 144 (no. 1):75–137, 2001. math.AG/9909015.

[21] R. P. Horja. Hypergeometric functions and mirror symmetry in toric varieties. 1999. math.AG/9912109.

[22] N. Katz. *Rigid Local Systems.* Number 139 in Annals of Mathematics Studies. Princeton University Press, 1996.

[23] S. Katz, A. Klemm, and C. Vafa. M-Theory, Topological Strings and Spinning Black Holes. *Adv.Theor.Math.Phys.*, 3:1445–1537, 1999.

[24] A. Klemm, M. Kreuzer, E. Rieger, and E. Scheidegger.. Topological String Amplitudes, Complete Intersection Calabi-Yau Spaces and Threshold Corrections. hep-th/0410018

[25] S. Katz, R. Schimmrigk, and A. Wißkirchen. The Calabi-Yau homepage. http://www.th.physik.uni-bonn.de/th/Supplements/cy.html.

[26] A. Klemm and S. Theisen. Considerations of one-modulus Calabi-Yau compactifications; Picard-Fuchs equations, Kähler potentials and mirror maps. *Nucl. Phys*, B 389:153–180, 1993. hep-th/9205041.

[27] M. Kontsevich. Homological algebra of mirror symmetry. In *Proceedings of the International Congress of Mathematicians (Zürich, 1994)*, volume 1, pages 120–139. Birkhäuser, 1995. alg-geom/9411018.

[28] M. Kontsevich. Talk. Rutgers University, Nov. 11 1996.

[29] M. Kontsevich and D. Zagier. Periods. In B. Engquist and W. Schmid, editors, *Mathematics Unlimited-2001 and beyond*, pages 771–808. Springer Verlag, 2000.

[30] S. Lefschetz. *L'Analysis Situs et la Géometrie Algébrique*. Gauthier-Villars, 1924.

[31] A. Libgober and J. Teitelbaum. Lines on Calabi-Yau complete intersections, mirror symmetry and Picard-Fuchs equations. *Int. Math. Res. Notices*, no.1:29–39, 1993.

[32] E. Looijenga. *Isolated Singular Points on Complete Intersections*, volume 77 of *London. Math. Soc. Lecture Note Series*. Cambridge University Press, 1984.

[33] D. Morrison. Picard-Fuchs equations and mirror maps for hypersurfaces. In S. T. Yau, editor, *Essays on mirror manifolds*, pages 241–264. Internat. Press, 1992.

[34] E. Rødland. The Pfaffian Calabi-Yau, its mirror, and their link to the Grassmannian $G(2,7)$. *Compositio Math.*, 122(no.2):135–149, 2000. math.AG/9801092.

[35] F. Rodriguez-Villegas. Hypergeometric families of Calabi-Yau manifolds. In N. Yui and J. Lewis, editors, *Calabi-Yau Varieties and Mirror Symmetry*, Fields Institute Communications, pages 223–232. Am. Math. Soc., 2003.

[36] Paul Seidel and Richard Thomas. Braid group actions on derived categories of coherent sheaves. 2000. math.AG/0001043.

[37] A. Strominger, S.-T. Yau, and E. Zaslow. Mirror symmetry is T-duality. *Nucl. Phys.*, B 479:243–259, 1996. hep-th/9606040.

[38] E. Tjøtta. *Rational Curves on the Space of Determinantal Nets of Conics*. PhD thesis, University of Bergen, 1997. math.AG/9802037.

[39] E. Tjøtta. Quantum cohomology of a Pfaffian Calabi-Yau variety: verifying mirror symmetry predictions. 1999. math.AG/9906119.

[40] F. Tonoli. Canonical surfaces in $\mathbb{P}^5$ and Calabi-Yau threefolds in $\mathbb{P}^6$. *J. Alg. Geom.*, 2004.

[41] C. van Enckevort and D. van Straten. The derived category for Calabi-Yaus in Grassmannians. In preparation.

FB 17 Mathematik und Informatik, Universität Mainz, Staudingerweg 9, 55099, Mainz, Germany

E-mail address: enckevor@mathematik.uni-mainz.de

FB 17 Mathematik und Informatik, Universität Mainz, Staudingerweg 9, 55099, Mainz, Germany

E-mail address: straten@mathematik.uni-mainz.de

AMS/IP Studies in Advanced Mathematics
Volume 38, 2006

Open string mirror maps from Picard–Fuchs equations

Brian Forbes

ABSTRACT. A definition of open string period integrals for noncompact Calabi–Yau manifolds is given. It is shown that the open string Picard–Fuchs operators, originally derived through physical considerations, follow from these period integrals. Also, we find that the natural extension to the compact case does not yield the expected results.

1. Introduction.

For some time now, mirror symmetry and its mathematical implications have provided unexpected connections between seemingly unrelated families of Calabi–Yau manifolds. In particular, closed string mirror symmetry, under which the complexified Kähler moduli space $\mathcal{X}$ of one Calabi–Yau family is locally isomorphic to the complex moduli space $\mathcal{Y}$ of another family, has proven to be a powerful tool of enumerative geometry [**5**],[**18**].

Closer to the present, there is the developing story of local mirror symmetry [**9**],[**4**],[**10**]. Here, the relevant space $\mathcal{X}$ may be described as the variation of complexified Kähler moduli of the canonical bundle K_S of a Fano surface S. The associated Gromov–Witten invariants have the (conjectural) interpretation as the number of holomorphic curves lying in the surface S. When compared with the compact case, the local periods take on a simplified form.

Finally, we come to the topic of interest, which is open string mirror symmetry [**1**],[**14**],[**16**]. Mathematically, this represents an extension of the moduli spaces considered in closed string mirror symmetry. We are now interested, on one side of the duality, in the moduli space of a pair (X, L), where we look at the complexified Kähler moduli space of X together with deformations of a Lagrangian submanifold $L \subset X$. Throughout, we will denote this moduli space as $(\mathcal{X}, \mathcal{L})$. On the other side, we consider the moduli space of (Y, C) with C a holomorphic curve in Y, in the sense of the complex structure moduli space of Y and normal deformations of C in X; this will be referred to as $(\mathcal{Y}, \mathcal{C})$. Then $(\mathcal{X}, \mathcal{L})$ and $(\mathcal{Y}, \mathcal{C})$ are supposed to be locally isomorphic, and the enumerative significance of this duality is given the

2000 *Mathematics Subject Classification.* Primary: 14N35, 14F25, 14J32, 32G20, 14D05.
Key words and phrases. noncompact Calabi–Yau manifold, open string, period, Picard–Fuchs operator.

interpretation of counting the number of holomorphic maps $f : D \to X$ such that $f|_{\partial D}(\partial D) \subset L$, where $D = \{z \in \mathbb{C} : |z| \leq 1\}$.

In this paper, we will give a definition of period integrals defining the moduli space $(\mathcal{Y}, \mathcal{C})$ in the noncompact case, and show that this is consistent with the physicists' open string PF operators [**14**]. Moreover, we attempt to extend this construction to include compact considerations, and show that this does not seem to be the right approach. Physically, this is due to the D-brane charge on C, which must be dealt with in a compact theory.

The organization of this paper is as follows. In section 2, we will review the noncompact pairs of spaces (X, L) and (Y, C) that are used in local open string mirror symmetry. Section 3 contains a definition of "open string period integrals" on $(\mathcal{Y}, \mathcal{C})$, which are then shown to reproduce the known Picard–Fuchs operators. Section 4 shows the failure of extending this idea to the compact case, and finally, in section 5 we give an example in detail.

Acknowledgements.
I would like to thank Wolfgang Lerche for patiently answering my questions about the superpotential computation. I also thank David Morrison and Matt Kerr for helpful conversations. Finally, thanks to my advisor, Kefeng Liu, for suggesting this direction, and to David Gieseker for support during the research.

2. Review of the spaces

2.1. Construction of $(\mathcal{X}, \mathcal{L})$. Although the primary focus of the present work is to define the moduli space $(\mathcal{Y}, \mathcal{C})$, it is instructive to first carefully review the mirror $(\mathcal{X}, \mathcal{L})$; moreover, the holomorphic discs that are "counted" by open string mirror symmetry actually lie in (X, L).

Following [**2**], we can define the spaces X as

$$X_r = \{(z_0, z_1, z_2, z_3) \in (\mathbb{C}^4 - Z) | \sum_{i=0}^{3} l_i |z_i|^2 = r\}/S^1, \tag{1}$$

where the S^1 action is

$$S^1 : z_i \longrightarrow e^{\sqrt{-1} l_i \theta} z_i, \quad i = 0, \ldots, 3,$$

and $r \in \mathbb{R}^+$. Also, Z is the singular set implied by the relation $\sum_{i=0}^{3} l_i |z_i|^2 = r$. For example, for the local $\mathbb{P}^2$ case, we have $l = (-3, 1, 1, 1)$, and this means that $Z = \{z_1 = z_2 = z_3 = 0\}$ here. Finally, we must impose the condition $\sum_{i=0}^{3} l_i = 0$, which is equivalent to $c_1(X_r) = 0$ (that is, so X_r satisfies the definition of a local Calabi-Yau manifold).

Notice also that r is the single Kähler parameter on this space. To see this, recall that in such a toric construction of X, the vector l corresponds to a 2-cycle $C_l \hookrightarrow X$. Then there is an element of cohomology ω_l dual to C_l, and we find that

$$\int_{C_l} \omega_l = r$$

when appropriately scaled. Hence, if t is the coordinate on the complexified Kähler moduli space of X, we may identify $Re(t) = r$.

In the interests of clarity, throughout the paper, we will consider spaces with only one Kähler modulus. The inclusion of additional parameters is in fact trivial [**14**].

Next, we should construct Lagrangian submanifolds L of X. These have been worked out previously [**1**], and are given by

$$L_{r,c} = X_r \cap \{\sum_i k_i^1 |z_i|^2 = c,\ \sum_i k_i^2 |z_i|^2 = 0,\ \sum_i arg(z_i) = 0\}.$$

The convention used is that $k^j = (k_0^j, \ldots, k_3^j)$. Again, we must insist that $\sum_{i=0}^3 k_i^j = 0$ for $j = 1, 2$, although this now has the interpretation of making L a Lagrangian submanifold of X.

Then, according to local open string mirror symmetry, there should be a family of spaces $(\mathcal{Y}, \mathcal{C})$, such that $Y \in \mathcal{Y}$ is Calabi–Yau and C is a holomorphic curve in Y, such that $(\mathcal{Y}, \mathcal{C}) \cong (\mathcal{X}, \mathcal{L})$. This isomorphism is expected to count genus 0 "open string Gromov–Witten invariants" on (X, L); these do not yet have a good mathematical description [**13**], [**6**], but their supposed interpretation is a count of holomorphic discs in X with boundary on L.

2.2. The mirror family (Y, C)**.** From the mirror constructions of e.g. [**9**],[**1**], we can immediately give a description of a mirror family Y for X, in the usual (closed string) sense of mirror symmetry. This is

$$Y_z = \{uv + \sum_{i=0}^{3} y_i = 0,\ \prod_{i=0}^{3} y_i^{l_i} = z,\ y_0 = 1\}, \tag{2}$$

where u, v are $\mathbb{C}$ variables and $y_i \in \mathbb{C}^*$. Also, the condition $y_0 = 1$ is of course arbitrary, and in practice we vary which y_i should be set to 1 based on convenience in studying the problem at hand.

One way to see that this is the right definition of a mirror Y to X is to look at the local period integrals for Y. So, first use the constraints $\prod_{i=0}^3 y_i^{l_i} = z, y_0 = 1$ on the equation $uv + \sum_{i=0}^3 y_i = 0$, and rewrite

$$Y_z = \{uv + f(z) = 0\},$$

where f is a function of e.g. y_1, y_2. Then we can define the period integrals on Y to be [**10**]

$$\Pi_\Gamma(z) = \int_\Gamma \frac{dudvdy_1dy_2/(y_1y_2)}{uv + f(z)} \tag{3}$$

Here we have $\Gamma \in H_4(\mathbb{C}^2 \times (\mathbb{C}^*)^2 - Y, \mathbb{Z})$. These period integrals will reproduce the enumerative information relevant to counting holomorphic curves in X. The reason for this is the same as in the usual formulation of Batyrev mirror symmetry [**3**], because of the use of l to define the polynomial for Y.

Next, we would like to see how mirror symmetry acts on the Lagrangian submanifolds L of X. In a sense, the mirror family $\mathcal{Y}$ (parameterized by z) of X was simply defined by exponentiation of the defining vector l. In the same way, we have but to exponentiate k^1, k^2 to get the mirror curve C to L. This yields

$$C_{z,w} = Y_z \cap \{\prod_{i=0}^{3} y_i^{k_i^1} = w,\ \prod_{i=0}^{3} y_i^{k_i^2} = 1\}. \tag{4}$$

In practice, the equation $\prod_{i=0}^{3} y_i^{k_i^1} = w$ will always take the form $y_i y_j^{-1} = w$, after accounting for the imposition of the constraint $y_0 = 1$.

Then, the moduli space $(\mathcal{Y}, \mathcal{C})$ is supposed to be locally isomorphic to $(\mathcal{X}, \mathcal{L})$. The next section will explore a way in which to make this correspondence precise.

3. Period integrals for $(\mathcal{Y}, \mathcal{C})$.

3.1. Periods and Picard–Fuchs operators for $\mathcal{Y}$. From the above, we have integrals

$$\Pi_\Gamma(z) = \int_\Gamma \frac{dudvdy_1dy_2/(y_1y_2)}{uv + f(z)} \tag{5}$$

which determine the complex structure of Y. Rather than directly integrating the Π_Γ, it is often easter to determine a set of Picard–Fuchs operators which annihilate these, and then write down a basis of solutions. Here, this process will be interpreted in a slightly different way from the conventional one, which will make the constructions of the following section natural.

First, looking back at (2), note that this can be written as

$$Y_z = \{uv + 1 + y_1 + y_2 + z(y_1^{l_1} y_2^{l_2})^{1/l_3} = 0\};$$

From here, define

$$Y_{a_0,\dots,a_3} = \{uv + a_0 + a_1y_1 + a_2y_2 + a_3(y_1^{l_1} y_2^{l_2})^{1/l_3} = uv + f(a, y) = 0\}.$$

This means that the $\mathbb{C}^*$ variables $(a_0, \dots, a_3)$ are considered to be homogeneous coordinates for the complex structure moduli space $\mathcal{Y}$. Then the period integrals

$$\Pi_\Gamma(a_0, \dots, a_3) = \int_\Gamma \frac{dudvdy_1dy_2/(y_1y_2)}{uv + a_0 + a_1y_1 + a_2y_2 + a_3(y_1^{l_1} y_2^{l_2})^{1/l_3}} \tag{6}$$

are annihilated by the GKZ differential operators, given by

$$\mathcal{L} = \prod_{l_i > 0} \partial_{a_i}^{l_i} - \prod_{l_i < 0} \partial_{a_i}^{-l_i} \tag{7}$$

and another operator. However, the addition of the second GKZ operator is equivalent to the statement that $\Pi_\Gamma(a_0, \dots, a_3)$ is actually a function of only one complex parameter, which we can identify as

$$z = \frac{\prod_{l_i>0} {a_i}^{l_i}}{\prod_{l_i<0} {a_i}^{-l_i}}. \tag{8}$$

Then, as is well known, we can reduce $\mathcal{L}$ via (8) to obtain a Picard–Fuchs differential operator

$$\mathcal{D} = \prod_{l_k>0}\prod_{n=0}^{l_k-1}(l_k\theta - n) - z\prod_{l_k<0}\prod_{n=0}^{-l_k-1}(l_k\theta - n). \tag{9}$$

As usual, $\theta = z\frac{d}{dz}$. Then the solution set of $\mathcal{D}f = 0$ is supposed to yield the periods of $\mathcal{Y}$.

3.2. Extension to $(\mathcal{Y}, \mathcal{C})$. It was proposed in [**16**] that the moduli space $(\mathcal{Y}, \mathcal{C})$ should be thought of as a variation of mixed Hodge structure of the relative cohomology $H^3(Y, C; \mathbb{C})$. Taking this as a starting point, note that there is an isomorphism $H^3(Y, C) \cong H^3_c(Y - C)$, where the subscript c denotes forms with compact support. Now, from the defining equations (5), since Y is given as a hypersurface, we are led to consider derivatives of the differential form

$$\frac{du dv dy_1 dy_2/(y_1y_2)}{uv + f(z)}$$

to determine the Picard–Fuchs operator, and hence the period integrals.

In the same spirit, from the equation of the curve C,

$$C_{z,w} = Y_z \cap \{\prod_{i=0}^{3} y_i^{k_i^1} = w,\ \prod_{i=0}^{3} y_i^{k_i^2} = 1\},$$

set

$$g(w) = \prod_{i=0}^{3} y_i^{k_i^1} - w,\quad h = \prod_{i=0}^{3} y_i^{k_i^2} - 1.$$

Then the natural differential form to look at, which lies in $H^3_c(Y - C)$, is

$$\frac{du dv dy_1 dy_2/(y_1y_2)}{(uv + f(z))g(w)h}. \tag{10}$$

This leads us to make the

Definition. *The relative period integrals on the moduli space $(\mathcal{Y}, \mathcal{C})$ are given as*

$$\Pi_\Gamma(z, w) = \int_\Gamma \frac{du dv dy_1 dy_2/(y_1y_2)}{(uv + f(z))g(w)h} \tag{11}$$

for $\Gamma \in H_4(\mathbb{C}^2 \times (\mathbb{C}^*)^2 - Y, Y - C, \mathbb{Z})$.

An important feature of these integrals is that, as g is independent of z, the usual Picard–Fuchs operator (9) still annihilates $\Pi_\Gamma(z, w)$. Thus, the period integrals for $\mathcal{Y}$ are a subset of those in the definition.

Next, it will be shown that this definition indeed reproduces the extended Picard–Fuchs system associated to the pair $(\mathcal{Y}, \mathcal{C})$ [**14**]. To this end, it is useful, as in the previous section, to first consider a different set of integrals which are annihilated by an extended GKZ system. Afterwards, the resulting GKZ operators will be reduced via canonically defined coordinates, and the result will be a relative Picard–Fuchs system consistent with that of the literature.

PROPOSITION 1. *The period integrals defined in (11) reproduce the known extended Picard–Fuchs differential operators defined as the moduli space $(\mathcal{Y}, \mathcal{C})$.*

Proof. It is useful to write the period integrals $\Pi_\Gamma(z,w)$ in a slightly different form, as was done for the periods on Y in the previous section. For simplicity, take the defining equation of g to be given as $g(w) = y_1 - w$. Then, set $g(b) = b_0 + b_1 y_1$, and consider the integrals

$$\Pi_\Gamma(a_0,\ldots,a_3,b_0,b_1) = \int_\Gamma \frac{dudvdy_1dy_2/(y_1y_2)}{(uv+f(a,y))g(b)h} =$$

$$\int_\Gamma \frac{dudvdy_1dy_2/(y_1y_2)}{(uv+a_0+a_1y_1+a_2y_2+a_3(y_1^{l_1}y_2^{l_2})^{1/l_3})(b_0+b_1y_1)h}. \tag{12}$$

This is, of course, inspired by the process used above when considering (6) in place of (5). The interpretation is that $(a_0,\ldots,a_3,b_0,b_1) \in (\mathbb{C}^*)^6$ are homogeneous coordinates for the moduli space $(\mathcal{Y},\mathcal{C})$.

From the explicit form (12), it is easy to see that $\Pi_\Gamma(a,b)$ is still annihilated by $\mathcal{L}$ (7), and one immediately has that

$$\mathcal{L}' = \partial_{a_0}\partial_{b_1} - \partial_{b_0}\partial_{a_1} \tag{13}$$

also sends $\Pi_\Gamma(a,b)$ to 0. So, take $\mathcal{L} = \mathcal{L}_z$, $\mathcal{L}' = \mathcal{L}_w$. There are then two canonically defined coordinates, based on these operators:

$$z = \frac{\prod_{l_i>0} a_i{}^{l_i}}{\prod_{l_i<0} a_i{}^{-l_i}}, \quad w = \frac{a_0b_1}{a_1b_0}. \tag{14}$$

The moduli space defined by (12) is larger than that of (6). This extended moduli space can be realized as a pair of vectors:

$$l_z = (l_0,\ldots,l_3,0,0), \quad l_w = (1,-1,0,0,-1,1). \tag{15}$$

These new vectors, in turn, are equivalent to the GKZ operators $\{\mathcal{L},\mathcal{L}'\}$. Hence, we may take $\{l_z,l_w\}$ as toric data for the moduli space of the pair $(\mathcal{Y},\mathcal{C})$. □

3.3. Extended Picard–Fuchs operators and solutions. The vectors (15) were found previously through physical considerations [**14**]. Thus, at this point the equivalence of the definition (11) to the physical calculations of [**14**],[**16**] has already been demonstrated. For completeness, we will present the extended Picard–Fuchs operators and solutions that follow from (15).

Let $z = z_1$, $w = z_2$, and change all other indices accordingly, so e.g. $l_z = l^1$. Writing the operators $\{\mathcal{L}_1,\mathcal{L}_2\}$ in the variables (14), we find

$$\mathcal{D}_k = \prod_{l_i^k>0}\prod_{n=0}^{l_i^k-1}(\sum_{k=1}^{2} l_i^k\theta_k - n) - z_k \prod_{l_i^k<0}\prod_{n=0}^{-l_i^k-1}(\sum_{k=1}^{2} l_i^k\theta_k - n). \tag{16}$$

The notation used here is $l^k = (l_0^k,\ldots,l_3^k)$. This is nothing other than the non-compact Picard–Fuchs system associated to the vectors $\{l^1,l^2\}$, so we can use the results of [**4**] to immediately write down a generating function for the solutions:

$$\omega_0(z,\rho) = \sum_{n\geq 0} c(n,\rho) z_1^{n_1+\rho_1} z_2^{n_2+\rho_2}, \tag{17}$$

where

$$c(n,\rho)^{-1} = \prod_i \Gamma(1 + \sum_k l_i^k(n_k + \rho_k)).$$

Then $\omega_0(z, 0)$ is constant, and there are two logarithmic solutions

$$t_i(z) = \partial_{\rho_i}\omega_0(z,\rho)|_{\rho=0}$$

providing the mirror map between $(\mathcal{X}, \mathcal{L})$ and $(\mathcal{Y}, \mathcal{C})$. According to [**16**], the function that is relevant for counting discs on $(\mathcal{X}, \mathcal{L})$ is

$$W(z_1, z_2) = \sum_{n\geq 0} (\partial^2_{\rho_2} c(n,\rho)|_{\rho=0}) z_1^{n_1} z_2^{n_2}$$

This is obtained by ignoring the logarithmic terms of $\partial^2_{\rho_2}\omega_0(z,\rho)|_{\rho=0}$.

Numerous examples have already been worked out [**16**] as evidence that, upon insertion of the inverse mirror map into W, the conjectural open Gromov–Witten invariants can be extracted, after accounting for the multiple cover formula [**13**]. Next, we would like to see whether the above considerations can be extended to the compact setting.

4. An attempt at the compact case.

From the mirror symmetry constructions of [**1**], it is evident that we cannot distinguish the compact and noncompact cases at the level of GKZ operators. That is to say, the GKZ system associated to the mirror family of the quintic (i.e., a zero section of $\mathcal{O}(5) \to \mathbb{P}^4$) is identical to that derived from the mirror family of the (complex dimension 5) noncompact Calabi–Yau space $\mathcal{O}(-5) \to \mathbb{P}^4$. This statement continues to hold true in the open string setting.

This means that the distinction of compact and noncompact geometry is made through the transition from GKZ to Picard–Fuchs operators. The reason for this is easy to explain: in the noncompact case, the differential form

$$\frac{du dv dy_1 dy_2/(y_1 y_2)}{uv + a_0 + a_1 y_1 + a_2 y_2 + a_3(y_1^{l_1} y_2^{l_2})^{1/l_3}}$$

is invariant when we pass to the quotient defined by the variable (14). However, in the compact case, we are working with a differential form

$$\Omega = \frac{\prod_i dy_i/y_i}{a_0 + \sum_{i>0} a_i y_i^{l_i}},$$

which has an additional $\mathbb{C}^*$ action on the denominator. Hence, we must modify the differential form as well as the GKZ operator in order to maintain invariance under this new $\mathbb{C}^*$ action:

$$\Omega \to a_0\Omega, \quad \mathcal{L} \to \mathcal{L}a_0^{-1}.$$

Naturally, the Picard–Fuchs operator arising from $\mathcal{L}a_0^{-1}$ will be different from that in (9).

From these considerations, as well as the mirror construction in [**1**], it is evident that the form

$$\Omega_1 = \frac{\prod_i dy_i/y_i}{(a_0 + \sum_{i>0} a_i y_i^{l_i})(b_0 + b_1 y_1)h}$$

will be the natural object of interest in the context of compact open string mirror symmetry. This now takes on the form from the complete intersection case [**11**], and it follows that invariance under the quotient given by (14) requires the modification

$$\Omega_1 \to \Omega_1 a_0 b_0, \quad \mathcal{L}_i \to \mathcal{L}_i a_0^{-1} b_0^{-1}.$$

Then we have new Picard–Fuchs operators

$$\mathcal{D}_k = \prod_{l_i^k>0} \prod_{n=0}^{l_i^k-1} (\sum_{k=1}^{2} l_i^k \theta_k - n) - z_k \prod_{l_i^k<0} \prod_{n=0}^{-l_i^k} (\sum_{k=1}^{2} l_i^k \theta_k - n)$$

and their associated solution set, which is given in full detail in [**11**]. The point now, though, is that the associated double logarithmic solution of $\{\mathcal{D}_1, \mathcal{D}_2\}$ will yield nontrivial numbers, after the insertion of the inverse mirror map. However, as was pointed out in [**1**], a holomorphic family of curves in a compact Calabi–Yau, such as that given by $\{b_0 + b_1 y_1 = 0\}$ in $\{a_0 + \sum_{i>0} a_i y_i^{l_i} = 0\}$, should have no double log solution. This is because such a family will have a vanishing disc number in the mirror. Thus, we cannot naively apply the techniques of [**16**] to compact situations.

5. An example.

Finally, it will be shown how the application of the above ideas reproduces the results of both [**6**] and [**14**], in the case of $K_{\mathbb{P}^2} = \mathcal{O}(-3) \to \mathbb{P}^2$.

First, we take $K_{\mathbb{P}^2} =$

$$X_r = \{(z_1, \ldots, z_4) \in \mathbb{C} - Z : |z_1|^2 + |z_2|^2 + |z_3|^2 - 3|z_4|^2 = r\}/S^1,$$

where $Z = \{z_1 = z_2 = z_3 = 0\}$, and the S^1 action is determined by the vector $l^1 = (1, 1, 1, -3)$. Also, consider the Lagrangian submanifolds

$$L_{r,c} = X_r \cap \{|z_1|^2 - |z_4|^2 = c, |z_2|^2 - |z_4|^2 = 0, \sum_i arg(z_i) = 0\},$$

$$L'_{r,-c} = X_r \cap \{|z_1|^2 - |z_4|^2 = -c, |z_2|^2 - |z_1|^2 = 0, \sum_i arg(z_i) = 0\}.$$

Here, $r, c \in \mathbb{R}^+$.

Then, as in previous sections, the moduli space $(\mathcal{Y}, \mathcal{C})$ can be specified by the toric data

$$l^1 = (1, 1, 1, -3, 0, 0), \quad l^2 = (1, 0, 0, -1, -1, 1),$$

while that of $(\mathcal{Y}, \mathcal{C}')$ is given as $\{l^1, -l^2\}$.

The vectors $\{l^1, l^2\}$ immediately give GKZ operators

$$\mathcal{L}_1 = \partial_{a_1}\partial_{a_2}\partial_{a_3} - \partial_{a_0}^3, \quad \mathcal{L}_2 = \partial_{a_1}\partial_{b_0} - \partial_{a_0}\partial_{b_1},$$

and these are the annihilators of the modified period vectors

$$\Pi_\Gamma(a,b) = \int_\Gamma \frac{dudv\frac{dy_1}{y_1}\frac{dy_2}{y_2}}{(uv + a_0 + a_1y_1 + a_2y_2 + a_3y_1^{-1}y_2^{-1})(b_0 + b_1y_1)(y_2 - 1)}.$$

These period vectors are, of course, arising from the mirror geometry to $(\mathcal{X}, \mathcal{L})$. This may be described by a pair $(Y, C) \in (\mathcal{Y}, \mathcal{C})$, where $\mathcal{Y}$ is the family of hypersurfaces

$$Y_{z_1} = \{(u, v, y_1, y_2) \in \mathbb{C}^2 \times (\mathbb{C}^*)^2 : uv + 1 + y_1 + y_2 + z_1y_1^{-1}y_2^{-1} = 0\}$$

together with the curves

$$C_{z_1,z_2} = Y_{z_1} \cap \{1 - z_2y_1 = 1 - y_2 = 0\}.$$

Then, the GKZ operators $\{\mathcal{L}_1, \mathcal{L}_2\}$ produce Picard–Fuchs operators

$$\mathcal{D}_1 = (\theta_1)^2(\theta_1 + \theta_2) + z_1(3\theta_1 + \theta_2)(3\theta_1 + \theta_2 + 1)(3\theta_1 + \theta_2 + 2),$$
$$\mathcal{D}_2 = (\theta_1 + \theta_2)\theta_2 - z_2(3\theta_1 + \theta_2)\theta_2.$$

The double logarithmic solution of the system $\{\mathcal{D}_1, \mathcal{D}_2\}$ is, modulo logarithmic terms,

$$W_{K_{\mathbb{P}^2}} = \sum_{n_1\geq 0, n_2\geq 1} \frac{(-1)^{n_1}(3n_1 + n_2 - 1)!}{n_2(n_1 + n_2)!(n_1!)^2} z_1^{n_1} z_2^{n_2}.$$

This is exactly the superpotential of [**6**], which was computed through explicit localization calculations on $(\mathcal{X}, \mathcal{L})$.

To reproduce the results of [**14**], use instead the moduli space of $(\mathcal{Y}, \mathcal{C}')$, which is defined by the vectors $\{l^1, -l^2\}$. Then there are new Picard–Fuchs operators $\{\mathcal{D}'_1, \mathcal{D}'_2\}$ which have a double log solution (again suppressing log terms)

$$W'_{K_{\mathbb{P}^2}} = \sum_{n_1\geq 0, n_2> n_1} \frac{(-1)^{n_1}(n_2 - n_1 - 1)!}{n_2(n_2 - 3n_1)!(n_1!)^2} z_1^{n_1} z_2^{n_2}.$$

This is the superpotential of [**14**].

References

[1] M. Aganagic and C. Vafa, *Mirror Symmetry, D- branes and Counting Holomorphic Discs*, hep-th/0012041.

[2] M. Aganagic, A. Klemm and C. Vafa, *Disc Instantons, Mirror Symmetry and the Duality Web*, Z. Naturforsch. A 57 (2002), no. 1-2, 1–28. hep-th/0105045.

[3] V. Batyrev, *Dual polyhedra and mirror symmetry for Calabi–Yau hypersurfaces in toric varieties.* J. Algebraic Geom. 3 (1994), no. 3, 493–535. alg-geom/9310003

[4] T.-M. Chiang, A. Klemm, S.-T. Yau and E. Zaslow, *Local mirror symmetry: calculations and interpretations.* Adv. Theor. Math. Phys. 3 (1999), no. 3, 495–565. hep-th/9903053

[5] D. Cox and S. Katz, *Mirror symmetry and algebraic geometry,* Mathematical Surveys and Monographs, 68. American Mathematical Society, Providence, RI, 1999

[6] T. Graber and E. Zaslow, *Open-string Gromov–Witten invariants: calculations and a mirror "theorem".* Orbifolds in mathematics and physics (Madison, WI, 2001), 107–121, Contemp. Math., 310, Amer. Math. Soc., Providence, RI, 2002. hep-th/0109075

[7] M. Green, J. Murre, and C. Voisin, *Algebraic cycles and Hodge theory.* Lectures given at the Second C.I.M.E. Session held in Torino, June 21–29, 1993. Edited by A. Albano and F. Bardelli.

[8] R. Hain and S. Zucker, *A guide to unipotent variations of mixed Hodge structure,* Lecture Notes in Math., 1246, Springer, Berlin, 1987.

[9] K. Hori and C. Vafa, *Mirror Symmetry.* hep-th/0002222

[10] S. Hosono, *Central charges, symplectic forms, and hypergeometric series in local mirror symmetry.* hep-th/0404043. In this volume.

[11] S. Hosono, A. Klemm, S. Theisen and S.-T. Yau, *Mirror symmetry, mirror map and applications to Calabi–Yau hypersurfaces.* Comm. Math. Phys. 167 (1995), no. 2, 301–350. hep-th/9308122

[12] S. Hosono, A. Klemm, S. Theisen and S.-T. Yau, *Mirror Symmetry, Mirror Map and Applications to Complete Intersection Calabi–Yau Spaces,* Nuclear Phys. B 433 (1995), no. 3, 501–552. hep-th/9406055.

[13] S. Katz and C.-C. Liu, *Enumerative geometry of stable maps with Lagrangian boundary conditions and multiple covers of the disc.* Adv. Theor. Math. Phys. 5 (2001), no. 1, 1–49. math.AG/0103074

[14] W. Lerche and P. Mayr, *On N=1 Mirror Symmetry for Open Type II Strings,* hep-th/0111113

[15] W. Lerche, P. Mayr and N. Warner, *Holomorphic N=1 Special Geometry of Open–Closed Type II Strings,* hep-th/0207259

[16] W. Lerche, P. Mayr and N. Warner, *N=1 Special Geometry, Mixed Hodge Variations and Toric Geometry,* hep-th/0208039.

[17] B. Lian, K. Liu and S.- T. Yau, *Mirror principle I.* Surveys in differential geometry: differential geometry inspired by string theory, 405–454. alg-geom/9712011

[18] D. Morrison, *Picard–Fuchs equations and mirror maps for hypersurfaces,* Essays on mirror manifolds, 241–264, Internat. Press, Hong Kong, 1992. hep-th/9111025.

[19] D. Orlov, *Triangulated Categories of Singularities and D-branes in Landau–Ginzburg Models.* math/0302304

[20] W. Schmid, *Variation of Hodge structure: the singularities of the period mapping.* Invent. Math. 22 (1973), 211–319

[21] J. Steenbrink and S. Zucker, *Variation of Mixed Hodge Structure 1,* Invent. Math 80 (1985), no.3, 489 - 542.

[22] C. Vafa, *Superstrings and topological strings at large N.* Strings, branes, and M-theory. J. Math. Phys. 42 (2001), no. 7, 2798–2817. hep-th/008142

Division of Mathematics, Graduate School of Science, Hokkaido University, Kita-ku, Sapporo, 060-0810, Japan

E-mail address: brian@math.sci.hokudai.ac.jp

Problems

During the problem session and through informal discussions, a number of problems were proposed. Some of the problems are compiled here by the editors for the sake of future discussions.

Problem 1: (Eckart Viehweg, Kang Zuo) Let M_h be the moduli scheme of polarized Calabi–Yau n-folds. Does there exists a compact curve C and a non-trivial morphism $\varphi : C \to M_h$, which is induced by a smooth family $f : X \to C$, such that C is a rigid Shimura curve, controlling the VHS $R^n f_* \mathbb{Q}_X$.

By taking the quotient of a special family of Abelian surfaces by the involution one finds examples for $n = 2$, i.e. for $K3$-surfaces. In "Families over curves with a strictly maximal Higgs field" [**18**], we show that there are no such families, for n odd, and that in general, the VHS $R^n f_* \mathbb{Q}_X$ is isometric to one, build up by the weight one VHS of the family of Abelian varieties, parameterized by C. So in a vague way, this problem is related to the question of whether such an isometry has some geometric meaning.

Problem 2: (Eckart Viehweg, Kang Zuo) Find ball quotients and Shimura varieties in the moduli scheme of polarized Calabi–Yau manifolds.

Problem 3: (Eckart Viehweg, Kang Zuo) Let $g : Z \to S \times S'$ be a smooth family of Calabi–Yau n-folds. In "Complex multiplication, Griffiths–Yukawa couplings, and rigidity for families of hypersurfaces" [**19**], we have shown that there exist polarized complex variation of Hodge structures $\mathbf{V}$ and $\mathbf{V}'$ on S and S', respectively, and a Hodge isometry $R^n g_* \mathbb{C}_Z \cong pr_1^* \mathbf{V} \otimes pr_2^* \mathbf{V}'$. Does such a decomposition exist for $R^n g_* K_Z$ where K is a number field, or perhaps even over $K = \mathbb{Q}$? And does the existence of such a decomposition have any geometric interpretation?

Problem 4: (James D. Lewis) Let X be an algebraic $K3$ surface, and consider the transcendental regulator

$$\Phi : \mathrm{CH}^2(X,1) \to \frac{H^{2,0}(X)^\vee}{H_2(X,\mathbb{Z})},$$

where $\mathrm{CH}^2(X,1)$ is Bloch's higher Chow group. Consider the subgroup V of classes in $\mathrm{CH}^2(X,1)$ obtained in the following way. Let $C \subset X$ be an irreducible nodal rational curve, $\sigma : \tilde{C} \overset{\approx}{\to} C$ its normalization, with points $P,\ Q \in \tilde{C}$ for which $\sigma(P) = \sigma(Q)$ is a nodal point on C. On $\tilde{C}$ there is a rational function $\tilde{f}$ for which $\mathrm{div}_{\tilde{C}}(\tilde{f}) = P - Q$. Since $\mathbb{C}(\tilde{C}) = \mathbb{C}(C)$, $\tilde{f}$ corresponds to a rational function f on C for which $\mathrm{div}_C(f) = 0$. Then $\xi := \{(f, C)\} \in \mathrm{CH}^2(X,1)$ defines a higher Chow cycle. If $\omega \in H^{2,0}(X)$, then the value $\Phi(\xi)(\omega)$ is described as follows. Consider $f : C \to \mathbb{P}^1$. Then one can argue that $f^{-1}[0, \infty]$ is a closed 1-cycle that bounds a real 2-dimensional membrane Γ. Then

$$\Phi(\xi)(\omega) = \int_\Gamma \omega \quad (\textit{modulo periods } H_2(X, \mathbb{Z})).$$

What can one say about the image $\Phi(V)$?

Problem 5: (Rolf Schimmrigk) Determine the relation between the conductor of a modular Calabi–Yau variety and the level of the corresponding modular form.

Background: Weil's important contribution to the Shimura–Taniyama conjecture was his observation of the relation between the conductor of the modular elliptic curve and the level of the corresponding modular form. Recently it has been shown that many Calabi–Yau varieties are modular in the sense that the Mellin transform of the Hasse–Weil L-functions of the variety are modular forms of weight four at some level N. In order to understand this modularity of Calabi–Yau varieties more systematically, it would be useful to have a generalization of Weil's conductor observation to higher-dimensional varieties, even if only conjecturally.

Problem 6: (Rolf Schimmrigk) Establish the relation between Kuga–Sato varieties and modular Calabi–Yau manifolds.

Background: Many Calabi–Yau threefolds have recently been shown to be modular. It has also been known for some time that cusp Hecke eigenforms of weight k admit a motivic interpretation in terms of the Kuga–Sato variety (Deligne, Scholl). We therefore encounter a situation which is rather different from the one encountered in the context of elliptic curves and abelian varieties, where it is known from Faltings' proof of the Tate conjecture that varieties over $\mathbb{Q}$ with the same L-function are isogenous. In the context of Calabi–Yau varieties we have two different varieties with the same modular form. This raises the question what precisely the relation is between such modular Calabi–Yau varieties and the corresponding Kuga–Sato variety associated to its modular form.

Problem 7: (Klaus Hulek) In [**8**] we discuss two rigid (birational) Calabi–Yau varieties called X_1 and X_9 and prove that they are both modular with the same associated modular form f_6 which is the unique normalized newform of weight 4 and level 6. By a conjecture of Tate there should be a correspondence between X_1 and X_9 which explains this fact. We can actually prove that X_1 and X_9 are not birational. It would be interesting to exhibit a correspondence as predicted by Tate's conjecture.

Remark: The Calabi–Yau variety X_1 is also birational to the Barth-Nieto quintic and to Verrill's variety associated to the root lattice A_3 (cf. [**9**] and [**15**]). The Barth–Nieto quintic has (birationally) a double cover $\tilde{X}_1$ which is again a Calabi–Yau variety with the same modular form. Here too X_1 and $\tilde{X}_1$ are not birational (they have different Euler numbers). But in this case the double cover gives the correspondence required by Tate's conjecture.

The following problem seems to have been posed by several people, and is posted here by Klaus Hulek.

Problem 8: (Hulek, et al.) For which Hecke eigenforms f of weight 4 does there exist a rigid Calabi–Yau variety X such that the L-series of X equals the Mellin transform of f (up to finitely many primes)?

Note: Hulek became aware of this problem when B. Mazur asked him this question in Oslo in June 2003. He was meanwhile informed by D. van Straten that Straten himself had asked this question to B. van Geemen on an earlier occasion.

Problem 9: (Andrey Todorov)

Introduction. Noriko Yui suggested to pose some problems about Calabi–Yau (CY) manifolds. The problems that I suggested are mainly connected to the construction of CY manifolds whose moduli space is a locally symmetric space. Up to now only one example is known, namely the Borcea–Voisin manifold. The problems suggested in this note are related to the conjecture of Oort–Andre–Mazur which states that the set of manifolds of CM types are dense in the moduli space if and only if the moduli space is a locally symmetric manifold. This conjecture is related to some questions concerning rational conformal field theories.

A. Locally Symmetric Spaces as Moduli of Polarized CY Manifolds

Problem 9A (1): Characterize all CY threefolds whose moduli spaces are locally symmetric spaces.

B. Gross in [**7**] classified all the symmetric domains that are also tube domains and over them one can construct a variation of Hodge structure of weight three with $\dim H^{3,0} = 1$.

Problem 9A (2): This is an open problem posed by B. Gross. Can one find a geometric realization of the variation of Hodge structure described in [**7**]?

Problem 9A (3): Show that the moduli space of CY threefolds that are double covers of $\mathbf{CP}^3$ ramified over eight planes in a general position is a locally symmetric space associated with $\mathrm{SU}(3,3)/\mathrm{S}(\mathrm{U}(3) \times \mathrm{U}(3))$. If the moduli space of CY manifolds that are double covers of $\mathbf{CP}^n$ ramified over $2n+2$ planes is a locally symmetric space then it should be $\mathrm{SU}(n,n)/\mathrm{S}(\mathrm{U}(n) \times \mathrm{U}(n))$.

One can show that the moduli space of a CY threefold is a locally symmetric space of rank greater than or equal to two if and only if the Yukawa coupling has no quantum corrections. It is a well known fact that any moduli space of CY manifolds that is one-dimensional is a locally symmetric space and the famous example of Candelas and coauthors shows that there exists a CY manifold whose moduli space is a one-dimensional locally symmetric space and there are quantum corrections. In the Candelas example the action of the mapping class group on the upper half plane is not arithmetic. It will be interesting to construct CY manifolds whose moduli space is one-dimensional and for which the action of the mapping class group is arithmetic on the upper half plane. I do not know how arithmetically this is related to the existence of quantum corrections to the Yukawa coupling.

Problem 9A (4): This problem was also discussed in [**2**]. It was stated that I. Dolgachev conjectured that the moduli space of CY manifolds that are double covers of $\mathbf{CP}^n$ ramified over $2n+2$ planes is the tube domain $S_n(\mathbf{C}) + \sqrt{-1}S_n^+(\mathbf{C})$, where $S_n(\mathbf{C})$ is the space of $n \times n$ Hermitian matrices and $S_n^+(\mathbf{C})$ is the space of positive Hermitian matrices.

The basis of proposing Problem 9A (4) is the following lemma:

Lemma: *Let $\mathbf{C}^{2n}$ be equipped with a Hermitian metric $\langle u, u\rangle$ with signature (n, n). Let $\mathbf{C}^{2n} = V \oplus \overline{V}$, where $\langle u, u\rangle$ when restricted to V is positive and on $\overline{V}$ is negative. Then $\wedge^n(V \oplus \overline{V})$ is a variation of Hodge Structures of weight n with $\dim_{\mathbf{C}} H^{n,0} = 1$. This variation of Hodge structures is parametrized by $SU(n, n)/S(U(n) \times U(n))$.*

B. CY Manifolds, Conic Singularities and M. Reid Conjecture

It is natural to ask if the generic point of the discriminant locus of an odd-dimensional CY manifold corresponds to a manifold with conic singularity. This is not true. Let us take double covers of $\mathbf{CP}^3$ ramified over eight planes in general position. After the resolution of singularities we will get a CY threefold. The discriminant locus corresponds to a double covering which is ramified over eight planes, three of them meeting in one point. So the generic point of the discriminant locus does not correspond to a threefold with conic singularities since the monodromy group around these points is finite. The monodromy group of a conic singularity is infinite.

Remark: The homological mirror conjecture predicts that CY threefolds should have always conic singularities. The example of double cover of $\mathbf{CP}^3$ ramified over eight planes in a general position shows that one needs to modify this part of the homological mirror conjecture.

Problem 9B (1): Suppose that M is a CY manifold whose moduli space is not a locally symmetric space. Is it true in this case that the generic point of the discriminant locus corresponds to a CY manifold with a conic singularity?

This problem is closely related to Miles Reid's conjecture that the moduli spaces of all CY threefolds are connected. So one can ask the following question:

Problem 9B (2): Is it true that for a CY threefold such that its moduli space is a locally symmetric space it follows that (a) the moduli space is contained in the discriminant locus of the moduli space of a CY manifold, and (b) the generic point of the discriminant locus corresponds to a manifold with a conic singularity?

C. The Analogue of the Dedekind Eta Function

In [**17**] we proved the following Theorem:

Theorem. *Let $\mathfrak{M}_L(M)$ be the moduli space of polarized CY manifolds. Let $\omega_{\mathcal{X}/\mathfrak{M}_L(M)}$ be the relative dualizing sheaf on $\mathfrak{M}_L(M)$. Let $det(\Delta_{\tau,1})$ be the regularized determinant of the Laplacian of the CY metric acting on $(0,1)$-forms. Then locally we have*

$$det(\Delta_{\tau,1}) = \langle\omega_\tau, \omega_\tau\rangle\, |\eta|^2,$$

where ω_τ is a family of holomorphic n-forms and η is a holomorphic function.

As a Corollary we get:

Corollary. *Let Γ_L be the subgroup in the mapping class group which preserves the polarization class. According to Sullivan and Kazhdan for CY manifolds the group $\Gamma_L/[\Gamma_L, \Gamma_L]$ is finite. (See [**4**].) Let $N = \#\Gamma_L/[\Gamma_L, \Gamma_L]$. Then there exists a* *section* η^N *of the line bundle* $\left(\omega^*_{\mathcal{X}/\mathfrak{M}_L(M)}\right)^{\otimes N}$ *such that*

$$det(\Delta_{\tau,1}) = \langle\omega_\tau, \omega_\tau\rangle\, |\eta|^2.$$

Let $\tau \in \mathfrak{M}_L(M)$. Then we know that τ corresponds to a CY threefold M_τ. Let us denote by ω_τ a non-zero holomorphic threeform on M_τ. Let $\beta \in H_3(M, \mathbf{Z})$, then we will write

$$\langle\tau, \beta\rangle := \int_\beta \omega_\tau.$$

Problem 9C (1): Can one find a product formula for the analogue of the Dedekind eta function of CY threefolds

$$\eta^N = \exp\left(2\pi\sqrt{-1}\,\langle\gamma, \tau\rangle\right) \times \prod_i \left(1 - \exp\left(2\pi\sqrt{-1}\,\langle\tau, \beta_i\rangle\right)\right),$$

around points of maximal degeneration, which would mean that around such points the monodromy operator has index of unipotency $n+1$, β_i are the vanishing invariant cycles of the monodromy operators of infinite order and $\tau = (\tau^1, \cdots, \tau^N)$ are the flat local coordinates? For a discussion of the product formulas for automorphic forms (see [**3**].)

Problem 9C (1) is closely related to paper [**1**], precisely, to the counting problem of elliptic curves on the CY threefold.

Problem 9C (2): Prove that $\det\Delta_{\tau,1}$ is bounded on the moduli space $\mathfrak{M}_L(M)$ of any CY manifold M.

This problem will follow directly if one can prove that the coefficients a_k for $k = -n, \cdots, 1$ of the short term asymptotic expansion

$$Tr\left(\exp\left(-t\Delta_{\tau,1}\right)\right) = \sum_{k=-n}^{1} \frac{a_k}{t^k} + a_0 + \cdots$$

are constants. We prove that a_0 is a constant if M is a CY. The solution of Problem 16 will show that the analogue of the Dedekind eta function η^N vanishes on the discriminant locus. This will imply that the section $\overline{\eta}^N$ constructed in [**17**] will be related to the algebraic discriminant as defined by Gelfand, Kapranov and Zelevinsky. The analogue of the Dedekind eta function for algebraic polarized K3 and Enriques surfaces was discussed in [**2**], [**10**], [**11**], [**12**], [**13**] and [**20**].

References

[1] Bershadsky, S. Cecotti, H. Ooguri and C. Vafa, *Kodaira–Spencer theory of gravity and exact results for quantum string amplitude*, Comm. Math. Phys. **165** (1994), 311-428.

[2] R. Borcherds, *The moduli of Enriques surfaces and the fake Monster Lie superalgebra,* Topology **35** (1996), 699-710.

[3] R. Borcherds, *Automorphic forms with singularities on Grassmanians,* Invent. Math. **132** (1998), No 3, 491-562.

[4] A. A. Kirillov and Cl. Delaroche, *Sur les relations entre l'espace dual d'un group et la struture de ses sous-groupes fermés (d'après D. A. Kazhdan)*, Bourbaki No. 343, 1967/68.

[5] S. Donaldson, *Infinite determinants, stable bundles and curvature,* Duke Math. J. **54** (1987), No.1, 231-248.

[6] P. Gilkey, *Invariance Theory, The Heat Equation, and the Atiyah-Singer Index Theorem,* Mathematics Lecture Series vol. **11**, Publish or Parish, Inc. Wilmington, Delaware (USA) 1984.
[7] B. Gross, *A remark on tube domains,* Math. Research Letters **1** (1994), 1-9.
[8] K. Hulek, and H. Verrill, *On modularity of rigid and non-rigid Calabi–Yau varieties associated to the root lattice* A_4, arXiv:math.AG/0304169. Nagoya J. Math. (to appear).
[9] K. Hulek, J. Spandaw, B. van Geemen and D. van Straten, *The modularity of the Barth–Nieto quintic and its relatives*, Adv. Geometry **1** (2001), 263–289.
[10] J. Jorgenson and A. Todorov, *A conjectural analogue of Dedekind's eta function for K3 surfaces,* Math. Research Letters **2** (1995), 359-360.
[11] J. Jorgenson and A. Todorov, *Analytic discriminant for manifolds with zero canonical class,* Manifolds and Geometry, eds. P de Bartolomeis, F. Tircerri and E. Vesantini, Symposia Mathematica **36** (1996), 223-260.
[12] J. Jorgenson and A. Todorov, *Enriques surfaces, analytic discriminants, and Borcherds's* Φ *function,* Com. Math. Phys **191** (1998), 249-264.
[13] J. Jorgenson and A. Todorov *Correction to "Enriques surfaces, analytic discriminants, and Borcherds's* Φ *function"*, Com. Math. Phys. **207** (1999), no. 2, 495–497.
[14] K. Liu, A. Todorov, Shing-Tung Yau and K. Zuo, *The analogue of the Shafarevich conjecture for CY manifolds I.,* preprint.
[15] M.-H. Saito and N. Yui, *The modularity conjecture for rigid Calabi–Yau threefolds over* $\mathbb{Q}$, J. Math. Kyoto Univ. **41** (2001), 403–419.
[16] A. Todorov, *Witten's geometric quantization of moduli of CY manifolds,* preprint 1999.
[17] A. Todorov, *Weil–Petersson volumes of the moduli of CY manifolds,* preprint.
[18] E. Viehweg and K. Zuo, *Families over curves with a strictly maximal Higgs field,* Asian J. Math. **7** (2003), no.4, 575–598.
[19] E. Viehweg and K. Zuo, *Complex multiplication, Griffiths–Yukawa couplings, and rigidity for families of hypersurfaces*, J. Alg. Geom. (2005), to appear.
[20] K. Yoshikawa, *Generalized Enriques surfaces and analytic torsion,* preprint (1998).
[21] S. T. Yau, *On the Ricci curvature of a compact Kähler manifold and the complex Monge-Ampère equation I*, Comm. Pure App. Math. **31** (1979), 339-411.

Titles in This Series

TITLES IN THIS SERIES

7 **D. A. Buell and J. T. Teitelbaum, Editors,** Computational Perspectives on Number Theory, 1998

6 **Harold Levine,** Partial Differential Equations, 1997

5 **Qi-keng Lu, Stephen S.-T. Yau, and Anatoly Libgober, Editors,** Singularities and Complex Geometry, 1997

4 **Vyjayanthi Chari and Ivan B. Penkov, Editors,** Modular Interfaces: Modular Lie Algebras, Quantum Groups, and Lie Superalgebras, 1997

3 **Xia-Xi Ding and Tai-Ping Liu, Editors,** Nonlinear Evolutionary Partial Differential Equations, 1997

2.2 **William H. Kazez, Editor,** Geometric Topology, 1997

2.1 **William H. Kazez, Editor,** Geometric Topology, 1997

1 **B. Greene and S.-T. Yau, Editors,** Mirror Symmetry II, 1997